Advances in Microcrystalline and Nanocrystalline Semiconductors–1996

MATERIALS RESEARCH SOCIETY
SYMPOSIUM PROCEEDINGS VOLUME 452

Advances in Microcrystalline and Nanocrystalline Semiconductors—1996

Symposium held December 2–6, 1996, Boston, Massachusetts, U.S.A.

EDITORS:

Robert W. Collins
The Pennsylvania State University
University Park, Pennsylvania, U.S.A.

Philippe M. Fauchet
University of Rochester
Rochester, New York, U.S.A.

Isamu Shimizu
Tokyo Institute of Technology
Yokohama, Japan

Jean-Claude Vial
Université Joseph Fourier
Grenoble, France

Toshikazu Shimada
Hitachi Ltd.
Tokyo, Japan

A. Paul Alivisatos
University of California, Berkeley
Berkeley, California, U.S.A.

PITTSBURGH, PENNSYLVANIA

Single article reprints from this publication are available through University Microfilms Inc., 300 North Zeeb Road, Ann Arbor, Michigan 48106

CODEN: MRSPDH

Published by:

Materials Research Society
9800 McKnight Road
Pittsburgh, Pennsylvania 15237
Telephone (412) 367-3003
Fax (412) 367-4373
Website: http://www.mrs.org/

Library of Congress Cataloging in Publication Data

Advances in microcrystalline and nanocrystalline seiconductors—1996: symposium held December 2–6, 1996, Boston, Massachusetts, U.S.A. / editors, Robert W. Collins, Philippe M. Fauchet, Isamu Shimizu, Jean-Claude Vial, Toshikazu Shimada, A. Paul Alivisatos
p.cm—(Materials Research Society symposium proceedings ; v. 452)
Includes bibliographical references and index.
ISBN 1-55899-356-8
1. Semiconductors—Congresses. 2. Nanostructure materials—Congresses. 3. Porous materials—Congresses. 4. Luminescence—Congresses. 5. Porous silicon—Congresses. 6. Thin film devices—Congresses. 7. Polycrystalline semiconductors—Congresses. I. Collins, Robert W. II. Fauchet, Philippe M. III. Shimizu, Isamu IV. Vial, Jean-Claude, V. Shimada, Toshikazu VI. Alivisatos, A. Paul VII. Materials Research Society symposium proceedings ; v. 452.
TK7871.85.A3545 1997 97-3594
537.6'226—dc21 CIP

Manufactured in the United States of America

CONTENTS

*Invited Paper

*Invited Paper

PART IV: SYNTHESIS AND PROPERTIES OF OTHER GROUP IV CLUSTERS, NANOCRYSTALS, AND NANOSTRUCTURES

*Invited Paper

PART VI: SYNTHESIS, SURFACES, AND STRUCTURE/PROPERTY RELATIONS OF LUMINESCENT POROUS SEMICONDUCTORS

*Invited Paper

*Invited Paper

PART IX: ELECTROLUMINESCENCE AND LIGHT-EMITTING DEVICES OF NANOCRYSTAL AND POROUS SILICON

PART X: PREPARATION AND PROPERTIES OF NANOCRYSTALLINE, MICROCRYSTALLINE, AND MICROSTRUCTURED THIN FILMS

*Invited Paper

PART XI: APPLICATIONS OF NANOCRYSTALLINE AND MICROCRYSTALLINE THIN FILMS

*Invited Paper

*Invited Paper

*Invited Paper

PREFACE

This volume includes the proceedings of Symposium Q "Advances in Microcrystalline and Nanocrystalline Semiconductors—1996," held during the 1996 MRS Fall Meeting in Boston, Massachusetts. This is the fourth in a series of MRS fall symposia devoted to research on the theory, preparation, characterization, and application of nanocrystalline, microcrystalline, and polycrystalline semiconductors. The first symposium was conducted in 1989 and yielded the volume entitled "Materials Issues in Microcrystalline Semiconductors," edited by P.M. Fauchet, K. Tanaka, and C.C. Tsai (MRS Symposium Proceedings Volume 164), which included about 60 papers. The second symposium was conducted in 1992 and yielded the volume entitiled "Microcrystalline Semiconductors: Materials Science and Devices," edited by P.M. Fauchet, C.C. Tsai, L.T. Canham, I. Shimizu, and Y. Aoyagi, (MRS Symposium Proceedings Volume 283), which included about 150 papers. The explosion of interest in nanocrystalline semiconductors, and the consequent expansion in the symposium between 1989 and 1992, was a result of the demonstration in 1990 by L.T. Canham that visible photoluminescence is emitted at room temperature from the nanostructures in porous silicon. Over the last five years, scientific and technological interest has continued undiminished. The third symposium was conducted in 1994 and yielded the volume entitled "Microcrystalline and Nanocrystalline Semiconductors," edited by R.W. Collins, C.C. Tsai, M. Hirose, F. Koch, and L. Brus, (MRS Symposium Proceedings Volume 358), which included about 60 papers. In the five-day span of the most recent symposium in 1996, 200 papers were presented in 25 oral and poster sessions, some 80% of which appear in this volume.

Symposium Q of the 1996 MRS Fall Meeting brought together scientists and engineers with diverse backgrounds to share their recent advances in (i) semiconductor nanocrystals and nanostructures, (ii) porous semiconductors, and (iii) nanocrystalline, microcrystalline, and polycrystalline semiconductor films. With the exception of the larger-grained microcrystalline and polycrystalline films, the unifying scientific theme underlying these research areas is the modification of the properties of semiconductors that results from electron confinement in nanometer-sized structures. The challenge in *all* of these fields, however, is to understand (and often to minimize, as in the case of the larger-grained films) the influence of structural features including surfaces, interfaces, grain boundaries, and intergranular phases on the often unique (as in the case of light-emitting Si) semiconducting properties of the grains themselves. A common technological application among semiconductor nanocrystals/nanostructures and porous semiconductors stems from their unique optical properties, specifically their efficient visible photoluminescence that persists at room temperature, owing to the confinement of the photoexcited charge carriers. As a result, electroluminescent devices with a wide variety of designs are being developed and optimized, and their integration with Si microelectronics has now been demonstrated. The challenges are to ensure that the crystallites are optimally-sized for visible light emission, and also that surfaces are optimally-oxidized to avoid nonradiative recombination and instability, while still allowing carrier transport in light-emitting diodes. An important application of the polycrystalline thin films is large-area electronics and photovoltaics on inexpensive glass substrates. In this case, the challenge is one of obtaining large crystallite sizes, while ensuring that grain boundaries are well-passivated. Common themes and challenges have provided a unique opportunity for fruitful interactions among the scientists and engineers active in these fields.

Materials advances presented during the symposium and described in this volume involve structures spanning more than five orders of magnitude in size, ranging from molecular clusters <1 nm in size to single-crystal regions in thin films >10 μm in size. At the smallest end of the scale, detailed descriptions of predicted and observed properties of Group IV molecular clusters are presented here. At the largest scale, novel excimer laser annealing approaches for thin Si films have been described that yield single-crystal grains large enough for fabrication of thin-film transistors within their boundaries. In a plenary session, excellent overviews of the current status of research in (i) microcrystalline and nanocrystalline semiconductor films, (ii) semiconductor nanocrystals, and (iii) porous semiconductors were provided by K. Tanaka (NAIR), L.E. Brus (Columbia University), and L.T. Canham (DRA-Malvern), respectively.

In the area of thin films, advanced techniques for characterizing the nucleation, growth, and structure of nanocrystalline Si films were reported, based on *in situ* scanning tunneling microscopy (K. Ikuta et al., NAIR). The motivation for better understanding and optimizing the growth of such films was provided through the demonstration of a 10.7%-efficient tandem solar cell made with a stable, all-microcrystalline Si bottom cell (H. Keppner et al., University of Neuchatel). In the area of nanocrystals, phase transition kinetics as a function of size are providing information on the intrinsic nucleation of solid-solid phase transformations, e.g., from four-coordinate wurtzite to six-coodinate rock-salt in CdSe (C.-C. Chen et al., University of California - Berkeley). In the future, this research may lead to approaches for chemical synthesis of a wide variey of nanocrystals in metastable phases. Other studies of the optical properties of CdSe nanocrystals clearly reveal the atomic-like nature of the light-emitting state. By eliminating spectral inhomogeneities, a factor of 50 reduction in the photoluminescence linewidth (to <120 μeV at 10 K -- resolution limited) in comparison with previous work was reported (S. Empedocles et al., MIT). In the area of porous silicon, materials advances reported during the symposium focused on improving control over the structure and enhancing the stability of room-temperature light-emitting material. The fact that such advances are meeting with success is demonstrated with the report of device integration of Si-based LEDs (Hirschman et al., University of Rochester). In addition, new applications of porous silicon continue to emerge in biotechnology (L.T. Canham et al.) and in passive optics such as Bragg reflectors, Fabry-Perot filters, rugate filters, and gratings (M. Thönissen et al., KFA-Jülich; G. Lérondel et al., Université de Joseph Fourier).

This is just a sampling of the wide-ranging, but interconnected topics covered by the microcrystalline/nanocrystalline semiconductor symposium and included in this volume. Owing to the challenging nature of the problems, the rapid advance of the field, and the importance of microcrystalline and nanocrystalline semiconductors in electronic and optoelectronic device applications, future symposia on this subject are expected to continue generating great interest. For this reason, we have adopted the title "Advances in Microcrystalline and Nanocrystalline Semiconductors - 1996," to indicate that the symposium and volume provide a snapshot in time of rapidly advancing research and applications. The volume is divided into twelve parts. In many cases, assigning a paper to a given part has been difficult, owing to the many interconnections of the three materials fields listed above. The plenary papers are collected in Part I, entitled *Current Status and Future Prospects of Research in Microcrystalline and Nanocrystalline Semiconductors*, and include the excellent overviews by K. Tanaka, L. Brus, and L.T. Canham.

The theme of the papers in Part II is the theory of semiconductor molecular clusters and nanocrystals confined in two or three dimensions (wires or dots, respectively), with an emphasis on calculating electronic band structure, density of electronic states, and optical properties. Studies on this topic are essential in order to develop a better understanding of many of the experimental observations presented in the proceeding parts of the volume. The themes of the papers in Part III concern luminescent Group IV clusters/nanocrystals and quantum wells, semiconductor systems confined in three and one dimensions, respectively. The papers in this part are sequenced primarily by the preparation technique, starting with an invited article by J. Budai and coworkers (Oak Ridge National Laboratory) on nanocrystals prepared by the ion implantation technique. After several articles of similar topic, the articles that follow describe nanocrystal and nanophase materials prepared by magnetron sputtering, pulsed laser ablation, spark processing, inert gas evaporation, and chemical synthesis. The later papers in Part III describe studies of Si quantum wells, including an invited article by D.J. Lockwood and coworkers (National Research Council of Canada), in which the wells are prepared by a variety of vapor deposition techniques, as well as studies of crystallites in thin films prepared by chemical vapor deposition. The themes of the papers in Part IV are similar to those of Part III, but the primary interest is in the electronic and vibrational states, self-assembly mechanisms, and thermoelectric properties of the Group IV nanostructures, rather than their photoluminescence characteristics. The papers in this part are sequenced by material, starting with an invited article by G.P. Lopinski and coworkers (Penn State University) on graphitic carbon nanocrystals, and progressing through Si, Ge, and $Si_{1-x}Ge_x$ alloy systems. The papers in Part V are concerned with the Group III-V, Group II-VI, and metal sulfide, iodide, and oxide nanocrystals, and are sequenced in that order. In the research described in the first several papers on III-V and II-VI nanostructures, the preparation method is vapor deposition including molecular beam epitaxy and chemical vapor deposition, whereas in most of the remaining papers, chemical or electrochemical deposition methods from solutions are employed.

Porous silicon provides the focus of Parts VI and VII. Part VI starts with papers that concentrate on the synthesis of porous silicon from various starting materials, including an invited article by J.-N. Chazalviel and coworkers (Ecole Polytechnique) which describes porous silicon from hydrogenated amorphous silicon films (a-Si:H). Most of the remaining papers in Part VI concern the process/property relationships in porous silicon, its structure and surface chemistry, and the role of the surfaces in various aspects of the photoluminescence process. Also included in Part VI are several papers that describe the deposition of polymers or other semiconductors into porous silicon. An article describing studies of porous SiC concludes Part VI. The papers in Part VII describe studies of the basic photoluminescence properties and mechanisms of porous Si, particularly focusing on the emitting nanocrystals within porous Si. Other papers in this part describe the optical properties and Raman spectra of porous Si. The emphases of the papers in Parts VIII and IX are the applications of nanocrystal and porous semiconductors, as well as the specific properties that relate to these applications, such as electronic transport. The applications in Part VIII include the non-light-emitting applications, starting with applications in biotechnology, as described in the initial invited article by L.T. Canham and co-workers. The subsequent series of papers in Part VIII present photochemical, photoelectronic, and photovoltaic properties and applications. The final papers describe passive optical applications including sensors, gratings, and interference filters. The first set of papers in Part IX describe light-emitting properties and applications of porous Si, including the electroluminescence process and the performance of light-emitting diodes. In addition, the integration of Si optoelectronics is demonstrated, and the performance of optical microcavities formed from porous Si is described at the end of Part IX.

Papers in Parts X-XII present research results on the nano-, micro-, and polycrystalline thin films, including their preparation, properties and applications. The distinguishing character of these studies from many in the earlier parts is that the intended applications are more often electronic or photovoltaic than light-emitting. Many of the materials and films described in earlier parts are prepared with a wide-gap intergranular phase designed to promote quantum confinement and passivate the surfaces, whereas the intergranular phase of the films in Parts X-XII is most often incidental and detrimental. Part X begins with papers that concentrate on the preparation and properties of micro/nanocrystalline silicon, including an invited one by F. Finger and coworkers (Forschungszentrum Jülich), continues with papers on the properties of these films, and concludes with those describing materials other than Si, such as ITO and PZT. Papers in Part XI describe the applications of the nano/microcrystalline films in devices. These include, in order of presentation, photovoltaic devices, thin-film transistors, photodiodes, light-emitting diodes, and photoconductors. An invited article by H. Keppner and coworkers highlights Part XI. Many papers in Part XII deal with the preparation, characterization, and application of crystalline Si films having large grain sizes, using inexpensive substrates such as glass. The initial papers in this part describe laser and thermal annealing methods for the preparation of these materials, including invited articles by K.S. Choi and M. Matsumura (Tokyo Institute of Technology) focusing on thin-film transistor applications and M. Tanaka and co-workers (Sanyo Electric) focusing on photovoltaic applications. Succeeding papers describe materials preparation by enhanced chemical vapor deposition (CVD) and low-pressure CVD techniques, and materials characterization. The final papers of the volume describe the preparation and properties of crystalline materials other than Si that exhibit novel properties with potentially important applications.

In concluding, the organizers wish to thank the many referees who dedicated their free time at the symposium in Boston to review all the articles in this volume. The authors also deserve praise for revising their articles within a very narrow time frame to ensure rapid publication. All invited speakers and session chairs should be commended for their essential contributions in making the symposium in Boston a success. Special thanks go to K. Tanaka, L. Brus, and L.T. Canham for their extra efforts in providing comprehensive surveys of prior research and anticipated future directions in the three focus areas of the conference. Our thanks also go to Jennifer Barker for her help in preparing and indexing the volume. Finally, the organizers are indebted to the sponsors below for their generous contributions in making this symposium possible:

Asahi Glass Company Limited
Hitachi Europe, Ltd.
Hitachi, Ltd.
KANEKA Corporation
Mitsui Toatsu Chemicals, Incorporated
SANYO Electric Company, Ltd.
Sharp Corporation
Xerox PARC

Robert W. Collins
Philippe M. Fauchet
Isamu Shimizu
Jean-Claude Vial
Toshikazu Shimada
A. Paul Alivisatos

January 1997

MATERIALS RESEARCH SOCIETY SYMPOSIUM PROCEEDINGS

Volume 420— Amorphous Silicon Technology—1996, M. Hack, E.A. Schiff, S. Wagner, R. Schropp, A. Matsuda 1996, ISBN: 1-55899-323-1

Volume 421— Compound Semiconductor Electronics and Photonics, R.J. Shul, S.J. Pearton, F. Ren, C-S. Wu, 1996, ISBN: 1-55899-324-X

Volume 422— Rare-Earth Doped Semiconductors II, S. Coffa, A. Polman, R.N. Schwartz, 1996, ISBN: 1-55899-325-8

Volume 423— III-Nitride, SiC, and Diamond Materials for Electronic Devices, D.K. Gaskill, C.D. Brandt, R.J. Nemanich, 1996, ISBN: 1-55899-326-6

Volume 424— Flat Panel Display Materials II, M. Hatalis, J. Kanicki, C.J. Summers, F. Funada, 1997, ISBN: 1-55899-327-4

Volume 425— Liquid Crystals for Advanced Technologies, T.J. Bunning, S.H. Chen, W. Hawthorne, T. Kajiyama, N. Koide, 1996, ISBN: 1-55899-328-2

Volume 426— Thin Films for Photovoltaic and Related Device Applications, D. Ginley, A. Catalano, H.W. Schock, C. Eberspacher, T.M. Peterson, T. Wada, 1996, ISBN: 1-55899-329-0

Volume 427— Advanced Metallization for Future ULSI, K.N. Tu, J.W. Mayer, J.M. Poate, L.J. Chen, 1996, ISBN: 1-55899-330-4

Volume 428— Materials Reliability in Microelectronics VI, W.F. Filter, J.J. Clement, A.S. Oates, R. Rosenberg, P.M. Lenahan, 1996, ISBN: 1-55899-331-2

Volume 429— Rapid Thermal and Integrated Processing V, J.C. Gelpey, M.C. Öztürk, R.P.S. Thakur, A.T. Fiory, F. Roozeboom, 1996, ISBN: 1-55899-332-0

Volume 430— Microwave Processing of Materials V, M.F. Iskander, J.O. Kiggans, Jr., J.Ch. Bolomey, 1996, ISBN: 1-55899-333-9

Volume 431— Microporous and Macroporous Materials, R.F. Lobo, J.S. Beck, S.L. Suib, D.R. Corbin, M.E. Davis, L.E. Iton, S.I. Zones, 1996, ISBN: 1-55899-334-7

Volume 432— Aqueous Chemistry and Geochemistry of Oxides, Oxyhydroxides, and Related Materials J.A. Voight, T.E. Wood, B.C. Bunker, W.H. Casey, L.J. Crossey, 1997, ISBN: 1-55899-335-5

Volume 433— Ferroelectric Thin Films V, S.B. Desu, R. Ramesh, B.A. Tuttle, R.E. Jones, I.K. Yoo, 1996, ISBN: 1-55899-336-3

Volume 434— Layered Materials for Structural Applications, J.J. Lewandowski, C.H. Ward, M.R. Jackson, W.H. Hunt, Jr., 1996, ISBN: 1-55899-337-1

Volume 435— Better Ceramics Through Chemistry VII—Organic/Inorganic Hybrid Materials, B.K. Coltrain, C. Sanchez, D.W. Schaefer, G.L. Wilkes, 1996, ISBN: 1-55899-338-X

Volume 436— Thin Films: Stresses and Mechanical Properties VI, W.W. Gerberich, H. Gao, J-E. Sundgren, S.P. Baker 1997, ISBN: 1-55899-339-8

Volume 437— Applications of Synchrotron Radiation to Materials Science III, L. Terminello, S. Mini, H. Ade, D.L. Perry, 1996, ISBN: 1-55899-340-1

Volume 438— Materials Modification and Synthesis by Ion Beam Processing, D.E. Alexander, N.W. Cheung, B. Park, W. Skorupa, 1997, ISBN: 1-55899-342-8

Volume 439— Microstructure Evolution During Irradiation, I.M. Robertson, G.S. Was, L.W. Hobbs, T. Diaz de la Rubia, 1997, ISBN: 1-55899-343-6

Volume 440— Structure and Evolution of Surfaces, R.C. Cammarata, E.H. Chason, T.L. Einstein, E.D. Williams, 1997, ISBN: 1-55899-344-4

Volume 441— Thin Films—Structure and Morphology, R.C. Cammarata, E.H. Chason, S.C. Moss, D. Ila, 1997, ISBN: 1-55899-345-2

Volume 442— Defects in Electronic Materials II, J. Michel, T.A. Kennedy, K. Wada, K. Thonke, 1997, ISBN: 1-55899-346-0

Volume 443— Low-Dielectric Constant Materials II, K. Uram, H. Treichel, A.C. Jones, A. Lagendijk, 1997, ISBN: 1-55899-347-9

Materials Research Society Symposium Proceedings

Volume 444— Materials for Mechanical and Optical Microsystems, M.L. Reed, M. Elwenspoek, S. Johansson, E. Obermeier, H. Fujita, Y. Uenishi, 1997, ISBN: 1-55899-348-7

Volume 445— Electronic Packaging Materials Science IX, P.S. Ho, S.K. Groothuis, K. Ishida, T. Wu, 1997, ISBN: 1-55899-349-5

Volume 446— Amorphous and Crystalline Insulating Thin Films—1996, W.L. Warren, J. Kanicki, R.A.B. Devine, M. Matsumura, S. Cristoloveanu, Y. Homma, 1997, ISBN: 1-55899-350-9

Volume 447— Environmental, Safety, and Health Issues in IC Production, R. Reif, A. Bowling, A. Tonti, M. Heyns, 1997, ISBN: 1-55899-351-7

Volume 448— Control of Semiconductor Surfaces and Interfaces, S.M. Prokes, O.J. Glembocki, S.K. Brierley, J.M. Woodall, J.M. Gibson, 1997, ISBN: 1-55899-352-5

Volume 449— III–V Nitrides, F.A. Ponce, T.D. Moustakas, I. Akasaki, B.A. Monemar, 1997, ISBN: 1-55899-353-3

Volume 450— Infrared Applications of Semiconductors—Materials, Processing and Devices, M.O. Manasreh, T.H. Myers, F.H. Julien, 1997, ISBN: 1-55899-354-1

Volume 451— Electrochemical Synthesis and Modification of Materials, S.G. Corcoran, P.C. Searson, T.P. Moffat, P.C. Andricacos, J.L. Deplancke, 1997, ISBN: 1-55899-355-X

Volume 452— Advances in Microcrystalline and Nanocrystalline Semiconductors—1996, R.W. Collins, P.M. Fauchet, I. Shimizu, J–C. Vial, T. Shimada, A.P. Alvisatos, 1997, ISBN: 1-55899-356-8

Volume 453— Solid–State Chemistry of Inorganic Materials, A. Jacobson, P. Davies, T. Vanderah, C. Torardi, 1997, ISBN: 1-55899-357-6

Volume 454— Advanced Catalytic Materials—1996, M.J. Ledoux, P.W. Lednor, D.A. Nagaki, L.T. Thompson, 1997, ISBN: 1-55899-358-4

Volume 455— Structure and Dynamics of Glasses and Glass Formers, C.A. Angell, T. Egami, J. Kieffer, U. Nienhaus, K.L. Ngai, 1997, ISBN: 1-55899-359-2

Volume 456— Recent Advances in Biomaterials and Biologically–Inspired Materials: Surfaces, Thin Films and Bulk, D.F. Williams, M. Spector, A. Bellare, 1997, ISBN: 1-55899-360-6

Volume 457— Nanophase and Nanocomposite Materials II, S. Komarneni, J.C. Parker, H.J. Wollenberger, 1997, ISBN: 1-55899-361-4

Volume 458— Interfacial Engineering for Optimized Properties, C.L. Briant, C.B. Carter, E.L. Hall, 1997, ISBN: 1-55899-362-2

Volume 459— Materials for Smart Systems II, E.P. George, R. Gotthardt, K. Otsuka, S. Trolier-McKinstry, M. Wun-Fogle, 1997, ISBN: 1-55899-363-0

Volume 460— High–Temperature Ordered Intermetallic Alloys VII, C.C. Koch, N.S. Stoloff, C.T. Liu, A. Wanner, 1997, ISBN: 1-55899-364-9

Volume 461— Morphological Control in Multiphase Polymer Mixtures, R.M. Briber, D.G. Peiffer, C.C. Han, 1997, ISBN: 1-55899-365-7

Volume 462— Materials Issues in Art and Archaeology V, P.B. Vandiver, J.R. Druzik, J. Merkel, J. Stewart, 1997, ISBN: 1-55899-366-5

Volume 463— Statistical Mechanics in Physics and Biology, D. Wirtz, T.C. Halsey, J. van Zanten, 1997, ISBN: 1-55899-367-3

Volume 464— Dynamics in Small Confining Systems III, J.M. Drake, J. Klafter, R. Kopelman, 1997, ISBN: 1-55899-368-1

Volume 465— Scientific Basis for Nuclear Waste Management XX, W.J. Gray, I.R. Triay, 1997, ISBN: 1-55899-369-X

Volume 466— Atomic Resolution Microscopy of Surfaces and Interfaces, D.J. Smith, R.J. Hamers, 1997, ISBN: 1-55899-370-3

Prior Materials Research Society Symposium Proceedings available by contacting Materials Research Society

Part I

Current Status and Future Prospects of Research in Microcrystalline and Nanocrystalline Semiconductors

RECENT PROGRESS IN MICROCRYSTALLINE SEMICONDUCTOR THIN FILMS

K. TANAKA
Joint Research Center for Atom Technology (JRCAT)
National Institute for Advanced Interdisciplinary Research, 1-1-4 Higashi, Tsukuba, Ibaraki 305,
tanaka@jrcat.or.jp

ABSTRACT

Nanocrystalline/microcrystalline thin films prepared at relatively low temperatures by plasma-enhanced chemical vapor deposition (PECVD), in particular hydrogenated microcrystalline Si films (μc-Si:H), have attracted an increasing attention not only as potential materials for thin film solar cells, but also as active layers in thin film transistor arrays for flat panel displays. This paper reviews recent progress in the investigation of these materials; preparation methods, structural and optical properties, and electronic transports. Emphasis is placed on the understanding of the growth mechanism of μc-Si:H films as well as the microscopic characterization of the film structure.

INTRODUCTION

What is a μc-Si:H film? - It is characterized as a structurally-inhomogeneous film, which consists of nanometer-size microcrystals embeded in amorphous tissue with some fraction of voids. Therefore, film properties vary significantly depending on a volume fraction (X_C) of Si microcrystals, their averaged grain size (δ), its distribution, nature of grain boundaries and amorphous tissue, a void fraction, a content (C_H) and bonding configuration of hydrogen atoms.

In the early stage of this field, Veprek and Marecek initiated a pioneering work in 1968, where they prepared polycrystalline Si and Ge films on glass substrates by a chemical transport method using a hydrogen plasma [1]. However, the actual growth of this field started ten years later, namely, after the initial success of *pn* control of amorphous Si (a-Si:H) in 1975 by plasma-enhanced chemical vapor deposition (PECVD) [2]. Actually, Usui and Kikuchi [3] and our group [4,5] independently found out doped as well as undoped μc-Si:H films in the course of the study of a-Si:H using PECVD. In this sense, this material was some sort of a byproduct of a-Si:H, but now, it has grown up into an independent and important research area as evidenced by a bunch of papers on μc-Si:H published for the recent ten years [6].

Why is this material so interesting? - Optical and transport properties are considerably different from those of bulk crystalline Si or amorphous Si:H, as is described later. From the viewpoints of materials science μc-Si:H belongs to a new type of materials, i.e., a mixed phase of microcrystals and amorphous tissue. Therefore, in spite of its complexity, a rich designability can be expected if the structure is microscopically controlled. This material has received a strong demand from the device application community as a potential material for thin film solar cells and flat panel displays [6]. PECVD guarantees a compatibility of this material with the a-Si:H technology.

In order to avoid too much diversity, I mainly pick up μc-Si:H in this review article as a representative of nanocrystalline/microcrystalline semiconductor thin films. Structural, optical and electronic transport properties will be reviewed with some emphasis on the growth mechanism of μc-Si:H because for the past decade much work has been done on this topic towards a better control of its microscopic structure.

PREPARATION TECHNIQUES

Different techniques have been employed for preparing nanocrystalline/microcrystalline semiconductor thin films so far ; chemical transport using a hydrogen plasma [1], plasma-enhanced chemical vapor deposition (PECVD) from H_2/SiH_4 gas mixture[3-5], reactive magnetron sputtering of c-Si by Ar/H_2 [7], reactive thermal CVD from GeH_4/Si_2H_6 [8], laser nucleation or crystallization from amorphous phase [9], solid-phase crystallization [10] etc. PECVD, among

Mat. Res. Soc. Symp. Proc. Vol. 452 © 1997 Materials Research Society

them, has been most extensively studied because of its advantages such as the capability of low-temperature processes in a range of 150 - 300°C and also the easiness to deposit a large area film, which is compatible with the a-Si:H technology.

There have been reported a lot of variations in PECVD so far. As for a reactor srtucture, in addition to conventional capacitively-coupled (diode)[5] and inductively-coupled reactors [3], a triode reactor has been used for controlling the amount of ionic species impinging on the surface [11] and/or selecting long-lifetime precursors from the SiH_4 plasma [12]. Very high frequency glow discharge (VHF-GD), in a frequency range of 50 - 120 MHz, has pruduced μc-Si:H films of larger-size grains with higher deposition rates than the conventional 13.56-MHz GD [13]. Higher deposition rates have also been achieved by electron cyclotron wave resonance (ECWR) for plasma excitation [14] and plasma-gun CVD [15]. SiF_nH_m or SiF_4/H_2 has been used as source gases with emphasis on chemical annealing in the film growth process [16]. Remote plasma [17] and hot filament [18] techniques emphasize soft reaction on the growing surface free from ion bombardment.

Those techniques and approaches have led to a significant improvement in film properties of μc-Si:H and also improved our understanding of the growth mechanisms. However, in all of the above preparation methods, excitation reactions of all the gases are coupled, and depositon reactions compete with other surface processes such as diffusion and desorption of adsorbates and etching. In this sense, the use of time-modulated gas flows enables us to separate the gas-phase reaction from the surface modification process. Actually, gas-flow modulation methods have been widely introduced for preparing μc-Si:H in a layer-by-layer (LBL) manner, and also for deeper understanding of the microscopic process of the film growth [19-22].

STRUCTURAL PROPERTIES

Macroscopic observations as well as microscopic characterizations have been extensively made on the structure of μc-Si:H films using a variety of techniques : X-ray diffraction [23], high-resolution TEM [13,24], atomic force microscopy (AFM) [25], scanning tunneling microscopy (STM) [12], spectroscopic ellipsometry [26,27], Raman scattering [28-32], infrared absorption [5, 33], proton NMR [34, 35] and ESR [36, 37]. All of those structural investigations show that μc-Si:H films are highly inhomogeneous, being composed of small crystallites with an average grain size δ of about 40 - 400 A, embedded in an amorphous matrix with some fraction of voids. The relative amount of crystalline phase contained in the film is described by the crystalline volume fraction X_c, which can vary from a few percent up to higher than 90% in fully-crystallized specimens. Hydrogen atoms, most of which are bonded with silicon atoms, are included in the film with the content C_H of about 3 - 15 at.% in a spatially-inhomogeneous manner, as is described below.

Morphology

Macroscopic inhomogeneity along the growth direction of μc-Si:H films has been directly observed by TEM [24] and STM [12], and also evidenced by spectroscopic ellipsometry [22] as well as the thickness dependence of the room temperature conductivity [24]. An initially-grown several-nm thickness of the μc-Si:H film is structurally amorphous with a lot of voids irrespective of substrate materials such as glass, SiO_2, a-Si:H or graphite and has, therefore, a less dense structure. Nucleation of nanocrystals occurs before the thickness reaches 5 nm for the case of the graphite substrate [12], and later, X_c keeps increasing as a result of the growth of μc nuclei and their coalescence until the film thickness exceeds a value of about 50 nm [22,24]. Namely, μc-Si:H films are structurally inhomogeneous at least up to an initial several-ten-nanometer thickness independent of what substrates they are deposited on.

Grain Size (δ)and Shape of Microcrystallites

The grain size δ of microcrystals is usually determined by X-ray or electron diffraction methods and lies in a range from about 4 nm to 40 nm in diameter. As is well known, the grain size of microcrystals in μc-Si:H films increases as the substrate temperature increases [38], and also as the plasma excitation frequency increases [39,13]. The grain size is also strongly correlated with the width and shift of the Raman peak of the μc-Si:H film. The Raman spectrum of crystalline Si is dominated by a sharp feature at 520 cm^{-1} while the spectrum of a-Si:H displays features which resemble the broadened density of vibrational states of c-Si because the momentum conservation rule is relaxed by the structural disorder. There have been several detailed analyses of the evolution of the Raman spectra as a function of crystallite size [28,29,31]. For the case of crystalline Si, as the crystalline domain size decreases, the Raman peak broadens and shifts to lower frequency [28,29,31]. Campbell and Fauchet calculated the width and shift of the TO Raman peak as functions of the microcrystal size for different shapes ; sphere, column and slab [31]. A good fit to experimental results was obtained in the result of the calculation for spherical crystals, which seems to be consistent with TEM images by Lucovsky et al.[40] and Otobe et al.[41], but a little bit different from the result of Y. He et al. where an elongated shape of microcrystals is concluded [42].

From a shift of the X-ray diffraction peaks corresponding to the (111) and (220) lines it has been concluded that the spacing of each lattice plane is elongated along the normal direction [23]. Iqbal and Veprek discussed the effect of the lattice expansion on the frequency shift of the TO Raman peak on the basis of the second order 2TA scattering peak, and concluded that the shift is not dominated by the expansion of the Si lattice but rather by the effect of finite crystal size [28].

Volume Fraction of Microcrystals (Xc)

The volume fraction of microcrystallites X_c has been determined by X-ray diffraction [28], Raman scattrering [30,32], or spectroscopic ellipsometry [26,27, 43], indicating that X_c varies from a few percent to nealy 100 % depending on deposition conditions as well as post-treatments such as thermal annealing [28] or hydrogen-plasma treatments [26,27].

Tsu et al. used Raman scattering to deduce X_c for the highly phosphorous-doped μc-Si:F:H alloys ; they determined X_c from the ratio of an integrated intensity of the TO Raman peak at 520 cm^{-1} (c-Si) to that of a total integration including the broad peak located at around 480 cm^{-1} (a-Si), using the experimentally-determined ratio ($y=0.88$) of the integrated Raman cross section for c-Si to a-Si [30]. However, as they pointed out, the value of y strongly depends on the grain size δ when δ becomes larger and, therefore, a care should be taken if one wants to employ Raman scattering spectra for determining the volume fraction of microcrystals. Nemanich et al. calculated the relationship between the Raman peak intensity ratio and the volume ratio of c-Si to a-Si as a function of the crystalline domain size [32]. According to their conclusions, because of the large difference in the optical absorption of the crystalline and amorphous domains, the relative fraction of the two regions cannot be deduced unless the domain size has been determined ; however, the model will break down below around 50-nm domain size and in the limit of very small domains y tends to unity [32].

Collins et al.[44] and Drevillon et al.[27] introduced real-time spectroscopic ellipsometry for characterizing thin films and surfaces with monolayer sensitivity during deposition, being followed by many groups in the field of μc-Si:H films [16, 26, 27]. This technique provides information not only on layer thicknesses of surface roughness layers as well as bulk layers but also on microstructures such as void volume fraction and X_c, and also various electronic structures [26]. Although the obtained information is essentially averaged on the basis of the Bruggeman effective medium approximation, the thickness of the surface roughness layer and its time evolution determined by this technique are consistent with the results of STM measurements [12].

Hydrogen in the Film

Infrared absorption spectra due to the Si-H stretching vibrational modes located around 2100 cm^{-1} show that the bonded-hydrogen content C_H of μc-Si:H films varies from a few atomic percent up to about 15 at.% depending on deposition conditions, and that hydrogen atoms are bonded to Si atoms mainly in a dihydride (SiH_2) configuration [5, 33]. It has been argued by Wagner et al. on the basis of the classical nucleation theory that hydrogen, swept out of the recrystallized volume, stabilizes the Si microcrystal nucleus by decorating its surface [45]. This statement is indirectly supported by a strong correlation between C_H and δ, namely, the grain size decreases with an increase of the bonded hydrogen content [38]. Overhof and Otte have also speculated from the electronic transport properties that hydrogen is tremendously efficient in μc-Si:H in removing defects at grain boundaries [46]. More direct evidence for the absence of hydrogen in the bulk of microcrystals has been obtained from the pulsed ESR experiments on μc-Si:H films [37]. Two different ESR signals have been observed in μc-Si:H; g=1.998 which has been ascribed to conduction electrons in the bulk of microcrystals, and g=2.005 coming from dangling bonds in amorphous tissue and/or grain boundaries, respectively [36]. Zhou et al. made an electron-spin-echo-envelope modulation (ESEEM) study of the pulsed ESR, from which a spatial relationship between each spin center and near-by nuclear spins was discussed. It is tentatively cocluded that hydrogen atoms are separated from conduction electrons with g=1.998 by at least 9 - 10 A while hydrogen is located at a distance of 4 - 5 A from dangling bonds with g=2.005 [37], indicating that hydrogen is almost absent in the bulk of Si microcrystallites.

Spatial distribution of hydrogen can be determined by ^1H NMR studies, and it is well known that the ^1H NMR consists of two components even in device-quality a-Si:H films, a broad Gaussian line with an FWHM of 20-30 kHz (clustered hydrogen) and a narrow Lorentzian line with an FWHM of 2-4 kHz (diluted hydrogen). Hayashi et al. performed detailed ^1H NMR studies on several different samples; a-Si:H(L) deposited from pure SiH_4 under a low rf power (0.025 W/cm^2), μc-Si:H from H_2-diluted SiH_4 under a high rf power (0.5 W/cm^2), and a-Si:H(H) from H_2-diluted SiH_4 under a higher rf power (1.5 W/cm^2). The results are shown in Fig. 1. The ^1H NMR spectrum of a-Si:H(L) is basically the same as the well-known spectrum of the device-quality a-Si:H, while other two spectra of μc-Si:H and a-Si:H(H) show a much narrower component down to 0.5-0.7 kHz of FWHM, which has been ascribed to moving hydrogen atoms in the network with respect to $(SiH_2)_n$ chain structures [34, 35]. Another characteristic feature common to μc-Si:H and a-Si:H(H), as shown in the figure, is the absence of a narrow line with a linewidth of about 4 kHz (diluted H). Consequently, micrscopic structure models for the above three samples have been proposed, as is schematically shown in Fig. 2 [34]. Moving hydrogen, perhaps related with $(SiH_2)_n$ chain in a low-density region, does exist in μc-Si:H as well as in a-Si:H films prepared under the high rf power.

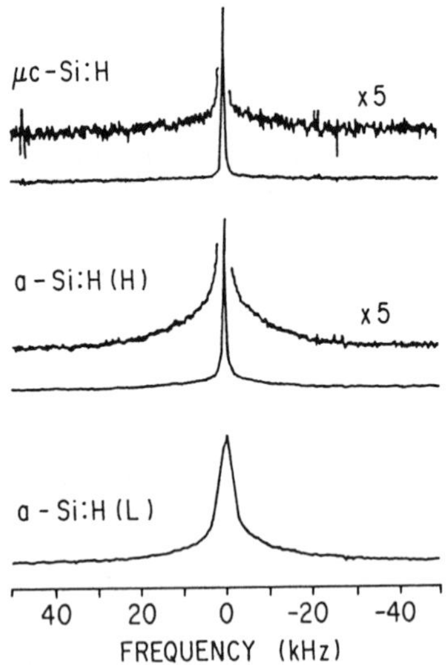

Fig. 1. ^1H NMR spectra for μc-Si:H and two different a-Si:H samples; a-Si:H(L) deposited under a low rf power and a-Si:H(H) under a very high rf power (after Hayashi et al.[35]).

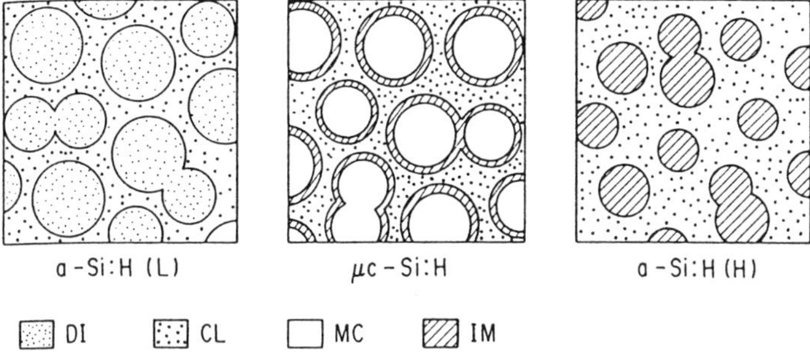

Fig. 2. Schematics of microscopic structure models for a-Si:H(L), a-Si:H(H) and μc-Si:H. DI and CL contain "diluted" and "clustered" hydrogen, respectively. MC means microcrystalline, and IM represents a region containing movable hydrogen (Tanaka and Hayashi [34]).

GROWTH MECHANISMS

As is described in the previous section, μc-Si:H films are highly inhomogeneous and their properties vary in a wide range depending on deposition condiitons. Obviously, deeper understanding of the growth mechanism of these materials will be needed for more precise control of structural, optical and electronic transport properties.

Several growth models have been proposed and discussed for μc-Si:H films grown via a PECVD process. One model describes the growth of the film in terms of a partial chemical equilibrium between deposition and etching of SiH_x species [47]. Preferential etching of amorphous phase may be categorized as one variation of the above concept [48]. Another model emphasizes the surface diffusion of precursors enhanced by hydrogen coverage of the growing surface, which causes the formation of microcrystallites [20,49]. In a more realistic model including a subsurface region H atoms diffuse into a growth zone of around several ten-nm thickness and mediate structural transformation from amorphous to crystalline phase [50]. Those models do not necessarily contradict with each other but, perhaps, one mechanism predominates under some particular conditions and others will stand out when the deposition conditions are changed. Actually, detailed experimental evidences were repoted supporting not only etching by atomic hydrogen [21,26] but also enhanced surface diffusion of radicals by hydrogen coverage [19,20]. In any models, however, hydrogen is a key element for the formation of μc-Si:H.

H atoms and H ions in Gas Phase

What is the role of H atoms and H ions (H^+) in gas phase? It is well known that a combination of high H_2-dilution of SiH_4 and high rf power in PECVD results in the formation of μc-Si:H films [38]. Gas-phase analyses by optical emission spectroscopy (OES) have indicated that a relative amount of atomic hydrogen with respect to Si-related radicals increases as an rf power increases [38,51]. So, phenomenologically, a large amount of hydrogen onto the growing surface favors the formation of μc-Si:H films [38]. Scheib et al. also have confirmed that the SiH_4 plasma by ECWR produces μc-Si:H when the OES intensity ratio of [SiH] to [Hα] is quite low [14]. A more microscopic role of atomic hydrogen will be decribed later.

On the other hand, H^+ ions have a tendency to prevent the nucleation of μc grains. As the amount of H ions reaching to the substrate increases, the size δ of microcrystallites in μc-Si:H films gradually decreases and, finally, the structure becomes amorphous [38]. Matsuda et al., using the triode reactor, intentionally controlled the grain size δ of μc-Si:H films in a wide range by changing the amount of impinging H ions without giving any disturbance to the main plasma [11]. Quite recently, Kondo et al. have unequivocally demonstrated that the crystallinity of μc-Si:H films is considerably affected by the bombardment of positive ions [52]. A μc-Si:H layer of 50 nm was deposited on the Al-evaporated-glass substrate where some parts of Al were floating and the others were electrically grounded. As shown in Fig.3, the crystallinity of the films on the grounded Al islands, evaluated by Raman scattering, is much better than that on the floating parts. The potential difference between the plasma and the substrate is enlarged for the insulating or floating substrate and thereby increases the kinetic energy of the impinging ions, which is the reason why the crystallinity is degraded [52].

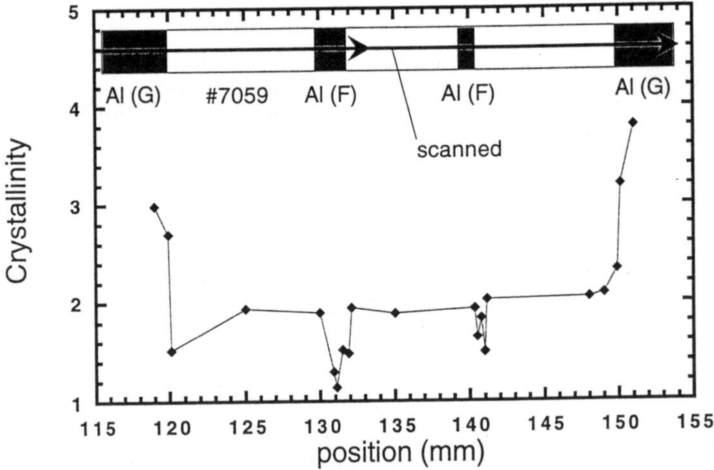

Fig. 3. Position dependence of the crystallinity for films of 50 nm on Al-evaporated glass substrate. The laser spot was scanned accross the Al islands. Al(G) and Al(F) represent an electrically-grounded island and a floated island, respectively. The crystallinity was estimated from Raman scattering spectra (after Kondo et al.[52]).

Substrate Effects

Persons et al. introduced the gas-flow modulation technique in PECVD and showed that time-modulated flow of silane into a hydrogen plasma allowed for substrate selective nucleation and enhanced crystallinity of microcrystalline silicon films [21,53,54]. It is concluded that more strongly bound nuclei are formed on crystalline substrates (sapphire vs. SiO_2) or those that readily form silicides (Mo vs. Al), and also that crystalline Si is an ideal substrate but the growth rate slowly decays with increasing hydrogen exposure, reflecting the ability of hydrogen to eliminate even bulk-like Si-Si bonds [53,54]. Ikuta et al. made STM and high-resolution TEM observations on PECVD-Si:H films on H-termunated Si(111) and plasma-oxidized SiO_2/Si(111) surfaces [55]. STM observation revealed that, on both surfaces, initial growth processes proceed through the nucleation-and-coalescence mechanism [12], although the structure of the Si:H film deposited on the ultrathin SiO_2 film is amorphous up to 50-nm thickness while μc-Si:H is epitaxially grown on the H-termunated Si(111) substrate in the initial stage [55].

Direct Conversion of Amorphous to Crystalline Phase

Conversion of the near-surface layer of a-Si:H to a μc-Si:H phase by the exposure to hydrogen plasma has been confirmed by several groups [21,22,26]. Nguyen et al. investigated the etching of a-Si:H films in thermally-generated (hot filament) atomic hydrogen in order to avoid the influence of ionic species, and observed the formation of ultrathin μc-Si:H films of relatively high density in the near-surface region using real-time spectroscopic ellipsometry [26]. During etching, the μc-Si:H surface layer is stable at 1.3 nm, representing an equilibrium between loss from the surface, and crystallization of deeper a-Si:H. Therefore, different from the direct nucleation of crystallites on the substrate, the above etching process can be used for the formation of thin μc-Si:H films on any substrate materials [26].

Roca i Cabarrocas et al. demonstrated also using real-time spectroscopic ellipsometry on the Layer-by-Layer (LBL) experiments that even a thick a-Si:H layer of 8.5 nm can be transformed into the crystalline phase, and that the nucleation of the crystalline phase takes place only after the initially dense a-Si:H film is converted to a porous a-Si:H by the hydrogen plasma treatment [22]. Similar phenomena have been observed by Yang et al. in the formation process of μc-Si:H films via the reactive magnetron sputtering technique [56].

What Happens in the Subsurface under Hydrogen Plasma?

Toyoshima et al. measured the hydrogen coverage of the growing surface of a-Si:H films under the SiH_4 plasma using real-time infrared reflection absorption spectroscopy [57]. The dominant bonding configuration of surface hydrogen changes from SiH_3 to SiH_2 and to SiH with increasing temperature but the absorption intensity of the hydrogen coverage is found to be compatible with the monolayer coverage scheme up to 380°C.

Recently, Beyer and Zastrow have discussed incorporation and kinetics of hydrogen during plasma post-hydrogenation and thermal treatment for a-Si:H and a-Ge:H films [58]. It has been demonstrated that the hydrogen diffusion coefficient and hydrogen solubility are much enhanced in a-Si:H during the hydrogen plasma treatment at a temperature lower than 400° C, which has been attributed to a plasma-induced rise of the surface hydrogen chemical potential [58]. Enhancement of the hydrogen diffusion coefficient under the hydrogen plasma was also observed by Santos et al [59].

Another new experiment using *in situ* ESR measurements has been made quite recently by Yamasaki et al [60] on a growing a-Si:H film placed in the ESR cavity. In this experiment the Si dangling-bond signal (g=2.0055) during and after deposition has been detected, in addition to the gas-phase ESR signals both of atomic hydrogen and SiH_x molecules. Furthermore, Si dangling-bond signal from the subsurface region under hydrogen plasma was also observed, being separated from the signal from the bulk region of the a-Si:H film by taking into account a difference in structural relaxation times between the subsurface and the bulk regions. Fig. 4 shows the Si dangling-bond ESR intensity as a function of elapsed time after stopping the film deposiiton [61]. The dangling-bond density gradually decreases due to a room-temperature relaxation of the network structure, but suddenly increases when the hydrogen plasma is ignited. As shown in the figure, this excess signal decays in a relatively short time after turning off the hydrogen plasma and the signal intensity returns to the initial level. It indicates that dangling bonds are created dynamically in the subsurface region when it is exposed to the hydrogen plasma because a top surface is basically covered by hydrogen. Dynamical concentration of dangling bonds in the subsurface induced by the hydrogen plasma was tentatively estimated to be of the order of 10^{19} cm^{-3} assuming a 1-nm thickness of the subsurface. Details will be published in the near future.

Consequently, hydrogen plasma produces dynamically a higher concentration of dangling bonds as well as hydrogen atoms with a higher diffusion coefficient in a subsurface region, resulting in a lower network connectivity due to the presence of high density of Si-H and Si dangling bonds and also a higher reaction rate of hydrogen with strained Si-Si bonds. In other words, weaker structural constraint and higher frequency factor can be expected for structural transformation from a dense to a porous a-Si:H network and subsequent nucleation of microcrystals. Similar pictures have been proposed by several groups [16,21,62].

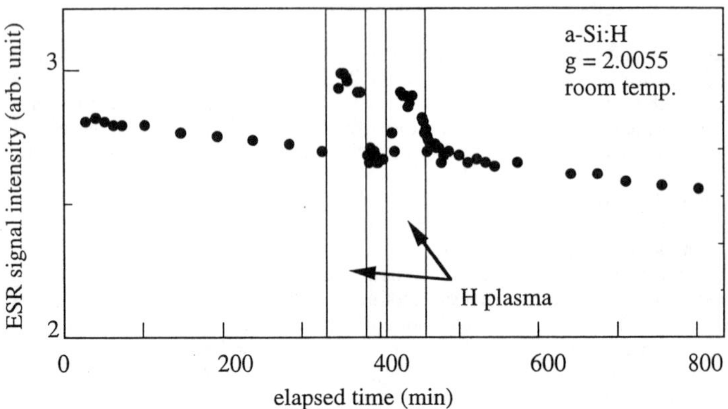

Fig. 4. ESR (g=2.0055) intensity as a function of elapsed time after stopping deposition. The intensity increases when the surface is exposed to the hydrogen plasma (after Yamasaki et al.[61]).

OPTICAL PROPERTIES

Optical absorption spectra of μc-Si:H films were reported by many groups, which were determined by optical transmission, photothermal deflection spectroscopy (PDS) and constant photocurrent method (CPM) [63-65]. PDS and CPM are very efficient for determining optical absorption coefficients in a range lower than 10^4 cm^{-1} although special care shoud be taken when CPM is employed. CPM assumes a constant $\eta\mu\tau$ product, but the photoconductivity of these materials shows a strong dependence on their film thickness due to the structural inhomogeneity along the growth directon and also the $\eta\mu\tau$-product is no longer constant below $h\nu$=1.4 eV [64].

Sub-bandgap absorption region of $h\nu < 1$ eV is mainly cotrolled by impurities and/or defects such as Si dangling bonds. Materials prepared by the high frequency GD technique at 70 kHz show a very low absorption coefficient down to 1 cm^{-1} at 0.8 eV [65]. Absorption coefficients in this photon energy range increases after hydrogen is evolved out of the film by thermal annealing perhaps due to creation of new dangling bonds [65].

Optical absorption in a bandgap energy range of μc-Si:H films (1 eV< $h\nu$ < 1.6 eV) is different from that of a-Si:H films and of c-Si. However, for the case of the material with a high volume fraction X_c of microcrystallites, its absorption spectrum $\alpha(h\,\nu)$ follows a square root dependence on photon energy, which is characteristic for single crystalline Si [64,65]. Therefore, this absorption edge can be attributed to the indirect optical transition with the assistance of phonons although the gap energy, obtained by extrapolation, is slightly shifted to a lower energy side compared to 1.08 eV of c-Si [65]. This apparent red shift of the edge might be ascribed to an enhanced optical absorption compared to c-Si, which was observed by several groups [63-65]. Internal light scattering effects which are generally present in inhomogeneous materials, lattice strain effect, and relaxation of the selection rule have been tentatively proposed as possible reasons for the enhanced optical absorption, but the quantum confinement effect may be excluded because the grain size of microcrystallites in the measured samples is around 200 A [65].

Absorption coefficients in the above-gap absorption region of $h\nu > 1.8$ eV is higher than in c-Si but considerably lower than in a-Si:H. However, a simple superposition of a c-Si phase and an a-Si:H phase according to each volume fraction failed to explain the experimental data [65]. This region is sensitive to the hydrogen content C_H and, actually, the absorption coefficient increases as hydrogen evolves out by thermal annealing [65]. The increased absorption in this energy range might be related with Si-H bonds located at the grain boundary.

Regarding quantum confinement effects in μc-Si:H films no systematic studies have been reported. According to theoretical works [66,67] , zero-phonon transitions, which are allowed due to the finite-size effect, are more important for a crystallite size smaller than 1.5-2.5 nm. In relation to this theoretical prediction, Nguyen et al. observed a systematic blue shift of the optical absorption edge in ultrathin μc-Si:H-cluster films in a thickness range of 0.6-1.25 nm, which has been ascribed to quantum confinement effects [68]. Liu et al. reported the room-temperature photoluminescence peaked at 1.8 eV from μc-Si:H films consisting of nanometer-size crystals, δ=2.8 nm with X_c=5-10 %, embedded in a large fraction of amorphous tissue [69]. However, a more detailed study will be needed for further discussions on quantum confinement effects.

ELECTRONIC TRANSPORT

Electronic transport properties of μc-Si:H films vary in a wide range depending strongly on their structural properties such as δ and X_c, and also on doping levels as well as temperature. For example, dark conductivities as well as Hall mobilities of undoped and P-doped specimens strongly depend on X_c [30,38], suggesting the existence of the percolation threshold [30,46,70]. Finger et al. reported that, in case X_c is close to unity, the Hall mobility of P-doped μc-Si:H films increases with an increase of δ [71]. The dark conductivity of heavily-doped samples shows a temperature dependence of thermal-activation type in a high-temperature range, namely, expressed by $\sigma_d = \sigma_0 \exp(-E_\sigma/kT)$, while σ_d is less sensitive to temperature in a low-temperature range and activated with not a single activation energy[71]. The room-temperature conductivity of phosphorus-doped μc-Si:H films increases drastically from 10^{-4} Scm^{-1} to 1 Scm^{-1} for a change of a doping level from 10^{-5} to 10^{-2} [5], indicating that the Fermi level is not pinned. Doping properties really demonstrate that defects at the grain boundaries are efficiently passivated by hydrogen [46].

It should be noted that transport data of these materials are considerably different from those of pure c-Si and a-Si:H in spite of the mixed-phase structure of both materials; for example, the double sign anomaly in the Hall mobility, which is characteristic for a-Si:H, is absent in μc-Si:H films, but the magnitude of the mobility is of the order of 1 $cm^2/Vsec$, being lower than that of c-Si at least by two orders of magnitude and comparable to the absolute value of the mobilty of doped a-Si:H [71]. Electron and hole drift mobilities of undoped μc-Si:H films are also in a range of 0.5-3 $cm^2/Vsec$ [65].

However, as far as undoped and lightly-doped μc-Si:H films are concerned, transport data can be decribed well within a homogeneous model in spite of their structural inhomogeneity. As is well known for a-Si:H films, the dark-conductivity prefactor σ_0 and dark-conductivity activation energy E_σ obey an empirical relationship which is called as the Meyer-Neldel rule (M-N rule), represented by $\sigma_0 = C \exp(E_\sigma/E_0)$ [72,73]. Data for undoped and doped a-Si:H can be fitted with the constants, $C = 2.5 \pm 1$ S/cm, and $E_0 = 0.12 \pm 0.02$ eV, respectively [74]. Also the observed prefactors and activation energies for undoped and lightly-doped μc-Si:H films were found to obey the M-N rule with the same constants as those for a-Si:H films [74,75].

In order to provide a theoretical basis for the empirical M-N relationship, Overhof et al. proposed a model in terms of the statistical shift of the Fermi energy, derived from the band-gap density of states of a-Si:H under the charge neutrality condition [72,76]. This model also predicts an "inverted" M-N relationship with a negative value of E_0 in case the Fermi level is moved deeply into the band-tail states, which corresponds to $E_\sigma < 0.2$ eV and does not occur for even the most heavily-doped a-Si:H. Recently, Kondo et al. observed the inverted M-N behavior with $E_0 = -15$ meV in undoped a-Si:H using a thin film transistor structure with Mg source/drain electrodes, which seems to support the above prediction [77].

In μc-Si:H films also, data for heavily-doped samples with $E_\sigma < 0.2$ eV display the inverted M-N relationship, with $E_0 = 305 \pm 10$ S/cm and a negative value of $E_0 = -20 \pm 5$ meV [74], which is close to the value of E_0 obtained for a-Si:H by Kondo et al. Therefore, phenomenologically, high-temperature dc conductivity of μc-Si:H films appears to be quite similar to that of a-Si:H samples in spite of distinct differences in other properties between the two. Lucovsky and Overhof have explained the inverted M-N behavior of heavily-doped μc-Si:H films by the statistical shift of the Fermi Energy assuming that the films are comprised of doped Si crystallites encapsulated in doped a-Si:H [74].

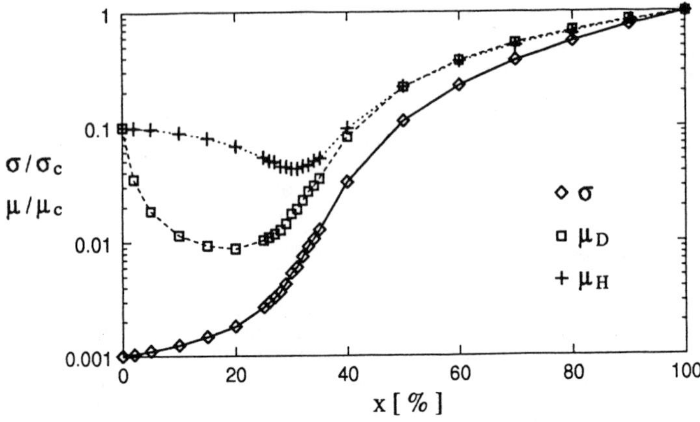

Fig. 5. Results of conductance network calculations : network conductivity σ, the mean drift mobility μ_D, and the Hall mobility μ_H for $\sigma_c = 1$ and $\sigma_a = 10^{-3}$ for a sixfold coordinated network (after Overhof and Otte [46]).

Overhof and Otte, quite recently, have made conductance network calculations for heavily-doped μc-Si:H films in order to explain a big difference in transport data between the highly-doped and lightly-doped cases; i.e., why the conductivity of heavily-doped samples is so sensitive to the crystallinity but less sensitive to temperature in comparison to lightly-doped ones [46]. The calculaton is based on "two-phase inhomogeneous model" consisting of highly conductive crystalline cubes and low-conductance amorphous cubes, and the conductivity of the model system was determined by averaging conductances calculated for a set of random numbers. They also calculated the Hall voltage in a similar way [46]. Fig. 5 shows the results of the calculation. Network coductivities σ calculated for $\sigma_c = 1$ and $\sigma_a = 10^{-3}$ for a sixfold coordinated network are plotted as a function of the volume fraction X_c of crystalline cubes, and the mean drift mobility μ_D and the Hall mobility μ_H are also shown.

As shown in the figure, σ changes steeply around at $X_c = 0.3$ that is close to the percolation threshold of this system and will show a more sharp transition if the ratio of the conductivity σ_c/σ_a is very large. Magnitudes of μ_D and μ_H near the percolation threshold are much lower than the microscopic mobility of c-Si, which is consistent with the experimental observations. The intricacies of the dominant current path and internal shorts may lead to a reduction of μ_D and μ_H, respectively, while the crystalline cubes that do not contribute much to the dc current in a range $X_c < 0.3$ will reduce the macroscopic drift mobility μ_D [46]. Consequently, this calculation demonstrates that sample inhomogeneity causes a reduction of the current by essentially geometrical constraints, resulting in a decrease in μ_D as well as μ_H that no longer reflects the microscopic mobility of highly conducting crystalline grains. More detailed and systematic studies are needed for further discussions.

CONCLUDING REMARKS

Microcrystalline semiconductor thin films are structurally inhomogeneous, and, as a result, not so easily characterized. However, much progress has been made in understanding the growth mechanism of μc-Si:H films during the past decade. Optical, electronic transport properties as well as their size effects have not yet been sufficiently investigated, although a strong demand for the materials does exist towards device applications such as solar cells and flat panel displays. At the Semiconductor Conference in Exeter in 1962, Sir Nevill Mott predicted that in the 1970's much of the excitement in solid state physics would come from challenges in technology. We believe this still remains true towards the 21st century.

ACKNOWLEDGMENTS

The author is grategul to Prof. H. Overhof (Paderborn) and Prof. S. Wagner (Princeton) for their valuable comments and discussions. He also would like to thank Drs. Roca i Cabarrocas(Palaiseau), N. Beck (Neuchatel), C. Malten(Julich), W. Beyer(Julich), H. Overhof for sending me preprints of their papers prior to publication. He gratefully acknowledges the collaboration of many colleagues at NAIR and ETL, particularly, S. Yamasaki, K. Ikuta, M. Kondo and A. Matsuda.

REFERENCES

1. S. Veprek and V. Marecek, Solid-State Electronics **11**, 683 (1968).

2. W.E. Spear and P.G. LeComber, Solid State Commun. **17**, 1193 (1975).

3. S. Usui and M. Kikuchi, J. Non-cryst. Solids **34**, 1 (1979).

4. A. Matsuda, S. Yamasaki, K. Nakagawa, H. Okushi, K. Tanaka, S. Iizima, M. Matsumura and H. Yamamoto, Jpn J. Appl. Phys. **19**, L305 (1980).

5. K. Tanaka, K. Nakagawa, A. Matsuda, M. Matsumura, H. Yamamoto, S. Yamasaki, H. Okushi and S. Iizima, Jpn J. Appl. Phys. **20**, Suppl. 20-1, 267 (1981).

6. Materials Issues in Microcrystalline Semiconductors, edited by P.M. Fauchet, K. Tanaka, and C.C. Tsai (Mater. Res. Soc. Proc. **164**, Pittsburgh, PA 1990); Amorphous & Microcrystalline Semiconductor Devices Vol.II : Materials and Device Physics, edited by J. Kanicki (Artech House, Norwood, MA 1992); Microcrystalline Semiconductors : Materials Science & Devices, edited by P.M. Fauchet, C.C. Tsai, L.T. Canham, I. Shimizu, and Y. Aoyagi (Mater. Res. Soc. Proc. **283**, Pittsburgh, PA 1993); Microcrystalline and Nanocrystalline Semiconductors, edited by R.W. Collins, C.C. Tsai, M. Hirose, F. Koch, and L. Brus (Mater. Res. Soc. Proc. **358**, Pittsburgh, PA 1995).

7. G.F. Feng, M. Katiyar, Y.H. Yang, J.R. Abelson and N. Maley in ref. [6] (Mater. Res. Soc. Proc. **283**, Pittsburgh, PA 1993), p. 501.

8. J. Hanna, T. Ohuchi and M. Yamamoto, J. Non-cryst. Solids **198-200**, 879 (1996).

9. D. Toet, S. Eitel, P.V. Santos and M. Heintze, J. Non-cryst. Solids **198-200**, 887 (1996).

10. M. Tanaka, S. Tsuge, S. Kiyama, S. Tsuda and S. Nakano (this volume).

11. A. Matsuda, K. Kumagai and K. Tanaka, Jpn J. Appl. Phys. **22**, L34 (1983).

12. K. Ikuta, Y. Toyoshima, S. Yamasaki, A. Matsuda and K. Tanaka, J. Non-cryst. Solids **198-200**, 863 (1996).

13. P. Hapke, F. Finger, M. Luysberg, R. Carius and H. Wagner, in ref. [6] (Mater. Res. Soc. Proc. **358**, Pittsburgh, PA 1995), p. 745 : F. Finger, R. Carius, P. Hapke, L. Houben, M. Luysberg and M. Tzolov (this volime).

14. M. Scheib, B. Schroder and H. Oechsner, J. Non-cryst. Solids **198-200**, 895 (1996).

15. N. Imajyo, J. Non-cryst. Solids **198-200**, 935 (1996).

16. Y. Miyamoto, J. Miita and I. Shimizu (this volume).

17. G. N. Parsons, D.V. Tsu and G. Lucovsky, J. Non-cryst. Solids **97&98**, 1375 (1987).

18. J.K. Rath, K.F. Feenstra, D. Ruff, H. Meiling and R. Schropp (this volume).

19. A. Asano, Appl. Phys. Lett. **56**, 533 (1990).

20. K. Nomoto, Y. Urano, J.L. Guizot, G. Ganguly and A. Matsuda, Jpn J. Appl. Phys. **29**, L1372 (1990).

21. G.N. Parsons, J.J. Boland and J.C. Tsang, Jpn J. Appl. Phys. **31**, 1943 (1992).

22. P. Roca i Cabarrocas, N. Layadi, B. Drevillon and I. Solomon, J. Non-cryst. Solids **198-200**, 871 (1996).

23. A. Matsuda, T. Yoshida, S. Yamasaki and K. Tanaka, Jpn J. Appl. Phys. **20**, L439 (1981).

24. P. Hapke, M. Luysberg, R. Carius, M. Tzolov, F. Finger and H. Wagner, in Proc. of POLYSE '95 (Gargano, Italy, 1995) (in press).

25. E. Baedet, J.E. Bouree, M. Cuniot, J. Dixmier, P. Elkaim, J. Le Duigou, A.R. Middya and J. Perrin, J. Non-cryst. Solids **198-200**, 867 (1996).

26. H.V. Nguyen, Ilsin An, R.W. Collins, Y. Lu, M. Wakagi and C.R. Wronski, Appl. Phys. Lett. **65**, 3335 (1994).

27. B. Drevillon, I. Solomon and M. Fang in ref. [6] (Mater. Res. Soc. Proc. **283**, Pittsburgh, PA 1993) p. 455.

28. Z. Iqbal and S. Veprek, J. Phys. C : Solid State Phys. **15**, 377 (1982).

29. H. Richter, Z.P. Wang and L. Ley, Solid State Commun. **39**, 625 (1981).

30. R. Tsu, J. Gonzalez-Hernandez, S.S. Chao, S.C. Lee and K. Tanaka, Appl. Phys. Lett. **40**, 534 (1982).

31. I.H. Campbell and P.M. Fauchet, Solid State Commun. **58**, 739 (1986).

32. R.J. Nemanich, E.C. Buehler, Y.M. Legrice, R.E. Shroder, G.N. Parsons, C. Wang, G. Lucovsky and J.B. Boyce, J. Non-cryst. Solids **114**, 813 (1989).

33. A. Hiraki, T. Imura, K. Mogi and M. Tashiro, J. De Physique C4 supplement, Tome 42, C4-277 (1981).

34. K. Tanaka and S. Hayashi in Glow-Discharge Hydrogenated Amorphous Silicon, edited by K. Tanaka (KTK, Tokyo / Kluwer, Dordrecht, 1989) p. 39-70.

35. S. Hayashi, K. Hayamizu, S. Yamasaki, A. Matsuda and K. Tanaka, Phys. Rev. **B35**, 4581 (1987): S. Hayashi, K. Hayamizu, S. Yamasaki, A. Matsuda and K. Tanaka, J. Appl. Phys. **56**, 2658 (1984).

36. C. Malten, F. Finger, P. Hapke, T. Kulessa, C. Walker, R. Carius, R. Fluckiger and H. Wagner in ref. [6] (Mater. Res. Soc. Proc. **358**, Pittsburgh, PA 1995) p. 757.

37. J.H. Zhou, S. Yamasaki, J. Isoya, K. Ikuta, M. Kondo, A. Matsuda and K. Tanaka (this volume).

38. A. Matsuda, J. Non-cryst. Solids **59-60**, 767 (1983).

39. W.E. Spear, G. Willeke, P.G. LeComber and A.G. Fitzgerald, J. De Physique C4 Supplement, Tome 42, C4-257 (1981).

40. G. Lucovsky, C. Wang, M.J. Williams, Y.L. Chen and D.M. Maher in ref. [6] (Mater. Res. Soc. Proc. **283**, Pittsburgh, PA 1993) p. 443.

41. M. Otobe and S. Oda in ref. [6] (Mater. Res. Soc. Proc. **283**, Pittsburgh, PA 1993) p. 519.

42. Y. He, Y. Chu, H. Lin and G. Qin in ref. [6] (Mater. Res. Soc. Proc. **283**, Pittsburgh, PA 1993) p. 537.

43. S. Hamma and P. Roca i Cabarrocas, Thin Solid Films (in press).

44. R.W. Collins and B.Y. Yang, J. Vac. Sci. Technol. **B7**, 1155 (1989).

45. S. Wagner, S.H. Wolff and J.M. Gibson in ref. [6] (Mater. Res. Soc. Proc. **164**, Pittsburgh, PA 1990) p. 161.

46. H. Overhof and M. Otte in Proc. of International Symposium on Condensed Matter Physics (ISCMP '96) (Varna, 1996) (in press).

47. S. Veprek, F.A. Sarrot, S. Rambert and E. Taglauer, J. Vac. Sci. Tech. **A7**, 2614 (1989).

48. M. Heintze, W. Westlake and P.V. Santos, J. Non-cryst. Solids **164-166**, 985 (1993).

49. A. Matsuda and T. Goto in ref. [6] (Mater. Res. Soc. Proc. **164**, Pittsburgh, PA 1990) p. 3.

50. K. Nakamura, K. Yoshino, S. Takeoka and I. Shimizu, Jpn J. Appl. phys. **34**, 442 (1995) and references therein.

51. A. Matsuda and K. Tanaka, Thin Solid Films **92**, 171 (1982).

52. M. Kondo, Y. Toyoshima, A. Matsuda and K. Ikuta, J. Appl. Phys. **80**, 6061 (1996).

53. G.N. Persons, Appl. Phys. Lett. **59**, 2546 (1991).

54. J.J. Boland and G.N. Persons, Science **256**, 1304 (1992).

55. K. Ikuta, J.W. Park, L.H.Kuo, T. Yasuda, S. Yamasaki and K. Tanaka (this volume).

56. Y.H. Yang, M. Katiyar, G.F. Feng, N. Maley and J.R.Abelson, Appl. Phys. Lett. **65**, 1769 (1994).

57. Y. Toyoshima, K. Arai, A. Matsuda and K. Tanaka, J. Non-cryst. Solids **137&138**, 765 (1991).

58. W. Beyer and U. Zastrow in Amorphous Silicon Technology - 1996, edited by A. Hack, E.A. Schiff, S. Wagner, A. Matsuda and R. Schropp (Mater. Res. Soc. Proc., Pittsburgh, PA 1996, in press).

59. P.V. Santos, N.M. Johnson and R.A. Street, Phys. Rev. Lett. **67**, 2686 (1991).

60. S. Yamasaki, T. Umeda, J. Isoya and K. Tanaka, Appl. Phys. Lett. (1997, in press).

61. unpublished data, S. Yamasaki et al.

62. I. Shimizu, J. Non-cryst. Solids **114**, 145 (1989).

63. H. Richter and L. Ley, J. Appl. Phys. **52**, 7281 (1981).

64. R. Krankenhagen, M. Schmidt, S. Grebner, M. Poscherieder, W. Henrion, I. Sieber, S. Koynov and R. Schwarz, J. Non-cryst. Solids **198-200**, 923 (1996).

65. N. Beck, J. Meier, J. Fric, Z, Remes, A. Poruba, R. Fluckiger, J. Pohl, A. Shar and M. Vanecek, J. Non-cryst. Solids **198-200**, 903 (1996): N. Beck, P. Torres, J. Fric, Z. Remes, A. Poruba, Ha Stuchlikova, A. Fejfar, N. Wyrsch, M. Vanecek, J. Kocka and A. Shar (this volume).

66. M.S. Hybertsen, Phys. Rev. Lett. **72**, 1514 (1994).

67. L.W. Wang and A. Zunger, J. Phys. Chem. **98**, 2158 (1994).

68. H.V. Nguyen, S. Kim, Y. Lu, M. Wakagi and R.W. Collins, J. Non-cryst. Solids **198-200**, 853 (1996).

69. X-N. Liu, X-W. Wu, X-M. Bao and Y-L. He, Appl. Phys. Lett. **64**, 220 (1994).

70. P. Hapke, F. Finger, R. Carius, H. Wagner, K. Prasad and R. Fluckiger, J.Non-cryst. Solids **164-166**, 981 (1993).

71. F. Finger,P. Hapke, M. Luysberg, R. Carius and H. Wagner, Appl. Phys. Lett. **65**, 2588 (1994).

72. H. Overhof and W. Beyer, Philos. Magazine B **47**, 377 (1983).

73. W. Meyer and H. Neldel, Z. Techn. Physik. **18**, 588 (1937).

74. G. Lucovsky and H. Overhof, J. Non-cryst. Solids **164-166**, 973 (1993).

75. G. Willeke in Amorphous & Microcrystalline Semiconductor Devices Vol.II : Materials and Device Physics, edited by J. Kanicki (Artech House, Norwood, MA 1992) p. 55.

76. H. Overhof and P. Thomas in Electronic Transport in Hydrogenated Amorphous Silicon , Springer Tracts in Modern Physics Vol. 114 (Springer Verlag, Heidelberg, 1989).

77. M. Kondo, Y. Chida and A. Matsuda, J. Non-cryst. Solids **198-200**, 178 (1996).

SEMICONDUCTOR NANOCRYSTALS: EXCITON QUANTUM MECHANICS, SINGLE NANOCRSYTAL LUMINESCENCE, AND METASTABLE HIGH PRESSURE PHASES

M.Nirmal and L. E. Brus
Department of Chemistry, Columbia University, New York, N.Y 10027

ABSTRACT

We review three areas where significant progress has recently occurred in our understanding of semiconductor nanocrystals. The first two involve luminescence properties of single and ensembles of Cadmium Selenide (CdSe) nanocrystallites (Quantum Dots) between 10 and 50 Å in radius. The size, magnetic field, and temporal dependence of emission from ensembles of nanocrystallites at cryogenic temperatures uncovers the fundamental mechanism of radiative recombination in these nanocrystals. Effective mass models that take into account the electron-hole exchange interaction can quantitatively account for observed luminescence Stokes shifts. Furthermore, the magnetic field dependence of luminescence lifetimes and longitudinal-optical (LO) phonon ratios demonstrate that the exciton ground state in these nanocrystals is optically passive ("dark exciton") with spin projection ±2. Picosecond time resolved measurements probe exciton relaxation into this level. Recent results on the spectroscopy of single CdSe nanocrystals at room temperature are also presented. Remarkably, emission from a single CdSe nanocrystal under C.W illumination is observed to turn on and off discretely (fluorescence intermittency) on a ~0.5s timescale. The excitation intensity dependence, and the influence of a passivating high band gap shell of Zinc Sulfide (ZnS) encapsulating the CdSe nanocrystal on the on/off times, suggest that this phenomenon is caused by photoionization. Finally, the third area originates in diamond anvil studies of the solid-solid phase transitions of nanocrystals under pressure. These studies show that a single nucleation event occurs per nanocrystal, and that as a consequence the nanocrystals change shape. The kinetic activation barrier increases with increasing size. Under suitable conditions nanocrystals in dense, six-coordinate high pressure phases may be metastable at STP.

A NANOCRYSTAL EXCITON QUANTUM MECHANICS

Semiconductor nanocrystals offer the opportunity to explore the evolution of bulk electronic and structural properties as the size of the system increases from the molecular scale. In addition, their strongly size-dependent optical properties render them attractive candidates as tunable light absorbers and emitters in optoelectronic devices such as light emitting diodes and quantum-dot lasers. New fabrication methods have enabled the synthesis of highly monodisperse ($\sigma_R < 4\%$) CdSe nanocrystals with radii tunable between 10 and 50 Å, which luminesce with high quantum yield (~0.1 to 0.9 at 10 K).[1] While the absorption spectra of these dots are now fairly well understood, the nature of the emitting state remains controversial. Even high quality samples exhibit unusually long radiative lifetimes ($\tau_R \sim 1\mu s$ at 10 K)[2,3] relative to the bulk exciton recombination time ($\tau_R \sim 1$ ns).[4] Since these radiative rates were expected to be comparable, band edge emission in II-VI nanocrystals was rationalized as a "surface effect" and assigned to the recombination of weakly overlapping surface-localized carriers.[2,3,5] However new magnetic field data implicates exciton spin dynamics in the recombination mechanism.[6] In CdSe nanocrystals, the electron-hole exchange interaction, enhanced by quantum confinement, strongly modifies the

Mat. Res. Soc. Symp. Proc. Vol. 452 © 1997 Materials Research Society

band edge exciton structure.[6,7] Exciton thermalization to the lowest state, which is optically forbidden, quantitatively explains the long lived emission, luminescence stokes shift, and the magnetic field dependence.

Nanocrystals of CdSe were produced from the pyrolysis of dimethylcadmium and tri-n-octyl phosphine selenide (TOPSe) in a solution of 50% tri-n-octyl phosphine (TOP) and 50% tri-n-octyl phosphine oxide (TOPO) according to the method of Murray et al.[1] This method produces CdSe nanocrystals with the surface Cd atoms passivated with TOPO and TOPSe molecules. The nanocrystals were studied dispersed in the growth medium (TOPO/TOP). Samples were placed in either an optical helium vapor flow or a cold finger cryostat to obtain low temperature spectra. Fluorescence spectra were recorded using a nanosecond Nd:YAG/dye laser system as the excitation source. The luminescence was dispersed through a 0.66 single spectrometer and detected by a time gated optical multichannel analyzer (OMA) to obtain spectra. Emission decays were recorded using a 500-MHz digitizing oscilloscope, a photomultiplier tube and a 0.75 m subtractive double spectrometer in order to eliminate scattered laser light. Magnetic field dependent studies were conducted using a superconducting magnet capable of attaining fields up to 10 Tesla. Picosecond time resolved measurements were conducted using a dye laser sync pumped by the third harmonic of a mode locked Nd:YAG laser followed by standard time correlated single photon counting detection (time resolution ~70 ps).

We begin with a brief description of the band edge exciton structure for CdSe quantum dots.[6,7] We extend effective mass models used previously to describe their absorption spectra. For spherical dots, the lowest state ($1S_{3/2}1S_e$), which has the electron and hole in 1S particle in a sphere orbitals, is eightfold degenerate in the spherical band approximation. However, anisotropy in the internal crystal structure and/or shape of the dots lifts this degeneracy. In our nanocrystals, which are prolate with a wurtzite crystal structure, the band edge state is split into two fourfold degenerate levels analogous to the bulk "A-B splitting". The net splitting is the sum of the internal crystal field and shape asymmetry components. The band edge state is further split by the electron-hole exchange interaction. Since this term is proportional to the spatial overlap between the electron and hole, an enhanced exchange effect is predicted in quantum dots due to the three dimensional confinement of the electron and hole. The anisotropy and exchange terms are included within the framework of perturbation theory. The initially eightfold degenerate band edge exciton ($1S_{3/2}1S_e$) is split into five states labeled by the total exciton angular momentum projection, $F_m=m_e+m_j$, where $m_{e(j)}$ is the electron(hole) spin projection.

The order of the levels for our nanocrystals is indicated Fig.1a. These states can be classified into two groups which converge to the bulk A and B excitons: $F_m=\pm2$, $\pm1^L$ and $F_m=0^L$, $\pm1^U$, 0^U, where the superscripts U and L distinguish upper and lower states with the same angular projection. As can be seen in Fig. 1a, the states fan out with decreasing size due to enhancement of the exchange term. The calculated exchange splitting between the ±2 and $\pm1^L$ states in the smallest dots (~12.5 meV) is two orders of magnitude larger than the corresponding bulk value (~0.13 meV). A similar enhancement has been suggested in porous silicon.[8] The oscillator strengths of the optically active levels are also strongly modified by the exchange term and change dramatically with size. Both the ±2 and 0^L levels are optically passive within the electric dipole approximation.

We find strong evidence for this band edge structure in our photoluminescence results.[6,7] In spite of the extremely narrow size distributions of our samples (σ_R<4%), the features seen in absorption exhibit significant inhomogeneous broadening due to the strong size dependence of the excitonic levels within the strong confinement regime. Excitation on the red edge of the sample absorption reduces residual sample inhomogeneities by optically selecting the "largest" crystallites in the distribution. In the resulting fluorescence line narrowing (FLN) spectrum, a well resolved

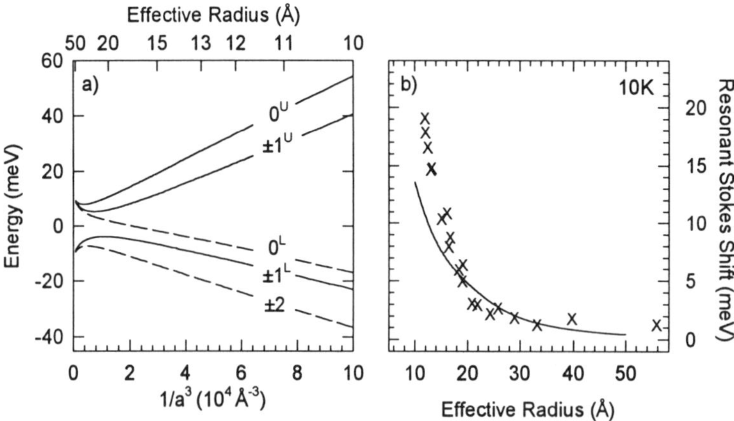

Fig. 1 a) Calculated band edge exciton structure vs $1/a^3$ for our CdSe quantum dots. b) Size dependence of the experimental Stokes shift (X) between the pump position and the peak of the zero LO phonon line compared with the calculated size dependent splitting between ±2 and $\pm1^L$ states from a).

LO phonon progression is observed. The Stokes shift between the excitation energy and the peak of the zero LO phonon line (ZPL) for nanocrystals between ~12 and 56 Å in radius is shown in Fig. 1b. This Stokes shift is strongly size dependent, ranging from ~20 to ~1 meV for this size series. According to the predicted band edge structure (Fig. 1a) excitation on the red edge of the absorption spectrum probes the lowest optically active level ($\pm1^L$). The luminescence is then Stokes shifted due to efficient relaxation to the band edge ±2 state. In Fig. 1b the data are compared with the calculated size dependent splitting between the $\pm1^L$ and ±2 states. We obtain good agreement with no adjustable parameters. In the smallest crystallites, however, the exchange splitting underestimates the observed FLN Stokes shift. This discrepancy may be due to an additional vibrational component since exciton-acoustic phonon coupling is predicted to scale as $1/a^2$.

The presence of an optically passive band edge state strongly affects the dynamics of electron-hole recombination.[6] In Fig. 2a we show magnetic field dependent luminescence decays for 12 Å radius nanocrystals between 0 and 10 T at 1.7 K. The emission comes primarily from the ±2 state, as thermalization processes are highly efficient relative to the radiative rates. The long μs emission at zero field is consistent with recombination from this weakly emitting state. With increasing magnetic field, the luminescence lifetime decreases due to an enhancement of the radiative rate. Within the strong confinement regime of quantum dots, the interaction with the magnetic field is well described as a molecular Zeeman effect and the field influences the mechanism of recombination primarily by mixing the various excitonic spin states. Since the wurtzite *c* axes of the nanocrystals in our samples are randomly oriented, the ±2 and $\pm1^L$ states are mixed in dots whose *c* axis is not aligned with the field. Semiclassically this can be understood in terms of exciton spin precession around the applied field. As a result of mixing, the ±2 state gains oscillator strength from the optically active $\pm1^{L,U}$ states, explaining the observed decrease in radiative lifetime with increasing magnetic field.

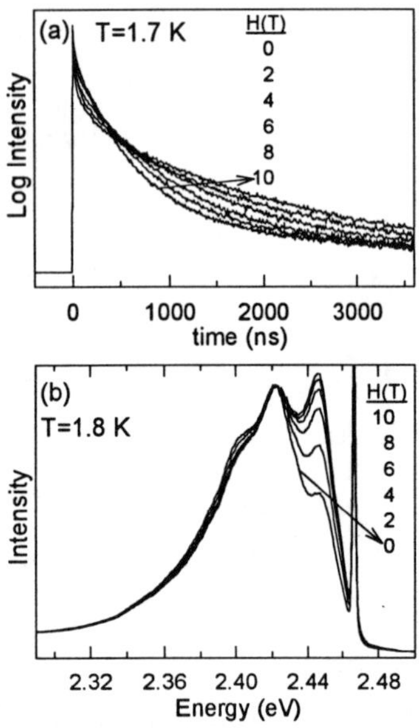

Fig.2 Magnetic Field dependence of (a) emission decays and (b) FLN spectra. The sharp feature at 2.467 eV, to the blue of the ZPL is the excitation laser which is included to mark the pump position. Experiments were carried out in the Faraday geometry (H||k)

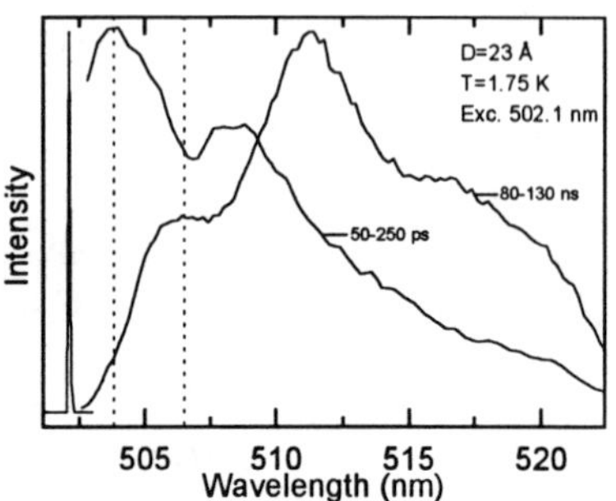

Fig.3 Time resolved FLN spectra for 23 Å diameter CdSe nanocrystals at 1.75 K corresponding to the time intervals 50-250 ps and 80-130 ns. The sharp peak at 502.1 nm is the excitation laser which is shown for reference.

The vibronic spectrum of the CdSe nanocrystals is also strongly influenced by the field.[6] Magnetic field dependent FLN spectra for 12 Å radius dots between 0 and 10 T at 1.7 K are shown in Fig.2b. Each spectrum is normalized to the one phonon lines for clarity. In isolation the ±2 state would have an infinite lifetime since the photon cannot carry an angular momentum of 2 within the electric dipole approximation. However the dark exciton can recombine via an LO phonon assisted momentum conserving transition. Spherical LO phonons with an orbital angular momentum of 1 or 2 are expected to participate in these transitions; the selection rules being determined by the coupling mechanism. Consequently at zero field higher LO phonon lines are strongly enhanced relative to the ZPL. With increasing magnetic field, the ±2 level gains optically active $\pm1^{L,U}$ character, diminishing the need for LO phonon assisted recombination. This explains the dramatic increase in the intensity of the ZPL, relative to the higher LO phonon replicas with increasing field.

Since thermalization processes out of the initially populated $\pm1^{L}$ state are highly efficient (τ_{th}~100 ps) relative to radiative recombination from this level (τ_R~10 ns), CW and nanosecond time resolved emission studies thus probe the ±2 "dark excitonic" state. Picosecond time resolved luminescence experiments however directly probe the emission from the $\pm1^{L}$ absorbing level.[9] Time resolved FLN spectra corresponding to time delays of 50-250 ps and 80-130 ns are shown in Fig. 3. There is a significant Stokes shift (~67 cm^{-1}) even in the 50-250 ps FLN spectrum. This might correspond to an initial fast vibrational relaxation which is beyond the experimental time resolution. Nonetheless, the emitting state at early time has significant $\pm1^{L}$ state character. Since this state is optically active, it does not require the participation of LO phonons for emission.

Consequently, even at 1.75 K the zero phonon line is stronger than the one phonon line (1PL). At long time delays (80-130 ns), after complete relaxation into the ±2 state, however, the 1PL intensity increases dramatically, indicating LO phonon assisted momentum conserving transitions.

B. ROOM TEMPERATURE SINGLE NANOCRYSTAL SPECTROSCOPY

Single nanocrystal measurements were conducted using an optical confocal microscope. Laser light was coupled through the microscope (Zeiss Axioscope) and focused onto the sample using a high numerical aperture oil immersion objective.[10] Sample preparation involves spin coating 1 drop of a ~5 nM solution of nanocrystals in a 0.5 % PVB/toluene solution onto a thin quartz flat which is then attached onto a piezo tube which generates the X-Y scanning motion. The fluorescence, excited by the diffraction limited laser spot (~0.38 μm), is collected by the objective (epi-illumination) and imaged onto a low noise, high quantum efficiency avalanche photo diode (APD). The digital output of the detector is fed into image acquisition and photon counting electronics. For spectroscopy, a fraction of the emission is split off into a spectrometer with a CCD camera. This permits simultaneous acquisition of fluorescence time traces (or lifetimes) and spectra.

In Fig.4a we show a fluorescence image of single 21 Å radius CdSe nanocrystals, embedded in a thin polyvinylbutyral film, at room temperature.[11] The streaks in the image are a consequence of raster scanning the sample across a diffraction limited laser spot, and arise due to the discrete turning on and off of the nanocrystal fluorescence signal while acquiring the image (~2 min). A fluorescence intensity versus time trace for a single nanocrystal under CW illumination is shown in Fig. 4b. The discrete turning on/off of the emission is strong evidence that the fluorescence originates from a single nanocrystal. The random overlapping of the on/off signals of many nanocrystals would obscure this phenomenon.

To determine whether this on/off phenomenon is light induced or occurs spontaneously we investigated the intensity dependence of the on/off times.[11] If the ons and offs were due to spontaneous transformations between an emitting and a 'dark' configuration, the on/off periods would be independent of excitation intensity.[12] On the other hand, for a light induced process where the 'dark' state is fed via a radiationless transition from the optically excited state, the 'on' times should decrease with increasing excitation intensity whereas the 'off' times would be determined by the spontaneous lifetime of the 'dark' state.[13] Fluorescence intensity versus time traces at two excitation intensities (Fig.5a) indicate that the average on time is inversely proportional to the excitation intensity while the average off time is intensity independent, suggesting a light-induced process.

This on/off behavior is consistent with a mechanism based on Auger ionization. In nanocrystals with two or more electron-hole pairs, the energy released from the annihilation of a pair may be transferred to the remaining carriers, one of which is preferentially ejected into the surrounding matrix depending on the conduction(valence) band offset at the nanocrystal interface and the electron(hole) confinement energy within the nanocrystal. In the resulting ionized nanocrystal, the Auger interaction, which is mediated by the coulomb repulsive potential between carriers, is strongly enhanced owing to the diminished screening relative to neutral nanocrystals. Subsequent photogenerated electron-hole pairs thus recombine primarily non-radiatively, transferring their energy to the resident carrier. This mechanism accounts for the decrease in the quantum efficiency of similar CdS nanocrystals embedded in glass on laser irradiation (photodarkening).[14] Photodarkened particles exhibit significantly shorter emission lifetimes and lower quantum yield compared to neutral nanocrystals.[15] The luminescence is then restored when

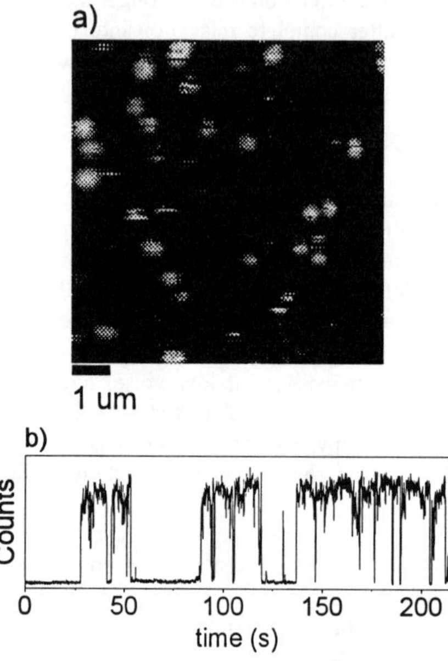

Fig.4 a) Image of a random field of single 21 Å radius CdSe nanocrystals (each with ~4 monolayers of ZnS on its surface) acquired by raster scanning the sample across a diffraction limited laser spot (wavelength 532 nm, full width at half maximum ~0.38 μm) and collecting the red shifted fluorescence onto an avalanche photodiode in an epi-illumination confocal geometry. This 160 x 160 pixel image (1 pixel=5ms) represents an 8μm x 8μm field of view. Sample preparation involves spin coating one drop of a ~5 nM solution of nanocrystals in a 0.5% polyvinylbutyral/toluene mixture onto a quartz cover slip. b) Fluorescence intensity versus time trace of a single ~21 Å radius CdSe nanocrystal from the same sample used in a) recorded using a multichannel scaler with a 40 ms sampling interval and an excitation intensity of ~0.52 kW/cm^2

the ejected carrier returns to neutralize the particle. At low flux (about three orders of magnitude lower than in our study) the Auger ionization rate was observed to scale as the square of the excitation intensity.[14] We observe that the average on period, which is determined by the rate of ionization, scales linearly with intensity. Furthermore, mode-locked pulsed laser excitation at the same average power but with over two orders of magnitude higher probability of generating two electron-hole pairs, yielded similar on times as were found with CW excitation. This suggests a two step ionization mechanism. The first step involves trapping of one of the carriers onto the nanocrystal surface followed by photoionization within the trap state lifetime (~5 μs). At low flux both processes should scale linearly with excitation intensity (I), the quantum yield of the ionization step being $[(I\sigma_p)/(h\nu)]/\{[(I\sigma_p)/(h\nu)]+\gamma_t\}$, where σ_p is the photoionization cross-section and γ_t is the relaxation rate from the intermediate trap state. Saturation of either step would lead to a net linear dependence. For instance, at an excitation rate of ~1 MHz, as in our study, a trap lifetime of ~5 μs and assuming an ionization yield of ~10-100%, the second step

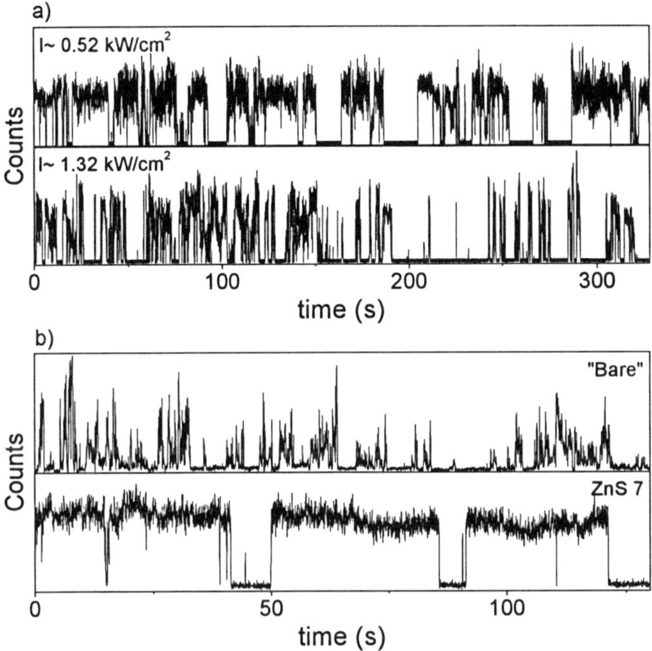

Fig.5 a) Comparison of fluorescence intensity versus time traces at excitation intensities (I) of ~0.52 and ~1.32 $kWcm^{-2}$ with a sampling interval of 10 ms. The average on/off times at the two excitation intensities are (I~0.52 $kWcm^{-2}$; $<\tau_{on}>$~0.97 s, $<\tau_{off}>$~0.44 s: I~1.32 $kWcm^{-2}$; $<\tau_{on}>$~0.32 s, $<\tau_{off}>$~0.43 s). Intensity dependent studies were carried out on CdSe nanocrystals with ~4 monolayers of ZnS on their surface. b) Fluorescence intensity versus time trace of a 'bare' nanocrystal compared with that of an overcoated one with a shell thickness of ~7 monolayers of ZnS (ZnS 7) at the same excitation intensity (I~0.7 $kWcm^{-2}$) and a sampling interval of 20 ms.

would approach saturation resulting in an effective three-level system with near linear dependence.

Based on this model the on/off times, determined by the photoionization/neutralization rates respectively, should be sensitive to changes at the nanocrystal-matrix interface. The nanocrystals as initially synthesized ('bare') have organic capping groups which passivate ~1/2 of all surface atoms. These particles can then be overcoated with a shell of ZnS, a higher bandgap material.[16] In Fig.5b, the fluorescence intensity versus time trace of a 'bare' particle is compared at the same excitation intensity with that of an overcoated one with ~7 monolayers of ZnS. ZnS has a bandgap ~2 eV higher than that of CdSe and so should serve as an effective barrier to ionization/neutralization and passivate potential surface trap sites. This is apparent in the dramatic increase in both the average on/off times for the ZnS overcoated nanocrystal relative to the 'bare' particle. To confirm this, we probed on/off times at the same excitation intensity as function of ZnS shell thickness. Consistent with photoionization/neutralization across a barrier, the average neutralization (off) time, increases, and the photoionization branching ratio, defined as the ionization probability per excitation, which depends on the average on period, decreases, with increasing shell thickness.

Although the nanocrystals are ionized and then neutralized reversibly, single nanocrystal spectra shift irreversibly to the blue (Fig.6a).[11] As the quantum confined energy levels vary as $1/\text{radius}^2$, this blue shift indicates a shrinkage of the effective radius as a function of time and is most probably due to surface photooxidation. This probably causes nanocrystals to photobleach permanently before we can accumulate accurate single dot statistics. But sometimes we find robust nanocrystals which we have been able to study for up to 40 minutes under constant illumination at room temperature. In Fig.6b we show the on/off histograms of one such nanocrystal. Both these distributions, which are also highly non-exponential, strongly suggest that even a single nanocrystal exhibits distributed kinetics, in sharp contrast to a single molecule, which usually has a single exponential intersystem crossing rate and triplet lifetime. A 21 Å radius nanocrystal consists of ~1400 atoms, of which ~25 % lie on the surface. A distribution of traps on the nanocrystal surface and within the surrounding amorphous matrix, could potentially lead to a range of ionization/neutralization rates.

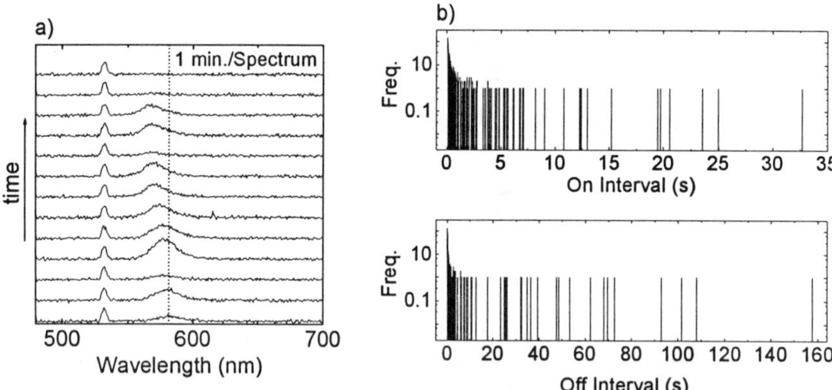

Fig.6 a) Fluorescence spectra (1 min. integration per spectrum) of a single nanocrystal as a function of time. The sharper feature to the blue (wavelength = 532 nm) is residual laser light which is included for reference. Fluctuations in fluorescence intensity relative to the laser are due to the discrete turning on and off of the emission during the experiment. Assuming an effective mass model for the quantum confined energy levels of the nanocrystal,[17] the observed blue-shift with time (~50 meV) corresponds to a 1.75 Å shrinkage in the radius representing ~2/3 of a monolayer of CdSe. The fluorescence signal vanishes permanently in the thirteenth minute. b) Histograms of on/off times for a single 'robust' nanocrystal probed for ~39 min. with a sampling interval of 40 ms and an excitation intensity of ~0.53 $kWcm^{-2}$

C. METASTABLE HIGH PRESSURE PHASES

Semiconductors such as CdSe and Si, are normally tetrahedral and are found in the diamond, wurtzite, and zinc blende phases. These materials also exhibit dense six-coordinate phases of rocksalt and beta-tin structure, that are stable thermodynamically only at high pressure. In the denser phases, these materials are electrically either semiconductors of smaller band gap, or metals. In bulk single crystals, nucleation of solid-solid phase transitions is apparently always caused by defects. Intrinsic nucleation is not well understood. The transformation kinetics show substantial hysteresis with pressure, suggesting the presence of large activation barriers.

Tolbert and Alivisatos have shown that study of phase transition kinetics in nanocrystals, as a function of size, provides insight into these processes.[18] Just one nucleation event per nanocrystal is observed: a single nanocrystal of one phase transforms into a single nanocrystal of the other phase. Thus the nanocrystal changes shape to reflect the distortion the unit cell undergoes upon phase change. Figure 7 is a schematic drawing of this expected shape change for the silicon diamond to beta-tin transition, for two different initial shapes. In this transformation, the Si unit cell c/a ratio undergoes a large change, from 1.414 in diamond to 0.55 in beta-tin. X-ray powder pattern data implying that such shape changes occur has been recently observed for 50 nm Si nanocrystals with a thin oxide surface layer.[19]

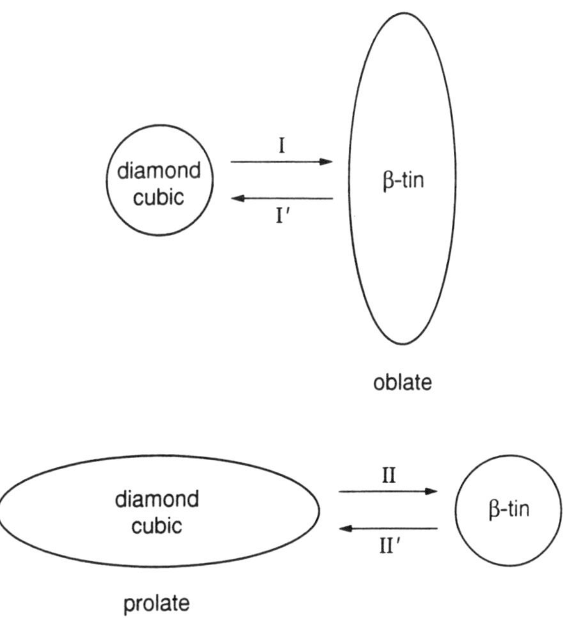

Figure 7

If one assumes that a nanocrystal thermally fluctuates from one phase into the other along the coordinate that simultaneously transforms all unit cells(a "bulk fluctuation"),[20] then a simple, activated unimolecular rate model can be formulated that quantitatively incorporates both surface and bulk energies. Figure 8 shows the energy surfaces for the beta-tin to diamond transition in Si, for the two different initial shapes. There is an activation barrier that depends upon both size and shape. For larger particles, the volume (size) contribution dominates, and the activation barrier increases proportional to the volume. For smaller particles, nearly round nanocrystals with lower surface energy are more stable than nanocrystals with larger surface areas. Yet, quantitative barrier estimates for sizes greater than 2 nm diameter show that such beta-tin nanocrystals should be metastable, with half-lives of years or more, for a wide range of shapes. These simple considerations suggest that nanocrystals in six-coordinate phases should be metastable at STP if they can be made defect free.

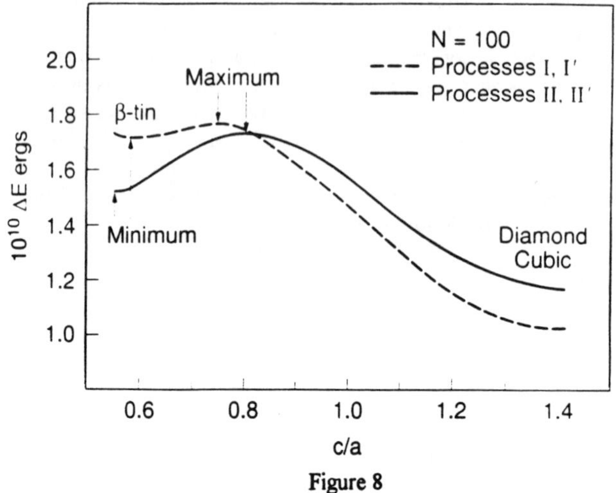

Figure 8

Metastable six-coordinate Si or CdSe nanocrystals experimentally revert to tetrahedral phases when pressure is released in the diamond anvil kinetic experiments. It may be that the sheer stress is too great. Recent experiments do show that the lifetimes do increase with increasing size, as predicted by the bulk fluctuation model.[21] Actually there are literature reports of metastable rocksalt AlN and GaN crystallites.[22] These ultrahard, large band gap semiconductors are thermodynamically stable at STP only in the wurtzite phase. These results suggest the possibility and hope that, some years into the future, chemical syntheses will be known for a wide variety of metastable nanocrystal phases.

CONCLUSIONS

The low temperature ensemble measurements indicate that the energetics and dynamics of the band edge emission from CdSe nanocrystals can be quantitatively understood in terms of the intrinsic band edge exciton structure. The electron-hole exchange interaction and anisotropy terms split the eightfold degenerate bulk band edge exciton into five levels, labeled by the total angular momentum projection F_m. Exciton thermalization into a dipole forbidden state with $F_m=\pm2$ resolves the issue of the long band edge luminescence lifetimes. The magnetic field dependence of the emission decays and LO phonon structure confirms the presence of this 'dark excitonic' state and reveals its luminescence properties. Remarkably, room temperature measurements on single CdSe nanocrystals reveal that under CW illumination, single nanocrystallites emit intermittently on a 0.5s timescale, an effect obscured in previous ensemble studies. While the abrupt turning off of the emission due to photoionization is rare(probability per excitation is $\sim10^{-6}$), it has a significant effect on luminescence owing to the unusually long neutralization times that follow ($\tau_n\sim0.5$s). This poses a potential limitation to optical devices based on CdSe nanocrystals for which fast response times and high luminescence yields are crucial. Approaches for modifying this on/off behavior could include the use of an inorganic shell material with lattice matching superior to ZnS.

ACKNOWLEDGEMENTS

Nanocrystal research at Columbia is partially supported by the Joint Services Electronics Program in the Columbia Radiation Laboratory. Low temperature ensemble measurements were conducted in collaboration with Prof. M. G. Bawendi at MIT, Cambridge, MA. Room temperature single nanocrystal studies were performed at Bell Laboratories, Lucent Technologies, Murray Hill, N.J in collaboration with J. K. Trautman, J. J. Macklin and T. D. Harris. Experimental work on Si nanocrystal phase transitons was conducted in collaboration with S. Tolbert, A. Herhold, and A. P. Alivisatos of the University of California at Berkeley. The simple model for nanocrystal phase transitions was developed in collaboration with F. S. Stillinger of Bell Laboratories. We deeply appreciate these collaborations. We also thank Al. L. Efros for collaboration and numerous stimulating discussions. We thank M. Kuno for providing Figure 1.

REFERENCES

1. C. B. Murray, D. J. Norris and M. G. Bawendi, J. Am. Chem. Soc. **11**, 8706 (1993).
2. M. G. Bawendi, P. J. Carroll, W. L. Wilson and L. E. Brus, J. Chem. Phys. **96**, 946 (1992).
3. M. Nirmal, C. B. Murray and M. G. Bawendi, Phys. Rev. B **50**, 2293 (1994).
4. C. H. Henry and K. Nassau, Phys. Rev. B **1**, 1628 (1970).
5. M. O'Neil, J. Marohn, and G. Maclendon, J. Phys. Chem. **94**, 4356 (1990); A. Hasselbrath, A. Eychmuller and H. Weller, Chem. Phys. Lett. **203**, 271 (1993).
6. M. Nirmal, D. J. Norris, M. Kuno, M. G. Bawendi, Al. L. Efros and M. Rosen, Phys. Rev. Lett. **75**, 3728 (1995).
7. Al. L. Efros, M. Rosen, M. Kuno, M. Nirmal, D. J. Norris and M. G. Bawendi, Phys. Rev. B **54**, 4843 (1996).
8. P. D. J. Calcott, et al., J. Phys. Condens. Matter **5**, L91 (1993).
9. M. Nirmal, To be Published.
10. J. J. Macklin, J. K. Trautman, T. D. Harris and L. E Brus, Science **272**, 255 (1996).
11. M. Nirmal, B. O. Dabbousi, M. G. Bawendi, J. J. Macklin, J. K. Trautman, T. D. Harris and L. Brus, Nature **383**, 802 (1996).
12. W. P. Ambrose, T. Basche and W. E. Moerner, J. Chem. Phys. **95**, 7150 (1991).
13. J. Bernard, L. Fleury, H. Talon and M. Orrit, J. Chem. Phys. **98**, 850 (1993).
14. D. I. Chepic et al., J. Lumin. **47**, 113 (1990).
15. P. Rossignol, D. Ricard, J. Lukasik and C. Flytzanis, J. Opt. Soc. Am. B **4**, 5 (1987).
16. M. A. Hines and P. Guyot-Sionnest, J. Phys. Chem. **100**, 468 (1996); A. R Kortan et al., J. Am. Chem. Soc. **112**, 1327 (1990).
17. D. J. Norris and M. G. Bawendi, Phys. Rev. B **53**, 16338 (1996).
18. S. Tolbert and A. P. Alivisatos, Annu. Rev. Phys. Chem. **46**, 595 (1995).
19. S. Tolbert, A. Herhold, L. Brus and A. P. Alivisatos, Phys. Rev. Lett. **76**, 4384 (1996).
20. L. Brus, J. Harkless and F. Stillinger, J. Am. Chem. Soc. **118**, 4834 (1996).
21. C. Chen and A. P Alivisatos, Private Communication.
22. H. Volldstadt, E. Ito, M. Akaishi, S. Akimoto and O. Fukunaga, Proc. Jpn. Acad. Ser. **B66**, 7 (1990); Y Xie, Y. Qian, W. Wang, S. Zhang and Y. Zhang, Science **272**, 1926 (1996).

POROUS SEMICONDUCTORS: A TUTORIAL REVIEW

L T Canham
DRA, St Andrews Road,
Malvern, Worcestershire. WR14 3PS, UK.

ABSTRACT

Porous semiconductors constitute a class of material that exhibit surprising properties, are quite easy to fabricate, but are however fragile, complex, and difficult to characterise. This tutorial review extracts specific topics from the large knowledge base now available on porous Si that are deemed relevant to other porous semiconductors beginning to receive study. It also highlights topics where controversy is now resolved, where problems remain, and where further effort could be focused.

I INTRODUCTION

Although porous semiconductors are strictly speaking, not a new class of material, it is only in the 1990's that a significant level of study has occurred. Over the last six years there have been > 1500 papers published with the focus primarily on visibly luminescent high porosity silicon (1). Given such a large body of data, and that work is now starting to appear on a range of other luminescent porous semiconductors (see table 1), it would seem timely to highlight some of the lessons learnt from porous Si work that are directly applicable to porous semiconductors in general.

TABLE 1 - INTEREST IN POROUS SEMICONDUCTORS

SEMICONDUCTOR RENDERED POROUS	No. of papers	RECENT REFERENCE
Si	> 1500	This MRS proceedings
SiC	> 13	MacMillan et at. J Appl Phys 80, 2412 (1996).
GaP	> 8	Meijerink et al. J Appl Phys 79, 9301 (1996).
Si_xGe_{1-x}	> 7	Schoisswohl et al. J Appl Phys 79, 9301 (1996).
Ge	> 4	Miyazaki et al. Thin Sol Films 255, 99 (1995).
GaAs	> 4	Schmuki et al. Appl Phys Lett 69, 1620 (1996).
InP	> 3	Kikuno et al. Jpn J Appl Phys 34, 177 (1995).
$CuInSe_2$	> 1	Lebedev et al. Tech Phys Lett 22, 467 (1996).

Mat. Res. Soc. Symp. Proc. Vol. 452 © 1997 Materials Research Society

The focus of this short, highly selective literature review is thus to summarise some of the significant advances made with porous Si over the last few years, but in addition the major issues and problems that are to some degree unique to porous semiconductors. Its aim is hence to attempt to guide new researchers in the field towards making better quality material, avoid some of the reproducibility problems that have arisen in the past, and target some of the key problems in the future. Table 2 illustrates the material complexity involved since both skeleton and pore microstructure influence properties, and can occur over widely varying length scales.

TABLE 2 DESCRIPTIONS OF A POROUS SEMICONDUCTOR

MAJOR STRUCTURAL FEATURES	EXAMPLES AND LABELS		EXAMPLES OF MEASUREMENT TECHNIQUES*
SKELETON:			
Crystallinity	eg poly, amorphous		TEM, XRD
Size distribution	eg gaussian, bimodal		SAXS, RSS
Morphology	eg columnar, fractal		TEM, SPM
Defects	eg extended, point		HRTEM, EPR
Strain	eg symmetry and sign		XRD
SKELETON SURFACE:			
Chemical composition	eg hydride, oxide		IBA, SIMS
Surface area	eg BET value		GADA, SAXS
Surface bonding	eg dihydride, hydroxyl		FTIR, XPS
Defects	eg dangling bond		EPR, PAS
POROSITY:			
Void content	0-30%	low porosity	GM, SE
	30-70%	medium porosity	GM, SE
	70-100%	high porosity	GM, SE
Size distribution	< 2 nm	microporous	FMC, NMR
	2-50 nm	mesoporous	TEM, GADA
	> 50 nm	macroporous	SEM, SOM
Morphology	eg columnar, branched		GADA, TEM

* GLOSSARY

TEM: Transmission Electron Microscopy, XRD: X-Ray Diffraction, SAXS: Small Angle X-ray Scattering, RSS: Raman Scattering Spectroscopy, SPM: Scanning Probe Microscopy, HRTEM: High Resolution Transmission Electron Microscopy, EPR: Electron Paramagnetic Resonance, IBA: Ion Beam Analysis, SIMS: Secondary Ion Mass Spectroscopy, GADA: Gas Adsorption Desorption Analysis, FTIRS: Fourier Transform Infrared Spectroscopy, XPS: X-ray Photoelectron Spectroscopy, PAS: Positron Annihilation Spectroscopy, GM: Gravimetric Data, SE: Spectroscopic Ellipsometry, FMC: Flow MicroCalorimetry, NMR: Nuclear Magnetic Resonance, SOM: Scanning Optical Microscopy.

II FABRICATION

A range of semiconductors can be made highly porous by anodization in HF based solutions. With regard formation mechanisms, we refer the reader to the comprehensive review of Smith and Collins (2) where the various models invoked are discussed in some detail. Several models address the manner in which the electrochemically induced etch pits ('pores') propagate into the semiconductor, but there remains a relatively poor understanding of why such pores nucleate on the surface in the first place. A recent model considers pore nucleation to arise from the instability of a planar Si-electrolyte interface to small perturbations (3), whilst other workers have suggested the involvement of specific point defects in the process (4). Much further experimental work is needed in this area.

The basic methods of rendering Si highly porous have also been briefly focused on, in more than one recent review (5,6). Anodization under galvanostatic conditions is generally the preferred approach for attaining a wide range of porosity and thickness. Careful design of the electrochemical cell is essential to maximise lateral uniformity of layers (6). In contrast, vertical inhomogeneity within a porous Si layer arises more from the nature of the electrochemical etching process itself, than the anodization equipment used. It is much harder to avoid, and indeed virtually all luminescent porous Si layers exhibit a finite porosity gradient with depth. The use of long anodization times, chemically aggressive electrolytes or light-assisted etching all promote negative porosity gradients, ie a porosity decreasing with increasing depth within the layer. On the other hand, electrolyte depletion within the pores can arise during etching with high current densities and acts to generate positive porosity gradients within such layers (7).

A number of significant modifications to the basic anodization process have been developed to either tailor desired properties, improve structural uniformity or generate more complex multilayer structures. Examples are shown in table 3.

TABLE 3 DEVELOPMENTS IN FABRICATING POROUS SILICON VIA ANODISATION

PROCESS MODIFICATION	OBJECTIVE	REFERENCE
PHOTOASSISTED ANODISATION	Improved tuneability of luminescence	Doan et al. J Phys Chem 97, 4503 (1993)
PERIODICALLY MODULATED ANODISATION	Multilayer structures	Berger et al. J Phys D 27, 1333 (1994)
MAGNETIC FIELD ASSISTED ANODISATION	Improved structural uniformity	Nakagawa et al. Appl Phys Lett 69, 3206 (1996)

Alternative electrochemical techniques to anodization, for creating visibly luminescent porous Si layers, have also been demonstrated such as stain-etching (8) and photo-induced synthesis (9). Such techniques are in principle simpler to implement, but have received much less study, and remain plagued by reproducibility problems.

A range of much cheaper forms of silicon than monocrystalline bulk wafers have also been rendered highly porous and visibly luminescent; namely poly Si (10), doped and undoped a-Si:H (11,12) and even metallurgical grade Si (13).

Although a wide range of semiconductors have also been made porous by similar techniques (see table 1), the realisation of useful structures for optoelectronic applications will probably require the development of alternative etch chemistries to those suitable for silicon. Key problem areas are non-stoichiometry induced by etching alloys or compound semiconductors and the need for excellent surface passivation.

III DRYING

The drying of a porous semiconductor should be considered an integral and important step in its fabrication process, but until recently was given little attention. Indeed, even to the present day, the vast majority of studies on porous Si allowed the electrolyte to evaporate out of the pores in a relatively uncontrolled fashion. For high porosity luminescent films this is unfortunately often a highly destructive process due to the enormous capillary forces exerted on the semiconductor skeleton (14). The porous Si literature is strewn with reports of layers fabricated with too high a porosity or thickness, that when dried become cracked, crazed or shrink to various degrees. The important message here is that the reproducibility of layer manufacture and properties can depend as much on how it is dried as on the anodization equipment and etch parameters used. Only recently have improved drying techniques been implemented. The first of these was supercritical drying (15) followed recently by pentane drying (16) and freeze drying (17). Controlled drying techniques should be made an essential part of the fabrication process for applications where high porosity films are required and the crystalline perfection of the semiconductor skeleton together with good structural integrity are major issues.

IV STABILISATION AND STORAGE

The stability of porous silicon structures remains a major issue for many of the potential application areas under development (see section VII). Freshly HF-etched layers 'age' in the sense that their metastable hydride surface reacts slowly with ambient air. This 'atmospheric impregnation' process induces contamination and partial oxidation of the Si skeleton and the structural and optical properties of a given layer continuously evolve with storage time from minutes to months (18). The speed and extent to which oxidation occurs depends upon many factors such as ambient air contamination and humidity levels, and the level of background illumination. It should also be noted that even when dried, freshly etched porous semiconductors can liberate significant levels of toxic vapours upon exposure to humid air (19). The hydrolysis of porous GaAs layers for example, could locally liberate ppm levels of the extremely toxic gas, arsine, due to their large internal surface area (see table 4).

TABLE 4 HYDRIDE EVOLUTION FROM POROUS SEMICONDUCTORS STORED IN AIR

POROUS SEMICONDUCTOR UNDERGOING HYDROLYSIS	ASSOCIATED TOXIC HYDRIDE	SHORT TERM (10 min) TLV EXPOSURE LIMIT [ppm]
Si, SiC	SiH_4	1.0
Ge, Si_xGe_{1-x}	GeH_4	0.6
GaAs, AlGaAs	AsH_3	- *
GaP, InP	PH_3	0.3
InSb, GaSb	SbH_3	0.3

* 40 hr week TLV is 0.05 ppm.

Not only do the properties of a highly porous semiconductor depend on the way the material is dried after etching, but also on the manner in which it is stored. A striking example of this is given by the blue-green luminescence band reported occasionally in heavily oxidised porous silicon (20). Significant blue output is in fact only observed after prolonged storage (21). It has been pointed out that in fact many high surface area materials stored in air for months exhibit such luminescence, and that contamination of the oxide phase is its likely origin (22). Even with bulk Si wafers, storage-induced surface contamination from containment vessels is a recognised issue (23). Most plastic wafer and sample containers, for example, contain complex molecules like phthalate esters and outgas over extended periods of time. Many such containers (eg, fluoroware cassettes) themselves contain organic chromophores that luminesce in the blue-green spectral range and could gradually impregnate a highly porous material.

For most applications such 'ageing effects are unacceptable and hence a number of techniques have been under development for stabilising porous Si. These can be grouped into four categories namely intentional oxidation, chemical derivitization, pore impregnation and layer 'capping' or encapsulation.

Most effort to date has been on controlled partial oxidation and a wide variety of treatments have been investigated. These include anodic oxidation (24), chemical oxidation (25), both dry and wet thermal oxidation (26,27), plasma assisted oxidation (28) and irradiation enhanced oxidation (29). All such oxidised structures can retain their desired luminescent properties and do exhibit improved stability in comparison to freshly etched material. An alternative approach which has the added attraction of not significantly consuming too much of the Si skeleton through oxidation, is that of surface derivitization, where Si-H_x bonds are replaced by, for example Si-CH_3 bonding (30). This has only quite recently begun to be utilised but preliminary results are very encouraging.

Nevertheless, it seems highly likely that such treatments alone will never render porous semiconductor structures sufficiently stable for many optoelectronic applications. Additional stabilisation steps will be needed, simply because with these aforementioned approaches, the structures are still porous and susceptible to atmospheric induced drift and contamination.

Such chemically modified structures need to be isolated from the environment either by an optically transparent capping layer or encapsulation technology. Relatively simple capping techniques have been shown to dramatically suppress some 'ageing' effects (31) but once again much further work is needed.

It is pore-filling techniques to generate 'nanocomposite' structures that perhaps offer the most realistic long-term solution to stability problems. UHV deposition of Ge into porous Si by CVD has provided the proof-of-principle that mesopores can be virtually completely filled (32). Alternative gas phase impregnation approaches (33) and chemical (34) and electrochemical techniques (35) for forming nanocomposites have also been reported. Although the infiltrated material often has to satisfy many criteria, it seems that such nanocomposites are needed to achieve the mechanical and thermal properties needed for many optoelectronic applications (see section VI). The field of porous semiconductor nanocomposites is still in its infancy, but is a very important area.

V LUMINESCENT PROPERTIES

This has been the major focus of a large amount of work world-wide over the last 6 years. Porous silicon based structures have been developed that can photoluminesce efficiently over a very wide spectral range (from the near infrared to the ultraviolet) (36). The general properties of the main luminescence bands (see table 5), together with the proposed mechanisms have been recently reviewed by the author (37).

TABLE 5 POROUS SILICON LUMINESCENCE BANDS

SPECTRAL RANGE	PEAK WAVELENGTH (nm)	LUMINESCENCE BAND LABEL	PL?	CL?	EL?
UV	~350	'UV band'	Yes	Yes	No
blue-green	~470	'F band'	Yes	Yes	No
blue-red	400-800	'S band'	Yes	Yes	Yes
near IR	1100-1500	'IR band'	Yes	No	No

The most important luminescence band, the 'S band' has received most attention, and has generated considerable controversy with regards its origin (37). Is this luminescence associated with quantum size effects, ie nanocrystalline silicon? What influence does the enormous surface area of the material have? The first question can now be answered with assurance. There are now in fact 3 independent spectroscopic observations that clearly link S band luminescence to the vibrational properties and bandstructure of monocrystalline Si.

These are:

(a) resonant photoexcitation data (38): TO and TA phonon satellites are observed at low temperatures with relative strengths and energies that match those observed in monocrystalline Si.

(b) photoluminescence polarisation data (39): Under excitation with linearly polarised light, the degree of polarisation memory at 6K reaches zero at exactly the bulk Si bandgap (1.17 eV) at that temperature.

(c) tuneable electroluminescence data (40): Voltage tuneable output with liquid electrolytic contacts under both cathodic and anodic bias demonstrates a 'polarisation gap' that exactly equals the Si bandgap at that temperature (1.1 eV at 300K).

Aside from the structural and compositional data that also renders many of the ~20 models reported in the literature unrealistic (37), such spectroscopic data, in my opinion, is quite conclusive in its implications. Similar 'spectroscopic signatures' of luminescence from other porous semiconductors, if they exist would also provide definitive insight into whether light originates from quantum confinement effects on semiconductor band structure.

Without such data it is often very difficult to determine the microscopic origin of a broad featureless luminescence band within a complex material system. Particular caution needs to be exercised with regards oxidized porous semiconductors. These are in reality two-phase materials and both PL and CL can originate from the oxide phase, rather than the remaining semiconductor skeleton. In the case of silicon, defective or contaminated silicon oxide is known to be capable of luminescing right across the visible and ultraviolet spectral range. Both the 'UV band' and 'F band' (see table 5) are only observed in oxidized material, in contrast with the 'S band' emission (37). The latter has been studied under a wide variety of conditions as listed in table 6.

TABLE 6 CONDITIONS FOR S BAND PL

CONDITION OF POROUS SILICON	MAJORITY CHEMICAL SPECIES
In-situ in HF during / after formation.	SiF_xH_y
Freshly etched in inert ambient	SiH_x
Chemically or anodically oxidized	SiO_xH_y
Stored in air for months to years	$SiO_xH_yC_z$
Rapid thermally oxidized at high temperatures	SiO_xH_y

With regards the role(s) of the internal surface, this question has to date only been partially answered. Much work on the photostability of the 'S band' luminescence (41) has been done since 1991. It is clear that a number of degradation and PL quenching mechanisms involve surface chemistry (42). In addition, not only PL efficiency, but also spectral position and decay kinetics can be influenced (43). Nevertheless, this luminescence band is observed under a wide variety of surface conditions (37), so if 'surface states' were to be involved with the radiative process directly, they would need to be impurity non-specific. In addition only shallow surface states with weak associated carrier localisation could be compatible with experimental evidence (38-40). On the weight of evidence to date it would seem the role of the surface is primarily through the introduction or modification of competing non-radiative processes.

One area where controversy does still exist however, regards the size and shape of the luminescent silicon nanostructure. An accurate quantitative relationship between wavelength of emission and size is still not available, in the author's opinion. Red emitting layers of 80% porosity were shown in 1991 to consist primarily of crystalline Si columns of fluctuating width around 2-3 nm (44). Graphic images of much higher porosity (> 95%) 'aerocrystals' (15) subsequently revealed the columnar nature in more detail. HRTEM studies have reported both 'crystalline clusters' and 'nanowires' in the 1-3 nm size regime (45, 46), and similar dimensions are obtained from small angle scattering techniques (47) and Raman scattering (48).

Recently however, the size and shape of luminescent regions of porous silicon have been extracted in a novel but indirect way from EXAFS data (49). It was concluded that visible emission (< 700 nm) arises from Si quantum dots, not quantum wires, and whose dimensions were < 1.5 nm for even deep red output. A subsequent EXAFS study has however been interpreted quite differently (50). Red, yellow and green emission was correlated with a nanowire model with mean crystalline core widths of 2.2, 1.9 and 1.3 nm respectively. These latter size estimates are also in better accord with most theoretical calculations of bandgap size (51). Much more work is needed in this area to clarify the origins of apparently conflicting data.

VI MECHANICAL AND THERMAL PROPERTIES

Until very recently, there was virtually no data available on either the mechanical or thermal properties of porous Si, which were nonetheless expected to differ dramatically from those of bulk Si, and could be critical constraints for a number of applications. The first such studies reveal a quadratic dependence of Young's modulus on relative density (52, 53) such that high porosity structures can exhibit values < 1GPa (53). Micro-hardness is also dramatically affected (34). Thermal conductivity data for different types of layer are also now available (54, 55) and typical values are shown in table 7.

TABLE 7 THERMAL CONDUCTIVITY OF POROUS SILICON AND RELATED MATERIALS

MATERIAL	POROSITY	THERMAL CONDUCTIVITY ($Wm^{-1}K^{-1}$)
Anodized p^+Si	45% (meso)	80
Anodized p^-Si	40% (micro)	1.2
Si	0	150
SiO_2	0	1.4
SiO_2 aerogel	> 90% (meso)	10^{-3}

The thermal properties of a layer are clearly highly dependent on skeleton microstructure as well as porosity. Stongly luminescent material will have thermal conductivity values well below 1 $Wm^{-1}K^{-1}$. Heat dissipation could thus in some circumstances be a limiting factor with regards device performance, unless a thermally conductive material is impregnated (see section IV) to improve heat-sinking.

TABLE 8 POTENTIAL APPLICATION AREAS OF POROUS SILICON

APPLICATION AREA AND EXAMPLE(S) OF ROLE OF PSi	KEY PROPERTY OF PSi	RECENT REFERENCE
OPTOELECTRONICS eg LED	efficient electroluminescence	Canham et al. Appl Surf Sci 102, 436 (1996).
eg field emitter	hot carrier emission	Boswell et al. J Vac Sci Techn B13, 437 (1995).
MICRO-OPTICS eg Fabry Perot filter	refractive index modulation	Berger et al. Thin Sol Films 255, 313 (1995).
eg photonic bandgap structures	regular macropore arrays	Gruning et al. Appl Phys Lett 68, 747 (1996).
eg all optical switches	highly non linear optical properties	Matsumoto et al. J Electrochem Soc 142, 3528 (1995).
ENERGY CONVERSION eg AR solar cell coating	low refractive index	Menna et al. Solar Energy Mat. and Sol. Cells 37, 13 (1995).
eg photoelectrochemical cells	photocorrosion shield	Mao et al. J Phys Chem 99, 3643 (1995).
ENVIRONMENTAL MONITORING eg gas sensing	ambient sensitive properties	Schechter et al. Anal. Chem. 67, 3727 (1995).
MICROELECTRONICS eg micro capacitors	high specific surface area	Lehman et al. Thin Sol Films 276, 138 (1996).
WAFER TECHNOLOGY eg wafer bonding for SOI	high etch selectivity	Sato et al. J Electrochem. Soc. 142, 3116 (1995).
eg heteroepitaxy	variable lattice parameter	Raiko et al. Diam Relat Mater 5, 1063 (1996)
MICROMACHINING eg thick sacrificial layers	highly controllable etch patterns	Bell et al. J Micromech. Microeng 6, 361 (1996).
BIOTECHNOLOGY eg tissue bonding	tuneable chemical reactivity	Canham et al. This MRS proceedings.
eg biosensors	enzyme immobilization	Laurell et al. Sens. Act. B31, 161 (1996).

VII POTENTIAL APPLICATION AREAS

Table 8 lists some of the perceived application areas for porous Si that have been proposed over the last few years. Attention to date has focused primarily on the luminescent optoelectronic properties of the material but there is potential in a number of diverse areas. Over the last 6 years porous Si structures have been shown to not only emit light efficiently but also be capable of guiding (56), modulating (57) and detecting light (58). However, in each case it is important to make comparisons with bulk Si devices to identify where exploitation of porosity is actually needed to achieve significant performance gains (59). In the case of light emitting diodes for example, the efficiency has improved dramatically over this period (60-62) to levels that are 10^4 those of bulk Si devices (59). In contrast bulk Si technology offers both fast and sensitive photodetection in the visible range, and it is not clear that porous Si offers major benefits (59). The motivations for striving towards Si-based optoelectronics have been described in detail (63) quite recently. It is important to stress that a major attraction of porous silicon is that it is indeed silicon, and a silicon structure that is VLSI compatible. This is exemplified by the recent integration of a porous Si LED with standard bipolar circuitry (64). This feature is also, in part, the motivation behind many of the applications listed in table 8. In addition it also explains why interest in porous Si currently greatly exceeds that of other porous semiconductors.

VIII CONCLUDING REMARKS

Porous Si has over the last few years proved to be a complex material that has aroused great excitement, interest and controversy. It is a material that could eventually help extend the functionality of Si technology into diverse areas such as optoelectronics (59) and biotechnology (65). These are both currently major areas of technological growth, for which bulk silicon's material properties alone restrict it to a passive role.

Traditionally, silicon has neither been regarded as an optoelectronic material nor a biomaterial. Porous Si may have initiated a gradual change in that viewpoint. The field of porous semiconductors is in its infancy and will no doubt yield more surprises in the future.

ACKNOWLEDGEMENTS

I would like to thank many DRA colleagues for close collaboration on porous Si over the last 6 years, especially Marc Beale, Doug Brumhead, Phil Calcott, Tim Cox, Tony Cullis, John Keen, Armando Loni, Keith Nash, Chris Pickering, Chris Reeves and Andy Simons. I am also grateful to the international porous and light emitting Si community that has grown over those years for many 'enlightening' discussions.

REFERENCES

[1] L T Canham. Appl Phys Lett 57, 1046 (1990).

[2] R L Smith, S D Collins. J Appl Phys 71, R1 (1992).

[3] Y Kang, J Jorre. J Electrochem Soc 140, 2836 (1993).

[4] P Allongue, C H de Villeneuve, L Pinsard, M C Bernard. Appl Phys Lett 67, 941 (1995).

[5] K H Jung, S Shih, D L Kwong. J Electrochem Soc 140, 3046 (1993).

[6] A Halimaoui in Porous Silicon Science and Technology edited by J C Vial and J Derrien. (Springer-Verlag, Berlin 1995) p33.

[7] M Thonissen, S Billat, M Kruger, H Luth, M G Berger, U Frotscher, U Rossow. J Appl Phys 80, 2990 (1996).

[8] R W Fathauer, T George, A Ksendzov, R P Vasquez. Appl Phys Lett 60, 995 (1992).

[9] N Noguchi, I Suemene. Appl Phys Lett 62, 1429 (1993).

[10] T Ueno, Y Akiba, T Shinohara, H Koyama, N Koshida, Y Tarvi. Jpn J Appl Phys 32, L5 (1993).

[11] E Bustarret, M Ligeon, L Ortega, Sol State Commun 83, 461 (1992).

[12] R B Wehrspoon, J N Chazalviel, F Ozanam, I Solomon. Phys Rev Lett 77, 1885 (1996).

[13] P Menna, Y S Tsuo, M M Al-Jassim, S E Asher, F J Pern, T F Ciszek. J Electrochem Soc, 143 L115 (1996).

[14] U Gruning, A Yelon. Thin Sol Films 255, 135 (1995).

[15] L T Canham, A G Cullis, C Pickering, O Dosser, T I Cox, T P Lynch. Nature 368, 133 (1994).

[16] O Belmont, D Bellet, Y Brechet. J Appl Phys 79, 7586 (1996).

[17] G Amato, N Brunetto. Mater Lett 26, 295 (1996).

[18] L T Canham, M R Houlton, W Y Leong, C Pickering, J M Keen. J Appl Phys 70, 422 (1991).

[19] L T Canham, S J Saunders, P B Heeley, A M Keir, T I Cox. Adv Mater 6, 865 (1994).

[20] D I Kovalev, I D Yarostietzkii, T Muschik, V Petrova-Koch, F Koch. Appl Phys Lett 64, 214 (1994).

[21] A Loni, A J Simons, P D J Calcott, L T Canham. J Appl Phys 77, 3557 (1995).

[22] L T Canham, A Loni, P D J Calcott, A J Simons, C Reeves, M R Houlton, J P Newey, K J Nash, T I Cox. Thin Solid Films 276, 112 (1996).

[23] K Saga, T Hattori. J Electrochem Soc 143, 3279 (1996).

[24] A Bsiesy, F Gaspard, R Herino, M Ligeon, F Muller, J C Oberlin. J Electrochem Soc 138, 3450 (1991).

[25] F Kozlowski, W Wagenseil, P Steiner, W Lang. Mat Res Soc Symp Vol 358, 677 (1995).

[26] V Petrova-Koch, T Muschik, A Kux, B K Meyer, F Koch, V Lehman. Appl Phys Lett 61, 943 (1992).

[27] H Chen, X Hou, G Li, F Zhang, M Yu, X Wang. J Appl Phys 79, 3282 (1996).

[28] P O Keefe, Y Aoyagi, S Komuro, T Kato, T Morikawa. Appl Phys Lett 66, 836 (1995).

[29] J S Fu, J C Mao, E Wu, Y Q Zia, B R Zhang, L Z Zhang, G C Qin, G S Wui, Y H Zhang. Appl Phys Lett 63, 1830 (1993).

[30] V M Dubin, C Viellard, F Ozanam, J N Chazalviel. Phys Stat Sol (b) 190, 47 (1995).

[31] T Giaddiu, K S Forcey, L G Earwaker, A Loni, L T Canham, A Halimaoui. J Phys D 29, 1580 (1996).

[32] A Halimaoui, Y Campidelli, P A Badoz, D Bensahel. J Appl Phys 78, 3428 (1995).

[33] C Ducso, N Q Khanh, Z Horvath, I Barsony, M Utriaven, S Lehto, M Nieminen, L Niinisto. J Electrochem Soc 143, 683 (1996).

[34] S P Duttagupta, X L Chen, S A Jerekhe, P M Fauchet. Sol State Commun 101, 33 (1997).

[35] F Ronkel, J W Schultze, R Arens-Fischer. Thin Sol Films 276, 40 (1996).

[36] P M Fauchet. J Lumin 70, 294 (1996).

[37] L T Canham. Phys Stat Sol (b) 190, 9 (1995).

[38] P D J Calcott, K J Nash, L T Canham, M J Kane, D Brumhead. J Lumin 57, 257 (1993).

[39] F Koch, D Kovalev, B Averboukh, G Polisski, M Ben Chorin. J Lumin 70, 320 (1996).

[40] A Bsiesy, J C Vial. J Lumin 70, 310 (1996).

[41] M A Tischler, R T Collins, J H Stathis, J C Tsang. Appl Phys Lett 60, 639 (1992).

[42] D L Fischer, A Gamboa, J Harper, J M Lauerhaus, M J Sailor. MRS Proc Vol 358, 507 (1995).

[43] I Mihalcescu, M Ligeon, F Muller, R Romestain, J C Vial. J Lumin 57, 111 (1993).

[44] A G Cullis, L T Canham. Nature 353, 335 (1991).

[45] I Berberzier, A Halimaoui. J Appl Phys 74, 542 (1993).

[46] A Albu-Yaron, S Bastide, D Bouchet, N Brun, C Collieux, C Levy-Clement. J Phys, France 4, 1181 (1994).

[47] M Binder, T Edelmann, T H Mitzger, J Peisl. Sol State Commun 100, 13 (1996).

[48] H Munder, C Andrzejak M G Berger, U Klemradt, H Luth, R Herino, M Ligeon. Thin Sol Films 221, 27 (1992).

[49] S Schuppler, S L Friedman, M A Marcus, D L Adler, Y H Xie, F M Ross, Y J Chabal, T D Harris, L E Brus, W L Brown, E E Chaban, P F Szajowski, S B Christman, P H Citrin. Phys Rev B 52, 4910 (1995).

[50] Q Zhang, S C Bayliss. J Appl Phys 79, 1351 (1996).

[51] B Delley, E F Steigmeier. Appl Phys Lett 67, 2370 (1995).

[52] R J M da Fonesca, J M Saurel, A Foucaran, J Camassel, E Massone, T Taliercio. J Mat Sci 30, 35 (1995).

[53] D Bellet, P Lamagnere, A Vincent, Y Brechet. J Appl Phys 80, 3772 (1996).

[54] G Amato, L Boarino, G Benedetto, R Spagnolo. Thin Sol Films 255, 111 (1995).

[55] W Lang, A Drost, P Steiner, H Sandmaier. MRS Proc Vol 358, 561 (1995).

[56] A Loni, L T Canham, M G Berger, R Arens Fischer, H Munder, H Luth, H Arrand, T M Benson. Thin Solid Films 276, 143 (1996).

[57] T Matsumoto, N Hasegawa, T Tamaki, K Ueda, T Futagi, H Mimura, Y Kanemitsu. Jpn J Appl Phys 33, L35 (1994).

[58] C Tsai, K H Li, J C Campbell, A Tasch. Appl Phys Lett 62, 2818 (1993).

[59] L T Canham, T I Cox, A Loni, A J Simons. Appl Surf Sci 102, 436 (1996).

[60] N Koshida, H Koyama. Appl Phys Lett 60, 347 (1992).

[61] P Steiner, F Kozlowski, W Lang. Appl Phys Lett 62, 2700 (1993).

[62] A Loni, A J Simons, T I Cox, P D J Calcott, L T Canham. Electron Lett 31, 1288 (1995).

[63] R A Soref. Proc IEEE 81, 1687 (1993).

[64] K D Hirschmann, L Tsybeskov, S P Duttagupta, P M Fauchet. Nature 384, 338 (1996).

[65] L T Canham. Adv Mater 7, 1033 (1995).

Part II

Theory of Semiconductor Clusters, Nanocrystals, and Porous Silicon

MOLECULAR DYNAMIC SIMULATIONS OF SEMICONDUCTOR CLUSTERS

GIRIJA S. DUBEY and **GODFREY GUMBS**
Department of Physics and Astronomy, Hunter College, City University of New York, 695 Park Avenue, New York, NY 10021

ABSTRACT

Molecular dynamics simulations are carried out for silicon clusters to analyze the effect of heating and cooling on their structural properties. The calculations are based on the Kaxiras and Pandey potential which has been derived from a microscopic calculation with the density functional method. Results for the cluster formation are presented for different rates of cooling and heating. Our simulations clearly show the dependence of the patterns on the cooling and heating rates as well as the way in which these clusters evolve. Our results indicate that the structures are most stable when the rate of loss of thermal energy is slow and the potential energy is lowest.

INTRODUCTION

For several years, there has been a considerable amount of interest and activity in understanding the atomic structure and stability of closed carbon clusters — the fullerenes such as C_{60}. The reason why so much attention has been given to this subject is due to the potential applications in materials science [1,2] The exact geometrical structure of small-to-medium size clusters cannot be determined directly but their sizes can be inferred from various experiments. For example, cluster mass distribution spectroscopy provides information about the most abundant magic cluster sizes, and this can be used to deduce theoretically the structure of the most stable clusters [3]. From a theoretical point of view, the difficulty in predicting the geometry of the structures and their sizes arises from the existence of multiple local minima in the potential-energy surface used in computer simulations. Therefore, one must perform full simulated annealing computer experiments in order to ensure that a global minimum and the most stable structure are indeed obtained. Clearly, the arrangement of the atoms in a cluster will play an important role in the assembly process, and thereby influence the properties of the resulting material. In order to develop a comprehensive microscopic theory of clusters, it is necessary to understand the basic physics underlying their structures in the medium-to-large cluster limit, as well as the dynamical properties of both isolated and several clusters fused together.

In this paper, our aim is a first step towards understanding the effects of cooling and heating on the stability of silicon clusters as well as identifying microscopic structural changes underlying the cooling and heating rate. The results presented here may be relevant for annealing phenomena and identifying the differences between the results for models developed under different theoretical and experimental methods of preparation and that experimental data on the chemical reactivities of clusters can also be applied to deduce the sizes of the most stable clusters [4,5]. To explain the above physical facts, we have carried out molecular dynamics (MD) simulations for silicon clusters using the potential of Kaxiras and Pandey [6,7]. This potential is based on a microscopic calculation with the density functional method and does not have any artificial features which could give rise to unreliable results. Our calculations can be helpful in testing the accuracy of interatomic potentials that would be suitable for classical MD simulations, cluster fusion and for the assembly of cluster materials.

Mat. Res. Soc. Symp. Proc. Vol. 452 © 1997 Materials Research Society

DESCRIPTION OF THE MODEL AND NUMERICAL RESULTS

The interparticle interaction potential energy V is the vital input to MD simulations. This N-body term is commonly expressed as a sum of one-body, two-body and three-body potentials with, for example,

$$V = \sum_{i,j} V_2(r_i, r_j) + \sum_{i,j,k} V_3(r_i, r_j, r_k) + \cdots , \tag{1}$$

where the terms V_2 and V_3 are often constructed on the basis of experimental information. The two-body or pairwise interactions among particles are frequently the most important interaction terms in MD simulations. The three body terms are usually introduced in simulations involving covalent semiconducting materials and for predicting their atomic structures.

We carried out the MD simulations using a classical interatomic potential containing both two- and three-body interaction terms. The two-body potential of Kaxiras and Pandey [7] is given by

$$V_{ij} = A_1 \exp\left(-\alpha_1 r_{ij}^2\right) - A_2 \exp\left(-\alpha_2 r_{ij}^2\right) , \tag{2}$$

where $A_1 = 57.316072$, $A_2 = 6.4373054$, $\alpha_1 = 0.82335230$, $\alpha_2 = 0.190619$. The three-body interaction is

$$V_3 = V_{ijk} + V_{jki} + V_{kij} , \tag{3}$$

with V_{ijk} defined by

$$V_{ijk} = \exp\left\{-\beta\left(r_{ij}^2 + r_{jk}^2\right)\right\} \left[Bg^2(\theta) + Dg^4(\theta)\right] , \tag{4}$$

where $g(\theta) = \left(\cos\theta + \frac{1}{3}\right)$ and θ is a polar angle.

In our MD simulations, we used the Nosé Hamiltonian $H_{\rm N}$ [8] with the potential defined in Eqs. (1) - (4) and

$$H_{\rm N} = V + \frac{K}{s^2} + \frac{p_s^2}{2Q} + 4k_{\rm B}T\ln(s) , \tag{5}$$

where K is the total kinetic energy, T is the temperature and Q plays the role of a thermal inertia parameter which controls the fluctuations in temperature; both Q and the time-dependent scaling factor s are free parameters. The momentum $p_s = Q\dot{s}$ [8]. The equations of motion were integrated over time steps $\Delta t = 0.05$ fs.

We took a 45-atom silicon cluster and equilibrated it at 4000 K in 2×10^4 time steps with $\Delta t = 0.05$ fs for different heating and cooling cycles to obtain the most stable structure of the cluster; the cluster was slowly cooled to 20 K. Then it was reheated to 4000 K again and cooled back to 20 K. This heating and cooling cycle was repeated several times, in order to obtain defect free structures. Our calculations showed that when the cooling/heating rate is sufficiently slow, some of the atoms left the surface of this cluster at various stages of the heating and cooling process, but some of them eventually came back to the cluster while others evaporated completely so that the end result could be a silicon cluster with a reduced number of atoms. This cluster did not evaporate any more despite further heating and cooling cycles. Figure 1(a) shows the structure of this Si_{39} cluster at 20 K for a rate of 2.0 K/fs. These results show that for this rate the Si_{39} cluster consists of three closed clusters fused together. Even more careful examination revealed that there are even smaller fused clusters within this larger cluster. We separated this larger cluster into the following groups consisting of 2, 4, 13, 19, 20 and 33 atom clusters. Careful examination of these clusters revealed that the 2, 4, 13, 19 and 20 atom clusters are all completely surface in the sense that there no atoms trapped in the interior of the cluster. The 13, 19 and 20 atom clusters form closed shell structures, are three-fold coordinated, and they form a chain-like network. The significance of these results can be understood through a detailed analysis and comparison with the experimental data for the sticking probability of silicon clusters with C_2H_4 and NH_3 reagents [4,5]. A very important observation which we made concerning the results in Refs. [4,5] is that all the clusters we obtained from our MD simulation are the least reactive clusters observed experimentally. As a follow-up, we built models consisting of clusters of atoms using as a matrix the surface of these clusters thereby forming π-bonded chain networks similar to the 2×1 reconstruction on the (111) surface of silicon. We also calculated the bond lengths and bond angles of 2- and 4-atom clusters obtained from our MD simulations and found that they are within 5% of the exact calculations of Raghavachari [9]. The good agreement between these results give us confidence that our cluster formations are reliable.

Figure 1(b) shows the structure of a cluster of Si atoms at 20 K subjected to a different rate of 20.0 K/fs of annealing still starting at 4000 K with 45 atoms and cooling down to 20 K. The diagram shows the cluster formation only after one cycle of cooling in which only four atoms have evaporated, leaving 41 atoms behind. Comparing Fig. 1(a) with Fig. 1(b), it is clear that the rates of cooling and heating as well as the repetition of the cycles to equilibrium have a nontrivial effect on the structure of the cluster. These calculations for identifying the stable structures are time-consuming (since the MD technique spends a considerable amount of time in the neighborhood of metastable states). This implies that the structure has an optimum configuration depending on the rate at which the temperature is changed. The structures are most stable when the rate of loss of thermal energy is slow and, in this case, the potential energy is lowest. The time scales of our simulation experiments are from 10 to 100 ps. In Fig. 1(b), there are isolated dangling bonds for the rate of 20.0 K/fs, whereas the slower rate in Fig. 1(a) shows the cluster forms a regular structure after many cycles are repeated whereby many of the atoms evaporate. Annealing causes the atoms to form bonds, illustrating the metastability of this configuration.

Our simulations show that if the annealing process is sufficiently slow then it produces well-defined, stable structures. When our quenching rate is rapid, we often found many atoms trapped in the interior of the cluster, thereby demonstrating the metastability of the cluster. There are more trapped atoms for rates larger than the rate of 20 K/fs. For the rate of 20 K/fs, there are many trapped atoms, thereby resulting in overcoordinated configurations. By reducing the cooling rate even further to 5 K/fs, we find that only a few atoms are trapped but that the group symmetry of the atoms in the cluster which is found in Fig. 1(a) at 2 K/fs is absent. This implies that the self-organization has an optimum configuration depending on the rate at which the temperature is reduced.

For the rate of 75.0 K/fs shown in Fig, 2, there are a few isolated bonds in the cluster but the surface atoms are still tightly bound since they survive the cooling and heating process. There are many atoms trapped inside the cluster clearly illustrating the metastability. At this high rate, all 45 atoms remain after many cycles are repeated. This is in strong contrast to Figs. 1(a) and 1(b) where some of the atoms have left the cluster. Even more careful examination reveals that some bonds are disconnected, there are a few isolated bonds, single atoms are split off, surface atoms are not tightly bound. There is no sign that a regular stable structure will be formed at 75 K/fs as we obtained in Fig. 1(a) for a rate of 2 K/fs. Since for a faster cooling rate, the duration of the MD period is short, it becomes more unlikely to end up with a regular structure. The rates we chose in our calculations are similar to those used in other numerical simulations [10]. At these rates, rapid solidification can be achieved experimentally, using plasma spray techniques, based on the specific heat and size of the cluster.

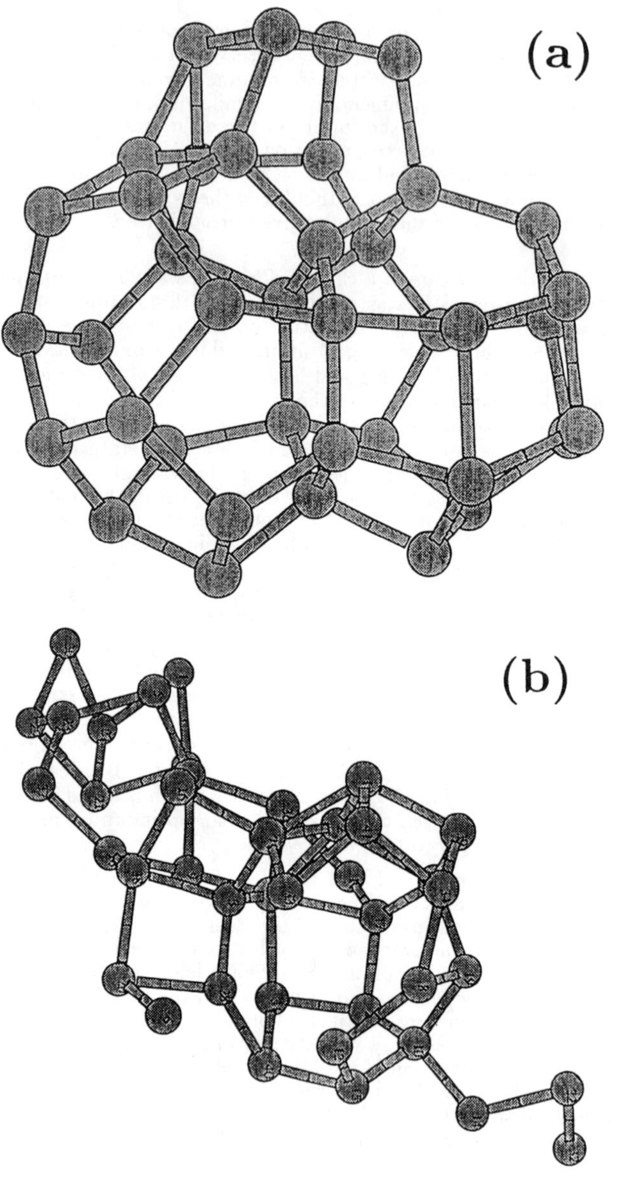

FIG. 1. (a) The structure of Si_{39} at 20 K from our MD simulations, for a heating/cooling rate of 2 K/fs, after many cycles were repeated and the structure equilibrated. (b) Starting with 45 Si atoms at 4000 K, we cooled the cluster down to 20 at a rate of 20 K/fs. Only once was this cooling process carried out.

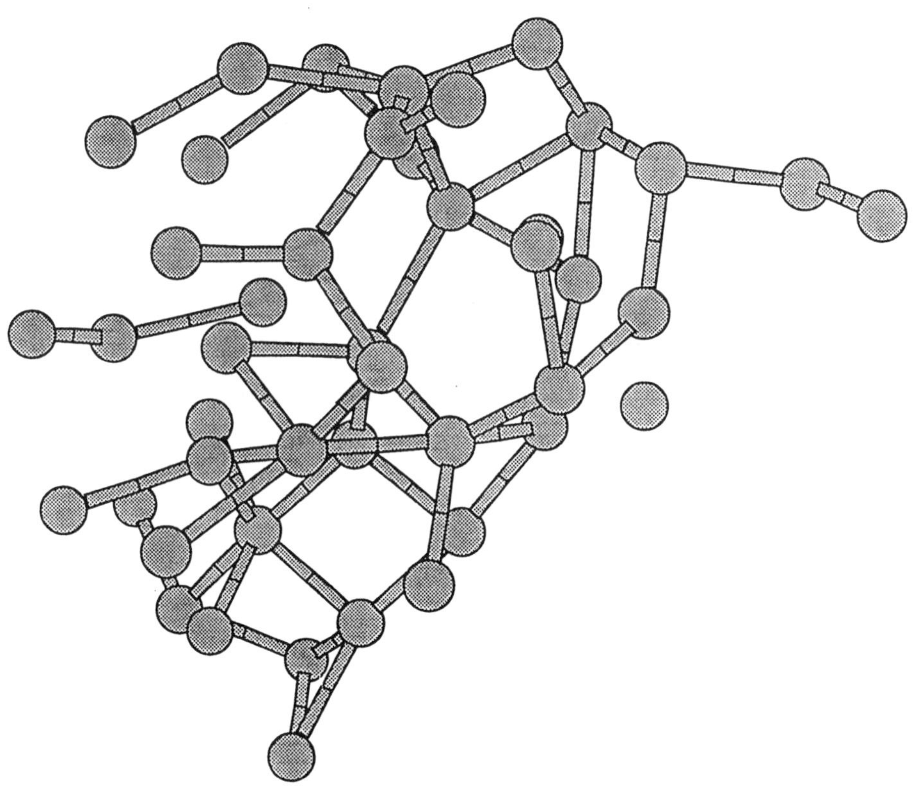

FIG. 2. Starting with 45 Si atoms at 4000 K, we cooled the cluster down to 20 at a rate of 75 K/fs. The structure equilibrated as shown at 20 K after many cycles were repeated but none of the atoms evaporated.

SUMMARY AND CONCLUSION

In summary, we studied the stability of silicon cluster as a function of the local-excitation processes and thermal annealing. The clusters are cooled down to 20 K, starting in the vapor state, and then reheated, with the heating and cooling cycle being repeated several times and at different rates to get rid of any defects and to see what effect the five different rates of cooling and heating have on the dynamics of the cluster formation. Our calculations showed that at various stages of the heating and cooling cycle, some of the atoms leave the cluster in an effort to minimize the surface energy of the structure, i.e., the structures are most stable when the rate of loss of thermal energy is slow and, in this case, the potential energy is lowest. The most stable structures are obtained when the rate of cooling is slow (~ 5 K/fs) and when the potential energy is a minimum. The lowest-energy structure we found is presented in Fig. 1(a). We have observed the generation of metastable pairs of bonds at the higher rates of cooling and heating; by further annealing, the weaker Si-Si bonds will disappear. One goal of our simulation is to determine what sort of structure is formed as we cool/heat these clusters. We also varied this rate to form the lowest-energy structure possible.

The MD simulation results in this paper make use of a very reliable potential to study the relationship between the stability and reactivity of semiconductor clusters to get a better understanding of the basic principles underlying the observed experimental behavior. We used the three-body potential of Kaxiras and Pandey [7] in our simulations. The experiments of Jarrold, et al. [4] and Smalley, et al. [5] on the reactivities of silicon clusters have clearly shown that small and intermediate sized silicon clusters exhibit size-dependent chemical reactivities. Other potentials give unrealistic and misleading results for the most stable and compact formations of clusters. The differences between the theoretical and experimental results are thus mainly due to the different model potentials used to simulate the structure.

ACKNOWLEDGMENTS

The author gratefully acknowledges the support in part from the City University of New York PSC-CUNY-BHE grant #666414.

1. H. W. Kroto, J. R. Heath, S. C. O'Brien, R. F. Curl, and R. E. Smalley, Nature **318**, 962 (1985).

2. W. Kratschmer, L. D. Lamb, K. Fostiroupoulos, and D. R. Huffman, Nature **347**, 354 (1990).

3. D. G. Pettifor, Solid State Physics **40**, 43 (1987).

4. M. F. Jarrold et al, J. Chem. Phys. **90**, 3615 (1989).

5. R. E. Smalley et al, J. Chem. Phys. **87**, 2397 (1987).

6. E. Kaxiras, Phys. Rev. Lett. **64**, 551 (1990).

7. E. Kaxiras and K. C. Pandey, Phys. Rev. B **38**, 12736 (1988).

8. S. Nosé, Mol. Phys. **52**, 255 (1984); J. Chem. Phys. **81**, 511 (1984).

9. K. Raghavachari, J. Chem. Phys. **84**, 5672 (1986).

10. N. Binggeli, J. L. Martins, and J. R. Chelikowsky, Phys. Rev. Lett. **68**, 2956 (1992).

ATOMIC AND ELECTRONIC PROPERTIES OF SMALL HYDROGENATED SILICON CLUSTERS: Si_6H_{2m} and $Si_6H^+_{2m+1}$

Takehide Miyazaki*,***, Ivan Stich**, Tsuyoshi Uda** and Kiyoyuki Terakura*
*JRCAT, National Institute for Advanced Interdisciplinary Research, **JRCAT, Angstrom Technology Partnership, ***Electrotechnical Laboratory, *,**1-1-4 Higashi, Tsukuba 305, Japan, ***1-1-4 Umezono, Tsukuba 305, Japan.

ABSTRACT

The atomic and electronic structures of Si_6H_{2m} and $Si_6H^+_{2m+1}$ clusters have been investigated in the framework of density-functional theory. For both neutral and ionized clusters we found the structure to belong to one of four distinct structural families. A molecular-orbital picture of hydrogenation is presented. From the calculated formation energies of these clusters, we infer the relative stability of the different structural families discussed.

INTRODUCTION

Due to their peculiar geometry, small silicon clusters (Si_n) with $n \sim 10$ have been subject of considerable interest in the past[1]. It has been well established that the Si_n clusters have compact structures which are quite different from the bulk crystal structure and that their electronic structure also dramatically deviates from that given by sp^3 hybridization in the crystal. However, because of the rapidly increasing complexity of the structures of larger clusters, a complete answer to fundamental questions such as the structure "evolution" of the clusters as a function of n has not been obtained yet[2].

The structure of hydrogenated silicon clusters (Si_nH_x) are even more complicated. The reason is that the stability of the Si_nH_x results from a delicate balance between two competing factors: the *global* stability of a "host" Si_n cluster and the *local* stabilization induced by attachment of H atoms and creation of sp^3-like configuration. We have recently performed *ab initio* optimization of neutral Si_6H_{2m} clusters as a typical example[3]. We found that the geometries of those clusters can be classified in terms of the arrangement of Si atoms in distinct structural families and demonstrated that the attachment of H atoms to the cluster proceeds in a systematic way.

In the present work we extend the *ab initio* study of the structure and energetics to the positively charged clusters, namely $Si_6H^+_{2m+1}$. We discuss the evolution of the structures of the Si_6 cluster due to hydrogenation in both neutral and ionized forms from an electronic-state viewpoint. We also discuss the relative stability of these clusters.

CALCULATION METHOD

The energetics and structural optimization are based on the density functional theory[4] augmented with the generalized gradient approximation[5] to the exchange-correlation energy. We used the DMol program[6], which uses the wavefunction expansion in the localized orbitals with double-numerical-plus-polarization (DNP) basis. For Si, the DNP

Mat. Res. Soc. Symp. Proc. Vol. 452 © 1997 Materials Research Society

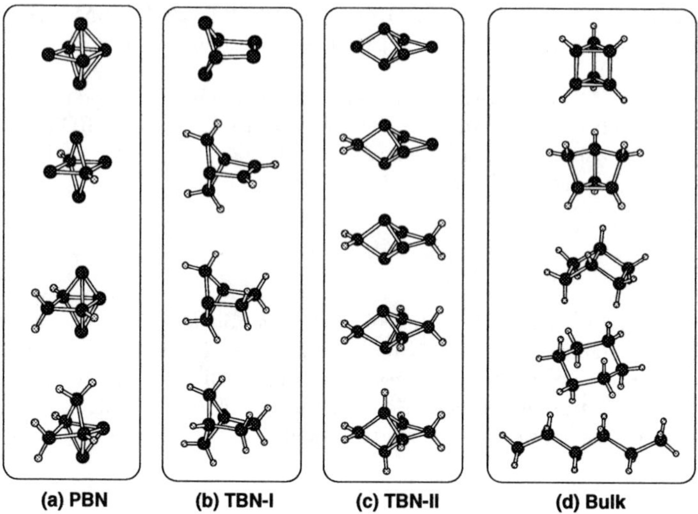

Figure 1: Calculated structures of Si_6H_{2m} clusters, classified into four geometry families in terms of the positions of Si atoms. Large, dark circles represent Si atoms, and small, bright ones are H atoms, respectively. The same convention applies also to Figs.2-4.

basis set contains all of the occupied orbitals of a neutral atom and the $3s$, $3p$ and $3d$ orbitals of an ion with a +2 charge. Occupancies of the $1s$ and $2s$ orbitals are frozen. For H, the basis is constructed of the $1s$ orbital of a neutral atom and the $1s$ and $2p$ orbitals of an ion with a +1.3 charge. Optimizations of the atomic coordinates were performed with the eigenvector-following method and continued until the residual atomic forces reduced below ~0.03 eV/Å.

RESULTS AND DISCUSSION

First we review the structure of the neutral Si_6H_{2m} clusters shown in Fig.1[3]. The most important feature is that the clusters can be classified into several distinct geometrical families according to the arrangement of Si atoms. We have found four different categories.

The first category (Fig.1(a)) is characterized by a compact structure of the Si atoms. Patterson and Messmer[7] called the structure of Si_6 in Fig.1(a) a polyhedral bonding network (PBN). The arrangement of Si atoms of Si_6 in the PBN is barely affected by hydrogenation, hence we classify the Si_6H_{2-6} clusters in Fig.1(a) into the PBN family.

The second and third families shown in Fig.1(b) and Fig.1(c) share some aspects of both compact and bulk-like configurations. Although the positions of Si atoms are very different from the diamond structure in the crystal, all of them are nearly tetrahedral when saturated with H atoms. Patterson and Messmer[7] call the structures of the Si_6 clusters in Fig.1(b) and Fig.1(c) tetrahedral bonding network (TBN).

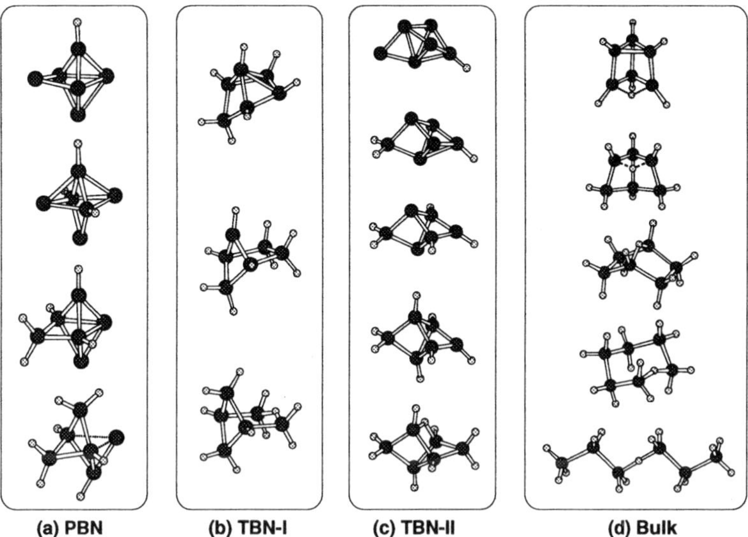

Figure 2: Calculated structures of $Si_6H^+_{2m+1}$ clusters, classified into four geometry families in terms of the positions of Si atoms.

The clusters shown in Fig.1(d) belong to a bulk-like structural family. At variance with first three categories (PBN, TBN-I and TBN-II), the connectivity among Si atoms changes at every step of the sequence.

Next we describe the structures of ionized clusters ($Si_6H^+_{2m+1}$). They were obtained by addition/removal of a H atom to/from the Si_6H_{2m}/Si_6H_{2m+2} clusters. We considered only the cases where the clusters have an even number of electrons, because experimental observations suggest considerable abundance of $Si_6H^+_{2m+1}$ relative to $Si_6H^+_{2m}$[8] . When adding an H atom to Si_6H_{2m}, a number of different attachment sites, which belong either to apexes, bond centers, inside of the n-membered rings (n = 3, 4, 5 and 6) or inside of the cluster cages, were investigated. On removal of an H atom from Si_6H_{2m+2}, all of the symmetrically irreducible cases were calculated. The obtained geometries of the ionized clusters (Fig.2) were essentially very similar to those of the neutral ones. They can also be classified into the same four structural families: PBN, TBN-I, TBN-II, and bulk-like frameworks of the Si atoms.

Now we present a molecular-orbital picture of hydrogenation of the Si_6 and Si_6^+ clusters. The structure of Si_6 in Fig.1(a) originates from a regular octahedron distorted by a compressive T_{1u} mode. As a result the symmetry of the structure is reduced from O_h to D_{4h} with an energy gain of 1.2eV and the triply degenerate HOMO of the regular octahedron (Fig.3(a)) is split into a doubly degenerate HOMO (Fig.3(b)) and a non-degenerate LUMO (Fig.3(c)) with energy gap of 2.2eV. The compactness of the geometry of Si_6 is caused by the "s-p separation"[1] in the valence electronic structure.

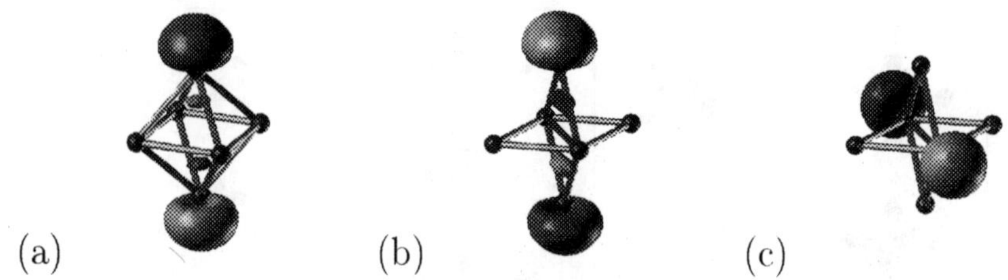

Figure 3: HOMO's of the Si_6 cluster with (a) O_h and (b) D_{4h} symmetries. The former is triply degenerate since there are two other equivalent MO's distributed in the two other orthogonal axes. The latter is doubly degenerate. Panel (c) illustrates LUMO of the Si_6 cluster with D_{4h} symmetry. Different signs of the wavefunction are illustrated in different colors.

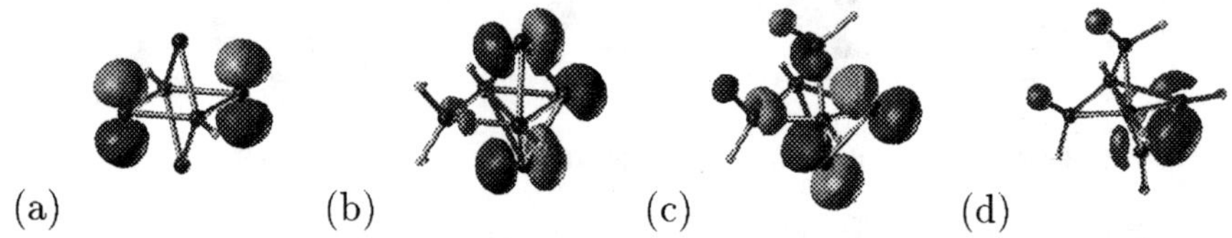

Figure 4: (a) Second LUMO of Si_6H_2, (b) LUMO of Si_6H_4, (c) second LUMO of Si_6H_6, and (d) second HOMO of Si_6H_8, respectively.

In the process of $Si_6H_{2m-2} + H_2 \rightarrow Si_6H_{2m}$, the two H atoms preferentially attach to the Si atoms where the LUMO wavefunction of Si_6H_{2m-2} has a large weight (for example, Si_6 (Fig.3(c)) + $H_2 \rightarrow Si_6H_2$ (Fig.4(a)) and Si_6H_4 (Fig.4(b)) + $H_2 \rightarrow Si_6H_6$ (Fig.4(c))). This is because the bonding state produced by the hybridization between H 1*s* and the LUMO may produce a large energy lowering. In cases where the overlap of the H 1*s* with the LUMO is small, the second LUMO will be used (for example, Si_6H_2 (Fig.4(a)) + $H_2 \rightarrow Si_6H_4$ (Fig.4(b)) and Si_6H_6 (Fig.4(c)) + $H_2 \rightarrow Si_6H_8$ (Fig.4(d))). The Si_6H^+ cluster in the PBN is obtained by attaching one H atom to the HOMO of the Si_6 cluster (Fig.3(b)). The hydrogenation sequence, $Si_6H^+_{2m-1} + H_2 \rightarrow Si_6H^+_{2m+1}$, proceeds in the same way.

We note that hydrogenation of the PBN Si_6 and Si_6^+ clusters does not necessarily recover the sp^3-like structure (*c.f.* Si_6H_2 (Fig.4(a))). This suggests the stability of the *s*-*p* separation in the valence electronic states in the PBN Si_6 cluster and explains why the skeletal framework of Si_6 in the PBN is not significantly affected by hydrogenation. Sequential hydrogenation of Si_6 in both the TBN-I and TBN-II proceeds in a similar manner as in the PBN.

In contrast, hydrogenation of the Si_6H_{2m} clusters in the "bulk" family is accompanied by change in the connectivity of the framework at every step of hydrogenation. Starting from hexasilabenzene (Si_6H_6), the sequential hydrogenation generates the linear chain of Si_6H_{14} by breaking a Si-Si bond of a specific pair of Si_6H_{2m-2} and then terminating the

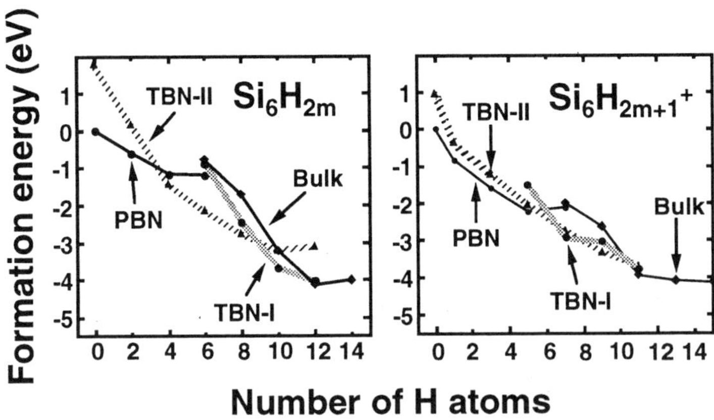

Figure 5: Formation energies of Si_6H_{2m} and $Si_6H_{2m+1}^+$ clusters.

broken bonds with two H atoms to obtain Si_6H_{2m}. The hydrogenation of the $Si_6H_{2m+1}^+$ clusters in this family is characterized by penetration of an H atom into the bond-center (BC) site of the clusters. The specific bonding in this Si-H-Si complex can be identified to be a three-centered bond[9].

Finally we turn to the calculation of the formation energies of the Si_6H_{2m} and $Si_6H_{2m+1}^+$ clusters (Fig.5). We have calculated the formation energy of the Si_6H_{2m} ($Si_6H_{2m+1}^+$) cluster, taking a system of the Si_6 (Si_6^+) cluster in the PBN plus m molecules ($m+\frac{1}{2}$ molecules) of H_2 as a reference. Zero-point energies of H atoms (0.134 eV/H for H_2 and 0.21 eV/H for H-Si bond [10]) were included in the calculation.

The changes in the formation energies as functions of the number of H atoms, x, are systematic. In order to see this, we first discuss the case of the neutral clusters. One can immediately see that the PBN clusters are the most stable in the low-coverage regime ($x \leq 2$) and that the H-saturated crystal fragments are stabilized in the high-coverage counterpart ($x \geq 12$). In the intermediate region ($4 \leq x \leq 10$), the TBN clusters are energetically favorable. It is important to note that the number of dangling bonds is minimized in the PBN, and that it becomes 8 and 10 in the TBN-II and TBN-I, respectively. In the "bulk" family, it increases to 14, the maximum number. Thus the systematic change in the relative stability of the Si_6H_{2m} clusters is a consequence of a balance between two competing factors, (1)minimization of the number of dangling bonds by removing as many sp^3-like configurations as possible at a given H-content, and (2)maximization of the energy gain due to hydrogenation by the recovery of as many sp^3-like arrangements as possible.

A similar trend is seen in the case of cation clusters in low- ($x \leq 3$) and high-coverage ($x \geq 11$) regimes, mainly because the principal part of the energetics is governed by the formation of a compact structure in the former and by saturation of the sp^3 configurations in the latter, respectively. However, the relative stability of the ionized clusters for medium coverages ($5 \leq x \leq 9$) is considerably different from the corresponding neutral case. In this regime, the two factors mentioned above may compete with each other with roughly equal strengths. Therefore changes in the charge state would induce large difference in the relative stability.

CONCLUSION

We have performed *ab initio* structure optimization and calculation of the formation energies of Si_6H_x and $Si_6H_x^+$ clusters. The cluster structures can be classified into at least four categories according to the structure of the Si framework. The categorization applies both to neutral and positively ionized clusters. Hydrogenation of silicon clusters proceeds either by saturation of the LUMO's with the character of dangling bonds or, where such localized LUMO's are not available, by creation of a three-centered bond. The calculated formation energy curves suggest that the relative stability in the medium-coverage regime is very sensitive to the charge state as well as the H-content.

ACKNOWLEDGMENT

One of the authors (TM) thanks T. Kanayama and M. Watanabe for fruitful discussions. TM also thanks K. Tanaka for his continuous encouragement. This work was partly supported by NEDO through the management of the Angstrom Technology Partnership (ATP).

References

[1] K. Raghavachari, Phase Trans. **24-26** (1990) 61.

[2] E. Kaxiras and K. Jackson, Phys. Rev. Lett. **71** (1993) 727.

[3] T. Miyazaki, I. Stich, T. Uda and K. Terakura, Mat. Res. Soc. Symp. Proc. **408** (1996) 533; T. Miyazaki, T. Uda, I. Stich and K. Terakura, Chem. Phys. Lett. **261** (1996) 346.

[4] P. Hohenberg and W. Kohn, Phys. Rev. **136** (1964) B864; W. Kohn and L.J. Sham, Phys. Rev. **140** (1965) A1133.

[5] A. D. Becke, Phys. Rev. A**38** (1988) 3098; J.P. Perdew, in "Electronic Structure of Solids '91", edited by P. Ziesche and E. Eschrig (Akademie Verlag, Berlin, 1991).

[6] *DMol User Guide*, October 1995 (Biosym/MSI, San Diego, 1995).

[7] C.H. Patterson and R.P. Messmer, Phys. Rev. B**42** (1990) 7530.

[8] T. Kayanama, private communication.

[9] C.G. Van de Walle, Physica B**170** (1991) 21.

[10] C.G. Van de Walle, Phys. Rev. B**49** (1994) 4579.

[11] H. Katagiri, Solid State Commun. **95** (1995) 143.

COMPARATIVE THEORETICAL STUDY OF AMORPHOUS AND CRYSTALLINE SILICON CLUSTERS

M. LANNOO, C. DELERUE,G. ALLAN,
Institut d'Electronique et de Microélectronique du Nord, Département Institut Supérieur d'Electronique du Nord, BP 69, 59652 Villeneuve d'Ascq Cedex, France, gal@isen.fr

ABSTRACT

The occurence of confinement effects for amorphous Si and Si:H clusters is investigated theoretically in an empirical tight binding treatment. The results show that one must consider three categories of states: strongly localized, weakly localized and delocalized. In all cases a substantial blue shift with reduction in size is obtained, of comparable magnitude for amorphous a-Si:H and crystalline clusters. These results are compared with available experimental data

INTRODUCTION

The recent observation of photoluminescence from porous silicon [1] raised the hope of realizing silicon optical devices. If the photoluminescence efficiency is very low in bulk silicon due to its indirect gap electronic structure, the situation improves in porous silicon as the electron wave vector is no longer a good quantum number [1-4]. However the radiative recombination rate remains low and might be a limitation to reach a high electroluminescence efficiency. A photoluminescence high efficiency has been obtained in amorphous silicon due to structural disorder [5]. Both effects (breakdown of the wave vector selection rule and structural disorder) if they are combined could lead to amorphous silicon-based integrated optoelectronics. Photoluminescence has been effectively observed in porous amorphous silicon [6-12]. The experimental results are rather similar to those obtained for porous crystalline silicon photoluminescence but are contradictory as regards the evidence of a confinement related blue shift. The aim of this paper is thus to provide some theoretical informations on these questions, particularly on the relative blue shift in amorphous and crystalline silicon clusters.

LOCALIZED VERSUS DELOCALIZED STATES

To calculate the electronic structure of amorphous silicon (a-Si) clusters, one has first to determine their atomic structure. The starting structure is obtained by isolating the atoms in a sphere of a given diameter in a 4096 atoms periodic cluster determined using the Wooten-Winer-Weaire (WWW) method [13-14], known to produce a correct radial distribution function. There is no dangling bond in this initial network. To simulate amorphous clusters, the center of the sphere is randomized. Dangling bonds occurring on atoms close to the surface of the sphere are saturated by hydrogen atoms. Even if the initial WWW atomic model was relaxed, the atoms in the sphere are no more in equilibrium positions since the boundary conditions are modified. We have thus relaxed the atomic positions in the sphere using a Keating potential [15] identical to the one used to build the initial periodic cluster. Two initial networks (with and without four-membered rings) have been used and give very close results. All the results in this paper have been obtained with the network with four-membered rings.

The electronic structure is calculated by the sp^3s^* empirical tight-binding method [16]. The Hamiltonian between silicon atoms includes interactions up to nearest neighbors. Associated with a R^{-n} dependence upon distance they produce reasonable deformation potentials. Tight-binding

Mat. Res. Soc. Symp. Proc. Vol. 452 © 1997 Materials Research Society

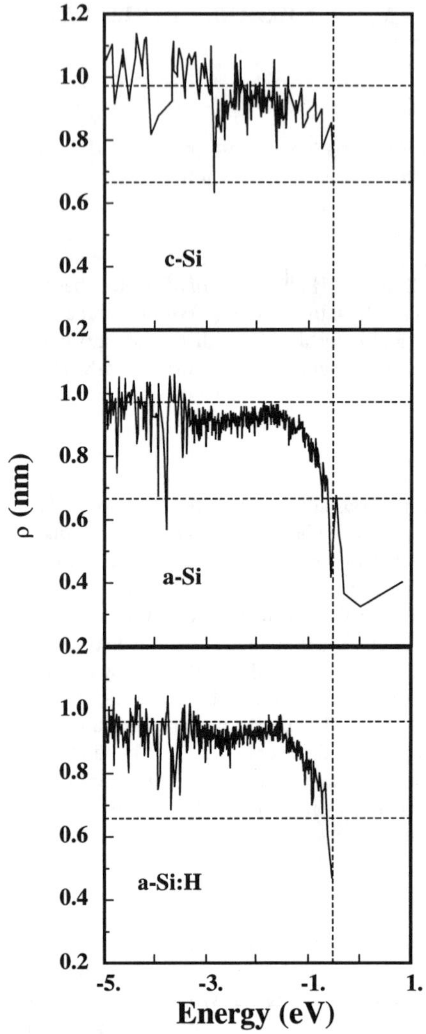

Fig. 1: Radius of cluster eigenstate as a function of the energy level. The clusters respectively have about 420 silicon atoms. Their diameter is close to 2.5 nm. The vertical dotted line shows the top of the crystalline cluster valence band. The horizontal lines give two limits: deep in the valence band, the radius is equal to 1.25 R for a uniform eigenstate and close to the gap, it is equal to 0.53 R for an effective mass state sin(kr)/r.

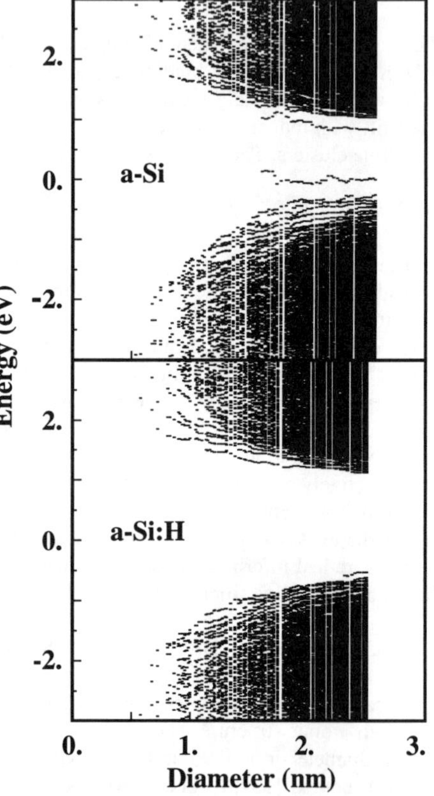

Fig.2: Confinement energy for an amorphous cluster as a function of its diameter for a-Si and a-Si:H. The center of the clusters are kept constant. Each column of points represents the energy levels of a cluster.

parameters between the silicon and hydrogen atoms have been obtained from Harrison's expressions [17]. Figure 1 compares the average radii $\rho\ (E_n)$ of the eigenstates of energy E_n for a crystalline cluster and an amorphous one. This is defined as the root mean square distance from the center of gravity of the distribution defined by the wave function. The diameters of the crystalline and amorphous clusters on Figure 1 are close to 2.5 nm (~420 silicon atoms in the clusters). One can see that the variation of the eigenstate radii as a function of the energy are very close for both these clusters and do not depend very much on the cluster atomic structure. However, in the energy range of the crystalline cluster gap, one finds for the amorphous cluster some levels which are more localized. Figure 1 also shows a similar plot for amorphous Si:H(a-Si:H) model clusters whose structure is obtained by passivating most of the strongly localized states. This was achieved by considering clusters of 2 nm size, varying the position of their center to explore the full 4096 atoms unit cell, identifying near gap states with an average root mean square extension in space smaller than 4 Å, removing the central Si atom and saturating the dangling bonds with H atoms. This is a somewhat empirical procedure but at the end it must produce a situation close to an optimized a-Si:H structure. Once this is done the atomic positions are again relaxed by use of a Keating model. Figure 1 shows that this results in a localization of the states much closer to what is obtained for crystallites, except in the vicinity of the gap.

Figure 2 presents typical results for the energy levels of a-Si clusters versus size and for a fixed center of the sphere. They show in a spectacular way the fact that the strongly localized states of the amorphous Si clusters are not shifted in energy by the reduction in cluster size. This is not true for the other states, more distant from the gap which experience a substantial shift.

For what follows we find it useful to classify these states into three categories:

- delocalized states, experiencing the full confinement effect as for c-Si
- strongly localized states with extension in space much smaller than the cluster diameter and energies deep in the gap, insensitive to the confinement effect and showing no blue shift
- weakly localized states with extension in space of the order of the cluster diameter and energies near the gap limits, subject to an intermediate blue shift.

Figure 2 also shows that all the states of our amorphous Si:H clusters are shifted with size and belong to the weakly localized or the delocalized category.

BLUE SHIFT OR NOT BLUE SHIFT

For luminescence the basic quantity of interest is the fundamental gap, i.e. the distance in energy between the HOMO (highest occupied molecular orbital) and the LUMO (lowest unoccupied molecular orbital). The statistical distribution of this quantity is plotted on Figure 3 for two cluster sizes, 2.2 and 1.2 nm. We find a substantial blue shift in both cases, more important for a-Si than for a-Si:H. Furthermore our larger a-Si clusters give rise to a two-peak distribution. This can be explained by the fact that a large quantity of the 2.2 nm clusters contain strongly localized states which give rise to a small HOMO-LUMO gap. This is no more the case of the smaller 1.2 nm a-Si clusters. The apparent blue shift in a-Si clusters has thus two origins: i) the varying proportion of clusters with strongly localized states and ii) the normal confinement effect on the other states. This is confirmed on the same figure by the a-Si:H clusters (dashed lines) which show only the second type behavior.

Figure 4 shows the average HOMO-LUMO gap versus size for clusters with randomly chosen center in the 4096 unit cell compared to the same quantity for c-Si clusters. As expected from the discussion on Figure 3 the lowest values correspond to a-Si while the blue shift for a-Si:H is surprisingly close to c-Si clusters. The second quantity of experimental interest is the radiative lifetime. This has been calculated along the lines of ref [18]. Figure 5 shows that for small clusters (1.2 nm) the results for a-Si and a-Si:H are comparable to c-Si, i.e. the lifetime is dominated by

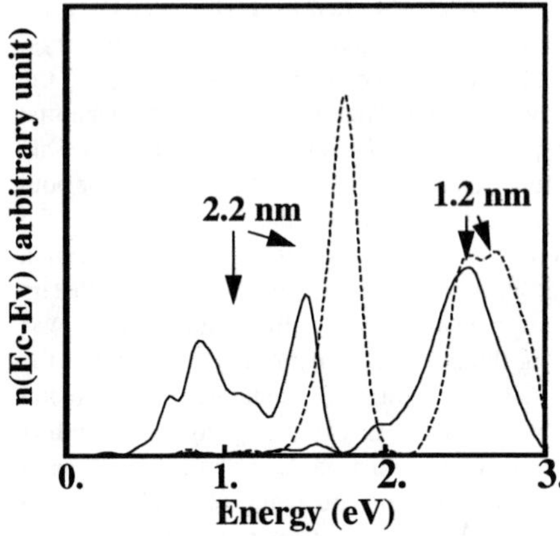

Fig. 3: Density of states n(E) for $E=E_{LUMO}-E_{HOMO}$ calculated for 200 clusters with 1.2 nm size and 2.2 nm size: a-Si (full line), a-Si:H (dotted line)

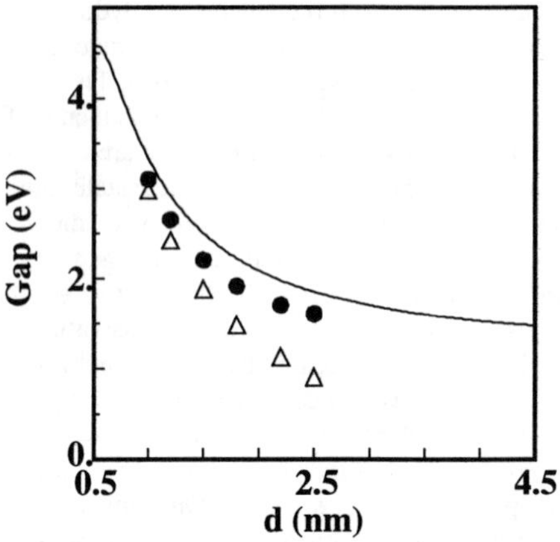

Fig. 4: Average HOMO-LUMO gap of amorphous clusters with randomized centers compared to crystallites: a-Si (△), a-Si:H (●), and c-Si (full line).

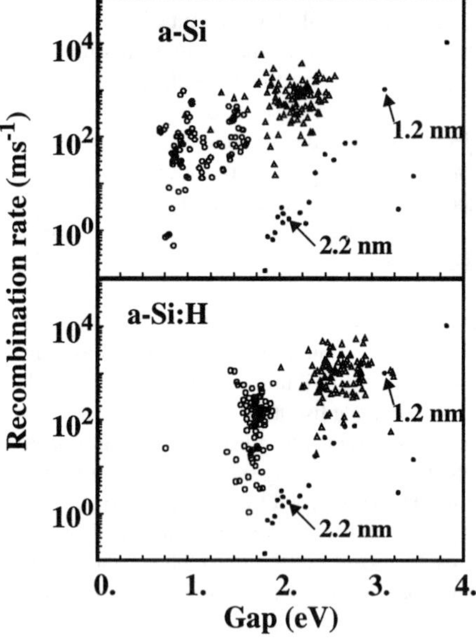

Fig. 5: Recombination rates amorphous silicon clusters with randomized centers and diameters equal to 1.2 nm (△) and 2.2 nm (○). The arrow points to the results obtained for nanocrystallites with diameters equal to 1.2 and 2.2 nm.

the breaking of selection rules. This is no longer true for the larger (2.2 nm) clusters for which the radiative lifetime of a-Si clusters (hydrogenated or not) is two orders of magnitude larger than for c-Si clusters due to disorder induced effects.

The main message one can extract from these calculations is that confinement effects do exist for a-Si and a-Si:H clusters at least in the range of sizes (1 to 2.5 nm) for which we have been able to perform calculations. In particular the predicted apparent blue shift is almost as important as for the c-Si clusters.

DISCUSSION

Comparison with experiment requires some comments about the models used to describe the atomic structure. The WWW model gives a good radial distribution function of a-Si and should thus provide a realistic description of its short range disorder [13-14]. Our a-Si:H model is more artificial but contains two essential features: suppression of deep gap states and correct order of magnitude of the hydrogen content (this should lead to an elastic mean free path comparable to the observed one ~ 1nm). What is absent from both models is the effect of long range fluctuations which cannot be produced by a 4096 atoms periodic cell. The effect of such fluctuations can be understood qualitatively by considering for instance that the regions of attractive potential will induce confinement effects in the conduction band whereas it simply shifts the valence band density of states to lower energies. The situation is symmetric in the regions of repulsive potential: confinement for hole valence states and shift of the conduction band. The net result is an overall blue shift of the states with the creation of exponential tails of weakly localized states in the confined regions in much the way as described by Ziman [21].

This description has some similarities with the model developed by Estes and al [20]. However in that work the tails are assumed to be strongly localized states which do not feel the effect of confinement. On the contrary in our case these states are built from the weakly localized or delocalized states of our a-Si:H model. They are located in the confined regions induced by the longer range attractive potential fluctuations and thus all become weakly localized states. The difference with Estes et al is that these states must experience a substantial blue shift which will add to the one already obtained by these authors and resulting from statistics on their spatial distribution.

Among the experimental data some show evidence of a blue shift with reduction in size [11-12] while others do not [8-9] or in some cases exhibit a small effect [8]. A common statement of most papers is that a-Si:H clusters should not be subject to quantum size effects due to the short mean free path, of order 1 nm. Our results conclude on the contrary that one should expect blue shifts with reduction in size comparable to what is obtained for c-Si clusters.

Although there are certainly several interpretations for the non observation of this effect in some cases we believe that a likely explanation is the formation of selftrapped excitons, proposed for bulk a-Si [5] and by us [21] for c-Si clusters. Indeed, in [21], we have been able to show that such selftrapped excitons are favored by the existence of local strains which are likely to be present in a-Si but should also strongly depend upon the method of preparation of the materials.

In conclusion we have calculated the confinement effect and radiative lifetime for amorphous silicon clusters and compared it to the corresponding crystallites. We find that the blue shift of a-Si:H and c-Si clusters is comparable. The radiative lifetimes are also comparable for small clusters (~ 1 nm) but become two orders of magnitude larger for the larger (~ 2.2 nm) a-Si:H clusters than for their crystalline counterparts.

ACKNOWLEDGMENTS

We would like to thank Drs B. R. Djordjevic, M. F. Thorpe, F. Wooten and D. Weaire for providing us the coordinates and neighbor tables of the silicon amorphous network we have used. The "Institut d'Electronique et de Microélectronique du Nord" is "Unité Mixte 9929 du Centre National de la Recherche Scientifique".

REFERENCES

1 L. T. Canham, Appl. Phys. Lett. **57**, 1046 (1990).
2 A. Halimaoui, C. Oules, G. Bomchil, A. Bsiesy, F. Gaspard, R. Herino, M. Ligeon, and F.Muller, Appl. Phys. Lett. **59**, 304 (1991).
3 A. Bsiesy, J. C. Vial, F. Gaspard, R. Hérino, M. Ligeon, F. Muller, R. Romestain, A. Wasiela, A. Maimaoui, and G. Bomchil, Surface Science **254**, 195 (1991).
4 V. Lehmann and U. Gosele, Appl. Phys. Lett. **58**, 856 (1991).
5 R. A. Street, Adv. Phys. **30**, 593 (1981).
6 E. Bustarret, M. Ligeon, and L. Ortega, Solid State Comm. **83**, 461 (1991).
7 E. Bustarret, E. Sauvain, M. Ligeon, and M. Rosenbauer, Thin Solid Films E-MRS 95.
8 R.B. Wehrspohn, J.-N. Chazalviel, F. Ozanam, and I. Solomon, Phys. Rev. Lett. **77**, 1885 (1996).
9 M. J. Estes, L. R. Hirsch, S. Wichart, and G. Moddel (To be published).
10 S. Lazarouk, S. Katsuba, N. Kazuchits, G. De Cesare, S. La Monica, G. Maiello, E. Proverbio, and A. Ferrari, "*Microcrystalline and Nanocrystalline Semiconductors*", L. Brus, M. Hirose, R.W. Collins, F. Koch, C.C. Tsai Eds. (Mat. Res. Soc. Proc. **358**, Pittsburgh, PA, 1995)
11 H.V. Nguyen, Y. Lu, S. Kim, M. Wakagi and R.W. Collins, Phys.Rev.Lett. **74**, 3880 (1995).
12 Z.H. Lu, D.J. Lockwood and J.-M. Baribeau, Nature **378**, 258 (1995); D. J. Lockwood, Z. H. Lu, and J.-M. Baribeau, Phys. Rev. Lett. **76**, 539 (1996).
13 F. Wooten, K. Winer, and D. Weaire, Phys. Rev. Lett. **54**, 1392 (1985); F. Wooten, and D. Weaire, Solid State Phys. **40**, 1 (1987).
14 B. R. Djordjevic, M.F. Thorpe, and F. Wooten, Physical Review B **52**, 5685 (1995).
15 P. N. Keating, Phys. Rev. **145**, 637 (1966).
16 P. Vogl, H. P. Hjalmarson, and J. D. Dow, J. Phys. Chem. Sol. **44**, 365 (1983).
17 W.A. Harrison, *Electronic Structure and the Properties of Solids* (Freeman, San Francisco, 1980).
18 J.-P. Proot, C. Delerue and G. Allan, Appl. Phys. Lett. **61**, 1948 (1992)
19 J. M. Ziman, Models of disorder, Cambridge University Press, p. 481 (1979)
20 M. J. Estes, and G. Moddel, Appl. Phys. Lett. **68**, 1814 (1996).
21 G. Allan, C. Delerue, and M. Lannoo, Phys. Rev. Lett. **76**, 3038 (1996).

AB-INITIO CALCULATION OF THE OPTICAL PROPERTIES OF SILICON QUANTUM WIRES

Stefano OSSICINI, M. BIAGINI, C. M. BERTONI, G. ROMA

INFM and Dipartimento di Fisica, Università di Modena, Via Campi 213/A, I-41100, Modena, Italy

O. BISI

INFM and Dipartimento di Fisica, Università di Trento,I-38050 Povo, Italy

ABSTRACT

We studied the effect of H, O passivation and inter-wire interaction on the optical properties of nanoscale Si wires. We find that wires with diameters as small as 10-25 Å are active in the visible range. Inter-wire interaction leads to the presence of localized states which lower the band gap energy. The presence of dangling bonds generates broad features in the infrared region. O-Si bonds reduce the absorption threshold. These results are important for the discussions concerning absorption and luminescence in porous Si.

INTRODUCTION

The optical absorption threshold and the visible luminescence (PL) in porous silicon (PS) show a large blue shift, as compared to bulk Si, which has been attributed to quantum confinement effects [1]. However the structure (wires [1,2] or dots [3]) and the size of Si crystallites [2,3] that are responsible for PL are controversial. The calculated band gaps for the proposed dimensions of the emitting nanoparticles are larger than the experimental PL energies [4]. Moreover it has been reported that PL occurs below the absorption gap [2,5]. Another important issue is related to the chemical composition of PS and to its effects on PL. To study these aspects we present results for the calculation of the optical properties of Si quantum wires (QW) extending our previous work [6]. The analysis has been performed on QW of different size to elucidate the role of the quantum size effect and has considered QW with totally or partially H-covered surfaces to understand the importance of surface passivation. We have also considered the presence of O-H complexes at the interfaces to show the role of O with respect to the PL properties.

THEORY

We use both the Linear Muffin Tin Orbitals (LMTO) method in the Atomic Sphere Approximation (ASA) and the norm-conserving pseudopotential (PP) method within the supercell technique. Exchange and correlation effects are described within the Density Functional Theory (DFT) in the Local Density Approximation (LDA). For the PP method we found that using a 10 Rydberg energy cutoff and sampling the Brillouin zone with the Γ-point only was enough to provide a value of the energy gap that converged within 2%. We have used the PP calculations in order to study the stability of the wire structures; the relaxation of these structures was done by minimizing the forces. The structures investigated consist of infinite wires of rectangular cross section with the axis along the [001] direction. We have considered QW of different cross section. In the larger

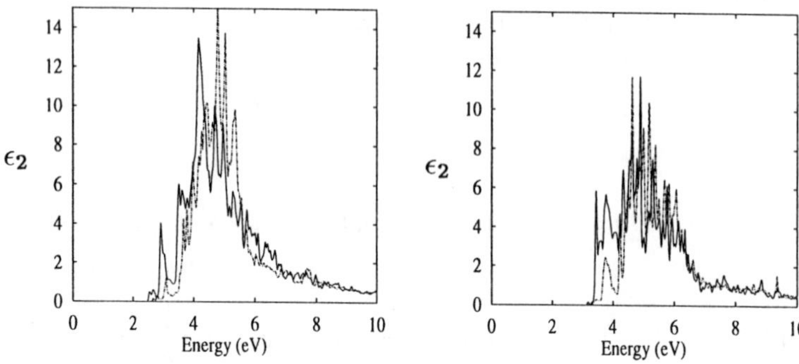

Figure 1: (a) Imaginary part of the dielectric function ϵ_2 for the 5x4 Si quantum wire, after a gaussian broadening of 0.02 eV. z polarization (solid line); average between x and y polarizations (dotdashed line). (b) The same for the 3x4 Si quantum wire

wire the cross section is delimited by zig-zag chains of 7 and 4 Si atoms,the edges being respectively 11.52 Å and 5.76 Å long. This structure is referred as 7x4. The smaller wire has a 3x4 structure of cross section 3.84 x 5.76 $Å^2$. The intermediate ones has a 5x4 structure (7.68 x 5.76 $Å^2$). In the fully passivated wire all the dangling bonds are saturated by H atoms. We use crystalline-like Si-Si distance and a Si-H distance of 1.64 Å [7]. By varying the size of the vacuum region we move from isolated, non interacting wire to structures, where strong wire-wire interaction is present. The structures describing the partial passivated QW are obtained by removing the H atoms, those describing O-Si interaction are obtained substituting some of the H atoms with a O-H complex, where the Si-O distance is 1.64 Å, and the O-H distance is 1.00 Å [8]. The calculation of the optical properties is performed through the evaluation of the dipole matrix elements within LMTO-ASA [6]. This allows to evaluate directly the imaginary part of the dielectric function ϵ_2 :

$$\epsilon_2^{\alpha}(\omega) = \frac{4\pi^2 e^2}{m^2\omega^2} \sum_{v,c} \frac{2}{V} \sum_{k_z} |<\psi_{c,k_z}|p_\alpha|\psi_{v,k_z}>|^2 \delta[E_c(k_z) - E_v(k_z) - \hbar\omega] \qquad (1)$$

It is possible to have informations about the PL processes looking at the features of ϵ_2, containing the various inter-band transitions, weighted by the momentum optical matrix element. The real part, $\epsilon_1(\omega)$, is obtained by a Kramers-Kronig transformation of $\epsilon_2(\omega)$ in which a tail of the same form used in Ref. 9 is used for energies greater than 12.0 eV.

RESULTS AND DISCUSSION

Quantum confinement.

The ϵ_2 functions of the fully H-passivated 5x4 and 3x4 QW are shown in panels (a) and (b) of Fig. 1 [6]. From the figure it is evident that the blue shift of the gap increases when the dimensions of the QW are reduced. The LDA gap of the 5x4 Si wire is 2.58 eV, while that of the 3x4 is 3.27 eV. Taking into account the self-energy correction to our DFT values of the gap (about 0.8 eV for QW of the dimensions considered in this study)

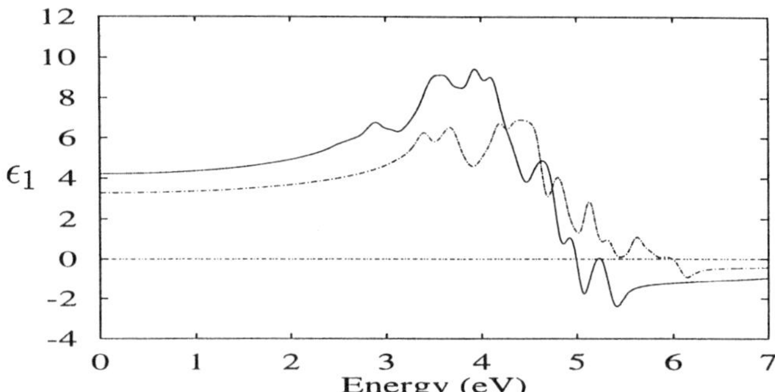

Figure 2: Real part of the dielectric function ϵ_1 for the 5x4 (solid line) and 3x4 (dotdashed line) Si quantum wires.

our results are not so far from the corresponding experimental results for absorption in PS: Zhang and Bayliss [2] found optical band gaps of 2.3, 2.6 and 3 eV for fresh PS samples, whose wire diameters have been estimated to be 22, 19 and 13 Å respectively. Furtherly, both Fig. 1(a) and 1(b) show well defined structures in the low energy side of the main peak of ϵ_2. These features, not present in bulk Si, are mainly related to the polarization parallel to the wire axis along z and are strongly size dependent, varying like the energy-gap. The dominant role of the z polarization transitions near the energy threshold shown in Fig. 1 is due to the one-dimensional structure of the wire. It is interesting to note that polarization studies of luminescent PS show that a high degree of polarization of luminescence can be generated [10]. Fig. 2 shows the results for the real part of the dielectric function $\epsilon_1(\omega)$ obtained via Kramers-Kronig transformation of the total $\epsilon_2(\omega)$; the oscillations at higher energies reflect those of ϵ_2 and are due to the quantum confinement effects. We note a strong reduction of the static dielectric constant $\epsilon_1(0)$ on going from the data for bulk Si (11.4) to 4.2 and 3.3 for our 5x4 and 3x4 Si QW. Koshida et al. [11] found a value of approximately 3 for a PS sample, whose porosity was about 70%.

Wire-wire interaction

The experimental evidence that in PS the emission occurs well below absorption (the difference ranges [5,2] from 0.2 eV to 0.6 eV) points towards the presence of some localized states acting as recombination-centers for the transition. Surprisingly when small crystalline particles are completely disconnected, as in the case of colloidal suspensions, no difference is observed between absorption and emission energy [12]. Looking in this direction we investigate the optical properties of interacting Si QW obtained by reducing the vacuum between the wires. Fig. 3 compares the LMTO results of the z-component of ϵ_2 (which is the dominant one [6]) for the interacting and non-interacting 5x4 wire. The role of the interactions between the wires is clear. The gap is strongly reduced (from 2.58 to 1.51 eV) by the presence of localized states. These states are strongly localized at the surface of the wires and their tails extend in the vacuum region and overlap the neighboring wires; they are inter-wire bonded states. We have investigated the stability of the

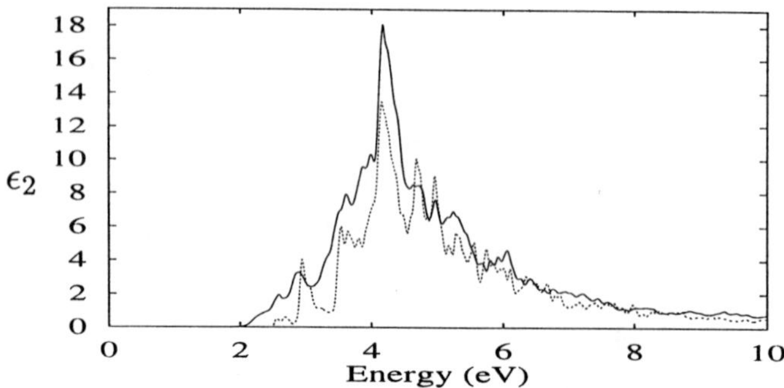

Figure 3: The z-component of ϵ_2, interacting 5x4 wire (solid line); non-interacting 5x4 wire (dotted line).

Table 1: Calculated energy gaps (eV) and inter-wire H-H and Si-Si distances (a.u.) for non-interacting (n-int.) and interacting (int.) 7x4 and 5x4 wires.

7x4	Energy Gap	H-H Distance	Si-Si Distance
n-int., unrel.	2.08	5.43	10.49
n-int., relaxed	2.23	5.70	10.57
int., unrel.	1.43	2.43	7.49
int., relaxed	1.76	3.23	7.94
5x4	Energy Gap	H-H Distance	Si-Si Distance
n-int., unrel.	2.41	5.43	10.49
n-int., relaxed	2.60	5.72	10.58
int., unrel.	1.45	2.43	7.49
int., relaxed	2.00	3.22	7.83

wire structure with respect to wire-wire interaction through PP calculations. The PP allows to optimize the geometrical structure through relaxation. Table 1 collects these results. For non-interacting wires the Si-Si bond lenghts in the wire remain unchanged, while the inter-wire H-H distance increases enhancing the gap of ~ 0.2 eV. These structures represent rather ideal crystalline Si wires. When the wires are forced to get closer the gap strongly reduces. This reduction (due to the interacting states in the gap) leads to a similar value of the gap for the different interacting wires we studied. Besides, the effect of the relaxation is now stronger, the most relevant being on the H atoms which repel each other. The Si atoms at the interface move consequently by ~ 0.3 a.u., while the subsurface and inner Si atoms remain in the same position. Whereas the values of the gap for the interacting QW before the relaxation are insensitive to the dimensions of the wires, the relaxation restores this dependence. For the 5x4 and 7x4 wire the gaps are now 2.00 eV and 1.76 eV in comparison to 2.60 eV and 2.23 eV before the switching on of the inter-wire interaction. These differences are of the same order of those between absorption and emission in PS [2].

H and O passivation.

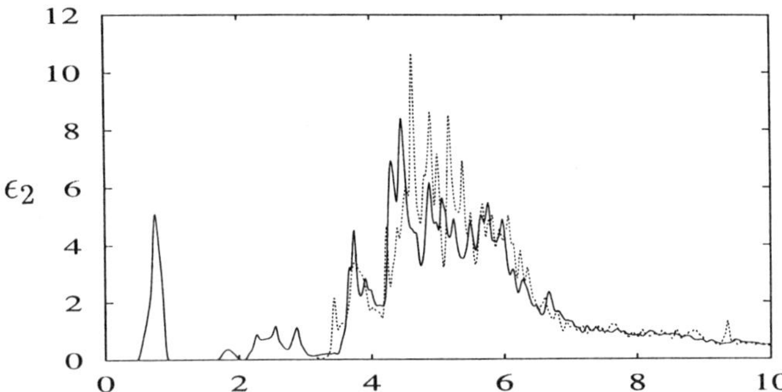

Figure 4: Imaginary part of the dielectric function ϵ_2 for Si wires, after a gaussian broadening of 0.02 eV. Partially passivated 3x4 wire (solid line); fully passivated 3x4 wire (dotted line).

The presence of dangling-bond states gives rise to drastic changes in the optical properties. The ϵ_2 of partially passivated 3x4 wires is shown in Fig. 4. For the partially covered wire we have removed 2 H atoms, over the 14 present in the unit cell, leading to a H vacancy concentration of 14%. The first structures in ϵ_2 (solid line), between $\sim$ 0.3 eV and $\sim$ 0.7 eV, correspond to the dangling bond-dangling bond transitions. These are intense transitions and dominate the spectrum. After a gap, we find the dangling bond-band state transitions, while the band state-band state transitions, characteristic of the fully passivated wire (dotted line), start only at $\sim$ 3.3 eV. It is evident that the ϵ_2 spectrum is entirely changed. Since we do not compute the emission spectra our analysis is only qualitative; nevertheless the result is clear: the presence of H vacancies strongly affects the luminescence properties. PL experiments show a broad infrared band between 0.7 and 1.1 eV in as prepared PS, showing the presence of some unsaturated dangling bonds at the surfaces of PS [13,14]. To understand the role of O we have substituted 4 H-Si bonds with 4 OH-Si bonds in the 3x4 Si wire. Fig. 5 compares the ϵ_2 for the wire covered partially with OH with respect to the fully H-covered ones. We see that the presence of O (solid lines) not only strongly reduces the energy gap, but also reduces the intensity of the lower energy side peaks in comparison with the H case (dotdashed line). This fact is confirmed by the experiments : in the first stage of oxidation the luminescence peak of as prepared PS splits into two components, the first, less intense, at lower energy, the second, more intense, at higher energy [14]. Our calculation suggests that the first component is related to the formation of Si-O-H bonds, while the second is probably related to the formation of SiO_2, which reduces the dimensions of the luminescent structures.

CONCLUSIONS

We have shown that the interaction between Si QW and the presence of different passivating agents at the surfaces of the QW strongly modify their optical properties. The comparison between our calculations and experimental results on PS provides support to the modeling of PS as formed by QW.

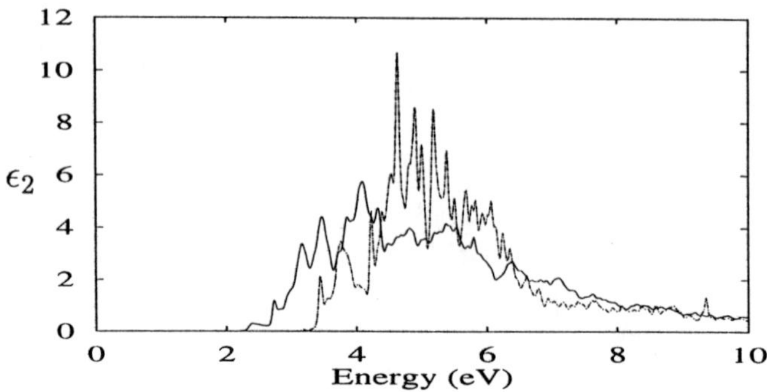

Figure 5: Imaginary part of the dielectric function ϵ_2 for Si wires, after a gaussian broadening of 0.02 eV. Partially OH passivated 3x4 wire (solid line); fully passivated 3x4 wire (dotdashed line).

ACKNOWLEDGEMENTS

This work was partially supported by Consiglio Nazionale delle Ricerche (CNR), Italy and has benefitted from collaborations within the Network on "Ab-initio (from electronic structure) calculation of complex processes in materials" (contract: ERBCHRXCT930369).

REFERENCES

1. L. T. Canham, Appl. Phys. Lett. **57**, 1046 (1990)
2. Q. Zhang, S. C. Bayliss, J. Appl. Phys. **79**, 1351 (1996)
3. S. Schuppler et al., Phys. Rev. Lett. **72**, 2648 (1994)
4. G. Allan, C. Delerue, and M. Lannoo, Phys. Rev. Lett. **76**, 2961 (1996)
5. A. Kux, and M. Ben Chorin, Phys. Rev. B **51**, 17535 (1995)
6. L. Dorigoni, O. Bisi, F. Bernardini, and S. Ossicini, Phys. Rev. B **53**, 4557 (1996). We observe that in this paper the ϵ_2 of the 5x4 and 3x4 wires (Fig. 6) were incorrectly plotted.
7. R. Jones, J. Phys. C **20**, 1271 (1987)
8. C. G. Van de Walle, and J. E. Northrup, Phys. Rev. Lett. **70**, 1116 (1993)
9. M. Alouani, L. Brey, and N. E. Christensen, Phys. Rev. B **37**, 1167 (1988)
10. H. Koyama, and N. Koshida, Phys. Rev. B **52**, 2649 (1995)
11. N. Koshida et al., Appl. Phys. Lett. **63**, 2774 (1993)
12. W. L. Wilson, P. F. Szajowski, and L. E. Brus, Science **262**, 1242 (1993)
13. A. Kux and D. Hoffman in *Optical Properties of Low Dimensional Silicon Structures*, edited by D. Bensahel, L.T. Canham, and S. Ossicini, Kluwer, Dordrecht (1993), p. 197
14. S. Gardelis and B. Hamilton, J. Appl. Phys. **76**, 5327 (1994)

A MICROSCOPIC MODEL FOR THE DIELECTRIC FUNCTION OF POROUS SILICON

M. CRUZ[†], M. R. BELTRAN, C. WANG, and J. TAGÜEÑA-MARTINEZ[‡],

Instituto de Investigaciones en Materiales, Universidad Nacional Autónoma de México, Apartado Postal 70-360, 04510, México, D.F., MEXICO.

ABSTRACT

Micro and nano-structures have opened a new area in materials research since they present interesting phenomena such as efficient luminescence and localization of carriers. An important example of these new materials is porous silicon (PS). It is considered that the quantum confinement is an essential cause of the opto-electronic properties of PS [1], thus microscopic analysis should be performed. We have developed a supercell model to study PS with a tight-binding Hamiltonian, where an sp^3s^* basis set is used. In an otherwise perfect silicon structure empty columns of atoms are produced and passivated with hydrogen atoms [2]. In this work we calculate the dielectric function and compare it against experimental data for bulk c-Si, ultrathin c-Si films and PS. We discuss the importance of considering the relaxation of the electron wavevector (**k**) conservation in order to include disorder effects in PS.

INTRODUCTION

Optical devices from indirect band gap semiconductors, such as crystalline silicon, are not as efficient as direct band gap semiconductors. However, in silicon nanocrystallites the confinement partially breaks the **k** vector selection rule and allows new radiative transitions even without phonon assistance [3]. Porous silicon (PS) is an example of nanostructured materials, whose fabrication is easy and inexpensive, and with a broad spectrum of applications [4], where efficient luminescence is observed.

There is a great interest in explaining the underlying mechanism of the light emission in PS from a microscopic point of view. In this work we study the optical properties of PS, through the dielectric function, which is related to the absorption coefficient and the refractive index.

All the quantum mechanical theoretical works, from the Hamiltonian's point of view, can be classified in two major categories: first principles and semi-empirical frameworks. In spite of the first principle methods success treating small systems, semi-empirical or tight-binding calculations are simple enough to be applied in large supercells with complex morphologies [2]. It would be worth mentioning that the use of phenomenological parameters includes many-body effects, which generally are neglected in a first principles Hamiltonian. In this work we will follow this last approach to calculate the interband transitions between valence and conduction states.

On the experimental side, the dielectric function of porous silicon has been measured [5] and it is quite different to that from the bulk crystalline one [6]. Recently a very complete study of the dielectric function of ultrathin crystalline silicon films (6$\AA$) has been performed [7]. Our calculations will be compared with all three mentioned measurements.

Mat. Res. Soc. Symp. Proc. Vol. 452 © 1997 Materials Research Society

In the next section we describe the microscopic supercell model. Following this we present and discuss the results obtained from the dielectric function. Finally, some conclusions are given.

THEORY

We start with a tight-binding Hamiltonian; the minimum basis capable of describing an indirect band gap along the X-direction is the sp^3s^* basis. We use P. Vogl, H.P. Hjalmarson and J. Dow's parameters [8], which reproduce an 1.1 eV gap in bulk crystalline silicon. Empty columns are produced removing columns of atoms within the supercell in the [001] direction.

As PS exhibits a very large surface mainly hydrogen passivated [1], we saturate the pore surface with hydrogen atoms The Si-H bond length is taken as 1.48 Å. The on-site energy of the H atom is considered to be -4.2 eV, since the free H atom energy level, -13.6 eV, is so close to the s-state energy level of a free Si atom, -13.55 eV [9], therefore the on-site energy of H is taken to be the same as that of silicon, as in [10]. The H-Si orbital interaction parameters are taken as $ss\sigma_{H-Si} = -4.075$ eV, $sp\sigma_{H-Si} = 4.00$ eV, which are obtained by fitting the energy levels of silane [11].

Apart from providing valuable information of the electronic behavior, the electronic structure calculations are a key factor for the study of optical properties through the oscillator strength analysis [12, 13]. We believe that this analysis could give insights into the controversy of having an indirect gap in PS as it is suggested by induced absorption experiments [4], even when high luminescence efficiency is observed. We start by defining the dimensionless interband oscillator strength following B. Koiller, R. Osorio and L. Falicov [12]:

$$f_{v,c} = \frac{2}{m}|\langle v|\mathbf{p}|c\rangle|^2/(E_c - E_v), \qquad (1)$$

where $|v\rangle$ and $|c\rangle$ are valence- and conduction-band eigenstates, respectively. In the tight-binding scheme eigenstates $|e\rangle = \sum_{i,\mu} a^e_{i,\mu}|i,\mu\rangle$, where i is the site index and μ identifies the orbital, then the dipole matrix in Eq. (1) can be expressed as

$$\langle v|\mathbf{p}|c\rangle = \sum_{i,j,\mu,\nu} a^v_{i,\mu}{}^* a^c_{j,\nu}\langle i\mu|\mathbf{p}|j\nu\rangle. \qquad (2)$$

The dipole matrix elements in Eq. (2) may be rewritten in terms of the Hamiltonian (H) and the position ($\mathbf{r}$) operators, using the commutation relation $\mathbf{p} = \frac{im}{\hbar}[H, \mathbf{r}]$,

$$\langle i\mu|\mathbf{p}|j\nu\rangle = \frac{im}{\hbar}\sum_{l,\lambda}(\langle i\mu|H|l\lambda\rangle\langle l\lambda|\mathbf{r}|j\nu\rangle - \langle i\mu|\mathbf{r}|l\lambda\rangle\langle l\lambda|H|j\nu\rangle). \qquad (3)$$

Since the polarizability of a free atom is much smaller than that of the corresponding semiconductor [14], Eq. (3) can be simplified as [12]:

$$\langle i\mu|\mathbf{p}|j\nu\rangle = \frac{im}{\hbar}\langle i\mu|H|j\nu\rangle\mathbf{d_{ij}}, \qquad (4)$$

where $\mathbf{d_{ij}} = \langle j\nu|\mathbf{r}|j\nu\rangle - \langle i\mu|\mathbf{r}|i\mu\rangle$ is the distance between the gravity centers of the orbitals μ and ν placed at atoms i and j, respectively, and it is independent on orbitals if the crystal field is symmetric. Notice that the contribution to the dipole matrix coming from two orbitals at the same atom is neglected.

The final stage in the calculation is to include all the interband transitions between the valence and the conduction band states, in order to obtain the dielectric function ($\epsilon = \epsilon_1 + i\epsilon_2$). When pores are introduced, we take also into account the oscillator strength from non-vertical transitions, considering that there is disorder in PS and therefore a relaxation of the electron $\mathbf{k}$-wavevector conservation should be included, *i. e.*, the existence of a random perturbative potential, which can always be expanded as a Fourier series, makes a broadening of the perturbed wavefunction in the $\mathbf{k}$-space. The imaginary part of the dielectric function is proportional to [9]

$$\epsilon_2 \propto \sum_{k,k'} f_{k,k'}\delta(\epsilon_{k'} - \epsilon_k - \hbar\omega), \tag{5}$$

where $f_{k,k'}$ are the oscillator strengths defined in equation (1), k and k' correspond to the valence and the conduction band states, respectively.

Optical experiments have been used widely to estimate the real electronic band structure of solids. In particular, the dielectric function measures the response to an external electromagnetic field, and its imaginary part is related with the absorption spectra.

RESULTS

We have considered transitions between states of the valence band and those of the conduction band, for x-direction polarized light. They are calculated in 8-atom supercells [2] with and without a 1-atom columnar pore saturated by hydrogen atoms. Figure 1a shows the dielectric function for crystalline silicon calculated with vertical interband transitions (solid line), *i.e.*, $\mathbf{k}_i^c = \mathbf{k}_f^v$, and it is compared with the experimental data reported in [6] (solid circles). The calculation has been performed by considering 205,379 $\mathbf{k}$-points in the first Brillouin zone. It can be seen that the theory gives reasonably well the energy range and the shape, in spite that no d-orbital is considered. It is important to notice that the absorption onset corresponds to the optical c-Si gap, which is much larger than the indirect band gap ($1.1eV$).

In figure 1b the calculated dielectric function for a supercell with a 1-atom columnar pore is shown, including non-vertical transitions, *i.e.*, $\mathbf{k}_i^c \neq \mathbf{k}_f^v$, to consider the disorder in PS which has been excluded by the supercell model. The calculation has been performed only for 729 $\mathbf{k}$-points in the first Brillouin zone, since adding non-vertical transitions is a very lengthy calculation. Notice that in the calculated dielectric function a tail appears in contrast with the c-Si case. This is due to the disorder, as occurred in a-Si [7]. In our calculation this shift is produced by the relaxation of the electron wavevector conservation allowing interband transitions between states close to the almost direct band gap of $2.43eV$, in order to introduce disorder effects. Furthermore, the optically active $\mathbf{k}$ space is enlarged by the quantum confinement, and then the dielectric function reveals the real band gap, instead of the optical band gap. Finally, notice that the porosity simulated by this supercell is 12.5% while the experimental data are obtained from 40% porosity samples [5].

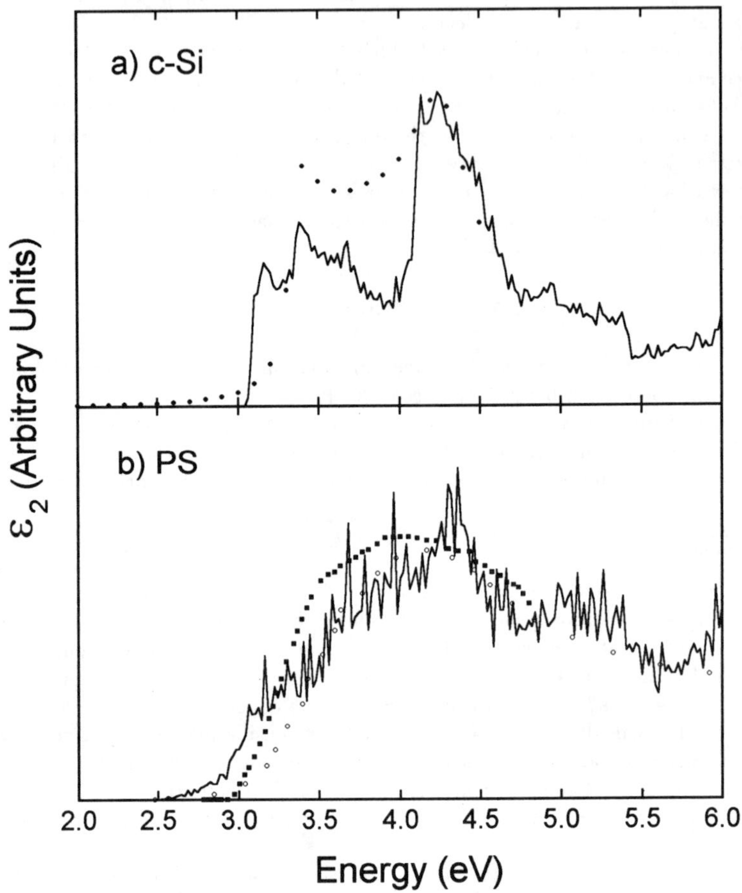

Figure 1: Imaginary part of the dielectric function calculated in an 8-atom supercell without (solid line in Fig. 1a) and with (solid line in Fig. 1b) a 1-atom columnar pore, considering interband transitions between valence and conduction band states, for light excitations polarized in the x-direction. The theoretical results are compared with experimental c-Si data (solid circles in Fig. 1a) [6], measured PS data (solid squares in Fig. 1b) [5] and thin $6\mathring{A}$ film results (open circles in Fig. 1b) [7].

CONCLUSIONS

We have shown that a simple microscopic quantum mechanical treatment, such as a phenomenological tight-binding technique, is capable of reproducing the essential features of the dielectric function of PS. Actually, from the comparison with the experimental data we can conclude that the tight-binding supercell model, when non-vertical transitions are included, gives the correct energy range and the shape of the dielectric response of PS. The need of introducing non-vertical interband transitions to take into account the PS disordered nature reveals an significant enlargement of the optically active **k** zone. This enlargement can be explained through the Heisenberg's uncertainty principle, due to the localization of the wavefunctions caused by the disorder of the shape and distribution of pores. Furthermore, this enlargement together with the trend observed towards a direct band gap [2], could reconcile the apparent controversy between the efficient luminescence of PS and its indirect gap observed experimentally.

It is interesting to notice the fact that a very thin c-Si film has similar dielectric function behavior to PS, since the quantum confinement is essential in both cases. As a first approximation we are describing the pores as columns and saturating the surface with hydrogen atoms. Clearly, both approximations can be extended to include other saturators, surface relaxation and amorphization these studies are currently in progress.

This work has been partially supported by projects DGAPA-IN104595, CONACyT-0205P-E9506, CONACyT-4229-E, and CRAY-UNAM-SC005096.

† Escuela Superior de Ingeniería Mecánica y Eléctrica - UC, IPN, México.

‡Centro de Investigación en Energía, UNAM, A.P. 34, C.P. 62580, Temixco, Mor., México.

References

[1] L.T. Canham, Appl. Phys. Lett. **57**, 1046 (1990); A.G. Cullis, and L.T. Canham, Nature **353**, 335 (1991); L.T. Canham, M.R. Houlton, W.Y. Leong, C. Pickering, and J.M. Keen, J. Appl. Phys. **70**, 422 (1991).

[2] M. Cruz, C. Wang, M. R. Beltrán, and J. Tagueña-Martínez, Phys. Rev. B **53**, 3827 (1996).

[3] C. Delerue, M. Lannoo, G. Allan, E. Martin, I. Mihalcescu, J.C. Vial, R. R. Romestain, F. Muller, and A. Bsiesy, Phys. Rev. Lett. **75**, 2228 (1995).

[4] P. Fauchet *Porous Silicon: Photoluminescent Devices* in Light Emission in Silicon ed. by D. Lockwood, to appear in the Semiconductors and Semimetals Series (Academic Press, New York, 1996).

[5] N. Koshida, H. Koyama, Y. Suda, Y. Yamamoto, M. Araki, T. Saito, and K. Sato, Appl. Phys. Lett. **63**, 2774 (1993).

[6] G.E. Jellison, Optical Materials **1**, 41 (1992).

[7] H.V. Nguyen, Y. Lu, S. Kim, M. Wakagi, and R.W. Collins, Phys. Rev. Lett. **74**, 3880 (1995).

[8] P. Vogl, H.P. Hjalmarson, and J.D. Dow, J. Phys. Chem. Solids **44**, 365 (1983).

[9] Walter A. Harrison, Electronic Structure and the Properties of Solids (Dover Pub., New York, 1989), p. 50 and p. 100.

[10] Shang Yuan Ren and John D. Dow, Phys. Rev. B **45**, 6492 (1992).

[11] Fu Huaxiang, Ye Ling, and Xide Xie, Phys. Rev. B **48**, 10978 (1993).

[12] B. Koiller, R. Osório, and L.M. Falicov, Phys. Rev. B **43**, 4170 (1991).

[13] A. Selloni, P. Marsella, and R. Del Sole, Phys. Rev. B **33**, 8885 (1986).

[14] L. Brey and C. Tejedor, Solid State Commun. **48**, 403 (1983).

THEORY OF PRESSURE EFFECTS ON SILICON NANOCRYSTALLITES

G. ALLAN, C. DELERUE, M. LANNOO
Institut d'Electronique et de Microélectronique du Nord, Département Institut Supérieur d'Electronique du Nord, BP 69, 59652 Villeneuve d'Ascq Cedex, France, gal@isen.fr

ABSTRACT

Pressure effects on silicon nanocrystallites are calculated using semi-empirical tight-binding and ab-initio local density calculations. Using the confinement model in porous silicon a red shift of the luminescence energy with increasing pressure is obtained. Quantum confinement in BC8 phase silicon nanocrystallites obtained after release of high pressure are also studied. It increases the cluster gap and also enhances the electron-hole radiative recombination rate.

INTRODUCTION

Pressure effects on the photoluminescence of porous silicon [1-8] or silicon nanocrystallites [9-10] have been measured by several groups. At low pressure, red shift as well as a blue shift of the luminescence energy is observed. Let us recall that a negative (-2.03 meV/kbar [11]) pressure coefficient of the indirect band gap is observed for bulk silicon whereas a positive one (0.65 mev/kbar [12]) is obtained for the direct gap. Higher values (10.2 mev/kbar and 5.1 mev/kbar) are measured respectively for the p-s gap at the Γ point and at the L point. The different shift to higher (blue shift) or lower (red shift) energy of the luminescence peak seems to be due to the used pressure medium [7]. An empirical pseudopotential calculation done for silicon quantum wires only gives a red shift of the band gap [13]. This result is interpreted in terms of wire conduction bands which can be expressed as a linear combination of the bulk conduction bands. In the next section, the band gap pressure coefficient of silicon nanocrystallites are calculated with an ab-initio local density calculation using the DMol code [14] which has already been applied with success to many different systems [15] and the empirical sp^3s^* Vogl tight-binding model [16].

High pressure-induced phase transitions have been observed [6-8-10]. After such phase transformations and upon release of the pressure, the BC8 phase recovered in bulk silicon [17] and in porous silicon [8] is not obtained for SiO_2 coated Si nanocrystals [10]. Contrary to silicon in the diamond phase, the valence band maximum and the conduction band minimum of silicon in the BC8 phase occur for the same H point in the Brillouin zone [18-19]. BC8 silicon is thus a "direct" gap material, so one might expect to get a higher radiative recombination rate of the electron-hole pair. In the last section, the electronic structure and the radiative recombination rate of BC8 phase nanocrystallites have been calculated using a non orthogonal tight-binding method. For small crystallites, the results are compared to those obtained with the local density calculation.

PRESSURE COEFFICIENT OF THE GAP IN SILICON NANOCRYSTALLITES

We have calculated the electronic structure of spherical silicon nanocrystallites with dangling bonds saturated by hydrogen atoms using ab-initio local density and semi-empirical tight-binding techniques. LDA is known to underestimate the semiconductor band gap. We have estimated for bulk silicon the variation of the self-energy correction with pressure using the approximated value given by V. Fiorentini, et al [20]. Its contribution to the bulk band gap pressure coefficient is very small and has been neglected for the clusters. In the tight-binging model, Harrison parameters [21] are used for the Si-H interactions. It is known that the sp^3s^* Vogl model gives a too narrow conduction band and that the cluster band gaps obtained with this model are smaller than the val-

Mat. Res. Soc. Symp. Proc. Vol. 452

ues obtained when the bulk conduction band is well described [23]. However, we have used this model because the s^* orbital interaction with the p states at the bottom of the conduction band gives a negative pressure coefficient for the bulk band gap. We expect that for a cluster, it will realize a good interpolation between the bulk states and the corresponding bulk pressure coefficients. We have assumed that the interatomic nearest-neighbor Slater-Koster tight-binding parameters used in the Vogl model vary as d^{-n_α} where α is for ssσ, spσ, ppσ, ppπ and s^*pσ. The n_α's have been fitted to the four bulk pressure coefficients listed above. We get $n_{ss\sigma} = 3.25$, $n_{sp\sigma} = 2.5$, $n_{pp\sigma} = 1.5$, $n_{pp\pi} = n_{pp\sigma}$ and $n_{s^*p\sigma} = 3$.

We have verified using the calculated total energy E_t from the LDA calculation that $\frac{\partial^2 E_t}{\partial V^2}$ is very close to the bulk value and the pressure coefficient $a = \frac{\partial E}{\partial P}$ (where E is the crystallite band gap) can be calculated as:

$$a = -\frac{V \partial E}{B \partial V} \tag{1}$$

where B is taken equal to the bulk value (978.8 kbar [22]). The variation with the cluster size is given on fig. 1. The results are slightly scattered for small clusters and one can see from the tight-binding calculation that it smoothly tends towards the bulk value when the diameter increases. For small clusters, the difference is comparable to the one between tight-binding and LDA bulk values. Recent measurements [9] give a=-0.4 and -0.6 mev/kbar for cluster size ~ 3 nm. This is in very good agreement with the tight-binding model which is in that case more reliable than the LDA calculation as it is fitted to bulk values. Due to the underestimated cluster gap obtained with the Vogl model, comparison with experiment which measures the photoluminescence energy and not the cluster size can only be made using for example the relation between the gap E_G and the cluster diameter we have obtained using a non orthogonal tight-binding model[26] and which is in agreement with empirical pseudopotential or first-principles LDA results:

$$E_G = \frac{2.29}{d^{1.39}} eV \tag{2}$$

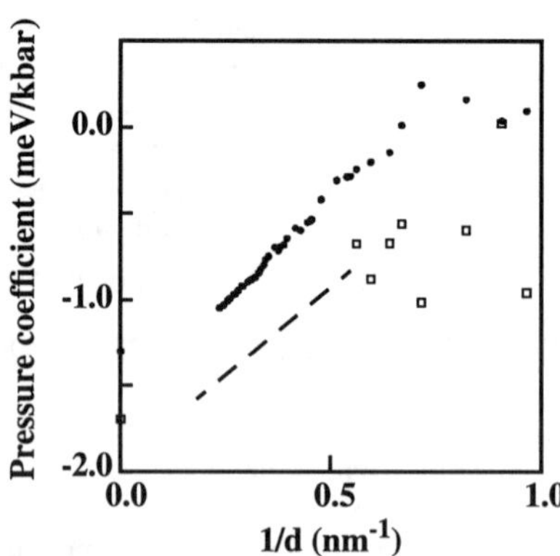

Fig 1: Band gap pressure coefficient of spherical nanocrystallites as a function of their diameter: LDA calculation [14] for small clusters (▫), sp^3s^* tight-binding model (•)[16]. The values for 1/D=0. are the bulk values. The dotted line is a guide foir the eyes between the LDA bulk and small clusters values.

In a confinement model, we get as for Si quantum wires negative pressure coefficients. This would not be the case if the luminescence comes from a surface defect as the self-trapped exciton we have previously studied [23-24]. In this case, the luminescence energy is directly related to the coupling between silicon hybrid orbitals. When the pressure is increased, the interatomic distance decreases and the coupling between the orbitals increase, which gives rise to a blue shift and a positive pressure coefficient

BC8 PHASE NANOCRYSTALLITES

The BC8 structure is metastable and is obtained after release of the pressure when the high-pressure β-tin phase (Si-II) is unloaded to ambient pressure [17]. The BC8 phase has a body-centered-cubic structure with 8 atoms in the primitive cell and a lattice parameter equal to 6.636 Å. Each silicon atom has a distorted tetrahedral environment with interatomic distances (2.30 and 2.39 Å) close to the diamond ones (2.35 Å)

It has been shown [25] that semiempirical methods can provide accurate results for cluster band gaps only if they give an excellent fit to the corresponding bulk band structure. Here we have chosen the non orthogonal tight-binding technique which has proved to be very efficient and accurate to calculate the electronic structure of silicon crystallites with the diamond structure [26]. The tight-binding overlap and hamiltonian have been obtained by a fit to a band structure we have calculated using norm conserving pseudopotential theory [27] within the local density approximation and an energy cutoff equal to 12. eV [28]. The conduction bands have been shifted by the same quantity (0.6 eV) as for Si in the diamond structure as LDA underestimates the bulk band gap. As in the diamond structure, we take into account interatomic parameters up to 0.5 nm.

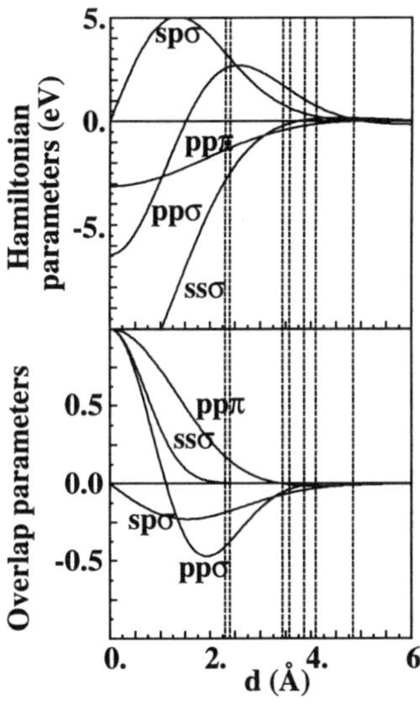

Fig. 2: Non orthogonal tight-binding parameters for Silicon in the BC8 phase. The vertical dotted lines indicate the positions of the nearest neighbors.

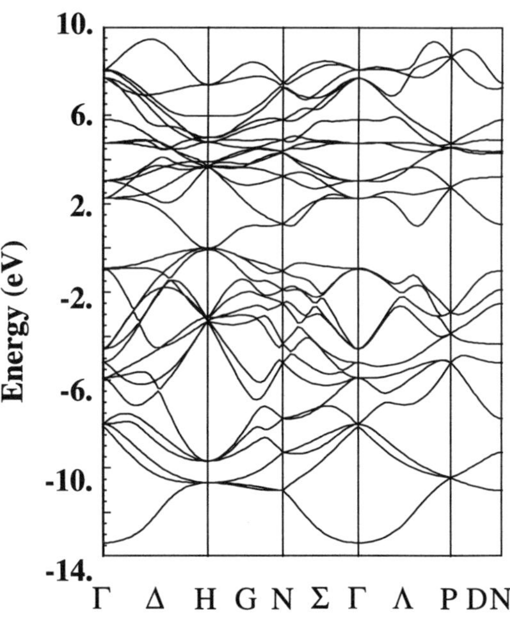

Fig. 3: Tight-binding band structure of BC8 silicon.

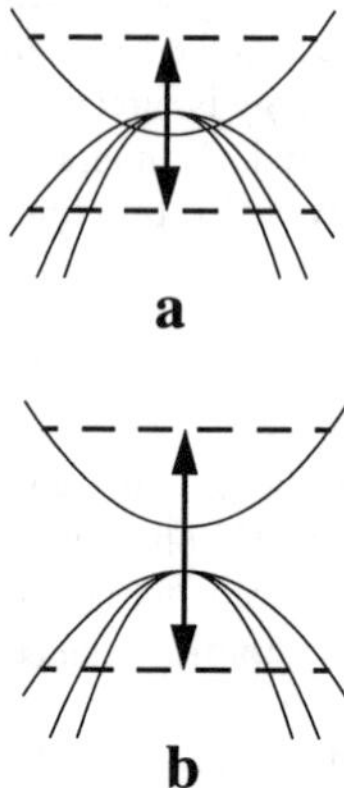

Fig. 4: Confinement effect in a cluster of a semimetal (a) and in a semiconductor (b) with equal valence (and conduction) effective masses.

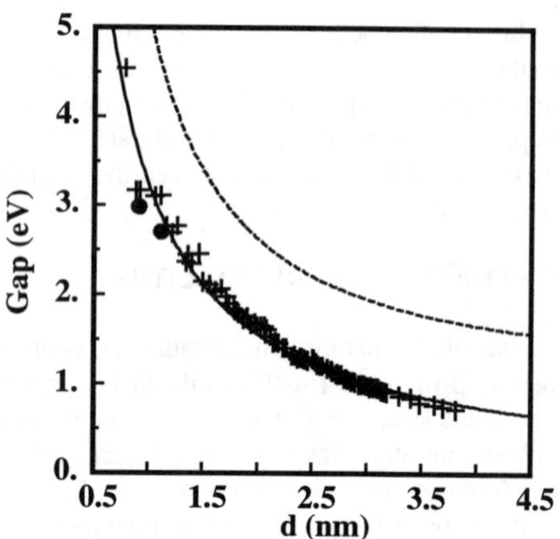

Fig. 5: Variation of the cluster band gap as a function of the cluster diameter for silicon in the diamond structure (dotted line) and in the BC8 phase: tight-binding model (crosses); LDA calculation (full dots). The full line is a mean least square fit to the calculated points.

This corresponds to the 7th nearest neighbors. In order to avoid a too large number of parameters to fit, we have taken an analytic expression of the parameters based on the overlap of gaussian orbitals [29]. The non orthogonal tight-binding parameters are shown on fig. 2. The corresponding band structure shown on fig. 3 agrees very well with an empirical pseudopotential result [18]. The only discrepancy appears at the H point and affects the gap value. The LDA calculation corrected by 0.6 eV and our tight-binding fit give a band gap close to zero whereas a value equal to 0.43 eV was obtained using empirical pseudopotentials. On fig. 4, in a simple effective mass model, one can see that this does not change the confinement energy, only the absolute value of the cluster gap is modified by a change of the bulk band gap if the valence (and conduction) effective mass remains constant.

As for silicon clusters in the diamond structure, we have considered spherical nanocrystallites with dangling bonds saturated by Hydrogen atoms. Fig. 5 shows the variation of the gap as a function of the cluster size for BC8 phase clusters compared to diamond phase clusters. The confinement effect and the blue shift are very similar for both structures. Here it varies like $3.22\ d^{-1.04}$, i.e. less rapidly than in a mass effective model ($\sim d^{-2}$). The only difference when one goes from the BC8 cluster to the diamond one with the same size comes from the bulk gap value which simply shifts the cluster gap energy. Thus in a quantum confinement model and for the cluster sizes observed in porous silicon, one would expect for porous BC8 and diamond silicon clusters quite different luminescence energies. For small clusters one can also remark on fig. 5 the good agreement between the values obtained by the non orthogonal tight-binding calculation and the results based on the ab-initio local density calculation using the DMol code [14].

As bulk BC8 silicon is a direct gap semiconductor, we expect a higher recombination rate than for silicon with the diamond structure. The radiative recombination time τ is given by the Fermi Golden rule [30]. Fig. 6 compares the variation of the recombination rate as a function of the cluster gap for the BC8 and the diamond structures. One can see that the BC8 phase recombination rate remains constant when the cluster size increases and the blue shift decreases. This is not the

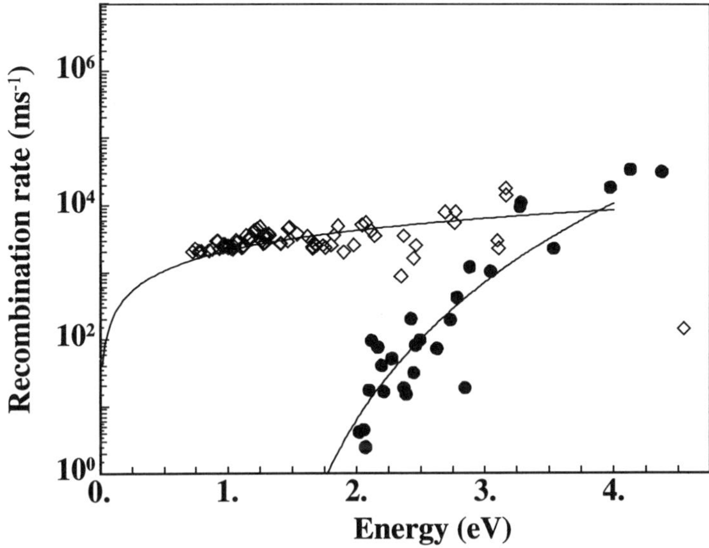

Fig. 6: Electron-hole recombination rates as a function of the gap energy for BC8 (◇) and diamond (●). The lines are least-square fits: $A*E^{6.6}$ for the diamond phase and A*E for the BC8 one.

case of diamondlike silicon: it rapidly decreases when the cluster size increases. A least square fit of the calculated recombination rate in the diamond case gives a $E^{6.6}$ variation which may be compared to the result of an effective mass model which varies as E^3 [31]. The difference comes from the fact that the fit is here done for large blue shift values where the effective mass model is no more valid. One must remark that the BC8 phase recombination rate remains quite high. It is of the order of a few μs^{-1}(i.e. below 2 eV more than 10^3 times larger than in the diamond phase), which is however lower than the result for GaAs (~ ns^{-1}) which is also a direct gap semiconductor. Nevertheless if one assumes a similar non radiative recombination rate for the BC8 and the diamond structures, the luminescence yield must be strongly improved for the BC8 structure as it is to first order proportional to this faster radiative recombination contribution.

CONCLUSION

We have studied two pressure effects on silicon nanocrystallites. First, the gap pressure coefficients remain negative in a confinement model of the luminescence. It increases when the cluster size decreases as the confinement mixes bulk states with positive pressure coefficients. The second effect is the possible occurrence of clusters with BC8 structure after release of the pressure. If the blue shift in these clusters is similar to the diamond one, the cluster gap is smaller due to a smaller bulk gap of silicon in the BC8 phase. However, the radiative recombination rate is improved but does not reach the GaAs one. This might nevertheless improves the electroluminescence efficiency as it enhances the recombination rate by a factor larger than 10^3. Such effects may be interesting in order to develop devices based on such a metastable structure.

REFERENCES

1 J. Camassel, E. Massone, S. Lyapin, J. Allegre, P. Vincente, A. Foucaran, A. raymond, and J.L. Robert, in *Proceeding of the 21st International Conference on the Physics of Semiconductors*, edited by Ping Jiang and Hou-Zhi Zhang (World Scientific, Singapore, 1992), p. 1463.

2 W. Zhou, H. Shen, J. F. Harvey, R.A. Lux, M. Dutta, F. Lu, C. H. Perry, R. Tsu, N. M. Kalhhoran, and F. Nawavar, Appl. Phys. Lett. **61**, 1435 (1992).

3 A. K. Sood, K. Jayaram, and D. V. S. Mathu, J. Appl. Phys. **72**, 4963 (1992).

4 X.-S. Zhao, P. D. Persans, J. Schroeder, and Y.-J. Wu, Mater. Res. Soc. Symp. **283**, 127 (1993).

5 N. Ookubo, Y. Matsuda, and N. Kuroda, Appl. Phys. Lett. **63**, 346 (1993).

6 J. M. Ryan, P.R. Wamsley, and K. L. Bray, Appl. Phys. Lett. **63**, 2260 (1993).

7 H. M. Cheong, P. Wickboldt, D. Pang, J. H. Chen, and W. Paul, Phys. Rev. B **52**, R11577 (1995).

8 J. Zeeman, M. Zigone, G. Martinez, and G. L. J. A. Rikken, J. Phys. Chem. Solids **56**, 655 (1995).

9 H. M. Cheong, W. Paul, S. P. Withrow, J. G. Zhu, J. D. Budai, C. W. White, and D. M. Hembree, Jr, Appl. Phys. Lett. **68**, 87 (1996).

10 BS. H. Tolbert, A. B. Herhold, L. E. Brus, and A. P. Alivisatos, Phys. Rev. Lett. **76**, 4384 (1996).

11 . Welber, C. K. Kim, M. Cardona, and S. Rodrigues, Solid State Comm. **17**, 1021 (1975).

12 E. Schmidt, and K. Vedam, Solid State Comm. **9**, 1187 (1971).

13 C.-Y. Yeh, S. B. Zhang, and A. Zunger, Appl. Phys. Lett. 64, 3545 (1994).

14 DMol User Guide, version 2.3.5. San Diego: Biosym Technologies, 1993.

15 B. Delley, J.Chem.Phys. **92**, 508 (1990).

16 P. Vogl, H. P. Hjalmarson, and J. D. Dow, J. Phys. Chem. Sol. **44**,3651 (983).

17 R. H. Wentorf, and J. S. Kasper, Science **139**, 338 (1963); J. S. Kasper, and S. M. Richards, Acta Crystallogr. **17**, 752 (1964).

18 J. D. Joannopoulos, and M. L. Cohen, Phys. Rev. B **7**, 2644 (1973)

19 M.T. Yin, Phys. Rev. B **30**, 1773 (1984).

20 V. Fiorentini, and A. Baldereschi, Phys. Rev. B **51**, 17196 (1995).

21 W. A. Harrison, *Electronic Structure and the Properties of Solids* (Freeman, San Francisco, 1980).

22 H. J. McSkimin, and P. Andreatch, Jr, J. Appl. Phys. **35**, 2161 (1964).

23 G. Allan, C. Delerue, and M. Lannoo "*Surface/interface and Stress Effects in Electronic Material Nanostructures*", S.M. Prokes, K.L. Wang, R.C. Cammarata, A. Christou Eds. (Mat. Res. Soc. Proc. **405**, Pittsburgh, PA, 1996) p. 235

24 G. Allan, C. Delerue, and M. Lannoo, Phys.Rev.Lett. **76**, 2961 (1996).

25 C. Delerue, M. Lannoo, and G. Allan, Phys. Rev. Lett. **76**, 3038 (1996).

26 J. P. Proot, C. Delerue, and G. Allan, Appl. Phys. Lett. **61**, 1948 (1992); C. Delerue, G. Allan, and M. Lannoo, Phys. Rev. B **48**, 11024 (1993).

27 N. Troullier, and J.L. Martins, Solid State Comm. **74**, 613 (1990).

28 Plane_wave User Guide, version 3.0.0. San Diego: Biosym Technologies, 1995.

29 G. Allan, C. Delerue, and M. Lannoo (To be published).

30 D. L. Dexter in *Solid State Physics*, edited by F. Seitz and D. Turnbull (Academic Press, New York, 1958), Vol. **6**, p. 360.

31 M. Hybertsen, , in *Light Emission from Silicon*, edited by S.S. Iyer, L.T. Canham, and R.T. Collins (Materials Research Society, Pittsburgh, 1992), p. 179.

THEORETICAL THERMAL CONDUCTIVITY OF POROUS SILICON: NONLINEAR BEHAVIOR

J. E. Lugo[+] J. A. de Río and J. Tagüeña-Martínez.
Centro de Investigación en Energía, Universidad Nacional Autónoma de México, A.P. 34, Temixco, Morelos, México.
[+]Facultad de Ciencias, Universidad Autónoma del Estado de Morelos, Av. Universidad 1001, 62210, Cuernavaca, Morelos, México.

ABSTRACT

The use of porous silicon (PS) in fabricating optoelectronic devices is in progress. However, the performance of such applications still needs to be improved. In particular, in solar cells heat must be dissipated to avoid a decay in their efficiency and one of the properties of PS that must be evaluated is the effective thermal conductivity. It is well known that the thermal conductivity of silicon is temperature dependent. Thus we cannot use a standard effective medium approach to obtain its effective thermal response. In this work, we extend the averaging volume and surface methods [1] to consider nonlinear effects in the effective transport coefficients. We model PS as composed of c-Si cylindrical columns covered by different overlayers (i.e. a-Si or SiO_2) immersed in another medium. In our model the effective thermal conductivity has an explicit dependence on the temperature gradient. We present a parametric analysis of the model, compare it with the c-Si behavior and evaluate the importance of the nonlinear contribution.

INTRODUCTION.

Real systems are heterogeneous, polycrystalline, amorphous, multicomponent or porous and the prediction of their properties presents an interesting basic challenge. Nevertheless, these derivations from homogeneity might be useful to produce devices. In particular, a nanocomposite material that has received special attention lately for its potential optoelectronic applications and cheap production process is porous silicon (PS) [2]. However, PS is a very fragile material and there is a search to introduce other material inside the pores to improve its mechanical properties, while maintaining its optical properties [2]. Its thermal properties, in particular the effective thermal conductivity, are also relevant to study heat dissipation.

Our previous works [1], [3] seem to indicate that, due to the relation between the characteristic length of the PS structure and the observation length, an effective medium theory is capable of describing the transport phenomena. Thus, in principle, we can obtain the effective thermal conductivity using one of these smoothing theories.

However, the non-linear character of the thermal conductivity of c-Si is well known, *i.e.* its thermal conductivity is temperature dependent. On the other hand, the materials used to give strength to PS are polymers, which also lead to a non-linear thermal conductivity. Clearly, it seems necessary to study the role of the non-linearity of the transport coefficients of the PS main components on the effective thermal conductivity.

One of the first assumptions of the effective medium theories is that the transport coefficients are constant. Few attempts exist in the literature to extend this type of calculation.

Mat. Res. Soc. Symp. Proc. Vol. 452 © 1997 Materials Research Society

Whitaker [4] studied the role of the temperature dependence of viscosity on the flow transport in porous media. Bergman analyzed the role of the non-linear dielectric function [5]. Previously, we have applied the volume averaging method to model PS as a three component system considering one of the components as a very thin overlayer [3]. In this work we extend our result to consider the non-linear contribution to the effective thermal conductivity.

In the next section we analyze the role of the nonlinear components of the thermal conductivity on the heat transport in a porous medium using an effective medium approach. In the third section we present the non-linear thermal conductivity tensor for PS modeled as a c-Si cylinder bundle covered by different overlayers and surrounded by another nonlinear material, for instant polymetylmethacrylate. Finally, we end the paper with some remarks and conclusions.

EFFECTIVE NONLINEAR PROPERTIES IN POROUS MEDIA.

Let us consider the general case of heat transport in a porous medium with a thin coverage. This problem is described by

$$\nabla \left[k^{\sigma}\left(T^{\sigma}\right)\nabla T^{\sigma}\right] = \frac{\partial u^{\sigma}}{\partial t} \quad in \quad V^{\sigma},$$

$$\nabla \left[k^{\beta}\left(T^{\beta}\right)\nabla T^{\beta}\right] = \frac{\partial u^{\beta}}{\partial t} \quad in \quad V^{\beta}, \tag{1}$$

$$k^{\sigma}\left(T^{\sigma}\right)\nabla T^{\sigma} n^{\sigma\beta} + k^{\beta}\left(T^{\beta}\right)\nabla T^{\beta} n^{\beta\sigma} = u^{\sigma\beta} \quad in \quad A^{\sigma\beta},$$

$$T^{\sigma} = T^{\beta} \quad en \quad A^{\sigma\beta},$$

with V^{σ} the first bulk component, V^{β} the second bulk component, $A^{\sigma\beta}$ the area between σ and β, $n^{\sigma\beta}$ a unitary normal vector, outgoing from interface, $u^{\sigma\beta}$ the surface source, $k^{\sigma}\left(T^{\sigma}\right)$ and $k^{\beta}\left(T^{\beta}\right)$ are the thermal conductivities for the σ and β components (in turn dependent on the local temperatures, T^{σ} and T^{β}), u^{σ} and u^{β} are the bulk sources. In the following we will use round brackets () to indicate variable dependence.

In general, averaging nonlinear terms is not an easy task. In this case, we need to perform the following average

$$\langle \nabla \left[k\left(T\right)T\right]\rangle = \frac{1}{V}\int \nabla \left[k\left(T\right)T\right]dV,$$

where $\langle\rangle$ indicates a spatial average.

It is clear that if the thermal conductivity could be evaluated based on the average temperature, the averaging procedure could be performed immediately; of course $\langle T^{\sigma}\rangle^{\sigma}$ is an effective quantity which varies on the macroscopic scale and therefore it is constant on the microscopic scale, within the averaging volume. In the mean field approach, it is considered that the local temperature may be expressed as

$$T^{\sigma} = \langle T^{\sigma}\rangle^{\sigma} + \tilde{T}^{\sigma}$$

where $\tilde{T}^{\sigma}$ is the local temperature deviation in the σ-component. In an effective medium approach $\tilde{T}^{\sigma}$ is considered to be a small quantity. With this definition, $\tilde{T}^{\sigma}$ has variations in the microscopic scale.

Then we can take a Taylor expansion of the $k^{\sigma}(T^{\sigma})$ around the average value $\langle T^{\sigma}\rangle^{\sigma}$in the σ component, namely

$$k^{l}\left(T^{l}\right)=k^{l}\left(\left\langle T^{l}\right\rangle^{l}+\tilde{T}^{l}\right)=k^{l}\left(\left\langle T^{l}\right\rangle^{l}\right)+\sum_{n=1}^{\infty}\frac{d^{n}k^{l}\left(\left\langle T^{l}\right\rangle^{l}\right)}{dT^{n}}\frac{\left[\tilde{T}^{l}\right]^{n}}{n!}. \tag{2}$$

In the following, we omit the explicit valuation of k^{σ} in $\langle T^{\sigma}\rangle^{\sigma}$ for clarity of notation. If we susbtitute eq. 2 and its analogue for the β-component and for the coverage in eq.1 and we use a constitutive equation for the local deviations, namely

$$\tilde{T}^{l}=g_{i}^{l}\partial_{i}\langle T\rangle \quad with \quad l=\sigma,\beta,\sigma\beta, \tag{3}$$

where g_{i}^{l} is a position dependent vector and represents the connection between the microscopic and the macroscopic domains, and ∂_{i} denotes the partial derivative with respect to the i coordinate. We can arrive at the effective non-linear conductivity tensor, given by[6]

$$\overleftrightarrow{k}_{efec}(\langle T\rangle)=\left[f^{\sigma}k^{\sigma}+f^{\beta}k^{\beta}+a_{v}k^{\sigma\beta}d\right]+\left\{\left[f^{\sigma}\left(\left\langle\sum_{n=1}^{\infty}\frac{d^{n}k^{\sigma}}{dT^{n}}\frac{\left[g_{j}^{\sigma}\partial_{j}\langle T\rangle\right]^{n}}{n!}\right\rangle^{\sigma}\right)\right.\right.$$

$$\left.+f^{\beta}\left(\left\langle\sum_{n=1}^{\infty}\frac{d^{n}k^{\beta}}{dT^{n}}\frac{\left[g_{j}^{\beta}\partial_{j}\langle T\rangle\right]^{n}}{n!}\right\rangle^{\beta}\right)+a_{v}\left(\left\langle\sum_{n=1}^{\infty}\frac{d^{n}k^{\sigma\beta}}{dT^{n}}\frac{\left[g_{j}^{\sigma}\partial_{j}\langle T\rangle\right]^{n}}{n!}d\right\rangle^{\sigma\beta}\right)\right]\delta_{ij}$$

$$\frac{1}{V}\int_{A^{\sigma\beta}}\left[-\left[k^{\sigma}+\sum_{n=1}^{\infty}\frac{d^{n}k^{\sigma}}{dT^{n}}\frac{\left[g_{j}^{\sigma}\partial_{j}\langle T\rangle\right]^{n}}{[n+1]!}\right]+\left[k^{\beta}+\sum_{n=1}^{\infty}\frac{d^{n}k^{\beta}}{dT^{n}}\frac{\left(g_{j}^{\sigma}\partial_{j}\langle T\rangle\right)^{n}}{(n+1)!}\right]+\right.$$

$$\left.\left.\left[k^{\sigma\beta}+\sum_{n=1}^{\infty}\frac{d^{n}k^{\sigma\beta}}{dT^{n}}\frac{\left[g_{j}^{\sigma}\partial_{j}\langle T\rangle\right]^{n}}{(n+1)!}\right]2Hd\right]n_{i}^{\beta\sigma}g_{j}^{\sigma}dA\right\}. \tag{4}$$

This is a general expression for a system composed by two bulk materials and one overlayer separating them. The effective tensor depends on the thermal conductivities of each component, the microstructure (vectors g_{j}^{l}), the gradient of the macroscopic average of the local variable and the effective thickness of the intercomponent surface d. We can derive a partial differential equation for g_{i}^{l}, which has to be solved in only one representative cell of porous media to obtain an explicit expression for the thermal conductivity tensor.

We use the simplest microgeometry to model PS [3], a periodic network of parallel cylinders, within Chang's cell[6]. The main restriction of this cell is that it is valid at high porosities, but this is the interesting case for PS. The next step is to find the local vector in our modeled microgeometry, and to calculate the two principal terms of the effective nonlinear tensor. One is the axial term $K_{\parallel}(\langle T\rangle)$ and the other is the transverse term $K_{\perp}(\langle T\rangle)$. The axial term is obtained if the tensor $\overleftrightarrow{k}_{efec}(\langle T\rangle)$ (eq. 4) is projected in the Z direction, which can be expressed as[6]

$$K_{\parallel}\left(\langle T\rangle\right)=\left[f^{\sigma}k^{\sigma}\left(\langle T\rangle\right)+f^{\beta}k^{\beta}\left(\langle T\rangle\right)+a_{v}k^{\sigma\beta}\left(\langle T\rangle\right)d\right]- \tag{5}$$

$$\left[\sum_{k=1}^{\infty}\frac{d^{2k}k^{\sigma}}{dT^{2k}}\frac{\left[\partial\langle T\rangle\right]^{2k}}{[2k]!}\frac{r_1^{2k}}{\delta^{2k}[k+1]}\prod_{l=1}^{k}\frac{2l-1}{2l}\right]-\left[\sum_{k=1}^{\infty}\frac{d^{2k}k^{\sigma\beta}}{dT^{2k}}\frac{\left[\partial\langle T\rangle\right]^{2k}}{n!}\frac{r_1^{2k}a_vd}{\delta^{2k}}\prod_{l=1}^{k}\frac{2l-1}{2l}\right]$$

$$\left[\sum_{k=1}^{\infty}\frac{d^{2k}k^{\beta}}{dT^{2k}}\frac{\left(\partial\langle T\rangle\right)^{2k}}{(2k)!}\frac{2}{\left[r_2^2-r_1^2\right]}\left[\frac{(1-P)}{P\delta}\right]^{2k}f\left(r_1,r_2,k\right)\prod_{l=1}^{k}\frac{2l-1}{2l}\right]$$

where $\delta=1+\frac{2\Delta}{P[1+\Lambda-\Delta]}$, $\Delta=\frac{k^{\beta}(\langle T\rangle)}{k^{\sigma}(\langle T\rangle)}$, $\Lambda=\frac{k^{\sigma\beta}(\langle T\rangle)d}{k^{\sigma}(\langle T\rangle)r_1}$, r_1, is the radius of the cylindrical column, P is the porosity, $a_v=\frac{2(1-P)}{r_1}$ the area per unit volume and

$$f\left(r_1,r_2,k\right)=\left\{\begin{array}{cc} r_2^4\ln\left(1+\frac{r_1}{r_2}\right)-2kr_2^3r_1 & si\quad k=1,\\ r_2^5r_1-4r_2^6\ln\left(1+\frac{r_1}{r_2}\right) & si\quad k=2,\\ r_2^{2k+1}r_1\left(1-2k\right) & si\quad k=3,4,5.... \end{array}\right\}.$$

We can see in the last equation a new ingredient: the effective thermal conductivity depends on the macroscopic temperature gradients, which do not appear in the linear case. Another interesting fact is that in the nonlinear contributions, the temperature gradients, appear only on even order terms. This is due to the assumed transverse homogeneity of PS.

If we project the tensor $\overleftrightarrow{k}_{efec}\left(\langle T\rangle\right)$ in the X direction, we obtain the transverse term, which can be written as

$$K_{\perp}\left(\langle T\rangle\right)=K_{\parallel}\left(\langle T\rangle\right)+k^{\sigma}\left(\langle T\rangle\right)\frac{[1-P][-1-\Lambda+\Delta]}{\delta}+$$

$$\sum_{k=1}^{\infty}\frac{2\left[\partial\langle T\rangle\right]^{2k}r_1^{2k}[1-P]}{[2k+1]!\delta^{2k+1}}\frac{d^{2k}\left[-k^{\sigma}\left(\langle T\rangle\right)+k^{\beta}\left(\langle T\rangle\right)+2Hdk^{\sigma\beta}\left(\langle T\rangle\right)\right]}{dT^{2k}}\prod_{l=1}^{k+1}\frac{2l-1}{2l}. \tag{6}$$

And once again, the only surviving terms in the derivatives are the even order ones and also again the macroscopic temperature gradients appear.

RESULTS

From eqs. 5 and 6 it is clear that the effective thermal conductivity can be written as

$$K(<T>)=K_L(<T>)+K_{NL}(<T>)$$

where the superscripts indicate the origin of the contribution, i.e. linear L or non-linear NL. The relative importance of these terms may be analyzed by dimensional inspection. The correction of the first non-linear term to the thermal conductivty (second order in the gradient of the average temperature) is

$$K_{NL}(<T>)=\frac{d^2k^{\sigma}}{dT^2}\frac{\left[\partial\langle T\rangle\right]^2}{8}\frac{r_1^2}{\delta^2}+...$$

The corresponding orders of each term may be estimated as

$$(\Delta T)^2 \sim \left(\frac{10K}{10^{-6}m}\right)^2 \sim 10^{14},$$

because the thickness of the porous layer is around $1\mu m$ and it is reasonable to assume a temperature difference in this layer of less than $10K$. The ratio of the thermal conductivities is much less than 1, then the value of δ is of the order of 1. Finally, the order of magnitude of the silicon column radius is $r_1 \sim 10^{-9}m$. Then, only if the following condition is satisfied

$$\frac{d^2k^\sigma}{dT^2} > \left\{\max\left(k^\sigma, k^\beta\right)\right\} 10^4$$

we would have an important nonlinear contribution. This condition is not satisfied for PS as one can see from the plots of the thermal conductivities for c-Si, a-Si:H and polymers or air. Therefore, the effective thermal conductivity of PS can be calculated using a standard mean field approximation without considering nonlinear contributions. Evaluating the thermal conductivities on the average temperature, we have

$$K_\parallel(\langle T\rangle) = \left[f^\sigma k^\sigma(\langle T\rangle) + f^\beta k^\beta(\langle T\rangle) + a_v k^{\sigma\beta}(\langle T\rangle)\, d\right]$$

and

$$K_\perp(\langle T\rangle) = K_\parallel(\langle T\rangle) + k^\sigma(\langle T\rangle)\frac{[1-P][-1-\Lambda+\Delta]}{\delta}, \tag{7}$$

where the surface contribution is considered. Using eq. 7 we calculated the effective thermal conducitivty of PS, and this is shown in fig. 1. The experimental c-Si result [7] is plotted for comparison together with the model including polymetylmethacrylate [8] and our result for PS with an "ideal" polymer that would increase the thermal conductivity. This "ideal" polymer should have a thermal conductivity comparable to that of c-Si, contrary to the polymetylmethacrylate which has a thermal response two order of magnitude smaller.

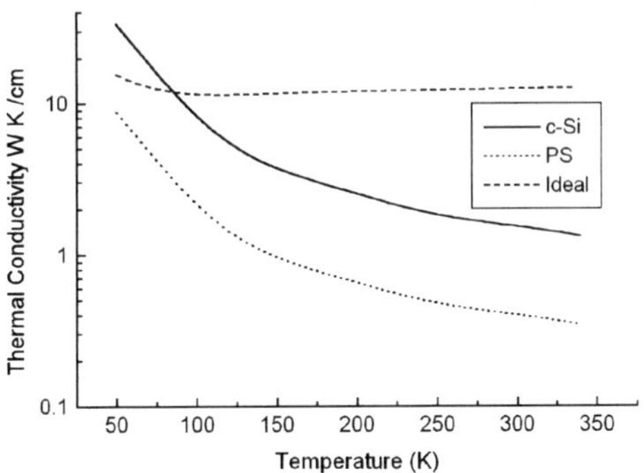

Figure. 1. Experimental thermal conductivity of c-Si compared with a calculation for our model of porous silicon including polymetylmethacrylate (PS) and an ideal polymer (Ideal) inside the pores.

It is necessary to mention that due to the low thermal conductivities of a-Si:H and SiO_2 the thermal behavior of these coverages has no relevant role in the effective thermal response, in contrast to the case of the optical behavior [3].

CONCLUSIONS.

We have analyzed the role of nonlinear contributions to the effective transport coefficients of nanocomposites within an effective medium framework, including surface effects. In the case of the thermal conductivity, nonlinear terms can be neglected if the component conductivities scale linearly with the temperature ($K(T) = a + bT$). Thus our PS model does not need to extend the effective medium approximation to include nonlinear terms.

We have calculated the effective thermal conductivity for our model of PS where a polymer is introduced inside the pores. To construct devices with PS with high thermal conductivity we need a polymer with a thermal conductivity near to that of c-Si. We have shown the result for this ideal polymer.

ACKNOWLEDGMENTS

We acknowledge the partial support of UNAM under projects IN108195 and IN104595 and of CONACYT 4229-E.

References

[1] J.E. Lugo, J.A. del Río, J. Tagüeña and J.A. Ochoa, Mat. Res. Soc. Symp. Proc. **358**, 43 (1995).

[2] P.M. Fauchet, Porous Silicon: Photoluminescent Devices in Light Emission in Silicon ed. D. Lockwood, Academic Press, (in press, 1997).

[3] J.A. del Río, J. Tagüeña and J.A. Ochoa, Solid State Comm. **87**, 541 (1993). J. Tagüeña, J.A. del Río and J.A. Ochoa, Solid State Comm. **90**, 411 (1994). J.E. Lugo, J.A. del Río and J. Tagüeña, J. Appl. Phys. (in press, 1997).

[4] S. Whitaker, Chem. Eng. Comm. **58**, 171 (1987).

[5] O. Levy and D.J. Bergman, Phys. Rev. B **46**, 7189 (1992).

[6] J. E. Lugo, J.A. del Río and J. Tagüeña, Transport in Porous Media (Submitted)

[7] C.M. Bhandari and D.M. Rowe, Thermal Conduction in Semiconductors, John Wiley (N.Y., 1988). L. Wieczorek, H.J. Goldsmith and G.L. Paul, Thermal Conductivity of Amorphous Films in Thermal Conductivity 20, eds. D.P.H. Hasselman and J.R. Thomas, Plenum (N.Y., 1989) p. 235.

[8] I.I. Perepechko, An Introduction to Polymer Physics, Mir (Moscow, 1981) pp. 131.

Part III

Synthesis and Properties of Luminescent Group IV Clusters, Nanocrystals, and Nanostructures

SYNTHESIS, OPTICAL PROPERTIES, AND MICROSTRUCTURE OF SEMICONDUCTOR NANOCRYSTALS FORMED BY ION IMPLANTATION

J. D. BUDAI, C.W. WHITE, S.P. WITHROW, R.A. ZUHR, AND J. G. ZHU
Oak Ridge National Laboratory, Oak Ridge, TN 37831-6030

ABSTRACT

High-dose ion implantation, followed by annealing, has been shown to provide a versatile technique for creating semiconductor nanocrystals encapsulated in the surface region of a substrate material. We have successfully formed nanocrystalline precipitates from groups **IV** (Si, Ge, SiGe), **III-V** (GaAs, InAs, GaP, InP, GaN), and **II-VI** (CdS, CdSe, CdS_xSe_{1-x}, CdTe, ZnS, ZnSe) in fused silica, Al_2O_3 and Si substrates. Representative examples will be presented in order to illustrate the synthesis, microstructure, and optical properties of the nanostructured composite systems. The optical spectra reveal blue-shifts in good agreement with theoretical estimates of size-dependent quantum-confinement energies of electrons and holes. When formed in crystalline substrates, the nanocrystal lattice structure and orientation can be reproducibly controlled by adjusting the implantation conditions.

INTRODUCTION

Recent advances in controlling and characterizing semiconductor nanocrystals and quantum dots has generated considerable interest in exploring new synthesis techniques and applications [1, 2]. Existing applications include commercially available optical filters [3], while promising potential applications range from quantum dot lasers to high-speed nonlinear optical switches. Several successful approaches have been developed for producing nanostructured materials, including arrested precipitation in solvents, controlling surface clusters during film growth, and lithographic etching. In addition, several research groups have initiated investigations into the formation of semiconductor nanocrystals using ion implantation [4 - 8]. In this approach, a supersaturated solid solution is produced by implantation into the surface region of a host material and nanocrystals are formed by precipitation either during high-temperature implantation or during subsequent thermal annealing. The formation and growth of precipitates in such a supersaturated solid solution follows a two-stage kinetic process described theoretically by Lifshitz and Slyozov (LS) [9, 10]. After an initial diffusion-limited nucleation stage, particle coarsening occurs by Ostwald ripening, where the larger particles grow at the expense of the smaller ones.

Much of the initial interest in ion-implanted semiconductor precipitates was focused on studies of the strong visible photoluminescence (PL) arising when indirect bandgap Si (or Ge) precipitates are formed by implantation in SiO_2 [5 - 8, 11]. This system is structurally related to porous silicon formed by electrochemistry, and the optical properties are also quite similar. The ion-implantation technique has since been extended to include the formation of compound semiconductor nanocrystals in SiO_2, Al_2O_3, and Si matrices [12, 13]. Such systems provide great opportunities for tailoring the composite materials properties since both direct and indirect compounds with a wide range of band gap energies are available. Table I

Mat. Res. Soc. Symp. Proc. Vol. 452 © 1997 Materials Research Society

displays the semiconductor nanocrystals which we have synthesized at ORNL by ion implantation into three technologically-important substrate materials.

Table 1. Semiconductor nanocrystals grown by ion implantation. Y indicates nanocrystals formed. N indicates nanocrystals not observed. No symbol indicates systems not yet investigated.

	SUBSTRATE		
Nanocrystal	SiO_2	Al_2O_3	Si
Si	Y	Y	-
Ge	Y	Y	
SiGe	Y	Y	
GaAs	Y	Y	Y
GaP	Y		Y
GaN	N	Y	N
InP	Y		Y
InAs	Y		Y
CdS	Y	Y	Y
CdSe	Y	Y	Y
$CdSe_{0.5}S_{0.5}$	Y		
CdTe			Y
$CdSe_{0.5}Te_{0.5}$			Y
ZnS	Y		
ZnSe	Y		

Ion implantation offers several advantages, as well as challenges, as a technique for synthesizing nanocrystals. First, ion implantation is versatile, since almost any ion can be implanted into any solid substrate with extreme chemical purity, including isotopic discrimination. In some cases, this versatility may in fact provide a unique route for incorporating precipitates in substrates with particular chemical sensitivity or extreme melting temperatures. Second, ions are embedded directly in the matrix by implantation and hence the crucial steps of nanocrystal encapsulation or passivation are typically integral parts of the synthesis. Third, implantation can be carried out through masks or with focused beams, and hence is capable of creating well-defined, high-resolution patterns of regions containing quantum dots. Fourth, the quenching of the kinetic energy for implanted ions is an intrinsically metastable process and hence presents opportunities for unique kinetic routes to nonequilibrium structures. Finally, since ion implantation represents an integral part of existing semiconductor fabrication technology, improvements in equipment capabilities will continue and it will be relatively straightforward to incorporate quantum dots created by ion implantation into electronic devices.

Considering the challenges, ion implantation often introduces structural damage into the substrate. As will be demonstrated in this paper, the ion damage itself sometimes provides useful changes in the host material. However in many cases, it must be controlled by high-temperature implantations or subsequent thermal annealing. In addition, since the implantation approach for creating nanocrystals relies on nucleation and Ostwald ripening of precipitates, the particles possess a range of sizes. Thus, applications requiring greater monodispersity will involve more complex thermal processing steps or advanced techniques such as writing individual quantum dots with finely focused ion beams [14].

EXPERIMENTAL

Semiconductor nanocrystals were formed by implantation of the constituent ions into several different host materials, including fused silica (Corning 7940 glass), silica grown by thermal oxidation on silicon wafers (SiO_2/Si), α-Al_2O_3(0001), and Si(001) wafers. Doses ranged up to $\sim 10^{17}/cm^2$ for each species, and the ion energies were chosen such that the ion concentration profiles were superimposed after implantation. As expected, the size of the nanocrystals increased when the substrate temperature was increased, either during implantation or during subsequent annealing in flowing Ar + 4% H_2. The implanted samples were characterized by Rutherford backscattering (2.3 MeV He ions), four-circle x-ray diffraction (Cu$K\alpha_1$), transmission electron microscopy (TEM), optical absorption, photoluminescence (PL), and Raman spectroscopy.

RESULTS AND DISCUSSION

I. OPTICAL PROPERTIES OF II-VI NANOCRYSTALS IN SiO_2

Silica glasses obviously represent an important class of optical materials and extensive efforts have been made throughout history to modify their properties through the addition of various forms of impurities or inclusions. We have succeeded in creating a wide range of elemental and compound semiconductor nanocrystals embedded in fused silica by implantation and thermal annealing. In general, precipitation in a glassy matrix results in the formation of spherical, randomly oriented, nanocrystals. Here, we present representative results for CdS, a direct bandgap II-VI compound formed in SiO_2. Semiconductor nanocrystals were formed by ion implantation of the constituent group II and group VI ions into bulk fused silica, or into a thermally oxidized silica layer grown on a silicon wafer.

Figure 1 shows x-ray diffraction θ-2θ scans obtained from samples implanted at room temperature with $1 \times 10^{17}/cm^2$ at a single energy for each of the group II and group VI ions, and subsequently annealed at 1000°C for one hour in an Ar/ 4% H_2 atmosphere. The measurements show that randomly oriented, CdSe and CdS precipitates are formed predominantly in the hexagonal wurtzite structure. The positions of the diffraction peaks for the Cd+0.5Se+0.5S sample lie intermediate between the pure compounds, indicating that the mixed compound $CdSe_{0.5}S_{0.5}$ has been formed. This conclusion is further supported by Raman spectra.

The optical properties of the CdS sample described above is shown in Figure 2, with the absorption and PL spectra shown on the left and right sides respectively. After annealing at 1000°C, the absorption spectra develops steep band-edge structure near the bulk value of the CdS bandgap (2.43 eV = 510 nm at room temperature). We attribute the higher energy peak in the PL spectra to bound excitonic recombination within the CdS nanocrystals and the broad, lower energy peak ($\lambda > 650$ nm) to surface-related interface states. Similar absorption features have been observed in doped bulk-glasses containing CdS nanoparticles formed during heat treatments [15], as well as in commercially available Cd(S,Se) glass filters [3, 16]. In addition, we have observed similar results for CdSe, ZnS, and ZnSe nanocrystals formed in SiO_2 by ion implantation. The broad, lower energy PL peak has been observed to strengthen significantly with temperatures above 45K in CdSeS samples [16], consistent with the relatively intense peak found in our room-temperature measurements. We note however, that unlike our observations, some previous studies of CdS nanoparticles in filter glasses failed to detect measurable PL near the direct bandgap energy [3]. This difference may be due to the high Cd and S concentrations and high-temperature heat treatments used in our work.

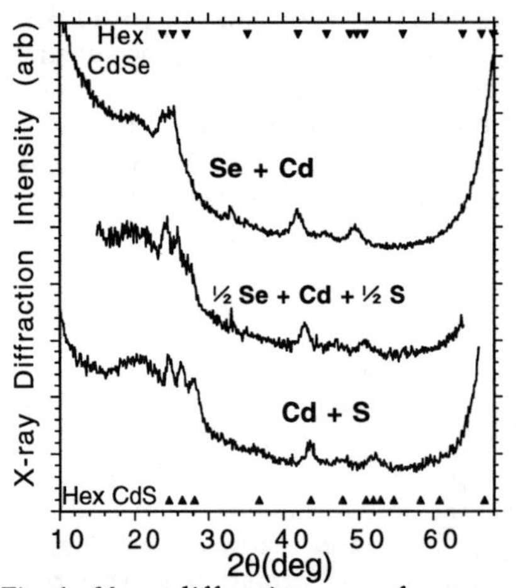

Fig. 1. X-ray diffraction scans from II-VI nanocrystals in a SiO_2 layer on Si. The implantation energies used were Cd(450 keV); Se(330 keV); S(164 keV).

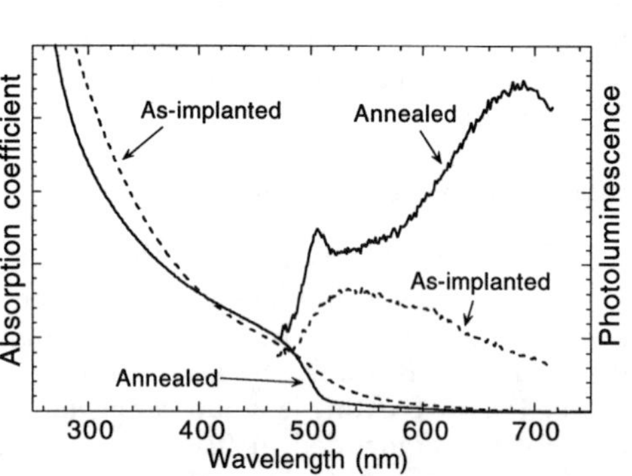

Fig. 2. Optical absorption (left side) and photoluminescence (right side) from a CdS in SiO_2 sample before and after annealing at 1000°C.

The results presented in Fig. 1 and Fig. 2 were obtained from samples synthesized using a single energy for each ion species, resulting an approximately Gaussian concentration depth profile for the implanted ions. After annealing, TEM images show relatively broad distributions of particle sizes, with the largest precipitates located near the peak position of the concentration profile, and the smallest in the tails. Thus, the optical results represent a convolution of spectra from particles with a wide range of sizes. In order to obtain more constant concentration profiles and hence more uniform particle-size distributions, samples were synthesized by implanting ions at three different energies (Cd ions at 35, 120, and 320 keV; S ions at 35, 60, and 110 keV). These energies yielded a relatively uniform concentration of ions in a ~1500Å surface layer. After high-temperature annealing, the late-stage particle-size distributions in these 'flat-profile' samples is controlled by Ostwald ripening [9, 10], where the average particle volume is proportional to the excess concentration.

Figure 3 shows the optical absorption coefficient, α, measured after annealing at 1000°C for one hour from four 'flat-profile' samples with different excess concentrations ranging from $4 \times 10^{20}/cm^3$ (~0.5 at %) to $6.3 \times 10^{21}/cm^3$ (~8 at %) for each of the implanted Cd and S ion species. As the concentration is decreased, the absorption threshold shifts towards shorter wavelength, as expected from quantum-size effects due to confinement of an electron and hole in nanocrystals smaller than the exciton Bohr diameter (~50Å for CdS). Our absorption spectra do not exhibit a distinct exciton peak as is often observed for semiconductor nanocrystal systems. This feature will be broadened by our relatively large size-distribution as well as the measurement temperature (RT), and our observations are consistent with previous findings [17] for S-rich Cd(Se,S) nanocrystals in doped SiO_2.

Since CdS is a direct band gap material, we can extract an experimental estimate of the gap, Eg, from the intercept value of a plot of $(\alpha E)^2$ vs. E, where E is the photon energy [18]. Applying this procedure to the data in Fig. 3 yields blue shifts, $\Delta Eg = Eg - Eg(bulk)$, of up to ~0.3 eV compared with the bulk CdS band gap of ~2.43 eV. A theoretical estimate of the confinement energy, ΔEg, as a function of a spherical particle diameter, d, using the effective-mass approximation has been given by Brus [19]:

$$\Delta Eg(d) = (2\hbar^2\pi^2/\mu d^2) - (3.6e^2/\varepsilon d) \qquad (1)$$

where $\hbar$ is Planck's constant, μ is the exciton reduced mass (~$0.154m_o$), e is the electron charge, and ε is the dielectric constant (~8.9). Inverting this theoretical expression, we have obtained predicted particle sizes for each of the four experimentally determined values of ΔEg. A tight-binding approximation calculation of $\Delta Eg(d)$ for CdS has been used to generate a curve similar in functional shape to equation (1), but quantitatively shifted [20].

Considering the expected dependence of the particle size on implantation dose, the Ostwald ripening process described by LS theory predicts an average particle volume, $(\pi d^3/6)$, which increases proportionately with the excess concentration, c_o. Thus $d \sim c_o^{1/3}$, and Fig. 4 shows the approximate particle diameters obtained from equation (1) plotted as a function of $(\text{concentration})^{1/3}$ for the four CdS samples. An approximately linear relation is observed, in spite of the fact that the LS treatment is strictly valid only in the limit of dilute concentrations where there are no interparticle diffusion interactions.

Note that the particle diameters displayed in Fig. 4 were estimated indirectly from the measured shifts of the absorption spectra. Direct TEM and x-ray diffraction measurements of the particle sizes for comparison are not yet available for these particular 'flat profile' CdS samples. However, we have measured the particle size distribution from the central region of a single energy CdS in SiO_2/Si sample, in which the peak concentration was calculated to be $\sim 8 \times 10^{21}/cm^3$. As discussed above, the highest concentration region for this sample possesses the largest nanocrystals and the particle size distribution in this region should be similar to that for the more uniform 'flat profile' sample with a similar excess concentration. The FWHM range of particle diameters for the central region of this sample is indicated on Fig. 4 with the midpoint marked by the open circle. This observation is in excellent qualitative agreement with the diameters extracted from the energy shifts, especially when one considers that the experimental optical properties are generally volume-weighted, and will be dominated by the largest nanocrystals. When the TEM distribution is weighted by the particle volumes, the midpoint is shifted to ~100Å, as indicated by an × in Fig. 4.

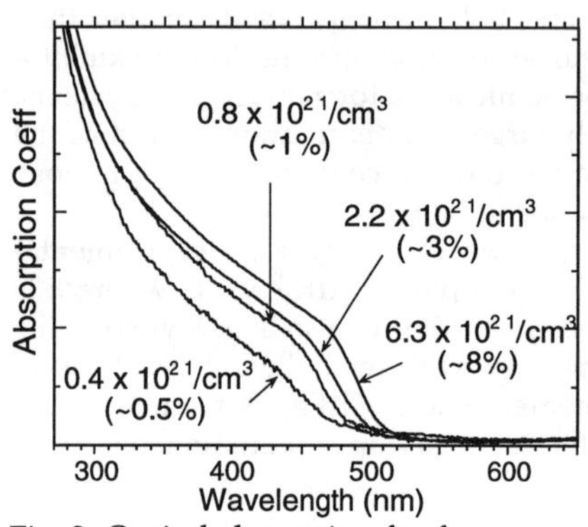

Fig. 3 Optical absorption for four samples implanted with Cd and S ions at multiple energies to produce uniform concentration profiles. The samples were all annealed at 1000°C for one hour.

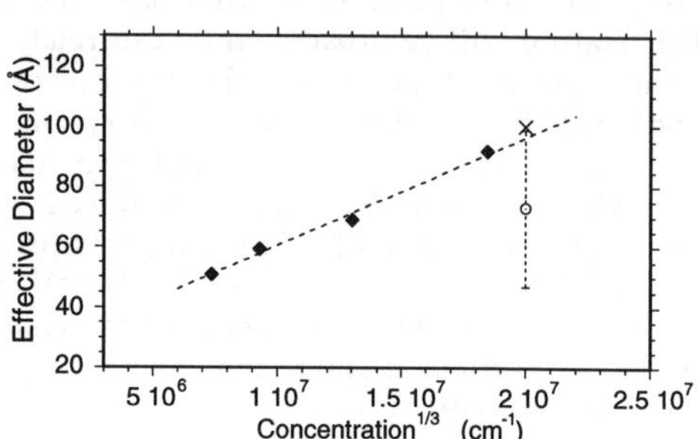

Fig. 4 The estimated CdS particle diameters calculated from equation (1) for different doses are marked by ◆. The open circle indicates the midpoint of a TEM measurement from a similar sample. The × indicates the TEM volume-weighted midpoint.

II. CONTROLLING NANOCRYSTALS IN Al_2O_3 USING ION DAMAGE

In contrast to the previous section, precipitation within a crystalline matrix generally results in the formation of crystallographically oriented precipitates, often with faceted shapes. We have previously reported the synthesis and preliminary optical properties of both elemental and compound semiconductor nanocrystals in α-Al_2O_3 [4, 12]. Here, we describe microstructural changes in semiconductor nanocrystals formed in Al_2O_3 associated with the accumulation of lattice damage in the substrate.

Figure 5 shows θ-2θ x-ray scans along the α-$Al_2O_3(000\ell)$ axis for six different combinations of ion species, after single-energy ion implantation and subsequent annealing, with a representative range of doses, substrate implantation-temperatures, and post-implant annealing temperatures: **Si**(400 keV, $6 \times 10^{17}/cm^2$, 650°C), anneal 1100°C; **Ge**(500 keV, $1.5 \times 10^{17}/cm^2$, RT), anneal 900°C; **Ge**(500 keV, $6 \times 10^{16}/cm^2$, 550°C) + **Si**(215 keV, $6 \times 10^{16}/cm^2$, 550°C), anneal 900°C; **As**(500 keV, $1 \times 10^{17}/cm^2$, LN_2) + **Ga**(470 keV, $1 \times 10^{17}/cm^2$, LN_2), anneal 1100°C; **Se**(330 keV, $4.3 \times 10^{16}/cm^2$, 600°C) + **Cd**(450 keV, $4.3 \times 10^{17}/cm^2$, 600°C), anneal 1000°C; and **Cd**(450 keV, $4.3 \times 10^{16}/cm^2$, 600°C) + **S**(215 keV, $4.3 \times 10^{16}/cm^2$, 600°C), anneal 1000°C.

These x-ray scans reveal that, in general, only a few diffraction peaks appear in addition to the intense substrate α-$Al_2O_3(0006)$ peak. The peak which appears in all scans in the range 2θ ~ 25°-30° is due to the precipitation of oriented semiconductor nanocrystals in the usual diamond cubic, hexagonal wurtzite, or cubic zincblende structures. These phases are structurally related, with an ABABAB plane-stacking sequence corresponding to the hexagonal (0001)⊥ structure and an ABCABC stacking sequence corresponding to the cubic (111)⊥ structures. The observed alignment of the hexagonal (0001) planes or the cubic (111) planes of the nanocrystals with the pseudo-hexagonal substrate (0001) planes is predicted from simple symmetry considerations. The intensity and width of this peak for particular implanted species depends on the number and size of the precipitates, and as expected, larger particles are produced by higher ion doses or higher temperatures (enhanced ripening).

In addition to the nanocrystal peaks, a series of diffraction peaks corresponding to an aligned γ-phase of Al_2O_3 is often observed. These peaks are labeled γ in Fig. 5 in the Ge and the Ga+As implanted samples. However, the presence or absence of the γ-phase x-ray peaks does not depend on the particular implanted ion species. Instead, the γ-phase is observed with any of the ion species whenever the substrate temperature during implantation is relatively low (typically room temperature or below) and the dose is sufficiently high. In such cases, the accumulated lattice-displacement damage due to the implanted ions is sufficient to amorphize the surface layer of the α-Al_2O_3 substrate. Subsequent thermal annealing is known to consist of a two-step recrystallization process [21-24]. An amorphous Al_2O_3 layer on sapphire recrystallizes during annealing first as an epitaxial, metastable γ-phase layer, and then transforms back to the original α-phase at a still higher temperature via layer-by-layer growth modes. The intermediate γ-phase consists of a cubic stacking sequence (ABCABC) variant of Al_2O_3 which has been hypothesized to be stabilized by the greater potential for defect accommodation in the γ-phase. The recrystallization into γ-Al_2O_3 occurs epitaxially with an orientation relation given by $\gamma(111)[1\bar{1}0] \parallel \alpha(0001)[10\bar{1}0]$ or the associated 60° rotated domains. Experimentally, we find that the amorphization dose and the regrowth kinetics depend strongly on the particular implanted impurities, but that the general microstructural evolution is the same for all implanted semiconductor species illustrated here.

The presence of intermediate γ-phase recrystallization in Al_2O_3 substrates leads to several interesting microstructural changes in the precipitated semiconductor nanocrystals, including changes in the orientation. Figures 6(a) and 6(b) show x-ray diffraction θ–2θ-scans taken from two α-Al_2O_3 samples implanted with the same energies and doses of Ge(500 keV, $6 \times 10^{16}/cm^2$) and Si(220 keV, $6 \times 10^{16}/cm^2$) and subsequently annealed for one hour at 1100°C. The only difference in the synthesis of these two samples was in the substrate temperature during implantation; for Fig 6(a) the substrate was held at an elevated temperature of 550°C, while for Fig. 6(b) the substrate was at room temperature (RT). The positions of the nanocrystal (111) peaks show that alloy SiGe nanocrystals (as opposed to separate Si or Ge nanocrystals) are formed in both samples with their diamond cubic (111) planes parallel to the substrate (0001) planes. In Fig. 6(a), although some lattice damage may be present, the elevated temperature of 550°C provides sufficient dynamic annealing that the substrate remains α-phase. Only the α-Al_2O_3(0006) and the SiGe(111) alloy peaks are observed. In contrast, the lattice damage in the RT-implanted sample accumulated to the point that the surface layer of the substrate was amorphized, as verified by Rutherford backscattering. Subsequent annealing produced γ-phase regrowth as shown in Fig 6(b).

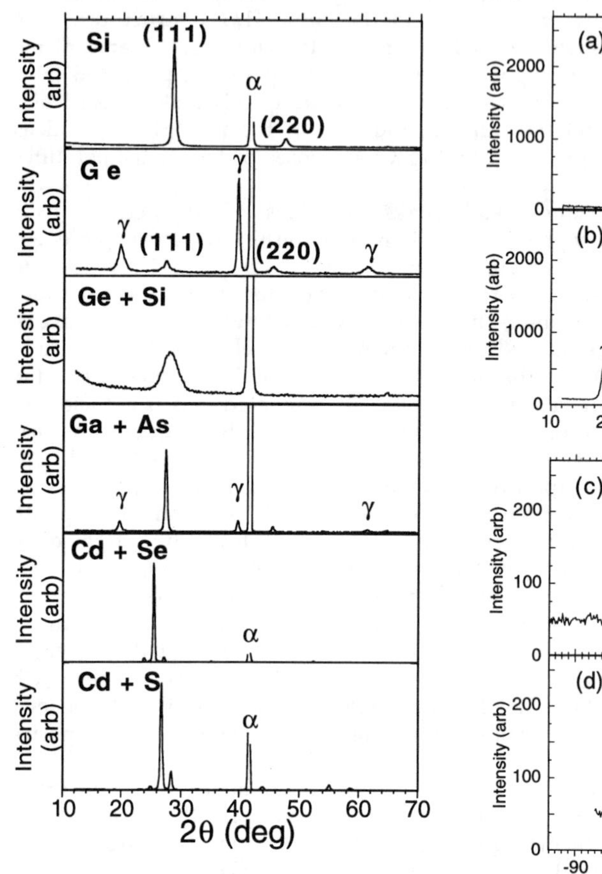

Figure 5 θ-2θ x-ray scans for α-Al_2O_3(0001) implanted with six different combinations of ions. Implant temperatures, energies and doses, and annealing conditions are given in the text.

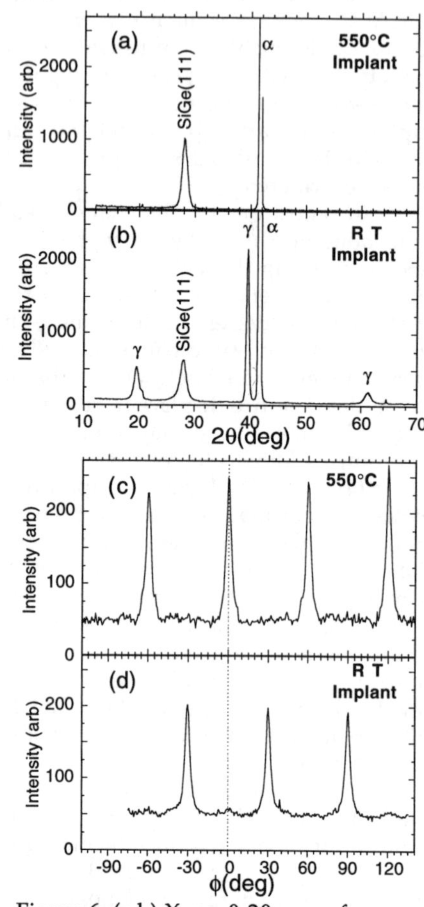

Figure 6 (a,b) X-ray θ-2θ scans for α-Al_2O_3 implanted with Ge + Si with substrate temperatures of (a) 550°C and (b) Room Temp during implantation. (c,d) X-ray φ scans showing nanocrystal orientation for the same samples: (c) 550°C and (d) Room Temp

The large microstructural difference in Al_2O_3 substrates implanted at different temperatures has a direct effect on the semiconductor nanocrystal orientation as illustrated in Figures 6(c) and 6(d). These figures show x-ray φ-scans using the SiGe{220} reflections, and indicate the in-plane nanocrystal orientations for the same samples used in Figures 6(a) and 6(b). For the sample implanted at 550°C, the in-plane orientation is described by SiGe$[1\bar{1}0] \parallel \alpha$-$Al_2O_3[11\bar{2}0]$ (and 60° rotated domains), whereas for the sample implanted at RT, the nanocrystals are rotated by ±30° away from this orientation. The orientations are distinct for these two samples because the SiGe nanocrystals were nucleated and grew while embedded in an α-Al_2O_3 matrix in one case, and were nucleated within a γ-Al_2O_3 matrix in the other. Although illustrated here with SiGe, this orientational dependence on substrate microstructure is a general phenomena and is observed for the other semiconductor nanocrystals. If samples containing γ-oriented nanocrystals (such as used in Fig 6(b) and 6(d)) are subsequently annealed at higher temperatures or longer times such that the γ-Al_2O_3 layer is transformed back to α-Al_2O_3 without melting the nanocrystals, then the nanocrystals remain and coarsen in the γ-orientation in which they were originally formed. Thus, it is possible to reproducibly create nanocrystals oriented in either of the two directions within an α-Al_2O_3 substrate.

CONCLUSIONS

We have synthesized a wide range of semiconductor nanocrystals encapsulated in SiO_2, α-Al_2O_3 and Si substrates using ion implantation and thermal annealing. The particle-size distribution is changed by controlling the processing conditions such as the ion dose and the annealing temperature. Nanocrystals are typically spherical in shape and randomly oriented when precipitated within an amorphous matrix such as SiO_2, and are faceted and oriented in crystalline hosts such as Al_2O_3 and Si. Strong visible photoluminescence has been observed from some systems, and the absorption spectra show quantum-confinement blue-shifts in good agreement with size-dependent theoretical estimates. Interesting microstructural effects related to substrate ion damage have been observed for nanocrystals formed in α-Al_2O_3. Damage accumulation in low-temperature implants produces an amorphous Al_2O_3 surface layer, which recrystallizes during subsequent thermal annealing as an epitaxial, metastable γ-phase Al_2O_3 layer. Nanocrystals which nucleate within γ-Al_2O_3 are found to be microstructurally distinct from those nucleated within α-Al_2O_3.

ACKNOWLEDGMENT

This research sponsored by the Division of Materials Sciences, U.S. Department of Energy under Contract No. DE-AC05-96OR22464 with Lockheed Martin Energy Research Corp.

REFERENCES

1. A.P. Alivisatos, Science **271**, 933 (1996).

2. Special issue: Chemistry of Materials, vol. **8**, no. 8 (1996).

3. N.F. Borrelli, D.W. Hall, H.J. Holland, and D.W. Smith, J. Appl. Phys. **61**, 5399 (1987).

4. C.W. White, J.D. Budai, S.P. Withrow, S.J. Pennycook, D.M. Hembree, D.S. Zhou, T. Vo-Dinh, and R.H. Magruder, *Mat. Res. Soc. Symp. Proc.* **316**, 487 (1994).

5. H.A. Atwater, K.V. Shcheglov, S.S. Wong, K.J. Vahala, R.C. Flagan, M.L. Brongersma, and A. Polman, *Mat. Res. Soc. Symp. Proc.* **316**, 409 (1994).

6. T. Shimizu-Iwayama, K. Fujita, S. Nakao, K. Saitoh, T. Fujita, and N. Itoh, J. Appl. Phys. **75**, 7779 (1994).

7. T. Komoda, J.P. Kelly, A. Nejim, K.P. Homewood, P.L.F. Hemment, and B.J. Sealy, *Mat. Res. Soc. Symp. Proc.* **358**, 163 (1995).

8. P. Mutti, G. Ghislotti, S. Bertoni, L. Bonoldi, G. Cerofolini, L. Meda, E. Grilli, and M. Guzzi, Appl. Phys. Lett. **66**, 851 (1995).

9. I.M. Lifshitz and V.V. Slyozov, J. Phys. Chem. Solids **19**, 35 (1961).

10. P.W. Voorhees, *Annu. Rev. Mater. Sci.* **22**, 197 (1992).

11. J.G. Zhu, C.W. White, J.D. Budai, S.P. Withrow, and Y. Chen, *Mat. Res. Soc. Symp. Proc.* **358**, 175 (1995).

12. C.W. White, J.D. Budai, J.G. Zhu, S.P. Withrow, R.A. Zuhr, D.M. Hembree, D.O. Henderson, A. Ueda, Y.S. Tung, R. Mu, and R.H. Magruder, J. Appl. Phys. **79**, 1876 (1996).

13. C.W. White, J.D. Budai, J.G. Zhu, S.P. Withrow, and M.J. Aziz, Appl. Phys. Lett. **68**, 2389 (1996).

14. L. Feldman, private communication.

15. V. Sukumar and R.H. Doremus, phys. stat. sol. (b) **179**, 307 (1993).

16. J. Warnock and D.D. Awschalom, Phys. Rev. B **32**, 5529 (1985).

17. Y. Fuyu and J. M. Parker, Materials Letters **6**, 233 (1988).

18. J.I. Pankove, Optical Processes in Semiconductors, Dover Publications, New York, 1971.

19. L. E. Brus, J. Chem. Phys. **80**, 4403 (1984).

20. P.E. Lippens and M. Lannoo, Phys. Rev. B **39**, 10935 (1989).

21. C.W. White, C.J. McHargue, P.S. Sklad, L.A. Boatner, and G.C. Farlow, Materials Science Reports **4**, 41 (1989).

22. C.W. White, L.A. Boatner, P.S. Sklad, C.J. McHargue, J. Rankin, G.C. Farlow, and M.J. Aziz, Nucl. Instr. Methods B **32**, 11 (1988).

23. N. Yu, P.C. McIntyre, M. Nastasi, and K.E. Sickafus, Phys. Rev. B **52**, 17518 (1995).

24 S. Cao, A.J. Pedraza, D.H. Lowndes, and L.F. Allard, Appl. Phys. Lett. **65**, 2940 (1994).

VISIBLE PHOTOLUMINESCENCE FROM Si ION-IMPLANTED AND THERMALLY ANNEALED SiO_2 FILMS

Y. Kanemitsu*, N. Shimizu*, S. Okamoto*, T. Komoda**, P. L. F. Hemment***, and B. J. Sealy***
*Institute of Physics, University of Tsukuba, Tsukuba, Ibaraki 305, Japan
**UK R&D Laboratory, Matsushita Electric Works Ltd., Guildford, Surrey, GU2 5YG, UK
***Department of Electronic and Electrical Engineering, University of Surrey, Guildford, Surrey, GU2 5XH, UK

ABSTRACT

We have experimentally studied the photoluminescence (PL) properties of Si clusters in SiO_2 glassy matrices. Si clusters in the SiO_2 matrices were fabricated by Si^+ ion implantation into SiO_2 glasses and then thermally annealed in forming gas. Broad PL peaks are observed in the visible spectral region at room temperature. Resonantly excited PL spectra indicate that the strong coupling of excitons and stretching vibrations of the Si-O bonds causes the broad luminescent spectra. It is concluded that the interaction between electronic and vibrational excitations controls the luminescent emission and the observed dynamics.

INTRODUCTION

The goal of achieving efficient visible luminescence from Si nanocrystals has stimulated considerable research in understanding the optical properties of group IV semiconductor nanocrystals [1]. A size reduction to a few nanometers is required for the observation of visible light emission from Si nanocrystals when the band structure is modified from that of bulk Si, which has an indirect gap of 1.1 eV. The large surface-to-volume ratios of Si nanocrystals are expected to enhance surface effects to the extent that the PL peak wavelength [2], the PL intensity [3] and the fine structures in the PL spectrum at low temperatures [4] will be modified by the surface chemistry of the Si nanocrystals, particularly with regard to the amounts of oxygen and hydrogen on the surfaces. Although there are many extensive studies concerning the origin of visible light emission, the mechanism of visible luminescence in Si nanocrystals and porous Si is still not clear.

In order to investigate the luminescence from Si nanostructures, we need to fabricate Si nanostructures with near identical and stable surfaces [5,6]. Silicon nanocrystals and clusters embedded in a SiO_2 matrix offer some advantages because SiO_2 is a well-characterized material known to passivate Si surfaces where the Si/SiO_2 system is fully compatible with Si technology. Si nanocrystals in the SiO_2 system have been produced by ion implantation and thermal annealing techniques [7-10]. Ion implantation can be used to create supersaturated Si solid solutions while the thermal annealing provides energy which drives the system to a two phase (Si/SiO_2) state. In this paper we discuss the PL properties of Si^+-implanted and annealed SiO_2 films on bulk Si and commercially available fused silica

Mat. Res. Soc. Symp. Proc. Vol. 452 © 1997 Materials Research Society

substrates. The excitation energy dependence of the PL spectra is presented and the data are interpreted in terms of strong coupling of excitonic and vibrational excitations.

EXPERIMENT

The substrates used in this work were thermally grown ~1 μ m SiO_2 thin films on bulk silicon substrate and commercially available fused silica glass of 1 mm thickness. All the samples used in these experiments were prepared by implanting at room temperature a dose of $2x10^{17}$ cm^{-2}, 200 or 400 keV $^{28}Si^+$ ions followed by lamp annealing in forming gas (10% H_2 + 90 % N_2) or N_2 gas for 3-180 minutes at 900-1300 ℃ [8].

The PL spectra were excited by using Cd-He, He-Ne, Ar and Ti:Al_2O_3 lasers. The samples were mounted in a cryostat and the measurement temperature was varied from 2 to 300 K. Time-resolved PL spectra, where the emission was stimulated by 2 ps pulses from a 380 nm laser line, were measured using a synchroscan streak camera. The spectral sensitivity was calibrated by using a tungsten standard lamp and the time resolution was 20 ps.

RESULTS AND DISCUSSION

The PL intensity of samples depends strongly on the annealing condition. Figure 1 summarizes PL spectra in samples (1 μ m SiO_2 thin films on bulk Si) treated by different annealing conditions. The as-implanted samples did not show efficient PL at room temperature, while annealed samples showed pronounced visible PL at room temperature. The PL intensity from samples annealed in forming gas were much higher than those annealed in N_2 gas. In all cases, the annealing of samples under forming gas enhances the PL intensity. However, no hydrogen related vibrational mode (Si-H modes) signals were clearly observed in the FTIR spectrum, although peaks due to Si-O bonds were observed. Thus the role of hydrogen during the annealing process is not clear from the FTIR spectrum but, on the other hand, there is a good correlation between the ESR signal intensity and the PL intensity where the PL intensity increases with a decrease of the ESR intensity [11]. Thermal annealing in forming gas reduces the number of nonradiative recombination centers and we speculate that hydrogen passivation reduces the number of nonradiative recombination centers.

Multipeak PL spectra from SiO_2 thin films on bulk Si were observed in the visible region. The interval of the interference pattern due to multiple reflection can be interpreted by the following equation:

$$2nd = (1/\lambda_2 - 1/\lambda_1)^{-1},$$

where n is the refractive index, d is the effective thickness of the oxide films, and λ_1 and λ_2 are the maximum peak wavelengths in the spectrum. We can change the effective thickness by changing the angle θ between the sample normal and the entrance slit for PL measurements. Figure 2 shows the angle dependence of the PL spectrum under 2.707 eV excitation at room temperature. From these experiments, we obtain n = 1.48. This value is larger than that of SiO_2 (n=1.46), and we speculate that this is because the composition of Si ion-implanted oxide films is below the stoichiometric value (SiO_x, x<2).

In order to discuss the PL mechanism in detail and to eliminate the interference effect, we used the thick fused silica glass of 1 mm thickness [12]. The samples were prepared by implanting at room temperature a dose of $2x10^{17}$ cm^{-2}, 200 keV $^{28}Si^{+}$ ions followed by lamp annealing in forming gas (10% H_2 + 90 % N_2) for 3 minutes at 900 ℃. Figure 3 shows typical PL spectra recorded at 18 K and 2 K using excitation photon energies of 3.814 eV (curve a), 2.707 eV (curve b), 2.540 eV (curve c), 2.409 eV (curve d), and 1.959 eV (curve e). A broad PL peak, which shifts to higher energy with increasing excitation energy, is evident in each case. The energy interval (Stokes shift) between the energy of excitation and first peak emission also increases monotonically. In each case, the spectra show some peaks which are in contrast to the featureless, Gaussian-like peaks recorded from these samples at room temperature [7].

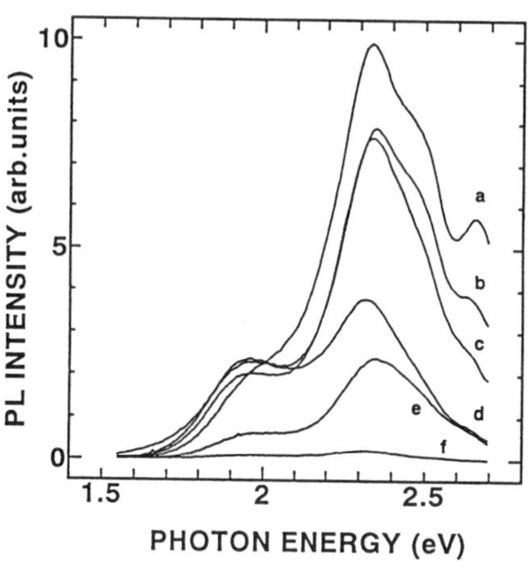

Figure 1 Photoluminescent spectra from Si^{+}-implanted (400 keV, 2 x 10^{17} cm^{-2}) and thermally annealed SiO_2 thin film on bulk silicon substrate. The thermal annealing conditions: (a) 180 min at 900 ℃ under forming gas (10 % H_2 + 90 % N_2), (b) 30 min at 900 ℃ under forming gas, (c) 3 min at 900 ℃ under forming gas, (d) 30 min at 1300 ℃ under N_2 gas, (e) 3 min at 900 ℃ under N_2 gas, and (f) as-implanted sample.

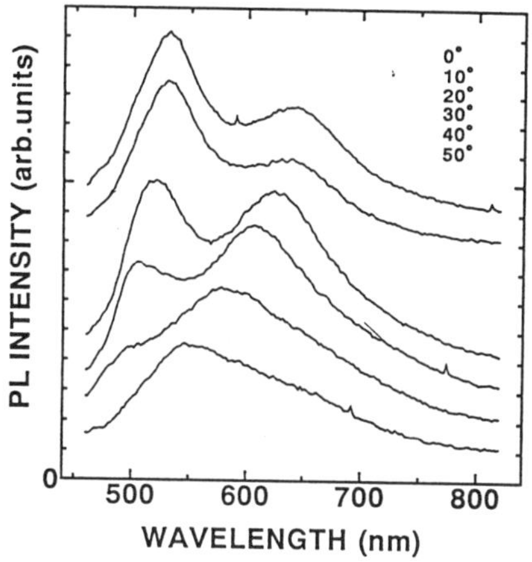

Figure 2 Photoluminescence spectra form Si^{+}-implanted (400 keV, 2 x 10^{17} cm^{-2}) and thermally annealed SiO_2 thin film on bulk silicon substrate as a function of the angle θ between the incident slit of the monochromator and the sample normal. The effective sample thickness is changed by changing the angle θ. The interference pattern in the luminescent spectra is due to multiple reflection.

Figure 4 shows, in more detail, spectra recorded at 18 K using excitation photon energies of 2.540 eV (curve a), 2.409 eV (curve b), 2.330 eV (curve c) where the zero on the abscissa scale corresponds to the excitation energy. A curve fitting program has been used to generate the Gaussian components, which are shown as dashed curves. The energy interval (Stokes shift) between the excitation energy and the highest calculated energy PL peak depends on the excitation energy and is appreciably larger than the exciton splitting energy [13,14]. We conclude that the large Stokes shift is mainly caused by strong exciton-phonon interactions or self-trapped exciton formation [15]. The experimental values of the spacing between the Gaussian peaks (horizontal arrows in the figure) are approximately 135-140 meV which is much larger than the reported energy of ~57 meV of TO phonons in crystalline silicon [4,16]. This large difference cannot be explained simply by the exciton-phonon coupling in the Si crystallites as excitons are highly delocalized within the nanocrystals and peak structures, seen in Fig. 4, , do not correspond to phonon spectra in bulk Si. Instead, this energy is almost equal to the local vibration energy of the Si-O-Si stretch mode (measured as ~ 1100 cm-1) [12]. The coupling of excitons and surface silicon oxide vibrations are expected to increase with localization of excitons in smaller dimensions [15].

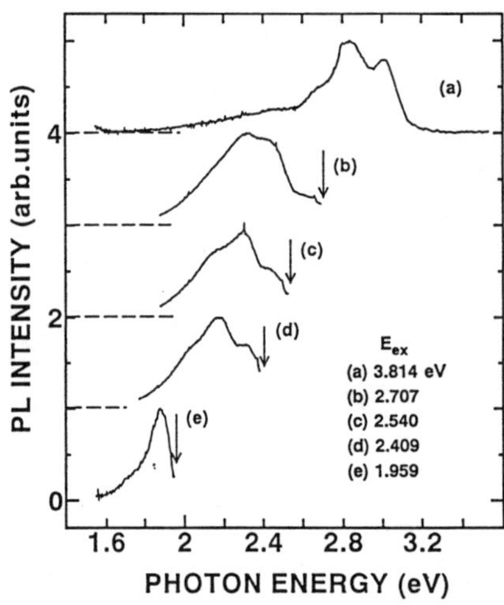

Figure 3 Photoluminescent spectra from Si^{+}-implanted and thermally annealed silica films stimulated by photon energies of (a) 3.814 eV at 18 K, (b) 2.707 eV at 18 K, (c) 2.540 eV at 18 K, (d) 2.409 eV at 18 K and (e) 1.959 eV at 2 K. The vertical arrows indicate the photon energies of the excitation radiation.

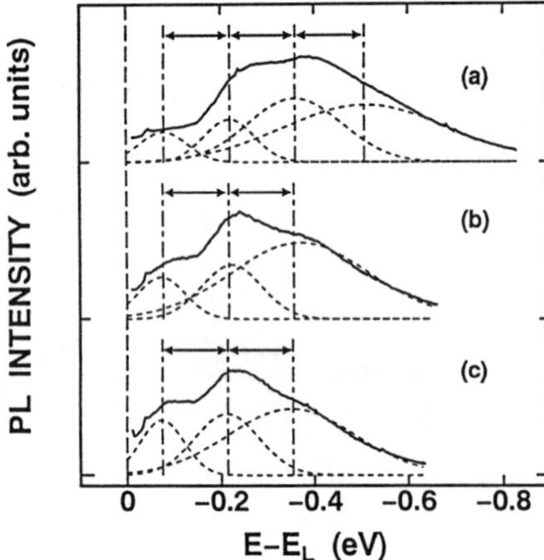

Figure 4 Photoluminescent spectra from Si+-implanted and thermally annealed silica samples at the low-energy side of the excitation energies at 18 K: (a) 2.540 eV, (b) 2.409 eV and (c) 2.330 eV, (d). The dotted lines show the Gaussian components of the experimental data. The peak energies are identified by the vertical broken lines.

These structures were clearly observed at high temperatures up to 250 K, although the phonon-related structures in porous Si are only observed at low temperatures below ~100 K [17]. Luminescent properties of the nanocrystals in these Si^+-implanted silica samples resemble, more closely, the emissions from molecules rather than a solid, where similar PL is usually observed in isolated small molecules at room temperature [18].

In porous Si, step-like phonon structures, due to TO phonons, are clearly observed at low temperatures (<70 K), by reducing the excitation laser energy below 2 eV (resonant excitation) [4,16]. However, these bulk Si phonon-related and step-like PL structures under resonance excitation are not observed in this work. Instead, the peak structures in the broad PL spectrum are clearly observed in Si clusters in SiO_2 glass under resonance excitations, as shown in Fig. 4. These peak structures are clear evidence that in smaller geometric structures the exciton and Si-O vibrational coupling dominate the luminescent process. Our results are in good agreement with the appearance of the peaks in the PL spectra if the phonon-assisted luminescent process determines the phonon structures [16]. Then, the peak structures in the PL spectrum show that the interaction between electronic and vibrational excitations is important in the generation of luminescence from small Si clusters. Since the Si-O bond is polar, the coupling of excitons and stretch vibrations of surface species increase with localization of excitons in smaller dimensions, through the Fröhlich interaction [19]. The relative strength between Fröhlich and deformation-potential contributions depends strongly on the size of nanocrystals and clusters [20]. In particular, in non-polar Si and Ge nanocrystals, the interaction strength between excitons and vibrations depends on either polar (Si-O bonding) or non-polar (Si-H bonding) surface structures. The Fröhlich interaction increases with decreases in the size of Si nanocrystals and clusters with polar bonding. The strong exciton-phonon coupling causes the self-trapping of excitons which are strongly bound in crystalline of very small dimensions [12,20]. Having access to resonantly excited PL spectra from Si nanostructures would reveal the nature of the visible luminescent mechanism in Si quantum structures and materials.

CONCLUSIONS

We have studied the dynamics and spectroscopy of Si^+-implanted SiO_2 and fused silica. Luminescent properties are understood in terms of the localization of excitons in very small dimensions. The coupling of excitons and polar Si-O-Si vibrations cause the broad luminescent spectra. This experiment is the first evidence that the interaction between excitons and Si-O-Si vibrations in the cluster-matrix interface dominate the luminescent properties of Si clusters and nanocrystals. The coupled states of excitons and vibrations are good elementary excitations in very small dimensions

ACKNOWLEDGMENTS

The authors would like to thank M. Kondo for his experimental assistance and discussion. This work was partly supported by a Grant-In-Aid for Scientific Research from the Ministry of Education, Science, Sports, and Culture of Japan, Matsushita Electric Works Ltd., and the University of Surrey.

REFERENCES

1. See, for example, Y. Kanemitsu, Phys. Rep. **263**, 1 (1995).
2. L. Tsybeskov and M. Fauchet, Appl. Phys. Lett. **64**, 1983 (1994).
3. J. C. Tsang, M. A. Tischler and R. T. Collins, Appl. Phys. Lett. **60**, 2279 (1992).
4. Y. Kanemitsu, Phys. Rev. B **53**, 13515 (1996); Y. Kanemitsu and S. Okamoto, to be published.
5. Y. Kanemitsu, T. Ogawa, K. Shiraishi, and K. Takeda, Phys. Rev. B **48**, 4883 (1993).
6. L. E. Brus, P. F. Szajowski, W. L. Wilson, T. D. Harris, S. Shuppler, and P. H. Citrin, J. Am. Chem. Soc. **117**, 2915 (1995).
7. T. Shimazu-Iwayama, S. Nakao, and K. Saitoh, Appl. Phys. Lett. **65**, 1814 (1994).
8. T. Komoda, J. Kelly, F. Cristiano, A. Nejim, P. L. F. Hemment, K. P. Homewood, R. Gwilliam, J. E. Mynard, and B. J. Sealy, Nucl. Inst. Meth. Phys. Res. B **96**, 387 (1995).
9. T. Fischer, V. Petrova-Koch, K. Shcheglov, M. S. Brandt, and F. Koch, Thin Solid Films **276**, 100 (1996).
10. H. A. Atwater, K. V. Scchglov, S. S. Wong, K. J. Vahala, R. C. Flagan, M. L. Brongersma, and A. Polma, Mater. Res. Soc. Symp. Proc. **316**, 409 (1994).
11. M. Kondo and Y. Kanemitsu, unpublished data.
12. Y. Kanemitsu, N. Shimizu, T. Komoda, P. L. F. Hemment, and B. J. Sealy, Phys. Rev. B **54**, R 14329 (1996).
13. E. Martin, C. Delerue, G. Allan, and M. Lannoo, Phys. Rev. B **50**, 18258 (1994).
14. T. Takagahara and K. Takeda, Phys. Rev. B **53**, R4205 (1996).
15. Y. Kanemitsu, S. Okamoto, M. Otobe, and S. Oda, to be published.
16. P. D. J. Calcott, K. J. Nash, L. T. Canham, M. J. Kane, and D. Brumhead, J. Phys. Condens. Matter. **5**, L91 (1993); J. Lumin. **57**, 257 (1993).
17. Y. Kanemitsu and S. Okamoto, to be published.
18. See, for example, M. Ueta, H. Kanzaki, K. Kobayashi, Y. Toyozawa, and E. Hanamura, *Excitonic Processes in Solids* (Springer-Velarg, Berlin 1986).
19. S. Okamoto and Y. Kanemitsu, Phys. Rev. B **54**, 16421 (1996).
20. See, for example, K. Inoue, A. Yamanaka, K. Toba, A. V. Baranov, A. A. Onuschchenko, and A. V. Fedorov, Phys. Rev. B. **54**, R8321 (1996); G. Scamarcio, V. Spagnolo, G. Ventrani, M. Lugara, and G. C. Righini, Phys. Rev. B **53**, R10489 (1996).

ANNEALING STUDIES OF VISIBLE LIGHT EMISSION FROM SILICON NANOCRYSTALS PRODUCED BY IMPLANTATION

G. Ghislotti [(a)], B. Nielsen [(a)], L. F. Di Mauro[(b)], B. Sheey [(b)], P. Mutti[(c)], A. Pifferi[(c)], P. Taroni [(c)], L. Valentini [(c)], F. Corni [(d)], R. Tonini [(d)]
(a)Department of Applied Science, Brookhaven National Laboratory, Upton, N.Y. 11973
(b)Department of Chemistry, Brookhaven National Laboratory, Upton, N.Y. 11973
(c)Dipartimento di Fisica, Politecnico di Milano, 20133 Milano Italy
(d)Dipartimento di Fisica, Università di Modena, Modena, Italy

ABSTRACT

The annealing behavior of silicon implanted SiO_2 layers is studied using continuous and time-gated photoluminescence (PL). Two PL emission bands are observed. A band centered at 560 nm is present in as implanted samples and it is still observed after 1000 °C annealing. The emission time is fast (0.2 -2 ns). A second band centered at 780 nm further increases when hydrogen annealing was performed. The emission time is long (1 μs - 0.3 ms).

INTRODUCTION

Ion implantation is a promising technique for producing silicon nanocrystals (Si_{nc}) in a dielectric matrix. The basic idea in this approach is to enrich by ion implantation the silicon concentration in a dielectric matrix. Subsequent thermal annealing can promote the formation of crystalline precipitates having nanometric dimensions. Ion implantation which is a well established technique in microelectronics, presents some extra advantages: the control of the depth and width of the region of Si_{nc} formation, and the possibility to implant different ions in different substrates.

Photoluminescence (PL) from Si_{nc} produced by ion implanation has been reported by several authors [1-4]. Shimizu-Iwayama et al. [2] presented data of room temperature luminescence from 1 Mev Si^+ implanted fused silica. Komoda et al. [3] observed a band centered at 600 nm, and a red shift that was interpreted as due to an increase in the size of Si_{nc}. The observation of near infrared and blue-green light emission in SiO_2 layers implanted at 1 x 10^{17} cm^{-2} and annealed at high temperature (T>1000 °C) has been reported by some of us [4- 6].

In spite of the many investigations, a complete study of the kinetics of formation of the two emission bands as a function of different preparation conditions, and of their emission times has not been performed. This study could provide a better understanding of the physical parameters which influence the luminescence mechanism. At the same time, this can also be of pratical interest to improve the PL intensity in view of possible optoelectronics applications. To address some of these issues, in this paper continuous and time-gated PL studies of Si^+ implanted SiO_2 layers were performed starting from as-implanted samples.

EXPERIMENT

Samples were prepared by implantation of 160 KeV $^{28}Si^+$ ions into 430 nm thick SiO_2 layers thermally grown on a (100) oriented p-type Si substrate. Fluences ranged from 3 x 10^{16}

Mat. Res. Soc. Symp. Proc. Vol. 452 © 1997 Materials Research Society

cm^{-2} to 3 x 10^{17} cm^{-2} . Vacuum annealing was performed in a high-vacuum chamber (10^{-7} torr) and the sample was mounted on a resistively heated tantalum foil. Gas annealing was done in a furnace using high purity gases. In the case of hydrogen annealing a 15% H_2 , 85% Ar mixture was used.

Photoluminescence measurements were performed using 464 nm excitation light from an excimer pumped dye laser. Emitted light was detected using an intensified charge coupled device (ICCD) detector. Nanosecond time-resolved spectroscopy was performed using a mode-locked krypton ion laser oscillating at 413 nm (see [7] for details on the experimental apparatus).

RESULTS AND DISCUSSION

Figure 1 represents photoluminescence (PL) spectra for samples implanted at different doses ($3x10^{16}$ - $3x10^{17}$ cm^{-2}) and annealed for 30 min. at 1000 °C. Two bands are observed: a green-blue band peaked at 560 nm, and a near-infrared one centered at 780 nm. This latter is observed only for an implantation dose higher than $1x10^{17}$ cm^{-2}, which corresponds to a Si excess in the implanted region higher than 5 % at. PL intensity at 560 nm and 780 nm for sample implanted at 2 x 10^{17} cm^{-2} after annealing at different temperatures for 30 min. is reported in fig. 2a. The intensity of the 560 nm band increases as a function of the annealing temperature up to 1000 °C. A band centered around 780 nm is detected only after annealing at 1000 °C was performed. The near-infrared band grows for increasing annealing time (see Fig. 2b).

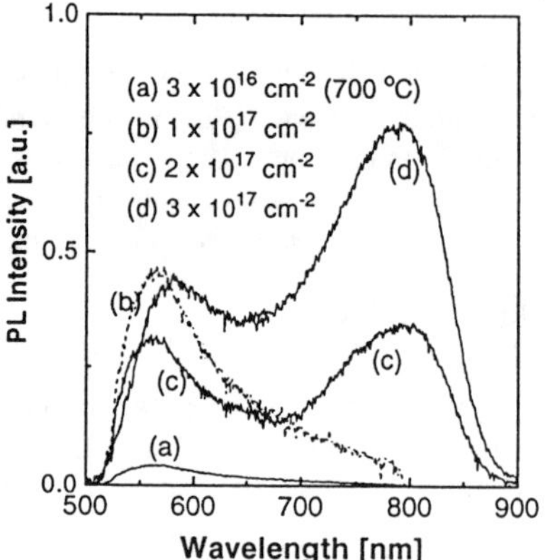

Fig.1. PL intensity for samples implanted at different doses and annealed 30 min. at 1000 °C.

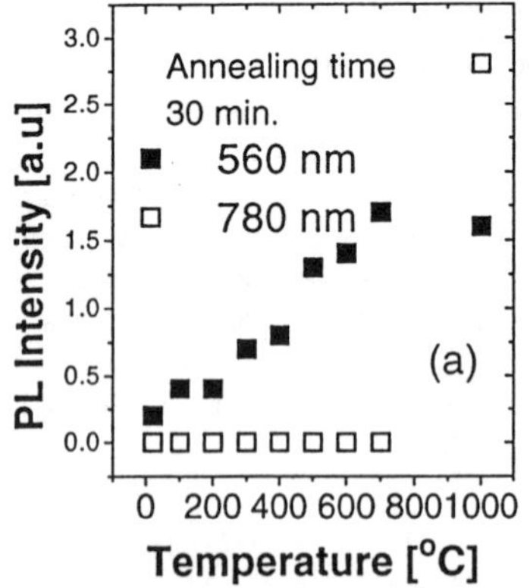

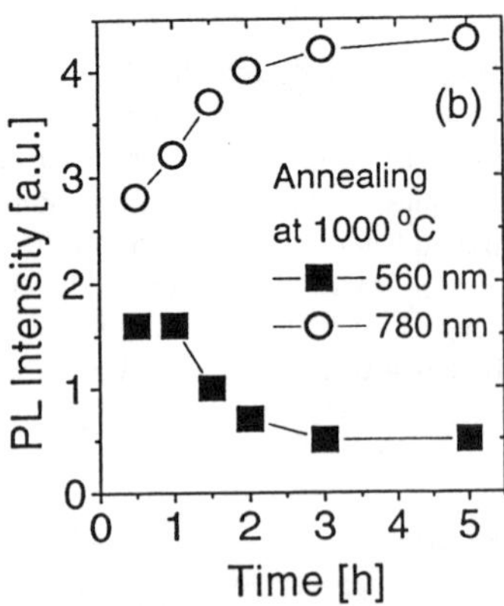

Fig. 2. Effect on PL intensity of annealing temperature (a), and time (b) for sample implanted at 2 x 10^{17} cm^{-2}.

To study the effect of hydrogen, a sample prepared by implanting at $2x10^{17}$ cm^{-2} fluence and annealed 30 min. at 1000 °C, was subsequently annealed in H_2 atmosphere at 500 °C. The intensity in the near-infrared spectral region increases after hydrogen annealing and saturates after about 3 hours annealing, whereas the PL intensity for the blue-green light remains almost the same (see Fig 3). The temporal evolution of the PL signal was studied using time-gated spectroscopy. In the case of sample implanted at 1 x 10^{17} cm^{-2} and annealed 30 min. in argon at 1000 °C, for very short time gates (0-0.5 ns, 2-5 ns) after the excitation pulse PL emission is peaked around 510 nm (see Fig. 4a). For the sample corresponding to an implantation dose of 2 x 10^{17} cm^{-2} and annealed under the same conditions, the fast component is shifted at a longer wavelength, while a longer component appears (Fig. 4b).

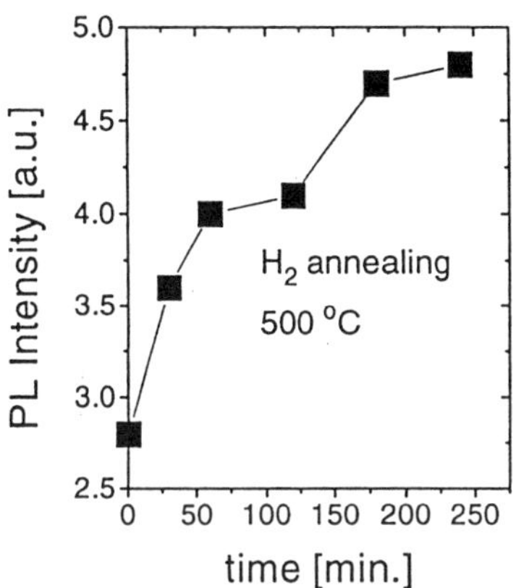

Fig. 3. Effect of hydrogen annealing on PL intensity at 780 nm.

Gated PL spectra for this sample on a longer time-scale are shown in Fig.5. For gates of 1 µs - 10µs the emission is centered at 750 nm. When the time gate is increased the PL emission is centered at a longer wavelength (800 nm when the gate is in the range 10 µs - 0.3 ms).

The observation of a green emission band in Si implanted SiO_2 layers has been reported by several authors [3,4-6]. Present data show that it is present in as implanted samples, it increases as a function of the annealing T up to 1000 °C, and the emission time is very fast. An emission around 600 nm and related to defects (twofold coordinated Si atoms bonded to two oxygen atoms) is observed in SiO_2 [8, 9]. However these defects are characterized by radiative times of the order of µs [9].

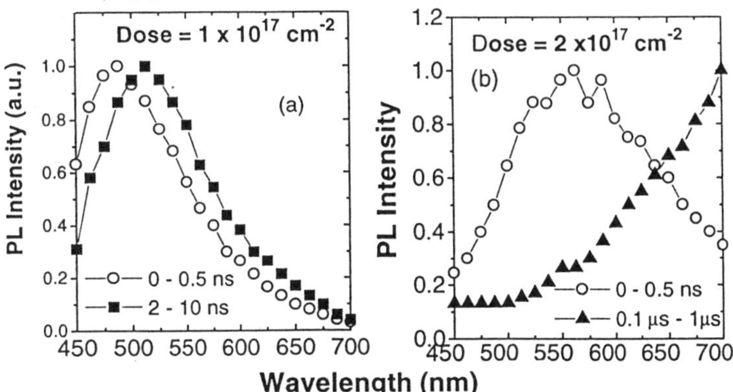

Fig.4 Nanosecond time-resolved gated spectra for samples implanted at 1 x 10^{17}(a), and 2 x 10^{17} cm^{-2} and annealed at 1000 °C.

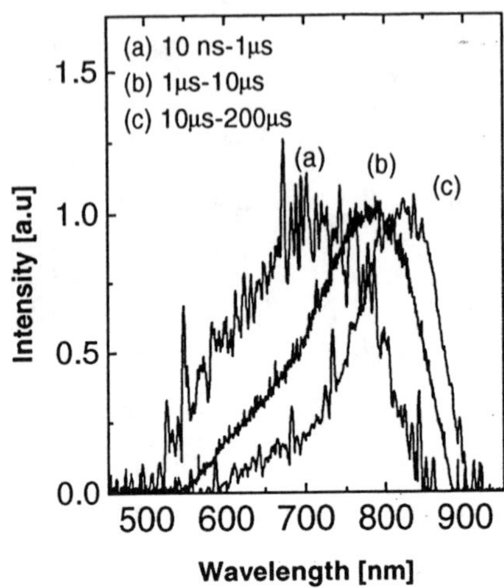

Fig.5 Gated spectra for sample corresponding to 2 x 10^{17} cm^{-2} dose for different time gates.

Previous annealing studies conducted on samples under investigation have shown that these defects anneal out above 700 °C [10]. The annealing behavior and the fast decay time seem to exclude that this band originates from the point defects mentioned above.

Fast lifetimes for silicon-based nanostructures are reported to be related with chain or ladder-like structures. Kanemitsu et al. [11] reported that linear or ladder structures of Si atoms can give a luminescence signal in the spectral region around 500 nm and they have radiative times of the order of ns, while cubic silicon clusters have radiative times of the order of ms [12]. Moreover Si clusters are more influenced by the presence of non-radiative centers at their surface [12, 13]. The behavior observed for the green light emission presents some similarities with results recently obtained by Augustine et al. [14] on amorphous silicon oxynitrides produced by plasma-enhanced chemical vapor deposition. They reported a fast (less than 10 ns) green emission which does not depend on annealing atmosphere. and which shows analogies with organosilane chemical compounds [15].

Present results could be therefore consistently explained assuming that after implantation in the Si-rich region corresponding to the end of range Si extended defects having such a molecular-like stucture (chain or ladder) are formed. These structures can act as radiative sites. Since the radiative times for the emission at 560 nm are fast, the luminescence intensity is not influenced by non-radiative processes which are usually characterized by capture rates of the order of ms [16]. This may explain why the intensity is not affected by hydrogen annealing. While based on annealing behavior both small silicon clusters and Si-chain or ladder structures can be responsible for the 560 nm band, the fast lifetime and the insensitivity to annealing atmosphere (argon, nitrogen, hydrogen) seem to indicate that the latter are the most probable candidates.

Near infrared light is observed only after annealing at high temperature and for implantation doses higher than 1 x 10^{17} cm^{-2}. After implantation a Si-rich SiO_2 (SiO_x , $x<2$) is produced, Si excess depending on the dose. If annealing temperature is raised to (or above) 1000 °C SiO_x becomes unstable [17, 18] and Si precipitates are known to be formed [19-21]. This is supported also by several TEM showing the presence of Si nanocrystals inside the SiO_2 matrix [3,4,]. Therefore the observation of a red band is related with the formation of Si precipitates (although not necessarily related to a recombination within the Si_{nc}). The observed annealing time behavior (see Fig. 2b), is consistent with a Ostwald ripening process (disappearence of smaller precipitates and formation of bigger ones).

Finally the observed emission times for the near-infrared light (from 1μs to 0.3 ms) are on a timescale comparable with that of non-radiative recombination processes. Centers like Si dangling bonds can represent important non-radiative channels. Lannoo et al. [13] showed that even few Si dangling bonds can completely quench the PL emission from a Si_{nc} . Since hydrogen is known to passivate the $Si\text{-}SiO_2$ interface [22, 23], the improvement in PL intensity, after hydrogen annealing (fig.3), can be explained assuming that passivation of Si dangling bonds at the $Si_{nc}\text{-}SiO_2$ interface is taking place.

CONCLUSIONS

Two emission bands are detected in SiO_2 layers implanted with Si^+ ions at doses of 3 x10^{16} cm^{-2} - 3 x 10^{17} cm^{-2}. The first centered at 560 nm is present in as-implanted samples, increases its intensity as a function of the annealing temperature, and it is characterized by fast emission (0.2-2 ns). A second band around 750 nm is detected after 1000 °C annealing was performed. The fast emission time and the annealing behavior suggest that the green emission is related to extended defects, while the near-infrared light is related to the presence of silicon precipitates.

REFERENCES

[1] H. A. Atwater, K. V. Shcheglov, S. S. Wong, K. J. Vahala, R. C. Flagan, M. L. Brongersma, A. Polman, Mat. Res. Soc. Symp. Proc. **316**, 409 (1994)

[2] T. Shimizu-Iwayama, K. Fujita, S. Nakao, K. Saitoh, T. Fujita, and N. Itoh, J. Appl. Phys. **75**, 7779 (1994)

[3] T. Komoda, J. Kelly, F. Cristiano, A. Nejim, P.L.F. Hemment, K.P. Homewood, R. Gwilliam, J.E. Mynard, and B.J. Sealy, Nucl Instr. and Methods B **96**, 387 (1995)

[4] P. Mutti, G. Ghislotti, S. Bertoni, L. Bonoldi, G.F. Cerofolini, L. Meda, E. Grilli, and M. Guzzi, Appl. Phys. Lett. **66**, 851 (1995)

[5] P. Mutti, G. Ghislotti, L. Meda, E. Grilli, M. Guzzi, L. Zanghieri, R. Cubeddu, A. Pifferi, P. Taroni, and A. Torricelli, Thin Solid Films **276**, 88 (1996)

[6] G. Ghislotti, B. Nielsen, P. Asoka-Kumar, K. G. Lynn, A. Gambhir, L. F. DiMauro, C. E. Bottani, J. Appl. Phys. **79**, 8660 (1996)

[7] R. Cubeddu, F. Docchio, W. Q. Liu, R. Ramponi, and P. Taroni, Rev. Sci. Instrum. **59**, 2254 (1988)

[8] L. Skuja, J. Non-Cryst. Solids **149** 77 (1992)

[9] L. Skujia, Solid State Commun. **84** 613 (1992)

[10] G.Ghislotti, B. Nielsen, P. Asoka-Kumar, K.G. Lynn, C. Szeles, C.E.Bottani, S.Bertoni, G.F.Cerofolini, and L. Meda, Thin Solid Films **276** (1996)

[11] Y. Kanemitsu, K. Suzuki, S. Kyushin, H. Matsumoto, Phys. Rev B **51**, 13103 (1995)

[12] Y. Kanemitsu, K. Suzuki, M. Kondo, S. Kyushin, H. Matsumoto, Phys. Rev. B **51**, 10666 (1995)

[13] M. Lannoo, C. Delerue, and G. Allan, J. of Lumin. **57**, 249 (1993)

[14] B.H. Augustine, E.A. Irene, Y.J. He, K.J. Price, L.E. McNeil, K.N. Christensen, and D.M. Maher, J. Appl. Phys. **78**, 4020 (1995)

[15]S. Kyushin, H. Matsumoto, Y. Kanemitsu, and M. Goto, J. Phys. Soc. Jpn. **63**, 46 (1988)

[16] D. Goguenheim and M. Lannoo, Phys. Rev. B **44**, 1724 (1991)

[17] D. Dong, E.A. Irene, and D.R Young, J.Electrochem. Soc. **125**, 819 (1978)

[18] F. Rochet, G. Dufar, H. Roulet, B. Pelloie, J. Perrière, E. Fogarassy, A. Slaoui, and M. Froment, Phys. Rev. B **37**, 6468 (1988)

[19] L.A. Nesbit, Appl. Phys. Lett. **46**, 38 (1985)

[20] C. Jaussand, J. Margail, J. Stoemenos, and M. Bruel, MRS Symp. Proc. **107**, 17 (1988)

[21] S. Mantl, Mater. Sci. Rep. **8**, 1 (1992)
[22] K.L. Brower, Phys. Rev. B **42**, 3444 (1990)
[23] M.L Reed, Semicond. Sci. Technol. **4**, 980 (1989)

IMPROVEMENT OF THE LUMINESCING BEHAVIOUR OF Si^+-IMPLANTED SiO_2 FILMS

T. SCHUSTER, T. DITTRICH, H. E. PORTEANU, T. FISCHER, E. HECHTL, V. PETROVA-KOCH, F. KOCH
Physik-Department E 16, Technische Universität München, D-85747 Garching, Germany

ABSTRACT

In previous work we reported on the observation of continuously tunable photoluminescence in Si^+-implanted SiO_2-films with moderate intensities. In this paper we demonstrate improved performance of such samples. The photoluminescence intensity increases abruptly up to two orders of magnitude when the anneal temperature is elevated to values higher than 1000...1100°C. This strong photoluminescence degrades less than that of porous silicon. Very fine tunability in the spectral range from 2.1 eV to 1.3 eV is achieved in samples implanted with a graded dose. In the analysis of the results we try to distinguish between the contributions of the Si-nanocrystals and of the oxide related defects.

INTRODUCTION

In the rush of papers dealing with light emitting silicon, where porous silicon is still the object attracting the most attention, more recently another photoluminescing system was introduced, namely Si^+-implanted SiO_2 [1-4]. The Si^+-implanted SiO_2 represents a very clean, MOS-compatible, quantitatively controllable, compact (absence of pores) system, which offers advantages for applications as well as for better academic understanding of the widely debated problem of the light emission in Si. Generally, the photoluminescence (PL) in the implanted layers is seen to appear after thermal annealing in inert gas atmosphere [1-4]. The Si precipitates in the SiO_2 are considered to be responsible for this behaviour, however the light emitting oxidic defects [5] can not be completely ruled out [1,4].

The formation of Si precipitates in SiO_2 can be understood in the frame of the well known theoretical work of oversaturated solid solutions [6], which was explored previously to understand, for example, the formation of CdS and CdSe in a glassy matrix [7].

In contrast to porous silicon, where the formation of very small Si crystallites is restricted by the fragility of the porous skeleton, very small Si clusters, containing only a couple of Si-atoms, can be formed in the implanted layers. As a result of the superior size reduction we observed and reported [4,8,9] superior tunability of the PL in the green/blue spectral region in comparison to porous silicon. However, the quantum efficiency of these samples was moderate in comparison to porous silicon. The decay times are in the nanosecond range for the PL in the green/blue region [9], while the PL decays with microseconds in the red/orange region at room temperature which is also typical for porous silicon [4].

The slow decay time of the PL of Si^+-implanted SiO_2 in the red/orange region can not be improved, in principle, for the given luminescence mechanism. In contrast, the quantum efficiency of the PL and its stability can be optimized within a given technological process. We used samples with graded implantation doses to get more detailed information about the dose

Mat. Res. Soc. Symp. Proc. Vol. 452

dependent tunability of the PL. Further we increased the annealing temperature to improve the passivation of the surfaces of the Si nanoparticles in the SiO_2 matrix.

EXPERIMENTAL

Ion implantation of Si^+ is performed into thermally grown SiO_2 layers (thickness 100 nm) on p-type Si(100) (resistivity 1 Ωcm). The implantation energy is chosen between 20 and 50 keV. The projected ranges are within the SiO_2 layer for these energies. The thermal anneal is carried out at temperatures between 300-1150°C and for times between 2 and 40 min in forming gas atmosphere. The PL is measured at room temperature in the spectral region 0.7-3 eV using a prism spectrometer and Ge and Si photodetectors and a photomultiplier tube. The PL is excited with the UV line of a HeCd laser (wavelength 325 nm, intensity about 500 mW/cm^2). Some of the samples are characterized by Fourier transform infrared spectroscopy (FTIR).

RESULTS AND DISCUSSIONS

Fig. 1 shows the normalized PL spectra of a sample with a graded dose (implantation energy 20 keV, dose in the maximum $5*10^{16}$ cm^{-2}, annealing time 30 min) for annealing temperatures below 600°C and for the annealing temperature of 1100°C. The PL spectrum of the bulk Si is also shown for comparison. The normalized PL spectra are independent of the dose for annealing temperatures below 600°C. The spectra are broad and peaked about 2.2 eV in this case. The PL spectra start to shift to lower energies for annealing temperatures higher than 600°C while the value of the red shift depends on the dose. A very fine tunability could be reached for annealing temperatures about 1100°C. The range of the tunability could be extended to lower peak energies (up to 1.3 eV) in comparison to our previous work.

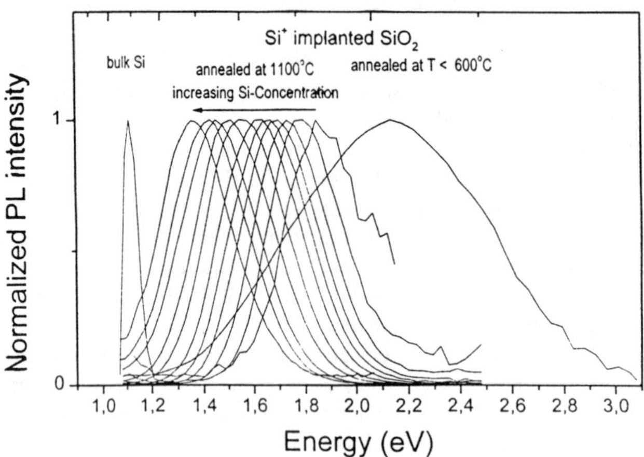

Fig.1: Normalized PL spectra of a sample with a graded dose (implantation energy 20 keV, dose in the maximum $5*10^{16}$ cm^{-2}, annealing time 30 min) for annealing temperatures below 600°C and for the annealing temperature of 1100°C. The PL spectrum of the bulk Si is also shown for comparison.

The annealing temperature is the major key for the improvement of PL of Si^+- implanted SiO_2 layers. This is demonstrated also for a homogeneously implanted sample (implantation energy 50 keV, dose $5*10^{16}$ cm^{-2}, annealing time 4 min) in fig. 2. In difference to fig.1 these PL spectra are not normalized. The PL peak position shifts continuously from 2.3 eV to 1.6 eV with increasing annealing temperature while the PL intensity is nearly constant over the whole temperature range up to 1100°C for the given sample. The PL intensity increases abruptly for annealing temperatures above 1000°C for the given annealing time but the PL peak position is not more shifting. This is likely to the constant PL at 1.6 eV which was observed in CVD processed Si nanocrystallites [10] or in thermally oxidized por-Si [11].

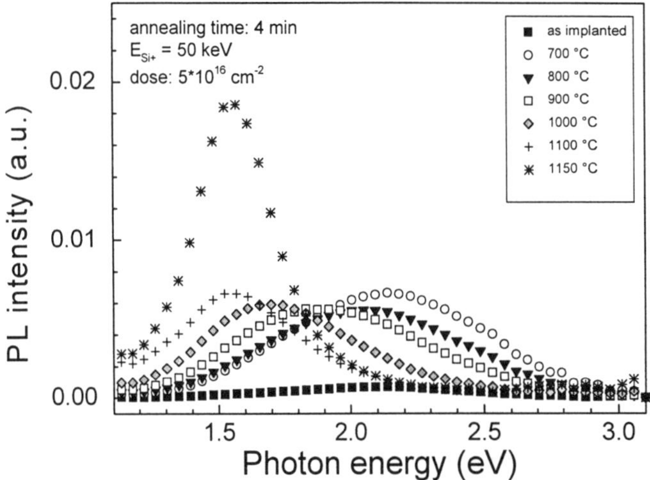

Fig.2: PL spectra of Si^+-implanted SiO_2 for isochronal annealing at 4 min. The implantation energy and dose are 50 keV and $5*10^{16}$ cm^{-2}, respectively.

The temperature dependence of the PL peak position for the samples shown in fig.2 is given in fig. 3. The PL peak position can be shifted continuously from 2.1 eV down to 1.6 eV in the temperature range from 700 to 1100 °C for this sample. The PL peak position is practically constant for lower annealing temperatures.

The dependence of the integrated PL amplitude on the annealing temperature for homogeneously Si^+-implanted SiO_2 (same implantation parameters as for figs.2 and 3) is shown in fig.4 in the case of isochronal annealing at 30 min. This dependence consists of three pronounced regions: (i) up to 600°C, (ii) from 650°C to 1000°C and (iii) from 1000°C to higher temperatures. The integrated PL intensity increases up to a constant value in region (i) and decreases by one order of magnitude between 600 and 650°C. The normalized PL spectra are unchanged in region (i). We relate the region (i) to PL at defects in the highly damaged SiO_2. The FTIR spectra are changing in this region from the damaged SiO_2 to the undamaged SiO_2. Consequently the increase of the PL intensity should be induced by the anneal of nonradiative defects in the SiO_2 layer. The integrated PL intensity increases up to a constant value also in

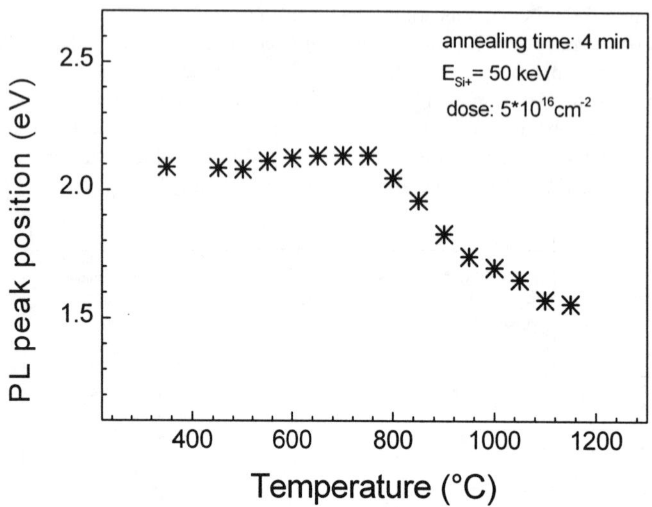

Fig.3: Dependence of the PL peak position on the annealing temperature for isochronal annealing at 4 min of Si^+-implanted SiO_2. The implantation energy and dose are 50 keV and $5*10^{16}$ cm^{-2}, respectively.

region (ii). We remark that the PL peak position is not constant for this region, it shifts to lower energies with increasing temperature similarely as shown in fig. 3. This region is related to the development of Si nanoparticles in the oversaturated SiO_2 matrix. The shapes of the nanoparticles increase with increasing temperature and therefore the quantum confinement of the electron wave function is decreasing. It is surprisingly that also in the case of isochronal annealing at 30 min the PL peak position could be shifted only up to 1.6 eV. Obviously, there are going on qualitative changes, probably in the surface region of the nanoparticles, which lead to effective radiative recombination channels via surface defects [11]. This effect is strongly increasing in region (iii). The PL intensity could be increased up to two orders of magnitude in comparison to the level at 1000°C in some cases. We will give more details on the origin of the three temperature regions in the near future [12].

Fig.5 shows the time dependence of the integrated PL intensity for the isothermal anneal at 1150°C of a homogeneously Si^+-implanted SiO_2 layer. The implantation parameters are the same as for figs.2-4. The integrated PL intensity increases up to about 8 min and remains nearly constant for annealing times up to about 30...35 min. The PL intensity decreases strongly for longer annealing times. The reason for this very strong decrase is not well understood at present. We guess that nonradiative recombination channels get more efficient at a given minimum of quantum confinement. The nearly constant value of the PL intensity opens a relatively wide technological window for producing highly efficient luminescent Si^+-implanted SiO_2 layers.

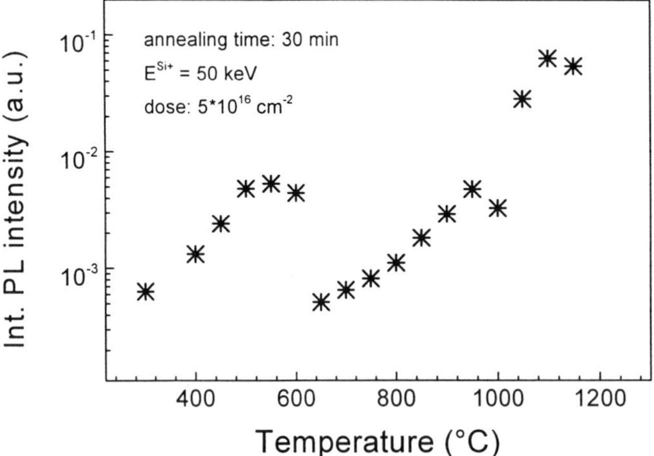

Fig.4: Dependence of the integrated PL intensity on the annealing temperature for isochronal annealing at 30 min of Si^+-implanted SiO_2. The implantation energy and dose are 50 keV and $5*10^{16}$ cm^{-2}, respectively.

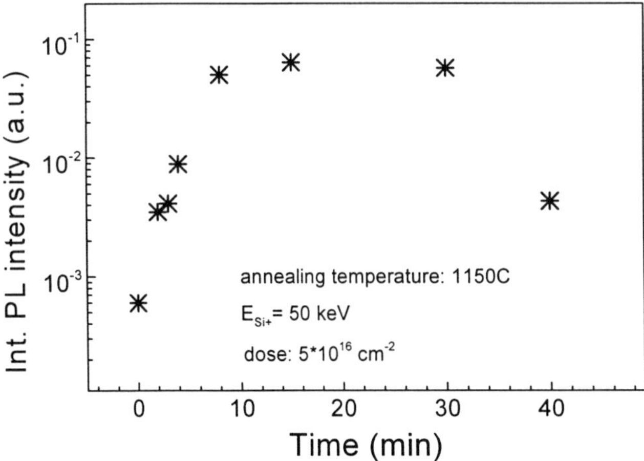

Fig.5: Time dependence of the integrated PL intensity for the isothermal anneal at 1150°C of a homogeneously Si^+-implanted SiO_2 layer. The implantation parameters are the same as for figs.2-4.

CONCLUSIONS

The PL of Si^+-implanted SiO_2 layers can be improved significantly by the use of a graded dose und by increasing the annealing temperature (higher than 1000°C). The PL peak position can be tuned very fine up to 1.3 eV. The PL intensity increases strongly at the high temperature anneal while the PL peak position is fixed at 1.6 eV.

REFERENCES

1. T. Shimizu-Iwayama, S. Nako, K. Saitoh, Appl. Phys. Lett. **65**, 1814 (1994).

2. H. A. Atwater, K. V. Sheglov, K. J. Vahala, R. C. Flayan, M. L. Brougersma, A. Polman, MRS Proc. **316**, 409 (1994).

3. P. Mutti, G. Ghislotti, S. Bertoni, L. Benoldi, G. F. Gerofilini, L. Meda, E. Grilliand, M. Guzzi, Appl. Phys. Lett. **66**, 851 (1995).

4. T. Fischer, V. Petrova-Koch, K Sheglov, M. S. Brandt, F. Koch, EMR Spring Meeting 1995.

5. D. L. Griscom, J. of Chemical Soc. of Japan **99**, 923 (1991).

6. I. M. Lifshitz, V. V. Zlesov, JETP **35**, 363 (1981).

7. A. I. Ekimov, A. A. Owushenko, JETP **34**, 363 (1981).

8. V. Petrova-Koch, T. Fischer, K. Sheglov, F. Koch, F-CS, Chicago, 1995

9. T. Fischer, Doctoral thesis, TU Munich, 1996.

10. M. Rückschloss, B. Landkammer, O. Ambacher, and S. Veprek, Mat. Res. Soc. Symp. Proc. Vol. **283**, 65 (1993).

11. Y. Kanemitsu, Phys. Rev. B **48**, 4883 (1993).

12. Th. Dittrich, T. Schuster, V. Petrova-Koch, H. Porteanu, E. Hechtl, F. Koch, to be submitted to J. Appl. Phys..

BLUE LUMINESCENCE FROM SiO_x FILMS CONTAINING Ge NANOCRYSTALS

M. ZACHARIAS*, R. WEIGAND, J. BLÄSING, and J. CHRISTEN
Institute of Experimental Physics, Otto-von-Guericke University,
PF4120, 39016 Magdeburg, Germany
* present address: Department of Electrical Engineering, University of Rochester, Rochester, N.Y. 14627, USA

We present a systematic analysis of SiO_x alloys films containing Ge nanocrystals prepared by dc magnetron sputtering. Increasing the sputtering power (50-175W) reduces the average Ge nanocrystal size exponentially down to ~2nm. Broad band photoluminescence spectra are observed in the visible at room temperature centered at ~3eV or/and ~2eV. Neither the 3eV nor the 2eV luminescence can be correlated to the change in the size. Excluding the quantum-confined origin, the presence of a luminescence center located in the inhomogeneous strain field of the Ge nanocrystal surface is discussed.

Introduction

Amorphous SiO_x alloys have been studied in the past mainly in order to understand vibrational modes, interface defects, impurity phenomena or the widening of the band gap. In recent years, great efforts have been made to investigate low-dimensional structures of indirect semiconductors. Since the discovery of strong visible room temperature luminescence from porous silicon[1] amazing progress has been made in the fabrication of electroluminescent devices using light-emitting porous silicon [2,3].
Superlattices and other low dimensional structures based on semiconductors like Si and Ge have been reported recently [4-6], in which quantum size effects are made responsible for the optical properties of these nanostructures. Ge nanocrystals fabricated by different techniques have attracted considerable attention[7-9]. Several approaches have been suggested for interpreting the photoluminescence and absorption spectra of quantum dots of indirect semiconductors in terms of quantum confinement. Starting from the simplified effective mass approximation with infinite potential barrier [10] then finite barrier [11] and anisotropic m_e* [12,13] the models which all used data from bulk semiconductors become more sophisticated. Some of the restrictions of the effective mass approximation for very small quantum dots are removed by the application of the effective bond-orbital model [14]. Commonly these calculations predict a quantum-size effect of Wannier excitons [15] easily detected in semiconductor nanocrystals as the high-energy shift of the interband absorption or of the luminescence peak.
On the other hand, defect states located at the interface between the quantum dot and the matrix or in the matrix have been considered [5,16]. Several defect mechanisms have been proposed to explain the absorption and luminescence spectra independent of size [17,18]. However, today the origin of the visible room temperature luminescence of the Ge nanocrystals is not unambiguously understood.
Much less is known about the precipitation of Ge nanoclusters in a SiO_x matrix and their crystallization by thermal annealing. Here we present results of a comprehensive

Mat. Res. Soc. Symp. Proc. Vol. 452

investigation of $Si_xGe_yO_z$ alloy films of different chemical composition prepared by dc magnetron sputtering. The as-prepared films show a precipitation of the amorphous Ge nanosize clusters. The cluster formation is determined by the preparation parameters such as sputtering power and number of Ge chips on the Si target.
Bright blue room temperature photoluminescence is observed from the films with a peak position at 410 nm for laser excitation at 325 nm. This seems to be in contrast to the peak position at 520 nm reported previously [7]. However, excitation of our films with the 441nm line of the HeCd laser results in the reported peak above 500 nm. Our results do not agree with a quantum confinement origin of the room temperature PL. We will discuss possible luminescence mechanisms including defects.

Experimental

Amorphous $Si_xGe_yO_z$ alloy films were deposited on (100) silicon substrate s using dc sputtering of a silicon target covered with up to 6 Ge chips in an Ar/water vapor atmosphere. The following parameters were varied in this study: the number of Ge pieces (1-6), the substrate temperature (100-200°C), the partial water pressure (0.05-0.2Pa), and the sputtering power (50-175W). This systematic variation of the sputtering parameters result in a wide range of as-prepared Ge cluster sizes (50nm down to 2nm) as well as in a variation of the chemical composition. The latter is investigated using Rutherford backscattering (RBS) providing an absolute error < 4at.% per constituent.
The cluster size distribution was obtained by analyzing the small angle X-ray scattering spectra (SAXS) of the thin amorphous films[19]. In a second step the transformation of the Ge-clusters into Ge nanocrystals is induced by thermal annealing at 800°C in an argon atmosphere. The formation of the Ge nanocrystals was studied with X-ray diffraction under grazing incidence. An X-ray goniometer with Seeman-Bohlin geometry was used for the thin film analysis. The grazing incidence angle was 1° and the divergence of the Soller collimator for the analysis of the interference field was 0.38°. The X-ray intensity was measured using a secondary monochromator and a scintillations counter tube. The resulting wide angle scattering intensities were analyzed with the Scherrer multiple peak method. Photoluminescence spectra were measured with the 325nm and the 441nm line of a HeCd laser, a 0.5m monochromator, and a N_2 cooled CCD matrix detector head with UV coating.

Results and discussion

An increase of the sputtering power from 50 to 175 W results in an increase of the deposition rate. As shown in fig. 1 the deposition rate increases from 0.2nm/s up to 1.2nm/s with the sputtering power. The higher substrate temperature results in a slight increase of the density of the films but shows no influence on the sputtering rate. The increase of the partial water pressure strongly changes the chemical composition of the ternary system.

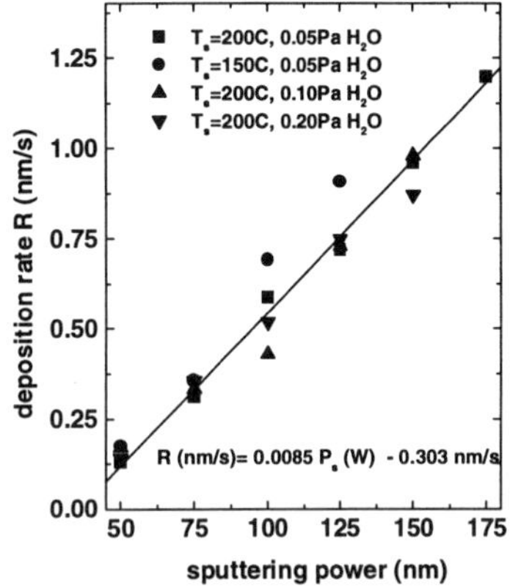

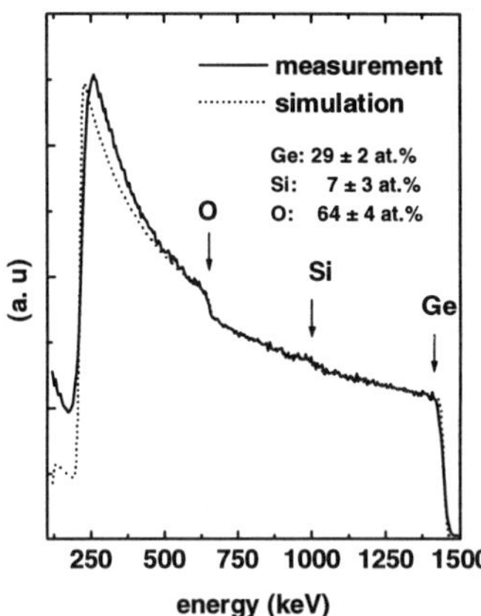

Fig.1: Deposition rate as a function of the sputtering conditions.

Fig.2: Investigation of the chemical composition using RBS analysis.

The oxygen content of the thin films is dependent on both the deposition rate and the partial water pressure. Fig. 2 demonstrates an example of the Rutherford backscattering investigation and simulation of the spectra. The films are stoichiometric for a partial water pressure of 0.2Pa. With an increase of the sputtering power from 50W to 150W, the oxygen content decreases slightly from 66at.% to 51at.% due to the limited number of oxygen atoms that can reach the growing film (water pressure 0.1Pa). With a water pressure of 0.05 Pa the oxygen content decreased from 60at.% to 30at.% when the sputtering power changes from 50W to 175W. By decreasing the sputtering power down to 50W, the Si target is more and more covered with an oxide layer. The low energy particle bombardment of the sputtering Ar ions can not destroy this layer, which decreases the sputtering rate of the Si target. It is well know that the sputtering yield of the germanium is larger than that of the silicon. The sputtering yield depends on the energy of the sputtering Ar ions and is 1.55 (Ge) compared to 1.0 (Si) for Ar ions of 1keV [20]. This may result in the observed drastic decrease of the Si content of the films when a low sputtering power is used in combination with a pressure water of 0.1 Pa or 0.2Pa. The preferred bonding of the oxygen to the silicon atom can be demonstrated with IR spectroscopy. The silicon atoms are always saturated with oxygen in an amorphous SiO_2 matrix for films sputtered with 0.1Pa. The germanium atoms are partially bonded to oxygen, too, when using a low sputtering power in combination with 0.1 or 0.2Pa partial water pressure.
The formation of the Ge nanocrystals in the SiO_x matrix is induced by annealing at 800°C. Further details of the X-ray analysis of the $Si_xGe_yO_z$ alloys and their interpretation proving the Ge nanocluster formation and the formation of the Ge nanocrystals with thermal annealing have been given previously [19, 21].

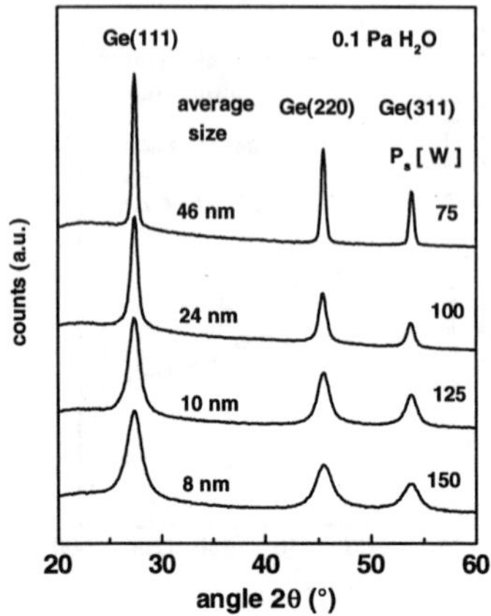

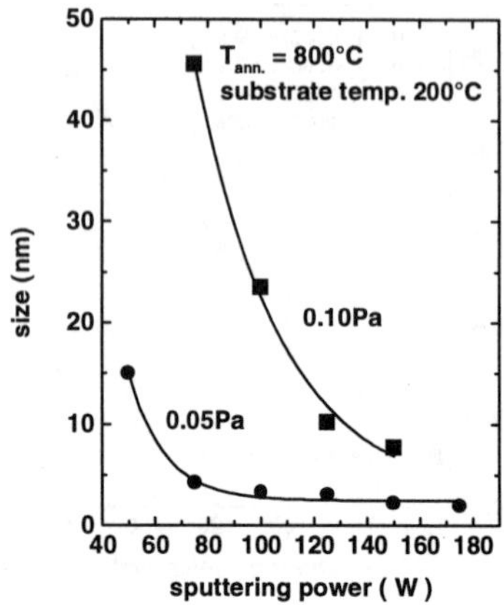

Fig.3: X-ray diffraction of selected samples prepared with various sputtering power.

Fig.4: Average size of the Ge nanocrystals as a function of the sputtering power.

Here we restrict the presentation to the X-ray diffraction spectra of the crystallized samples sputtered with 0.1Pa water (fig. 3). The crystallization of the Ge nanoclusters (precipitated in the films) is indicated by the splitting of the (220) and (311) reflections. Fig. 4 demonstrates the decrease of the average crystal size with the sputtering power. An increase of the sputtering power results in a decrease of the average nanocrystal size down to 2nm. The fitting curves in fig. 4 are a single exponential decay. The increased oxygen content in the sputtering atmosphere is correlated with a larger Ge/Si ratio of the films, as discussed above, and therefore with larger Ge nanocrystals. Annealing near the melting point of Ge results in a doubling of the average nanocrystal size due to recrystallization and merging of neighboring nanocrystals.

The samples do not luminescence in the as-prepared amorphous state. After crystallization the samples show a broad blue photoluminescence (PL) centered ~3eV under excitation at 325nm. However, excitation at 441nm results in a PL peak position between 540 and 645nm for all samples. The latter is consistent with data reported previously[7,8]. Fig. 5 demonstrates the PL for two samples with an average crystal size of ~45nm (a), (b) and ~7nm (c), (d) excited with 325nm and 441nm. The occurrence of the visible room temperature PL is correlated with the formation of the nanocrystals. However, we can not confirm a systematic shift to higher energies with decreasing size of the Ge nanocrystals. Without going in details, we combine in fig. 6 the observed two peak positions of a large set of samples to the confinement energy as a function of the size calculated using the equation given in ref. 10. The strong size-relation of the confinement energy is not correlated with either of the two PL-peaks.

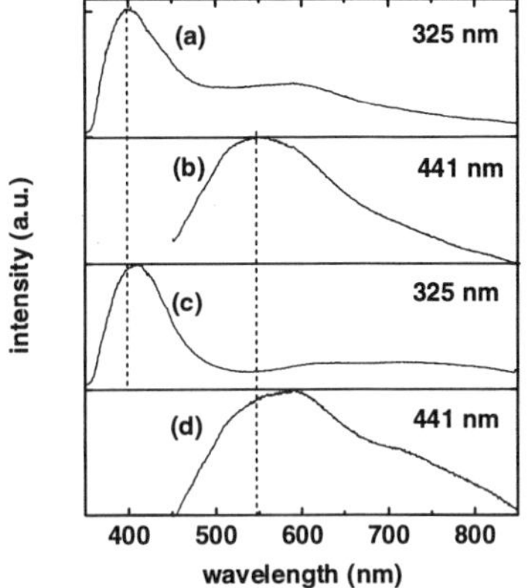

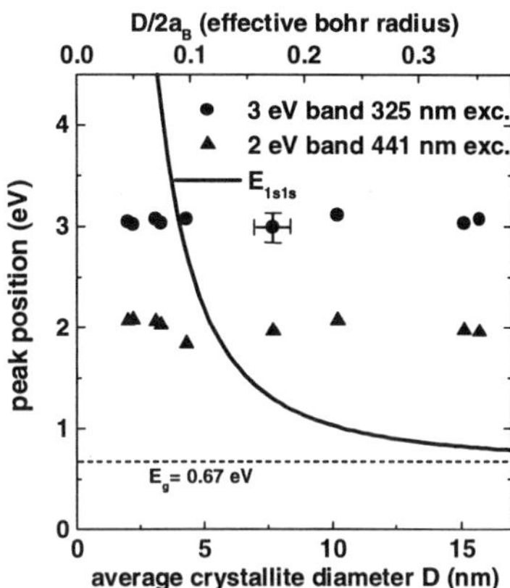

Fig.5: Effect of the excitation wavelength on the photoluminescence. Average size of the selected sample (a), (b) 10nm, and (c), (d) 46 nm.

Fig.6: Peak position of both PL bands of our sample in comparison with the predictions of the quantum confinement model.

Excluding the quantum size effect as the origin of the visible luminescence, we suggest that defects may contribute to both visible luminescence lines. However, the presence of the PL is clearly correlated to the presence of the Ge crystals, or may be to specific modifications of the crystal surface or the surrounding matrix of the nanocrystals. The crystallization of the Ge clusters is associated with cleaning and reconstruction processes. A further segregation results from a densification of the Ge-phase and out-diffusion of oxygen to the grain surface. Highly disturbed bonds located in the inhomogeneous strain field of the Ge surface may be contribute to the defect related luminescence. Further investigations such as electron paramagnetic resonance (EPR) or time resolved measurements will bring support to the defect mechanism.

Conclusion

We studied the precipitation behavior, chemical composition, and size of the Ge crystals over a broad range of the sputtering conditions. Using these results, we could control the size of the Ge nanocrystals from 50nm down to 2nm. The properties of photoluminescence did not shown a strong correlation with the change in size as expected for quantum confinement. These PL properties seemed to be more consistent with the properties of special defects located at the surfaces of the nanocrystals.

Acknowledgments

The authors would like to acknowledge Dr. Wendler (University Jena/Germany) for RBS analysis of the films, M. Schmidt, and P. Kohlert (University Magdeburg) for preparation of the films and performing the X-ray measurements, and P. Fauchet for helpful discussion. Financial support by the Deutsche Forschungsgemeinschaft and the Kultusministerium of Sachsen-Anhalt is gratefully acknowledged.

References

1 L.T. Canham, Appl. Phys. Lett. 57, 1046 (1990).
2 L. Tsybeskov, S.D. Duttagupta, K.D. Hirschman, P.M. Fauchet, Appl. Phys. Lett. 68, 2058 (1996).
3 K.D. Hirschman, L. Tsybeskov, S.D. Duttagupta, P.M. Fauchet, Nature 384, 338 (1996)
4 D.J. Lockwood, Z.H. Lu and J.-M.Baribeau, Phys.Rev.Lett.76, 539 (1996).
5 K.S. Min, K.V. Shcheglov, C.M. Yang, H.A. Atwater, M.L. Brogersma and A. Polman, Appl.Phys.Lett.68, 2511 (1996)
6 B. Delley and E.F. Steigmeier, Phys.Rev. B47, 1397 (1993)
7 Y. Maeda, Phys. Rev. 51 (1995) 1658.
8 Y. Kanemitsu, H. Uto. Y. Masumoto, Y. Maeda, Appl. Phys. Lett. 61, 2187 (1992).
9 D.C. Paine, C. Caragianis, S. Shigesato, Appl. Phys. Lett. 60, 2286 (1992).
10 Y. Kayanuma, Phys.Rev. B38, 9797 (1988).
11 D.B. Tran Thoai, Y.Z. Hu and S.W. Koch, Phys.Rev. B42, 11261 (1990).
12 T. Takagahara and K. Takeda, Phys.Rev. B46, 15578 (1992).
13 T. Richard P. Lefebvre, H. Mathieu and J. Allegre, Phys.Rev. B53, 7287 (1996).
14 G.T. Einevoll, Phys.Rev. B45, 3410 (1992).
15 U. Woggon and S.V. Gaponenko, phys.stat.sol.(b) 189, 285 (1995).
16 F. Koch, V. Petrova-Koch, T. Muschik, A. Nikolov and V. Gavrilenko, MRS Symp. Proc.283, 197 (1992).
17 H. Hosono, Y. Abe, D.L. Kinser, R.A. Weeks, K. Muta and H. Kawazoe , Phys.Rev. B46, 11445 (1992).
18 J. Nishii, K. Fukumi, H. Yamanaka, K. Kawamura, H. Hosono and H. Kawazoe, Phys.Rev.B52, 1661 (1995).
19 M. Zacharias, J. Bläsing, M. Löhmann, J. Christen, Thin Solid Films, 278, 32 (1996).
20 H. Oechner, Appl. Phys. 8, 185 (1975).
21 M. Zacharias, J. Bläsing, J. Christen, U. Wendt, J. of Non-Cryst. Solids, 198-200, 919 (1996).

CHARACTERIZATION OF SiGe ALLOY NANOCRYSTALLITES PREPARED BY PULSED LASER ABLATION IN INERT GAS AMBIENT

YUKA YAMADA*, TAKAAKI ORII**, and TAKEHITO YOSHIDA*
*Opto-Electro Mechanics Research Laboratory, Matsushita Research Institute Tokyo, Inc., 3-10-1 Higashimita, Tama-ku, Kawasaki 214, Japan, yyamada@mrit.mei.co.jp
**Institute of Applied Physics, University of Tsukuba, Tsukuba, Ibaraki 305, Japan

ABSTRACT

We report nanometer-sized silicon germanium (SiGe) alloy crystallites prepared by excimer laser ablation in constant-pressure inert gas. Size distribution of the Si_xGe_{1-x} ultrafine particles decreases with decreasing x under fixed conditions of deposition such as ambient gas pressure. Raman scattering spectra of the deposited SiGe ultrafine particles show three peaks intrinsic to crystalline SiGe alloys, and the linewidths of these peaks broaden due to the reduced size of the crystallites. Furthermore, a visible photoluminescence (PL) band with a peak at around 2.2 eV is obtained at room temperature after an annealing process.

INTRODUCTION

In the past several years, extensive studies have been carried out on nanometer-sized silicon (Si) and germanium (Ge) structures since visible photoluminescence (PL) spectra were observed [1-4]. In particular, the strong PL from porous Si has attracted much attention and much research has been undertaken to clarify its origin [5]. Furthermore, SiGe alloys have the possibility of band structure engineering by controlling the composition [6], and there have been several reports on optical properties of porous SiGe [7,8]. When we discuss the optical properties of nanoscale group IV materials as one of the quantum confinement effects, it is significant to adopt another approach using the nanoscale "spherical" structures which have well-controlled size and surface chemical structure. For this purpose, pulsed laser ablation into a gas ambient has been applied to the preparation of Si ultrafine particles [9-12]. Werwa et al.[9] reported that the minimum diameter of Si ultrafine particles was about 2 nm in laser ablation into a pulsed inert gas. Yoshida et al.[11] reported that the size distribution of Si nanocrystallites was effectively controlled by varying the ambient gas pressure. Furthermore, pulsed laser ablation has the potential for use in the deposition of complex materials with congruent transfer of the target composition [13], and SiGe thin films have been prepared using this method [14]. However, there have been few studies where alloy nanocrystallites are obtained using laser ablation in a gas ambient.

In this work, we adopt the laser ablation method combined with the constant-pressure inert gas evaporation system. A sintered mixture of Si and Ge powders is used as a target. We characterize structures and optical properties of SiGe alloy nanocrystallites prepared by laser ablation in an inert gas.

EXPERIMENT

In order to deposit Si and Ge, which have different melting points and vapor pressures, at the same time by laser ablation, an appropriate target should be a mixture of Si and Ge with high purity. Thus, sintered SiGe targets were fabricated as follows. Si and Ge powders with the size of μm order and high purity of 6N were mixed without binder and sintered using a hot press system

Mat. Res. Soc. Symp. Proc. Vol. 452 © 1997 Materials Research Society

in an inert gas. Then, we could obtain Si_xGe_{1-x} ($x = 0.18, 0.82$) sintered targets of 2-inch diameter with purity over 4N. The alloy composition x of the sintered target was determined using inductively coupled plasma optical emission spectroscopy (ICPS).

SiGe ultrafine particles were prepared using pulsed laser ablation of the sintered SiGe target in an inert gas ambient at constant reduced pressure. A schematic diagram of the laser ablation apparatus used in this study is shown in Fig. 1. After the vacuum chamber was evacuated to 1.0×10^{-6} Pa, helium (He) gas was introduced into the chamber and was kept at a constant pressure (333 Pa) using a differential evacuation system. An argon-fluoride (ArF) excimer laser (λ: 193 nm, energy density: 1.0 J/(cm^2·pulse), pulse duration: 12 ns, repetition rate: 10 Hz) beam was focused onto a 3×1 mm^2 rectangular spot at the surface of the sintered SiGe target of 2-inch diameter. Then, a plume of ejected species was created and extended almost perpendicular to the target surface. The target was rotated at 8 rpm. As described above, the size of the mixed powders of the sintered target was of μm order, which is much smaller than the spot size of the irradiated laser beam. Thus, the ejected species can have a mixed composition. A deposition substrate was located at a distance of 7 mm normal to the target. The species ejected from the target condensed during flight in a gas ambient and then were deposited as ultrafine particles on the substrate [11]. The substrate was kept at room temperature during the deposition process. Finally, the deposited substrates were annealed in dry nitrogen (N_2) at 800℃ for 10 min.

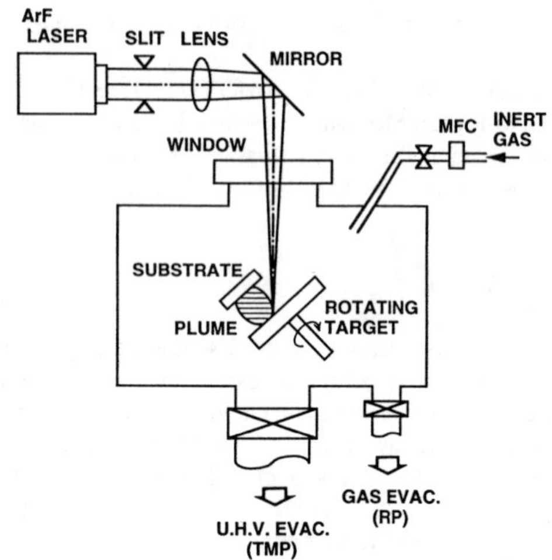

Fig. 1. Schematic diagram of preparation system of nanometer-sized SiGe ultrafine particles using ArF excimer laser ablation in constant-pressure inert gas.

Scanning electron microscope (SEM) observation was employed to evaluate the particle size of the deposits. The average composition of the deposited particles was evaluated using ICPS measurement. Raman scattering spectra were measured with a filtered single monochromator (Jasco TRS-600) system in a backscattering geometry. Excitation was provided by the 514.5 nm Ar^+ ion laser line. A CCD system (Photometrics TK512CB) was used as a detector. To prevent damage of samples by the laser and/or a change of the Raman spectrum during measurements, the laser power was chosen to be less than 4 mW. In the Raman scattering measurement, fused quartz was used for the deposition substrate. Furthermore, PL measurements were performed at room temperature using a single polychromator (Jasco CT-25C) system with a cooled MOS linear-image sensor unit (Hamamatsu C4834-02) as a detector. An Ar^+ ion laser (λ: 488 nm, power: 10 mW) was used as an excitation light source. The spectral sensitivity of the measurement system was calibrated using a halogen standard lamp.

RESULTS

The as-deposited Si_xGe_{1-x} ultrafine particles were observed using SEM to clarify the

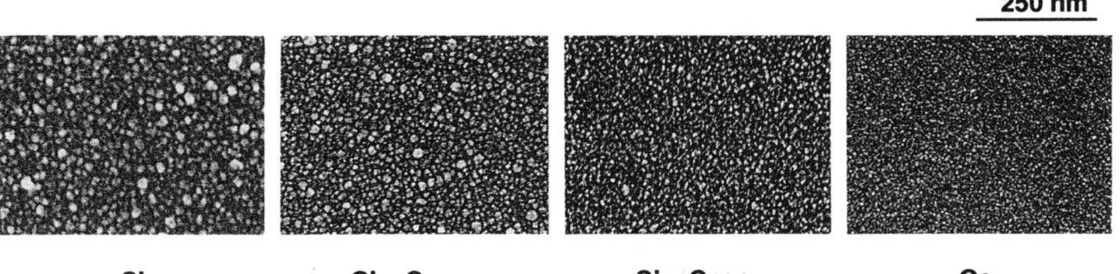

Fig. 2. SEM micrographs of as-deposited Si_xGe_{1-x} ultrafine particles. The average size of the ultrafine particles decreases with decreasing composition x.

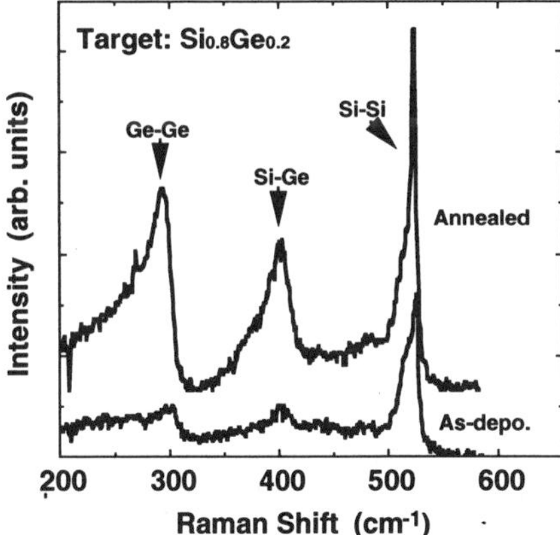

Fig. 3. Raman spectra of $Si_{0.8}Ge_{0.2}$ alloy nanocrystallites. Three Raman peaks intrinsic to crystalline SiGe alloys are observed.

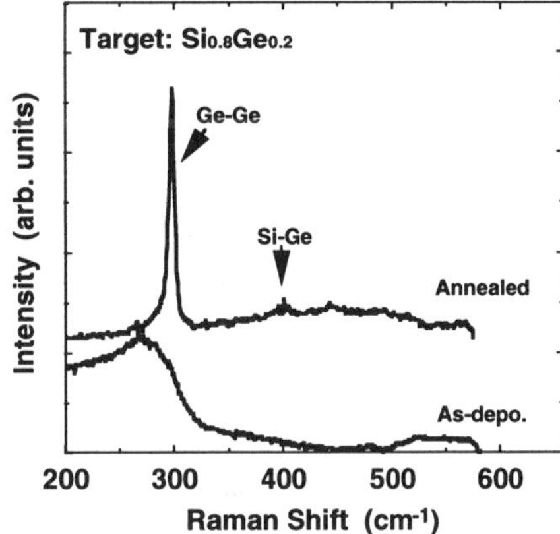

Fig. 4. Raman spectra of $Si_{0.2}Ge_{0.8}$ alloy nanocrystallites. Ge-Ge and Si-Ge peaks are observed after annealing.

dependence of the particle size on the composition x. These results are shown in Fig. 2. Nanometer-sized ultrafine particles are observed. Furthermore, the average size decreases with decreasing x under the fixed process conditions.

The average composition of the ultrafine particles deposited on the Al_2O_3 substrate using the $Si_{0.82}Ge_{0.18}$ target was evaluated by ICPS. The total ratio (at.%) of Si:Ge of the deposited ultrafine particles was obtained as 78.9:21.1. From this result, it is revealed that our pulsed laser ablation process can transfer as a whole the composition of the target to the deposits.

Figure 3 shows Raman scattering spectra of the $Si_{0.8}Ge_{0.2}$ ultrafine particles before and after annealing. Three Raman peaks ascribed to the vibrations of Ge-Ge, Si-Ge and Si-Si pairs are observed at 292, 400 and 520 cm^{-1}, respectively. These peaks are intrinsic to the crystalline SiGe alloy system [15]. The intensities of these peaks increase after annealing. This result shows that crystallization progressed after the annealing process.

Figure 4 shows Raman scattering spectra of the $Si_{0.2}Ge_{0.8}$ ultrafine particles before and after

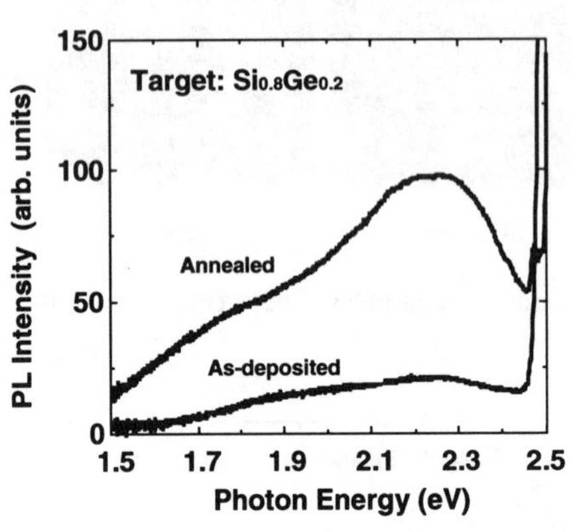

Fig. 5. PL spectra from the $Si_{0.8}Ge_{0.2}$ alloy nanocrystallites. A green PL band appears at around 2.2 eV with a low energy shoulder.

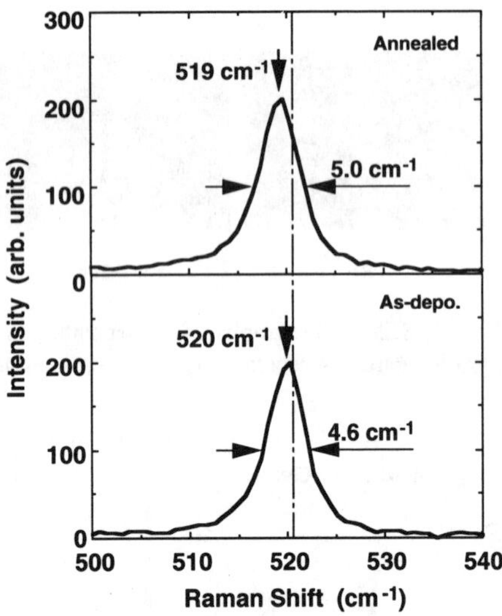

Fig. 6. Raman spectra of Si nanocrystallites. Dash-dotted line indicates the peak position of the bulk single-crystalline Si. FWHM of the bulk Si was 3.7 cm^{-1} in this measurement.

annealing. A broad peak appears at 200-300 cm^{-1} before annealing, which is associated with amorphous Ge [4]. After annealing, Ge-Ge and Si-Ge peaks are observed at 298 and 400 cm^{-1}, respectively [15]. The Si-Si peak cannot be observed. As a result, the Ge-rich ultrafine particles, which are amorphous-like before annealing, are crystallized after the annealing process.

Figure 5 shows the PL spectrum from the $Si_{0.8}Ge_{0.2}$ nanocrystallites before and after annealing. A green PL band appears at around 2.2 eV with a low energy shoulder. The PL intensity increases after annealing.

DISCUSSION

There have been several reports on the study of the dependence of the Raman scattering spectra on the composition x of bulk Si_xGe_{1-x} alloys [15,16]. The frequencies and linewidths of three Raman peaks ascribed to vibrations of Ge-Ge, Si-Ge, and Si-Si pairs strongly depend on the composition x. The frequencies of these peaks in Figs. 3 and 4 correspond well to those of the bulk $Si_{0.8}Ge_{0.2}$ and $Si_{0.2}Ge_{0.8}$ alloys reported in Ref. 15. This result shows that the composition is maintained in the SiGe alloy nanocrystallites deposited by pulsed laser ablation in an inert gas.

However, the linewidths of the three Raman peaks in our experiment are about 2 times broader than that of the bulk alloys in Ref. 15. According to a spatial correlation model, finite size effects relax the q-vector selection rule and this relaxation of the momentum conservation leads to a broadening and downshift of the Raman spectrum [17]. To clarify the finite size effects shown in the Raman spectrum, Raman scattering measurements were also performed for Si nanocrystallites. Figure 6 shows Raman scattering spectra of the Si nanocrystallites before and after annealing [18]. The dash-dotted line indicates the peak position of the bulk single-crystalline Si. The Raman peak of the bulk Si was symmetric and had a linewidth of 3.7 cm^{-1} (FWHM) in this measurement. It was found that the Si nanocrystallites have a broader and lower frequency peak in comparison with the bulk Si. The average diameter of the as-deposited Si nanocrystallites in Fig. 6 was estimated to be about 11 nm using the strong phonon confinement model [19]. This result is in good agreement with the result obtained from high-resolution transmission electron microscopy (TEM) [18]. Furthermore, the annealing process causes a much broader shape and lower frequency, as shown in Fig. 6. This result suggests that the annealing process caused a size reduction of the core of the Si nanocrystallite.

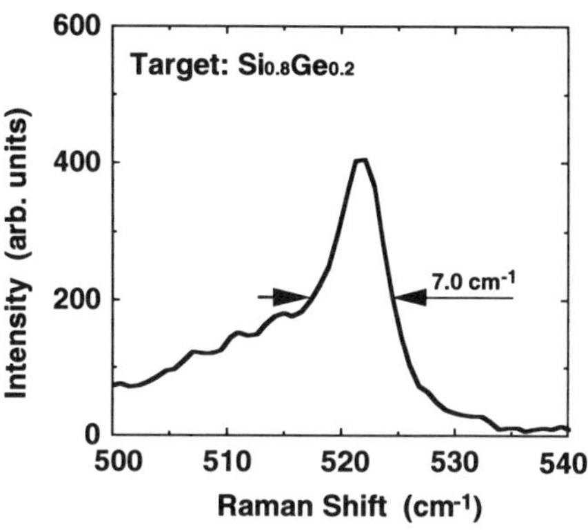

Fig. 7. Raman peak ascribed to Si-Si vibration of the $Si_{0.8}Ge_{0.2}$ alloy nanocrystallites after annealing in Fig. 3. FWHM broadens to 7.0 cm^{-1}.

The broadened linewidths of the Raman peaks in SiGe alloy nanocrystallites must also be caused by the reduction of the particle size. Figure 7 shows the Si-Si peak of the $Si_{0.8}Ge_{0.2}$ alloy nanocrystallites after annealing in Fig. 3. The linewidth broadens to 7.0 cm^{-1}, especially to lower frequencies compared to the Si nanocrystallites. This result cannot be explained by only the reduced size of the SiGe nanocrystallites. The inclusion of Ge as impurities in Si should increase lattice disorder which would result in the relaxation of the q-vector selection rule. Furthermore, Okada and Iijima indicated the existence of the strong stress in oxidized Si particles from the study of the oxidation property [20]. This stress can exist in the SiGe nanocrystallites and cause the strain effects, which also results in the broadening of the Raman peaks. Advanced study is necessary to analyze the lattice distortion, for example using the electron or x-ray diffraction.

Consequently, Raman scattering spectra of Si_xGe_{1-x} alloy nanocrystallites depend on the composition x. The peak shifts of nanocrystalline and bulk SiGe alloys can be explained by the same mechanism. Furthermore, the linewidths broaden to lower frequencies compared to the bulk alloys. This property is considered to be derived from the synergistic effect of the finite size confinement and the lattice disorder.

A green PL band around 2.2 eV with a low energy shoulder was clearly observed from SiGe alloy nanocrystallites after annealing at room temperature. Similar PL spectra were obtained from Si nanocrystallites and the origin was associated with a quantum confinement effect [12]. However, we don't have enough experimental data to identify the relation between the SiGe alloy structure and the PL. Further study is in progress.

CONCLUSIONS

SiGe alloy nanocrystallites have been obtained using excimer laser ablation of sintered SiGe targets in a constant-pressure inert gas. The deposited Si_xGe_{1-x} nanocrystallites have the same alloy composition x as the target. The Raman peaks intrinsic to crystalline alloys have broadened to lower frequencies compared with those of bulk alloys. This should be explained in terms of the synergistic effect of the finite size confinement and the lattice disorder. Furthermore, a visible PL band has been clearly observed at room temperature after the annealing process.

ACKNOWLEDGMENTS

The authors would like to express their thanks to Professor T. Arai (Ishinomaki Sensyu Univ.) and Professor S. Onari (Univ. of Tsukuba) for very stimulating discussion concerning optical properties and spectroscopy of mesoscopic materials. Thanks are also due to J. Ikeda and G. Oda for encouragement throughout this work.

REFERENCES

1. S. Furukawa and T. Miyasato, Jpn. J. Appl. Phys. **27**, L2207 (1988).
2. H. Takagi, H. Ogawa, Y. Yamazaki, A. Ishizaki, and T. Nakagiri, Appl. Phys. Lett. **56**, 2379 (1990).
3. L. T. Canham, Appl. Phys. Lett. **57**, 1046 (1990).
4. Y. Maeda, N. Tsukamoto, Y. Yazawa, Y. Kanemitsu, and Y. Masumoto, Appl. Phys. Lett. **59**, 3168 (1991).
5. Porous Silicon, edited by Z. C. Feng and R. Tsu (World Scientific, Singapore, 1994).
6. R. Braunstein, A. R. Moore, and F. Herman, Phys. Rev. **109**, 695 (1958) .
7. S. Gardelis, J. S. Rimmer, P. Dawson, B. Hamilton, R. A. Kubiak, T. E. Whall, and E. H. C. Parker, Appl. Phys. Lett. **59**, 2118 (1991).
8. A. Ksendzov, R. W. Fathauer, T. George, W. T. Pike, R. P. Vasquez, and A. P. Taylor, Appl. Phys. Lett. **63**, 200 (1993).
9. E. Werwa, A. A. Seraphin, L. A. Chiu, Chuxin Zhou, and K. D. Kolenbrander, Appl. Phys. Lett. **64**, 1821 (1994).
10. T. Ohyanagi, A. Miyashita, K. Murakami, and O. Yoda, Jpn. J. Appl. Phys. **33**, 2586 (1994).
11. T. Yoshida, S. Takeyama, Y. Yamada, and K. Mutoh, Appl. Phys. Lett. **68**, 1772 (1996).
12. Y. Yamada, T. Orii, I. Umezu, S. Takeyama, and T. Yoshida, Jpn. J. Appl. Phys. **35**, 1361 (1996).
13. Pulsed Laser Deposition of Thin Films, edited by D. B. Chrisey and G. K. Hubler (Wiley, New York, 1994).
14. F. Antoni, E. Fogarassy, C. Fuchs, J. J. Grob, B. Prevot, and J. P. Stoquert, Appl. Phys. Lett. **67**, 2072 (1995).
15. M. A. Renucci, J. B. Renucci, and M. Cardona, in Light Scattering in Solids, edited by M. Balkanski (Flammarion, Paris, 1971), p. 326.
16. W. J. Brya, Solid State Commun. **12**, 253 (1973).
17. H. Richter, Z. P. Wang, and L. Ley, Solid State Commun. **39**, 625 (1981).
18. T. Yoshida, Y. Yamada, S. Takeyama, T. Orii, I. Umezu, and Y. Makita, Proc. SPIE **2888**, 6 (1996).
19. I. H. Campbell and P. M. Fauchet, Solid State Commun. **58**, 739 (1986).
20. R. Okada and S. Iijima, Appl. Phys. Lett. **58**, 1662 (1991).

EXCITATION INTENSITY AND TEMPERATURE DEPENDENT PHOTOLUMINESCENCE BEHAVIOR OF SILICON NANOPARTICLES

E. WERWA, A.A. SERAPHIN, AND K.D. KOLENBRANDER
Department of Materials Science and Engineering,
Massachusetts Institute of Technology
Cambridge, MA 02139, werwa@mit.edu, aseraphi@ida.org, kdk@mit.edu

ABSTRACT

The luminescence properties of silicon nanoparticles have been studied as a function of the excitation light intensity, the temporal nature of the excitation source, and of sample temperature. The excitation intensity dependence of the luminescence was found to depend strongly on the temporal nature of the excitation source. Under high intensity excitation from a pulsed 355 nm source, the photoluminescence (PL) intensity saturates and the peak PL wavelength shifts to the blue at room temperature. This behavior persists at reduced temperature. In contrast, under high intensity excitation using a cw 488 nm source at room temperature, the PL intensity saturates but does not shift in wavelength. At reduced temperatures, there is no saturation of luminescence intensity with high intensity cw excitation. These differences indicate that photogenerated carrier recombination occurs via different pathways depending on the temporal profile of the excitation, with cw excited samples following the expected Auger pathway while pulsed samples exhibit a state filling mechanism. Auger models for the pulsed behavior are found to be inconsistent with the experimental data. The temperature dependence of the PL from a pulsed excited sample for a constant excitation intensity was also monitored. The variation of the peak emission wavelength was found to be similar in magnitude to that observed for amorphous silicon, suggesting that structural disorder may play a role in the luminescence. The change in emission intensity was fairly weak, indicating enhanced carrier confinement, as would be expected in a quantum confined system.

INTRODUCTION

There has been much interest in the photoluminescence behavior of silicon nanostructures. Several potential mechanisms for the light emission have been put forth, with the two most popular being excitonic recombination in quantum confined semiconductor nanoparticles[1] and recombination in semiconductor particles through some undefined surface states.[2,3] Much of this work has been performed on porous silicon, synthesized by etching of bulk silicon wafers to produce a pore network within which there have been shown to be nanometer sized particles.[4] Unfortunately, the intricate morphology of porous silicon has made it difficult to identify the luminescence mechanism, leading to the various theories of emission.

We have chosen to study silicon nanoparticles synthesized by pulsed laser ablation supersonic expansion in order to circumvent the problems of the porous silicon morphology. This will allow us to shed light on the behavior of the actual silicon particles, rather than particles whose behavior may be masked by the porous silicon structure as a whole. In order to examine the luminescence mechanism active in the particles, we have chosen to study the excitation intensity dependence of the luminescence using both pulsed and continuous excitation sources. These experiments tailor the excitation conditions to favor different recombination mechanisms, which permits the study of all possible controlling mechanisms PL.

Mat. Res. Soc. Symp. Proc. Vol. 452

EXPERIMENTAL

Silicon nanoparticle thin films were synthesized using a pulsed laser ablation supersonic expansion technique, the details of which are reported elsewhere.[5] Briefly, a doubled Nd:YAG laser (λ = 532 nm, 6 mJ/pulse, 7 nsec pulse width, 20 Hz repetition rate) was focused to a 1 mm diameter spot at the surface of a polycrystalline silicon rod (99.9999% pure), producing a plasma. A pulsed valve gated the flow of a pressurized He carrier gas over the ablation spot and carried the ablated Si through an interaction channel. The interaction between the silicon plasma and the He carrier gas nucleates silicon nanoparticles, which are then carried by the gas into a vacuum system (base pressure 10^{-7} torr). Films are deposited by placing the substrate, here Teflon or silicon, into the path of the nanoparticles as they travel through the vacuum chamber.

The photoluminescence experiments performed herein used two separate excitation light sources. Pulsed excitation at λ=355 nm was achieved using a 3x Nd:YAG laser, operating at 20 Hz with a 7 nsec pulse width. Continuous wave (cw) excitation was supplied using the λ=488 nm line of an Ar^+ laser. In both cases, the illuminated spot size was ~1cm^2, which remained constant throughout the studies. The excitation intensity used was varied for both sources as indicated in the discussion of the experimental results. The luminescence spectra were measured using a 0.275 m spectrometer coupled with a silicon photodiode array cooled to -20° C and were corrected for detector and spectrometer response. PL characterization was performed in a cryostat, which allowed the temperature of the samples to be varied between 4K and 300K.

RESULTS AND DISCUSSION

To clarify the nature of the luminescence recombination pathway, we have studied the behavior of the PL emission from silicon nanoparticles at high excitation intensity. For a quantum confined system under these conditions, both Auger recombination and state filling mechanisms must be present and will compete to control the observed luminescence behavior. The relative importance of each mechanism will depend upon the kinetic conditions under which the carrier generation and recombination occurs. These conditions may be tailored to promote one pathway relative to the other by varying the rate at which exciting photons are incident upon the nanoparticles under study. Following a brief description of the Auger and state filling mechanisms, we show below evidence of both mechanisms at work in our silicon nanoparticles.

As the light intensity incident upon a sample of silicon nanoparticles increases, multiple excitations are created in some particles. When Auger recombination kinetics dominate, these multiple excitations (one electron-hole pair plus any additional carriers) will give rise to nonradiative Auger recombination. If additional incident photons are absorbed by a nanoparticle at a rate slower than existing electron-hole pairs are consumed through this nonradiative path, little or no additional emission intensity will be observed.[6] Auger-limited emission has been observed by researchers studying porous silicon using cw excitation sources.[7] Auger lifetimes of 0.1 to 100 nsec have been calculated for silicon nanoparticles.[8]

The state-filling mechanism is exclusive to highly excited quantum confined particles. Under intense excitation, the first excited state can become saturated and transitions from the higher excited states are observed, resulting in the emission of higher energy light.[9] The resultant luminescence saturation and blue-shift can occur only when photons are absorbed at a rate greater than energy is released through some other path. The energy difference resulting from saturation can be estimated as the separation between the energy levels in a spherical potential well,[10] which is ~0.1 to 0.2 eV for the Si particles in the size range of interest.

State-filling behavior has been observed in other quantum dot systems. Saturation of the luminescence intensity from the first excited state with increased excitation intensity and the emergence of intensity from the second excited state has been observed in InGaAs,[11] InAs,[12] InP[13] and GaInP[14] quantum dots, CdSe nanoparticles in SiO_2,[15] and porous silicon in PMMA.[16]

In order to preferentially promote one pathway we must vary the rate at which exciting photons are made available to the silicon nanoparticles. The state-filling mechanism will be active only when sufficient time for the Auger processes is not allowed, and to access this we will need to use an intense, short-pulse excitation laser. Conversely, the Auger pathway will be optimized through the use of a cw excitation source. Also, the Auger recombination rate is known for bulk semiconductors to decrease with decreasing temperature.[17] A temperature dependence study under the two excitation conditions can thus confirm the dominance of one mechanism relative to the other.

Figure 1 shows a series of PL spectra measured at 300K as a function of excitation intensity from a cw λ = 488 nm source. Under these ideal conditions for observing Auger recombination dominated behavior, the emission intensity saturates uniformly at high excitation intensity. Figure 2 shows a similar series of PL spectra measured at 300K, this time using a

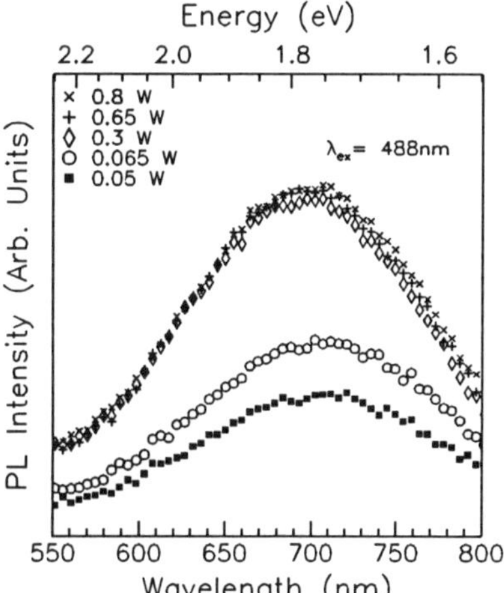

Figure 1. Excitation intensity dependence of emission using a cw 488 nm excitation source. Note the uniform saturation of the PL intensity at all wavelengths.

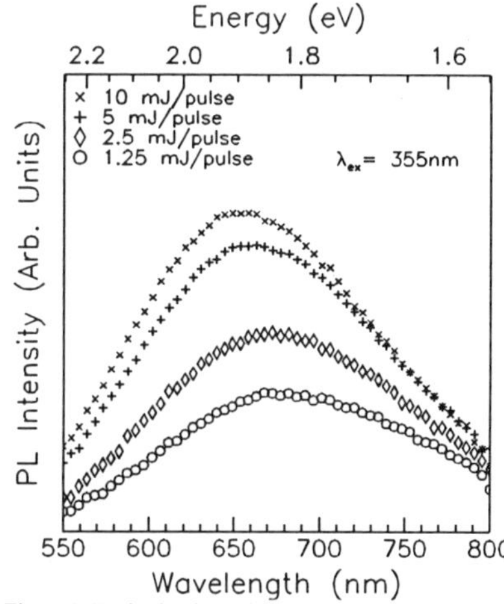

Figure 2. Excitation intensity dependence of emission using a pulsed 355 nm source. Note the blueshift at high excitation intensity.

pulsed λ = 355 nm source. Here, the peak emission wavelength shifts significantly to shorter wavelengths with increasing emission intensity. The peak shift results from the rapidly saturating PL behavior at longer wavelengths relative to the shorter. There are two possible explanations for this behavior. First, one might consider that the spectral shift is a result of the different luminescence lifetimes for the high and low energy portions of the spectrum. This difference, which has been observed experimentally for porous silicon[18] and silicon nanoparticles,[19] could mean that particles emitting in the blue (small particles with short lifetimes) might undergo multiple absorption and emission events during the course of an excitation pulse without ever experiencing two excitations simultaneously. In contrast, the long lifetime large particles could absorb multiple photons before ever emitting, which would lead to Auger recombination. The net result of this would be a blue shift of the spectrum. A second explanation uses a simple quantum confinement framework: the intensity saturation and blue shift of the emission are the necessary consequences of a state-filling mechanism at work on a distribution of nanoparticle sizes. Both of these mechanisms are proposed by the authors of Reference 6 to explain similar behavior from porous silicon. They opt for an Auger model but do not explain the blue shift.

A study of the temperature dependence of the excitation intensity behavior helps to clarify these assignments. Figure 3 shows the PL emission intensity at $\lambda = 600$ nm as a function of excitation intensity at 300 and 150 K for the cw excitation scheme. Figure 4 is a plot of the variation of the integrated PL intensity as a function of temperature for a variety of excitation intensities. From these plots, we see that the emission intensity at room temperature saturates at low intensity, while it increases linearly with increasing intensity at low temperature, reflecting

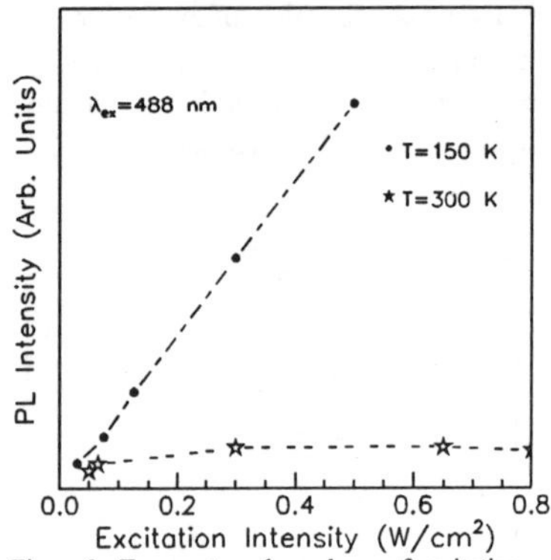

Figure 3. Temperature dependence of excitation intensity dependent luminescence at 600 nm. Unlike at room temperature, the cw sample shows no saturation at reduced temperature.

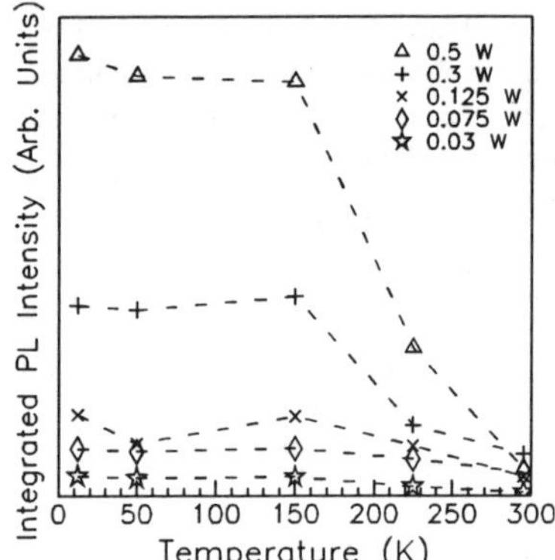

Figure 4. Integrated PL intensity vs. temperature for different excitation intensities from 488 cw laser. Emission intensity saturates at high excitation intensity at high temperature, but not at low temperature.

the temperature dependence of the Auger recombination rate. Figure 5 shows a series of PL spectra for a sample excited with the pulsed source at different excitation intensities. For the differential lifetime model to hold under the experimental conditions of Figure 5, it is necessary that the lifetime for the blue end emission be shorter than that for the red end emission. Since it is known that the lifetime of emission becomes independent of wavelength in porous silicon[20,21,22,23] and silicon nanoparticles[18] as the temperature is decreased, if lifetime differential were to lead to our observed behavior at room temperature, the behavior should not persist as the temperature is decreased, and the entire spectrum should show saturation. This is not the case in our material, which indicates that differential lifetime is not the source of our behavior. The state filling mechanism, however, should not show any temperature dependence, which is consistent with what we observe experimentally.

We have also studied the temperature dependence of the PL emission excited by the pulsed 355 nm source at a constant excitation intensity, as shown in Figure 6 for an excitation intensity of 1.25 mJ/pulse. The PL intensity increases monotonically as the temperature decreases and the peak emission wavelength shifts to the blue. The weak temperature dependence of the emission intensity is evidence the enhanced exciton binding energy predicted for a quantum confined system[24] and is consistent with observations in CdS quantum dots.[25]

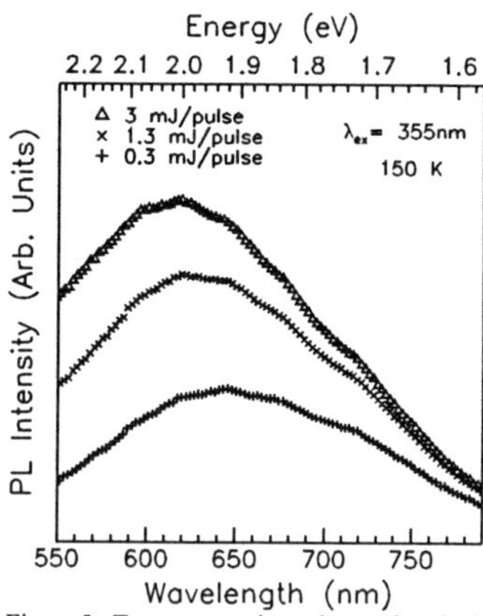

Figure 5. Temperature dependence of excitation intensity behavior for pulsed source. Spectra show the same trends as observed at room temperature.

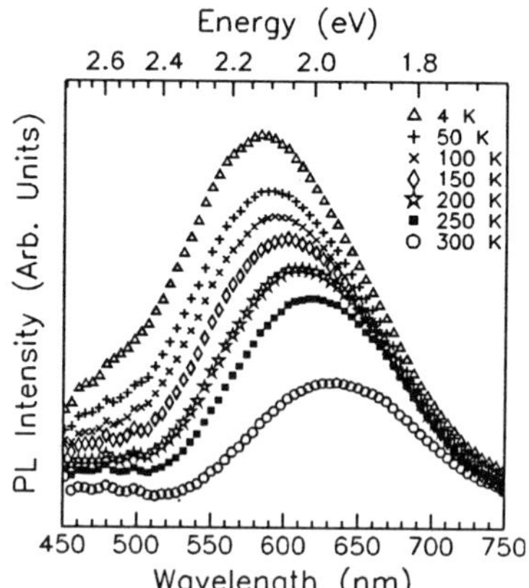

Figure 6. PL spectrum of a single sample measured at a series of temperatures. Sample was excited with a pulsed 355 nm laser, excitation intensity 1.25 mJ/pulse.

The magnitude of the energy shift of the PL peak is approximately 225 meV. This can be compared to the peak shift observed in bulk silicon of approximately 50 meV and that observed for amorphous silicon, 250-300 meV.[26] The behavior in the silicon nanoparticle system is more similar to amorphous silicon than bulk silicon, suggesting that structural disorder similar to that in amorphous silicon may play a role in the emission. This is consistent with the observation by Lockwood et al. that amorphous silicon quantum well systems exhibit the same kind of luminescence as our silicon nanoparticles.[27] It is also consistent with PL decay results on silicon nanoparticles[28] and optical studies of porous silicon[2] that indicate disorder. We are currently developing a modification to our synthesis apparatus which will allow us to deposit only particles which we know must be structurally disordered in order to study this further.

CONCLUSION

We have shown that the recombination pathway of photoexcited carriers in silicon nanoparticles, as evaluated by the photoluminescence behavior, is determined by the temporal nature of the excitation source. Upon increasing the excitation intensity using a pulsed 355 nm source, the PL intensity saturates and the peak wavelength shifts to the blue, at both room temperature and reduced temperature. Increasing the excitation intensity from a cw 488 nm source results in a PL intensity saturation without a spectral shift, and at low temperatures this saturation behavior disappears. The 488 cw behavior is consistent with an Auger recombination mechanism, while the 355 pulsed behavior can be explained using a simple quantum confinement framework where the observed saturation and blue shift of the emission are the necessary consequences of a state-filling mechanism at work on a distribution of nanoparticle sizes. We have also shown that the emission spectrum of a pulsed 355 nm excited sample blue shifts and increases in intensity as the temperature is reduced. The magnitude of the intensity change suggests an enhanced exciton binding energy, which is expected in a quantum confined system. The magnitude of the energy shift of the PL peak is closer to that observed in amorphous silicon than bulk silicon, suggesting that structural disorder plays a role in the luminescence. We are currently developing experiments to probe this very interesting area of study.

ACKNOWLEDGMENTS

We thank T.A. Burr for assistance with some of the temperature dependent photoluminescence. Authors E.W, A.A.S., and K.D.K. appreciate support from AT&T Bell Laboratories, the Department of Defense, and the National Science Foundation, respectively.

REFERENCES

1. L. T. Canham, Appl. Phys. Lett. **57,** 1046-1048 (1990).
2. F. Koch, V. Petrova-Koch, and T. Muschik, J. Lumin. **57,** 271 (1993).
3. P. M. Fauchet, E. Ettedgui, A. Raisanen, L. J. Brillson, F. Seiferth, S. K. Kurinec, Y. Gao, C. Peng, and L. Tsybeskov in Silicon-Based Optoelectronic Materials, edited by R.T. Collins, M.A. Tischler, G. Abstreiter, and M.L. Thewalt (Mater. Res. Soc. Proc. **298**, Pittsburgh, PA, 1993) pp. 271-276.
4. A. G. Cullis and L. T. Canham, Nature **353,** 335-338 (1991).
5. E. Werwa, A.A. Seraphin, LA. Chiu, C. Zhou, and K.D. Kolenbrander, Appl. Phys. Lett. **64** 1821 (1994).
6. I. Mihalcescu, J.C. Vial, A. Bsiesy, F. Muller, R. Romestain, E. Martin, C. Delerue, M. Lannoo, and G. Allan, Phys. Rev. B **51**, 17605 (1995).
7. M. Koos, I. Pocsik, and E.B. Vazsonyi, Appl. Phys. Lett. **62**, 1797 (1993).
8. C. Delerue, M. Lannoo, G. Allan, E. Martin, I. Mihalcescu, J.C.Vial, R. Romestain, F. Muller, and A. Bsiesy, Phys. Rev. Lett. **75**, 2228 (1995).
9. S. Schmitt-Rink, D.A.B. Miller, and D.S. Chemla, Phys. Rev. B **35**, 8113 (1987).
10. L.E. Brus, J. Chem. Phys. **80**, 4403 (1984).
11. S. Fafard, R. Leon, D. Leonard, J.L. Merz, and P.M. Petroff, Phys. Rev. B **52**, 5752 (1995).
12. S. Fafard, Z. Wasilweski, J. McCaffrey, S. Raymond, and S. Charbonneau, Appl. Phys. Lett. **68**, 991 (1996).
13. D. Hessman, P. Castrillo, M.-E. Pistol, C. Pryor, and L. Samuelson, Appl. Phys. Lett. **69**, 749 (1996).
14. M. Sopanen, M. Taskinen, H. Lipsanen, and J. Ahopelto, Appl. Phys. Lett. **69**, 3393 (1996).
15. A.I. Ekimov, F. Hache, M.C. Schanne-Klein, D. Richard, C. Flytzanis, I.A. Kudryavtsev, T.V. Yazeva, A.V. Rodina, Al. L. Efros, J. Opt. Soc. Am. B **10**, 100 (1993).
16. S. Guha, G. Hendershot, D. Peebles, P. Steiner, F. Kozlowski, and W. Lang, Appl. Phys. Lett. **64**, 613 (1994).
17. A. R. Beattie and P.T. Landsberg, Proc. Royal Soc. London **249**, 16 (1959).
18. H. Koyama, T. Ozaki, and N. Koshida, Phys. Rev. B **52**, R11561 (1995).
19. Y. Kanemitsu, Phys. Rev. B. **49**, 16845 (1994).
20. N. Ookubo, N. Hamada, and S. Sawada, Sol. St. Comm. **92**, 369 (1994).
21. K.J. Nash, P.D.J. Calcott, L.T. Canham, M.J. Kane, and D. Brumhead, J. Lumin. **60&61**, 297 (1994).
22. G.W. 't Hooft, Y.A.R.R. Kessener, G.L.J.A. Rikken, and A.H.J. Venhuizen, Appl. Phys. Lett. **61**, 234 (1992).
23. S. Finkbeiner and J. Weber, Thin Solid Films **255**, 254 (1995).
24. T. Takagahara and K. Takeda, Phys. Rev. B. **46**, 15578 (1992).
25. A. Eychmüller, A. Hasselbarth, L. Katskias, and H. Weller, J. Lumin. **48&49**, 745 (1991).
26. R. A. Street, Adv. Phys. **30**, 593 (1981).
27. D.J. Lockwood, Z. H. Lu, and J. M. Baribeau, Phys. Rev. Lett. **76,** 539 (1996).
28. Y. Kanemitsu, Phys. Rev. B **53**, 13515 (1996).

NANO-STRUCTURED SILICON-BASED FILMS WITH VISIBLE LIGHT EMISSION SYNTHESIZED BY LASER ABLATION

T. MAKIMURA, Y. KUNII, N. ONO and K. MURAKAMI
Institute of Materials Science, University of Tsukuba, Tsukuba, Ibaraki 305, Japan, makimura@mat.ims.tsukuba.ac.jp

ABSTRACT

Applying laser ablation technique, we have synthesized two types of SiO_2 films that include nanometer-sized Si particles. One is synthesized by alternative deposition of Si nanoparticles layers and SiO_2 layers. The synthesized film exhibits red photoluminescence (PL) with a peak energy below 1.5 eV. The other is synthesized by annealing at 1000°C of SiO_x films, which are formed by laser ablation in diluted O_2 gas. We find that there is a narrow range of composition for efficient red PL. Based on the experimental results, we tentatively discuss a possible model for the origin of the red PL.

INTRODUCTION

Nanostructured Si-based materials can exhibit bright light emission in the visible wavelength range [1–8], although crystalline Si has a 1.1-eV indirect band gap. The structure responsible for the light emission is still controversial partly because most of the materials contain undesirable chemical impurities such as hydrogen and fluorine. For further investigation, it is important to develop a clean synthesis method utilizing laser ablation [2, 3] and annealing of SiO_x films [4]. It is shown that nanometer-sized Si particles can be synthesized by ablating a Si target by laser light in He gas and collecting the final products on a substrate [2]. Shimizu-Iwayama *et al.* showed that Si nanocrystallites with visible photoluminescence (PL) precipitate in a SiO_2 matrix when Si^+-implanted SiO_2 films are annealed [4].

In the present work, applying the laser ablation technique, we have synthesized two types of SiO_2 films including Si nanoparticles. One was synthesized by alternative deposition of Si nanoparticle layers and SiO_2 layers. The other was synthesized by annealing of SiO_x films deposited in diluted O_2 gas. These films exhibit red PL with a peak energy below 1.5 eV. SiO_x films with particular compositions are found to exhibit efficient PL. Based on the experimental results, we propose that the red PL is due to excitons with large lattice distortion, localized at the interfaces of the precipitated Si core and the SiO_2 matrix.

Si NANOPARTICLE LAYERS ISOLATED BY SIO_2 LAYERS

The laser ablation was performed by focusing a Q-switched Nd:YAG laser beam (λ=532 nm. FWHM=10 ns, 10 J/cm^2) on Si targets in He gas filled in a vacuum chamber. The chamber was pumped down to 2×10^{-6} Torr before filling with the gas. The synthesized Si particles deposited on Si substrates were observed using an atomic force microscope (AFM). The AFM has space resolution of 0.05 nm in height and 20 nm in the direction parallel to the sample surface. For the AFM measurement, the Si target was ablated 50 times

Mat. Res. Soc. Symp. Proc. Vol. 452 © 1997 Materials Research Society

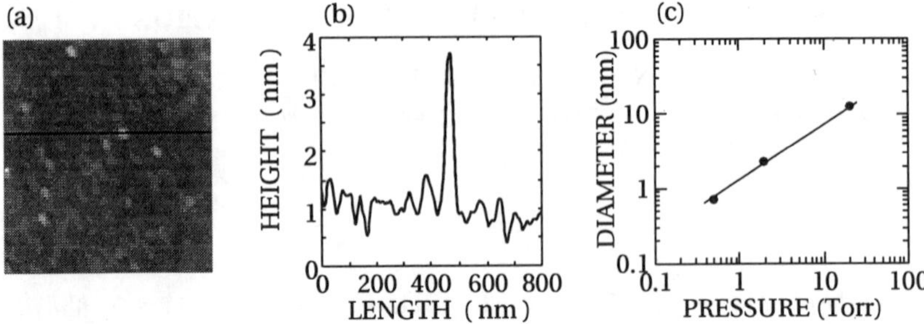

Figure 1: (a) An image of final products observed by an atomic force microscope. The products were synthesized by laser ablation of a Si target in He gas. (b) A section of the image along a straight line indicated in Fig. 1 (a). (c) Maximum of the size distribution as a function of gas pressure in which the particles are synthesized.

and the products were deposited on a Si substrate placed 2 cm above the target. For the PL measurement, the particle layers were embedded between SiO_2 layers as follows. The particles were deposited on a Si substrate by 1500 shots of laser ablation in the 2-Torr He gas. The thickness of the particle layer is estimated to be 50 nm. Then, a SiO_2 layer was deposited over the particles by 3000 shots of laser ablation of a Si target in 50-mTorr O_2 gas. Note that the use of O_2 gas causes an oxidation of the surface of the particles [8]. These sequences were repeated 7 times. We confirmed the deposition of a SiO_2 film in the O_2 gas by soft X-ray absorption spectroscopy. For this spectroscopy, the film was formed on a polycarbonate film. The details of the *in situ* spectroscopy are given elsewhere [5, 9, 10]. The deposited film was taken out of the chamber and annealed in flowing O_2 gas at 800°C. PL was measured in a photon energy range from 1.5 eV to 2.5 eV using a cw Ar^+ laser (457.9 nm, 30 W/cm^2) at room temperature [6]. The sensitivity of the spectrometer at each wavelength was corrected using a standard lamp.

Figure 1(a) shows a typical AFM image of the final products synthesized in a 2-Torr He gas. Several particles are seen in the image. To estimate a size of each particle, we analyzed a section along a straight line, as illustrated in Fig. 1(a). A typical section is shown in Fig. 1(b). This particle has a size of 2.7 nm. Sizes obtained in this way have a distribution of 2.5 ± 1 nm. The maxima of the distributions for particles synthesized in the gas with various gas pressures is shown in Fig. 1(c). The size distributions clearly depend on the ambient gas pressure. This result suggests that the formation of the particles is determined by the dynamics of the laser-ablated Si species in the gas phase. The details of the dynamics have been reported previously [6].

Figure 2(a) shows PL spectrum of Si particles embedded between SiO_2 layers. The synthesized film exhibits PL with a broad band width and a peak below 1.5 eV. It should be emphasized that the Si particles showing visible PL can be principally synthesized without containing any chemical impurities such as hydrogen, fluorine and so on, except for oxygen.

Si NANOCRYSTALS PRECIPITATED IN A SIO_2 MATRIX

The SiO_x films were formed by laser ablation of Si targets in diluted O_2 gas. The ablation was performed by irradiation with the Nd:YAG laser light for 9000 times. Soft X-ray absorption spectra were measured for analysis of the deposited films. The films were formed in vacuum or in 1% O_2 gas diluted with He gas with total pressures ranging up to 2 Torr. For PL measurement, products were formed on Si substrates placed 2 cm away from the Si targets. The ablation was performed under two conditions: (A) The energy densities of laser light was from 7 J/cm^2 to 20 J/cm^2 and the ambient gas was 1% O_2 gas diluted with He gas at a total pressure of 500 mTorr. (B) The energy density was 10 J/cm^2 and ambient gas was O_2 gas diluted with Ar gas at a total pressure of 200 mTorr. The concentration of the O_2 gas was changed from 1% to 10%. Ar gas was used because the mass of an Ar atom is close to that of an O_2 molecule. It is known that the dynamics of the laser-ablated species is strongly affected by mass of ambient gas [11]. After depositing the films, the substrates were taken out of the chamber into the air and were annealed at 1000°C in flowing Ar gas for 30 minutes. The PL from the synthesized films was measured in the same way described above.

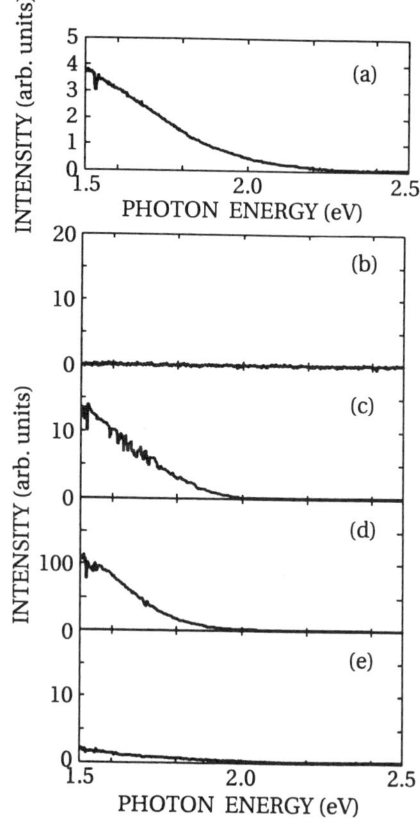

Figure 2: (a) Photoluminescence spectrum of nanometer-sized Si particle layers embedded between SiO_2 layers. The particles were synthesized in 2-Torr He gas. (b)–(e) Photoluminescence spectra of nanocrystals precipitated in SiO_2 films. These films were synthesized by annealing of SiO_x films prepared by laser ablation of Si targets in diluted O_2 gas using YAG laser light at (b) 7 J/cm^2, (c) 10 J/cm^2, (d) 13 J/cm^2, (e) 20 J/cm^2.

Figure 3 shows the X-ray absorption spectra for the films deposited in the diluted O_2 gas at total pressures of (a) 2 Torr, (b) 500 mTorr, (c) 200 mTorr, (d) 50 mTorr and (e) 10 mTorr, and (f) in vacuum. Spectra (a) and (f) in Fig. 3 coincide with those of SiO_2 and Si, respectively [5, 9]. Spectra (b)–(e) in Fig. 3 can be represented by superposition of spectra of Si and SiO_2. These results indicate that the O_2 gas with higher pressure results in formation of SiO_x films with larger x and that almost SiO_2 films are formed at pressures higher than 200 mTorr. It is found that the composition can be easily controlled by the partial pressure of the O_2 gas.

Figures 2(b)–2(e) show PL spectra of the films deposited under condition *A*. The energy densities of the laser light were (b) 7 J/cm^2, (c) 10 J/cm^2, (d) 13 J/cm^2 and (e) 20 J/cm^2. All of the films exhibit red PL with a peak below 1.5 eV (red PL) except for the film formed at 7 J/cm^2. The PL spectra (c)–(e) almost coincide with spectrum (a) in Fig. 2. Figure 4(a) shows the PL intensity at 1.5 eV as a function of the energy density. There is a limited range of the energy density appropriate for efficient red PL. The same PL was observed for films formed under condition *B*. Figure 4(b) shows the intensity of the PL after annealing as a function of O_2 concentration. As is clearly seen in Fig. 4(b), a limited range of partial pressure yields efficient PL.

The dependence shown in Fig. 4(a) and 4(b) can be explained in terms of the composition x in SiO_x and size of Si nanocrystallites precipitated in a SiO_2 matrix during annealing. The composition clearly depend on the partial pressure of the O_2 gas, as shown in Fig. 3. Therefore, under condition *B*, the composition would depend on the concentration of O_2 gas. Under condition *A*, the composition would also depend on the energy density because relative deposition rate of the laser-ablated Si atoms increases as the energy density increases. By annealing the SiO_x film with larger composition x, smaller nanocrystallites would precipitate in a SiO_2 matrix. Therefore, it can be concluded that Si crystallites with particular sizes exhibit the red PL efficiently.

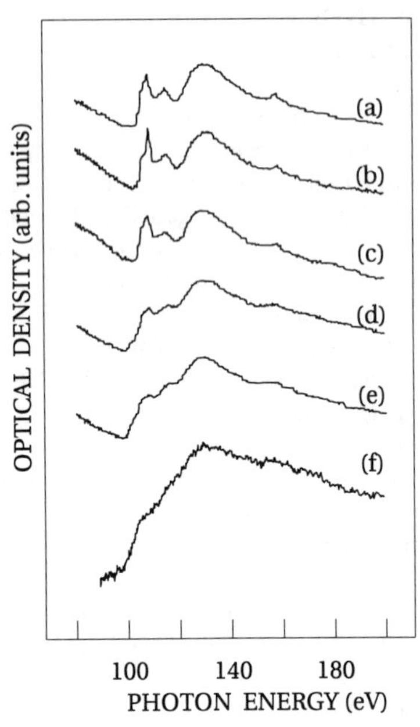

Figure 3: *In situ* soft X-ray absorption spectra of films synthesized by laser ablation. The ablation was performed by irradiation of Si targets with YAG laser light in 1% O_2 gas diluted with He gas. The films were deposited on polycarbonate films in the gas with total pressures of (a) 2 Torr, (b) 500 mTorr, (c) 200 mTorr, (d) 50 mTorr and (e) 10 mTorr and (f) in vacuum.

DISCUSSION

It is reasonable to introduce a model for the origin of the red PL other than electron-hole (*e*-*h*) recombinations in the Si cores or the SiO_2 matrix, for the following reasons. The binding energy of the excitons in the core is expected to be of the order of 10 meV. Therefore luminescence due to excitons in the core would result in PL with a peak energy almost equal to the band gap of the core. In this point of view, PL with peaks higher than 1.5 eV would be observed when the composition of the SiO_x films were changed. We could, however, observe only PL with peaks below 1.5 eV. Furthermore, the peak energies are reported to be much lower than the band gap [7]. On the other hand, excitons in silica are localized to an extent of the order of the lattice constant, with large lattice distortion,

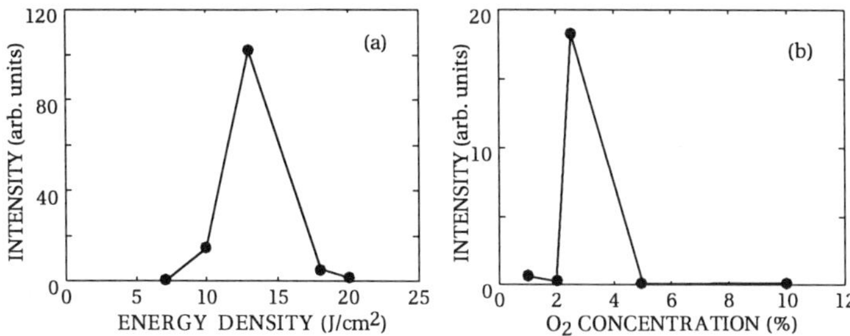

Figure 4: (a) The intensity of the photoluminescence at 1.5 eV as a function of the energy density of the laser light. (b) The intensity of photoluminescence from SiO_x films as a function of the concentration of O_2 gas in which the films were formed.

resulting in PL at 2.2–2.7 eV [12]. In addition, silica has point defects that exhibit PL in the visible range [13]. These localized *e-h* pairs have PL bands, with widths of the order of 0.1 eV due to electron-lattice interactions, with large oscillator strength. In spite of these suitable features, however, the peak energies are much higher than that of the observed red PL. Therefore, other models are required. Kanemitsu *et al.* proposed that the red PL is due to an exciton localized at the interface region between the Si core and the SiO_2 matrix [8].

In the interface region, there exist Si atoms bonded to both Si atoms and O atoms. At such sites, the excitons are expected to have strong interactions with the lattice, as excitons in silica and ionic crystals do [14]. As a possible model, we tentatively propose that the origin of the red PL is excitons localized at the interface between the Si core and the SiO_2 matrix, to an extent of the order of the lattice constant, with large lattice distortion [5]. In our framework, the experimental results are interpreted as follows. 1) The bandwidth of more than 0.1 eV is a result of strong electron-lattice interaction. 2) The bright PL originates from localized sites. At such sites, oscillator strength is expected to be large. 3) The peak energy of the red PL is lower than those of excitons in silica. A probable origin is excitons localized at neighborhood of a Si atoms at the interface, bonded to both Si atoms and O atoms. The lowered peak energy may be due to coexistence of Si-Si bonds and Si-O bonds. 4) The PL intensity depends on the composition, which is related to the core size. The Si cores absorb excitation light resulting in generation of *e-h* pairs. The pairs relax into more stable states at the interfaces if the cores are small enough to have band gaps wider than the energy separations of the localized states. If the cores are too small, the band gap is wider than the photon energy for the excitation, resulting in no generation of *e-h* pairs. Consequently, efficient PL requires Si cores with particular sizes. Determination of the peak energy of the PL should be required for further investigation.

CONCLUSIONS

We have synthesized two types of SiO_2 films with Si nanoparticles by laser ablation. One was synthesized by alternative deposition of Si nanoparticle layers and SiO_2 layers.

The synthesized films exhibit red PL with a peak energy below 1.5 eV. The other type of films were SiO_2 films in which Si nanocrystallites precipitate. These films were synthesized by annealing at 1000°C of SiO_x films, which were formed by laser ablation of Si targets in diluted O_2 gas. These films also exhibit the red PL. We found that SiO_x films with particular compositions exhibit efficient PL after annealing. This result suggests that the particular size of the nanocrystallites exhibit efficient PL because the sizes would depend on the composition. Taking these experimental results into account, we tentatively propose that the red PL is due to excitons with large lattice distortion, localized to an extent of lattice constant, at the interfaces between the precipitated Si core and the SiO_2 matrix.

ACKNOWLEDGMENTS

The authors would like to thank Prof. J. Ohnari for helping with the photoluminescence experiment. This work was supported partly by Grant-in-Aid for Scientific Researchers from the Ministry of Education, Science and Culture of Japan, and partly by the contract research from the NEDO.

References

[1] L. T. Canham, Appl. Phys. Lett. **57**, 1046 (1990).

[2] E. Werwa, A. A. Seraphin, L. A. Chiu, C. Zhou and K. D. Kolenbrander, Appl. Phys. Lett. **64**, 1821 (1994).

[3] T. Yamada, T. Orii, I. Umezu, S. Takeyama and T. Yoshida, Jpn. J. Appl. Phys. **35**, 1361 (1996).

[4] T. Shimizu-Iwayama, Y. Terao, A. Kamiya M. Takeda, S. Nakao and K. Saitoh, Nanostructured Materials **5**, 307 (1995).

[5] T. Makimura, Y. Kunii, N. Ono and K. Murakami, Jpn. J. Appl. Phys. to be published.

[6] T. Makimura, Y. Kunii and K. Murakami, Jpn. J. Appl. Phys. **35**, 4780 (1996).

[7] Y. Kanemitsu: *Optical Properties of Low-Dimensional Materials*, eds. T. Ogawa and Y. Kanemitsu (World Scientific, Singapore, 1995) Chap. 5, p. 258.

[8] Y. Kanemitsu, T. Ogawa, K. Shiraishi and K. Takeda, Phys. Rev. **B48**, 4883 (1993).

[9] T. Makimura and K. Murakami, Appl. Surf. Sci. **96–98**, 242 (1996).

[10] T. Makimura, T. Sakuramoto and K. Murakami, Jpn. J. Appl. Phys. **35**, L735 (1996).

[11] D. B. Geohegan, *Laser Ablation of Electronic Materials –Basic Mechanism and Applications–* eds. E. Fogarassy and S. Lazare (Elsevier, North Holland, 1992) p. 73.

[12] K. Tanimura, C. Itoh and N. Itoh: J. Phys. C **21**, 1869 (1988).

[13] D. L. Griscom: The Centennial Memorial Issue of The Ceramic Society of Japan **99**, 923 (1991).

[14] K. S. Song and R. T. Williams: *The Self-Trapped Excitons* (Springer-Verlag, Berlin, 1993).

LUMINESCENCE PROPERTIES OF SILICON NANOCRYSTALS

SHOUTIAN LI, STUART J. SILVERS and M. SAMY El-SHALL*
Department of Chemistry, Virginia Commonwealth University
Richmond, VA 23284-2006

ABSTRACT

Weblike aggregates of coalesced Si nanocrystals are produced by a laser vaporization - controlled condensation technique. SEM micrographs show particles with ~ 10 nm diameters but the Raman shift suggests the presence of particles as small as ~ 4 nm. FTIR of the freshly prepared particles shows weak peaks due to the stretching, bending and rocking vibrations of the Si-O-Si bonds indicating the presence of a surface oxidized layer SiO_x ($x<2$).

The particles show luminescence properties that are similar to those of porous Si and Si nanoparticles produced by other techniques. The nanoparticles do not luminesce unless, by exposure to air, they acquire the SiO_x passivated coating. They show a short-lived blue emission characteristic of the SiO_2 coating and a biexponential longer-lived red emission. The short lifetime component of the red emission, about 12 μs, does not depend on emission wavelength. The longer-lived component has a lifetime that ranges from 90 to over 130 μs (at 300 K), increasing with emission wavelength. The results are consistent with the quantum confinement mechanism as the source of the red photoluminescence.

INTRODUCTION

The discovery that porous and Si nanocrystalline emit visible light with a high quantum yield has stimulated interest in the synthesis of Si nanocrystals which are believed to be the luminescent centers in porous silicon [1-7]. Since the structure, size distribution, morphology and surface composition of the nanocrystals depend to a large extent on the method of preparation, it is important to compare the physical and optical properties of Si nanocrystals prepared by different synthesis techniques. In this paper we describe the application of our new technique, Laser Vaporization Controlled Condensation (LVCC) [8-11], to the synthesis of Si nanocrystals.

Among various synthesis method, laser vaporization of metals has the advantage of eliminating the need for high temperatures and for chemical precursors. By coupling laser vaporization with controlled condensation from the supersaturated vapor, it is possible to achieve a good control over the particles' size and the aggregate state [8-11]. The present work is focused on the photoluminescence of the Si nanocrystals prepared by the LVCC method .

EXPERIMENTAL

The Si nanocrystals used in these experiments are prepared in a modified, upward thermal diffusion cloud chamber by the LVCC method [11]. The Si vapor is generated by pulsed laser vaporization using the second harmonic (532 nm) of a Nd-YAG laser (15-30 mJ/pulse, 10^{-8} s pulse). The large temperature gradient between the bottom and top plates results in a steady convection current which can be enhanced by using a heavy carrier gas such as Ar under high pressure conditions (10^3 torr). The role of convection in the experiments is to remove the small particles away from the nucleation zone (once condensed out of the vapor phase) before they grow into larger particles. Glass slides or metal wafers can be attached to the top plate when it is desired to examine the morphology of as-deposited particles. No particles are found on any other place in the chamber except on the top plate.

Mat. Res. Soc. Symp. Proc. Vol. 452 © 1997 Materials Research Society

RESULTS AND DISCUSSION

The Si nanocrystals appear as a yellow powder. The SEM micrographs of the as - deposited particles on glass substrates reveal highly organized weblike structures characterized by micropores with wall thicknesses of 10 - 20 nm. The weblike morphology is similar to that of the silica nanoparticles prepared by the same method using O_2/Ar or O_2/He gas mixtures [10-11]. The weblike structure with strings of aggregated Si nanocrystals is shown in Figure 1. The invidual particles are spherical and uniform in size, about 10 nm in diameter and are connected in a weblike structure.

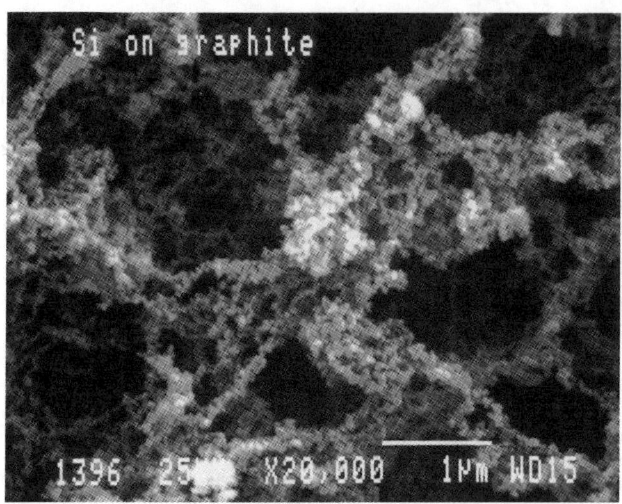

Figure 1. SEM of the weblike agglomeration of the surface-oxidized Si nanocrystals.

The FTIR of the as-deposited sample (Fig. 2-a) shows a broad absorption band at about 1080 cm^{-1} accompanied by weak bands at 956 cm^{-1}, 880 cm^{-1} and 460 cm^{-1}. We assign the observed IR peaks to a surface oxide layer of SiO_x that apparently forms after the particles are removed from the reaction chamber. The main band at 1080 cm^{-1} is due to Si-O-Si stretching and is usually observed in thin films of amorphous silicon oxide (SiO_x with x varying from 0-2) [12,13]. Annealing the particles at higher temperatures results in the appearance of a weak band at 800 cm^{-1} and an increase in both the broad band at 1080 cm^{-1} and the 460 cm^{-1} band. These changes are attributed to developing the oxide layer (SiO_x) by both increasing its oxygen content (i.e. x increases toward 2) and by increasing its thickness through the slow oxidation of the Si core.

The Raman spectrum, displayed in Figure 2-b, shows a sharp peak at about 510 cm^{-1}, close to the Raman allowed optical phonon characteristic of microcrystalline silicon at 520 cm^{-1} [14]. The downshift of this band in our Si nanoparticles is attributed to size and strain effects [15]. For example, Raman shifts of 517 cm^{-1}, 515 cm^{-1}, 507 cm^{-1}, and 501 cm^{-1} were correlated to particle sizes of 7 nm, 4.8 nm, 2.5 nm and ~ 1.6 nm, respectively [16]. Based on the Raman shift of our sample, the average particle size can be estimated as ~ 4 nm. This is considerably smaller than the particle size shown by the SEM micrographs and this suggests that particles smaller than the SEM is able to detect may be present in our sample.

The crystallinity of the particles is verified by the X-Ray diffraction spectrum which shows crystalline Si lines at scattering angles of 28°, 47° and 56°. Such crystalline lines are not present in the X-Ray diffraction pattern of silica nanoparticles prepared in the presence of O_2 [10,11].

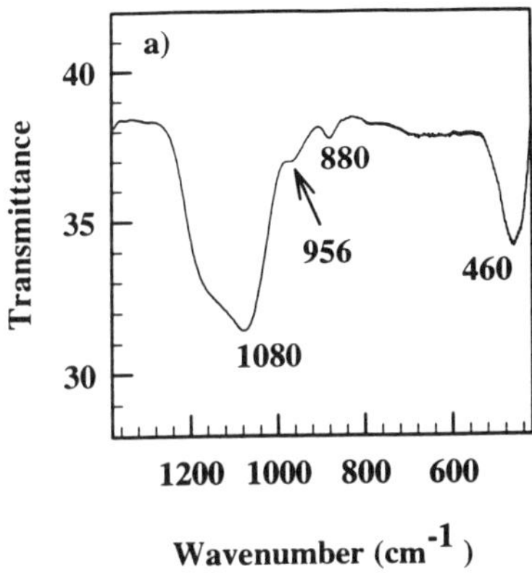

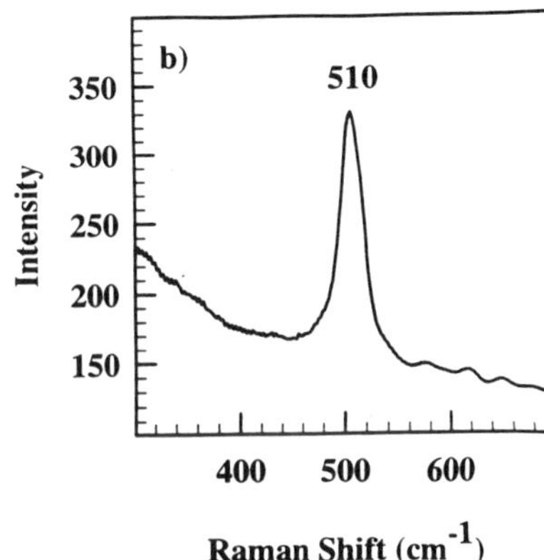

Figure 2 (a) FTIR of the as deposited Si nanocrystals, **(b)** Raman spectrum of the surface-oxidized Si nanocrystals.

The dispersed luminescence spectra of the Si nanoparticles excited by the 363.8 and 514.5 nm Ar ion cw laser lines are shown in Figure 3. In both spectra a broad emission is seen in the red region. This emission can be seen with the naked eye in the presence of normal room light. The red emission curve can be fit to a Gaussian shape. For the 363.8 nm excitation the width (FWHM) is 184 nm and the maximum is at 740 nm; for the 514.5 nm excitation the width is 190 nm and the maximum is at 760 nm. The 20 nm blue shift in the emission maximum of the 363.8 nm excitation may arise as a result of excitation of higher energy states which are not accessible with longer wavelength excitations. With the quantum confinement interpretation [5,7], this suggests that different sizes of nanocrystals are simultaneously present in our sample. Accordingly, the longer wavelength excitation selects larger particles which in turn emit further to the red than the smaller particles selected by the shorter wavelength excitation.

The insert in Fig. 3 shows a weak blue emission feature that is observed when 363.8 nm excitation is used. This feature peaks at 450 nm and appears similar to the blue emission observed when SiO_2 nanoparticles are excited. This blue emission is more pronounced when pulsed excitation is used as shown below.

Dispersed luminescence spectra obtained with 340 nm pulsed laser excitation are shown in Figure 4. The spectra differ by the position of the boxcar gate, which ranges from 0 delay with respect to the laser excitation pulse to 40 μs delay. The 0 delay spectrum enhances the blue emission component, since the lifetime associated with it is short (less than 20 ns) compared to the lifetimes of the red emission. The shape of the blue emission appears similar to that from SiO_2 nanoparticles and it is probably due to the oxidized surface layer of the Si nanoparticles [10,11].

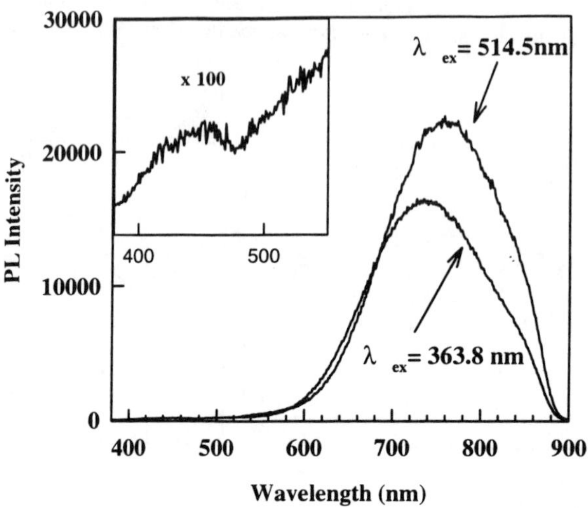

Figure 3. Dispersed emission of the Si nanocrystals excited by the 363.8 and 514.5 nm Ar ion laser.

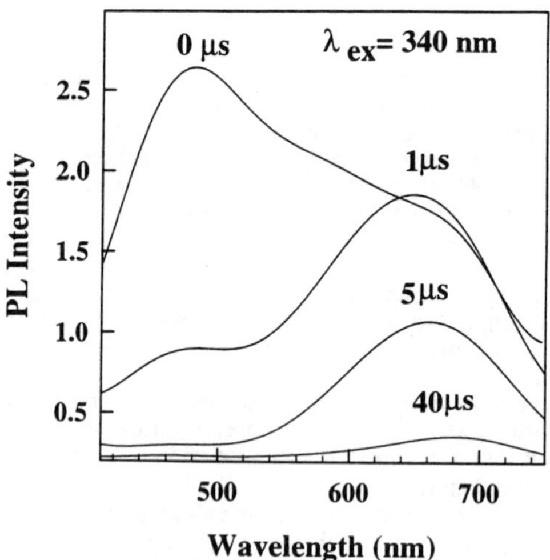

Figure 4. Dispersed emission of the Si nanocrystals obtained with 340 nm pulsed laser excitation.

The photoluminescence decays are multiexponential, but clearly have both long and short components. They can be fit to the biexponential form

$$I(t) = I^{\circ}_{s} \exp(-t/\tau_{s}) + I^{\circ}_{l} \exp(-t/\tau_{l})$$

where the subscripts 's' and 'l' refer to the short and long-lived components respectively. A decay at 740 nm is shown in Figure 5-a together with the biexponential fit. The short lifetime, about 12 μs, does not depend on emission wavelength; the long lifetime ranges from 90 to over 130 μs, increasing with emission wavelength. The dependence of τ_l on emission

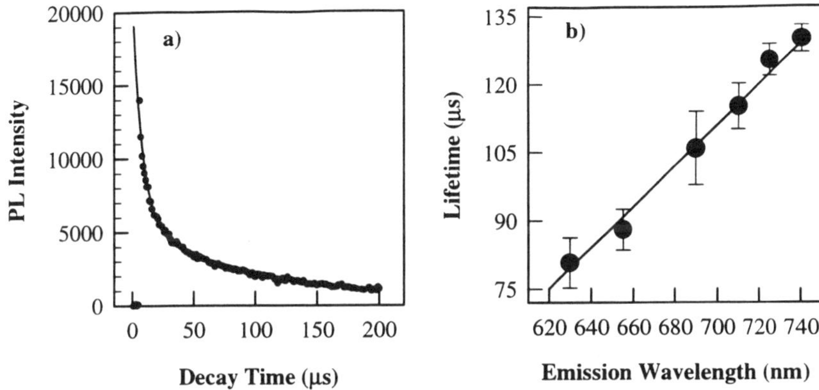

Figure 5 (a) Decay of the PL from the Si nanocrystals at 740 nm (solid line represents the biexponential fit), **(b)** Dependence of the lifetime of the longer-lived component of the red emission on the emission wavelength.

The features associated with the red emission are similar to those observed for porous Si and from other Si nanocrystals [1-7]. The features we observe for the blue PL are similar to those reported by Kanemitsu using surface-oxidized Si crystallites prepared by laser breakdown of silane [6,17]. Kanemitsu proposed a model in which the red PL originates from a surface localized state while the blue PL originates from the nanocrystal core [6,17]. According to this model, the shift of the red PL to longer wavelength with time delay is due to the relaxation of the carriers or excitons to lower energy states.

In view of the many different mechanisms proposed to explain the visible PL from porous Si and Si nanocrystals [1-7], we note that many of the observed features associated with the red PL from our particles can be explained by the quantum confinement mechanism. For example, the failure to observe the red PL from the freshly prepared particles and the long annealing time in ambient air required before emission can be observed, can be explained by the need to provide an efficient surface passivation which permits radiative recombination in the crystallites by removing competitive nonradiative relaxation at the surface states. Also, the shift of the PL towards shorter wavelength by increasing excitation energy is explained by the selective excitation of the smaller particles which emit more to the blue. Finally, the increase in the lifetime of the longer-lived component of the red emission with emission wavelength is consistent with larger particles having longer lifetimes. In addition, we note that the features associated with the fast-decay blue emission can't be explained with the confinement mechanism. For example, this emission does not shift with the excitation energy and its short lifetime does not depend on the emission wavelength. Moreover, this blue emission is similar to that observed from silica particles and several models related to surface defects and surface adsorbed OH groups have been proposed to explain it [10,18].

CONCLUSIONS

A method which combines laser vaporization of metal targets with controlled condensation in a diffusion cloud chamber is used to synthesize Si nanoparticles. Our particles were prepared by laser vaporization of Si and the method does not involve hydrogen. The nanoparticles aggregate into a novel weblike porous microstructure. The particles show luminescence properties that are similar to those of porous Si and Si nanocrystals produced by other techniques. The nanocrystals do not luminesce unless, by exposure to air, they acquire the SiO_x passivated coating. They show a short-lived blue emission characteristic of the SiO_2 coating and a biexponential longer-lived red emission. The short lifetime component of the red emission, about 12 μs, does not depend on emission wavelength. The longer-lived component has a lifetime that ranges from 80 to over 130 μs (at 300 K), increasing with emission wavelength. The results are consistent with the quantum confinement mechanism as the source of the red photoluminescence.

ACKNOWLEDGMENT

The authors gratefully acknowledge financial support from the NASA Microgravity Materials Science Program (Grant NAG8-1276). We thank Professor Jim Terner (VCU) for the Raman measurements.

REFERENCES

*. To whom inquiries should be addressed.
1. L. T. Canham, Appl. Phys. Lett., **57**, 1046 (1990).
2. A. G. Cullis and L. T. Canham, Nature **353**, 335 (1991).
3. L. E. Brus, Nature **353**, 301 (1991).
4. H. Takagi, H. Ogawa, Y. Yamazaki, A. Ishizaki and T. Nakagiri, Appl. Phys. Lett. **256**, 117 (1992).
5. W. L. Wilson, P. F. Szajowski and L. E. Brus, Science **262**, 1242 (1993).
6. Y. Kanemitsu, T. Ogawa, K. Shiraishi and K.Takeda, Phys. Rev B. **48**, 4883 (1993).
7. L. E. Brus, P. F. Szajowski, W. L. Wilson, T. D. Harris, S. Schuppler and P. H. Citrin, J. Am. Chem. Soc. **117**, 2915 (1995).
8. M. S. El-Shall, W. Slack, W. Vann, D. Kane and D. Hanley, J. Phys. Chem. **98**, 3067 (1994).
9. M. S. El-Shall, D. Graiver, U. Pernisz and M. I.Baraton, NanoStructured Materials **6**, 297 (1995).
10. M. S. El-Shall, S. Li, T. Turkki, D. Graiver, U. C. Pernisz and M. I. Baraton, J. Phys. Chem. **99**, 17805 (1995).
11. M. S. El-Shall, S. Li, D. Graiver and U. C. Pernisz in "Nanotechnology: Molecularly Designed Materials", edited by G. M. Chow and K. E. Gonsalves (ACS Symposium Series 622, Washington DC, 1996), Chapter 5, PP. 79-99.
12. Nakamura, M.; Mochizuki; Usami, K. Solid State Comm. **50** (12), 1079 (1984).
13. S. Hayashi, S. Tanimoto and K. Yamanoto, J. Appl. Phys. **68**, 5300 (1990).
14. T. Okada, T. Iwaki, K. Yamamoto, H. Kasahara and K. Abe, Solid State Comm. **49** (8), 809 (1984).
15. R. Tsu, H. Shen and M. Dutta, Appl. Phys. Lett. **60**, 1112 (1992).
16. S. M. Prokes in "Nanomaterials: Synthesis, Properties and Applications" edited by A. S. Edelstein and R. C. Cammarata (Institute of Physics Publishing, Bristol, 1996), pp. 349-457 ; S. M. Prokes, Bull. Am. Phys. Soc. **38**, 157 (1993); I. H. Campbell and P. M. Fauchet, Solid State Commun. **58** (10), 739 (1986).
17. Y. Kanemitsu, Phys. Rev. B. **49**, 16845 (1994).
18. H. Tamura, M. Ruckschloss, T. Wirschem, S. Veprek, Appl. Phys. Lett. **65**, 1537 (1994) ; H. Morisaki, H. Hashimoto, F. W. Ping, H. Nozawa and H. Ono, J. Appl. Phys. **74**, 2977 (1993).

Ferromagnetic Properties of Spark-Processed Photoluminescing Silicon

J. Hack, M.H. Ludwig, W. Geerts, R.E. Hummel
Materials Science and Engineering Department, University of Florida, Gainesville, FL. 32611.

ABSTRACT

Magnetic properties of photoluminescing spark-processed silicon (sp-Si) have been investigated for the first time. Contrary to the diamagnetic signal known for bulk silicon, sp-Si displays a paramagnetic resonance as well as a ferromagnetic hysterisis loop. The paramagnetic resonance was studied using an EPR system and showed a high concentration of at least two distinct paramagnetic centers. One center can be eliminated by annealing in Ultra-High Purity nitrogen for 30 minutes at 600 °C. Measurements utilizing a SQUID magnetometer revealed that sp-Si displays ferromagnetic ordering with a saturization magnetization occuring at low fields. This is attributed to the high density of paramagnetic centers. Temperature dependent measurements were performed to establish possible links between magnetic properties and the luminescence of sp-Si.

INTRODUCTION

In previous investigations, strong photoluminescence (PL) was found in non-luminescing materials such as silicon, after spark processing.[1] The method of spark processing conducted on silicon (sp-Si) is carried out by exposing *c*-Si to a pulsating plasma-based discharge process. The directed energy causes the silicon to be flash evaporated. As the sp-Si settles back down onto the substrate wafer, it is rapidly quenched and forms a highly disordered structure. Depending on the atmospheric conditions utilized during spark processing, different luminescing peak emissions can be seen. For example, when sparked in an oxygen environment, a red luminescence (1.9 eV) is produced. Stagnant air yields a bright green luminescence, peaking at 2.2-2.4 eV, while passing compressed dry air across the sample during spark processing creates a material with highly intense UV/blue (3.2 eV) luminescence.[2] The intensity of the luminescence is of the same order of magnitude as that of porous silicon (PS).

This study demonstrates that spark processing also produces a ferromagnetic material from a diamagnetic substance such as Si. This novel preparation technique has significant ramifications from both an application, as well as a purely scientific point of view.

Thin amorphous magnetic layered materials are of widespread interest at this time. By altering the morphology of a material, it is hoped that the magnetic properties can be tailored to suit a particular application. These materials can then be utilized in such diverse disciplines as sensors and micro-machines. Typically, innate ferromagnetic materials are used as precursors and various amorphization techniques are performed to enhance the magnetic properties. This paper discusses how spark processing can be used to affect the magnetic properties of Si.

To understand the defect structure of sp-Si, Electron Paramagnetic Resonance (EPR) studies were conducted. Due to its high sensitivity for spin-active defect centers and its specific sensitivity to the local crystallographic structure of the defect center, EPR is an excellent technique for resolving the atomic nature of defects. [3] EPR can detect the presence, concentration and local environment of a given defect. Since considerable research has been conducted on silicon to characterize defects in this material, [3,4,5] comparisons of defects found

Mat. Res. Soc. Symp. Proc. Vol. 452

in sp-Si and those known in the literature for Si are possible.

Magnetic susceptibility measurements were conducted to investigate interactions of the defects created during spark processing. Using a Superconducting Quantum Interference Device (SQUID) with an accuracy better than $1x10^{-8}$ emu. [6]

EXPERIMENTAL PROCEDURE

Commercial, p-doped silicon wafers having a purity of $5.5x10^{15}$ cm^{-3} were spark-processed by applying a unidirectional high voltage (15kV) discharge between an anode tip and the silicon wafer acting as the cathode. To exclude ferromagnetic contamination, the anode tip was chosen to be from the same silicon wafer as the substrate. A single-pulse spark event had a duration of about 10 ns. The repetition frequency was 16kHz with the off time between two spark events lasting about 60 μs. Moreover, fabrication was done in an isolated environment to avoid possible contamination from ferromagnetic elements.

For magnetic susceptibility measurements, long Si strips 10.0 by 0.64 cm, were fashioned to eliminate possible stray fields seen by the magnetometer. Spark-processing was performed for 6 days. This led to a processed area which extended completely through the wafer. Specifically, a processed area of 0.5 cm in diameter on the top portion and 0.2 cm on the bottom of the wafer was observed. For grounding of the wafer, the ends of the long strips were adhered to an aluminum plate using a conductive, black polymer tape. The ends of the strips were cut off after fabrication to ensure no contamination. Moreover, susceptibility measurements of the conductive polymer were performed to check for any possible ferromagnetic response. None was found. The susceptibility measurements were conducted in a SQUID magnetometer at temperatures ranging from 4.3 to 300 K and fields varying from ±200 to ±50,000 Gauss. Annealing of the specimens were performed in UHP nitrogen (99.999% purity).

Samples for EPR measurements were produced similar to those above with some exceptions. All wafers were cut to 3 mm^2. Sp-Si samples were then spark-processed for 2 hours. They were produced using stagnant air, flowing air, and in a bell jar back filled with oxygen (99.99% purity). For the oxygen-prepared sample, a bell jar was evacuated and reflushed five times to minimize the presence of unwanted residual gases or contaminants. Again, samples were annealed in UHP nitrogen for 30 minutes. EPR experiments were conducted at room temperature. The samples were mounted on quartz rods to fix their position. Measurements were performed both parallel and perpendicular to the applied magnetic field.

Silicon wafers used in the luminescence experiments were cut to 1 cm^2 squares. The spark processing was initiated for two hours under differing atmospheric conditions (see above). Annealing of sp-Si occurred again in UHP nitrogen flowing across the samples for 30 minutes. The PL measurements were performed in air within less than 5 min. after heat treatment. Excitation was performed using a HeCd laser having a lasing wavelength of 325 nm. A long-pass filter was inserted between the sample and the monochromator to block scattered laser light having a 50% transmittance at 348 nm.

RESULTS

Magnetic susceptibility

Magnetic susceptibility measurements conducted on bulk silicon at 4.3 K revealed diamagnetism consistent with the literature.[6] As seen in Fig. 1, the silicon displays the typical negative slope in an M vs. H diagram characteristic for a diamagnetic material. In contrast, sp-Si measured under similar conditions reveals significantly altered magnetic properties. Specifically, as seen in Fig. 1, sp-Si displays a hysterisis loop as known from soft ferromagnetic materials.[7] A remnant magnetization at zero field is clearly visible concomitant with a positive slope.

To elucidate the saturization magnetization of sp-Si, the magnetic field was increased to 50,000 Gauss, the limit of the SQUID magnetometer. The as-prepared sp-Si still does not completely saturate at 4.5 K, although the curve shows a trend of approaching a zero slope, see Fig. 1 insert.

In order to further investigate this behavior, a series of anneals in UHP nitrogen at 200 °C intervals and subsequent measurements at 4.5 K was conducted on sp-Si. As one can deduce from Figs. 3 (a) and (b) the remanent magnetization and the largest magnetization attained at 500 Gauss decreases with successively increasing annealing temperatures and approaches the silicon value above 600 °C. Thus, sp-Si is initially a soft ferromagnetic material while, upon annealing, passes through a paramagnetic, and finally, converts to a diamagnetic state consistent with bulk Si, as shown in Fig 3b.

Moreover, it is noteworthy to state that essentially the same behavior is observed when the magnetization measurements are conducted at room temperature.

It is interesting to note, that when measured at low fields, i.e. <2000 Gauss, sp-Si displays the characteristic ferromagnetic hysteresis loop. However, at very large fields, >10,000 Gauss, the material does not reveal a hysteresis loop. Moreover, the same sp-Si sample which was initially exposed to high magnetic fields and then subsequently measured at low magnetic fields, showed no hysterisis. This suggests that high magnetic fields permanently quench ferromagnetism in sp-Si.

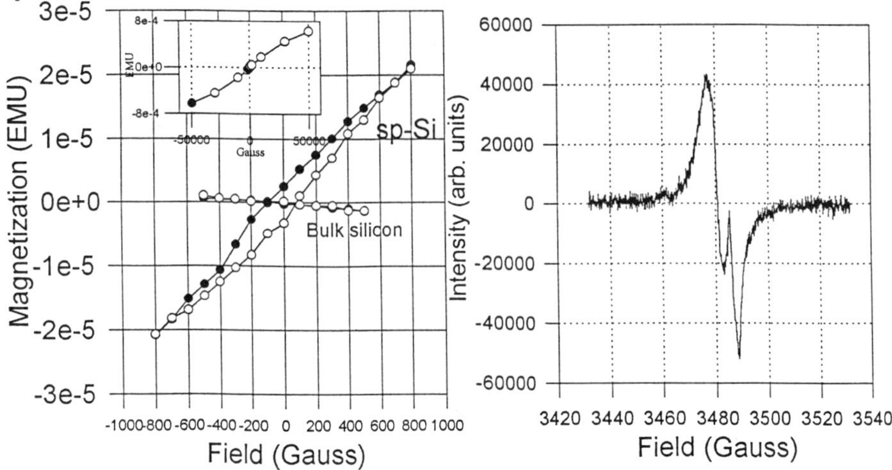

Fig. 1 Magnetization of sp-Si and of virgin Si as a function of applied magnetic field measured at 4.3 K. The insert shows sp-Si at substantially higher fields.

Fig. 2 Electron Paramagnetic Resonance spectrum of sp-Si measured at room temperature.

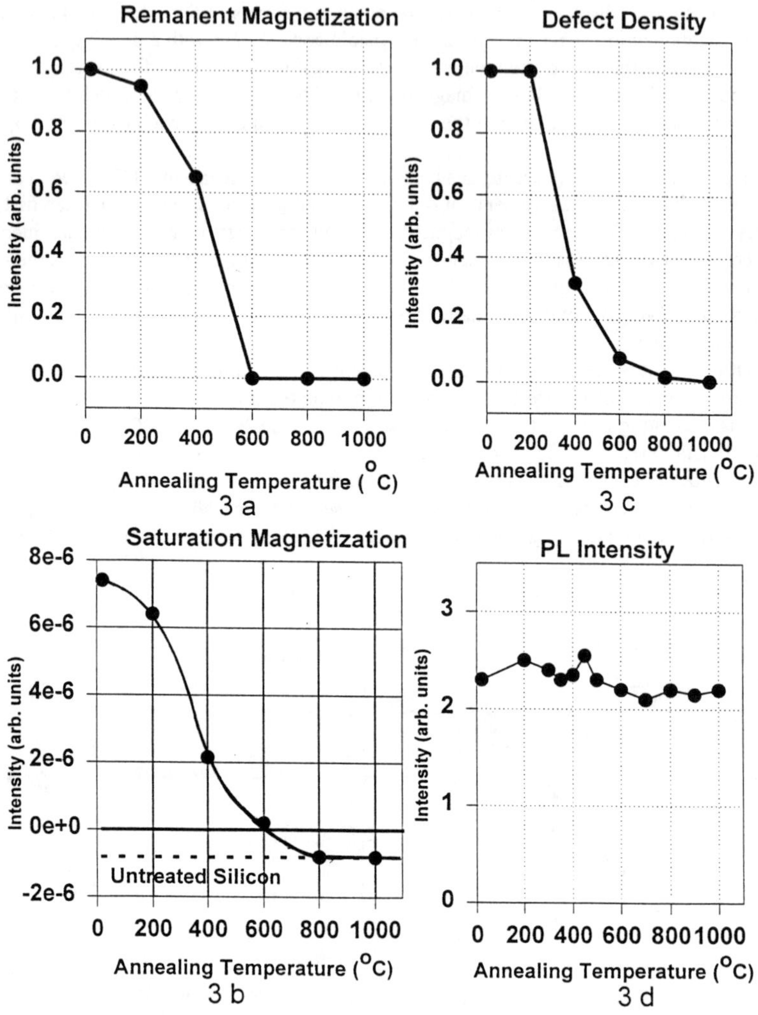

Fig. 3 Annealing behavior of certain physical properties of sp-Si as indicated on the individual graphs.

Electron Paramagnetic Resonance

First, a bulk Si wafer was measured in the EPR system at room temperature using an x-band microwave energy of 9.7 GHz. No paramagnetic centers were observed within the detection limit of the system. Next, sp-Si was measured under identical conditions. A strong double peak structure was found, see Fig. 2. To determine the number of paramagnetic centers in sp-Si, comparison with a known material was conducted. A concentration of 3.8×10^{20} unpaired spins/cm^3 was calculated for sp-Si.

Following the initial EPR measurement of as-prepared sp-Si, an annealing experiment similar to that for magnetic susceptibility was conducted. After the first anneal, there is only a slight decrease in the defect concentration. The 400 °C anneal however, displays a large decrease in paramagnetic centers. By 600 °C, the second peak is extinguished, and the defect concentration has decreased by an order of magnitude. The 800 °C and 1000 °C anneal further reduce the defect concentration. These results are shown in Fig. 3c.

Photoluminescence

In contrast to the magnetic properties of sp-Si, photoluminescence is essentially not affected by annealing, as seen in Fig. 3d.

DISCUSSION

The results of this study show a strong correlation between the paramagnetic defects produced during spark processing, and the ferromagnetic properties of the resultant material. As the defect concentration is reduced by an order of magnitude through annealing, the ferromagnetic properties are destroyed within the same temperature range. This is consistent with experiments conducted on irradiated a-Si.[8] Further annealing does not eliminate all paramagnetic centers. A point is reached however, where the paramagnetic centers in sp-Si are overwhelmed by diamagnetic centers present in the material. This leads to sp-Si attaining diamagnetism similar to bulk silicon above the 800 °C anneal. It is suggested that dangling bonds produced during spark processing combine with atoms in the material, most probably either oxygen, or silicon atoms, to reduce the number of unpaired electrons.

It is generally accepted that, ferromagnetism is the result of exchange interactions between unpaired electrons in the *d* or *f* shells of a substance. A material such as silicon, has no electrons occupying the *d* shell, so ferromagnetism cannot be a result of unpaired electrons residing in this shell. It is proposed, however, that the ferromagnetic properties of sp-Si are the result of exchange interaction occurring between the high concentration of paramagnetic centers present. This large concentration of unpaired spins as seen in the EPR experiments, forces the paramagnetic centers close together. It is believed that since exchange interaction depends on the proximity of unpaired electrons, the ferromagnetism as seen in sp-Si is a result of coupling of the large number of unpaired electrons in the *p* shell.

This proposition is substantiated by annealing experiments. Reducing the number of unpaired spins in sp-Si reduces the defect density. As the defect density is lowered, the distance between any two defects is expanded and the probability for exchange interaction to occur is greatly reduced. Thus at some critical distance, ferromagnetism is highly improbable since unpaired spins are too far apart to initiate coupling. The 600 °C anneal apparently reduces the defect density to the point where the unpaired electrons are just beyond this critical distance. At higher annealing temperatures, the magnetic behavior of sp-Si is similar to that of bulk Si. At this point, the concentration of paramagnetic centers is low enough so that the overall magnetic properties are governed by the diamagnetic centers present.

The annealing experiments also lead to the conclusion that the observed ferromagnetism is not due to contaminants in sp-Si. Had impurities been the cause of ferromagnetism, annealing would not have destroyed the hysterisis. Contaminants would still be present even after annealing, and would continue to yield hysteresis.

By comparison, there is no apparent correlation between the magnetic properties and photoluminescence. It is evident that annealing destroys the paramagnetic centers but does not affect the PL.

Conclusions

This study has shown that, spark processing transforms silicon, a diamagnetic material, into a ferromagnetic substance. There is a large concentration of paramagnetic centers in the sp-Si (3.8×10^{20} cm^{-3}). After a 600 °C anneal, the ferromagnetism is destroyed and the material becomes diamagnetic as pure Si. There is no correlation between the magnetic properties and photoluminescence. The ferromagnetic phenomenon is thought to result from an exchange interaction occurring from the close proximity of the unpaired electron spins produced during spark processing.

REFERENCES

1. Hummel, R.E., Chang, S.S., Applied Phys. Lett. 61, pg. 1965, 1992.

2. Ludwig, M. H., Review of Materials Science, 1996.

3. Stesmans, J. B. et. al., Surface Science, 141, pgs. 255-284, 1984.

4. Griscom, D.L., The Centennial Memorial Issue of the Ceramic Society of Japan, 99, pgs. 923-942, 1991.

5. Piondexter, E. H., et.al., J. Appl. Phys., 52, pg. 879, 1981.

6. McElfresh, M., Fundamentals of Magnetism and Magnetic Measurements, Quantum Design, 1994.

7. Hummel, R. E., Electronic Properties of Materials 2nd Edition, Springer-Verlag, 1992.

8. Khokhlov, A. F., et. al., JEPT Lett., Vol. 24, pg. 212, 1976.

9. Laiho, R., et.al., Journal of Luminescence, 57, pg. 197, 1993.

10. Gerasimenko, N. N., Soviet Physics-Semiconductors, Vol. 5, No. 9, pg. 1487, 1972.

MULTICOLOR-EFFECTS OF LUMINESCING, NANOSTRUCTURED SILICON AFTER SPARK-PROCESSING IN PURE AND COMPOSITE GASES

M.H. LUDWIG, A. AUGUSTIN, R.E. HUMMEL
University of Florida, Department of Materials Science and Engineering, P.O. Box 116400, Gainesville, FL 32611-6400, USA

ABSTRACT

Spark-processing has been shown recently to generate a silicon-based substance which intensely photoluminesces in the UV/blue or green spectral range, depending on whether the preparation was performed in flowing or stagnant air, respectively. This study reports about significant differences for the radiative properties of spark-processed silicon after preparations in oxygen, nitrogen, and mixtures of both gases. Whereas there is essentially no photoluminescence (PL) after processing in pure nitrogen, an orange PL band at 1.89 eV occurs after preparation in pure oxygen. The orange PL degrades with time and when exposed to laser light. Furthermore, it switches immediately to a more intense blue emission centered at 2.61 eV when the sample is subjected to lower gas pressures. A similar 2.6 eV PL is observed after heat treatments up to 350°C. If, however, spark-processing of Si is carried out in nitrogen and oxygen is added, a green PL arises, centered at 2.35 eV. Moreover, for a volume ratio of about 1:1 N_2/O_2 a PL band at 3.22 eV emerges, superimposing the green emission. Both bands are about two orders of magnitude more intense than the orange PL, when prepared under otherwise identical processing conditions.

INTRODUCTION

Directing pulsating plasma discharges to Si modifies macro- and microstructure and generates a substance which strongly photoluminesces in several bands in the visible spectrum [1][2]. The process of plasma-based discharges is called spark-processing. Spark-processed silicon (sp-Si) is prepared by exposing crystalline Si to pulsating discharges whereby the Si substrate serves as cathode and a tipshaped counterelectrode as anode. Depending on the preparation conditions, the photoluminescence (PL) of sp-Si peaks in the UV/blue (380-400 nm / 3.26-3.1 eV) or green (520-580 nm / 2.38-2.14 eV) spectral range when prepared in flowing or stagnant air, respectively [3]. Emissions of both PL bands are very intense and thermally stable up to annealing temperatures of 850°C for the UV/blue and up to 1100°C for the green band [4]. Furthermore, PL of sp-Si decays very rapidly with lifetimes in the 5 to 10 ns range [5]. Thermal stability and fast decay rule out similarities with chemically etched, porous silicon [4] or other luminescent Si-based substances such as a-Si:H [6] or Si-O-H derivatives [7].

It has been demonstrated by Raman studies combined with high resolution electron microscopy that spark-processing of Si generates a substance which contains Si nanocrystallites imbedded within a predominantly amorphous SiO_2 matrix (Ref. [8] and ref. therein). The sizes of these Si particles

Mat. Res. Soc. Symp. Proc. Vol. 452 © 1997 Materials Research Society

range between 3 to 20 nm. The surrounding matrix is formed mainly by SiO_2 with additions of nitrogen (up to 7 at%) when prepared in air. Present understanding of the radiative mechanism within sp-Si does not allow to assign the luminescence to a quantum size effect, transitions via interface states, or specific defect structures. The inevitably formed silicon oxide or oxynitride may support emissions via defect or band tail states [9]. Also, alloys of silicon oxide or silicon oxynitride may provide a matrix with an energetic gap wide enough to shift possible radiative transitions into the visible spectrum when compared to Si.

To decide whether the PL mechanism of sp-Si is related to specific chemical agents, either related to silicon oxynitrides or contaminations (e.g., from the processing atmosphere or the counter-electrode), spark-processing of Si is performed in pure oxygen, pure nitrogen, or mixtures of both in this study. The results verify that the UV/blue and the green PL of sp-Si, obtained after spark-processing in flowing and stagnant air, can be duplicated by conducting the process in certain mixtures of oxygen/nitrogen gases. However, the PL spectra differ considerably from those of sp-Si prepared in air when spark-processing is performed in pure oxygen or pure nitrogen. Whereas preparation in pure oxygen causes a relatively weak and rapidly degrading PL centered at 1.89 eV (650 nm), essentially no emission can be detected after spark-processing of Si in pure nitrogen. These findings are in striking contrast to those of Steigmeier et al. [10] who reported about nearly identical emissions after spark-processing in different gases.

EXPERIMENT

Spark-processed Si was prepared in a vacuum chamber which was evacuated, baked out, and flushed several times prior processing with oxygen (purity >99.99%), nitrogen (purity >99.99%), or mixtures of both gases. During preparation, the gas pressure was held constant at 1 bar. All Si specimens were cut from a single-crystalline, (111)-oriented wafer (n-type doped by As to an extent of 2×10^{19} cm^{-3}). Spark-processing was performed by applying unidirectional high voltage (15 kV) discharges between a tungsten tip (anode) and a Si specimen (cathode). A single spark event lasted about 10 ns and was repeated with a frequency of 16 kHz for an overall processing time of 30 min for each sample preparation. PL spectra were recorded by exciting the sample with the 325 nm line of a continuous wave HeCd laser at a power density of ~0.4 mW cm^{-2}. All spectra were corrected for the spectral response of the detection system which consisted of a 0.33 m single-grating monochromator and a cooled GaAs photomultiplier.

RESULTS AND DISCUSSION

Figure 1 provides a comparison of normalized room-temperature PL spectra for sp-Si after processing in oxygen, nitrogen and various ratios of both gases. The most intense emission was found to peak at 3.22 eV (385±10 nm) if the processing atmosphere contains an equal ratio of N_2 and O_2 (curve (a) in Fig. 1). This PL band is identical in peak energy, shape, FWHM and intensity to the one which is observed after preparation under flowing air conditions; that is, by directing a stream of compressed, dry air toward the discharges between the electrodes. (Detailed descriptions of preparation conditions and optical properties for spark-processing in air can be found elsewhere.[8])

Surprisingly, increasing the nitrogen content above a volume ratio of ~3:1 as well as reducing it below ~1:3 cause the same broad PL band (curve (c) in Fig. 1) centered around 2.36 eV (525 nm). These emissions are identical to those found after spark-processing in stagnant air. The green PL intensity can be considerably enhanced when the specimen is heated during spark-processing.

The peak positions for both PL bands stay essentially fixed and do not shift with processing time or temperature. For certain conditions, both bands (UV/blue and green peaks) can be observed for the same sample. It is however, important to stress that no gradual peak shift or transition occurs.

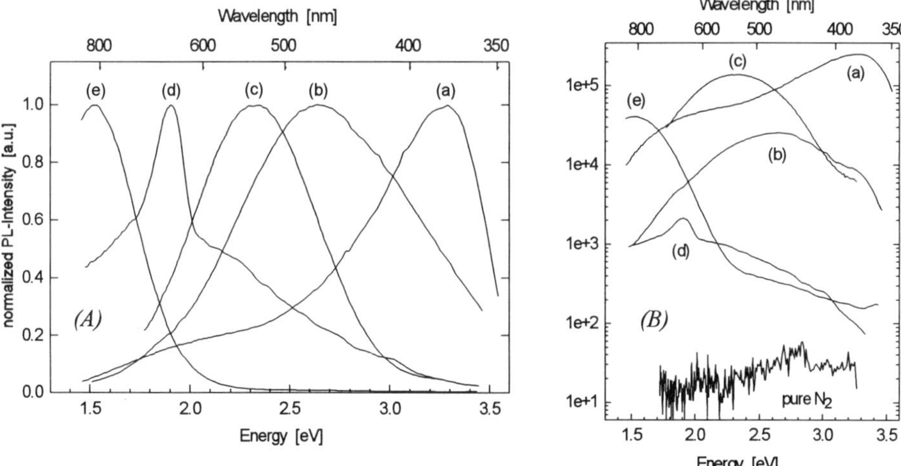

Fig. 1: *(A)* Normalized PL spectra for spark-processed Si, prepared (a) in a 1:1 oxygen/nitrogen mixture or under flowing air, (b) in pure oxygen and subjected to low pressure, (c) in nitrogen/oxygen atmospheres with volume ratios of >3:1 or <1:3 or under stagnant air conditions, (d) in pure oxygen, and (e) in oxygen and annealed in oxygen at 850°C.
(B) Comparison of PL intensities for the bands (a) to (e). All samples were processed for 30 min and measured under identical conditions (note the logarithmic scale).

If, however, the nitrogen content is more and more enlarged, the PL intensity decreases significantly and diminishes eventually. Under very stringent conditions, by avoiding any contaminations by oxygen in the processing chamber, the reaction still causes a surface erosion. However, green PL is no longer discernable. That is, spark-processing of Si in pure nitrogen generates a non-luminescent material. Only weak PL near 2.7 eV (460 nm) appears at room temperature after storage in air. The latter PL is probably caused by an oxidation of the air-exposed rough surface during the PL measurement. Emissions around 2.7 eV are a common feature of oxygen-vacancy defects in SiO_2 [11].

Processing in pure oxygen leads to a completely different spectrum as displayed by curve (d) in Fig. 1. Superimposed on a broad background luminescence, a PL peak appears at 1.89 eV (~650 nm). This emission can be easily distinguished as orange luminescence by the naked eye even though it is about two orders of magnitude lower in intensity when compared with the green or UV/blue PL bands. Such a 1.9-eV PL has been repeatedly ascribed to radiative transitions via non-bridging oxygen hole centers [12]. Apparently, these structures may develop in sp-Si as a result of the rapid quenching between flash-evaporations initiated by the spark discharge. However, the 1.9-eV PL peak of sp-Si is highly unstable with time and under UV illumination and bears more similarities to the radiative relaxation of excited oxygen atoms (O ^{1}D to O ^{3}P transition) originating from the dissociation of ozone molecules [13]. Generally, ozone is generated during spark-processing to a large extent, and may therefore become a constituent of the matrix, either as interstitial molecule or by being trapped in microcavities.

The instability of the 1.9-eV band shows again when the specimen is subjected to lower pressure. Under such conditions, a more intense blue emission (curve (b) in Fig. 1) emerges, centered at 2.61 eV (475 nm). The same blue PL band is also observed, when the sample is annealed up to 350 °C instead of stored in vacuum. At higher annealing temperatures (above 400 °C), the peak changes again, peaking now around 1.55 eV (800 nm). The PL spectrum is shown by curve (e) in Fig. 1.

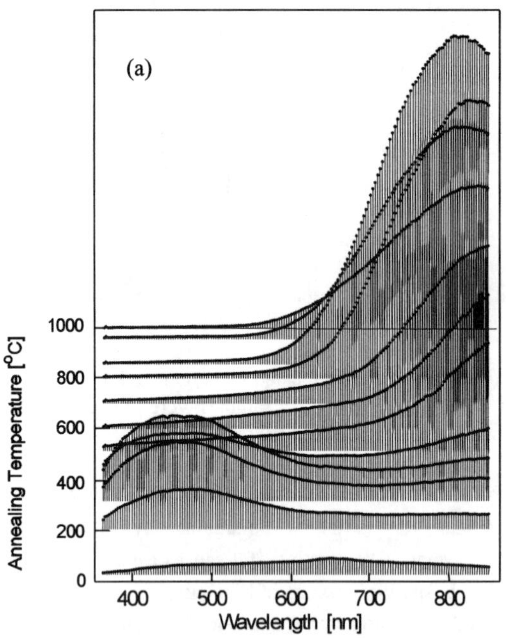

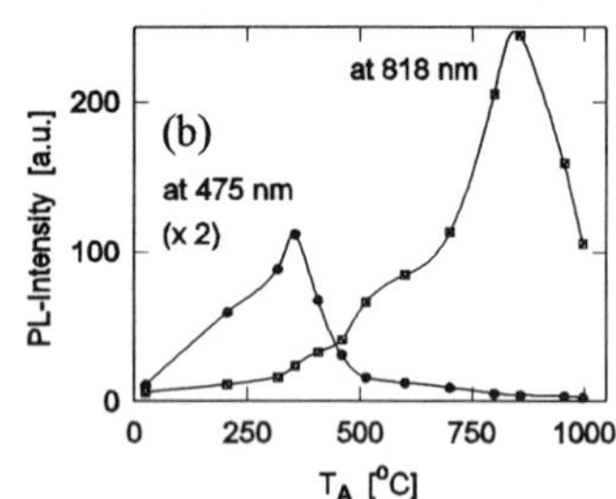

Fig. 2: (a) 3D display of a series of PL spectra taken at room temperature on sp-Si which was prepared in pure oxygen and successively annealed in oxygen for 5 min at each step. (b) Variation of PL intensities with annealing temperature T_A for the indicated wavelengths.

Fig. 2 depicts the changes of PL characteristics with annealing temperature for sp-Si prepared in pure oxygen. The initial 1.9-eV peak of the as-prepared specimen becomes rapidly dwarfed by the blue PL band centered at 2.6 eV (475 nm). At annealing temperatures (T_A) above 500°C the blue luminescence is not longer discernable and the spectrum is then dominated by a broad and intense PL in the infrared range. As shown in more detail in Fig. 2b, the PL intensity at 475 nm increases up to T_A's of 350°C and drops considerably at higher temperatures. Such an intensity vs. T_A dependence shows similarities with that of silanol groups as reported by Tamura *et al.* [14]. FTIR spectra, taken from sp-Si prepared in oxygen and after several annealings in oxygen, revealed, however, that the most prominent Si-OH vibrational modes were absent. Only the HSi-O absorption line was recorded, though essentially unchanged over the temperature range employed. It can thus be argued that oxygen deficiency centers, as proposed above, may be responsible for the blue emission band. The IR band, having highest intensities after annealings at 900°C, carries features which are similar to the IR luminescence found in strongly oxidized por-Si [15].

Interestingly enough, the 2.6-eV blue PL shows a completely different temperature dependence when compared with the UV/blue and green bands of sp-Si. The latter display a well known temperature activated quenching of intensity with rising temperature [16] (see Fig. 3). For the blue PL an inverse behavior was observed. The PL intensity first increases with temperature, having a maximum between 200 and 300 K, and lowers at higher temperatures again. A similar unusual characteristic has been reported for the 2.7-eV band in SiO_2 and assigned to radiative transitions via oxygen deficiency centers [17]. For comparison, Fig. 3 includes also the temperature dependence of the blue cathodoluminescence (CL) band which can be found in all sp-Si specimens, independent of the particular preparation conditions [18].

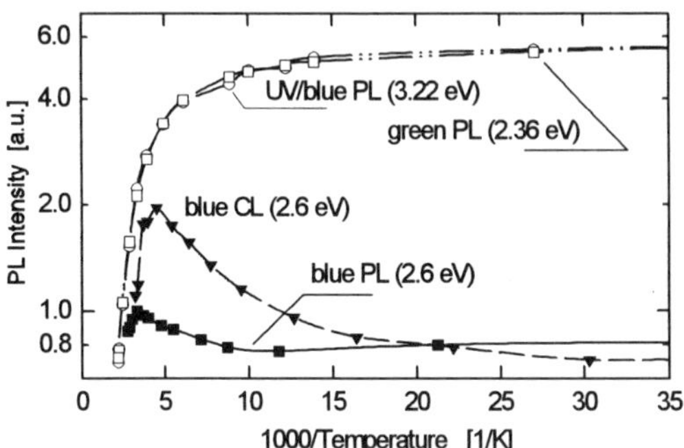

Fig. 3: PL intensities vs. temperature for the UV/blue (3.22 eV), green (2.36 eV), and for the blue (2.61 eV) emission bands of sp-Si. For comparison, the temperature dependence of the blue CL band is added (from ref. [18]).

CONCLUSIONS

In summary, five different PL bands can be distinguished in sp-Si when the preparation is performed in air, oxygen, nitrogen, or mixtures of both gases. Each of those PL bands depends in peak position and intensity on the composition of the processing gas. More specifically, the volume ratios between N_2 and O_2 apparently determines which of the most intense PL bands dominates, the one with UV/blue emission at 3.22 eV or the green band centered at 2.4 eV. Both gaseous components are, however, necessary and probably essential constituents of the luminescing center. It is, therefore, more likely that the radiative center is related to an effect caused by the coimplantation of nitrogen and oxygen ions. Preparations in pure O_2 or in pure N_2 causes completely different PL bands, which are unstable and much less intense, or not luminescent at all, respectively. Furthermore, the distinctive temperature dependence for the 2.6-eV PL and CL band indicates that the emission process is not related to that of the UV/blue and green PL bands.

The above described experiments demonstrate that the intense UV/blue and green PL bands found for sp-Si prepared in air cannot be related to radiative transitions in silicon oxides or silicon nitrides because these bands are absent when spark-processing was performed in pure oxygen or nitrogen, respectively. Moreover, possible contaminations by the counterelectrode alone cannot account for the observed multi-color effects in spark-processed Si.

The authors are indebted to NSF-DMR and to AFOSR for financial assistance and to Wacker Chemitronic for providing Si wafers. One of the authors (M.H.L.) gratefully acknowledges the financial support by a Habilitationsstipendium of the Deutsche Forschungsgemeinschaft.

REFERENCES

1. R. E. Hummel and S-S. Chang, Appl. Phys. **61**, 1965 (1992)
2. D. Rüter and W. Bauhofer, J. Luminescence **57**, 19 (1993)
3. M. H. Ludwig, A. Augustin, R. E. Hummel and Th. Gross, J. Appl. Phys. **80**, 5318 (1996)
4. R. E. Hummel, M. H. Ludwig, J. Hack and S.-S. Chang, Sol. State Commun. **96**, 683 (1995)
5. R. E. Hummel, M. H. Ludwig, S.-S. Chang, P. M. Fauchet, Ju. V. Vandyshev, and L. Tsybeskov, Sol. State Commun. **95**, 553 (1995)
6. D. J. Wolford, J. A. Reimer and B. A. Scott, Appl. Phys. Lett. **42**, 369 (1983)
7. M. S. Brandt, H. D. Fuchs, M. Stutzmann and M. Cardona, Sol. State Commun. **81**, 307 (1992)
8. M. H. Ludwig in Handbook of Optical Properties, Vol. II: Optics of Small Particles, Interfaces and Surfaces, CRC Press, Boca Raton, 1996, chapter 5
9. D. L. Griscom. J. Ceramic Soc. of Jap. **99**, 923 (1991)
10. E. F. Steigmeier, H. Auderset, B. Delley and R. Morf, J. Luminescence **57**, 9 (1993)
11. R. Tohmon, Y. Shimogaichi, H. Mizumo, Y. Ohki, K. Nagasawa and Y. Hama, Phys. Rev. Lett. **62**, 1388 (1989)
12. L. Skuja, Sol. State Commun. **84**, 613 (1992)
13. K. Awazu and H. Kawazoe, J. Appl. Phys. **68**, 3584 (1990)
14. H. Tamura, M. Rückschloss, T. Wirschen and S. Veprek, Appl. Phys. Lett. **65**, 1537 (1994)
15. Y. Kanemitsu, H. Uto, Y. Matsumoto, T. Matsumoto, T. Futagi and H. Mimura, Phys. Rev. B **48**, 2827 (1993)
16. M. H. Ludwig, R. E. Hummel, A. Augustin, J. Hack and J. Menniger, Appl. Phys. Lett. **67**, 2542 (1995)
17. L.-S. Liao, X.-M. Bao, X.-Q. Zheng, N. S. Li and N.-B. Min, Appl. Phys. Lett. **68**, 850 (1996)
18. M. H. Ludwig, J. Menniger and R. E. Hummel, J. Phys.: Condens. Matter 7, 9081 (1995)

PHOTOLUMINESCENCE CHARACTERISTICS OF HF-TREATED SILICON NANOCRYSTALS

S. NOZAKI, S. SATO, H. ONO and H. MORISAKI
Department of Communications and Systems, The University of Electro-Communications, 1-5-1 Chofugaoka, Chofu-shi, Tokyo 182, Japan, nozaki@cas.uec.ac.jp

ABSTRACT

Si nanocrystals were deposited in a helium atmosphere by the gas-evaporation technique. Their average size is 3.5 nm, much smaller than those of the Si nanocrystals deposited in an argon atmosphere. The PL spectra of the as-deposited and the HF-treated Si nanocrystals were compared. A great increase in the PL intensity of the HF-treated Si nanocrystals is attributed to the hydrogen passivation of Si surface dangling bonds. A good correlation between the amount of Si-O bonds and the PL intensity suggests that the oxygen-passivation of dangling bonds is required for the red-band PL. The PL spectra of the HF-treated Si nanocrystals resemble those of porous Si and clearly indicate that the HF-treated Si nanocrystals well simulate the porous Si.

INTRODUCTION

Recently there has been great interest in the visible-light emission from various forms of silicon (Si) nanostructures such as porous Si, Si nanocrystals embedded in SiO_2 and Si nanocrystals deposited by the gas-evaporation technique. Although the luminescence mechanism of Si nanostructures is still a subject of debate at present, it is clear that the surface plays an important role in the luminescence process because of a large surface-to-volume ratio. Kumar et al. calculated the bandgap of Si nanostructures whose surface dangling bonds were terminated by H, F, or O atoms and concluded that the oxygen surface termination of the Si nanostructures was one of the requisite conditions for obtaining the luminescence in the red region [1]. According to the reports on porous Si, the photoluminescence (PL) intensity was proportional to the absorption in the Si-H mode [2], and the PL which was quenched by annealing could be recovered by dipping in HF solution [3]. However, understanding the mechanism of visible light emission from porous Si is further complicated by presence of amorphous Si which may surround Si nanocrystals and many chemical species induced by the anodization process. Furthermore, even if Mimura et al. found a correlation between H or O atoms and the luminescence spectra [4], it is difficult to determine whether these atoms terminated dangling bonds on the surface of Si nanocrystals or the surface of a porous Si layer. In this paper, we propose Si nanocrystals whose surface was hydrogen-passivated by dipping in HF solution as a model to simulate porous Si and deduce the mechanism of visible-light emission from porous Si from the study of the luminescence from such a simpler system.

EXPERIMENTAL

The Si nanocrystals were deposited on (100) p-type Si substrates by the gas-evaporation technique. Although the apparatus was the same as that reported previously [5, 6], pure helium gas was used as an ambient gas during evaporation of Si in order to obtain smaller-size Si nanocrystals. The He pressure in the chamber was kept at 10 Torr during the deposition. Some samples were immersed in a 46 % HF solution for 3 sec immediately after the preparation of Si

Mat. Res. Soc. Symp. Proc. Vol. 452

nanocrystals by the gas-evaporation technique. After this HF treatment, the samples were blown dry in N_2 flow. The chemical analyses were made by x-ray photoelectron spectroscopy (XPS) and Fourier-transform infrared spectroscopy (FTIR) with an ESCA-K1 (Shimadzu) and a Spectrum 2000 (Perkin Elmer), respectively. The PL measurements were carried out at room temperature with the 313 nm line selected from the incident light of a mercury lamp as an excitation source using an optical band-pass filter.

RESULTS AND DISCUSSION

The average size of the Si nanocrystals estimated from the width of the x-ray diffraction peaks is 3.5 nm, which is consistent with the size determined by transmission electron microscopy (TEM). The size and its distribution are much smaller and narrower, respectively, than those of the Si nanocrystals deposited in an argon (Ar) atmosphere [5], as expected from a study of metal particles deposited in Ar, He and xenon (Xe) atmospheres [7].

Figures 1 (a) and (b) show the Si2p XPS spectra of the samples before and after the HF treatment, respectively. The depth profiles of the spectra were measured after sputter-etching by Ar ion bombardment (2 kV, 20 mA) for various periods of time as indicated in the figures. It is noted that the as-deposited Si nanocrystals contain a large amount of native oxide, while the sample which was dipped in the HF solution shows no trace of oxide. However, it should be pointed out that the Si nanocrystals deposited in an Ar atmosphere contain a larger amount of native oxide and are hardly left after the HF treatment. Less natural oxidation for a He atmosphere can be understood by the experimental finding that the oxidation rate of the smaller-sized Si nanocrystals was less [8].

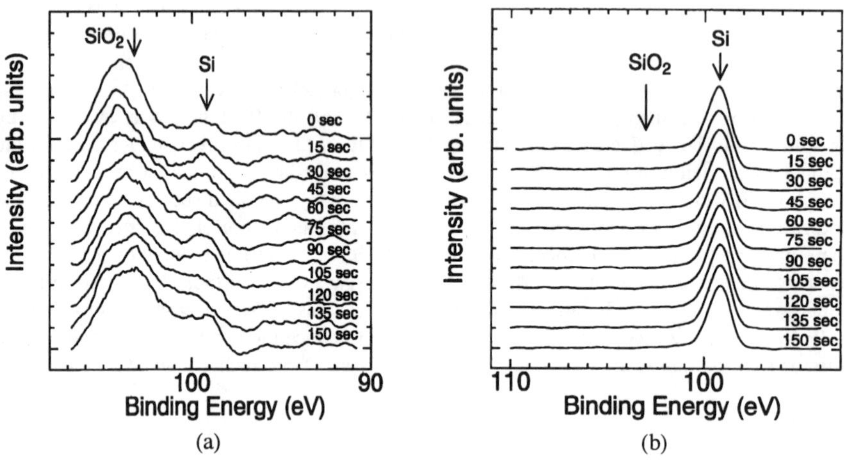

Fig. 1. Si2p XPS spectra of (a) the as-deposited and (b) HF-treated Si nanocrystals. In (b) the measurement was carried out immediately after the HF treatment. The XPS spectra were collected after sputter-etching for various periods of time as shown in each spectrum.

The FTIR spectra of the Si nanocrystals measured after the HF treatment are shown in Fig. 2. Prior to the measurements, the samples were left in the air for various periods of time. Both SiH_n - and SiO-related peaks are extremely small immediately after the HF treatment (spectrum (a)) and gradually grow with time. The intensities of the SiH_2 scissors and Si-H stretching modes at 910 and 2100 cm^{-1}, respectively, saturate 30 min after the HF treatment (spectrum (b)), while that of the Si-O-Si asymmetric stretching mode at 1068 cm^{-1} continues to increase even 60 min after the HF treatment (spectra (c) and (d)). A small amount of H atoms at the surface present immediately after the HF treatment is rather surprising, and it suggests that the passivation of Si-dangling bonds by H atoms becomes significant only after removal of HF vapor and water from the Si nanocrystals or redistribution of H atoms to the dangling bonds [9].

Figure 3 shows the PL spectra of the Si nanocrystals before and 10 min after the HF treatment. As discussed later, the PL intensity is extremely low immediately after the HF treatment and then gradually increases with time. Although the PL peak energies are almost the same in both spectra, the spectral width becomes considerably narrower after the HF treatment. It is rather difficult to compare both intensities quantitatively, because the thickness of the Si-nanocrystal layer is reduced to less than one hundredth by dipping in the HF solution. Nevertheless, if we consider that the as-measured PL intensity of the HF-treated sample is 2-20 times stronger than that of the as-deposited one, it can be said that the PL intensity is increased at least by two orders of magnitude by the HF treatment.

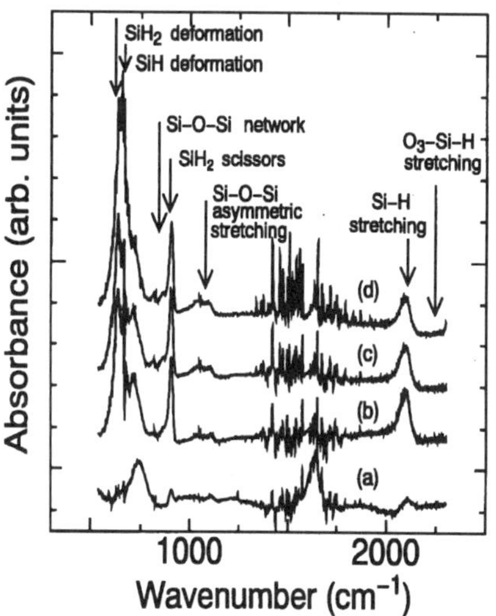

Fig. 2. FTIR spectra of the HF-treated Si nanocrystals, which were exposed to the air for various periods of time: (a) 0 min, (b) 30 min, (c) 60 min, and (d) 90 min.

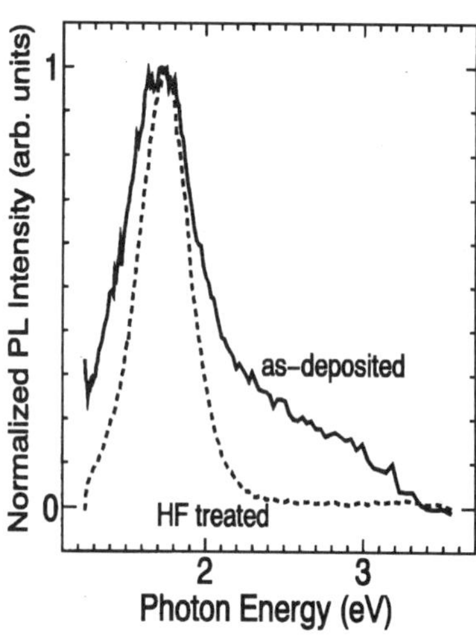

Fig. 3. PL spectra of the as-deposited and HF-treated Si nanocrystals. The PL spectrum of the latter sample was collected 10 min after the HF treatment.

Both the PL peak intensity and PL peak energy increase with the elapsed time after the HF treatment as shown in Fig. 4. As mentioned earlier, it takes time for the sample to become luminescent, and the intensity gradually increases with time. Since the PL peak energy does not show a significant change after the HF treatment, the origins of the luminescence before and after the HF treatment are considered to be the same. From the evidence that there is no hydrogen atoms present before the HF treatment, it is clear that hydrogen passivation is not required for the luminescence. However, the increased intensities of the SiH_2 scissors and Si-H stretching modes result in an increase of PL peak intensity. For an elapsed time after the HF treatment longer than 30 min, the PL intensity still continues to increase, while the amount of SiH_2 and Si-H bonds does not show a significant change. In contrast to hydrogen passivation, the intensity of Si-O-Si asymmetric stretching mode continues to increase, and the amount of Si-O-Si bonds seems to exhibit a better correlation with the PL intensity. As theoretically predicted by Kumar et al., it can be concluded that the oxygen passivation of the surface of Si nanocrystals is one of the requisite condition to obtain luminescence in the red region, while the hydrogen passivation removes the Si dangling bonds which may act as nonradiative recombination centers.

A blue shift of the PL peak energy similar to one shown in Fig. 4 was also observed in porous Si exposed to the ambient air by Maruyama and Ohtani [10], who attributed the shift to room-temperature oxidation. In addition to a blue shift of the PL peak energy in the red region,

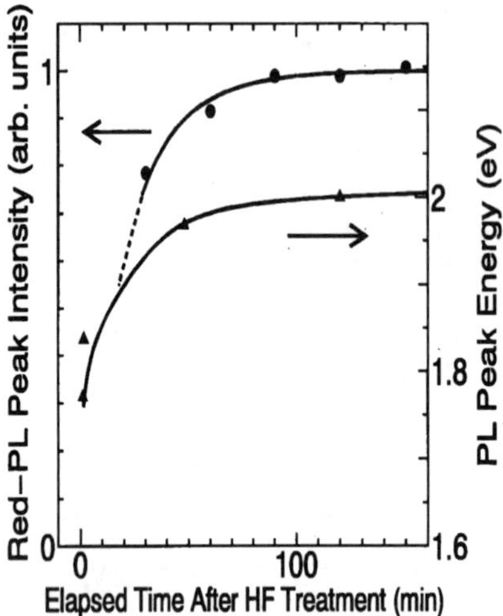

Fig. 4. Change of the peak intensity and peak energy in the red-band PL with air-exposure time after the HF treatment.

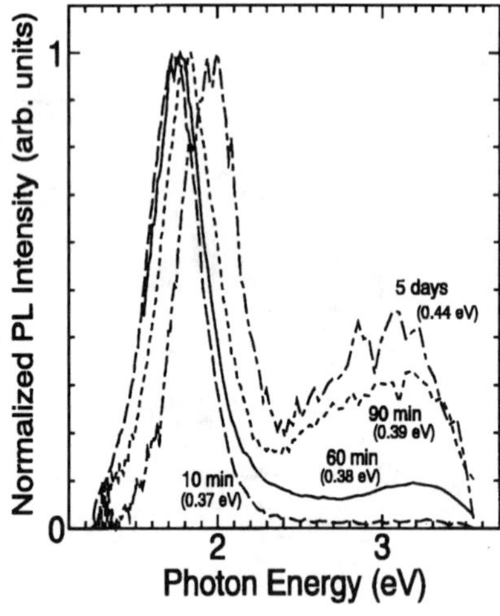

Fig. 5. PL spectra of the HF-treated Si nanocrystals which were exposed to the air for various periods of time. The numbers in the parentheses are FWHM's of the peak in the red-band PL.

the blue-band PL emerges 60 min after the HF treatment and grows slowly with the air-exposure time, as shown in Fig. 5. A similar blue-band PL was observed in the oxidized porous Si [4, 11]. Both red- and blue-band PL spectra observed in the HF-treated Si nanocrystals are similar to those in porous Si. A similarity in the luminescence spectra of the HF-treated Si nanocrystals and porous Si suggests that the Si nanocrystals passivated with H and O atoms in the porous Si are responsible for the visible light luminescence in porous Si, and that the HF-treated Si nanocrystals well simulate porous Si. The surface modification by H and O atoms yields red- or blue-band PL in the Si nanocrystals and porous Si.

CONCLUSIONS

The Si nanocrystals deposited in a He atmosphere by the gas-evaporation technique yield sizes much smaller than those deposited in an Ar atmosphere. Some of the samples were dipped in HF solution, and their PL spectra were compared with those before the HF treatment. The PL intensity is increased by more than 100 times after hydrogen passivation by the HF treatment. However, the same PL peak energy before and after the HF treatment suggests that the hydrogen passivation of Si nanocrystals is not required. A good correlation between the amount of Si-O bonds and the PL peak intensity supports the importance of oxygen at the surface in observing the red-band luminescence from Si nanocrystals. The hydrogen passivation of the surface appears to terminate Si dangling bonds, which may act as nonradiative recombination centers and contribute to the increased PL intensity in the red band. The blue-band PL emerges after long air-exposure of the HF-treated Si nanocrystals. These red- and blue-band PL spectra are similar to those observed in the porous Si. The roles of H and O atoms in the luminescence process discussed for the HF-treated Si nanocrystals can be applied to porous Si. It is deduced from the study that the luminescence from porous Si is also attributed to passivation of the surface of Si nanocrystals but not of the porous layer with H and O atoms.

ACKNOWLEDGMENTS

The authors thank Professor N. Ito of the University of Osaka prefecture for the X-ray diffraction measurements. This study was partly supported by research grants from the Ogasawara Foundation for the Promotion of Science and Engineering, the Mechanical Industry Development and Assistance Foundation and the Yazaki Science Promotion Foundation, the Special Coordination Funds for Promoting Science and Technology entitled "Research on New Material Development by Nanospace Lab" and by a Grant-in-Aid for Scientific Research, Ministry of Education, Science and Culture.

REFERENCES

1. R. Kumar, Y. Kitoh, K. Shigematsu and K. Hara, Jpn. J. Appl. Phys. **33**, 909 (1994).

2. A. I. Belogorokhov, V. A. Karavanskii, L. I. Belogorokhova and A. N. Obraztsov, Semiconductors **28**, 800 (1994).

3. S. M. Prokes, O. J. Glembocki, V. M. Bermudez and R. Kaplan, Phys. Rev. B **45**, 13788 (1992).

4. H. Mimura, T. Futagi, T. Matsumoto, T. Nakamura and Y. Kanemitsu, Jpn. J. Appl. Phys. **33**, 586 (1994).

5. H. Morisaki, F. W. Ping, H. Ono, and K. Yazawa, J. Appl. Phys. **40**, 1869 (1991).

6. S. Nozaki, S. Sato, H. Ono and H. Morisaki, Mat. Res. Soc. Symp. Proc. **351**, 399 (1994).

7. N. Wada, Jpn. J. Appl. Phys. **7**, 1287 (1968).

8. R. Okada and S. Iijima, Appl. Phys. Lett. **58**, 1662 (1991).

9. L. Tsybeskov and P. M. Fauchet, Appl. Phys. Lett. **64**, 1983 (1994).

10. T. Maruyama and S. Ohtani, Appl. Phys. Lett. **65**, 1346 (1994).

11. A. J. Kontkiewicz et al., Appl. Phys. Lett. **65**, 1436 (1994).

UV AND BLUE PHOTOLUMINESCENCE FROM SILICON NANOCOLLOIDS

Shingo Iwasaki and Keisaku Kimura
Department of Material Science, Faculty of Science, Himeji Institute of Technology, 1479-1, Kamigori, Hyogo, 678-12 Japan

ABSTRACT

We have prepared silicon nanocolloids by trapping silicon nanocrystallites produced by a gas evaporation method into an organic liquid. Blue photoluminescence (PL) at a peak maximum around 480 nm from silicon nanocolloids has been observed at room temperature. The external quantum efficiency of blue PL is 0.16 %. A room temperature UV PL has been observed by dropping alkaline solids such as sodium hydroxide (NaOH), potassium hydroxide (KOH) and calcium hydroxide ($Ca(OH)_2$) into silicon nanocolloids. The mechanism of blue and UV PL from silicon nanocolloids is discussed.

INTRODUCTION

It has been reported that oxidized silicon nanoparticles embedded in SiO_2 matrix display a blue shift with decreasing size [1-3]. To the contrary, the chemical vapor deposited (CVD) films made of nanocrystalline silicon do not show any shift in the PL spectrum upon varying the particle size in the range 2-8 nm [4]. Hence it seems that the quantum size effect on the optical properties of porous silicon or nanocrystalline silicon is not fully understood theoretically and experimentally, although it has been suggested that the surface structure of porous silicon also plays an important role in visible PL [5]. A highly oxidized porous silicon or nanocrystalline sample shows, in addition to red-yellow luminescence (1.4-1.7 eV), distinct blue-green luminescence in the photon energy range of about 2.4-2.7 eV [6-9]. The origin of this blue-green PL seems no more clear than that of red-yellow PL. The crystalline silicon core may participate in this PL, because a further oxidation of silicon crystallites decreases the intensity of blue-green PL in accordance with the decreasing number of silicon cores [5]. On the other hand, it has been ascribed to structural defects in the SiO_2 layer in some reports [10,11], because similar blue-green PL also appears when certain commercial silica glasses are heated in a reducing gas and treated with water [10].

In general, the optical properties of small particles are complicated because they are a function of the dielectric constant of the surrounding media or the substrate, and also of the interparticle interaction [12,13]. Thus the preparation of well-isolated particles is strongly required in order to elucidate the origin of visible PL from silicon nanostructures. Recently, Littau et al. prepared silicon nanocrystals dispersed in ethylene glycol by gas-phase pyrolysis of disilane, and observed a significant shift of red PL depending on the particle size [14]. They also showed that the intensity of red PL of the colloidal silicon is enhanced by treatment with acidic water. Recently we have prepared colloidal silicon nanocrystallites by a gas evaporation method, and observed blue PL at room temperature [15]. The observed strong blue PL may be ascribed to a radiative recombination of excitons at a trap site near the interfacial region between the silicon core and the amorphous SiO_x layer.

It is reported in this paper that not only blue PL but also UV PL have been observed at room temperature by dropping alkaline solids (NaOH, KOH and $Ca(OH)_2$) into silicon nanocolloidal dispersions. We will discuss the mechanism of blue and UV PL from

Mat. Res. Soc. Symp. Proc. Vol. 452

silicon nanocolloids at room temperature.

EXPERIMENTAL

Silicon nanocolloid was prepared by means of gas evaporation and matrix isolation techniques [16,17]. We used He gas (99.9999 %) as an inert gas atmosphere and methanol or 2-propanol (99.9 %) as an organic liquid. A Pyrex-glass chamber was evacuated to about 2.0 $\times 10^{-6}$ Torr with a turbomolecular pump and then filled with methanol or 2-propanol vapor which had previously been degassed several times by a freeze and thaw technique. The methanol or 2-propanol vapor was frozen onto the inner wall of the glass chamber, the outer wall of which was refrigerated with liquid nitrogen, and then the glass chamber was filled with He gas after re-evacuation. Under refrigeration, silicon nanocrystallites were prepared by the resistant heating technique where small pieces of silicon with a purity of 99.9999 % were heated in a tungsten crucible covered with alumina followed by sublimation into He gas atmosphere and the formed nanocrystallites adhered to the surface of the frozen methanol or 2-propanol. As it returned to room temperature, silicon nanocrystallites were recovered as flocculants in an organic liquid, which could be dispersed with ultrasonic irradiation. The prepared silicon nanocolloid was yellow in color. We prepared five samples (Ⅰ to Ⅴ) of silicon nanocolloid with different particle sizes by varying the pressure of the He gas from 0.5 to 5 Torr. The concentration of the colloidal solution determined by measuring the weight of the evaporated silicon nanocrystallites in some of the solutions varied from 1.1 to 2.5 mg/ml. In order to normalize the concentration, the initial solutions were diluted with methanol or 2-propanol to 1.0 mg/ml for photoluminescence and 0.13 mg/ml for UV-visible absorption measurements.

The size of silicon nanocrystallites was determined from direct observation by transmission electron microscopy (TEM) using a Hitachi H-8100 system operated at 200 kV. Samples for TEM observation were prepared by dropping silicon nanocolloids onto amorphous carbon film deposited on a copper sheet mesh and leaving it to dry in air.

The chemical analysis of nanocrystallites was conducted utilizing an energy dispersive X-ray (EDX) unit attached to the TEM. The IR spectra were measured using a Horiba FT-210 system immediately after the silicon nanocolloid was dropped onto an NaCl substrate and dried in Ar gas atmosphere. The UV-visible absorption spectra of the colloidal samples were measured with a Hitachi U-3210 spectrophotometer. The photoluminescence spectra were measured with a Hitachi F-3010 fluorescence spectrophotometer excited by a 150 W xenon lamp, where the excitation energy was set at the range of ultraviolet. The external quantum efficiency was measured with fluorescence spectrophotometer by a relative measurement using Fluorescein as a standard material with the quantum efficiency of 97 %.

Several alkaline solids were tested as sodium hydroxide, purity of 96 %, potassium hydroxide, 85 % and calcium hydroxide, 99.9 %. We have used pH values of the sample solutions as a measure of alkaline concentration because there is a positive correlation between pH reading and the amount of alkaline solid added.

RESULTS AND DISCUSSION

A TEM image of the prepared nanocrystallites (sample Ⅰ) is shown in Fig.1. In this micrograph, each nanocrystallite is consistent with the (2 0 0) plane of the diamond structure. Sharp rings in the electron-diffraction pattern also indicated that the nanocrystallite is the silicon nanocrystallite with diamond structure. The average diameter of the silicon nanocrystallite was found to be 3.7 nm by measuring the number of nanocrystallites of 150 samples as a function of size in the TEM image. We prepared several samples with the average diameter of 3.7 nm, 4.8 nm (sample Ⅱ), 6.2 nm (sample Ⅲ), 7.3 nm (sample Ⅳ) and 9.8 nm (sample Ⅴ).

In the IR spectra of fresh and one-day-old silicon colloidal dispersions, broad bands due to the Si-O-Si asymmetric stretching mode changed from 1018 to 1060 cm^{-1}. If we assume the location of this peak to be indicative of the degree of oxidation of SiO_x [18], the values of x are estimated to be x = 1.2 for the fresh nanocrystallites and x = 1.7 for the one-day-old ones. This means that the surface oxidation of silicon nanocrystallites proceeded within one day even if silicon nanocrystallites were kept in an organic liquid at room temperature. The degrees of oxidation of samples Ⅱ, Ⅲ, Ⅳ and Ⅴ were almost the same as that of sample Ⅰ. This means that upper most surface of the particle was mainly oxidized.

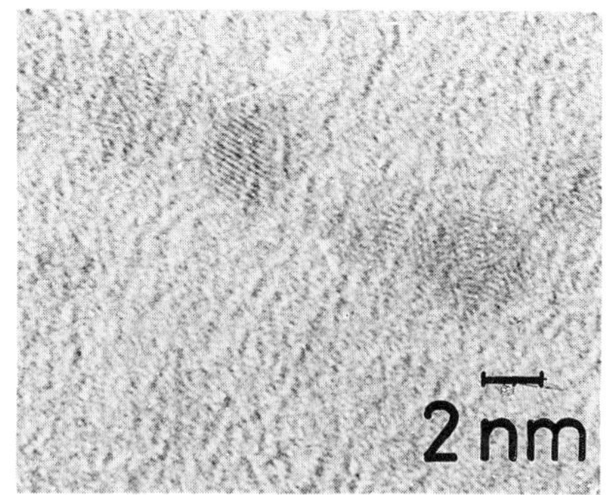

Fig.1 TEM image of silicon nanocrystallites

In the UV-visible absorption spectra, the absorption edge shifts to high energy with the decrease of the average diameter from 9.8 to 3.7 nm, which is explained in terms of a quantum size effect. The spectra of the one-day-old samples were identical to those of fresh samples. This result implies that the size of nanocrystallites does not decreases remarkably due to the spontaneous surface oxidation in an organic liquid at room temperature. This suggests that the electronic state of the silicon nanocrystallites is different from bulk silicon because it is known that the bulk surface is readily oxidized to form a few nm oxidized layer. The weak tail from about 370 nm across the visible range, which is the origin of the yellow color, may be attributed to indirect gap transitions.

The photoluminescence (PL) spectrum of fresh sample Ⅰ is shown in Fig.2. Since the excitation energy was 4.4 eV (wavelength 280 nm), the secondary diffraction of the incident light appears at about 560 nm. The broad PL band of the silicon nanocolloid appears at about 2.6 eV (wavelength 480 nm). We can confirm that it is not a secondary diffraction of Raman scattered light, because the peak position or the spectral profile of the PL band did not change when the excitation energy of the incident light was varied from 4.1 to 5.0 eV (wavelength 250 nm). This PL was so strong that the bright blue luminescent light from the sample was visible with naked eye even at room

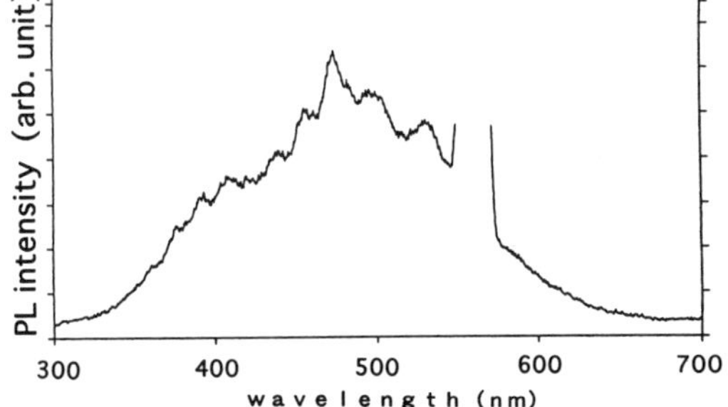

Fig.2 The photoluminescence spectrum of silicon nanocolloid (fresh sample Ⅰ) under 280 nm excitation at room temperature.

temperature. In contrast to the results reported by Littau et al. [14], no red-yellow PL was perceptible in the observed spectra. Since their preparation method was different from ours, it may be due to the difference in the microscopic structure near the surface of the the silicon nanocrystallites.

In the PL spectrum of the one-day-old silicon nanocolloid, we have found that the PL intensity at 2.6 eV (wavelength 480 nm) is further increased to about twice as that of the fresh one and that the external quantum efficiency is increased from 0.06 % to 0.16 %. Since the results of the IR measurement show the increase of the degree of oxidation of the SiO_x layer in the old samples, the enhancement of blue PL is likely to be caused by the modification of the surface structure of the silicon nanocrystallites. The PL spectral profiles and the peak position of blue PL of samples Ⅱ, Ⅲ, Ⅳ and V have been found to be almost unchanged from that of sample Ⅰ, while the PL intensity is reduced upon increasing nanocrystallite size. Figure 3 summarizes the dependence of the PL intensity as a function of the aging of the sample and the average size of silicon nanocrystallites. The size dependence of the PL intensity is primarily explained in terms of the increasing surface area ratio of smaller nanocrystallites. The size-dependent shift of the UV-visible absorption edge implies that the electronic excitation is dominated by the electronic state of the silicon nanocrystallite affected by a quantum confinement, while the luminescence process seems to have a local nature bound to the

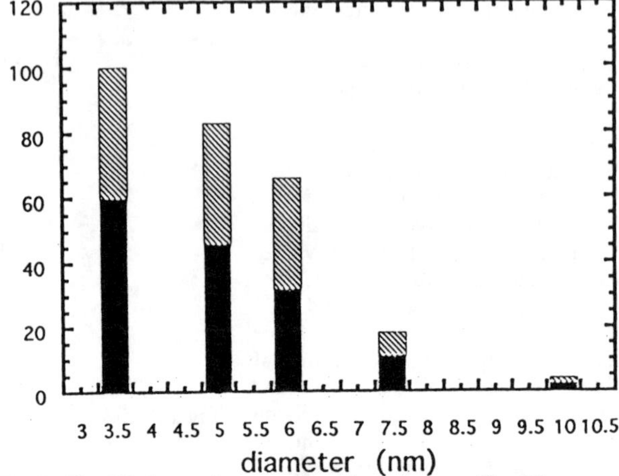

Fig.3 The PL intensity as a function of size. The PL intensity of the fresh sample is shown by solid column and that of the one-day-old sample is shown by hatched column.

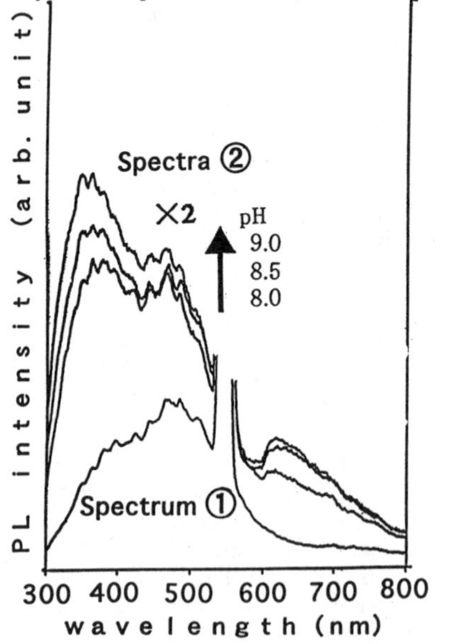

Fig.4 The photoluminescence spectra of silicon of non-alkaline treated sample (Spectrum ①; pH7.0) and treated samples (Spectra ②).

interfarcial region between the silicon nanocrystallite and the amorphous SiO_x layer. Hence blue PL may also be ascribed to a radiative recombination of excitons at a localization site near the interfacial region, as has been inferred for the mechanism of the red-yellow PL [5][19].

	fresh sample		spontaneous oxidation		alkaline sample
Quantum efficiency (%)	0.06	→	0.16	→	0.45

Scheme I The variation of the external quantum efficiency as a function of sample treatment.

Accordingly, we have dropped alkaline solids into silicon nanocolloid in order to chemically modify the electronic state of interfacial region. We have observed not only blue PL but also UV PL upon alkaline treatment at room temperature. However, acidic treatment caused no spectral shift. The PL spectrum of one-day-old sample I modified with potassium hydroxide is shown in Fig.4. Since the excitation energy was 4.6 eV (wavelength 270 nm), the secondary diffraction of the incident light appears at about 540 nm. The broad PL band of the silicon nanocolloid appears at about 3.55 eV (wavelength 350 nm) and 2.6 eV (wavelength 480 nm), the latter is coincided with that of non-treated sample. The intensity of broad band of UV PL rises up with the increase in quantity of potassium hydroxide, while the intensity of blue PL unchanged. We have found that the external quantum efficiency is increased from 0.16 % to 0.45 % at pH 9.0. Scheme I summarizes a variation of the external quantum efficiency as a function of sample treatments for sample I. We have confirmed that cation does not affect UV PL, because the peak position or the spectral profile of the PL band hardly changed at all in dropping sodium hydroxide or calcium hydroxide instead of potassium hydroxide. We have obtained the IR spectra of samples after the alkaline treatment, where broad band due to vibrational frequencies of OH groups including the SiOH group are located between 3400 cm^{-1} and 3750 cm^{-1}. The UV PL process seems to correlate with the SiOH bond or chemisorbed OH on the surface of silicon nanocrystallites. Hence UV PL may also be ascribed to a radiative recombination of exicitons at a localizated site on the surface SiOH region as well as blue PL.

CONCLUSION

Silicon nanocolloids were prepared by trapping silicon nanocrystallites produced by inert gas evaporation into an organic liquid. Strong blue and UV photoluminescence from silicon nanocolloids have been observed at room temperature. The photo absorption process seems to be dominated by the electronic state of silicon nanocrystallite cores. The origin of blue PL seems to be ascribed to a radiative recombination of excitons at a localization site near the interfacial region between the silicon nanocrystallite and the amorphous SiO_x layer, while the origin of UV PL seems to be ascribed to that on the surface SiOH region.

ACKNOWLEDGEMENT

The anthors' thanks are due to a Grant-in-Aid for Scientific Research from the Ministry of Education, Science, Sports and Culture, Japan and to a financial support from Hosokawa Powder Technology Foundation.

REFERENCES

1. H. Takagi, H. Ogawa, Y. Yamazaki, A. Ishizaki and T. Nakagiri, Appl. Phys. Lett. **56**, 2379 (1990)
2. Y. Osaka, K. Tsunetomo, F. Toyomura, H. Myoren and K. Kohno, Jpn. J. Appl. Phys. **31**, L365 (1992)
3. H. Morisaki, H. Hashimoto, F. W. Ping, H. Nozawa and H. Ono, J. Appl. Phys. **74**, 297 (1993)
4. M. Ruckschloss, B. Landkammer and S. Veprek, Appl. Phys. Lett. **63**, 1474 (1993)
5. Y. Kanemitsu, T. Futagi, T. Matsumoto and H. Mimura, Phys. Rev. **B49**, 14732 (1994)
6. M. Ruckschloss, O. Ambacher and S. Veprek, J. Lumin. **57**, 1 (1993)
7. H. Tamura, M. Ruckschloss, T. Wirschem and S. Veprek, Appl. Phys. Lett. 65, 1537 (1994)
8. D. I. Kovalev, I. D. Yaroshetzkii, T. Muschik, V. Petrova-Koch and F. Koch, Appl. Phys. Lett. **64**, 214 (1994)
9. T. Ito, T. Ohta and A. Hiraki, Jpn. J. Appl. Phys. **31**, L1 (1992)
10. H. Tamura, M. Ruckschloss, T. Wirschem and S. Veprek, Appl. Phys. Lett. **65**, 1537 (1994)
11. K. Kohno, Y. Osaka, F. Toyomura and H. Katayama, Jpn. J. Appl. Phys. **33**, 6616 (1994)
12. T. Yamaguchi, S. Yoshida and A. Kinbara, Thin Solid Films **21**, 173 (1974)
13. H. Ishikawa, T. Ida and K. Kimura, Surf. Rev. and Lett. **3**, 1153 (1996)
14. K. A. Littau, P. J. Szajowski, A. J. Muller and L. E. Brus, J. Phys. Chem. **97**, 1224 (1993)
15. S. Iwasaki, T. Ida and K. Kimura, Jpn. J. Appl. Phys. **35**, L551 (1996)
16. N. Wada and M. Ichikawa, J. Appl. Phys. **15**, 755 (1976)
17. K. Kimura and S. Bandow, Bull. Chem. Soc. Jpn. **56**, 3578 (1983)
18. S. Hayashi, S. Kawata, H. M. Kim and K. Yamamoto, Jpn. J. Appl. Phys. **32**, 4870 (1993)
19. Y. Kanemitsu, T. Ogawa, K. Shiraishi and K. Takeda, Phys. Rev. **B48**, 4883 (1993)

SOFT X-RAY EMISSION STUDIES OF THE ELECTRONIC STRUCTURE IN SILICON NANOCLUSTERS

T. VAN BUUREN*, L. N. DINH*, L.L.CHASE*, W. J. SIEKHAUS*, I. JIMENEZ*, L.J. TERMINELLO*, M. GRUSH **, T.A. CALLCOTT**, J.A. CARLISLE †
* Chemistry and Materials Science Department, Lawrence Livermore National Laboratory, Livermore, CA , 94556
** Department of Physics, University of Tennessee, Knoxville, TN 37996
† Department of Physics, Virginia Commonwealth University, Richmond, VA 23284-2000

ABSTRACT

Density of states changes in the valence and conduction band of silicon nanoclusters were monitored using soft x-ray emission and absorption spectroscopy as a function of cluster size. A progressive increase in the valence band edge toward lower energy is found for clusters with decreasing diameters. A similar but smaller shift is observed in the near-edge x-ray absorption data of the silicon nanoclusters.

INTRODUCTION

There are two main models for explaining the visible luminescence found in porous silicon. The first model " quantum confinement " is based on the idea that the restricted size of the nanometer scale silicon particles alters the band structure relative to bulk silicon [1]. In the competing model a surface layer or surface defect is responsible for the visible luminescence properties [2,3]. Recently it has been proposed that a combination of both models is needed to describe the optical properties found in porous silicon [4]. In order to shed more light on this matter, it seems useful to first examine more controllable systems, such as silicon nanocrystals with well defined sizes. In this paper, we show how Si nanoclusters with a narrow size dispersion can be synthesized in a well-controlled environment. The electronic structure of the silicon nanoclusters is then investigated using soft x-ray emission (SXE) and x-ray absorption spectroscopies. A prediction of the quantum confinement model is that the energies of the valence band (VB) and conduction band (CB) are shifted relative to the bands of bulk silicon. In order to overcome the charging and surface sensitivity problems of the electron spectroscopies we have used a method based on soft x-ray emission. SXE is explicitly a bulk-sensitive probe of the electronic structure. Photons penetrate many atomic layers deep into the material, as compared to a few atomic layers for electrons.

For the determination of the occupied valence electronic states we have measured the spectral distribution of the energy-resolved soft x-ray fluorescence (SXF) radiation generated in the transitions of the Si valence band electrons to the Si $2p_{3/2}$ core level. The core hole is selectively excited using monochromatized synchrotron radiation. Due to the photon in, photon out process the experiment is not affected by sample charging. With selective excitation it is furthermore feasible to separate the emission from Si atoms with and without oxygen neighbors, due to the chemical shift induced by the oxygen atoms.

Near-edge x-ray absorption fine structure spectroscopy (NEXAFS) measured with total photon yield can give complimentary information about the unoccupied states. This method is also bulk sensitive and is element specific since the incident x-ray photons excite electrons from selected core levels. The NEXAFS spectra were acquired by measuring the total Si L_{2-3} emission yield with the same detector used for the fluorescence. SXF and NEXAFS measurements were performed on silicon nanoclusters synthesized in-situ on beamline 8.0 at the Advanced Light Sources. The beamline 8.0 and the fluorescence spectrometer are described in detail elsewhere [5].

SAMPLE PREPARATION

Silicon nanoclusters were synthesized by thermal vaporization of Si in an inert buffer gas [6]. The synthesis was conducted in a chamber with a base pressure of < 1 x 10^{-8} Torr. Argon was leaked into the chamber up to the desired pressure. Pre-melted and outgassed Si inside a carbon boat, which was fitted inside a W basket, was resistively heated to a few hundred degrees above its melting point. Growth temperature was estimated with an optical pyrometer. The

Mat. Res. Soc. Symp. Proc. Vol. 452

substrate used to collect the silicon nanocluster was a (111) oriented Ge wafer with a native oxide layer, mounted two inches directly above the evaporation boat. A shutter was placed between the evaporation boat and the substrate to control the Si arrival fluences. The size of the Si clusters was varied by increasing (or decreasing) the Si source temperature or the pressure of the Ar buffer gas. After synthesis, the Ar buffer gas was pumped out of the chamber and the as prepared silicon clusters were passivated by exposure to atomic hydrogen for 5 minutes. The atomic hydrogen source was molecular hydrogen at $1x10^{-5}$ Torr passed through a very fine tungsten mesh at 2000°C. After synthesis the samples were transferred into the ultra high vacuum (UHV) analysis chamber at the ALS beamline 8.0 without exposure to air.

To measure the size of the clusters a highly oriented pyrolitic graphite (HOPG) substrate was used as a witness sample for each deposition on Ge. The basal plane of graphite was chosen as a substrate because all the carbon bonds in this plane are satisfied, also the surface is atomically flat making it ideal for subsequent scanning tunneling microscope (STM) or atomic force microscope (AFM) measurements of clusters size. Characterization of the size and morphology of the gsynthesized material was done in situ using a STM and ex-situ using atomic force microscopy. In Fig. 1(a) we show an AFM image of silicon clusters deposited on HOPG by evaporation of Si at 1700°C in an argon buffer gas of 112 mTorr. We find the silicon nanoclusters gather at the step edges on the graphite surface or assemble into snowflake like superclusters. STM on one of the superclusters shows that each snowflake like structures is made up of many individual nanoclusters as reported earlier [4]. X-ray diffraction and high resolution transmission electron microscopy work

Fig.1(a)

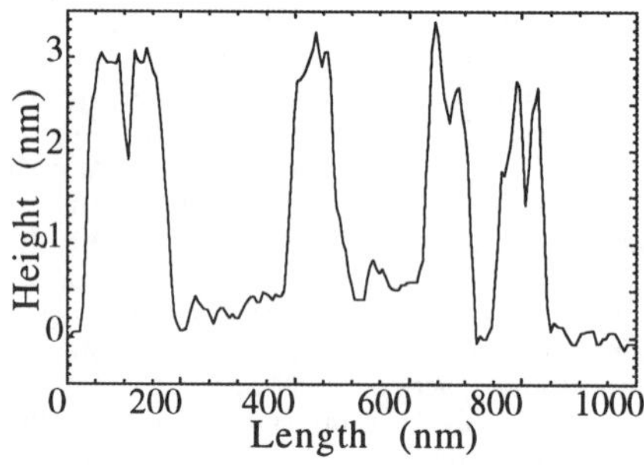

Fig.1(b)

Fig. 1(a). AFM image of silicon nanoclusters on the basal plane of HOPG formed by evaporation of Si at 1700°C in an argon buffer gas of 112 mTorr. Note how the nanoclusters gather at the step edges on the graphite surface or assemble into snowflake like superclusters. In Fig. 1(b) we shown the height cross-section along the black line on the AFM image.

has revealed that the silicon nanoclusters formed in this manner are crystalline and approximately spherical in shape. As shown in Fig. 1(b), the diameter of the nanoclusters is determined from the z scale of the AFM image, which is calibrated with a set of known standards. The average diameter of the clusters shown in Fig. 1 is 3 nm, with distribution in size that is approximately 20% of the average size. The distribution shape has an asymmetric bell shape with a tail toward larger diameters. This is characteristic of the log-normal distribution found in ultrafine metal particles formed by evaporation in a reduced atmosphere of an inert gas [6]. Since the AFM is done in air, the silicon clusters will oxidize.

RESULTS

Two Si cluster samples were grown by evaporation of Si at 1700°C in an argon buffer gas of 40 and 60 mTorr and deposited on a Ge substrate. AFM measurements after spectroscopic characterization show that the average diameter of the clusters is 1.6 nm in the 40 mTorr sample and 2.0 nm in the 60 mTorr sample. In Fig.2 we show the NEXAFS spectra of the $L_{2,3}$ edge absorption for bulk silicon and the two silicon nanocluster samples. The bulk silicon sample was oriented perpendicular to the incident photon flux and grazing incidence to the fluorescence detector to minimize the effects of self absorption in the total fluorescence yield NEXAFS spectra. The $L_{2,3}$ -edge of the cluster samples is shifted to higher energy relative to the bulk silicon by 1.4 eV for the 2.0 nm clusters and 2.2 eV for the 1.6 nm clusters, in agreement with quantum confinement which raises the energy of the bottom of the conduction band as the nanocluster particle size is decreased. In addition the well defined double threshold behavior associated with the 0.6 eV splitting of the Si 2p core level is less pronounced in the L-edge of the nanoclusters. We also note that the onset of the absorption edge in the clusters is not as sharp as that in the bulk silicon. We attribute these features to a distribution of quantum shifts caused by the variation of particle size within the sample. This effect has been modeled by the absorption edge of bulk silicon shifted in energy to simulate the average quantum shift and broadened by convolution of a Gaussian determined by the size distribution measured in the AFM images [7].

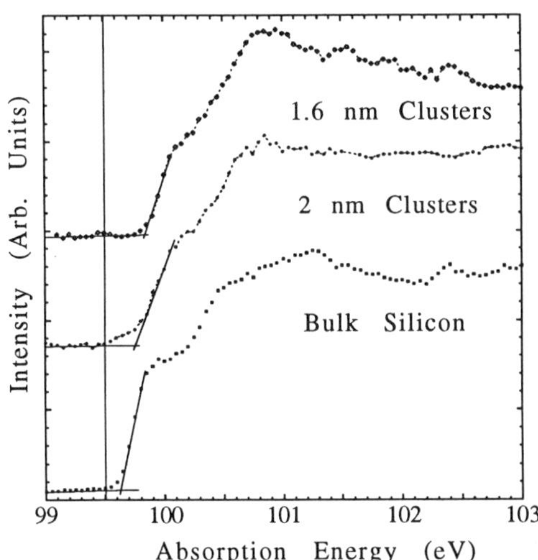

Fig. 2. Fluorescence yield x-ray absorption spectra at the silicon $L_{2,3}$ edges in bulk silicon, of silicon nanoclusters with an average size of 2.0 nm and 1.6 nm. The solid line indicates the extrapolation of the L_3 edge to baseline in order to determine the CB edge position.

We note that the L-edge absorption of the Si nanoclusters is very similar to that found in porous silicon [8,9]. Changes in the electronic structure of the VB were monitored by SXF spectroscopy on the same samples that there previously investigated by NEXAFS. In the SXF process a valence electron fills the core vacancy previously generated by the absorption of a photon. The generated fluorescence photon was analyzed in a spherical grating Rowland spectrometer. In all spectra presented the excitation energy was chosen to be below the Si L_2 absorption edge as determined from the NEXAFS spectra. Therefore, the fluorescence spectrum is generated by transitions from the valence band to the silicon $2p_{3/2}$ core hole. The SXF technique has an advantage over photoemission because it is insensitive to sample charging and is a bulk probe due to the large photon mean free path (~0.1 micron). By tuning the excitation energy below the Si L-edge absorption threshold of SiO_x we selectively investigate the electronic structure of the Si nanocluster not any surface oxide or substrate feature.

In Fig. 3 we show the SXF spectra for bulk silicon and the same two cluster samples as shown in Fig. 2. The SXF spectra were excited at 100 eV for the bulk silicon and the 1.6 nm clusters and 100.2 eV for the 2nm clusters. The bulk silicon SXF spectrum exhibits the three characteristic peaks, one at 89 eV associated with low-lying 3s states, another due to a density of states (DOS) maximum at 91.5 eV with strong s-p hybridization and a high DOS at 96 eV which is dominated by p-type states [9-11]. The intense peak at approximately 100 eV is due to reflected light from the undulator beamline into the spectrometer. The corresponding peak in the bulk Si spectrum is very weak as the reflectivity of the polished Si wafer is low in the non-specular position. Using these reflection peaks we reference the emission energy to the excitation energy and the NEXAFS spectra.

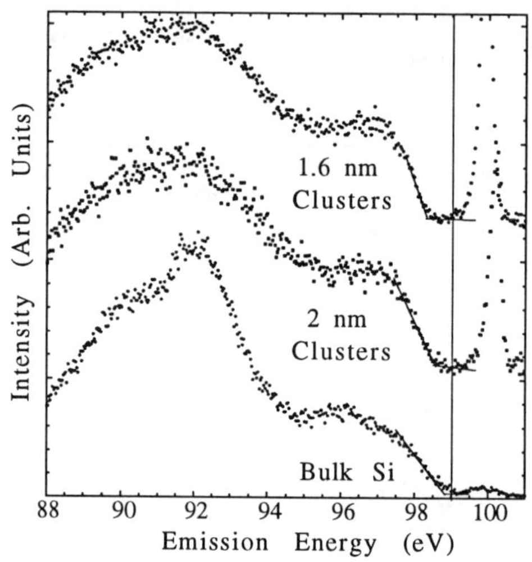

Fig. 3. The L_3 emission spectrum of bulk silicon and the 2.0 and 1.6 nm diameter cluster samples. The excitation energy is 100 eV for the bulk silicon and for the 1.6 nm clusters and 100.2 eV for the 2nm clusters. This corresponds to the L_3 edge measured in the absorption spectra so the L_2 edge is suppressed. The solid lines indicate the extrapolation of the high energy cut-off to the baseline in order to determine the VB edge position.

The overall appearance of the nanocluster spectra is different than that of the bulk silicon. In the nanocluster spectra the valence band is shifted to lower emission energy by 0.30 eV and 0.51 eV for the 2.0 nm and 1.6 nm clusters respectively. The VB edge is much sharper in the cluster samples than in the bulk Si opposite to what is observed in the x-ray absorption edge. Also the shape of the spectra is flattened from a peak like feature in the bulk Si to a plateau in the silicon nanoclusters. This effect has been observed in the SXF spectra of porous silicon and has been attributed to changes in the DOS at the VB edge due to confinement effects in the Si crystallites which alter the bulk electronic band structure [11].

The pronounced peak at 91.5 eV is not observed in the SXF spectra for the silicon nanoclusters. In fact the spectra looks more like what is observed for amorphous silicon [10]. Yet x-ray diffraction and TEM studies of these clusters have shown they are crystalline in nature [4]. It is interesting to note that the peak at 91.5 is observed in the SXF spectra of hydrogenated porous silicon [9,11].

It can be seen that the smaller silicon nanoclusters exhibit a larger VB shift as well as a larger conduction band shift. The VB shifts are larger than the shifts in the CB in accordance with theory and are similar to recent photoemission data on porous silicon although the ratio of the conduction band shift to valence band shift is slightly larger than that found in the photoemission study [8]. The observed band edge shifts are in qualitative agreement with the increasing luminescence blue shifts with decreasing particle size found in silicon nanoclusters with an oxide passivation [4].

In conclusion, we were able to investigate the unoccupied and occupied electronic states in silicon nanoclusters combining total fluorescence yield NEXAFS and selectively excited SXF spectroscopy. We observe shifts in the both the CB and VB edges indicating quantum size effects in the band structure of the nanoclusters. The onset of the absorption edge progressively broadens with increased confinement due to a distribution in cluster size. A steeper VB edge is observed in the nanoclusters as compared to bulk Si due to a change in the DOS at the VB edge.

ACKNOWLEDGEMENTS

We would like to thank Randy Hill for his technical assistance. This work was supported by the Director, Office of Energy Research, Office of Basic Energy Sciences, Chemical Sciences Division of the U.S. Department of Energy, LBNL under Contract No. DE-AC03-76SF00098 and LLNL under Contract No. W-7405-ENG-48, this work was done at the ALS, which is supported by the Department of Energy (Division of Materials Sciences and Division of Chemical Sciences of Basic Energy Sciences) under Contract No. DE-AC02-76CH0016.

REFERENCES

1. T. Canham, Appl. Phys. Lett. **57**, 1046 (1990).
2. F. Koch, V. Petrova-Koch, and T. Muschik, J. Lumin. **57**, 271 (1993).
3. S.M. Prokes, and O.J. Glembocki, Mater. Chem. Phys. **35**, 1 (1993).
4. L.N. Dinh, L.L. Chase, M. Balooch, W.J. Siekhaus, F. Wooten, Phys. Rev. B **54,** 5029 (1996).
5. J.J. Jia, T.A. Callcott, J. Yurkas, A.W. Ellis, F.J. Himpsel, M.G. Samant, D.L. Ederer, J.A. Carlisle, E.A. Hudson, L.J. Terminello, D.K. Shuh, and R.C.C. Perera, Rev. Sci. Instrum. **66**, 1394 (1995).
6. C. G. Granqvist, R.A. Buhrman, J. Appl. Phys. **47**, 2200 (1976).
7. T. van Buuren, L.N. Dinh, L.L. Chase to be published.
8. T. van Buuren, T. Tiedje, J.R. Dahn, and B.M. Way, Appl. Phys. Lett. **63**, 2911 (1993).
9. S. Eisebitt, J. Luning, J.-E. Rubensson, T. van Buuren, S.N. Patitsas, T. Tiedje, M. Berger, R. Arens-Fisher, S. Frohnhoff and W. Eberhardt, Solid State Comm. **97**, 549 (1996).
10. K.E. Miyano, D.L. Ederer, T.A. Callcott, W.L. O'Brien, J.J. Jia, L.Zhou, Q.-Y. Dong, Y.Ma, J.C. Woicik and D.R. Mueller, Phys. Rev B **48**,1918 (1993).
11. S. Eisibitt, S.N. Patitsas, T. Tiedje, T. van Buuren, J. Luning, J.-E. Rubensson and W. Eberhardt , submitted to Euro. Phys. Lett.

COMPARATIVE OPTICAL STUDIES OF CHEMICALLY SYNTHESIZED SILICON NANOCRYSTALS

GILDARDO R. DELGADO[1,2], HOWARD W. H. LEE[1], SUSAN M. KAUZLARICH[3], AND RICHARD A. BLEY[3]

[1]Lawrence Livermore National Laboratory, Photonics Group, P.O. Box 808, L-174, Livermore CA 94551

[2]University of California-Davis, Department of Applied Physics, Livermore, CA

[2]University of California-Davis, Department of Chemistry, Davis, CA

ABSTRACT

We studied the optical and electronic properties of silicon nanocrystals derived from two distinct fabrication procedures. One technique uses a controlled chemical reaction. In the other case, silicon nanocrystals are produced by ultrasonic fracturing of porous silicon layers. We report on the photoluminescence, photoluminescence excitation, and absorption spectroscopy of various size distributions derived from these techniques. We compare the different optical properties of silicon nanocrystals made this way and contrast them with that observed in porous silicon. Our results emphasize the dominant role of surface states in these systems as manifested by the different surface passivation layers present in these different fabrication techniques. Experimental absorption measurements are compared to theoretical calculations with good agreement. Our results provide compelling evidence for quantum confinement in both types of Si nanocrystals. Our results also indicate that the blue emission from very small Si nanocrystals corresponds to the bandedge emission, while the red emission arises from traps.

INTRODUCTION

Numerous models for the luminescence mechanism in porous silicon (PSi) and in monodispersed nanocrystalline systems have been proposed and have been described in various review articles [1-3]. Many of these models have been refuted in some form or another. Though a complete understanding of the mechanism for efficient light emission from Si nanoparticles is still lacking, the consensus has leaned strongly toward quantum confinement as being at least partly responsible. The important role of surface states as recombination centers in the luminescence process has also been proposed [4].

A number of methods have been used to prepare Si nanocrystals including thermal pyrolysis [5], evaporation and laser ablation into an inert atmosphere [6-9], and high pressure solution phase synthesis [10]. Most methods create a wide range of sizes and structures including the technique of ultrasonic fracturing of PSi [11]. The technique that has produced the most uniform sized Si nanocrystals reported to date involves a high temperature pyrolysis in which the Si particles are collected as an ethylene glycol colloid [12]. Different nanocrystalline systems yield different optical results due in part to the different size distributions and fabrication techniques involved. The origins of the observed luminescence have been attributed to may different sources such as quantum confinement, oxygen-related defects, polysilanes, siloxenes, SiH_x species, amorphous Si [1, 2].

We investigated the light emitting mechanism in order to understand how quantum confinement, surface states, and passivation affect the luminescence of different sized silicon nanocrystals. We studied monodispersed Si nanocrystals that were fabricated via two distinct techniques and compared their optical properties. One method involved ultrasonic fracturing of the nanostructured PSi to give a colloidal suspension in various solvents [11]. The second method involves a chemical synthetic route [13], again giving a colloidal suspension of Si nanocrystals. These two nanocrystalline Si systems provide two distinctly different surface passivation schemes as well as different particle sizes and size distributions. This will enable a critical comparison of their optical and electronic properties, and provide insights into the properties of nanocrystalline Si.

Mat. Res. Soc. Symp. Proc. Vol. 452 © 1997 Materials Research Society

The PSi technique gives a surface largely passivated by an oxide layer, while the synthetic route provides an organic (methyl and methoxy groups) surface passivation layer.

We performed various optical studies on these nanocrystalline Si systems, which include absorption, photoluminescence (PL), and photoluminescence excitation (PLE) spectroscopy. The experimental results are compared to theoretical models of nanocrystalline Si. Our analysis provides compelling evidence for quantum confinement in these two systems. Finally, our results indicate that the blue emission from these Si nanocrystals corresponds to the bandedge emission, while the red emission results from trapping and radiative relaxation of the initial excitation generated by bandedge absorption [14].

EXPERIMENT

Porous silicon is made by anodic electrochemical etching of silicon wafers in hydrofluoric acid solutions [1-3]. The nanostructured surface of the PSi was mechanically scraped off the wafer and place in appropriate solvents. A colloidal suspension of silicon nanocrystals in various solvents was prepared by ultrasonicating the PSi remnants for several days. The final colloidal suspension was centrifuged and filtered through 0.45 μm filters.

Colloidal suspensions of Si nanocrystals derived via a chemical synthetic route were prepared through a controlled low temperature phase chemical reaction. This technique is described in greater detail in Ref.13 and will not be considered here. Several major advantages were gain through this synthetic procedure. Smaller sized Si nanocrystals were achieved and surface termination with different organic groups was possible.

Colloidal suspensions of both nanocrystalline systems in various solvents, such as hexane, methanol and toluene, all gave identical behavior thus indicating that the solvent does not interfere with the mechanism for light emission.

RESULTS AND DISCUSSION

PL spectra from PSi gave a broad red peak centered at 680 nm. The PL from the PSi is consistent with other optical studies on PSi. However, no blue emission is observed even after prolonged periods (5 months) of exposure to ambient conditions.

After processing the PSi to give monodispersed nanocrystals in colloidal suspensions, the broad red peak disappeared and a different broad red peak at lower energies appeared along with an intense blue emission [14]. Similar spectral features are observed for the synthetic nanocrystals.

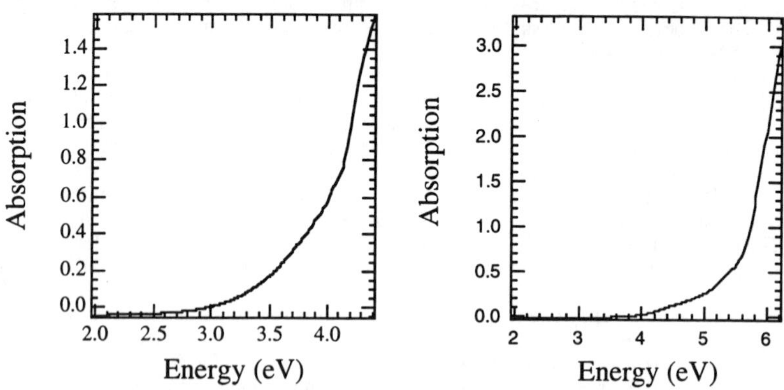

Figure 1. Absorption spectrum of Si nanocrystals derived from PSi.

Figure 2. Absorption spectrum of chemically synthesized Si nanocrystals.

Absorption spectra for both types of Si nanocrystals are smooth and nearly structureless and are shown in figures 1 and 2. Theoretical models by Delerue et al. [15] and by Wang and

Zunger [16] calculate absorption spectra for nanocrystals of a specific size. The calculated absorption edge shifts to higher energies due specifically to quantum confinement effects. In contrast, our samples consist of Si nanocrystals with a distribution of particle sizes. Nevertheless, the experimental absorption edge reflects the characteristics of the largest particles. Therefore, comparison of the theoretical and experimental absorption edges provides an upper limit on the size distribution of the samples.

The experimental absorption edges for both of our Si nanocrystal systems are in good agreement with the calculated absorption edge for the appropriate particle size. Based on the calculations by Delerue et al.[15], we find that our experimental absorption edge for nanocrystals derived from PSi agrees well with the calculation for 3.86 nm diameter nanocrystals. This would indicate that the nanocrystals derived from PSi have diameters less than 3.86 nm. This result is consistent with Transmission Electron Microscopy (TEM) data. In contrast, the experimental absorption edge for synthetic Si nanocrystals is shifted to higher energies from that calculated by Delerue et al. for 1.56 nm diameter nanocrystals. This would indicate that our synthetic nanocrystal sample has a diameter less than 1.56 nm [15]. Atomic force microscopy (AFM) confirms this and shows the presence of nanocrystals with diameter of 1.5 nm. Thus, based on comparisons of the experimental and theoretical absorption edge, we conclude that the Si nanocrystals derived from PSi are larger on average than those fabricated synthetically. This is supported by TEM and AFM data.

The PL and PLE spectra also reflect quantum size effects for both types of Si nanocrystals. A comparison of the PL and PLE spectra of both types of nanocrystals is shown in figures 3 and 4, respectively, and they support our conclusion that the synthetic nanocrystals are smaller than the nanocrystals derived from PSi. The PL peaks (red and blue) and the PLE peaks from the synthetic nanocrystals are blue-shifted from that seen from nanocrystals derived from PSi. Both of these results are consistent with quantum confinement effects. In addition, the width of the PLE and PL spectra of the synthetic nanocrystals are narrower, indicating that the synthetic nanocrystals have a smaller size distribution.

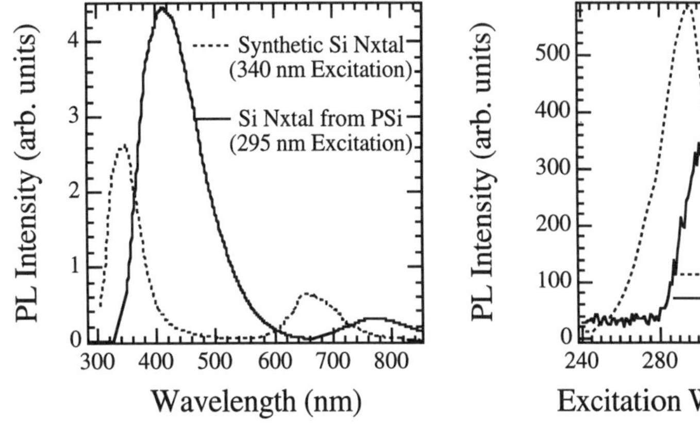

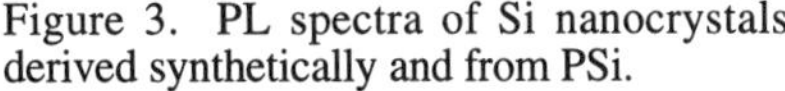

Figure 3. PL spectra of Si nanocrystals derived synthetically and from PSi.

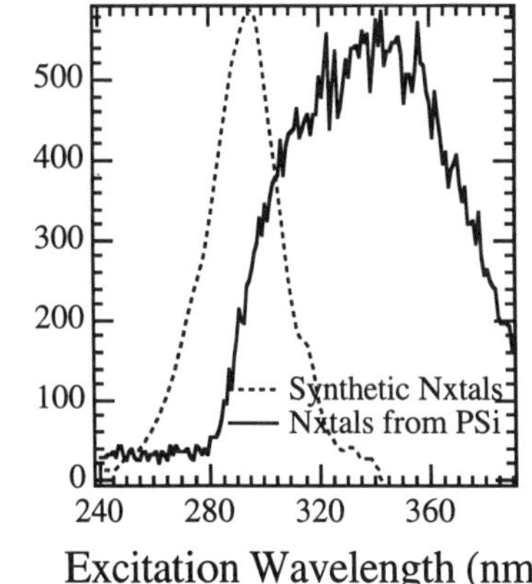

Figure 4. PLE spectra of Si nanocrystals derived synthetically and from PSi.

Figures 5 and 6 show the PLE spectra for the red and blue emission peaks from both types of nanocrystals. It is clear from the figure that the PLE spectra for the red and blue peaks are nearly identical for the synthetic nanocrystals, and are very similar for the nanocrystals derived from PSi. This strongly indicates that the blue and red emission peaks originate from the same absorption process [14].

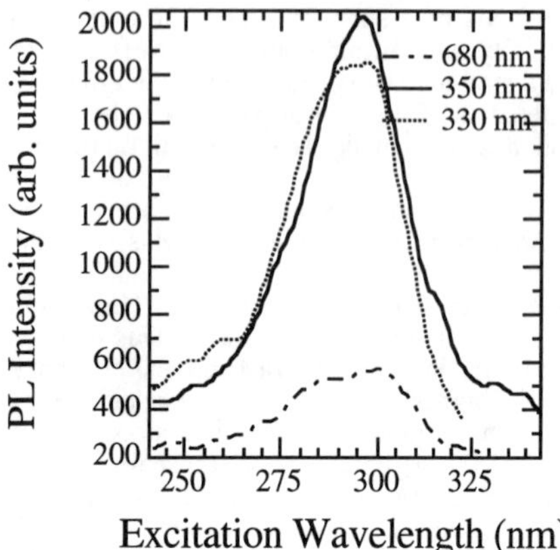

Figure 5. PLE spectra of synthetic Si nanocrystals.

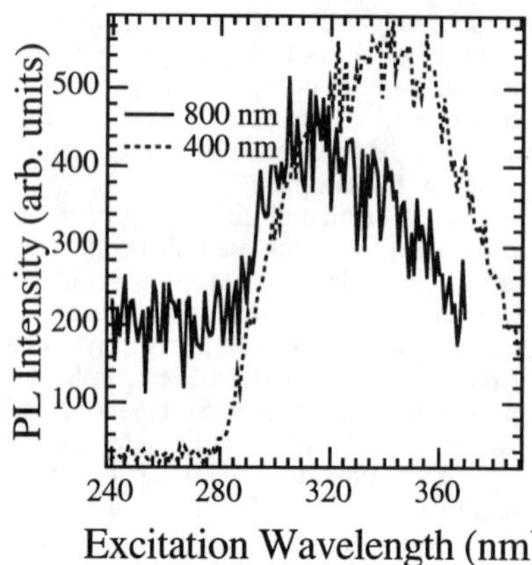

Figure 6. PLE spectra of Si nanocrystals derived from PSi.

We also performed photoluminescence size-selective spectroscopy, which is a variation of site-selective spectroscopy or Fluorescence Line-Narrowing. Quantum confinement dictates that the absorption and PL from quantum confined systems occur at energies that depend on the particle size. In size-selective spectroscopy on a sample with a distribution of particle sizes, a narrow band excitation selects a narrow distribution of particle sizes from the broader distribution. The resultant PL consequently reveals information on this particular size of particles. One manifestation of quantum confinement in size-selective spectroscopy would be a PL peak that spectrally shifts as the excitation energy is changed [14].

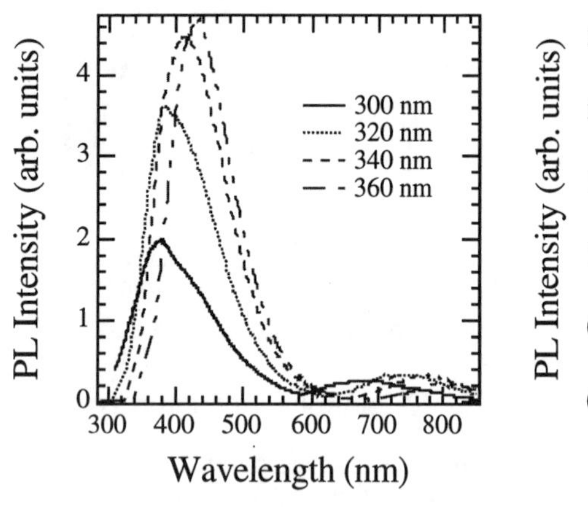

Figure 7. Size selective spectroscopy on Si nanocrystals derived from PSi.

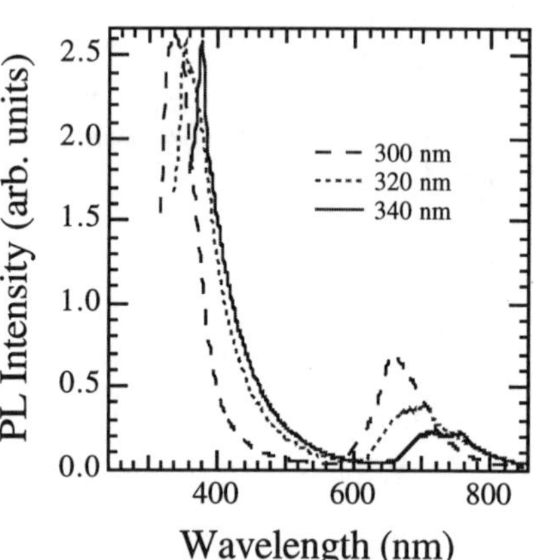

Figure 8. Size selective spectroscopy on chemically synthesized Si nanocrystals.

Figures 7 and 8 show the results of size selective spectroscopy on both types of Si nanocrystals. The red and blue PL peaks from both types of nanocrystals shift uniformly when the excitation wavelength is changed. For example, the PL shifts for the blue peaks ranged from 378-432 nm when excited from 300-360 nm for the nanocrystals derived from PSi, and from 340-378 nm when excited from 300-340 nm for the synthetic nanocrystals. In addition, each spectrum represents PL from nanocrystals with a narrowed size distribution. In spite of the fact that the two types of nanocrystals were fabricated with distinctly different techniques and have very different types of surface passivation, both yielded the same behavior with size selective spectroscopy. This strongly confirms that the presence of quantum confinement is responsible for the red and blue emission in both types of nanocrystals [14]. In addition, the extremes of the spectral shifts define the regime for quantum confinement and the particle size distribution. Size selective spectroscopy would therefore indicate that the Si nanocrystals derived form PSi have a larger size distribution than that made synthetically, in agreement with our earlier results.

The origin of the blue and red emission peaks also remains an important issue, and Si nanocrystals from different fabrication schemes provides useful insights into this problem. There are many models for the origin of the blue peak [17]. One possibility involves bandedge emission that results from interband recombination. In cases where the surface is passivated by an oxide layer, another possibility involves emission from Si oxides. Since the bandgap of SiO_2 is $\geq$ 8 eV, interband recombination in SiO_2 is unlikely to account for the blue emission. However, nonstoichiometric sub-oxides, SiO_x, have a bandgap of $\geq$ 3-4 eV for x ~ 1.4-1.6. In addition, defects in Si oxides are known to emit in the red and blue region. Surface states arising from Si-O bonds at the surface or Si-oxide interface have also been proposed as recombination centers for the light emission. Finally, hydroxyl groups adsorbed on structural defects in the SiO_2 matrix has also been suggested.

Our results indicate that the blue emission peak in our samples is not due to Si oxides or oxygen-related defects [14]. The blue peak is observed to shift monotonically in size selective spectroscopy when excited in the spectral range where quantum confined Si nanocrystals absorb. Si oxides should not show spectral shifts under these conditions. In general, the blue emission from Si oxides and oxygen-related defects occurs at a variety of different wavelengths and requires higher energy excitation. More importantly, the Si nanocrystals derived synthetically and from PSi behave very similarly in spite of having very different surface passivation schemes. This strongly suggests that the optical behavior we observe does not depend on the type of surface passivation and, in particular, does not originate from oxide effects but rather from the Si core.

A comparison of the calculated bandedge and the experimental absorption edge provides meaningful information on the origin of the blue emission. Calculations by Delerue et al. [18] show that the bandedge occurs at 1.67 eV for 3.86 nm diameter nanocrystals. The Si nanocrystals derived from PSi have diameters less than 3.86 nm and the experimental absorption edge is observed near 3.0 eV. Though our smaller diameter nanocrystals should have slightly larger bandedges, this difference of ~ 1.33 eV between the calculated bandedge and the experimental absorption edge indicates that the oscillator strength near the bandedge is still very small for this particle size.

In contrast, Delerue et al. calculate a bandedge of 3.45 eV for 1.56 nm diameter Si nanocrystals. The observed absorption edge of the synthetic nanocrystals, which have diameters $\leq$ 1.5 nm, is also ~3.45 eV. This shows that the oscillator strength near the bandedge for small nanocrystals of this size is already large. The obvious trend indicated here is that the bandedge approaches the absorption edge for very small nanocrystals. The utility of this trend is that the bandedge for very small Si nanocrystals can be determined by the absorption edge.

This trend can be applied to the two types of Si nanocrystals. The nanocrystals derived from PSi are less than 3.86 nm in diameter, indicating that the bandedge and the absorption edge should differ, but not by much. The observed absorption edge is ~3.02 eV. This would suggest that the blue emission is the bandedge emission. Ascribing the red emission to the bandedge would imply an unrealistically large Stokes shift. The small size of the synthetic nanocrystals provides more definitive conclusions. The synthetic nanocrystals are $\leq$ 1.5 nm in diameter, and thus their bandedge and absorption edge should be the same. The observed absorption edge is ~3.54 eV. The caluculated bandedge is 3.45 eV for 1.56 nm diameter nanocrystals. Again, this result indicates that the blue emission corresponds to the bandedge emission [14]. This conclusion is further supported by our earlier result that the blue emission arises from quantum confinement effects, indicating that it originates from the Si nanocrystalline core.

Consequently, the red emission cannot be the bandedge emission, as interpreted by many reserachers [14]. Ascribing the red emission to the bandedge would indicate an unrealistically large Stokes shift of almost a couple of eV. Delerue et al. have calculated a Stokes shift of ~ 28 meV for Si nanocrystals of 2 nm in diameter [18]. We propose that the large red shift from the bandedge of the red emission indicates trapping and radiative relaxation of the initial excitation generated by bandedge absorption [14]. Interestingly, our size selective spectroscopy results show that the red emission also arises from quantum confinement and our PLE spectroscopy results indicate that the red and blue emission originate from the same source. This further indicates that the trap states shift similarly as quantum confined states. This may be reasonable for very small nanocrystals for which these conclusions apply. The wave functions for trap states in very small nanocrystals have a large component of the nanocrystal core wave function. In essence, the surface states may "overlap" the core, and differences between the surface and the core are less defined.

CONCLUSIONS

We have studied the optical properties of Si nanocrystals derived synthetically and from PSi. Both types of nanocrystals show clear quantum confinement effects. The chemically synthesized Si nanocrystals have smaller diameters and a narrower size distribution. Our results also indicate that the blue emission we observe corresponds to the Stokes shifted bandedge emission, while the red emission arises from traps [14]. These trap states shift similarly as quantum confined states and are likely due to the surface.

ACKNOWLEDGMENT

This work was performed under the auspices of the U. S. Department of Energy by Lawrence Livermore National Laboratory under contract No. W-7405-ENG-48.

REFERENCES

1. See, for example, Mat. Res. Soc. Symp. Proc. 298, (1993) or J. Lumin., **57**, (1993).
2. Porous Silicon, Z. C. Feng and R. Tsu eds., (World Scientific Publishing Company, River Edge, New Jersey, 1994).
3. Porous Silicon Science and Technology, J.C. Vial and J. Derrien eds., (Springer-Verlag, New York, 1995).
4. F. Koch, V. Petrova-Koch, and T. Muschik, J. of Lumin., **57**, 271(1993).
5. J. J. Wu, R. C. Flagan, J. Appl. Phys., **61**, 1365 (1987).
6. S. Hayashi, S. Tanimoto, K. Yamamoto, J. Appl. Phys., **68**, 5300 (1990).
7. R. Okada, S. Ijima, Appl. Phys. Lett., **58**, 1662 (1991).
8. S. Ijima, Jpn. J. Appl. Phys., **26**, 357 (1987).
9. Y. Saito, J. Cryst. Growth, **47**, 61 (1979).
10. J. R. Heath, Science, **258**, 1131 (1992).
11. J. L. Heinrich, C. L. Curtis, G. M. Credo, K. L. Kavanagh, M. J. Sailor, Science, **255**, 66 (1992).
12. K. A. Littau, P. J. Szajowshki, A. J. Muller, A. R. Kortan, L. E. Brus, J. Phys. Chem. **97**, 1224 (1993).
13. R. A. Bley and S. M. Kauzlarich, *J. Am. Chem. Soc.*, in press, (1996).
14. G. R. Delgado and H. W. H. Lee, Phys. Rev. Lett, (submitted).
15. C. Dellerue, G. Allan, M. Lannoo, Phys. Rev. B. **48**,11024 (1993).
16. L.W. Wang, A. Zunger, Phys. Rev. Lett. **73**,1039 (1994).
17. See, for example, D.L. Griscom, J. Ceram. Soc. Jpn. 99, 923 (1991).
18. C. Dellerue, G. Allan, E. Martin and M. Lannoo, Porous Silicon Science and Technology, J.C. Vial and J. Derrien eds., (Springer-Verlag, New York, 1995), page 91.

VISIBLE LUMINESCENCE IN Si/SiO_2 SUPERLATTICES

D.J. LOCKWOOD, J.-M. BARIBEAU, P.D. GRANT, H.J. LABBÉ, Z.H. LU, J. STAPLEDON and B.T. SULLIVAN
Institute for Microstructural Sciences, National Research Council, Ottawa, Canada K1A 0R6

ABSTRACT

Amorphous Si/SiO_2 superlattices with periodicities from 2 to 5 nm have been grown on (100) Si wafers by several different techniques: molecular beam epitaxy, magnetron sputtering, and plasma enhanced chemical vapor deposition (PECVD). With the first two methods little or no hydrogen was incorporated during growth and visible photoluminescence (PL) was obtained at wavelengths from 520 to 800 nm. The shift in the PL peak position with Si layer thickness is consistent with quantum confined emission. Annealing the sputtered superlattices at temperatures up to 1100°C produced a very bright red PL that is similar in intensity to that found in porous Si. The PL was also considerably enhanced by deposition on aluminum-coated glass substrates. For large numbers of periods (e.g., 425) the PL was strongly modulated in intensity owing to optical interference within the superlattice. Similar quantum-confined PL was also observed in the PECVD grown superlattices, where the amorphous Si layers were heavily hydrogenated. The blue–red cathodoluminescence observed from sputtered superlattices is due primarily to defects in the SiO_2 layers.

INTRODUCTION

The fabrication of light-emitting Si based materials using band structure engineering (e.g., $Si_{1-x}Ge_x$ and $Si_{1-x-y}Ge_xC_y$ alloying and zone folding in Si/Ge atomic layer superlattices) or by synthesizing crystalline Si (c-Si) nanoparticles has become a very active area of research [1–3]. By modifying the c-Si environment it is hoped to induce a more efficient luminescence than that found in indirect-gap bulk c-Si and also to shift the emission from the infrared (~ 1.1 eV) into the visible region. Such modified materials could then be incorporated into the devices required by the optoelectronics industry using well known Si processing technology [4]. Efforts in this general area have been intensified by the discovery of bright visible light emission in porous silicon (π-Si) formed by electrochemical etching of c-Si [5]. Although evidence of quantum confinement effects in π-Si has been obtained from optical absorption, the recombination mechanism is poorly understood at present [6–9]. This is because the surface and bulk structure of π-Si is difficult to control, and it was recognized early on that more uniform and reproducible nanoscale Si structures were required to quantify the luminescence mechanism(s) [10]. Considerable efforts are now in progress to produce quantum wires and dots from c-Si and c-$Si_{1-x}Ge_x$, as has been done for GaAs-based systems, but the available lithographic and chemical etching techniques have not been able to produce small enough nanostructures to observe substantial confinement effects [11–13], until very recently [14]. In c-Si, dimensions much less than the free-exciton Bohr radius of ~ 5 nm are required [7] for such effects.

Another approach is to use crystal growth techniques such as planar epitaxy, quantum wire formation along wafer edge steps, and self-organized dot growth [15] to fabricate, in a controllable fashion, quantum confined structures. Silicon molecular beam epitaxy (MBE) is now a mature technology and quantum well materials can be engineered with atomic layer precision at the thick-

Mat. Res. Soc. Symp. Proc. Vol. 452

ness required. It is thus now possible to prepare nanometer thick Si layers intercalated with a wide band gap material to produce the required electron confinement. Such semiconductor/insulator superlattices employing hydrogenated amorphous Si (a-Si:H) and a-SiN_x:H were first demonstrated in 1983 by Abeles and Tiedje [16]. Since then, there has been considerable work aimed at understanding the optical properties of these amorphous semiconductor layered structures (e.g., see Refs. 17–25). More recently, room temperature luminescence has been reported from a-Si/SiO_2 [26], a-Ge:H/a-SiN_x:H [27], c-Si:H/a-Si:H [28], polycrystalline Si/CaF_2 [29,30], recrystallized a-Si:H/a-SiN_x:H [31] and recrystallized a-Ge:H/a-SiN_x:H [32] superlattices, but no convincing evidence for quantum confinement induced emission was obtained with the possible exception of Refs. 30 and 31.

Here we report on the visible light emitting properties of Si/SiO_2 superlattices grown on Si(001). These regular periodic structures are free of the size distribution and contamination effects that complicate the optical properties of Si nanoparticles and π-Si. In the earlier work of Zayats *et al.* [26], superlattices were prepared by magnetron sputtering of SiO_2 and Si in an argon atmosphere. Some evidence was obtained of confinement shifted luminescence from very thin a-Si layers, but only under pulsed laser excitation. The analysis of the optical emission was complicated by the presence of additional features due to recombination across the a-Si/SiO_2 interfaces and within the SiO_2 layers. More recently, we have prepared Si/SiO_2 superlattices under quite different growth conditions using MBE, magnetron sputtering (MS), and plasma enhanced chemical vapor deposition (PECVD) techniques [33,34]. For these superlattices, the Si layer thickness dependence of the optical emission when correlated with the conduction and valence band shifts provides the first direct evidence of light emission due to quantum confinement in Si nanostructures [35–38].

EXPERIMENT

MBE growth

The a-Si/SiO_2 superlattices were grown at room temperature on lightly phosphorous-doped n-type (001) Si wafers. The Si wafer was first submitted to a 600 s exposure to ultraviolet (UV) ozone before introduction into the vacuum chamber. This UV-ozone treatment is a rate-limited oxidation process that produces a ~ 1 nm thick carbon-free oxide layer [39]. A thin Si layer was next deposited at a 0.1 nm/s rate on the oxidized wafer by MBE in a VG Semicon V80 system. The wafer was then taken out of the ultra-high vacuum chamber and submitted to another 600 s exposure to UV ozone (the wafer was typically exposed to air for ~ 100 s before and after the oxidation treatment). This procedure was repeated to create a six period a-Si/SiO_2 superlattice.

MS growth

The a-Si/SiO_2 superlattices were fabricated using an automated radio-frequency MS deposition system developed at the National Research Council of Canada [40]. The layers were deposited as follows: after reaching a base pressure of 4–6 x 10^{-7} Torr, argon gas was flowed in and the Si target was pre-sputtered for ten minutes. A Si layer was then deposited at a rate of ~ 0.025 nm/s after which the substrate was rotated away from the Si target. Oxygen gas was then introduced into the chamber to create a sufficiently intense atomic oxygen plasma to partially oxidize the Si layer [41]. The oxidation time was typically 100 s. After the SiO_2 layer was formed, the O_2 flow was stopped and the Si target was again pre-sputtered for a short period of time. This process was then repeated for the desired number of periods (100–525). Various substrates were used including (001) electronic grade Si, quartz and cover glass.

PECVD growth

The PECVD samples were grown on (001) Si substrates in a Plasma-Therm 730 system operating at a base pressure of 1.5 x 10^{-5} Torr and with a substrate temperature of 350°C. The a-Si:H layers were deposited at a rate of 3.0 nm/min with a process pressure of 0.9 Torr, an RF power of 50 W, and flow rates of 800 sccm/min for He and 200 sccm/min for SiH_4. The SiO_2 layers were deposited at a rate of 3.0 nm/min with a process pressure of 1.5 Torr, an RF power of 20 W, and flow rates of 4, 900, and 45 sccm/min for SiH_4, He, and N_2O, respectively. Twenty period superlattices were grown with constant SiO_2 layer thickness and varying a-Si:H layer thickness. The first and capping layers were always SiO_2.

Optical measurements

The optical properties of the Si/SiO_2 superlattices were investigated at room temperature using photoluminescence (PL) [37], optical absorption [34], and cathodoluminescence (CL) techniques. The CL measurements were performed in a small stainless steel vacuum chamber pumped by a Balzers TCP-380 turbo pump. Measurements were started when the pressure was below 6x10^{-6} Torr. The electron source was a Kimball Physics EFG-7F electron flood gun with a Kimball Physics EGPS-7H power supply. The experiments were performed at 760 eV electron energy with the beam focused on a spot approximately 0.5 mm in diameter. The excitation energy was selected after visual observation of a number of samples. It was found that 760 eV was sufficient to produce visible luminescence and the luminescence did not age rapidly. The maximum emission current available from the electron gun, 0.1 mA, was used on some samples to produce more signal, but generally was reduced for samples with strong luminescence. The sample current was not measured. However, based on past experience, it can be expected that about 50% of the emitted current is flowing in the sample. The electron beam was incident on the sample at about 8° from normal, while the emitted light was collected at an angle of 45° from normal.

RESULTS AND DISCUSSION

Photoluminescence

Visible to near-infrared wavelength PL was observed in all superlattices with Si layer thickness $d < 3$ nm no matter which growth method was used [34,37]. The PL of a number of a-Si/SiO_2 superlattices grown by MS is shown, for example, in Fig. 1. All the superlattices had 100 periods and a SiO_2 layer thickness of ~ 1.2 nm (increasing the SiO_2 layer thickness did not affect the results). The PL peak wavelength shifts to longer wavelengths as the period increases and the peak intensity rises and falls again, as found in MBE samples [37]. The same PL was observed for a-Si/SiO_2 superlattices deposited on Si, quartz or cover glass substrates and even on "roughened" Si and cover glass substrates. Hence, no special substrate or smoothness is required to observe the PL from a-Si/SiO_2 superlattices. The MBE and MS grown layers were shown to be chemically clean and essentially hydrogen free [33,34] and no PL was seen in the 460 to 850 nm wavelength range for thick a-Si and SiO_2 reference layers. There was no noticeable deterioration in the PL over periods of several months.

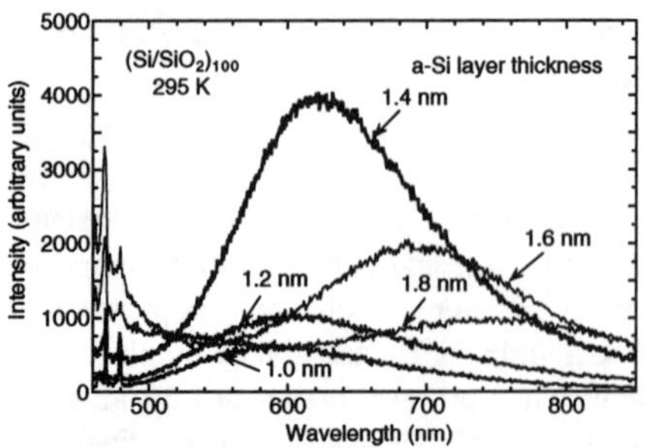

Fig. 1. Room-temperature PL from MS-grown $(a\text{-}Si/SiO_2)_{100}$ superlattices of different a-Si layer thicknesses (from Ref. 34).

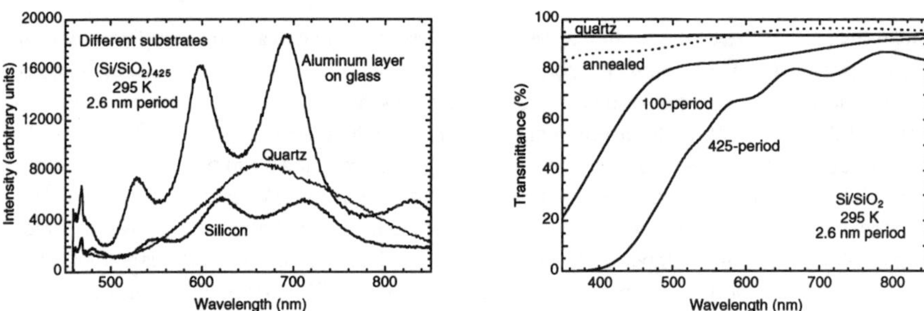

Fig. 2. (a) Room-temperature PL from an MS-grown $(a\text{-}Si/SiO_2)_{425}$ superlattice with a 2.6 nm periodicity deposited on different substrates. (b) Optical transmittance of $a\text{-}Si/SiO_2$ superlattices with a 2.6 nm periodicity grown by MS on quartz substrates; uncoated substrate, 100 period superlattice, 100 period superlattice annealed in air at 1100°C for 1/2 h, and 425 period superlattice (from Ref. 34).

Figure 2 (a) shows the PL obtained from a 425 period superlattice deposited simultaneously on Si, quartz and Al-coated glass substrates. For the Si and Al-coated substrates, there is a pronounced modulation of the PL intensity, which can also be observed in transmission (see Fig. 2 (b)). Since the periodicity of the superlattice, ~ 2.6 nm, is small compared to the wavelength of visible light, an average refractive index, n_{SL}, can be assigned to the superlattice material. Fitting the transmittance curve, and assuming a metric thickness $t_m \approx 1100$ nm, then $n_{SL} \approx 1.8$ at 650 nm giving an optical thickness $t_o \approx 2000$ nm. From this an intensity modulation period $T \approx 0.25\ \mu m^{-1}$ can be calculated for the 425-period superlattice, which is in good agreement with the observed PL modulation periods. This confirms that the PL peaks arise from optical interference of the emitted light within the superlattice structure. The lack of a significant PL modulation for the 425-period superlattice deposited on quartz can be explained by the low reflectance, ~ 1%, between the superlattice and the quartz substrate.

Rapid thermal annealing of an MBE grown superlattice with $d = 2.81$ nm at high temperatures (~ 1000°C) in nitrogen has a dramatic effect on the PL [36], as can be seen in Fig. 3. The PL peak

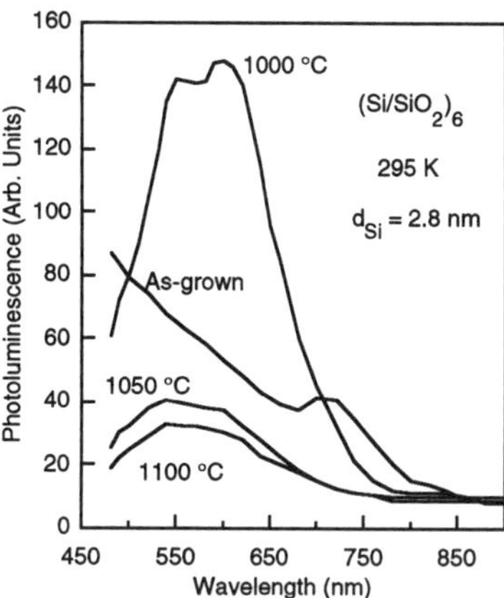

Fig. 3. Room-temperature PL from an MBE-grown (a-Si/SiO_2)$_6$ superlattice with a 3.7 nm periodicity before and after rapid thermal annealing for 30 s in nitrogen at different temperatures (from Ref. 36).

wavelength shifts from the far red (730 nm) to yellow-green (570 nm) and the peak intensity grows significantly. Raman and x-ray photoelectron spectroscopy (XPS) show that the a-Si layers relax towards a more ordered structure after annealing, but the layers do not yet become ordered c-Si [36]. Also, the a-Si layers become thinner with annealing due to Si-O interdiffusion across the Si/SiO_2 interfaces [36]. Hence annealing at high temperature mimics the effect of growing superlattices with commensurately thinner a-Si layers and has a similar effect on the PL.

Slow annealing of an MS grown superlattice in air resulted in a quite different behavior, as can be seen from Fig. 4 (a). The PL after annealing comprises two components: one shifts to shorter wavelength with increasing anneal temperature while the other emerges at longer wavelength and continually grows in intensity. The peak shifting to shorter wavelengths appears to have the same origin as the PL in the as-grown sample and the shift is caused by the thinning of the a-Si layers on annealing. The red wavelength emission dominates after the 1100°C air anneal and is most likely due to a stable oxygen-related defect center created by annealing [42]. Such non-bridging oxygen hole center clusters have been shown to produce the same red PL in π-Si and silica glasses [42–44]. Figure 4 (b) shows the PL obtained from an MS grown superlattice before and after a 1100°C anneal in air for 0.5 hour. Annealing the superlattice resulted in approximately a 40 times increase in the PL intensity, which is comparable to that observed from π-Si samples [45] and is readily seen in a brightly lit room.

The PECVD grown samples also give evidence of quantum confinement effects in the shift of the PL peak position with a-Si:H layer thickness, as can be seen in Fig. 5. These PECVD results correlate very well with those obtained from the MBE grown superlattices. However, the PL peak shifts to the red in the PECVD grown superlattices after rapid thermal annealing in nitrogen, which is contrary to the MBE case. This suggests that the PL emission in PECVD grown material is due to recombination in the band tails and/or is related to defects within the a-Si:H layers [19]. Our results indicate that PECVD is not a good way to grow quantum-confined light emitting a-Si layers.

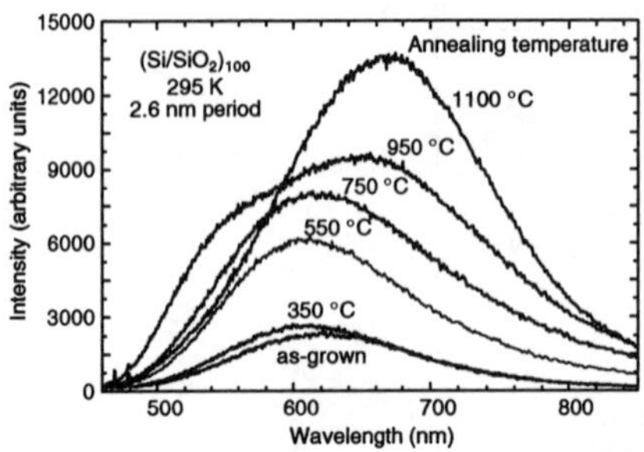

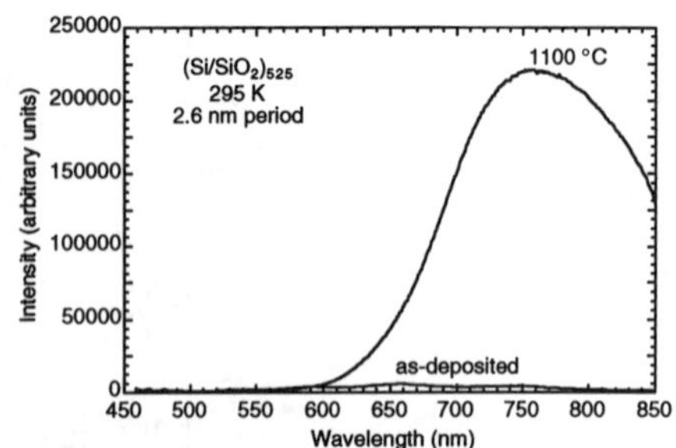

Fig. 4. Room-temperature PL of MS-grown a-Si/SiO_2 superlattices with a 2.6 nm periodicity: (a) 100 period superlattice before and after annealing for 1/2 h in air at different temperatures and (b) 525 period superlattice before and after annealing for 1/2 h in air at 1100°C. All spectra were recorded under the same conditions (after Ref. 34).

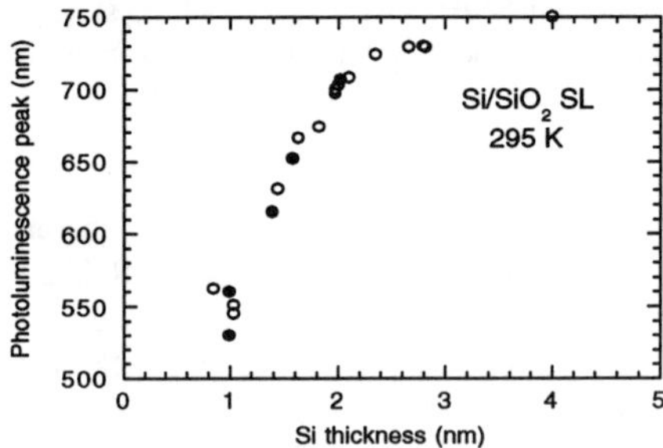

Fig. 5. Dependence of the room-temperature PL peak wavelength on a-Si layer thickness for MBE (o) and PECVD (•) grown a-Si/SiO_2 superlattices.

Cathodoluminescence

The CL spectra obtained from a series of annealed samples is shown in Fig. 6. These samples are the same as those whose PL is given in Fig. 4. The CL spectra are representative of results obtained from other as-grown and annealed samples, where generally a green CL was visible to the eye. Curve resolving the as-grown sample spectrum revealed CL peaks at 1.47, 1.92, 1.95, 2.00, ~ 2.3, 2.65 and 2.95 eV. These peaks vary considerably in intensity with annealing temperature (see Fig. 6), but, apart from the ~ 2.3 eV band, do not shift in frequency. The ~ 2.3 eV peak shifts smoothly up in frequency from 2.3 to 2.5 eV on annealing up to 1100°C. Also shown in Fig. 6 is the CL spectrum of a thick SiO_2 reference layer. Here, there is a dominant peak at 2.67 eV with a weak peak at 1.94 eV. Comparison with other work on CL in a-SiO_2 [46] shows that with the exception of the ~ 2.3 eV band all these peaks are associated with various defects such as the nonbridging oxygen hole center (1.9 eV) and self trapped exciton (2.6–2.7 eV) in the a-SiO_2. Some of these defects are induced by irradiation with the 760 eV electron beam [46]. The differences in intensity on annealing and on comparison with the a-SiO_2 reference layer reflect the changes in the density

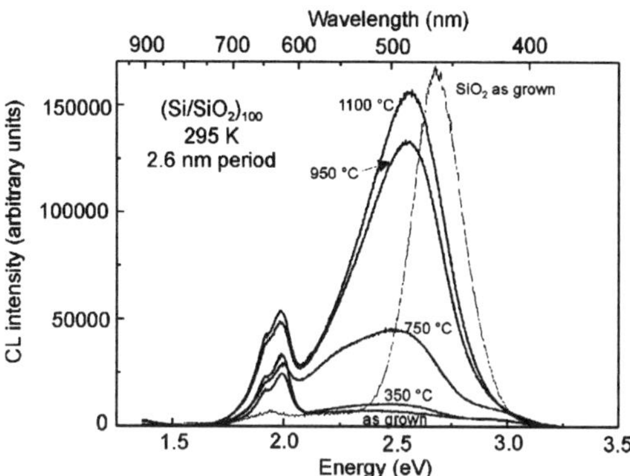

Fig. 6. Room-temperature CL of an MS-grown $(a\text{-}Si/SiO_2)_{100}$ superlattice with 2.6 nm periodicity before and after annealing for 1/2 h in air at different temperatures. Also shown is a reference SiO_2 layer spectrum.

of the different defects. Only the ~ 2.3 eV band does not fit into this pattern. There has been one report of a defect band at 2.28 eV due to a-SiO_2 outgrowths [46] that lies close in energy, but the shift up in energy on annealing is not characteristic of a-SiO_2 defects. Thus it is likely that the ~ 2.3 eV (~ 540 nm) band arises from CL in the a-Si layers and that the increase in energy on annealing is due to an increased quantum confinement, as was found for the PL. However, the CL from the as-grown sample occurs at a higher energy than that of the PL (2.0 eV or 620 nm), which is still a puzzle at present.

Optical Absorption

The optical transmittance of these a-Si/SiO_2 superlattices is important if they are to be useful as light-emitting devices. Figure 2(b) shows the transmittance of an as-grown and 1100°C air-annealed 100-period superlattice deposited on a quartz substrate and that of the 425-period superlattice discussed earlier. Increasing the number of periods from 100 to 425 in the as-grown superlattices results in only a 10% decrease in the transmittance for wavelengths above 600 nm. Below 600 nm, the transmittance starts to decrease and at the PL excitation wavelength of 458 nm it is 75% and 18% for the 100 and 425 period superlattices, respectively. Hence, the number of periods could be yet further increased and still have the excitation light penetrate all the layers and most of the emitted light being transmitted. After annealing in air, there is a significant increase in the transmittance of the 100 period superlattice indicating that the a-Si layers are nearly fully oxidized.

Quantum Confinement

The variation in PL peak energy with Si layer thickness for MBE grown samples is shown in Fig. 7, which demonstrates the pronounced blue shift of the PL with decreasing Si layer thickness. This blue shift of the PL and its increasing intensity with decreasing Si layer thickness is indicative of quantum confinement effects [47–53]. According to effective mass theory and assuming infinite

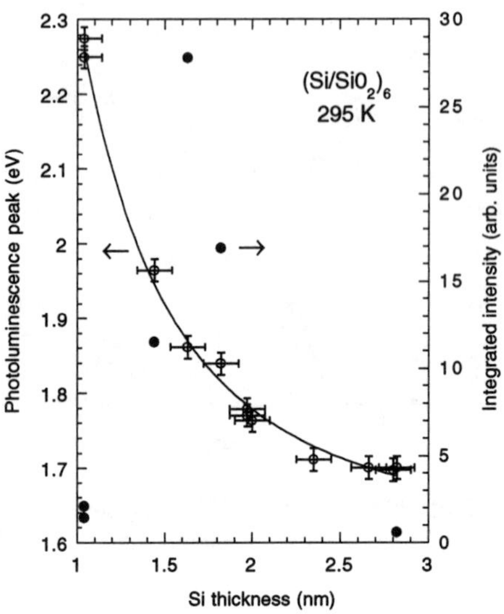

Fig. 7. The room-temperature PL peak energy (o) and integrated intensity (•) in MBE-grown (a-Si/SiO$_2$)$_6$ superlattices as a function of the a-Si layer thickness. The solid line is the fit to effective mass theory (from Ref. 37).

potential barriers, which is a reasonable approximation since wide-gap (9 eV) SiO_2 barriers are used, the energy gap E for one-dimensionally confined Si should vary as

$$E = E_g + \frac{\pi^2 \hbar^2}{2d^2}\left(\frac{1}{m_e^*} + \frac{1}{m_h^*}\right) \tag{1}$$

where E_g is the bulk material energy gap and m_e^* and m_h^* are the electron and hole effective masses. A least squares fit to the data given in Fig. 6 produces an excellent fit with $E(\text{eV}) = 1.60 + 0.72\, d^{-2}$. The fitted E_g of 1.60 ± 0.01 eV is larger than that of c-Si (1.12 eV at 295 K), but is in excellent agreement with that of bulk a-Si (1.5–1.6 eV at 295 K) [54] and indicates direct band-to-band recombination. The fitted confinement parameter of 0.72 ± 0.02 eV/nm^2 indicates $m_e^* \approx m_h^* \approx 1$. These effective masses are reassuringly close to those of c-Si ($m_e^* = 1.18$ and $m_h^* = 0.81$ at 300 K [55]), but carry a different meaning since they are parameters representing an average effect in an irregular lattice. For comparison, in the a-Si:H/a-Si_3N_4:H system, Miyazaki *et al.* found $m_e^* = 0.6$ from resonant tunneling experiments [20] while Hattori *et al.* reported an m_e^* less than 0.48 for ultrathin single layers of a-Si:H [24].

Confirmation of the direct nature of the gap comes from measurements of the a-Si conduction and valence band edges using x-ray techniques [35]. The density of states of the occupied valence band can be probed by XPS. Figure 8 compares two valence band maximum (VBM) XPS spectra: The relative VBM is determined from a linear extrapolation of the steepest portion of the curve. The density of states of the unoccupied conduction band can be studied by Si L-edge x-ray absorption spectroscopy (XANES). In these measurements, the occupied core shell $p_{3/2,1/2}$ electrons are

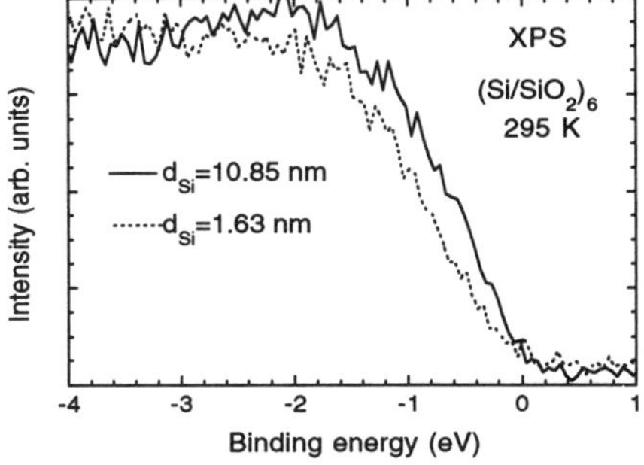

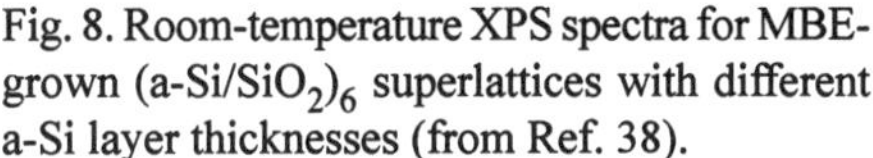

Fig. 8. Room-temperature XPS spectra for MBE-grown (a-Si/SiO_2)$_6$ superlattices with different a-Si layer thicknesses (from Ref. 38).

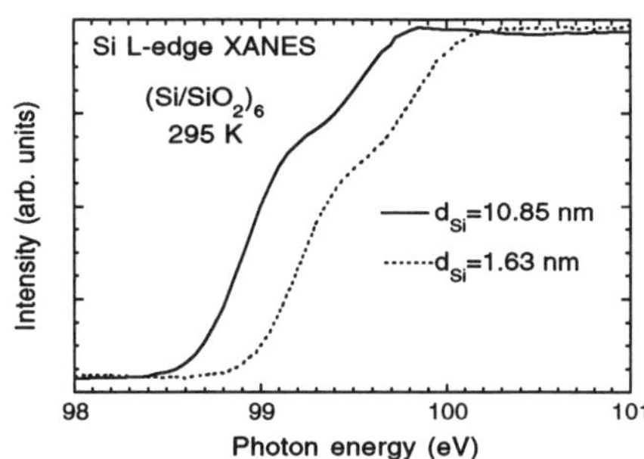

Fig. 9. Room-temperature XANES spectra for MBE-grown (a-Si/SiO_2)$_6$ superlattices with different a-Si layer thicknesses (from Ref. 38; see also Ref. 35).

excited to the empty conduction band and the absorption threshold is related to the conduction band minimum (CBM) s-like electronic states owing to the dipole selection rule. The absolute CBM position may not be determined accurately, but the relative CBM shift can be measured precisely. Figure 9 shows experimental results for the same two samples as given in Fig. 8. These spectra clearly show the shift up in energy of the absorption edge with decreasing a-Si layer thickness.

The results obtained for these CBM and VBM shifts are shown in Fig. 10, where it can be seen that the CBM shifts up from the bulk by ~ 0.3 eV for $d = 1.5$ nm whereas from XPS data the VBM shifts down slightly by ~ 0.1 eV for the same diameter. Thus the bandgap widens by ~ 0.4 eV for $d = 1.5$ nm, in good agreement with the PL result of ~ 0.3 eV (see Fig. 7). The direct relationship between the CBM shift and the PL peak energy is demonstrated in Fig. 11. A straight line is a reasonable fit to the data with a slope of 1.1 ± 0.2, which indicates direct recombination for the PL. The zero CBM shift intercept is at -1.8 ± 0.3 eV, which is in good agreement with the band gap of 1.6 eV determined from the PL dependence on Si layer thickness. Figure 12 shows a similar plot to that of Fig. 11, but this time the band gap widening determined by adding the VBM and CBM shifts is compared with the PL results. Again a straight line fit is a good representation of the data, with a slope of 1.7 ± 0.2 and an intercept of -0.02 ± 0.02 eV. The slope has increased from the CBM case due to the inclusion of the VBM shift, but the intercept is within error the expected value of zero electron volts. The VBM shift is more difficult to measure and the error bars are thus much larger than for the CBM case. Allowing for this, a slope of one is of the right order for the relationship between the band gap and the PL peak energy.

Taken together these PL and x-ray results strikingly demonstrate the effect of confinement in widening the a-Si band gap and confirm the direct band-to-band recombination mechanism for the PL.

CONCLUSIONS

In conclusion, convincing evidence of quantum confined luminescence due to band-to-band recombination has been obtained for the first time in a Si based nanostructure. The bright photoluminescence obtained from as-grown and annealed a-Si/SiO_2 superlattices offers interesting pros-

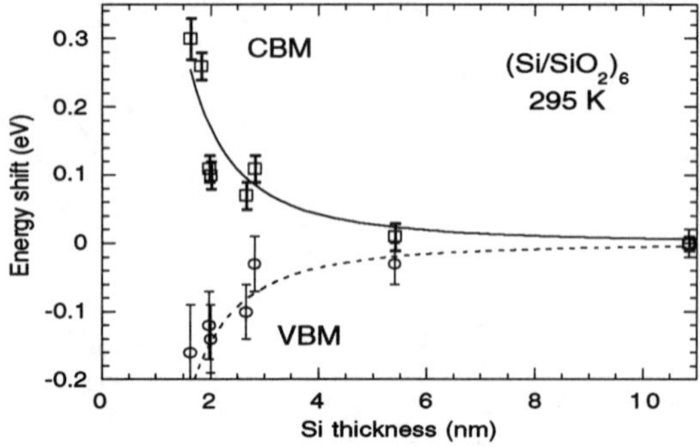

Fig. 10. The shifts in conduction (CBM) and valence (VBM) bands at room temperature for MBE-grown (a-Si/SiO_2)$_6$ superlattices as a function of a-Si layer thickness. The solid and broken lines are the fits to effective mass theory for the CBM [E(eV) = (0.67 ± 0.07) d^{-2}] and VBM [E(eV) = (-0.47± 0.04) d^{-2}] shifts, respectively (from Ref. 38; see also Ref. 35).

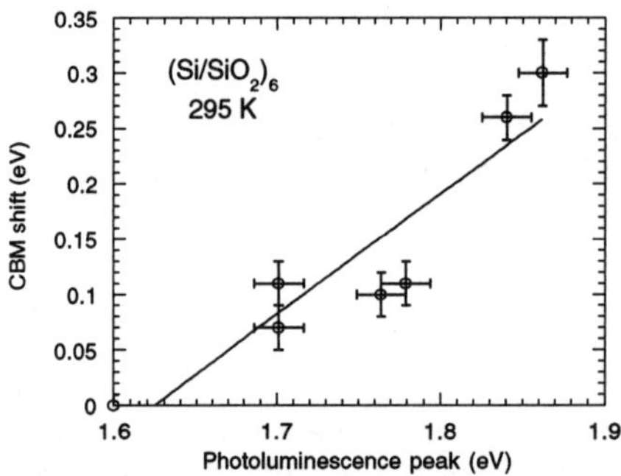

Fig. 11. The CBM shift in a-Si layers versus the PL peak position in MBE-grown (a-Si/SiO_2)$_6$ superlattices at room temperature. The solid line is the best linear fit (from Ref. 38).

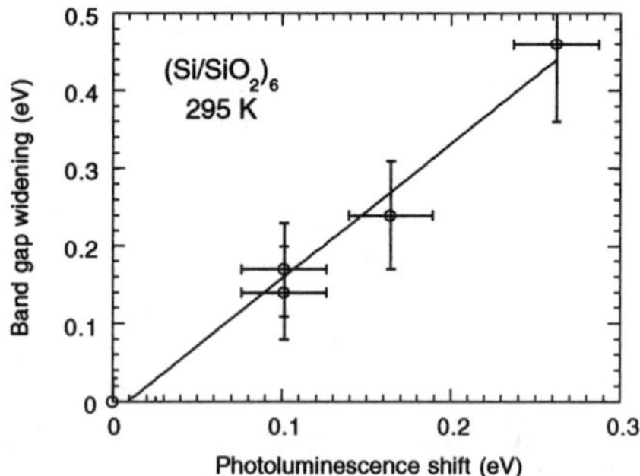

Fig. 12. The band gap widening in a-Si layers versus the PL peak position shift from the bulk value of 1.60 eV for MBE-grown (a-Si/SiO_2)$_6$ superlattices at room temperature. The solid line is the best linear fit (from Ref. 38).

pects for the fabrication of a Si-based light emitter that can be tuned from 500 to 800 nm by varying the a-Si layer thickness and/or the annealing conditions, all using available vacuum deposition technology and standard Si wafer processing techniques. The next important step is to develop light emitting diodes based on such a-Si/SiO_2 superlattices.

ACKNOWLEDGMENTS

We are grateful for the assistance of our colleagues G.C. Aers, M. Davies, M.W. Denhoff, J.P. McCaffrey, C. Montcalm, L.M. Sewrey, and H.T. Tran in this work.

REFERENCES

1. S.S. Iyer and Y.-H. Xie, Science **260**, 40 (1993).
2. R.A. Soref, J. Vac. Sci. Tech. A **14**, 913 (1996).
3. D.J. Lockwood, Editor, *Light Emission in Silicon* (Academic, Orlando, 1997).
4. R.A. Soref, Proc. IEEE **81**, 1687 (1993).
5. L.T. Canham, Appl. Phys. Lett. **57**, 1046 (1990).
6. D.C. Bensahel, L.T. Canham and S. Ossicini, Editors, *Optical Properties of Low Dimensional Silicon Structures* (Kluwer, Dordrecht, 1993).
7. D.J. Lockwood, Solid State Commun. **92**, 101 (1994).
8. Z.C. Feng and R. Tsu, Editors, *Porous Silicon* (World Scientific, Singapore, 1994).
9. Y. Kanemitsu, Physics Reports **263**, 1 (1995).
10. D.J. Lockwood, G.C. Aers, L.B. Allard, B. Bryskiewicz, S. Charbonneau, D.C. Houghton, J.P. McCaffrey, and A. Wang, Can. J. Phys. **70**, 1184 (1992).
11. Y.S. Tang, C.D.W. Wilkinson, C.M. Sotomayor Torres, D.W. Smith, T.E. Whall and E.H.C. Parker, Superlatt. Microstruct. **12**, 535 (1992).
12. A.S. Chu, S.H. Zaidi and S.R.J. Brueck, Appl. Phys. Lett. **63**, 905 (1993).
13. Y.S. Tang, C.M. Sotomayor Torres, S. Nilsson, B. Dietrich, W. Kissinger, T.E. Whall, E.H.C. Parker, W.-X. Ni, G.V. Hansson, H. Presting and H. Kibbel, J. Electron. Mater. **25**, 287 (1996) and references therein.
14. A.G. Nassiopoulos, S. Grigoropoulos and D. Papadimitriou, in *Advanced Luminescent Materials*, edited by D.J. Lockwood, P.M. Fauchet, N. Koshida and S.R.J. Brueck (Electrochemical Society, Pennington, 1996), p. 296.
15. H. Sunamura, N. Usami, Y. Shiraki and S. Fukatsu, Appl. Phys. Lett. **66**, 3024 (1995).
16. B. Abeles and T. Tiedje, Phys. Rev. Lett. **51**, 2003 (1983).
17. B. Abeles and T. Tiedje, Semicond. Semimetals **21**, Part C, 407 (1984).
18. T. Tiedje, B. Abeles and B.G. Brookes, Phys. Rev. Lett. **54**, 2545 (1985).
19. B.A. Wilson, C.M. Taylor and J.P. Harbison, Phys. Rev. B **34**, 8733 (1986).
20. S. Miyazaki, Y. Ihara and M. Hirose, Phys. Rev. Lett. **59**, 125 (1987).
21. L. Yang and B. Abeles, Appl. Phys. Lett. **51**, 264 (1987).
22. S. Kalem, Phys. Rev. B **37**, 8837 (1988).
23. K. Hattori, T. Mori, H. Okamoto and Y. Hamakawa, Phys. Rev. Lett. **60**, 825 (1988).
24. K. Hattori, T. Mori, H. Okamoto and Y. Hamakawa, Appl. Phys. Lett. **53**, 2170 (1988).
25. L. Yang, B. Abeles, W. Eberhardt, H. Stasiewski and D. Sondericker, Phys. Rev. B **39**, 3801 (1989).
26. A.V. Zayats, Yu.A. Repeyev, D.N. Nikogosyan and E.A. Vinogradov, J. Lumin. **52**, 335 (1992).
27. J. Jiang, K. Chen, X. Huang, Z. Li and D. Feng, Solid State Commun. **92**, 227 (1994).
28. S. Tong, X. Liu and X. Bao, Appl. Phys. Lett. **66**, 469 (1995).
29. F. Arnaud d'Avitaya, L. Vervoort, F. Bassani, S. Ossicini, A. Fasolino and F. Bernardini, Euro. Phys. Lett. **31**, 25 (1995).
30. L. Vervoort, F. Bassani, I. Mihalcescu, J.C. Vial and F. Arnaud d'Avitaya, Phys. Stat. Sol. (b) **190**, 123 (1995).
31. D.A. Grützmacher, E.F. Steigmeier, H. Auderset, R. Morf, B. Delley and R. Wessicken, Mat. Res. Soc. Symp. Proc. **358**, 833 (1995).

32. P. Wickboldt, D. Pang, J.H. Chen, H.M. Cheong and W. Paul, J. Non-Crystal. Solids **198-200**, 813 (1996).
33. J.-M. Baribeau, D.J. Lockwood and Z.H. Lu, Mat. Res. Soc. Symp. Proc. **382**, 259 (1995).
34. B.T. Sullivan, D.J. Lockwood, H.J. Labbé and Z.H. Lu, Appl. Phys. Lett. **69**, 3149 (1996).
35. Z.H. Lu, D.J. Lockwood and J.-M. Baribeau, Nature **378**, 258 (1995).
36. Z.H. Lu, D.J. Lockwood and J.-M. Baribeau, Solid State Electron. **40**, 197 (1996).
37. D.J. Lockwood, Z.H. Lu and J.-M. Baribeau, Phys. Rev. Lett. **76**, 539 (1996).
38. D.J. Lockwood, J.-M. Baribeau and Z.H. Lu, in *Advanced Luminescent Materials*, edited by D.J. Lockwood, P.M. Fauchet, N. Koshida and S.R.J. Brueck (Electrochemical Society, Pennington, 1996), p. 296.
39. J.R. Vig, in *Handbook of Semiconductor Wafer Cleaning Technology: Science, Technology, and Applications*, edited by W. Kern (Noyes, Park Ridge, 1993), p. 233.
40. B.T. Sullivan and J.-A. Dobrowolski, Appl. Opt. **32**, 2351 (1993).
41. B.T. Sullivan and K.L. Byrt, Appl. Opt. **34**, 5684 (1995).
42. S.M. Prokes and W.E. Carlos, J. Appl. Phys. **78**, 2671 (1995).
43. L. Skuja, Solid State Commun. **84**, 613 (1992).
44. S. Munekuni, T. Yamanaka, Y. Shimogaichi, R. Tohmon, Y. Ohki, K. Nagasawa and Y. Hama, J. Appl. Phys. **68**, 1212 (1990).
45. D.J. Lockwood and A.G. Wang, Solid State Commun. **94**, 905 (1995).
46. M.A. Stevens Kalceff and M.R. Phillips, Phys. Rev. B **52**, 3122 (1995).
47. A.J. Read, R.J. Needs, K.J. Nash, L.T. Canham, P.D.J. Calcott and A. Qtiesh, Phys. Rev. Lett. **69**, 1232 (1992).
48. F. Buda, J. Kohanoff, and M. Parrinello, Phys. Rev. Lett. **69**, 1272 (1992).
49. T. Ohno, K. Shiraishi, and T. Ogawa, Phys. Rev. Lett. **69**, 2400 (1992).
50. C.G. Van de Walle and J.E. Northrup, Phys. Rev. Lett. **70**, 1116 (1993).
51. X. Wang, D. Huang, L. Ye, M. Yang, P. Hao, H. Fu, X. Hou, and X. Xie, Phys. Rev. Lett. **71**, 1262 (1993).
52. M.S. Hybertsen, Phys. Rev. Lett. **72**, 1514 (1994).
53. A. Zunger and L.-W. Wang, Appl. Surf. Sci. **102**, 350 (1996).
54. *Properties of Amorphous Silicon, 2nd Ed.*, (IEE, London, 1989), pp. 269-286.
55. H.D. Barber, Solid State Electron. **10**, 1039 (1967).

PHOTOLUMINESCENCE MECHANISM OF SILICON QUANTUM DOTS AND WELLS

Y. KANEMITSU and S. OKAMOTO
Institute of Physics, University of Tsukuba, Tsukuba, Ibaraki 305, Japan

ABSTRACT

We discuss the mechanism of efficient photoluminescence (PL) from Si quantum dots and wells. Luminescence properties of SiO_2-capped Si nanocrystals are different from those of H-terminated Si nanocrystals, but are very similar to those of Si quantum wells sandwiched by SiO_2 layers. The size-dependence of PL properties and resonantly excited PL spectra of SiO_2-capped Si dots and wells indicate that excitons are localized near the interface between the crystalline Si core and surface oxide layer, and the strong coupling of electronic and vibrational excitations causes the broad PL spectrum. The exciton localization plays an essential role in the efficient luminescence process in nanoscale Si/SiO_2 systems.

INTRODUCTION

The discovery of efficient visible photoluminescence (PL) from Si [1] Ge [2], and SiC [3] nanocrystals has stimulated considerable efforts in understanding optical properties of indirect-gap group IV semiconductor nanocrystals [4]. In semiconductor nanocrystals, quantum confinement effects would play an essential role in optical transitions. However, experimental observation of the PL wavelengths from surface-oxidized Si nanocrystals does not show the dependence on size which is expected from simple quantum confinement [4,5]. The mechanism of visible PL from Si nanostructures is still under discussion.

Well-characterized Si nanostructures are desirable for the understanding of the PL mechanism of porous Si and Si nanocrystals. In particular, we need to study the optical properties of Si nanostructures with near identical and stable surface structures. In this paper, we report luminescence properties of SiO_2-capped Si nanocrystals and single Si quantum wells sandwiched by SiO_2 layers. The understanding of the electronic structures of nanoscale c-Si/SiO_2 systems is a challenging task in condensed matter physics and an inevitable step toward realization of high-performance semiconductor devices.

SIZE DEPENDENCE OF THE LUMINESCENCE PEAK ENERGY

Zero-Dimensional Si Nanocrystals

Surface-oxidized Si nanocrystals were produced by laser-breakdown of SiH_4 gas and plasma decomposition of SiH_4 gas. Details of the preparation and characterization of these nanocrystallite samples are given in Ref. [6] and [7]. The surface of the as-prepared Si nanocrystals was terminated by hydrogen atoms. Si nanocrystals with SiO_2 surface layers were formed after thermal oxidation. The size of crystalline Si core was varied from 2.5 to 12 nm. The porous Si layers were formed by electrochemical etching of p-type c-Si wafers (3.5～4.5 Ω cm) or n-type c-Si wafers (1～5 Ω cm) under light illumination [8]. For n-type substrates, light illumination during electrochemical etching is needed for the formation of the porous layer,

Mat. Res. Soc. Symp. Proc. Vol. 452

because holes are necessary for the electrochemical etching process of Si. The Si nanocrystal size can be controlled by changing illumination wavelength.

Figure 1 summarizes the size dependence of the observed PL peak energies in surface-oxidized Si nanocrystals [1,6,9,10] and as-prepared porous Si [8,11]. The size dependence of the calculated band-gap energy in Si nanocrystals is also plotted in the figure [12-14]. The observed PL peak energy in as-prepared porous Si is sensitive to the nanocrystal size, and the observed size-dependence is consistent with the theoretical calculations, as shown in Fig.1. However, after prolonged air exposure, the PL spectra of any of the samples are similar to each other and appear in the red spectral region [15]. The surface of as-prepared Si nanocrystals is mainly covered with hydrogen during etching, but after air exposure the surface is naturally oxidized. The surface oxidation of nanocrystals changes the PL wavelength. The luminescence spectrum and dynamics of Si nanocrystals are very sensitive to surface structures of Si nanocrystals, particularly with regard to the amount of oxygen and hydrogen on the surfaces.

The size dependence of the PL peak energy in surface-oxidized Si nanocrystals is different from that in H-terminated Si nanocrystals. In smaller nanocrystals, the PL energy in surface-oxidized Si nanocrystals is much lower than that in H-terminated Si nanocrystals. Furthermore, there is a large different between the PL peak energy and the theoretical band gap energy in small oxidized nanocrystals. These results suggest that excitons relax from the higher-energy absorption state to the lower-energy emission states in oxidized Si nanocrystals. In particular, in small nanocrystal exhibiting efficient PL in the visible spectral region, the exciton localization (the disorder-induced exciton localization and extrinsic self-trapping of excitons) is important in the electronic process from the light absorption to light emission [16].

Si Quantum Wells

The Si single quantum wells were formed on SIMOX (separation by implanted oxygen) wafers. The detailed fabrication methods and TEM (transmission electron microscopy) images were shown in Ref. 17. The 2D Si layer was sandwiched between thin surface SiO_2 and thick buried SiO_2 layers. Good crystalline quality in the Si layers was confirmed by the lattice image of the TEM and the roughness of the Si layers was about a few monolayer. The asymmetric PL spectra were observed in the red and infrared spectral region, and can be fitted by two Gaussian bands, the weak PL band (denoted as Q) and the strong PL band (denoted as I) [18]. The PL peak energy of the I band is almost independent of the well thickness. In contrast, the peak energy of the Q band shifts to higher energy with a decrease of the Si well thickness. Thickness dependence of the PL peak energies of the I and the Q bands is plotted in Fig. 2. The peak energy of the main I band does not depend on the well thickness. Similar size-independence of the PL peak energy has been observed in SiO_2-capped Si nanocrystals (see, Fig. 1). Experimental observation of the size-insensitive PL in both 0D and 2D Si/SiO_2 systems suggests that main luminescence originates from excitons localized at the interface between c-Si and SiO_2 layers.

Theoretical calculations of the thickness dependence of the band gap energy for 2D Si wells [19,20] are denoted by the lines in Fig. 2. The Q band peak energy is roughly consistent with theoretical calculations based on the quantum confinement model. The results imply that the Q band is caused by radiative recombination in the Si well. The size-dependence of the PL peak energy in c-Si/SiO_2 quantum wells is different from that in a-Si/SiO_2 quantum wells [21]. This is mainly because the coherent length of the wavefunctions in c-Si is different from that in a-Si.

Moreover, we speculate that the atomic configurations at the interface between c-Si and SiO_2 are different from those between a-Si and SiO_2. The strains at the interface, the interface roughness, and the compositional fluctuation of the interface region affect the electronic structures of very thin 2D wells.

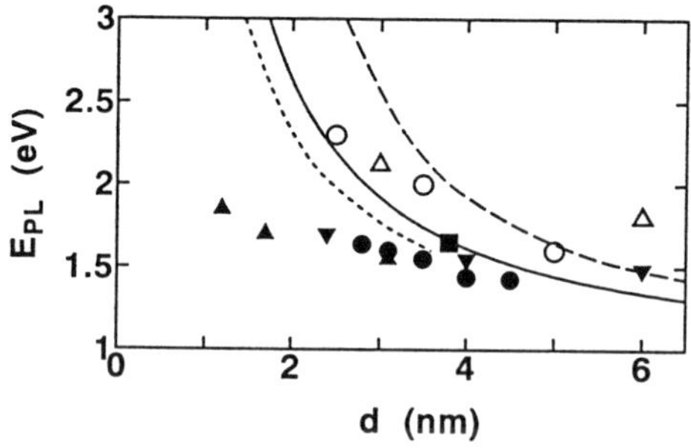

Figure 1. Size dependence of the photoluminescence peak energy in Si nanocrystals. The photoluminescence peak energies in surface-oxidized Si nanocrystals are reproduced from Ref. 1 (●), Ref. 6 (■), Ref. 9 (▲), and Ref. 10 (▼) and those in H-terminated Si nanocrystals are reproduced from Ref. 8 (○) and Ref. 11(△). Theoretical curves for the band-gap energy are also shown from broken (Ref. 12), solid (Ref. 13), and dashed (Ref. 14) lines.

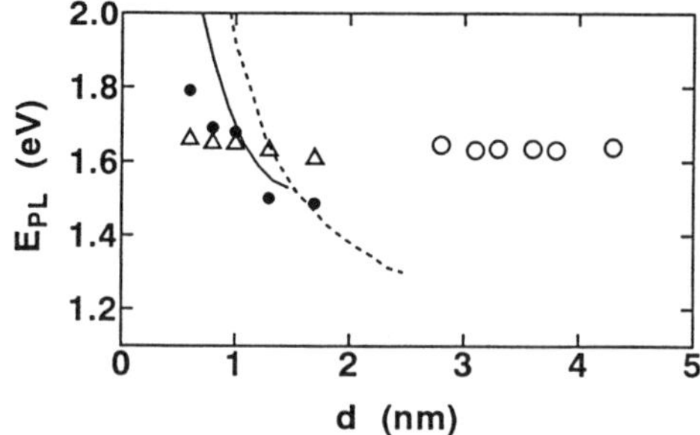

Figure 2. Thickness dependence of the PL peak energy of the Q (●)and I (△)bands. The PL peak energies in thick Si wells (○)are reproduced from Ref. 17. The solid (Ref. 19) and dotted (Ref. 20) lines show the calculated band-gap energy of 2D Si wells.

RESONANTLY EXCITED LUMINESCENCE SPECTRA

0D Si nanocrystals and 2D Si wells show broad PL spectra in the infrared and visible region. Under UV and blue laser excitation, these samples show featureless and broad PL spectrum. For excitation energies within the PL band of samples, we observed well-resolved fine structures in luminescence spectra at low temperatures. By reducing the excitation energy below 2 eV, step-like phonon structures due to TO phonons are clearly observed in as-prepared prepared porous Si. These step-like structures are more clearly observed with a decrease of the excitation photon energy. A spacing between the steps can be explained by ~57 meV TO-phonons in crystalline Si [22]. On the other hand, the phonon-related structures are not observed in oxidized Si nanocrystals. Figure 3 summarizes PL spectra in oxidized Si nanocrystals and porous Si at the low-energy side of the laser line at 2 K under the same experimental conditions. No fine structures in the PL spectrum of surface-oxidized Si nanocrystals suggest that a potential fluctuation in surface states in oxidized Si nanocrystals is much larger than that of H-terminated Si nanocrystals [23]. The experimental observations (Fig. 1 and Fig. 3) rule out the suggestion [9] that the band-to-band luminescence occurs even in the 1-2 nm nanocrystals with SiO_2 surface layers. In very small nanocrystals exhibiting efficient PL, the localization of excitons plays an essential role in controlling the radiative recombination process. The disordered potential in the interface causes the exciton localization. The H-terminated Si nanocrystal shows the crystalline nature, while the oxidized Si nanocrystals show the disorder nature.

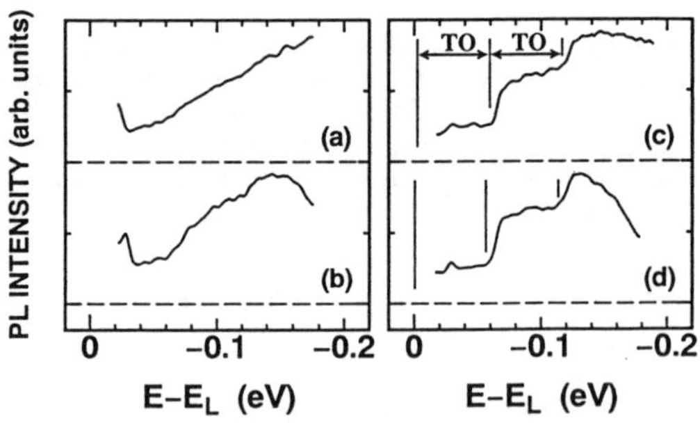

Figure 3. Resonantly excited photoluminescence spectra of surface-oxidized Si nanocrystals [(a) and (b)] and as-prepared porous Si [(c) and (d)] at the low-energy side of the laser lines of 1.724 eV [(a) and (c)] and 1.658 eV [(b) and (d)] at 2 K.

We measured PL spectra under excitation at energies within the PL band in 2D quantum wells and the degree of linear polarization of luminescence, $\rho = (I_{\parallel} - I_{\perp})/(I_{\parallel} + I_{\perp})$, where $I_{\parallel}$ ($I_{\perp}$) is the intensity of the PL polarized parallel (perpendicular) to the polarization of the excitation laser light and the polarization of the excitation laser is parallel to the unconfinement direction in the cleavage plane. The PL polarization memory suggests the Q-band PL originating from optically anisotropic region such as the 2D quantum wells. When the Q band is resonantly excited, satellite PL structures similar to Fig. 3 are observed. From the comparison with porous Si data, we speculate that the satellite structures in the 2D Si are due to the deformation potential coupling between excitons and TO phonons. Then, TO-phonon related fine structures in the PL spectrum and the PL polarization memory show that the Q band PL is related to the size-quantized states of 2D wells. On the other hand, TO-phonon related structures are not observed when the I band is excited. The peak energy of the broad PL band is shifted to 140 meV from the laser excitation energy. This energy approximately corresponds to the vibrational energy of the Si-O-Si stretch mode ($\sim$1100 cm^{-1}). The coupling of localized excitons and local vibrations of silicon oxide species causes new structures in the PL spectrum. Since the Si-O bond is polar, the coupling of excitons and vibrations increases with localization of excitons in smaller dimensions [24]. Resonantly excited PL spectroscopy also shows that the Q band is due to the radiative recombination of excitons in the 2D Si wells, while the I band is due to the recombination at the interface state between the c-Si well and the SiO_2 layer.

INTERIOR AND INTERFACE STATES

Surface-oxidized Si nanocrystals and naturally oxidized porous Si exhibit the stable visible luminescence in the red spectral region. The PL peak wavelength is not sensitive to the crystallite size. Instead, the PL spectrum and dynamics are sensitive to the surface chemistry. Luminescence properties of surface-oxidized Si nanocrystals are different from those of H-terminated Si nanocrystals. Oxidized Si nanocrystals and naturally oxidized porous Si show the size-insensitive red PL. We then propose a three-region model for oxidized nanocrystals composed of (a) a crystalline core, (b) the interface region, and (c) an outer oxide layer or glass matrix [6]. In this model, the interfacial region plays the most important role in the stable and red PL process in small nanocrystals where the energy gap of the crystalline core state is higher than that of the interface states. In Si/SiO_2 junctions, the lattice mismatch is very large. Moreover, spherical curvature of the interface, strains between c-Si and SiO_2, roughness and compositional fluctuation at the interface cause a disorder potential at the interface. The Si nanocrystals with a disorder potential of interface states show complicated luminescence properties. In very thin 2D and 0D Si quantum structures, the band-gap energies of c-Si quantum wells and dots are blueshifted due to the quantum confinement effect. Then, the interface states which are not observed in the bulk crystalline Si and SiO_2 system would appear in low-dimensional nanostructures. Because the interface states between c-Si and SiO_2 layer plays the most essential role in PL process in 0D and 2D Si/SiO_2 systems, the luminescence properties of 0D Si nanocrystals are similar to those of 2D Si wells.

On the other hand, in large Si nanocrystals, the band gap energy of the crystalline core state is lower than the $\sim$1.65 eV interface state. The radiative recombination of excitons would occur in the crystalline core state similar to H-terminated Si nanocrystals. We can observe the phonon-related fine structures under IR laser excitation. Moreover, since excitons can move in 2D wells, luminescence properties due to the delocalized exciton is more clearly observed in 2D well structures rather than 0D nanocrystals. The size-sensitive PL in H-terminated Si nanocrystals comes from mostly delocalized states of the Si crystalline core, while the size-insensitive PL in oxidized Si nanocrystals comes from oxygen-modified localized states.

CONCLUSIONS

We have studied PL properties of 0D Si dots and 2D Si quantum wells. In oxidized 0D and 2D Si nanostructures, excitons are localized near the interface between c-Si and SiO_2 layers and the strong coupling of excitons and vibrations causes the broad PL spectrum. PL properties of hydrogen-terminated Si nanocrystals are quite different from those of surface-oxidized Si nanocrystals. In nonpolar semiconductor nanocrystals, the polar bonding at the surface plays an active role in luminescence processes in nanometer dimensions.

ACKNOWLEDGMENTS

The authors would like to thank Dr. K. Shiraishi, Dr. H. Kageshima, Dr. Y. Takahashi, and Dr. K. Murase of NTT Basic Research Laboratories and Dr. M. Otobe and Prof. S. Oda of Tokyo Institute of Technology for discussions.

REFERENCES

1. H. Takagi, H. Ogawa, Y. Yamazaki, A. Ishizaki, and T. Nakagiri, Appl. Phys. Lett. **56**, 2349 (1990).
2. Y. Maeda, N. Tsukamoto, Y. Yazawa, Y. Kanemitsu, and Y. Masumoto, Appl. Phys. Lett. **59**, 3168 (1992).
3. T. Matsumoto, J. Takahashi, T. Tamaki, T. Futagi, H. Mimura, and Y. Kanemitsu, Appl. Phys. Lett. **64**, 226 (1994).
4. Y. Kanemitsu, Phys. Rep. **263**, 1 (1995).
5. C. Delerue, M. Lannoo, and G. Allan, Phys. Rev. Lett. **76**, 3038 (1996); N. A. Hill and K. B. Whaley, *ibid* **76**, 3039 (1996).
6. Y. Kanemitsu, T. Ogawa, K. Shiraishi, and K. Takeda, Phys. Rev. B **48**, 4883 (1993).
7. M. Otobe, T. Kanai, T. Ifuku, H. Yajima, and S. Oda, J. Non-Cryst. Solids **198-200**, 875 (1996).
8. H. Mimura, T. Matsumoto, and Y. Kanemitsu, Mater. Res. Soc. Symp. Proc. **358**, 635 (1995).
9. S. Schuppler S. L. Friedman, M. A. Marcus, D. L. Adler, Y. H. Xie, R. M. Ross, Y. J.Chabal, T. D. Harris, L. E. Brus, W. L. Brown, E. E. Chaban, P. F. Szajowski, S. B. Christman, and P. H. Citrin, Phys. Rev. B **52**, 4910 (1995).
10. Y. Kanemitsu, S. Okamoto, M. Otobe, and S. Oda, to be published.
11. K. Ito, S. Ohyama, Y. Uehara, and S. Ushioda, Appl. Phys. Lett. **67**, 2536 (1995).
12. T. Takagahara and K. Takeda, Phys. Rev. B **46**, 15578 (1992).
13. P. Proot, C. Delerue, and G. Allan, Appl. Phys. Lett. **61**, 1948 (1992).
14. L. W. Wang and A. Zunger, J. Chem. Phys. **100**, 2394 (1994).
15. Y. Kanemitsu, H. Uto, Y. Masumoto, T. Matsumoto, T. Futagi, and H. Mimura, Phys. Rev. B **48**, 2827 (1993).
16. Y. Kanemitsu, N. Shimizu, T. Komoda, P. L. F. Hemment, and B. J. Sealy, Phys. Rev. B **54**, R14329 (1996).
17. Y. Takahashi, T. Furuta, Y. Ono, T. Ishiyama, and M. Tabe, Jpn. J. Appl. Phys. **34**, 950 (1995).
18. S. Okamoto and Y. Kanemitsu, to be published.
19. H. Kageshima, Surf. Sci. **357/358**, 312 (1996).
20. S. B. Zhang and A. Zunger, Appl. Phys. Lett. **63**, 1399 (1993).
21. D. J. Lockwood, Z. H. Lu, and J. M. Baribeau, Phys. Rev. Lett. **76**, 539 (1996).
22. P. D. J. Calcott, K. J. Nash, L. T. Canham, M. J. Kane, and D. Brumhead, J. Phys. Condens. Matter. **5**, L91 (1993).
23. Y. Kanemitsu, Phys. Rev. B **53**, 13515 (1996).
24. S. Okamoto and Y. Kanemitsu, Phys. Rev. B **54**, 16421 (1996).

POSSIBLE MECHANISM OF THE 30-100 PICOSECONDS FAST AND EFFICIENT PHOTOLUMINESCENCE FROM nc-Si/a-SiO_2 DOPED WITH TRANSITION METALS

S. VEPREK[*], Th. WIRSCHEM[*], J. DIAN[*], S. PERNA[*], R. MERICA[*] M.G.J. VEPREK-HEIJMAN[*], V. PERINA[**], M. FUSS and X. LIN[***]

[*] Institute for Chemistry of Inorganic Materials, Technical University Munich, Lichtenbergstrasse 4, D-85747 Garching/Munich, Germany

[**] Institute of Nuclear Physics, CZ-25068 Rez u Prahy, Czech Republic

[***] Institute for Radiochemistry, Technical University Munich, Walther-Meissner-Str. 3, D-85747 Garching, Germany

ABSTRACT

The nc-Si/a-SiO_2 composite thin films doped with tungsten show very fast and efficient photoluminescence (PL). In order to obtain insight into the PL mechanism we have performed a comparative study with other metals. The results lend support to the suggested mechanism which includes the photogeneration of charge carriers due to efficient absorption of the excitation UV light in the silicon nanocrystals followed by energy transfer to the W^{n+} radiative center from which the light emission occurs.

INTRODUCTION

The usually-observed yellow-red photoluminescence from porous (PS) and nanocrystalline silicon is relatively slow with a decay time $\tau_{1/2}$ of tens of microseconds which renders it relatively uninteresting technologically. After elucidating the mechanism of this PL from nc-Si/a-SiO_2 thin films we have suggested using fast inorganic dyes in order to obtain efficient PL from such a material which could be sufficiently fast to enable, e.g. optical chip-to-chip communication [1]. More recently we reported that the same PL is obtained from spark processed silicon when a tungsten electrode is used [2]. In this paper we shall present new data on the doping with other transition metals which support the suggested mechanism.

The charge transfer transitions are optically allowed and therefore very fast. In the case of tungsten complexes they occur in the blue-violet spectral region where also substoichiometric WO_{3-x} shows an intense absorption called "Berlin blue" [3,4]. Various tungstates also show efficient PL which is, however relatively slow with $\tau_{1/2}$ of the order of 0.1 μs [5-7] to 10 μs [8,9] at room temperature. The idea underlying our work was based on the expectation that isolated W^{n+} ions incorporated in the nc-Si/a-SiO_2 composites close to the Si nanocrystals could facilitate similar fast transitions as observed in the molecular complexes. In its highest oxidation state 6, all valence electrons of tungsten are localized at the ligands which is the reason for the observed charge transfer spectra. However, doping of nanocrystals can also result in an enhancement of the PL efficiency and shortening of the decay time as shown, for example, for the case of doping of ZnS nanocrystals with manganese [10,11]. The decay time of the PL from erbium doped Si is also very slow. The intrinsic life time of about 1 ms can be shortened only by non-radiative recombinations which lead to a corresponding decrease of the PL efficiency. For these reasons the W-doped nc-Si represents an attractive approach towards the development of efficient and fast optoelectronic devices.

Because the W^{n+}-related transitions are expected to occur in the photon energy range of 2.5-3.0 eV, the band gap of the Si nanocrystals should be equal or larger in order to facilitate the

Mat. Res. Soc. Symp. Proc. Vol. 452

desirable energy transfer. Figure 1 shows the theoretical calculations of the dependence of the band gap on the crystallite size by Delley and Steigmeier [12] and by Delerue et al. [13]. It is well known that the PL from nc-Si/a-SiO_2 films shows no blue shift with decreasing crystallite size and that the shift observed in PS or isolated Si-nanocrystals is much smaller than the calculated increase of the band gap. This is illustrated for our films in Fig. 1 as well [15]. Recently we have measured the increase of the band gap by the excitation spectra of the red PL [14]. In such a case, the intensity of the red PL at a constant energy (here the peak energy of 1.5 eV) is measured as a function of the energy of the excitation light E_{ex}. Because the crystallite sizes have a certain distribution which is difficult to determine, the onset of the PL excitation spectra is somewhat broadened. Therefore we use the energy corresponding to a 20% increase of the PL intensity I_{PL} with respect to the maximum (see symbols Δ in Fig. 1) and the value obtained by extrapolation of the dependence of $(I_{PL} \cdot E_{ex})^{1/2}$ on E_{ex} to $I_{PL}=0$, which is linear for an indirect band gap (see full symbols in Fig. 1). One can see that a very good agreement with the theoretical calculations is obtained in both cases. Thus, for the W^{n+}-stimulated PL one has to use nc-Si/a-SiO_2 films with an average crystallite size of ≤ 2 nm.

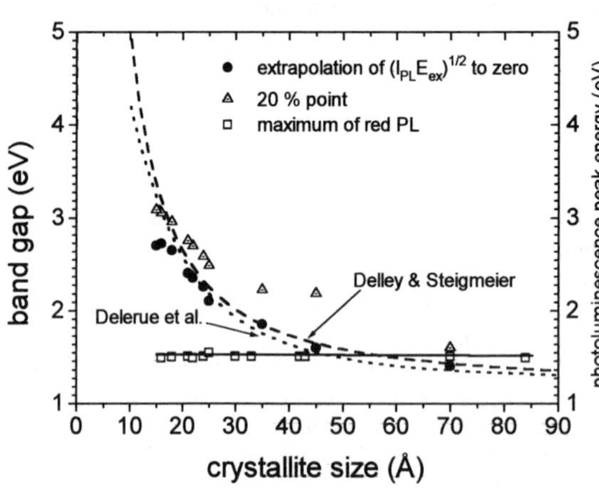

Fig. 1: Dependence of the band gap of nc-Si on the crystallite size calculated by Delley and Steigmeier [12] and by Delerue et al [13] and experimentally determined from the PL excitation spectra [14]. For comparison, the peak position of the broad PL band which remains unchanged [15] is shown as well.

EXPERIMENTAL

The sample preparation was described in our earlier papers [1,2,15-17]. Thus only a brief summary is given here. The nc-Si films of desirable crystallite size were prepared either by chemical transport of silicon in hydrogen glow discharge or by depositing a-Si followed by its controlled recrystallization. Afterwards, the surfaces of the crystallites are passivated by oxidation of the grain boundaries. The oxidation also allows us to control the average separation between the crystallites which is very important for obtaining an efficient red PL [18]. A subsequent annealing in forming gas further decreases the concentration of dangling bonds (which act as non-radiative recombination centers) thus resulting in a further increase of the red PL. Afterwards, the films are doped with tungsten by evaporating a desirable amount of WO_3 at the surface and diffusion of the tungsten into the films at about 870°C in forming gas. This preparation procedure is rather tricky. Because of lack of space available here we must refer to a subsequent paper for further details.

In the course of these studies we have realized a close similarity between the PL spectra from our W^{n+}-doped nc-Si/a-SiO_2 films and those reported for spark processed silicon [19-24]. Therefore we have used an arrangement and procedure as reported by Hummel et al. [19,20]. The spark discharge was maintained between a silicon wafer and a sharp counter-electrode made of the chosen material, such as W, Mo, Cr, Mn, Fe, Co, Ni, Cu, Au, Pt, Ag and Si by means of a Tesla coil (Edwards ST 4 Spark Tester).

The measurements of the PL spectral distribution, its polarization and decay time were described elsewhere [2,18]. The Rutherford backscattering (RBS) measurements of the metal content and depth distribution was done either with He^+ or H^+ ions [25] and evaluated by means of the Gisa 3 PC program [26] using the cross sections reported in [27,28].

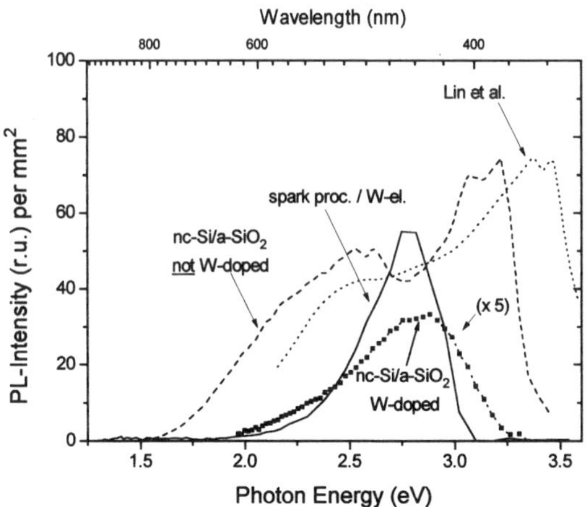

Fig. 2: PL spectra of a W-doped nc-Si/SiO_2 sample prepared by plasma CVD, a spark processed sample with W-counter-electrode and an undoped nc-Si/SiO_2 sample which was stored in air for one year prior to the oxidation. The latter is similar to a PL spectra reported by Lin et al. [34] and green PL from a sample oxidized in air. The measured PL intensities were normalized to a sample area of 1 mm^2.

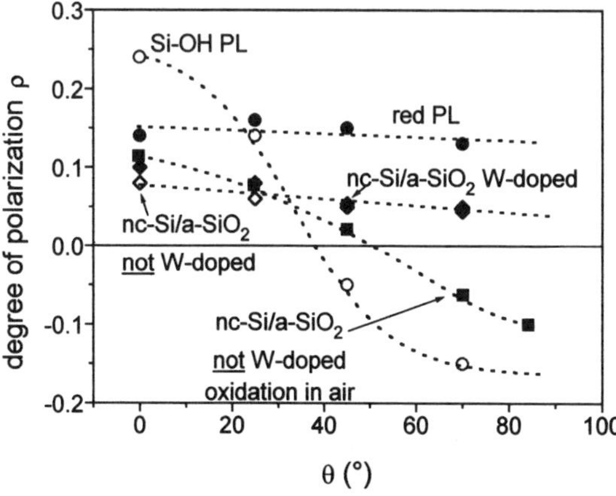

Fig. 3: Dependence of the degree of polarization on the angle between the electric vector of the incident light and the z-axis normal to the plane. For experimental details see [33,35]

RESULTS

Figure 2a shows as an example PL spectra from plasma deposited nc-Si/SiO_2 films doped with W^{n+}; spark processed Si using a tungsten counter-electrode; an undoped, heavily oxidized nc-Si film which has been oxidized after one year storage in air; and a similar PL spectrum recently reported by Lin et al. [34] and attributed to nanocrystalline silicon. The green PL observed after oxidation in air is due to defects in SiO_2. One notices that the PL spectra from the undoped sample which was stored in air for a long time is very similar to that reported by Lin et al and quite different from the PL of the W-doped samples. The dependence of the degree of polarization of the PL on the angle of the polarization vector with respect to z-axis (see [33,35]) is shown in Fig. 3. The absence of any dependence in the case of the tungsten doped samples indicates an energy transfer between the absorbing and emitting center whereas the dependence of the degree of polarization on θ is typical for the case when the same dipole absorbs and emits the light [36].

During the spark processing with a tungsten electrode, the blue PL appears visible to the naked eye already after about one hour. For comparison with the other metals all samples were processed for about 5 hours. In all cases the concentration and depth distribution of the respective metal was measured by means of RBS and found in the range between about $5x10^{18}$ and $5x10^{19}$ metal atoms/cm^3 and relatively uniform in depth. These data are not shown here for lack of space and we refer to a full length paper to follow.

Samples processed with chromium and molybdenum (which belong to the same VI B group as tungsten) showed PL with the same spectral distribution as those doped with tungsten (Fig. 2) but with a 7 - 10 times smaller intensity. On the other hand, no PL could be detected even with orders of magnitude higher sensitivity for samples processed with the other metals: Mn, Fe, Co, Ni, Au, Pt and Ag. These metals were chosen because their complexes show a large variety of color and because some of them might be contaminants of the Si samples during the spark processing.

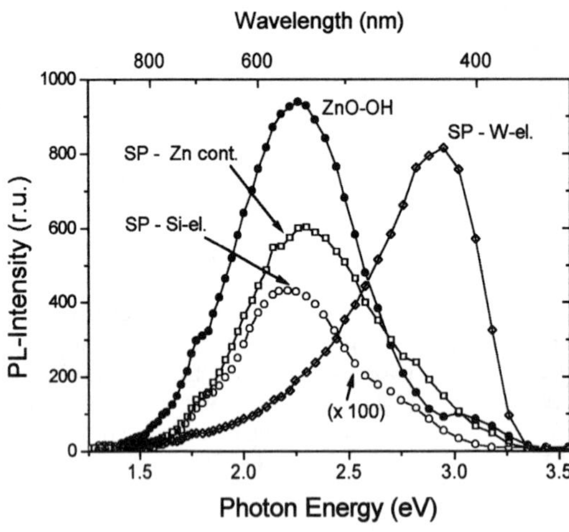

Fig. 4: PL spectra of spark processed Si wafers with a Si-counter-electrode: Open squares: a sample with Zn impurities measured by RBS; open circles: a sample in which the impurities were just at the detection limit of the RBS. Full symbols: ZnO-OH sample from [33]. The PL from W-doped sample with a maximum at 2.8 eV is indicated for comparison as well.

Processing of the Si wafer with a Si counter-electrode resulted in a PL with a spectral maximum at 2.25 eV but about 30 times smaller intensity than that from the tungsten doped sample. As the RBS indicated that there might be some metallic impurities in this sample, we have investigated another sample processed in a similar way and showing a much stronger PL with a maximum at 2.25 eV. The PL spectra of these two samples are shown in Fig. 4 in comparison with the PL from tungsten processed Si and from hydrated ZnO [33]. The similarity between the latter PL and those from spark processed Si with the Si-counter-electrode is obvious. The RBS analysis clearly reveals the presence of zinc in the range of 10^{18} Zn-atoms/cm^3. Thus, the green PL observed from samples processed with the Si-counter-electrode are due to Zn impurities which are introduced into the sample during the processing. In order to assure that the PL observed from the samples doped with tungsten, chromium and molybdenum is not due to some other impurities, we have done neutron activation analysis of a series of representative samples which showed visible PL. With the exception of Si dopants, such as B, Sb, As and the metallic impurities intentionally introduced, the concentration of all other impurities was below 200 ppb.

We have also verified that no detectable PL could be seen when spark processing pure SiO_2 with a W-counter-electrode for a period up to 12 hours.

CONCLUSIONS

All results available so far (those published earlier in [1] and those reported here) are in agreement with the suggested mechanism according to which the ≤ 2 nm small silicon crystallites efficiently absorb the excitation light with the formation of electron-hole pairs. This is followed by energy transfer to a radiative Me^{n+} center (Me being tungsten, molybdenum or chromium) from which emission of the PL quantum occurs. Details of this process need further clarification. We can also exclude other impurities as being responsible for the observed blue PL. Furthermore, the fact that even 12 hours of spark processing of SiO_2 with a tungsten electrode did not yield any detectable PL shows that the presence of silicon nanocrystals is necessary, and that the observed PL is not due to defects in SiO_2. However, this mechanism still cannot be considered as proven with certainty and further, very careful investigations are needed because many organosilicon compounds show efficient PL in that region. We have recently found fairly efficient PL from silsesquioxanes $R_8(SiO_{1.5})_8$ (R is an organo ligand such as butyl, phenyl, methyl etc.) in the spectral range between about 2.3 and 4 eV, the position of the maximum being dependent on the nature of the ligands. It is very unlikely that such impurities were responsible for the PL reported in this paper because of the care to assure cleanliness of the processing and because of the fact that no PL was observed with many other metal dopings. Zinc impurities are obviously responsible for the PL with a maximum at 2.25 eV when a Si-counter-electrode is used. This PL is very weak or not observable if the microsparking is carefully minimized or avoided.

We should like to thank Deutsche Forschungsgemeinschaft (German Research Foundation) for financial support of this work. J.D. thanks the Alexander-von-Humboldt-Stiftung for financial support.

REFERENCES

[1] S. Veprek, M. Rückschloss, Th. Wirschem, B. Landkammer, M. Fuss and Y. Lin, Appl. Phys. Lett. **67**, 2215 (1995).

[2] S. Veprek, Th. Wirschem, J. Dian, S. Perna, R. Merica, M.G.J. Veprek-Heijman, R. Heinecke and V. Perina, European Mater. Res. Soc., Spring Meeting, June 1996, Strasbourg, France, in press.

[3] F.A. Cotton and G. Wilkinson, Advanced Inorganic Chemistry, J. Wiley, New York 1974.

[4] N.N. Greenwood and A. Earnshaw, Chemistry of Elements, Pergamon Press, Oxford 1984.

[5] P. Bräunlich, K. Reiber and A. Scharmann, Zeitschrift f. Physik **183**, 431 (1965).
[6] M. Becker and A. Scharmann, Z. f. Naturforschung **29a**, 1060 (1974).
[7] Gmelins Handbuch der Anorg. Chem., Wolfram Ergänzungsband **B4,** Springer-Verlag, Berlin 1980.
[8] M.J. Treadaway and R. Powell, J. Chem. Phys. **61**, 4003 (1974).
[9] M.J. Treadeway and R. Powell, Phys. Rev. B **11**, 862 (1975).
[10] R.N. Bhargava, D. Gallager, X. Hong and A. Nurmikko, Phys. Rev. Lett. **72**, 416 (1994).
[11] Y.L. Soo, Z.H. Ming, S.W. Huang, Y.H. Kao, P.N. Bhargawa and D. Gallager, Phys. Rev. B **50**, 7602 (1994).
[12] B. Delley and E.F. Steigmeier, Phys. Rev. **B 4**7,1397(1993) and Appl. Phys. Lett. **67**, 2370 (1995).
[13] C. Delerue, M. Lannoo and G. Allan, J. Lumin. **57**, 249 (1993).
[14] S. Veprek and Th. Wirschem, in: Handbook of Optical Properties, Vol.II: Optics of Small Particles, Interfaces and Surfaces, ed. R.E. Hummel, CRC Press, Boca Raton 1996 (in press).
[15] M. Rückschloss, O. Ambacher and S. Veprek, J. Lumin. **57**, 1 (1993).
[16] M. Rückschloss, B. Landkammer and S. Veprek, Appl. Phys. Lett. **63**, 1474 (1993).
[17] S. Veprek, M. Rückschloss, B. Landkammer and O. Ambacher, Mater. Res. Soc. Proc. **298**, 117 (1993).
[18] S. Veprek, H. Tamura, M. Rückschloss and Th. Wirschem, Mater. Res. Soc. Symp. Proc. **345**, 311 (1994).
[19] R.E. Hummel and S.-S. Chang, Appl. Phys. Lett. **61**, 1965 (1992).
[20] M.H. Ludwig, R.E. Hummel and M. Stora, Thin Solid Films **255**, 103 (1994).
[21] R.E. Hummel, M.H. Ludwig and S.-S. Chang, Solid State Commun. **93**, 237 (1995).
[22] R.E. Hummel, M.H. Ludwig, J. Hack and S.-S. Chang, Solid State Commun. **96,** 683 (1995).
[23] M.H. Ludwig, J. Menninger and R.E. Hummel, J. Phys.: Condens. Matter. 7, 9081 (1995).
[24] R.E. Hummel, M.H. Ludwig and S.-S. Chang, Mater. Res. Soc. Proc. **358**, 151 (1995).
[25] W.-K. Chu, J.W. Mayer and M.A. Nicolet, Backscattering Spectrometry, Academic Press, New York 1978.
[26] J. Saarliathi and E. Rauhala, Nucl. Instr. & Methods in Phys. Res. **B 64**, 734 (1992).
[27] R. Salamonovic, Nuclear Instr. & Methods in Phys. Res. **B 82**, 1 (1993).
[28] M. Lumoajärvi, E. Rauhala and M. Hautala, Nucl. Instr. & Methods in Phys. Res. **B 9,** 255 (1985).
[29] D.L. Griscom, J. Ceramic Soc. Japan **99**, 923 (1991).
[30] H. Nishikawa, T. Shiroyama, R. Nakamura, Y. Ohki, K. Nagasawa and Y. Hama, Phys. Rev. **B 45**, 586 (1992).
[31] A. Aneda, G. Bongiovani, M. Cannas, F. Congiu, A. Mura and M. Martini, J. Appl. Phys. **74**, 6993 (1993).
[32] J.H. Sathis and M.A. Kastner, Phys. Rev. **B 35**, 2972 (1987).
[33] M. Rückschloss, Th. Wirschem, H. Tamura, G. Ruhl, J. Oswald and S. Veprek, J. Lumin. **63**, 279 (1995).
[34] J. Lin, G.O. Yao, J.Q. Duan and G.G. Qin, Solid State Commun. **97**, 221 (1996).
[35] S. Veprek, Th. Wirschem, M. Rückschloss, H. Tamura and J. Oswald, Mater. Res. Soc. Symp. Proc. **358**, 99 (1995).
[36] A.C. Albrecht, J. Molec. Spectr. **6**, 84 (1961).
[37] K.J. Laidler, Chemical Kinetics, 3rd ed., Harper & Row, New York 1987.

PREPARATION AND CHARACTERIZATION OF SILICON NANOCRYSTALS IN A SiO_2 MATRIX AND STUDY OF SUBOXIDE STABILITY

B.J. HINDS, A. BANERJEE†, R.S. JOHNSON, and G. LUCOVSKY
Department of Physics, North Carolina State University, Raleigh, NC 27695-8202

ABSTRACT

The kinetics of the decomposition of silicon suboxides (SiO_x, x<2) to Si_c + SiO_2 was studied as a function of composition and post-deposition annealing. Amorphous hydrogenated SiO_x films (0.8<x<1.4) were deposited by remote plasma enhanced chemical vapor deposition (RPECVD) and rapid thermal annealed (RTA) at temperatures of 500-1000°C. By monitoring the Si-O infra-red (IR) bond-stretch mode frequency, it was found that at temperatures below 850°C, or at a oxygen poor composition near $SiO_{0.8}$, the decomposition reaction only proceeded to a metastable form of $SiO_{1.6}$ + Si. Characterization by Raman and spectroscopic ellipsometry confirm similar trends. Cross sectional transmission electron microscopy (TEM) confirms that Si nanocrystals (Si_{nc}) are formed with anneals at 900°C (30 sec). As deposited suboxides show band edge photoluminescence at 1.6 eV which disappears upon annealing at 900°C, indicating a sharp suboxide free interface between Si_{nc} and SiO_2 matrix.

INTRODUCTION

The interface between SiO_2 and Si is the subject of important investigations primarily due to its dominating effects in MOSFET performance. As the dimensions of devices are reduced to gate oxide thicknesses of less than 30Å, control of the interface becomes crucial. The Si/SiO_2 interface is well known to be stoichiometric SiO_2 within 4-6Å of the interface where suboxides, SiO_x (x<2) are seen in x-ray photoelectron spectroscopy (XPS) studies [1,2]. The amount and nature of the suboxides is highly dependent on surface preparation, annealing temperatures, and growth temperatures [1]. Interestingly, SiO is thermodynamically unstable below 1173°C [3], pyrophoric and readily decomposes as in the unbalanced Eqn. (1) [3-5]. To date the decomposition kinetics of SiO has not been studied, primarily due to the ambient reactivity of the non H-terminated Si dangling bonds.

$$SiO_x \longrightarrow SiO_2 + Si \qquad (1)$$

A method to produce H-termination stabilized Si suboxides is the chemical vapor deposition (CVD) of oxygen doped a-Si, also known as semi-insulating polysilicon (SIPOS). Early reports show that these SIPOS films decompose into Si_{nc} and amorphous SiO_2 matrix [6] as expected in Eqn. (1). Transmission electron microscopy (TEM) and x-ray diffraction (XRD) analysis show Si_{nc} size increasing with annealing temperature and Si content [6]. Due to the relatively high growth temperature of 650°C, the material initially contains Si_{nc} before heat treatments [7-9]. Further studies have more thoroughly examined the properties of SIPOS with various heating treatments by XRD [10], TEM [6, 11-13], auger electron spectroscopy (AES) [9], XPS [6,8], and various optical techniques including infra-red (IR), Raman and ellipsometry [11,14,15]. However these studies did not closely examine the temperature region which ultra-thin MOSFET gate dielectric growth is important (700-900°C), investigate the stability or kinetics of the decomposition reaction, or probe the defect structure of SiO_2/Si_{nc} interface.

Remote plasma enhanced chemical vapor deposition (RPECVD) has been demonstrated to grow high quality gate dielectrics [16] and can readily grow *amorphous* SiO_x [17]. An advantage over the SIPOS method is that this amorphous SiO_x has no initial Si_{nc} nucleation sites, facilitating the study of SiO_x stability. Recent work has shown that these SiO_x films form Si_{nc} upon annealing at 900°C and lose the ability for suboxide photoluminescence (PL) at energies < 2eV, indicating a sharp Si_{nc}/SiO_2 interface [18,19]. The object of this research is to closely examine the

† Present address: Texas Instrument Incorporated, MS 944, Dallas TX 75243

stability of RPECVD SiO_x over the technologically important temperature ranges of 700-900°C. Characterization includes IR, Raman, Ellipsometry, PL, TEM and XRD.

EXPERIMENTAL

A RPECVD reactor previously described [20] was employed to grow thin films SiO_x ($0.8<x<1.4$). 100 sccm He with 0.05 to 0.2% O_2 was passed through the RF plasma coil tube with 33W power at 13.56 MHz. 10 sccm of 10% SiH_4/He was introduced through a diffusion ring 20 cm downstream from plasma source. Total pressure was maintained at 300mTorr by a feedback controlled throttle. The Si(100) substrate was positioned 23 cm from plasma source was heated to 210°C and biased at -65V. Film thickness were typically 1700Å (6Å/min). The suboxide films were then annealed *ex-situ* in an AG associates rapid thermal annealing furnace under 1atm of 5% O_2/Ar or Ar for times of 30 sec. to 15 min. Fourier Transform-IR measurements were performed with a Nicolet 750 spectrometer. Spectra are subtracted from a reference sample cut from the same Si(100) single side orange peel polished wafer. Raman measurements were performed with a Ar^+ laser at 514nm, ISA U-1000 double diffractometer, and photomultiplier detector. Samples for Raman spectra were deposited on Ge(111) substrate under identical processing conditions. Ellipsometry data were collected with benchtop rotating analyzer system previously described [21]. PL data measurements utilized double grating 0.85m monochromator and 1 W Ar^+ laser at 488nm (2.54 eV). A liquid N_2 Joule-Thompson cryostat cooled the sample to 80K. TEM cross sections were prepared by standard methods and viewed with a Topcon 002B microscope. XRD was collected on Rigaku diffractometer using $Cu_{k\alpha}$ radiation with single quartz monochromator. Instrument resolution was 0.15° and calibrated to Si powder standards.

RESULTS AND DISCUSSION

An effective method to monitor the degree of suboxide character is to observe the IR Si-O bond stretching frequency ($\nu_{Si\text{-}O}$). For stoichiometric SiO_2 $\nu_{Si\text{-}O}$ is ~1075cm^{-1} and linearly decreases to about 940 cm^{-1} with decreasing oxygen content [17]. Thus as the suboxide decomposes as in equation (1), the observed Si-O stretch would increase. For anneals over 850°C $\nu_{Si\text{-}O}$ is seen to increase to near 1070 cm^{-1} as is seen in Figure 1. There is a time dependence for the decomposition reaction, with a rate that increases with temperature. Reaction completion seen in the order several minutes at 850°C while at 900°C only 30 sec. is required. Most notable is that the decomposition reaction does not go to completion for annealing temperatures below 850°C, implying a kinetically limited stability of Si suboxides below this temperature. To examine this trend further, other compositions of suboxide were annealed by RTA at various temperatures and $\nu_{Si\text{-}O}$ monitored by IR as shown in Figure 2. Each data point shows $\nu_{Si\text{-}O}$ for maximum reaction at the given temperature. Fig. 2 show that with higher initial O concentration allows for a more facile progression to completion (Si + SiO_2), while at a low O concentrations ($SiO_{0.8}$) the decomposition reaction cannot go to completion even at relatively high annealing temperatures of 1000°C. The highest $\nu_{Si\text{-}O}$ of 1044 cm^{-1} (from an initial composition

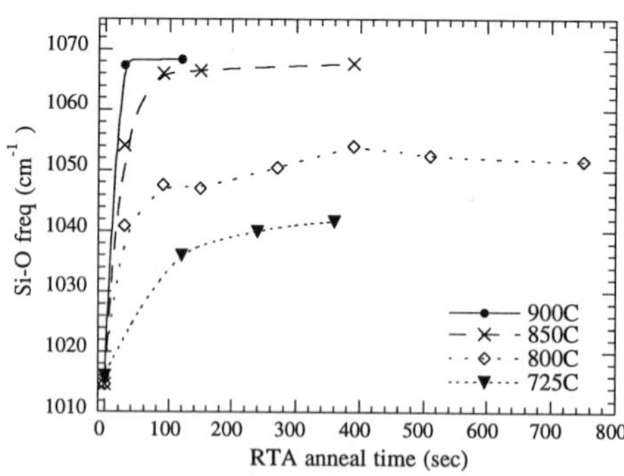

Figure 1. FTIR Si-O stretch freq. as a function of time at various annealing temperatures for initial $SiO_{1.0}$ thin films deposited by RPECVD. Curve shown as guide to eye.

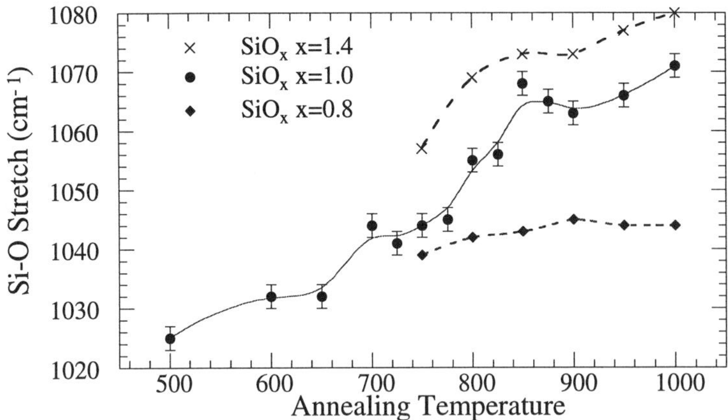

Figure 2. Si-O stretch freq. as a function of annealing temperature for RPCVD suboxides of various initial O concentrations. Annealing times are long enough for reaction completion. Error bars are $\pm 2cm^{-1}$ for x=1.4 and 0.8 but are omitted for clarity, and the curves are guides to the eye.

of $SiO_{0.8}$) corresponds roughly to $SiO_{1.6}$ for the suboxide matrix after annealing.

The most interesting trend is seen for the $SiO_{1.0}$ sample (solid line Figure 2) which shows three very distinct regions of maximum Si-O stretch from 500-1000°C. The first indicates a low level of decomposition, and the shift from $1015cm^{-1}$ (as deposited) to $1023cm^{-1}$ (500°C anneal) occurs simultaneously with the loss of H-passivation, (e.g., the loss of Si-H bond stretching mode at $2200cm^{-1}$). The second region from 700-800°C, the suboxide decomposition only proceeds to ~1044 cm^{-1} (~$SiO_{1.6}$). In the final region, temperatures greater than 850°C, the reaction proceeds to completion. Hence it is of great interest that at temperatures below 850°C, an intermediate metastable state of Si suboxide (~$SiO_{1.6}$) exists.

As a complement to FT-IR measurements, Figure 3. shows the Raman spectrum of $SiO_{1.0}$ films as deposited, annealed at 750°C (Si + $SiO_{1.6}$), and annealed at 900°C (Si+SiO_2). The Raman techniques is more sensitive to Si-Si vibrational modes and shows the progression of amorphous Si (broad peak at $480cm^{-1}$) in the as deposited and 750°C RTA case, to distinct polysilicon peak at $518cm^{-1}$ for the 900°C RTA case. The Raman spectrum display several important points. First is that at the intermediate temperature (750°C) the suboxide decomposition reaction does not go to completion form crystalline Si which is consistent with FT-IR observations. Second is that at reaction completion (900°C) crystalline Si is formed. However, of interest is the broad a-Si feature at ~480cm-1 which is similarly seen in SIPOS heat treatment studies [14], which may indicate a large amount of disorder in the Si_{nc}. The broad peak shows little polarization dependence, hence is unlikely to be attributed to SiO_2. Third the location of the c-Si peak is shifted from $520cm^{-1}$ to $518cm^{-1}$ which has been seen for nanoparticle sized (~100nm) polysilicon [22,23]. Fourth, in the polarized Raman spectra we are unable to detect a chemically order $SiO_{1.5}$ species (Si bonded to one Si and three O) which had been seen analogous GeS_x system [24]. However the weak signal from Si-Si-O bonds cannot be resolved above the background, thus cannot be discounted at this time. The Raman data supports the trends seen in FT-IR of suboxide decomposing into SiO_2 + Si_{nc}, with the kinetics being hindered at temperatures less that 900°C

Using spectroscopic ellipsometry (SE) at energies above the direct bandgap of Si (3.6eV) it is possible to distinguish between c-Si, a-Si, and SiO_2 and model the relative amounts using the effective medium theory. Figure 4. shows the SE spectra of the imaginary part of the complex dielectric function ($\varepsilon = \varepsilon' + i\varepsilon''$) for samples of $SiO_{0.84}$, $SiO_{1.0}$, $SiO_{1.4}$ annealed at 750 and 900°C. For the Si-rich samples, the characteristic two peak feature of c-Si can be seen at both

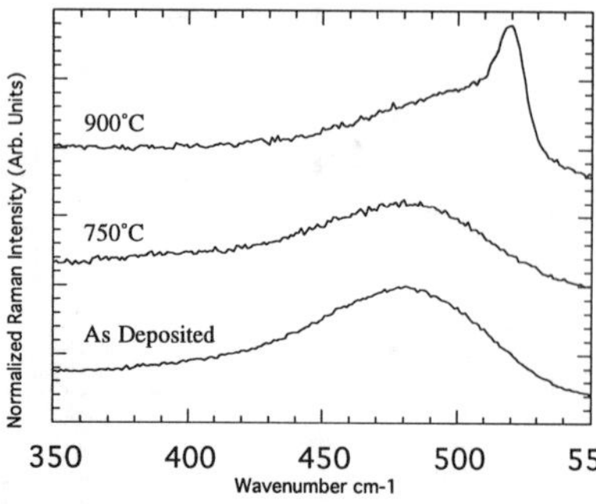

Figure 3. HH polarized Raman spectra of $SiO_{1.0}$ film on Ge substrate as deposited, and after 750°C (4 min.) and 900°C (30 sec.) anneal.

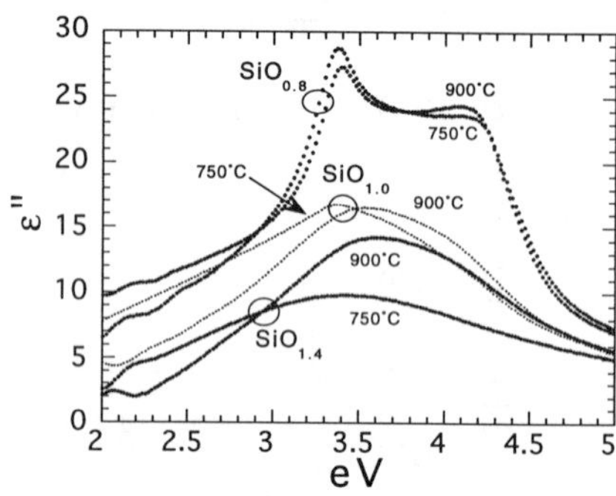

Figure 4. Ellipsometric dielectric loss (ε") as a function of photon energy for suboxides (x = 0.8, 1.0, 1.4) annealed at 750 and 900°C.

temperatures, indicating a dominant amount of polysilicon. At higher annealing temperature, the sample becomes more crystalline (large crystallite size) as seen by the increase of the peak at ~4.2eV. At lower Si content ($SiO_{1.0}$ and $SiO_{1.4}$) the film is of much more a-Si nature. A shoulder near 4.2eV is indicative of nano crystals (40-100Å) size effect [25], which is seen for $SiO_{1.0}$ and $SiO_{1.4}$ films annealed at 900°C. This is consistent with Raman Si_{nc} observations of Figure 3. The quantitative analysis of SE data for nanocrystalline materials requires knowledge of composition or size distribution thus is not presented here.

Figure 5. High resolution cross-sectional TEM micrograph of $SiO_{1.4}$ thin film after anneal at 900°C (30sec) showing Si_{nc} of ~50Å dimensions. Electron diffraction (not shown) confirms Si crystallinity.

Direct observation of the structure of annealed suboxides films can be best seen in the cross-sectional TEM photograph shown in Figure 5 [19]. As is readily seen there are regions of Si_{nc} (~50Å dimensions) within an amorphous SiO_2 matrix. Electron diffraction confirms the nanocrystals to be c-Si, thus showing reaction (1) to be appropriate, i.e. amorphous SiO_x decomposes into c-Si and SiO_2. The size distribution of Si_{nc} was examined by XRD with use of the Scherrer equation over the range of annealing temperatures. No systematic trend in Si_{nc} size vs. annealing temperature could be determined with sizes ranging through 5-30nm (instrument resolution). It is experimentally difficult to collect such data on thin polycrystalline films and larger crystallites tend to dominate the peak width. Annealing studies of SIPOS derived suboxides showed a systematic increase in Si_{nc} size with annealing temperature[6], however SIPOS films have

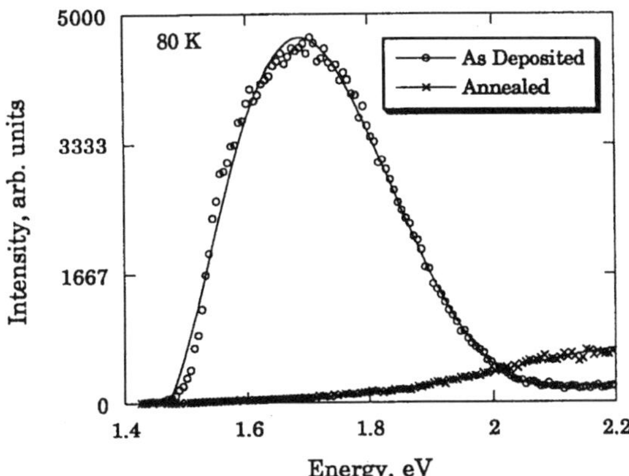

Figure 6. Photoluminescence measurement of $SiO_{1.4}$ as deposited and annealed at 900°C (30sec), which shows disappearance of tailstate transition associated with suboxides at 1.6eV. Measurements are made at 80K

inherent polycrystalline nucleation sites. These amorphous RPECVD derived suboxides have a less controlled nucleation process, possibly accounting for the lack of correlation in Si_{nc} size with annealing temperature.

The possible effect Si_{nc} size may have on the Si-O stretch frequency is worth mentioning, since if the distance between Si_{nc} becomes ~7Å or less, the oxide may be inherently suboxide in nature due to the proximity of bulk Si-Si bonds. One can estimate the SiO_2 thickness by using the known molar volume fraction of Si and SiO_2 and approximating the geometry as a Si_{nc} cube surrounded by SiO_2. An initial $SiO_{0.8}$ suboxide (decomposing to $0.6Si_{nc}$ + $0.4SiO_2$) with Si_{nc} about 100Å in size would require a SiO_2 thickness on the order ~35Å, thick enough to have bulk properties. Thus it is unlikely for the $SiO_{0.8}$ samples that the maximum Si-O stretch of 1044cm^{-1} is due solely to a size effect. Si_{nc} size would have to be on the order of 20Å for the effect to become dominate while Raman, ellipsometry, and XRD show Si_{nc} to be >>100Å. This simple cubic nanocrystal geometry argument is a lower limit for SiO_2 thickness, as a spherical Si_{nc} in SiO_2 matrix would have larger average SiO_2 thickness.

PL has been used to examine the interface structure between Si_{nc} and SiO_2 in this study. Figure 6. shows the PL of SiO_x for as deposited and RTA annealed at 900°C. The peak at 1.7eV for the unannealed film is characteristic for suboxide tail state transitions [26]. The disappearance of this PL at 1.7eV upon annealing strongly suggests a suboxide free interface between Si_{nc} and SiO_2 matrix. There are numerous studies examining the PL of Si_{nc} derived from the heat treatment of SiO_x films (prepared by Si^+ ion implanted SiO_2 or co-sputtering) [27-30]. PL at 1.7eV is seen for anneals at 1100°C however this may be attributed to forming defects from the formation of SiO as in the reverse of equation (1). At lower annealing temperatures (< 1000°C), only PL at ~2.4 eV is seen without the contribution of the 1.7eV suboxide PL.

CONCLUSIONS

The kinetics of the decomposition of SiO_x has been studied for a variety of suboxide compositions over a range of annealing temperatures. Most importantly, we found by monitoring the IR Si-O bond stretching frequency, that below 850°C the decomposition reaction of $SiO_{1.0}$ does not go to completion but to a metastable state of $SiO_{1.6}$. For an initial composition $SiO_{0.8}$, this metastable state of $SiO_{1.6}$ was stable to annealing temperatures as high as 1000°C. Raman, ellipsometry, XRD, and TEM support the observation of the decomposition of SiO_x to SiO_2 + Si_{nc} with the reaction being kinetically hindered at lower temperatures. TEM observes Si_{nc} with sizes on the order of 30Å for $SiO_{1.0}$ annealed at 900°C. XRD analysis shows no systematic dependence of Si_{nc} size on temperature as is seen in SIPOS films which have inherent nucleation sites. Photoluminescence shows a sharp elimination of the peak at 1.7eV upon annealing, which is indicative of a sharp interface between Si_{nc} and SiO_2 which lacks suboxides. The implications for MOSFET gate oxides is significant since we show that interfacial Si-suboxides can be removed by the otherwise kinetically limited decomposition of suboxides with a short anneal at temperatures

>850°C. This decomposition of suboxide is strikingly coincidental with the observation that an RTA at 900°C (30sec.) is required for optimal MOS capacitor performance (D_{it}~$2x10^{10}/cm^2$)[16] with RPECVD deposited gate dielectrics.

ACKNOWLEDGMENTS

The authors would like to thank Steve Ellis and Prof. Robert Nemanich for aid in the collection of Raman data, Prof. Dave Aspnes for aid in collection and discussion of ellipsometry data, K. Christiansen and D. Maher for the TEM sample preparation and imaging, and K. Price and Prof. L. McNeil for PL measurements. Funding was provided by SRC and ONR.

REFERENCES

[1] F.J. Grunthaner and P.J. Grunthaner, Mater. Science Reports **1**, 65 (1986).
[2] T. Hatttori and T. Suzuki, Appl. Phys. Lett. **43**, 470 (1983).
[3] L. Brewer and R. K. Edwards, J. Electrochemical Soc. **58**, 351, (1954).
[4] M.V. Coleman and D.J.D. Thomas, Phys. Stat. Sol. **22**, 593 (1967).
[5] G.W. Brady, J. Electrochemical Soc. **63**, 1119 (1959).
[6] M. Hamasaki, T. Adachi, S. Wakayama, and M. Kikuchi, J. Appl. Phys. **49**, 3987 (1978).
[7] J.H. Thomas III, and A.M. Goodman, J. Electrochemical Soc. **126**, 1766 (1979).
[8] J. Liday, S. Tomek, and J. Breza, Appl. Surf. Sci. **99**, 9 (1996).
[9] J.R. Shallenberger, J. Vac. Sci. Technol. A **14**, 693 (1996).
[10] B. Greenberg and T. Marshall, J. Electrochemical Soc. **135**, 2295 (1979).
[11] G. Kragler, H. Bender, G. Willeke, E. Bucher, and J. Vanhellemont, Appl. Phys. A **58**, 77 (1994).
[12] M. Catalano, M.J. Kim, R.W. Carpenter, K. Das Chowdhury, and J. Wong, J. Mater. Res. **8**, 2893 (1993).
[13] J. Wong, D.A. Jefferson, T.G. Sparrow, J.M. Thomas, R.H. Milne, A. Howie, and E.F. Koch, Appl. Phys. Lett. **48**, 65 (1986).
[14] G. Compagnini, S. Lombardo, R. Reitano, and S.U. Campisano, J. Mater. Res. **10**, 885 (1995).
[15] P Bruesch, Th. Stockmeier, F. Stucki, P.A. Buffat, and J.K.N. Linder, J. Appl. Phys. **73**, 7690 (1993).
[16] G. Lucovsky, D. R. Lee, S. V. Hattangady, H. Niimii, Z. Jing, C. Parker, and J. R. Hauser, Jpn. J. Appl. Phys. **34**, 6827 (1995).
[17] P.G. Pai, S.S. Chao, Y. Takagi, and G. Lucovsky, J. Vac. Sci. Tech. A **4**, 689 (1986).
[18] A. Banerjee and G. Lucovsky, in Amorphous Silicon Technology-1996, edited by M. Hack, R. Schropp, E.A. Schiff, A. Matsuda, and S. Wagner (Mater. Res. Soc. Proc. **420**, Pittsburgh, PA, 1996) in press.
[19] A. Banerjee, Ph.D. thesis, North Carolina State University (1996).
[20] D.V. Tsu, G.N. Parsons, G. Lucovsky, and M.W. Watkins, J. Vac, Sci. Technol. A **7**, 1115 (1989).
[21] D.E. Aspnes and A. A. Studna, Appl. Opt. **14**, 220 (1975).
[22] Z. Iqbal, S. Vebrek, A.P. Webb, and P. Capezzuto, Solid State Comm. **37**, 993 (1981).
[23] P.M. Fauchet, and I.H. Cambell, Critical Reviews in Solid State and Mater. Sci. **14**, S79 (1988).
[24] G. Lucovsky, F.L. Galeener, R.H. Geils, and R.C. Keezer, in The Structure of Non-Crystalline Materials, edited by P. H. Gaskell (Taylor & Francis LTD, London, 1977) p 127-130.
[25] D. E. Aspnes, S. M. Kelso, C.G. Olson, and D.W. Lynch, Phys. Rev. Lett. **48**, 1863 (1982).
[26] M.A. Paesler, D.A. Anderson, E.C. Freeman, G. Moddel, and W. Paul, Phys. Rev. Lett. **41**, 1492 (1978).
[27] Y. Osaka, K. Tsunetomo, F. Toyomura, H. Myoren, and K. Kohno, Jpn. J. Appl. Phys. **31**, L365 (1992).
[28] T. Shimizu-Iwayama, K. Fujita, S. Nakao, K. Saitoh, T. Fujita, and N. Itoh, J. Appl. Phys. **75**, 7779 (1994)
[29] T. Shimizu-Iwayama, T. Niimi, S. Nakao, K. Saitoh, T. Fujita, and N. Itoh, J. Phys.: Conden. Matter **5**, L375 (1993).
[30] S. Hayashi, T. Nagareda, Y. Kanzawa, and K. Yamamoto, Jpn. J. Appl. Phys. **32**, 3840 (1993).

Part IV

Synthesis and Properties of Other Group IV Clusters, Nanocrystals, and Nanostructures

Electronic States of Nanocrystalline Carbon

G.P. Lopinski*, V.I. Merkulov, and J.S. Lannin
Dept. of Physics, Penn State University, University Park, PA 16802

Abstract

Electron energy loss spectroscopy (EELS) has been used to investigate the electronic states of isolated, nanocrystalline carbon particles. Small carbon nanocrystals were prepared via sputter deposition onto SiO_2 substrates, followed by annealing to 700C. The structure and size distribution of the particles have been characterized by Raman scattering, Auger electron spectroscopy and electron microscopy. EELS observations indicate that a semimetal to semiconductor transition occurs for particles smaller than 1nm. In addition, hydrogen adsorption is found to significantly affect the electronic states of these particles, indicating that both finite size and dangling bond effects modify the properties of small carbon nanocrystallites.

Introduction

The structure and properties of small carbon particles vary widely with the conditions of formation, yielding molecular clusters, amorphous networks and disordered nanocrystallites. Nanocrystalline carbon (nc-C) particles, consisting of two-dimensional graphite-like fragments, have been considered as important constituents of diverse systems, ranging from interstellar dust to arc generated soots and vacuum deposited thin films. These objects present an unique opportunity to observe finite size effects in a π electron system, in contrast to nanostructures based on other group IV elements such as Si and Ge which only exhibit σ-bonding interactions. Theoretical calculations for carbon have indicated that the electronic structure is quite sensitive to medium range correlations [1, 2, 3]. For two dimensional graphitic fragments, tight binding calculations [1, 2] predict a loss of semimetallic character as the fragment size is reduced although such a transition has not yet been observed experimentally. Small nanocrystalline particles will also have a significant number of undercoordinated edge atoms. Passivation of these dangling bond sites by adsorption of atomic hydrogen is expected to strongly influence the properties of the nc-C particles. For example, planar molecules consisting of several fused hexagonal carbon rings with edges terminated by hydrogen, known as PAHs (polycyclic aromatic hydrocarbons), exhibit substantial gaps of order 2.5-4eV[4, 5].

In this paper we report size dependent modifications of the electronic states of isolated nc-C particles supported on SiO_2 substrates. Electron energy loss spectroscopy has been used to observe the predicted semimetal-semiconductor transition as the particle size is reduced. Hydrogenation of the dangling bonds is also found to strongly affect the electronic states. These results demonstrate that both finite size and the increasing

Mat. Res. Soc. Symp. Proc. Vol. 452 © 1997 Materials Research Society

importance of dangling bonds influence the properties of nanoscale carbon particles. Furthermore the present work also demonstrates the utility of EELS as a powerful probe of the properties of nanocrystalline solids.

Sample Preparation and Characterization

In order to study the properties of isolated nanocrystalline carbon particles it was necessary to choose a substrate which interacts weakly with carbon. Deposition onto such a surface is expected to result in the growth of three-dimensional clusters with properties only slightly perturbed by interaction with the substrate. These considerations motivated the choice of SiO_2 as a substrate for the present work.

The experiments were carried out in a multichamber UHV system with a base pressure of $2x10^{-10}$ Torr. In situ Raman scattering was performed employing 514.5 nm excitation and a Spex triplemate spectrometer with multichannel detection. An LK2000 spectrometer with rotatable analyzer was used for the EELS measurements. The typical resolution, determined by the FWHM of the elastic peak, was typically $\sim$ 20 meV for the electronic loss spectra reported here but could be improved to $\sim$7 meV for vibrational measurements. A VSW HA5000 hemispherical analyzer was employed for Auger and XPS measurements.

The SiO_2 films were prepared in situ via reactive sputtering onto Ag covered Si substrates. Cleanliness and stoichiometry of the substrate films were verified by EELS, Auger and XPS. Ultrathin island structures of amorphous carbon were formed by dc magnetron sputtering (225 W, 5 mtorr Ar) at room temperature. The clusters were subsequently crystallized by annealing to 650-700 C, using interference enhanced Raman scattering [6, 7] to monitor the crystallization. While the as-deposited amorphous clusters show a single broad peak centered at 1500 cm^{-1}, the annealed particles exhibit an additional feature at $\sim$ 1350 cm^{-1} characteristic of the nanocrystalline state [8, 9]. Changes in the Raman spectra with film thickness indicate modifications of the phonon density of states for the smallest particles, discussed in detail elsewhere [10].

Auger electron spectroscopy was used to monitor the carbon deposition and cluster growth. The signal from a monolayer of C_{60} on Ag was used to calibrate the intensity. In addition, the carbon deposition rate was determined to be $\sim$ 6 Å/minute from profilometry studies on thick films. The initial uptake of carbon was found to increase superlinearly with deposition time, indicating a sticking probability that increases with carbon coverage. This is consistent with a low sticking coefficient on bare SiO_2 which increases as clusters are nucleated at defect sites. Figure 1 shows Auger intensities as a function of the amount of carbon deposited, expressed in terms of an equivalent film thickness. The increase of the carbon intensity as well as the attenuation of the substrate silicon and oxygen features differs significantly from that expected for layer by layer growth. The observed behavior is consistent, however, with three-dimensional growth of roughly rectangular or cylindrical particles which begin to coalesce at a thickness of $\sim$20 Å.

Imaging of nanocrystallites of light elements such as carbon is usually hampered by poor contrast relative to the supporting substrate. However, ex-situ transmission electron microscopy (STEM) using a novel dark field, thin annular detector method [11, 12] was used to image the nc-C particles. Contrast was observed when the detector was set to observe diffraction corresponding to the 3.4 Å interplanar spacing between

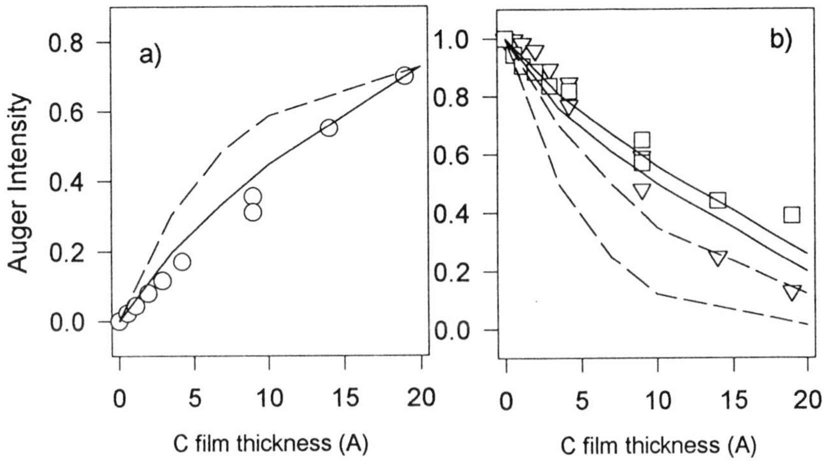

Figure 1: Auger intensities for a) carbon 271 eV (circles), b) silcon 92 eV (triangles) and 510 eV oxygen (squares) as a function of carbon film thickness. The lines indicate calculations for layer by layer growth (dashed) and the nucleation and growth of cylindrical clusters (solid). Escape depths of 5 Å 7 Å and 10 Å were used for silicon, carbon and oxygen respectively.

graphitic sheets. Images of this (002) diffraction facilitated estimates of the particle size. Measurements on a 11 Å thick film yield an average particle size of 10±2 Å. Size estimates for a 19 Å thick film are consistent with the dimensions scaling as $t^{1/3}$, where t is the film thickness, as expected for three-dimensional growth.

Based on the Raman, Auger and STEM observations we conclude that the particles obtained here consist of two-dimensional graphite-like fragments, randomly oriented with respect to neighboring planes in a cluster as well as relative to the plane of the substrate as shown schematically in Figure 2. In addition to the sp^2 bonded atoms within the interior of the planes, each particle has a substantial fraction of two-fold bonded edge atoms.

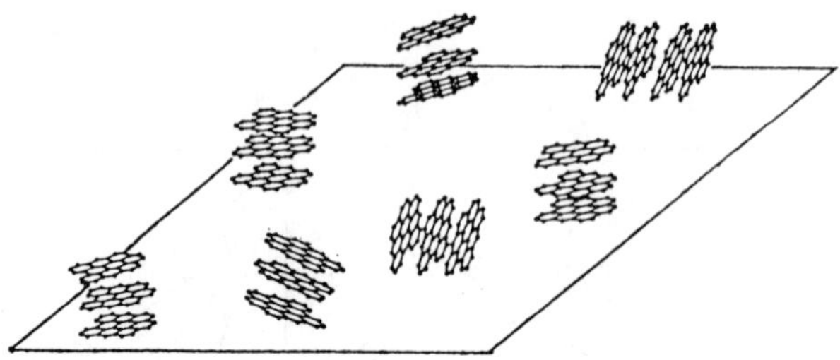

Figure 2: Schematic of the structure of nc-C particles on SiO_2, consistent with the Raman, Auger and electron microscopy measurements. The clusters shown above correspond roughly to the smallest particles studied here obtained for deposition of 4 Å of carbon.

EELS Results and Discussion

Electronic properties were investigated using electron energy loss spectroscopy (EELS) in reflection. Within the dipole scattering approximation, the EELS response is proportional to $Im\{-1/(\epsilon(\omega)+1)\}$ where $\epsilon(\omega)$ is the dielectric function of the sample [13]. This method has the sensitivity to observe electronic losses at submonolayer coverages and has been employed to provide information on the surface and near surface optical properties of semiconductor surfaces [13, 14, 15]. As indicated in Figure 3, large changes are observed in the EELS spectra as a function of film thickness. Since the bare SiO_2 substrate exhibits only weak scattering below 9 eV, the observed spectra are characteristic of the nc-C particles. Also shown for reference is the spectrum from a continuous thicker film, deposited at elevated temperatures (with a Raman spectrum characteristic of bulk glassy carbon (g-C) with 30-40 Å domain size [8, 9]), which can be satisfactorily accounted for using bulk optical constants of g-C [16].

As the film thickness, and consequently the average particle size, is reduced, the spectra deviate from the form expected for ultrathin films of g-C on SiO_2. In particular, the ~5.7 eV feature, associated with $\pi - \pi^*$ transitions and sometimes referred to as a π-plasmon, is found to broaden and shift slightly (~0.4 eV) to lower energy for the larger isolated particles (>11.5 Å thickness). For thinner films there is a dramatic change in the form of the EELS spectra with a substantial intensity reduction and broadening of the ~5.7 eV feature. Concurrent with this change are variations in the spectra at lower loss energies which indicate the formation of a gap in the electronic density of states. The spectra for the larger particles exhibit large low energy scattering, appearing as an extended tail on the elastic peak, which disappears for thicknesses below 11 Å. This large low energy tail has also been observed in EELS studies of highly oriented pyrolitic graphite (HOPG) and attributed to low energy interband transitions characteristic of

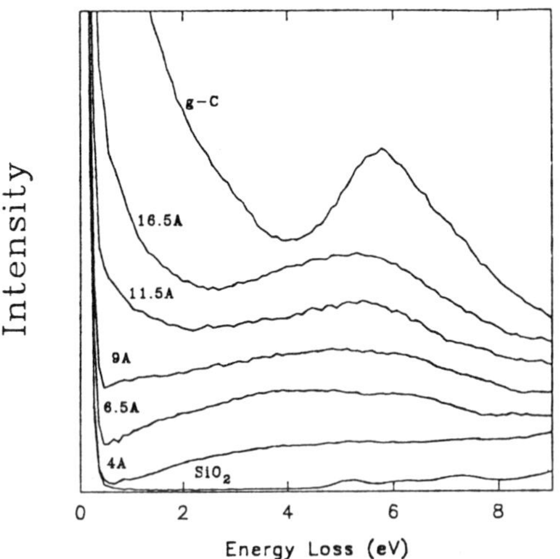

Figure 3: EELS spectra of ultrathin nc-C films for different amounts of carbon deposited. Spectra of bare SiO_2 and of a thick (200 Å) "glassy" carbon film are shown for comparison.

a semimetal [17]. The disappearance of this tail is indicative of the loss of metallic character and a transition to a semiconducting state.

For ultrathin films, where the effective probe depth of the electrons exceeds the film thickness, it is important to distinguish between intrinsic and extrinsic size effects on the EELS spectrum. In this case the response function becomes $Im\{-1/(\epsilon'(\omega,t,E_i)+1)\}$, where $\epsilon'(\omega,t,E_i)$ is an effective dielectric function which depends on the film thickness and incident energy (E_i) and can be calculated from the dielectric functions of the film and substrate using a two-layer model [13]. However, calculated EELS spectra for ultrathin nc-C films, using optical constants of g-C [16] cannot account for the observed spectra. While the calculations do indicate some broadening of the ~5.7 eV peak as the film thickness is reduced, these changes are considerably smaller than observed in Fig. 3 and no shift of this feature is obtained. In addition, the low energy tail is present in the calculated spectra at all thicknesses. Therefore the observed changes with film thickness can largely be attributed to an intrinsic size dependence of the dielectric function for small carbon particles.

The STEM measurements allow an estimate of the number of carbon atoms within the graphite-like fragments at the semimetal-semiconductor transition. Detailed measurements on a 11.5Å thick film yield an average particle size of 10±2 Å which approximately corresponds to the smallest metallic particles. From the average particle size of ~ 10 Å and assuming an interplanar spacing of 3.4 Å, these particles consist of 3-4 graphitic layers. If the in-plane particle dimensions are similar to the perpendicular dimension measured in STEM, each two dimensional fragment will contain of order 60 atoms, corresponding to 20-25 hexagonal rings.

In order to determine the onset of electronic transitions and study its variation with thickness the loss function was extracted from the data of Fig. 3 by subtracting the SiO_2 spectrum, removing contributions of the elastic peak and substrate vibrational modes, and multiplying the result by energy. In the vicinity of the gap, where the refractive index $n(\omega)$ is much larger than the extinction coefficient $k(\omega)$, it is readily shown that the EELS response is proportional to $k(\omega)$. If $n(\omega)$ varies weakly with energy as is usually the case near the gap, the form of $k(\omega)$ and hence the absorption coefficient $\alpha = 4\pi k/\lambda$, can be extracted from the loss function as recently demonstrated [18]. As in amorphous semiconductors it is useful to consider the form of $(\alpha E)^{1/2}$, as the intercept yields an estimate of the optical or Tauc gap. In amorphous solids, the Tauc gap is often used a measure of the pseudogap of the system and typically corresponds to absorption coefficients in the range of 10^3-10^4. Tauc plots for different film thicknesses are shown in Fig. 4. For the 9 Å film the Tauc gap is found to be ~0.3 eV. The gap is seen to increase with decreasing film thickness up to a value of ~0.7 eV for the thinnest (4 Å) film studied to date. Figure 4 also indicates that the Tauc form holds only over a limited range. The tailing of the absorption to lower energies is similar to the Urbach behavior of amorphous semiconductors, where it is associated with structural disorder. In the nc-C fragments possible structural disorder effects due to bond angle strain, as well as a distribution of cluster sizes, may both contribute to this Urbach behavior.

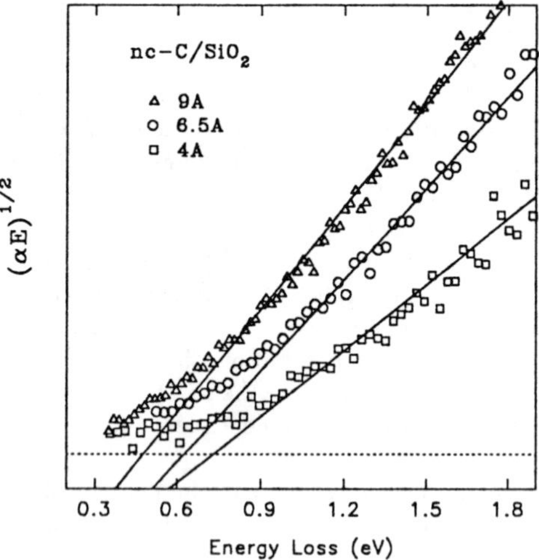

Figure 4: Tauc plots derived from EELS spectra for different film thicknesses.

The observation of a semimetal-semiconductor transition for small carbon particles is in agreement with theoretical calculations of Robertson[1, 2] who calculated the electronic density of states for fused six-fold rings of carbon, within the tight binding approximation. An infinite number of rings, corresponding to a single graphitic plane, results in a zero gap semiconductor. However, for a finite number of rings the calculation indicates the opening of a gap, increasing in magnitude with a decreasing number of rings. For a compact cluster of 18 rings, the gap is predicted to be 0.4 eV, in good agreement with experimental observations near the transition. The loss spectra in Fig. 3 also indicate substantial changes in the higher energy losses which accompany the transition to semiconducting behavior. While the calculations of Robertson indicate some changes in the π bands as the gap opens, the predicted changes in the joint density of states do not appear to account for the broadening of the 5.7 eV loss band. Given the simple nature of the tight binding calculations, additional theoretical calculations utilizing ab initio quantum calculation methods and considering possible structural relaxations and dangling bond effects would be useful here.

The large fraction of two-fold bonded edge atoms (see Fig. 2) with their associated dangling bonds suggests that the electronic structure of the particles can be modified by hydrogen adsorption. The formation of C-H bonds at the edge sites is expected to eliminate dangling bond states as well as modify adjacent π bonding and antibonding states of the nanocrystalline particles. Figure 5 shows Tauc plots for hydrogenated nc-C samples of 4 Å and 19 Å film thicknesses. The nc-C particles were exposed to

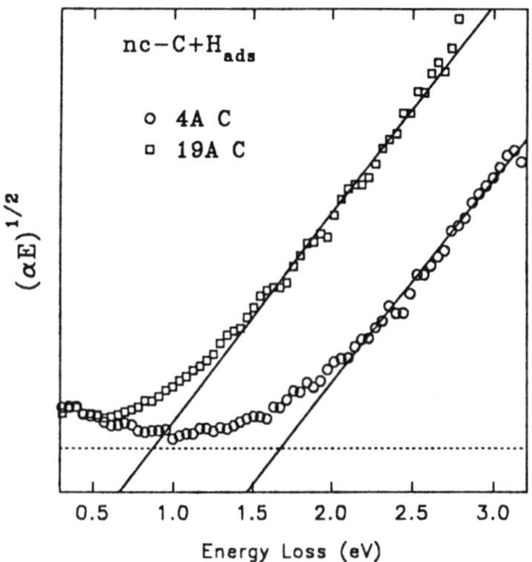

Figure 5: Tauc plots for hydrogenated nc-C particles obtained from 4 Å and 19 Å thick films.

atomic H using the hot filament of a Kaufmann source to dissociate H_2. Adsorption was monitored via observation of C-H vibrational modes (~2900-3000cm^{-1}) in the HREELS spectra. For the 4 Å film, hydrogenation increases the gap to ~1.7 eV from ~0.7 eV (see Fig. 4). This gap value is approaching that of the larger PAH molecules, such as ovalene ($C_{32}H_{14}$), in which the lowest energy electronic transition is at 2.7eV [5]. In the case of the 19 Å film, which is initially semimetallic (Fig. 3), hydrogen adsorption induces a gap of 0.7 eV. The hydrogenated particles also exhibit a broad background in Raman scattering studies with 514.5nm excitation, due to luminescence extending down to shorter wavelengths.

In addition to terrestrial interest in the properties of small carbon particles, these entities have also been suggested to be important components in interstellar dust. PAH molecules have been investigated extensively as possible dust candidates with some degree of agreement between laboratory infrared data and astronomical observations[19, 20]. Thermal balance calculations of uv adsorption and ir emission processes have suggested, however, that three dimensional structures composed of PAH units with ~50 carbon atoms, larger than those currently studied in laboratory environments, may be required [19, 20, 21]. Hydrogenated nc-C structures of the type studied here may serve as possible models for interstellar dust particles as they exhibit the required optical gap and broad luminescence possibly related to the extended red emission observed in the interstellar medium [22, 23]. Since the smallest clusters studied here are slightly larger than the PAH molecules studied to date, the isolated particles of the present work appear to serve as a useful bridge between these molecules and larger soot and glassy carbon particles considered previously. Further work is in progress to characterize the vibrational modes of these nc-C particles for comparison with infrared spectra of PAH molecules and interstellar infrared emission spectra. Wether or not these particles turn out to be true dust analogs, the results indicate that calculations of emission and absorption of small graphitic clusters (less than ~2nm) must include an intrinsic size dependence of the optical constants.

Conclusions

In summary, size dependent modifications of the electronic states have been observed for isolated nanocrystalline carbon particles. Electron energy loss spectra indicate a transition from semimetallic to semiconducting behavior for clusters with a size of approximately 1nm, in good agreement with theoretical predictions. Hydrogenation of the dangling bonds further modifies the electronic properties, increasing the magnitude of the gap for the smaller semiconducting clusters and opening a gap in the case of the larger semi-metallic particles. These results indicate the importance of both finite size and dangling bond effects in determining the properties of nanocrystalline carbon. Electron energy loss spectroscopy has also been shown to be a useful probe for studies of nanocrystalline materials.

Acknowledgements

This work was supported by NSF Grant DMR 92-24165 and USDOE Grant DE-FG02-84ER45095. We thank Prof. J.M. Cowley for the electron microscopy measurements.

* Present address; Steacie Institute for Molecular Sciences, National Research Council, 100 Sussex Dr., Ottawa, Ontario, Canada K1A 0R6.

References

[1] J. Robertson, Adv. Phys. **35**, 317 (1986).

[2] J. Robertson and E.P. O'Reilly, Phys. Rev. **B35**, 2946 (1987).

[3] U. Stephan, Th. Frauenheim, P. Blaudeck, and G. Jungnickel, Phys. Rev. **49**, 1489 (1994).

[4] I. Nakhimovsky, M. Lamotte, and J. Joussot-Dubien, Handbook of Low Temperature Electronic Spectra of Polycyclic Aromatic Hydrocarbons, Elsevier, Amsterdam, 1989.

[5] S. Leach in Polycyclic Aromatic Hydrocarbons and Astrophysics, edited by A. Leger et al., Reidel, Dordrecht, 1987.

[6] G.A.N. Connell, R.J. Nemanich, and C.C. Tsai, Appl. Phys. Lett. **36**, 31 (1980).

[7] W.S. Bacsa and J.S. Lannin, Appl. Phys. Lett. **61**, 2116 (1992).

[8] R.J. Nemanich and S.A. Solin, Phys. Rev. **B20**, 392 (1979).

[9] F. Li and J.S. Lannin, Appl. Phys. Lett. **61**, 2116 (1992).

[10] V.I. Merkulov, J.S. Lannin and J.M. Cowley, *in this volume.*

[11] J.M. Cowley, Ultramicroscopy **49**, 4 (1993).

[12] J.M. Cowley, V.I. Merkulov and J.S. Lannin, Ultramicroscopy, in press.

[13] H.L. Ibach and D.L. Mills, Electron Energy Loss Spectroscopy and Surface Vibrations, Academic Press, New York, 1982.

[14] H. Froitzheim, H. Ibach, and D.L. Mills, Phys. Rev. **B11**, 4980 (1975).

[15] G.P. Lopinski, J.R. Fox, J.S. Lannin, F.S. Flack, and N. Samarth, Surf. Sci. **L355**, 355 (1996).

[16] H.J. Hagemmann, W. Gudat, and C. Kunz, J. Opt. Soc. Amer. **65**, 742 (1975).

[17] R.E. Palmer, J.F. Annett, and R.F. Willis, Phys. Rev. Lett. **58**, 2490 (1987).

[18] G.P. Lopinski and J.S. Lannin, Appl. Phys. Lett., in press.

[19] A. Leger in Experiments on Cosmic Dust Analogues, edited by E. Bussoletti et al., Kluwer, Dordrecht, 1988.

[20] J.L. Puget and A. Leger, Ann. Rev. Astron. Astrophys. **27**, 161 (1989).

[21] K. Selgren in Interstellar Dust, edited by Allamondola and Tielens, Kluwer, Dordrecht, 1988.

[22] W.W. Duley and D.A. Williams, Mon. Not. Royal Astron. Soc. **247**, 647 (1990).

[23] W.W. Duley, Astron. Journal **445**, 240 (1995).

STRUCTURE AND VIBRATIONAL PROPERTIES OF ISOLATED CARBON NANOCRYSTALLITES

V.I. MERKULOV, J.S. LANNIN
Dept. of Physics, Penn State University, University Park, PA 16802
J.M. COWLEY
Dept. of Physics and Astronomy, Arizona State University, Tempe, AZ 85287-1504

ABSTRACT

Small isolated nanocrystallites of carbon have been prepared on amorphous SiO_2 and studied by interference enhanced Raman scattering (IERS) and scanning transmission electron microscopy (STEM). A new mode of dark field STEM, using a thin annular detector, has allowed imaging of 1-2 nm graphite-like particles using (002) diffraction. For such small particles, the Raman spectra provide the first evidence for changes in the phonon density of states of a nanocrystalline system.

INTRODUCTION

As crystallites are decreased in size to the nanoscale level their electronic and vibrational properties will be modified as surface and quantum size effects play an increasing role. Experimental studies to date have focused on modification of the electronic states of nanocrystalline semiconductors, such as CdSe or Si, wherein substantial changes in the optical gap have been attributed to quantum size effects [1,2]. However, the influence of size effects on the phonon states of very small, isolated nanocrystallites <2nm in size has been relatively unexplored. Theoretical studies of small Si nanocrystallites, for example, predict substantial changes in Raman scattering and infrared absorption [3].

With decreasing particle size Raman scattering is modified due to relaxation of crystal momentum conservation. This results in disorder-induced scattering from states away from the center of the Brillouin zone[4]. For very small nanocrystallites modifications of the bonding and electronic density of states due to finite size effects and surface perturbations should also result in changes in the phonon density of states relative to a bulk crystal. While such changes are possible, they have not been observed to date in nanocrystallites. In small isolated amorphous Ge particles surface effects on the phonon spectrum have been noted using IERS [5].

Previous Raman scattering studies of small nanocrystallites of sizes below ~3nm have been confined to either nonisolated particles or particles within a matrix or with chemically bound ligands. The Raman spectra of these systems exhibit shifts and broadening of the allowed q=0 modes but overall are often weakly modified relative to bulk solids. This may be a consequence of weak sensitivity of the phonons to size effects. One exception is nanocrystalline carbon (nc-C). Here finite size, disorder effects have induced scattering from optic modes at the Brillouin zone boundary for 3-20nm particles [6,7]. For relatively smaller, <2nm, particles of nc-C prepared by crystallization of continuous films of amorphous C (a-C) additional disorder induced Raman scattering is observed at both low and high frequencies [8]. Extensive electron microscope studies of different forms of carbon-based materials have suggested that hexagonal rings constitute the basic structural unit [9]. Relatively random correlations between graphite-like planes in nc-C imply the absence of three dimensional order. Structural order is two dimensional in character as noted in radial distribution function (rdf) studies of 4nm size particles in bulk, "glassy" nc-C samples [10]. In contrast, true a-C has an rdf which implies few ordered hexagonal units [11]. Of basic interest for very small, isolated nanocrystallites of C are differences and similarities with a-C

Mat. Res. Soc. Symp. Proc. Vol. 452 © 1997 Materials Research Society

and the role of the surface (dangling bonds) and quantum size effects in the modification of the vibrational and electronic states .

EXPERIMENT

In order to study very small nanocrystallites without surface bound molecules or atoms it is necessary to prepare particles that are isolated on a weekly interacting substrate. With this geometry *in situ* Raman scattering enhancement methods may be utilized to study ultrathin films [12] to obtain information on finite size effects. In the present study bilayer interference enhanced Raman scattering (IERS) [13] is employed. Thin films of a-C were deposited at ~20C by dc magnetron sputtering at an Ar pressure of 5mTorr in an ultrahigh vacuum chamber with a base pressure of 2 x 10^{-10} Torr. Substrates were composed of a thin film of a-SiO_2 on a thick layer of Ag. Thickness of SiO_2 was chosen to yield bilayer IERS. Nanocrystallites of C were prepared from a-C by UHV crystallization at ~670C. Film thicknesses were calibrated using Auger analysis relative to a monolayer of C_{60} on Ag. Auger analysis of the C and SiO_2 core lines were consistent with island growth and the absence of appreciable coalescence for the films studied. Raman scattering measurements were performed *in situ* at 20C with 514.5nm Ar laser excitation using a Spex Triplemate Spectrometer with a Mepsicron multichannel detector. These measurements confirmed the conversion of a-C particles to nc-C. High resolution specular EELS measurements demonstrated the absence of any substantial H bound to the nc-C particles.

The observation by electron microscopy of very small isolated nanocrystallites of light elements such as C is limited by contrast effects relative to the supporting substrate. In addition, the absence of three dimensional order implies that (hkl) diffraction is constrained and only (002) reflections are strong. Conventional methods of bright-field transmission electron microscopy (TEM) are ineffective because the information concerning the nanocrystallites is lost in the noise from the amorphous support. The dark field TEM modes and the usual dark-field STEM modes, using a broad annual detector to collect all the intensity scattered outside the incident beam cone, are not effective because the signals are approximately proportional to the density of scattering matter [14] and the signal from the support is noisy. A new mode of dark-field STEM imaging, employing a thin annular detector (the TADDF mode) has recently been explored [15,16] and found to provide good contrast for even light atom amorphous or nanocrystalline films on amorphous supports by selecting the intensity at particular radii in the diffraction pattern for which there are characteristic scattering maxima.

For STEM measurements thin layers of ~5-6nm of SiO_2 were sputter-deposited in high vacuum on Cu or Mo TEM grids initially covered on one side with a photolithography material that was removed after SiO_2 deposition. Preparation of small C nanocrystallites was done in the same fashion as that for the Raman scattering studies described above.

Dark-field STEM images were obtained using a thin annular detector in the HB-5 STEM instrument from VG Microscopes, modified by the addition of a high-resolution objective lens pole-piece, two post-specimen lenses and a detector system which includes three interchangeable annular detectors and a YAG phosphor screen. The thin annular detector used had a ratio of outside to inside radii of about 1.1. An objective aperture was chosen to give a beam convergence at the specimen of 7.4 mrad so that the incident beam probe at the specimen for 100keV electrons had a diameter at half-height of ~0.7nm, which was approximately the dark-field STEM image resolution. By using the post specimen lenses, the scattering angle of electrons collected by the detector could be varied through angles corresponding to real-space spacing varying from about 1 nm to 0.05nm.

RESULTS

For specimens of ~1.1nm of nc-C on ~6nm of a-SiO_2 the TAD was set to collect the spacing of d=0.34nm, which corresponds to the inter-layer spacing of graphitic sheets. The image showed numerous white spots with diameters ranging from 1nm to 3nm (Fig. 1(a)) suggesting the presence of carbon nanocrystals. For similar settings of the TAD, films of pure a-SiO_2 showed only occasional smaller bright spots having diameters less than 1 nm (Fig. 1(b)).

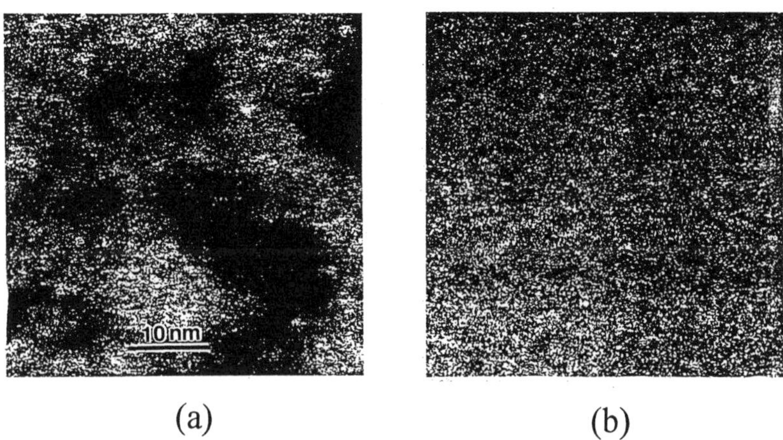

Fig.1 TAD, dark field STEM for (a) 1.1nm of nc-C on a-SiO_2 , (b) pure a-SiO_2. Dark regions on the scale of 10-20nm are attributed to SiO_2 thickness variations.

From the distribution of larger bright spots an estimated diameter of ~1.2±0.2nm was obtained. Deconvolution with the beam diameter gives crystallite diameters of 1±0.2nm. For an apparent crystallite diameter of L (nm), the number of graphitic planes of C atoms may be estimated as (L/0.34) + 1, which suggests an average of 4 or 5 such planes for Fig. 1(a). The Laue conditions for diffraction from (002) planes, with inter-layer spacing of 0.34nm, are somewhat relaxed for such small crystals so that spots of appreciable intensity in the dark-field image may be produced for about 10% of possible orientations. The density of white spots in Fig. 1(a) is therefore consistent with the assumption that all of the island film of a-C has been converted to nanocrystalline carbon particles.

A second method of using the TAD to detect nanocrystals is to use the STEM post-specimen lenses to enlarge the central spot of the diffraction pattern so that it is almost as large as the inner diameter of the annular detector. Then the TAD picks up the small-angle scattering around the central spot due to the finite size of the particles. Further results using this approach and a theoretical discussion of the expected variation of contrast with experimental parameters will be given elsewhere [17].

The high frequency, *in situ* Raman spectra are shown before and after crystallization of a-C in Fig. 2. The spectra, which are primarily due to stretching modes, are shown for $\omega > 1200 cm^{-1}$ for a 0.65nm thick film. The distinct two features, the peak and the higher frequency shoulder of the nanocrystalline spectrum (Fig.2(b)) qualitatively differ from the single broad a-C peak (Fig. 2(a)).

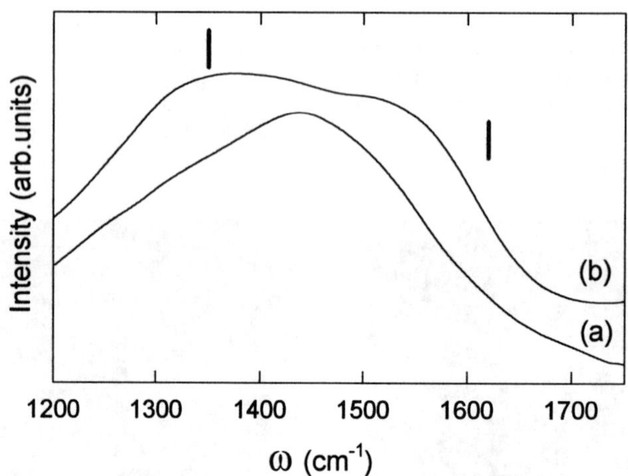

Fig. 2 Comparison of the high frequency IERS spectra of a 6.5 A thick film of C before (a) and after (b) formation of nc-C. Vertical bars indicate density of states peaks for larger nanocrystallites.

As STEM measurements suggest an average particle size L=8.5A for this film, the spectra demonstrate that even for such small particles structural differences between amorphous and nanocrystalline particles are observable in the Raman function response. This is consistent with theoretical calculations for ~1nm size Si and Ge modes [3].

The quite small average particle size for the spectra of Fig.2 suggests that the spectra represent a coupling parameter weighted phonon density of states. For the peak at ~1370 cm^{-1} a small shift of ~15cm^{-1} is noted for this disorder induced peak relative to 3nm and larger nc-C particles [7]. For the higher frequency shoulder at ~1520-1530cm^{-1} a shift down by ~90cm^{-1} from the second order derived overtone density of states of 2.5-3nm particles is observed. The vertical bars in Fig. 2 indicate the peak positions of the optic band density of states for these larger particles which are similar to those of graphite. The frequency shift noted for the 6.5 A thick film may thus be associated with the changes in the phonon density of states for the small, isolated particles. Lower resolution, inelastic neutron scattering [18] of graphite suggest that the higher frequency feature in Fig. 2 for nc-C has an enhanced Raman coupling parameter relative to the density of states. This may, in part, be due to π band resonant enhancement effects. The increased breadth of the nc-C peak, along with a shallower minimum may be due to disorder effects combined with some inhomogeneous broadening due to a distribution of particle sizes.

In contrast to the high frequency spectra, more substantial changes in the low frequency Raman spectra with the film thickness are shown in Fig. 3. Two peaks in the spectra, which are associated with density of states features of graphite derived bands, are found to shift. The sharper, low frequency feature has a large shift of ~68cm^{-1}, or 14%, to higher frequency. In contrast, the higher frequency peak exhibits a smaller shift to lower frequency along with the broadening. These two broad low frequency modes of graphite are related to bond bending modes with displacements perpendicular to the hexagonal plane [19]. With the formation of dangling bonds on the edge of C layers, assuming no reconstruction, the reduction in local coordination as well as possible changes in near edge atom force constants may result in changes in the phonon

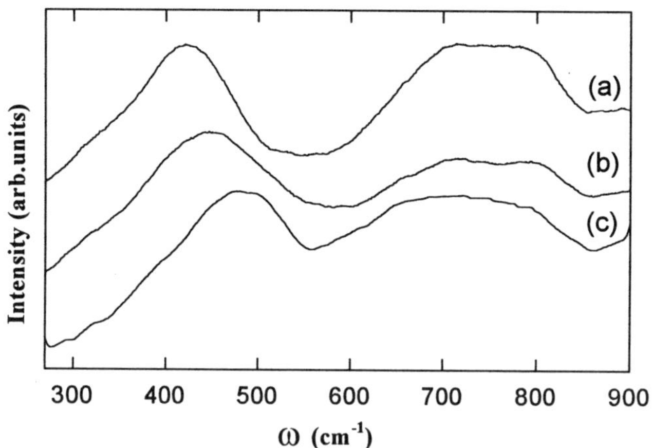

Fig. 3 Comparison of the lower frequency IERS spectra of nc-C for 16.5A (a), 9A (b) and 6.5A (c) thick films

spectrum. In addition, 'interior' atoms with graphite-like, local structure may have their phonon density of states modified by quantum size effects. Preliminary electron energy loss spectra suggest changes in the electronic states of small nanocrystallites.

CONCLUSION

In summary, a new method for imaging very small, light atom crystallites using a thin annular detector has been applied to study nc-C. The TADDF method offers important possibilities for detecting and characterizing nanocrystals even for the unfavorable circumstance of the presence of a relatively thick film of amorphous supporting material. The observation of distinct (002) diffraction as well as the form of the taken *in situ* Raman spectra have demonstrated two dimensional, graphitic-like order in crystallized a-C particles whose sizes were between 1-2nm. The form of the IERS demonstrates substantial changes, particularly at low frequencies, in the phonon density of states due to finite size effects associated with either or both near-surface atom bonding or quantum size effects. Future studies of the effects of atomic H adsorption will be useful along with theoretical modeling to distinguish these effects.

ACKNOWLEDGMENT

This work was supported at Penn State University by NSF Grant DMR 92-24165 and at Arizona State University by NSF support for the Center for High Resolution Electron Microscopy. The assistance of G.P. Lopinski and J.R. Fox in Auger analysis is gratefully acknowledged.

REFERENCES

1. M.L. Steigerwald et al. J. Am. Chem. Soc. **110**, 3046 (1988)
2. S. Schlupper et al. Phys. Rev. Lett. **72**, 2648 (1994)

3. R. Alben, D. Weaire, J.E. Smith, Jr., and M.H. Brodsky, Phys. Rev. B**11**, 2271(1975)
4. J.S Lannin in Semiconductors and Semimetals, Vol.21(B) ed. J.I. Pankove (Academic Press, 1984), p. 159
5. J.A. Fortner and J.S. Lannin, Surf. Sci. **254**, 251 (1991)
6. F. Tuinstra and J.L. Koenig, J. Chem. Phys. **53**, 1126 (1970)
7. R.J. Nemanich and S.A. Solin, Phys. Rev. B**20**, 392 (1979)
8. F. Li and J.S. Lannin, Appl. Phys. Lett. **61**, 2116 (1992)
9. A. Oberlin in Chemistry and Physics of Carbon, Vol.22 ed. P.A. Thrower (M. Dekker, N.Y., 1989), p. 1
10. D.F.R Mildner and J.M. Carpenter, Proc. 5th Int. Conf. Amorphous and Liquid Semiconductors, ed. J. Stuke and W. Brenig (Taylor and Francis, London, 1974) p. 463
11. F. Li and J.S. Lannin, Phys. Rev. Lett. **65**, 1905 (1990)
12. G.A.N. Connell, R.J. Nemanich and C.C. Tsai, Appl. Phys. Lett. **36**, 31 (1980)
13. W.S. Bacsa and J.S. Lannin, Appl. Phys. Lett. **61**, 19 (1992)
14. A.V. Crewe, J. Wall and J.P. Langmore, Science **168**, 1338 (1970)
15. J.M. Cowley, Ultramicroscopy **49**, 4 (1993)
16. R.-J. Liu and J.M. Cowley, J. Microscopy Soc. Amer. (1996) (in press)
17. J.M. Cowley, V.I. Merkulov and J.S. Lannin, Ultramicroscopy (in press)
18. W.A. Kamitakahara, J.S. Lannin, R.L. Cappelletti, J.R.D. Copley and F. Li, Physica B **180**, 709 (1992)
19. C.Z. Wang and K.M. Ho (private communication)

RAMAN SCATTERING INVESTIGATION OF SUPERCONDUCTIVITY IN Si_{46} CLATHRATES

S.L. Fang*, L. Grigorian*, A.M. Rao*, P. C. Eklund*, G. Dresselhaus**, M. S. Dresselhaus**, H. Kawaji***, S. Yamanaka***
*Department of Physics and Astronomy, University of Kentucky, Lexington, KY 40506, USA
**Massachusetts Institute of Technology, Cambridge, MA 02139, USA
***Department of Applied Chemistry, Faculty of Engineering, Hiroshima University, Higashi-Hiroshima 724, Japan

ABSTRACT

Room temperature Raman scattering spectra are reported for the type II superconductors $M_xBa_ySi_{46}$ (M=Na, K) which were recently shown to exhibit T_c's ~ 3.5 K. The spectra are compared to those of other Si_{46}-clathrates which exhibit normal metallic behavior down to 2K. Thirteen of the twenty first-order Raman frequencies predicted by group theory have been detected, and the frequencies are found to be sensitive to the particular dopants. The Raman linewidths observed for the $M_xBa_ySi_{46}$ system are comparable to those observed for Na_8Si_{46} and K_7Si_{46}. The data, taken collectively, suggest that the line broadening in the metallic Si-clathrates is due to important contributions from both the electron-phonon interaction as well as to a random filling of the Si cages.

INTRODUCTION

In this paper we report the first Raman scattering results on the vibrational modes of Si-clathrates which are constructed from face-sharing Si polyhedra as shown in Fig. 1. This Raman study was motivated by recent reports by Yamanaka and coworkers of type II superconductivity (T_c ~ 3.5 K) in $M_xBa_ySi_{46}$ (M=Na, K)[1,2]. Soon after this discovery, Saito *et al.* [3] calculated the band structure for the ideal stoichiometry $Na_2Ba_6Si_{46}$ in which all eight cages per formula unit were filled (Fig. 1). They found a strong hybridization between the Si_{46} bands and a barium orbital which yields a very high density of states at the Fermi level, $N(\varepsilon_F)$ ~48 states/eV. Consistent with the BCS theory of superconductivity, they proposed that this electronic effect is important for the superconductivity in these particular Si_{46} clathrates [3].

So far, little experimental information on the electronic and phonon dynamics of these materials is known, and the effects of doping on the lattice dynamics have, as yet, not been explored. The purpose of this work is to investigate possible correlations between the Raman scattering spectra and the superconducting properties of the Si clathrate samples.

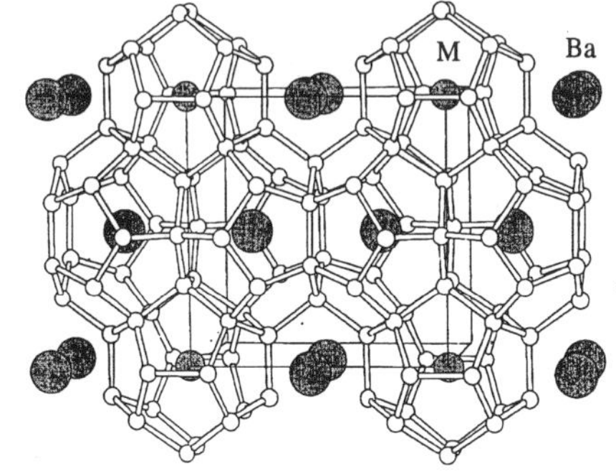

Fig.1. Simple cubic structure of $M_2Ba_6Si_{46}$. Only the Si polyhedra for the front face are shown.

Silicon clathrate compounds M_xSi_{46} and M_xSi_{136} (M=Na, K, Rb, and Cs) were first synthesized in 1965 and studied extensively by Hagenmüller and coworkers [4,5]. The crystal structures are built from polyhedral Si cages. Two distinct polyhedra are found in the

Mat. Res. Soc. Symp. Proc. Vol. 452

Si_{46} clathrates: a dodecahedron (Si_{20}) with 12 pentagonal faces, and a tetrakaidecahedron (Si_{24}) with 12 pentagonal and 2 opposing hexagonal faces. As shown in Fig.1, these polyhedra share faces forming a three dimensional isotropic sp^3 covalent network. The M_xSi_{46} clathrate compounds form metallic line compounds, where x = 8, 7, 5 for Na, K, Rb, respectively [5]. A second family of clathrates, M_xSi_{136}, can also be formed in which x can be varied over a wide range for a particular choice of M. The system has been shown experimentally to be generally metallic for $x>10$ and semiconducting for $x<10$ [5]. The structure for the Si_{136} clathrates is similar but distinct from the Si_{46} clathrates, involving a periodic lattice of Si_{20} and Si_{28} cages.

EXPERIMENT

The preparation of Ba-doped Si_{46} clathrates has been described in detail elsewhere [2,6]. Briefly, two Zintl phases of the silicides $BaSi_2$ and MSi (M=Na or K), were mixed stoichiometrically and placed in a Ta tube which was vacuum-sealed in a stainless steel cylinder. The mixture was heated at 600°C for several days to obtain a third metal silicide Na_2BaSi_4. The silicide was then evacuated at elevated temperature (500°C) under dynamic vacuum. As a result, part of the M atoms were removed, and the Si_{46} clathrate containing M and Ba in the cages is formed. X-ray structural analysis (Rietveld) was performed and the results indicate that Ba atoms occupy the cavities of the larger Si_{24} cages and M atoms reside mainly inside the smaller Si_{20} cages [2,6]. Some cages could be empty, depending on the conditions of sample preparation. Chemical analysis indicated that the samples used for this work are $Na_{0.2}Ba_{5.6}Si_{46}$ and $K_{2.9}Ba_{4.9}Si_{46}$. Both samples were characterized by x-ray powder diffraction, electrical resistivity measurements, and temperature-dependent magnetization measurements. The T_c for the two samples studied here are 3.5 K ($Na_{0.2}Ba_{5.6}Si_{46}$) and 3.2 K ($K_{2.9}Ba_{4.9}Si_{46}$). For comparison, we also collected the Raman scattering spectra of semiconducting Na_xSi_{136} and metallic Na_8Si_{46} and K_7Si_{46} samples. These samples were prepared at the University of Kentucky as follows. First, silicon powder was loaded with a large excess of Na (or K) in a stainless steel ampoule sealed in vacuum in a quartz tube and heated to 600°C. This resulted in the formation of the compound NaSi (KSi) and elemental Na (K). This mixture was heated under high dynamic vacuum at about 330°C for a period of several days, decomposing the silicides and yielding Si_{46} or Si_{136}, depending on the decomposition temperature. Rietveld analysis shows that the polycrystalline powders exhibited structures consistent with those reported by Kasper *et al.* [4].

Raman scattering experiments were carried out on various pelletized silicide powders using the 5145-Å line of an argon ion laser in the Brewster backscattering geometry. Using a mortar and pestle, the Si_{46} clathrate samples were mixed with 20 wt % KBr and then pressed into small pellets. A cylindrical lens was employed to create an illuminated stripe (0.1×2 mm^2) on the sample surface. Low laser flux (20 mW/mm^2) was used in order to prevent damage to the samples, and a 0.46m, f/5 single grating spectrometer (Jobin-Yvon HR460) equipped with a Charge-Coupled-Device (CCD) detector cooled by liquid nitrogen was used to collect the Raman spectra.

RESULTS AND DISCUSSION

The Si_{46} lattice (Fig.1) is a simple cubic (sc) lattice with a space group symmetry Pm3n [1]. In the figure, the specific structure is for $M_2Ba_6Si_{46}$, where the Ba atoms occupy the larger Si_{24} cages [1]. A group theoretical analysis of the Raman- and IR-allowed modes for the ideal composition $M_2Ba_6Si_{46}$ was carried out. The optically active modes are: $13T_{1u}(3) + 3\ A_{1g}(1) + 8E_g\ (2) + 9T_{2g}(3)$, where the ungerade (u) and gerade (g) modes are IR- and Raman-active, respectively. One of the 13, triply degenerate T_{1u} modes contains the three ω=0 acoustic modes. The A_{1g} modes are polarized, and the E_g and T_{2g} modes are unpolarized in isotropic well-ordered samples. One of the $8E_g$ modes is associated with Ba displacements and would be expected at low frequencies. The $9T_{2g}$ modes have one mode at very low frequency associated with Ba displacements and 8 modes associated with Si displacements which should be found at relatively much higher frequencies. No Raman lines were found experimentally to be strongly polarized, and we attribute this to a breakdown in symmetry from random, rather than regular, periodic doping of the Si cages. Raman scattering may be much more sensitive to this form of disorder in the Si-clathrates than X-ray powder diffraction.

The room-temperature Raman spectra for the 3.5 K superconductors $Na_{0.2}Ba_{5.6}Si_{46}$, $K_{2.9}Ba_{4.9}Si_{46}$ and the metallic line compounds Na_8Si_{46} and K_7Si_{46} are shown in Figs. 2a and 2b, respectively. In Fig. 2a, the results of a Lorentzian line shape analysis appear above, and the individual Lorentzians below, the experimental spectrum. In all cases, for clarity, a smooth polynomial background has been subtracted from the data. First, we compare the T ~ 300 K Raman spectra for the superconducting Ba-containing samples (Fig. 2a). Note that the positions of the Raman bands are sensitive to the particular alkali metal dopant (i.e., K or Na). Twenty Raman frequencies are predicted by group theory, and 10 Raman bands

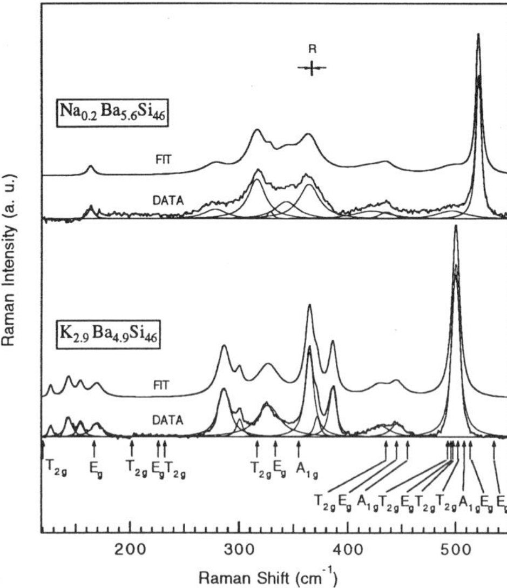

Fig. 2a. Raman spectra of $Na_{0.2}Ba_{5.6}Si_{46}$ and $K_{2.9}Ba_{4.9}Si_{46}$ at T = 300 K.

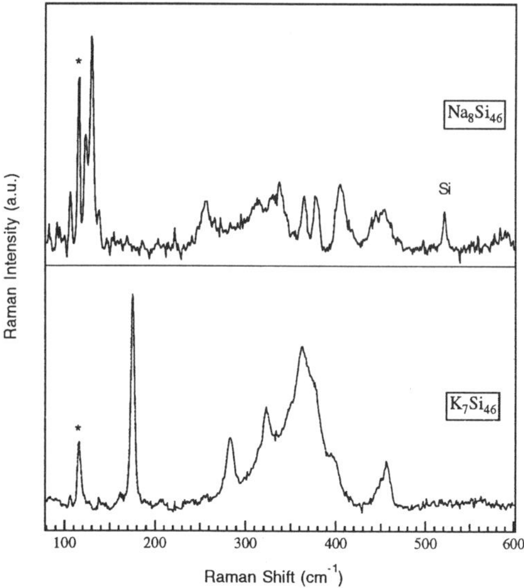

Fig. 2b. Raman spectra of Na_8Si_{46} and K_7Si_{46} at T = 300 K. Lines marked with * are plasma lines from laser.

are observed for $Na_{0.2}Ba_{5.6}Si_{46}$ and 13 for $K_{2.9}Ba_{4.9}Si_{46}$. These bands are somewhat broad; they undoubtedly contain some unresolved Raman lines. The strongest bands in these two spectra appear close in frequency to strong lines observed in possible minority phases, i.e., silicides and diamond Si. For example, in the $Ca_{1-x}La_xSi_2$ silicides, a strong line has been reported in the range 510-515 cm^{-1}, exhibiting a weak x-dependence [7], and in diamond silicon a single, Raman-allowed line is found at 520 cm^{-1}. Therefore, the assignment of the strong band at 500.0 cm^{-1} for $K_{2.9}Ba_{4.9}Si_{46}$ as well as the strong band at 520.5 cm^{-1} for $Na_{0.2}Ba_{5.6}Si_{46}$ to a Si-clathrate mode(s) is questionable. In support of this view is the *absence* of strong bands in the range 510-520 cm^{-1} in the Na_8Si_{46} and K_7Si_{46} spectra (Fig. 2b). The weak band at 520 cm^{-1} in the Na_8Si_{46} is assigned to a small amount of diamond Si in the sample. It is also interesting to note that this strong line near 520 cm^{-1} is absent in the Raman spectra of semiconducting Na_3Si_{136} [8].

The results of a Lorentzian lineshape analysis (frequency and linewidth) for the Raman spectra of the Ba-doped M-clathrates shown in Fig. 2a are listed in Table I, and it should be kept in mind that unresolved modes may have been mistakenly identified with a single band. Previously, using a semi-empirical force constant model, Alben *et al.* calculated the $q=0$ Raman frequencies of a non-relaxed Si_{46} clathrate structure with *empty* cages, i.e., the "undoped" lattice [9]. Since we observe very noticeable, dopant-dependent changes in the Raman frequencies of these superconducting clathrates, improvements over this empty lattice model are now required. As shown in Fig. 2a, the frequencies calculated by Alben *et al.* (indicated by arrows) are not found to be in very good agreement with the experimental data. A more sophisticated theory, including structural relaxation, charge transfer between dopant atoms and the Si cages, and Si-M interactions will be necessary to quantitatively understand our data. As can be seen in Fig. 2a, the two spectra are quite similar, albeit the Raman lines in the $Na_{0.2}Ba_{5.6}Si_{46}$ spectrum are, in many cases, noticeably up-shifted relative to their spectral counterparts in $K_{2.9}Ba_{4.9}Si_{46}$. The shift is not rigid for every mode, as shown in Table I. The question is whether or not this shift, or the broader linewidths tell us anything conclusive about the superconducting pairing mechanism. We discuss this below.

Table I. Raman mode frequencies (ω), widths (Γ), and deviation ($\Delta\omega$) for $Na_{0.2}Ba_{5.6}Si_{46}$ and $K_{2.9}Ba_{4.9}Si_{46}$.

$Na_{0.2}Ba_{5.6}Si_{46}$		$K_{2.9}Ba_{4.9}Si_{46}$		
ω (cm^{-1})	Γ (cm^{-1})	ω (cm^{-1})	Γ (cm^{-1})	$\Delta\omega$
		127.1	7.0	
		143.4	7.2	
		154.4	6.8	
163.5	6.8	169.6	12.6	-6.1
278.0	24.8			
316.0	19.0	286.3	12.8	29.7
328.4	5.8	300.5	6.6	27.9
343.1	25.2			
364.1	22.4	326.7	25.2	37.4
		364.7	9.0	
		371.7	6.2	
		386. 2	7.8	
421.6	41.2	428.9	22.2	-7.3
435.3	16.0	445.5	14.6	-10.2
495.8	32.8			
520.7	5.8	499.8	8.4	20.9

The doping-dependent differences in frequency that we observe for the Raman bands in $Na_{0.2}Ba_{5.6}Si_{46}$ and their counterparts in $K_{2.9}Ba_{4.9}Si_4$ (Fig. 2a) may be due to several factors. First, the Si-M interaction may be responsible, as the mass of K is almost a factor of two larger than that of Na. Second, there is a doping dependence of the Si-Si bondlengths as observed through the lattice constant (10.273 Å for $K_{2.9}Ba_{4.9}Si_{46}$ and 10.261 Å for $Na_{0.2}Ba_{5.6}Si_{46}$). This lattice expansion should weaken the Si-Si bonds and therefor contribute to the observed doping-induced red-shift in the Raman spectrum. The atomic radius of K (2.27 Å) is larger than that of Na (1.54 Å), suggesting a stronger charge transfer between the M atoms and the Si cages in $K_xBa_ySi_{46}$. As a result, a higher concentration of electrons would occupy antibonding states of the Si network and K-doping would then be expected to weaken the Si-Si bonds. Better theoretical calculations are clearly needed to help understand the doping-dependence of the vibrational mode frequencies.

Another intriguing difference in the room-temperature Raman spectra of the superconducting clathrates is the larger linewidths observed for several Raman bands in the $Na_{0.2}Ba_{5.6}Si_{46}$ spectrum over their counterparts in the $K_{2.9}Ba_{4.9}Si_{46}$ spectrum (see Table I). This might be attributed to an increased electron-phonon coupling which decreases the phonon lifetime and increases the Raman linewidths. However, two factors argue against this interpretation for the linebroadening: (1) the T_c's of these two materials are comparable, and (2) comparable Raman linewidths are observed in the non-superconducting Na_8Si_{46} and K_7Si_{46} samples. We feel that an extrinsic mechanism, such as doping disorder must also play an important role in the line broadening.

In the samples considered here, x-ray diffraction data reveal no significant difference in the diffraction peak width and these data can not identify that a particular sample exhibits more (or less) crystalline order. However, we believe that Raman scattering, rather than x-ray diffraction, may be the more sensitive probe for doping disorder in the Si clathrates. Random doping, i.e., either undoped or empty cages, or M- doping the larger Si_{24} cages rather than smaller Si_{20} cages is expected to induce considerable local strain in the Si superstructure. This should lead to a disorder-induced, line broadening of Si vibrational modes. *It should be remembered that a significant doping difference occurs in the two superconducting clathrate samples studied here.* The measured sample stoichiometries indicate that the Si_{20} cages in the $Na_{0.2}Ba_{5.6}Si_{46}$ sample are only 1/10th full, whereas they are completely full in the $K_{2.9}Ba_{4.9}Si_{46}$ sample. Furthermore, as the K-doped sample has 2.9 K atoms per formula unit, presumably the excess 0.9 K ions per formula unit reside in the larger Si_{24} cages, in competition with the Ba dopant. Therefore, the $Na_{0.2}Ba_{5.6}Si_{46}$ sample is expected to exhibit a stronger doping disorder line broadening contribution than found in the $K_{2.9}Ba_{4.9}Si_{46}$, in agreement with experiment. This line of reasoning can be applied to the metallic line compounds Na_8Si_{46} and K_7Si_{46}. In Na_8Si_{46}, all the cages are full, and we expect electron-phonon and phonon-phonon broadening components. However, in K_7Si_{46} one out of eight cages is empty. Therefore, in addition to the these two broadening mecahnisms, we anticipate a third, important, disorder-induced mechanism in K_7Si_{46}. This proposal is consistent with the spectra shown in Fig. 2b, where the Raman linewidths in the Na_8Si_{46} spectrum are, overall, noticeably narrower than in the K_7Si_{46} spectrum.

CONCLUSIONS

Thirteen of the twenty first-order Raman frequencies for a Si_{46} clathrate have been detected for the first time. Normal-state Raman scattering data on the superconducting $M_xBa_ySi_{46}$ system cannot support a significantly enhanced electron-phonon coupling constant in these materials. The Raman linewidths in these samples, however, have significant contributions from doping disorder.

Many of the mode frequencies in the K-doped Ba_ySi_{46} sample were found to be downshifted relative to their counterparts in the Na-doped Ba_ySi_{46} sample. We have proposed various mechanisms for this observation. Further theoretical calculations on the phonon modes in these novel materials are needed to understand the doping-induced effects observed in this study.

ACKNOWLEDGEMENTS

This work was supported by NSF # OSR-94-52895 (S.L.F., L.G., and P.C.E.) and NSF # DMR-95-10093 (G.D. and M.S.D.).

REFERENCES

1. H. Kawaji, H. Horie, S. Yamanaka and M. Ishikawa, Phys. Rev. Lett. **74**, p. 1,427 (1995).

2. S. Yamanaka, H.O. Horie, H. Kawaji and M. Ishikawa, *Eur*. J. Solid State Inorg. Chem. t. **32**, p. 799 (1995).

3. S. Saito and S. Oshiyama, Phys. Rev. B **51**, p. 2,628 (1995).

4. J.S. Kasper, P. Hagenmüller, M. Pouchard and C. Cros, Science **150**, p. 1,713 (1965).

5. C. Cros, M. Pouchard and P. J. Hagenmüller, Solid State Chem. **2**, p. 570 (1970).

6. S. Yamanaka, H.O.Horie, H.Nakano and M. Ishikawa, Fullerene Science & Technology **3**, p. 21 (1995).

7. H. Nakano and S. Yamanaka, J. Solid State Chem. **108**, p. 260 (1994).

8. S.L. Fang, L. Grigorian, P.C. Eklund, Paper in preparation.

9. R. Alben, D. Weaire, J. E. Smith, Jr., and M. H. Brodsky, Phys. Rev. B **11**, p. 2,271 (1975).

THE CHARACTERISTICS AND OXIDATION OF VAPOR - LIQUID - SOLID GROWN Si NANOWIRES

J. WESTWATER, D.P. GOSAIN, S. TOMIYA, Y. HIRANO, AND S. USUI

Sony Research Center, 174 Fujitsuka-cho, Hodogaya-Ku,
Yokohama 240, Japan, jwestw@src.sony.co.jp

H. RUDA

Department of Metallurgy and Materials Science, University of Toronto,
184 College Street, Toronto, Canada M5S 3E4

ABSTRACT

The Vapor - Liquid - Solid (VLS) technique allows the growth of high aspect ratio Si wires. The Si nanowires formed by this technique can be thinned down by oxidation. This approach allows the formation of very thin Si cores which may be used to research the properties of Si nanostructures. In this work the growth and oxidation of these wires is characterized.

In the growth a very thin layer of Au is deposited on a Si (111) surface, silane gas is introduced into the chamber as the Si source gas and the temperature is raised to 300 - 600°C. Initially a catalytically active Au surface phase leads to the growth of a defective epitaxial Si layer. As Au / Si molten alloy balls nucleate and grow in size to approach the threshold size for VLS wire growth, which is determined by the Gibbs - Thomson effect, the epitaxial layer growth rate decreases and a transition to Si nanowire growth occurs. The morphology and width of the wires is strongly dependent on the growth temperature and pressure. At low pressure and high temperature relatively thick well-formed wires grow straight up from the substrate surface along the [111] direction. As the temperature is decreased and the pressure is increased thinner wires (as thin as 10 nm) grow which tend to exhibit growth defects. A light oxidation yields Si cores which are of the order of 5 nm in diameter.

INTRODUCTION

The Vapor - Liquid - Solid (VLS) wire (whisker) growth mechanism was studied in the 1960s on the basis of its novelty as a crystal growth phenomenon [1,2]. Well-formed high aspect ratio sub 100 nm diameter (nano) wires were grown via this method [3].

The current interest in the physics and possible applications of nanowires and the difficulty in fabricating well-formed structures using conventional techniques, such as electron beam lithography and reactive ion etching, has led to some renewed interest in VLS growth. Hiruma *et al* have studied the optical properties of VLS - grown GaAs nanowires and identified quantum confinement effects [4].

It has recently been demonstrated that growth of Si wires as thin as 10 nm can be achieved using silane as the Si source gas in the VLS reaction [5]. In the case of Si wires further thinning can be achieved by oxidation. This is a better means of thinning than simple etching because the thin Si cores are supported by the surrounding SiO_2 and thus maintain some mechanical strength. Furthermore, the interface between the Si and the surrounding SiO_2 should allow good confinement of the carriers in the Si core.

In the current work the nucleation, growth and oxidation of Au / silane VLS nanowires is described. Our results show that the Si wires can be oxidized to yield Si cores as thin as 5 nm

Mat. Res. Soc. Symp. Proc. Vol. 452

in diameter.

EXPERIMENTAL DETAILS

Rectangular samples (10 mm x 45 mm) were cut from polished n-type Si (111) wafers with ρ = 0.4 - 4 Ohm cm. The samples were cleaned in acetone, etched using HNO_3 and HF solutions and then placed in the reaction chamber which was pumped down to 5 x 10^{-8} Torr. Au was deposited from a W filament source onto the Si which was held at room temperature. The Au thickness, which was monitored using a calibrated quartz crystal monitor, was held constant during these experiments at 0.6 nm. Following Au deposition silane gas diluted to 10 % in He was introduced into the chamber. The silane flow rate was 40 SCCM and the total pressure was varied in the range 0.1 - 10 Torr. The sample was then heated to the reaction temperature by flowing direct current along the sample long axis. The temperature was measured using an optical pyrometer and a thermocouple. The temperature during reaction was in the range 320 - 600°C. Oxidation was carried out by backfilling the chamber with oxygen to 500 Torr after wire growth and heating the samples to temperatures in the range 700 - 875°C. The samples were evaluated using secondary electron microscopy (SEM) and transmission electron microscopy (TEM). TEM samples for oxidation studies were prepared by scraping the oxidized wires off the substrate onto a TEM observation grid.

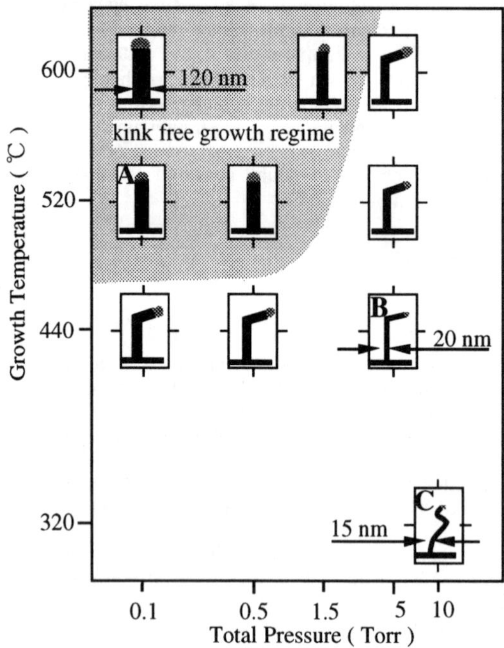

Figure 1 : Si Nanowire Growth Modes.

RESULTS AND DISCUSSION

Nanowire Growth

Wires grown under different pressure and temperature conditions grow at different rates and exhibit different morphology. The growth modes (dependence of the wire morphology and average diameter on temperature and pressure) of the wires observed under different reaction conditions are summarized in figure 1. Figures 2 (a), (b) and (c) show SEM images of wires grown under the conditions indicated by A, B and C, respectively in figure 1. Different growth times were used in each case. At high temperature and low pressure the wires grow straight up from the substrate along the [111] direction. As the silane pressure increases the wires become thinner because the silane chemical potential is higher at higher pressure which allows thinner wires to grow (the Gibbs - Thomson effect). In addition, as the pressure is raised and / or the reaction temperature is lowered, growth defects such as kinking tend to occur. In figure 2 (b) the wires switch growth direction spontaneously during

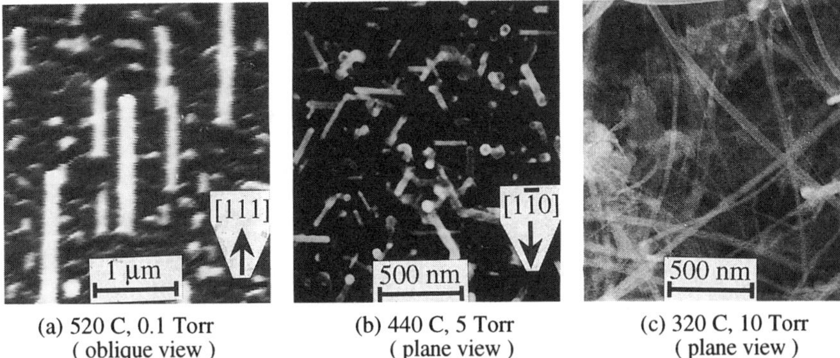

(a) 520 C, 0.1 Torr (oblique view)

(b) 440 C, 5 Torr (plane view)

(c) 320 C, 10 Torr (plane view)

Figure 2 : Dependence of the Wire Width and Morphology on Growth Temperature and Pressure. (a), (b) and (c) correspond to A, B and C in Figure 1, respectively.

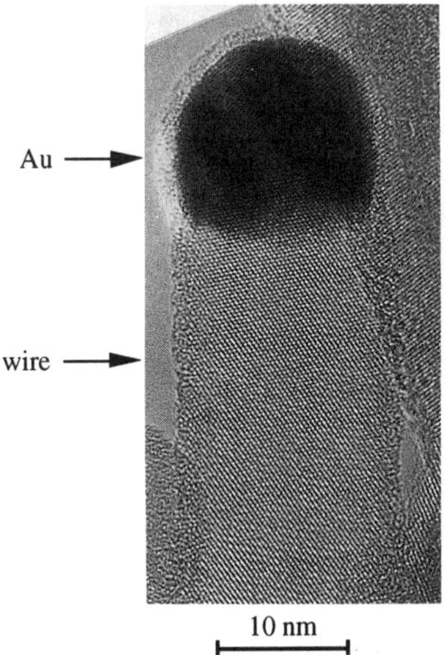

Figure 3 : TEM Lattice Image of Si Nanowire.

growth. They switch from the initial [111] growth direction to one of the other available {111} directions. Thus in the plane view SEM image three sets of parallel wires can be seen. In figure 2 (c) the wires twist and turn apparently randomly. The thinnest wires grown under these conditions are of the order of 10 nm wide. The TEM lattice image of a 16 nm wide wire shown in figure 3 shows that the wires are single crystals.

Nanowire Nucleation

The growth time for the wires shown in figure 2 (a) was 1 hour. The SEM images in figure 4 show the extent of the growth after 5 minutes and 20 minutes under the same conditions. The wire nucleation does not occur immediately; there is no evidence of wire growth during the first 5 minutes of the reaction. The Si surface does, however, become very rough during the first stages of the reaction. Wires have clearly nucleated after 20 minutes of growth, and after 1 hour wires which are about 1 µm long can be seen. Cross-sectional TEM images for samples with reaction times of 5 and 60 minutes (ie prior to and after wire nucleation) are shown in figures 5 (a) and (b). No wires are visible after 5 minutes of growth although Au balls up to about 15 nm in diameter can be seen on the top of a layer of Si which is of the order of 15 nm thick. In figure 5 (b) the Si wires have grown on top of the Si layer. The layer thickness is now of the order of 30 nm. The rate of growth of the Si layer is high at the beginning of the reaction and then slows down afterwards.

When the Au is deposited at room temperature it forms a thin Au / Si amorphous alloy layer at the substrate surface [6]. When the sample temperature is raised in the silane ambient molten Au / Si alloy balls nucleate and the Au on the surface tends to gather at the balls. Most of

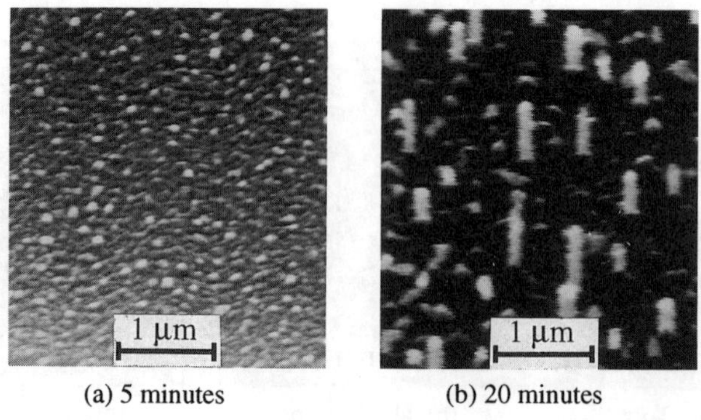

(a) 5 minutes (b) 20 minutes

Figure 4 : Time Dependence of Growth (oblique view).

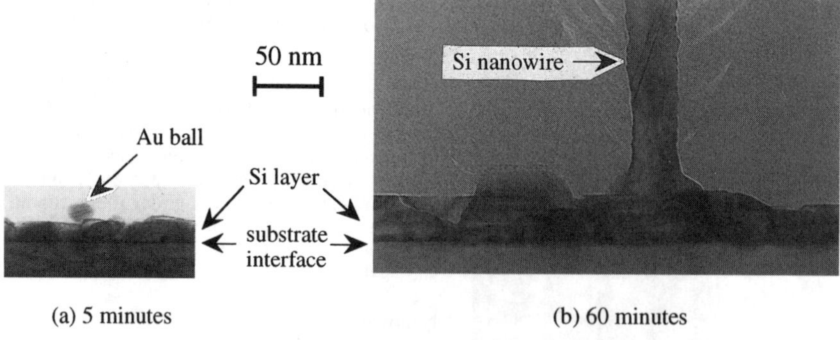

(a) 5 minutes (b) 60 minutes

Figure 5 : Time Dependence of Growth : TEM

the Si layer growth occurs prior to and during this nucleation phase when the Au is distributed across the Si layer surface in significant quantities. The Au surface phase therefore plays a catalytic role because the rate of Si layer growth greatly exceeds the rate of growth of a Si film on a bare Si surface under similar temperature and pressure conditions. The alloy balls grow by diffusion of Au at the surface and agglomeration of the liquid alloy balls themselves until they exceed the lower limit for wire growth which is determined by the Gibbs - Thomson effect [3]. Si nanowire growth then commences. The TEM images show that the Si layer is a defective epitaxial layer with crystallinity which is inferior to that of the Si nanowires. There was evidence of twin defects and stacking faults in the diffraction patterns of the Si layers.

Oxidation

The Au ball at the tip of the wire plays a role in the wire oxidation. Figure 6 shows a bright field TEM image of a wire which was grown under the same conditions as the wires shown in figure 2 (a) and then oxidized at 875°C for 2 hours. The Au ball is no longer at the tip of the wire. The Au / Si molten alloy ball appears to mediate in the oxidation of the Si wire, traveling along the wire, converting Si to SiO_2 as it does so.

In the oxygen ambient the wires are heated well beyond the Au / Si eutectic temperature (363°C [7]). If an oxide crust forms on the Au / Si molten alloy tip as the oxygen removes Si from the melt, Si from the wire must then be dissolved to re-establish the eutectic composition.

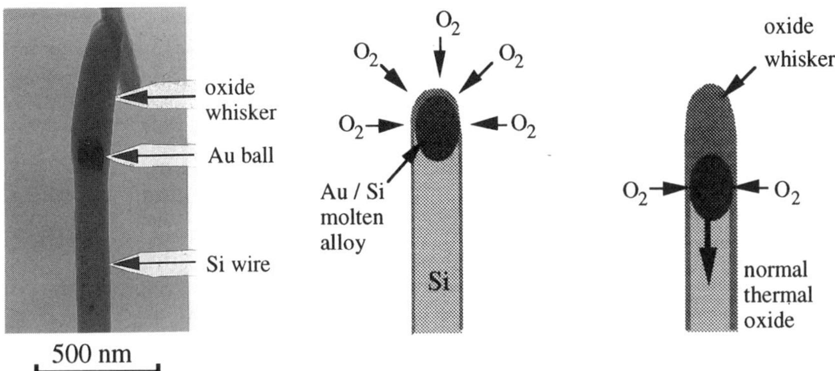

Figure 6 : Bright Field TEM Image of Oxidized Wire Showing the Oxide Whisker.

(a) Introduction of Oxygen - Formation of Oxide Crust. (b) After Extended Oxidation - Growth of Oxide Whisker.

Figure 7 : Proposed Mechanism of Oxide Whisker Growth.

In this way the Au / Si molten alloy can mediate in the oxidation of the Si wire. As the crust on the tip of the wire thickens the alloy ball apparently does not become starved of oxygen because experiments to investigate how this reaction proceeds with time showed that the Au ball continues to travel until it reaches the substrate surface and then comes to rest there. It is probable that oxygen is supplied to the ball through the side of the wire. Figure 7 illustrates this possible mechanism.

In order to investigate the nanowire lateral oxidation, wires were grown under the same conditions as the wires shown in figure 2 (c) and then oxidized at 700°C for 6 hours. The Au mediated oxidation mechanism described above was also active under these conditions. However, after 6 hours of oxidation the Au ball had only moved a relatively short distance along these wires, which were several µm in length. The dark field TEM image in figure 8 shows that the Si cores which remain after the oxidation are about 5 nm in diameter. The wires appear to be discontinuous. This is because the wires are crossed by other wires which cannot be seen in the dark field image. Close observation reveals that there is a weak diffraction signal at the points where the wires appear to be broken which suggests that the wires are indeed obstructed.

Figure 8 : Dark Field TEM Image of Oxidized Si Nanowires.

It is difficult to draw conclusions from these results alone about the lateral oxidation rate because of the distribution of initial wire sizes. Stresses caused by the volume expansion associated with oxidation are expected to affect the oxidation rate in structures with such small radius of curvature. These stresses are particularly relevant in the case of low temperature oxidation when stress relieving viscous flow of oxide is negligible. According to Kao *et al* [8], oxidation of wires with diameters of the order of 40 nm or less at low temperature is likely to be controlled by the rate of the interface reaction, which slows significantly due to the stress, rather than the rate of diffusion of oxidants through the oxide.

CONCLUSIONS

The nucleation, growth and oxidation of Si nanowires via the silane / Au VLS reaction has been characterized. Prior to the growth of Si nanowires a catalytically active Au surface phase leads to the growth of a defective epitaxial Si layer. As Au / Si molten alloy balls nucleate and grow in size to approach the threshold size for VLS wire growth, the layer growth rate decreases and a transition to Si nanowire growth occurs.

The morphology and width of the wires is strongly dependent on the growth temperature and pressure. At low pressure and high temperature relatively thick well-formed wires grow straight up from the substrate surface along the [111] direction. As the temperature is decreased and the pressure is increased thinner wires grow which tend to exhibit growth defects.

When the wires were oxidized the Au / Si molten alloy at the wire tip mediated in the growth of oxide whiskers. Si cores as thin as 5 nm were obtained by the lateral oxidation.

ACKNOWLEDGEMENTS

The authors would like to thank C. Edirishinge, L. Sidha, S. Zukotyhski, and M. Nakagoe for useful discussions and experimental support. The support of H. Ohki and T. Yamada is gratefully acknowledged.

REFERENCES

1. R. S. Wagner and W. C. Ellis, Appl. Phys. Lett., **4**, 89 (1964).
2. R. S. Wagner and C. J. Doherty, J. Electro. Chem. Soc., **115**, 93, (1968).
3. E. I. Givargizov, J. Cryst. Growth, **31**, 20 (1975).
4. K. Hiruma, M. Yazawa, T. Katsuyama, K. Ogawa, K. Haraguchi, M. Koguchi, and H. Kakibayashi, J. Appl. Phys., **77**, 447 (1995).
5. J. Westwater, D. P. Gosain, S. Tomiya, S. Usui, and H. Ruda, submitted to J. Vac. Sci. Tech.
6. G. Lelay, Surface Science, **132**, 169, (1983).
7. Binary Phase Diagrams , edited by H. Okamoto, P. R. Subramanian and L. Kacprzak (W. W. Scott, 1992), Vol. 1, 2nd edition.
8. D. B. Kao, J. P. McVittie, W. D. Nix, K. C. Saraswat, IIIE Trans. Elect. Dev., **34**, no. 5, 1008, (1987).

SELF-ASSEMBLING FORMATION OF SILICON QUANTUM DOTS BY LOW PRESSURE CHEMICAL VAPOR DEPOSITION

K. NAKAGAWA, M. FUKUDA, S. MIYAZAKI AND M. HIROSE
Department of Electrical Engineering, Hiroshima University
Higashi-Hiroshima 739, Japan

ABSTRACT

The growth of Si dots on thermally-grown SiO_2/c-Si from a thermal decomposition of pure silane has been systematically studied in the temperature range from 500 to 650°C. It has been suggested that the Si dot height and dot diameter on as-grown SiO_2 are rate-limited by the cohesive action of adsorbed precursors and the thermal decomposition of silane on Si nucleation sites, respectively. The nucleation site on as-grown SiO_2 is likely to be generated by the thermal dissociation of surface Si-O bonds. It has been also found that in Si dot formation on OH-terminated SiO_2 surface the nucleation density is dramatically enhanced and consequently the dot size and its distribution become small. This implies that surface OH bonds can provide the nucleation sites.

INTRODUCTION

Nanometer-sized silicon structures have increasingly attracted much attention because their unique physical properties associated with quantum mechanical effects lead us to develop novel Si-based functional devices such as resonant tunneling devices [1-3], one-dimensional transport devices [4] and single electron tunneling devices [5,6]. A crucial issue for the room temperature operation of such Si quantum-effect devices is to fabricate well-defined Si structures with a feature size below 3 nm as predicted from the quantization energy or coulombic charging energy. In preparing Si quantum dots without defects and damage, a self-assembling process is thought to be suitable to achieve a good uniformity in size and a high areal density. Recently, it was reported that Si nanocrystallites can be spontaneously grown on SiO_2 by controlling the early stages of low-pressure chemical vapor deposition (LPCVD) [3,7] and exhibit a blue shift of the optical absorption edge which is attributable to the quantum confinement effect [7]. In addition, we have demonstrated room temperature resonant tunneling through double barrier structures consisting of a single Si quantum dot sandwiched with 1 nm-thick SiO_2 [3].

In this paper, we focused on the formation mechanism of Si dots on thermally-grown SiO_2/c-Si from a thermal decomposition of pure silane. The size distribution and the nucleation density of Si dots have been evaluated as a function of growth temperature by using atomic force microscopy (AFM) and transmission electron microscopy (TEM). The Si dot formation on HF-treated SiO_2 has been compared with the case on as-grown SiO_2 to reveal the influence of chemical bonding features of the oxide surface on the growth kinetics.

EXPERIMENTAL

Czochralski-grown Si(100) and (111) wafers were used as substrates in this study. After conventional RCA cleaning steps, Si(111) wafers were immersed in 40% NH_4F for 5 min at room temperature to obtain an atomically-flat monohydride-terminated surface and Si(100) wafers were treated with a 4.5% HF solution just to remove the chemical oxide, followed by a pure water rinse to eliminate residual fluorine atoms from the surface. Subsequently, oxides with thicknesses of 2~10 nm were grown at a temperature of 900 or 1000°C in dry O_2. Some of SiO_2/c-Si so prepared

Mat. Res. Soc. Symp. Proc. Vol. 452

were slightly etched back by dipping into a 0.1% HF solution to change the surface chemical bonding features. Si nanocrystallites were deposited on as-prepared or HF-treated SiO_2/c-Si by the thermal decomposition of pure monosilane in the temperature range from 500 to 650°C. During the deposition, the gas pressure was maintained at 0.2 Torr. AFM observations were carried out in air using a Au-coated Si_3N_4 probe to assess the nanocrystallite size. High resolution TEM images were taken for nanocrystallites deposited near a substrate edge in the cross sectional view and for samples whose substrates were removed locally by etching in a KOH solution.

RESULTS AND DISCUSSION

The formation of cap-shaped Si nanocrystals was confirmed at deposition temperatures above 525°C by high resolution TEM observations as shown in Fig. 1. Notice that the lattice images for Si(111) planes were always observed at a tilt by ~30 degrees from the substrate surface normal. This indicates that the growth of Si dots proceeds so that a Si(100) plane becomes parallel to the substrate surface to minimize the surface free energy. The Si dot growth was also evidenced from the change in the X-ray excited Si2p spectrum measured at a photoelectron take-off angle of 5~30°. AFM images taken for Si dots grown on atomically flat SiO_2/c-Si(111), whose surface morphology maintains an initial biatomic step structure of Si(111), clearly show that Si nucleation occurs randomly irrespective of the atomic scale roughness of the SiO_2 surface as represented in Fig. 2. Indeed, there is no significant difference between the Si dot formation on atomically flat SiO_2/c-Si(111) and on rough SiO_2/Si(100).

Figure 3 shows an Arrhenius plot of the areal Si-dot density on as-grown SiO_2 evaluated by AFM observations. In the early stages of the dot formation, as the deposition temperature rises from 550 to 570°C, the dot density is dramatically increased with an activation energy of ~5.4 eV until reaching a value of $2x10^{11}$ cm^{-2}, at which point it tends to be saturated because of the coalescence growth as confirmed by TEM observations. Considering that the Si-O bond energy is estimated to be 4.85 eV from the atomization energy of SiO_2 [8], the Si-O bond breaking might play a role in the creation of nucleation sites. The size distribution of obtained Si dots evaluated from AFM images can be fitted to a log-normal function [9] as indicated in Fig. 4. It is obvious that higher temperature deposition provides a wider size distribution.

To get a clear insight into the growth mechanism, the average dot diameter or dot height was determined at each deposition temperature by fitting a measured size distribution to a log-normal

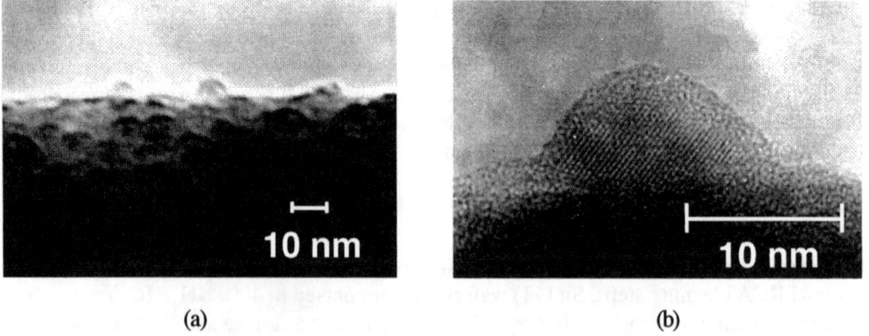

Fig. 1. TEM micrographs of a sample after 36 sec deposition at 600°C on as-grown SiO_2 which were taken at a tilt cross sectional view with a low resolution (a) and at a cross section with a high resolution (b).

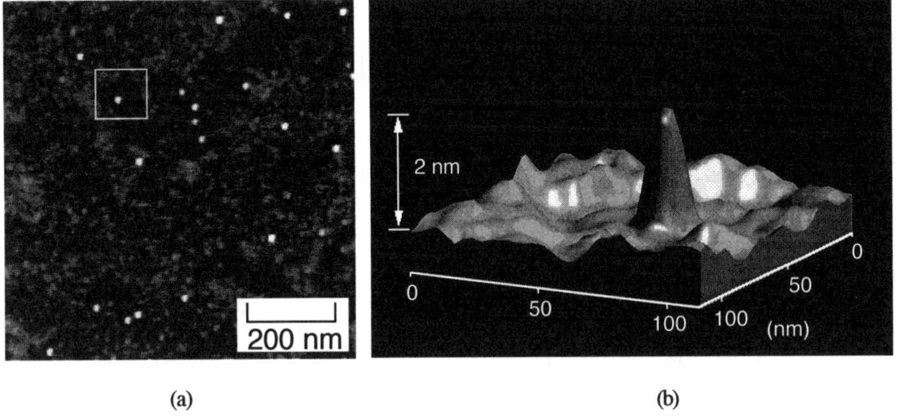

(a) (b)

Fig. 2 AFM images of Si dots formed at 550°C for 12 sec on atomically-flat as-grown SiO_2/Si(111). An open squared region in (a) is magnified in the three-dimensional image (b).

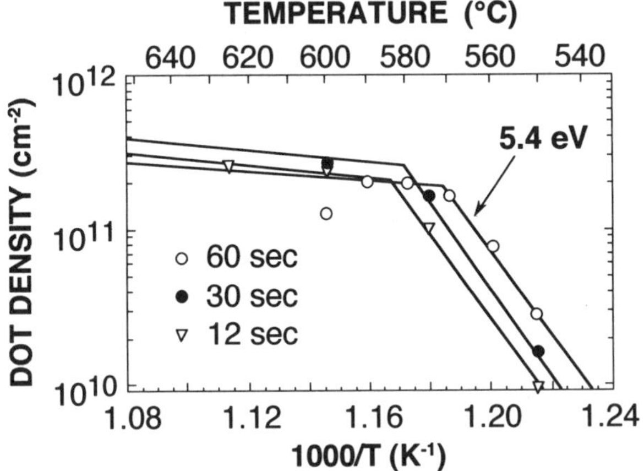

Fig.3 Deposition temperature dependence of areal Si-dot density on as-grown SiO_2. The deposition time was varied from 12 to 60 sec.

function, and the activation energy of dot formation was assessed from the Arrhenius plot as shown in Fig. 5. The dot diameter increase with temperature has an activation energy of 0.82 eV which corresponds to the SiH_4 decomposition energy on c-Si [10]. On the other hand, the activation energy for the dot height is found to be 2.2 eV, being almost identical to the Si cohesive energy of 2.14 eV [11] or 2.45 eV [12]. These results imply that the dot diameter is controlled by the thermal decomposition process of SiH_4 on the Si nucleation sites created on SiO_2 and the height is rate-limited by the cohesive action of adsorbed precursors. In addition, it is likely that reactive species diffusing on the surface mainly contribute to the dot growth as described later. In a thermodynamic concept for nucleation, it is energetically favorable for clusters exceeding a critical size to grow

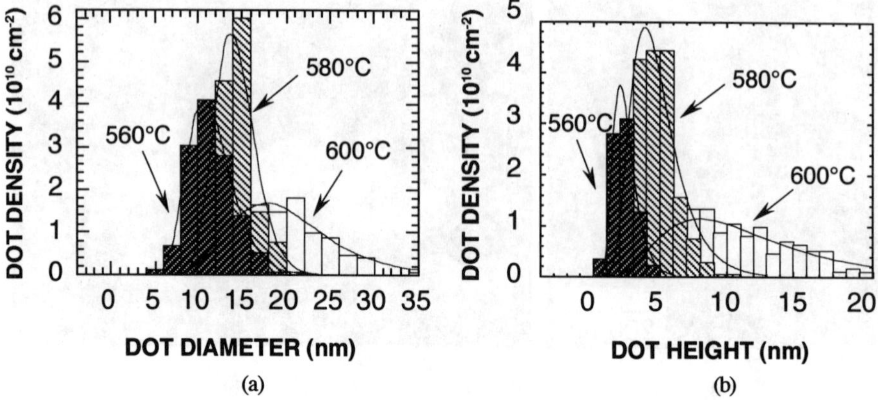

Fig. 4 Distributions of Si-dot diameters (a) and of dot heights (b) obtained by AFM images for samples prepared at different deposition temperatures for 60 sec. The solid lines denote log-normal functions well-fitted to the measured distributions.

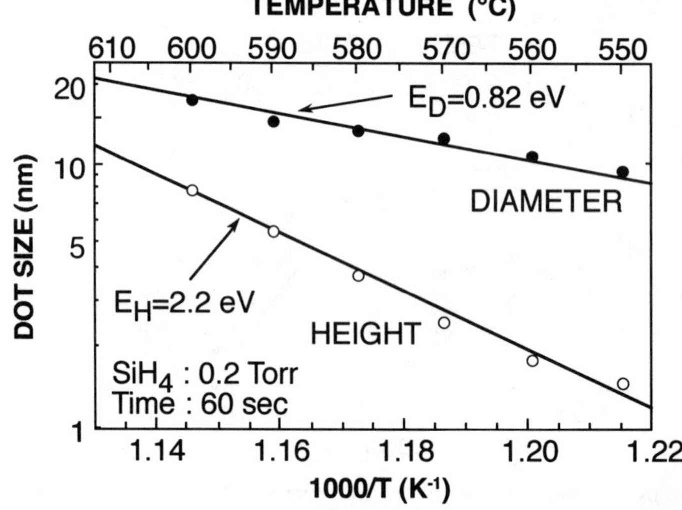

Fig. 5 Deposition temperature dependences of average Si-dot diameter and dot height on as-grown SiO_2. The deposition time was 60 sec.

further while for clusters smaller than the critical size to decay. According to a simple nucleation theory [13], the critical radius R_c for a stable Si cluster can be estimated by using the following equation: $R_c = 2\gamma\upsilon/kT\ln\alpha$, where γ is the surface free energy, υ the atomic volume, $kT\ln\alpha$ the supersaturation function giving the gain in free energy, and α the ratio of SiH_4 pressure to the equilibrium vapor pressure of Si. The estimated value of R_c = 0.23 nm is slightly smaller than the size of crystalline silicon unit cell. For more detailed discussion, the potential energy difference between clusters and adsorbed precursors and the interfacial energy between clusters and SiO_2 are needed although we do not have their reliable values at present.

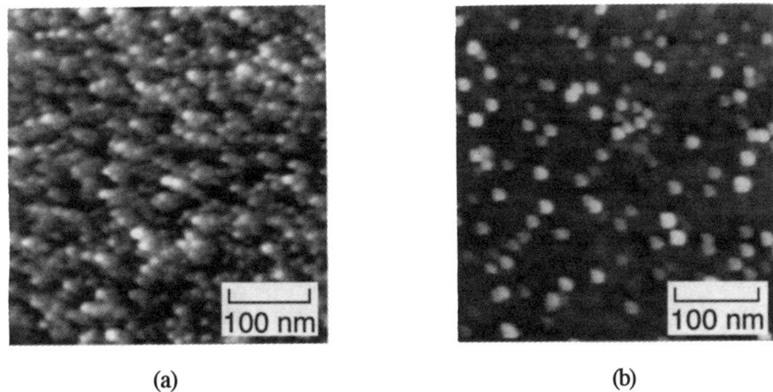

Fig. 6 AFM images of Si dots formed at 565°C for 60 sec on SiO_2 surfaces with (a) and without HF treatment (b).

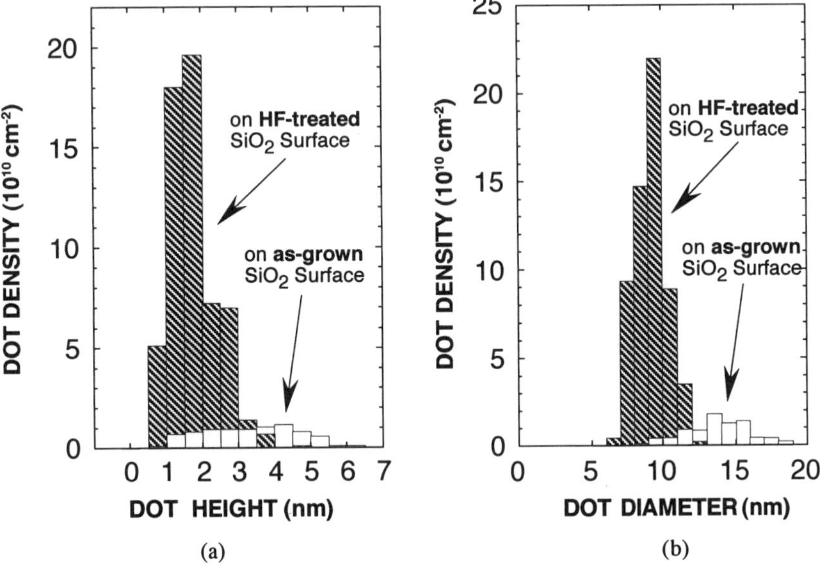

Fig.7 Distributions of Si-dot heights (a) and dot diameters (b) for samples shown in Fig. 6.

Significant enhancement of the areal Si-dot density and resultant decrease in the dot size are caused by treating the SiO_2 surface in a 0.1% HF solution before LPCVD as seen in Figs. 6 and 7. It should be noted that the size distribution becomes appreciably narrower compared with the case on as-grown SiO_2 as indicated in Fig. 7. This implies the uniform nucleation on the HF-treated SiO_2 surface. The Fourier transform infrared attenuated total reflection (FT-IR-ATR) spectrum of the HF-treated SiO_2 exhibits absorption bands centered at ~3450 and ~3250 cm^{-1}, indicating that the surface is terminated by OH bonds. Taking into account the fact that the surface silanol groups (≡Si-OH) are not stable at temperatures above 500°C, it is likely that they can act as nucleation sites for reactive species such as SiH_2. The result of Fig. 6 can be interpreted in terms of a model in which the surface diffusion length of reactive species is reduced by OH termination of the SiO_2 surface. Providing that the ratio of square root of dot density on as-grown SiO_2 to that on HF-treated SiO_2 is equal to the reciprocal ratio of the surface diffusion length, the activation energy for surface diffusion of reactive species is increased by 0.15 eV at 565°C by means of the HF treatment.

CONCLUSIONS

The formation of nanometer-sized single-crystalline Si dots on thermally grown SiO_2 has been demonstrated by LPCVD using pure silane. The Si-dot diameter and dot height are controlled by the SiH_4 decomposition and the Si cohesive process, respectively. The dilute HF treatment of the SiO_2 surface enables us to prepare uniformly fine Si dots as a result of a significant increase in the nucleation density.

ACKNOWLEDGEMENTS

This work has been supported in part by Grants-in-aid for General Scientific Research B (No.07455142) from the Ministry of Education, Science, Sports and Culture and for the Core Research for Evolutional Science and Technology (CREST) from the Japan Science and Technology Corporation (JST).

REFERENCES

[1] M. Hirose, M. Morita and Y. Osaka, Jpn. J. Appl. Phys. **16**, Suppl. 16-1, 561 (1977).

[2] K. Yuki, Y. Hirai, K. Morimoto, K. Inoue, M. Niwa and J. Yasui, Jpn. J. Appl. Phys. **34**, 860 (1995).

[3] M. Fukuda, K. Nakagawa, S. Miyazaki and M. Hirose, Extended Abstract of 1996 Intern. Conf. on Solid State Devices and Materials (Yokohama, 1996) p. 175.

[4] K. Morimoto, Y. Hirai, K. Yuki and K. Morita, Jpn. J. Appl. Phys. **35**, 853 (1996).

[5] K. Yano, T. Ishii, T. Hashimoto, T. Kobayashi, F. Murai and K. Seki, IEEE Transactions on Electron Devices **41**, 1628 (1994).

[6] S. Tiwari, F. Rana, H. Hanafi, A. Hartstein, E. F. Crabbè and K. Chan, Appl. Phys. Lett. **68**, 1377 (1996).

[7] A. Nakajima, Y. Sugita, K. Kawamura, H. Tomita and N. Yokoyama, Jpn. J. Appl. Phys. **35**, L189 (1996).

[8] R. T. Sanderson in Chemical Bonds and Bond Energy, edited by E. M. Loebl (Academic Press Inc., New York, 1976) p.135 .

[9] R. R. Irani and C. F. Callis in Particle Size Measurement: Interpretation and Application, (Wiley, New York, 1963)

[10] B. A. Joyce, R. R. Brandley and G. R. Booker, Philoso. Mag. **15**, 1167 (1967).

[11] J. C. Philips in Bonds and Bands in Semiconductors, edited by A. M. Alper, J. L. Margrave and A. S. Nowick (Academic Press Inc., New York, 1973) p. 52.

[12] W. A. Harrison in Electrnic Structure and the Properties of Solids -The Physics of the Chemical Bond, edited by P. Renz and K. Sargent (W. H. Freeman and Company, San Fransisco, 1980) p.171.

[13] D. W. Jones, E. Kaldis, A. I. Mlavsky, D. W. Shaw and G. A. Wolff in Crystal Growth Theory and Techniques, edited by C. H. L. Goodman (Plenum Press, New York, 1974) p. 55

RAMAN SPECTROSCOPY OF GE NANOCRYSTALS GROWN BY SELF-ORGANIZATION PROCESSES

A. STELLA*, C. E. BOTTANI**, P. CHEYSSAC***, R. KOFMAN***, P. MILANI****, P. TOGNINI*.
*Istituto Nazionale per la Fisica della Materia, Dipartimento di Fisica "A. Volta", Università di Pavia, Via Bassi 6, 27100 Pavia, Italy
**Istituto Nazionale per la Fisica della Materia, Dipartimento di Ingegneria Nucleare, Politecnico di Milano, Via Ponzio 34/3, 20133 Milano, Italy
***Laboratoire de Physique de la Matière Condensée, URA 190, Université de Nice Sophia Antipolis, Nice Cedex, France
****Istituto Nazionale per la Fisica della Materia, Dipartimento di Fisica, Università di Milano, Via Celoria 16, 20133 Milano, Italy

ABSTRACT

We report Raman spectroscopy measurements on Ge nanocrystals with average radii ranging from about 65 Å down to 10 Å (with a size dispersion lower than 20 %).

Ge has been deposited by UHV evaporation on an amorphous substrate, kept at such a temperature as to produce the Ge nanodroplets nucleation in the liquid phase.

A nanocrystalline size dependence of the Raman spectra has been observed and explained in the framework of a phonon confinement model. We have observed the softening of the TO Raman peak predicted by the theory when the dimensions of the particles are decreased. Moreover the observed inhomogeneous broadening of the Raman lines has been correlated with the size distribution of the particles in the samples.

Our results provide a characterization of Ge nanoparticles exhibiting a good crystalline nature, down to about 10 Å, and in conditions of substantial absence of perturbations of the environment.

INTRODUCTION

The investigation of the properties of metal and semiconductor clusters embedded in dielectric matrices presents several aspects of high interest both for basic research and technological applications [1],[2].

Due to size reduction down to dimensions of the order of one nanometer, significant deviations from bulk behaviour take place: in particular, quantum confinement gives rise for instance to a large discretization of the electronic levels and to a relevant blueshift of structures and singularities in the density of states [3]. The nanometer size affects rather strongly the thermodynamic properties (size dependence of melting temperature and latent heat of fusion [4]) and produces sizeable nonlinear optical effects [5].

In view of these considerations, it appears to be crucial, for the development of the field, to master the growth techniques for sample preparation in order to tailor, to a large extent, their physical behaviour. High priority must be given to the possibility of varying the size in a wide range, to restrict the size dispersion to a very narrow range and to achieve regular and reproducible shapes.

Mat. Res. Soc. Symp. Proc. Vol. 452

Several growth techniques have been developed for Ge nanoparticles, each one with advantages and limitations: synthesis in zeolite [6], ultraviolet-assisted oxidation of Si-Ge layers [7], hydrothermal oxidation of Si_xGe_{1-x},[8], inorganic solution phase synthesis [9], reduction in glassy matrix [10]. In this paper we present significant aspects of a vapour condensation growth technique and the we relate them to size and matrix dependence of the nanocrystals Raman response.

SAMPLE CHARACTERISTICS

The nanoparticles are grown in ultra high vacuum in the Volmer-Weber mode: they are produced by a self-organization of vapour condensing on a partially wetting substrate [11]. This allows to obtain samples with nanoparticles characterized by a relatively low size dispersion and a uniform shape. Moreover there is the possibility of varying the sizes in a wide range, i.e. more than one order of magnitude. Due to the high temperature of the substrate, the deposit consists of liquid droplets.

The conditions of nanocrystals nucleation are based on the surface energy balance between the interfaces involved in the process. The fulfilled relation is [11]:

$$\Gamma_{SV} - \Gamma_{LV} < \Gamma_{SL} < \Gamma_{SV} + \Gamma_{LV} \qquad (1)$$

where the Γ_{SV}, Γ_{LV} and Γ_{SL} are the free energies per unit area of the substrate solid-vapour (sv), deposit liquid-vapour (lv) and substrate solid-deposit liquid (sl) interfaces.

The surface tensions play also a primary role in determing the shape of the nanoparticles, which result to be truncated spheres. The volume of the particles can be determined from their projected area, knowing the contact angle (about 106°). Their size distribution is bimodal, with a bell-shaped part due to the coalescence and a tail caused by renucleation and growth of small particles. The transition from two-dimensional to three-dimensional structures happens in the early stages of growth.

The amorphous embedding matrix and the growth at the liquid state minimize any stress or strain due to interaction of the nanoparticles with the environment.

The method here summarized has been applied to different materials: Sn, Pb, Ga, Ge; in this work we intend to study the size effects on the Raman spectrum of Ge quantum dots. We present the investigations performed on two different series of samples.

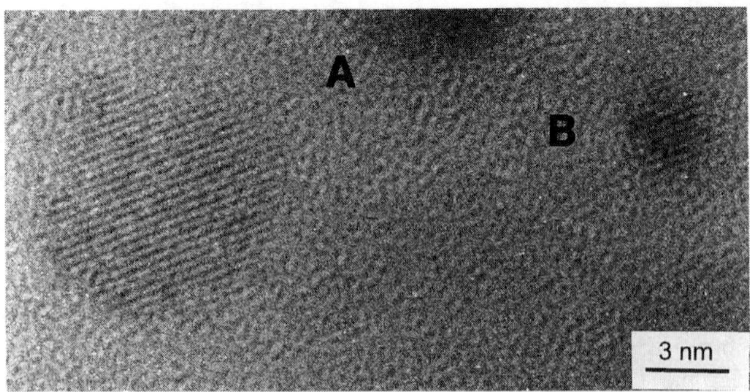

Fig. 1 HREM image of the Ge nanocrystals

In the first one the nanocrystals are evaporated on a sapphire substrate and embedded in amorphous Al_2O_3. The average sizes, obtained by Transmission Electro Microscopy, are between 25 and 130 Å [12]. They are proportional to the Ge quantity deposited in each layer.

The size range in the second series is considerably narrower, but, in this case, the conditions of growth have been different from the previous series: in particular the substrate (spectrosil), and the nucleation temperature (which is sligthly higher) have been varied.

In order to increase the optical response of the samples, the Ge/amorphous matrix evaporations have been repeated five times.

The corresponding transmission spectra are reported in fig. 2; the two main absorption peaks are due to the E_1 and E_2 spectral structures. Their presence, and their position very close to that in the bulk, is a good check of the crystallinity of the nanoparticles which has been directly inspected by electron microscopy in the high resolurtion mode [12]. Quantum confinement effects on these structures, related to their origin inside the Brillouin Zone, have been previously discussed [12].

Sample	Nanoparticles average radius (Å)	Number of layers	Substrate	Series
Ge13	13	5	sapphire	1
Ge15	15	5	sapphire	1
Ge48	48	5	sapphire	1
Ge65	63	5	sapphire	1
Ge*	17-47	5	spectrosil	2

Tab. 1: growth characteristics of the samples investigated in this work (Ge*: bimodal distribution).

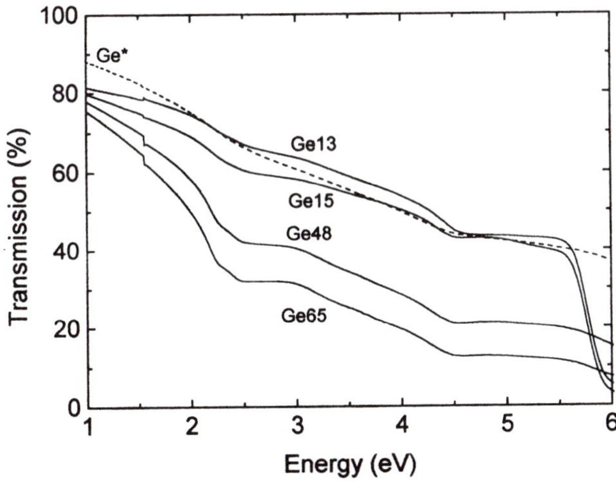

Fig.2 Transmission spectra at 77K. Solid lines: series 1 samples; dashed line: series 2 sample.

RAMAN INVESTIGATION

Raman spectra are in principle expected to be quite sensitive not only to quantum confinement but also to disorder, strain or in general to interface peculiarities and structural characteristics of the systems [13].

They were recorded at room temperature in a backscattering geometry using a Jobin--Yvon T64000 triple-grating spectrometer with a liquid--nitrogen cooled CCD camera. Spectral resolution was about 1 cm^{-1}. The excitation wavelength was λ= 514.5 nm.

In figure 3 we show the dependence of the Raman spectrum on particle size for the series 1 samples. For bulk crystalline Ge the Raman spectrum can be assigned to the $\mathbf{q} \cong \mathbf{0}$, Γ_5^+ optic phonon exhibiting a Lorentzian line at 300 cm^{-1} with an intrinsic width of about 3 cm^{-1} [14].

The spectrum of Ge65 already starts showing a shift and an incipient asymmetry in the low frequency tail and a FWHM of 5.8 cm^{-1}. As the size decreases we observe an inhomogeneous broadening and a softening of the peak. The evolution of the peak shape has been simulated with a model developed to take into account both phonon confinement [13],[15] and size distribution of the particles.

The $\mathbf{q} \cong \mathbf{0}$ selection rule is a consequence of the full translational symmetry of a crystalline sample. In the presence of finite size effects, e.g. partial phonon confinement in nanoparticles, the Raman intensity can be written as a superposition of Lorentzians centered at $\omega = \omega(\mathbf{q})$ (the dispersion relation of optical phonons) with q in the first Brillouin zone. This produces a symmetric broadening and red shift of the Raman line, which can be computed as in ref. [16], taking fully into account the effective distribution of particle radii.

We note that Raman spectra show significant deviations from the bulk features even for larger size particles.

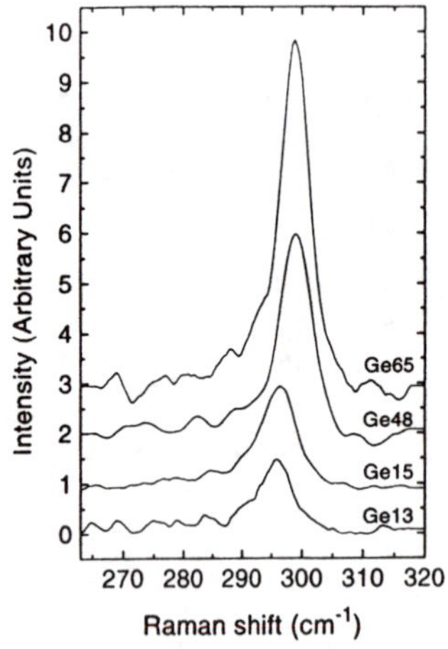

Fig. 3 Raman spectra of Ge nanocrystals (series 1).

The Raman shift and broadening can be associated to quantum confinement rather than to matrix interaction. Our results can be compared with those of ref. [9] concerning Raman spectra of colloidal Ge particles in the size range 200 - 80 Å, where the observed shift and broadening is in agreement with the theory. On the other hand, different authors [17] observed the expected broadening with no frequency shift in particles having a mean diameter of 20 Å embedded in SiO_2 matrices; however the strain due to the matrix-cluster interaction and the incertitude on the structure of the dots, and on their mass distribution, make any comparison with the theory very difficult.

In order to further investigate the influence of the matrix on the Raman spectra of Ge nanoparticles we performed Raman spectroscopy on a sample of the series 2. In fig. 4 we show the the Raman signal of Ge nanoparticles after subtraction of the signal due to the substrate; it has been fitted with two Lorentzian lines according to ref. [16]. From the goodness of this fit we confirm the bimodal nature of the size distribution centered at 47 Å and 17 Å respectively. This bimodal distribution is reminiscent of the growing method.

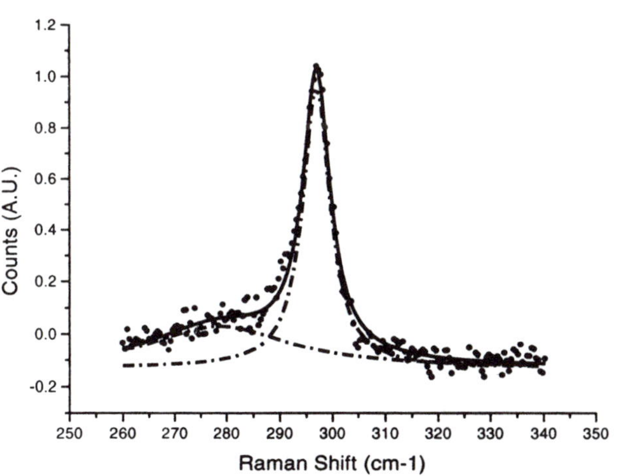

Fig. 4: Raman spectra of the sample Ge* (series 2).

CONCLUSIONS

We have shown that Raman spectroscopy is a valuable tool for the characterization of nanoparticles embedded in a matrix. Phonon confinement theory can be used to evaluate the size distribution of the particles. Our results suggest that the growth method here described minimizes the particle-matrix interaction thus favoring the creation of particles having the shape of truncated spheres with a high degree of crystallinity even for very small radii.

REFERENCES

1. C.B. Murray, C.R. Kagar and M.G. Bawendi, Science **270**, 1335 (1995).

2. A.P.Alivisatos, Science **271**, 933 (1996).

3. L.Banyai and S.W.Koch, Semiconductor Quantum Dots (World Scientific, 1993).

4. A.Stella, P.Cheyssac and R.Kofman in Science and Technology of Thin Films (World Scientific, 1995).

5. F. Hache, D.Ricard, C.Flytzanis, U.Kreibig, Appl. Phys. A **47**, 347 (1988).

6. H. Miguez, V. Fornes, F. Meseguer, F. Marquez, C. Lopez, Appl. Phys. Lett. **69**, 2347 (1996).

7. V. Craciun, C. Boulmer-Leborgne, E. Nicholis, I. Boyd, Appl. Phys. Lett. **69**, 1506 (1996).

8. D. Paine, C.Caragianis, T.Kim, Y.Shigesato, T.Ishahara, Appl. Phys. Lett. **62**, 2842 (1996).

9. J.R. Health, J.J: Shiang and A.P. Alivisatos, J. Chem. Phys. **101**, 1607 (1994).

10. Y.Maeda, Phys. Rev. B **51**, 1658 (1995).

11. E. Sondergard, R. Kofman, P. Cheyssac, A. Stella, Surf. Sci. **364**, 467 (1996).

12. P.Tognini, L.C.Andreani, M.Geddo, A.Stella, P.Cheyssac, R.Kofman and A.Migliori, Phys. Rev. B **53**, 6992 (1996).

13. P.M. Fauchet and I.H.Cambell, Critical Reviews in Solid State and Material Science **14**, S79 (1988).

14. J.Gonzalez-Hernandez, G.H.Azerbayejani, R.Tsu and F.H.Pollak, Appl. Phys. Lett. **47**, 1350 (1985).

15. R.J.Nemanich and S.A. Solin, Phys. Rev. B **20**, 392 (1979).

16. C.E. Bottani, C. Mantini, P. Milani, M. Manfredini, A. Stella, P. Tognini, P. Cheyssac, R. Kofman, Appl. Phys. Lett. **69**, 2409 (1996).

17. M.Fujii, S.Hayashi and K.Yamamoto, Jpn. J. Appl. Phys. **30**, 687 (1991).

GROWTH OF GERMANIUM ON POROUS SILICON (001)

W.H. THOMPSON *, Z. YAMANI *, H.M. NAYFEH *+, M.-A. HASAN ***, J.E. GREENE **, M.H. NAYFEH *

* Department of Physics, University of Illinois at Urbana-Champaign, Urbana, Illinois USA

** Department of Materials Science, Coordinated Science Laboratory, University of Illinois at Urbana-Champaign, Urbana, Illinois 61801

*** C.C. Cameron Applied Research Center and Department of Electrical Engineering, University of North Carolina, Charlotte, NC 28223

ABSTRACT

The surface morphology of Ge grown on Si (001) and porous Si(001) by molecular beam epitaxy at 380 °C is examined using atomic force microscopy (AFM). For layer thicknesses of 30 nm, the surface shows islanding while still maintaining some of the underlying roughness of the surface of porous Si. For thicknesses in the 100 nm range, the surface roughness is not visible, but the islanding persists. Unlike the case of silicon where islands tend to merge and nearly disappear as the thickness of the deposited layer rises, we observe on the porous layer the persistence of the islands with no merging even for macroscopic thicknesses as large as 0.73 microns.

INTRODUCTION

The discovery of strong, visible photoluminescence in porous silicon was a surprise due to its indirect band gap [1, 2]. This raises the possibility of incorporating an opto-electronic device on silicon which would eliminate many of the problems associated with the compound semiconductors that have been developed to create opto-electronic devices. They are expensive and not easily integrated with silicon due to the difference in their lattice spacing.

Heteroepitaxial growth of metals, insulators and semiconductors and their compounds on silicon substrates is a rapidly expanding field of research in Si molecular beam epitaxy (MBE). In particular, the growth of device-quality material on Si is of great interest for several reasons, including strong incentive to utilize large diameter, structurally robust, and inexpensive Si wafers as substrates, advantages in integrating the high-speed and optical properties of the III - V devices with Si very large scale integration (VLSI) technology. The interest is also driven by the anticipated band gap engineering, quantum confinement of carriers, and the incorporation of direct band gap materials such as GaAs in well established Si microdevice technology.

Recently, growth of a high quality epitaxial Ge film as an intermediate layer prior to subsequent device quality growth of GaAs and the other III-V compounds has been suggested as a means for alleviating many of the problems of growing on Si that are encountered due to lattice mismatch among other things [3]. In the hope of combining optically active porous silicon with GaAs, we have studied the growth of Ge on porous silicon

We compare the surface topography of epitaxial growth of Ge on porous Si(001) to growth on Si(001) using atomic force microscopy (AFM), and examine the dependence of the surface topography on the thickness of the germanium layer. For nanometer scale thicknesses, the surface topography shows islanding while reflecting some of the underlying roughness of the surface of porous Si. For thickness in the 100 nm range the roughness is not visible but the islanding persists. Unlike the case of silicon where islands tend to merge and nearly disappear, as

Mat. Res. Soc. Symp. Proc. Vol. 452

the thickness of the deposited layer rises, we observe the persistence of the islands with no merging even for macroscopic thickness as large as 0.73 microns.

EXPERIMENT

The silicon samples used were p-type boron doped Si (001) of 1-2 ohm-cm wafers supplied by Virginia Semiconductor. The samples were cleaned ultrasonically in an acetone bath for five minutes, then a methanol bath for five minutes and immersed in a solution of HF:H_2O:ethanol of ratio 1:1:4 for five minutes to remove the native oxide layer and to passivate the surface. The samples were then laterally anodized [4] at ~ 0.5 mA/cm^2 in an electrochemical cell containing the same solution of HF:H_2O:ethanol with continuous stirring of the solution. The cathode was a platinum wire ~ 0.5 cm from the sample, which was held by a Pt pad. The samples were anodized for 15 minutes in the dark. At this point the sample is porous but not luminescent. If luminescent samples are required then an additional period of anodization for 10 minutes under ultraviolet (UV) illumination is carried out. After etching with the additional UV step, uniform orange PL was visible on the surfaces of the samples when they were illuminated with UV light. Samples contained both a porous and an unetched region. After viewing, samples were immediately placed in a vacuum desiccator to minimize oxidation.

Prior to being placed in the MBE chamber, the samples were rinsed in HF. Once in vacuum, the samples were annealed at 550 - 600 °C to for 25 min. to desorb hydrogen. The samples were not observed at this time. Annealing between 300 to 600 C does induce a slight change in the microstructure, so some surface modifaction occured at this point[5]. The Ge was grown at a rate of 1.02 Å/s at 380 °C. The samples were removed from the growth chamber and transported to the AFM facility. AFM data was taken in air on a Topometrix Explorer AFM in contact mode.

RESULTS

Figure 1 shows the AFM image of a surface of silicon (100) after a layer of 30 nm thickness of germanium is grown on it. The image shows islands that are approximately 100 nm - 150 nm in size with no large clusters but less frequently we have some of 20 nm size. It is clear that the growth is not in a 2D mode but rather a three dimensional island mode. In this mode, the system minimizes its built-in strain energy by undergoing strain relaxation through island formation.

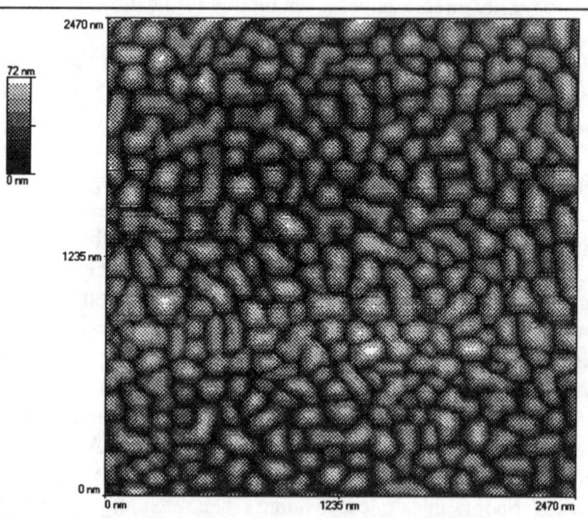

Fig. 1) 30 nm of Ge on Si (001). The scan size is 2.5 μ x 2.5 μ. The maximum height is 72 nm.

Figure 2 shows also the AFM image of a surface of porous silicon after a layer of 30 nm thickness of germanium is

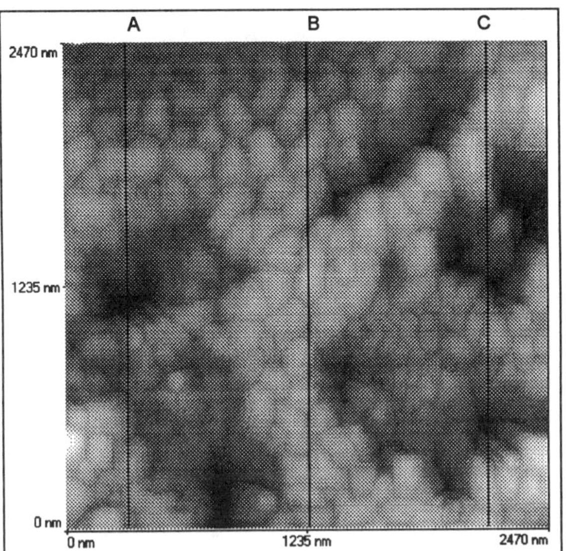

Fig. 2) 30 nm of Ge on por Si (001). The scan size is 2.5 μ x 2.5 μ. The line profiles for lines A, B, and C are shown in fig. 1c).

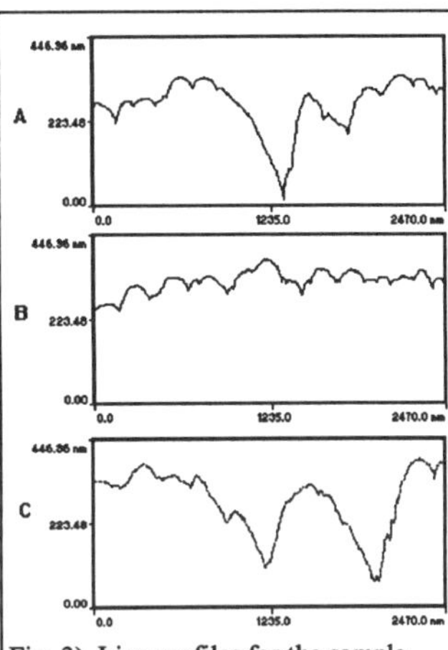
Fig. 3) Line profiles for the sample shown in fig. 1b). The maximum height is 447 nm.

grown. The image shows islands that are fairly homogeneous just as in the unetched case but reflecting some of the underlying roughness of the surface of porous Si. However the islands are not as rectangular. The islands are approximately 100 nm - 150 nm in size with no large clusters but less frequently we have some of 20 nm size. We show in figure 3 three line profiles, two of which go through some of the rough craters in the underlying surface and one that misses all craters. These profiles show that the thickness of the Ge layer is not sufficient to fill some of the large craters or holes normally found in porous silicon.

We studied the changes that occur as the thickness of the Ge layer increases on both the porous and the unetched substrates. We grew Ge layers of 98 nm, 214 nm, 428 nm, and 734 nm thickness. On both the porous and unetched substrates we see a clear progression with increasing thickness. On the porous region the islands grow in size from 50 nm to 200 nm while the shape changes from roughly rectangular to irregular. On the unetched substrate, the islands begin to fuse together and this

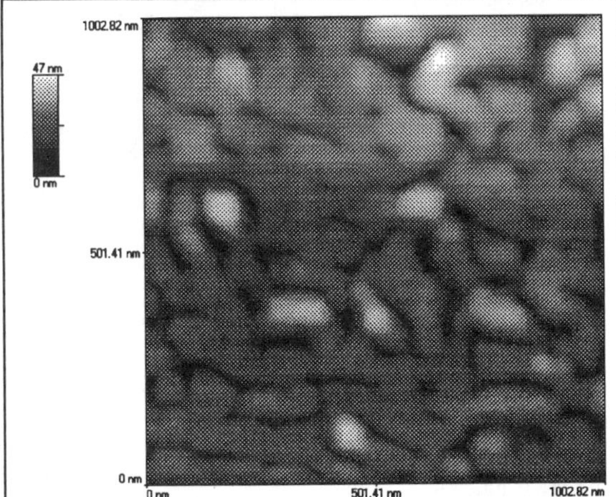

Fig. 4) 214 nm Ge on unetched Si (001) The scan size is 1μ x 1μ and the maximum height is 47 nm.

trend continues until the surface appears to be fairly smooth.

Figures 4 and 5 show the surface of 214 nm of Ge on nonporous and porous Si respectively. On the nonporous substrate, the islands are combining, yet distinct valleys can be seen between the merging islands, and a good deal of height variation exists on the surface. Growth on the porous substrate yielded islands of a fairly uniform size, approximately 100 nm wide. The islands are well differentiated and roughly rectangular in shape. No pattern seems present in the island placement.

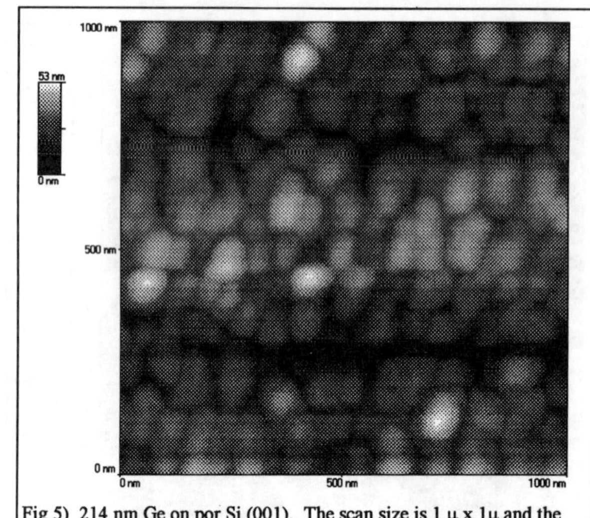

Fig 5) 214 nm Ge on por Si (001). The scan size is 1 μ x 1μ and the maximum height is 53 nm.

These growth patterns changed as the layer thickness increased. The extreme case is shown in figures 6 and 7. These samples have a layer of Ge approximately 0.73 microns thick. As seen on the unetched region, distinct islands are nonexistent. They have merged together, forming an extremely smooth surface when compared to growth on the porous substrate. The islands on the porous region are of varying sizes, the largest being roughly 200 nm wide. The island shapes are irregular, but the individual islands still remain well defined.

The obvious explanation for the persistence of islanding would seem to be the rough surface topography of the porous layer. So by changing the topography, we should be able to change the growth. Therefore, we also grew Ge on a nonluminescent porous sample. This sample was fabricated in a similar manner as the luminescent samples except that the sample is not exposed to UV for the last 10 minutes. SEM images show that the nonluminescent porous layer is composed of isolated pores, i.e. pores with diameters smaller than the interspacing, hence the structure is much less complex than luminescent porous layers where the pores have a larger

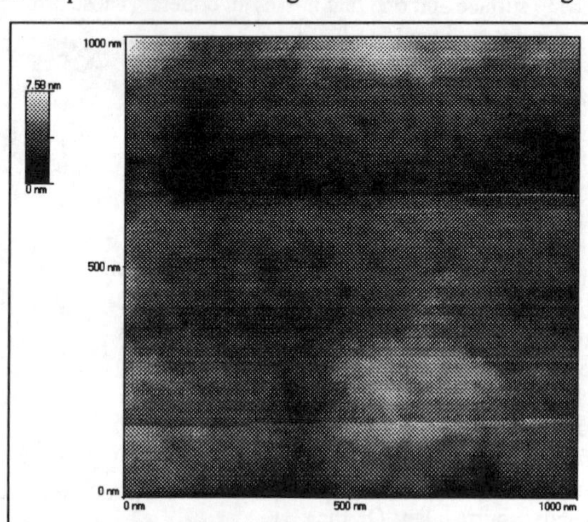

Fig. 6) 730 nm of Ge on Si (001). The scan size is 1 μ x 1 μ and the maximum height is 13 nm.

diameter and are interconnected forming a Si skeleton [6]. The nonluminescent sample should be significantly less rough than the luminescent samples. A 214 nm layer of Ge was grown on this sample at the same temperature as those discussed above. As Fig 8 shows, the islands are roughly of the same shape and size seen on the luminescent counterpart. Therefore it seems unlikely that the surface roughness is causing the persistence of islanding.

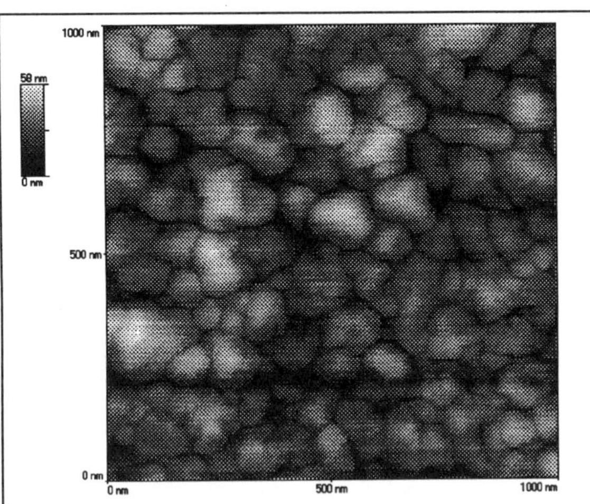

Fig. 7) 730 nm of Ge on porous Si (001). The scan size is 1 μ x 1 μ and the maximum height is 74 nm.

It is currently thought that the islanding of Ge on crystalline Si is the result of strain and not dislocations [7]. The islands on c-Si disappear because of the formation of strain relieving defects [8]. If we assume that the islands on porous Si are due to a similar stress mechanism then the relieving mechanism must be counteracted by the porous layer. So either the strain relieving defects are not forming, or they can't relieve enough strain to get relaxed bulk growth. Since the luminescent and the nonluminescent samples have different surface topologies, topology alone does not seem to be a good candidate to cause islanding. It might be worthwhile to examine chemical similarities. FTIR data show that both luminescent and nonluminescent porous silicon have similar chemical bonds [9]. It is possible that the Si bonds on the surface have been altered compared to the unetched Si. This could be affecting the growth. Or possibly it could be some contaminants left over from the etching process that were not successfully removed in preparation. Another strong candidate for causing islanding is the strain inherent in porous Si [10]. And if indeed a strain gradient exists [11], then it may not be possible to avoid island growth.

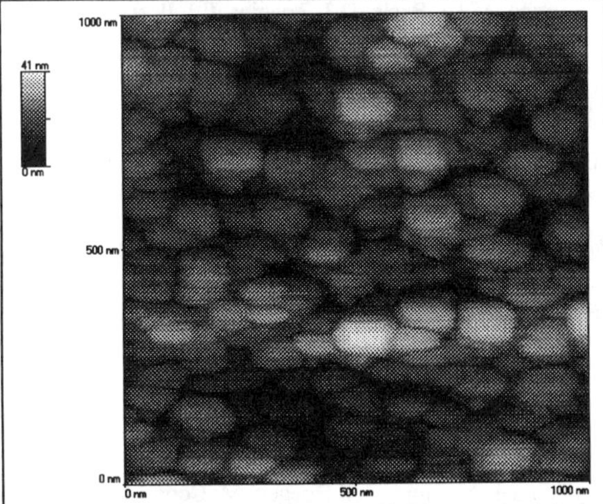

Fig. 8) 214 nm Ge on nonluminescent por Si. The scan size is 1 μ x 1 μ and the maximum height is 41 nm.

CONCLUSION

We have studied the MBE growth of Ge on porous Si (001) at various thicknesses. The islands merge as the thickness increases on a nonporous substrate, yet they remain distinct on the porous substrate. Also, the persistence of islanding does not seem to be related to the initial porosity of the substrate, as seen in the comparison between growth on the nonluminescent and luminescent substrates. Instead, the islanding appears to be dependent on the thickness of the Ge layer. If porous Si is to be used in an electrooptical device, such defects must be avoided. Future work will involve determining the cause of the persistence and seeing if it can be avoided.

ACKNOWLEDGMENTS

This work has been supported by grant NSF INT 93-07794.

+ present address: Massachusetts Institute of Technology, Department of Electrical Engineering, Cambridge, MA 02139.

REFERENCES

1. C. Pickering, M.I.J. Beale, D.J. Robbins, P.J. Pearson, and R. Greef, J. Phys. C. **17**, 6535 (1984).

2. L.T. Canham, Appl. Phys. Lett. **57**, 1046 (1990).

3. G. Bajor, K.C. Cadien, M.A. Ray, J.E. Greene, and P.S. Vijaykumar, Appl. Phys. Lett. **40**, 696 (1982).

4. K.H. Jung, S. Shih, T.Y. Hsieh, D.L. Kwong, and T.L. Lin, Appl. Phys. Lett. **59**, 3264 (1991).

5. R. Herino, A. Perio, K. Barla, and G. Bomchil, Mater. Letters **2**, 519 (1984).

6. D.K. Andsager, Ph.D. thesis, University of Illinois, Department of Physics, Urbana, IL, 1994.

7. H.J. Osten, Phys. Stat. Sol. A **145**, 235 (1994).

8. F.K. LeGoues, M. Horn-Von Hoegen, M. Copel, and R. M. Tromp, Phys. Rev. B **44**, 12894 (1991).

9. J. Hilliard, D. Andsager, L.H. AbuHassan, H. M. Nayfeh, M. H. Nayfeh, J. Appl. Phys. **76**, 2423 (1994).

10. K. Barla, R. Herino, G. Bomchil, J. C. Pfister, and A. Freund, J. Cryst. Growth **68**, 727 (1984).

11. E. Koppensteiner, A. Schuh, G. Bauer V. Holy, D. Bellet, and G. Dolino, Appl. Phys. Lett. **65**, 1504 (1994).

EFFECT OF QUANTUM-WELL STRUCTURES ON THE THERMOELECTRIC FIGURE OF MERIT IN THE $Si/Si_{1-x}Ge_x$ SYSTEM

X. Sun*, M. S. Dresselhaus*, K. L. Wang**, M. O. Tanner**
*Department of Physics, Massachusetts Institute of Technology, Cambridge, MA 02139
**Department of Electrical Engineering, University of California, Los Angeles, CA 90024

ABSTRACT

The $Si/Si_{1-x}Ge_x$ quantum well system is attractive for high temperature thermoelectric applications and for demonstration of proof-of-principle for enhanced thermoelectric figure of merit Z, since the interfaces and carrier densities can be well controlled in this system. We report theoretical calculations for Z in this system, based on which $Si/Si_{1-x}Ge_x$ quantum-well structures were grown by molecular-beam epitaxy. Thermoelectric and other transport measurements were made, indicating that an increase in Z over bulk values is possible through quantum confinement effects in the $Si/Si_{1-x}Ge_x$ quantum-well structures.

INTRODUCTION

Recently, it has been shown theoretically [1] that it may be possible to increase the thermoelectric figure of merit (Z), defined by [2]

$$Z = \frac{S^2\sigma}{\kappa}, \tag{1}$$

where S is the thermoelectric power (Seebeck Coefficient), σ is the electrical conductivity and κ is the thermal conductivity, of certain materials by preparing them in the form of two-dimensional quantum-well structures. This has already been demonstrated experimentally [3] using $PbTe/Pb_{1-x}Eu_xTe$ multiple-quantum-well structures grown by molecular-beam epitaxy.

In bulk form, $Si_{1-x}Ge_x$ is a promising thermoelectric material for high temperature applications [4, 5, 6], and has been used in radio-isotope thermoelectric generators (RTGs) on satellites and spacecraft for compositions of about $Si_{0.7}Ge_{0.3}$ operating at $\sim 1000\,K$ [7]. Because of the large amount of expertise and information available on this system regarding the materials science of fabricating $Si_{1-x}Ge_x/Si$ quantum wells, it is an interesting system for the demonstration of both proof-of-principle and high performance thermoelectric devices operating at high temperatures.

THEORETICAL MODELING

For a material to be a good thermoelectric cooler, it must have a high thermoelectric figure of merit, Z. In order to achieve a high Z, one requires a high thermoelectric power S, a high electrical conductivity σ, and a low thermal conductivity κ. In general, it is difficult to increase Z because a modification to any one of the three parameters S, σ, or κ, adversely affects the other transport coefficients, so that the resulting Z does not vary significantly. Currently, the materials with the highest Z are Bi_2Te_3 alloys such as $Bi_{0.5}Sb_{1.5}Te_3$, with $ZT \simeq 1.0$ at $T = 300\,K$ [8].

Mat. Res. Soc. Symp. Proc. Vol. 452

The effect on Z of using materials in two-dimensional (2D) structures, such as 2D multiple-quantum-well (MQW) structures, has been studied earlier [1] and it was shown theoretically that this approach could yield a significant increase in Z over the bulk value as the quantum-well width is decreased. The proposed increase in the power factor $S^2\sigma$ arises mainly from the enhancement of the density of electron states per unit volume (near the Fermi level) that occurs for small quantum well widths. Further increase in Z is possible through the reduction of thermal conductivity, κ, resulting from enhanced phonon scattering at the interfaces between the quantum wells and barriers.

Let the quantum well be parallel to the x-y plane and the current flow in the x direction. In the $Si/Si_{1-x}Ge_x$ quantum well structures we have investigated, the samples are grown along the [100] direction and the transport measurements are performed along the principal directions of each of the six ellipsoids in the conduction band. By assuming that the conduction band is parabolic and that the electrons occupy only the lowest ($n=1$) sub-band of the quantum well, the electronic dispersion relation is

$$\mathcal{E}(k_x, k_y) = \frac{\hbar^2 k_x^2}{2m_x} + \frac{\hbar^2 k_y^2}{2m_y} + \frac{\hbar^2 \pi^2}{2m_z a^2}, \tag{2}$$

where a is the width of the quantum well. Using

$$Z_{2D} = \frac{S^2\sigma}{\kappa_e + \kappa_{ph}}, \tag{3}$$

for the figure of merit for a two-dimensional system, where κ_e and κ_{ph} are the electronic thermal conductivity and the phonon thermal conductivity, respectively, we get [1]

$$Z_{2D}T = \frac{\left(\frac{2F_1}{F_0} - \zeta^*\right)^2 F_0}{\frac{1}{B} + 3F_2 - \frac{4F_1^2}{F_0}}, \tag{4}$$

where the Fermi-Dirac function F_i is given by

$$F_i = F_i(\zeta^*) = \int_0^\infty \frac{x^i\,dx}{e^{(x-\zeta^*)} + 1}, \tag{5}$$

and $\zeta^* = \zeta/k_BT$ is the reduced chemical potential relative to the edge of the first sub-band. The value of B in Eq. (4) is given by

$$B = \frac{1}{2\pi a}\left(\frac{2k_BT}{\hbar^2}\right)\frac{k_B^2 T\tau\alpha}{\kappa_{ph}}, \tag{6}$$

where in the case of Si quantum wells

$$\alpha = 2\left(\frac{m_{\parallel}}{m_{\perp}}\right)^{\frac{1}{2}} + 2\left(\frac{m_{\perp}}{m_{\parallel}}\right)^{\frac{1}{2}} + 2, \tag{7}$$

accounting for the anisotropy of the six different ellipsoids in the Si conduction band, in which $m_\perp$ and $m_\parallel$ are, respectively, the transverse and longitudinal effective mass components of electrons in the conduction band, and τ is the relaxation time which was approximated by [9]

$$\tau = \frac{3\mu_n}{e\left(\frac{1}{m_\parallel} + \frac{2}{m_\perp}\right)}, \tag{8}$$

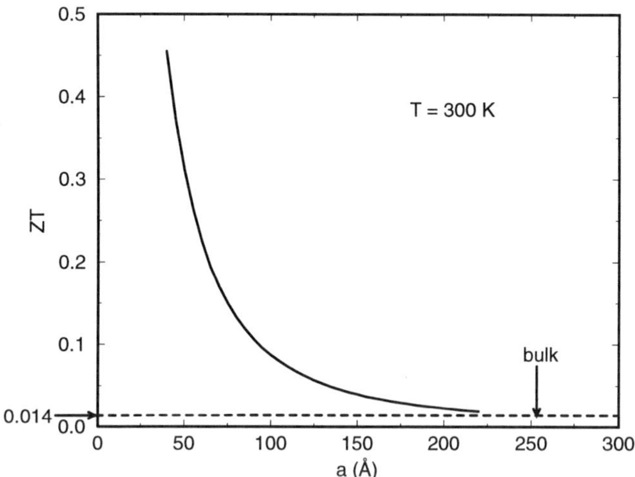

Figure 1: The calculated $Z_{2D}T(\zeta^*)$ vs layer thickness a at room temperature for a Si quantum well. The dashed line indicates the ZT for bulk Si.

where μ_n is the electron mobility.

The value B in Eq. (6) is determined by the intrinsic properties of Si and the width of the quantum well. For a given value of B, the reduced chemical potential ζ^* in Eq. (4) may be optimized to yield the maximum value of $Z_{2D}T$ within the quantum well. In the 2D case, ζ^* may be varied both by doping and by changing the layer thickness a. This extra degree of freedom provides a new approach for increasing $Z_{2D}T$ above the value characteristic of bulk materials.

Si crystallizes in the diamond structure. The conduction band is characterized by six equivalent minima along the $\langle 100 \rangle$-axes of the Brillouin zone. The surfaces of constant energy are ellipsoids of revolution with their major axes along $\langle 100 \rangle$. The transverse and longitudinal effective masses are $m_\perp = 0.1905\, m_0$ and $m_\parallel = 0.9163\, m_0$. The electron mobility at room temperature for bulk Si is $\mu_n = 1447\,\mathrm{cm^2\,V^{-1}\,s^{-1}}$. The bulk phonon thermal conductivity is $\kappa_{ph} = 1.313\,\mathrm{W\,cm^{-1}\,K^{-1}}$ at 300 K [10].

In a quantum-well structure, since phonons can scatter off the interfaces, the phonon thermal conductivity may be reduced relative to the bulk value which is given by

$$\kappa_{ph} = \frac{1}{3} C_v v l, \tag{9}$$

where l is the phonon mean free path, C_v is the lattice heat capacity, and v is the velocity of sound in the material. For Si, $C_v = 1.658\,\mathrm{J\,K^{-1}\,cm^{-3}}$ and $v = 8.4332{\times}10^5$ cm/s, giving a value of $l = 282$ Å. If the layer thickness a is greater than 282 Å, then layering does not seriously affect the mean free path l, and κ_{ph} should then be similar to its bulk value. However, if a is less than 282 Å, then l and κ_{ph} are limited by phonon scattering off the interfaces and a rough estimate for κ_{ph} is obtained by setting $l = a$ and using Eq. (9).

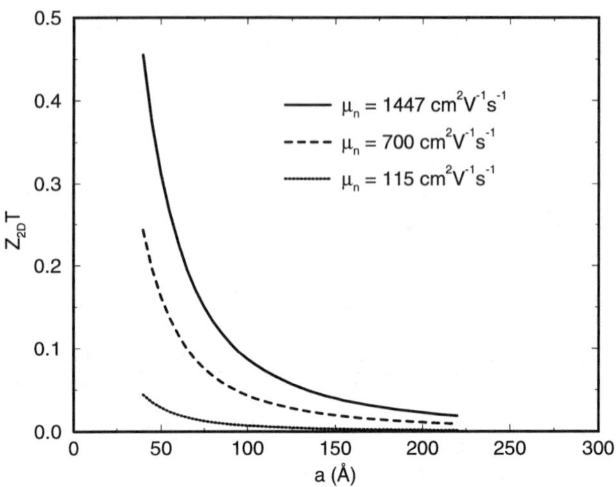

Figure 2: The calculated optimal $Z_{2D}T(\zeta^*)$ vs layer thickness a at room temperature for a Si quantum well with various electron mobilities. Thus $Z_{2D}T$ has a strong dependence on the carrier mobility.

The calculated $Z_{2D}T(\zeta^*)$ as a function of a for Si quantum wells at room temperature is shown in Fig. 1, together with a dashed line indicating the room temperature 3D figure of merit of $Z_{3D}T = 0.014$. The quantum well width in the calculations is less than the phonon mean free path l, and therefore $\kappa_{ph} = \frac{1}{3}C_v v a$ is used to estimate the phonon contribution to the thermal conductivity for small quantum well widths. The results show a significant increase in the thermoelectric figure of merit for a quantum well width a below 100 Å.

Our calculations further show that the optimal thermoelectric figure of merit $Z_{2D}T$ for the $Si/Si_{1-x}Ge_x$ superlattice structure has a strong dependence on the carrier mobility, as shown in Fig. 2. However, $Z_{2D}T$ is not very sensitive to a variation in the carrier concentration in the range of 10^{18}–$10^{19}\,cm^{-3}$. Although the dependence of the mobility on the carrier concentration can be neglected in the case of modulation doping, the optimized carrier concentration indicated by the theoretical modeling ($\sim 2 \times 10^{19}\,cm^{-3}$) is higher than what can be achieved in the $Si/Si_{1-x}Ge_x$ quantum well system. Therefore it is advantageous to lower the carrier concentration below the theoretically optimized value in order to get high quality samples with good thermoelectric performance.

EXPERIMENTAL STUDY

The Si quantum well sample was grown in a Perkin Elmer 430S MBE chamber. A protective oxide resulting from the chemical cleaning was desorbed by an anneal at 900 °C for 10 minutes. A 3000 Å Si buffer layer was first grown, followed by a 4500 Å $Si_{0.7}Ge_{0.3}$ layer which formed the substrate. The thickness and Ge fraction of this layer ensure nearly complete relaxation of the film used for the thermoelectricity study. On this substrate, a 100 Å Si quantum well was then grown, followed by a 50 Å $Si_{0.7}Ge_{0.3}$ spacer layer and a 200 Å $Si_{0.7}Ge_{0.3}$ Sb-doped carrier supply layer. The substrate temperature was held at 600 °C except for the growth of the supply layer, where the substrate temperature was reduced to 350 °C because of the reduced sticking coefficient of Sb at higher temperatures. Subsequent to the growth of the cap layer, the wafer was ramped to 600 °C for improving the film quality.

Hall measurements were performed at 300 K and 77 K, showing that the sample has mobilities of $\sim 700\,\mathrm{cm^2\,V^{-1}\,s^{-1}}$ and $\sim 4000\,\mathrm{cm^2\,V^{-1}\,s^{-1}}$ and carrier densities of $\sim 4\times10^{18}\,\mathrm{cm^{-3}}$ and $\sim 2\times10^{18}\,\mathrm{cm^{-3}}$, at 300 K and 77 K, respectively. The resistivity of the sample was measured using a four-point probe method. Ohmic contacts were made with an In alloy. Chromel-Gold/Iron thermocouples were used in measuring the Seebeck coefficient.

Some preliminary experimental measurements of the resistivity and the Seebeck coefficient suggest that the power factor ($S^2\sigma$) at room temperature may be significantly higher than that of bulk Si, suggesting a room temperature figure of merit significantly higher than that of bulk Si. The enhancement in ZT in a superlattice may be even larger because the phonon scattering due to the surface roughness and imperfections give rise to diffuse scattering which is expected to reduce the thermal conductivity significantly. The reduction of thermal conductivity in a 2D system, part of which can be accounted by Eq. (9), can further increase the figure of merit perhaps to a useful value for application to thermoelectric devices. In fact, a giant reduction by almost two orders of magnitude in the thermal conductivity has been reported for a Si membrane with only a decrease by a factor of two in the carrier mobility [11]. Thus, further detailed thermoelectricity studies on the $Si/Si_{1-x}Ge_x$ system should address the optimization of ZT rather than $S^2\sigma$.

Since $Si/Si_{1-x}Ge_x$ is a system aimed at high temperature operation (up to 1000 K), it will be interesting to look at the thermoelectric performance of $Si_{1-x}Ge_x$ quantum well systems at high temperatures. Since the power factor generally increases with increasing T above room temperature [4, 5], the power factor as well as the thermoelectric figure of merit for the quantum well are expected to show even greater enhancements above room temperature. Detailed thermoelectric measurements on the $Si/Si_{1-x}Ge_x$ system at 300 K and at elevated temperatures are now in progress.

CONCLUSION

We have done theoretical modeling and a preliminary experimental investigation of the thermoelectric figure of merit for $Si/Si_{1-x}Ge_x$ quantum well structures. Calculations suggest that an increase in the thermoelectric figure of merit over bulk values in the $Si/Si_{1-x}Ge_x$ quantum well system may be possible at room temperature. Further enhancement in $Z_{2D}T$ for $Si/Si_{1-x}Ge_x$ quantum well structures is expected at elevated temperature through both quantum confinement and phonon interface scattering effects.

ACKNOWLEDGMENTS

The authors would like to thank Drs. G. Dresselhaus, T. C. Harman and Professor Gang Chen for valuable discussions. The MIT authors gratefully acknowledge support by Navy under Contract No. N00167-92-K-0052. The work at UCLA was in part supported by AFOSR under URI.

REFERENCES

[1] L. D. Hicks and M. S. Dresselhaus, Phys. Rev. B **47**, 12727 (1993).

[2] H. J. Goldsmid, *Thermoelectric Refrigeration* (Plenum Press, New York, 1964).

[3] L. D. Hicks, T. C. Harman, X. Sun, and M. S. Dresselhaus, Phys. Rev. B **53**, R10493 (1996).

[4] G. A. Slack and M. A. Hussain, J. Appl. Phys. **70**, 2694 (1991).

[5] C. B. Vining, in *CRC Handbook of Thermoelectrics*, edited by D. M. Rowe (CRC Press, New York, 1995) p. 329.

[6] G. D. Mahan, in *Solid State Physics*, edited by H. Ehrenreich and F. Spaepen (Academic Press, to be published).

[7] C. Wood, Rep. Prog. Phys. **51**, 459 (1988).

[8] H. J. Goldsmid, *Electronic Refrigeration* (Pion, London, 1986).

[9] S. Krishnamurthy, A. Sher, and A.-B. Chen, Phys. Rev. B **33**, 1026 (1986).

[10] *Landolt-Börnstein Numerical Data and Functional Relationships in Science and Technology,* New Series, edited by K.-H. Hellwege (Springer-Verlag, Berlin, 1982), Vol. 17a, pp. 43–87.

[11] X. Y. Zheng, M. Chen, and K. L. Wang, in *Proceedings of the seventh International Symposium on MEMS*, Atlanta, GA, Nov. 18-20, 1996 (to be published).

Part V

Groups III-V and II-VI, and Metal Sulfide, Iodide, and Oxide Nanocrystals and Nanostructures

OPTICAL PROPERTIES OF NANO CRYSTALLINE InP IN OPAL 3-DIMENSIONAL GRATINGS

N.P. Johnson*, S. G. Romanov**, V. Butko**, H. Yates***, M.E. Pemble***, C.M. Sotomayor Torres****
*Department of Electronics and Electrical Engineering, University of Glasgow, Glasgow, G12 8LT, UK. Email: n.johnson@elec.gla.ac.uk
**A.F. Ioffe Physical Technical Institute, St. Petersburg, 194021, Russia
***Department of Chemistry, University of Salford, Manchester, UK.
****Department of Electronics, University of Wuppertal, Germany

ABSTRACT

Opal consists of a three dimensional array of silica balls with diameter in the range 150 -350 nm and acts as a 3-D grating. Within the microcrystals (a few hundred microns) of opal there is typically 5% variation in ball diameter arranged in a fcc structure. Using a specially modified MOCVD process InP can be grown within the voids between the touching silica balls. These interconnecting voids are approximately 1/5 and 2/5 the silica sphere diameter. Raman spectroscopy confirms the crystallinity of the InP for two samples loaded with 2.5 volume % of InP. The LO phonon is shifted to the red by 10 cm^{-1} from the bulk value of 345 cm^{-1}, while the TO phonon is shifted 22 cm^{-1} to the blue from the bulk value of 303.7 cm^{-1}. This shift is attributed to strain in the nanocrystals.

The half width of the reflectance is dependent upon InP loading with the 2.5 vol.% samples giving 0.19 eV and the lower loaded 0.4 vol. % giving 0.17 eV for samples with the same ball diameter. Photoluminescence emission near 1.7 eV is attributed to confined InP compared with a bulk value of 1.4 eV. The energy of the emission is determined by the combined contribution of the laterally confined InP and its interaction with the opal matrix.

INTRODUCTION

Opal is a natural grating for visible light consisting of regularly spaced balls of silica. These refracting properties give precious opal value as a gem stone. The refracting properties of opal alone (bare opal) are insufficient to produce a complete stop band in all directions for light transmission. Material with a complete stop band or Photonic Band Gap (PBG) is potentially useful for zero threshold current lasers and other optoelectronic devices. Very high efficiency is possible because a complete PBG will not allow any mode to propagate; however, an irregularity subsequently introduced within the PBG will allow only one mode of emission to propagate. Full PBGs have been demonstrated in the microwave region [1] using wire meshes and in colloidal liquids with polystyrene balls [2]. A solid system such as opal would have obvious processing advantages and could operate in the visible region .

The silica balls comprising opal are arranged in the fcc configuration. Theoretical calculations [3] show that this arrangement and the small refractive index contrast between silica and air do not allow a complete photonic gap. However, between the touching silica balls an interconnecting array of voids may be filled with a suitable material. InP has a high refractive index, and as a semiconductor is an optical gain medium. Here we report on the optical properties of indium phosphide grown within the 3-dimensional opal gratings.

Mat. Res. Soc. Symp. Proc. Vol. 452 © 1997 Materials Research Society

EXPERIMENTAL DETAILS

Samples and growth method

Samples of synthesised opal were impregnated with InP using a specially modified MOCVD process [4]. Firstly, trimethylindium was added in a flow of hydrogen at 52° C in a standard MOCVD atmospheric pressure reactor. Secondly phosphine was passed through the reactor for several hours at 350° C to decompose the hydride. For higher loadings of InP the above cycle was repeated. The opal had ball diameters in the range of 189 to 304 nm, measured by SEM with microcrystallites ~100 µm in size and were in the form of polished plates, 0.5 mm thick and area of 25 mm^2. Two samples were loaded to 2.5 volume % and the remainder low loaded samples contained 0.4 volume % of InP measured as by electron-microprobe analysis.

Experimental techniques

Raman spectra were measured in near backscattering configuration using 514.5 nm excitation from an Ar^+ laser and charge coupled device detector. Photoluminescence spectra were measured by 457.9 nm excitation from an Ar^+ laser and collected by a double monochromator and photomultiplier tube. All spectra were corrected for spectral response. Reflectance spectra were taken using a single spectrometer and integrating sphere.

RESULTS AND DISCUSSION

Raman spectra

We first examined the Raman spectra opal-InP. Raman spectra could only be obtained for the highly loaded samples. One sample (blue in colour) contained a range of ball sizes from 189 to 232 nm and is shown as the solid trace in fig. 1. The dashed line shows a sample (dark green in colour) with uniform ball size of 227 nm. The peaks are coincident with the solid line but are much better defined. The shift in position of the LO and TO phonon peaks from bulk values is notable. The LO phonon shifts to lower frequency in a similar manner to InP structurally confined in other matrices [5]. For comparison, a Raman spectrum of InP grown in etched chrysotile asbestos is shown in fig. 2. This has the form of long (~1 cm) spiral roll of SiO_2 -MgO planes with a 2-15 nm channel [6]. Etching the asbestos removes the MgO component leaving a pure silica framework with ~100 times more surface area [7]. The shift to lower frequencies of the LO phonon is indicative of tensile stress[8]. The TO phonon for opal-InP however shifts in the opposite direction of the InP-asbestos system (and other wire like structures) this is attributed to compressive strain in the opal system due to the extra structural confinement.

Photoluminescence

Strong photoluminescence is observed for the 217 nm diameter sample shown in fig. 3. When empty, defects from the opal dominate around 2.4 eV. InP growth produces a new peak at 1.7 eV which is substantially shifted from the bulk value of InP (1.4 eV) and dominates at low temperature. Near room temperature the peaks intensities are comparable and the InP peak shifts with the change in bandgap with temperature to lower energy [4]. An estimate of the size of the particles based on the exciton energy is 8.5 nm for a shift of 0.3 eV. However the nominal thickness of the InP layer is around 1-2 monolayers or less than 1 nm, and would produce an even greater shift. The reduction in energy shift is attributed to stress as observed in Raman spectra. Thus the overall energy shift is determined by a combination of confinement and stress from the opal matrix.

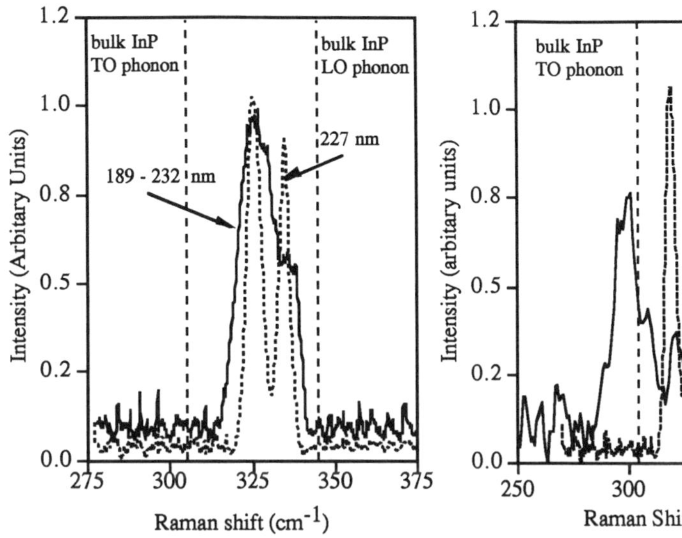

Figure 1 Raman scattering of high loading InP in opal with a well defined diameter (227 nm) and a range of diameters (189-232 nm) of the silica balls.

Figure 2 Comparison of high loaded InP in opal (227 nm, dashed line as in figure 1) with InP in etched chysotile asbestos wires (solid line).

Reflectance Spectra

The reflectance spectra for bare opal and opal filled with InP is shown in figure 4. It is surprising that the deposition of a very small amount of InP has such a strong effect on the reflectance spectrum; this is because of the high value of refractive index for InP (3.5), and a likely non-linear dependence on composition, which determines the overall composite refractive index.

Reflectance spectra of several opal-InP samples are shown in figure 5. The position of the reflectance peak correlates with ball diameter and is shown in the inset in figure 5. The error bars represent the variation of ball diameter within one sample. This behaviour is consistent with Bragg reflection. By using the gradient of the line (0.47) relating the wavelength to the ball diameter and assuming the reflections are dominated by silica balls in the (111) plane we may estimate the effective refractive index from the simple Bragg formula, $\lambda = 2\ d\ n\ \mathrm{Sin}\theta$. This gives a disappointingly low value of n= 1.23 assuming the scattering takes place at $\sqrt{3}/2$ of the silica ball diameter. According to reference [9] the entire grating should be taken in to account to achieve an effective refractive index (silica, air and InP). Recently [10] it has been shown that porosity in the silica ball which is, inaccessible to the InP filling, effectively lowers the refractive index for silica to 1.35. Two samples with the same ball size (227 nm) and different loading show that the higher loaded sample is approximately the same but broadened to lower energy. This corresponds to an increase of half width from 0.17 to 0.19 eV for an increase of 0.4 to 2.5 % volume of InP, and is consistent with a higher refractive index contrast. For a full photonic gap the ratio of energy half width to spectral position must exceed 17% [4]. By increasing the loading for the D=227 nm samples this was increased from 5.9 to 6.5 % with the maximum ratio for the D=220-260 nm sample at 7.3 %.

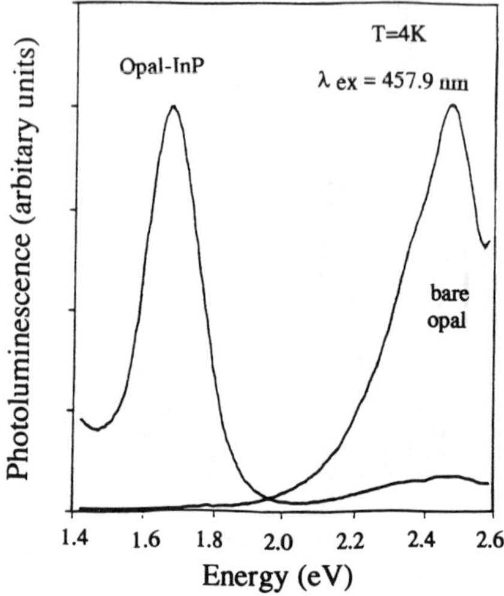

Figure 3 Photoluminescence of bare and InP in opal, D=217 nm.

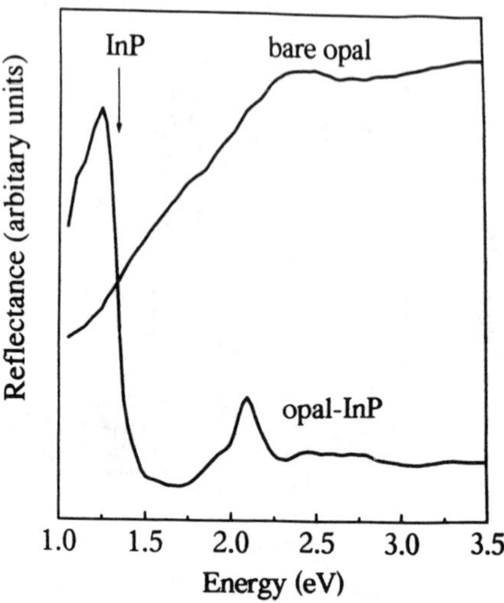

Figure 4 Reflectance spectra of bare opal and InP in opal, D=217 nm

A further observation of the reflectance spectra in fig. 5 reveals that the absorption edge around 1.5 eV depends on loading and order of the silica balls. The two lower loaded samples of well defined ball diameter have the sharpest absorption at lower energy. The high loaded well defined sample has the absorption shifted to higher energy, while the two samples with a range of ball diameters occupy an intermediate position.

Returning to figures 3 and 4, the photoluminescence (PL) emission is seen to occur between the InP absorption and the Bragg reflection peak (or partial stop band). Thus there is a resonance between the shifted PL and region of the grating allowing propagation of light.

We wish to note that the refractive index calculated from the position of the reflectance peak in fig. 4 is much higher at n=1.55 for the D=217 nm sample, compared with the value calculated from the line in the inset of fig. 5. The reason for the difference is not clear.

CONCLUSION

We have examined the optical properties of InP in opal. Raman spectra show a shift in the TO phonon in the opposite direction to wire like structural confinement. This is attributed to additional compressive stress. The photoluminescence properties show strong confinement of the emission from InP compared to the bulk. Reflectance spectra show correlation of the reflectance maximum with silica ball size and that higher loading increases the width of the partial photonic band gap. These properties will help us design future 3 D gratings for potential PBG structures.

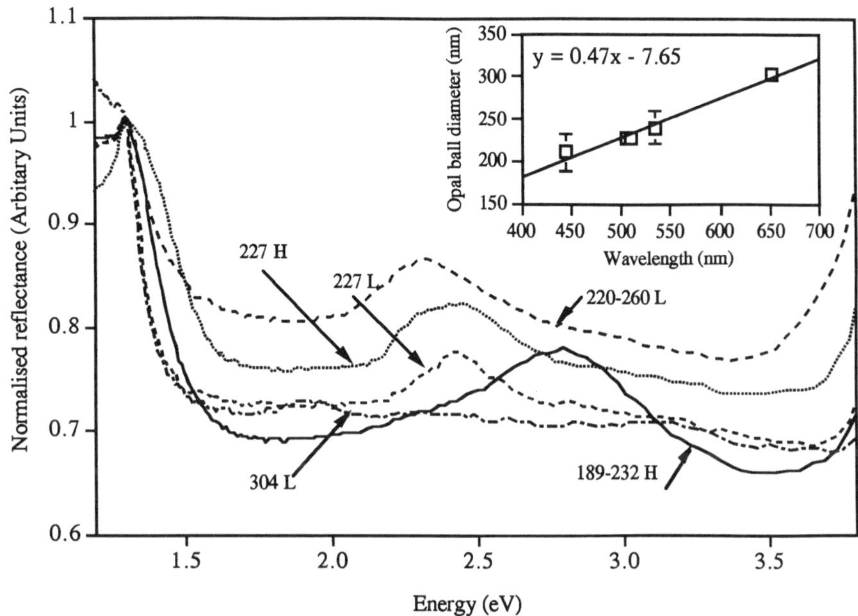

Figure 5 Normalised reflectance of InP in opal: variation of ball diameter and loading. Ball diameter given in nm, H=high loading, 2.5 %; L= low loading, 0.4 %. The inset shows the wavelength of the opal -InP reflectance maximum versus opal ball diameter.

ACKNOWLEDGEMENTS

This work was partially supported by the Russian Foundation for Basic Research grant no. 96- 02-17963 and UK EPSRC grant no GR/J90718. SGR acknowledges the support from Leverhulme Trust grant no. F/179/AK. The authors wish to thank Mr A Ross and Mr G McCulloch for technical support.

REFERENCES

1. see, for example, review articles in J. Optical Soc. America B **10** (1993).
2. R. Williams and R.S. Crandall, Phys. Lett. A **48** 225 (1974).
3. E. Yablonovitch, J. Modern Optics **41** 173 (1994).
4. H. M. Yates, W. R .Flavell, M.E. Pemble, N. P. Johnson, S G Romanov and C M Sotomayor Torres. J. Crystal Growth, **170** pp 611-615 (1996).

5. S.G. Romanov, N.P. Johnson, C.M. Sotomayor Torres, H.M. Yates, J. Agger, M.E. Pemble, M.W. Anderson, A.R. Peaker, V. Butko. in *Quantum Confinement: Quantum wires and dots*, Eds S Badyopadhyay, M. M. Cahay, P. J. Leburton and M. Razegui, Electrochemical Society Proceedings Volume PV 95-17, Pennington, USA pp 14-30 (1996).

6. V.N. Bogomolov and Yu. A. Kumzerov, JETP Lett. **21** 434 (1975).

7. V. P. Petranovskii (private communication).

8. S. Perkowitz, *Optical Characterization of Semiconductors Academic Press,* London (1993).

9. W.L. Vos, R. Sprik, A. Blaaderen, A. Imhof, A. Lagendijk and G.H. Wegdam, Submitted to Appl. Phys. Letters.

10. V.N. Bogomolov, D.V. Kurdyikov, A.V. Prokofiev, S.M. Samoilovich, JETP Lett. **63** 496 (1996).

LIMITS AND PROPERTIES OF SIZE QUANTIZATION EFFECTS IN InAs SELF ASSEMBLED QUANTUM DOTS

K.H. SCHMIDT*, G. MEDEIROS-RIBEIRO**, M. CHENG***, P.M. PETROFF****
*Werkstoffe der Elektrotechnik, Ruhr-Universität Bochum, D-44780 Bochum,
email: schmidt@lwe.ruhr-uni-bochum.de
**Hewlett Packard Co., 3500 Deer Creek Rd., Bldg. 26, Palo Alto, CA 94304-1392
***M/A-COM, Microelectronics Division, 100 Chelmsford, St. Lowell, MA 01853-3294
****QUEST and Materials Department, University of California, Santa Barbara, CA 93106

ABSTRACT

In this paper we report on the limits and properties of size quantization effects in InAs self assembled quantum dots (QDs). Size, density and character of the InAs islands are investigated by transmission electron microscopy. The electronic and optical properties of the islands in the coherent and dislocated growth regime are studied using capacitance, photoluminescence, photovoltage and photocurrent spectroscopy. In the data measured with the different techniques, the change in dot size and density as well as the transition from coherent to dislocated island growth is clearly observable. An increasing QD size causes a red shift in the energetic position of the QD features while the density of the islands is reflected in the intensity of the QD signal. The decrease in intensity at high InAs coverage is attributed to dislocated island formation.

INTRODUCTION

Recently quantum dots have been the subject of intense research [1-9] since they promise new devices with improved performance compared to applications based on higher dimensional physics [10-12]. The Stranski-Krastanow growth mode is one tool nature provides for producing zero dimensional systems in a quite simple way [13]. It can be applied to a number of systems like Si/Ge [14], InP/GaAs [15], InSb,GaSb,AlSb/GaAs [16] and InAs/GaAs [17]. This paper is focused on the InAs/GaAs system. In the Stranski-Krastanow growth mode a coherently strained InAs monolayer (ML) covers the GaAs substrate at the initial stage of the growth. The two dimensionally grown layer is called wetting layer (WL). With further InAs deposition small coherently strained InAs islands form on top of the WL due to the 7% lattice mismatch between InAs and GaAs. The size and density of these islands increase with InAs coverage up to a limit above which misfit dislocations are incorporated and more and more coherent islands transform into dislocated ones. Those islands grow without restriction [17].

In this paper we report on limits and properties of size quantization effects in InAs self assembled QDs using a number of different techniques such as transmission electron microscopy (TEM), capacitance-voltage (C-V), photoluminescence (PL), photovoltage (PV) and photocurrent (PC) spectroscopy.

Our TEM micrographs as well as the measured data reflect all three phases of the Stranski-Krastanow growth described above. In the TEM images changes in the dot density and the island character (coherent or dislocated) in respect to the InAs coverage are clearly observable. Since strain contrast is also present in the TEM images, it is very difficult to get precise size information [18]. Electrical and optical spectroscopy is much better suited for such investigations.

At low InAs coverage the QD signal in our measured spectra is weak due to the low dot density. When more InAs is deposited the related increase in density and size of the coherently strained QDs results in a higher intensity and in a shift of the QD signal to lower energies. Finally,

Mat. Res. Soc. Symp. Proc. Vol. 452

the energetic position as well as the line shape of the QD spectra is no longer affected by a change in InAs coverage but the intensity of the QD signal decreases. We attribute this effect to the onset of dislocated island formation accompanied by an increase of non radiative recombination centers.

EXPERIMENT

We have grown two samples by MBE under an As pressure of $7x10^{-5}$ Torr. All material except the InAs was deposited at 600°C with rotating wafers. The InAs was grown at 530°C. Due to an asymmetric position of the In cell in the MBE chamber and a fixed wafer position, we achieved a change of the InAs coverage across the wafer.

Sample 1 was designed for capacitance spectroscopy. 40 periods of a AlAs/GaAs (2nm/2nm) superlattice and 200nm intrinsic GaAs compensate the surface roughness of the semiinsulating (100) substrate. It follows a 20nm thick Si-doped GaAs layer ($n \approx 10^{18} cm^{-3}$). 30nm intrinsic GaAs separates the InAs system (WL and QDs) from the back contact. The deposition of InAs was stopped immediately after the RHEED pattern changed from streaky to spotty indicating the onset of the island formation. The QDs were covered with 5nm GaAs, 7 periods of an AlAs/GaAs (2nm/2nm) short period superlattice and finally with a 10nm thick GaAs cap layer. The bias is applied between a circular Schottky gate (100μm diameter, Au on top of NiCr) and a Ni/AuGe ohmic back contact. Details of the growth are described elsewhere [19].

Sample 2 is a pin-structure grown on a (100) semiinsulated GaAs substrate. On top of a 1.5μm n-doped GaAs ($n \approx 10^{18} cm^{-3}$) buffer layer, we have grown 40 periods of a n-doped GaAs/AlAs (1.5nm/1.5nm) short period superlattice followed by 30nm i-GaAs and the InAs QD system. In order to observe the dislocated island formation, we stopped the InAs deposition at ≈2ML. 100nm intrinsic GaAs cap the dot system. The p-region consists of 40 periods of a GaAs/AlAs (1.5nm/1.5nm) short period superlattice and a cap layer of 10 nm GaAs. At the edge of a 100μm square mesa structure we defined a ohmic Zn/Au p-contact. A Au/Ge pad forms the ohmic n-contact at the bottom of the mesa.

We used a 5210 EG&G dual phase Lock-In amplifier in our C-V experiments. In order to measure the capacitance at T=4.2K, a small AC bias of 5mV was added to a variable DC bias. The modulation frequency was low enough (≈30Hz) to ensure a carrier population equilibrium during each cycle. For our PL studies we used an Ar^+ laser. The luminescence signal was detected by a cooled Ge detector. In our PV and PC experiments we illuminated our sample with a Tungsten Halogen lamp (100W) dispersed by a monochromator. The chop frequency was ≈200Hz. The modulated PV and the PC signal was measured with a SRS 530 Lock-In amplifier.

RESULTS

Coherent island regime

The coherent island formation is studied on sample 1 with C-V, PL and PV spectroscopy. Fig.1 depicts typical C-V and PL spectra measured at different positions on the wafer. In both figures the measured data show a strong dependence on the InAs coverage (i.e. position on the wafer). In C-V trace A of Fig.1a only a weak shoulder at U≈0.7V describes the tunneling of electrons from the back contact into the ground state of small QDs. At higher forward bias the electrons tunnel into the two dimensional WL and cause the strong change in the capacitance at U≈0.8V . With increasing InAs coverage (A→E) the size and the density of the InAs QDs increases. Thus, the electronic structure in the QDs changes which results in a red shift

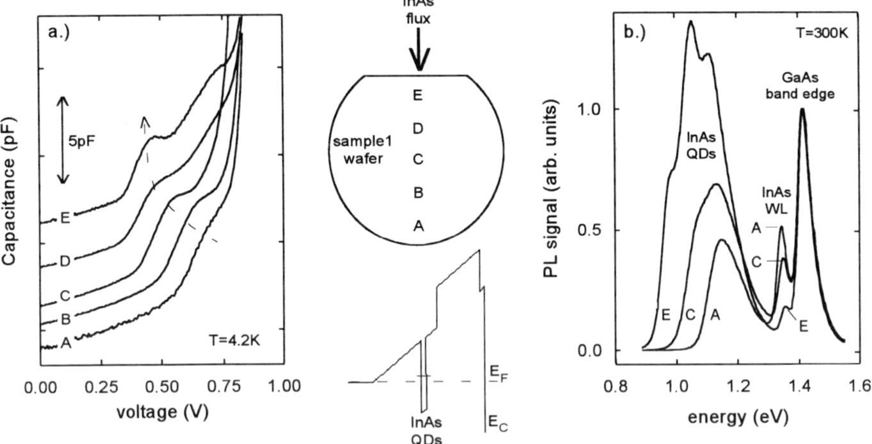

Fig.1: a.) C-V spectra of sample 1 measured at different positions on the wafer (A-E). The modulation frequency was F=30Hz and the modulation amplitude was ΔU=5mV. An offset is added for clarity. The capacitance at U=0V is C≈65pF. b.) PL spectra pumped with P≈10W/mm^2 of a focused Ar$^+$-laser. The spectra are measured at wafer positions A,C,E. The data are normalized to the GaAs band edge signal. The diagram between Fig.a and b sketches the positions on the wafer (A-E) and the conduction band structure (E_C) of the sample. E_F marks the Fermi level when no bias is applied to the structure.

of the QD ground state transition (trace A-E, dashed arrow). An additional structure appears below the WL feature in trace D and E reflecting the tunneling into the first excited electron state of the QDs. In small dots size quantization effects shift this electronic level into the WL continuum.

Along with an increase of the InAs coverage the onset of the QD PL shows also a strong red shift reflecting the growth of the QDs in size (Fig.1b). At the same time the QD signal becomes more intense since additional QDs form. More carriers are able to relax down from the WL ground state into the QDs and the WL signal decreases. In addition the full width at half maximum of the QD size distribution of trace E is reduced compared to trace A. This is a strong hint that more and more QDs reach their coherent size limit. We also took PV spectra at the same positions on the wafer (Fig.2). Since we illuminated the sample from the back side only the light with an energy below the GaAs band edge reached the field region of our Shottky diode and was able to produce a PV signal. Thus, the PV signal disappears above ≈1.4eV. Absorption from the WL or the QD takes place below the GaAs band gap energy. In this case the electric field separates the electron hole pairs which results in a PV signal (inset of Fig.2). As already discussed in Fig.1 the QD signal between 0.9eV and 1.3eV shows a clear red shift reflecting the increasing QD size (i.e. InAs coverage). The decrease of the WL feature at high InAs coverage can be explained by a slight blue shift which is observable in the PL-data of Fig.1b. It might be caused by enhanced In/Ga intermixing or higher strain in the WL at higher dot densities.

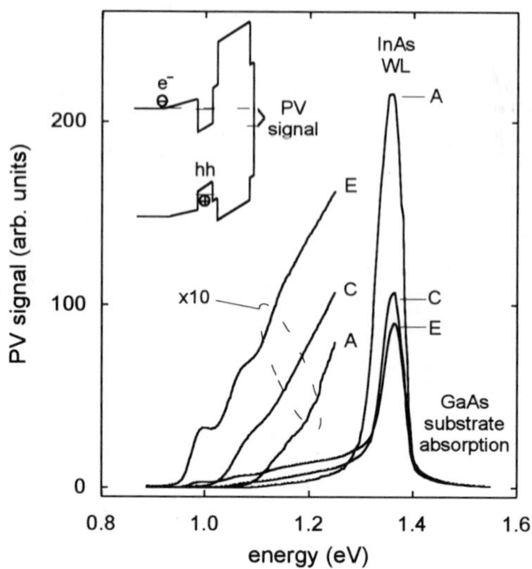

Fig.2: Room temperature photovoltage spectra of sample 1 measured at wafer positions A,C,E. Light of a tungsten halogen lamp dispersed by a monochromator was used for illumination of the sample from the back side through the GaAs substrate. The data are normalized to the signal above the GaAs band edge and multiplied by a factor of 10 for energies below 1.25eV. The inset depicts the band structure of our sample under illumination.

Dislocated island formation

In order to study the effects of dislocated island formation on the optical properties of our QDs we have investigated sample 2. In the cross section as well as in the plan view TEM micrographs taken from a position with low InAs coverage no dislocated islands can be observed. However, on a position with high InAs coverage coherent and dislocated islands are clearly visible (inset of Fig.3a). The coherent islands are surrounded with a homogenous strain field in contrast to dislocated ones. In some of the dislocated islands the misfit dislocation starts from the island and extends along the growth direction to the surface of the sample (inset of Fig. 3a). The onset of dislocated island formation is also clearly reflected in the intensity of the spatial dependent PL signal. In the part of the wafer where the size and the density of the coherent islands increases with the InAs coverage the PL signal of the QDs shows a similar behavior as already discussed in Fig.1b. When the size and the density of the coherent QDs have reached their limits dislocated island formation starts. Those islands act as non radiative recombination centers. Since less coherent islands luminesce while the excitation density per dot remains roughly constant, the intensity of the coherent QD luminescence decreases but the line shape of the PL-signal does not change significantly (Fig.3a). The energetic position of the QD transitions does no longer depend on the amount of InAs deposited since the coherent islands reached their size limit. In order to grow bigger, they have to incorporate a misfit dislocation.

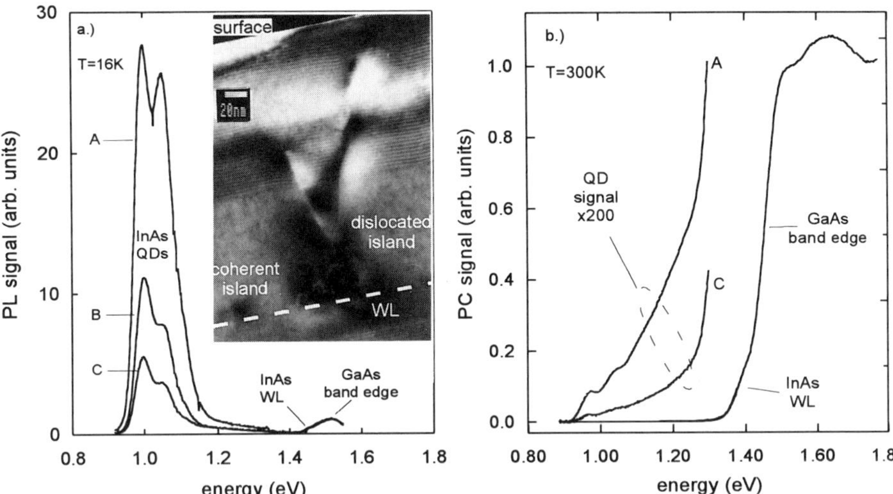

Fig.3: PL (a) and PC (b) spectra of sample 2 measured at different positions on the wafer (A→C: increasing InAs coverage). The data are normalized to the GaAs band edge. An Ar^+-laser with a pump intensity of $10mW/mm^2$ was used in the PL experiments. In order to measure the PC signal we illuminated sample 2 from the front (in contrast to Fig.3). Thus, Franz-Keldysh oscillations can be observed above the GaAs band edge. The decrease in the intensity of the QD features in Fig.a.) and b.) is attributed to dislocated island formation. The inset of Fig.a shows a cross section TEM image taken at position C (i.e. high InAs coverage).

A similar effect is also observable in the PC spectra of Fig.3b. This figure compares the PC signal of a device with and without dislocated islands. The signal is normalized to the GaAs band edge. Like in the PV spectra of Fig.2 the PC traces reflect mostly the absorption of the sample. For energies between 0.95eV and 1.3eV the QD signal shows a steady increase reflecting the density of states and the size distribution of the InAs QDs. The small shoulder at E≈1.4eV represents the absorption into the WL. In contrast to Fig.2 the p-i-n structure was illuminated from the front. Thus, Franz-Keldysh oscillations caused by the internal field in the p-i-n structure are clearly observable above the GaAs band edge. According to the PL spectra of Fig.3a, the QD signal of trace C (coherent and dislocated islands) is reduced compared to the one of trace A (only coherent islands). Again, the electron and hole pairs created in a dislocated island recombine non radiatively. The one in coherent islands are separated by the intrinsic electric field of the p-i-n diode and produce a PC signal.

CONCLUSION

In conclusion we have investigated the limits of size quantization effects in InAs self assembled QDs with different experimental techniques. Changes in InAs coverage result in various InAs QD sizes and densities which are clearly observed by TEM. We have studied the electronic and optical properties of our samples in respect to the InAs coverage with C-V, PL, PV and PC spectroscopy. All spectra reflect size quantization effects due to the change of the nominal InAs layer thickness. In the coherent island regime an increasing InAs coverage results in bigger QDs which causes a clear shift of the QD features to lower energies and higher excited states appear. In addition the intensity of the QD signal increases reflecting an increasing dot density. Above the limit where dislocated island formation starts, the lineshape of the PL and PC signal remains constant while the intensity of the signal decreases with further amount of InAs deposited. We explain this effect by an increasing number of dislocated islands and by non radiative recombination paths in such islands.

ACKNOWLEDGMENT

The authors wish to acknowledge the financial support by QUEST, the Alexander von Humboldt Foundation (KHS) and the NSF-STC center (DMR No. 91-20007, GMR). One of the authors (KHS) wishes to thank Prof. U. Kunze for fruitful discussions and funding.

REFERENCES

1 C.B. Murray, D.J. Norris, and M.G. Bawendi, J. Am. Chem. Soc. **115**, 8706 (1993)
2 G.S. Solomon, J.A. Trerza, A.F. Marshall, and J.S. Harris Jr., Phys. Rev. Lett. **76**, 952 (1996)
3 Q. Xie, A. Madhukar, P. Chen, and N.P. Kobayashi, Phys. Rev. Lett. **75**, 2542 (1995)
4 M. Sopanen, H. Lipsanen and J. Ahopelto, Appl. Phys. Lett. **66**, 2364 (1995)
5 A. Wojs, P. Hawrylak, S. Fafard, and L. Jacak, Phys. Rev. B **54**, 5604 (1996)
6 M. Grundmann, O. Stier, and D. Bimberg, Phys. Rev. B **52**, 11969 (1995)
7 J.-Y. Marzin, J.M. Gerard, A. Izrael, D. Barrier, and G. Bastard, Phys. Rev. Lett. **73**, 716 (1994)
8 H. Drexler, D. Leonard, W. Hansen, J.P. Kotthaus, and P.M. Petroff, Phys. Rev. Lett. **73**, 2252 (1994)
9 S. Fafard, R. Leon, D. Leonard, J.L. Merz, and P.M. Petroff, Supertatt. Microstruct. **16**, 303 (1994)
10 N. Yokoyama, S. Muto, K. Imamura, M. Takatsu, T. Mori, Y. Sugiyama, Y. Sakuma, H. Nakao, and T. Adachihara, Solid State Electron. **40**, 505 (1996)
11 H. Shoji, K. Mukai, N. Ohtsuka, M. Sugawara, T. Uchida, and H. Ishikawa, IEEE Phot. Techn. Lett. 7, 1385 (1995)
12 G. Yusa, H. Sakaki, Electronics Lett. **32**, 491 (1996)
13 I.N. Stranski and Von L. Krastanow, Akad. Wiss. Lit. Mainz Math.-Natur. Kl. IIb **146**, 797 (1939)
14 R. Apetz, L. Vescan, A. Hartmann, C. Dieker, and H. Lüth, Appl. Phys. Lett. **66**, 445 (1995)
15 N. Carlsson, W. Seifert, A. Petterson, P. Castrillo, M.-E. Pistol, and L. Samuelson, Appl. Phys. Lett. **65**, 3093 (1994)
16 B. R. Bennett, R. Magno, and B.V. Shanabrook, Appl. Phys. Lett. 68, 505 (1996)
17 D. Leonard, K. Pond, and P. M. Petroff, Phys. Rev. B **50**, 11687 (1994)
18 D. Leonard, M. Krishnamurthy, S. Fafard, J.L. Merz, and P.M. Petroff, J. Vac. Sci. Technol. B **12**, 1063 (1994)
19 G. Medeiros-Ribeiro, D. Leonard, and P.M. Petroff, Appl. Phys. Lett. **66**, 1767 (1995)

DETERMINATION OF OPTICAL PROPERTIES OF THE MICRO-FACETTED InGaAs QUANTUM WELLS AND QUANTUM WIRES USING MAGNETOPHOTOLUMINESCENCE

Sung-Bock Kim*, Jeong-Rae Ro*, and El-Hang Lee*
*Research Department, Electronics and Telecommunications Research Institute,
Yusong P. O. Box 106, Taejon, 305-600, Korea

ABSTRACT

We report optical properties of the micro-facetted InGaAs quantum wells and quantum wires on non-planar substrates employing magnetophotoluminescence (MPL). The InGaAs/GaAs structures were grown by chemical beam epitaxy on V-groove patterned GaAs substrates. In the presence of a magnetic field of 18 *T*, the diamagnetic shifts of exciton ground states of the (001)- and side-QWLs are $\Delta E = 15.6$ and 10.3 meV, respectively. In MPL of the facetted microstructure, we found that the different diamagnetic shifts strongly depend on the magnitude of the effective magnetic field as well as the quantum confinement. From comparing the intensities and full widths at half maximum, we easily found that side-QWLs are of higher quality than (001)-QWLs. We also fabricated InGaAs/GaAs quantum wires with a size of about 200 Å x (500~600) Å. By fitting the diamagnetic shifts ($\Delta E = 9.5$ meV) of the exciton ground state with the calculated results of a variational method, we estimated that the reduced mass of the exciton is approximately 0.052 m_e.

1. INTRODUCTION

The use of non-planar substrates has recently drawn a great deal of attention for the possibility of obtaining quantum well (QWL), quantum wire (QWR) and quantum dot (QD) structures having very smooth interfaces and better optical and electrical properties [1, 2]. The reduced dimensionality is expected to improve many physical properties of semiconductors and semiconductor base devices such as high performance devices [3] and opto-electronic and photonic devices of lasers [4] and waveguides [5].

On the other hand, epitaxial growth on non-(001)-oriented planar substrates which are far from (001) substrates has been tried for superior properties such as low threshold current, low compensation and better surface morphology [2, 6-8]. In the case of (111) substrates, a higher peak-to-valley current ratio in resonant tunneling devices and a higher luminescence intensity than (001) substrates have been reported [9, 10]. To achieve better crystal growth properties, (113) substrates have been used. The ideal (113) surface is non-polar with equal densities of single and double dangling bonds, i.e., an equal number of steps and terraces, and this is an average of the (001) and (111) surfaces [11].

The evaluation of optical properties of the various facetted epilayers is very important in developing an integration technique for electronic and photonic devices on a single substrate. Although a variety of measurements have been employed to determine the electronic states and optical properties, it was difficult to investigate the optical characteristics of the various facetted micro-structures on a single substrate.

In this paper, we have reported the optical properties of the micro-facetted InGaAs

Mat. Res. Soc. Symp. Proc. Vol. 452 © 1997 Materials Research Society

QWLs and QWRs grown on V-grooved substrates using magnetophotoluminescence (MPL). Due to the dependence of the quantum confinement states on the magnitude of applied magnetic field, we simply estimate the optical properties in the facetted QWLs by varying the direction of applied external magnetic field. Using the variational method of the magnetoexciton ground state energy, we also determine the reduced mass of exciton in QWRs.

2. EXPERIMENTS

InGaAs/GaAs facetted microstructures and QWRs were grown on V-groove patterned GaAs substrates with a (001)-(223) plane system in a conventional chemical beam epitaxy (CBE) system which consisted of an ultrahigh vacuum growth chamber, a load-lock chamber, and a source gas control system. The group III and V source gases were introduced through automatic pressure-controlled leak valves without the use of a carrier gas. In this study, we used trimethylindium (TMIn) and triethylgallium (TEGa) as metalorganic sources together with unprecracked monoethylarsine (MEAs) as the group V gas.

The V-grooves were fabricated on semi-insulating GaAs(100) substrates by standard photolithography and a wet chemical etching process at room temperature. The V-grooves were formed along the $[0\bar{1}1]$ direction having (223) sidewalls and (001) top- and bottom planes and the periodicity of the V-grooves was about 5 μm. The patterned wafers were degreased by organic solutions. Prior to growth, the V-grooved substrates were sequentially annealed at 150 °C and 600 °C in order to remove the moisture and the oxide layer from the surface, respectively. The gas pressure in the chamber during growth was in a range of $10^{-5} \sim 10^{-4}$ Torr depending upon the hydride gas treatment. The detailed configuration of the system used in this work was described in the previous works [12, 13].

To study the optical characteristics of InGaAs/GaAs microstructures in detail, low temperature MPL was employed. In low temperature MPL investigations, single fiber optics was used to couple the light in and out of the sample mounted in a superconducting magnet. A magnetic field of up to 18 Tesla was applied in the direction parallel to the <001> axis, and the PL was dispersed by a 1/4 meter monochromator and detected with a charge coupled device (CCD) array. The layer thickness and cross-sectional profile were also examined by scanning electron microscope (SEM) and transmission electron microscopy (TEM).

To determine the exciton reduced mass of the exciton ground state from the measured diamagnetic shifts in the presence of magnetic fields, we calculated the magnetoexciton ground state in a QWLs using a variational method. In the envelope-function approximation, ignoring the valence band mixing effect, the trial wave function of the exciton ground state was expressed in terms of subband wave functions in the <110> direction and the product of two-dimensional hydrogen and oscillator-like wave functions with two variational parameters for the in-plane component [14, 15].

3. RESULTS AND DISCUSSION

3. 1. InGaAs/GaAs facetted microstructure

InGaAs/GaAs QWLs were grown on the (223)-(001) facetted microstructure which consisted of (001) planes and (223) facets. Fig. 1(a) shows a schematic diagram of InGaAs/GaAs

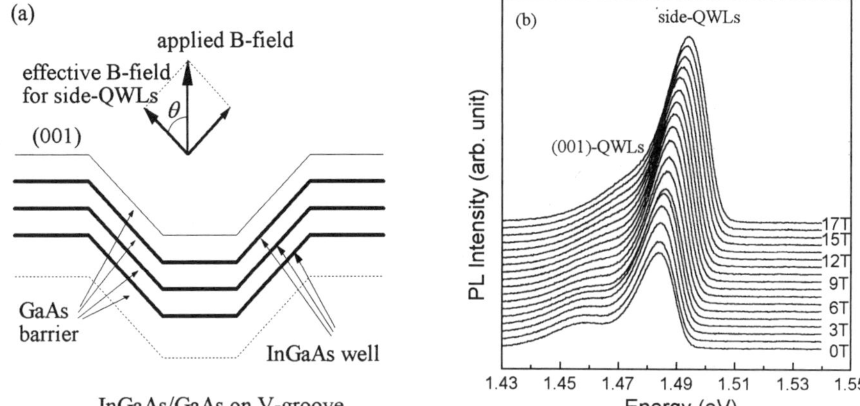

Fig. 1 (a) Schematic diagram of facetted InGaAs/GaAs QWLs on a V-grooved GaAs substrate. The effective magnetic field for side-QWLs was indicated by the arrow. (b) The typical MPL spectra of the InGaAs/GaAs QWLs illustrated in (a).

facetted QWLs studies in this work. The thickness of the InGaAs active layers between GaAs barriers was about 60 Å on the (001) top plane. From low temperature (T = 4.2 K) PL measurements at zero magnetic field, we observed two distinct peaks at 1.462 and 1.483 eV, which are associated with the exciton ground states (e1-h1) of the top-QWLs and the (223) side-QWLs, respectively. The origin of PL peaks was confirmed by the diamagnetic shifts of the PL

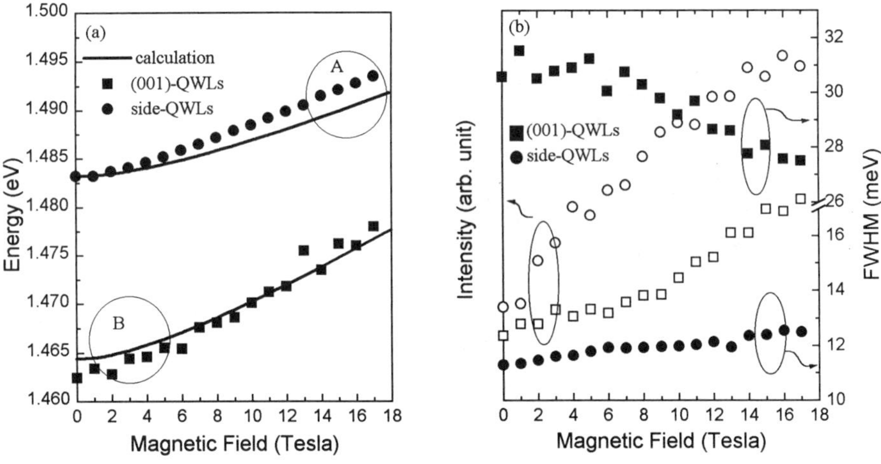

Fig. 2 (a) The solid symbols denote the diamagnetic shifts of the (001)- and side-QWLs while the solid lines represent calculated results. (b) Magnetic field dependent PL intensity and FWHM of (001)- and side-QWLs.

transitions in the presence of magnetic fields up to B = 18 Tesla uniformly applied parallel to the <001> direction, $\vec{B} = B\hat{z}$. Fig. 1(b) shows typical experimental recordings of MPL spectra of InGaAs/GaAs facetted QWLs. As shown in Fig. 1(b), both PL peak positions of the (001) top- and (223) side-QWLs shift to higher energy side with increasing magnetic field. The diamagnetic shift of the top-QWLs' peak is larger than that of side-QWLs' peak. Fig. 2(a) shows the dependence of the energy changes of the magnetoexciton ground states of (001)- and side-QWLs on magnetic field. In Fig. 2, two peaks show a considerable difference in the diamagnetic shift; the one at 1.462 eV shows $\Delta E = 15.6$ meV, while the other at 1.483 eV shows $\Delta E = 10.3$ meV, respectively. The solid lines represent the calculated results using the variational method. We found that the different diamagnetic shifts of the magnetoexciton ground states strongly depend on the level of effective magnetic field corresponding to the angle between the facet and applied magnetic field as well as quantum confinement. The deviation of the A region of high magnetic field in Fig. 2(a) can be understood upon consideration of the perpendicular component of magnetic field, while in the B region, the calculated result does not coincide with experimental results due to excitons bound to impurities and defects induced by the etching process. When comparing the intensities and full widths at half maximum (FWHM) of the spectra in Fig. 2(b), the peak of side-QWLs seems much stronger and sharper. The intensities of PL for both (001)- and side-QWLs increase with increasing magnetic field resulting from the magnetic confinement. On the other hand the FWHM of (001)-QWLs decreases with increasing field because the exciton bound to impurities or defects radiate due to increasing magnetic field. These results clearly show that InGaAs side-QWLs are higher quality than (001)-QWLs. Hence, magnetophotoluminescence is a simple means to identify the origin of the exciton transition of the micro-facetted quantum structure on a single substrate.

3. 2 InGaAs/GaAs QWRs

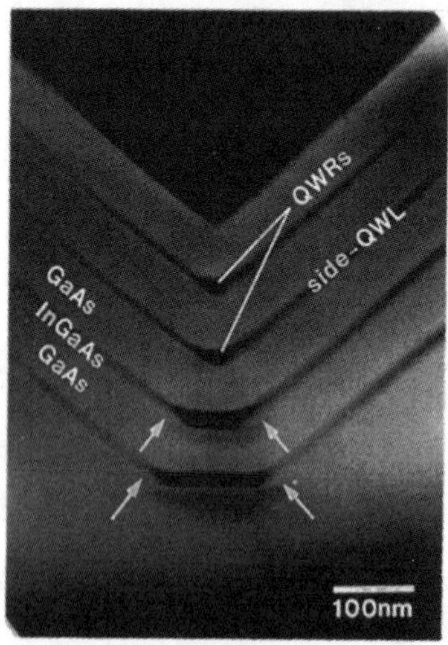

Fig. 3. Cross-sectional TEM image of vertically stacked InGaAs/GaAs QWRs with sizes of 200 Å x (500~600) Å at the bottom of the V-grooves (upper two images). The lower two images clearly show the resharpening effect of GaAs layers.

Fig. 3 shows a cross-sectional TEM photograph of stacked multiple InGaAs/GaAs QWRs using the resharpening effect of a GaAs layer at a growth temperature of 450 °C. As shown in the upper two images of Fig. 3, the crescent-shaped InGaAs QWRs were formed with sizes of about 200 Å x (500~600) Å. To achieve the vertically stacked QWRs, we used the V-grooves with (223) planes instead of (111)A planes. The (223) plane has a greater number of arsenic dangling bonds than the (111)A plane and can effectively suppress Ga migration, resulting in the enhancement of the GaAs growth rate on the (223) plane. If the GaAs layer grows thick enough, the concave parts of the GaAs become sharper and sharper at the bottom of the V-grooves. This result is significantly different from the crescent-shaped GaAs QWRs where the GaAs layer shows a fast planarization of the V-groove by creating a new (001) surface at the bottom due to the fast surface migration of Ga atoms on the (111)A sidewall. However, in this paper, we suggest the possibility of fabricating stacked InGaAs QWRs based on GaAs barriers.

Fig. 4 shows MPL spectra of InGaAs QWRs at the bottom along with the entire spectra illustrated in Fig. 3. In the inset in Fig. 4, we plotted the entire PL spectrum of InGaAs/GaAs on the V-grooves without applying the magnetic field. Although the FWHMs of these luminescence peaks have somewhat large values of about 20~30meV due to the compositional disordering and the interface roughness, three distinct peaks related with side-QWLs, top-QWLs, and QWRs are observed. In the presence of a magnetic field of 18 T, the diamagnetic shift of the magnetoexciton ground state in QWRs is about ΔE = 9.5 meV. From the measured diamagnetic shifts, we calculated the magnetoexciton ground state using a variational method. In our calculation, the effect of edges of QWRs was not taken into account because the width of QWRs was greater than the exciton Bohr radius which is less than 150 Å. By comparison of the calculated result with the experimental data, we estimated that the reduced mass of exciton is approximately 0.052m_e.

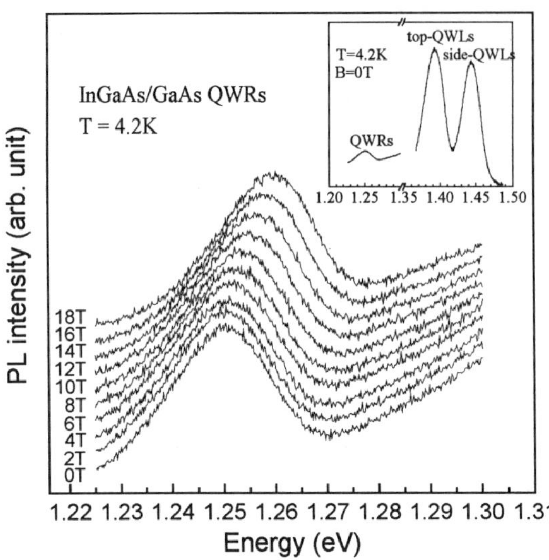

Fig. 4. Typical MPL spectra of InGaAs QWRs. Each spectrum was measured at the magnetic field indicated on the left-hand side of the spectra. The entire PL spectrum of the same sample at zero magnetic field is shown in the inset.

4. CONCLUSION

In summary, the optical properties of InGaAs micro-facetted QWLs and QWRs have been studied by using MPL. Because the diamagnetic shifts strongly depend on the effective magnetic field (parallel component to the growth direction), we clearly assigned the origin of the peaks radiated from facetted QWLs. By comparing intensities and FWHMs of these peaks, we easily determined the optical properties in facetted QWLs. In stacked QWRs with a size of 200 Å x (500~600) Å, the transition energy of the ground state shows a diamagnetic shift of 9.5 meV. The exciton reduced mass was considered as an adjustable parameter, and the best value, $\mu = 0.052m_e$, was chosen from the comparison of the calculated result of the variational method for the magnetoexciton ground state energy and the experimental data.

ACKNOWLEDGMENTS

The authors would like to thank Dr. K. S. Lee for their help with the MPL analysis. This work was partially supported by the Ministry of Information and Communications, Korea.

REFERENCES

1. S. Shimomura, A. Wakejima, A. Adachi, Y. Okamoto, N. Sano, K. Murase, and S. Hiyamizu, Jpn. J. Appl. Phys. **32**, L1728 (1993).
2. T. Hayakawa, K. Takahashi, M. Kondo, T. Suyama, S. Yamamoto, and T. Hijikata, Phys. Rev. Lett. **25**, 349 (1988).
3. K. Imamura, N. Yokoyama, T. Ohnishi, S. Suzuki, K. Nakai, H. Nishi, and A. Shibatomi, Jpn. J. Appl. Phys. **23**, L342 (1984).
4. B. Rose, D. Remienes, V. Hornung, and D. Robein, J. Cryst. Growth, **107**, 850 (1991).
5. J. B. D. Soole, H. Schumacher, H. P. LeBlanc, R. Bhat, and M. A. Koza, Appl. Phys. Lett. **56**, 1518 (1990).
6. H. Sugawara, K. Itaya, H. Nozaki, and G. Hatakoshi, Appl. Phys. Lett. **61**, 1775 (1991).
7. G. H. Olsen, T. J. Zamerowski,and F. Z. Hawrylo, J. Cryst. Growth, **59**, 654 (1982).
8. S. S. Bose, B. Lee, M. H. Kim, G. E. Stillman, and W. J. Wang, J. Appl. Phys. **63**, 743 (1988).
9. L. F. Luo, R. Beresford, W. I. Wang, and E. E. Mendez, Appl. Phys. Lett. **54**, 213 (1989).
10. T. Hayakawa, M. Kondo, T. Suyama, K. Takahashi, S. Yamamoto, and T. Hijikata, Jpn. J. Appl. Phys. **26**, L302 (1987).
11. D. J. Chadi, Phys. Rev. B29, 786 (1984).
12. S. J. Park, J. R. Ro, J. K. Sim, and E. H. Lee, ETRI J. **16**, 1 (1994).
13. S. B. Kim, S. J. Park, J. R. Ro, and E. H. Lee, J. Cryst. Growth, **164**, 356 (1996).
14. K. S. Lee, Y. Aoyagi, and T. Sugano, Phys. Rev. B**46**, 10267 (1992).
15. K. S. Lee and E. H. Lee, ETRI J. **17**, 13 (1996).

RAMAN SCATTERING AND PHOTOLUMINESCENCE OF SPONTANEOUSLY ORDERED $Ga_{0.5}In_{0.5}P$ ALLOY

G. H. LI*, Z. X. LIU*, H. X. HAN*, Z. P. WANG*, J. R. DONG** AND Z. G. WANG**

*National Laboratory for Superlattices and Microstructures, Institute of Semiconductors, Chinese Academy of Sciences, Beijing, 100083, China

**Laboratory for Material Scientific, Institute of Semiconductors, Chinese Academy of Sciences, Beijing 100083, China

ABSTRACT

Samples of the spontaneously ordered $Ga_{0.5}In_{0.5}P$ alloys were grown by the MOCVD method on [001]-oriented GaAs substrates. The thickness of the epitaxal layer is about 2 μm. Raman scattering and photoluminescence spectra have been measured at room temperature. The result from photoluminescence measurement indicates that the direct-band gaps of the spontaneously ordered samples are lower than that of the disordered sample. Three scattering peaks have been observed in the Raman spectra, corresponding to the GaP-like LO, InP-like LO and InP-like TO modes in the alloys, respectively. The frequencies of the GaP- and InP-like LO modes increase with the decrease of the band-gap of the ordered alloys. It is related to the formation of the $(GaP)_1/(InP)_1$ monolayer superlattice along [111] direction in the ordered alloys. The polarization properties of the ordered alloys are similar to those of the bulk III-V semiconductors with the zinc-blende structure.

INTRODUCTION

The $Ga_{0.5}In_{0.5}P$ alloy, lattice matched to the GaAs substrate, has attracted considerable interest in recent years for its applications in optoelectronic devices [1-4]. It has been found that $Ga_{0.5}In_{0.5}P$ alloys grown on [001]-oriented GaAs substrates by the metalorganic chemical vapor deposition (MOCVD) under the appropriate growth conditions will display spontaneously long-range order in some degree. And the degree of ordering can be controlled by the variations in the growth parameters. A striking difference is that the band gap of the ordered alloy is smaller than that of the disordered alloy grown by liquid phase epitaxy [5-8]. In general, the band gap reduction is about 50-100 meV. In fact, in the ordered alloy, the Ga and In atoms occupy the alternate (111) atomic planes of group III sublattice forming a $(GaP)_1/(InP)_1$ monolayer superlattice (MSL) along the [111] direction. The theoretical calculations by Wei et al. [9] indicated that the formation of $(GaP)_1/(InP)_1$ MSL along [111] direction leads to the folding of L valley to Γ point in the Brillouin zone, and the repulsion between Γ valley and folded L valley results in a reduction of the band gap.

Raman scattering spectra of the disordered alloy of $Ga_xIn_{1-x}P$ have been extensively studied [10-15]. In the early reports, the infrared and Raman spectra of $Ga_xIn_{1-x}P$ alloys display a one-mode behavior, however, it is difficult to explain the additional weak structures between LO and TO modes. In the later reports [13-15], it was found that the frequency of the gallium local mode in InP is slightly higher than that of InP LO mode, so that they attributed to this alloy system a modified two-mode behavior. As to Raman scattering of the spontaneously ordered $Ga_{0.5}In_{0.5}P$ alloy, in the earlier works [5,6,16,17], the GaP- and InP-like LO phonon peaks have been observed at about 380 and 360 cm^{-1}, respectively. Their lineshapes have been analyzed and found that the lineshape is correlated with the energy of photoluminescence peak. In the recent reports [18,19], the new modes have been found and assigned to the phonon modes related to the ordered phase.

In this paper, we present the measured results of room temperature Raman scattering and photoluminescence (PL) of three spontaneously ordered $Ga_{0.5}In_{0.5}P$ alloy samples and one

Mat. Res. Soc. Symp. Proc. Vol. 452

disordered $Ga_{0.5}In_{0.5}P$ alloy sample. It is found that the frequencies of GaP- and InP-like LO modes increase with the decrease of the band-gap of the ordered alloys. It is related to the formation of the $(GaP)_1/(InP)_1$ monolayer superlattice along [111] direction in the ordered alloys. The polarization properties of the ordered alloys is similar to those of the bulk III-V semiconductors with the zinc-blende structure.

SAMPLES AND EXPERIMENT

The undoped $Ga_{0.5}In_{0.5}P$ epilayers were grown on semi-insulating GaAs substrates as described elsewhere [20]. Sample A and B were grown by the atmospheric pressure MOCVD on [001]-oriented GaAs substrate. Sample C was grown by the low-pressure MOCVD on [001]-oriented GaAs substrate. Sample D was grown by the gas-source molecular beam epitaxy on [111]-oriented GaAs substrate. The growth parameters of the four samples are listed in Table I. The epilayer thickness is about 2 μm. The Ga compositions in epilayers were determined by the double crystal X-ray diffraction and are also listed in Table I. For x = 0.516, the lattice constant of the $Ga_xIn_{1-x}P$ ternary mixed crystal is matched with that of the GaAs substrate.

Table I. The growth parameter of samples and energy (E_g) of band gap

Sample	Substrate	Temperature	V/III ratio	x	E_g (eV)
A	(100)GaAs	650°C	70	0.496	1.788
B	(100)GaAs	610°C	70	0.508	1.811
C	(100)GaAs	705°C	200	0.525	1.860
D	(111)GaAs	500°C		0.514	1.903

PL measurements were performed at room temperature. Fig.1 shows the PL spectra of three ordered samples, A, B and C, (in solid lines) and one disordered sample D (in dashed line). Only one peak was observed in each curve in Fig.1, corresponding to the direct transition from the Γ valley of the conduction band to the valence band. For convenience, we will take the energies of PL peaks as the corresponding energies of direct band gap of the alloys at room temperature, which are also listed in the Table I. The energies of band gap in the three ordered samples are lower than that of the disordered sample by 115, 92 and 43 meV, respectively.

Raman spectra were measured in the backscattering geometry at room temperature. The excitation light is the 514.5 nm line of a model SP 165-09 Ar ion laser. The power focused on the samples is about 50 mW. The Raman scattering signals were analyzed by a model JY-HRD-2 double-grating monochromator and then detected with a cooled RCA C31034A GaAs photomultiplier and PAR 1140 photon counting system. A

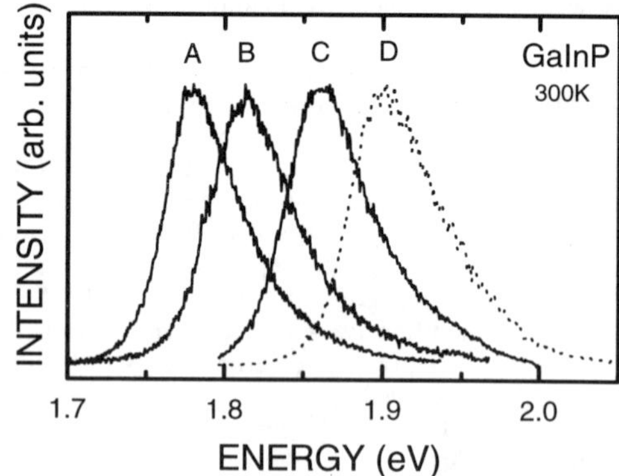

Fig.1. PL spectra of the ordered (solid lines) and disordered (dashed lines) $Ga_{0.5}In_{0.5}P$ alloys at room temperature.

polarization rotator was used to change the polarization direction of the incident laser beam. And a polarization analyzer was adapted behind the collecting lens of scattering light for the polarization analysis.

RESULTS AND DISCUSSION

Fig.2 shows the Raman spectra of the ordered sample A at room temperature. Three Raman scattering peaks were observed at 380, 360 and 330 cm^{-1}, respectively. They were assigned to the GaP-like LO mode, InP-like LO mode and InP-like TO mode [18,19], respectively. The TO mode should be forbidden on [001]-oriented substrate, but the disorder-activated InP-like TO mode was observed. However, the weak GaP-like TO mode was between the strong GaP and InP- like LO modes so that it is too difficult to observe

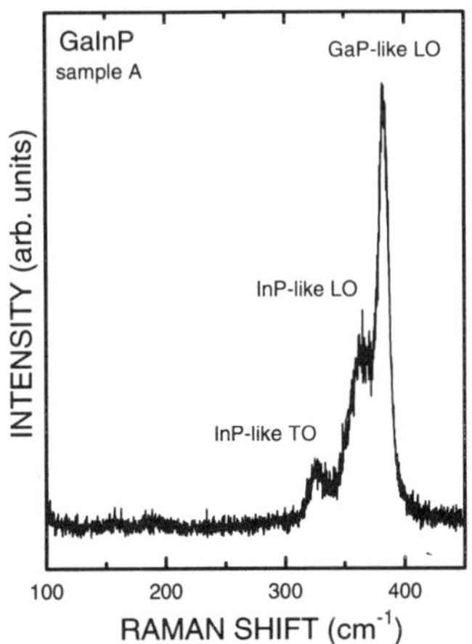

Fig.2. Raman scattering spectra of sample A of the ordered alloy at room temperature.

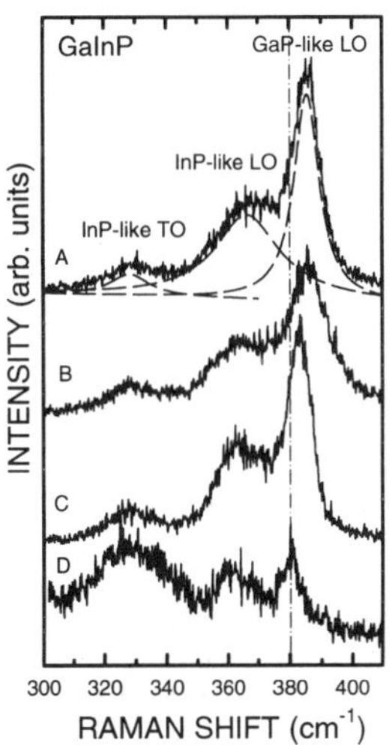

Fig.3. Raman scattering spectra of the ordered alloy A, B, C and disordered alloy D, the dashed curves is the fitting result.

Fig.3 shows the Raman scattering spectra in the range of 300-400 cm^{-1} of the three ordered samples and one disordered sample, their growth and structure parameters are shown in Table I. Three scattering peaks are observed in all four samples. The shape of spectra curves in three ordered sample are same essentially. However, the InP-like TO mode is the strongest peak in the three peaks of the disordered sample since it was grown on [111]-oriented GaAs substrate, in this case, the TO mode is Raman-active. The frequencies of the three Raman peaks were obtained from the fitting for spectra curves by three peaks with Lorentian lineshape. The fitting result for curve A is shown as dashed line in Fig.3. The fitting parameters are listed in Table II.

The frequency of the GaP-like LO mode in the disordered sample obtained from Fig.3 is 380.1(5) cm^{-1}, it is in good agreement with the results of $Ga_{0.5}In_{0.5}P$ mixed crystal in references [12,13,15,21]. However, the frequencies of the GaP-like LO mode in the three ordered samples are larger than that of the disordered sample by 3-6 cm^{-1}. Fig.4 shows the dependences of the frequencies of the GaP-like and InP-like LO modes on the band gap of the samples. The solid line in Fig.4 is the fitting result by the least-squares method. From Fig.4, the frequencies of the GaP-like LO modes increase with the decrease of the band gap energies of the ordered alloys. We suppose that it is related to the formation of $(GaP)_1/(InP)_1$ monolayer superlattice along [111] direction in the ordered alloys.

Table II. The frequency values of Raman peaks (in cm^{-1})

Sample	GaP-like LO mode	InP-like LO mode	InP-like TO mode
A	386.0(5)	366.3(5)	328(1)
B	385.9(5)	363.1(5)	328(1)
C	383.0(5)	363.8(5)	327(1)
D	380.1(5)	362.4(5)	328.4(5)

The LO confined modes on the [001] oriented- GaAs/AlAs short period superlattices have been extensively investigated by Raman scattering [22-24]. Because the optical phonon branches of the bulk GaAs and AlAs phonon dispersion curves are separated in frequency, the optical phonon modes can not propagate through the entire superlattice structures and hence become the confined modes localized in the GaAs or AlAs layers, respectively. In $(GaAs)_n/(AlAs)_n$ superlattices (n is the number of monolayer), there are n GaAs-like LO modes and n AlAs-like LO modes confined in the GaAs and AlAs layers, respectively. The frequency of the mth GaAs LO mode is equal to that of the corresponding value at the effective wave vectors $q_m = 2\pi m/(n+1)a_0$ on the bulk GaAs LO phonon dispersion curve. For [111] oriented GaAs/AlAs superlattices, the effective wave vectors are $q_m = \pi m/(n+1)a_0$ [25]. Because the optical phonon branches of the bulk GaP and InP phonon dispersion curves are also separated by a gap, in the ideal $(GaP)_1/(InP)_1$ monolayer superlattice along [111] direction, there are one GaP-like LO mode and one InP-like LO mode confined in the GaP and InP layers, respectively. Their frequencies are equal to the values at the center of the corresponding bulk LO phonon dispersion curves along Γ-L direction. For the bulk GaP, LO(Γ) = 403 cm^{-1} [26] and LO(L) = 390 cm^{-1} [27]. Thus, the frequency of the GaP-like LO_1 confined mode in the ideal $(GaP)_1/(InP)_1$ monolayer superlattice along [111] direction should be about 395 cm^{-1}. In fact, the frequencies of the GaP-like LO_1 confined modes in [001] and [211]-oriented $(GaP)_1/(InP)_1$ monolayer superlattices are between 393 and 396 cm^{-1} from Raman scattering measurement at 300 K [28]. However, the frequency of GaP-like LO mode in the $Ga_{0.5}In_{0.5}P$ disordered alloy is about 380 cm^{-1}, which is lower than the former by 15 cm^{-1}. In fact, the ordered samples are only ordered partly at the different degree, they should be between two critical cases : the disorder alloy and the ideal monolayer superlattice. That is to say, there are two effects in the ordered $Ga_{0.5}In_{0.5}P$ alloys : the mixed crystal effect and the confinement effect of LO phonons. The higher the ordered degree, that is to trend to the ideal monolayer superlattice, the lower the band gap energy and the higher the frequency of GaP-like LO mode. Thus, the frequency of GaP-like LO mode and the band gap energy in the spontaneously ordered $Ga_{0.5}In_{0.5}P$ alloys can be a measured criterion of the ordering degree. The similar case has been observed in the InP-like LO modes as shown in Fig.4.

Fig.5 shows the Raman spectra of the ordered sample_B in the different scattering configurations, where z // [001], x′ // [100], y′ // [010], x // [1 $\bar{1}$ 0] and y // [110]. From Fig.5, LO modes are Raman active in the $z(x'y')\bar{z}$ and $z(yy)\bar{z}$ scattering configuration, and Raman inactive in the $z(y'y')\bar{z}$ and $z(xy)\bar{z}$ scattering configuration. The polarization selection rule is the same with that of the zinc-blende structure. The similar results were observed in A and C samples. We have reported that there are two effects in the ultra-short period $(GaAs)_n/(AlAs)_n$ superlattices : the confinement effect of LO phonons and the alloy effect, and the smaller n, the stronger the alloy effect so that Raman spectra of $(GaAs)_1/(AlAs)_1$ monolayer superlattices are quite similar with those of the disordered $Al_{0.5}Ga_{0.5}As$ alloy [29]. In Fig.5, the

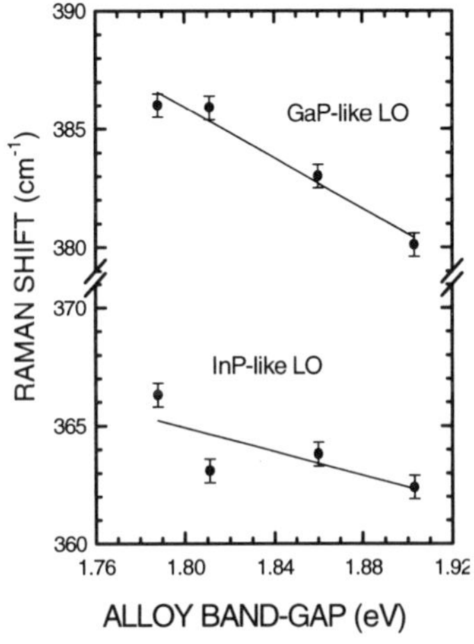

Fig.4. The dependence of the GaP-like LO mode frequency on the energy of band gap of the alloy.

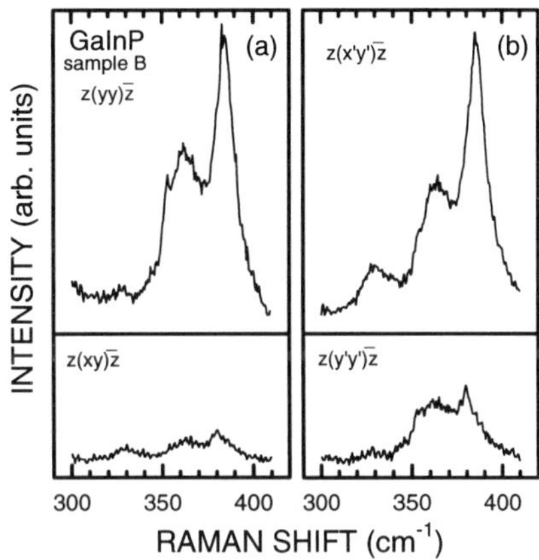

Fig.5. Raman spectra of the ordered alloy B at different scattering configuration, where z // [001], x′// [100], y′// [010], x // [1 $\bar{1}$0], y // [110].

weak LO modes were observed in the $z(y'y')\bar{z}$ and $z(xy)\bar{z}$ scattering configuration, which can be attributed to the relaxation of the polarization selection rule due to the alloy disordering.

CONCLUSIONS

Raman scattering and photoluminescence of three ordered samples and one disordered sample have been measured at room temperature. It is found that the frequencies of the GaP- and InP-like LO modes increase with the decrease of the band-gap of the ordered alloys. It is related to the formation of the $(GaP)_1/(InP)_1$ superlattice along [111] direction in the ordered alloys. In general, there are two effect in the partly ordered $Ga_{0.5}In_{0.5}P$ alloys : the mixed crystal effect and the confinement effect of the optical phonons. The frequency of the GaP-like LO mode can be a measured criterion of the ordered degree. The polarization properties of the ordered alloys is similar to those of the bulk III-V semiconductors with the zinc-blende structure.

ACKNOWLEDGMENTS

This work was partly supported by National Science Foundation of China. Authors are grateful to D. C. Lu and D. Z. Sun for the assistance in the sample growth.

REFERENCES

1. H. Fujii, Y. Ueno, A. Gomyo, K. Endo and T.Suzuki, Appl. Phys. Lett. **61**, 1959 (1992).
2. K. H. Huang, J. G. Yu, C. P. Kuo, R. M. Fletcher, T. D. Osentowski, L. J. Stinson, M. G. Craford and T.Suzuki, Appl. Phys. Lett. **61**, 1045 (1992).
3. K. A. Bertness, S. R. Kurtz, D. J. Friedman, A. E. Kibbler, C. Kramer and J. M. Olson, Appl. Phys. Lett. **65**, 989 (1994).
4. R. P. Schneider, E. D. Jones, and D. M. Follstaedt, Appl. Phys. Lett. **65**, 587 (1994).
5. T. Suzuki, A. Gomyo, S. Iijima, K. Kobayashi, S. Kawata, I. Hino, and T. Yuasa, Jpn. J. Appl. Phys. **27**, 2098 (1988)
6. M. Kondow, H. Kakibayashi, S. Minagawa, Y. Inoue, T. Nishino and Y. Hamakawa, Appl. Phys. Lett. **53**, 2053 (1988).
7. A. Mascarenhas, S. Kurtz, A. Kibbler and J. M. Olson, Phys. Rev. Lett. **63**, 2108 (1989).
8. S. R. Kurtz, J. Appl. Phys. **74**, 4130 (1993).
9. S. H. Wei and A. Zunger, Appl. Phys. Lett. **56**, 662 (1990).
10. G. Lucovsky, M. H. Brodsky, M. F. Chen, R. J. Chicotka, and A. T. Ward, Phys. Rev. **B4**, 1945 (1971).
11. R. Beserman, C. Hirlimann, M. Balkanski, and J. Chevallier, Solid State Commun. **20**, 485 (1976) .
12. E. Jahne, W. Pilz, M. Giehler, and L. Hildish, Phys. Stat. Sol. **(b) 91**, 155 (1979).
13. E. Bedel, R. Carles, G. Landa, and J. B. Renucci, Rev. Phys. Appl. **19**, 17 (1984).
14. B. Jusserand and S. Slempkes, Solid State Commun. **49**, 95 (1984).
15. T. Kato, T. Matsumoto and T. Ishida, Jpn. J. Appl. Phys. **27,** 983 (1988).
16. M. Kondow and S. Minagawa, J. Appl. Phys. **64**, 793 (1988).
17. M. Kubo, M. Mannoh, and T. Narusawa, J. Appl. Phys. **66**, 3767(1989).
18. F. Alsina, N. Mestres, J. Pascual, C. Geng, P. Ernsa and F. Scholz, Phys. Rev. **B53**, 12994 (1996).
19. A. M. Mintairov and V. G. Melehin, Semicond. Sci. Technol., **11**, 904 (1996).
20. J. R. Dong, X. L. Liu, D. C. Lu, D. Wang and Z. G. Wang, Chin. J. Semicond. to be published.
21. H. Lee, D. Biswas, M. V. Klein, H. Morkoc, D. E. Aspnes, B. D. Choe, J. Kim, and C. O. Griffiths, J. Appl. Phys. **75**, 5040 (1994).
22. K. Huang, B. F. Zhu, and H. Tang, Phys. Rev. **B41**, 5825 (1990).
23. A. K. Sood, J. Menendez, M. Cardona, and K. Ploog, Phys. Rev. Lett. **54**, 2111 (1985).
24. Z. P. Wang, D. S. Jiang, and K. Ploog, Solid State Commun. **65**, 661 (1988).
25. F. Calle, D. J. Mowbray, D. W. Niles, M. Cardona, J. M. Calleja and K. Ploog, Phys. Rev. **B43**, 5904 (1991).
26. A. Mooradian and G. B. Wright, Solid State Commun. **4**, 431 (1966).
27 Landolt-Bornstein Vol. 17a, edited by O. Madelung (Springer, Heiderberg, 1982), p.209.
28.R. M. Abdelouhab, R.Braunstein, M. A. Rao and H. Kroemer, Phys. Rev. **B39**, 5857 (1989).
29. Z. P. Wang, H. X. Han, G. H. Li, D. S. Jiang and K. Ploog, Phys. Rev. **B43,** 12650 (1991).

Growth and Characterization of $Zn_{1-x}Mn_xSe_{1-y}S_y$ Epilayers and Related Heterostructures

F. S. Flack and N. Samarth
Department of Physics, Pennsylvania State University, University Park, PA, 16801

F. Semendy
SEDD, The Army Research Laboratory, Fort Belvoir, VA 22060

Abstract

We describe the growth of (Zn,Mn)(S,Se) epitaxial layers of high structural quality on (100) GaAs substrates. Double crystal x-ray diffraction (DCXRD) measurements indicate that quaternary epilayers nearly lattice-matched with GaAs are characterized by DCXRD curves with a full width at half maximum in the range 30 - 60 arc seconds. Photoluminescence (PL) spectroscopy is employed to map the variation of the energy gap of the quartenary alloys over a wide range of alloy compositions. Finally, temperature dependent PL is used to examine the viability of (Zn,Mn)(S,Se) alloys as confining layers for ZnSe and (Zn,Cd)Se quantum wells. While efficient exciton confinement is demonstrated through the observation of robust PL from such quantum wells up to high temperatures, the renormalization of the quartenary band gap by spin fluctuations leads to a rapid decrease in confinement with increase in temperature.

Introduction

The epitaxial growth of heterostructures derived from the wide bandgap II-VI magnetic semiconductor (MS) alloys has received substantial attention for several years primarily because of the fundamentally interesting magneto-optical phenomena (e.g. spin superlattice formation and a giant quantum confined Faraday effect) that result from the exchange interaction between electrons in extended band states and in localized 3d states [1-4]. A parallel outcome of the growth of such heterostructures is that they extend the possibilities of lattice constants and bandgaps available within the context of opto-electronic applications of II-VI heterostructures at short visible wavelengths [5]. For instance, at a superficial level, the (Zn,Mn)(S,Se) alloy system [6-9] offers an interesting alternative to the better investigated (Zn,Mg)(S,Se) buffer layers [5] currently used to confine the optically active region of these devices. Like Mg, the addition of Mn into a II-VI semiconductor such as ZnSe opens up the bandgap and increases the lattice constant. However, the presence of Mn also introduces additional effects such as an absorption band in the vicinity of 2.3 eV as well as a renormalization of the bandgap by spin fluctuations. While such phenomena raise interesting fundamental questions, their influence on purported device applications of MS needs to be carefully considered. This paper discusses our research into the (Zn,Mn)(S,Se) material system and related heterostructures. We first discuss our efforts to lattice-match (Zn,Mn)(S,Se) to GaAs substrates in an effort to produce buffer layers of higher structural integrity than than the

Mat. Res. Soc. Symp. Proc. Vol. 452 © 1997 Materials Research Society

ZnSe buffer layers typically employed for MS quantum structures. The successful achievement of this goal is important for the fabricaton of high quality MS heterostructures for sophisticated magneto-optical studies. Next, we discuss the growth and optical characterization of larger regions of the quaternary phase diagram and explore the potential of this material to act as a confining layer for II-VI quantum wells (QWs).

Sample growth is carried out on (100) GaAs substrates in a cryoshrouded MBE chamber using Zn, Mn, Se, and ZnS sources of 6N, 5N, 5N and 5N purity, respectively, evaporated from standard effusion cells. *In situ* reflective high energy electron diffraction (RHEED) at 12 keV monitors the surface during growth. Background pressures in the chamber during growth are typically $\sim 10^{-10}$ torr. Growth rates of the various materials grown are measured using RHEED intensity oscillations and sample compositions of calibration epilayers are determined by electron probe microscopy. Structural characterization is carried out using double-crystal x-ray diffraction measurements. The steady state photoluminescence (PL) of these samples is measured using lock-in techniques in a 0.4 m spectrometer equipped with a cooled PMT over the temperature range 4K - 300K in a continuous flow optical cryostat using the 325 nm line of a HeCd laser with an incident intensity of about 1 W/cm^2. The temperature of the sample during PL measurements is measured using a Rh-Fe thrmocouple heat sunk to the sample stage.

We first discuss the growth and characterization of $Zn_{1-x}Mn_xS_ySe_{1-y}$ epilayers of 1-4 microns thickness in the composition range $0<x<0.09$, $0<y<0.36$. Substrate temperatures during growth are typically in the range 150-300° C. Due to the relatively high vapor pressure, significant incorporation of sulfur does not occur unless the substrate temperature is lower than ~250° C. This lowering of substrate temperature necessitates careful adjustment of other growth parameters to obtain high quality layers. A rough idea of layer quality is obtained by observing the RHEED pattern which is typically streaky and unreconstructed. An increased roughening of the surface is observed with increasing sulfur content. We speculate that the rough surface growth stems from the decreased surface mobility of the adatoms at these lower growth temperatures.

Figure 1 below shows DCXRD scans from two of the quartenary epilayers. The relative position of the diffraction peak with respect to that from GaAs shows that these epilayers have ~ 0.05% lattice mismatch with the substrate. Due to the fact that this mismatch is much smaller than that of ZnSe on GaAs, we find a significant improvement in the structural quality of the epilayers, with half widths in the range of 35 - 65 arcseconds. For comparison, the typical widths of rocking curves reported by most researchers for ZnSe epilayers of similar thickness grown on (100) GaAs lies in the range 150 - 200 arcseconds. More recently, significant improvements in the structural quality of ZnSe epilayers has been reported in work that has paid careful attention to the inital nucleation of ZnSe growth on GaAs, resulting in DCXRD widths of ~ 27 arc seconds [10].

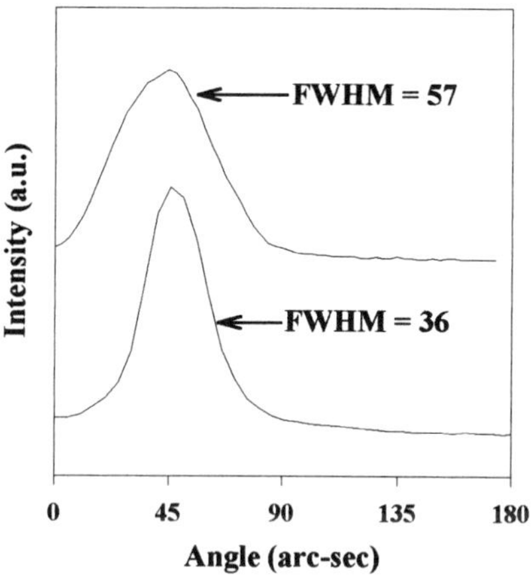

Fig. 1. DCXRD measurements for two (Zn,Mn)(S,Se) epilayers. The compositions of the two samples are: $Zn_{0.89}Mn_{0.11}S_{0.13}Se_{0.87}$ (upper curve) and $Zn_{0.93}Mn_{0.07}S_{0.11}Se_{0.89}$ (lower curve).

The low temperature energy gaps of the quartenary alloys are estimated from the near band-edge PL and range from near that of ZnSe (2.8 eV) to as high as 3.1 eV. A summary of the variation of the band gap with alloy composition is shown in Fig. 2. We note that these estimates are probably lower than the actual band gap since it is likely that the PL from the quartenary alloy epilayers originates from the recombination of excitons localized in a disorder induced tail to the density of states. Judging from the widths of the PL spectra, Stokes' shifts of up to 30 meV are likely to be encountered. Nonetheless, the

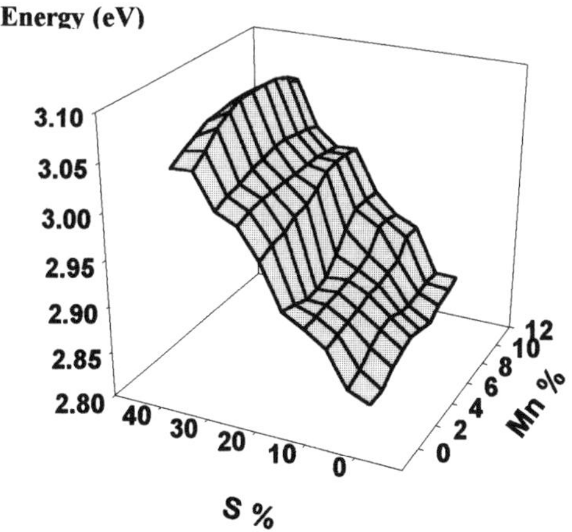

Fig. 2. Variation in band gap with both sulfur and manganese content for the samples studied. The energy gap vs. composition surface is produced by estimating the band gap from PL studies of 21 samples of (Zn,Mn)(S,Se) of different compositions. Note that the variation of the band gap with sulfur content qualitatively follows standard expectations for a semiconductor alloy while there is little change in the band gap with manganese content.

measurement of PL does provide a useful and convenient qualitative measure of the variation of the energy gap with alloy composition. A more careful study of the energy gap using absorption and/or reflectivity is currently in progress

Naively, one might anticipate that the energy gap variation with the quartenary alloy composition may be estimated from the energy gaps of the binary end points: ZnSe (2.8 eV), MnSe (~3.2 eV), ZnS (3.8 eV) and MnS (~3.7 eV). However, as can be seen in the figure above, the energy gap of the quartenary alloy does not vary much with Mn concentration in the composition range studied here. This is rooted in the same physics that leads to the unusual nonmonotonic relationship between x and E_g in $Zn_{1-x}Mn_xSe$ alloys which show a small minimum for low x [11]. This "bowing" of the energy gap in MS alloys arises because of the strong *sp-d* exchange interaction between conduction and valence band states and the localized 3d Mn states. When the *s-pd* exchange is included as a perturbation in the standard crystal Hamiltonan, the band gap has second-order corrections that involve the averages of two-spin correlation functions. These corrections can be reformulated in terms of the magnetic susceptibility and essentially result in an energy gap which varies with concentration x and temperature T as [11]:

$$E_g\ (x) = E_o\,(T)\ + ax - b\chi(x)T\,,$$

where a and b can be treated as empirical fitting parameters, χ is the magnetic susceptibility, and T is temperature and $E_O\,(T)$ describes the temperature variation of the band-gap due to non-magnetic effects
(the Varshni dependence [12]). The magnetic susceptibility has a Curie-Weiss form and is a nonlinear function of concentration, exhibiting a maximum at a Mn concentration $x \sim 0.1$ due to the anti-ferromagnetic coupling between the Mn spins. At very low values of x, a large number of Mn spins are isolated and the susceptibility increases with x. As the concentration increases, the Mn spins begin to compensate for one another, negating their contribution to the susceptibility. Therefore, the negative term is largest at small x, resulting in very little variation of E_g with x. For this reason, the Mn concentration has a minimal contribution to the band gap for the composition range examined in this study. Further, the spin-dependent contribution to the band gap strongly affects the temperature variation of E_g.. An important feature of this unusual variation of bandgap with temperature is that the energy gap of a MS alloy decreases with temperature more rapidly than the non-magnetic II-VI host lattice.

The realization that the energy gap of (Zn,Mn)(S,Se) alloys red shifts with increasing temperature at a faster rate than that of ZnSe or Zn(S,Se) raises a cautionary note about the use of this material for short wavelength applications. In order to examine the efficiency of (Zn,Mn)(S,Se) epilayers as confining barriers, we have embedded quantum wells of ZnSe and (Zn,Cd)Se within the epitaxial quartenary layers.

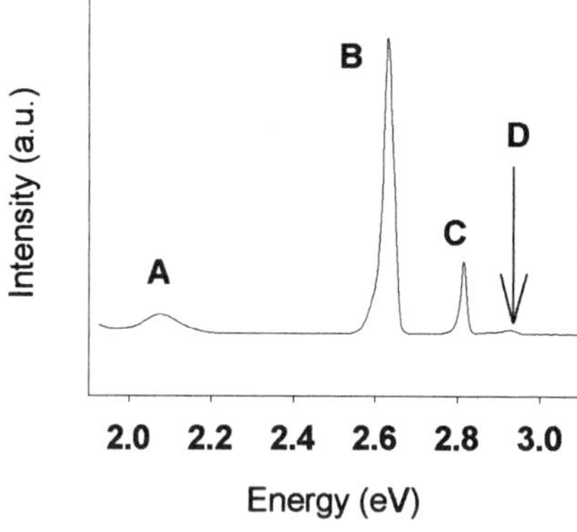

Fig 3. PL emission spectrum at 4 K taken from a $Zn_{0.82}Mn_{0.18}S_{0.20}Se_{0.80}$ epilayer containing a 50 Å ZnSe QW and a 65Å $Zn_{0.75}Cd_{0.25}Se$ QW.

Fig. 3 above shows a typical PL spectrum from a sample containing two isolated, single QWs of 5 nm
ZnSe and 6.5 nm (Zn,Cd)Se. We estimate the conduction (valence) band offsets to be roughly 20 meV (100 meV) and 150 meV (150 meV) for the ZnSe QW and the (Zn,Cd)Se QW, respectively.

The spectrum shows 4 distinct features:

(a) A broad, orange (E_{PL} ~ 2.1 eV) emission from the Mn^{2+} intraionic $^4T_1 \rightarrow {}^6A_1$ transitions.
(b) A strong emission in the vicinity of 2.6 eV from the (Zn,Cd)Se QW;
(c) A strong emission in the vicinity of 2.8 eV from the ZnSe QW;
(d) A weak near band edge PL from the quaternary epilayer at 2.9 eV

The strong PL from the QWs remains quite efficient up to room temperature but also shows a significant inhomogeneous broadening of ~20 meV and ~ 40 meV for the ZnSe and (Zn,Cd)Se QWs, respectively. This should be compared to the typical 5-10 meV PL linewidths obtained for similar QWs in a ZnSe host lattice. The presence of low-energy tails in the PL spectra are indicative of exciton localization in fluctuations of the confining potential due to interdiffusion at the well interfaces.

Variation in PL spectra with temperature provides additional information about these structures. Studies have been performed on several samples at temperatures up to 300K. In order for this material to be useful as a confining region for optoelectronic applications, it must continue to act as a barrier to the optically active region at high temperatures. As shown in Figure 4, the increased rate of bandgap closure

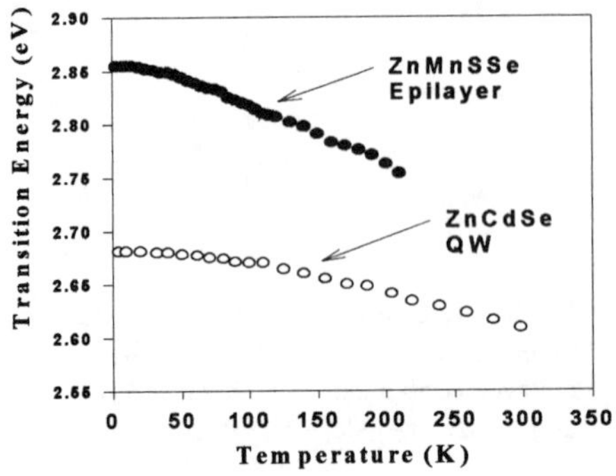

Fig 4. Temperature dependence of the energy position of the PL from a (Zn,Mn)(S,Se) epilayer containing a 8 nm (Zn,Cd)Se QW. Note the differing rates of band gap closure with increasing temperature..

with temperature due to the *sp-d* exchange interactions results in a barrier band gap that decreases with temperature more rapidly than that of the quantum well material. The consequent decrease in quantum confinement creates a rapid redshift of the luminescence and increased thermalization of the well excitons. Despite the reduced confinement, the PL from these samples is quite strong, even up to room temperature. The PL from the shallower ZnSe quantum wells, however, does tend to vanish at higher temperatures (~200K) as the excitons are thermalized from the well. At room temperature, the remaining signal is from the barrier and the deeper (Zn,Cd)Se quantum well, showing that as long as the barrier is high enough, (Zn,Mn)(S,Se) performs quite effectively as a confining material

Acknowledgements

This work was supported by the Army Research Office under grants DAAH04-94-G-0394 and DAAH04-95-I-0202.

References

1. D. D. Awschalom and N. Samarth, in *Optics of Semiconductor Nanostructures*, eds. F. Hennenberger, S. Schmitt-Rink and E. O. Gobel (Akademie Verlag, Berlin, 1993) and references therein.
2. N. Dai *et al.* Phys. Rev. Lett. **67**, 3824 (1991); W. C. Chou *et al.*ibid. 3820 (1991).
3. S. A. Crooker *et al.*, Phys. Rev. Lett. **75**, 508 (1995); S. A. Crooker, J. J. Baumberg, F. Flack, N. Samarth and D. D. Awschalom, Phys. Rev. Lett. **77**, 2814 (1996).
4. A review of early work on MS heterostructures may be found in R. L. Gunshor, L. A. Kolodziejski, A. V. Nurmikko and N. Otsuka, Ann. Rev. Mat. Sci. **18**, 325 (1988).
5. G. F. Neumark, R. M. Park and J. M. DePuydt, Physics Today **47**, 26 (1994).
6. T. Karasawa, *et al.* J. Crystal Growth **32**, No. 11B, L1657 (1993).
7. Y. P. Chen *et al.* in *Opto-electronic Materials: Ordering, Composition Modulation and Self-assembled Structures,* eds. E. D. Jones, A. Mascarenhas and P. M. Petroff (MRS Symposia Proceedings, Vol 417, 1996), p. 337.
8. M. Pessa *et. al.* Phys. Stat. Sol. (b) **187**, 337 (1995);
9. J. W. Hutchins *et al.* J. Crystal Growth **159**, 50 (1996).
10. C. C. Kim and S. Sivananthan, Phys. Rev. B **53**, 1475 (1996).
11. R. B. Bylsma *et al.*, Phys. Rev. B **33**, 8207 (1986).
12. Y. P. Varshni, Physica **34**, 149 (1967).

The Fabrication of Semiconductor Nanostructure Arrays on a Silicon Substrate Using an Anodized Aluminum Template

S.P. MCGINNIS *, J.N. CLEARY **, B. DAS*
** Department of Electrical and Computer Engineering*
*** Department of Chemical Engineering*
West Virginia University, P.O. Box 6104, Morgantown, WV 26506

ABSTRACT

We have developed a nanogrowth technology for the fabrication of periodic arrays of semiconductor nanostructures on silicon that is currently being investigated for silicon based x-ray detectors. The semiconductor nanostructures are formed by chemical synthesis in pores of a template created by the anodization of aluminum on a silicon substrate. The use of the silicon substrate allows greater control over the aluminum thin film properties, better *in situ* monitoring of the pore formation process, and the direct integration of nanostructure arrays with conventional silicon technology.

INTRODUCTION

Semiconductor nanostructure arrays are of great scientific and technical interest due to the strong non-linear and electro-optic effects that occur due to carrier confinement in three dimensions [1-3]. Room temperature operation of these arrays requires that the energy separation due to quantization be greater than that due to thermal broadening. This restricts the size of the quantum dot to less than 20 nm for most semiconductor materials. Traditional nanofabrication technologies are not suitable for the fabrication of QD arrays of this size due to lithographic limitations, fabrication induced damage and prohibitive cost [4]. This has led to the development of a number of nanogrowth techniques, where the semiconductor material is synthesized in the size and shape of a quantum dot. However, most techniques lack the desired control over both the quantum dot size and the array periodicity [5].

We have reported earlier on the use of a template to guide the nanogrowth process as a means to control both the QD size distribution and the array periodicity [6,7]. The template is formed by the anodization of aluminum, which forms a two dimensional hexagonal lacework of Al_2O_3 cells with uniform tubular pores. This structure was first characterized by Keller, Hunter and Robinson [8] as a close-packed array of columnar hexagonal cells each containing a central pore normal to the substrate surface. The pore diameter and density depend on the type, pH and temperature of the acid and the anodizing current density. By varying these parameters, the pore diameter can be precisely controlled over a range of 5-100 nm in diameter with a pore density of 10^{10} -10^{11} pores/cm^2.

The properties of porous anodized aluminum have recently been used for the fabrication of metallic nanowires [9], and the deposition of Fe nanoparticles [10], on aluminum substrates. However, since silicon is the dominant material for integrated circuit

Mat. Res. Soc. Symp. Proc. Vol. 452 © 1997 Materials Research Society

(IC) technology, it is desirable to fabricate the nanostructure arrays on a silicon substrate to allow for optoelectronic integration with silicon. In addition, fabrication on a silicon substrate allows the direct use of the well-developed silicon IC process technology.

This paper describes the fabrication process for the formation of nanostructure arrays on a silicon substrate, the results of fabrication, and the use of this technology towards a silicon based x-ray detector for digital radiography applications.

FABRICATION AND RESULTS

The semiconductor nanostructure arrays are fabricated on a p-type silicon wafer with 0.3 Ω-cm resistivity. The substrate is first cleaned using a standard silicon cleaning process. Next, a 1 μm Al film is sputter deposited on the back side of the wafer and annealed at 450° C for 30 minutes. Then, a 500 nm 99.999% pure aluminum film is sputter deposited on the polished side of the wafer. This is annealed at 400° C for 30 minutes in a nitrogen atmosphere to ensure good adhesion of aluminum to the substrate. The top layer was then anodized in 15% sulfuric acid for 1 minute at a constant current ranging between 10 to 40 mA/cm^2. A schematic of the anodization apparatus is shown in Figure 1.

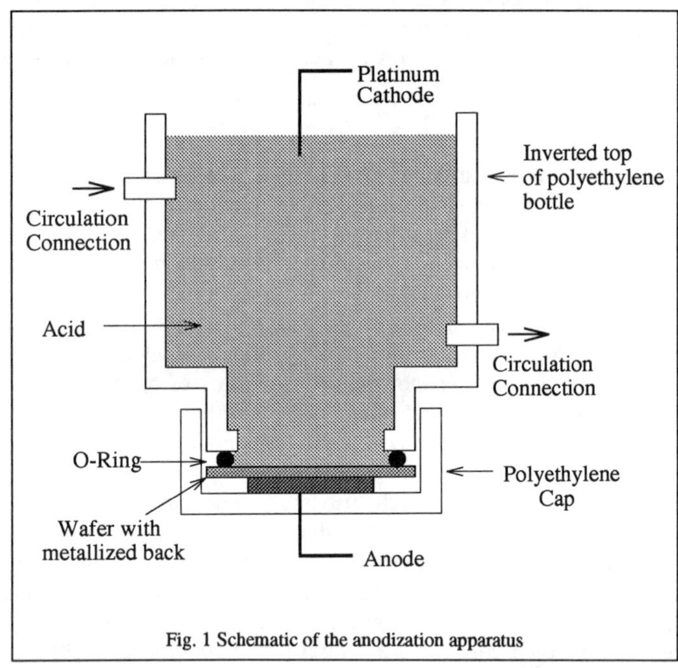

Fig. 1 Schematic of the anodization apparatus

A sample potential-time plot for a 500 nm aluminum film during anodization is shown in Figure 2, which can be correlated with the pore formation mechanism in the aluminum film. Initially, a thin non-porous barrier layer of aluminum oxide is formed on the top of the aluminum film, which corresponds with the initial increase in potential. As anodization continues, an array of pores develops on the barrier layer, which grow in diameter until reaching a dimension determined by the anodization conditions alone. During pore diameter growth, the potential decreases. This is followed by a pore propagation period during which the fixed diameter pores propagate vertically into the aluminum thin film. During this period the potential remains constant. Finally, when the pores contact the silicon surface, a rapid increase in potential with complex features is observed. These features are believed to result from the presence of multiple alloy layers at the silicon/aluminum interface.

Since the thickness of the aluminum film is accurately known, and the time from the beginning to the end of pore formation can be determined from the potential profile, the pore formation rate can be precisely determined in this system. This is illustrated in Figure 3 where the pore formation rate is plotted as a function of anodization current density. Therefore both the pore diameter, which is solely determined by the anodization conditions, and the pore depth, which can be determined from the linear pore formation rate, can be precisely controlled.

The semiconductor quantum dot array is fabricated based on a unique property of AC anodization that sulfide (S^{2-}) ions are deposited within the pores during the cathode half-cycle. Since sulfide ions are extremely reactive, it is possible to fabricate sulfide compound semiconductors (i.e., ZnS, CdS, etc.) by supplying the appropriate cation (i.e., Zn, Cd, etc.). This is accomplished by immersing the anodized substrates in a solution of cadmium chloride (for CdS), or zinc chloride (for ZnS) and boiling for 10 minutes. The substrates are then removed, rinsed in DI water, and blown dry in nitrogen.

Preliminary energy dispersive x-ray analysis data for the synthesis of ZnS and CdS are shown in Figure 4. This data shows the presence of elemental sulfur as well as elemental

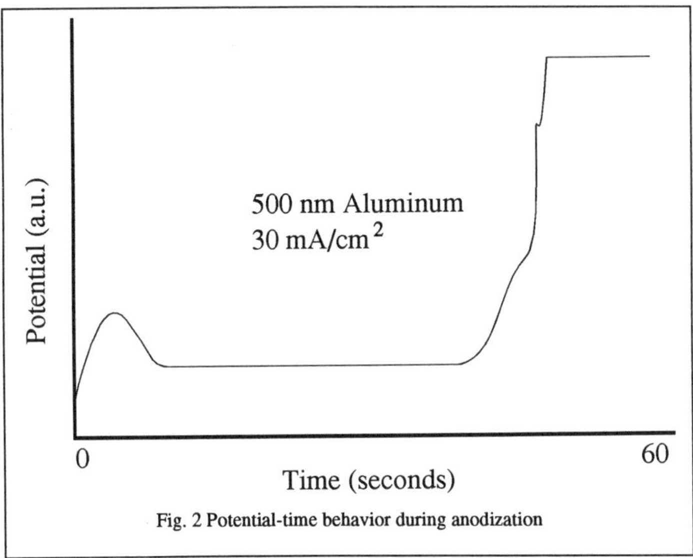

Fig. 2 Potential-time behavior during anodization

Zn and Cd in the appropriate samples. The presence of copper is due to impurities in the salt solutions. More detailed analysis is needed to determine the ionic states of these materials as well as their structures. It is important to point out here that the crystal structures of these materials may be difficult to control, thus these materials may not be suitable for applications where crystalline quality is important (e.g., laser diodes). However these materials will find wide applications as optical modulators and detectors.

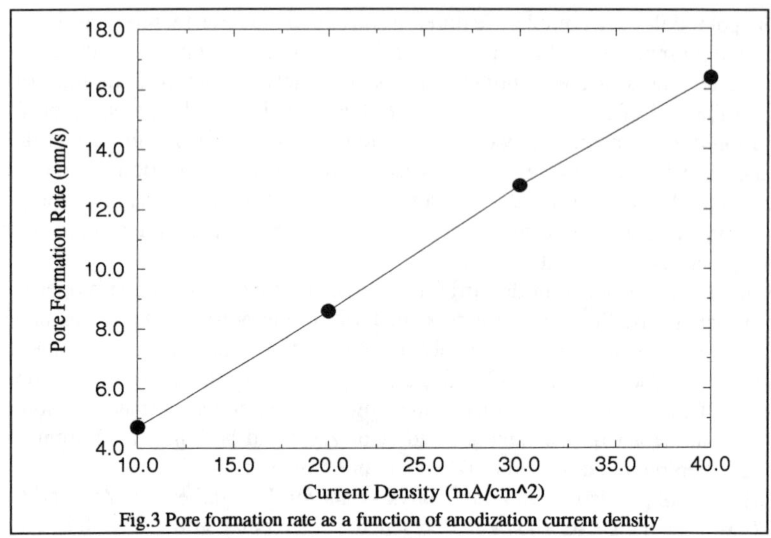

Fig.3 Pore formation rate as a function of anodization current density

After the semiconductor quantum dots have been synthesized, they are encapsulated in aluminum hydroxide. Aluminum hydroxide is optically transparent, electrically insulating, and corrosion resistant. The encapsulation step is accomplished by boiling the substrate in DI water for 30 minutes.

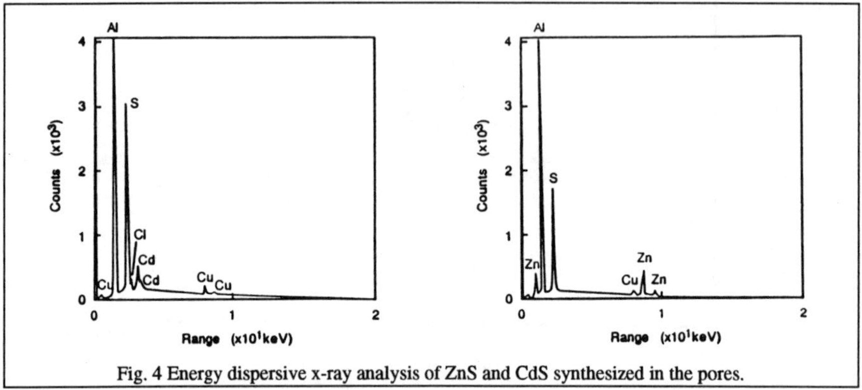

Fig. 4 Energy dispersive x-ray analysis of ZnS and CdS synthesized in the pores.

APPLICATION

The development of new medical imaging techniques such as digital radiography requires the development of sensitive x-ray detectors that can be easily integrated with silicon technology [11]. Current scintillation detectors used for this purpose such as bulk silicon, CdS, and ZnS have a low detection sensitivity primarily due to the relatively inefficient recombination of generated electrons and holes. However, photoluminescence studies of Mn^{2+} doped ZnS nanocrystals have shown a substantial increase in both the speed and efficiency of luminescent recombination compared to the bulk material [12]. Since there is a strong correlation between the photoluminescence and scintillation characteristics of a material [13], this indicates that doped semiconductor nanostructures have the potential to be used as very fast and efficient scintillators.

The nanostructure arrays discussed here are currently being investigated for use as scintillators for a silicon based x-ray detector for digital radiography applications. This technology is attractive since it allows direct integration with silicon based optical detectors and allows precise control over the size of the QDs and flexibility in their composition. Since the scintillation properties of the nanostructure arrays are strongly dependent upon the size and composition of the QDs, this allows the ability to engineer an optimum scintillator for silicon digital radiography.

CONCLUSIONS

A technology for the fabrication of semiconductor nanostructure arrays on a silicon substrate was presented. This technology allows precise control over the size of the QDs in the array, and flexibility in their composition. Due to these characteristics, this technology is being investigated as a means to fabricate high efficiency scintillators for silicon based medical imaging.

ACKNOWLEDGMENTS

This work was supported by the National Science Foundation's grant #ECS-9521729.

REFERENCES

1. D.A.B. Miller, D.S. Chemla, T.C. Damen, T.H. Wood, C.A. Burrus, A.C. Gossard and W. Wigemann, IEEE J. Quantum Electronics, **QE-21**, 1462 (1985).
2. T. Takagahara, Surface Science, **267**, 310 (1992).
3. H. Rossmann, A. Schulzgen, F. Henneberger and M. Muller, Phys. Stat. Sol. B **159**, 287 (1990).
4. L. Brus, J. Phys. Chem., **90**, 2555 (1986).
5. L. Brus and N. Peyghambarian, OE Reports (1993).
6. B. Das, G. Banerjee and A.E. Miller, Electronic Materials Conference, Boulder, CO, June 22-24 (1994).
7. S.P. McGinnis and B. Das, IEEE Cornell Conference, Ithaca, NY, August (1995).
8. F. Keller, M.S. Hunter, and D.L. Robinson, J. Electrochem. Soc., **100**, 411 (1953).
9. D. AlMawlawi, N. Coombs and M. Moskovits, J. Appl. Phys., **70**, 4421 (1991).
10. D. AlMawlawi, C.Z. Liu and M. Moskovits, J. Matter. Res., **9**, 1014 (1994).
11. L.E. Antonuk, Photonics Spectra, June, 108 (1995).
12. R.N. Bhargava, D. Gallagher, X. Hong, and A. Nurmikko, Phys. Rev. Lett., **72**, 416 (1994).
13. E. Fenyves and O. Haiman, The Physical Principles of Nuclear Radiation Measurements, (Academic Press, New York, 1969), p. 144.

EFFECTS OF Mn^{2+} DISTRIBUTION CONTROLLED BY CARBOXYLIC ACIDS ON PHOTOLUMINESCENCE INTENSITY OF NANOSIZED ZnS:Mn PARTICLES

T. ISOBE, T. IGARASHI, M. SENNA
Department of Applied Chemistry, Keio University, 3-14-1 Hiyoshi, Kohoku-ku, Yokohama 223, Japan. isobe@applc.keio.ac.jp

ABSTRACT

Addition of methacrylic acid (MA) during preparation of ZnS doped with Mn^{2+} (ZnS:Mn) increased the photoluminescence (PL) due to $^4T_1$-$^6A_1$ transition of Mn^{2+}. According to X-ray fluorescence analysis and electron paramagnetic resonance spectroscopy, ion exchange between Zn^{2+} and Mn^{2+} through a preferential dissolution of Mn^{2+} was promoted by acidic additives. This caused that Mn^{2+} ions were isolatedly incorporated into ZnS. The X-ray photoelectron spectra show that the intensity of S $2p_{3/2}$ peak due to S^{6+} increased relative to that of S^{2-} by virtue of carboxylic groups. The intensities of PL peaks at 450 and 580 nm, corresponding to polymethacrylic acid and Mn^{2+}, respectively, increased after heating at 80°C for 1 week. We conclude that MA plays important roles on selective leaching to increase the amount of isolated Mn^{2+} ions, chemical interaction between ZnS:Mn and MA, and energy transfer to Mn^{2+}, leading to the increase in PL intensity.

INTRODUCTION

The PL intensity of a nanosize phosphor is generally believed to be lower than that of a micronsize phosphor, since the increase in the ratio of surface to volume results in the increase in defects, i.e., PL killers, localized near surface. Bhargava et al., however, suggests that quantum efficiency increases with decreasing the particle size less than 10 nm [1].

ZnS:Mn is well-known to be the orange phosphor. Mn^{2+} cannot easily be incorporated into ZnS by a conventional chemical process, such as coprecipitation. ZnS:Mn is, therefore, prepared by thermal diffusion of Mn^{2+} into ZnS, where nanosize particles cannot be produced because of the thermal growth of particles. We proposed the chemical process to prepare 2~3nm ZnS:Mn powdered sample by adding MA [2]. The PL intensity of this sample was 5 times higher than that of a ca. 0.5 μm commercial sample. The purpose of this study is to elucidate the roles of MA on preparation and PL processes of ZnS:Mn.

EXPERIMENT

Preparation of ZnS:Mn Particles

Two methanol solutions, i.e., 150 cm^3 of 0.133 mol dm^{-3} zinc acetate and 25 cm^3 of $8x10^{-3}$ mol dm^{-3} manganese acetate were mixed and then put into 200 cm^3 of 0.4 mol dm^{-3} sodium sulfide solution with a volume ratio, water/methal=1/1, to obtain a ZnS:Mn colloidal suspension. After stirring for 20 min, CH_2=C(CH_3)COOH (methacrylic acid, MA), CH_2=C(CH_3)$COOCH_3$ (methyl methacrylate, MMA) or HCl was added into the ZnS:Mn suspension obtained above, where MA and MMA were used without removal of the polymerization inhibitor. A precipitate separated by centrifugation at $8x10^3$ rpm was dried at 50°C for 24 h to obtain a ZnS:Mn powdered sample. Each sample name is given in table I.

Mat. Res. Soc. Symp. Proc. Vol. 452 © 1997 Materials Research Society

Table I List of sample names.

species	additive amount (mol)	volume (cm^3)	sample name
none	0	0	ZS
MA	0.12	10	ZSMA1
	0.35	30	ZSMA3
	0.58	50	ZSMA5
MMA	0.28	30	ZSMMA3
HCl[a)]	0.24	20	ZSHCl2

a) concentration: 12.1 M

Characterization of ZnS:Mn Particles

Contents of Mn and Zn in the supernatant solution during preparation were measured by X-ray fluorescence analysis (XFA) (Rigaku, system 3080E2). An Mn content in a ZnS:Mn sample was also measured by XFA, where 0.1 g of sample was dissolved in 10 cm^3 of 12.1 M HCl aqueous solution.

An Mn^{2+} distribution in a ZnS:Mn sample was examined by electron paramagnetic resonance (EPR) absorption spectroscopy (JEOL, JES-RE3X) [3,4]. A second derivative spectrum was obtained at 9.5 GHz, where the modulation frequency was 100 kHz, the modulation width 0.32 mT, and the microwave power 2.0 mW. Chemical states of ZnS:Mn and additives were examined by Fourier transform infrared (FT-IR) absorption spectroscopy (Bio-Rad, FTS175) and X-ray photoelectron spectroscopy (XPS, JEOL, JPS-90SX). A photoluminescence spectrum of the sample excited by 350 nm light was measured at intervals of 1 nm by spectrophotofluorometer (Shimadzu, RF-1500). The PL intensity of each sample is expressed as the percentage relative to the intensity of the sample ZSMA5 at 580 nm.

RESULTS

Change in the PL Intensity by Additives

Fig. 1 shows the PL spectra of the samples. The PL intensity at 580 nm for the sample ZSMA3 was about twice larger than that of the sample ZS. In contrast, the PL intensity of the sample ZSMMA3 was less than that of the sample ZS. On the other hand, the addition of the inorganic acid, HCl, increased PL intensity by a factor of ~1.5.

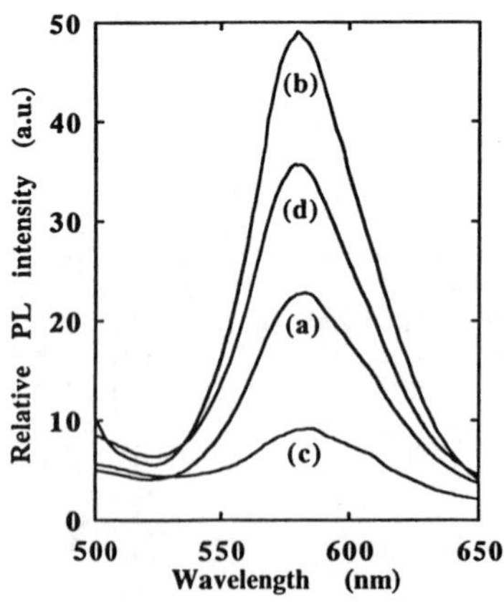

Fig. 1
PL spectra of the samples excited by 350 nm light. (a) ZS, (b) ZSMA3, (c) ZSMMA3, (d) ZSHCl2.

Contents and Distribution of Mn^{2+}

The sharp sextet hyperfine structure due to Mn^{2+} was observed in the EPR spectra of the samples ZSMA3 and ZSHCl2, as shown in curves (b) and (d) of fig. 2, respectively. In contrast, the resolution of the hyperfine structure was lower and the baseline tilted for the samples ZS and ZSMMA3, as shown in curves (a) and (c) of fig. 2, respectively. The large broad line represents the clustered Mn^{2+} ions with superexchange interaction [4]. The EPR results, therefore, indicate that Mn^{2+} ions of the samples ZSMA3 and ZSHCl2 were isolatedly incorporated into ZnS with the aid of acidic additives.

The change in the contents of Mn and Zn in the supernatant solution separated from ZnS:Mn particles is plotted as a function of the MA amount and is shown in fig. 3. The Mn content was 2~3 times higher than the Zn content, although ZnS:Mn was prepared at the Mn/Zn molar ratio, 1/100. Enhanced preferential dissolution of Mn by addition of MA reduced the Mn content in dry ZnS:Mn powders, as also shown in fig. 3.

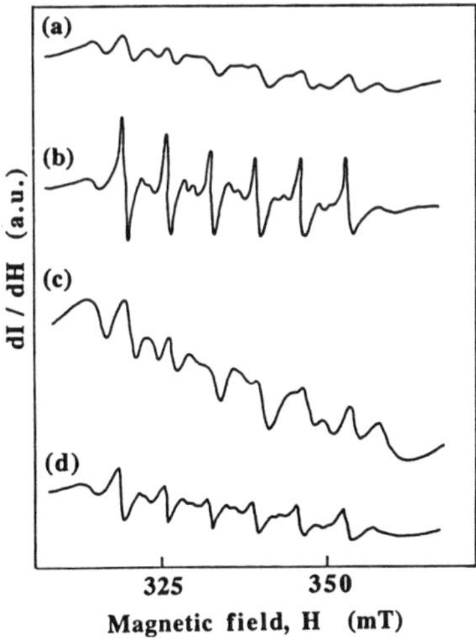

Fig. 2
EPR spectra of ZnS:Mn powders. (a) ZS, (b) ZSMA3, (c) ZSMMA3, (d) ZSHCl2. (I: intensity, H: magnetic field)

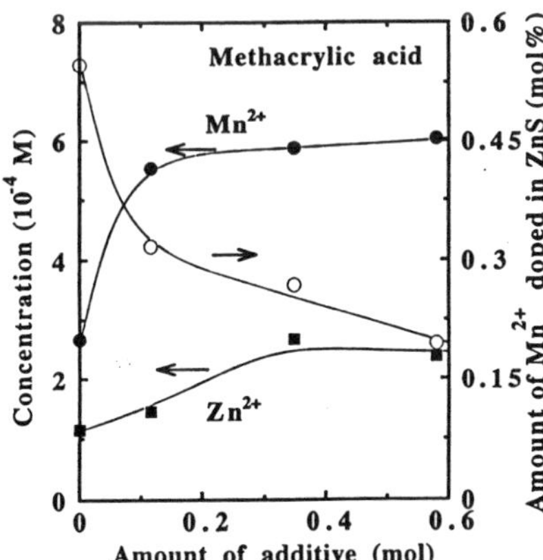

Fig. 3
Change in the contents of Mn and Zn in the supernatant solution and in ZnS:Mn powders with the MA amounts.

Furthermore, another sample was prepared by adding mixed colloidal suspensions of ZnS and MnS with 0.35 mol MA. The sharp sextet peaks due to Mn^{2+} were observed in the EPR spectrum of this sample, while not for the sample prepared from mixed suspensions of ZnS and MnS without MA, as shown in fig. 4. This indicates that Mn^{2+} ions were isolatedly incorporated in ZnS in the former.

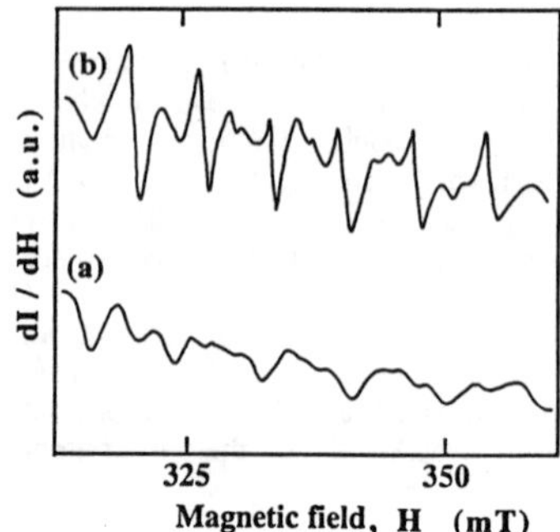

Fig. 4
EPR spectra of the powdered samples which were prepared by mixing colloidal suspensions of ZnS and MnS (a) without and (b) with 0.35 mol MA, and then dried at 50°C for 24 h.
(I: intensity, H: magnetic field)

Chemical States of Methacrylic Acids Attached on Particles

According to the thermal analysis, the weight loss, accompanied by exothermic DTA peak at around 450°C, was observed by heating the samples prepared by adding MA. This indicates that the organic species remained adsorbed on ZnS:Mn particles even after drying at 50°C for 24 h.

Fig. 5 shows the XPS spectra of S $2p_{3/2}$ for the samples ZS and ZSMA5. Two peaks were observed at 162 and 168.5 eV, respectively. The former is assigned to S^{2-}, the latter to S^{6+} on the basis of the report [5]. The relative intensity at 168.5 eV to 162 eV was higher for the sample ZSMA5 rather than for the sample ZS. This relative intensity of the sample ZSMA5 decreased after etching by Ar^{+} for 10 sec. A new IR absorption band due to the stretching vibration of S-O bond was observed at 805 cm^{-1} in the spectrum of the sample ZSMA5 (see fig. 6).

Effects of Methacrylic Acid on PL Intensity

In order to elucidate the role of MA on PL intensity, we attempted to remove MA of the sample ZSMA1 by Soxhlet's extractor in boiled methanol for 24 h. Removal of MA from ZnS:Mn particles was confirmed by IR spectroscopy. As a result of Soxhlet's extraction, the PL intensity decreased by ca. 65%, although the Mn content decreased by only ca. 20%.

When the samples ZSMA5 was additionally heated in the closed vessel at 80°C for 1 week, the PL peaks at 580 and 450 nm increased, as shown in fig. 7. The PL peak at 450 nm was not observed in the PL spectrum of the samples ZSHCl2. Fig. 8 shows the change in the PL spectra of ZSMA5, heated at 80°C for 1 week, with a nominal Mn content which represents the molar percent of Mn/(Mn+Zn) in the mixed methanol solution at the stage of preparation. With increasing the Mn content, the PL peak at 580 nm increased and the peak at 450 nm decreased, although the organic content, ca. 45 wt%, was kept constant for all these samples.

DISCUSSION

Homogenization of Mn^{2+} Distribution in ZnS:Mn by Carboxylic Acid

The dissolution of Zn and Mn occurs by organic or inorganic acids, such as methacrylic acid and HCl. The elemental analysis of supernatant solution reveals that Mn^{2+} is selectively leached by methacrylic acid. Subsequently, ion exchange between Zn^{2+} and Mn^{2+} is promoted through repeated dissolution and precipitation. This causes incorporation and isolation of Mn^{2+} in ZnS and enhances homogenization of Mn^{2+} distribution, as confirmed by sharp sextet EPR peaks due to Mn^{2+}. This isolation of Mn^{2+} ions, in turn, results in the increase in the PL intensity at 580

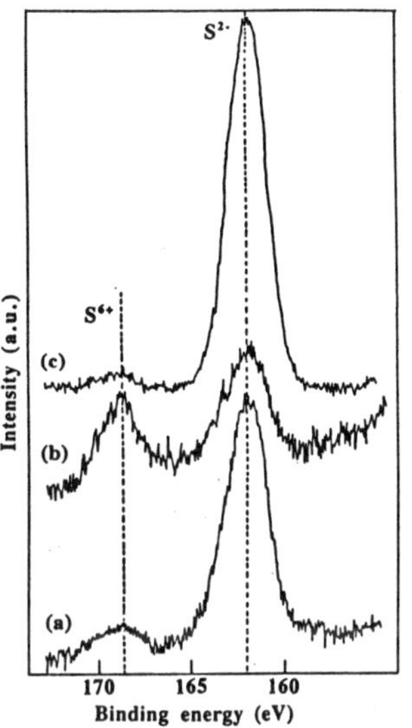

Fig. 5
XPS spectra of ZnS:Mn powders. (a) ZS, (b) ZSMA5 , (c) ZSMA5 etched by Ar^+ for 10 sec.

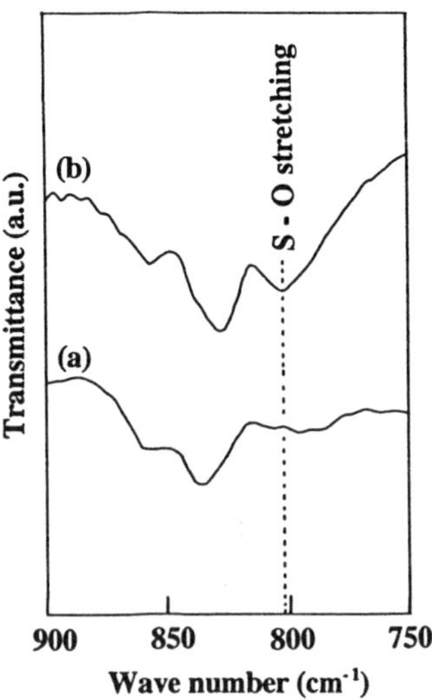

Fig. 6
FT-IR absorption spectra of ZnS:Mn powders. (a) ZS, (b) ZSMA5.

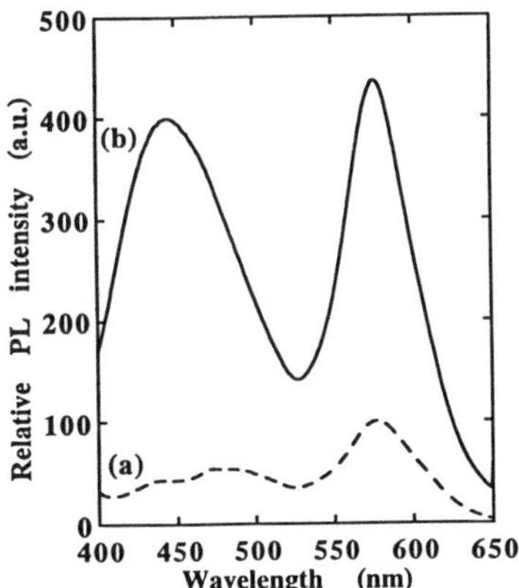

Fig. 7
PL spectra of the sample ZSMA5 (a) without and (b) with heating at 80°C for 1 week.

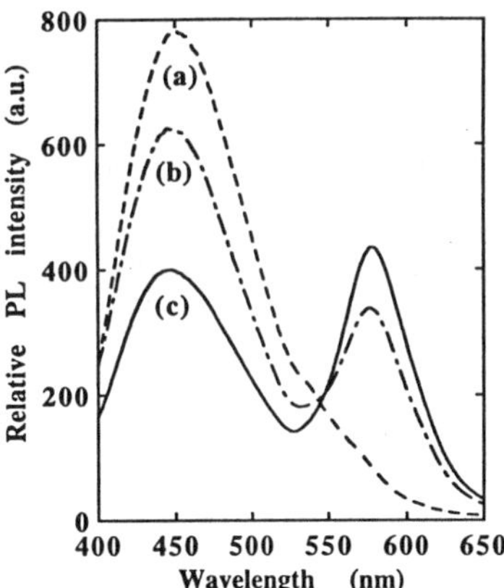

Fig. 8
Change in PL spectra of the ZnS:Mn powders with the nominal Mn contents. (a) 0 mol%, (b) 0.4 mol%, (c) 1.0 mol%. All ZnS:Mn powders with different Mn contents were prepared by adding 0.58 mol MA, and heated at 80°C for 1 week.

nm, since the increased amount of isolated Mn^{2+} detectable by EPR is known to increase the PL intensity [3]. By comparison of methacrylic acid and methyl methacrylate, it is also verified that a carboxylic group plays a significant role on homogenization of Mn^{2+} distribution in ZnS:Mn via the above-mentioned mechanism.

Effects of Carboxylic Acid on Photoluminescence

The existence of MA additive on ZnS:Mn particles was confirmed by TG-DTA and XPS. Furthermore, chemical interaction between sulfur of ZnS:Mn and oxygen of carboxylic acid in MA was recognized by the appearance of XPS peak due to S^{6+} and IR absorption due to S-O vibration only for the ZnS:Mn sample with MA. Therefore, MA additive serves as passivating of surface of nanosize particles and quantum confinement effects by coating with organic species, as pointed out in refs. [1,6].

Additional heating at 80°C for 1 week promotes polymerization of MA adsorbed on ZnS:Mn particles, leading to the increase in the PL peak at 450 nm due to polymethacrylic acid. At the same time, the PL intensity at 580 nm increased. The PL intensity at 450 nm for the samples with a constant amount of MA decreased with increasing a nominal Mn content. It is strongly suspected that energy transfer occurs from MA to ZnS:Mn. Further work on optical properties, involving quantum confinement and energy transfer mechanism, will be done in near future.

CONCLUSIONS

Mn^{2+} ions was isolatedly incorporated into ZnS particles of 2~3 nm through ion exchange between Zn^{2+} and Mn^{2+}, as confirmed by EPR. This kind of homogenization of Mn^{2+} distribution, being one of the important keys to increase the PL intensity of nanosize ZnS:Mn particles, was achieved by adding a carboxylic acid, e.g., methacrylic acid, during preparation from methanol solutions of Zn and Mn.

ACKNOWLEDGMENTS

Authors thank Prof. H. Hirashima and Dr. H. Imai for helping PL measurement and Dr. K. Fujimoto for valuable discussion.

REFERENCES

[1] R.N. Bhargava, D. Gallagher and T. Welker, J. Lumin. **60&61**, 275 (1994).
[2] I. Yu, T. Isobe and M. Senna, J. Phys. Chem. Solids **54**, 373 (1996).
[3] I. Yu, T. Isobe, M. Senna and S. Takahashi, Mater. Sci. Eng. **B38**, 177 (1996).
[4] I. Yu and M. Senna, Appl. Phys. Lett. **66**, 424 (1995).
[5] P. Baláz, Z. Bastl, J. Briancin, I. Ebert and J. Lipka, J. Mater. Sci. **27**, 653 (1992).
[6] D. Gallagher, W.E. Heady, J.M. Racz and R.N. Bhargava, J. Mater. Res. **10**, 870 (1995).

SYNTHESIS OF Mn^{2+} DOPED CdS NANOCRYSTALS EMBEDDED IN A SOL-GEL SILICA MATRIX: CHARACTERIZATION OF THE LUMINESCENCE ACTIVATOR

G. COUNIO, S. ESNOUF, T. GACOIN, P. BARBOUX, A. HOFSTAETTER*, J.-P. BOILOT
Laboratoire de Physique de la Matière Condensée, C.N.R.S. U.R.A.1254D, Ecole Polytechnique, 91128 Palaiseau, France.
*1. Physics Institute, University of Giessen, D 35392 Giessen, Germany.

ABSTRACT

Mn^{2+}-doped CdS nanocrystals (1.2 to 2.4 nm in diameter) embedded in organic-inorganic silica xerogels have been synthesized. Extensive studies (EXAFS, ESR and ENDOR) allow us to localize the ions responsible for the bright luminescence observed in such materials (quantum yield of 7%). The average number of Mn^{2+} per nanocrystal is in the 0.2-0.8 range, and the emission arises from an energy transfer from surface trapped carriers to Mn^{2+} ions.

INTRODUCTION

The progressive transition from bulk to molecular-like behavior shown by nanometer sized semiconductor nanocrystals has been extensively investigated and is now well understood in terms of quantum confinement [1]. These studies have singled out the major importance of the surface of such particles, which cannot be considered as a simple excised fragment of the bulk compound. The incorporation of Mn^{2+} ions as dopants in sulfides is of special interest since doped nanocrystals show a bright photoluminescence due to the $^4T_1 \rightarrow {}^6A_1$ Mn^{2+} transition. The first part of this paper briefly describes the synthesis of CdS:Mn/silica nanocomposites and the second part is dedicated to the characterization of the Mn^{2+} ions in terms of concentration and spatial location inside the nanoparticles. This information leads us to give an interpretation of the recombination process responsible for this high luminescence efficiency.

EXPERIMENTAL

The process we used for the synthesis of CdS:Mn/silica nanocomposites follows two general steps: capped nanocrystals are first prepared as a powder, and then dispersed in a solution of hydrolyzed alkoxides whose condensation will give the final matrix. Nanometer sized nanoparticles with a narrow size distribution are obtained within inverted micelles from the AOT/water/heptane system, where AOT stands for the di-2 ethylhexyl sulfosuccinate surfactant molecule; [AOT] = 0.5 $mol.L^{-1}$ and $[H_2O]$ = 2.5 $mol.L^{-1}$. Mixing a solution containing metallic ions ($[Cd(NO_3)_2]$ = 0.20 $mol.L^{-1}$ and $[Mn(NO_3)_2]$ = 0.12 $mol.L^{-1}$) and one containing sulfur ions ($[Na_2S]$ = 0.38 $mol.L^{-1}$) dissolved in the water pools yields a colloidal suspension of Mn^{2+} doped CdS nanocrystals. Then, the chemical grafting of thiophenol at the surface of the particles leads to the flocculation of the CdS:Mn colloid [2]. The precipitate is separated by centrifugation and washed with petroleum ether to eliminate residual AOT. The nanocrystals are then washed in an aqueous solution acidified with HCl down to pH 2.5 to dissolve MnS like by-products of the synthesis. This process leads to a pure thiophenol capped CdS:Mn nanoparticle powder. It can

Mat. Res. Soc. Symp. Proc. Vol. 452 © 1997 Materials Research Society

be dispersed in pyridine to give a transparent yellow orange colloid.

The size of the nanocrystals is determined by comparing the shift of the UV-visible absorption band with the correlation proposed by Wang and Herron [3]. The mean size of the particles is found to fluctuate between 1.2 and 2.4 nm, if we assume that this correlation is not affected by the presence of Mn^{2+} dopants. This approximation is justified by the low dopant concentration we measure.

The nanocrystals are incorporated in a transparent organic-inorganic silica matrix synthesized using the sol-gel chemistry [4]: CdS:Mn particles are dispersed in a solution of hydrolyzed methyltriethoxysilane $CH_3Si(OEt)_3$. The presence of a non hydrolyzable alkyl group provides matrices with good optical and mechanical qualities, and improves the stability of the nanocrystals towards photodegradation. After the sol-gel transition, the careful drying of the gel at ambient temperature gives the final material either as transparent disks of a few centimeters in diameter or as thin films previously deposited onto various substrates (glass, ITO ...).

CHARACTERIZATION OF CdS:Mn NANOPARTICLES

EXAFS, ESR and ENDOR studies were performed on powders of pure capped nanoparticles to determine the Mn^{2+} concentration and also to probe the environment of the manganese ions inside the nanocrystals.

EXAFS study of manganese ions in CdS:Mn nanocrystals

This experiment was conducted using the DCI ring at the LURE synchrotron radiation facility (Orsay, France). Low dopant concentration forbids simple absorption measurements, and EXAFS spectra have to be measured as X-ray fluorescence. Data were acquired on the EXAFS III spectrometer equipped with a Si (111) double crystal monochromator. Spectra were recorded between 6500 and 7200 eV with 2 eV steps. The EXAFS oscillations were extracted using a conventional software treatment [5]. The Fourier transform of these oscillations yielded the pseudo-radial distribution function, whose simulation (Fig. 1) allows us to identify the manganese in a tetrahedral site, surrounded by four sulfur atoms. This environment is consistent with high resolution transmission electron microscopy studies which have shown that the CdS:Mn nanocrystals have the same zinc blende structure as their undoped counterparts [11].

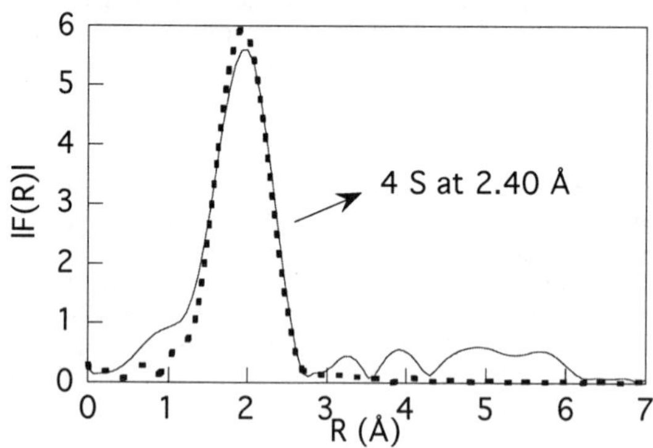

Fig. 1: Pseudo-radial Distribution Function (straight line) and its simulation (dotted line) from an EXAFS spectrum of CdS:Mn nanocrystals.

ESR characterization of CdS:Mn nanocrystals

X-band ESR spectra between 4.2 and 300K were collected on a BRUKER ER200D instrument equipped with an OXFORD continuous flow cryostat. Mn^{2+} concentrations were measured by comparing the intensities of the signals to a standard sample of copper sulfate pentahydrate powder. The number of particles per gram was determined by careful drying and weighing of a colloidal suspension of known optical density. It was checked that no significant change is observed in the ESR spectra for nanoparticles assembled as a powder or dispersed in a solvent or a gel matrix, and for microwave frequencies of 9 and 35GHz.

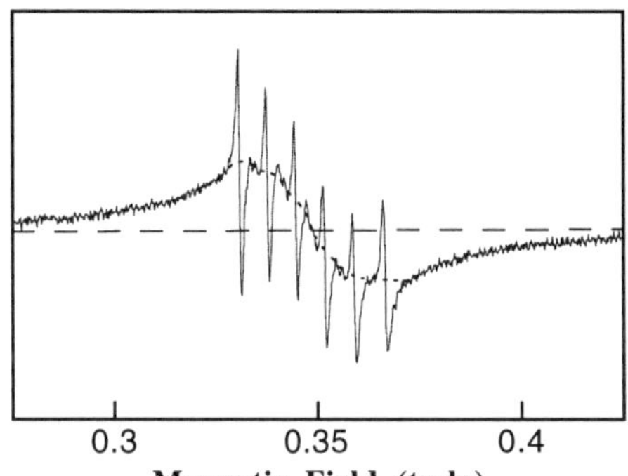

Fig. 2: ESR spectrum of CdS:Mn particles
Microwave frequency ν=9.43 GHz, microwave power P=10mW. Spectrum recorded at 25K.

The theory of ESR of Mn^{2+} in cubic hosts such as zinc blende has already been largely developed [6]. In this case, the hyperfine interaction reduces itself to its isotropic part, and the spectrum is split into 6 evenly spaced lines. Moreover, since the Mn^{2+} ion is in the ^6S configuration which has the g parameter of free electrons, broadening due to crystal field distributions can only be measured through hyperfine interactions.

A detailed analysis of the ESR spectra taken at 25K (Fig. 2) allows us to distinguish two contributions whose intensities follow a Curie law with temperature:

- signal (I): This is composed of 6 sharp and intense lines and of intermediate weaker bands, which correspond respectively to allowed (Δm_I=0) and forbidden (Δm_I=±1) hyperfine transitions between the Zeeman sublevels (m_s=±1/2). The nuclear hyperfine splitting can be directly determined (198 MHz) and is characteristic of Mn^{2+} ions in a tetrahedral crystal field. The Mn^{2+} ions contributing to this signal represent about 0.01 ion per nanoparticle.

We notice that Mn^{2+} ions do not occupy purely cubic centers, since in such centers the probability of forbidden transitions vanishes. The similarity between this spectrum and those observed in strained single crystals suggests disorder in the nanocrystals. This disorder is not healed by annealing at 230°C for 1 hour and arises from the finite size of the crystallites: the nanocrystals are so small that the cubic symmetry is always broken.

- signal (II): This is a broad signal whose broadening (1000 MHz) is inhomogeneous. Depending on the sample, the Mn^{2+} ions contributing to this signal represent between 0.4 and 0.8 ion per nanoparticle. Therefore, it is representative of the majority of the Mn^{2+} ions. EXAFS results allows us to interpret this signal as ESR transitions corresponding to Mn^{2+} ions occupying tetrahedral centers in a disordered material.

In disordered materials, non central transitions are always unresolved due to a broad distribution of zero-field splitting parameters. Therefore, this signal is an envelope of central transitions. Dipolar interactions and crystal field distribution are the two usual origins for the broadening of the ESR signal of transition metal ions [6,7]. However, the manganese concentrations measured and the similarity between ESR spectra performed on nanocrystal powders and on diluted colloidal suspensions prevent dipolar couplings. Since the Q band and the X band spectra are similar, we suggest that the main source of broadening for the signal (II) is a distribution of hyperfine interactions. This distribution may arise from isolated Mn^{2+} ions located inside the nanocrystal but close enough to its surface to affect their electronic density. If the Mn^{2+} ions are located in the two or three last atomic layers of the nanoparticle, the polarity of the Mn-S bound, and consequently the hyperfine constant, undergoes large fluctuations which are responsible for the broadening of signal (II).

ENDOR study of CdS:Mn nanocrystals

ENDOR spectra were collected at 4.2K on a BRUKER ESP300E instrument. Our aim was to specify which nuclei are interacting with the two ESR signals we have previously identified. To provide us with a local surface probe, 4-fluorothiophenol was used in place of thiophenol as a surface capping molecule. Two different ENDOR spectra were recorded (Fig. 3–a and 3-b): We chose to saturate the ESR transition corresponding to $m_I=1/2$ (line A), therefore probing both signals (I) and (II), and its high field shoulder (line B), thus only probing signal (II).

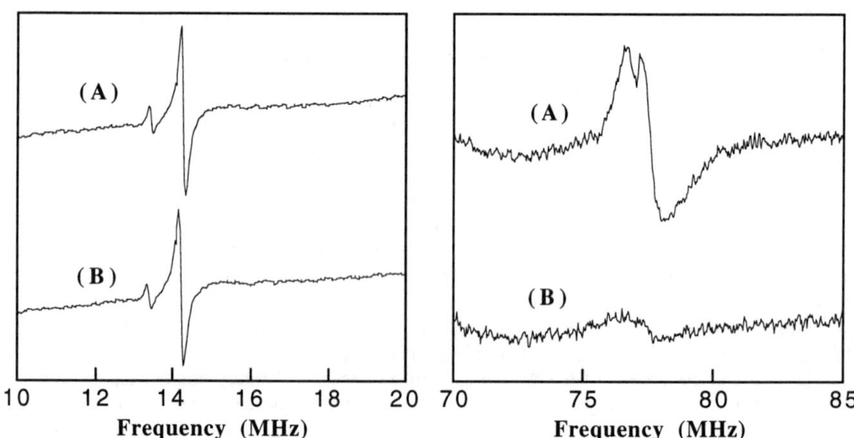

Fig. 3-a: 10 - 20 MHz range of the ENDOR spectra of CdS:Mn nanocrystals.

Fig. 3-b: 70 - 85 MHz range of the ENDOR spectra of CdS:Mn nanocrystals.

With a magnetic field of about 0.35 T, two different ranges of the spectra are of particular interest: below 20 MHz and between 70 and 120 MHz. In the first range (Fig. 3-a), we observe signals centered around the magnetic resonance of ^1H and ^{19}F (respectively 13.3 and 14 MHz). They are related to a distant ENDOR effect [8]: the interacting nuclei are located in the vicinity of the paramagnetic electrons, but too far to split the ENDOR lines with a hyperfine dipolar coupling. This excludes Mn^{2+} ions linked to thiophenol grafted at the surface of the nanocrystals.

In the second range, we observe the ENDOR spectrum of the manganese. It is composed of four lines whose frequencies are: $\nu=|A/2 \pm 3Q \pm \nu_n|$, where $|A|=198$ MHz is the nuclear hyperfine frequency, $|Q|\approx 7$ MHz is the nuclear quadrupolar coupling and $\nu_n=3.5$ MHz is the nuclear Larmor frequency of the manganese. Figure 3-b presents the low frequency part of the spectra. As the ESR saturation shifts from the center of the $m_I=1/2$ (A) transition to its high field shoulder (B) , the ENDOR band becomes less intense and finally vanishes in the baseline.

Broad distributions of Q and A constants may explain these results. This supports the conclusion that the Mn^{2+} ions in the crystallites have a large distribution of environments.

We conclude that the Mn^{2+} ions located inside the nanocrystals occupy strongly distorted sites, characterized by a broad hyperfine constant distribution which may be explained by the proximity between Mn^{2+} ions and the surface of the nanoparticles.

PHOTOLUMINESCENCE OF CdS:Mn/SILICA XEROGELS

Photoluminescence and excitation spectra were performed on CdS:Mn nanocrystals dispersed in acetone and silica xerogels at room temperature using a HITACHI F4500 spectrophotometer. The photoluminescence excitation spectrum (Fig. 4) follows the absorption of the sample, and does not significantly change with the emission wavelength. The luminescence spectrum is dominated by a yellow band peaking at 2.16 eV, characteristic of the Mn^{2+} $^4T_1 \rightarrow ^6A_1$ transition. Its radiative lifetime is found to be 1.7 ms. This band is superimposed on a broad weaker band which is attributed to the surface recombination commonly observed in the case of undoped CdS nanoparticles [9]. The lifetime of this second band is in the microsecond time scale [10]. It is not clear yet if both signals originate from the same nanocrystals through a competitive mechanism, or if they arise from two populations of particles such as Mn^{2+}-doped and undoped nanocrystals.

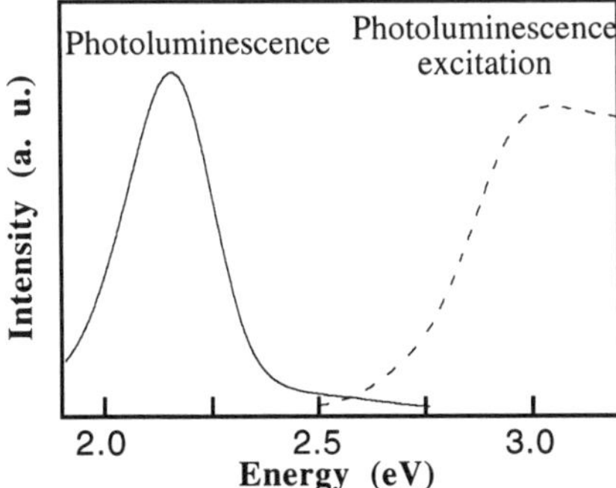

Fig. 4: Photoluminescence and excitation spectra of CdS:Mn nanoparticles embedded in a silica matrix.

Quantum yield evaluations were made by comparing the intensity of the luminescence band with standards of known absorption and efficiency (Rhodamine 6G in acetone for liquid samples and Er-doped fluoride glasses for solid samples). Typical results were 5% in solution and 7% after the incorporation of the nanocrystals in a methyltriethoxysilane-based xerogel.

It is clear that the surface of the nanoparticles plays an important role in the radiative recombination process that leads to the Mn^{2+} $^4T_1 \rightarrow ^6A_1$ transition [11]. In bulk Mn^{2+} doped II-VI semiconductors, it is generally admitted that a strong luminescence requires the presence of so-called sensitizers such as Cl, In, Ga, Ag or Cu [12]. The Mn^{2+} emission is thought to occur after a resonant energy transfer from defect centers associated with these impurities. In our case, we suggest that the role played by these defects in the bulk compounds is now played by the surface of the nanoparticles: the Mn^{2+} emission is then assumed to occur after an energy transfer from excited carriers trapped at the surface of the nanocrystals to Mn^{2+} ions.

The high quantum yield we measured suggests a high transfer efficiency between surface states and the Mn^{2+} ions. Indeed, the assumption that the majority of the Mn^{2+} ions located inside the nanocrystals are very close to the surface may explain the efficiency of the energy transfer between the sensitizer defect and the luminescence activator.

CONCLUSION

We have described a process for the synthesis of CdS:Mn nanoparticles embedded in silica xerogels. The nanocrystals are obtained in inverted micelles through a simple precipitation of CdS in presence of an excess of manganese ions. EXAFS, ESR and ENDOR were used to characterize the luminescent center: the average number of Mn^{2+} per particle ranges between 0.2 and 0.8, and the majority of them is located in distorted sites. We interpret this result in terms of proximity between the ion and the surface of the nanocrystal. The bright Mn^{2+} luminescence observed in CdS:Mn/silica xerogels compounds is believed to occur after a resonant energy transfer between charge carriers trapped at the surface of the particles and manganese ions.

REFERENCES

1 L. E. Brus, Appl. Phys. A53, 465 (1991).

2 M. L. Steigerwald, A. P. Alivisatos, J. M. Gibson, T. D. Harris, R. Kortan, A. J. Muller, A. M. Thayer, T. M. Duncan, D. C. Douglas, L. E. Brus, J. Am. Chem. Soc. 110, 3046 (1988).

3 Y. Wang, N. Herron, Phys. Rev. B, 42-11, 7253 (1990).

4 T. Gacoin, C. Train, F. Chaput, J.-P. Boilot, P. Aubert, M. Gandais, Y. Wang and A. Lecomte, SPIE Sol-gel optics II, 1758, 565 (1992).

5 A. Michalowicz in EXAFS pour le Mac, Logiciels pour la Chimie, Société française de Chimie, Paris (1991).

6 T. B. Allen, J. Chem. Phys. 43, 3.820 (1965).

7 J. Kreissl, W. Gehlhoff, Phys. Stat. Sol. (a), 81, 701 (1984).

8 S. Geschwind in Hyperfine interactions, edited by A. J. Freeman and R. B. Frankel, Academic Press, 226 (1967).

9 A. Hässelbarth, A. Eychmüller, H. Weller, Chem. Phys. Lett. 203-2, 271 (1993).

10 M.A. Chamarro, V. Voliotis, R. Grousson, P. Lavallard, T. Gacoin, G. Counio, J.-P. Boilot and R. Cases, J. of Crystal Growth, 159, 853 (1996).

11 G. Counio, S. Esnouf, T. Gacoin and J.-P. Boilot, accepted in J. Phys. Chem.

12 H.C. Froelich, J. Opt. Soc. Am. 43, 320 (1953).

ELECTRON AND HOLE TRAPPING DYNAMICS IN SEMICONDUCTOR NANOCRYSTALS: FEMTOSECOND NONLINEAR TRANSMISSION AND PHOTOLUMINESCENCE STUDY

V. KLIMOV, D. McBRANCH
Chemical Sciences and Technology Division, CST-6, MS-J585,
Los Alamos National Laboratory, Los Alamos, NM 87544, klimov@lanl.gov

ABSTRACT

Application of two complementary femtosecond techniques (time-resolved nonlinear transmission and photoluminescence up-conversion) allows us to observe separately the electron and the hole relaxation paths in CdS nanocrystals. The obtained data indicate that hole relaxation channels are different at low and high pump fluences which is attributed to an Auger-process-assisted hole trapping at surface/interface states activated at high excitation intensities.

INTRODUCTION

The nonlinear optical and luminescent properties of semiconductor nanocrystals (NC's) are significantly affected by carrier dynamics. Carrier trapping(localization) [1-3] and a nonradiative Auger process [4] are believed to play a major role in the early stages of carrier relaxation, resulting in ultrafast dynamics measured in femtosecond pump-probe, photoluminescence (PL) up-conversion, and photon-echo experiments [1-5]. The mechanisms for carrier localization, and in particular the role of the surface/interface states, in the trapping process have not yet been clarified. In particular, it is still open to question which of the carriers (an electron or a hole) is trapped first (see, *e.g.*, Refs. 6 and 7) and what are the rates of the respective trapping processes.

In this paper, we report studies on the ultrafast carrier trapping in CdS NC's performed using two different femtosecond techniques: PL up-conversion and time-resolved nonlinear transmission. These techniques provide complementary information on carrier dynamics and allow us to separately observe electron and hole relaxation paths. The obtained data clearly indicate that hole relaxation channels are distinctly different at low and high pump levels, leading to a strong difference in optical nonlinearities observed in quasi-equilibrium at long times after excitation. Experimental data and modeling show that this difference is due to an Auger-process-assisted trapping of the hole at surface/interface states that leads to efficient charge separation and generation of a dc-electric field, which modifies the nonlinear-optical response at high pump levels.

EXPERIMENTAL

CdS NC's of ~4 nm radius were grown in a glass matrix by a secondary heat treatment method [8]. Time-resolved nonlinear transmission was measured using a femtosecond pump-probe experiment. The samples were excited at 3.1 eV by frequency-doubled 100-fs pulses from a regeneratively amplified mode-locked Ti-sapphire laser (Clark-MXR NJA-4/CPA-1000). The transmission of the excited sample is probed by delayed pulses of a femtosecond continuum generated in a 1-mm-thick sapphire plate. As a measure of transmission changes we use the differential transmission (DT) defined as follows: $DT = (T - T_0)/T_0 = \Delta T/T_0$, where T_0 and T are transmissions in the absence and in the presence of the pump, respectively. Time-resolved DT was measured using either a CCD coupled to a spectrometer (time-resolved DT spectra) or in a single-wavelength configuration with a phase-sensitive detection (DT time-transients) (for experimental details see Ref. 2).

PL dynamics were measured using a femtosecond up-conversion set-up [1]. The luminescence from the sample was mixed in a nonlinear β-barium borate crystal with variably delayed femtosecond pulses at a fundamental frequency of the Ti:sapphire laser to generate a sum-frequency signal in the near-UV spectral region. The up-converted PL was dispersed by a monochromator and was finally detected by a cooled photomultiplier. All measurements reported below were performed at room temperature.

CARRIER TRAPPING DYNAMICS AT LOW PUMP INTENSITIES

First we examine the time-resolved DT at low pump fluences $w_p < 0.5$ mJ cm $^{-2}$ where the PL and DT dynamics are independent of the pump intensity. This corresponds to excitation of less than one electron-hole (e-h) pair per NC on average and allows us to avoid the saturation of trap states and fast Auger recombination.

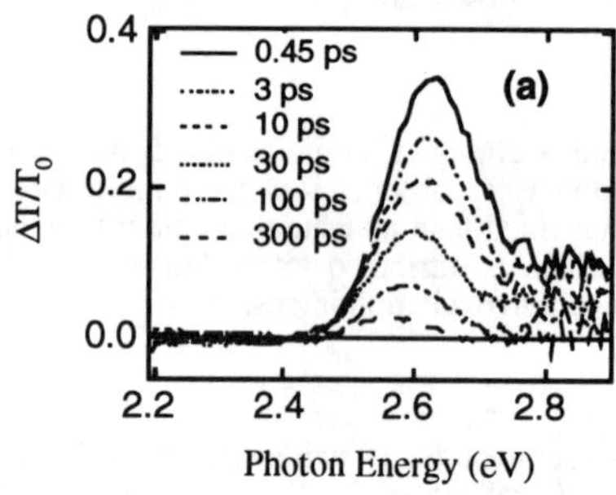

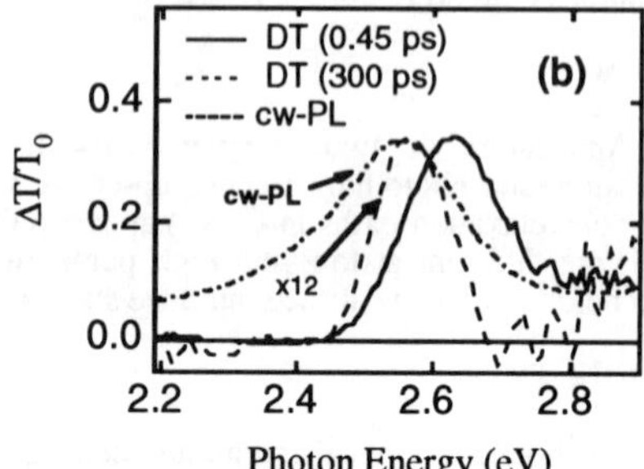

FIG. 1. (a) DT spectra recorded at different delay times after excitation at a pump level $w_p = 0.45$ mJ cm^{-2}, below the threshold for nonlinear recombination processes. (b) Normalized DT spectra measured at $\Delta t = 0.45$ ps (solid line) and 300 ps (dashed line) ($w_p = 0.45$ mJ cm^{-2}) in comparison to the cw-PL spectrum.

DT spectra recorded in this pump range are shown in Figs. 1(a) and (b). At early times after excitation, the DT spectra are dominated by an intense bleaching band at 2.63 eV arising from state-filling-induced saturation of the lowest optical transition [$1S(e)-1S_{3/2}(h)$] between electron and hole quantized states (for notation of quantum-confined states and optical transitions in NC's see, *e. g.*, Ref. 9). The decay of the DT band is almost exponential with a time constant of about 30 ps [see Fig. 2 (squares)]. Due to the large difference between electron and hole masses in CdS ($m_h/m_e \sim 6$), the bleaching of the lowest transition is dominated by the electron population [10]. Therefore, the initial DT dynamics can be attributed to the depopulation of the lowest quantized state in the conduction band due to electron trapping. The trapping states responsible for this process are likely the same as those that give rise to the so-called deep-trap emission [11] located in our samples at ~1.9 eV [2]. This explanation is consistent with direct measurements of buildup dynamics of the deep-trap PL which yield a time-constant of ~30 ps [12].

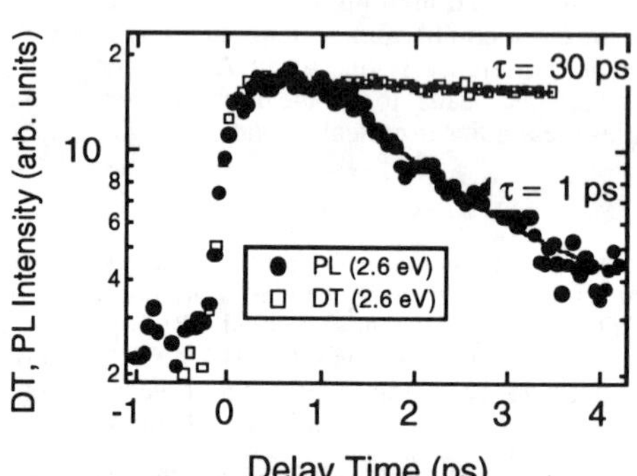

FIG. 2. Single-wavelength ($\hbar\omega = 2.6$ eV) PL (solid circles) and DT (open squares) dynamics in CdS NC's. Solid lines show exponential fits.

As nonlinear transmission is dominated by the electron population, the initial hole dynamics are almost "invisible" in DT. To resolve the hole dynamics, we make use of a complementary femtosecond technique–PL up-conversion. At early times after excitation, the PL spectra are dominated by the band located at 2.63 eV (the lowest optical interband transition between quantized states) [1]. In contrast to the relatively slow DT decay, the PL relaxation is extremely fast and occurs on the ps time scale. The drastic difference in early-time PL and DT dynamics is pronounced in single wavelength time-transients, as shown in Fig. 2. The initial decay of the PL occurs on a much faster time scale than that of the DT. Exponential fits to the experimental data yields 1 and 30 ps relaxation times for the first stages of PL and DT dynamics, respectively. The strong difference in the early-time DT and PL dynamics suggests that the carrier decay is nongeminate and that the electrons and holes have different relaxation paths. In contrast to DT, the PL intensity is determined by the product of the electron and hole occupation numbers and therefore is significantly affected by the dynamics of both carriers. Since DT at 2.6 eV [the 1S(e)–$1S_{3/2}$(h) transition] decays relatively slowly, the fast 1-ps relaxation of the PL intensity can only be attributed to a change in the hole occupation numbers due to trapping at localized states. The magnitude of the low-energy shift of the PL band measured in Ref. 1 indicates that this state has a depth of about 70 meV. The transition coupling this state to the lowest quantized electon state dominates the cw-emission, as well as DT measured at long times after excitation [see Fig. 1(b)].

CARRIER DYNAMICS AT HIGH PUMP-POWERS: AUGER-PROCESS-INDUCED CHARGE SEPARATION

At pump levels above ~0.5 mJ cm^{-2}, the DT dynamics begin to show pump-intensity dependence (see Fig. 3). In contrast to the low-pump-intensity data, at high pump fluences,

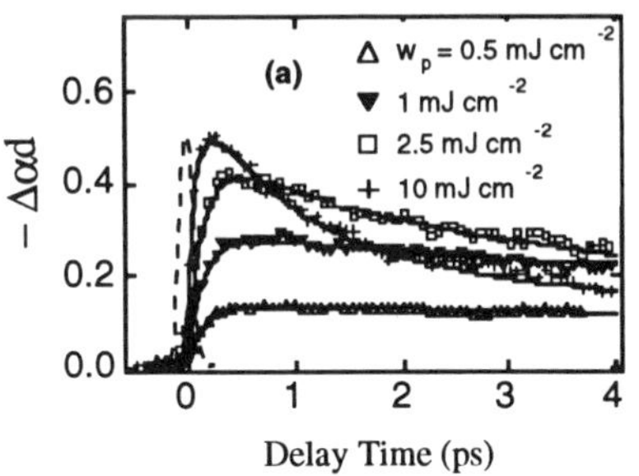

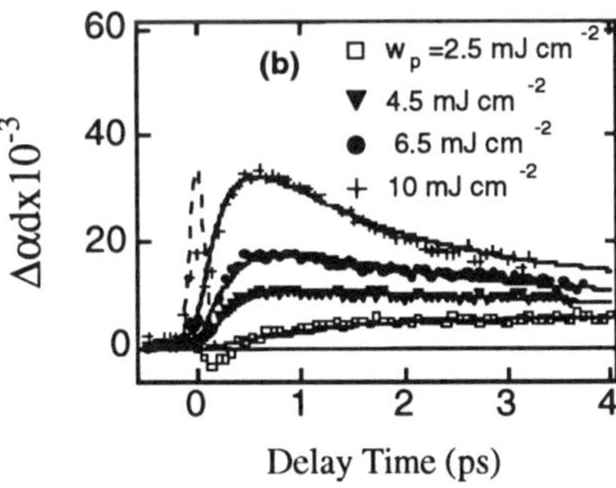

FIG. 3. Pump-dependent DT dynamics (symbols) recorded at (a) 2.6 eV (bleaching) and (b) 2.4 eV (increased absorption) along with fits (lines) calculated using a model of the Auger-process-assisted trapping. The dashed line shows the pump-probe cross-correlation function.

the bleaching maximum at 2.63 eV exhibits a much faster initial decay on the subpicosecond time scale. Another feature of the high-intensity DT is a new band of increased absorption located below the main bleaching band. The relative contribution from the induced absorption increases with increasing time delay until both features (bleaching and increased absorption) become essentially equivalent in magnitude. The shape of the DT spectra measured at high pump levels and at long times after excitation is drasticallydifferent from that measured at low pump fluences (see Fig. 4), although in both cases the average number of e-h pairs per NC is well below one. Instead of a single bleaching band, the high-pump-intensity spectra show a derivative-like feature which is a clear signature of a dc-electric-field-induced shift of the optical transition (dc-Stark effect) [13]. Similar transient absorption features were previously observed in NC's in the case of nonlinearities arising from carrier-trapping-induced charge separation [13] and the biexciton effect [14]. The shape of DT at high pump fluences clearly indicates that the transit ion responsible for the nonlinearity couples unoccupied states (see analysis of DT for competing contributions from the state filling and the transition shift in Ref. 14). Interestingly, the hole state involved in this transition is occupied at low pump intensities which is seen as a pronounced bleaching at 2.56 eV dominating DT at long times after excitation [see Figs. 1(b) and 4 (squares)]. This shows that the hole relaxation path changes at high pump fluences. Instead of being trapped at a shallow state with a relatively extended wavefunction, the hole most likely gets trapped at a state with much stronger

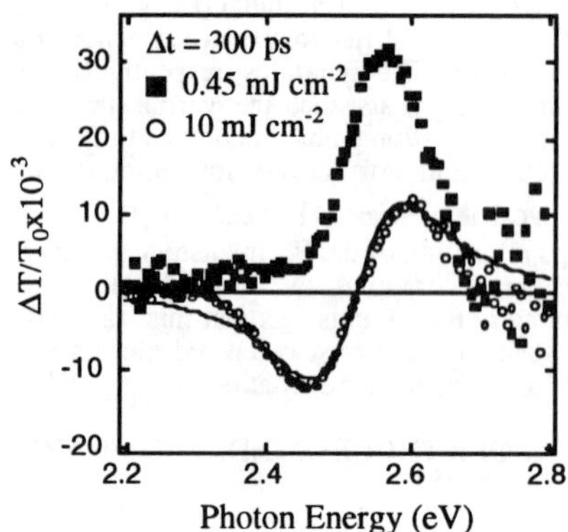

FIG. 4. Comparison of DT spectra recorded at 300 ps after excitation at low (solid squares) and high (open circles) pump intensities. A solid line shows the fit of the high-pump-intensity spectrum to a model spectrum calculated for a dc-electric-field-induced shift of the optical transition centered at E_0 = 2.546 eV with broadening factor Γ = 110 meV.

localization. Due to the reduced overlap of the hole and the electron wave functions and/or a large low-energy shift of the corresponding transition (possibly below the observation range), this state is not marked by a pronounced feature in DT spectra. However, DT gets affected by new trapped states via the dc-electric field generated as a result of the hole trapping. As the new hole relaxation channel is associated with an efficient charge separation, the hole traps activated at high pump fluences are most likely surface/interface related.

To get more information on carrier dynamics, we have performed a careful comparison of DT time transients measured at 2.4 eV (induced absorption) and 2.6 eV (bleaching) (see Fig. 3). The transients recorded at w_p = 2.5 mJ cm^{-2} (slightly above the threshold for observation of the induced absorption feature) show that the buildup of the signal at 2.4 eV is complementary to the decay of DT at 2.6 eV (see transients shown by open squares in Fig. 3); both are characterized by a time constant of about 2.5 ps. The complementary behavior of the bleaching and the increased absorption is preserved at higher pump fluences. Increasing the pump intensity leads to a faster decay of the bleaching band [Fig. 5(a)] which is accompanied by faster buildup dynamics of the induced absorption [Fig. 5(b)].

To understand the mechanism of a pump-dependent decay rate of the bleaching band and associated complementary buildup dynamics of the induced absorption feature, we have measured the pump-intensity dependence of DT at 2.4 and 2.6 eV at a fixed delay time (Δt = 1 ps) between pump and probe pulses (Fig. 5, symbols). At pump levels below ~2 mJ cm^{-2}, the bleaching increases almost linearly with pump intensity. A linear fit to the low-intensity data in the logarithmic plot gives a slope p = 0.9 as shown by the dashed line in Fig. 5. The induced absorption is characterized by a much steeper rise with initial slope of 3.1 (solid line in Fig. 5). At pump fluences above 2 mJ cm^{-2}, both curves show saturation. Interestingly, the bleaching signal saturates at a level $|\Delta\alpha/\alpha_0| \cong 0.7$, that is below 1 expected for a simple Fermi blocking of the transition with one unoccupied state. This shows that the observed saturation is caused by a mechanism different from Fermi blocking, as discussed below.

The complementary dynamics measured at 2.6 and 2.4 eV (Fig. 3) as well as the data on the pump dependences in Fig. 5 clearly indicate that the growth of induced absorption is associated with a cubic decay of the bleaching band. This type of nonlinear decay is a signature of Auger processes which have been previously reported as playing an important role in carrier recombination in highly-excited NC's [4, 15]. The Auger decay involves three particles, two of which recombine, with the energy transfer to the third particle (an electron or a hole). The nonlinear transmission data show that this is a hole relaxation path which is different at high pump fluences. From this, we deduce that thc Auger processes play the role of a re-excitation mechanism which promotes holes high enough in energy allowing them to overcome a NC-boundary-related potential barrier and to reach interface/surface-related states. Auger-process-assisted hole trapping at the surface leads to charge separation and the generation of strong local fields with anassociated dc-Stark shift seen in DT.

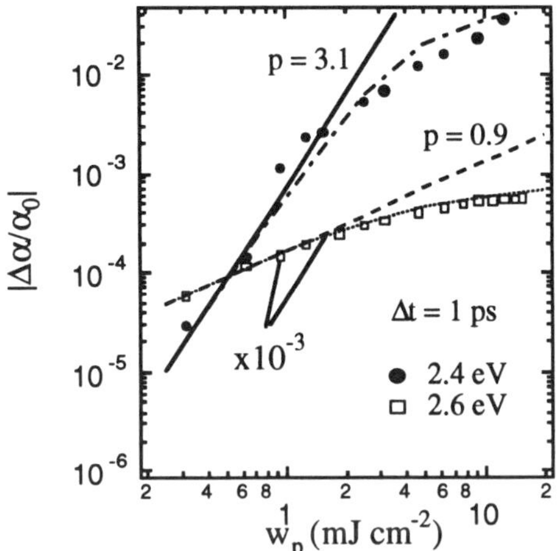

FIG. 5. Pump-intensity dependence of increased absorption at 2.4 eV (solid circles) and bleaching at 2.6 eV (open squares) measured at 1 ps after excitation in comparison to modeling results shown by dotted and dashed-dotted lines. Dashed and solid lines show pump-power dependences with logarithmic slopes 0.9 and 3.1, respectively.

Using the model of the Auger-process-assisted trapping we were able to describe very closely the complementary behaviour of time transients measured at 2.6 and 2.4 eV (lines in Fig. 3) as well as the pump-dependence of DT (dotted and dashed-dotted lines in Fig. 5). The proposed model explains very well not only the slopes measured for the bleaching ($p \cong 1$) and the increased absorption ($p \cong 3$) at low pump levels but also the saturation of both at high pump intensities. This shows that the saturation observed experimentally occurs entirely due to the Auger recombination which effectively limits the occupation of the lowest quantized states at a level below the threshold for the Fermi blocking. As a result of the modeling, we obtain the magnitude of the Auger constant C: $C^{-1} V_0^2 = 2 - 6$ ps (V_0 is the volume of a single NC).

Finally, we address a peculiarity in the behavior of the bleaching time transients observed at high intensities and manifested as intersections of traces recorded at different pump fluences [see Fig. 3(a)]. This effect is entirely reversible and, therefore, cannot be attributed to a sample degradation (photodarkening) [16]. However, this behavior cannot be described by the proposed model without an additional assumption about the pump dependent oscillator strength of the optical transition at 2.6 eV. To fit accurately not only dynamics but also the magnitude of the bleaching, we have assumed that at high pump levels (above ~1 mJ cm^{-2}) the oscillator strength starts to decrease with increasing pump intensity. The reduction factor amounts to about 2.5 at w_p ~ 10 mJ cm^{-2}. This effect can be attributed to the influence of the electric field generated as a result of the hole surface trapping, leading to a modification of the symmetry of the electron and hole wavefunctions [17].

CONCLUSIONS

We performed femtosecond studies of ultrafast carrier dynamics in CdS NC's. Applying methods of time-resolved PL and DT, we were able to separate the electron and hole relaxation paths following photoexcitation with 100-fs pulses. In particular, we have demonstrated that of the two carriers (an electron and a hole) the hole gets trapped first into a localized state with a time-constant ~1ps. In contrast, electron trapping is a much slower process occuring on the time scale of tens of picoseconds.

We have observed a strong difference in nonlinear optical response at low and high pump intensities, indicating a change in the dominant hole relaxation channel as a result of nonlinear interactions in the system of photoexcited carriers. The nonlinear transmission measured at high pump fluences has a clear signature of a dc-electric-field-induced Stark shift of optical transitions, indicative of charge separation due to carrier surface trapping. A careful analysis of the nonlinear transmission time-transients and pump dependences shows that this charge separation is caused by Auger processes acting as a re-excitation mechanism which promotes holes high enough in energy to allow them to overcome the NC-boundary-related potential barrier and to reach interface/surface-related states.

ACKNOWLEDGMENTS

This research was supported by Los Alamos Directed Research and Development funds, under the auspices of the U.S. Department of Energy.

REFERENCES

1. V. Klimov, P. Haring-Bolivar, and H. Kurz, Phys. Rev. B **53**, 1463 (1996).

2. V. Klimov, P. Haring-Bolivar, H. Kurz, and V. Karavanskii, Superlattices and Microstructures **20**, 395 (1996).

3. M. C. Nuss, W. Zinth, and W. Kaiser, Appl. Phys. Lett. **49**, 1717 (1986).

4. J. Z. Zhang, R. H. O'Neil, and T. W. Roberti, Appl. Phys. Lett. **64**, 1989 (1994).

5. R. W. Schoenlein, D. M. Mittleman, J. J. Shiang, A. P. Alivisatos, and C. V. Shank, Phys. Rev. Lett. **70**, 1014 (1993).

6. M. O'Neil, J. Marahn, and G. McLendon, J. Phys. Chem. **94**, 4356 (1990).

7. M. G. Bawendi, P. J. Carroll, W. L. Wilson, and L. E. Brus, J. Chem. Phys. **96**, 946 (1992).

8. N. F. Borelli, D. W. Hall, H. J. Holland, and D. W. Smith, J. Appl. Phys. **61**, 5399 (1987).

9. A. I. Ekimov, F. Hache, M. C. Schanne-Klein, D. Ricard, C. Flytzanis, I. A. Kudryavtsev, T. V. Yazeva, A. V. Rodina, and Al. L. Efros, J. Opt. Soc. Am. B **10**, 100 (1993).

10. V. Klimov and V. Karavanskii, Phys. Rev. B **54**, 8087 (1996).

11. N. Chesnoy, T. D. Harris, R. Hull, and L. E. Brus, J. Phys. Chem. **90**, 3393 (1986).

12. M. O'Neil, J. Marohn, and G. McLendon, Chem. Phys. Lett. **168**, 208 (1990).

13. D. J. Norris, A. Sacra, C. B. Murray, and M. G. Bawendi, Phys. Rev. Lett. **72**, 2612 (1994).

14. V. Klimov, S. Hunsche, and H. Kurz, Phys. Rev. B **50**, 8110 (1994).

15. M. Ghanassi, M. C. Schanne-Klein, F. Hache, A. I. Ekimov, D. Ricard, and C. Flytzanis, Appl. Phys. Lett. **62**, 78 (1993).

16. M. Tomita and M. Matsuoka, J. Opt. Soc. Am. B **7**, 1198 (1990).

17. Y. Z. Hu, M. Lindberg, and S. W. Koch, Phys. Rev. B **42**, 1713 (1990).

SURFACE DERIVATIZATION OF NANOCRYSTALLINE CdSe SEMICONDUCTORS

J.-K. LEE, M. KUNO, M.G. BAWENDI
Department of Chemistry, Massachusetts Institute of Technology, Cambridge, MA 02139

ABSTRACT

CdSe quantum dots (QDs) with several different ligands were prepared by ligand exchange reactions of trioctylphosphine oxide (TOPO)/ trioctylphosphine selenide (TOPSe) passivated QDs with an excess of new ligands, such as pyridine, 4-picoline, thiophenol, 4-(trifluoromethyl)-thiophenol, and tris(2-ethylhexyl)phosphate. We find that 85~90% of ligands passivating the surface of the QDs are the newly introduced capping species and 10~15% are native TOPO/TOPSe ligands, reflecting the incomplete exchange of the QD surface. Thermal gravimetric analysis (TGA) and quantitative proton NMR measurements show that on average only 30% of surface cadmium atoms are passivated by TOPO/TOPSe. Approximately 1/3 of these native ligands occupy sites too strongly bound to be replaced by ligand exchange reactions.

INTRODUCTION

Stable nanocrystallites (or QDs) have been prepared by stabilizing the surface with various organic ligands.[1-4] Usually these organic ligands are introduced during the preparation of QDs and affect many properties of QDs such as their solubility and stability. Ligand exchange reaction is one of the simplest methods to make high quality QDs with various surface ligands, because in most cases the size of the QD is only controllable using a very limited numbers of capping ligands. Although there have been several reports on the ligand exchange reaction of the QD surface[4,5], precise characterization and quantitative analyses of the ligand exchange reaction have not been explored. We derivatized the surface CdSe QD to evaluate the influence of surface ligands on the optical and electronic properties of the QDs. We report our results of ligand exchange reactions on the QD surface and quantitative analyses of surface ligands.

EXPERIMENT

We synthesized CdSe QDs according to the literature method.[4] The QDs produced by this method differ from CdSe QDs prepared by many other synthetic methods in that they luminescence strongly at the band edge (quantum yields 10~20% at 300K and 10~90% at 10K) and show little deep trap emission. To obtain nearly monodisperse samples we follow the growth of the QDs with size selective precipitation. This procedure involves precipitating out of solution the largest QDs present in the initial growth mixture. By repeatedly performing this procedure on the recovered material it is possible to obtain distribution with standard deviations $(\sigma) \leq 5\%$.

Surface Exchanged QDs

To modify the surface of size selected CdSe QDs with different ligands we employed the following procedure. For a surface modification with pyridine (or 4-picoline) size selected QDs passivated with TOPO/TOPSe were recovered as powder and redispersed in a small volume (typically 5~10 mL) of neat pyridine (or 4-picoline). After overnight stirring the QDs were

Mat. Res. Soc. Symp. Proc. Vol. 452

precipitated from solution by adding an excess of n-hexane. The resulting suspension was then centrifuged and the recovered precipitate dispersed in neat pyridine (or 4-picoline). The solution was left to stir for 5 hours whereupon the dots were precipitated, redispersed in pyridine (or 4-picoline) as above and left to stir for an additional 5 hours. The QDs were later precipitated from solution and dried under N_2.

To passivate the surface of QDs with thiophenol, size selected QDs were recovered as a powder by drying the QD precipitate under a continuous flow of N_2. The isolated nanocrystallites were then brought into a glove box where they were dissolved in about 10 mL of thiophenol. The mixture was left to stir for a day inside the glove box. The QDs were recovered as precipitate by adding an excess of n-hexane to the solution. Dilute solutions of 4-(trifluoromethyl)thiophenol in CH_3Cl were used to exchange the surface with 4-(trifluoromethyl)thiophenol. In each case the isolated QDs were dissolved in 3 mL of CH_3Cl and an estimated 20 fold excess of 4-(trifluoromethyl)thiophenol was added to the CH_3Cl solution. The mixture was left to stir for 4 days in a glove box. At the end of the exchange reaction, the QDs were recovered by precipitation with methanol.

Control Experiments

Various control experiments were conducted to monitor the extent of ligand exchange in our reactions. QDs were prepared in several different ways from the standard preparation described above. This was done to evaluate the extent of surface passivation by TOPO and TOPSe. Clusters were prepared with an excess of TOPO over TOP (trioctylphosphine). This involved excluding TOP in our injection solution. QDs were also prepared exclusively in TOPO by replacing our TOPSe precursor with bis(trimethylsilyl)selenium, $(TMS)_2Se$, and excluding TOP in the injection solution.

RESULTS

^{1}H NMR Measurements of Surface Ligands

In our ligand exchange experiments we found ourselves unable to completely change the surface of the QDs. This was confirmed by ^{1}H NMR experiments. The NMR spectra of QDs passivated with thiophenol, 4-(trifluoromethyl)thiophenol, pyridine and 4-picoline all show residual alkyl protons upfield from the aromatic resonances characteristic of these ligands. Based on the integrated intensities of the proton peaks we find that 85~90% of the ligands passivating the surface of the QDs are the newly introduced capping species. The remaining 10~15% are native capping ligands. Similar results have been seen by Katari et al. in CdSe nanocrystallites exchanged with pyridine.[5]

To overcome this problem we varied the length and number of times the QDs were exposed to the new capping species. In all instances we found that neither length nor time had a significant effect on the ligand exchange reactions as 10 to 15% of the total surface species remained unaltered. Several control experiments were carried out to determine the reason for the incomplete exchange reactions. Of interest was the identity of the tightly bound surface species; whether it was TOPO, TOP, TOPSe or some impurity in our TOPO source such as phosphinic acid. The first control experiment involved preparing the QDs using an excess of TOPO. The nanocrystallites prepared in this manner were subjected to ligand exchange with thiophenol. The exchange reaction showed that 91% of the surface species consisted of the newly introduced cap, in agreement with previous results. A second control experiment involved preparing the QDs exclusively in TOPO. This meant excluding TOP in our injection solution and replacing our TOPSe precursor with $(TMS)_2Se$. The exchange reaction of this material with thiophenol was also incomplete and showed a 95% exchange.

Both experiments suggested that TOPO was specially coordinated to the surface of the QD but did not exclude the possibility that TOPSe or TOP was also tightly bound. To eliminate the possibility that the special surface species was an impurity from our technical grade TOPO we prepared QDs according to the method described in ref [5]. This procedure uses higher purity TOPO (99% Aldrich) and minimizes the potential contribution from a host of impurities such as phosphinic acid. (See ref. [6] for a careful investigation of common impurities found in TOPO) The exchange reaction in this case was also incomplete (86%) and suggested that TOPO, possibly TOPSe/TOP, remains strongly bound to the surface even in the presence of a large excess of coordinating ligands.

^{31}P NMR Measurements of Surface Ligands

Several ^{31}P NMR experiments were conducted to give us more insight into the nature of the tightly bound surface species. The proton decoupled ^{31}P NMR of normally prepared QDs in deuterated chloroform is shown in Figure 1a. Most prominent in the spectra is a broad feature centered between 20 to 30 ppm. It appears to be composed of three peaks and is neighbored by two smaller peaks on either side of its center. These three peaks along with a small feature centered at -20 ppm indicate the existence of at least four distinct chemical environments for the ^{31}P nuclei on the surface of the QD. Several of these peaks have been seen before in the solid state 31P NMR of CdSe QDs.[7]

We attribute the broadness of the NMR features to the inhomogeneous distribution of magnetic environments found on the QD surface, not to the equilibrium between free and bound ligands. Measuring the proton coupled ^{31}P NMR had little effect on the observed peak widths. This result in conjuction with structural studies indicating a disordered or reconstructed surface

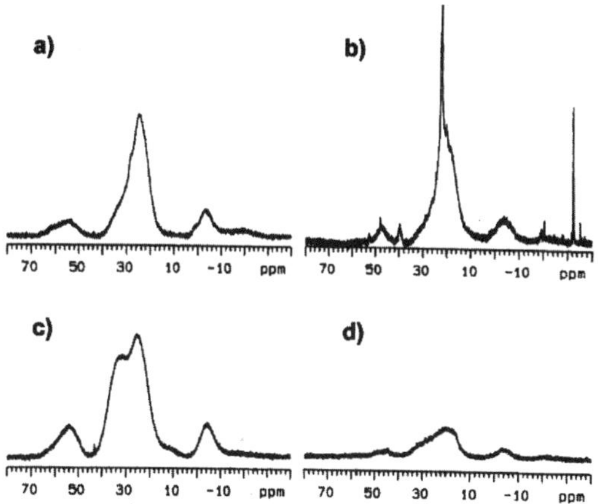

Fig.1 ^{31}P NMR of (a) normally prepared QDs in $CDCl_3$, (b) normally prepared QDs in py-d_5, (c) QDs made exclusively in TOPO using $(TMS)_2Se$ instead of TOPSe as our selenium precursor (in $CDCl_3$), (d) pyridine capped QDs in py-d_5.

supports the notion that the spectral broadness is due to the multiple environments found on the QD surface.[7-9]

To help assign the features in the QD ^{31}P spectra we measured the ^{31}P NMR of TOPO, TOP and TOPSe in several deuterated solvents. TOPO has a peak around 50 ppm in chloroform; TOPSe has a peak at 30 ppm in benzene and TOP has a resonance at -30 ppm in chloroform. One puzzling feature of the QD ^{31}P NMR which becomes apparent after comparing the various peak positions is that a number of phosphorous resonances appear to have shifted upfield rather than downfield. This is inconsistent with deshielding effects normally observed when a phosphine oxide is bound to an electrophile.[10,11] If we assume that the broad feature between 20 and 30 ppm arises from TOPO/TOPSe bound to the QD surface, this implies that the surface shields rather than deshields the phosphorous nuclei moving its resonance to higher energies. Possible explanations for this effect may lie in the presence of Se lone pairs on the surface of the QD or in the high electron density of the nanocrystallite which can screen the ^{31}P nuclei from external magnetic fields.

The ^{31}P NMR of TOPO/TOPSe passivated QDs in deuterated pyridine (Figure 1b) shows that the broad peak centered at 25 ppm splits into at least two components revealing a sharp peak at 22 ppm superimposed on top of a broader peak centered at 25 ppm bordered by two neighboring peaks. The spectrum also shows that the small feature at -20 ppm diminishes in intensity and is replaced by a sharp peak at -32 ppm which we interpret as free TOP in solution. If we monitor the ^{31}P NMR with time we observe that the narrow peak at 22 ppm diminishes in intensity and is followed by the appearance of a sharp peak at 47 ppm. Based on a comparison to known peak positions we interpret this feature as free TOPO in solution. These results suggest that an excess of pyridine can successfully displace both TOP and TOPO from the surface of the QD. We tentatively assign the small feature at -20 ppm in the QD chloroform spectra as TOP loosely bound to surface selenium atoms and also assign the narrow feature at 22 ppm in the QD pyridine spectra as TOPO bound to surface cadmium atoms. The identity of the broad central feature accompanied by its two neighboring peaks is still unknown at this point.

To further clarify the identity of the various phosphorous resonances, we measured the ^{31}P NMR of QDs made exclusively in TOPO. The spectrum is shown in Figure 1c. As with the first chloroform spectra, we observe a broad feature centered at 20-30 ppm bordered by two neighboring peaks. There are differences between spectrum 1c and 1a. The broad central feature in this case appears to be composed of two peaks, not three. In addition, the small peak at -20 ppm is absent. Since no TOP or TOPSe was introduced at any point in the sample preparation, the appearance of the exclusive TOPO spectra supports our initial assignment that the peak at -20 ppm in the chloroform spectra is due to TOP bound to surface selenium atoms. The presence of the broad central feature with symmetric neighbors also suggests that these peaks are TOPO related and the absence of a third component to the central feature implies that this contribution arises from TOPSe bound to surface cadmium atoms.

The ^{31}P NMR of normally prepared QDs which have been extensively washed with pyridine is shown in Figure 1d. It indicates that even after extensive exposure to a new coordinating solvent there are still native ligands which remain tightly bound to the QD surface. Although the signal to noise is poor we identify the dominant central peak at 20-30 ppm. If our assignments are correct, the presence of this peak suggests that there are sites on the QD surface which coordinate strongly to TOPO and possibly TOPSe.

Quantitative Analysis of Surface Ligands

To give a measure of the amount of TOPO/TOPSe on the surface of the QD before and after ligand exchange we performed thermal gravimetric analysis (TGA) measurements in conjunction with quantitative proton NMR measurements. Table I summarizes our results and indicates that on average only 30% of surface cadmium atoms are passivated by TOPO/TOPSe. Exposing the QDs to an excess of pyridine replaces approximately 2/3 of the initial TOPO/TOPSe ligands leaving 1/3 of the sites unchanged. At the same time the absolute coverage of surface cadmium sites appears to increase dramatically, from roughly 30% to 90%. These results are consistent with NMR measurements indicating that 85 to 90% of the ligands on the surface are the new capping species and that roughly 10 to 15% are the original TOPO/TOPSe molecules. The amount of unchanged Cd sites can also be expressed in terms of the amount of residual TOPO left on the surface of the QD after ligand exchange, 38% (from TGA) compared to 40% from NMR measurements (see Table I).

Table I. Quantitative Analysis for the ligand Coverage in CdSe QDs

	TOPO capped QD (r = 18.4Å)	Py capped QD (r = 18.4Å)	
Sample wt.	11.772 mg	10.248 mg	
wt. loss[a]	2.75 mg (23% loss due to TOPO)	1.015 mg (10% due to Py)	0.676 mg (6.6% due to TOPO)
Mole loss	5.4 μmol of TOPO	12.8 μmol of Py	1.8 μmol of TOPO
CdSe core wt.	9.647 mg (50.4 μmol)	8.557 mg (44.7 μmol)	
Estimated moles of surface Cd[b]	18.6 μmol	16.5 μmol	
Surface Cd coverage by TGA[c]	5.37/18.6 = 0.29 (29% due to TOPO)	12.8/16.5 = 0.78 (78% due to Py)	1.8/16.5 = 0.11 (11% due to TOPO)
NMR integrated area of TOPO aliphatic protons[d]	10.0	3.99	

TGA Summary	11/29 = 38% TOPO passivation of all possible Cd sites
NMR Summary	3.99/10.0 = 40% TOPO passivation of all possible Cd sites

a) The weight loss is correlated with the removal of TOPO and pyridine through independent TGA/Mass Spec. measurements. However, we cannot distinguish TOPO from TOP in the mass spectra and therefore assume that the majority species is TOPO.

b) The number of surface cadmiums was calculated by comparing the number of CdSe units on the surface to the total number of CdSe units in a 18.4Å radius QD (233). The number of CdSe units in a 15.78Å radius QD, obtained by subtracting the CdSe bond length (2.62Å) from the original radius (18.4Å), is 147. Subtracting this number from the total gives 86 which corresponds to the number of CdSe units on the surface of the QD.

c) Assuming that TOPO does not occupy bridging positions over the Cd.

d) Aliphatic proton peaks were integrated using CH2Cl2 as an internal standard. The concentration of each sample was normalized using the intensities of the first absorption peak at 540 nm.

CONCLUSIONS

We find that on average only 30% of surface cadmiums in our CdSe nanocrystallites are passivated by TOPO/TOPSe. Although the absolute coverage of the surface increases from roughly 30% to 90%, after ligand exchange with small ligands such as pyridine, 4-picoline, thiophenol and 4-(trifluoromethyl)thiophenol, approximately 1/3 of initial TOPO/TOPSe ligands remain bound to surface Cds. At this point we can only speculate that the unremoved TOPO/TOPSe occupy special sites related to selenium vacancies on the surface which leave behind empty voids of exposed cadmium atoms. Bonding by TOPO (TOPSe) could then take the form of bridging positions over several cadmium atoms which would account for the stronger bonding strengths observed. Further work is needed to clarify the true surface morphology of the QDs.

ACKNOWLEDGMENT

We thank Dr. Andrew McGhie at U. Penn. for conducting TGA/Mass Spec. measurements on our samples. MK is grateful to the Corning Foundation for a pre-doctoral fellowship. MGB thanks the Lucille and David Packer Foundation as well as the Alfred P. Sloan Foundation for fellowships. This work was supported in part by the NSF-MRSEC program (NSF-DMR-4-00034) and NSF (DMR-91-57491).

REFERENCES

[1] I.G. Dance, A. Choy, and M.L. Scudder, J. Am. Chem. Soc. **106**, 6285 (1984).

[2] M.L. Steigerwald and L.E. Brus, Acc. Chem. Res. **23**, 183 (1990).

[3] G. Schmid, Chem. Rev. **92**, 1709 (1992).

[4] C.B. Murray, D.J. Norris, and M.G. Bawendi, J. Am. Chem. Soc. **115**, 8706 (1993).

[5] J.E. Bowen Katari, V.L. Colvin, and A.P. Alivisatos, J. Phys. Chem. **98**, 4109 (1994).

[6] M. Koloskt, J. Vialle, and T. Cotel, J. of Chromatogr. **299**, 436 (1984).

[7] L.R. Becerra, C.B. Murray, R.G. Griffin, and M.G. Bawendi, J. Chem. Phys. **100**, 3297 (1994).

[8] M.A. Marcus, L.E. Brus, C.B. Murray, M.G. Bawendi, A. Prasad, and A.P. Alivisatos, Nanostruc. Mater. **1**, 323 (1992).

[9] J.R. Sachleben, E.W. Wooten, L. Emsley, A. Pines, V.L. Colvin, and A.P. Alivisatos, Chem. Phys. Lett. **198**, 431 (1992).

[10] P.A.W. Dean and M.K. Hughes, Can. J. Chem. **58**, 180 (1980)

[11] G.E. Maciel and R.V. James, Inorg. Chem. **3**,1650 (1964).

RAMAN SCATTERING AND PHOTOLUMINESCENCE MEASUREMENTS ON II-VI SEMICONDUCTOR NANOCRYSTALS AS A FUNCTION OF PRESSURE AND PARTICLE SIZE

John Schroeder, Mierie Lee, Markus R. Silvestri, Lih-Wen Hwang and Peter D. Persans
Department of Physics, Applied Physics, and Astronomy, Rensselaer Polytechnic Institute, Troy, NY 12180-3590

ABSTRACT

Pressure tuned photoluminescence and Raman scattering were employed to study the structural phase transition from wurtzite to rocksalt structure of CdS_xSe_{1-x} nanocrystals embedded in a glass matrix as a composite material and in the colloidal form. In both the composite and the colloidal form the increased phase stability of the wurtzite to rocksalt structure is reversible and size dependent. In contrast the phase transition in bulk semiconductors of cadmium-sulfide is sharp but for the nanocrystals it is broad and occurs at higher pressures. A size dependent study of the phase stability was conducted and a correlation between size and increase in phase stability was definitely established. For the nanocrystals in the composite form the difference in surface tension in both phases allowed for a model that predicted the tendencies observed in optical spectra. For the colloidal nanocrystals a model based on surface tension and volume effects (defects) gave good agreement in explaining the observations. The above results are discussed with respect to surface tension and deep level defects and their pressure dependence.

INTRODUCTION

II-VI nanoparticles can be crystalline with wurtzite, zincblend, or rocksalt structure: or some disordered phase related to one of these structures. The electronic properties depend upon phase, and, conversely, optical measurements provide a probe of structure. The variety of phases observed due to variation of pressure is quite different from that observed by varying temperature. Pressure leads to more closely packed structures and increased interatomic energies.

The phase stability of nanocrystallites may be affected by several size-dependent factors. Firstly, the interface (or surface) energy becomes increasingly important as size is decreased simply because the number of unit cells on the crystal surface increases relative to the number of unit cells embedded inside the crystal. The relative number of molecular orbitals associated with the surface also must increase, chemical termination becomes increasingly important. Secondly, electronic confinement may destabilize one phase relative to another if the effective masses or volumes differ. Thirdly, defects which might be unimportant in the bulk might become important due to strain, interface effects or different growth conditions in nanocrystalline systems. In addition, differential thermal expansion between the crystallites and their host may be important in composites.

Solid phase-solid phase structural transitions fall into two classes [1]. The first type of transition is where the unit cells of the two phases have different topology and the atoms must reconstruct a new lattice. The second type of phase transition is one in which the unit cell is deformed but the topology remains the same. The details of how one structure transforms into the other may be very important in determining the shape and faceting of the final transformed crystallite [2-5].

Many studies have found changes in the thermo-dynamic properties of II-VI semiconductor crystals with nanometer dimensions (quantum dots) such as reduction of the melting temperature

Mat. Res. Soc. Symp. Proc. Vol. 452 © 1997 Materials Research Society

[6,7] or an increase in the structural phase transition pressure point [2,8-13]. The stability of such crystals can be controlled by an external parameter such as their sizes. Tolbert *et al.* [8,9] conducted a series of studies on nearly monodisperse CdSe quantum dots in a colloidal suspension and confirmed a size dependence which was attributed to surface tension. However, similar size dependent studies have not been performed on quantum dots embedded in a solid glass matrix, a composite which has promising potential for technical applications [14]. The diffusion grown quantum dots in a glass matrix pose in many ways a different problem with respect to phase stability. Strains, a large size distribution, and a small number concentration of quantum dots complicate the picture.

It is the goal of this paper to examine these effects in more detail. A model is used that can account for the observed enhanced phase stability of the quantum dots. This model, which we had developed previously, is based on the difference of the surface tension between the wurtzite and rocksalt phases [15,16]. We then examine quantum dots not in the composite form but as a colloid and will propose a model that can account for their observed behavior.

EXPERIMENT

As samples we used Schott filters RG630 and OG570. These filters are composite materials based on borosilicate glass containing CdS_xSe_{1-x} crystallites. Their compositions were determined through the relative shift between the S-like and Se-like peak positions in the Raman spectra [17] and their sizes by TEM [18] or by x-ray diffraction through the Debye-Scherrer broadening. The composition of RG630 and samples grown from its melt [18] is $CdS_{0.44}Se_{0.56}$ and their sizes are given in the legend of Fig. 5.

The colloidal samples were prepared by the injection of 20 ml of 9.1×10^{-3} M $CdSO_4$ aqueous solution into 100 ml of rapidly stirred 5×10^{-3} M Na_2S aqueous solution at room temperature. In this way we obtain fresh colloid particles that are about 10 nm in diameter [11] and can be suspended for more than one week as aqueous solutions. The fresh colloids are about 10 nanometer in diameter as determined from TEM measurements [18].

For the pressure measurements the composite quantum dot samples were polished down to ~ 60 μm thickness and then loaded into a diamond anvil pressure cell, which is described in detail elsewhere [11]. A 4:1 methanol-ethanol mixture served as a pressure medium. For the colloids we used water instead of methanol-ethanol as pressure medium. The pressures were measured with the standard ruby fluorescence technique [19]. The shape of the ruby lines and their spacing did not alter with pressure indicating that hydrostatic pressure was well maintained throughout all our experiments.

All optical experiments were carried out at a 135° scattering geometry at room temperature unless otherwise noted. Depending on the sample used, the 457.9 nm, 488.0 nm, or 514.5 nm lines of the Ar^+ laser were used. The photoluminescence spectra were recorded with a double monochromator (Spex-1403) with slits set to a resolution of about 2 cm^{-1}. A cooled photomultiplier (RCA 31034A) served as a detector from which the counts were stored in a computer for further processing.

RESULTS AND DISCUSSIONS

Photoluminescence spectra of a C6, a $CdS_{0.44}Se_{0.56}$ sample of 6.2 nm size, grown from the melt of RG630 is shown in Fig. 1. The strong band edge luminescence peaks in the energy range of 2.0 - 2.2 eV are clearly visible at pressures far beyond the bulk structural phase transition pressure point (~ 30 kbar at room temperature). In CdS the high pressure rocksalt phase has an indirect

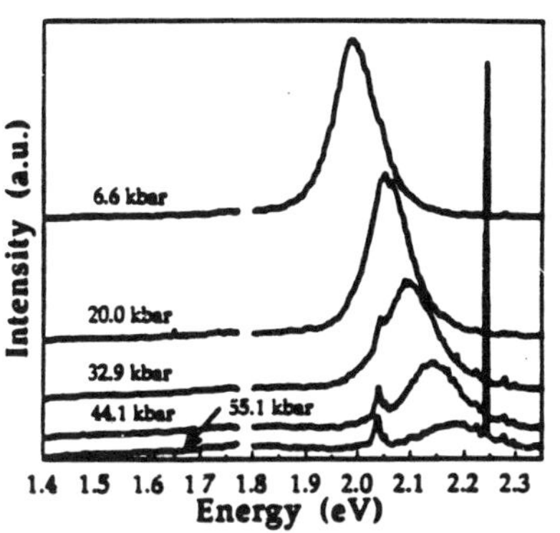

Fig. 1 Typical photoluminescence spectra of sample C6 at different pressures at room temperature. The spike around 2.25 eV is due to the diamond anvils and the emerging peak at around 2.05 eV is from the pressure transmitting medium. Note the emergence of a broad deep level defect at high pressures.

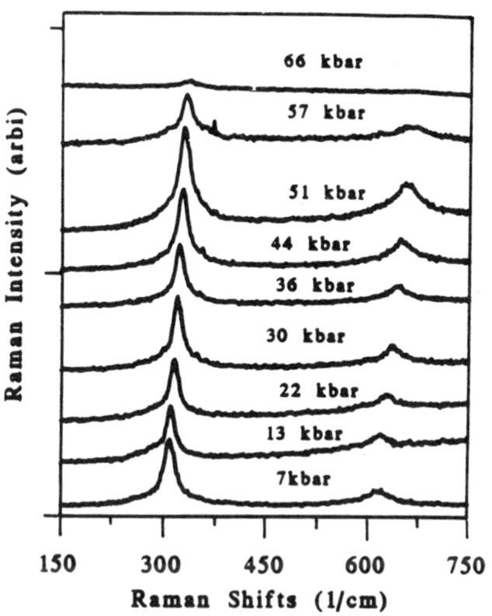

Fig. 2 The Raman spectra of 1 month aged CdS colloids at different pressures.

band gap of about 1.5 eV and a similar reduction is also expected for CdSe; thus, the band edge luminescences at high pressure clearly indicate that quantum dots remain in the low pressure wurtzite structure. This experimental observation is further supported by the Raman spectra: Raman lines typical for the wurtzite structure persist up to the same high pressures [15]. However, for quantitative estimates, the Raman lines are difficult to interpret since they are also subject to resonance effects [20]. Fig. 2 shows typical Raman spectra for aged CdS colloids (one month) at various pressures. Both the LO and 2LO phonons are observed at pressures as high as 66 kbar. The size of these colloids is about 10 nm. The optical phonon energy increases with increasing pressure. The pressure coefficient for LO is 0.44 cm^{-1}/ kbar which is in good agreements with the bulk value of 0.50 cm^{-1}/ kbar [21].

The phase transition pressure point for the colloidal quantum dots was determined by observing the drastic decrease of the Raman intensity going to zero as the wurtzite phase changes to the rock salt phase. This is illustrated in Fig. 3 where we plot the Raman intensity of the 1LO line and 2LO line as a function of pressure. Note that the Raman intensity increases up to about 43 kbar and then drops to negligible intensity beyond the 66 kbar pressure point. The dispersion of the data points may come about from the inhomogeneity of the particle sizes for the colloidal sediment. At pressures below the bulk phase transition point (~27 kbar) the intensity tends to increase with pressure. Beyond 60 kbar the Raman signal drops off markedly, indicating that a structural phase transition is taking place. Here the wurtzite phase is Raman active while the rocksalt phase is not Raman active due to centro-symmetry.

Typical photoluminescence spectra as a function of pressure are given in Fig. 4 for the colloidal samples (one month). Here peak 1 is a defect peak and peak 2 represents the band edge. The spectra consist of two emission bands with peaks at 1.73 eV (peak 1) and at 2.44 eV (peak 2). Both photoluminescence peaks exhibit a strong pressure dependence in the energy shift and in intensity. Notice that the photoluminescence intensity is enhanced for a different pressure than the Raman intensity. The difference exists due to competing effects in the Raman lines, namely structural changes and resonance effects take place [20]. Pressure does not only change the resonating level and coupling strength, but also affects other variables such as dipole moments in the Raman cross section [20].

The relative intensities of the band edge luminescence for quantum dots of four different average sizes are given in Fig. 5. In order to compare the different samples, the intensities were normalized to the average intensity of measurements that were taken below the bulk phase transition pressure point. Increased phase stability with decreasing size is observable. A number of effects may account for this phenomenon. The most important are: effects from the glass matrix, strains from defects, changes in the electronic energy due to confinement, and surface tension. The first three effects contribute energetically too little to the total free energy of the nanocrystalline semiconductor composite [15]. Surface tension however may certainly be the mechanism that brings about the enhanced high pressure phase stability in the nanocrystalline composites. If the surface tension in the rocksalt phase is larger than in the wurtzite phase, the Gibbs free energy will not be minimum, due to its additional energy term, and this energy must be overcome in the phase transition. The result is a shift in the phase transition pressure to higher values. Such a model has been constructed [15] and it does give a qualitative explanation to the observed pressure enhanced phase stability of nanocrystalline semiconductor in a glass composite.

The elevated phase transition pressure for the colloids as opposed to the nanocrystalline semiconductor composite suggests also an enhancement of the structural phase stability. To account for this elevated phase transition pressure in the colloidal quantum dots, a model simulating a high concentration of defects (volume effect) and a second model simulating surface effects are put forth.

A model to take account of the volume effects postulates the simultaneous existence of a solid particle and a static liquid particle [22]. Each quantum dot is simulated by the coexistence of a solid particle and a glass-like (below T_g) high viscosity liquid particle. The interaction between them is assumed to be weak. The solid particle has the same crystal structure as the bulk material. The liquid particles here stand for a random three-dimensional network as is postulated for glass and it also implies a certain ease of diffusion of atoms. The ratio of the size of the solid particle to the liquid particle may be scaled by the size of the quantum dot.

Details of such a model and its calculations are given elsewhere [16]. By considering the disorder effects in the colloid nanocrystalline material one may construct the free energy curves relative to the ground state of bulk CdS at ambient pressure (see Fig. 6). From this graph it is seen that the phase transition activation energy for colloidal CdS is higher than that of bulk CdS. However, when the pressure is released the activation energy for colloidal CdS is smaller than its bulk value. This implies that for colloidal CdS the transition from the wurtzite phase to the rocksalt phase occurs at higher pressure than in the bulk CdS and in the reverse case the bulk CdS needs more energy to recover its ambient pressure structure.

If the entropy due to the defects is constant as a function of volume, the Gibbs free energy curve of the colloids would have the same shape as that of the bulk CdS. If this is true the behavior of the phase transition curve of colloidal and bulk CdS should be the same. But our results show that colloidal CdS has an elevated phase transition stability, see Figs 2-4; and does not have a hysteresis [16] as observed in bulk CdS. As a consequence, as shown in Fig.6, for colloidal CdS to overcome the energy barrier much more energy and higher external pressure is needed than in the bulk case and this explains why higher pressure for the colloids is required. On the other hand, if the hysteresis is not observed upon releasing the pressure [16], this can simply be explained by the lower energy barrier for the colloids. Hence, entropy variation may be responsible for the novel properties reported here. Therefore, the elevated phase transition pressure for colloidal CdS may be due to the migration of the defects.

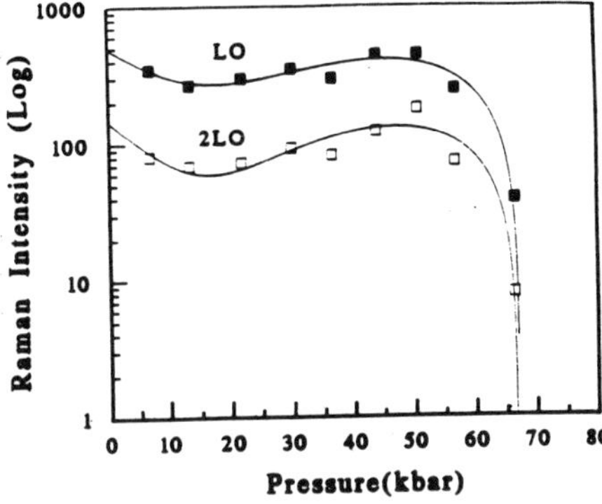

Fig. 3 The intensity of the LO and 2LO Raman peaks for 1 month aged CdS colloids at different pressures.

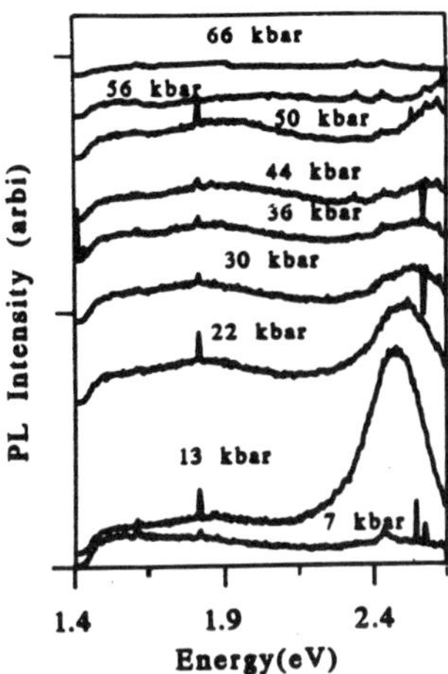

Fig. 4 The photoluminescence spectra for 1 month aged CdS colloids at different pressures. The spikes emerging at higher pressures are due to the diamond anvils (around 2.25 eV) and CdS Raman peaks (around 2.60 eV).

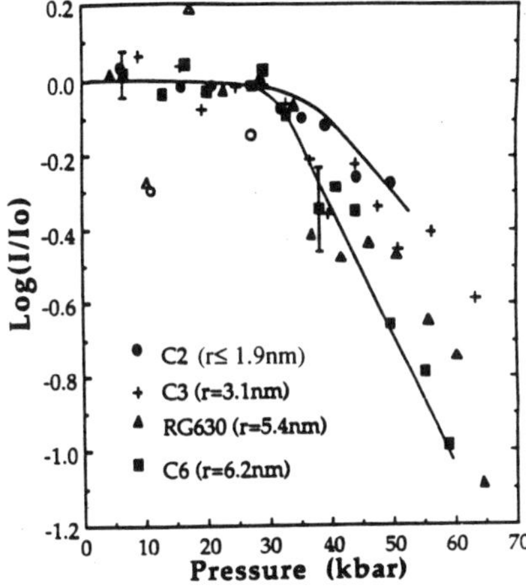

Fig. 5 Phase stability for different average sized quantum dots. The luminescence intensities were normalized to the average intensities below 30 kbar. Open figures were considered as outliers. The lines are only guides for the eye to appreciate the change in phase stability.

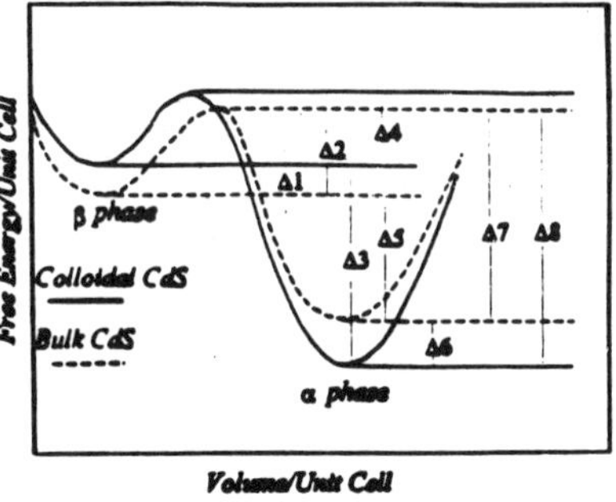

Fig. 6 Schematic of colloidal and bulk CdS free energy as a function of volume. Free energy difference: Δ1, plastic deformation for bulk; Δ2 (Δ4), energy barrier for bulk (colloids) after releasing pressure; Δ3 (Δ5), driving force for colloids; Δ6, entropy arises from amorphous property of colloids; Δ7 (Δ 8), energy barrier for bulk (colloids) when increasing pressure.

SUMMARY AND CONCLUSION

From photoluminescence and Raman measurements we determined that the structural phase transition due to pressure in the nanocrystalline systems is vastly different from the same bulk material. A size dependent study on nanocrystalline semiconductor composites showed a correlation between size and increase in phase stability. For the nanocrystalline semiconductor composite a model, based on surface tension between the semiconductor nanocrystallite and the surrounding glass matrix, allowed us to predict the observed behavior. For the colloidal nanocrystalline material a somewhat different explanation was given. That is volume effects (defects) and surface tension are both important contributors.

REFERENCES

1. K.A. Muller and H. Thomas, Structural phase transitions-I, Springer Verlag, Berlin, 1981.
2. M. Haase and A.P. Alivisatos, J. Phys. Chem., **96**, p. 6756 (1992).
3. S.H. Tolbert and A.P. Alivisatos, Z. Phys. D, **26**, p. 56 (1993).
4. S.H. Tolbert and A.P. Alivisatos, J. Chem. Phys., **102**, p. 4642 (1995).
5. A.P. Alivisatos, Bull. Mat. Res.Soc., **8**, p. 23 (1995).
6. A.N. Goldstein, C.M. Echer, and A.P. Alivisatos, Science, **256**, p. 1425 (1992).
7. P. Buffat and J.P. Borel, Phys. Rev. A, **13**, p. 2287 (1976).
8. S.H. Tolbert and A.P. Alivisatos, Science, **265**, p. 373 (1994).
9. S.H. Tolbert, A.B. Herhold, C.S. Johnsson, and A. P. Alivisatos, Phys. Rev. Lett., **73**, p. 3266 (1994).
10. A.P. Alivisatos, T.D. Harris, L.E. Brus, and A. Jayaraman, J. Chem. Phys., **89**, p.5979 (1988).
11. X.S. Zhao, J. Schroeder, P.D. Persans, and T.G. Bilodeau, Phys. Rev. B, **43**, p. 12580 (1991).
12. X.S. Zhao, J. Schroeder, M.R. Silvestri, T.G. Bilodeau and P.D. Persans, in Clusters and Cluster Assembled Materials, ed. R.S. Averback, J. Bernholc, and D.L. Nelson, MRS Symposia Proceedings No. 206, (Materials Research Society, 1991), p. 151.
13. J. Schroeder et al., in Chemical Processes in Inorganic Materials: Metal and Semiconductor Clusters and Colloids, ed. P.D. Parsens, J.S. Bradley, R.R. Chianelli and G. Schmid, MRS Symposia Proceedings No. 272, (Materials Research Society, 1990), p. 251.
14. N.M. Lawendy and R.L. MacDonald, J. Opt. Soc. Am. B, **8**, p. 1307 (1991).
15. M.R. Silvestri and J. Schroeder, J. of Phys. Condensed Matter, **7**, p. 8519 (1995).
16. J. Schroeder, L.W. Hwang, M.R. Silvestri, Mierie Lee and P.D. Persans, J. Non-Crystalline Solids, **203**, p.217 (1996).
17. P.D. Persans et al., in Materials Issues in Microcrystalline Semiconductors, ed. P.M. Fauchet, K. Tanaka, and C.C. Tsai, MRS Symposia Proceedings No. 164, (Materials Research Society, 1990), p. 105.
18. G. Mei, S. Carpenter, L.E. Felton, and P.D. Persans, J. Opt. Soc. Am. B, **9**, p. 1394 (1992).
19. J. Schroeder and G. Kennendy, Encyclopedia of Physics, ed. R. Lerner and G. Trigg (VCH Press,1990), p. 1315.
20. M.R. Silvestri and J. Schroeder, Phys. Rev. B, **50**, p. 15108 (1994).
21. S.S. Mitra, O. Brafman, W.B. Daniels and W.D. Crawford, Phys. Rev., **186**, p. 942 (1969), O. Brafman and S.S. Mitra, Proc. 2nd Inter. Conf. on light scattering in Solid, ed. M. Balkanski, Flammarion, Paris, 1971, p.284.
22. S. Valkealahti and M. Manninen, Z. Phys. D., **26**, p. 255 (1993).

SPECTRAL DIFFUSION OF ULTRA-NARROW FLUORESCENCE SPECTRA IN SINGLE QUANTUM DOTS

S.A. EMPEDOCLES, D.J. NORRIS and M.G. BAWENDI
Chemistry Department, Massachusetts Institute of Technology, 77 Massachusetts Ave. Cambridge, Ma. 02138

ABSTRACT

We collect and spectrally resolve photoluminescence from single CdSe nanocrystallite quantum dots. By eliminating spectral inhomogeneities, we reveal resolution limited linewidths < 120μev at 10K. These lines are more than fifty times narrower than what has previously been reported using ensemble measurements. Light driven spectral diffusion is seen as a form of power broadening and may be the cause of surprisingly broad linewidths at room temperature. In addition, we see no evidence of excited state emission or coupling to acoustic phonons.

INTRODUCTION

Low dimensional materials such as quantum wells (QWs) have had a profound impact in the field of semiconductor physics, yielding numerous fundamental observations as well as important optoelectronic applications. Due to the reduced number of thermally accessible states, low dimensional structures display significantly enhanced efficiencies in many devices relative to bulk materials. Quantum dots (QDs), the zero-dimensional analogs of QWs, represent the ultimate in semiconductor based quantum confined systems.[1-12] These structures, often referred to as artificial atoms, are predicted to have discrete, atomic-like energy levels, and a spectrum of ultra-narrow transitions that is tunable with dot size.[1-3] Potential applications for QDs include light emitting diodes, lasers, and other optoelectronic devices.[4]

Nanocrystallite QDs synthesized as colloids are particularly flexible building blocks for quantum dot heterostructures. They can be produced in macroscopic quantities with a high degree of reproducibility and control, diameters that are tunable during synthesis (~15-100Å) in a narrow size distribution (<5% rms), and with a uniform shape.[5] Nanocrystallite QDs can be incorporated into a variety of polymers as well as thin films of bulk semiconductors.[6] They can also be manipulated into close packed glassy thin films[7], ordered three dimensional superlattices (colloidal crystals)[8] or linked to form QD molecules.[9]

An inherent complication in the study of ensembles of dots is the loss of spectral information resulting from structural and environmental inhomogeneities.[10] While the discrete nature of the spectrum has been verified for CdSe quantum dots[2], transition linewidths appear significantly broader than expected. This is true even when size selective optical techniques are used to extract homogeneous linewidths.[2,10] Previous attempts to obtain luminescence spectra from single nanocrystallite QDs using two photon microscopy have also failed to reveal narrow features.[11] As a result, broad linewidths were thought to be an inherent limitation of colloidal QDs, hindering their study and restricting their potential use in QD heterostructures. These dots were thought to be inherently different from the more traditional QD structures fabricated using epitaxial growth techniques which, individually, do show ultra-narrow luminescence linewidths.[12]

Since its introduction, single molecule spectroscopy has been successful in extracting new microscopic properties from ensemble systems.[13,14] In this report we use far-field microscopy to image and obtain ultra-narrow single dot luminescence (SDL) spectra from single CdSe

Mat. Res. Soc. Symp. Proc. Vol. 452

nanocrystallites at 10K. The elimination of spectral inhomogeneities reveals new spectral phenomenon including light driven spectral diffusion which is explored in further detail.

EXPERIMENT

Single QD images and spectra were taken using a far-field epi-fluorescence microscope, described elsewhere.[15] Two types of dots were used in this study. The first, referred to as "standard" dots were synthesized using the method of Ref. 5. The second referred to as "overcoated" dots were prepared in the same manner but with the addition of a final layer of ZnS.[16] A ZnS or ZnSe overcoating has been shown to protect dots from physical deterioration during processing[6] as well as enhance QY.[17] Both types of dots are coated with an organic surface layer which effects solubility and passivates non-radiative surface relaxation pathways.[5]

Samples were prepared by spin-coating a 500Å layer of poly(methyl methacrylate) onto a crystalline quartz substrate followed by a layer of QDs at extreme dilution in hexane. The concentration of dots was chosen to produce a lateral dispersion of < 1 dot/μm^2 in order to facilitate spatial resolution using far-field optics. The sample was then placed in a liquid helium cryostat and cooled to 10K.

RESULTS

A typical image of standard dots at 10K can be seen in figure 1a. Each dark spot corresponds to the fluorescence from a single dot. As can be seen, individual dots fluoresce with different intensities within the same image, the brightest producing > 500counts/sec (~10^5 photons emitted/sec, assuming an isotropic source). Differences in fluorescence intensity are attributed to variations in the quality of the organic passivating layer on individual dots.

On this time scale, individual dots are occasionally seen to turn off and then back on again during consecutive scans. These "off" periods of darkness can be very short (on the order of 1 sec) or as long as 10 min in some cases. This on/off behavior has been shown to be intensity dependent and is thought to be the result of QD photoionization.[18] It is also strong evidence that we are observing single QDs. If the individual spots in figure 1a were the result of aggregates of QDs, we would expect to see a gradual dimming as individual dots turned on and

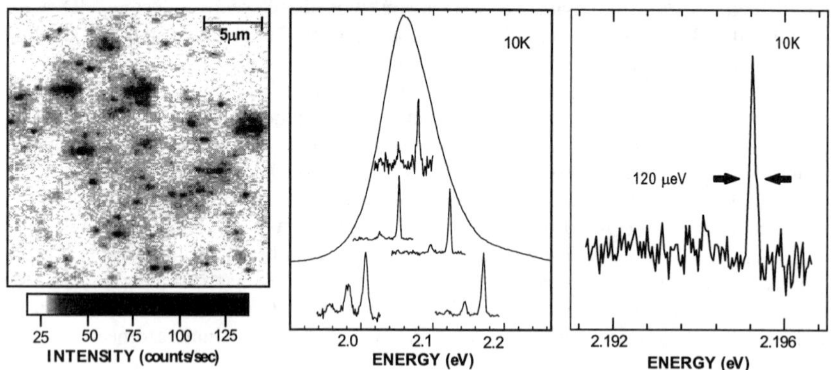

Figure 1 (a). Image of standard dots at 10K with 1 sec integration time and 150W/cm^2 excitation intensity. In this image, we detect ~5% of the total photons emitted (assuming and isotropic source) with a spatial resolution of ~0.5μm. (b) Several SDF spectra with the corresponding ensemble spectrum from that sample. (c) Ultra-narrow SDF spectrum taken at 25W/cm^2 excitation intensity.

off within the ensemble. The fact that individual spots show an all on/all off behavior strongly suggests that we are observing single QDs.

Spatial isolation of individual dots within the image field is followed by spectral dispersion of the fluorescence signal. Figure 1b shows a representative sample of SDF spectra along with the full ensemble fluorescence from the same sample. As can be seen, SDF spectra show a significantly reduced linewidth relative to the ensemble spectrum. Individual SDF spectra show qualitatively similar characteristics (a phonon progression with average coupling comparable to that measured using ensemble techniques[15] and no observable emission from higher excited states). In addition, the distributed intensity of the SDF spectra within a sample can be used to reproduce the intensity distribution of the ensemble spectrum.

While the spectra displayed in figure 1b are significantly narrower than what can be obtained using ensemble techniques, the observed linewidths are broadened by a form of power broadening. At sufficiently low excitation intensities, this broadening can be reduced to reveal resolution limited linewidths as narrow as 120μeV at 10K (figure 1c).[19] As these lines are resolution limited, we can only report an upper bound for the true homogeneous linewidth. However, even at this resolution, the observed linewidths are significantly narrower than kT, suggesting the absence of coupling to low energy acoustic phonons.

In these experiments, we observe a wide range of linewidths (0.1-10meV) for dots under different conditions. Differences occur between dots at a single intensity and within the same dot over a range of intensities.

Observation of a SDF spectrum over many minutes reveals that the peak emission energy changes with time, similar to the spectral diffusion observed in single molecules. Spectral diffusion in these nanocrystallites is highly power dependent, with some dots at higher excitation intensities (2.5kW/cm^2) shifting as much as 80meV over several minutes. The rate and extent of spectral diffusion varies from dot to dot.

Spectral diffusion provides additional evidence that we are taking spectra from single QDs. Observation of several SDF spectra at once reveals that individual spectra shift independently. As this is the case, it would be expected that multiple, overlapping SDF spectra would split over time. This is not observed.

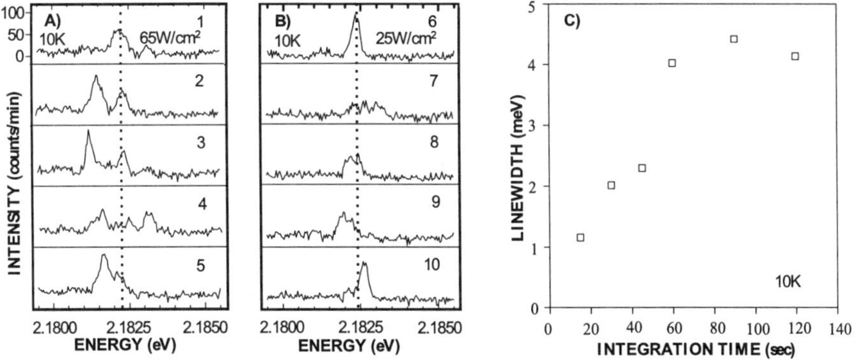

Figure 2 (a) Five consecutive one minute spectra of a single overcoated dot at 65W/cm^2 excitation intensity and 10K. (b) The next five consecutive minutes of the same dot at 25W/cm^2. (c) Plot of linewidth (ZPL) of a single dot as a function of integration time at constant intensity and 10K.

Figures 2a and b show a series of high resolution spectra of a SDF zero phonon line (ZPL) over several minutes at two different excitation intensities. These spectra show a narrow peak shifting over a range of ~2meV over several minutes, often changing energy during a single scan. As the excitation intensity is reduced, the rate and extent of spectral shifting decreases. At lower resolution and/or longer integration times, this shifting results in a form of power broadening due to the sampling of many spectral position during a single integration time. Figure 2c plots the average linewidth of a single dot at constant excitation intensity over a range of integration times. We see an increase in linewidth at longer times, with saturation occurring for times > 1 minute. Similar to single molecule spectroscopy and spectral hole burning, the linewidth of a SDF spectra is dependent on the time over which the measurement is made.[14,20] At this intensity, different dots yield different saturation linewidths.

At higher excitation intensities, much larger spectral shifts are observed. Figure 3a shows 16 consecutive minutes of a low resolution SDF spectra taken at 2.5kW/cm^2. This progression shows a large, red shift over time, accompanied by a decrease in fluorescence intensity and an increase in phonon coupling. This shift is reversible, with a complete recovery of fluorescence intensity and phonon coupling as the spectra begins to shift back to higher energies.

These characteristics are consistent with a stark effect.[21] Low temperature stark studies show a red shift of emission with a corresponding decrease in band edge emission and an increase in LO phonon coupling as electric field strength is increased. Spectral diffusion may be the result of changing local electric fields. Room temperature studies of single CdSe QDs suggest the occurrence of QD photoionization, resulting in periods of darkness (no photons emitted).[18] These dark periods are the result of very rapid Auger relaxation of an exciton within a charged QD (one with an unpaired charge carrier). Auger relaxation occurs when a second electron-hole pair is created within the charged particle and can then relax non-radiatively, transferring its energy to the remaining charge carrier. The remaining, highly excited lone carrier then releases its energy through rapid vibrational relaxation, leaving a charged (dark) particle in a catalytic-like cycle. The QD remains dark until the core is neutralized by either the return of the ejected charge carrier or by the ejection of the remaining carrier. The probability of

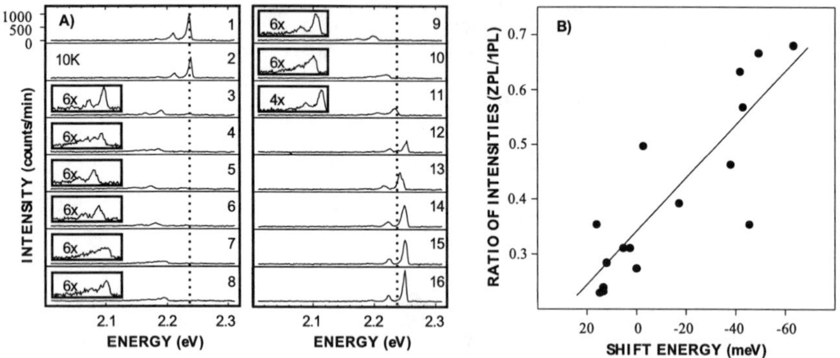

Figure 3 (a) 16 consecutive 1 minute spectra of a single standard dot taken at 2.5 kW/cm^2 excitation intensity. Insets show magnification of the y-axis by the indicated amounts. (b) Plot of the ratio of intensities of the 1 phonon line/ZPL vs the shift position of the ZPL relative to the initial position. This data is taken directly from fig. 3a.

the latter case is greatly increased by this highly energetic Auger process during subsequent excitation events.

In the event of bi-ionization, the product is a neutral (emitting) QD in the presence of a potentially large electric field resulting from the nearby localized charge carriers. The magnitude of this field is dependent on the positions of the electron and hole, relative to the QD core. Each bi-ionizing event or recombination of trapped carriers should then be accompanied by a corresponding large Stark shift.

Theory predicts and experiments confirm that emission energy and phonon coupling both vary as the square of the electric field across the QD core.[21] The observed linear relation between phonon coupling and shift magnitude in figure 3b is consistent with the proposed Stark mechanism. Variations in organic surface coverage would explain differences in the frequency of carrier trapping, resulting in particle to particle variations in the extent and rate of the large spectral shifts.

Once trapped, ejected charge carriers may relocalize by relaxing into different local trap states either though thermal or light induced pathways. The result is a redistribution of charge around the QD core yielding local electric fields that vary with time. These smaller, more continuous field changes may result in the observed small spectral shifts seen in figure 2a and b. As multiple bi-ionizing events occur, shifting the spectrum to lower energies, the large number of trapped carriers would be expected to produce the increased line broadening observed in figure3a (minutes 3-10).

While spectral diffusion is clearly evident in these low temperature studies, it may also play an important role in the room temperature spectra of single QDs. Room temperature SDF spectra reveal asymmetric, featureless, broad peaks with a peakwidth ~50-70meV (figure 4a). This broadening at room temperature has been attributed to thermal broadening of the spectral lines. However, the low temperature data suggest the absence of coupling to low energy acoustic phonons, which would limit the observed effects of thermal broadening. Instead, the broad lineshape may be the result of very rapid, thermally induced spectral diffusion. The range of energies covered by the largest spectral shifts at 10K and the asymmetry of the intensity distribution seen in figure 3a are consistent with the observed room temperature lineshape. In addition, we see a qualitatively similar dependence of peakwidth on integration time as was seen at 10K (figure 4b), consistent with the effects of spectral diffusion.

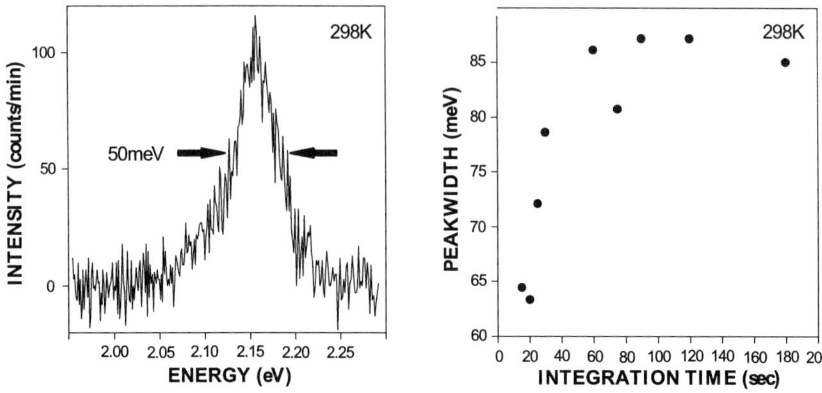

Figure 4 (a) Room temperature SDF spectrum of an overcoated dot at 150W/cm^2 excitation intensity. (b) Plot of peakwidth vs integration time for a single dot at room temperature.

CONCLUSIONS

We have presented evidence of the zero-dimensional nature of the emitting state in CdSe nanocrystallite QDs, reinforcing the description of these structures as "artificial atoms". We observe light driven spectral diffusion and suggest a self induced Stark mechanism to describe the effect. Spectral diffusion has been shown to result in homogeneous line-broadening at 10K and may also be the cause of the broad SDF lineshapes observed at room temperature.

ACKNOWLEDGEMENTS

We thank Manoj Nirmal, Jay Trautman, Louis Brus, and Alexander Efros for invaluable conversations and advice, and Bashir Dabbousi for the synthesis of overcoated dots. S.A.E. and D.J.N benefited from DoD and NSF fellowships respectively. M.G.B. thanks the David and Lucille Packard Foundation and the Sloan Foundation for Fellowships. This research was funded in part by NSF grants DMR-91-57491 and by the NSF-MRSEC program (DMR-94-00034). We also thank the MIT Harrison Spectroscopy Laboratory (NSF-CHE-93-04251) for support and use of its facilities.

REFERENCES

[1] Al.L. Efros and A.L. Efros, Sov. Phys. Semicond. **16**(7), 772 (1982). L.E. Brus, J. Chem. Phys. **80**, 4403 (1984).
[2] D.J. Norris and M.G. Bawendi, Phys. Rev. B **53**, 16338 (1996). D.J. Norris *et al.*, Phys. Rev. B **53**, 16347 (1996).
[3] Al.L. Efros *et al*, Phys. Rev. B **54**(7), 1 (1996).
[4] Weisbuch, J. of Cryst. Growth **138**, 776 (1994).
[5] C.B. Murray, D.J. Norris, and M.G. Bawendi, J. Am. Chem. Soc. **115**(19), 8706 (1993).
[6] M. Danek, K.F. Jensen, C.B. Murray, and M.G. Bawendi, J. Cryst. Growth **145**, 714 (1994); Chem. Mater. **8**(1), 173 (1996).
[7] C.R. Kagan, C.B. Murray, M. Nirmal, and M.G. Bawendi, Phys. Rev. Lett. **76**(9), 1517 (1996).
[8] C.B. Murray, C.R. Kagan, and M.G. Bawendi, Science **270**, 1335 (1995).
[9] A.P. Alivisatos *et al*, Nature **382**, 609 (1996).
[10] A.P. Alivisatos *et al*, J. Chem. Phys. **90**(7), 3463 (1989).
[11] S. A. Blanton, A. Dehestani, P. Lin, and P. Guyot-Sionnest, Chem. Phys. Lett **229**, 317 (1994).
[12] R. Leon, P.M. Petroff, D. Leonard, and S. Fafard, Science **267**, 1966 (1995).
[13] W.E. Moerner, Science **265**, 46 (1994) and references therein.
[14] J.K. Trautman, J.J. Macklin, L.E. Brus, and E. Betzig, Nature **369**, 40 (1994).
[15] S.A. Empedocles, D.J. Norris and M.G. Bawendi, Phys. Rev. Lett. **77**(18), 3873 (1996).
[16] B.O. Dabbousi *et al*, (to be published).
[17] M.A. Hines and P. Guyot-Sionnest, J. Phys. Chem. **100**(2), 468 (1996).
[18] M. Nirmal *et al.*, Nature **383**, 802 (1996).
[19] Individual linewidths at any given intensity vary, with most SDF spectra dropping below the background before reaching the resolution limit. As a result, the narrowest lines are observed in overcoated dots. This is most likely a result of higher fluorescence efficiency and not inherently narrower linewidths.
[20] M. Berg *et al*, Chem. Phys. Lett. **139**(1), 66 (1987).
[21] A. Sacra, "Stark Spectroscopy of CdSe Nanocrystallites", Ph.D. Thesis (Dept. of Chemistry, Massachusetts Institute of Technology, Cambridge, Ma. 1996).

ELECTRONIC STRUCTURE OF O-D EXCITON GROUND STATE IN CdSe NANOCRYSTALS

M. CHAMARRO*, M. DIB*, C. GOURDON*, P. LAVALLARD*, O. LUBLINSKAYA** and A. I. EKIMOV**
*Universités Paris VI-VII, Groupe de Physique des Solides, 75251Paris France
**A.F. Ioffe Physical-Technical Institute, 194021 St Petersburg, Russian Federation

ABSTRACT

We present results on photoluminescence excitation spectra (PLE) of wurtzite CdSe nanocrystals (NCs) embedded in a glass with effective radii in the range 15-35 Å. Information on the near band-gap absorption of an assembly of NCs is obtained by selecting a narrow energy range in the inhomogeneously broadened photoluminescence band. The size and shape dependence of the lowest exciton states are calculated for slightly non-spherical wurtzite NCs. The experimental results are in good agreement with the theoretical predictions when both shape and size dispersions are taken into account.

INTRODUCTION

The determination of semiconductor NC electronic structure has been the center of interest of a considerable number of theoretical works in the last ten years. In the strong confinement regime, the NCs electronic states are calculated separately for electrons and holes within the framework of the effective mass approximation[1,2]. Confined electron states are well described by one conduction band while the complexity of the valence states has to be considered for confined hole states.

The NC eigenstates have an excitonic nature and are obtained by taking into account the electron-hole Coulomb interaction. The first effect of this interaction consists in a red shift of the eight-fold degenerate lowest electron-hole pair state. A second effect is the lifting of the degeneracy of the lowest exciton state due to the electron-hole short-range exchange interaction. The exciton lowest state is split into two states: the optically forbidden state and the optically active state, at higher energy. This exchange splitting is extremely small in bulk semiconductors, but is strongly enhanced for NCs in the strong confinement regime[3,4] and should be treated in perturbation theory on the same level as the crystal field and the effect of shape anisotropy in non-spherical wurtzite NCs[5]. The combination of all these effects gives a complicated fine structure of exciton states that has to be compared carefully to experimental spectral studies.

Nowadays high quality CdSe semiconductor NCs with dimensions smaller than the bulk exciton Bohr radius are obtained by using different fabrication methods[6,7]. However, an inherent problem remains: the inhomogeneous broadening of the optical transitions which makes difficult the size dependence study. To reveal single size optical properties, spectroscopic techniques as hole burning, excitation of photoluminescence (PL) with a narrow spectral width in the low-energy side of the first absorption band and PLE experiments with detection of PL in the blue side of the PL band have been used.

Studies of the absorption spectra by hole burning[8] and PLE experiments[9,10] show that the multiple valence band model predicts with a good approximation the energy of the excited electronic levels. On the other hand, the nature of the long-lived and red shifted luminescent state observed in high quality CdSe NCs has remained for a long time a controversial question[11]. This state was finally identified as the optically forbidden exciton state and the enhancement of the electron hole interaction by quantum confinement was clearly demonstrated [12,13]

In this paper, we study the fine electronic structure of the lowest exciton state in CdSe NCs grown in a oversaturated solid solution. Selective spectroscopic technique is used to investigate the size dependence of near band-gap absorption. The characteristics of real samples are taken into account in order to calculate the size and shape dependence of the split states of the lowest exciton. The theoretical predictions are compared with the experimental results.

Mat. Res. Soc. Symp. Proc. Vol. 452 © 1997 Materials Research Society

EXPERIMENTAL

The CdSe NCs are embedded in a silicate or phosphide glass. They are obtained by diffusion-controlled phase decomposition of an oversaturated solid solution. The average radius (R_m= 15Å, 17Å, 19Å, 27Å) of the NCs was determined by small angle X-ray scattering (SAXS) and corrected to take into account the correlation of SAXS results with high resolution transmission electron microscopy data Slightly non-spherical wurtzite NCs without evident structural defects were observed by the latter method[14]. The shape of the NCs studied was found either spherical or slightly oblate. Let us call a (b) the length of the short (long) axis, respectively, and $\mu = (a-b) / (b^2 a)^{1/3}$ the ellipticity parameter. The observed oblate NCs have an ellipticity $\mu = -0.15$. Unfortunately we have not enough images to make a good statistics of the dispersion of ellipticity in a large assembly of NCs. In the following, the effective radius will be defined as the radius R of the sphere, which has the same volume as the ellipsoid , $R^3 = b^2 a$.

RESULTS and DISCUSSION

The electronic structure of the lowest electron-hole correlated pair states depends very much on the specific characteristics of considered NCs (size, shape, crystalline structure, chemical composition...). The confinement energy of electrons and holes in an infinitely deep spherical potential has been calculated by Efros for NCs of cubic crystalline structure[15]. The hole confinement energy depends on the ratio ß of the light and heavy masses. The wavefunction of the lowest electron-hole pair states $|F,F_z,S,S_z>$ is written as the tensorial product of the electron wavefunction $|1S_{1/2},S_z>$ with $S_z=\pm 1/2$ and the hole wavefunction $|1S_F,F_z>$ with F=3/2, $F_z=\pm 3/2, \pm 1/2$.

In order to describe the fine structure of lowest electron-hole correlated states in the studied samples, we have added three perturbations terms having the same order of magnitude. These three terms are: the anisotropy of the crystal field related to the wurtzite crystalline structure, the ellipticity of the infinitely deep potential of confinement and the Coulomb short-range electron-hole exchange interaction[5]. To determine the eigenstates and eigenfunctions of the total Hamiltonian we diagonalize the (8 x 8) matrix representing the whole perturbation Hamiltonian in the sub-space of the degenerate electron-hole pair states. The eigenstates are labelled by F_z+S_z, the projection of the exciton angular momentum J^{exc}=S+F along the crystal axis[16]. They are obtained as a linear combination of the |3/2,Fz,1/2,Sz> NCs states. $|\pm 2>$, $|\pm 1^L>$ and $|\pm 1^U>$ are doubly degenerate states,while $|0^L>$ and $|0^U>$ are non-degenerate. The eigenvalues are written as:

$$E_{\pm 2} = \alpha - \gamma - \frac{3\kappa}{4} \qquad (1.1)$$

$$E_{\pm 1^{U,L}} = \frac{\kappa}{4} \pm \sqrt{(\alpha - \gamma + \kappa)^2 + (\gamma - \alpha)\kappa} \qquad (1.2)$$

$$E_{0^{U,L}} = -\alpha + \gamma + (\tfrac{1}{4} \pm 1)\kappa \qquad (1.3)$$

where superscript L and U mean lower and upper states and correspond respectively to the minus and plus signs in the right-hand expressions. α, γ and κ correspond to the ellipticity, crystal-field and exchange perturbation Hamiltonians, respectively. The expressions of these parameters as a function of NC radius and ellipticity are:

$$\alpha = -\mu u(\beta)\frac{\hbar^2 k(\beta)^2}{2m_A R^2}, \; \gamma = \frac{v(\beta)\delta}{2}, \; \kappa = 4.\chi(\beta)\left(\frac{a_B}{R}\right)^3 E^{Bulk}_{exch} \qquad (2)$$

where the functions u(ß), v(ß) and χ(ß) have been defined by Efros[5]

m_A is the heavy hole mass, δ is the bulk A-B splitting (25 meV), a_B is the bulk exciton Bohr radius (54 Å) and E^{Bulk}_{exch} is the electron-hole exchange energy (0.12 meV) in bulk wurtzite CdSe.

For ß=0.3 we find $\alpha = \frac{5.34\mu\hbar^2}{2m_A R^2}$, $\gamma = 0.47\delta$ and $\kappa = 3.09\left(\frac{a_B}{R}\right)^3 E_{exch}^{Bulk}$.

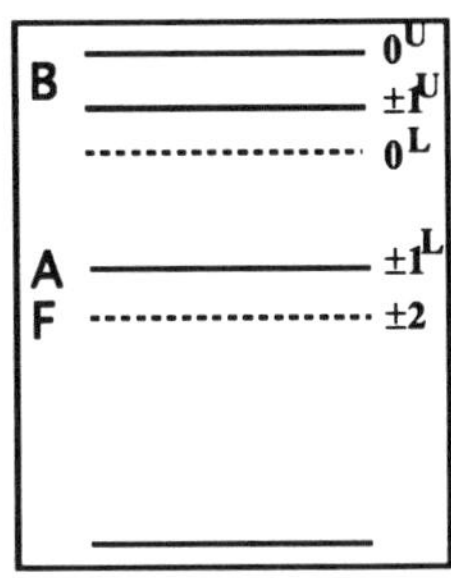

Fig.1 Scheme of NC exciton states.

Fig.1 shows a schematic representation of NC exciton states. Fig.2a shows the calculated energy differences $\Delta_{\pm1^L} = E_{\pm1^L} - E_{\pm2}$, $\Delta_{\pm1^U} = E_{\pm1^U} - E_{\pm2}$ and $\Delta_{0^U} = E_{0^U} - E_{\pm2}$ as a function of the effective NC radius for an ellipticity μ= -0.15. Fig.3 shows $\Delta_{\pm1^U}$, $\Delta_{\pm1^L}$ and Δ_{0^U} as a function of the ellipticity for an effective NC radius equal to 19 Å. We observe that the energy difference $\Delta_{\pm1^L}$ almost does not depend on the ellipticity value but depends on the NC size and increases with decreasing size (Fig.2a). The energy differences $\Delta_{\pm1^U}$ and Δ_{0^U} also increase with decreasing NC size. They depend on the ellipticity and increase with decreasing ellipticity.

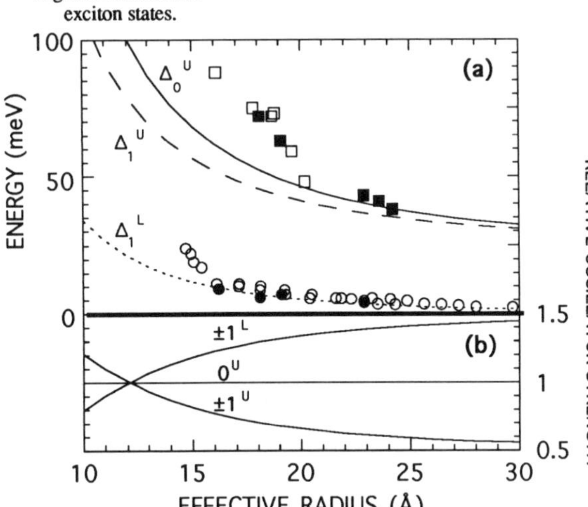

Fig. 2 (a) Size dependence of the calculated energy differences $\Delta_{\pm1^L}$, $\Delta_{\pm1^U}$ and Δ_{0^U} for a fixed ellipticity μ = -0.15. Experimental data are shown by circles and squares, open symbols: difference between the F-line and the excitation energy in the A-states (○) or B-states (❑) from PL data, filled symbols: difference between the detection energy and the A-band (●) or the B-band (■) from PLE data.
2 (b) Calculated oscillator strength of the excitonic states for a fixed ellipticity μ = -0.15

The oscillator strength of the optical transition for the exciton states is proportional to the square optical matrix element: $\left|\left\langle 0|\vec{e}\vec{p}|F_z + S_z{}^{L,U}\right\rangle\right|^2$ where |0> is the ground state, $\vec{e}$ is the polarization vector of the absorbed light, $\vec{p}$ is the exciton momentum operator. The |±2> and |0^L> states are optically forbidden. The |0^U> state can be excited with polarized light with polarization vector parallel to the c-axis of wurtzite NC and its oscillator strength is independent of NC size and ellipticity. The |$\pm1^L$> and |$\pm1^U$> states can be created with polarized light with polarization vector perpendicular to the NC c-axis and the oscillator strength depends on NC size and ellipticity. In order to obtain the relative intensities of the three optically allowed transitions one needs to average over the random distribution of orientation of NCs crystalline axes and to take into account that the |$\pm1^L$> and |$\pm1^U$> states are doubly degenerate but that |0^U> is a non-degenerate state. The dependence on size and ellipticity of the relative averaged oscillator strength of optically allowed transitions to the the |$\pm1^U$> and |$\pm1^L$> states can be written as:

$$f_{U,L} = \frac{3}{2\left(1+d_{U,L}^2\right)}\cdot\left(1-\frac{d_{U,L}}{\sqrt{3}}\right)^2 \quad \text{with } d_{U,L} = \frac{2\alpha-2\gamma+\kappa\mp2\sqrt{(\alpha-\gamma+\kappa)^2+(\gamma-\alpha)\kappa}}{\kappa\sqrt{3}}. \qquad (3)$$

In Fig.2b these expressions are plotted as a function of the NC effective size for a fixed ellipticity (μ= -0.15). We see that the excitation probability of the $|\pm1^L\rangle$ exciton state decreases with decreasing size while the excitation probability of the $|\pm1^U\rangle$ exciton state increases.

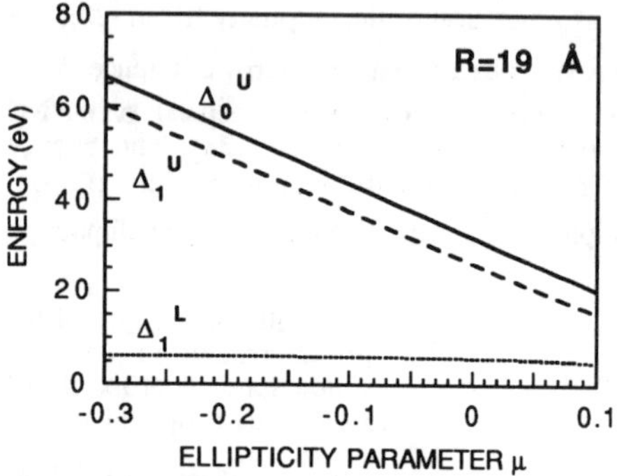

Fig. 3 Shape dependence of the calculated energy of excitonic transitions relative to the energy of $|\pm2\rangle$ excitonic transition for a fixed NC radius of 19 Å.

Fig.4 (a) shows selectively excited PL (left hand) and selectively detected PLE (right hand) spectra obtained at 5K with the same energy of excitation and detection, respectively. These two spectra have been obtained for the sample containing NCs of mean radius 19Å. In this sample the maximum of the first absorption is at 2.219 eV and the maximum of the PL band of the entire distribution is at 2.172 eV. Selectively excited PL has been studied in more details in a previous paper[12]. It consists in a red-shifted band (F-band) and its two phonon replica. It has been demonstrated that the F-band arises from the recombination of the forbidden exciton state which is identified as the $|\pm2\rangle$ state.

The shape of the PLE spectra depends on the detection energy selected in the PL spectrum. Thus, the PLE spectra detected at 2.134 eV and 2.157 eV show broader bands than the PLE spectra detected at higher energies. Three bands, denoted A, A' and B, are seen in the PLE spectrum of fig.4(a). The A-band is always narrower than the B-band. The A-band originates from creation of excitons in the $|\pm1^L\rangle$ states that relax to the $|\pm2\rangle$ states and emit light. The B-band mainly originates from creation of excitons in the $|\pm1^U\rangle$ and $|0^U\rangle$ states which relax to the $|\pm2\rangle$ states. Fig.2a shows the energy difference between the position of the B-band and the detection energy and between the position of the A-band and the detection energy as obtained from selectively detected PLE spectra. The figure also shows the energy difference between the excitation energy and the F-band energy obtained from the selectively excited PL spectra. The energy differences determined experimentally are in clear disagreement with the calculated energy differences for spherical NCs (not shown here). They are however, close to the calculated energy differences in oblate NCs.

For the A-band, there is a good agreement between the experimental data and the calculated energy difference $\Delta_{\pm1^L}$ with ellipticity μ=-0.15. For the B-band, the experimental energy differences are larger than the calculated ones. We explain this discrepancy by the contribution of phonon replicas to the detected luminescence. The energy of the $|\pm2\rangle$ states is determined by a couple of parameters (R, μ). This energy does not vary for a whole set of NCs with slightly different values of the couple (R, μ). In PLE experiment, one detects the F-line from one set of NCs. On the other hand, one detects at the same energy the phonon replicas of the F-line due to other sets of NCs. PL intensity results from the superposition of the F-line from set 1, the F-1LO line from set 2, the F-2LO line from set 3 etc. Each set gives a different spectral contribution to the total PLE specrum. The relative importance of these contributions depends on the population of the corresponding set of NCs and on the exciton-LO phonon coupling strength. Fig.4 (b) shows the calculated PLE spectrum of set 1 of NCs for a detection energy at 2.212 eV and a Huang-Rhys parameter of 0.1. The absorption spectrum is usually a mirror image of the emission spectrum. However, in the case of CdSe NCs, the emitting state is a forbidden exciton state. Therefore the ratio of the LO phonon replica to the zero phonon line is larger in emission than in absorption. We used a Huang-Rhys parameter of 0.8 for emission. The PLE spectrum calculated by taking into account the contribution of sets 1, 2 and 3 is shown in Fig.4 (a) superimposed onto the experimental one. One clearly sees that including the contribution of sets 2 and 3 is of prime importance to fit the spectral position of the B-line in the experimental PLE spectrum. The intensity of the A'-line is also well represented. This line corresponds mainly to the absorption band of the $|\pm1^L\rangle$ states of NCs of set 2 whose F-1LO line is observed at the detection energy.

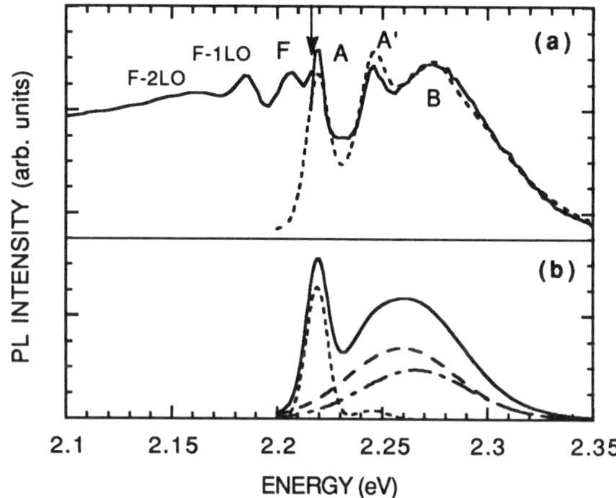

Fig 4 (a) Selectively excited PL spectrum and selectively detected PLE spectrum obtained at 5 K for a sample containing NCs of average size 19 Å. The energy of excitation or detection at 2.212 eV is indicated by the arrow. The calculated PLE spectrum is shown by the dotted line.
(b) PLE spectrum of one set of NCs (see text) with contributions of transitions to the $|\pm1^L\rangle$ states (dotted line), $|\pm1^U\rangle$ states (dashed line) and $|0^U\rangle$ state (dashed-dotted line).

The relative oscillator strength used to simulate the PLE spectra is 0.6 for the transition to the $|\pm1^L\rangle$ states and 1.4 for the transition to the $|\pm1^U\rangle$ states. These values disagree with the calculated ones for NCs of radius R around 19 Å and ellipticity μ in the range 0.10-0.15 (fig.2b) Recently Norris *et al*[17] have reported a similar result for prolate CdSe NCs obtained by wet chemical synthesis. The discrepancy may be due to several reasons. First, we have calculated PLE spectra from absorption spectra neglecting the fact that relaxation from the different excited states to the $|\pm2\rangle$ emitting states may not be fully efficient. Second, according to Goupalov *et al* [18], there is a contribution of the long-range exchange interaction to the exchange splitting. Using a larger value of the exchange term and a smaller value of the ellipticity parameter μ, it is possible to obtain a larger oscillator strength for the transition to the $|\pm1^U\rangle$ states than for the transition to the $|\pm1^L\rangle$ states for R around 19 Å, in better agreement with the experimental results.

Let us finally discuss the width of the A- and B-bands. Fig.4 (a) shows that the full width at half maximum (FWHM) of the B-band is much larger than the FWHM of the A-band. this can be explained by the dispersion of the NC ellipticity. At a given energy one detects PL light from NCs characterized by different couples (R, μ). Let us consider for example that the dispersion of ellipticity is $\Delta\mu = \pm 0.2$. The size dispersion of the NCs which contribute to the selected luminescence light is then a fraction of angstrom. As the energy difference $\Delta_{\pm1^L}$ does not depend on ellipticity but only on the NC effective size, the width of the A band is small, about 1meV in our example. In fact, the line is broadened to 4 meV by the experimental resolution. On the other hand the energy differences $\Delta_{\pm1^U}$ and Δ_{0^U} depend not only on NC size but also on their ellipticity. From the variation of $\Delta_{\pm1^U}$ and Δ_{0^U} with ellipticity parameter μ, (see fig. 3) one deduces that the width of the B band is as large as 60 meV.

CONCLUSION

The fine structure of the excitonic levels in CdSe NCs has been established by a careful comparison of the PLE spectra with theory. The excitonic levels are calculated by taking into account the crystal field anisotropy, the electron-hole exchange interaction and the non-spherical confining potential for slightly oblate wurtzite NCs. The calculated PLE spectra are found to be in good agreement with the experimental ones when on tales into account phonon assisted transition both in emission and absorption. The NCs are ellipsoidal. The strong difference between the widths of the A and B absorption bands is well explained by a different dependence of the energy of the corresponding states on the NC shape.

REFERENCES

[1] T. Richard, P; Lefebvre, H. Mathieu and J. Allègre, Phys. Rev. B **53**, 7287, (1996).and ref. therein

[2] J. B. Xia, J. Lumin. **70,** 120, (1996) and ref. therein.

[3] Takagahara, Phys. Rev. B**47**, 4569 (1993)

[4] M. Nirmal, D. J. Norris, M. Kuno and M.G. Bawendi, Al. L. Efros and M. Rosen, Phys. Rev. Lett,75, 3728,(1995)

[5] Al. L. Efros, M. Rosen, M. Kuno, M. Nirmal, D. J. Norris and M. Bawendi, Phys. Rev. B **54**, 4843, (1996).

[6]C.B. Murray, D. B. Norris and M. G. Bawendi, J. Am. Chem. Soc. **115**, 8706, (1993). J. E.Bowen Katari, V.L.Colvin and A.P. Alivisatos, J. Phys. Chem. **98**, 4109, (1994)

[7] A. Ekimov, J. Lum. **70**, 1, (1996)

[8] D. J. Norris, A. Sacra, C. B. Murray and M. G. Bawendi, Phys. Rev. Lett. **72**, 2612, (1994)

[9] A. I. Ekimov, Phys. Scr. **39**, 217, (1991).

[10] D . J. Norris and M. G. Bawendi, Phys. Rev. B **53**, 16338, (1996)

[11] M. G. Bawendi, W.L. Wilson, L. Rothberg, P. J. Carroll, T.M. Jedju, M.L. Stegerwald and L. E. Brus, Phys. Rev. Lett **65**,1623 (1990).

[12]M. Chamarro, C. Gourdon, P. Lavallard and A.I. Ekimov, Jpn. J. Appl. Phys. **34,** Suppl. 34-1, 12 (1995). M. Chamarro, C. Goudon, P. Lavallard; O Lublinskaya and A. I. Ekimov, Phys. Rev. B**53**, 1336,(1996)

[13] M. Nirmal, D. J. Norris, M. Kuno and M.G. Bawendi, Phys. Rev. Lett,**75**, 3728,(1995)

[14] M. Gandais, private communication.

[15] Al. Efros, Phys. Rev. B **46**, 7448, (1992).

[16] M. Chamarro, C. Gourdon and P. Lavallard, J. Lum. **70**, 222, (1996)

[17] D. J. Norris, Al. L Efros, M. Rosen and M. G. Bawendi. Phys. Rev. B **53**, 16347, (1996).

[18] S.V.Goupalov, E.L.Ivchenco and A.V.Kavokin:Superlattices and Microstructure.(to be published)

THE BAND EDGE LUMINESCENCE OF SURFACE MODIFIED CDSE NANOCRYSTALLITES

M. KUNO, J. K. LEE, B. O. DABBOUSI, F. V. MIKULEC, M. G. BAWENDI
Department of Chemistry, Massachusetts Institute of Technology, Cambridge, MA. 02139, mkkuno@mit.edu

ABSTRACT

We study the band edge luminescence of CdSe nanocrystallites to determine the origin of this emission. Previous studies have attributed the band edge emission to the recombination of photo-generated carriers trapped in localized surface states. Recently a number of "dark exciton" theories have been proposed which explain the luminescence in terms of recombination through internal core states. To address this issue we modify the surface of CdSe nanocrystallites with a number of organic/inorganic ligands and monitor the effect this has on the energetics of the resonant and non-resonant band edge luminescence. Our results for nanocrystallites passivated with trioctylphosphine oxide (TOPO), ZnS, 4-Picoline, 4-(trifluoromethyl)thiophenol, and tris(2-ethylhexyl)phosphate are in agreement with a dark exciton description of the band edge luminescence.

INTRODUCTION

With the recent development of high quality CdSe nanocrystallites or quantum dots (QDs) there have been numerous studies on the linear optical properties of this system.[1] Various theories have been developed that explain changes in the absorption spectrum with decreasing size. These effects arise from the quantum confinement of the excitation to the physical dimensions of the nanocrystallite. While the linear absorption is mostly understood, less is known about the luminescence. For instance, the band edge emission shows an unusually long radiative lifetime (~1μs at 10K) compared to the nanosecond lifetimes seen in the bulk. In addition, the band edge luminescence displays a prominent, size dependent, redshift from the peak of the first absorption feature or $1S_{3/2}1S_e$ transition.

To explain these observations the band edge luminescence has often been attributed to the recombination of surface localized carriers.[2,3] The role of surface related states in the luminescence is a distinct possibility because of the large surface to volume ratios in these structures. Surface reconstruction and the presence of defects potentially create "trap" states with energies within the band gap of the material. These trap states could subsequently act to localize photo-generated carriers and would account for the long lifetimes as well as the redshift of the emission.

While these luminescence results are consistent with proposed surface models they can also be explained with models which do not invoke the role of surface states.[4-7] Instead the luminescence is explained in terms of recombination through internal core states. We address the issue of whether the band edge emission is surface related by systematically altering the surface of CdSe nanocrystallites with a variety of organic and inorganic ligands. We perform photoluminescence (PL) measurements in conjunction with fluorescence line narrowing (FLN) experiments to see if changing the surface affects the energetics of the radiative recombination. Our results are found to be in good agreement with an intrinsic exciton description of the band edge luminescence.

Mat. Res. Soc. Symp. Proc. Vol. 452 © 1997 Materials Research Society

EXPERIMENT

We prepare nearly monodisperse CdSe nanocrystallites using the procedure described in ref. 8. This technique, based on the pyrolysis of organometallic precursors, yields nanocrystallites of tunable size which are slightly prolate and have surfaces passivated by trioctylphosphine oxide (TOPO) and trioctylphosphine selenide (TOPSe). The resulting nanocrystallites luminesce strongly at the band edge (QY~10-20% at room temperature, 10-90% at 10K) and show very little deep trap emission.

To chemically modify the surface of these nanocrystallites, we repeatedly expose the QDs to a large excess of new coordinating ligands. We perform ligand exchange with ZnS, 4-picoline, 4-(trifluoromethyl)thiophenol and tris(2-ethylhexyl)phosphate. More details about the exchange procedure, including characterization of the resulting material, can be found in ref. 9.

Absorption, photoluminescence (PL) and fluorescence line narrowing (FLN) experiments were performed on the same optical setup. Transmitted light from a 300W Hg-Xe lamp was used in the absorption experiment. The same light was also passed through a spectrometer to provide excitation for PL measurements. Typically, samples were excited far to the blue of their first absorption feature with spectrally broad light (~50nm FWHM). An optical multichannel analyzer (OMA) and 0.33m spectrometer was used to detect the transmitted/emitted light. In the FLN experiment, the output of a Q switched Nd:YAG/dye laser system excited samples on the red edge of their first absorption feature. The laser intensity was attenuated to ensure a linear response of the luminescence with the excitation. Emitted light was collected and dispersed with the spectrometer/OMA combination with the OMA gated to reduce scattered laser light.

RESULTS

Figure 1a presents the low temperature (10K) absorption and luminescence spectra of ~22Å radius QDs passivated with (a) TOPO/TOPSe, (b) ZnS, (c) 4-(trifluoromethyl)thiophenol, and (d) 4-picoline. The solid line is the absorption spectrum of each sample. The dotted line is the accompanying band edge luminescence. Figure 1b shows the FLN spectrum (solid lines) obtained when we excite each sample on the red edge of its low temperature absorption. The resulting line narrowed spectrum shows a well resolved longitudinal optical (LO) phonon progression and is characteristic of the emission from a narrow subset of QDs present in the residual size distribution of the sample. The dotted line is the laser excitation included for reference purposes.

In Figure 1a the band edge luminescence is redshifted from the peak of the first absorption feature. This shift, which we refer to as the non-resonant Stokes shift, is highly size dependent and ranges from 20 meV at large sizes to approximately 100 meV at small sizes. Similarly, in Figure 1b, the peak of the zero phonon line is redshifted with respect to the laser excitation. This shift, referred to as the resonant

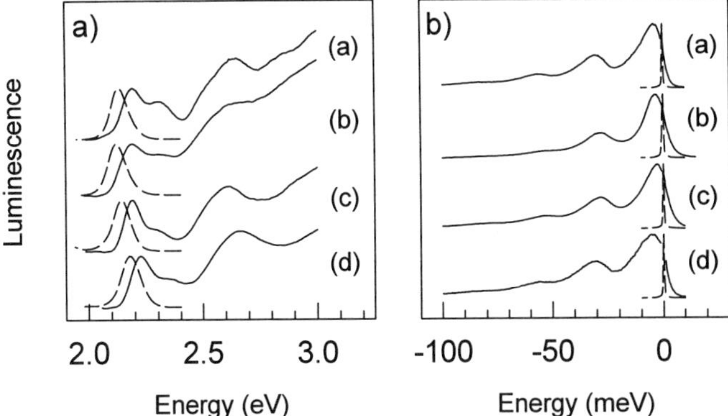

Figure 1. a) Low temperature absorption (solid line) and PL (dashed line) of 22Å QDs. b) Accompanying fluorescence line narrowed spectra (solid lines) with laser excitation (dashed lines) included for reference puroposes.
(a) QDs with TOPO/TOPSe, (b) ZnS, (c) 4-(trifluoromethyl)thiophenol, (d) 4-picoline

Stokes shift, ranges from 2 meV at large sizes to approximately 20 meV at small sizes. A discrepancy, however, exists between these two experimentally observed shifts. Assuming a single absorbing and single emitting state, it should be possible to reproduce the PL from the FLN spectra. This is not the case. Figure 2 shows the situation for 18Å radius QDs. The experimental absorption and PL spectra are shown in solid lines. Convoluting the line narrowed spectrum with the intrinsic size distribution of the sample ($\sigma \sim 5\%$) produces the dashed line. Although the shape and width of the PL are well matched, the predicted non-resonant Stokes shift is much smaller (74 meV vs. 12 meV). The band edge luminescence is therefore anomalous because it requires too large a Huang-Rhys S parameter to be explained by simple electron-phonon coupling.

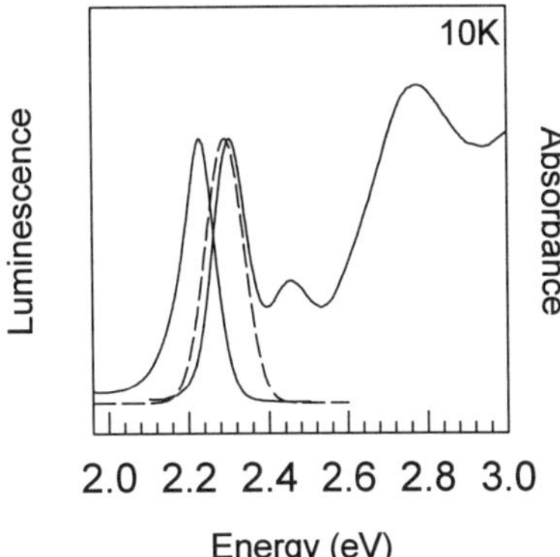

Figure 2. Low temperature absorption and PL spectra of 18Å radius QDs (solid lines). The dashed line is the result of convoluting the FLN spectrum with the intrinsic size distribution of the sample. The predicted non-resonant shift (12 meV) is smaller than the measured value (74 meV).

To explain this discrepancy we turn to recent theory considering effects such as the hexagonal lattice structure, QD asymmetry, and the electron-hole exchange interaction which lift the eightfold degeneracy of the $1S_{3/2}1S_e$ transition.[7] The theory predicts the presence of fine structure within the first optical transition of these QDs. It also postulates that the exciton ground state (or luminescing state) is an optically passive "dark" exciton. Figure 3a shows the predicted fine structure. States are labeled with the value of their angular momentum projection along the "c" axis of the nanocrystallite. Optically active (passive) states are shown in the solid (dashed) lines. In terms of the theory, exciting on the red edge of the first absorption maxima probes the lowest optically active $\pm1^L$ state. Non-radiative relaxation to the $\pm2^L$ dark exciton follows. Radiative recombination to the ground state is then achieved through a phonon assisted or nuclear/paramagnetic spin flip transition. The difference in energy between the $\pm1^L$ and $\pm2^L$ states is the resonant Stokes shift. Figure 3b compares experiment to theory. There is agreement at large sizes but as the size of the QD decreases the theory begins to increasingly underestimate the resonant Stokes shift. This may be related to the presence of acoustic phonons and is not considered by the theory.

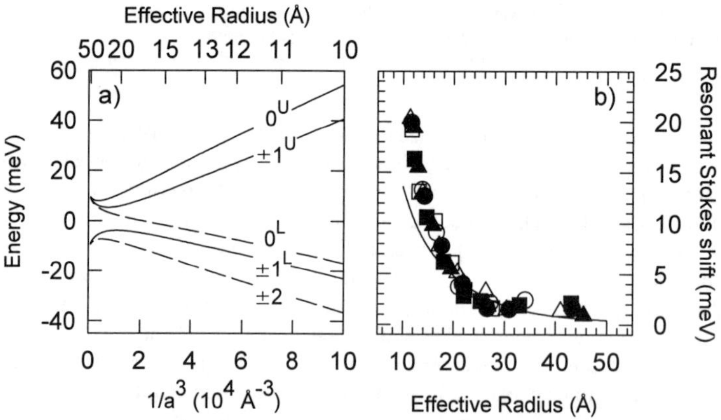

Figure 3. (a) Theoretical prediction of the size dependent band edge exciton fine structure. States are labeled by their angular momentum projection along the "c" axis of the nanocrystallites. Optically active (passive) states are denoted by solid (dashed) lines. The superscripts "U" and "L" refer to upper and lower to distinguish states with the same angular momentum projection. (b) Resonant Stokes shift plotted against the mean effective radius of each sample. The theory is shown as the solid line. (solid triangles) QDs with TOPO/TOPSe, (open squares) ZnS, (solid circles) TOPO/TOPSe in BuOH, (open triangles) 4-picoline, (open circles) tris(2 ethylhexyl)phosphate, (solid squares) 4-trifluoromethyl)thiophenol

The data also shows the insensitivity of the resonant shift to the nature of the passivated surface. Figure 4 plots the resonant Stokes shift against excitation energy for QDs passivated with TOPO/TOPSe, ZnS, 4-picoline, 4-(trifluoromethyl)thiophenol and tris(2-ethylhexyl)phosphate. We graph the data in this way because of difficulties in determining the exact size that we probe when exciting on the red edge of each sample's absorption. Moving the excitation changes the value of the measured resonant Stokes shift but Figure 4 shows this to be consistent with exciting larger or smaller nanocrystallites present in the residual size distribution of each sample. The reproducibility of the shifts strongly suggests that the luminescence is unrelated to the surface.

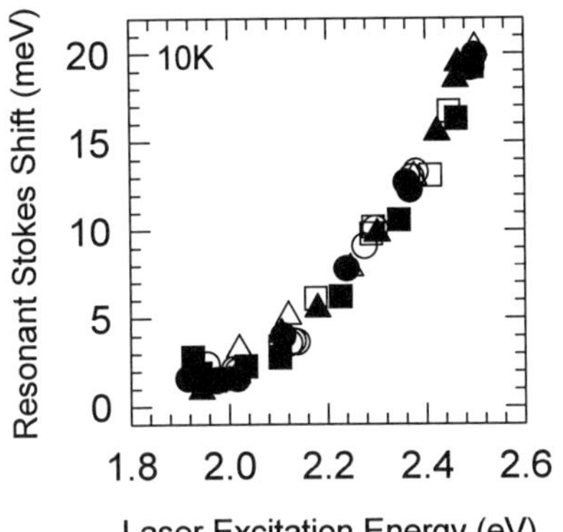

Figure 4. Resonant Stokes shift plotted against laser excitation energy. (solid triangles) QDs with TOPO/TOPSe, (open squares) ZnS, (solid circles) TOPO/TOPSe in BuOH, (open triangles) 4-picoline, (open circles) tris(2 ethylhexyl)phosphate, (solid squares) 4-trifluoromethyl)thiophenol

To explain the non-resonant Stokes shift we consider the presence of three optically active fine structure states in the first transition. The absorption maxima occurs at the energetic mean of these three states when their oscillator strengths and the sample size distribution are taken into account. The non-resonant Stokes shift is described as the difference in energy between the mean energy of the three optically states and the exciton ground state. Figure 5 shows a comparison of experiment and theory at 10K for all instances of surface modification. There is reasonable correlation between the theory (dashed line) and the measured non-resonant Stokes shifts from surface modified QDs (symbols). The discrepancy may stem from a tendency of the model to underestimate the splitting and oscillator strengths of the various fine structure states.

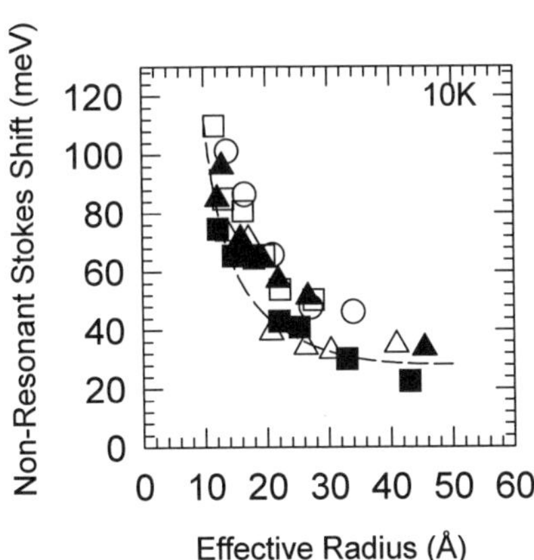

Figure 5. Non-resonant Stokes shift plotted against the mean effective radius of each sample. The theory is the dashed line. It considers the role of phonons, a 10% size distribution and assumes that all emission originates from the dark exciton ground state. (solid triangles) QDs with TOPO/TOPSe, (open squares) ZnS, (solid circles) TOPO/TOPSe in BuOH, (open triangles) 4-picoline, (open circles) tris(2 ethylhexyl)phosphate, (solid squares) 4-trifluoromethyl)thiophenol

CONCLUSIONS

We have studied the luminescence of surface modified CdSe QDs. Results from PL and FLN experiments show that the energetics of the luminescence have little correlation with the nature of the nanocrystallite surface. QDs passivated with TOPO/TOPSe, ZnS, 4-picoline, 4-(trifluoromethyl)thiophenol and tris(2-ethylhexyl)phosphate show essentially the same resonant and non-resonant Stokes shift. The good agreement between experiment and theory suggests that the band edge luminescence from CdSe QDs can be understood in terms of an intrinsic exciton model.

ACKNOWLEDGMENTS

We thank Al. L. Efros, and D. J. Norris for helpful discussions during the course of this work. MK thanks the Corning Foundation for a pre-doctoral fellowship. BOD benefited from a pre-doctoral fellowship from the Saudi Aramco Corporation. MGB thanks the Lucille and David Packard Foundation as well as the Alfred P. Sloan Foundation for fellowships. This work was supported in part by the NSF-MRSEC program (NSF-DMR-4-00034) and NSF (DMR-91-57491).

REFERENCES

[1] A. I. Ekimov, F. Hache, M. C. Schanne-Klein, D. Ricard, C. Flytzanis, I. A. Kudryavtsev, T. V. Yazeva, A. V. Rodina, and Al. L. Efros, J. Opt. Soc. Am. B **10**, 100 (1993). And references within.

[2] A. Eychmüller, A. Hässelbarth, L. Katsikas, and H. Weller, Ber Bunsenges. Phys. Chem. **95**, 79 (1991).

[3] M. G. Bawendi, W. L. Wilson, L. Rothberg, P. J. Carroll, T. M. Jedju, M. L. Steigerwald, and L. E. Brus, Phys. Rev. Lett. **65**, 1623 (1990).

[4] S. Nomura, Y. Segawa, and T. Kobayashi, Phys. Rev. B **49**, 13571 (1994).

[5] M. Chamarro, C. Gourdon, P. Lavallard, and A. I. Ekimov, Jpn. J. Appl. Phys. **34**, Suppl. 34-1, 12 (1995).

[6] P. D. J. Calcott, K. J. Nash, L. T. Canham, M. J. Kane, and D. Brumhead, J. Luminescence, **57**, 257 (1993).

[7] Al. L. Efros, M. Rosen, M. Kuno, M. Nirmal, D. J. Norris, and M. G. Bawendi, Phys. Rev. B **54**, 4843 (1996).

[8] C. B. Murray, D. J. Norris, and M. G. Bawendi, J. Am. Chem. Soc. **115**, 8706 (1993).

[9] J. K. Lee, M. Kuno, M. G. Bawendi, (published in this volume)

STRUCTURAL INVESTIGATIONS OF COLLOIDAL SEMICONDUCTOR NANOCRYSTAL HETEROSTRUCTURES: FACETING AND EPITAXY

A.V. KADAVANICH, A. MEWS, S.H. TOLBERT*, X. PENG, M.C. SCHLAMP, J.C. LEE, A.P. ALIVISATOS
University of California, Department of Chemistry, Box 101, Latimer Hall, Berkeley, CA 94720-1460
*currently: University of California, Department of Chemistry, Los Angeles, CA 90065

ABSTRACT

We use High Resolution Transmission Microscopy (HRTEM) to study CdSe, CdS/HgS/CdS quantum-dot quantum well (QDQW), and CdSe/CdS core-shell nanocrystals, grown by wet-chemical techniques. The nanocrystals have faceted Wulff polyhedron shapes.

In addition to HRTEM, we employ multi-wavelength anomalous dispersion (MAD) x-ray diffraction. We use computer modeling to help interpret the experimental data.

Growth of a heterogeneous phase proceeds epitaxially preserving the overall shape and point-group symmetry of the original seed nanocrystals, both for wurtzite (CdSe/CdS) and zincblende (CdS/HgS/CdS) type structures.

Recently prepared InAs nanocrystals also show evidence of faceting as observed by HRTEM and may lend themselves equally well to epitaxy.

INTRODUCTION

Colloidal semiconductor nanocrystals, being a soluble chemical reagent, distinguish themselves from other quantum dot systems by the relative ease of preparation and handling,. However, structural details such as morphology and surface chemistry are poorly defined. In particular the surfaces are incompletely passivated by the organic ligands commonly used to confer solubility. This leads to a large number of dangling bonds and other defect states on the nanocrystal surface which adversely affect the luminescence properties of these potentially useful optoelectronic materials. To this end it is desirable to epitaxially passivate nanocrystals using an inorganic material, which can then be independently treated to achieve solubility.

We present structural evidence that the syntheses of QDQW nanocrystals and CdSe/CdS core-shell nanocrystals do in fact proceed by epitaxy

EXPERIMENT

Synthesis

Nanocrystal syntheses for CdSe [1,2] and CdS/HgS/CdS[3] have been published elsewhere. The synthesis of CdSe/CdS core-shells is to be published and closely follows that for plain CdSe. In brief, CdSe nanocrystals are dissolved in pyridine and heated to ~115°C. Trimethyl cadmium and atomic sulfur dispersed in tributyl phosphine solvent are rapidly injected and allowed to grow on the CdSe seeds. InAs nanocrystals are synthesized by the method of Guzelian.[4]

HRTEM

For HRTEM studies, CdSe and CdSe/CdS samples were deposited from toluene solution onto standard 3mm TEM grids coated with a thin (100-500Å) film of amorphous carbon. Excess solvent was wicked away with absorbent tissue and the remainder was allowed to evaporate.

QDQW samples dissolved in methanol were prepared similarly except the carbon films were pretreated by a glow discharge in Ar gas at 10mtorr to improve adhesion of the nanocrystals. We studied samples from each step of the synthesis: as-prepared CdS nanocrystals, CdS/HgS shell nanocrystals, and the final QDQW nanocrystals.

InAs samples deposited from toluene solution were embedded by evaporating a layer of carbon (~150Å) onto the grid after depositing the sample solution. This improves the stability inside the microscope. Without this treatment, InAs nanocrystals degrade rapidly under the influence of the electron beam.

Mat. Res. Soc. Symp. Proc. Vol. 452 © 1997 Materials Research Society

HRTEM was performed on a TopCon EM002B microscope with Ultra-High Resolution pole piece and a JEOL 200CX both at 200kV. Point-to-point resolutions are 1.8Å and 2.2Å respectively. For InAs the microscope was operated at 120kV to minimize beam damage. Micrographs were obtained under black atom contrast conditions near Scherzer defocus. Nanocrystals are randomly oriented so no tilt was used. Different orientations of nanocrystals are obtained by selecting different nanocrystals to image.

MAD

The basic principles of anomalous dispersion experiments and theory are discussed in the literature.[5-7] For MAD the QDQW solutions were placed onto kapton tape, the solvent allowed to evaporate and the precipitate then covered with additional kapton tape. Diffraction patterns were collected at Beamline 10-2 of the Stanford Synchrotron Radiation Laboratory (SSRL) in transmission geometry (Debye-Scherrer patterns) with image plate detectors.

Diffraction patterns were collected 10eV and 100eV below the Hg L-edge (12291eV). The ring patterns obtained were radially averaged using a custom program to yield intensity versus scattering vector plots with a resolution of $0.005Å^{-1}$ in reciprocal space. Plots were normalized to the incoherent background intensity between diffraction peaks. After interpolation to equalize the x-axes, the plots are subtracted to give difference plots. Such difference plots give a measure of the coherence between the Hg atoms and all other atoms in the structure. Thus they are partial diffraction patterns, analogous to the partial scattering factors of reference [7].

Computer Modeling

HRTEM images were simulated on an IBM Powerstation model 550 using Biosym Insight II 3.00 software. Structural models for the simulations were created in Insight II as well.

RESULTS

QDQW Nanocrystals

A HRTEM micrograph of a CdS nanocrystal is shown in Figure 1a. The nanocrystals is approximately triangular in projection. From the fringe spacing and pattern it can be determined that the crystallite is aligned along the zincblende [110] axis. The contrast decreases from the apex of the triangle towards the base, indicating a decrease in thickness. These features are reproduced by an image simulation of a tetrahedral zincblende CdS nanocrystal, terminated with Cd (111) surfaces, Figure 1b. For the sample investigated, 79 nanocrystals were oriented such that their shapes could be determined. 71% of the initial CdS crystallites had tetrahedral shape, with an average edge length of 50Å (17% standard deviation) obtained by counting lattice fringes. The remainder had shapes approaching octahedral morphology with (111) surface termination or were severely faulted, resulting in low-symmetry shapes with indiscernible surface structures.

Similar tetrahedral shapes were observed after HgS was deposited on such CdS nanocrystals, however only in 64% yield. Faulted structures were more prominent in this sample.

Adding the final shell of CdS still yielded tetrahedral symmetry in 63% of the 117 nanocrystals studied. An example is shown in Figure 1c. Nearly 20% of tetrahedral nanocrystals (12% of the total) exhibited rotation twin faults around the tetrahedral core, resulting in images similar to Figure 1d. The lobes around the core are zincblende structure, but are in a twin relationship to the core. This also causes lobes to be crystallographically mismatched with each other. There is a line of increased contrast at the interface between the core and one of the lobes, as indicated by the arrow in the figure.

No evidence of amorphous material was found in any of the samples. Also, there were no indications of point defects in the crystals, albeit such defects are difficult to detect in HRTEM.

MAD was performed on QDQW's with HgS layer thicknesses of 1, 2, and 3 monolayers (ML) of HgS. The general features are the same but 1 ML samples have too much noise to show any detail. The Diffraction pattern for a 3 ML sample is presented in Figure 2. The MAD difference pattern obtained shows peaks in positions matching those in the normal diffraction pattern. However, the peak intensities relative to the 111 peak are different. The 200 and 222 peaks are enhanced by over a factor of 2. The 311 and 400 peaks show weaker enhancements.

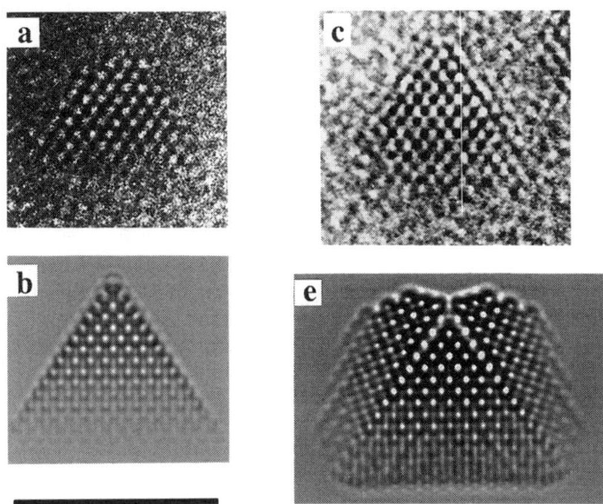

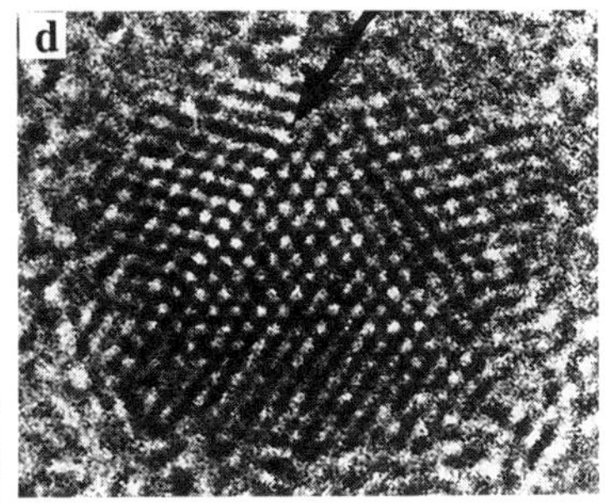

Figure 1: HRTEM images. The bar in the lower left indicates 50Å length. All crystallites are imaged in [110] projection. Dark areas are atomic positions, the white spots are channels in the crystal lattice. a) CdS nanocrystal, b) image simulation of a tetrahedral zincblende CdS nanocrystal, c) QDQW with tetrahedral morphology, d) QDQW with tetrahedral core and twin-faulted "lobes". The arrow indicates the line of increased contrast due to the presence of the HgS layer at the interface. e) image simulation for a QDQW with a single layer of HgS surrounding a tetrahedral CdS core, and with CdS "lobes" outside the HgS.

<u>CdSe/CdS Core Shell Nanocrystals</u>

Figure 3 shows HRTEM micrographs of wurtzite structure core (CdSe only) and core-shell (CdSe/CdS) nanocrystals. The cores (a,b) facet along (001) and (100) surfaces as previously observed for these CdSe nanocrystals. [8] The core-shells (c,d) retain this overall morphology. For the sample studied the average core diameter was 37Å (16% standard deviation) in the <001> direction (along the unique wurtzite c-axis) and 32Å (11%) perpendicular to it. On average each core had 1.9 ± 0.9 faults along the c-axis. Core-shells had mean diameters of 54Å (15%) along <001> and 52Å (16%) perpendicular to it. This corresponds to 2-3 ML of CdS coating the core. The stacking fault density is 2.1 ± 1.9 per crystallite. Again there are no signs of amorphous material or point defects. There is a slight dropoff in contrast

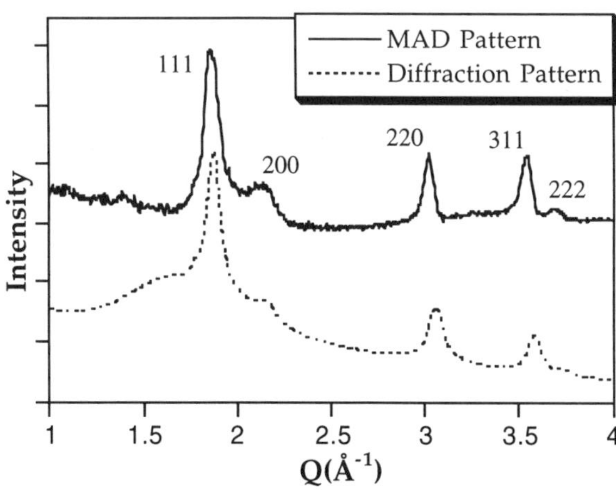

Figure 2:
Raw diffraction pattern 10eV below the Hg L-edge (dotted line) and MAD pattern (difference of 10eV below and 100eV below) for a 3ML QDQW sample. The peak positions in the MAD pattern correspond to a zincblende lattice with the same lattice constant as CdS and the peaks are indexed appropriately. The intensities are scaled to give equal height in the 111 peak and the patterns are offset vertically for clarity.
The horizontal axis is distance in reciprocal space (independent of x-ray energy).

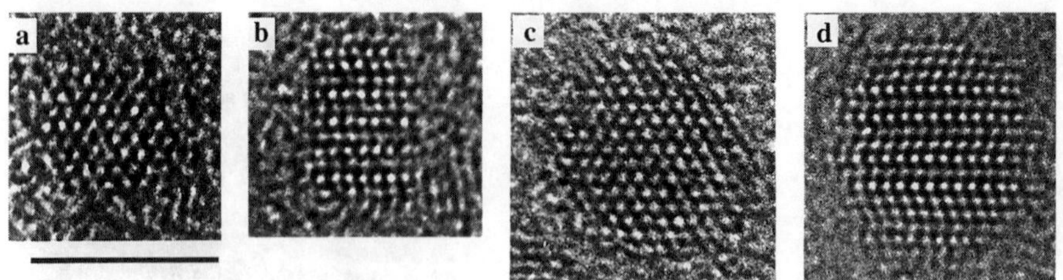

Figure 3:
a) CdSe imaged along the [001] axis, exhibiting facets along (100) planes, b) CdSe imaged along [010] showing (001) top and bottom facets and (100) side facets, c) core-shell imaged along [001] with (100) facets, d) core-shell imaged along [010] showin (001) facets and (100) facets perturbed by twin faults running through the crystallite. The length bar is 50Å.

over the last 2-3 layers of the crystallite, which is more apparent in a line profile, Figure 4. Occasionally, nanocrystals had zincblende lobes similar to those in the QDQW's.

InAs Nanocrystals

Figure 5 shows HRTEM micrographs of zincblende InAs nanocrystals in [110] projection. The small crystallite (a) exhibits (111) facets and is consistent with an octahedral shape. The larger nanocrystal has some (111) facets but the overall morphology is less defined.

CONCLUSIONS

QDQW Nanocrystals

The slow growth from aqueous, ionic reagents under dilute conditions favors the Wulff shape, as redissolution of deposited ions provides a means for mass transport that can anneal out high energy surfaces. A crystallite completely terminated in (111) surfaces is the expected outcome. However, in the zincblende structure there are in fact two different (111) surfaces. One is entirely terminated in Cd atoms, the other in S atoms. The expected Wulff shape is an octahedron, with alternating Cd and S surfaces, which gives an overall point group symmetry T_d. If, however, one of the surfaces is made energetically more favorable than the other, the Wulff shape distorts toward a tetrahedron, as observed in the HRTEM. The driving force for this is likely the presence of the anionic polyphosphate ligands coating the nanocrystals, which would stabilize Cd surfaces at the expense of S surfaces. Thus the CdS nanocrystals would have predominantly Cd (111) surfaces. This notion is further supported by the fact that it was shown in the original synthesis[3] that the growth of HgS proceeds by stoichiometric replacement of surface Cd. This mechanism itself is corroborated by the HRTEM which shows no change in size upon growth of the HgS layer.

The growth process itself introduces more defects as only 64% of CdS/HgS nanocrystals

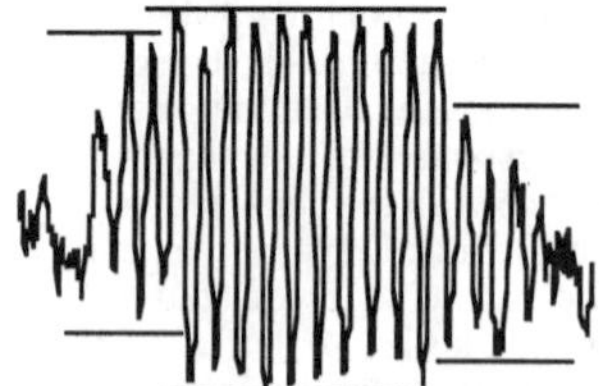

Figure 4:
Intensity profile along a horizontal row in Figure 3d. The horizontal lines indicate how the maximum intensity varies across the scan.

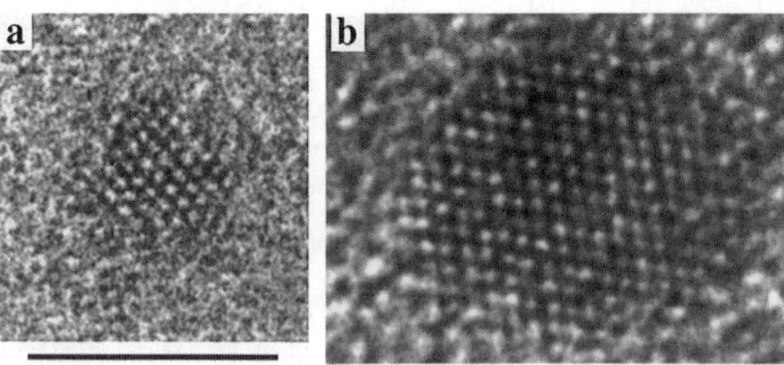

Figure 5:
In As nanocrystals in [110] projection. The length bar is 50Å

are tetrahedral, compared to 71% of the initial CdS nanocrystals. Only 51% preserve the full tetrahedral shape through the entire synthesis of the QDQW, with another 12% comprising the lobed tetrahedral structures. However, the epitaxial process seems quite efficient as no signs of amorphous material or misfit dislocations are found, even in badly misshaped nanocrystals. Rather the main defect structures detected are rotation twin faults of the type seen in Figure 1d. This is a very low energy defect as no bonds are broken or strained. Rather, this is only a single layer of wurtzite stacking in the otherwise zincblende crystal and hence only affects third-nearest neighbor and higher distances. Such defects are not surprising, given the low growth temperature (room temperature). Given the small lattice mismatch of 0.7% for the CdS and HgS lattices, high quality epitaxy could be expected.

The line of increased contrast at the core-lobe interface seen in Figure 1d is indicative of the HgS layer in the structure. Contrast in HRTEM increases with increasing electron scattering power of the material observed. HgS thus should yield more contrast than CdS. The computer simulation of a HRTEM image for a model lobed structure with 1 ML of HgS reproduces this contrast change at the interface, Figure 1e. The same model with the Hg atoms replaced by Cd does not show this contrast. The process of forming the lobes helps to expose the HgS layer while keeping it passivated. In projection, the HgS is then unobstructed by CdS, unlike the case of the perfect tetrahedral shell structure. For the latter, the computer simulation indicates a weak contrast change, but it is likely below the noise level in the micrographs and has not been observed in the experiment. Other lobed crystallites were observed that did reproduce the enhanced contrast at the interface.

The MAD experiment was undertaken to overcome the limited sampling in the HRTEM experiment. In principle, MAD can be used to obtain structural information of a subset of atoms in a structure, similar to the information obtainable by EXAFS. MAD gets additional information by exploiting the crystalline nature of the material. However, we are limited in having to use powdered samples rather than single crystals.

The MAD pattern in Figure 2 is conceptually a partial diffraction pattern measuring diffraction for all scattering vectors whose structure factor contains at least one Hg atom. The peaks observed correspond to a zincblende lattice and the peak positions match those of the normal diffraction pattern for the QDQW's. The background is nearly flat. This means that the majority of Hg atoms occupy zincblende lattice sites, matched to the CdS lattice. Additionally, the peak intensities, relative to the 111 peak are larger in the MAD pattern compared to the normal diffraction for all peaks observed. In the case of the 200 and 222 peaks, the increase is more than a factor of two. This is more than would be expected from replacing the Cd atomic scattering factor in the expression for the zincblende structure factor by an average Cd/Hg scattering factor as would be appropriate for a random alloy. For a range of alloy compositions up to 30% Hg, the intensity change never exceeds 5%. This indicates that there are coherences outside the zincblende unit cell that affect the structure factor of the total crystallite. Albeit it is not possible to make conclusive statements regarding the exact distribution of the Hg, based on the present level of analysis. We are currently in the process of developing computer code to simulate the MAD pattern. Potentially it will be possible to distinguish different shell arrangements, for instance tetrahedral versus spherical, using the computer models.

Core-shell Nanocrystals

The high crystallinity and the appearance of facets point towards an epitaxial process as the growth mode. The change in size is consistent with 2-3 ML of CdS grown on the outside of the CdSe core. There is no apparent change in aspect ratio, which supports the hypothesis that growth takes place uniformly. Again the main defects are rotation twin faults, leading to polytypism. There appears to be a slight increase in the stacking fault density after shell growth, but the error limits are currently too large and more data is required to substantiate that claim. At present it should be noted that when stacking faults occur, they traverse the entire crystallite, including the core. Such faults are probably present in the cores and are propagated during shell growth. Another point is that in wurtzite systems, point group C_{3v}, the presence of a single rotation twin fault does not affect the point group of the crystallite and hence the overall shape is rather insensitive to the presence of such faults. This contrasts the case of zincblende structures, where a single layer of wurtzite stacking immediately lowers the symmetry from T_d to C_{3v} or less. Hence wurtzite nanocrystals are less likely to develop lobes.

The dropoff in contrast at the last 2-3 layers in the micrographs of Figure 3, and shown in the line scan Figure 4, is suggestive evidence for the presence of CdS in the shell as opposed to the more strongly scattering CdSe in the core. To confirm this more directly we are currently pursuing the possibility of EELS-filtered imaging to extract sulfur elemental maps. Nevertheless, optical absorption and x-ray photoelectron data support the shell model, and photoluminescence yields indicated excellent electronic passivation in these systems. A manuscript detailing these results is in preparation.

InAs nanocrystals

The nanocrystals depicted show faceting along (111) planes as expected. There is less tendency towards a tetrahedral morphology which we ascribe to the non-ionic trioctyl phosphine oxide ligands used to passivate the surfaces. Indeed octahedral shape of the small crystallite in Figure 5a may be more representative for this sample. These observations are still rather preliminary. However, these nanocrystals III-V nanocrystals are now approaching the level of quality that CdSe samples have enjoyed for several years, and hence open up the possibility of studying epitaxy in III-V nanocrystal systems. This should provide for interesting comparisons to bulk III-V epitaxial systems.

ACKNOWLEDGEMENTS

We thank A. Bienenstock and S. Brennan for helpful discussions on MAD. The HRTEM was performed at the National Center for Electron Microscopy, Lawrence Berkeley Laboratory. The computer modeling was done at the Computer Graphics Lab of the College of Chemistry at UC Berkeley. This work was partially supported by the Department of Energy and the Molecular Design Institute.

REFERENCES

[1]Janet E. Bowen Katari, V. L. Colvin, and A. Paul Alivisatos, "X-ray Photoelectron Spectroscopy of CdSe Nanocrystals with Applications to Studies of the Nanocrystal Surface," Journal of Physical Chemistry 98 (15), 4109-17 (1994).

[2]Chris B. Murray, D. J. Norris, and Moungi G. Bawendi, "Synthesis and Characterization Of Nearly Monodisperse CdE (E = S, Se, Te) Semiconductor Nanocrystallites," Journal Of the American Chemical Society **115** (19), 8706-8715 (1993).

[3]Alf. Mews, A. Eychmuller, M. Giersig, D.Schooss, Horst Weller, "Preparation, Characterization, and Photophysics Of the Quantum Dot Quantum Well System CdS/HgS/CdS," Journal Of Physical Chemistry **98** (3), 934-941 (1994).

[4]A. A. Guzelian, U. Banin, A. V. Kadavanich *et al.*, "Colloidal chemical synthesis and characterization of InAs nanocrystal quantum dots," Applied Physics Letters **69** (10), 1432-4 (1996).

[5]A. Bienenstock, "Applications of synchrotron radiation," presented at the Proceedings of the 1979 Particle Accelerator Conference Accelerator Engineering and Technology, San Francisco, CA, USA, 1979 (unpublished).

[6]J. Kortright, W. Warburton, and A. Bienenstock, "Anomalous X-ray scattering and its relationship to EXAFS," presented at the Proceedings of the International Conference, Frascati, Italy, 1982 (unpublished).

[7]O. Lyon and J. P. Simon, "Anisotropy of partial structure factors in an unmixed Cu-Ni-Fe single crystal studied by anomalous small-angle X-ray scattering," Journal of Physics: Condensed Matter **4** (28), 6073-86 (1992).

[8]Joseph J. Shiang, Andreas V. Kadavanich, Robert K. Grubbs, A. Paul Alivisatos, "Symmetry of Annealed Wurtzite CdSe Nanocrystals: Assignment to the C_{3v} Point Group," Journal of Physical Chemistry **99** (48), 17417-17422 (1995).

SYNTHESIS AND CHARACTERIZATION OF HIGHLY LUMINESCENT (CdSe)ZnS QUANTUM DOTS

F.V. MIKULEC*, B.O. DABBOUSI*, J. RODRIGUEZ-VIEJO**, J.R. HEINE***, H. MATTOUSSI***, K.F. JENSEN**, M.G. BAWENDI*
*Department of Chemistry, Massachusetts Institute of Technology, Cambridge, MA 02139, mgb@mit.edu
**Department of Chemical Engineering, Massachusetts Institute of Technology, Cambridge, MA 02139
***Department of Materials Science and Engineering, Massachusetts Institute of Technology, Cambridge, MA 02139

ABSTRACT

We report the synthesis of a series of nearly monodisperse ZnS-overcoated CdSe quantum dots whose room temperature photoluminescence quantum yield approaches 50%. This spectrally narrow (FWHM < 40nm) band edge luminescence spans the visible region from blue to red light. We use a two-step synthesis based on the high temperature decomposition of organometallic precursors in a coordinating solvent. We characterize our composite quantum dots using optical spectroscopies, wavelength dispersive x-ray spectroscopy, x-ray photoelectron spectroscopy, high resolution transmission electron microscopy, small angle x-ray scattering, and wide angle x-ray diffraction. The data indicate that samples displaying the highest quantum yield are those which have just achieved a closed shell of ZnS encapsulating the CdSe core. As more ZnS is added, the quantum yield decreases somewhat, possibly due to the many defects present in larger ZnS shells.

INTRODUCTION

Semiconductor nanocrystallites (quantum dots) whose radii are smaller than the bulk exciton Bohr radius exhibit interesting optical and structural properties[1]. Quantum confinement in all three dimensions of both the electron and hole leads to an increase in the effective band gap of the material with decreasing crystallite size. Consequently, both the absorption and emission of these quantum dots shift to the blue (higher energies) as the size of the dots gets smaller. Numerous optical and structural studies have been carried out on such systems in order to better understand the evolution of a material from the molecular level to the bulk[2,3].

Core-shell type composite quantum dots, in which the interior core of the dots is composed of one material while the exterior shell is made up of another, exhibit properties which make them interesting from both an experimental and a practical point of view. While the optical and structural properties of quantum dots of uniform composition have been studied extensively, composite quantum dots are not as well characterized. Examples of core-shell type quantum dot structures in the literature include $Cd(OH)_2$ on CdS[4], ZnS on CdSe and the inverse structure[5], CdS/HgS/CdS quantum dot quantum wells[6] and ZnSe on CdSe[7]. Composite quantum dots can potentially be useful for the fabrication of electrooptic devies, since the core of the dots can be chosen for a specific optical property while the external shell can be selected to protect the internal dot from its surrounding environment.

Mat. Res. Soc. Symp. Proc. Vol. 452

Very recently Hines *et al.*[8] reported making (CdSe)ZnS nanocrystallites that photoluminesce with a quantum yield of 50% at 530nm. We use a modification of their synthetic procedure and that of Danek *et al.*[7] to synthesize high quantum yield (30%-50%) core-shell (CdSe)ZnS nanocrystallites of various sizes with narrow band edge luminescence spanning most of the visible spectrum from 470nm to 625nm. Such materials should improve the stability of elecroluminescent devices (LED's) based on CdSe dots and aid in the production of thin-film CdSe/ZnS quantum dot composites which exhibit cathodoluminescence[9] and which may be useful for alternating-current thin-film light-emitting devices (ACTFLED's).

EXPERIMENTAL

Synthesis of CdSe and (CdSe)ZnS

Nearly monodisperse ($\sigma < 5\%$) CdSe quantum dots ranging in size from 20Å to 55Å in diameter are synthesized via the pyrolysis of dimethyl cadmium and trioctylphospine selenide in trioctylphosphine oxide (TOPO), as described previously[3]. The dots are purified using size-selective precipitation, ~0.1mmol CdSe is added to ~5g dry TOPO, and heated to 180°C. Diethyl zinc and hexamethyl disilithiane (1:1 ratio) are dissolved in ~3mL of trioctylphosphine and added dropwise with stirring to the hot CdSe/TOPO solution. The amount of zinc and sulfur precursors added depends upon the size of the CdSe core and the desired ZnS shell thickness. Slow addition (3-5 minutes total) of the zinc and sulfur precursors ensures that ZnS grows exclusively on the CdSe surfaces, instead of nucleating discrete ZnS particles. After the addition is complete the mixture is cooled to 90 °C and left stirring for several hours. The overcoated dots are then purified by size-selective precipitation.

Characterization

Optical Spectroscopy: UV-Visible absorption spectra were acquired on an HP 8452 diode array spectrophotometer. The dots were dispersed in hexane in a 1cm quartz cuvette. The photoluminescence spectra were taken on a SPEX fluorolog 2 spectrometer in front face collection mode. The quantum yields were obtained by comparing the integrated emission of the dots in solution to the emission of a solution of rhodamine 590 or rhodamine 640 of identical optical density at the excitation wavelength.

Wavelength Dispersive X-Ray Spectroscopy: A JEOL SEM 733 Electron Microprobe was used to determine the chemical composition of the composite quantum dots using both Energy Dispersive X-Ray (EDX) and Wavelength Dispersive X-Ray (WDS) spectroscopies. 1μm thick films of (CdSe)ZnS quantum dots were cast from pyridine solution onto silicon wafers and then coated with a thin layer of amorphous carbon to prevent charging.

X-Ray Photoelectron Spectroscopy: XPS was performed using a Physical Electronics 5200C equipped with a dual X-ray anode (Mg and Al) and a Concentric Hemispherical analyzer (CHA). All samples were exchanged with pyridine and spin coated at high speed onto Si substrates forming a thin layer several monolayers thick.

High Resolution Transmission Electron Microscopy: A Topcon EM002B transmission electron microscope (TEM) was operated at 200kV to obtain high resolution images of individual quantum dots. The samples were prepared by placing one drop from a dilute solution of dots in octane onto a copper grid supporting a thin film of amorphous carbon and then wicking off the remaining solvent after 30 seconds.

Small and Wide Angle X-Ray Scattering (SAXS and WAXS): X-ray powder diffraction patterns were measured on a Rigaku 300 rotating anode diffractometer operating in the Bragg configuration using CuKα radiation. SAXS samples were made by dissolving quantum dots and poly(vinyl butyral) in toluene. This solution is cast onto a silicon wafer, giving a dilute thin film of nanocrystallites (~10% by weight) dispersed in a polymer matrix. WAXS samples were made by evaporation of a concentrated pyridine solution of dots on silicon wafers.

RESULTS AND DISCUSSION

The noteworthy aspects of our synthetic route to (CdSe)ZnS composite nanocrystallites are (1) the ability to make a range of sizes of CdSe and (2) the retention – at least for thin ZnS shells -- of the small initial size distribution by the overcoated dots. The former property allows

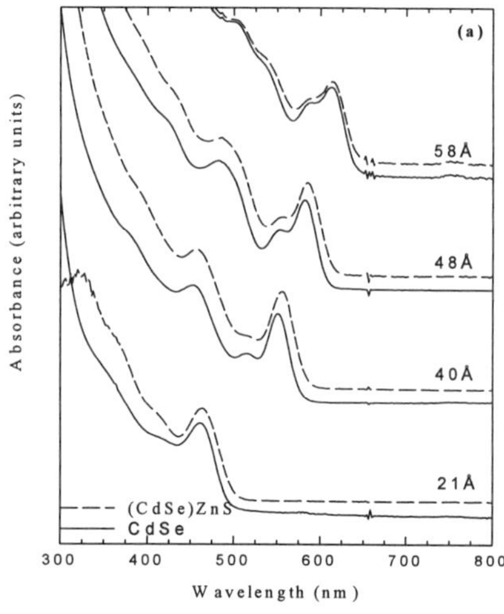

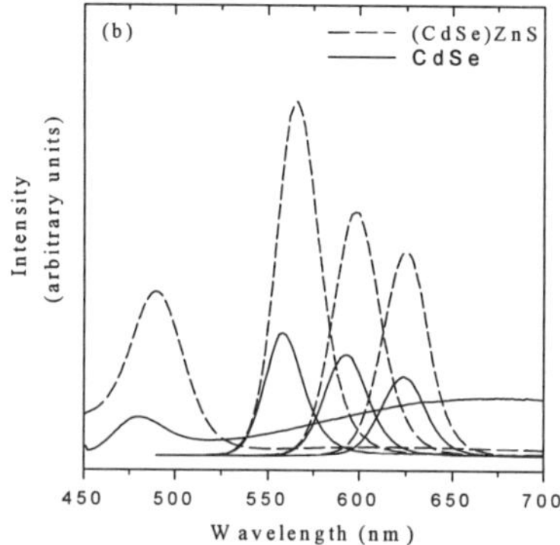

Figure 1. Room temperature absorption (a) and photoluminescence (b) spectra of bare and ZnS overcoated CdSe quantum dots in hexane. Sizes in 1a refer to the major diameter of the CdSe core. (CdSe)ZnS samples have 1-3 monolayers of ZnS.

for the synthesis of composite nanocrystallites which emit blue through red light. Figure 1a displays the absorption spectra of dilute hexane solutions of four CdSe quantum dot sizes both before and after ZnS overcoating. Separate control experiments have shown that the size and size distribution of the CdSe core do not change when those nanocrystallites are subjected to the same conditions experienced during the overcoating process, but without addition of zinc and sulfur precursors. Therefore, we believe that the small red shift of the overcoated dots' spectra is due to a slight leakage of the exciton into the ZnS shell. The shape of the absorption curve, an indication of the quality of the sample's size distribution, remains largely unchanged, confirming that the monodispersity of the CdSe core is retained. Figure 1b shows the corresponding PL of these samples. The PL of the overcoated dots is red-shifted from the bare dots, with the magnitude of the shifts increasing for smaller sizes. Experiments investigating this size-dependence are in progress. The room temperature quantum yield increases from 5-15% to 30-50%. Concentrated hexane solutions of (CdSe)ZnS nanocrystallites can be stored in air for at least 6 months without a noticable decrease in quantum yield.

To determine the effect of the size of the ZnS shell on the optical properties we synthesized a large amount of CdSe quantum dots (42Å diameter), divided that quantity into 5

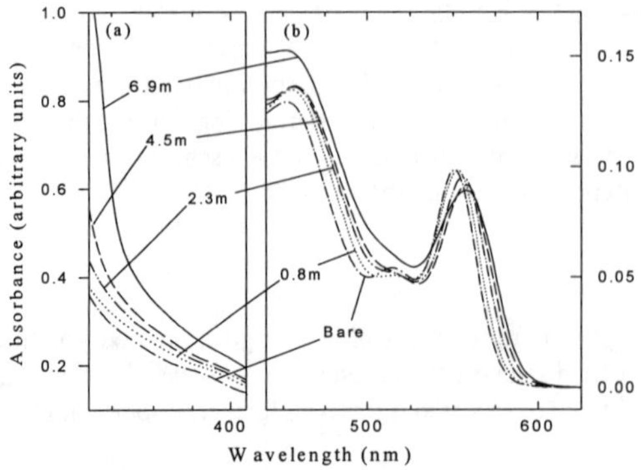

Figure 2. Room temperature absorption spectra of bare and ZnS overcoated CdSe quantum dots in hexane. (a) Detail of the high energy region. (b) Detail of the band edge absorption. The spectra are plotted so that the areas under the first absorption feature are equal.

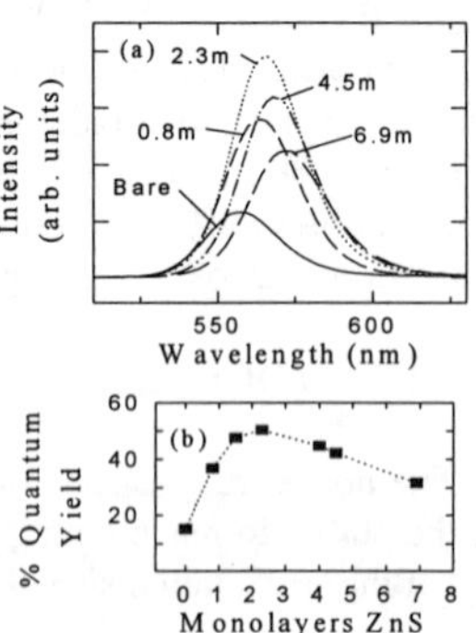

Figure 3. (a) Room temperature photoluminescence of Bare and ZnS overcoated CdSe nanocrysatllites. (b) Room temperature quantum yield as a function of ZnS coverage.

parts, and overcoated those samples with increasing amounts of ZnS. WDS provided the samples' Zn/Cd ratios, from which we calculated a nominal value for the ZnS coverage (assuming spherical particles and using 2.3Å as the width of 1 monolayer). All data presented below is from this particular sample series.

Figure 2a displays the high energy region of the room temperature absorption of the bare and overcoated nanocrystallites; spectra have been normalized to the first absorption feature. The slight increase in the UV absorption with increasing shell thickness confirms that we actually are adding ZnS onto the surface of our dots. Figure 2b shows a small red shift and a broadening of the spectra, both of which increase as the overlayer thickness increases. This behavior is consistent with the expected leakage of the exciton into the ZnS shell. The PL of this sample series is shown in Figure 3a. In a plot of quantum yield versus the number of ZnS monolayers, there are two important features: a maximum in the quantum yield at ~2 monolayers and then a gradual decrease in the quantum yield as the coverage increases.

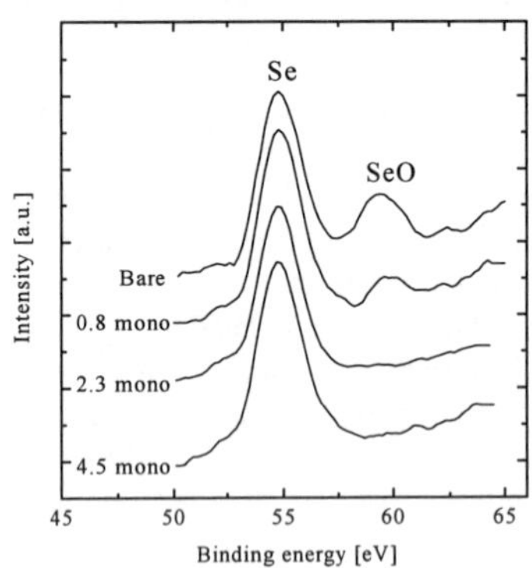

Figure 4. X-ray photoelectron spectra of bare and ZnS overcoated CdSe quantum dot films after 80 hrs of air exposure.

In order to understand these two features, we investigate the (CdSe)ZnS surfaces. XPS data (Figure 4) of oxidized nanocrystallite films show that when the ZnS shell thickness is ~2 monolayers or more there is no peak corresponding to oxidized selenium[10]. All the surface selenium sites have been passivated, suggesting that by 2 monolayers the ZnS shell is complete. The onset of a complete shell correlates with the maximum in the quantum yield graph. Histograms of sizes measured from HRTEM micrographs (Figure 5) show an increase in size and also in size distribution for the entire (CdSe)ZnS nanocrystallite. Additionally, the

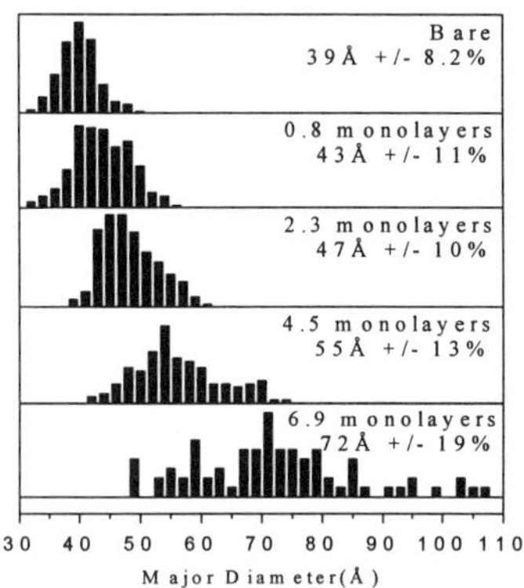

Figure 5. Histograms of sizes measured from HRTEM images of bare and ZnS overcoated CdSe quantum dots. Each graph represents at least 150 measurements.

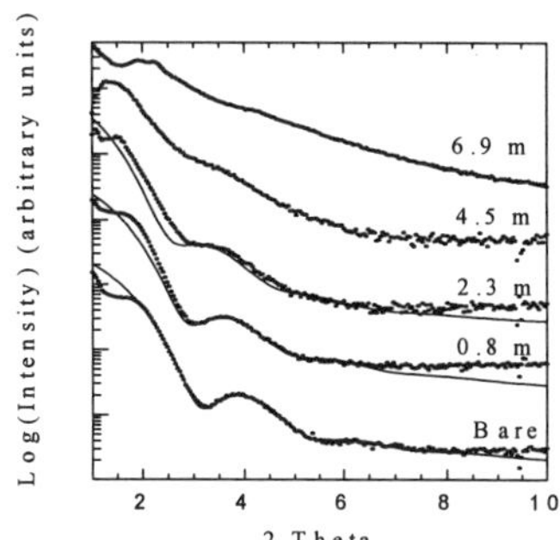

Figure 6. SAXS of bare and ZnS overcoated CdSe quantum dots dispersed in polymer films. Circles represent experimental data, lines the corresponding fits (major diameter, aspect ratio): bare (42Å +/- 10%, 1.1), 0.8 monolayers (46Å +/- 13%, ~1.2), 2.3 monolayers (50Å +/- 19%, ~1.2).

biggest (CdSe)ZnS dots (the 4.5 and 6.9 monolayer samples) display a rather large distribution of shapes; whereas the the ZnS overlayer is relatively uniform in the 0.8 and 2.3 monolayer samples, it is highly non-uniform in the dots with a thicker ZnS shell. SAXS (Figure 6) also provided information about the size, size distribution, and shape of the dots. The shift of the patterns to smaller angle with increasing ZnS coverage is consistent with increasing particle size. Also, loss of definition in the ringing pattern implies an increase in sample size distribution and/or shape distribution. We model the SAXS data by approximating the nanocrystallites as ellipsoidal fragments of the bulk (CdSe or ZnS) lattice, including a size distribution and a shape distribution (varying aspect ratio), then calculating the resulting pattern using the discrete form of the Debye equation[3]. These theoretical fits assume an unchanging CdSe core. Only the first 3 patterns had enough structure to be fit confidently.

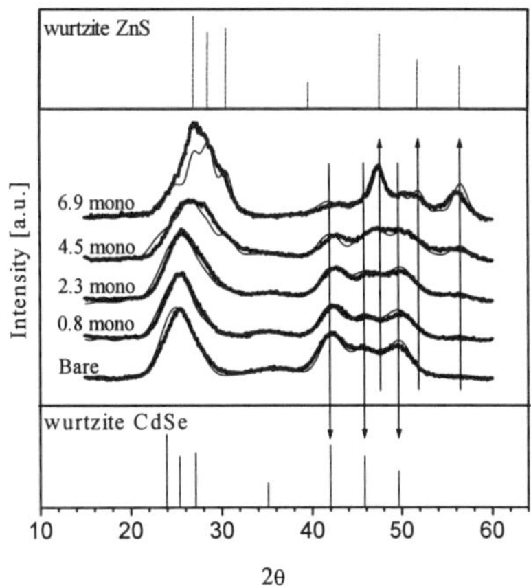

Figure 7. WAXS of bare and ZnS overcoated CdSe quantum dots. Experimental data (dots) and the corresponding fits (solid lines) are described in text.

The weak scattering intensity of the ZnS made studying the shell's internal structure by HRTEM very difficult. However, the WAXS data (Figure 7) clearly show a shift from a pattern which is dominated by diffraction from a wurtzite CdSe structure to a pattern whose features come mainly from wurtzite ZnS having the lattice parameter of bulk ZnS. This shift can occur only if the shell has the lattice parameter of bulk wurtzite ZnS. We model the data as in the SAXS case, neglecting the size distribution (it has little to no effect on the resulting fit), but including 1 randomly placed stacking fault in the CdSe core and 3 in the ZnS shell. The high number of stacking faults needed to model the shell suggests that it contains many defects. Such defects could explain why the quantum yield decreases as more ZnS layers are added to the dot.

CONCLUSION

We have reported a two-step synthesis of highly monodisperse ZnS overcoated CdSe quantum dots. This method allows for the production of a range of sizes of both the CdSe core and also the ZnS shell. Absorption spectroscopy displays that the CdSe core size is unchanged by the overcoating process. XPS results show that the composites with the highest quantum yields are those which have just achieved a complete shell of ZnS. TEM and SAXS give the size, size distribution, and shape of the particles. Fitting of the WAXS patterns reveals that the ZnS shell is wurtzite in structure, has the lattice parameter of bulk ZnS, and contains many defects. These defects may be the cause of the decrease in quantum yield with increasing ZnS shell thickness. Additional experiments are necessary to confirm this hypothesis.

REFERENCES

1. L.E. Brus, Appl. Phys. A **53**, 465 (1991).
2. D.J. Norris and M.G. Bawendi, Phys. Rev. B **53**, 16338 (1996).
3. C.B. Murray, D.J. Norris, M.G. Bawendi, J. Am. Chem. Soc. **115**, 8706 (1993).
4. L. Spanhel, M. Haase, H. Weller, A. Henglein, J. Am. Chem. Soc. **109**, 5649 (1987).
5. A.R. Kortan, R. Hull, R.L. Opila, M.G. Bawendi, M.L. Steigerwald, P.J. Carroll, L.E. Brus, J. Am. Chem. Soc. **112**, 1327 (1990).
6. A. Mews, A. Eychmuller, M. Giersig, D. Schooss, H. Weller, J. Phys. Chem. **98**, 934 (1994).
7. M. Danek, K.F. Jensen, C.B. Murray, M.G. Bawendi, Chem. Mater. **8**, 173 (1996).
8. M.A. Hines, P. Guyot-Sionnest, J. Phys. Chem. **100**, 468 (1996).
9. K.F. Jensen, J. Rodriguez-Viejo, B.O. Dabbousi, J.R. Heine, H. Mattoussi, J. Michel, M.G. Bawendi, published in this volume.
10. J.E. Bowen Katari, V.L. Colvin, A.P. Alivisatos, J. Phys. Chem. **98**, 4109 (1994).

CATHODOLUMINESCENCE OF CdSe/ZnS QUANTUM DOT COMPOSITES

J.RODRIGUEZ-VIEJO[*], J.R.HEINE[$], B.O.DABBOUSI[#], H.MATTOUSSI[$], J.MICHEL[$], M.G.BAWENDI[#], K.F.JENSEN[*$].

Massachusetts Institute of Technology, Departments of Chemistry[#], Chemical Engineering[*] and Materials Science and Engineering[$], Cambridge, Massachusetts 02139.

ABSTRACT

We report cathodoluminescence and photoluminescence spectra originating from ZnS overcoated CdSe nanocrystals of 33 and 42 Å diameter imbedded in a ZnS matrix. The thin film quantum dot composites were synthesized by electrospray organometallic chemical vapor deposition. The cathodoluminescence intensity depends on the crystallinity of the ZnS matrix and on the voltage and current density applied. Electron beam irradiation caused a decrease of the luminescence which may be explained by electron ionization of the quantum dots. Quenching of the cathodoluminescence at low temperatures is attributed to a more efficient trapping of the ionized electrons in deep traps inside the ZnS matrix.

INTRODUCTION

Semiconductor crystallites which are small compared to the exciton Bohr radius exhibit unique optical properties due to confinement of the electronic excitations [1]. The optical properties, such as absorption and emission, can be tuned across the visible by changing the size of the nanocrystals. Highly monodisperse CdSe quantum dots organically passivated with trioctylphosphines have quantum yields between 10 and 15% [2]. Solution chemistry growth of a thin ZnS overlayer on the CdSe nanoparticles raises the quantum yield to values between 30 and 50% for particles in the size range from 20 to 60 Å in diameter [3,4].

The incorporation of high luminescence quantum dots into an inorganic matrix of higher bandgap, provides a wide spectral window for exploring the quantum size effect and the ability of electrical integration of the composites into optoelectronic devices. Danek et al. [5] have already shown the feasibility of such composites by Electrospray Organometallic Chemical Vapor Deposition (ES-OMCVD). Films of CdSe/ZnSe have been grown by this method, but their luminescence characteristics were too low to be useful in optoelectronic devices [6].

In this work, we demonstrate excellent optical properties of quantum dot composites formed by ZnS overcoated CdSe quantum dots embedded in a ZnS matrix grown by OMCVD. In particular, we show cathodoluminescence from these composites and characterize the behavior of the CL intensity during irradiation and at low temperatures.

Mat. Res. Soc. Symp. Proc. Vol. 452

EXPERIMENTAL

Nearly monodisperse, highly luminescent ZnS overcoated CdSe quantum dots were prepared as described elsewhere [2]. The quantum yield of the 33 and 42 Å diameter tri-n-octylphosphine oxide/tri-n-octylphosphine (TOPO/TOP) cap nanocrystals used in this work was 45%. Pyridine is a solvent compatible with the OMCVD process of ZnS. Therefore, the TOPO/TOP cap was exchanged with pyridine by floculation with hexane and redispersion in pyridine. Afterwards, the dots were mixed with acetonitrile (1:2) to achieve stable operation of the electrospray. The thin film composites have been synthesized according to the technique developed by Danek et al. [4]. In this method the dots were injected into a capillary electrode at a rate of 15 μl/min. The electrospray atomization was carried out at 4kV (dc) imposed between the capillary and a ring counterelectrode, and the dots were carried to the reactor by a H_2 stream of 1slm. H_2S and diethylzinc (DEZn) were used as precursors to form a ZnS thin film. Flow rates of 25 and 2.5 μmol/min were used respectively. The composites were grown on glass and Si substrates at temperatures between 80 and 250^0C. The pressure was maintained at 600 Torr.

Photoluminescence measurements were carried out at room temperature in a Spex Fluorolog-2 spectrometer with front face collection. CL was performed at room or low temperature in a scanning electron microscope (SEM) equipped with an Oxford Mono-CL spectrometer.

RESULTS AND DISCUSSION

Figure 1 shows PL and CL spectra of quantum dot composites containing 33 and 42 Å diameter CdSe/ZnS QDs. The PL QY of the quantum dot composites was determined by comparing the emission to Rhodamine 640 in methanol. A background correction to subtract the contribution from the ZnS matrix in the absorption spectra was done for each of the films. Absorption of the excitation light was measured with a Cary spectrometer. QY values around 10-15% were typically obtained. The surface roughness of the films (2000-4000 Å) prevented us from determining a more accurate number. The red shift of the near band edge peak with increasing the size of the dots indicates that the CL and PL emission is generated from the quantum confined excitons inside the QDs. The similarity between PL and CL shows that in both cases light emission comes from the same state in the QDs. The bandwidth of the PL spectra are relatively narrow (30-40 nm). CL spectra are red shifted with respect to PL and have broader bandwidths (40-50 nm). This behavior is attributed to a quantum confined Stark effect produced by the electric field generated inside the thin films during electron irradiation.

The CL intensity depends strongly on the acceleration voltage and current density of the electron beam, as can be seen in Figure 2. The increase in CL signal with both acceleration voltage and current density is related to the generation of secondary electrons in the ZnS matrix. The energy required to produced an electron/hole pair is approximately three times the band gap, which for the CdSe QDs considered in this work is about 6 eV. The formation of e^-/h pairs is not uniform, as deduced from energy dissipation curves calculated using Monte Carlo methods [7], and the majority of the excitons are generated near the surface within a relatively small radius. Thus, the saturation of CL with voltage could be explained in terms of the small thickness (0.5-1 μm) of our films. Increasing the voltage shifts the maximum efficiency of the e^-/h pair generation towards the substrate, leading to a saturation of the CL intensity.

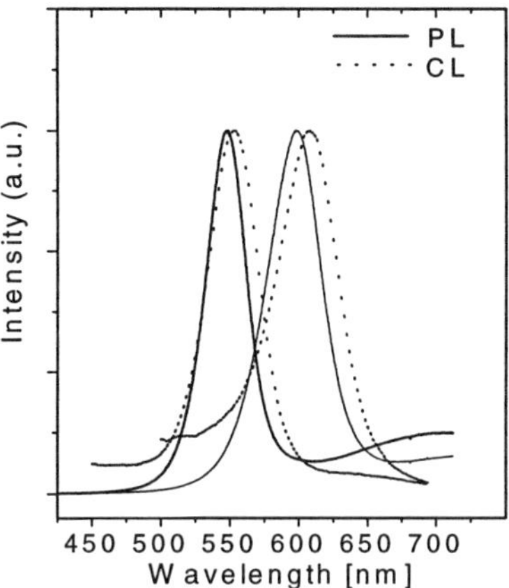

Figure 1. Photoluminescence and cathodoluminescence (dot lines) spectra for thin film quantum dot composites containing 33 and 42 A diameter dots. CL spectra were recorded at 30 kV, 20 nA. PL excitation was at 480 nm.

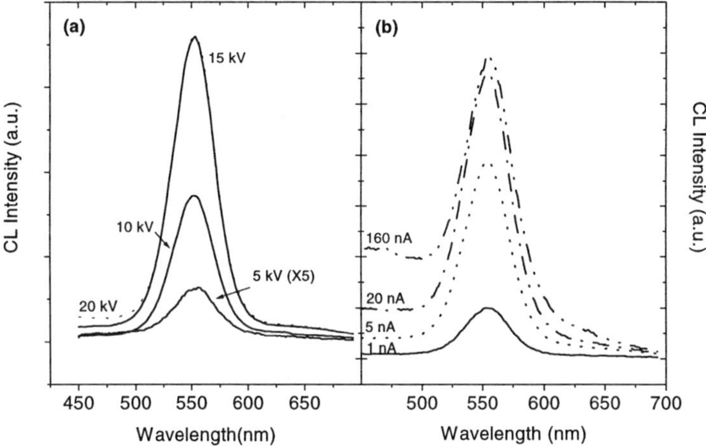

Figure 2. Room-temperature cathodoluminescence spectra of a thin film composite containing 33 A diameter dots at different applied voltages (a) and currents (b).

Figure 3 shows the degradation of the CL signal over time at different values of the acceleration voltage and current density. The degradation is more important at high voltages and currents. This process can not be attributed to thermal effects arising from power dissipation of the electrons. Thermal conductivity calculations show that under the given experimental conditions the temperature increases by a maximum of 1-5°C. On the other hand, the composites are stable up to temperatures around 250°C, which is the processing temperature. The decrease in CL intensity is usually characterized by a fast exponential decay (inset of figure 3), followed by a slow multiexponential decay at longer times. The CL degradation is faster in poorly crystallized samples, grown at 100°C, than in polycrystalline films grown at temperatures around 250°C. The steps in Figure 3 correspond to time intervals without electron beam irradiation. At room temperature, time intervals up to 100 min were done without observing any recovery of the CL intensity. However, the luminescence from a sample irradiated at 30kV, 40 nA for 30 min was partially recovered after annealing it for 3 h at 100°C (figure 4). We have also performed CL measurements at low temperatures (up to 5 K). A strong quenching of the luminescence was observed at 5 K and the CL curves showed faster decays as the temperature decreased.

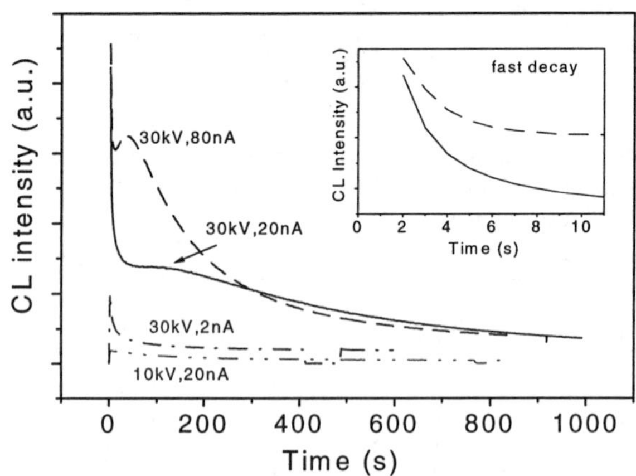

Figure 3. Decay time curves of a thin film composite containing 42 A diameter dots. (Inset) Fast decay at 30kV, 80nA and 30kV, 20 nA.

The experimental results presented so far can be rationalized through the presence of electron ionization of the quantum dots. Photoionization has been observed in semiconductor doped glasses (SDG) during laser irradiation [8,9]. In this case, as the energy of the incident photons is not enough to produce direct ejection of the electrons from the quantum dots, the main source of ionization in the SDG occurs through Auger recombination. In this process the creation of two electron-hole pairs inside a quantum dot results in the ejection of one of the electrons, leaving a positively charged dot. In cathodoluminescence, the ionization process may be more efficient because many of the secondary electrons involved in the electron-hole pair generation inside the QDs have enough energy to induce direct electron ionization. Increasing the current or voltage also increases the number and energy of secondary electrons and therefore the probability of direct ionization or the generation of more than one e^-/h^+ pair in each dot. The ejection of an electron through an Auger process also leaves behind a positively charged dot, which is dark from the point of view of optical transitions. These processes lead to a degradation of the optical properties of these semiconductor structures. The ejected electrons are captured by deep traps in the semiconductor matrix. When the sample was heated, electrons returned to ionized QDs recombining with the holes and neutralizing the QDs. The quenching of the cathodoluminescence at low temperatures is also consistent with this picture, because at low temperature the capture of the electrons in deep traps is more efficient.

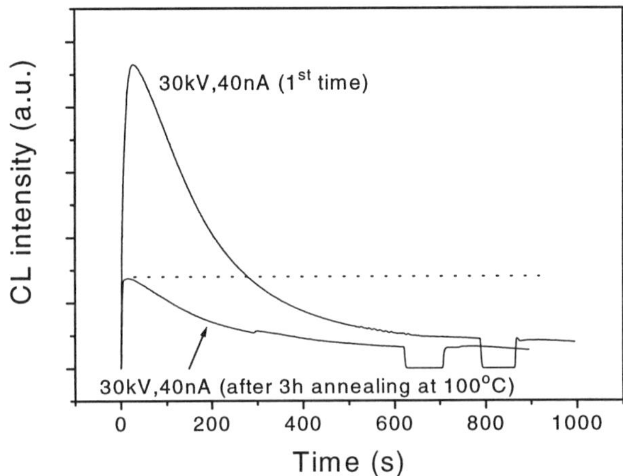

Figure 4. Partial recovery of the cathodoluminescence of a sample irradiated at 30 kV, 40 Na during 30 min and subsequently annealed for 3h at 100°C.

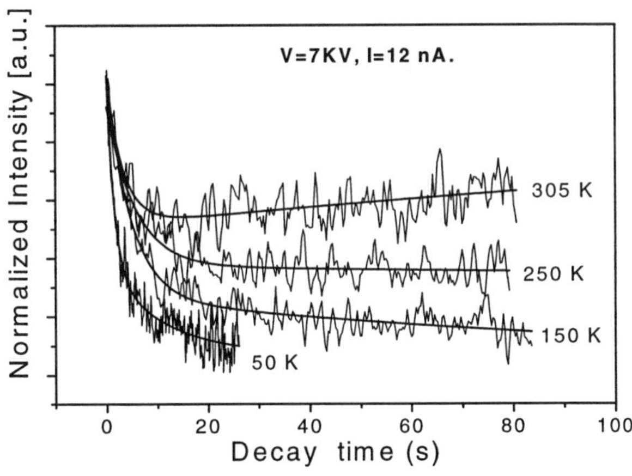

Figure 5. Influence of temperature on the degradation of the cathodoluminescence over time.

CONCLUSIONS

We have reported the CL and PL of 33 and 42 Å diameter ZnS overcoated CdSe nanocrystals embedded in polycrystalline ZnS matrices. CL and PL originates from the same state in the quantum dots. The degradation of the CL is attributed to electron ionization through direct ionization or Auger recombination followed by trapping of the electrons in the semiconductor matrix. Annealing at temperatures around 100°C produced a significant recovery of the CL signal, showing that the darkening of the quantum dot composites after electron irradiation is not irreversible. At low temperatures the efficient trapping of the ejected electrons resulted in the quenching of the cathodoluminescence.

Acknowledgments

JRV and BOD thank the *Direccio General de Recerca* from Catalonia and the Saudi Aramco respectively, for fellowships.

References

1. L.Brus, Appl. Phys. **A53**, 465 (1991).
2. C.B.Murray, D.J.Norris, M.G.Bawendi, J. Amer. Chem. Soc. **115**, 8706 (1993).
3. M.A.Hines, P.Guyot-Sionnest, J. Phys. Chem. **100**, 468 (1996).
4. B.O.Dabbousi et al. To be published.
5. M.Danek, K.F.Jensen, C.B.Murray, M.G.Bawendi, Appl. Phys. Lett. **65**, 2795 (1994).
6. M.Danek, K.F.Jensen, C.B.Murray, M.G.Bawendi, Chem. of Mater. **8**, 173 (1996).
7. T.E.Everhart and P.H.Hoff, J. Appl. Phys. **42,** 5837 (1971).
8. D.J.Chepic, Al.L.Efros, A.I.Ekimov, M.G.Ivanov, V.A.Kharchenko, I.A.Kudriaevtsev, T.V.Yazeva, J. Lumin. **47,** 113 (1990).
9. P.Roussignol, D.Ricard, J.Lukasik, C.Flytzanis, J. Opt. Soc. Am. **B 4,** 5 (1987)

SYNTHESIS AND OPTICAL PROPERTIES OF MOS_2 NANOCLUSTERS

J. P. WILCOXON, P. P. NEWCOMER AND G. A. SAMARA
Sandia National Laboratories, Albuquerque, New Mexico, USA 87185

ABSTRACT

Highly crystalline nanoclusters of MoS_2 were synthesized and their optical absorption and photoluminescence spectra were investigated. Key results include: (1) strong quantum confinement effects with decreasing size; (2) preservation of the quasiparticle (or excitonic) nature of the optical response for clusters down to ~2.5 nm in size which are only two unit cells thick; (3) demonstration that 3-D confinement produces energy shifts which are over an order of magnitude larger than those due to 1-D confinement; (4) observation of large increases in the spin-orbit splittings at the top of the valence band at the K and M points of the Brillouin zone with decreasing cluster size; and (5) observation of photoluminescence due to both direct and surface recombination.

INTRODUCTION

This paper deals with the synthesis and optical properties of size-selected nanoclusters of the layered compound, molybdenum disulfide (MoS_2). Bulk MoS_2 has been studied extensively because it approximates a 2-D system. Understanding how this 2-D character influences the synthesis, structure and electronic properties of nanoclusters of these materials represents a serious challenge. On the applied side there is a need to develop semiconductors for use as photocatalysts for solar fuel production and detoxification of chemical waste. In such applications the semiconductor spectrum must be resistant to photochemical degradation and must have a bandgap that is matched to the solar spectrum. Early work[2] has suggested that bulk MoS_2 is a good, stable candidate for solar photoelectrochemical applications, and subsequent bandstructure calculations[3] have shown that this stability derives from the fact that the relevant state at the top of the valence band is an antibonding state. While the stability of MoS_2 against photocorrosion is attractive, a limitation to its potential as solar a photocatalyst is the fact that in bulk form it is a black solid with indirect bandgap in the near infrared (~1.0 eV)[1,3] and its valence and conduction band energies are not sufficiently large to allow desired chemistry (e.g., water splitting). This limitation can, in principle, be overcome by tailoring the electronic structure by forming clusters. Space limitations restrict this paper to the presentations of few of the results. A more detailed account will appear elsewhere.[4]

EXPERIMENTAL DETAILS

The clusters were grown inside inverse micellar cages in non-aqueous solvents.[4,5] They are formed by first dissolving a molybdenum (IV) halide salt inside the cages and then combining this solution with another inverse micellar solution containing a sulfiding agent (e.g., metal sulfide or H_2S). The cluster size is varied by using different sized micellar cages. Electron diffraction and x-ray diffraction show the expected bulk hexagonal structure,[4,6] and high resolution TEM fringe imaging reveals that clusters down to ~3 nm are highly crystalline with no defects.[4] Cluster size was determined by TEM and estimated from dynamic light scattering (which measures the hydrodynamic radius) and from the chromatographic elution time in

Mat. Res. Soc. Symp. Proc. Vol. 452

calibrated columns (for the smallest clusters). As grown cluster samples were purified using high pressure liquid chromatography.

RESULTS AND DISCUSSION

Optical Absorption of MoS_2

Figure 1 shows the absorption spectra of different MoS_2 cluster samples and compares these spectra to those of two bulk crystalline samples taken from Refs. 7 and 8. The spectrum of bulk MoS_2 consists of a series of absorption thresholds the first of which corresponds to weak absorption in the near IR at ~1040 nm (~1.2 eV) associated with an indirect gap between Γ and the middle of the Brillouin zone between Γ and K. The second threshold occurs at ~700 nm (~1.8 eV) and is associated with a direct transition at the K point.[3] The two peaks (A_1 and B_1) on the short wavelength (λ) side of this threshold are the first (n=1) members of two excitonic Rydberg series corresponding to the transitions K4→K5 and K1→K5, respectively, whose energy separation (~0.20 eV) is due to the spin-orbit splitting of the top of the valence band at K.[3,7] A third threshold at ~500 nm (~2.5 eV) is due to a direct transition from deep in the valence band to the conduction band. Excitonic features (C and D) are also associated with this transition. A fourth threshold at ~350 nm (~3.5 eV) is also due to transitions from deep in the valence band. Features on the high energy side of this threshold (e.g., E in Figure 1) are seen in thin samples as illustrated in curves 2′ and 2″.[7,9]

The cluster results in Fig. 1 show that the main absorption features in this spectral region remain essentially unaltered, but exhibit large blue shifts with decreasing cluster size. Because our measurements were made on dilute solutions (10^{-4} to 10^{-3}) molar, and because of the inherent weak absorption characteristic of indirect transitions, the signal-to-noise ratio for the cluster samples did not allow measurements in the region of the indirect gap. Nevertheless, the shape of the cluster spectra on the short λ side of this gap suggests that the gap remains indirect. Figure 1 shows the evolution of the room temperature absorption spectrum with decreasing size reveals a one-to-one correlation between the bulk and cluster spectra.[4]

The resolved features in the spectra of the purified samples in this figure make it clear that the bulk-like excitonic features are retained to smaller cluster sizes than originally thought.[6] The earlier spectra of unpurified samples suggested that a bulk-like to molecule-like crossover occurs in the 3 to 2.5 nm size range. The present results on purified samples suggest that this expected crossover occurs below 2.5 nm.[4] This is a remarkable result which indicates that clusters which are only two unit cells thick have enough densities of states to reproduce bulk-like absorption properties, a feature undoubtedly related to the 2-D nature of MoS_2.

A variety of theoretical models have been used to describe size quantization in semiconductor clusters.[1] They include effective mass (EM) treatments, semi-empirical tight binding treatments and recent empirical pseudo-potential methods. Despite their shortcomings, the EM models contain some of the essential physics and provide a qualitative framework for examining experimental data and their trends. In terms of these models, the shift of the absorption edge, E(R), of a cluster of radius R relative to that of the bulk is to a reasonable approximation proportional to $1/2\mu R^2$, where μ is the reduced mass of the exciton.[1] A plot of E(R) vs. $1/R^2$ should then yield a straight-line with slope $\propto 1/(2\mu)$. Figure 2 shows such a plot for the various spectral features in the absorption spectrum of MoS_2. The curves show the now familiar deviations from linearity for small clusters, highlighting the inadequacy of EM theory in

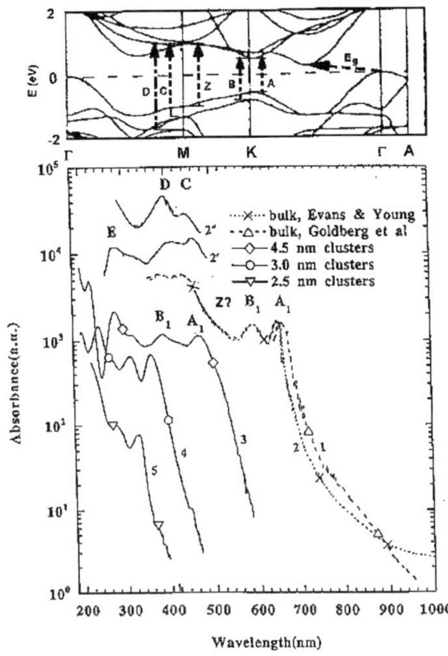

Fig. 1
Optical absorption spectra of two bulk crystalline samples (curve 1 synthetic crystal, Ref. 8; curve 2 natural crystals, Ref. 7) showing various absorption features discussed in the text. Curves 2′ and 2″ (from Ref. 9) provide better resolution of the high energy features. For comparison, long wavelength spectra for three cluster sizes are shown. The top of the figure is a portion of the band structure of MoS_2 (taken from Ref. 3) showing the proposed transitions which correspond to the various features in the optical absorption spectra.

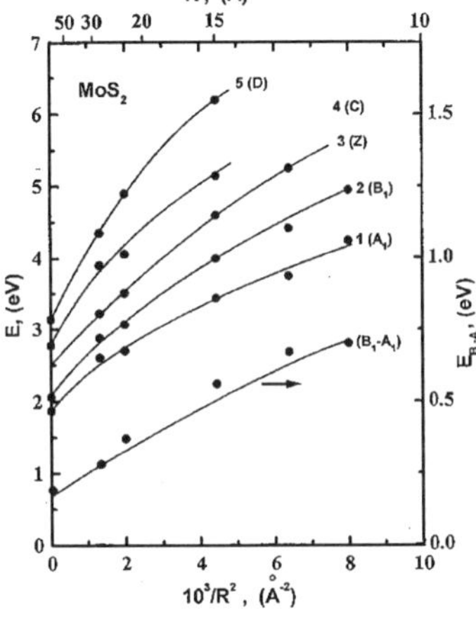

Fig. 2
Shifts of the various features observed in the optical absorption spectra of MoS_2 clusters. Also shown (right ordinate) is the increase in the spin-orbit splitting of the A_1 and B_1 excitonic peaks.

this regime. For the larger clusters the initial slope in Fig. 2 yields for the A_1 exciton μ in the range (0.1-0.2) m_0, where m_0 is the free-electron mass. This value of μ is comparable to the bulk value of 0.18 m_0 for the A exciton deduced from optical measurements.[7] Figure 2 also shows the influence of cluster size on the separation of the excitonic peaks A_1 and B_1, i.e., $E_{B_1-A_1}$ which is due to spin-orbit splitting at the top of the valence band at the K point. This splitting increases from 0.19 eV in the bulk to 0.67 eV for the 2.5 nm clusters.

Consistent with their being weakly bound (i.e., delocalized) Wannier excitonic peaks, the bulk excitonic Bohr radius (r_B) for the n=1 Rydberg states of A and B is r_B=2.0 nm.[7,9] Thus, our

4.5 nm (diameter) clusters are only slightly larger than the size of the exciton in the bulk, and the 3.0 nm and 2.5 nm clusters are smaller. Theoretical considerations[1] suggest that when $R/r_B > 4$, the quasiparticle characteristics of the excitons are preserved, whereas for $R/r_B < 2$ these characteristics are lost leading to individual confinement of the electron and hole which then behave like independent particles. Clearly, the data in Fig. 2 show that for MoS_2 clusters the quasiparticle nature of the A and B excitons (as well as the C-D excitons) is preserved to much smaller cluster sizes. No specific assignment of peaks C and D in bulk material have been made; however, using Coehoorn et al's[3] band structure results, we assign[4] these peaks to direct transitions from the spin-orbit-split valence band to the conduction band at the M point (see Fig. 1). The spin-orbit splitting of the top two valence band states at M in the bulk is about twice as large as that at K (0.37 vs. 0.19 eV), and the data in Fig. 2 clearly show that quantum confinement has a strong influence on this splitting. For the 3 nm clusters $\Delta_{D\text{-}C} \approx 1.05$ eV. The large increases in the spin-orbit splittings with decreasing cluster size reflect the influence of quantum confinement on the split-off energy levels and to some degree changes in the degree of hybridization among the various Mo and S atomic orbitals. Understanding these changes requires detailed electronic structure calculations.

Photoluminescence of MoS_2 Nanoclusters

In our earlier work on unpurified MoS_2 clusters[6], the photoluminescence (PL) was dominated by surface recombination. Chromatographic purification has led to significant improvement in the quality of the PL spectra and to the observation of both band edge (direct) as well as surface e-h recombination. Figure 3 shows the room temperature absorption and PL spectra of a 3.0 nm cluster sample. For excitation at λ=450 nm, which is seen to be a slightly longer λ than the first absorption threshold at ~420 nm, a single PL peak centered at 520 nm is observed. Clearly this PL is associated with surface recombination. Excitation at 348 nm (i.e., near the A_1 excitonic peak), however, leads to a PL spectrum consisting of two overlapping peaks. The more intense peak is centered at 420 nm which is seen to be the first absorption threshold for these clusters, and is attributed to direct band (K) edge recombination. Decomposition of the PL spectrum into its two components reveals the second peak which is centered at ~520 nm. This much less intense peak is at the same λ as the PL observed for 450 nm excitation, again pointing to its surface recombination character. For excitation at 225 nm (not shown), the PL spectrum consists of two overlapping peaks. The more intense PL is centered at 354 nm which is near the absorption threshold for the B_1 excitonic peak and thus represents direct band-edge recombination. The less intense peak is centered at 420 nm which is again the first absorption threshold for these clusters. For excitation at 425 nm, i.e., just above the first absorption threshold (not shown), the main PL is again due to surface recombination and is centered at 520 nm.[4] In an earlier collaboration[10], we studied the PL emission kinetics of as-prepared ~3.0 nm and ~4.5 nm clusters. These samples exhibited weak PL peaks close to the first excitonic thresholds (A_1 peak). The PL decay times were nonexponential having components with time constants ranging from <100 ps to >10 ns. The small values of these time constants are consistent with direct e-h recombination as found in our data on purified samples.

1-D vs. 3-D Quantum Confinement

The above results have clearly illustrated strong three- dimensional (3-D) quantum confinement effects. Because the chalcogen-metal-chalcogen layers in MoS_2 and its Mo and W

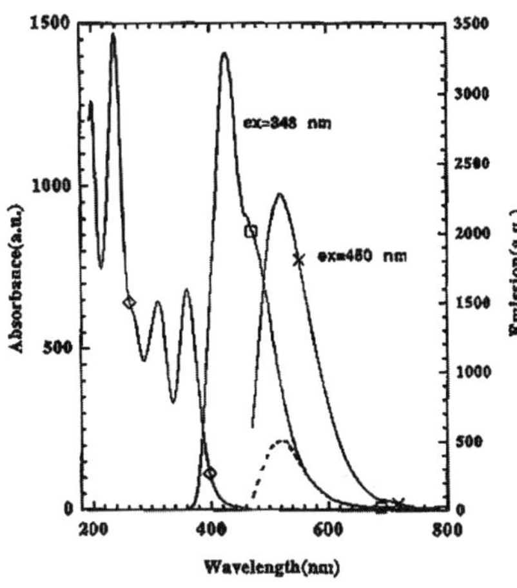

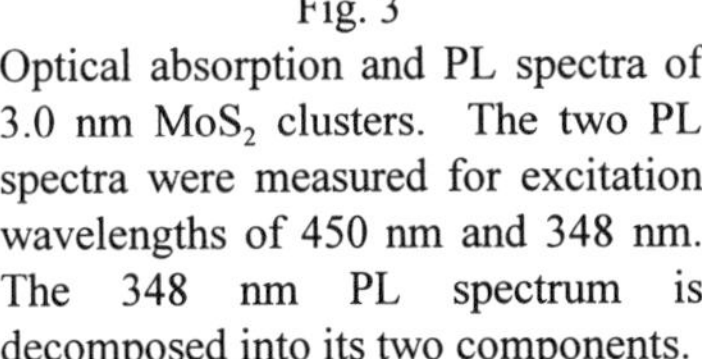
Fig. 3
Optical absorption and PL spectra of 3.0 nm MoS_2 clusters. The two PL spectra were measured for excitation wavelengths of 450 nm and 348 nm. The 348 nm PL spectrum is decomposed into its two components.

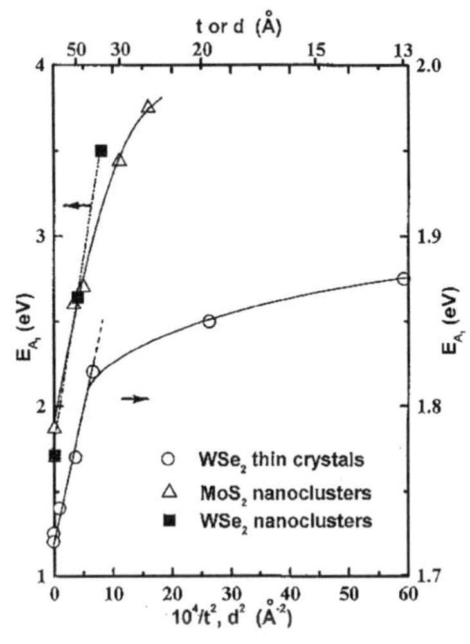

Fig. 4
Influence of dimensionality on the strength of quantum confinement of A_1 exciton in WSe_2 and MoS_2. Shown is the large difference between 1-D (thin crystals) and 3-D (clusters confinement).

isomorphs are held together by weak van der Waals forces, it is easy to cleave bulk crystals parallel to the layer planes. Remarkably, it has been possible to cleave bulk samples down to thicknesses of ~13Å (which corresponds to 1 unit cell thick) and measure the effects of quantum confinement (one-dimensional, 1-D, in this case) on their optical properties.

Consadori and Frindt[11] investigated the effects of crystal thickness (t) down to 13Å on the A_1 excitonic peak of WSe_2 at 77K. Their results, E_{A_1} vs. $1/t^2$ (Fig. 4), effective mass-like behavior down to t≈40Å, but deviate strongly at smaller thicknesses. The total shift in E_{A_1} between thick and 13Å thin crystals is, however, only 0.15 eV. For comparison, we have also plotted our results (3-D confinement) for MoS_2 and WSe_2 as E_{A_1} vs. $1/d^2$, where d is cluster diameter. For WSe_2 we have data on only two cluster sizes, (d=50Å and 35Å), but clearly the trend in the WSe_2 data is the same as for the more extensive MoS_2 data. Note that our data are read against the left ordinate and Consadori and Frindt's thin crystal data are read against the right ordinate. The comparison reveals a dramatic difference between 1-D and 3-D confinement. For sizes down to 40Å, the shift in E_{A_1} is over an order of magnitude larger for 3-D than for 1-D confinement. The difference becomes even larger for smaller sizes. The large magnitude of the difference is undoubtedly related to the quasi 2-D structural properties of these materials. Furthermore, the very large difference between the behavior of the 13Å (1 unit cell thick) thin crystal and our smallest clusters (d≈25Å) clearly demonstrates that lateral confinement of the carriers within the X-M-X layers is much stronger than confinement in the direction perpendicular to the layers, again emphasizing the 2-D character.

CONCLUDING REMARKS

We have produced high crystalline quality nanoclusters of MoS_2 and demonstrated great tailorability of its optical absorption and photoluminescence properties with change in cluster size -- features that bode well for its potential as a solar photocatalyst. MoS_2 exhibits strong 3-D quantum confinement effects which reflect the 2-D nature of the structure and bonding. The quasiparticle (or excitonic) nature of the optical response is preserved down to clusters ~2.5 nm in size (~2 unit cells thick) and there are large increases in the spin-orbit splittings at the top of the valence band at the K and M points of the BZ with decreasing cluster size. Calculations of the electronic structure as function of cluster size are needed to more fully understand these features.

ACKNOWLEDGMENTS

The excellent technical assistance of Sharon Craft is greatly appreciated. This work was supported by the Division of Materials Sciences, Office of Basic Energy Sciences, U.S. Department of Energy under Contract DE-AC04-94AL85000.

REFERENCES

1. See A. D. Yoffe, Adv. in Phys. 42, 173 (1993) and references therein for a recent review..

2. H. Tributsch, Z. Naturforsch 32a, 972 (1977).

3. R. Coehoorn, C. Haas, J. Dijkstra, C. J. F. Flipse, R. A. deGroot and A. Wold, Phys. Rev. B., 35, 6195 (1987), 35, 6203 (1987).

4. J. P. Wilcoxon, P. P. Newcomer, and G. A. Samara, J. Appl. Phys. (submitted).

5. J. P. Wilcoxon, DOE patent #5,147,841, issued Sep. 15, 1992.

6. J. P. Wilcoxon and G. A. Samara, Phys Rev. B 51, 7299, (1995).

7. B. L. Evans and P. A. Young, Proc. Roy. Soc. A., 284, 402 (1965).

8. A. M. Goldberg, A. R. Beal, F. A. Levy and E. A. Davis, Phil. Mag. 32, 367 (1975).

9. R. F. Frindt and A. D. Yoffe, Proc. Roy. Soc. A., 273, 69 (1963).

10. F. Parsapour, D. F. Kelley, S. Craft and J. P. Wilcoxon, J. Chem. Phys. 104, 1 (1996).

11. F. Consadori and R. F. Frindt, Phys. Rev. B 2, 4893 (1970).

THE INVESTIGATION OF DONOR AND ACCEPTOR STATES IN THE NANOPARTICLES OF PbI_2 LAYERED SEMICONDUCTOR

E. LIFSHITZ *
* Permanent address: Department of Chemistry and Solid State Institute, Technion, Haifa 32000, Israel, ssefrat@tx.technion.ac.il

ABSTRACT

The nanoparticles of PbI_2 layered semiconductor, embedded in SiO_2 films, were prepared by the sol-gel method. The low-temperature luminescence of PbI_2 nanoparticles consists of a series of exciton lines in the region of ~2.5eV and additional donor-acceptor recombination luminescent bands at lower energies, centered at 2.44eV (Green) and 2.07 eV (red), respectively. This paper reports the investigation of the nonexcitonic transitions, associated with of the Green and Red bands. The correlation between donor-acceptor recombination emission processes and lattice imperfections was examine, utilizing optically detected magnetic resonance (ODMR) spectroscopy. The results identified the following imperfection sites of both the Green and Red bands: an acceptor site associates with an isotropic Lead vacancy defect, $[V^-]Pb^{2+}$ and a donor site, associate with an anisotropic Iodine vacancy, $[V^0]_{Iodine}$. However, the results pointed on differences in relaxation processes of these bands. The relaxation processes correspond to the existence of competitive nonradiative recombination, spin-lattice relaxation and thermalization among the spin substrates. Thus, the results suggest that the Red band corresponds to stoichiometric defects at the surface, while the Green band corresponds to the same stoichiometric defects at the interior part of the nanoparticles.

INTRODUCTION

Lead iodide, PbI_2, is a direct band gap semiconductor with a layered structure. Compounds exhibiting a structure of this kind possess strong intralayer bonding and only weak, so called, van der Waals interlayer interactions. The band-edge optical properties of PbI_2 single crystals have been studied extensively for about two decades [1]. However, there are only few studies of these properties in the corresponding nanometer-sized particles (nanoparticles) [2,3].

The low-temperature photoluminescence (PL) spectrum of PbI_2 nanoparticles consists of a series of exciton lines in the region of ~2.5 eV and additional broad luminescent bands at lower energies centered at 2.44 eV (Green) and 2.07 eV (Red), respectively. Although it has been shown that the absorption and photoluminescence spectra of nanoparticles exhibit similar properties to those of the bulk [2,3], several differences arise for the following reasons: (1) When the particle radius approaches the size of the exciton Bohr radius (with bulk crystallographic structure), size quantization takes place [4]. The latter is pronounced as a blue shift in the optical transitions, (2) Due to the large ratio difference between surface/bulk atoms in the epitaxial and nanoparticle samples, there may be a proportionally larger amount of surface defects. These are capable of acting as traps for photogenerated excitons, holes or electrons and may alter the recombination processes.

Recently we have reported the study of the exciton properties in the nanoparticles of PbI_2 [4]. The latter showed unique quantum size effect in which the size confinement permitted the creation of an acceptor-like exciton. The present work reports the investigation of the

Mat. Res. Soc. Symp. Proc. Vol. 452

nonexcitonic recombination processes in PbI_2 nanoparticles. The research focuses on the identification of donor and acceptor sites that give rise to the Green and Red luminescence bands appearing below the exciton manifold. Moreover, the investigation emphasizes the dynamic properties of these nonexcitonic transitions, such as trapping into nonradiative sites, spin-lattice relaxation and thermalization effects. The research utilizes Optically Detected Magnetic Resonance Spectroscopy (ODMR) in combination with the conventional luminescence technique.

EXPERIMENTAL

Materials

The nanoparticles of PbI_2, embedded in SiO_2, films were prepared by a sol-gel technique according to the procedure described in reference 4. It should be noted that the synthesis of PbI_2 nanoparticles may be accompanied by the formation of several by-products (e.g., iodides and oxoiodides complexes). However, several control experiments, described in reference 5 and 6, have excluded their existence. Thus, the optical transitions discussed in the present document correspond only to the PbI_2 nanoparticles. The mean particle size and distribution of the PbI_2 crystallites were determined using Transmission Electron Microscope (TEM) while the crystal structure of the particles was determined by XRD technique.

Instrumental

The ODMR spectra were obtained by measuring the change in luminescence intensity, ΔI_{PL} (or its circular component), induced by a magnetic resonance event at the excited state. Thus, ΔI_{PL} was plotted versus the strength of the external magnetic field, H_0, leading to magnetic resonance-like spectra. The samples were immersed in a cryogenic dewar (2K) centered within a High-Q resonance cavity, coupled to an audio modulated microwave (mw) source (9 GHz) and surrounded by a superconducting magnet. The ODMR signal was detected in either of the following configurations: (a) in a direction, k_{em}, parallel to the external magnetic field ($k_{em}||H_0$, Faraday configuration) or (b) in a direction perpendicular to H_0 ($k_{em} \perp H_0$, Voight configuration). In the Faraday configuration, the change in the circular polarization component of the luminescence intensity was detected. The ODMR spectra were recorded at different laser excitation powers, audio modulation frequencies and microwave power output.

RESULTS

The photoluminescence (PL) spectrum of PbI_2 nanoparticles with mean particle radii of 6 nm is shown in figure 1a. It comprises an exciton envelope (FE and BE) and two broad bands, centered at 2.44 eV (Green) and 2.07 eV (Red), respectively. Qualitatively, the latter spectrum resembles the corresponding hexagonal bulk single crystal. However, the exciton band in the nanoparticles is shifted by about 40 meV, with respect to the bulk. The latter is a manifestation of the quantum size effect, as discussed in great detail elsewhere [3]. It should be indicated that post annealing treatment of the samples under iodine atmosphere caused quenching of the Red band and narrowing of the Green band as shown in figure 1b. Thus, the nonexcitonic luminescence bands are associated with stoichiometric defects.

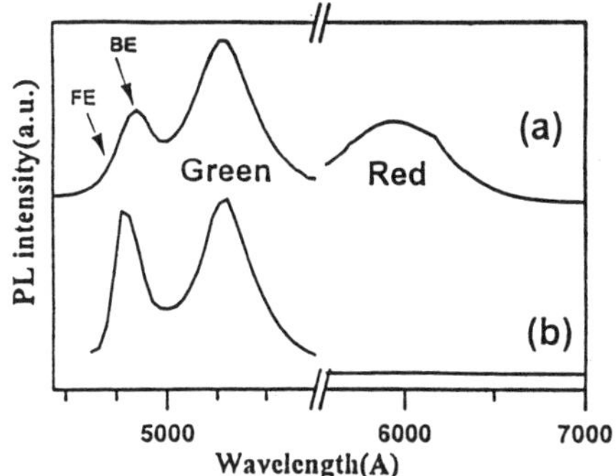

Fig. 1: Representative PL spectra of PbI_2 nanoparticles, before (a) and after (b) annealing with Iodine vapors. These spectra were recorded at 1.4K and excited with 457.9 nm (2.706 eV) Ar^+ laser, at laser power of 200 mW/cm^2.

Typical ODMR spectra of the PbI_2, monitored at the Red and Green luminescence bands, are shown in figure 2a and 2b, respectively. The latter spectra were recorded either at the Voight or Faraday configurations with laser excitation power of 200 mW/cm^2. Both spectra consist of three resonance signals, labeled I, II and III. The resonance signals monitored at the Green band are associated with quenching of the luminescence intensity (negative signals), while the I and III resonance signals monitored at the Red band correspond to enhancement of the luminescence intensity (positive signals). Moreover, resonance II is relatively narrow with a full width half maximum (FWHM) of 20 Gauss and a g-factor of 2.05. Resonance I is rather broad (FWHM ≈ 650 Gauss) with g-factor of 2.30. Resonance III in figure 2 also appears to be broad, with a g factor of about 1.73. Figure 2c show an ODMR spectrum or PbI_2 epitaxial film, monitored either at the Green or Red band. The latter was added to this document in order to assist in the forthcoming discussion with reference to the identification of the defect sites. Spectrum 2c showed strong dependence on the relative orientation between the c-crystallographic axis and the direction of the external magnetic field [5]. This dependence showed that resonance II is isotropic, while resonance I and III show anisotropic behavior. However, this angle dependence was not pronounced in the ODMR of the nanoparticles due to their random orientation within the SiO_2 medium.

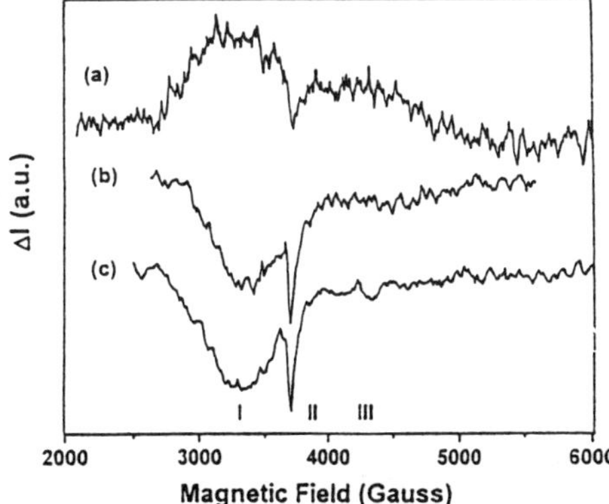

Fig.2. Representative ODMR spectra, obtained under laser excitation of 200 mW/cm^2: (a) monitored at the nanoparticles' Red band, (b) monitored at the nanoparticles' Green band, (c) monitored either at the Green or at the Red band of an epitaxial film.

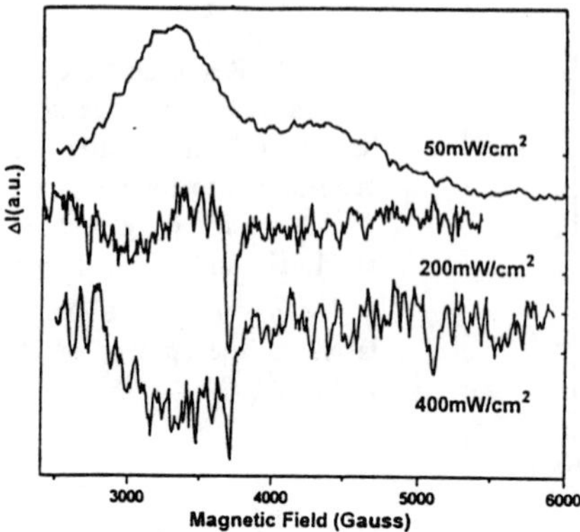

Fig.3. Representative ODMR spectra of PbI_2 nanoparticles, monitored at the Green band and recorded at various laser excitation powers, as indicated in the figure.

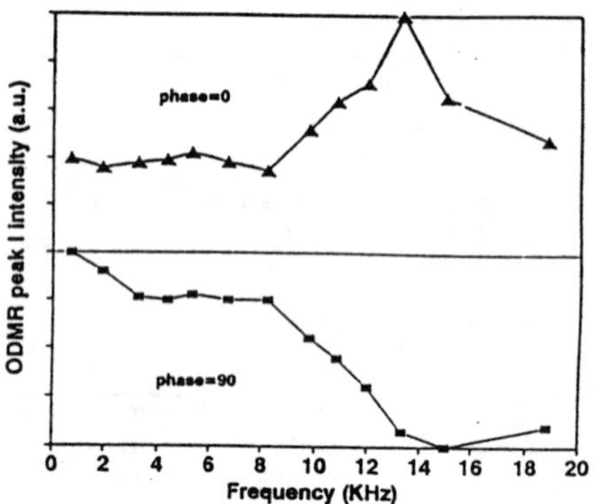

Fig.4. Plot of resonance I's integrated intensity versus the mw audio frequency, when monitored at the nanoparticles's Green band. Triangles correspond to the signal detected with lock-in at phase=0°, while squares correspond to the lock-in at phase=90°.

Figure 3 shows representative ODMR spectra of the Green band recorded under different laser excitation powers. While resonance II seems to appear as a negative ODMR signal in the studied laser power range, the signals I and III altered from negative, to derivative-like, to positive, with decreasing laser excitation power.

In general, the ODMR spectra were monitored by modulating the mw power at audio frequency and collecting the modulated luminescence by a lock-in technique. Thus, the ODMR of PbI_2 nanoparticles were monitored with audio modulation frequencies in the range 30 Hz to 17 kHz and lock-in detected either at phase=0° or phase=90°. Representative resonance amplitude dependence on the audio frequency is shown in figure 4.

DISCUSSION

The identification of the donor an acceptor sites

The dependence of the ODMR spectra on the experimental conditions (laser power, audio modulation) indicates that resonance II behaves substantially different from those of the other resonance signals. This suggests that I and III are associated with the same site, while resonance II corresponds to a separate site. In such a case, the photogenerated electrons and holes are trapped at different defect sites with weak e-h exchange interaction. Moreover, the observance of an identical spectrum when monitored in the Faraday and Voight configurations suggests that each one of the trapped carriers have a spin quantum number M_S=1/2, while the recombination corresponds to the singlet e-h pair. The relevant spin levels are drawn in figure 5. The solid arrows correspond to the spin resonance transitions of each carrier, while the dashed arrows represent the e-h recombination transitions.

The pronounced change of the Green and Red band intensities, upon Iodine annealing, suggests that these bands are associated with stoichiometric defects. The consideration of external impurities, such as O^{-2} and I_3^- in a substitutional or interstitial position , can be excluded for the following reasons: the existence of O^{-2} can be eliminated by carrying the growth and post annealing treatment, either in vacuum or in inert conditions. The integration of I_3^- complex is avoided due to its large size.

It was shown in figure 2a and 2b that the resonance signals, I, II and III appeared when detected either in the Green or in the Red bands of the PbI_2 nanoparticles. The latter observance suggests that the considered emission processes are sharing common states, one is radiative and the other is nonradiative (labeled n_e, and n_h in figure 5). The contradiction between quenching and enhancement effects are associated with dynamic processes that will be explained in the forthcoming discussion.

In the present case the possible acceptor sites associated with either a Lead vacancy or an Iodine interstitial. The latter may be excluded as the iodine atom with its relatively large atomic radius, will not occupy interstitial sites. Previously, the ESR studies of PbI_2 reported by Verwey showed the existence of relatively narrow resonance signal with g=2.03 associated with Lead vacancies [6]. The properties of the latter resonance are nearly identical to those of resonance II in the present case. Moreover, the Thermal Stimulated Current measurements, carried by De Blasi et. al. [7] showed that Lead vacancy acceptor state is located at 0.59eV above the valance band. Thus, the deepness of this defect state suggests that Lead vacancy acts as a nonradiative center. The metal vacancy at the ground state, $[V^{-2}]_{Pb^{+2}}$, is assessed to capture a hole at the excited state. Thus, trapping of one hole creates $[V^-]_{Pb^{+2}}$ center, that is paramagnetic with $(M_S)_h$= ±1/2. The trapping of two holes may create a double acceptor site, $[V^0]_{Pb}$, that is diamagnetic. It is anticipated that the nonradiative acceptor site in the present model corresponds to $[V^-]_{Pb^{+2}}$, while the radiative acceptor site either of the Green or Red bands corresponds to the double acceptor site $[V^0]_{Pb}$. Moreover, previous studies discussed the correlation of the Red band with metal vacancy defects at the external surfaces of the samples [8]. A hole trapped at the center of the Lead vacancy has a centrosymmetric property. Indeed, the hole exhibits isotropic behavior when its orientation is examined with respect to the direction of the external magnetic field in the oriented epitaxial films.

The experimental observations suggest that resonances I and III correspond to a single donor site. These resonances are relatively broad and, in the epit axial film exhibit anisotropic behavior. Identification of the donor site is based on analysis of the ODMR angle dependence of the corresponding epitaxial film (reported in reference 5). This analysis included simulation of the resonance lineshape and the utilization of a phenomenological spin Hamiltonian. The analysis showed that the broadening is due to an overlapping of resonance signals, associated with magnetically inequivalent sites and to an unresolved hyperfine structure, originating from the interaction with the nearest neighboring Iodine unclear spins (I=5/2. 100% abundance). Because of the volatile nature of the Iodine element, we assumed that the latter is the most probable donor site. The paramagnetic Iodine vacancy donor site is labeled $[V^0]_{Iodine}$ in the excited state, after the trapping of the photogenerated electron.

The mechanism of recombination and the dynamic properties of the excited states

The appearance of negative or positive resonance signals in the ODMR spectra is associated with the following considerations: (1) thermalization effect, (2) existence of competing nonradiative process. When the e-h radiative recombination process is much faster than the spin-lattice relaxation time, $\tau << T_1$, then thermalization does not take place, resulting in positive ODMR signals. However, when the recombination process occurs on a time scale that is about equivalent of the spin-lattice relaxation, $\tau \cong T_1$, thermalization will occur and the ODMR will exhibit both positive and negative signals. Thus, the change of sign of resonance I with increasing excitation power (fig. 3) is associated with a transfer from thermalized to an unthermalized situation. However, the quenching effect of the entire ODMR spectrum, recorded with excitation >200 mW/cm^2 corresponds to the existence of competitive nonradiative e-h recombination process. The coupling mechanism between the radiative and noradiative process and its influence on the magnetic resonance transitions can be distinguished by following the dependence of the ODMR spectra on the mw audio modulation frequency. The coupling can be via the ground state or coupling at the excited. The ODMR dependence shown in figure 4, as a typical characteristic to a case in which the radiative and nonradiative transitions are coupled at the excited state, by sharing a common donor state [9].

The above discussion suggests that the Green and Red bands are both associated with e-h recombination, trapped on Iodine and Lead vacancies, respectively. The Iodine forms a common radiative state, while the Lead creates a single nonradiative acceptor state. As discussed above, the radiative acceptor of the Green band can be associated with the Lead double acceptor at the interior parts of the nanoparticles, while the radiative acceptor of the Red band may be a similar Lead double acceptor, located at the surface of the particles.

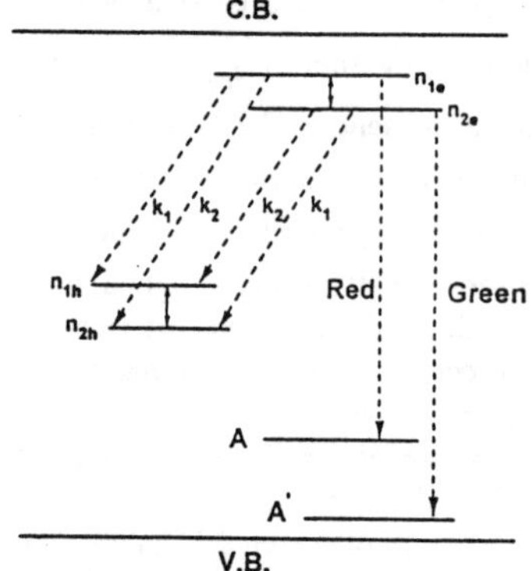

Fig. 5. The electron-hole spin and electronic energy level diagram

1 M.S. Skolnik, D. Bimberg, Phys. Rev. B, **18**, 7080 (1978).
2 S. Takeyama, K. Watanabe, T. Komatsu, N. Miura, Proceed. 20th International Conf. Phys. Semicond. Thessaloniki, Greece, August, 1 (1990).
3 E. Lifshitz, M. Yassen, L. Bykov, I. Dag, J. Phys. Chem., **98**, 1459 (1994) and references within.
4 A.I. Ekimov, Al.L. Efros, A.A. Onushchenko, Solid State Commun. **56**, 921 (1985).
5 E. Lifshitz, L. Bykov and M. Yassen. J. Phys. Chem., 99, 15262 (1995).
6 J. Arends, J.F. Verwey, Phys. Stat. Solidi, **23**, 137 (1967).
7 C. De Blasi, S. Galassini, C. Manfredotti, G. Micocci, L. Ruggiero and A. Tepore, Solid State Commun., **25**, 149 (1978).
8 I. Baltog, I. Piticu, M. Constantinescu, C. Ghita, L. Ghita, Phys. Stat. Solidi (a), 52, 103 (1979).
9 J. Shinar, A.V. Smith, J. Partee, O. Amir, P.A. Lane, X. Wei, Z.V. Vardeny and Y. Yoshino, Phys. Rev. Lett, in press.

OPTICAL PROPERTIES OF CUPROUS OXIDE NANOCRYSTALS

P.J. RODNEY,[a,f] M.I. FREEDHOFF,[b] A.P. MARCHETTI,[c,f] G.L. MCLENDON,[d] P.M. FAUCHET[e,f]

[a]Institute of Optics and Laboratory for Laser Energetics, University of Rochester, Rochester, NY 14627, prod@lle.rochester.edu

[b]MRS/OSA 1996-97 Congressional Fellow, Office of Rep. Edward Markey, Washington, DC

[c]Imaging Research and Advanced Development, Eastman Kodak Company, Rochester, NY 14652-4708.

[d]Department of Chemistry, Princeton University, Princeton, NJ 08544.

[e]Department of Electrical Engineering and Laboratory for Laser Energetics, University of Rochester, Rochester, NY 14627

[f]NSF Center for Photoinduced Charge Transfer, Department of Chemistry, University of Rochester, Rochester, NY 14627.

ABSTRACT

Optical properties of size restricted Cu_2O obtained through aqueous and non aqueous preparations are compared with those of commercially available bulk crystals. One method of synthesis involved using polyvinyl alcohol as a restraining agent in an aqueous preparation to produce nanocrystals having diameters with a mean of 11 nm and a standard deviation of 6 nm. Low-temperature spectroscopic studies indicate size restriction effects are manifest by a decrease in exciton luminescence and a 3.0 to 9.0 meV blue shift in 0,0 transition. No noticeable changes in the dynamics were observed for the nanocrystals, indicating no alteration or relaxation of the selection rules for the direct forbidden transition.

INTRODUCTION

Cu_2O represents a class of semiconductors with symmetry forbidden lowest energy band-to-band transitions. The effect of quantum confinement on the forbidden nature of this transition is of fundamental interest. Band structure calculations show that the lowest energy band-to-band transition is direct [1] at the Γ point of the Brillouin zone as depicted in Fig. 1. At the zone center of Cu_2O the valence band is largely comprised of copper 3d orbitals and has Γ_7^+ symmetry while the conduction band is made up of copper 4s orbitals [2] and has symmetry Γ_6^+ (following the relativistic notation of Stolz [3]). An optical transition between these bands would be s $\rightarrow$ d in an atomic sense, making it parity forbidden. Our investigations focus on the lowest energy 1S exciton whose transition dipole matrix element is identically zero [4], although the transition is quadrupole allowed. Because most quantum confinement studies have focused on either direct allowed or indirect band gap materials, a study of size restriction effects on this direct forbidden semiconductor may indicate to what extent parity selection rules are affected by size restriction.

Cu_2O is a well studied material both as a single-crystal and as polycrystalline film [3-13]. Recent studies include picosecond quantum beat spectroscopy of the lowest exciton state [3], growth of nanocrystals of Cu_2O from copper clusters [11], and thin film growth for device applications [13].

Cu_2O is the material in which exciton absorption lines forming a Rydberg series were first observed [12]. These spectra helped in development of the concept of Wannier or large radius excitons. Wannier excitons can be described using a scaled hydrogenic model with the effective mass approximation [14,15]. When the Coulombic interaction between the electron and hole is taken into account, the energies, E_n, referenced to the appropriate band-gap energy are

$$E_n = E_g - \frac{2\mu e^4\pi^2}{h^2\varepsilon^2 n^2} \quad (1)$$

Mat. Res. Soc. Symp. Proc. Vol. 452 © 1997 Materials Research Society

where Eg is the band gap (2.17 eV for the lowest energy exciton series) [3], μ is the reduced exciton mass (0.365 m_o) [16], e is the electron charge, h is Planck's constant, ε is the static dielectric constant (7.11) [16], and n is the principal quantum number. It has been shown that for $n \geq 2$, this model accurately predicts the energies of the Rydberg series of the yellow exciton when the static dielectric constant is used [17, 18], but fails for the $n = 1$ transition unless the high-frequency dielectric constant of 6.46 is used. The failure of Eq. 1 for the $n = 1$ transition has been attributed to the small exciton radius, which suggests a more localized exciton where the use of the static dielectric constant would be inaccurate [3, 14]. This work will focus on quantum confinement effects in the lowest energy forbidden $n = 1$ excitonic transition.

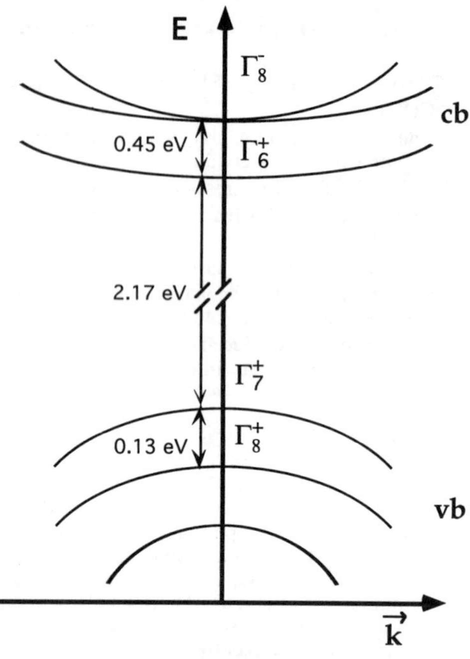

FIG. 1 Cu_2O band structure near the Γ point of the Brillouin zone. Valence and conduction bands are labeled as vb and cb respectively. From [4].

EXPERIMENTAL

Bulk Cu_2O powder was obtained from J. T. Baker and annealed in N_2 under different combinations of temperature and time with the most highly structured photoluminescence spectra resulting from an anneal schedule of 500 °C for 17 hrs. The method used to grow Cu_2O microcrystals followed that of reference 19 with the modification of using ascorbic acid as a reducing agent instead of glucose.

In order to make nanocrystalline samples, the above preparation was further modified in a number of ways. One approach, following work done with AgBr [20], involved the creation of reverse micelles by introducing the aqueous reactants into an organic solution (heptane) containing a small amount of surfactant, dioctlsulfosuccinate sodium (AOT). In another approach nanocrystals dispersed in toluene were produced by modifying a preparation by Zou [21]. This dispersion was created by vigorously mixing toluene with aqueous prepared Cu_2O and a small amount of surfactant in a separation funnel.

A third approach involved the addition of a small amount of polyvinyl alcohol (PVA) in water immediately after the addition of ascorbic acid. The polymer acted as a restraining agent, and some size control was achieved by varying the time at which the polymer was added. Transmission electron micrographs indicated that samples grown using this method were polydisperse, with a mean diameter of 11 nm and a standard deviation of 6 nm. This size metric can be deceiving as micrographs indicate that there is a noticeable population of crystallites as large as 50 nm. Quantifying size restriction effects would be difficult since each sized crystallite will undergo a transition at a slightly different energy which results in significant inhomogeneous broadening in both absorption and luminescence.

Excitation spectra were recorded using an Ar-ion pumped tunable dye laser (Rhodamine 6G) operating at 30 mW. The rest of the low-temperature spectroscopic apparatus used to measure photoluminescence and lifetimes has been described previously [22, 23].

RESULTS AND DISCUSSION

Fig. 2(a) shows the low-temperature photoluminescence spectra from bulk and nanocrystalline Cu_2O samples. A) and B) label spectra from the heat-treated bulk and aqueous prepared bulk samples respectively. Luminescence from the nanocrystals are labeled by preparation method: C) reverse micelles, D) PVA restrained, and E) toluene dispersion. The sharp peaks in the heat treated bulk and reverse micelles spectra in the region between 600-640 nm correspond to previously catalogued free and bound exciton transitions [4]. The high resolution spectrum of the heat treated bulk material in Fig. 2(b) more clearly reveals these transitions. The broad peaks in the spectra of all samples are probably due to donor acceptor recombination. The exciton emission in all but the heat treated sample was suppressed. This behavior strongly differs from that of the indirect gap semiconductor AgBr where the luminescence arising from free and weakly trapped excitons in quantum confined AgBr is enhanced [22, 23]. The emission suppression in Cu_2O is attributed to the incorporation of lattice irregularities into the crystallites from impurities, from CuO, and from hydrated species both in the bulk and on the surface. This is in contrast to quantum confined AgBr where iodide, an exciton trap, is eliminated by impurity exclusion [22, 23].

Band-edge excitation spectra of nanocrystalline Cu_2O reveal exciton transitions otherwise unobservable in emission. Fig. 3(a) shows the excitation spectra for bulk and nanocrystalline Cu_2O samples.

The excitation spectrum of the heat treated Cu_2O sample (trace A) reveals the phonon-assisted absorption band edge located at 606 nm in good agreement with literature values [7]. The two sets of narrow triplets, centered at 606 and 610 nm respectively, in the high resolution excitation spectrum, Fig. 3(b), are due to the n = 1 exciton and a phonon replica at 606 nm. The two triplets are separated by 102 cm^{-1}, closely matching the ground state phonon energies of 105-110 cm^{-1} given in the literature [2, 7, 24]. The multiplicity of the lines is attributed to the removal of the degeneracy of the n = 1 transition [2] by strain splitting [6] likely caused by the fast cooling the Cu_2O sample experienced following heat treatment. The origin of the degeneracy of this

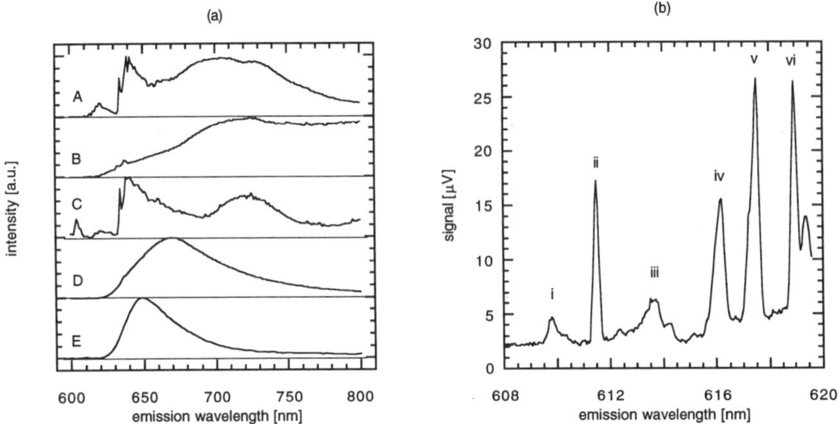

FIG. 2 Low-temperature photoluminescence spectra, T = 6 K, excitation 514 nm. (a) Spectra from bulk Cu_2O crystals: A) heat treated bulk, B) aqueous prepared microcrystals, and Cu_2O nanocrystals: C) reverse micelles, D) PVA restrained, and E) toluene dispersion. (b) High resolution spectra of heat treated bulk Cu_2O revealing: Ar laser plasma lines ii) and v); and photoluminescence from i) the orthoexciton, iii) the orthoexciton accompanied by creation of a Γ_3^- phonon, and iv) the paraexciton accompanied by creation of a Γ_5^- phonon.

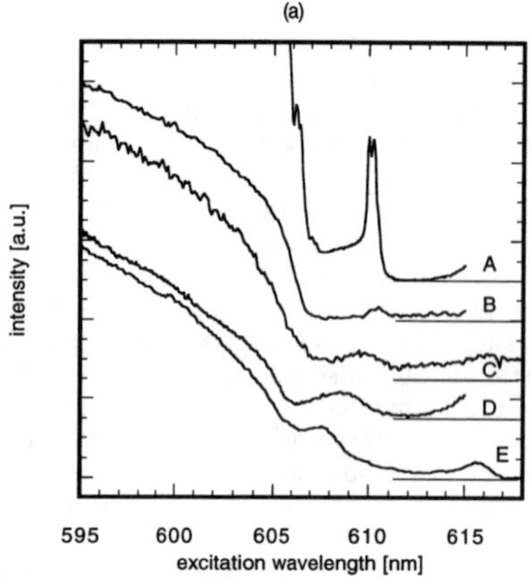

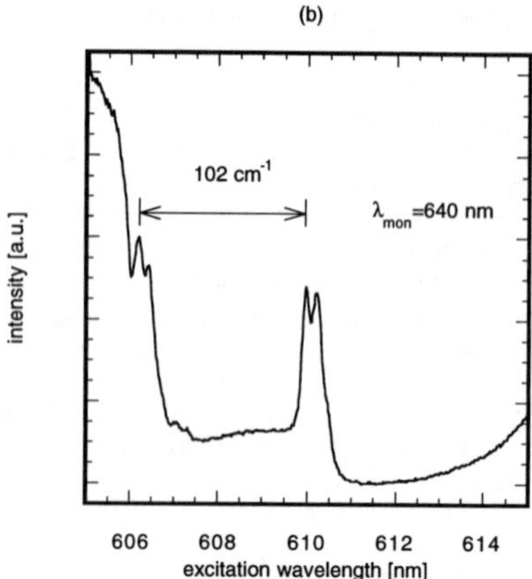

FIG. 3 Low-temperature excitation spectra, T = 6 K. (a) Low resolution spectra corresponding to Cu_2O materials labeled as in Fig. 2(a) with luminescence monitored at A) 640 nm, B) 720 nm, C) 640 nm, D) 640 nm, and E) 650 nm revealing a blue shift of the 0,0 transition (b) High resolution scan of heat treated bulk Cu_2O revealing the orthoexciton ca 610 nm and a phonon replica ca 606 nm.

transition can be explained using group theory. In group-theoretical terms the symmetry of the 1S exciton in Cu_2O can be expressed as a direct product of the irreducible representations which correspond to the symmetries of the valence band, conduction band, and the 1S excitonic envelope function, respectively, Γ_7^+, Γ_6^+, and Γ_1^+ . The direct product

$$\Gamma_{ex,1S} = \Gamma_7^+ \otimes \Gamma_6^+ \otimes \Gamma_1^+ = \Gamma_2^+(\text{orthoexciton}) + \Gamma_5^+(\text{paraexciton}). \qquad (2)$$

is reduced to a linear combination of excitonic states with symmetries Γ_5^+ and Γ_2^+ which are referred to as the orthoexciton and paraexciton respectively. The former is three-fold degenerate while the latter is nondegenerate. The strain in the heat treated sample removes the three-fold degeneracy of the orthoexciton. Additionally, transitions into the nondegenerate Γ_2^+ state, which are normally forbidden in all orders, are observed in the photoluminescence of the heat treated sample in Fig. 2(b) because the strain has altered the symmetry of the crystal.

Band-edge features of the aqueous prepared bulk, spectrum B in Fig. 3(a), closely follow those of the heat treated bulk, spectrum A in Fig. 3(a), establishing a connection between Cu_2O crystals prepared by different methods. This connection allows comparison of the optical behavior of the nanocrystals with that of the bulk because all of the nanocrystalline preparations used some form of the aqueous preparation as a precursor.

The n = 1 excitonic transition and the phonon-assisted absorption edge are clearly blue-shifted in the nanocrystalline samples, spectra C, D, and E. However, these features are less well defined because of inhomogenous broadening.

As a consequence of inhomogeneous broadening, quantitative modeling would be inaccurate. However, a qualitative picture can be obtained by comparing the blue shift predicted using a confined hydrogen atom model with the experimentally observed 'average' blue shift. In the model the exciton is treated as a hydrogen atom confined in an impenetrable spherical box for which exist exact and very good approximate calculations of the energy shift as a function of box size [25, 26]. The energy shifts for the exciton are obtained by scaling the results for the hydrogen atom to the exciton. The experimental blue shift for the nanocrystalline samples range between 3.0 and 9.0 meV from the change in energy of the n = 1 exciton absorption. Using the mean particle size given above and the exciton binding energy of 98 meV [16], the confined hydrogen atom model predicts a blue shift of $\approx$1 meV which has the correct order of magnitude.

The peaks ca 615 nm, especially evident in the excitation spectra of the nanocrystalline samples, are not likely related to phonon assisted exciton absorption, but rather are the result of surface defects, traps, or impurities such as hydrated species at the surface and in the interior. This is because at low temperatures the population of phonons present in the crystal lattice with sufficient energy to assist a low energy photon in creating an exciton is extremely small.

Finally, no noticeable change in the dynamics is observed when the photoluminescence lifetimes of the bulk and nanocrystalline samples are compared as seen in Fig. 4. The similar lifetimes are indicative that the selection rules for the direct forbidden transition are not affected by quantum confinement for the nanocrystals produced.

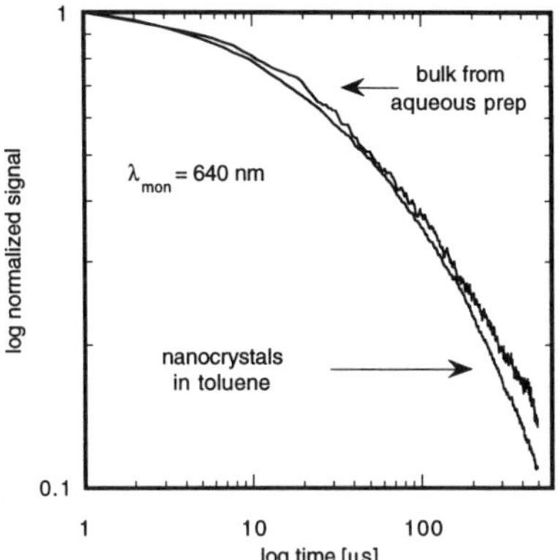

FIG. 4 Comparison of the exciton photoluminescence lifetimes in bulk and nanocrystalline Cu_2O.

CONCLUSIONS

The first study of the low-temperature optical properties of Cu_2O nanocrystals has been made and quantum confinement effects were observed as blue shifts in the energy of the lowest n = 1 excitonic transition. Size restriction, however, did not seem to cause any substantial relaxation in the selection rules as the n = 1 transition remains very weak in excitation and the photoluminescence lifetime does not change. In contrast with AgBr, a typical indirect gap semiconductor, quantum size effects do not result in an elimination of other radiative pathways which would lead to an increase in the luminescence yield. The low energy peaks evident in the excitation spectra of the nanocrystalline samples are probably related to surface states or a result of the method of synthesis. The size distributions of the samples are quite broad, and in order to obtain a better understanding of the photophysics of Cu_2O nanocrystals, methods of synthesis that result in narrower size distributions must be developed.

ACKNOWLEDGMENTS

We gratefully acknowledge NSF Science and Technology Center grant # CHE-9120001 which supported this work.

REFERENCES

[1] J. P. Dahl and A. C. Swittendick, J. Phys. Chem. **28**, 931 (1966).
[2] R. J. Elliott, Phys. Rev. **124**, 340 (1961).
[3] H. Stolz, *Time-Resolved Light Scattering from Excitons* (Springer-Verlag, Berlin, 1994), Ch. 7.
[4] V. T. Agekyan, Phys. Stat. Sol. (a), **43**, 11 (1977).
[5] V. Langer, H. Stolz and W. von der Osten, Phys. Rev. B, **51**, 2103 (1995).
[6] E. F. Gross and A. A. Kaplyanskii, Sov. Phys. Solid State, **2**, 1518 (1961).
[7] A. Compann and H. Z. Cummins, Phys. Rev. B, **6**, 4753 (1972).
[8] M. A. Washington, A. Z. Genack, H. Z. Cummins, R. H. Bruce, A. Compaan and R. A. Forman, Phys. Rev. B, **15**, 2145 (1977).
[9] J. L. Deiss, A. Daunois and S. Nikitine, Sol. State Commun. **8**, 521 (1970).
[10] E. F. Gross, Sov. Phys. Usp. **5**, 195 (1962).
[11] A. Kellersohn, E. Knözinger, W. Langel and M. Giersig, Adv. Mater. **7**, 652 (1995), and references therein.
[12] M. Hayashi and K. Katsuki, J. Phys. Soc. Japan, **7**, 599 (1952).

[13] I. Grozdanov, Mat. Lett. **19**, 281 (1994).
[14] R. S. Knox, *Theory of Excitons*, (Academic Press, New York, 1963), pp. 37-59.
[15] W. Kohn in *Solid State Physics Vol 5*, edited by F. Seitz and D. Turnbull (Academic Press, New York, 1957), pp. 257-320.
[16] A. Goltzené and C. Schwab, in *Landolt-Börnstein, New Series Vol 17e*, edited by O. Madelung (Springer, Berlin, 1982), pp. 144-151.
[17] P. W. Baumeister, Phys. Rev. **121**, 359 (1961).
[18] R. A. Forman, W. S. Brower and H. S. Parker, Phys. Lett. **36A,** 395, (1971).
[19] P. McFadyen and E. Matijevic, J. Coll. Interface Sci. **44**, 95 (1973).
[20] W. Chen, J. M. Rehm, C. Meyers, M.I. Freedhoff, A. Marchetti, G. McLendon, Mol. Cryst. Liq. Cryst. **252**, 79, (1994).
[21] B. Zou, X. Zheng, G. Tang, G Zhang, and W. Chen, Phys. Lett. A, **182**, 130 (1993).
[22] K. P. Johansson, A. P. Marchetti and G. L. McLendon, J. Phys. Chem. **96**, 2873 (1992).
[23] M. I. Freedhoff, A. P. Marchetti and G. L. McLendon in *Microcrystalline and Nanocrystalline Semiconductors*, edited by R. W. Collins, C. C. Tsai, M. Hirose, F. Koch, and L. Brus (Mater. Res. Soc. Proc. **358**, Pittsburgh, PA, 1995) pp. 301-306 and references therein.
[24] Y. Petroff, J. de Physique, **35**, C3-277 (1974).
[25] J. L. Marin and S. A. Cruz, Am. J. Phys. **59**, 931 (1991).
[26] J. L. Marin and S. A. Cruz, J. Phys. B: Mol. Opt. Phys. **24**, 2899 (1991).

PREPARATION, CHARACTERIZATION AND OPTICAL PROPERTIES OF ZINC OXIDE NANOPARTICLES

SHOUTIAN LI, STUART J. SILVERS and M. SAMY EL-SHALL*
Department of Chemistry, Virginia Commonwealth University
Richmond, VA 23284-2006

ABSTRACT

ZnO nanoparticles are produced by the laser vaporization-controlled condensation technique. These particles are connected in a web-like agglomeration. Their properties are compared to those of ZnO nanoparticles produced by solution sol-gel and reverse micelle techniques. All particles have the bulk crystal structure and show quantum size effects in absorption and emission. They show emissions that consist of a blue bandgap feature with a sub-nanosecond lifetime and a green feature with multiexponential lifetime decays. Emission from the stearate coated particles produced by the reversed micelle method is particularly intense.

INTRODUCTION

Nanoparticles have been the focus of many recent studies because their finite size and large surface to volume ratio can result in novel properties not exhibited by the bulk material [1-3]. Such properties are both intrinsically interesting and have important applications in areas such as microelectronics, catalysis, and optical communications.

Zinc oxide, a semiconductor, has been prepared as nanoparticles in transparent colloidal solutions by Bahnemann et al. [4], Henglein and co-workers [5,6], Spanhel and Anderson [7] and others. These particles exhibit quantum size effects [8,9]: their bandgap absorption and emission are blue-shifted with respect to bulk ZnO and their visible emission, in the 500 nm region, shows wavelength and lifetime dependence on size. Surface alterations cause noticeable photophysical changes.

Here ZnO nanoparticles are prepared by the laser vaporization-controlled condensation technique (LVCC) and by sol-gel and reversed micelle methods. The properties of these particles, produced by different methods, are compared.

EXPERIMENTAL

Preparation of ZnO Nanoparticles

(1) By the Laser Vaporization-Controlled Condensation Technique. A modified, upward thermal diffusion cloud chamber, shown in Figure 1, is used to prepare ZnO nanoparticles [10-12]. The chamber is filled to 800 Torr total pressure with a known mixture of O_2 and He gases. The top plate is cooled to about 150 K with circulating liquid nitrogen. A Zn metal rod (Aldrich) is set on the bottom plate which is maintained at a temperature higher than the upper one (300-400 K). The 532 nm second harmonic of a pulsed Nd:YAG laser is used to evaporate the target. Zn atoms/clusters react with O_2 to form ZnO clusters which are carried by convection to the nucleation zone. The particles grow in the nucleation zone and are deposited on the top plate.

(2) By the Sol-Gel Method. ZnO nanoparticles in methanol are prepared by the method of Henglein and co-workers [5,6]. A 10 ml sample of 0.1 M $Zn(OAc)_2$ (Aldrich) in methanol is diluted with 190 ml of methanol. To this is added 4 ml of 10 M NaOH under strong stirring.

(3) By a Reversed Micelle Technique. ZnO nanoparticles, coated with stearate, are prepared in a water-in-oil microemulsion [13,14]. In this method, 200 ml of a 0.001 M zinc stearate - toluene solution are mixed with 0.4 ml H_2O and 0.4 ml 10 M NaOH solution.

Mat. Res. Soc. Symp. Proc. Vol. 452

The mixtures are stirred for two hours and then refluxed for one half hour. After distillation to reduce the amount of solvent, a clear colloidal sol results. A powdered sample is obtained by removing all the solvents.

Characterization Methods

X-ray diffraction patterns from powder samples are obtained with a Ragaku spectrometer. UV-visible absorption spectra are recorded with a Shimadzu UV-265 spectrometer. FTIR spectra are measured with a Perkin-Elmer FT-IR 1600 spectrometer. Photoluminescence studies use 340 and 266 nm excitation. The 340 nm excitation is the frequency doubled output of a Lambda-Physik LPD-3002 dye laser, pumped by the 532 nm second harmonic of a Nd:YAG laser. The 266 nm excitation is the fourth harmonic of the Nd:YAG laser. The sample luminescence is dispersed by a SPEX 1 m spectrometer fitted with an EMI 9558 (S20 photocathode) photomultiplier. The photocathode sensitivity cuts off before 900 nm and spectra are not corrected for the wavelength dependence of the detection system. A boxcar averager is used for dispersed spectra which are recorded by computer. Time-resolved decays are averaged with a LeCroy 9450 oscilloscope and fitted by a computer.

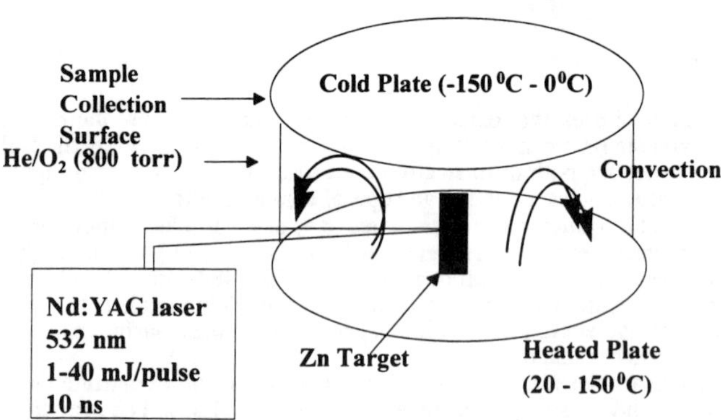

Figure 1. Experimental set-up for the preparation of ZnO nanoparticles by the Laser Vaporization-Controlled Condensation (LVCC) Technique.

RESULTS AND DISCUSSION

The X-ray diffraction patterns of ZnO powder samples obtained by the three different synthetic techniques are identical in peak positions and relative intensities as shown in Figure 2. They are also the same as the pattern for bulk, crystalline ZnO. Thus the nanoparticles have crystalline structures identical to the bulk structure, which is of the hexagonal, wurzite type. The peak intensities, which correspond to the degree of crystallinity, are strongest for the particles made in methanol and weakest for those coated with stearate. Particle diameters, estimated from peak widths, are about 5 nm for those made from a one-day-old methanol solution and about 6.5 nm and 6.8 nm for those made by the laser and reversed micelle techniques, respectively. These sizes are consistent with those obtained from the absorption spectra.

FTIR results indicate that the phonon spectra of the ZnO nanoparticles are the same as bulk ZnO. Stearic acid vibrational features were observed in the coated sample prepared by reversed micelle method.

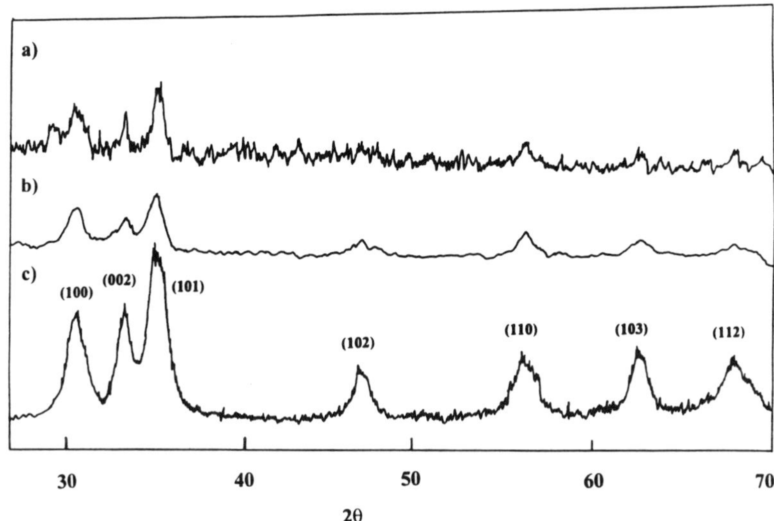

Figure 2. X-ray diffraction patterns of ZnO nanoparticles prepared by (a) Reversed Micelle, (b) Laser Vaporization and (c) Sol-Gel methods.

Bandgap Absorption and Particle Size

Bandgap absorption occurs when semiconductor electrons are excited from the valence band to the conduction band. The onset of this absorption occurs at the bandgap energy, E_g, which for bulk ZnO is 3.3 eV or 380 nm. For ZnO nanoparticles the electron levels become discrete and E_g increases: this increase correlates with decreasing particle size. The bandgap energy for a sample is calculated from the absorption coefficients in the onset region [15,16] and then the shift in bandgap from bulk ZnO is used to estimate particle size [17].

The bandgap and size of ZnO particles produced by the LVCC technique depend on the O_2 partial pressure in the chamber. The bandgap wavelength changes from 370 to 380 nm as the O_2 partial pressure increases from 20 to 300 Torr. From 300 to 800 Torr O_2 partial pressure, the bandgap wavelength does not change. Figure 3a shows absorption spectra (in methanol) for nanoparticles prepared at two different O_2 pressures. These results indicate that the smallest ZnO particles produced by the LVCC method are made at low O_2 pressures. In this method the Zn atoms produced by the laser react with oxygen and are carried by convection through the nucleation and growth zones to the cold upper plate where the ZnO nanoparticles are deposited as a web-like agglomeration.

One minute after zinc acetate is mixed with NaOH in the sol-gel preparation method, a bandgap absorption is seen with an onset at 328 nm as shown in Figure 3b. This shift of 52 nm from the bulk onset corresponds to particles of about 4 nm diameter. With time the particles grow until at 15 minutes after mixing, the growth is nearly complete. The bandgap wavelength shifts to 350 nm and the corresponding particle diameter is about 5.5 nm.

In the reversed micelle method, particle formation occurs in the nanoscale environment generated by surfactant zinc stearate molecules, water and toluene. The particle size is dependent on the $[OH^-]/[Zn^{2+}]$ ratio. The bandgap wavelength shifts from 373 nm to 365 nm when the ratio increases from 10 to 40; the particle size decreases with increasing OH^- concentration. Figure 3c shows absorption spectra at three different OH^- concentrations. By comparing bandgap shifts, it appears that the stearate coated ZnO particles produced by this method are comparable in size to the ZnO particles made by the LVCC method, but are larger than those made in methanol by the sol-gel technique.

When the ZnO nanoparticles, produced by the sol-gel technique, are irradiated for two hours with 266 nm laser radiation, a blue shift in the bandgap absorption occurs. This blue shift persists up to one day after illumination. It has been previously reported for ZnO and other colloids [5,6] and may be due to laser-induced dissolution of ZnO particles which results in smaller particles that absorb further to the blue.

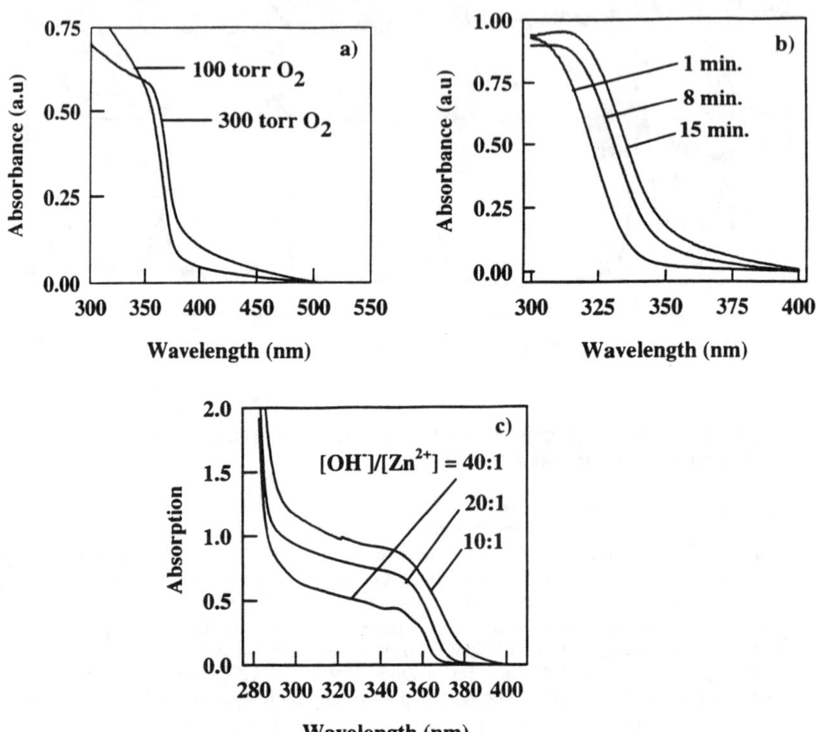

Figure 3. Absorption spectra of ZnO nanoparticles : (a) (in methanol) prepared by the LVCC method at two different O_2 pressures, (b) (in methanol) prepared by the sol-gel method and (c) (in toluene) prepared by the reversed micelle method.

Photoluminescence Spectra

Bulk ZnO shows two broad luminescence peaks: one, attributed to bandgap emission, is at 380 nm and the other is at 500 nm. The nanoparticle samples all show these two emissions when excited with 266 nm light, but the absolute and relative intensities and the positions of these two bands vary with the method of particle preparation. For the laser prepared particles, the blue maximum at 384 nm is less intense than the green one at 515 nm. For the stearate coated particles, the blue peak is again at 384 nm but is now relatively stronger than the green one which is red-shifted to 540 nm. The absolute intensity of the emission from stearate coated particles is about ten times greater than that from the laser prepared ones. The freshly prepared colloidal ZnO sol is turbid and shows only green emission; the blue emission apparently is reabsorbed by the sample. After the sol ages for four hours, weak blue emission appears; this emission then grows more intense and shifts to the blue. After aging for twenty four hours, the blue peak, now at 335 nm, is more intense than the green one.

Different explanations have been proposed for the green emission. Henglein and co-workers [5,6] have attributed it to surface anion vacancies, while Bahnemann and co-workers [4] have explained it by the tunneling of surface-bound electrons to pre-existing, trapped holes. The trapped state explanation better fits our alkaline colloids and stearate coated particles which are unlikely to have surface anion vacancies. The intense green emission from the stearate coated particles is similar to the strong red emission that arises when Si nanoparticles are passivated [18]. In both cases non-radiative surface relaxation may be reduced by surface modification.

Photoluminescence Lifetimes

The blue bandgap emission has a picosecond lifetime [4,5] which cannot be measured by our nanosecond lasers. The green emission has been shown to be multiexponential [19], with at least two components shorter than our 10 ns laser pulses. Our wavelength selected time decays can be fit to a biexponential form where the fast component includes all components with lifetimes less than 10 ns and the slow component lifetime depends on emission wavelength. For our particles, the fast component amplitude is much greater ($\sim 10^3$ times) than the slow component amplitude; the integrated intensities for the two components, however, are more comparable.

The slow component lifetime at an emission wavelength of 520 nm is 0.85 μs for the laser prepared ZnO particles and 1.42 μs for the stearate coated particles. The longer lifetime for the coated particles may reflect the ability of the coating to reduce non-radiative relaxation. The slow component lifetime increases with emission wavelength. For the coated particles, this lifetime increases from 1.35 μs at 480 nm to 1.50 μs at 600 nm. Figure 4a shows a series of emission spectra from laser prepared particles; the spectra have different boxcar delay times. A similar series is shown in Figure 4b for stearate coated particles. The increase in slow component lifetime with emission wavelength is shown by the red shift in the green maximum with increasing boxcar delay. This red shift, similar to that seen for CdS nanoparticles [20], can be explained by the trapped electron-trapped hole model. In this model, larger particles with longer electron-hole distances, emit at longer wavelengths with slower rates. This also explains the 10 nm red shift that is observed when the excitation wavelength for laser prepared particles is changed to 340 nm from 266 nm. The longer wavelength excitation preferentially excites larger particles which then preferentially emit to the red.

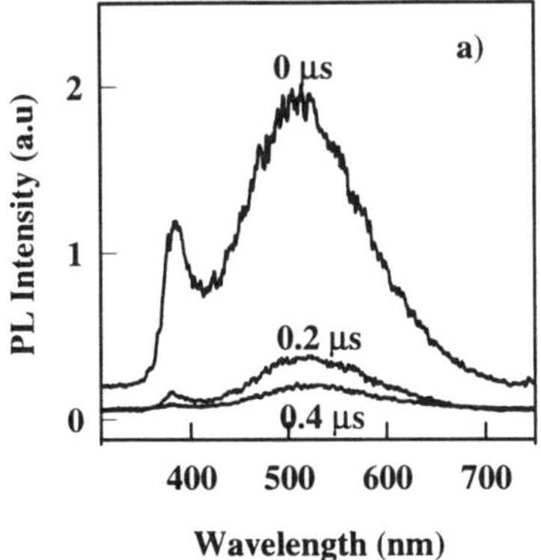

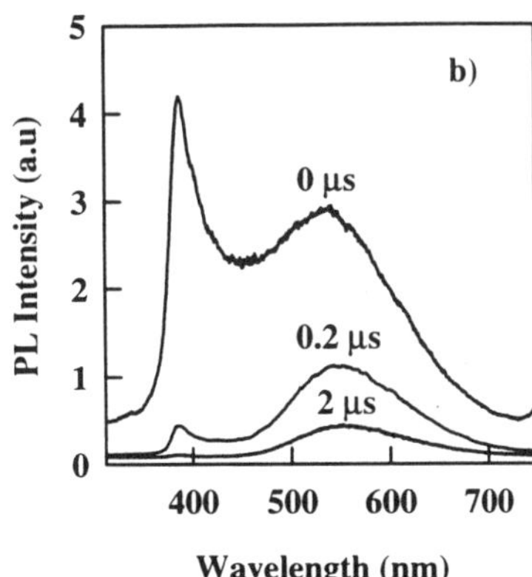

Figure 4. Dispersed emission of the ZnO nanoparticles excited by 266 nm. Nanoparticles prepared by (a) Laser Vaporization and (b) Reversed Micelle.

CONCLUSIONS

In this work ZnO nanoparticles produced by the laser vaporization-controlled condensation technique are compared to those produced in solution by sol-gel and reversed micelle techniques. The laser produced particles form a web-like agglomeration and, like the solution produced particles, show quantum size effects and interesting photoluminescence properties. The particles produced by the sol-gel method are smallest, but all show bandgap shifts and size-dependent dispersed and time-resolved emission spectra. Of particular interest is the intense luminescence from the stearate coated particles produced by the reversed micelle method; the surface coating may inhibit non-radiative relaxation processes.

ACKNOWLEDGMENT

The authors gratefully acknowledge financial support from the NASA Microgravity Materials Science Program (Grant NAG8-1276).

REFERENCES

*. To whom inquiries should be addressed.
1. P. V. Kamat, Chem. Rev. **93**, 267 (1993).
2. A. Henglein, Chem. Rev. **89**, 1861 (1989).
3. M. R. Hoffmann, S. T. Martin, W. Choi, and D. W. Bahnemann, Chem. Rev. **95**, 69 (1995).
4. D. W. Bahnemann, C. Kormann, and M. R. Hoffmann, J. Phys. Chem. **91**, 3789 (1987).
5. U. Koch, A. Fojtik, H. Weller, and A. Henglein, Chem. Phys. Lett., **122**, 507 (1985).
6. M. Haase, H. Weller, and A. Henglein, J. Phys. Chem. **92**, 482 (1988).
7. L. Spanel and M.J. Anderson, J. Am. Chem. Soc. **113**, 2826 (1991).
8. L. E. Brus, J. Chem. Phys. **80**, 4403 (1984).
9. L. Brus, J. Phys. Chem. **90**, 2555 (1986).
10. M. S. El-Shall, W. Slack, W. Vann, D. Kane and D. Hanley, J. Phys. Chem. **98**, 3067 (1994).
11. M. S. El-Shall, S. Li, T. Turkki, D. Graiver, U.C. Pernisz, and M. A. Baraton, J. Phys. Chem. **99**, 17805 (1995).
12. M.S. El-Shall, S. Li, D. Graiver, and U.C. Pernisz in "Nanotechnology - Molecularly Designed Materials", edited by G. M. Chow and K. E. Gonsalves (ACS Symposium Series **622**, American Chemical Society, Washington, DC 1996), p.79.
13. E. Joselevich and I. Willner, J. Phys. Chem. **98**, 7628 (1994).
14. S. Li, B. S. Zou, Y. Yang, L. Xiao, T. Li, J. Zhao, S. Huang, and J. Yu, J. Chem. J. Chin. Univ. **13**, 1597 (1992).
15. A. J. Hoffman, G. Mills, H. Yee, and M.R. Hoffmann, J. Phys. Chem. **96**, 5546 (1992).
16. A. Hagfeldt and M. Gratzel, Chem. Rev. **95**, 49 (1995).
17. Y. Wang and N. Herron, Phys. Rev. B **42**, 7253 (1990).
18. L. Brus, P. F. Szajowski, W. L.Wilson, T. D. Harris, S. Schuppler, and S. Citrin, J. Am. Chem. Soc. **117**, 2915 (1995).
19. P. V. Kamat and B. Patrick, J. Phys. Chem. **96**, 6329 (1992).
20. N. Chestnoy, T. D. Harris, R. Hull, and L. E. Brus, J. Phys. Chem. **90**, 3393 (1986).

THE STRUCTURE AND TEXTURE OF Y_2O_3:Tb NANOCRYSTALS

J. Taylor, M. Libera, E. Goldburt*, and R. Bhargava*
Dept. MS&E, Stevens Institute of Technology, Hoboken NJ 07030
*Nanocrystals Technology, Briarcliff Manor, NY 10510

ABSTRACT

This paper presents the results of microstructural studies of terbium-doped yttria quantum nanocrystals by imaging and diffraction in a transmission electron microscope. The nanocrystals are typically found in clusters of varying size. Electron energy-loss spectroscopy confirms that the particles consist of yttrium and oxygen. Analysis of selected-area electron diffraction patterns shows that the nanocrystals largely have the cubic structure found in bulk yttria. These patterns furthermore suggest that within each cluster the nanocrystals align themselves with a preferred orientation relative to the incident electron beam as well as to each other.

INTRODUCTION

Nanocrystalline structures are of interest because of their attractive luminescent properties. They are being widely developed for potential use in display technology and other optoelectric device applications. This class of materials includes porous silicon [1] and semiconductor quantum nanocrystals [2,3]. Doped quantum nanocrystals (DNCs) are particularly attractive, because the dominant electron-hole recombination process can be effectively controlled by the dopant [4]. The organometallic synthesis of Mn-doped nanocrystalline ZnS with high luminescent efficiency and ultrafast decay time has been achieved by Bhargava et al. [5,6].

The synthesis of doped quantum nanocrystals has recently been extended to the yttria (Y_2O_3) system doped with Tb or Eu emitting green or red light, respectively [7]. A correlation between DNC particle size and luminescent efficiency was established and described earlier [8]. The particle sizes were determined using high-resolution transmission electron microscopy (TEM) in order to make real-space estimates of particle diameters. When the average diameter decreases from 9.1 nm to 4.4 nm, the photoluminescence efficiency increases by a factor of four. This result was attributed to quantum-confinement effects. The present work concentrates on quantifying the crystal structure and microstructural properties of the DNC Y_2O_3 studied previously [8]. Deviations from standard bulk structure are often found in both structural and optoelectronic materials systems when the crystal size is decreased to nanometer length scales due to the increased significance of broken symmetries at crystal surfaces. The bulk form of yttria is cubic [9]. This work uses electron diffraction to confirm that DNC yttria is also cubic and explore the behavior of these DNC nanocrystals in the form of aggregated clusters.

Mat. Res. Soc. Symp. Proc. Vol. 452 © 1997 Materials Research Society

EXPERIMENTAL PROCEDURE

Three different samples of doped nanocrystalline (DNC) Y_2O_3:Tb phosphors were prepared by room-temperature organometallic synthesis [8]. Specimens for TEM imaging and diffraction were prepared from aqueous solutions of colloidal phosphors. Each solution was sonicated for 10-30 minutes. A drop of solution was placed on a holey carbon TEM grid and allowed to dry.

Both high-resolution and bright-field image data were collected using a 300 keV Philips CM30 SuperTwin TEM with a LaB_6 source. Its point resolution is 0.19 nm. Selected-area diffraction (SAD) patterns were collected from specimen areas ~200-300 nm in diameter. Approximately 15 patterns were collected from each DNC specimen. These were digitized by scanning negatives or enlarged prints and analyzed using Gatan's Digital Micrograph software running on a Power Macintosh. The patterns and the analysis procedure were calibrated using reflections from nanocrystalline gold evaporated onto the surfaces of the DNC TEM specimens. Experimental electron-diffraction data were compared to JCPDS cardfile data [9] and to electron-diffraction patterns calculated using the EMS software system [10]. Focused-probe electron energy-loss spectroscopy (EELS) was done using a 200 keV Philips CM20 TEM/STEM. This instrument has a Schottky field-emission electron source and a Gatan 666 parallel EELS spectrometer.

RESULTS AND DISCUSSION

The individual DNC particles are almost always found as clusters ranging in size from ~100 nm to ~50 μm in diameter. Figure 1 shows a bright-field image from a portion of one such cluster. The light and dark contrast corresponds to individual crystals satisfying different Bragg conditions. Near the cluster edges the nanocrystals are arranged almost in monolayer fashion, but the

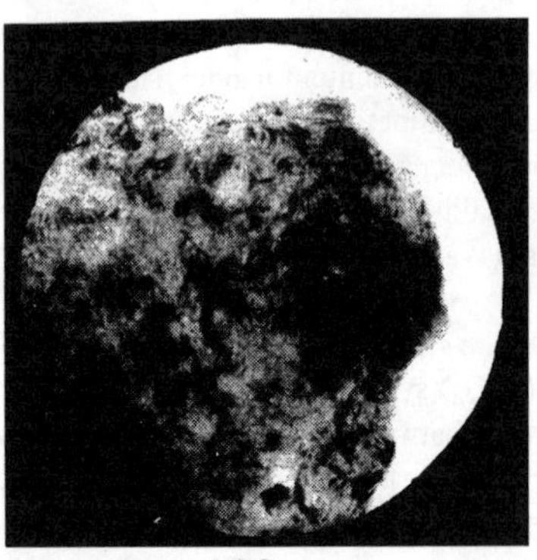

Figure 1 - Bright-field TEM image of part of a Y_2O_3:Tb nanocrystal cluster.

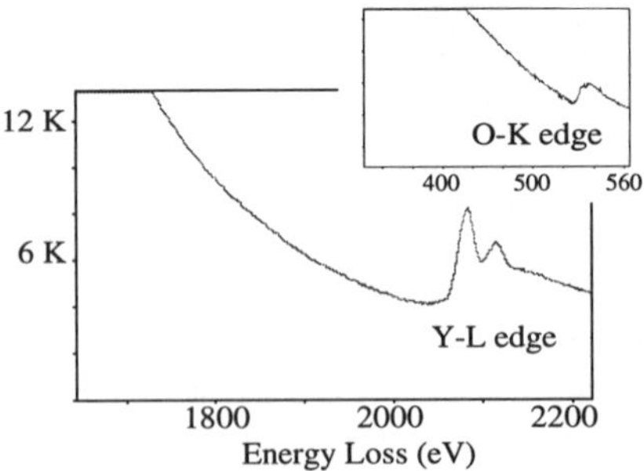

Figure 2 - Electron energy-loss spectra showing oxygen K-edge and yttrium L-edge peaks at 532eV and 2080eV, respectively.

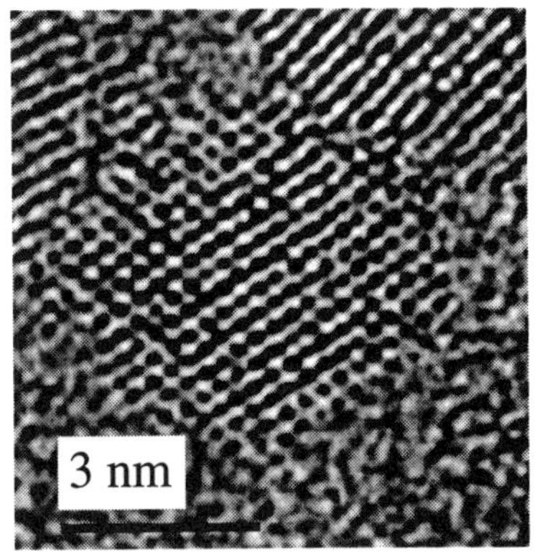

Figure 3 - High-resolution image of a typical Y_2O_3:Tb nanocrystal showing faceted surfaces.

thickness rapidly increases away from the edge. Interpretation of image data from areas with more than one or two DNC layers is difficult. Increased sonication tends to break the clusters into smaller sizes and enhances the effectiveness of TEM imaging and diffraction. Electron energy-loss spectroscopy confirmed that the principal elemental species in these nanocrystals are in fact yttrium and oxygen. Figure 2 presents is a typical result. There is an oxygen K-edge at 532 eV on a strong carbon K-edge (284 eV) background due to the support film. The yttrium L-edge begins at 2080 eV. No effort was made to quantify the composition. Figure 3 shows a high-resolution image of a typical single nanocrystal lying on the carbon support film. This crystal shows facets typical of these and other nanocrystalline materials [11].

The cubic form of yttria presents a rich array of Bragg reflections in a diffraction experiment. A simulation of the electron diffraction behavior under kinematical conditions assuming a specimen consisting of randomly-oriented yttria grains is presented in Figure 4. The diffracting angles for 300 keV electrons are indicated by the position of the vertical lines, and the relative intensities of each reflection are represented by the height of each vertical line. Only reflections with an intensity greater than a few percent of the strongest reflection - <222> - are included. This pattern shows that a number of intense reflections (<222>; <404>; <226>) as well as groups of several closely-spaced reflections should be present in specimens of randomly-oriented cubic yttria grains.

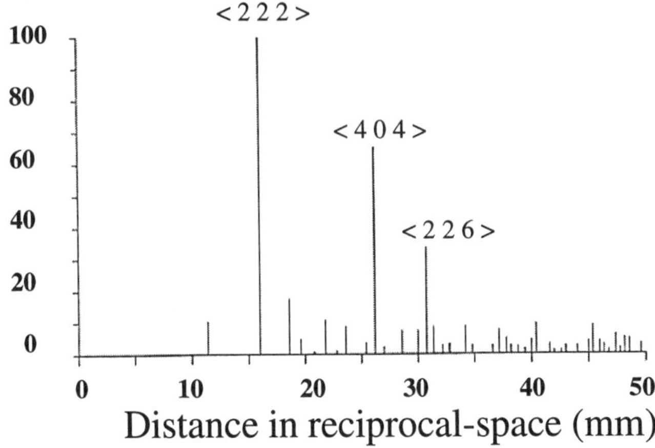

Figure 4 - Calculated electron-diffraction pattern for cubic yttria assuming conditions of kinematical scattering from a large population of randomly-oriented grains.

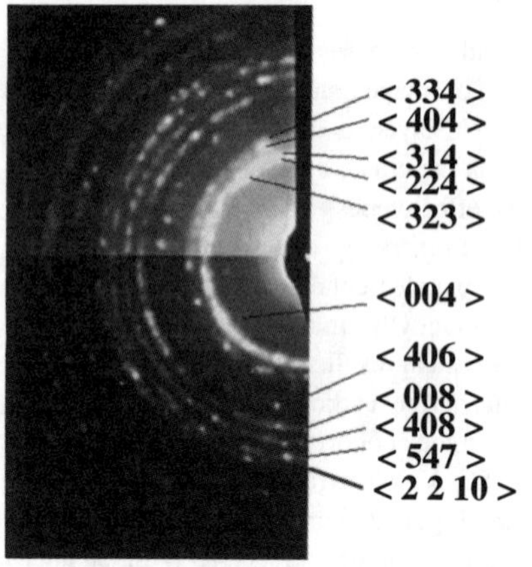

Figure 5 - Type A selected-area electron diffraction pattern from the cluster in fig. 1. Different exposure times were used to record each of the two quadrants.

Figure 5 is a composite which combines two quadrants from identical patterns recorded for different photographic exposure times in the microscope. This pattern is typical of the majority of patterns collected from the three different DNC specimens studied and is labeled the type A pattern. There is a large number of reflections extending to large angles in reciprocal space. Table I summarizes the positions of the most prominent experimental diffraction peaks and compares these to the standard data from the JCPDS cardfile. A large fraction of the reflections are absent in the experimental data. This can in part be attributed to experimental error. Reflections at high scattering angles are typically faint over the background due to thermal diffuse scattering. Distinctions between closely-spaced reflections are often difficult to make, particularly at high angles where there are many high-index planes. Many of the high-index planes also have interplanar spacings differing by only 0.02 A or less. Nevertheless, the experimental data are consistent with the cubic structure and are sufficient to distinguish it, for example, from hexagonal yttria.

Perhaps more striking then the presence of lines agreeing with the cubic structure is that there is a significant number of cubic reflections missing. Most prominent among these is the <222> peak. This would have the highest diffracted intensity in a sample of randomly-oriented crystals (fig. 3). The <224> reflection is not the strongest in the experimental pattern (fig. 5), and the <226> reflection is missing altogether. These would be the second and third most intense lines, respectively, in a sample of randomly-oriented crystals. While the specific reflections present or absent vary between the various SAD patterns analyzed from the three specimens studied here, the Type A

Table I - JCPDS cardfile data for cubic Y_2O_3 (partial list) and experimental d spacings (fig. 5)

plane < hkl >	interplanar spacing (A)	experimental spacing (A)
112	4.70	---
222	3.06	---
004	2.65	2.64
114	2.50	---
204	2.37	---
323	2.26	2.21
224	2.17	2.12
314	2.08	2.03
215	1.94	---
404	1.88	1.87
334	1.82	1.80
325	1.72	---
116	1.72	---
415	1.64	---
226	1.60	---
316	1.56	---
444	1.53	---
406	1.47	---

patterns all are qualitatively similar. In particular, the <222> reflection is often missing, and, when present, it is not the most intense peak. Instead, the <314> reflection is consistently observed, and experimentally it almost always has the highest intensity.

The absence of peaks and the deviations in their intensity from standard powder-pattern crystallographic data indicates that the nanocrystals contained in each cluster are textured and share some common crystallographic orientations. Texture is often observed in macroscopic processes associated with metal deformation and annealing (eg. 12), and thin-film growth (eg. 13), because deformation and crystal-growth processes tend to occur along specific crystallographic habits, typically close-packed planes and directions. Texture in nanocrystal CdSe has been reported by Bawendi et al. [14] who showed that it can occur to an impressive degree in self-organized clusters of CdSe nanocrystals where the relative crystallographic orientations are maintained over length scales orders of magnitude larger than the nanocrystal particle size. The origin of such behavior clearly rests in the minimization of some form of interfacial energy associated with nanocrystal/nanocrystal and nanocrystal/ambient interfaces.

The texture manifests itself in figure 5 as the absence of Debye-Scherrer rings associated with particular Bragg reflections. This means that the crystallographic planes associated with the missing reflections on average are not parallel to the incident electron-beam direction. In contrast, the fact that <314> Debye-Scherrer rings are consistently seen in the Type A patterns and are always intense indicates that this family of planes is aligned roughly parallel to the incident beam direction. The present research has also occasionally observed a second form of texture where diffracted intensity is present as arcs rather than as full Debye-Scherrer rings. This indicates that there are families of planes parallel to the incident beam whose orientations are also correlated over a finite range of directions perpendicular to the incident beam. This observation suggests some sort of self-organizing effect, though to a lesser extent, similar to that reported by Bawendi et al. (15).

A second and qualitatively different pattern from that presented in figure 5 was observed in approximately 20% of the patterns collected in this research. A typical example is presented in figure 6. This pattern is labeled the Type B pattern. It has far fewer reflections than the Type A patterns (eg. fig. 5). While the majority of the reflections agree with cubic lines, at least one strong reflection consistently does not. The crystallographic structure associated with the Type B pattern has not yet been established. The fact that the Debye-Scherrer rings are almost fully continuous suggests, however, that the

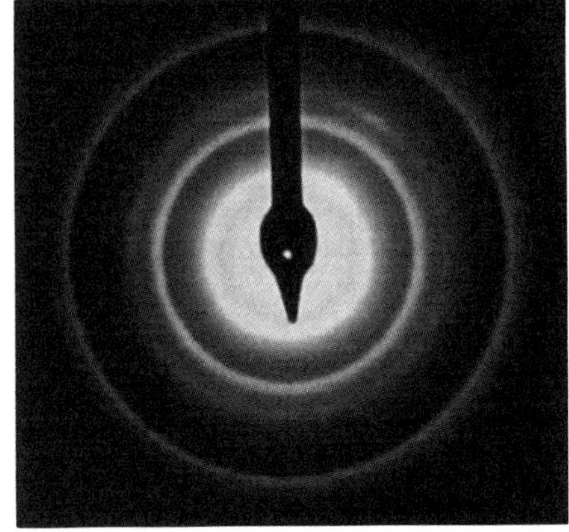

Figure 6 - Selected-area electron-diffraction pattern from the Type B structure.

particle size is smaller than that associated with the Type A patterns. Further work is ongoing to establish the structure associated with the Type B patterns.

CONCLUSION

This research has used electron diffraction, imaging, and spectroscopy to study the structure of Tb-doped yttria quantum nanocrystals. These nanocrystals tend to form clusters of varying size. Electron diffraction indicates that the nanocrystals within each cluster are textured and share preferred orientations presumably due to self-organizing surface-energy effects. Electron diffraction confirms that the nanocrystals predominantly have the cubic structure associated with bulk yttria though a minority fraction of the clusters display diffraction behavior which can not be indexed purely as cubic.

ACKNOWLEDGMENT

The authors are grateful for partial support from DARPA (contract number 61333-96-C-0035) and to the Stevens Institute of Technology for an Institute Research Initiation award.

REFERENCES

1. L.T. Canham, Appl. Phys. Lett. **57**, 1046 (1990).
2. L. Brus, J. Phys. Chem. **90**, 2555 (1986).
3. A. P. Alivisatos, Science **271**, 933 (1996).
4. R. Bhargava, Proc. Int. Conf. Solid State Devices and Matls., Yokohama, pp. 58-59 (1994).
5. R. Bhargava, D. Gallagher, X. Hong, and A. Nurmikko, Phys. Rev. Lett. **72**, 416 (1994).
6. D. Gallagher, W. Heady, J. Racz, and R. Bhargava, J. Mater. Res. **10**, 870 (1995).
7. E.T. Goldburt and R.N. Bhargava, "Tb doped nanocrystalline phosphors," in press, the Journal of the Society for Information Display.
8. E.T. Goldburt, B. Kulkarni, R. Bhargava, J. Taylor, and M. Libera, MRS Symp. Proc. Vol 424, *Flat Panel Display Materials*, ed. M. Hatilis, F. Funada, J. Kanicki, and C. Summers, in press.
9. Joint Committee on Powder Diffraction Standards (JCPDS) cardfile number 41-1105.
10. P.A. Stadelmann, Ultramicroscopy **21**, 131(1987).
11. A. P. Alivisatos, Materials Research Society Bulletin 20, 23 (1995).
12. C. Barrett and T. Massalski, *Structure of Metals* 2nd ed. (Pergamon, Oxford 1980) p. 541.
13. J.A. Thornton, Ann. Rev. Mater. Sci. **7**, 239 (1977).
14. C.B. Murray, C.R. Kagan, and M.G. Bawendi, Science **270**, 1335(1995).

Part VI

Synthesis, Surfaces, and Structure/Property Relations of Luminescent Porous Semiconductors

POROUS SILICON FROM HYDROGENATED AMORPHOUS SILICON: COMPARISON WITH CRYSTALLINE POROUS SILICON

J.-N. CHAZALVIEL, R.B. WEHRSPOHN, I. SOLOMON and F. OZANAM
Laboratoire de Physique de la Matière Condensée, CNRS-École Polytechnique,
91128 Palaiseau-Cedex, France

ABSTRACT

Device-grade, boron-doped amorphous hydrogenated silicon can be made microporous by anodization in ethanoic HF. The thickness of the porous layer is limited by an instability due to the high resistivity of the material. Amorphous porous silicon exhibits strong room-temperature photoluminescence around 1.5 eV even in samples containing a high density of non-radiative recombination centers. This demonstrates the presence of a spatial confinement effect, as opposed to quantum confinement effect for crystalline porous silicon. The temperature dependence of the luminescence intensity is also accounted for on the same grounds.

INTRODUCTION

The origin of the bright room-temperature photoluminescence of porous silicon has been a very controversial matter these last few years [1]. An intriguing experimental fact was the observation by Bustarret et al. that amorphous hydrogenated silicon (a-Si:H) can be made porous and then exhibits a room-temperature luminescence rather similar with that from crystalline porous silicon [2-4]. Strong luminescence in bulk a-Si:H is known to occur at temperatures lower than ~100 K. However, since it arises from localized band-tail states, the observed change in luminescence upon making the material porous appeared rather puzzling, and especially hard to explain in terms of quantum-confinement effects. Meanwhile, several other groups tried to make a-Si porous and were unable to obtain luminescent material, in apparent contradiction with the results of Bustarret [5-6]. It was just recently that luminescent material was obtained from porous a-Si by two independent groups [7-9]. The obtained results also shed new light on the physics of crystalline porous silicon itself.

In the following, we will report on our results on the preparation and study of amorphous porous silicon (aPS). The paper will be organized in two main sections, respectively devoted to the formation and the luminescent properties of aPS. In each case, the discussion will try to highlight the interest of these data with respect to crystalline porous silicon (cPS).

FORMATION OF AMORPHOUS POROUS SILICON

Experimental

a-Si:H was produced by rf-plasma decomposition of pure silane SiH_4 in a capacitively-coupled reactor at 13.56 MHz. Normal deposition conditions are: total pressure 50 mTorr, gas flow rate 2 l/h, rf power density 0.1 W/cm^2, substrate temperature 250°C. These parameters produce a homogeneous amorphous film, far from the conditions of production of micro-crystalline material [10]. The amorphous deposited material is of device grade, with a gap defined by E_{04}=1.90 eV and a density of states at Fermi level of about 5×10^{15} cm^{-3}eV^{-1} for the undoped material [11]. Boron doping was obtained by injecting a small amount of an admixture of diborane gas and hydrogen (3% of B_2H_6 in H_2) in the plasma. The films were deposited on optically polished stainless steel substrates of dimensions 25×75×1 mm^3. In order to provide a good ohmic contact with the substrate, a very thin film of heavily boron-doped a-Si:H (1000 Å, 2% B) was first deposited on the stainless steel. Resistivity of the obtained material was checked by sandwich measurements using evaporated aluminum pads. As shown in Fig.1, doping results in a decrease of the material resistivity from over 10^6 to 10^4 Ωcm for our highest doped material.

The a-Si:H layers were anodized in an electrochemical cell made of a polytrifluorochloroethylene container, with an Ag/AgCl reference electrode and a platinum counter electrode. The a-Si:H sample on its stainless-steel substrate was pressed with an O-

Mat. Res. Soc. Symp. Proc. Vol. 452

ring seal (12 mm inner diameter) against a circular aperture in the bottom of the cell. The geometry of the cell and counter electrode insured a good current homogeneity over the a-Si:H surface. Anodization was performed at constant current density in ethanoic HF solution, and the electrode potential (relative to the reference electrode) was monitored as a function of time during the anodization. The preparation space was explored by varying four parameters: a-Si:H doping (from 0.1% to 3% B), HF concentration (from 5% to 40%), current density (from 5 to 50 mA/cm^2) and anodization time (up to final peeling-off of the a-Si:H layer).

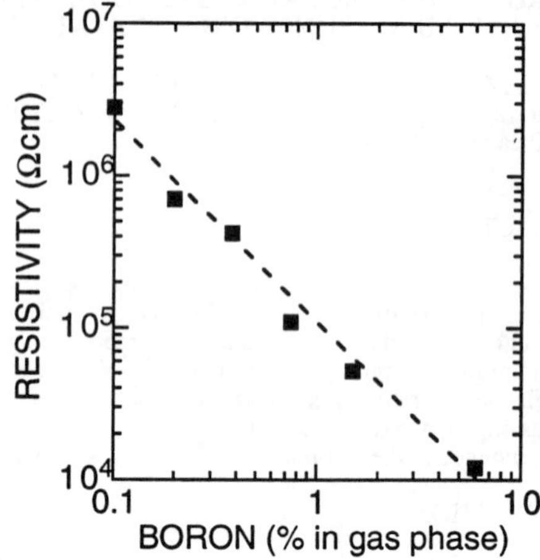

***Fig.1**: Electrical resistivity of our a-Si:H starting material.*

The thickness of the obtained porous layers was determined using VIS/IR reflectance spectrophotometry [12], before and after anodization. An accurate measurement (±0.01 μm) of the porous layer thickness was deduced by difference.

Results: a study of the anodization process

Figure 2 shows a typical variation of potential versus time during constant-current anodization of a-Si:H. The large value of the initial potential is clearly associated with the ohmic drop across the a-Si:H layer: for a resistivity $\rho_{Si}=10^6$ Ωcm, a thickness of only d=1 μm and a typical current density J=10 mA/cm^2, the ohmic drop ρJd reaches 1 V, which is several orders of magnitude larger than that for the case of a typical c-Si wafer. As anodization proceeds, potential evolves in a complex manner. The potential/time curve can be conveniently described as the succession of four regions, labelled A-D in Fig.2.

In region A, the potential decreases linearly with increasing time. This decrease is clearly related to formation of a porous layer of thickness Δ. Since the resistivity of the electrolyte is much smaller than that of a-Si:H, the ohmic drop is now $\rho_{Si}J(d-\Delta)$ instead of $\rho_{Si}Jd$. The slope dV/dt is indeed found to closely match $-\rho_{Si}J(d\Delta/dt)$, with Δ determined as stated above.

In region B, the potential decreases dramatically to a value of ~0.4 V, comparable to that found for c-Si. However, there is no significant variation of Δ during this stage. Next, the potential remains constant for some time (region C), then increases to ~1 V (region D), which is the value observed when stainless steel is anodized in the

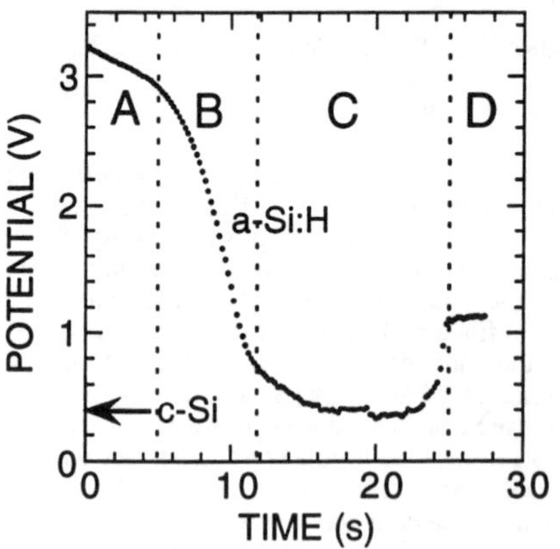

***Fig.2**: Typical potential/time curve during anodization of a-Si:H. Electrolyte: 25% ethanoic HF. Current density: 10 mA/cm^2. a-Si:H layer: 2μm, 0.2% boron.*

same electrolyte. The whole a-Si:H layer is seen to peel-off around the transition between region C and region D.

Keeping in mind that the resistivity of the electrolyte is much smaller than that of a-Si:H, these observations can be accounted for as follows. The fast decrease of the potential in region B corresponds to a progressive short-circuiting of the a-Si:H layer by fast growing channels of mesoscopic size. When the channels reach the back side of the a-Si:H film, the anodization process goes on laterally in the highly-doped contacting layer, with negligible ohmic drop. When the contacting layer is wholly made porous, the potential can rise up to a value so that the stainless-steel substrate is anodized, and the a-Si:H layer peels off.

This scenario has been verified by repeating the same experiments on a-Si:H films deposited on different substrates. When the anodization potential of the substrate was lower than that of silicon (e.g., titanium substrate), peeling off of the layer occurred right at the end of region B. For c-Si substrates, the potential/time curves were the same as in Fig.2, except for the expected disappearance of region D, and the samples could be cleaved and observed by SEM. Typical pictures have been shown elsewhere [8]. The fast-growing pipes are clearly seen. Their typical diameter is from 200 to 400 nm, and their typical density is 1 per μm^2.

Theoretical model: interface instability and channel growth

The instability associated with channel growth can be simply understood as a "Laplacian instability" related to the difference in the resistivities of the electrolyte ρ_{el} and of the a-Si:H layer ρ_{Si} . The onset of channel growth can be modelled by a linear stability analysis à la Mullins and Sekerka [13]. Let us assume that the initial growth leads to the formation of a uniform porous layer, and let us perturb the growing front by a small sinewave deformation $\delta\cos qx$ (see Fig.3). Then, if the interface impedance is neglected, and if the thicknesses d, Δ are taken very large, the distribution of potential $\phi(x,y)$ is just

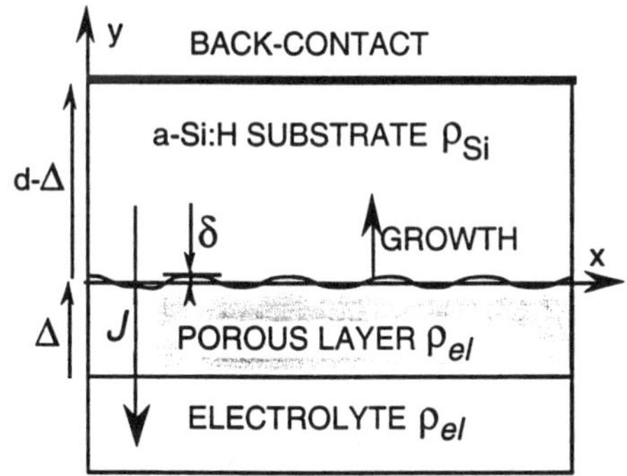

Fig.3: Calculation of the growth instability of aPS.

$$\phi = \rho_{el} J\, y + \gamma \rho_{el} J\, \delta\, e^{qy} \cos qx \quad (y<\delta \cos qx) \tag{1}$$

$$\phi = \rho_{Si} J\, y - \gamma \rho_{Si} J\, \delta\, e^{-qy} \cos qx \quad (y>\delta \cos qx) \tag{2}$$

where J is the mean current density (here J>0) and $\gamma=(\rho_{Si}-\rho_{el})/(\rho_{Si}+\rho_{el})$. [Eqs.(1-2) are obtained upon stating that ϕ satisfies Laplace equation in both media, and matching the potential and the current density at the boundary $y = \delta \cos qx$. Terms of second order in δ have been dropped]. The current density at the interface is then

$$J_y = \left[-\frac{1}{\rho_{el}}\frac{\partial\phi}{\partial y}\right]_{\delta \cos qx - 0} = \left[-\frac{1}{\rho_{Si}}\frac{\partial\phi}{\partial y}\right]_{\delta \cos qx + 0} \approx -J - \gamma J q \delta \cos qx \tag{3}$$

The growth of the porous layer is ruled by Faraday's law, hence the increase of Δ is governed by the constant term J, while the increase of δ is governed by the sinewave term, hence the deformation of the growth front evolves according to

$$\frac{d\delta}{d\Delta} = \frac{\partial\delta/\partial t}{\partial\Delta/\partial t} = \gamma q \delta \tag{4}$$

The amplitude of the front deformation evolves exponentially with the layer thickness, with a characteristic exponent

$$\alpha = \gamma q = \frac{\rho_{Si} - \rho_{el}}{\rho_{Si} + \rho_{el}} q \tag{5}$$

If $\alpha>0$, the front is unstable. In this simple model, such is the case as soon as the resistivity of the semiconductor is larger than that of the electrolyte. The model can be further refined, in order to take into account the finite thicknesses of the a-Si:H layer and of the porous layer, the interface impedance, a possible back-contact series resistance, and also the fact that the resistivity of the porous layer is not exactly that of the electrolyte but actually depends on the porosity. However, all these effects result in quantitative changes but no qualitative changes, and the criterion for instability always appears to be given by the condition $\rho_{Si} > \rho_{el}$.

This simple linear analysis does not allow for making predictions about the morphology of the front (e.g., in regime B of Fig.2). However, it does give the clue to the *origin* of the formation of channels and the limitation of porous layer thickness on a-Si:H. Furthermore, this theoretical result is also of relevance to cPS formation. It has been known that the front of PS layers exhibits a roughness which increases from p^+ to p [14]. There is little doubt that any source of noise (poor stirring of the electrolyte near the bottom of the pores, potential distribution due to the irregular nature of PS on the microscopic scale, or to ionized impurities in the space-charge layer, ...) will be much more efficiently damped for a high-conductance substrate (p^+) than for a p-substrate of typical resistivity 10 Ωcm. Interestingly, the resistivity of usual electrolytes (15% to 50% aqueous or ethanoic HF) falls in the range 10-100 Ωcm, i.e., barely higher than that of typical p-Si substrates. As a complementary experiment, we have anodized low-doped p-Si samples, of resistivities 100 and 1000 Ωcm, so that the instability condition be fulfilled. Above a few micrometers, the porous layer became black, while the potential exhibited a variation of the same kind as regime B in Fig.2 (though of a smaller amplitude, due to the lower resistivity of the substrate, as compared to a-Si:H). The obtained samples were cleaved and investigated by SEM. A typical result has been shown in ref.[8]. An array of macropores is seen below the microporous layer, as expected from fulfilment of the instability condition.

These observations demonstrate that the simple ratio ρ_{Si}/ρ_{el}, which has been largely overlooked in the preparation of PS from c-Si, has profound effects upon PS morphology. For example, temperature has sometimes been found to be a critical parameter in PS preparation [15]. Since increasing temperature makes ρ_{Si} increase and ρ_{el} decrease, the effect of temperature may proceed, at least partially, through the change of ρ_{Si}/ρ_{el}.

How to increase aPS thickness

The instability of aPS growth is a serious limitation to the growth of layers of thickness larger than a few thousand Å: when channels start growing, there is no significant thickening of the microporous layer. In order to increase the maximum thickness Δ_{max} of aPS, it would be of high interest to make the ratio ρ_{Si}/ρ_{el} as small as possible. Several strategies may be considered in this respect.

The resistivity of the a-Si:H substrate may be decreased by increasing its doping. This does result in an increase of Δ_{max}, as seen from the results summarized in Fig.4. This incidentally explains why the first reports of luminescent aPS were obtained from a highly-doped material [2-4]. However, increasing the doping can hardly bring the resistivity below $\sim 10^4$ Ωcm, and this is realized at the cost of introducing a large concentration of non-radiative recombination centers in the material. This fixes a practical limit to using this strategy.

The resistivity of the electrolyte may be increased in various ways. We have gained a factor of ~2 on Δ_{max} by using HF:H_2O:EtOH solutions in the ratio (25:11:64) instead of (25:45:30) (i.e., just increasing ethanol content with fixed HF concentration). Still higher resistivities may be obtained by using a heavy alcohol such as tert-butyl alcohol instead of ethanol. However, a porous layer is then hardly formed, because the electropolishing regime appears at lower current densities. The difficulty in this strategy is to keep the chemical conditions (i.e., dissolution rate of the fluosilicate ion) compatible with PS formation.

Although much hope can be put in the use of non-aqueous electrolytes which can exhibit very high resistivities, finding an appropriate system still represents a big challenge.

An artificial but very efficient way of decreasing the effective resistivity of the a-Si:H substrate is using carrier injection from its backcontact ("Space-Charge Limited Current") [11]. In contrast to illumination which creates carriers of both types, and may result in etching of the already formed PS layer [16], injection from the backcontact is a very "clean" manner of mimicking a low-resistivity substrate. The use of this technique has been reported for anodization of n-type c-Si epilayers [17]. The conditions for injection in a-Si:H can be found in [11]. A high carrier concentration can be achieved at the electrochemical interface if the a-Si:H layer is taken sufficiently thin and if the concentration of bandgap states is low enough. The latter condition is best realized for an intrinsic sample (density of states in the gap $<10^{16}$ $cm^{-3}eV^{-1}$). Using a current density of 10 mA/cm^2, we have been able to anodize a 0.2 µm layer of intrinsic a-Si:H in 25% HF and to make the whole layer porous. This is much larger than the value of Δ_{max} expected for such a highly-resistive material. This technique appears most promising for increasing aPS layer thickness.

Characterization of aPS and comparison with cPS

Due to the small thickness of the aPS layers, a gravimetric determination of the porosity is not possible. However, if the dissolution is assumed to be divalent, an order of magnitude for the porosity can be obtained from the measurements of the layer thickness and of the electric charge consumed. The results are shown in Fig.4. The trends are seen to be quite similar to those for c-Si. This confirms the overall similarity of the electrochemistry of a-Si:H and c-Si, already apparent from voltammetry in dilute HF [8].

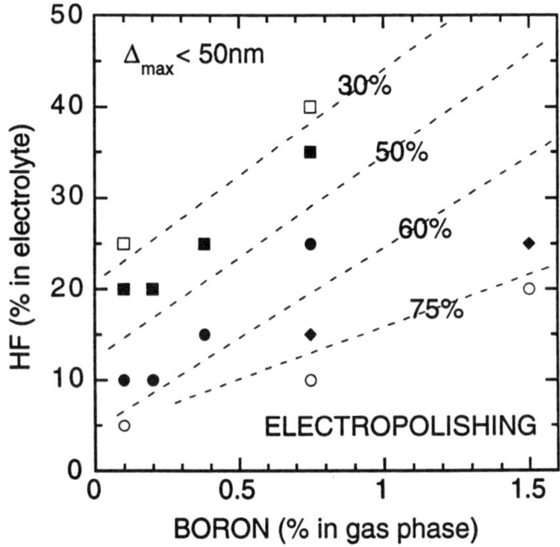

***Fig.4:** Map of aPS preparation, as a function of substrate doping and HF concentration (with 30% ethanol in the electrolyte). Current density 10 mA/cm². The dotted lines are isoporosity curves (porosity indicated on the lines). The shape of the points indicates the maximum obtainable thickness: ■ 0.1 µm, ● 0.2 µm, ◆ 0.4 µm, ❍ and ❐ not measurable.*

Further evidence for the similarity of aPS and cPS can be found from infrared vibrational spectroscopy. Fig.5 shows IR spectra of freshly prepared aPS, obtained in normal-incidence transmission from samples formed from a-Si:H deposited on a c-Si substrate. The formation of porous silicon results in a loss of Si-H bonds (the band at 2000 cm^{-1}), corresponding to a loss of a-Si:H material, and the appearance of surface SiH_x groups (the band at 2100 cm^{-1}). No signal associated with Si-O bonds is found. This confirms that the surface of freshly prepared aPS is hydrogenated, just as that of cPS. The magnitude of the SiH_x signal is of the same order of magnitude as for a cPS layer of same thickness, which indicates that the two materials have comparable specific surface areas.

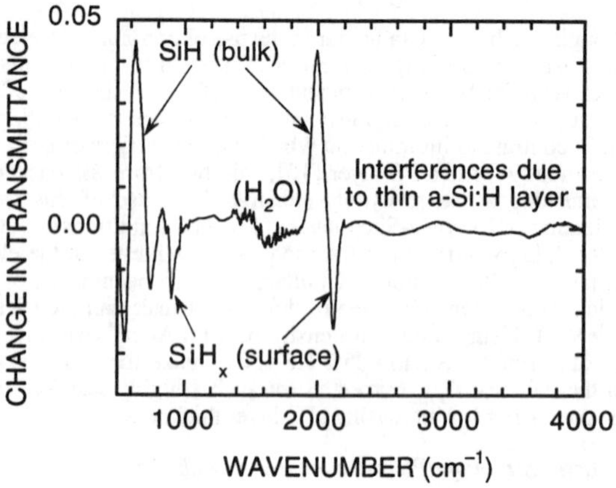

Fig.5: IR transmission spectrum of an aPS layer formed from a-Si:H on c-Si substrate. Reference is same sample before anodization. a-Si:H: 2 μm, 1.5% boron. aPS: 0.4 μm (25% HF, 10 mA/cm²). Notice the loss of Si-H bonds from a-Si:H (positive peaks), the appearance of surface SiH_x groups (negative peaks) and the absence of detectable Si-O bonds.

PHOTOLUMINESCENCE OF aPS: SPATIAL VERSUS QUANTUM CONFINEMENT

Experimental

Photoluminescence (PL) of aPS has been investigated as a function of temperature for layers formed under various conditions. The sample was mounted on a copper plate in a primary vacuum (10^{-3} Torr, liquid-nitrogen trapped). The temperature of the plate was varied from 77 K to 400 K, using liquid-nitrogen flow and/or Joule heating. Optical access to the sample was provided through a SiO_2 window. A krypton-ion laser (337 nm or 476 nm, 2 mm spot diameter) was used for excitation, and PL was analyzed by a monochromator coupled with a cooled CCD detector and an optical multichannel analyzer.

In-situ photoluminescence of aPS in HF was also investigated, in order to study the effect of photoelectrochemical etching on aPS luminescence, and compare the results with those obtained on cPS. This experiment was performed at room temperature, in a PTFCE cell with a CaF_2 window, where the porous layer was formed and left standing during the PL experiment. More details on this cell can be found elsewhere [16].

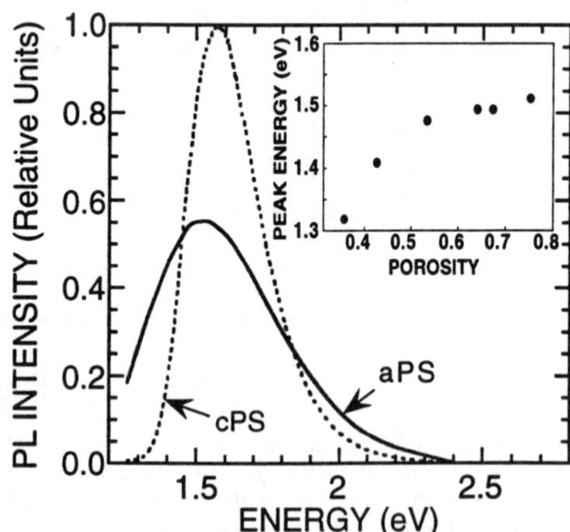

Fig.6: Typical room-temperature PL spectra of aPS and cPS. The inset shows the variation of PL peak energy as a function of aPS porosity (determined as explained in the text).

Results

Typical PL spectra are shown in Fig.6 and the temperature dependence of the integrated PL intensity is shown in Fig.7. In agreement with Bustarret [2-4], the energy of aPS luminescence is seen to lie in the same range as that of cPS. The integrated intensity, after scaling to layer thickness, is also comparable to that of cPS, though generally somewhat lower. In the same way as for cPS, the samples of highest porosity give the best room-temperature PL, an effect which can be traced back to the temperature dependence, which gets more pronounced as the porosity decreases (see Fig.7). A most intriguing point, however, is the marked insensitivity of the PL energy to the preparation conditions: in contrast to the case of cPS, the PL of aPS invariably exhibits its maximum in the 1.3-1.5 eV range (see inset in Fig.6). In order to make this point clear, we have investigated the PL of aPS in HF solution.

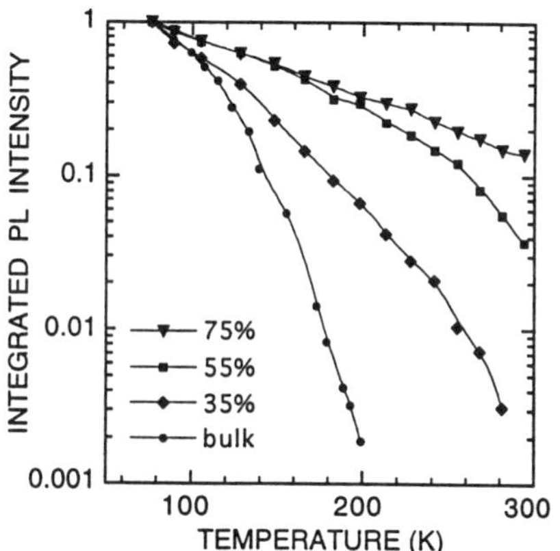

Fig.7: Temperature dependence of PL integrated intensity for aPS of different porosity (●: undoped bulk a-Si:H; other symbols: aPS of different porosities, see figure).

When cPS is left in HF solution while its PL is monitored, a *huge blue shifting* of the PL is observed with time, typically from 1.5 to 2.3 eV [16] and possibly even higher [18]. At the same time, the luminescence intensity *increases*, stays at a high level while at high energy, and finally decreases when the layer is destroyed. These effects are readily understood in the framework of quantum-confinement models: illumination induces photoelectrochemical etching, and the nanosized structures become thinner, resulting in an increased efficiency and a blue shifting of the luminescence [16]. However, the same experiment, performed on aPS, just results in a *decrease* of the luminescence intensity, attributable to a loss of material, *without any sizable blue shifting* (see Fig.8). This experiment is crucial in that it demonstrates the essentially different character of the luminescence of aPS and cPS. In the following, we will try to understand the luminescence mechanism in aPS. We will first address the problem of the luminescence intensity, then that of the luminescence energy.

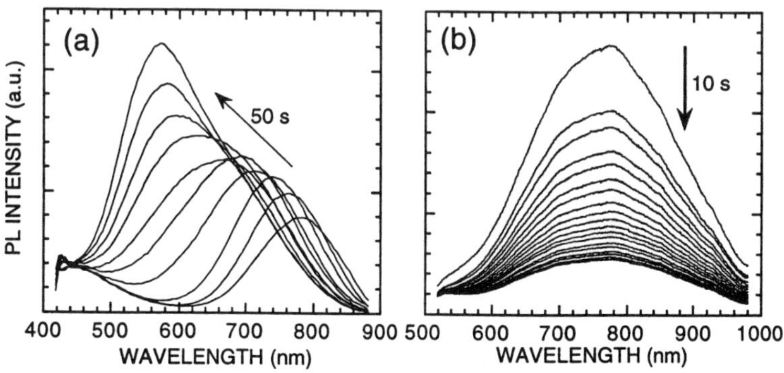

Fig.8: *Evolution of the in-situ PL of cPS (a) and aPS (b) in ethanoic HF solution. Excitation 337 nm, 1.5 mW/cm^2 (a) and 20 mW/cm^2 (b). The direction of change and interval between two spectra are indicated in the figure. Note that a large blue shifting is only present for cPS.*

Discussion: luminescence intensity

Let us first summarize what is known about the PL of bulk a-Si:H [10]. Bulk a-Si:H is a good luminescent material at low temperatures (T<100 K). Radiative recombination takes place through localized near-gap states: when an electron-hole pair is created, the electron and hole undergo fast ($\sim 10^{-12}$ s) relaxation to these near-gap "bandtail" states, so that they stay close to each other, and recombine radiatively at a rate of $\sim 10^3$ s^{-1}. However, this recombination path is in competition with non-radiative capture by midgap states. The practical capture range of such centers is determined by the distance where the radiative and non-radiative paths have equal probability. In a-Si:H, this length is $R_c \approx 12$ nm. Bulk a-Si:H at low temperature may then be regarded as a radiative medium with "dark" spots of radius R_c surrounding the non-radiative centers. The quantum PL efficiency η_0 is then given by the probability of photocreating the electron-hole pair in a "bright" region, i.e.,

$$\eta_0 = \exp\left(-\frac{4}{3}\pi R_c^3 N\right) \tag{6}$$

where N is the concentration of non radiative centers.

When temperature is increased, an electron-hole pair created in a bright region can either dissociate or diffuse and reach the dark spots. This results in a decrease of the PL ("thermal quenching"). The experimental variation can be fitted as [19]

$$\eta(T) = \frac{A}{1 + B\exp(T/T_0)} \tag{7}$$

At first glance, a simple Arrhenius behavior of $\eta_0 - \eta$ would have been expected. In the high-temperature limit, Eq.(7) can be regarded as resulting from an exponential distribution of activation energies. In a simple model, the characteristic temperature T_0 can be related to the width of the distribution: the pairs awaiting for recombination are supposed to be distributed in energy according to an exponential law

$$f(E) \propto \exp(E/E_u) \tag{8}$$

Such a distribution might result from the exponential shape of the band-edge density of states ("Urbach tail"). At a temperature T, the pairs near the top of the distribution will escape to a dark spot before radiative recombination can take place. The intensity of the radiative recombination is then given by counting the pairs below a demarcation energy E_d, such that the escape time $\nu_0^{-1}\exp[(E_m - E)/k_BT]$ equals the radiative recombination time τ_r. Here ν_0 is a characteristic attempt-to-escape frequency and E_m is mobility-edge energy. This gives

$$\frac{\eta(T)}{\eta_0} = \exp[-k_BT\ln(\nu_0\tau_r)/E_u] \tag{9}$$

in typical agreement with the observed behavior. The energy $k_BT_0\ln(\nu_0\tau_r)$ identifies with E_u, the characteristic energy of the joint density of states of the band tails, which can be accessed, e.g., by optical absorption measurements.

For the porous material, the temperature dependence of the PL intensity can still be described by Eq.(7), but the characteristic temperature T_0 appears larger than in bulk a-Si:H. This result is very similar to that found for amorphous silicon "superlattices" [20-21]. Although this is still a matter of debate [22], it has been interpreted as evidence that the bandtail density of states is affected by quantum confinement. In the framework of such an interpretation, the fair fits obtained with Eq.(7) would suggest that the exponential bandtail shape is preserved, and only its logarithmic slope E_u^{-1} is decreased. One might actually expect that conduction states will be affected by quantum confinement, whereas deep bandtail states will be little affected [23]. This would lead to a spreading of the bandtail, and to an increase in the activation energy for particle diffusion [24]. However, from a theoretical standpoint, it is not clear whether E_u^{-1} could be decreased by *over a factor of 3*, the exponential shape of the density of states being preserved. Especially, the deepest states, which dominate the room-temperature luminescence, should hardly be affected by confinement, since their localization radius is smaller than typical particle size, and a dependence more complex than Eq.(7) would probably be expected.

The above arguments cast some doubts on an explanation of aPS luminescence in terms of a pure quantum-confinement picture. Furthermore, highly-doped a-Si:H, which is *not luminescent at low temperature* ($NR_c^3 >> 1$), becomes *strongly luminescent when made porous*. This suggests that the photocarriers are prevented from reaching the non-radiative centers by a *spatial* (rather than *quantum*) confinement effect. Such an approach was recently attempted by Estes and Moddel [25]. At low temperature, in a nanostructured material (spheres, wires, or sheets), the probability for a photocarrier of encountering a nonradiative center is now given by the volume V_c of the intersect of the "dark sphere" with the nanostructure. When the characteristic size of the nanostructure is smaller than R_c , one has $V_c < {}^4/_3 \pi R_c^3$, hence the nonradiative recombination probability is decreased and the quantum efficiency is increased as compared to the bulk case. Furthermore, surface recombination has been included in the picture. At first sight, one might fear that surface recombination would have a dramatic effect on a nanostructured material, due to the very high specific surface area. However, it turns out that surface recombination is not exceedingly critical (except for sheets), because only a limited area is included within a length R_c from a photocarrier. In practice, surface recombination velocities up to 10^3 cm/s (which is far worse than achievable on HF-rinsed Si surfaces [26]) were found to little affect the photoluminescence of nanostructures of size lower than R_c . However, the above results only hold at low temperature. Estes and Moddel did not address the temperature dependence of the PL. In order to discuss room-temperature data, they just assumed that the effect of finite temperature in a-Si:H amounts to replacing the capture range R_c of the non-radiative centers by an effective value R'_c that they obtained by replacing η_0 by $\eta(T)$ in Eq.(6). However, this appears as an oversimplified treatment. In the standard model of thermal quenching of the PL, confinement is not expected to affect the escape probability of a pair in the band tail, hence $\eta(T)$ for aPS *should just be proportional* to that for a-Si:H, in clear contradiction with the experimental results [27].

In a reasonable model of thermal quenching, it is clear that the PL of small isolated particles ($a < R_c$) should be *temperature independent*, since only those particles with no non-radiative center will ever luminesce, and any motion of the photocarriers inside such a particle cannot lead to PL quenching. Hereafter we will attempt an improved description of thermal motion of the photocarriers and thermal quenching of PL in a-Si:H and aPS. We will assume that an electron-hole pair *diffuses as a whole*. In the presence of such a process, the probability p of radiative recombination of a photocreated pair *depends upon its distance* r to the *nearest* dark spot. The quantum efficiency of the sample is then the average of p over all possible locations for the photocreated pair, which can be expressed as

$$\eta(T) = \int_0^\infty g(r)\, p(r)\, dr \tag{10}$$

where $g(r)$ is the probability distribution of r. In the bulk material, one has

$$g(r) = 4\pi\,(r+R_c)^2 N \exp\left[-\frac{4}{3}\pi(r+R_c)^3 N\right] \tag{11}$$

Now we need a model for $p(r)$. The standard treatment of thermal quenching assumes that a *single* thermal excitation event is sufficient for driving the pair to a non-radiative center. However, retrapping in a bandtail state should occur in a time of the order of $\tau \sim 10^{-12}$ s. The estimated path length during such a time is of the order of $\lambda = (D_0\tau)^{1/2} \sim 30$ Å (here we take a typical value $D_0 \sim 0.1$ cm^2/Vs for the diffusion constant above the mobility gap and, for the sake of simplicity, we assume that λ is independent of T). This may be far smaller than the distance to the nearest non-radiative center and a second chance for efficient trapping and radiative recombination is again present at this stage. Several detrapping-retrapping sequences may then be possible before non-radiative recombination can occur (see Fig.9). The problem is actually that of a random walk with a trapping probability $1/n \sim \exp(-T/T_0)$ at each step. In the absence of encounter of a non-radiative recombination center, the pair will be efficiently trapped after $\sim n$ jumps, i.e., at a *typical distance* $\lambda n^{1/2}$ from the place where it was created. *Pairs created at a distance* $r > \lambda n^{1/2}$ *from a dark spot will then recombine radiatively* (p=1). If r is smaller than $\lambda n^{1/2}$, the dark spot may be reached after $\sim r^2/\lambda^2$ jumps, and the probability of radiative recombination for this pair is reduced to $p(r)=r^2/n\lambda^2$. In the dark-spot picture, this corresponds to appearance of a "grey" zone around the spots. The width $\lambda n^{1/2}$ of this grey zone increases with temperature, and its "brightness" profile is described by $r^2/n\lambda^2$. At low temperature, $p(r)$ is unity, and using Eq.(11) into Eq.(10) leads, as expected, to Eq.(6). In the opposite limit $T >> T_0$, $p(r) \sim (r/\lambda)^2 \exp(-T/T_0)$ is obtained, and an exponential behavior of η as a function of T/T_0 [similar to Eq.(9)] is found. Notice however that a preexponential factor appears, which may be appreciably larger than unity.

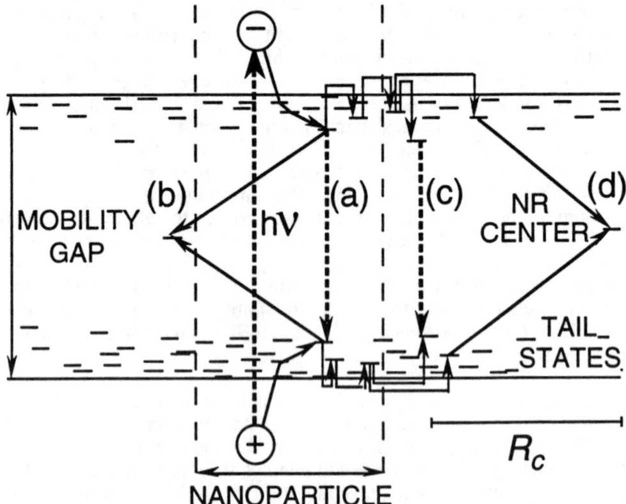

Fig.9: *Scheme of the recombination pathways in a-Si:H. At low temperature: radiative recombination from a bandtail state (a), or capture by a non-radiative mid-gap center (b). At high temperature: diffusion by thermal activation to the mobility edge, followed by radiative recombination from a lower-energy tail state (c), or (more probable) capture by a non-radiative center (d). Spatial confinement in a nanoparticle hinders processes (b-d) and favors direct radiative recombination (a).*

In the porous material, $g(r)$ is modified if the characteristic size a of the structures is smaller than R_c . At zero temperature, η_0 is just the integral of $g(r)$. In the limit $a << R_c$, the result for 2D structures (planar layers) and 1D structures (cylindrical wires) can be compared with Eq.(6) and to that for 0D structures (spherical particles):

$$\eta_{0,2D} = \exp\left(-\pi a R_c^2 N\right) \qquad (a = \text{layer thickness}) \qquad (12)$$

$$\eta_{0,1D} = \exp\left(-2\pi a^2 R_c N\right) \qquad (a = \text{wire radius}) \qquad (13)$$

$$\eta_{0,0D} = \exp\left(-\frac{4}{3}\pi a^3 N\right) \qquad (a = \text{sphere radius}) \qquad (14)$$

The beneficial effect of confinement in small-sized structures ($a << R_c$) is obvious and accounts for the strong low-temperature luminescence of PS made from highly boron-doped a-Si:H. It arises from the increase of the typical distance r_{ds} between nearby dark spots as dimensionality is decreased. At high temperatures, a law $\eta \propto \exp(-T/T_0)$ is invariably obtained, but the crossover temperature, given by r_{ds}~$\lambda n^{1/2}$, is found to be of the order of $T_0 \ln(N^{-2/3}\lambda^{-2})$ for bulk a-Si:H, $T_0 \ln(N^{-1}a^{-1}\lambda^{-2})$ for 2D structures and $T_0 \ln(N^{-2}a^{-4}\lambda^{-2})$ for 1D structures and thus pushes in all cases the crossover to higher temperatures. Note that η is temperature-independent for small isolated 0D structures. If the size and/or dimensionality of the structures is distributed, the exponential regime should not appear above a sharp crossover temperature, but rather a progressive transition is to be expected. This could easily account for a reduced slope of the $\ln(\eta)$ vs. T plots when a-Si:H is made porous.

In conclusion, spatial confinement appears as a necessary ingredient in order to understand the strong room-temperature luminescence of aPS, especially when the starting material is not luminescent, even at low temperature. This effect appears able to account for all the variations in luminescence intensity, at least qualitatively. However, possible quantum effects on the shape of the band tails might also be involved. A more quantitative analysis of the experimental data, together with quantum-theoretical calculations of the electron states, might help in assessing the possible role of such effects.

Discussion: luminescence energy

The luminescence energy of aPS has been found to exhibit weak changes depending upon the preparation conditions. Although the effect is small and hardly larger than the experimental uncertainties (think that the shift is at most 0.2 eV and the FWHM width of the luminescence band is about 0.5 eV), its presence is ascertained by related effects: i) the PL band of bulk a-Si:H is centered at ≈1.4 eV at low temperature, and red-shifts to ≈1.2 eV at room temperature (with a very weak intensity); ii) aPS samples with a room-temperature PL centered at 1.3 eV exhibit a blue shift to 1.5 eV upon lowering the temperature, whereas samples with a room-temperature luminescence centered at 1.5 eV hardly exhibit any blue shift upon cooling. All these observations are consistent with the statement that *the energy of the luminescence decreases as the volume explored by the electron-hole pair before recombination is larger.* This can be simply understood as a statistical effect: the states for recombination are located in the band tail. If the electron-hole pair is confined in a small volume, the total number of available states in the band tail becomes small, and the lowest energy accessible to the pair before recombination is not so low as in the bulk material (see Fig.9). If the bandtail joint density of states is of the form $N_c \exp(E/E_u)$, the energy dependence of the PL is expected to be [28]

$$h\nu = h\nu_0 - E_u \ln(N_c E_u \mathcal{V}) \tag{15}$$

where $h\nu_0$ is the luminescence energy for an electron-hole pair at the upper edge of the band tails, N_c is a characteristic band density of states, and $\mathcal{V}$ is the volume accessible to the pair. This simple scheme predicts a logarithmic dependence of $h\nu$ upon $\mathcal{V}$. It provides a simple explanation for the observed behavior in the bulk material [$\mathcal{V}$~$\lambda^3 n^{3/2}$] as well as in nanostructures [e.g., $\mathcal{V}$~$a^2\lambda n^{1/2}$ for wires], and it makes clear the connections between the two systems: when the material is made porous, the smallness of a induces a blue shift of the luminescence, but the smaller sensitivity to n reduces the temperature dependence of $h\nu$.

However, Lockwood et al. [29] recently found that the photoluminescence of 2D-confined non-hydrogenated a-Si appears to follow the predictions from quantum-confinement models (intended for crystalline semiconductors). The apparent conflict of these data with ours, as well as with most of the literature on a-Si thin films [20-21], probably stems from the different nature of the investigated materials: characteristic size, dimensionality, hydrogen content (hydrogen is known to decrease the density of midgap states by five orders of magnitude).

Our data of PL energy versus porosity are strikingly different from those obtained with cPS, in spite of the overall similarity in material characteristics. This clearly excludes simple quantum confinement models of the $1/d^2$ type which apply fairly well to cPS.

CONCLUSION

We have shown that device-grade amorphous hydrogenated silicon can be made porous and strongly luminescent at room temperature. Growth of the porous layer is limited by an instability, arising from the high resistivity of the material. The same instability appears to occur on crystalline silicon, when its resistivity exceeds that of the electrolyte. The luminescence of aPS appears to be of a nature essentially different from that of cPS. The variations of the luminescence intensity with temperature and the very weak variations of the luminescence energy are well accounted for by the *spatial* confinement of the photocarriers in the porous structure. By contrast, the large variations of the luminescence energy of cPS demonstrate that *quantum* confinement effects are an essential ingredient for understanding the luminescence of this material.

REFERENCES

1. See, e.g., L.T. Canham, Phys. Status Solidi (b) **190**, 9 (1995) and references therein.
2. E. Bustarret, M. Ligeon and L. Ortega, Solid State Commun. **83**, 461 (1992).
3. E. Bustarret, J.-C. Bruyère, F. Muller and M. Ligeon, MRS Symp. Proc. **283**, 39 (1993).
4. E. Bustarret, M. Ligeon and M. Rosenbauer, Phys. Status Solidi (b) **190**, 111 (1995).
5. K.H. Jung, S. Shih, D.L. Kwong, C.C. Cho and B.E. Gnade, Appl. Phys. Lett. **61**, 2467 (1992).
6. A.I. Yakimov, N.P. Stepina, A.V. Dvurechenskii and L.A. Scherbakova, Physica B **205**, 298 (1995).
7. M.J. Estes, L.R. Hirsch, S. Wichart and G. Moddel, MRS Spring Meeting 1996 (Symp. A).
8. R.B. Wehrspohn, J.-N. Chazalviel, F. Ozanam and I. Solomon, EMRS Meeting (Strasbourg, June 1996) (to be published in Thin Solid Films).
9. R.B. Wehrspohn, J.-N. Chazalviel, F. Ozanam and I. Solomon, Phys. Rev. Lett. **77**, 1885 (1996).
10. R.A. Street, *Hydrogenated Amorphous Silicon* (Cambridge Solid State Science Series, Cambridge, 1991) and references therein.
11. I. Solomon, R. Benferhat and H. Tran-Quoc, Phys. Rev. B **30**, 3422 (1984).
12. R. Swanepoel, J. Phys. E: Sci. Instrum. **16**, 1214 (1983).
13. W.W. Mullins and R.F. Sekerka, J. Appl. Phys. **35**, 444 (1964).
14. G. Lérondel, R. Romestain, F. Madéore and F. Muller, Thin Solid Films **276**, 80 (1996).
15. L.T. Canham, private communication.
16. F. Ozanam, J.-N. Chazalviel and R.B. Wehrspohn, EMRS Meeting (Strasbourg, June 1996) (to be published in Thin Solid Films).
17. T. Unagami and K. Kato, Jpn. J. Appl. Phys. **16**, 165 (1977).
18. H. Mizuno, H. Koyama and N. Koshida, Thin Solid Films (to be published).
19. R.W. Collins, M.A. Paesler and W. Paul, Solid State Commun. **34**, 833 (1980).
20. B. Abeles and T. Tiedje, Phys. Rev. Lett. **51**, 2003 (1983).
21. S. Miyazaki and M. Hirose, in *Amorphous and Microcrystalline Semiconductor Devices: Optoelectronic Devices*, ed. by J. Kanicki (Artech, Boston, 1991) chap.5, pp. 180-183.
22. S. A. Koehler and H. Fritzsche, MRS Spring Meeting 1996 (Symp. A).
23. G. Allan, C. Delerue and M. Lannoo, MRS Fall Meeting 1996 (Symp. Q).
24. F. Yonezawa and F. Satoh, Phil. Mag. B **60**, 109 (1989).
25. M.J. Estes and G. Moddel, Appl. Phys. Lett. **68**, 1814 (1996).
26. E. Yablonovitch, D.L. Allara, C.C. Chang, T. Gmitter and T.B. Bright, Phys. Rev. Lett. **57**, 249 (1986).
27. R.B. Wehrspohn et al. (to be published).
28. L.R. Tessler and I. Solomon, Phys. Rev. B **52**, 10962 (1995).
29. D.J. Lockwood, Z.H. Lu and J.-M. Baribeau, Phys. Rev. Lett. **76**, 539 (1996).

CHARACTERIZATION OF LIGHT EMITTING POROUS POLYCRYSTALLINE SILICON FILMS

M.C. POON *, P.G. HAN *, J.K.O. SIN *, H. WONG ** , P.K. KO *
* Department of Electrical & Electronic Engineering, Hong Kong University of Science & Technology, Clear Water Bay, Hong Kong
** Department of Electronic Engineering, City University of Hong Kong, Kowloon Tong, Hong Kong

ABSTRACT

Polycrystalline silicon (poly-Si) thin films (~700nm) were deposited by LPCVD, doped with 950°C phosphorous diffusion, and rendered porous by anodization and stain etching. From x-ray photoelectron spectroscopy, poly-Si films have atomic concentration of C(1s):O(1s):Si(2p) = 6%:15%:79%. However, porous poly-Si (PPS) films with weak photoluminescence (PL) have C:O:Si of 20%:38%:42%. For PPS films with strong PL, C:O:Si is 11%:38%:51%. From micro-Raman, scattered spectra for 632nm laser source has peak at 735nm and full wave half maximum (FWHM) of 76nm, and is similar to the PL spectra excited by 400nm uv laser source. High resolution transmission electron microscopy (TEM) study shows that PPS film is of complex structure and composes of numerous Si nano-crystals (1~10nm) surrounded by amorphous materials.

INTRODUCTION

Efficient light emission from porous silicon (Si) has drawn considerable interest since it was first reported by Canham[1]. In recent years, visible photoluminescence (PL) has also been found to be emitted from porous Si (PS) layers formed by anodized or stain-etched polycrystalline Si (poly-Si) [2-12]

The new luminescent porous poly-Si (PPS) film can be formed on Si or non-Si substrates such as oxide, metal or glass, and hence has numerous novel practical applications in future Si-based opto-electronic devices, such as sensors, photodetectors and displays. We have reported recently new nanostructures and formation mechanisms of luminescent PPS films using high resolution SEM and AFM studies[12] . In this article, we present more characterization results on the nanostructure and composition of this novel material.

EXPERIMENT

Undoped poly-Si thin films with a thickness of 670 nm were deposited onto <100> 5-10 Ω-cm p-type crystalline-Si (c-Si) substrates. Poly-Si films of around 450 nm in thickness were also deposited onto 95 nm-thick thermal oxide or 200 nm-thick low pressure chemical vapor deposition (LPCVD) nitride layers formed on Si substrates. The poly-Si films were formed by thermal decomposition of silane gas (SiH_4) in a LPCVD reactor at 625°C for 50 min. The flow rate and pressure of silane were 150 sccm and 300 mTorr, respectively, and the poly-Si deposition rate was 10 nm /min. Phosphorus was then diffused into the samples at 950°C for 20 min to form 12 Ω/sq n-

Mat. Res. Soc. Symp. Proc. Vol. 452 © 1997 Materials Research Society

type poly-Si films. To form the PPS layers, poly-Si films were anodized in dilute HF solution (49%HF :ethanol (C_2H_5OH)=1:1) with a Pt electrode at a current density of 10 mA/cm^2 for 5-10 min in room light and at room temperature. P-type Si wafers were also anodized under the same conditions for comparison. Poly-Si films were also stain etched in HF:HNO_3:DI 1:3:5 solution for 0.5-2 min in room light and at room temperature. The incubation time for the stain-etched PPS film was about 20-30 sec. For passivation tests, some PPS samples were further boiled in nitric acid (HNO_3) for 5-10 min. The poly-Si and PPS films were then studied using a Philips CM20 transmission electron microscope (TEM), Physical Electronics 5600 x-ray photoelectron spectroscope (XPS), and Renishaw 3000 micro-Raman and micro-PL scattering spectrometer with 632 nm laser excitation source. PL spectrum was also measured using SPEX 1403 PL spectrometer with innova 400 nm uv laser source.

RESULTS AND DISCUSSION

X-ray Photoelectron Spectroscopy (XPS)

Fig.1 shows the typical XPS spectrum of the anodized PPS films. Table 1 compares the x-ray diffraction (XRD) peaks [12] and XPS structural composition of the as deposited poly-Si films with no PL, and the PPS films with weak PL and strong PL. Results show that in poly-Si and anodized PPS films, there are high concentration of oxide and carbon residues. Poly-Si films have atomic concentration of C(1s):O(1s):Si(2p) = 6%:15%:79%. However, porous poly-Si (PPS) films with weak PL have C:O:Si of 20%:38%:42%. For PPS films with strong PL, C:O:Si is 11%:38%:51%. Intensity of PL increases with the increase in C and O concentrations. The C atoms might be due mainly to the ethanol (C_2H_5OH) in the electrolyte, and O atoms might be related to the water (H_2O), ethanol, or siloxene complex ($Si_6O_3H_6$)[13] in the PPS films. The relationship among the XRD peaks, XPS atomic composition and the PL intensity are currently being studied in our laboratory.

Table 1: Structural comparison of different PPS layers

Samples	As deposited Poly-Si thin films	PPS thin films with weak PL	PPS thin films with strong PL
XRD peaks [12]	(200), (220), (111)	(200), (220)	(200) or no peak
C:O:Si	6%:15%:79%	20%:38%:42%	11%:38%:51%

Figure 1. X-ray Photoelectron Spectrum of the PPS films

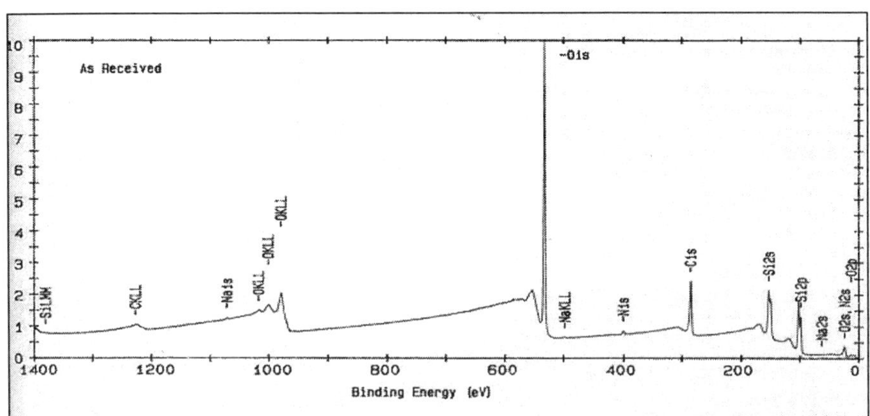

Micro-Raman and PL Spectroscopy

Fig.2a shows the micro-Raman scattering and micro-PL spectrum. The spectrum has a Raman scattering peak of 515 cm^{-1} , and a broad PL peak at about 735 nm with a full wave at half maximum (FWHM) of 76 nm. For the PL spectrum (Fig.2b), the peak is around 690 nm and the FWHM is about 100 nm. The different PL spectra obtained using two different excitation sources might be due to the inhomogenous PPS layer. According to the light-emitting mechanism of semiconductor, when a semiconductor is exposed to light with energy greater than the energy gap, excess electron-hole pairs are created in the material, which subsequently recombined to give photon emission. Photons with higher energy will be strongly absorbed and penetrate only a thin surface layer of the material. As we have obtained a peak of 690 nm for greater energy photon excitation (400 nm laser) and a peak of 735 nm for smaller energy excitation (632 nm laser), the results suggest that the energy gap in the PPS layer close to the surface is wider than that in the deeper layer close to the substrate. The surface PPS layer hence has smaller crystallites from quantum size theory[1] and the PPS film has a non-uniform structure.

Figure 2a. Micro-Raman and PL scattering spectra. Y axis is intensity in arbitrary unit.

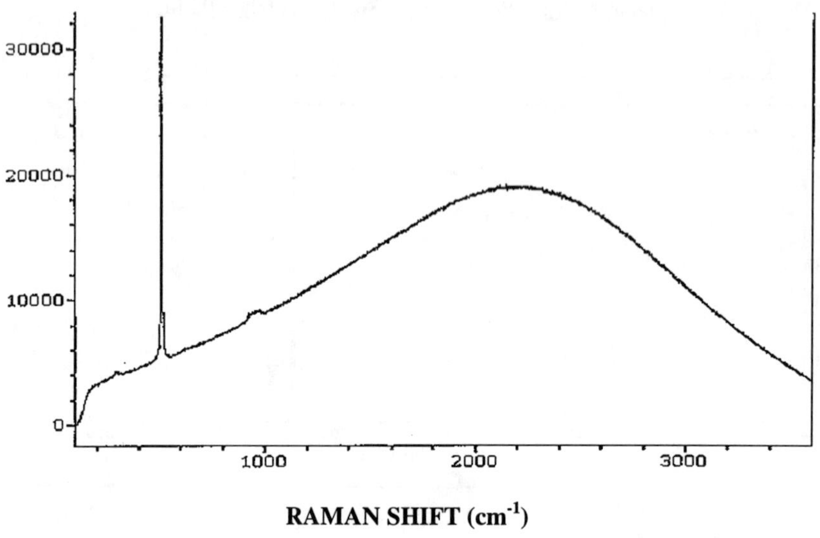

Figure 2b. Typical PL emission spectrum of the PPS films. Y axis is intensity in arbitrary unit.

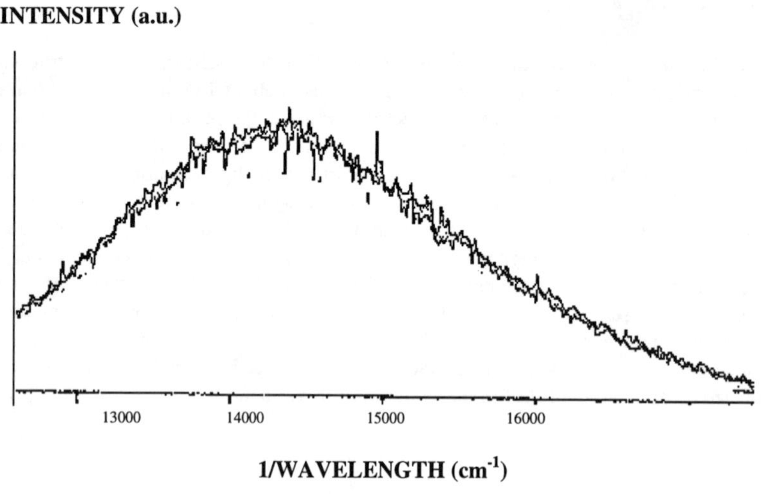

High resolution TEM study

Fig.3 shows the high resolution TEM micrograph of the cross-section of anodized n-type poly-Si thin films deposited on p-type c-Si substrate. Time for anodization is 10 min and current density is 10mA/cm^2. From the image, the etched layer is composed of two layers: a PPS layer formed from the top poly-Si layer (left side of micrograph), and a PS layer formed from the c-Si substrate (right side of micrograph). The PPS film is of complex structure and composes of numerous Si nano-crystals (1~10nm) surrounded by amorphous materials. Contrarily, the PS layer (~300nm) composes mainly of larger nano-crystals (~2-20nm) embedded in less amorphous materials. The atomic concentration of O:Si, which were obtained from the analytical TEM system, are 59% : 41% for the PPS layer and 44% :56% for the PS layer. If assume that most of the oxygen atoms are used to oxidize the silicon atoms, in the PPS layer, 59% O atoms will consume about 30% Si atoms to form silicon dioxide and there are about 11% crystal Si atoms. Contrarily, in the PS layer, 44% O atoms will consume about 22% Si atoms to form silicon dioxide and there are about 34% crystal Si atoms.

Fig.3 TEM of the interface of a PPS layer and the over-etched PS substrate

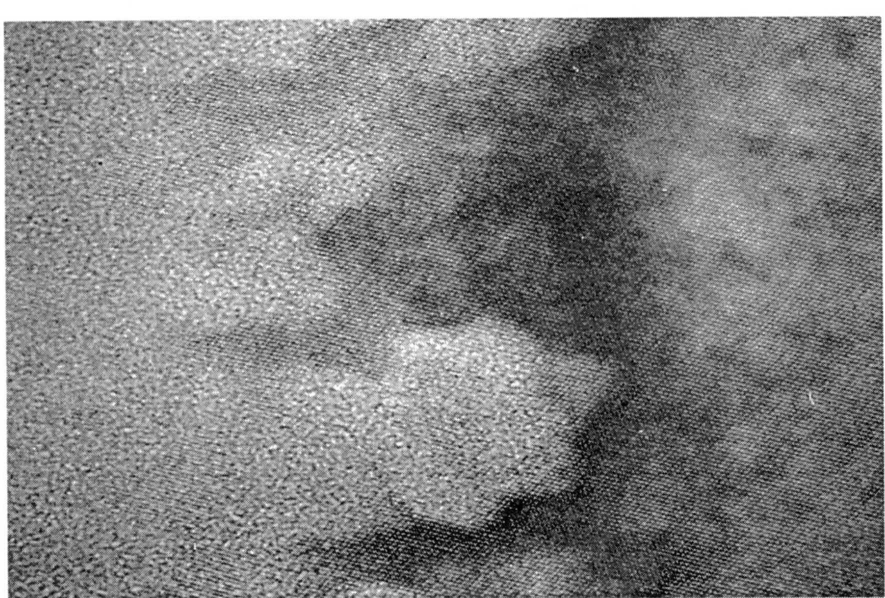

CONCLUSIONS

In summary, we obtained new results with respect to the nanostructures, photo-luminescence and crystallinity in porous poly-Si films formed by anodization or stain etching of phosphorus-doped poly-Si films deposited by low pressure chemical vapor deposition onto different substrates. The light emission may be related to both the Si nano-crystallites mainly in <200> orientation and oxygen-rich amorphous Si materials surrounding the nano-crystals.

ACKNOWLEDGMENTS

We would like to thank Dr. K.K. Fung and Dr. N. Wang for their helps on the samples analysis by transmission electron microscope. We also wish to thank Professor Ping Ko, Dr. Jack Lau, Professor Hoi Kwok, Dr. Philip Chan and Dr. Man Wong for their valuable comments, and to Mr. T.M. Kwok for his help in the PL measurement. This work was partially supported by grant number HKUST 691/95E of the RGC of Hong Kong.

REFERENCES:

1. L.T. Canham, Appl. Phys. Lett. **57**, 1406 (1990).
2. P. Guyader, P. Joubert, M. Guendouz, and M. Sarret, Appl. Phys. Lett. **65**, 1787 (1994).
3. N.M. Kalkhoran, F. Namavar, and H.P. Maruska, Appl. Phys.Lett. **63**, 2661 (1993).
4. K. Higa, T. Asano and T. Miyssato, Jpn. J. Appl. Phys. **33**, 1733 (1990).
5. T. Ueno, Y. Akiba, T. Shinohara, H. Koyama, N. Koshida and Y. Tarui, Jpn. J. Appl. Phys. **32**, L5 (1993).
6. A.J. Steckl, J. Xu and H.C. Mogul, Appl. Phys. Lett. **62**, 2111 (1993).
7. Y. Kolic, R. Gauthier, M.A. Garcia Perez, A. Sibai, J.C. Dupuy, P. Pinard, R.M.'Ghieth and H. Maaref, Thin Solid Films **255**, 159 (1995).
8. P. Joubert, A. Abouliatim, P. Guyader, D. Briand, B. Lambert and M. Guendouz, Thin Solid Films **255**, 96 (1995).
9. K.H. Jung, S. Shih, D.L. Kwong, C.C. Cho and B.E. Gnade, Appl. Phys. Lett. **61**, 2467 (1992).
10. E. Bustarret, M. Ligeon, J.C. Bruyere, F. Muller, R. Herino, F. Gaspard, L. Ortega and M. Stutzmann, Appl Phys. Lett. **61**, 1552 (1992).
11. R. Czaputa, R. Fritzl and A. Popitsch, Thin Solid Films **255**, 212-215, (1995).
12. P. G. Han, M C Poon, P K Ko and J K O Sin, J. Vac. Sci. Technol. B 14(2), Mar/Apr 1996
13. M. S. Brandt, H.D. Fuchs, M.Stutzmann, J. Weber and M. Cardona, Solid state Commun., **81**, 307, (1992).

TEM STUDY OF POROUS SILICON FABRICATED FROM N- AND P-TYPE DOPED POLYCRYSTALLINE FILMS

L. HAJI, Y. Le THOMAS, F. CHANE CHE LAI and P. JOUBERT
Groupe de Microélectronique et Visualisation, Université de Rennes 1;
IUT de Lannion, Boîte Postale 150; 22302 Lannion Cedex, France

ABSTRACT

The formation of porous silicon (PS) from *n/p*, n^+/p and p^+/n structures carried from polycrystalline silicon films (poly-Si) deposited on single crystal silicon (c-Si) substrates was studied by cross-sectional transmission electron microscopy. Our results clearly show that the pore formation in such structures involve the extended defects of the poly-Si film. The role played by these defects depends on the doping type and level, and on whether the anodization is performed under illumination or not.

INTRODUCTION

Recent interest in porous silicon (PS) results from its potential use as a new material for optoelectronic device realisation. This interest is stimulated by its strong visible phototoluminescence at room temperature [1]. The feasibility of PS-based light emitting diodes (LED) have been firstly demonstrated with Schottky type structure where a metal is deposited on top of the PS layer [2-3]. But the poor performance of these devices could be due to inefficient carrier injection. New generation LED's are based on silicon p^+/n or n^+/p homojunctions which are post-processing anodized [4-9]. The heavily doped top layer is usually obtained by using ion implantation [4-8], but *in-situ* doped polycrystalline silicon (poly-Si) films deposited by Low Pressure Chemical Vapor Deposition (LPCVD) is an alternative solution [9]. In this case, a burried layer of luminescent porous silicon has to be obtained by electrochemical etching through the poly-Si film.

We report in this paper results for porous formation in *n/p*, n^+/p and p^+/n structures carried from polycrystalline silicon films deposited on single crystal Si substrates. We investigate the microstructure of porous silicon by cross-sectional transmission electron microscopy (XTEM). We study the role played by defects (grain-boundaries, twins...) of the poly-Si on the pore formation and the effect of illumination used during anodization.

EXPERIMENTAL

Poly-Si were deposited by LPCVD on (100)-oriented single Si substrates. The films were *in-situ* doped by using a gaseous mixture of silane and phosphine or diborane [10]. Amorphous silicon films were deposited at 550 °C and were subsequently thermally crystallized at 600 °C for 12 hours. The resulting thickness of the layer is of about 0.3 µm. Backside ohmic contact is achieved by evaporation of 200 nm thick Al film and subsequently sintered.

The porous silicon was formed by electrochemical etching in an ethanoic HF solution (50% HF: C_2H_5OH: H_2O = 1:2:1) at a constant current density of 18 mA/cm^2. The anodization was performed either in the dark or under illumination using a halogen lamp (8 mW/cm^2). In table 1 we report the characteristics of the studied structures and the anodization conditions.

The specimens for cross-sectional TEM observations were prepared by using mechanical polishing and ion-milling techniques. The observations were performed using a JEOL 1200EX.

Mat. Res. Soc. Symp. Proc. Vol. 452 © 1997 Materials Research Society

Table 1: Characteristics of the studied samples and their anodization conditions.

Name of sample	NP1	NP2	NP3	N^+P	P^+N
Poly-Si type	*n*	*n*	*n*	*n+*	*p+*
Poly-Si resistivity (Ω.cm)	0.6	0.6	0.6	2×10^{-3}	3×10^{-3}
Substrate type	*p*	*p*	*p*	*p*	*n*
Substrate resistivity (Ω.cm)	1-5	1-5	1-5	1-5	2-5
Anodization time (s)	90	20	20	90	60
Illumination during anodization	no	no	yes	no	yes

RESULTS AND DISCUSSION

n POLY-Si / p-Si STRUCTURE:

In a first step we describe the porous morphologies in such a structure in the case where the anodization was done in dark. Figure 1 is a XTEM micrograph of the sample NP1 (see table 1) showing the porous microstructure of the poly-Si layer and c-Si substrate. The top layer is partially transformed into porous poly-Si (PPS) material but with a low porosity. Large areas in the poly-Si layer remain unattacked. These regions are visible near the interface and around the extended defects of the grains(see arrows in Fig.1). Although the top n-type layer was not completely transformed, we notice that the pore formation has advanced into the underlying p-type substrate. The porous Si layer (PSL) formed in the substrate exhibit a very rough interface with the non attacked substrate. This undulation of the anodization front in the substrate is due to an enhanced etch rate at some points located at the original poly-Si/substrate interface. The PSL has a nanoporous structure, with pore orientations radiated from these local points. The bright contrast which appear in the micrograph at these points indicates a high porosity material. We have observed, that these singular points correspond to twins running throughout the grains.

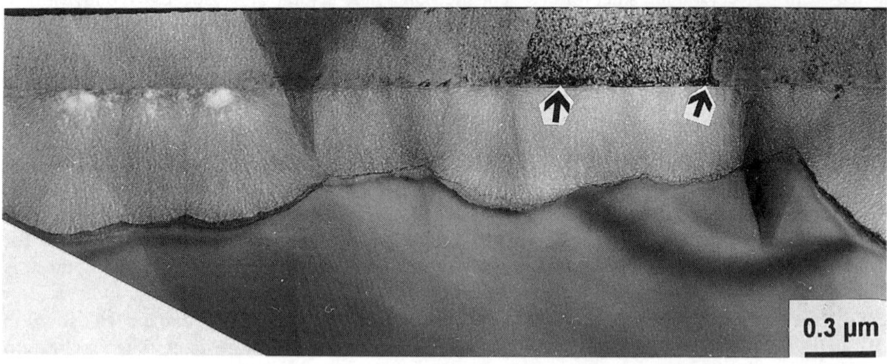

Figure 1. XTEM micrograph of sample NP1 anodized in the dark for 90 s.

Anodization at reduced etching time (sample NP2) may give information about the early stage of the pore formation. Figure 2 shows the result for an anodization time of 20 s. The microstructure of the PPS is quite similar to that described before. It consists of primary pores located within the twinned areas of the grains and secondary tiny pores having slight ramifications. Large unattacked regions are clearly observed inside the poly-Si grains as well as at the poly-Si/substrate interface (see arrow in Fig.2). Figure 3 is a HRXTEM view of a crystalline zone near the interface. We can notice that the unattacked region is bounded by micro-twins. The difference between the microstructures of the pores in the film and substrate is clearly evidenced. Also shown in figure 2 the heteregeneous character of the pores formation in the underlying substrate. Large cavities having various sizes, filled with nanoporous Si, appear. The center of these cavities, from which the PS propagates, corresponds to twins.

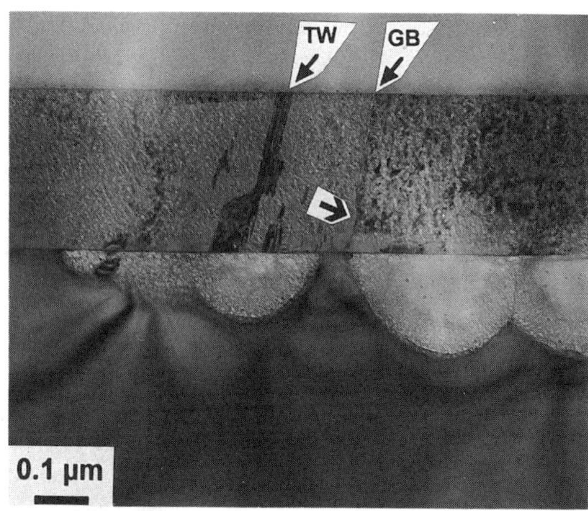

Figure 2. XTEM micrograph of sample NP2 anodized in the dark for 20 s. Primary pores are located at twinned areas (TW); unattacked areas (arrowed) around grain-boundaries (GB). Notice the thin secondary pores within the grains.

Figure 3. HRXTEM showing details around unattacked area near the film/substrate interface (sample NP2).

In order to get better understanding of the role played by the poly-Si defects during the porous formation, the same structure was anodized under illumination. Figure 4 shows the result for 20 s etching (sample NP3). The microstructure of the PPS and PS are quite similar to that observed previously for the anodization in the dark (fig.2). Large unattacked crystalline areas within the top poly-Si layer are still observed. The difference is that the starting points of PS formation in the substrate are now localized at grain boundaries termination instead of twins.

During the anodization, the current densities around electrically active defects are higher than in defect-free areas, consequently preferential etching has been observed [11]. From our results, we have noticed that the etching rate is higher within the grain boundaries when the anodization is enhanced by light induced photocurrent. On the other hand, intra-grain defects are etched faster in the dark. However, in both cases the preferential etching around poly-Si defects induces local enhanced current densities at the points where the electrolyte reaches the interface. Thus, a fast dissolution within the substrate is observed which results in a non uniform PS layer. In a previous work [12], we have pointed out this preferential etching of the grain boundaries of poly-Si during the anodization of n/n$^+$ structures under illumination.

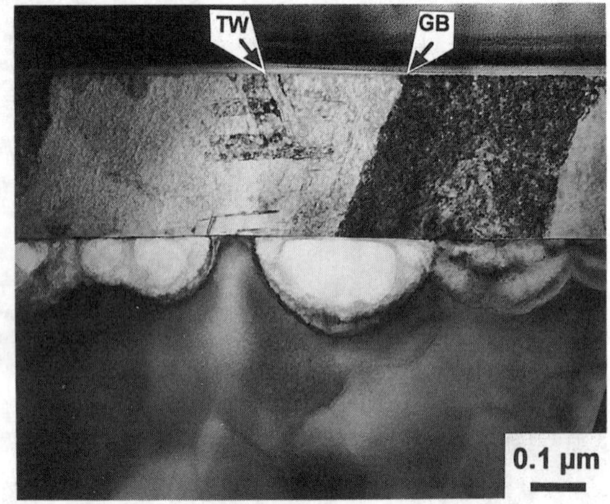

Figure 4. XTEM micrograph of sample NP3 anodized under illumination for 20 s. Primary pores are now located at grain-boundaries (GB).

n$^+$ POLY-Si / p-Si STRUCTURE:

Anodization for 90 seconds in the dark of degenerately doped n$^+$-type poly-Si (sample N$^+$P) results in the formation of porous materials which are very uniform in visual appearance. An XTEM image corresponding to this sample is shown in figure 5. The poly-Si is almost completely transformed into PPS. Its microstructure is now more homogeneous. Similarly to the *n/p* structure, the porous Si formed in the p-type substrate (PSL) is heteregeneous in thickness and porosity. As previously noticed, the pores have grown very fast along the twins leading to a local current density enhancement as soon as the electrolyte has reached the interface. Although not shown here, observations on identical samples anodized in the same conditions but under illumination have shown very similar porous morphology. However, as for *n/p* structure the roughness of the anodization front in the substrate is due to grain boundaries while in the dark, twins prevail.

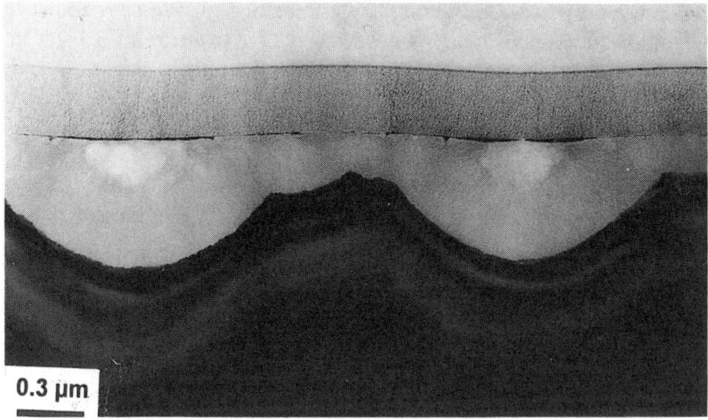

Figure 5. XTEM micrograph of sample $N^{+}P$ anodized in the dark for 90 s.

p^{+} POLY-Si / n-Si STRUCTURE:

The anodization of such a structure is quite different from the previous cases, since the *p/n* junction is now reverse biased and thus essentially no current flows. To avoid this problem, the anodization was then performed under illumination. Figure 6 shows a XTEM view of the sample anodized for 60 seconds. At the top of the poly-Si film, a thin porous layer of about 100 nm thick was formed. The anodization front within the poly-Si clearly delimits the PPS from the remaining unattacked film. In the defect-free regions of the film one should expect a uniform photocarrier generation and thus a uniform thickness of the PPS layer. On the opposite, around an extended defect the dissolution rate may be different which results in a local thickness inhomogeneity. This is clearly evidenced in Fig.6. The PPS layer consists of large pores having a columnar microstructure and running nearly perpendicular to the interface, except at the twinned area where the dissolution is faster.

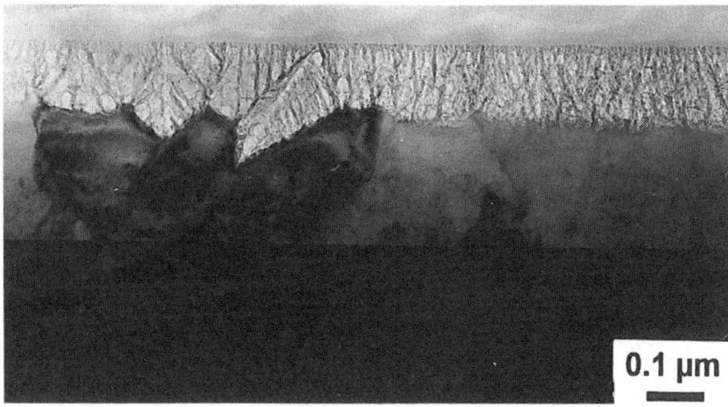

Figure 6. XTEM micrograph of sample $P^{+}N$ anodized under illumination for 60 s. Notice that the PPS is only formed at the top of the film

Furthermore, in contrast with *n/p* structure, no pore formation in the substrate was observed, as long as the poly-Si was not completely transformed. We have found that in this case a very long anodization time (> 10 min) is needed in order to form a PS layer in the substrate. However, the PS formed within the substrate was more uniform in thickness and did not show any variations in porosity.

CONCLUSION

The microstructure of porous Si formed in poly-Si / c-Si structures has been investigated in detail by cross-section TEM. It is found that the mecanisms of PS formation involve grain boundaries as well as intra-grain defects (twins and micro-twins). During the anodization under illumination, primary pores are formed along the grain boundaries of the poly-Si film due to a preferential etching; While in the dark the preferential etching is observed around the twins. On the other hand, pore formation is dependent on the substrate doping type irrespective of poly-Si layer *p*-type substrate may promote fast etching of the top layer and more particularly around twins and grain-boundaries.

ACKNOWLEDGMENTS

The authors gratefully acknowledge D. Briand, K. Kis-Sion and M. Sarret for the preparation of the poly-Si films.

REFERENCES

1. L.T Canham, Appl. Phys. Lett. **57**, 1046, (1990).
2. A. Richter, P. Steiner, F. Kozlowsky and W. Lang, IEEE Electron. Device Lett. **12**, 691 (1991).
3. N. Koshida and H. Koyama, Appl. Phys. Lett. **60**, 192, (1992).
4. P. Steiner, F. Kozlowsky and W. Lang, Appl. Phys. Lett. **62**, 2700, (1993).
5. A. Loni, A. J. Simons, T.I. Cox, P.D.J. Calcott and L.T. Canham, Electron. Lett. **31**, 1288 (1995).
6. J. Linros and N. Lalic, Appl. Phys. Lett. **66**, 3048, (1995).
7. Tsykeskov, S. P Duttagupta, K. D. Hirschman and M. Fauchet, Appl. Phys. Lett. **68**, 2058, (1996).
9. F. Chane Ché Laï, C. Beau, D. Briand and P. Joubert, Appl. Surf. Sci. **102**, 399 (1996).
10. M. Sarret, A. Liba, F. Le Bihan, P. Joubert, and B. Fortin, J. Appl. Phys. **76**, 5492 (1994).
11. H. Föll, Appl. Phys. A, **53**, 8, (1991).
12. P. Guyader, P. Joubert, M. Guendouz, C. Beau and M. Sarret, Appl. Phys. Lett. **65**, 1787, (1995).

FORMATION OF BURIED POROUS SILICON STRUCTURE BY HYDROGEN PLASMA IMMERSION ION IMPLANTATION

Z. Fan*, Paul K. Chu*, X. Lu**, S. S. K. Iyer**, N. W. Cheung**
* Department of Physics & Materials Science, City University of Hong Kong, Tat Chee Avenue, Kowloon, Hong Kong
** Plasma Assisted materials Processing Laboratory, Department of Electrical Engineering and Computer Sciences, University of California at Berkeley, CA 94720

ABSTRACT

Plasma Immersion Ion Implantation (PIII) excels in several areas over conventional ion implantation, for example, higher dose, shorter implantation time, and lower overall cost. The technique can be used to fabricate buried porous silicon. In our experiment, hydrogen is implanted into Si by PIII at 5-30kV to form underlying porous silicon (PS) which emits light at an energy higher than the Si bandgap. The optical properties of the PS samples as measured by photoluminescence are quite good. The PIII technique therefore offers an alternative means to fabricate buried porous silicon structures which can potentially be used to fabricate optoelectronic devices in silicon.

INTRODUCTION

In the last few years, strong room-temperature light emission from electrochemically etched porous silicon (PS) [1] has attracted a lot of attention. The technology has the potential to integrate optoelectronic and microelectronic devices on a single Si wafer. Unfortunately, the chemically etched PS process is difficult and not compatible with traditional Si fabrication technology because the HF solution which is the anodic electrolyte [1] causes a rough surface and leaves surface impurities after the etching process [2]. Furthermore, the mechanism of light emission from PS is still unclear.

In this paper, we introduce an alternative way to prepare "clean" PS. Our method avoids contamination from acid and has higher compatibility with Si-base IC fabrication [3]. Hydrogen is implanted into a crystalline Si wafer by plasma immersion ion implantation (PIII) to generate innumerable nano-sized bubbles under the surface [4]. When the silicon walls between these bubbles are thin enough to allow quantum effects to take place, silicon will be transformed from an indirect bandgap material to a direct bandgap one and give out light at an energy higher than the silicon bandgap. We call it buried porous silicon (BPS) since its structure is similar to ordinary porous silicon.

PIII is a novel ion implantation technology that was first introduced by Conrad and co-workers at University of Wisconsin [5]. It has many advantages when compared to a conventional implanter, especially high throughput and low cost. In PIII, the whole wafer is immersed in a plasma and pulsed or DC biased. Positive ions are implanted normally into the wafer, and the implantation time is consequently independent of the wafer size. It is thus a

Mat. Res. Soc. Symp. Proc. Vol. 452 © 1997 Materials Research Society

promising method to produce low-cost large-area Si-based display divices and silicon-on-insulator materials[6,7].

EXPERIMENT

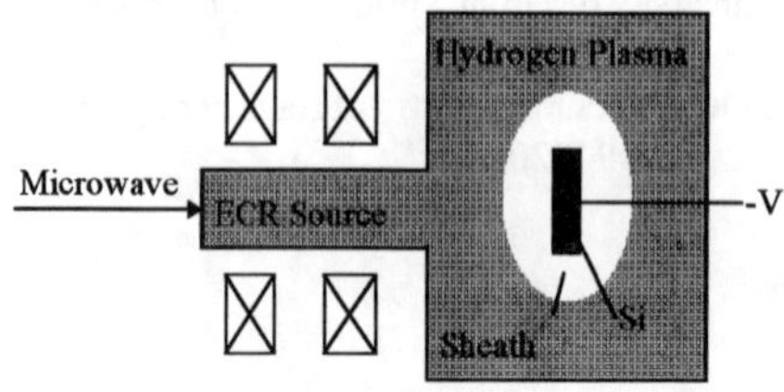

Fig.1 *Schematic of hydrogen plasma Immersion ion implantation (PIII) process*

P-type Czochralski grown (100) Si wafers with resistivity from 1 to 50 Ω-cm were used in our experiments. The PIII chamber was pumped down to $2x10^{-6}τ$ before starting the implantation. A negative dc bias, from -5kV to -30kV, was applied to the Si wafer that was immersed in a hydrogen plasma created by an ECR (Electron Cyclotron Resonance) source. Positive ions were accelerated cross the sheath by this dc voltage and implanted into the Si wafer (Fig. 1). The hydrogen dose was varied from $3x10^{14}cm^{-2}$ to $4x10^{16}cm^{-2}$ by adjusting the implantation time and gas pressure in the chamber. In this work, our attention was focused on the specimen implanted at -5kV for 10 seconds in a 0.5mτ hydrogen ambient because this sample emitted the stronger light.

Photoluminescence (PL) spectra were taken using a standard piece of apparatus. The samples were cooled in liquid nitrogen to 77K. The exciting light source was an argon laser, and both the 514.5 nm and 488 nm lines were tried. The PL output was collected by a monochromator and detected by a 700K type photomultiplier (PMT) which was sensitive to wavelength ranging from 400 nm to 1200 nm. The signal was amplified by a lock-in amplifier and the spectrum was recorded by a computer.

RESULTS AND DISCUSSION

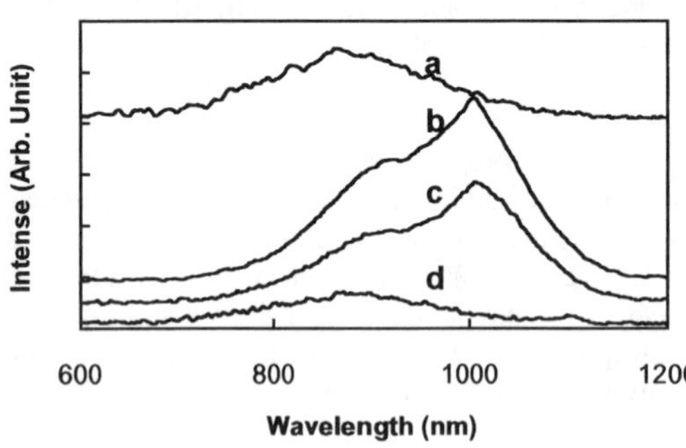

Fig. 2 *The PL spectra of pure Si (curve d) and hydrogen implanted samples: (a) 670℃ 2 hours annealing, (b) 325℃ 1 hour annealing, and (c) as implanted.*

The PL spectrum in Fig. 2 was acquired at an excitation wavelength of 514.5 nm. There are two broad fluorescent bands in the as implanted sample: at 1010 nm and 890 nm (curve c in Fig. 2). By comparing to the PL spectrum of pure Si wafer acquired under the same conditions, the 890 nm peak is determined to be instrumental background. The small peak at 1100 nm in the pure Si PL spectrum corresponds to the band to band emission of Si. Its energy is equivalent to the Si indirect bandgap. Thus the band located at 1010 nm is believed to originate from the BPS. The full-width half-maximum (FWHM) of this band is about 0.2 eV. The BPS emission band can still be observed after annealing at 325°C for 1 hour, but almost disappears after annealing at 670°C for 2 hours. As suggested by D. Bisero et al.[8], this is due to the coarsening of the bubbles during the annealing process as well as the loss of hydrogen from the silicon wafer when the annealing temperature exceeds 400°C. The implanted hydrogen can passivate nonradiative recombination centers stemming from defects and dangling bonds and consequently increase the efficiency of light emission.

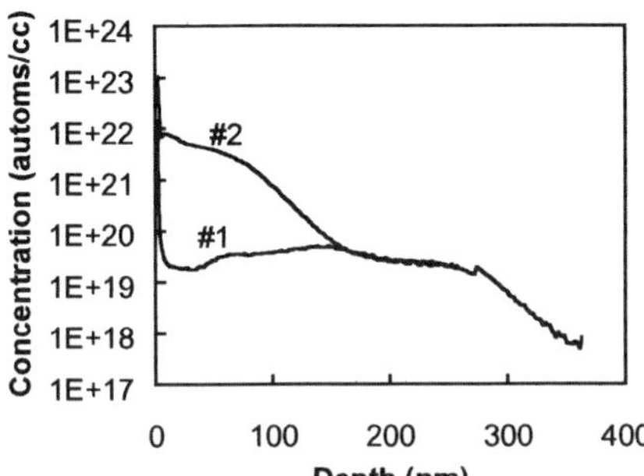

Fig.3 *The SIMS hydrogen profile of two samples implanted at -5kV with the gas pressure of 0.5mτ.*

Fig. 3 shows the SIMS results of two samples implanted under similar conditions but differing in implantation time: #1 for one minute and #2 (the same sample described above) for ten seconds. In general, the implantation dose increases with implantation time. However, based on the SIMS results, the hydrogen doses in samples#1 and #2 are $3.2x10^{14}cm^{-2}$ and $3.8x10^{16}cm^{-2}$, respectively. We attribute this reverse trend to the heat created during PIII. Because there is no sample cooling in our instrument, the wafer temperature goes up significantly during PIII as a large number of ions impact the wafer at same time. By observing the color of the wafers during the implantation experiment, the substrate temperature is believed to exceed 500°C after 1 minute. After the plasma source and the high voltage power supplier are shut down, the wafer remains hot for quite some time. The wafer is usually left in the PIII chamber in a vacuum of $2x10^{-6}$ τ until it is allowed to reach close to room temperature before it is taken out. This waiting process which can take up to an hour is thus similar to thermal annealing in vacuum. More hydrogen diffusion is thus expected for sample #1 because of the higher temperature, whereas the short implantation time and subsequently the lower substrate temperature for sample #2 result in a higher retained hydrogen dose in the sample. This phenomenon is supported by the PL data. Sample #2 shows strong luminescence at 77K while that in sample #1 is undetectable. In addition, the hydrogen dose in sample #1 is not sufficient to passivate the nonradiative recombination centers. Therefore, the possibility of excited carriers recombining through the radiative path is low.

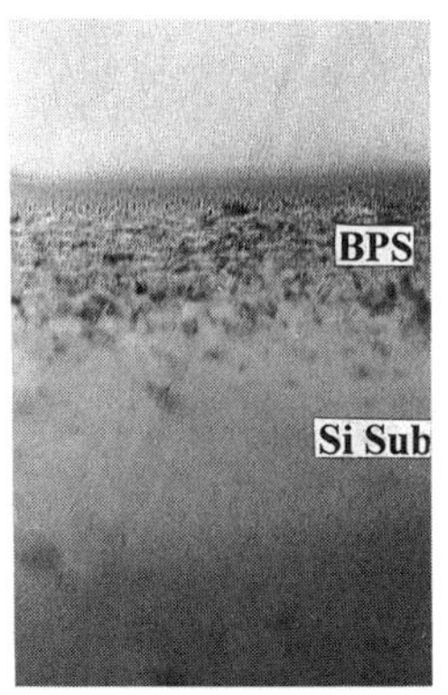

Fig. 4 *XTEM Picture of BPS*

Pavesi et al. [9] showed no luminescence from the as implanted samples, and a salient fluorescence band was observed only after annealing at 200°C or higher. However, our as implanted samples show strong fluorescence in 77K. This is believed to be caused by the unintentional "annealing process" from the high ion flux during our PIII experiment as explained above.

The XTEM image in Fig. 4 reveals a large number of bubbles under the surface. The projected range of BPS is about 50nm deep. The Si absorption coefficient of photons at an energy of 1.5eV is $0.78x10^3$ cm^{-1} [10]. Therefore, the emitted light from the underlying BPS layer penetrates the thin overlay unattenuated. Even if the light energy shifts to 2.5eV (blue), the transmission loss will still be minimal and 91% of the fluorescent light can be transmitted. This has important ramification if indeed both photonic and microelectronic devices can be fabricated in the same substrate. Our BPS materials are also believed to be more stable than the electrochemically etched counterparts because they are protected by the overlayer.

CONCLUSION

Silicon implanted by hydrogen PIII shows obvious light emission at an energy higher than the Si bandgap. This is observed for the as implanted samples and after low temperature annealing. The PL peak disappears after annealing at a high temperature. The SIMS profiles confirm the relationship between luminescence and the amount of implanted (and retained) hydrogen. Since absorption by silicon is minimal at the fluorescent wavelength, BPS produced by PIII can potentially be used to fabricate integrated photonic / electrical devices.

ACKNOWLEDGMENTS

We acknowledge the Department of Materials Science, Fudan University for the PL experiment and Mr. Song Liu and Ms. Zhiguang Gu for experimental assistance. The work is supported by the City University of Hong Kong Strategic Grants 7000621 and the Hong Kong RGC grants 9040220 and 8730005.

REFERENCES

1. L. T. Canham, et al., Appl. Phys. Lett. **57**, 1046 (1990).
2. L. T. Canham, M. R. Houlton, W. Y. Leong, C. Pickering, and M. J. Keen, J. Appl. Phys. **70**, 422 (1992).
3. R. Siegele, G. C. Weatherly, H. K. Haugen, and D. J. Lockwood, Appl. Phys. Lett. **66**, 1319 (1995).
4. J. Min, P. K. Chu, X. Lu, S. S. K. Iyer, and N. W. Cheung, to be published in Thin Solid Films.
5. R. Conrad, J. L. Radtke, R. A. Dodd, F. J. Worzala, and N. C. Tran, J. Appl. Phys. **62**, 4591 (1987).
6. P. K. Chu, N. W. Cheung, and C. Chan, Semicond. Int. **6**, 165 (1996)
7. J. B. Liu, S. S. K. Iyer, C. M. Hu, N. W. Cheung, R. Grousky, J. Min, and P. Chu, Appl. Phys. Lett. **67**, 2361 (1995)
8. D. Bisero, F. Corni, C. Nobili, R. Tonini, G. Ottaviani, C. Mazzoleni, and L. Pavesi, Appl. Phys. Lett. **67**, 3447 (1995).
9. L. Pavesi, G. Giebel, R. Tonni, F. Corni, C. Nobili, and G. Ottaviani, Appl. Phys. Lett. **65**, 454 (1994).
10. O. Madelung, Semiconductors Group IV Elements and III-V Compounds, (Springer-Verlag, Berlin, 1991), p. 20.

DEPTH GRADIENTS IN POROUS SILICON: HOW TO MEASURE THEM AND HOW TO AVOID THEM

M.Thönissen, M.G. Berger, W. Theiß *, S. Hilbrich *, M. Krüger, R. Arens-Fischer, S. Billat, G. Lerondel **, H. Lüth
KFA Jülich, ISI, D-52425 Jülich, Germany
* RWTH Aachen, I. Physik. Inst., D-52056 Aachen, Germany
** Université J. Fourier, B.P. 87, F-38402 St. Martin d'Hères cedex France

ABSTRACT

For the formation of single layers and layer systems the depth homogeneity of porous silicon (PS) plays a key role. We have measured qualitatively and quantitatively structural gradients in p-PS layers with increasing layer thicknesses by fitting reflectance spectra in the infrared and by measuring thickness oscillations in the reflectance during the formation of PS layers. These results allow us to give recipes for the formation of PS layers with a homogenous optical thickness in depth.

INTRODUCTION

The anodic etch rate r (and therefore the thickness of a layer) is related to the porosity P (and therefore to the refractive index) via the equation

$$r(j) / j = A_r / [P(j,c)\ \nu(j)\ e\ N\ \rho] \quad (1),$$

where ν is the valence, ρ the density of silicon, A_r the relative atomic mass of silicon, N the molar number, j the current density and c the HF concentration [1]. Changes in this etch rate and therefore in the thickness of single layers as well as changes in the microstructure with depth degrade the performance of PS interference filters [2]. Both effects depend on the thicknesses, the anodization current densities and the doping levels [2,3,4] of the layers in the filter structure. They are caused by chemical etching [5,6] and changes in the HF concentration with depth [4,7]. The resulting inhomogeneities of the optical thickness of single layers as well as layer systems with more complex filter structures [2] lead to changes of the designed filter frequency and have to be avoided. In this paper we have determined the effects of chemical etching and HF concentration changes with depth on single layers. For a qualitative estimation of these influences in the first part of this paper the optical thicknesses of layers above and below layers with different thicknesses are determined by a fit of the reflectance spectra of these layer systems using effective medium theory [8]. In the second part we investigate oscillations in the reflectance during the formation of PS layers caused by the increasing layer thickness. From these results we are able to give a functional description of the changes in the optical thickness with depth. By using these results, an on line correction of these changes by a variation of the anodization current density is possible as it will be shown below.

EXPERIMENT

To study the changes of the etch rate and microstructure with depth we have investigated PS layers formed on p^- (70 mΩcm) and p^+-doped (10 mΩcm) substrates (orientation (100)) using

Mat. Res. Soc. Symp. Proc. Vol. 452 © 1997 Materials Research Society

electrolyte compositions of (HF (50%) : C_2H_5OH) (1:1), (1:0.9) and (1:1.1). Before the etching process the samples were cleaned in an ultrasonic bath using propanol and thereafter they were rinsed in deionised water. In order to analyse changes in the porosity of a probe layer (200mA/cm², 8s ≈ 1.1µm), filling layers were etched below and above the test layers with a current density of 100mA/cm² whereby the etching times of these layers are varied from 12s (≈1µm) up to 240s (≈20µm). The resulting layer system was analysed by a Perkin-Elmer λ2 reflectance spectrometer (9000cm^{-1}-50000cm^{-1}). The spectra were fitted using Looyenga [9] effective medium theory as described in [8]. For the on-line measurements of the thickness oscillations we have used a laser diode operating at 785nm (power <3mW). In order to avoid effects of sample illumination [10] the maximum power was reduced to 0.3mW. The reflectance was measured by a photodiode (0.45A/W, 400-1200nm) as a function of etching time.

RESULTS

The qualitative influences of chemical etching during the anodization and HF concentration changes are analysed by measuring the reflectance of a system consisting of a probe layer and a filling layer as described in the experimental section. In order to separate the two effects we have prepared different layer systems: The first one consists of a filling layer on top of the probe layer. By the variation of the thickness of the filling layer from 1µm up to 20µm we are able to investigate an influence of changes in the HF concentration with depth on the formation of the probe layer for different thicknesses of the filling layers and different electrolyte compositions. A typical spectrum and the fit of this spectrum is shown in Fig. 1. It can be seen that the theoretical and experimental curves are in very good agreement.

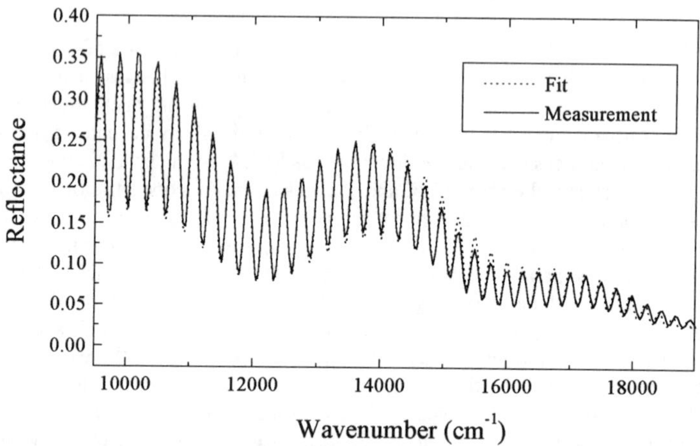

Fig. 1: Experiment (solid line) and fit (dotted line) of the reflectance spectrum of a p-PS layer consisting of a 10µm filling layer (100mA/cm²) with a 1.1µm probe layer (200mA/cm²) below. The electrolyte composition was (HF (50%) : C_2H_5OH) (1:1). The fit was performed using effective medium theory (Looyenga model).

In order to investigate the influence of chemical etching during anodization we used layer systems consisting of a probe layer and a filling layer which is below the probe layer. So during the formation of the filling layers the probe layer is exposed to the electrolyte. The results of the fits for different electrolyte compositions are given in Fig. 2. As it could be deduced from Fig. 2, the changes in the optical thickness caused by changes in the electrolyte composition are much larger than the changes caused by the chemical etching. Both effects strongly depend on the electrolyte composition itself. The overall effect for filling layers of low thicknesses can be described in the following way: For lower HF concentration the porosity of the probe layer is larger and the optical thickness lower, which is consistent with known results [11]. With increasing thickness of the filling layer this behaviour changes. For the layer system etched with the lowest HF concentration a higher optical thickness is observed. In fact this means a higher optical thickness in combination with a lower HF concentration. With respect to the known relationship of HF concentration and porosity [11] only two possible reasons can be given for this: The etch rate must increase drastically or the composition of the electrolyte has changed in depth. The first reason can be excluded because REM measurements of the thicknesses of these samples for different HF concentrations show deviations of about 4%-6% from the values found by the fit. The second reason of changes in the electrolyte is different from changes in HF concentration with depth. Here the composition of Ethanol, HF and water may cause the crossing in the functional dependence of optical thickness and thickness of filling layer. For further investigations layers etched with changed HF concentrations have to be investigated, where different HF concentrations are produced by increasing or decreasing the amount of water in the electrolyte.

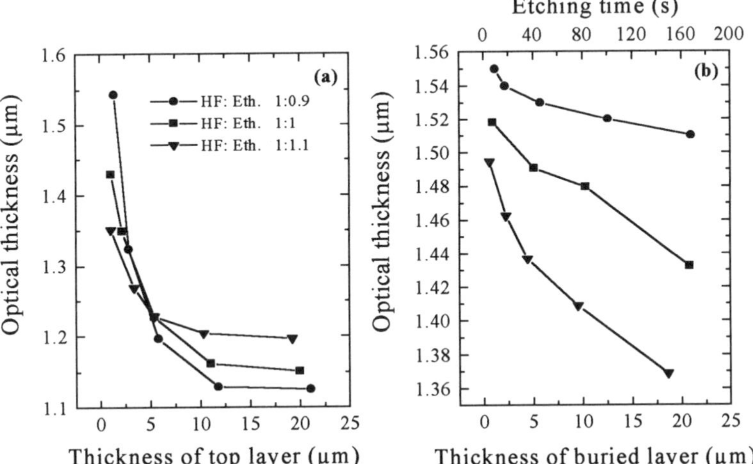

Fig. 2: (a) Optical thickness of the probe layer as a function of the thickness of the filling layer on the top of the probe layer for different electrolyte compositions given in the legend. The optical thickness was calculated by multiplying the refractive index at 12000cm^{-1} and the geometrical thickness of the probe layers taken from the fit. (b) Optical thickness of the probe layers as a function of the thickness (etching time) of the buried filling layer.

For the quantitative investigations we have analysed the time evolution of the thickness interferences during the formation of p^+ porous silicon layers in order to get the quantitative changes in the porous structure: For a fixed wavelength the reflectance of samples recorded during the formation exhibits oscillations due to the increasing layer thickness with increasing etching time. Because of the non-uniform porosity and a non-uniform etch rate the frequency of the oscillations changes with time. We have investigated these changes for different current densities and doping levels. In Fig. 3 typical oscillations as a function of time are shown. Each three neighbouring peaks are fitted by a sine function in order to get the oscillation frequency for a certain range of time. From these data we are able to determine the frequency of oscillations as a function of the etching time for various current densities. In Fig. 4 (a) these frequencies normalised to the current density are plotted as a function of the layer thickness using the etch rate for each current density and each doping level [1,12,13]. These normalised frequencies are a measure for r_{opt}/j from equation (1), where $r_{opt} = r \cdot n(P)$ is the optical etch rate and n(P) the refractive index of the system silicon/electrolyte. According to this equation, r_{opt}/j is proportional to the quotient n(P)/P(j,c), which should be constant with depth in order to get a homogenous layer structure. This can only be achieved by adjusting the anodization current j with depth. The results in Fig. 4(a) demonstrate a recipe for how to vary the current density with thickness (etching time): We start with a fixed current density (e. g. 200mA/cm^2, see point 1 in Fig. 4 (a)). From this point we draw a horizontal line of fixed oscillation frequency through all other curves (in Fig. 4 (a) the ones plotted for 120 and 100mA/cm²). The crossing with these other curves (marked as point 2 and 3 in the figure) gives the depth at which the current density of the crossed curve has to be applied to get the same ratio n(P)/P(j,c). This functional dependence for all applied current densities is plotted in Fig. 4(b). It can be seen that for example the highest applied current density at each doping level has to be reduced to half its initial value at a depth of 10μm to get a homogenous microstructure with depth. Using these results we are now able to form homogenous layers by adjusting the current density with depth as it is shown in Fig. 5.

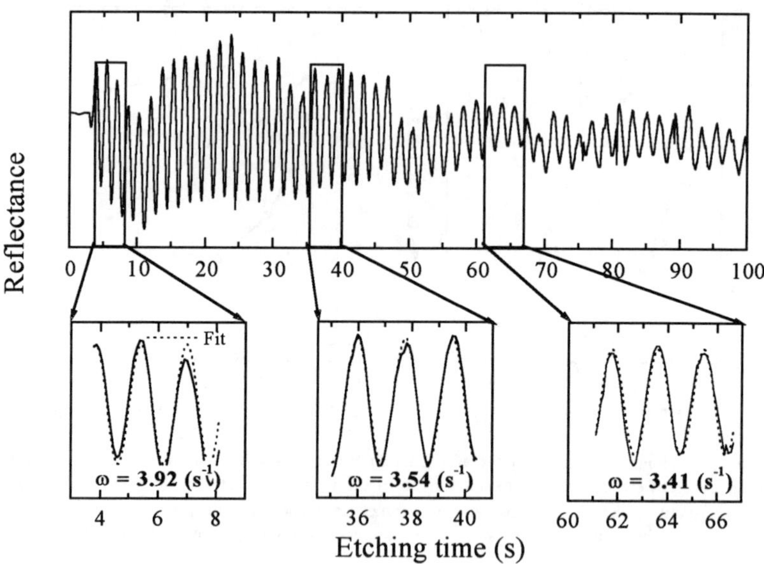

Fig. 3: In-situ oscillations in the reflectance of a p-PS layer etched with a current density of 80 mA/cm². Each three neighbouring peaks were fitted by a sine function in order to determine the oscillation frequency with depth.

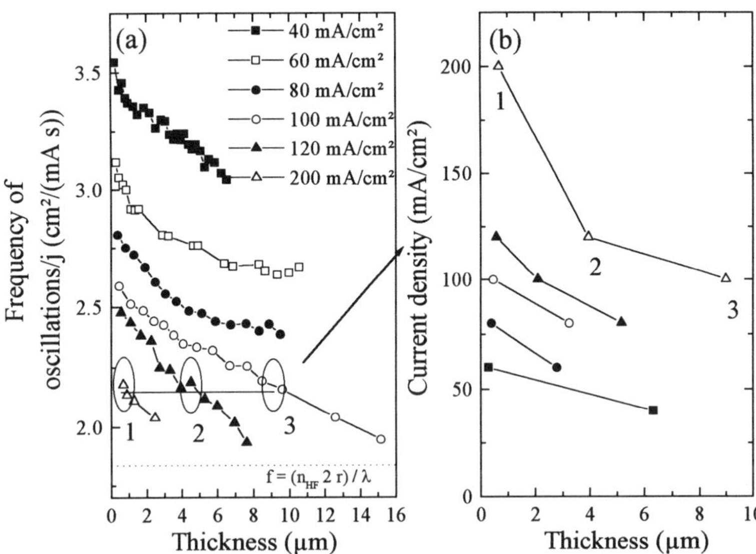

Fig. 4: (a) Frequency of oscillations for PS on p-doped substrates determined for the current densities given in the legend. The frequency is normalised for the current density. The dashed horizontal line in the bottom of the figure gives the calculated porosity in the case of peeling off the PS layer from the substrate. The derivation of (b) from (a) is explained in the text. In (b) a recipe for changing the current density with depth is given to form homogenous layers.

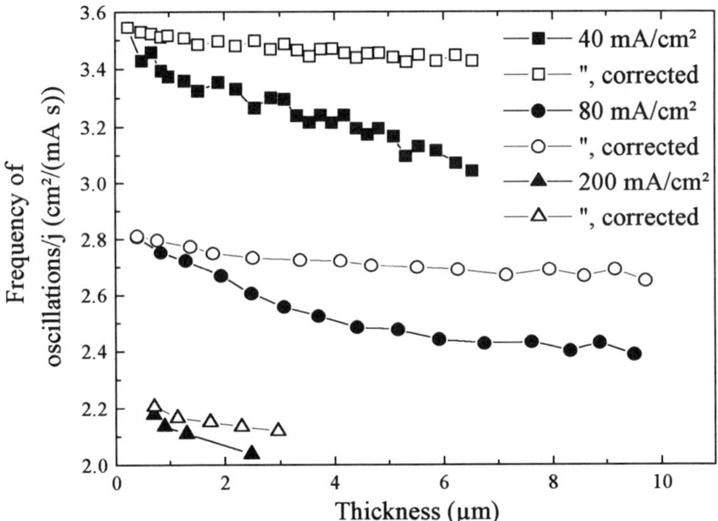

Fig. 5: Comparison of the frequencies of thickness oscillation with and without correction of current density with depth given by the recipe in Fig. 4 (b). Changes in the optical thickness can be reduced drastically by this correction.

CONCLUSIONS

We have shown by measurements of the beat of oscillations of different two-layer systems, that effects of chemical etching are lower than diffusion problems in the electrolyte. But also changes in the composition of the electrolyte seem to play a role. By in situ measurements of thickness oscillations in the reflectance we are able to give a recipe for changing the current densities with depth in order to get a homogenous optical thickness in depth.

REFERENCES

1. M.G. Berger, PhD thesis „Poröses Silicium für die Mikrooptik: Herstellung, Mikrostruktur und optische Eigenschaften von Einzelschichten und Schichtsystemen", RWTH Aachen, (1996)

2. M.G. Berger, R. Arens-Fischer, M. Thönissen, S. Hilbrich, W. Theiß, M. Krüger, S. Billat, P Grosse, H. Lüth, Thin Solid Films, accepted

3. S. Billat, M. Thönissen, M. Krüger, M.G. Berger, H. Lüth, Thin Solid Films, accepted

4. M. Thönissen, S. Billat, U. Frotscher, U. Rossow, M. Krüger, H. Lüth, M.G. Berger, J. Appl. Phys. 80(5), (1996), 2990-2993

5. D.R. Turner, J. Electrochem. Soc., 107, 810, (1960)

6. S.M. Hu, D.R. Kerr, J. Electrochem. Soc., 14, 414, (1967)

7. V. Lehmann, J. Electrochem. Soc., 140 (10), 2836, (1993)

8. W. Theiß, The use of effective medium theories in optical spectroscopy, R. Helbig (editor), Advances in Solid State Physics, 33, p. 149, Vieweg

9. Looyenga, Physica 31, 401 (1965)

10. M. Thönissen, M.G. Berger, R. Arens-Fischer, M. Krüger, O. Glück, H. Lüth, W. Theiß, M. Wernke, Thin Solid Films, 276 (1996), 21-24

11. A. Halimaoui, Surf. Sci. Lett., 306, L550 (1994)

12. M.G. Berger, S. Frohnhoff, W. Theiß, U. Rossow, H. Münder, Porous silicon science and technology, ISBN 3-540-58936-8, Springer (1994), p. 345

13. M.G. Berger, C. Dieker, M. Thönissen, L. Vescan, H. Lüth, H. Münder, W. Theiß, M. Wernke, P. Grosse, J. Phys. D: Appl. Phys., 27, 1333, (1994)

X-RAY DIFFRACTION AND REFLECTIVITY STUDIES OF THIN POROUS SILICON LAYERS

D. Buttard[*], G. Dolino[*], D. Bellet[*] and T. Baumbach[**]

[*]Laboratoire de Spectrométrie Physique. Université J. Fourier, CNRS (UMR 5588)

B.P 87, 38402 Saint-Martin-d'Hères Cedex - FRANCE, buttard@grenet.fr

[**]Fraunhofer Institut, Zerstörungsfreie Prüfverfahren, Dresden - GERMANY.

ABSTRACT

High resolution X-ray diffraction and reflectivity have been used for the structural characterization of thin porous silicon layers of p and p^+ doping type. Thin porous silicon layers studied either by diffraction or reflectivity, in the range of 10-1000 nm, exhibit several thickness fringes, corresponding to a lateral homogeneity of the layer thickness. The comparison between the experimental results with simulations enables one to deduce structural information relative to the porosity, thickness, lattice parameter as well as interface thickness. For p^+ type samples a double fringe system was observed, showing the existence of a surface film probably at the porous silicon layer top surface.

INTRODUCTION

In recent years, there has been a considerable interest in silicon nanostructures, especially in porous silicon (PS), which exhibits visible luminescence [1]. Many research efforts have been devoted on both fundamental and applied aspects of this material [2]. While the optical properties have been investigated the most, the structural ones are also of current interest. The etching of p and p^+ type porous silicon wafers under well defined conditions leads to the formation of a pore network of nanometer size. Then this material, with a specific area of a few 100 m^2/cm^3, exhibits very large surface effects, which is important either for optoelectronic devices, or for applications in other fields (gas sensor, biomaterials...).

A porous silicon layer (at least for p and p^+ doping types) is a nearly perfect single crystal, as revealed by high resolution X-ray diffraction [3]. This technique has been used to investigate the strains of thick PS layers after an anodic oxidation [4] or in the presence of a liquid or a vapour in the pore network [5]. However thin porous silicon layers have also interesting properties and could also give some hints about the PS formation mechanism, which is still an open problem. High resolution X-ray diffraction studies of thin PS layer and superlattices were reported previously [6,7], with the presence respectively of thickness fringes and satellites, showing the good crystalline quality of such structures.

In this paper we report a study of thin PS layers, by using X-ray diffraction and reflectivity. These two techniques allow a non-destructive determination of several parameters of a porous layer (thickness, porosity, lattice parameter, interface between the PS layer and the substrate). The last parameter is of great importance since it is known for instance that any roughness between the PS layer and the substrate seems to be an important loss factor for optical guides [8].

Mat. Res. Soc. Symp. Proc. Vol. 452

EXPERIMENTAL CONDITIONS

We have used (001) Si wafers, positively doped, etched in darkness by an electrochemical process in order to produce a porous silicon (PS) layer at the top surface of the wafer. The electrolyte was composed of $HF:C_2H_5OH:H_2O$ (1:2:1) and the value of the applied current density leads to different porosity values. In this paper we report some results obtained on p^+ type samples (with resistivity $\rho \approx 0.01$ Ωcm), of 36% porosity (p^+(36%)) and 60% porosity (p^+(60%)), and on p type samples ($\rho \approx$ 4-6 Ωcm) of 65% porosity (p (65%)). The thickness layers (which depends linearly on the anodisation time) were in the 10 nm - 1000 nm range.

The X-ray intensity diffracted by the substrate and the porous layer, were obtained with the $CuK_{\alpha 1}$ radiation of a Philips MRD apparatus [4]. We used the symmetrical diffraction of the (004) reflection in order to be sensitive to the perpendicular lattice misfit. The porous silicon layers have still the properties of nearly perfect single crystal, giving rise to coherent crystal truncation rods through all reciprocal lattice point. Specular X-ray reflectivity investigates the truncation rod through the origin of the reciprocal space. This method allows to investigate the depth profile of the mean electron density. Both methods can be performed by the same experimental geometry using $\omega/2\theta$ scans which correspond to a displacement along the normal of the sample surface.

RESULTS AND DISCUSSION

Diffraction measurements

The four reflection monochromator and the two reflection analyzer of the MRD apparatus allow a high resolution analysis of the sample structure in the reciprocal space. The introduction of an analyzer allows to separate the coherent diffracted intensity along the crystal truncation rods from the diffuse scattered intensity due to the pore structure.

The $\omega/2\theta$ scans obtained along the (004) truncation rod, for a p^+ (60%) sample and for a p (65%) sample are reported on Fig. 1.

-p^+ (60%): On the experimental curve (dotted line of Fig. 1a), one observes the Bragg peak of the substrate (S), and the Bragg peak of the thin porous silicon layer (P). The substrate peak width is meanly given by the resolution function of the multiple crystal arrangement. The width of the layer peak is proportional to the inverse of the layer thickness, which is in our case around 520 nm. The elastic expansion of the layers caused by surface effects [4] leads to a perpendicular misfit with respect to the substrate ($\Delta a/a = 8.2 \times 10^{-4}$) which produces the angular shift of the layer Bragg-peak from the substrate peak. Additionally to the Bragg peaks, one observes several thickness oscillations. From their period one obtains the layer thickness. The high number of oscillations gives evidence for the lateral homogeneity of the sample and the layer thickness. A previous study [7] showed a similar result for low porosity p^+ samples. To fit the experimental curves one used the dynamical theory of X-ray diffraction, which includes refraction and multiple scattering effects [9]. In our model a layer is described by three parameters, the layer thickness (t), the porosity (P) and the lattice strain ($\Delta a/a$) (The calculated values are given in table I). Keeping the simple model of a single porous layer structure on a substrate, the peak and the oscillations positions can be reproduced, but the fringe intensity differs from the measured one. Moreover, the progressive damping of the thickness fringe amplitude for higher oscillation orders indicates a gradual interface similar to the graduated heterotransition in compound semiconductors. By simulating the intensities, assuming for simplicity a linear gradient of the porosity and of the related lattice misfit at the PS/substrate interface, we find h = 15 nm for the thickness of the interface region which is of the same order of magnitude as the one calculated for the p$^+$ (36%) sample [7].

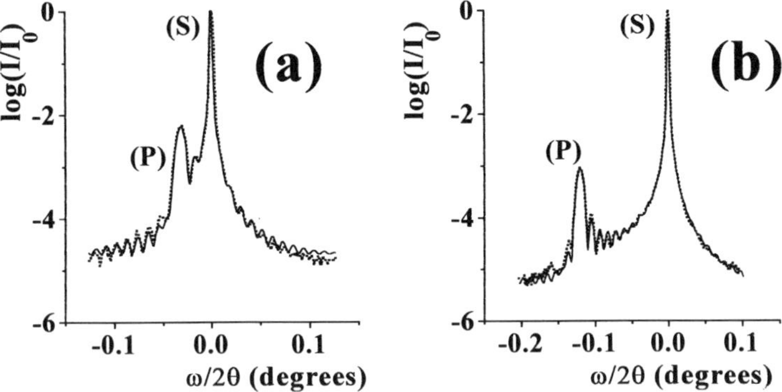

Figure 1: Experimental high resolution X-ray diffraction rocking curves (dotted line) and dynamical theory simulations (solid line) around the (004) reflection:
(a) for a p$^+$ (60%) type sample.
(b) for a p (65%) type sample.

-p (65%): The experimental results (dotted line of Fig. 1b), show also two Bragg diffraction peaks (S) and (P). The lattice mismatch ($\Delta a/a = 3.0\times10^{-3}$) is larger than for p$^+$ type sample. Thickness fringes are also observed on the experimental curve, but with a small number of oscillations than for the p$^+$ type sample maybe due to a poor quality of the interfaces or to thickness fluctuations in the plane of the sample. The dynamical theory simulation (solid line) is in rather good agreement with the experimental data. In this case the heterotransition width is

Table I: Parameters of the porous layers used in the dynamical theory simulations for the 5 different samples plotted in Fig. 1 and Fig. 2 (Thickness (t), Porosity (P), lattice misfit ($\Delta a/a$), thickness of the heterotransition (h), PS/substrate roughness (σ)). The indexes IF, PS and SF correspond respectively to the PS/substrate interface, to the porous silicon layer and to the surface film. On the left column are reported the expected values of porosity and thickness extrapolated from the usual data of thick porous silicon layers.

Sample type	Diffraction
p 65% **500nm**	t=500 nm P=84% h=10 nm $\Delta a/a=3.0\times10^{-3}$
p$^+$ 60% **500nm**	t=520 nm P=66 % h=15 nm $\Delta a/a=8.2\times10^{-4}$
p$^+$ 36% **300nm** (from[7])	t=299nm P=35% h=15 nm $\Delta a/a=4.6\times10^{-4}$

Sample type	Reflectivity
p 65% **70nm**	t_{PS}=82 nm P_{PS}=70% t_{IF}=5 nm P_{IF}=27%
p$^+$ 60% **150nm**	t_{SF}= 22 nm P_{SF}=45% t_{PS}=173 nm P_{PS}=58% t_{IF}=16 nm P_{IF}=30%
p$^+$ 36% **100nm**	t_{SF}=44 nm P_{SF}=15 % t_{PS}=63 nm P_{PS}=45 % σ_{IF}=15 nm

smaller than for p^+ type samples.

It is worth noting that for both p^+ samples, the calculated curves have been obtained with only four parameters. Moreover one can note that the calculated porosity is often larger than the expected one. This is probably due to the presence of large strain in some crystallites which reduce the coherent diffraction, and then leads to a systematic increase of the calculated porosity.

Reflectivity measurements

In order to limit the parasitic effect of the diffuse scattering, two slits of 0.1 mm width have been used, one after the monochromator, and the other one in front of the detector. Moreover owing to the finite width of the sample (≈ 7 mm) the experimental curves have been normalised by dividing the reflected intensity by a factor $\sin(\theta)$. In the inset of Fig. 2a the characteristic reflectivity curve of a bulk silicon crystal is plotted: the critical angle θ_{Si} (≈ 0.22°) at the end of the total reflection plateau is related to the silicon density, while the decrease slope is related to the surface roughness (≈ 2nm) for a virgin wafer [10].

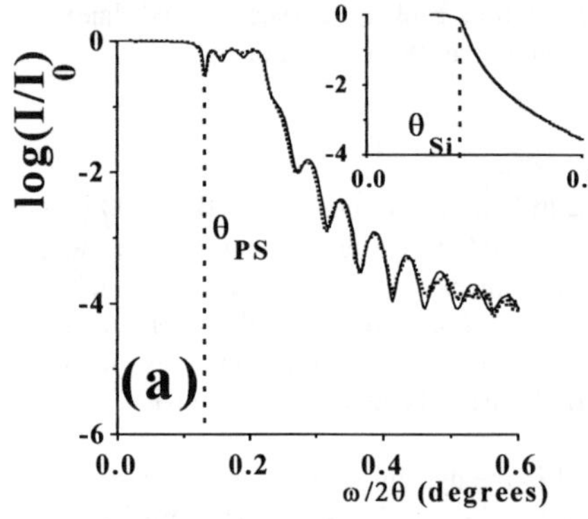

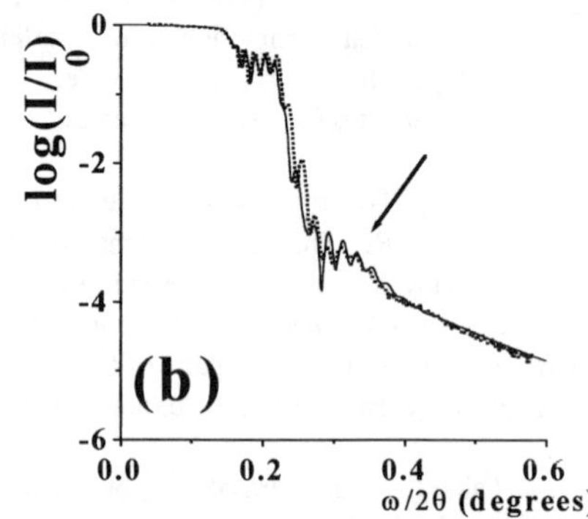

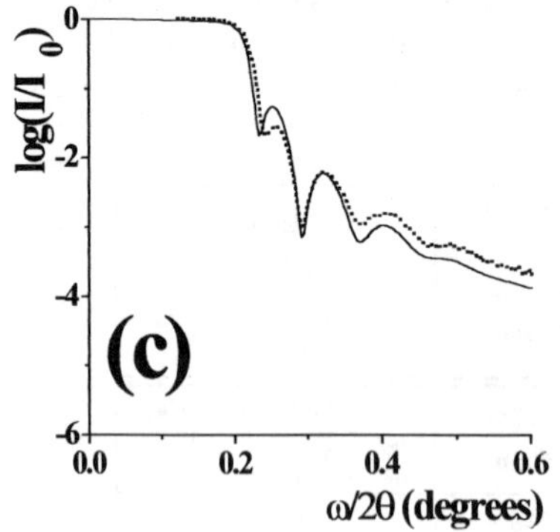

Figure 2: Experimental specular reflected intensity (dotted line) and simulated curve (solid line) for grazing incidence angle:
(a) for a p (65%) sample.
(b) for a p^+ (60%) sample showing a double fringe system; the arrow indicates the main fringe coming from the surface film.
(c) for a p^+ (36%) type sample: only the fringes produced by the surface film are observed in spite of the presence below of a 44 nm porous layer.

P(65%): The experimental reflectivity curve (dotted line) of the specular reflected intensity of a p (65%) sample is plotted on Fig. 2a. With ω increasing from 0 to 0.6°, one observes several features:

-for $\omega < \theta_{PS}$ there is a total reflection, (θ_{PS} is the critical angle of the porous layer, smaller than θ_{Si} as the porous layer is less dense than bulk Si).

-for $\theta_{PS} < \omega < \theta_{Si}$, the X-ray beam goes into the porous layer but is reflected by the bulk substrate. The value of θ_{PS} leads to a good estimation of the layer density.

-for $\omega > \theta_{Si}$ the reflected intensity strongly decreases because the X-rays penetrate into the substrate. The observed fringes are due to the interferences between the beam reflected by the (air/PS) interface and by the (PS/ bulk silicon) interface and the fringe period is determined by the thickness of the PS layer. The oscillation decay after a few periods indicates a substantial roughness at one or both interfaces. This experiment was simulated (solid line) by using a reflectivity routine in the dynamical simulation program. In order to have the good agreement of Fig. 2a, it is necessary to introduce a very thin porous layer (IF) at the (PS/substrate) interface (table I). It is also necessary to take into account the roughnesses of interfaces. The roughness of the top surface with a value of 2.8 nm corresponds approximately to the initial roughness of the bulk silicon substrate, whereas the internal interface roughness is in agreement with the value estimated by Chason et al [11] in a previous X-ray reflectivity study of p type porous silicon. Studies of p (65%) type samples for thicknesses going from 10 nm up to 130 nm have also been performed, and reveal the linearity of the layer thickness versus the anodisation time which is also in agreement with [11].

P^+60%: The first results on a p^+ (60%) type sample are presented on Fig. 2b, showing unexpectedly the superposition of two fringes systems:

-one system of narrow fringes corresponds to the expected porous layer.

-another system with a larger period, which corresponds to a thinner layer probably at the top surface of the sample. The solid line of Fig. 2b corresponds to a simulation obtained with three layers from bottom to top: an interfacial layer with the substrate (IF), the main porous silicon layer (PS) and a surface film (SF). The calculated parameters are given in table I. Morover the surface film is maybe inhomogenous according to the observation of only one period of the broad fringes.

p^+ type samples of various porosities: A systematic study as a function of the porosity, reveals that the observation of surface film fringes depends on the porosity value: for the higher porosities ($P \geq 70\%$) one can only observe the interference system of the porous layer, whereas for lower porosities ($P < 50\%$) one observed mostly the broad fringes of the thin surface film. Indeed for the p^+ (36%) sample of Fig. 2c, only the well defined fringes produced by a surface film of about 44 nm thick are visible. Then the reflected intensity appears to be no more sensitive to the presence of the usual porous layer (which still exists as shown by measurements with a profilometer). The fact that the porous layer does not influence the reflectivity curve is probably due to the large roughnesses of the (PS/substrate) interface and surface film. Actually the reflectivity results are more sensitive to the interfaces than diffraction: this explains why thickness fringes of PS layer are observed in diffraction and no more in reflectivity. Indeed the solid line of Fig. 2c was obtained with a roughness σ_{IF} of about 15 nm at the PS/substrate interface. This value corresponds to the value of the heterotransition estimated on a p^+ 36% sample by the X-ray diffraction dynamical theory (see table I). The simulation shows that the surface film porosity is quite different from the porous layer.

For several p^+(36%) samples, with thickness increasing from 10 nm to 1000 nm, only the surface film fringes are observed with parameters similar to those given in table I.

Furthermore a study on a p^+(36%) sample using the grazing incident diffraction at the European Synchrotron Radiation Facilities (Grenoble-France) have been performed: diffraction measurements along the <2*kl*> direction around the (200) reflection have been performed and reveal several thickness fringes. The thickness estimated with the period of this system fringes is around 50 nm corresponding to the surface film. This observation in grazing diffraction gives also evidence of the single crystal nature of the surface film.

It is interesting to remember that in 1980, Unagami [12] reported the existence of a surface film above the porous layer in p^+ type samples: a maximum film thickness of 200 nm was observed in p^+(35%) samples obtained from (111) wafers of 0.02 Ωcm resistivity. Although these observations were quite detailed, it seems that only few other observations of this film have been reported. Our X-ray results are probably due to the presence of a similar surface film.

CONCLUSION

X-ray diffraction and reflectivity measurements on thin porous silicon layers (of p and p^+ type) are reported showing the good lateral homogeneity of the layer thickness. The X-ray diffraction appears to be more sensitive to the porous layer volume and gives direct evidence that thin porous silicon layers are single crystal. The X-ray reflectivity gives more accurate information about surface and interfaces. These two non destructive techniques appear to be complementary, especially when the experimental results are compared with simulations. This enables one to get a quantitative analysis of parameters such as porosity, thickness, lattice parameter and the interfaces heterotransition. The present investigation show for instance that p^+ type samples exhibit a film layer at the top of the porous silicon layer which does not exist for p type samples.

ACKNOWLEDGEMENTS

We thank warmly F. Muller for his fruitful discussions and G. Martin-Granel for his help during the X-ray measurements.

REFERENCES

1. L.T. Canham, Appl. Phys. Lett. **57**, 1046 (1990).
2. Eur. Mat. Res. Soc. Symp. Proc. Thin Solid Film vol. **276** (1996); Mat. Res. Soc. Proc. Symp. vol. **358** (1995).
3. D. Bellet and G. Dolino, Thin Solid Films **276**, 1 (1996).
4. D. Buttard, D. Bellet and G. Dolino, J. Appl. Phys. **79,** 8060 (1996).
5. G. Dolino, D.Bellet and C. Faivre, Phys. Rev. B (December 1996), to be published.
6. A.A. Lomov, D. Bellet and G. Dolino, Phys. Stat. Sol. (b) **190**, 219 (1995).
7. D. Buttard, D. Bellet and T. Baumbach, Thin Solid Films **276**, 69 (1996).
8. A. Loni, L.T. Canham, M.G. Berger, R. Arens-Fisher, H. Munder, H. Luth, H.F. Arrand, T.M. Benson, Thin Solid Films **276**, 143 (1996).
9. T. Baumbach, H.-G. Brühl, H. Rhan and U. Pietsch, J. Appl. Crystallogr. **21**, 386 (1988).
10. A. Schalchli, J.J. Benattar and C. Licoppe, Eur. Phys. Lett. **26**, 271 (1994).
11. E. Chason, T.R. Guilinger, M.J. Kelly, T.J. Headley and A.J. Howard, Mat. Res. Soc. Symp. Proc. **358**, 321 (1995).
12. T. Unagami, J. Electrochem. Soc. **127**, 476 (1980).

NUCLEAR MAGNETIC RESONANCE (NMR) OF POROUS SILICON

W.K. CHANG, M.Y. LIAO AND K. K. GLEASON
Department of Chemical Engineering, MIT, Cambridge MA 02139, kkgleasn@mit.edu

ABSTRACT

Porous silicon (PS) was characterized by ^{1}H, ^{19}F and ^{29}Si solid-state nuclear magnetic resonance (NMR). On freshly prepared samples, hydrogen contents were between 3 x 10^{14} and 3 x 10^{15} per cm^2 of PS surface area, while fluorine concentrations were below the detection limit. Cross-polarization (CP) was used to selectively observe the ^{29}Si near the hydrogen passivation. The features of the ^{29}Si NMR spectra are assigned as $(O)_2(Si)Si\text{-}H$ (-50 ppm), $(O)_3Si\text{-}H$ (-84 ppm), $(Si)_3Si\text{-}H$ (-91 ppm), $(Si)_2Si\text{-}H_2$ (-102 ppm) and $(O)_4Si$ (-109 ppm). Changes resulting from low temperature annealing in air and an HF soak were observed by both NMR and infrared spectroscopy. The ^{29}Si NMR line widths for PS fall between those for crystalline silicon and those for amorphous hydrogenated silicon films (a-Si:H), suggesting disorder near the PS surface is intermediate between these extremes. However, comparison of the isotropic chemical shift values shows that the bonding in the disordered regions of PS differs from that found in a-Si:H. In addition, the sharp ^{29}Si NMR resonance observed in the bulk single crystal starting material can not be resolved in the spectra of PS. Thus, well-ordered silicon nanocrystallites in the PS are either several bond-lengths removed from hydrogen or comprise only a small fraction of the PS layer.

INTRODUCTION

Studying porous silicon passivation presents several challenges. Standard surface sensitive techniques will observe the top wafer surface rather than the pore walls. Depth profiling will detect the average PS composition but will not favor the detection of passivation species near the pore wall. In addition, PS can age upon exposure to ambient air, altering its surface passivation.[1]

NMR is often not considered to be a surface characterization technique and has not previously been used to study PS. The high surface area of PS allows the inherently low sensitivity of the NMR technique to be overcome. NMR generally does not discriminate between the surface and the bulk. Indeed, because the number of nuclei in the bulk of a solid is generally vastly greater than the number at the surface, NMR spectra often reflect bulk structure, However, the concentration of ^{1}H and ^{19}F in bulk silicon is well below the detection limit of NMR spectroscopy. Thus, direct polarization of ^{1}H and ^{19}F NMR reflects only the PS passivation.

Cross-polarization (CP) can be used to favor the detection of ^{29}Si near ^{1}H or ^{19}F.[2] The CP transfer rate is generally proportional to the inverse sixth power of the ^{1}H to ^{29}Si internuclear distance. However, the efficiency of polarization will be reduced if the ^{1}H-^{29}Si dipole couplings are averaged by molecular motion. In addition, CP enhances the sensitivity of low gyromagnetic ratio nuclei (such as ^{29}Si) having slow spin-lattice relaxation rates. Magic-angle spinning (MAS) is often combined with CP to further enhance NMR sensitivity and resolution via line narrowing.[3]

In this work, the assignment of the features of the ^{29}Si NMR spectra of PS are addressed. Changing the passivation of the as-prepared PS by low temperature annealing in air and by an HF soak aid in this assignment. For this work, a thick, high surface area, photoluminescent PS layer will be used. The presence of hydrogen only on the surface has also allowed the passivation of PS to be extensively studied by Fourier-transform infrared spectroscopy (FTIR) [4,5,6,7,8,9,10] and may provide insight into the ^{29}Si NMR results. Thus, complementary FTIR characterization will be performed immediately after the NMR measurements.

Mat. Res. Soc. Symp. Proc. Vol. 452

EXPERIMENTAL

In preparation for the CP experiments, ^{1}H and ^{19}F direct polarization spectra were acquired on several freshly prepared PS samples. By using rectangular sections of PS layers intact on their silicon wafer substrate, background signals were minimized.[11]

For ^{29}Si NMR, porous Si was made from 4" diameter (100) p-type Si wafers with resisitivities of 1-10 Ω cm, a current density of 20 mA/cm^2, and an anodization time of 90 minutes. The 30 wt% HF electrolyte was prepared by diluting 48 wt% aqueous HF with ethanol. A few seconds of increased current density at the end of the anodization produced delamination [12] of the PS layer which was collected by filtration for MAS. After anodization, the PS was washed with distilled water. About two months elapsed between the growth and the initial characterization. Upon completion of these measurements, the as-prepared PS sample was divided into two parts. The first one was oxidized in air at ~ 200° C for 4 hours. After NMR and FTIR measurements were taken, this first part was further oxidized for about 60 hours at the same temperature, and the corresponding changes in NMR and FTIR were compared. The second part of the sample was reduced by soaking it in HF for 10 minutes, followed by drying in air without being washed by water. NMR measurements were taken immediately after the HF treatment in order to minimize any signal originating from the oxidation of the sample in air. FTIR of the same sample was taken after the NMR characterization.

All NMR spectra were obtained using a home-built solid-state NMR spectrometer with a ^{29}Si resonance frequency of 53.64 MHz. For CP-MAS NMR, the PS samples were spun at a frequency ~ 2900 Hz. The 90° pulse time was 4 μs, the cross polarization 3 ms and the dipolar decoupling time was 10 ms. Also, since the recycle delay (40 s) between signal averages is determined by the recovery of proton magnetization, the shorter T_1 constants of the proton resonance allow for more signal averages in a given time, also leading to improved sensitivity. All ^{29}Si NMR spectra are plotted in ppm from the tetramethylsilane (TMS) reference.

RESULTS

The as-prepared PS layer used for ^{29}Si for was 110 μm thick and had a Brunauer-Emmett-Teller (BET) surface area of 893 m^2/g. The Barret-Joyner-Halenda (BJH) desorption method revealed a monodisperse pore size distribution with an average diameter 42 Å. The photoluminescence spectrum showed a single feature centered at ~700 nm, having a FWHM of approximately 150 nm.

Proton NMR of freshly prepared samples measured H concentrations ranging from $3x10^{14}$ to $3x10^{15}$ per cm^2 of PS surface area. The density of surface atoms on the different crystallographic faces of silicon is between $6x10^{14}$ and $1x10^{15}$ per cm^2 of surface area. [13] Thus, the hydrogen concentration range is equivalent to 0.5 to 3 monolayers and is sufficient for cross-polarizing nearby ^{29}Si nuclei. In contrast, the F content of all samples was found to be below our NMR detection limit. Thus, ^{19}F cross-polarization was not attempted.

The ^{1}H NMR of all PS examined showed a single peak with no resolved features having a full width at half maximum (FWHM) of ~3 kHz. If hydrogen directly terminated exposed planes of crystalline silicon, a much broader line would be anticipated. For example, a fully hydrogen passivated Si(100) surface would have a 41 kHz FWHM as calculated by the a Van-Vleck second-moment calculation for dipolar broadening. The narrower observed linewidth indicates the presence of motional averaging and/or increased interproton spacing. Water adsorbed in the pores could account for the narrow line width. However, little intensity was observed from ~3000 cm^{-1} to ~ 3700 cm^{-1} in the FTIR spectra, indicating low concentrations of OH and the absorbed water.[5,10] The absence of C-H stretching modes at ~2800 cm^{-1} showed contamination by adsorbed hydrocarbons was not detected. Motional averaging involving the hydrogen containing species identified by FTIR seems unlikely. In order to increase average interproton spacing while maintaining the same overall areal concentration, hydrogen would need to be distributed several atomic layers deep into the PS wall, This hypothesis is consistent with the idea of a passivation "tissue" rather than a well-defined surface passivation layer. One model which has been proposed for the structure of PS is that nanocrystals are surrounded by passivation "tissue" containing hydrogen and/or oxygen.[14] The presence of silicon nanocrystallites in PS has been supported by X-ray diffraction.[15]

Fig. 1 compares the FTIR Si-H stretch region for the as-prepared, reduced and oxidized PS. In the as-prepared PS (Fig. 1b), four absorptions are clearly identifiable: $(Si)_3Si\text{-}H$ (2090 cm^{-1}), $(Si)_2Si\text{-}H_2$ (2116 cm^{-1}), $(O)_2(Si)Si\text{-}H$ (2195 cm^{-1}) and $(O)_3Si\text{-}H$ (2248 cm^{-1}).[4,5] As expected, the first two become more dominant upon reduction (Fig. 1a), while the latter two are relatively more intense after oxidation (Fig. 1c). As expected, the $(O)_3Si\text{-}H$ mode becomes more intense relative to the $(O)_2(Si)Si\text{-}H$ as annealing time increases from 4 to 60 hours. Both $(O)(Si)_2Si\text{-}H$ and $(Si)Si\text{-}H_3$ stretch modes appear at ~ 2134 cm^{-1}.[5,6,7]

The ^{29}Si CP-MAS NMR spectra of the as-prepared, oxidized and reduced PS are also shown in Fig. 1. The as-prepared PS spectrum (Fig. 1e) contains two main regions. First, there is a relatively sharp feature centered at ~ -50 ppm, with FWHM ~ 6 ppm. The second region, ranging from ~ -80 ppm to ~ -115 ppm. Both regions may represent several overlapping peaks. Also indicated is the -81 ppm position of ^{29}Si NMR resonance from the single crystal starting material, acquired by direct polarization MAS. As expected for a highly ordered material, the single crystalline Si peak was sharp (FWHM< 2 ppm). Upon reduction, the ^{29}Si NMR spectrum of PS (Fig. 1d), the first region at ~ -50 ppm is no longer detected. In addition, the line shape of the remaining region changes. Shoulders, at approximately -84 ppm and -109 ppm, observed in the as-prepared material (Fig. 1e), were both diminished by the HF soak. The opposite intensity changes occur upon 4 hours of oxidation, as can be seen in Fig. 1f. The features at centered approximately -50, -84 and -109 ppm predominate after oxidation while the intensity in between the two shoulders declines. After 60 hours of oxidation, the same features are present with a slight increase in the proportion of the feature at -84 ppm. The additional treatment also degrades the signal-to-noise of the spectrum.

Standard CP was combined with Polarization Inversion (CPPI) to further help resolve and assign the components of the ^{29}Si MAS NMR spectra.[16] Following an ordinary CP period, a second CP period is applied, having a phase of 180° with respect to the first. As this inversion period proceeds, the intensity of a given peak will become less positive until zero intensity is observed. After this crossover point, the intensity will become increasingly negative. The inversion rate of the magnetization reflects the strength of the heteronuclear dipolar coupling between ^1H and ^{29}Si. In the absence of motional averaging, the shorter the $^1H\text{-}^{29}Si$ internuclear distance, the faster the expected inversion rate of the corresponding resonance.

The zero-intensity cross-over point occurs after only 240 µs of inverse polarization at approximately -50, -84, and -102 ppm. This rapid inversion is expected for short $^1H\text{-}^{29}Si$ internuclear distances. The component at ~ -109 ppm gives the slowest inversion, requiring ~960 µs before the zero intensity crossover point is observed. The slow inversion rate suggests hydrogen is not directly bonded to the Si in this species. Intermediate inversion rates are obtained for magnetization lying between -90 and -100 ppm, with null intensity observed at ~480 µs of inversion time.

Table I summarizes the NMR assignments discussed below. Comparison of the FTIR and NMR spectra (Fig. 1) shows that the reduced species present in the FTIR, $(Si)_3Si\text{-}H$ and $(Si)_2Si\text{-}H_2$, must appear in the NMR spectra at less than -80 ppm. The CPPI experiments suggest resonances centered at ~ -91 and ~ -102 ppm. The CPPI inversion is faster at ~ -102 ppm, suggesting its assignment as the dihydride rather than the monohydride. Also, the upfield (more negative) shift from crystalline Si (~ -81 ppm) and its magnitude (~ -10 ppm per H) are consistent with those observed for the small reference compounds. For $(H_3Si)_3SiH$ and $(H_3Si)_2SiH_2$, σ is -96.0 and -116.5 ppm, respectively.[17] A difference of 15 ppm between Ph_3SiH and Ph_2SiH_2 has been reported.[18]

The reduced PS ^{29}Si NMR spectrum (Fig. 1d) shows no evidence of the sharp bulk Si resonance at -81 ppm. This suggests any remaining well-ordered nanocrystalline regions are

Table I. FTIR and ^{29}Si NMR Assignments

	Configuration	FTIR (cm^{-1})	σ (ppm)	CPPI null time(ms)
Reduced Surface	$(Si)_3$ Si-H	2090	-91	480
	$(Si)_2$ Si-H_2	2116	-102	240
Oxidized Surface	$(O)_2$ (Si) Si-H	2195	-50	240
	$(O)_3$ Si-H	2248	-84	240
	$(O)_4$ Si	--	-109	960

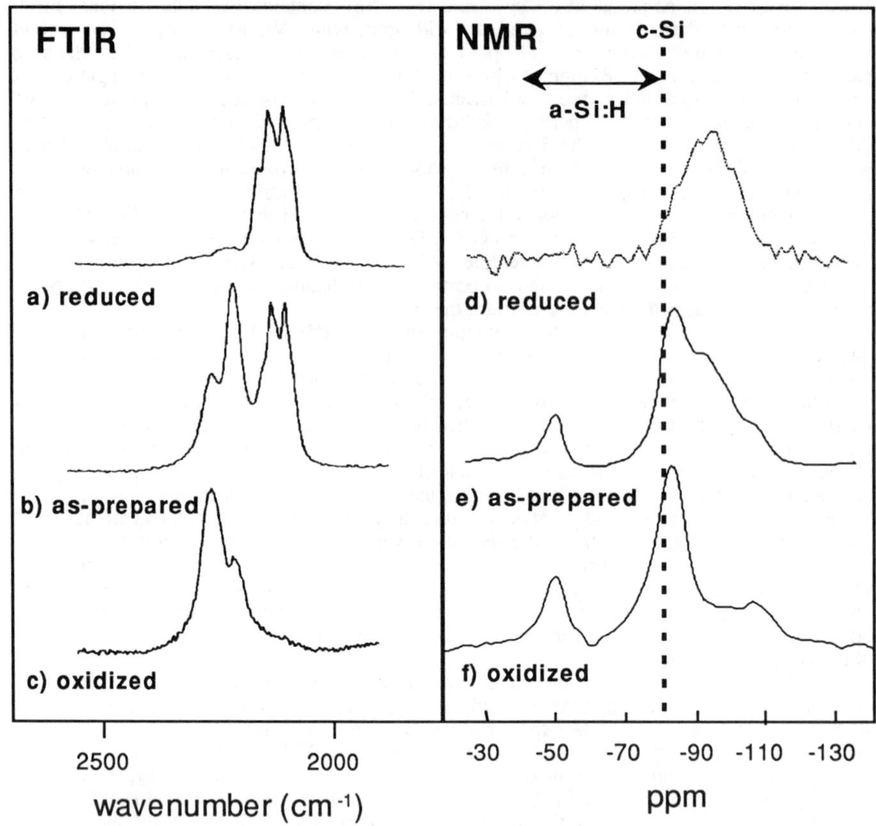

Figure 1. The FTIR Si-H stretch region (a-c) and ^{29}Si CP-MAS NMR (d-f) for reduced, as-prepared and oxidized porous silicon.

several bond lengths removed from hydrogen passivation or that such nanocrystalline regions comprise only a small fraction of the photoluminescent PS layer. In support of the latter hypothesis, strong luminescence has been observed in a PS sample which showed no evidence of crystallinity in its transmission electron diffraction pattern.[19]

The oxidized NMR spectrum (Fig. 1f) also shows no evidence of sharp crystalline signal, although, in this case, a sharp -81 ppm peak could be obscured by overlapping features. Based on its slow inversion rate and presence upon oxidation (Table 1), the peak at ~ -109 ppm is assigned as SiO_2 units which are close to the hydrogenated surface. This agrees with the chemical shift assignment for SiO_2 in silica gel and bulk glass.[20,21]

In the oxidized PS, there is low intensity in the 90-100 ppm NMR region where $(O)_2(Si)Si-(OH)_2$, $(O)_3Si-OH$ appear in silica gel. This is consistent with the low hydroxyl content indicated in the FTIR spectra at >3000 cm^{-1}. The $(O)_2(Si)Si-H$ (2195 cm^{-1}) and $(O)_3Si-H$ (2248 cm^{-1}) FTIR features must correspond to the two remaining NMR resonances at -50 and -84 ppm. Assuming, as for other silicates [20], that the more negative chemical shift should be associated with more nearest neighbor oxygens, the -84 ppm peak is assigned as $(O)_3Si-H$. In support of this assignment, both the 2248 cm^{-1} FTIR peak and the -84 ppm NMR peak become relatively more intense upon further oxidation.

The NMR (Fig. 1e) and FTIR (Fig. 1b) of the as-prepared PS shows both reduced and oxidized species, revealing that the as-prepared surface is partially oxidized. The reduced species most likely exist when the PS is removed from the HF electrolyte. The observed partial oxidation of the surface probably results from the subsequent water rinsing and air drying followed by exposure to room temperature air. When no water rinsing follows the post-anodization HF dip and the exposure to air is limited to a few hours, little surface oxidation is observed by NMR (Fig. 1e) or IR (Fig. 1a).

While the Si-H assignments in PS (-91 ppm for $(Si)_3Si-H$ and -102 for $(Si)_2Si-H_2$,) fall upfield of crystalline silicon, amorphous hydrogenated silicon films (a-Si:H) spectra show shifts in the opposite direction.[22,23,24,25] Variation in both hydrogen content (~5-50 at.%) and local order contribute to the wide rate of ^{29}Si isotropic chemical shifts for a-Si:H. For example, the -38 ppm shift of as-deposited a-Si:H film becomes increasingly negative with annealing until a recrystallization peak was observed at -79.9 ppm.[25] Annealing reduces both hydrogen content and disorder in a-Si:H. In PS, the absence of Si-H peaks downfield from c-Si demonstrates PS does not contain amorphous hydrogenated silicon regions comparable to those found in deposited films. This is a significant observation, as the luminescence behavior of PS has been discussed in relationship to the optical behavior of a-Si:H.[14]

In a-Si:H films, the ^{29}Si line widths are quite broad (30 to 100 ppm), even upon applying MAS. In part, this reflects the multiplicity of bonding environments, including $(Si)_4Si$, $(Si)_3Si-H$, $(Si)_2Si-H_2$ and $(Si)SiH_3$. In addition, considerable charge transfer can arise from disorder. [25] A distribution of bond angles and bond lengths will cause the chemical shift to be distributed over a range of values. While fast MAS spinning can average many line broadening interactions, it does not reduce chemical shift dispersion arising from disorder.[22]

The ^{29}Si NMR CP-MAS line widths for PS (~6 ppm) fall between those for crystalline silicon and those for a-Si:H, suggesting the disorder at the PS surface is intermediate between these extremes. This agrees with the observation that the FTIR vibrations at ~2100 cm^{-1} are much broader in PS than for hydrogenated crystalline Si surfaces.[26]

CONCLUSIONS

We have shown that solid-state NMR spectroscopy can characterize the bonding environments on PS surfaces. Surface concentrations equivalent to 0.5 to 3 monolayers of hydrogen were found by ^1H NMR. However, the relatively narrow ^1H dipolar linewidth (3 kHz FWHM) suggests that the hydrogen is distributed through several atomic layers of passivation "tissue" rather than forming a well-defined surface passivation layer. Fluorine concentrations were below the ^{19}F NMR detection limit. The ^1H nuclei were used to selectively cross-polarize ^{29}Si nuclei. For ^{29}Si NMR, isotropic chemical shift assignments were made of $(O_2)(Si)Si-H$ (-

50 ppm), $(O)_3$Si-H (-84 ppm), $(Si)_3$Si-H (-91 ppm), $(Si)_2$Si-H (-102 ppm) and $(O)_4$Si (-109 ppm).

Many of the observations made by the NMR measurements confirm those made by earlier infrared spectroscopy studies. However, two new observation were made via the NMR characterization. First, even though some disorder is apparent at the PS surface, no regions with NMR chemical shifts comparable to those in hydrogenated a-Si:H films were observed. Second, well-ordered silicon nano-crystallites were not observed by cross-polarization from hydrogen. This suggests that either that the crystallites are several bond-lengths removed from hydrogen (even for an HF treated surface) or nanocrystallites comprise only a small fraction of the PS layer.

ACKNOWLEDGMENTS

We would like to thank Professors Jackie Ying and Kurt Kolenbrander for generously allowing use of their FTIR and PL equipment, respectively. The support of the NSF under contracts DMR-9321224 and CTS-9411656 is gratefully acknowledged.

REFERENCES

1. T. Ito and A. Hiraki, J. Lumin. **57**, 331 (1993).
2. G. E. Maciel "The Characterization of Silica Surfaces by NMR." in Encyclopedia of NMR, edited by D.M. Grant and R.K. Harris (Wiley, New York, 1996).
3. C. A. Fyfe, Solid-state NMR for Chemists, (CFC Press, Guelph, Canada, 1983).
4. A. Borghesi,G.G. Guizzetti, A. Sassella, O. Bisi, and L. Pavesi, Solid State Comm. **89**, 615 (1994).
5. Z.C. Feng, A.T.S. Wee, and K.L. Tan, J. Phys. D: Appl. Phys. **27**, 1968 (1994).
6. É.B.Vázsonyi, M. Koós, G. Jalsovszky, I. Pócsik, Thin Solid Films **255**, 121 (1995).
7. P. O'Keefe and Y. Aoyagi, Appl. Phys. Lett. **66**, 836 (1995).
8. A. Borghesi, S. Sassella, B. Pivac, L. Pavesi, Solid State Comm. **87**, 1 (1993).
9. Y. Ogata, H. Niki, T. Sakka, and M. Iwasaki, J. Electrochem. Soc. **142**, 195 (1995).
10 P. Gupta, A.C. Dillon, A.S. Bracker, and S.M. George Surf. Sci. **245**, 360 (1991).
11. D.H. Levy and K.K. Gleason, J. Vac. Sci. Technol. A **11**, 195 (1993).
12. N. Ookubo, Y. Matsuda, and N. Kuroda, Appl. Phys. Lett. **63**, 346 (1993).
13. S. Wolf and R.N. Tauber, Silicon Processing for the VLSI Era: Volume 1: Process Technology (Lattice Press: Sunset Beach, California, 1996), p. 647.
14. F. Koch, Mat. Res. Soc. Symp. Proc. **298**, 319 (1983).
15. S. Schuppler, et al. Phys. Rev. Lett. **72**, 2648 (1994).
16. N.J. Zumbulyadis, Chem. Phys. **86**, 1162 (1987).
17. S. Hayashi, K. Hayamizu, A. Yamasaki, A. Matsuda, and K. Tanaka, Phys. Rev. B **35**, 4581 (1987).
18. E.A. Williams and J.D. Cargiollia, Ann. Rep. NMR Spectros. **9**, 221 (1962).
19. K.H. Jung, S. Shih, D.L. Kwong, C.C. Cho, and B.E. Gnade, Appl. Phys. Lett. **61**, 2467 (1992).
20. G. Engelhardt and H. Koller in Solid-State NMR II, Inorganic Matter, vol.. 31, edited by B. Blümlich (Springer-Verlag, Berlin, 1994), pp. 1-31.
21. X.-D. Cong, R.J. Kirkpatrick, and S. Diamond, Cem. Concr. Res., **23**, 811 (1993).
22. J.A. Reimer, P. Dubois Murphy, B.C. Gerstein, and J.C. Knights, J. Chem. Phys. **74**, 1501 (1981).
23. F.R. Jeffrey, P. Dubois Murphy and B.C. Gerstein, Phys. Rev. B **23**, 2099 (1981).
24. W.K. Cheung and M.A. Petrich, M.A. J. Appl. Phys. **73**, 3237 (1993).
25. W.-L. Shao, J. Shinar, B.C. Gerstein, F. Li, J.S. Lannin, Phys. Rev. B **41**, 9491 (1990).
26. G.S. Higashi, Y.J. Chabel, G.W. Trucks, K. Raghavachari. Appl. Phys. Lett. **56**, 656 (1990).

OPTICAL PROPERTIES OF DEUTERIUM TERMINATED POROUS SILICON

Takahiro Matsumoto*, Yasuaki Masumoto*, and Nobuyoshi Koshida**

*Single Quantum Dot Project, ERATO, Japan Science and Technology Corporation, 5-9-9 Tokodai, Tsukuba 300-26, Japan, tomato@sqdp.trc-net.co.jp

**Division of Electronic and Information Engineering, Tokyo University of Agriculture and Technology, Koganei, Tokyo 184, Japan

ABSTRACT

We have studied the optical properties of deuterium-terminated porous silicon. The photoluminescence spectrum was different from that of usual hydrogen-terminated porous Si despite porous Si showing both the same structure and the same absorption spectrum. These results indicate that the surface vibration of terminated atoms couples to the quantum confined states.

INTRODUCTION

Porous silicon formed by electrochemical anodization has attracted a lot of interest recently because it exhibits strong visible photoluminescence at room temperature [1]. Although a great deal of effort has been made to elucidate the origin of the photoluminescence, much of the uncertainty still remains due to the difficulty of physical characterization in nanometer structure of porous Si.

Isotopes are powerful tools for searching molecular terminations, kinetics of atoms or molecules, or the interaction between nuclei and extra nuclear electrons. Especially, hydrogen and its isotope are widely used to study the surface structure [2] and the diffusion of hydrogen in crystalline Si [3]. A huge amount of terminating hydrogen is supplied from hydrofluoric acid (HF) in porous Si [4,5], therefore by using isotope electrolyte such as deutrofluoric (DF) acid, the surface terminations can be changed from hydrogen to deuterium.

In this paper, we report the optical properties of isotope terminated porous Si. We have fabricated deuterium-terminated porous Si (D-PS) using DF-ethanolD6 solution ($DF:C_2D_5OD:D_2O=1:1:2$). We have characterized the D-PS together with the usual hydrogen-terminated porous Si (H-PS) by Raman spectroscopy and transmission electron microscope (TEM) analysis. In spite of the same nanometer structure and the same optical absorption spectrum for both samples, D-PS shows shorter PL than that of H-PS. These results can be analyzed by considering both the bandgap upshift due to the quantum size effect and the energy reduction due to the coupling of the confined carriers to the surface vibration of terminated atoms on nanocrystals.

Mat. Res. Soc. Symp. Proc. Vol. 452

EXPERIMENTAL DETAILS AND RESULTS

Hydrogen-terminated porous Si (H-PS), and deuterium-terminated porous Si (D-PS) were formed by electrochemical anodizations. The H-PS was fabricated in the dark to avoid oxidation using HF-ethanol solution ($HF:H_2O:C_2H_5OH=1:1:2$) by applying positive bias to a p-type 3-5 Ωcm Si substrate at various current densities in the range from 20 mA/cm^2 to 100 mA/cm^2. (Samples fabricated with high current densities show wide bandgap energy [6].) The D-PS were also fabricated under the same conditions except for the use of DF-ethanol-D6 solution ($DF:C_2D_5OD:D_2O=1:1:2$). Both the porous layers (H-PS and D-PS) have the same thickness such as 20 μm for 20 mA/cm^2 and 30 min anodization. The microstructures and the surface chemistries of these porous Si films were studied using high resolution transmission electron microscope (TEM) analysis, Raman spectroscopy, and Fourier transform infrared (FTIR) measurements [7]. Our TEM observations indicate that the statistically averaged diameters of H-PS and D-PS were 3.1 nm and 3.3 nm, which agrees well with the estimation from broad (15 cm^{-1}) and downshift (6 cm^{-1}) Raman spectra of these samples. The surface chemistries of H-PS and D-PS were characterized by the stretching- and the scissor-modes of absorption peaks. In the H-PS samples, absorbance due to the stretching mode of Si-H species at around 2100 cm^{-1} and the scissor modes of Si-H_2 at around 910 cm^{-1} can be observed. In the D-PS samples, both the stretching modes shift to around 1530 cm^{-1} and the scissor modes of Si-D_2 to 657 cm^{-1} due to deuterium termination. As there is no Si hydride-peak observed in D-PS, the surface of Si nanocrystals can be regarded as completely deuterated by the DF-ethanolD6 electrolyte.

Figure 1 shows the optical absorption and the photoluminescence (PL) spectra of H-PS and D-PS samples at room temperature. The PL spectra of these porous samples were measured in a vacuum using 325-nm excitation light from a He-Cd laser. The calibration of the spectral sensitivity was performed using a standard lamp. Despite the absorption spectra of both PS samples being the same, the peak wavelengths of the PL spectra were different; 800 nm for H-PS and 760 nm for D-PS. The PL intensities were almost the same for H-PS and D-PS samples. (The number of dangling bonds of H-PS was almost equal to that of D-PS samples, which was confirmed by electron spin resonance experiments.) With increasing dissolution current density, both H-PS and D-PS show higher energy shift of the absorption and the PL spectra. Figure 2 shows the relation between the bandgap energy E_g determined from the absorption spectra [8] and the PL peak energy E_p of H-PS (solid circle) and D-PS (open square) dissolved with various current densities in the range from 20 mA/cm^2 to 100 mA/cm^2. At the same bandgap energy, the PL peak energy of D-PS is always higher than that of H-PS as shown in Fig. 2.

The remarkable point of D-PS samples is that replacing hydrogen with deuterium greatly reduces the PL degradation of porous Si. Figure 3 shows a comparative time-dependent degradation of PL intensity for H-PS and D-PS. As shown in this figure, the PL stability improves by factors of 50 by replacing hydrogen with deuterium. This feature is very promising for the fabrication of optoelectronic devices from the point of the compatibility of carrier injections with the device stability. The reason for the improvement of PL stability is not clearly understood,

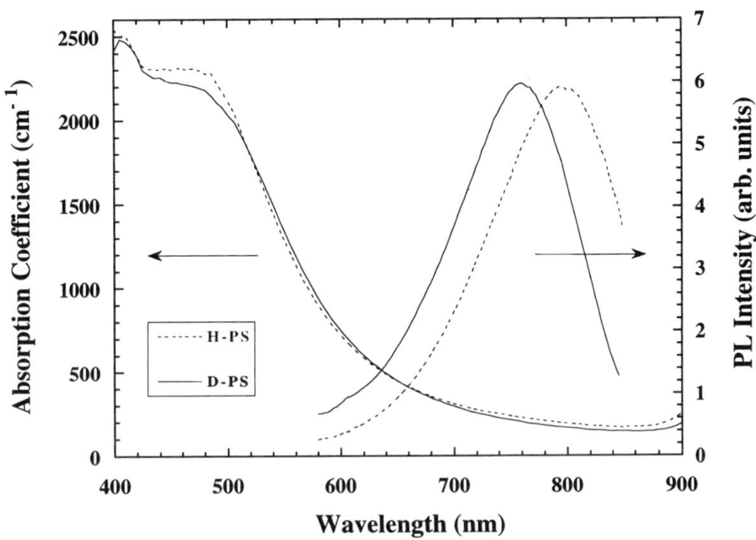

Fig. 1 Optical absorption and photoluminescence spectra of H-PS and D-PS.

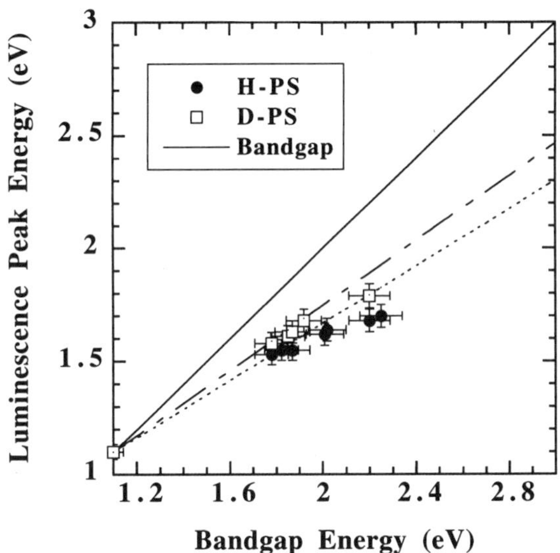

Fig. 2 PL peak energy-plots as a function of the bandgap energies in H-PS (solid circle) and D-PS (open square). These experimental plots are theoretically fitted with the chain line (H-PS) and the broken line (D-PS). (The solid line shows the bandgap energy.)

however; this could be related to the difference of zero point energy between hydrogen and deuterium or to the mass dependence of tunneling of the nuclei [3,9].

DISCUSSION AND ANALYSIS

Despite H-PS and D-PS having the same nanostructure and the same bandgap energy, the observed peak energies of PL are different. This cannot be explained by the simple quantum confinement effects, because the electronic energy states related to coordinate Si atoms, such as the bandgap of the nanocrystal, exciton confinement energies, or the defect states, are expected to be the same for both hydrogen- and deuterium-surface passivations. Furthermore, the electronic states of molecules are also the same for H- and D-terminations [10]. However, the above results show that the surface terminated species around Si nanocrystal crucially modify the electronic states of nanocrystals.

These results can be quantitatively analyzed by considering both the bandgap upshift due to the quantum confinement effect and the energy relaxation due to the coupling of the confined carriers to the vibration of the surface-terminated species around Si nanocrystals. As longitudinal motion can only couple with conduction electron, we consider stretching modes which deform the surface of Si nanocrystals. We adopt a Hamiltonian with deformation potential as,

$$\hat{H} = -\frac{\hbar^2}{2m^*}\partial_z\partial_z + \sum_q \hbar\omega_s\, b_q^+ b_q + \sum_q \left[V_q b_q e^{iqz} + V_q^* b_q^+ e^{-iqz} \right] , \tag{1}$$

where ω_S is the stretching frequency of the surface-terminated atoms on Si nanocrystals, m* is electron mass, and V_q is the one dimensional deformation potential.

The energy of the coupling state can be calculated by applying Lee-Low-Pines transformations [11], and it turns out to be

$$E = E_{gq} - \alpha\hbar\,\omega_S \ , \tag{2}$$

where E_{gq} is the bandgap energy of confined electrons, and α is the coupling constant. Up-shifted levels due to the quantum size effect (E_{gq}) are reduced by $\alpha\hbar\,\omega_S$. The energy downshift depends on the stretching frequency of the surface terminated atoms. Therefore the PL peak energy is different for hydrogen- and deuterium-terminated porous Si in spite of the same absorption spectrum. The above energy reduction indicates that the PL energy upshift is smaller than what is expected from the theoretical calculations, which is clearly shown by Schuppler et al. [12].

Generally, the coupling constant α becomes large as the confinement becomes strong. To calculate the PL peak energy as a function of the bandgap energy, here, we assume the parameter α to be linearly proportional to the upshift energy as

$$\alpha = \eta\cdot\left(E_{gq} - E_{gb}\right) , \tag{3}$$

where η is a parameter to fit both PL peak energies and E_{gb} is the bulk bandgap energy. Using Eq. (2) and (3), we obtain the PL peak energies of H-PS (chain line) and D-PS (broken line) by selecting η equal to 1.4 as shown in Fig. 2. The fitting of the PL peak energies using Eq. (3) well explains the low energy behavior. However, to fit the PL peak energies of the higher energy side, the addition of higher order terms to Eq. (3) seems to be necessary.

CONCLUSION

We have fabricated isotope-terminated porous Si, and have compared the structure and the optical properties with the usual hydrogen-terminated porous Si. Replacement of hydrogen with deuterium results in the substantial reduction of the PL degradation of porous Si. The electrochemical anodization using DF electrolyte is promising not only for the study of the surface effects of nanometer sized crystallites but also for the application of various devices such as optoelectronic integrated circuits. Important problems related to the determination of the size dependence of the coupling constant in nanocrystalline system and the high PL efficiency of porous Si at room temperature remain unresolved.

ACKNOWLEDGMENTS

The authors gratefully acknowledge S. Nair, E. Tokunaga, M. Tanaka, M. Sugisaki, H. Ren, and N. Matsura, for their useful advises.

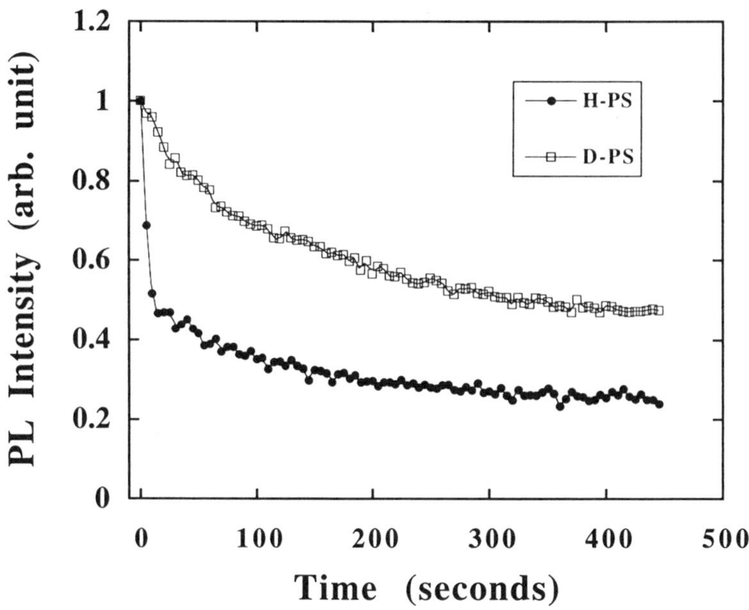

Fig. 3 Comparative time-dependent degradation of PL for H-PS and D-PS.

REFERENCES

[1] L. T. Canham, Appl. Phys. Lett. **57**, 1046 (1990).

[2] Y. J. Chabal and K. Raghavachari, Phys. Rev. Lett. **53**, 282 (1984).

[3] Hydrogen in Semiconductors, edited by J. I. Pankove and N. M. Johnson (Academic Press, New York, 1991).

[4] P. Gupta, V. L. Colvin, and S. M. George, Phys. Rev. B **37**, 8234 (1988).

[5] R. W. Hardeman, M. I. J. Beale, D. B. Gasson, J. M. Keen, C. Pickering, and D. J. Robbins, Surf. Sci. **152/153**, 1051 (1985).

[6] V. Lehmann, B. Jobst, T. Muschik, A. Kux, and V. Petrova-Koch, Jpn. J. Appl. Phys. **32**, 2095 (1993).

[7] T. Matsumoto, Y. Masumoto, and N. Koshida, to be published.

[8] Absorption coefficient $\alpha = 400\ cm^{-1}$ is used to determine the bandgap. The detailed discussion is given in Y. H. Xie et al., Phys. Rev. B **49**, 5386, (1994).

[9] R. Ramirez and C. P. Herrero, Phys. Rev. Lett. **73**, 126 (1994).

[10] C. J. Karlsson, F. Owman, E. Landemark, Y. -C. Chao, P. Mårtensson, and R. I. G. Uhrberg, Phys. Rev. Lett. **72**, 4145 (1994).

[11] T. D. Lee, F. E. Low, and D. Pines, Phys. Rev. **90**, 297 (1953).

[12] S. Schuppler, S. L. Friedman, M. A. Marcus, D. L. Adler, Y. -H. Xie, F. M. Ross, T. D. Harris, W. L. Brown, Y. J. Chabel, L. E. Brus, and P. H. Citrin, Phys. Rev. Lett. **72**, 2648 (1994).

ROLES OF SURFACE TERMINATION IN PHOTOLUMINESCENCE MECHANISMS OF POROUS SI

Y. SUDA, K. OBATA, A. KUMAGAI AND N. KOSHIDA
Faculty of Technology, Tokyo University of Agriculture and Technology,
2-24-16 Naka-cho, Koganei, Tokyo 184, Japan, sudayos@cc.tuat.ac.jp

ABSTRACT

The relationship between the oxidation states and PL properties and the effects of H/O termination exchange on the PL properties of PS have been investigated using synchrotron radiation photoemission spectroscopy (SR-PES), Auger electron spectroscopy (AES), Fourier transform infrared (FTIR), and photoluminescence (PL) techniques. The energy band gap, the peak energy and FWHM of the PL spectrum are almost unchanged by the oxidation process and by the H/O termination exchange. After the oxidation, the PL peak intensity decreased, suggesting the creation of nonradiative centers. In the H/O termination exchange experiment, the PL peak intensity decreased by more than 65 % upon annealing. However, it recovered the initial PL intensity by oxygen exposure. These results suggest that the surface termination itself functions to eliminate the nonradiative centers without depending on the termination species of hydrogen or oxygen, and that the skeletal structure of PS crystallites is important in the PL mechanisms.

INTRODUCTION

An understanding of the mechanism of visible luminescence in porous Si (PS) has become important from the viewpoint of its application to optoelectronic devices. We have investigated the surface structure and surface electronic states of as-anodized porous Si using synchrotron radiation photoemission spectroscopy (SR-PES), Auger electron spectroscopy (AES) and photoluminescence (PL) techniques [1-4]. The data indicate that there is almost no trace of oxygen on the surface of PS and its surface is covered with hydrides. However, our Fourier transform infrared (FTIR) study has also indicated no relationship between the amount of the surface hydrides and the PL intensity [5]. Thus, these results exclude hydrogen- and oxygen-related models [6-9] for the PL origin. In addition to these results, the results of our band gap and energy band diagram analyses, using the SR-PES data, also support a bulk-effect model including a quantum confinement model [10]. However, the relationship between the surface electronic states and PL mechanisms has not been fully understood yet. In this paper, we demonstrate the results of H/O termination exchange experiments together with the results of oxygen dosing experiments and discuss the relationship between the bulk-effect model and roles of the surface termination in the PL mechanisms.

EXPERIMENTAL

Porous Si was formed by anodizing p-type 2~6 ohm-cm Si(001) wafers in a mixture of 55 wt % aqueous HF solution and ethanol at an anodization current density of 100 mA/cm^2 for 30 s. This sample preparation was completed in a dark room to prevent oxidation [1].

SR-PES measurements were carried out using a 0.4 GeV electron storage ring at the Synchrotron Radiation Laboratory of the Institute for Solid State Physics, at the University of Tokyo. Photoemitted electron energies were analyzed using a double-pass cylindrical mirror analyzer contained in an ultrahigh-vacuum (UHV) chamber with a base pressure of $< 5 \times 10^{-11}$ Torr. Using this photoemission system, valence band spectra, Si 2p core spectra, and total yield spectra were obtained. Both AES and PL measurements were carried out using a separate UHV chamber from that used in the SR-PES system. The base pressure of the AES chamber was $< 2 \times 10^{-9}$ Torr. AES electron energies were analyzed using a cylindrical mirror

Mat. Res. Soc. Symp. Proc. Vol. 452 © 1997 Materials Research Society

analyzer, and differentiated AES spectra were measured with a modulation voltage of 10 V_{p-p}. PL spectra were measured using 325 nm He-Cd laser light for oxygen dosing experiments and 333.6 - 363.8 nm Ar^+ ion laser light for H/O exchange experiments.

Oxygen exposure was carried out separately in the SR-PES or the AES chamber using the same ionization gun. In this oxygen dosing experiment, PS samples were exposed to oxygen excited by electron impact at room temperature. The flux density at the surface was determined by calculating the angular distribution of the effused gas using Clausing's relation [11].

To avoid unexpected oxidation, PS samples were kept in pure ethanol before the SR-PES, AES and PL measurements. They were placed in the chambers while wet and then the chambers were evacuated [1]. FTIR measurements were carried out in a nitrogen atmosphere.

RESULTS AND DISCUSSION

Effects of Oxygen Dosing on Electronic States and PL Properties of PS

Measured Si 2p core spectra were decomposed into the Si $2p_{3/2}$ and Si $2p_{1/2}$ spin-orbit partner lines [12,13]. Figure 1 shows Si $2p_{3/2}$ core spectra (by dots) obtained from as-anodized porous Si and after oxygen exposure. To investigate the oxidation states, the Si $2p_{3/2}$ core spectra were further decomposed into several components. Since oxygen is not apparent on the surface of as-anodized porous Si which is mainly covered with mono- and di-hydride as reported in our previous papers [1-3], the Si $2p_{3/2}$ spectrum of as-anodized PS was decomposed into bulk Si denoted by Si^0, $\equiv$Si-H denoted by Si-H, and =Si-H_2 components [14]. Upon oxygen exposure up to a dose of 3 x 10^{17} cm^{-2}, the signal for the dihydride component almost disappears. On the basis of this result, the Si 2p spectra of the oxygen-exposed PS samples were decomposed into bulk Si (Si^0) and Si-H components, and five components in intermediate oxidation states, i.e., Si_2O (Si^{1+}), SiO (Si^{2+}), Si_2O_3 (Si^{3+}), SiO_2* and SiO_2 (Si^{4+}) [12,13]. The SiO_2* component has been reported by Niwano at al. for thin Si(001) oxides [12]. In the case of PS, better curve fitting has also been obtained with the SiO_2* component than without it, particularly, in the initial stage of oxidation up to an oxygen dose of 7 x 10^{17} cm^{-2}. Each of the distribution functions of these components were a Lorentzian convoluted by a Gaussian. Precise decomposition procedures have been described in our previous paper [4].

Fitting curves using these models are also shown in Fig.1 by solid lines. Good curve fitting has been achieved, as indicated in Fig.1. Peak intensities of the decomposed components as a function of the oxygen dose are shown in Fig.2. As the oxygen dose increases, Si suboxides develop and SiO_2 peak intensity gradually increases. After an excited-oxygen dose of 71 x 10^{17} cm^{-2}, the most intense surface oxide is SiO_2; the surface oxides include oxides in the intermediate oxidation states.

PL peak energies and PL peak intensities corresponding to the results of the SR-PES experiments are shown in Fig.3. As the oxygen dose increases, the PL peak intensity decreases, however, the peak energy and FWHM of its PL spectra are almost unchanged from the initial value for as-anodized PS. The corresponding energy band gaps estimated using the SR-PES results of valence band spectra, Si 2p core spectra, and total yield spectra are also shown in Fig.3. The method for estimating the energy band gap has been described in detail in our previous paper [1]. The energy band gap is also unchanged from the initial value for as-anodized PS.

In the PL measurements, a 325-nm laser beam was led to the PS sample surfaces at an angle of ~ 30° to the surface plane. Considering this incident angle and the absorption coefficient of PS for 325 nm light [15], the detection depth for the PL measurements has been estimated to be ~250 Å. In the SR-PES experiments, a SR light beam was incident on the PS sample surfaces at 45° to the surface plane. Since the PS surface has columnar and convex structures whose diameters are on the order of 20 ~ 100 Å [16,17], photoelectrons emitted from within a depth range of the same order are detected by measuring the light beam incident angle. Thus, the detection depths in the SR-PES and PL measurements are comparable. In the FTIR measurements for as-anodized and oxygen-exposed PS, hydrogen-related absorption signal intensity decreased by 34 % and oxygen-related absorption signal intensity increased

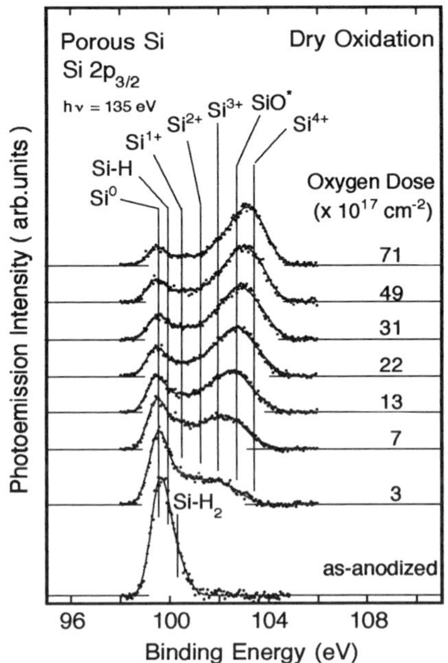

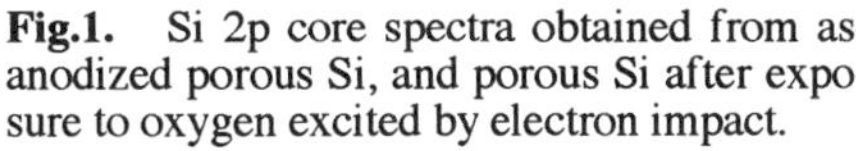

Fig.1. Si 2p core spectra obtained from as-anodized porous Si, and porous Si after exposure to oxygen excited by electron impact.

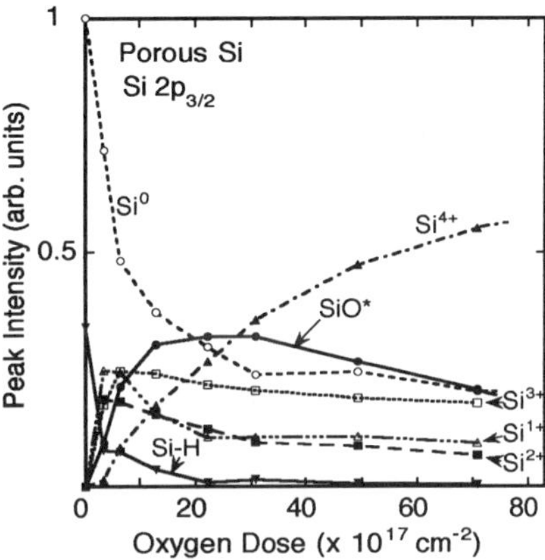

Fig.2. Peak intensities of decomposed monohydride and Si oxide components obtained from as-anodized porous Si, and porous Si after exposure to oxygen excited by electron impact.

by 61 % after an oxygen dose of 62 x 10^{17} cm^{-2}. Thus, a considerable amount of hydrogen desorbs even from the deep region, at least 34 % of the PS thickness, and oxidation proceeds to a comparable region in depth. Measured thicknesses of the PS layers were ~ 2.5 ± 0.5 μm. These results suggest that the oxidation depth is beyond the detection limit of the PL and SR-PES measurements.

Although the degree of oxidation of PS by oxygen dose varies along the depth of a PS layer and regions closer to the surface in PS are expected to be more oxidized, the energy band gap for PS, PL Peak energies, and FWHMs and relative shapes of the PL spectra remained almost unchanged from the initial value or state throughout the entire oxidation process. The result indicates that the band gap and the PL energy are insensitive to oxidation states or the degree of oxidation under the oxidation conditions adopted in this work. Thus, the SR-PES results are consistent with the PL results, implying that the estimated energy band gap traces the PL peak energy. This analysis is also supported by the previously obtained results [1] that the estimated band gaps obtained for as-anodized PS samples formed by anodization with different anodization currents and for PS samples after chemical etching with the same anodization solution under light exposure well trace the corresponding PL peak energies.

Since the energy positions and relative shapes of the PL spectra, which are associated with the transition states, were almost unchanged, the decrease in the PL intensity corresponds to the decrease in the number of photoluminescent elements. This decrease in the PL intensity is probably due to an increase in the number of nonradiative surface states produced by the intermediate oxidation states or via the oxidation process. Although the surface structures and the surface termination structures of Si crystallites in PS were modified through reaction with excited oxygen, the energy band gap and the PL peak energy or the luminescent transition energy, which are related to the PL mechanisms, were unchanged. These results suggest that the connection between the origin of PL and the surface termination structure of PS is weak and the skeletal structure of PS is important in the PL mechanisms.

Effects of H/O Termination Exchange on PL Properties of PS

In H/O termination exchange experiments, a part of hydrogen was desorbed from an as-anodized PS surface by thermal annealing at 200 °C for 300 s and its surface was exposed to excited oxygen using the AES chamber. The degree of oxidation was estimated by AES analysis. Two experimental procedures were carried out to confirm the reproducibility of the experimental results. In the first procedure, AES and PL measurements were carried out for as-annealed PS after annealing (sample1). In the second procedure, oxygen exposure was carried out after annealing without doing AES and PL measurements (sample2). PL peak energies and intensities as a function of O (KLL) to Si (LVV) Auger signal intensity ratio (I_O/I_{Si}) obtained in these experiments are shown in Fig. 4. The results in both the cases are very similar. The results indicate that the PL peak intensity sharply decreased due to the annealing, and that then it increased proportionally to the I_O/I_{Si} Auger signal intensity ratio. Finally, it recovered the initial peak intensity. However, the PL peak energy was almost constant without depending on the change in the PL peak intensity. Since the PL intensity well responds to the oxygen dosing, the PL detection region in the PS layer was almost entirely affected by the exposed oxygen. The sharp decrease in the PL peak intensity upon annealing is probably due to creation of nonradiative centers resulting from hydrogen desorption. Nonradiative centers

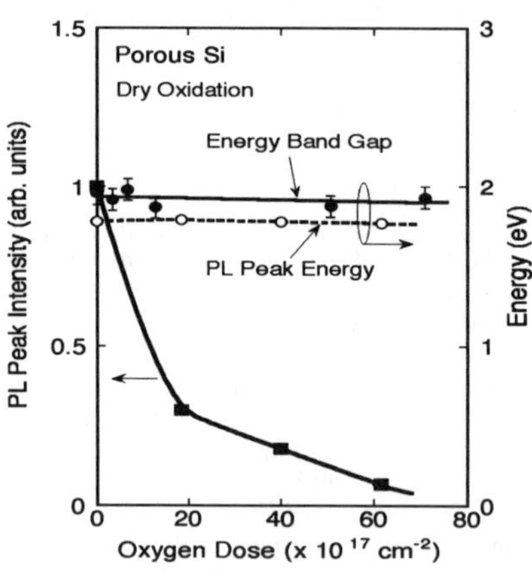

Fig.3. Energy band gaps, PL peak energies, and PL peak intensities obtained from as-anodized porous Si, and porous Si after exposure to oxygen excited by electron impact.

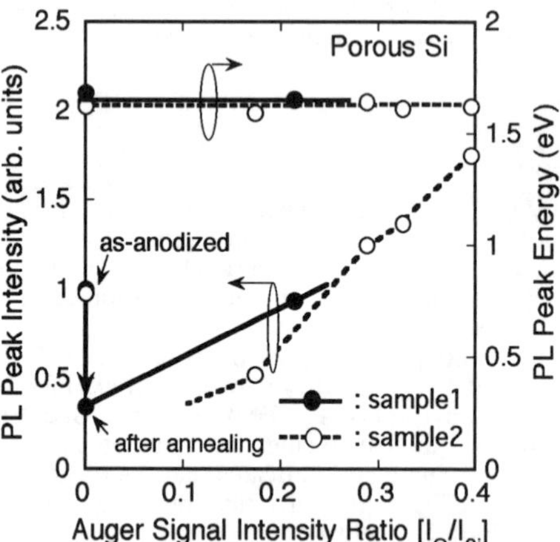

Fig.4. PL peak energies and PL peak intensities as a function of O (KLL) to Si (LVV) Auger signal intensity ratio (I_O/I_{Si}) obtained from as-anodized porous Si, porous Si after annealing at 200 °C for 300 s, and porous Si after annealing at 200 °C for 300 s followed by oxygen exposure.

are thought to be dangling bonds [18]. In the range of annealing temperature of ~ 200 °C, the value of which is used in this study, hydrogen mainly desorbs from a dihydride phase [19] leaving dimer bonds. Using a preexponential factor and an activation energy for hydrogen desorption from a monohydride phase [19], an increase in the number of dangling bonds resulting from hydrogen desorption from a monohydride phase has been estimated to be fairly small compared to the decrease in the PL intensity as shown in Fig. 4. These analyses suggest that desorption of a very small fraction of hydrogen atoms, which cover one luminescent crystallite, destroys its luminescent properties. From the relationship between the band gap and the PL peak energy discussed in the former section, the results shown in Fig. 4 suggest that the band gap of PS remains almost unchanged throughout the H/O exchange experiments. The vibration of atoms terminating on the surface of a PS crystallite has been calculated to lower the band gap of a PS crystallite with no termination atoms [20]. However, since the amount of the desorbed hydrogen has been estimated to be fairly small compared to the total amount of the terminated hydrogen, the vibration effect on the band gap may be small. Upon annealing, the PL peak intensity decreased by more than 65 % from the initial intensity. However, it recovered the initial peak intensity with the PL peak energy, which relates to the luminescent transition energy, being constant. The results suggest that surface termination itself is important to make the luminescent crystallites luminous without depending on the termination species of hydrogen or oxygen, and that the skeletal structure of PS crystallites is important in the PL mechanisms. Recent theoretical calculations [21,22] also support this explanation. In those calculations using quantum dots and wires, wave functions converge in the interior of the skeleton, suggesting that the electronic states of PS do not depend on the surface termination species.

CONCLUSIONS

In conclusion, the relationship between the oxidation states and PL properties and the effects of H/O termination exchange on the PL properties of PS have been investigated. The energy band gap, the peak energy and FWHM of the PL spectrum are almost unchanged by the oxidation process and by the H/O termination exchange. By the oxidation, the PL peak intensity decreased, suggesting creation of nonradiative centers. In the H/O termination exchange experiment, the PL peak intensity decreased by more than 65 % upon annealing. However, it recovered the initial PL intensity by oxygen exposure. These results suggest that surface termination itself is important to make the luminescent crystallites luminous without depending on the termination species of hydrogen or oxygen, and that the skeletal structure of PS crystallites is important in the PL mechanisms. Many researchers have reported to date that after PS is heavily oxidized, its PL blue-shifts [23,24]. This result is not inconsistent with the results of this work in terms that oxygen termination does not affect the radiative transition-states unless it modifies the size and the structure of the skeleton of luminescent crystallites.

ACKNOWLEDGMENTS

The support of the staff of the Synchrotron Radiation Laboratory of the Institute for Solid State Physics, the University of Tokyo, is greatly appreciated. This work is supported in part by a Grant-in-Aid from the Ministry of Education, Science, Sports and Culture of Japan.

REFERENCES

1. Y. Suda, T. Ban, T. Koizumi, H. Koyama, Y. Tezuka, S. Shin and N. Koshida, Jpn. J. Appl. Phys. **33**, 581 (1994).

2. Y. Suda and N. Koshida, Denki Kagaku **63**, 892 (1995) [in Japanese].

3. Y. Suda, T. Koizumi, K. Obata, Y. Tezuka, S. Shin and N. Koshida, J. Electrochem. Soc. **143**, 2502 (1996).

4. T. Koizumi, K. Obata, Y. Tezuka, S. Shin, N. Koshida and Y. Suda, Jpn. J. Appl. Phys. **35**, L803 (1996).

5. T. Ban, T. Koizumi, S. Haba, N. Koshida and Y. Suda, Jpn. J. Appl. Phys. **33**, 5603 (1994).

6. M. A. Tischler, R. T. Collins, J. H. Stathis and J. C. Tsang, Appl. Phys. Lett. **60**, 639 (1992).

7. C. Tsai, K.-H. Li, J. Sarathy, S. Shih, J. C. Campbell, B. K. Hance and J. M. White, Appl. Phys. Lett. **59**, 2814 (1991).

8. E. Bustarret, M. Ligeon and L. Ortega, Solid State Commun. **83**, 461 (1992).

9. M. S. Brandt, H. D. Fuchs, M. Stutzmann, J. Weber and M. Cardona, Solid State Commun. **81**, 307 (1992).

10. L. T. Canham, Phys. World **5**, 41 (1992).

11. P. Clausing, Z. Physics. **60**, 471 (1930).

12. M. Niwano, H. Katakura, Y. Takakuwa and N. Miyamoto, J. Appl. Phys. **68**, 5576 (1990).

13. F. J. Himpsel, F. R. McFeely, A. Taleb-Ibrahimi, J. A. Yarmoff and G. Hollinger, Phys. Rev. **B38**, 6084 (1988).

14. L. Ley, J. Reichardt and R. L. Johnson, Phys. Rev. Lett. **49**, 1664 (1982).

15. N. Koshida, H. Koyama, Y. Suda, Y. Yamamoto, M. Araki, T. Saito, K. Sato, N. Sata and S. Shin, Appl. Phys. Lett. **63**, 2774 (1993).

16. T. George, M. S. Anderson, W. T. Pike, T. L. Lin, R. W. Fathauer, K. H. Jung and D. L. Kwong, Appl. Phys. Lett. **60**, 2359 (1992).

17. V. P. Parkhutik, J. M. Albella, J. M. Martinez-Duart, J. M. Gómez-Rodríguez, A. M. Baró and V. I. Shershulsky, Appl. Phys. Lett. **62**, 366 (1993).

18. C. Delerue, G. Allan and M. Lannoo, Phys. Rev. **B48**, 11024 (1993).

19. P. Gupta, V. L. Colvin and S. M. George, Phys. Rev. **B37**, 8234 (1988).

20. T. Matsumoto, Y. Masumoto, S. Nakashima, H. Mimura and N. Koshida, in Ext. Abs. '96 Int. Conf. on Solid State Devices and Materials (Japan Soc. Appl. Phys., Tokyo, 1996) pp.709-711.

21. L. -W. Wang and A. Zunger, J. Phys. Chem. **98**, 2158 (1994).

22. C. -Y. Yeh, S. B. Zhang and A. Zunger, Phys. Rev. **B50**, 14405 (1994).

23. V. Petrova-Koch, T. Muschik, A. Kux, B. K. Meyer, F. Koch and V. Lehmann, Appl. Phys. Lett. **61**, 943 (1992).

24. S. Shih, C. Tsai, K.-H. Li, K. H. Jung, J. C. Campbell and D. L. Kwong, Appl. Phys. Lett. **60**, 633 (1992).

LUMINESCENCE AND SURFACE-STATE CHARACTERISTICS IN P-TYPE POROUS SILICON

A. RAMIREZ PORRAS,* O. RESTO,* S.Z. WEISZ,* Y. GOLDSTEIN,** A. MANY,** AND E. SAVIR**
*Department of Physics, University of Puerto Rico, Rio Piedras, PR 00931
**Racah Institute of Physics, The Hebrew University, Jerusalem 91904, Israel

ABSTRACT

Pulse measurements on the porous-Si/electrolyte system are employed to determine the surface effective area and the surface-state density at various stages of the anodization process used to produce the porous material. Such measurements were combined with studies of the photoluminescence spectra. These spectra were found to shift progressively to the blue as a function of anodization time. The luminescence *intensity* increases initially with anodization time, reaches a maximum and then decreases with further anodization. The surface state density, on the other hand, increases with anodization time from an initial value of $\sim 2\times10^{12}$ cm^{-2} for the virgin surface to $\sim 10^{13}$ cm^{-2} for the anodized surface. This value is attained already after ~2 min anodization and upon further anodization remains fairly constant. In parallel, the effective surface area increases by a factor of 10 - 30. This behavior is markedly different from the one observed previously for n-type porous Si.

INTRODUCTION

Porous silicon,[1-4] (PS) obtained by electrochemical etching procedures applied to crystalline Si surfaces, exhibits high luminescence efficiencies in the visible range. It is quite clear now that the visible luminescence originates from the band-gap enlargement due to quantum confinement.[4] At the same time, the reasons for the high-efficiency luminescence are still somewhat under debate.[4,5] It was suggested that it is the amorphous[6] or microcrystalline[7] nature of the porous Si that is responsible for the phenomenon, or that the formation of silicon compounds such as siloxene ($Si_6O_3H_6$) or species of Si-H, Si-O and Si-F bonds are involved in the luminescence.[8] One way of gaining further insight into the luminescence process is to carry out a variety of measurements on samples of different porosity. To that end we have employed combined studies of the luminescence spectrum, the surface-state density and the effective surface area of the porous surface. Such studies were carried out at different stages of the anodization process and thus for different morphologies of the porous surface. The luminescence spectra were measured by conventional methods. The surface state characteristics and the effective surface area were determined by pulse measurements[9] on the PS/electrolyte system. This system is particularly suitable since a capacitative contact to the terrain of the porous surface is best achieved by an electrolyte, and it was successfully used[10] to investigate n-type PS. There we found[10] a strong correlation between the surface-state density near the conduction band edge and the luminescence intensity. In this paper we present similar measurements on *p-type* porous Si and we compare the results with those obtained on n-type material.

EXPERIMENTAL

The starting material was high-grade p-type silicon wafers of resistivity in the range 20 – 50 Ωcm. A p^+ layer was formed by diffusing metallic Al into one of the faces to obtain an ohmic

Mat. Res. Soc. Symp. Proc. Vol. 452 © 1997 Materials Research Society

contact. The sample was attached to a cylindrical Teflon cell via a Kalrez O-ring, the sample constituting the bottom of the cell, with its free surface facing upwards. The samples were etched in 20% HF. In order to prepare the porous surface,[4] a solution of HF, ethanol and water (1:1:2) was poured into the cell. A platinum electrode was immersed in the solution and a spring contact was attached to the p^+ contact. The anodization of the Si surface was carried out with a current density of 100 mA/cm^2.

The luminescence of the PS was excited by a 10 mW He-Cd laser beam (λ = 442 nm). The luminescence spectra at different stages of the anodization process were measured by a Control Development spectrometer. The samples were rinsed and dried before the measurement.

The electronic characteristics of the PS/electrolyte interface were studied at different stages of the anodization process, starting from the "virgin" surface and up to an anodization time of 20 minutes. To that end, the anodizing solution was replaced after each anodization stage by an electrolyte; an aqueous solution of KCl. The measurement technique applied to the semiconductor/electrolyte (S/E) interface has been described elsewhere,[9] and will be reviewed only briefly here. A short voltage pulse of duration T = 20 μsec, applied between the Pt electrode and the sample's p^+ contact, is used to charge up the interface region. The voltage drop across these electrodes, measured just after the termination of the pulse (T + dT), represents to a very good approximation the change δV_s in barrier height across the semiconductor space-charge layer induced by the applied pulse. If an insulating layer, such as an oxide, is present at the semiconductor surface, the measured voltage drop is $\delta V_s + \delta V_g$, where δV_g is the drop across the insulating layer. Obviously, $\delta V_g = Q_{tot}/C_g$, where Q_{tot} is the total charge density induced at the surface and C_g is the "geometric" capacitance of the insulating layer (per cm^2). Q_{tot} is obtained from the voltage V_c developed across a large series capacitor C, again at the termination of the pulse. Pulses of varying amplitude are applied singly, one per data point taken. In this manner possible damage to the porous surface is minimized.

In general, Q_{tot} is made up of three components:

$$Q_{tot} = \delta Q_{sc} + \delta Q_{ss} + Q_L \; , \tag{1}$$

where δQ_{sc} is the change in the free space-charge density, δQ_{ss} is the change in surface-state charge density, and Q_L is the charge density that has leaked across the interface due to imperfect blocking of the S/E interface. In order to determine each component of Q_{tot}, the platinum electrode is shorted to ground by an electronic switch at T + dT, where dT is very short (0.1 – 0.2 μsec), just sufficient to permit accurate readings of δV_s and V_c right after the termination of the pulse. At this point, charge redistribution between C and the S/E interface begins to take place. In the first stage, the free charge δQ_{sc} and its equal counterpart in C discharge relatively fast through the low resistance of the sample and the electrolyte. The decay constant associated with this process is typically several microseconds. As a result, V_c decays to the value $\delta Q_{ss}/C$, δQ_{ss} being the charge remaining in C after the fast decay process. Thereafter, V_c decays to zero usually much more slowly, as charge trapped in the surface states by the charging pulse are thermally re-emitted into the conduction band, in the case of n-type semiconductor, or valence band, in the case of p-type semiconductor. The decay time is larger the farther away the surface states are located energetically from the relevant band edge and the lower the temperature.[11] If charge leakage exists, V_c does not decay to zero but to the value Q_L/C. Subsequently it remains practically constant since the leaked charge has been lost from the interface and the remaining charge Q_L on C cannot be dissipated. This behavior enables the separate determination of the three components in Eq. (1), all as functions of δV_s. In what follows, we shall express these components in terms of hole surface densities: $\delta P_s = \delta Q_{sc}/q$ and $\delta P_{ss} = \delta Q_{ss}/q$, where q is the absolute magnitude of the electronic charge.

In most cases a space-charge layer already exists at the semiconductor surface, before applying any bias. It is characterized by an equilibrium barrier height V_{so} and an equilibrium surface hole density P_{so}. If surface states are present, there may also be an equilibrium density

P_{sso} of "occupied" surface states. V_{so} and P_{so} can be determined quite accurately from measurements in the depletion range.[12] The entire plots of P_s and P_{ss} *vs.* V_s can then be constructed by using the relations $V_s = V_{so} + \delta V_s$, $P_s = P_{so} + \delta P_s$, and $P_{ss} = P_{sso} + \delta P_{ss}$. So much so in the absence of an insulating buffer layer (such as an oxide) at the semiconductor surface. If such a layer is present, the as-measured barrier height, i. e., the measured voltage drop between the Pt electrode and the p^+ contact just after the pulse termination, yields V_s+V_g, where V_g is the voltage drop across the insulating layer.

RESULTS AND DISCUSSION

In Fig. 1 we present typical photoluminescence spectra of p-type PS surfaces prepared by anodization. The different anodization times are marked on the spectra. We notice that the luminescence intensity at the beginning increases with anodization time, attains a maximum and then decreases. This behavior is illustrated by the higher curve in Fig. 2 and is quite similar to that observed[10] previously for n-type PS. The lower curve in Fig. 2 shows the appreciable blue shift of the spectra, suggesting that, on the average, the porous structure gets finer with anodization time.

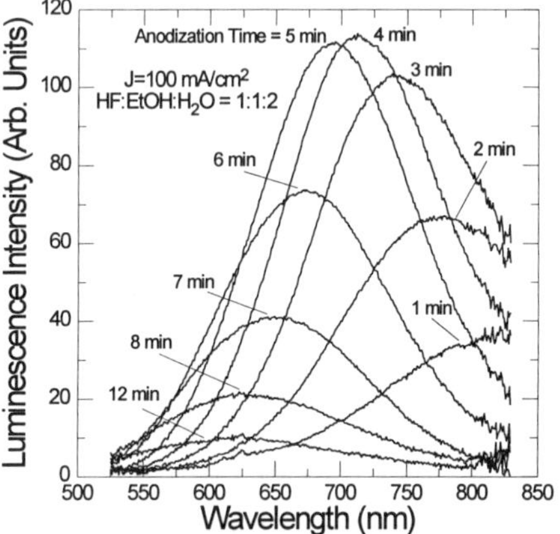

Fig.1. Photoluminescence spectra for various anodization times, as marked.

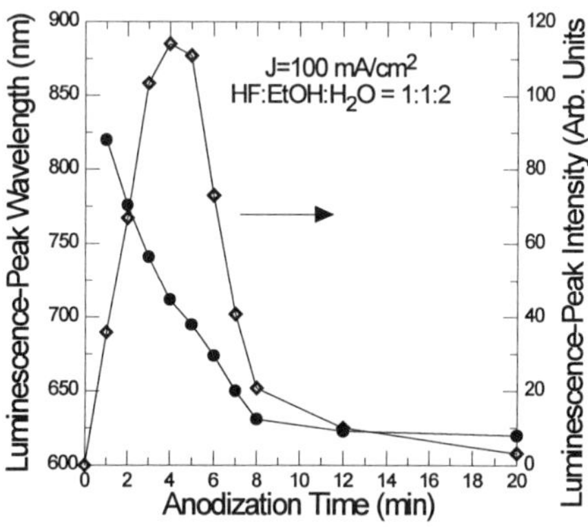

Fig. 2. Peak-photoluminescence intensity and wavelength as functions of anodization time for the sample in Fig. 1.

Typical results of the free surface-hole density P_s (diamonds) and the density of occupied surface states P_{ss} (stars) against the as-measured barrier height V_s+V_g as obtained for an etched virgin silicon surface, are displayed in the semilog plot of Fig. 3. In the depletion range, P_s is negative but, because of the logarithmic scale used, the plot is that of $-P_s$. The light curve, in the accumulation range, labeled $C_g = \infty$, represents the theoretical dependence of P_s on V_s for a buffer-free surface ($C_g = \infty$, $V_g = 0$), as derived from a solution of Poisson's equation for the value of the hole bulk concentration p_b marked in the figure. It is seen that this curve does not account well for the data in the accumulation range. The best fit, represented by the bold curve, labeled $C_g = 4.5\ \mu F/cm^2$, was obtained by assuming the presence of an insulating buffer layer of $C_g = 4.5\ \mu F/cm^2$, corresponding to a thickness of ~4 A. Most probably this layer consists of an oxide, but a small contribution of the series Helmholtz layer cannot be ruled out. Again, this behavior is similar to that found on n-type PS. Turning now to the surface-state hole occupancy P_{ss}, it is seen to rise slowly from a low value of $\sim 10^{11}\ cm^{-2}$ at $V_s+V_g \approx -0.5$ V up to $\sim 10^{14}\ cm^{-2}$

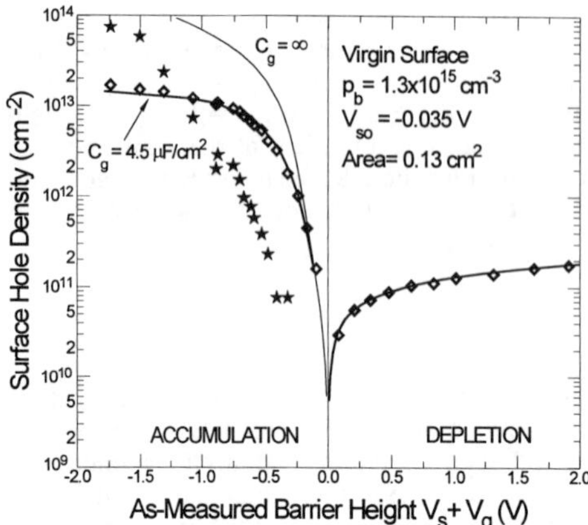

Fig. 3. Free surface hole density P_s (diamonds) and density of occupied surface states P_{ss} (stars) *vs.* the as-measured barrier height V_s+V_g for a virgin Si surface. The light and bold curves are theoretical plots of P_s as explained in the text.

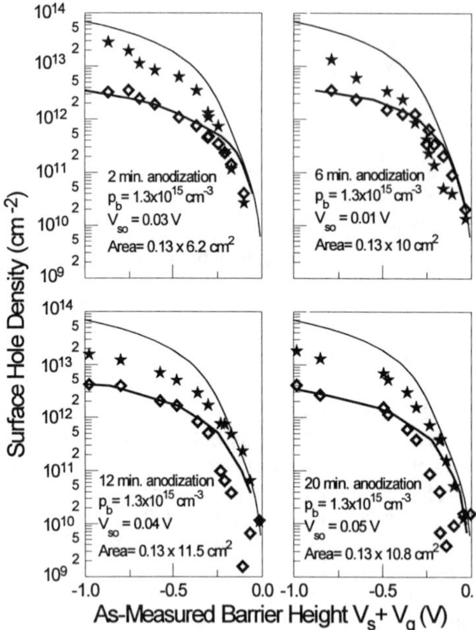

Fig. 4. P_s (diamonds) and P_{ss} (stars) *vs.* V_s+V_g for four anodized, porous Si surfaces. The light and bold curves are theoretical plots of P_s.

at $V_s+V_g \approx -2$ V. No saturation value for the surface-state density could be reached. This is quite different than the behavior we found[10] on n-type PS, where the surface-state density saturated at 10^{12} cm^{-2}. Because of this lack of saturation, we shall use the values of P_{ss} at $V_s+V_g \approx -1$ V to compare surface state occupancies for different anodization times. For the case of the virgin surface, this value is $\sim 2x10^{12}$ cm^{-2}.

Results of P_s and P_{ss}, similar to those in Fig. 3, for four porous surfaces are presented in Fig. 4. These results were obtained after the sample of Fig. 3 has been anodized for different times, as marked in the figure. Since our aim is to compare the surface-state densities, we show only the accumulation range here. Because of the increase of the effective surface area, the highest surface potential barriers (for holes) attained were around -1 volt. The curves in the figure are theoretical, calculated for the same C_g values as in Fig. 3. We notice again that for all four porous surfaces, the surface-state density (stars) increases monotonously with the potential barrier through the whole region shown and does not exhibit signs of saturation. As mentioned above, we choose for comparison the values of P_{ss} at ~-1 volt. These values in the figure are scattered around $2x10^{13}$ cm^{-2}, about an order of magnitude higher than on the virgin surface. However, the interesting thing is that this surface-state density remains fairly constant with anodization time, very much different from the results found[10] for n-type PS. A behavior similar to that of the surface states was observed also for the effective surface area. In Fig. 5 the surface area factor, i. e. the ratio of the effective area to the area of the virgin sample, for two typical samples is plotted against the anodization time. The area factors were derived from measurements of the type shown in Figs. 3 and 4 in the strong accumulation range. The results for the two samples are qualitatively the same; the area factor increases with anodization time till it reaches a saturation. The saturation values observed varied from sample to

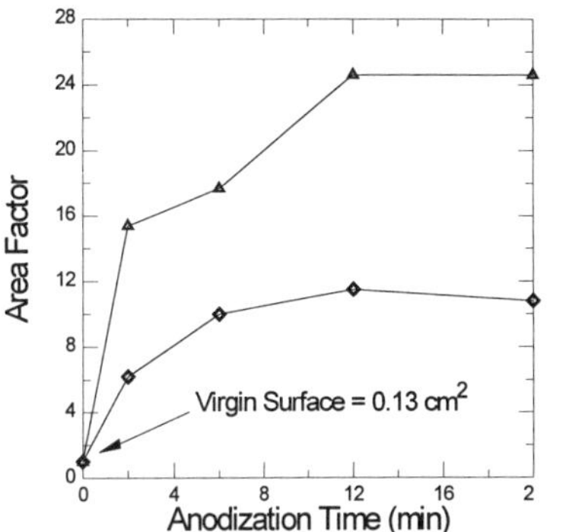

Fig. 5. Surface area factor *vs.* anodization time for two samples.

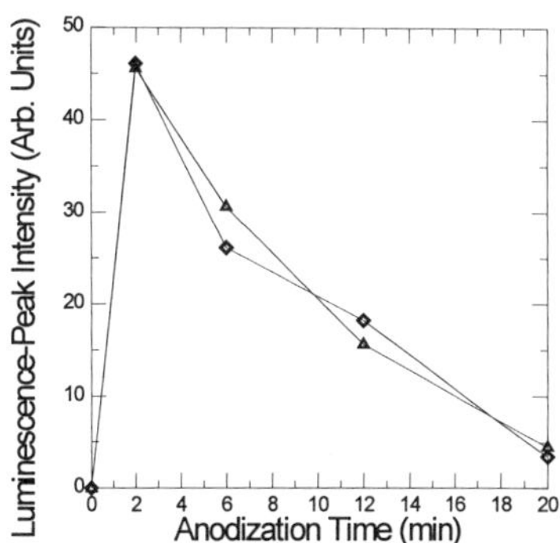

Fig. 6. Photoluminescence-peak intensity *vs.* anodization time for two samples.

sample between 10 and 30.

In parallel with the surface-state density, we measured the photoluminescence spectrum for each anodization time. These measurements were performed after the measurements on the PS/electrolyte system. In Fig. 6 the luminescence- peak intensity is plotted against the anodization time for the same two samples as in Fig. 5. We notice that the maximum luminescence peak obtains already after 2 min anodization and then the luminescence decreases upon further anodization. In Fig. 7 we plot the value of the luminescence-peak wavelength as a function of anodization time. The peak shifts to the blue upon anodization close to 200 nm from its value at 2 min anodization.

Comparing the results of Figs. 6 and 7 to those in Fig. 2 we notice that they both show comparable blue shifts, but there is a discrepancy in the anodization time needed to attain maximum luminescence. To check the influence of the KCl electrolyte used in the electronic measurements, we measured the spectra on PS samples that after anodization were immersed for 10 min into a KCl electrolyte. The results are shown in Fig. 8 and, for comparison, we also re-plotted the corresponding curve from Fig. 1. We notice that the KCl treatment does not shift the maximum to a different anodization time, however, it does lower the luminescence intensity. This latter is probably due to adsorption of some species, from the electrolyte. Thus we ascribe the

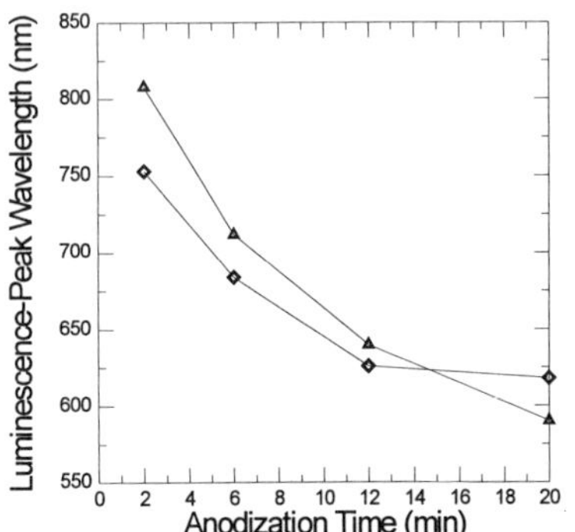

Fig. 7. Photoluminescence-peak wavelength *vs.* anodization time for two samples.

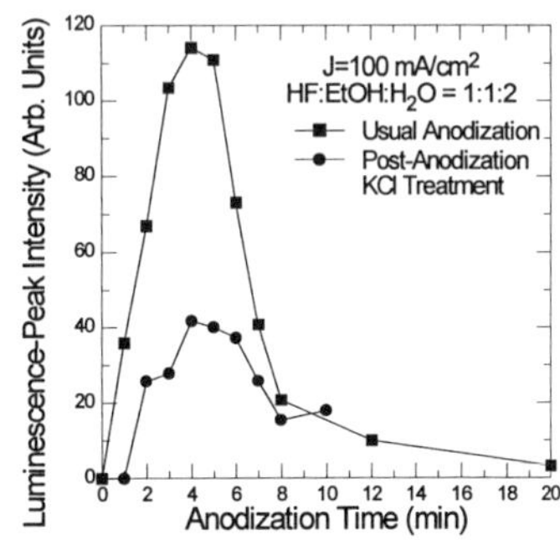

Fig. 8. Photoluminescence-peak intensity for an untreated and for a KCl treated sample *vs.* anodization time.

shift of the anodization time for maximum luminescence, to possible surface damage due to the application of the voltage pulses.

CONCLUSION

P-type PS behaves quite differently from that of n-type PS. While the photoluminescence intensity exhibits a pronounced maximum with anodization time, the effective surface area and the surface state density appear to reach a more or less constant value as a function of anodization time. This is in contrast to n-type PS, where a close correlation between the effective surface area, the luminescence intensity and the surface-state density was found.[10] The discrepancy between the two results can be reconciled once we realize that in both cases the surface states are measured under accumulation conditions. The surface states involved are then those near the majority carrier band edge; namely, near the conduction band for the n-type material and near the valence band for the p-type material. Thus our present results suggest that the surface states near the valence-band edge are not involved in the luminescence process.

ACKNOWLEDGMENT

The work in Israel was supported by the Volkswagen Foundation Hannover, Germany, and by the Ministry of Science, Israel, within the framework of the infrastructure-applied physics support project. The work in Puerto Rico was supported by NSF EPSCoR Grant No. OSR-94-52893, by NASA grant No. NCCW-0088 and U.S. Army Research Office grant No. DAAHO4-96-1-0405.

REFERENCES

1. L.T. Canham, Appl. Phys. Lett. **57**, 1046 (1990).
2. I. Amato, Science **252**, 922 (1991).
3. A.G. Cullis and L.T. Canham, Nature **353**, 335 (1991).
4. For the recent developments in this field see: Z.C. Feng and R. Tsu, *Porous Silicon* (World Scientific, Singapore, 1994).
5. F. Koch in *Silicon Based Optoelectronic Materials,* edited by R.T. Collins, M.A. Tischler, G. Abstreiter, and M.L. Thewalt (Mater. Res. Soc. Proc. **298**, Pittsburgh, PA, 1993) pp. 319-324.
6. R.P. Vasquez, R.W. Fathauer, T. George, A. Ksendzov and T.L. Lin, Appl. Phys. Lett. **60**, 1004 (1992).
7. J.S. Foresi and T.D. Moustakas in *Light Emission from Silicon,* edited by S.S. Iyer, L.T. Canham, and R.T. Collins (Mater. Res. Soc. Proc. **256**, Pittsburgh, PA, 1992) pp.77-82.
8. M.S. Brandt, H.D. Fuchs, M. Stutzmann, J. Weber and M. Cardona, Solid State Commun. **81**, 307 (1992).
9. M. Wolovelsky, J. Levy, Y. Goldstein, A. Many, S.Z. Weisz and O. Resto, Surf. Sci. **171**, 442 (1986).
10. S.Z. Weisz , J. Avalos, M. Gomez, A. Many, Y. Goldstein, and E. Savir in *Defect- and Impurity-Engineered Semiconductors and Devices,* edited by S. Ashok, J. Chevallier, I. Akasaki, N.M. Johnson, and B.L. Sopori (Mater. Res. Soc. Proc. **378**, Pittsburgh, PA, 1995) pp. 899-904.
11. A. Many, Y. Goldstein, and N.B. Grover, *Semiconductor Surfaces* (North-Holland, Amsterdam, 1971).
12. S.Z. Weisz, J. Penalbert, A. Many, S. Trokman, and Y. Goldstein, J. Phys. Chem. Solids **51**, 1067 (1990).

ELECTROCHEMICAL AND CHEMICAL DEPOSITION OF II-VI SEMICONDUCTORS IN POROUS SILICON

R. HERINO *, M. GROS-JEAN *, L. MONTES *, D. LINCOT **
* Laboratoire de Spectrométrie Physique, Université Joseph Fourier de Grenoble 1 - CNRS (UMR 5588), BP 87, 38402 St Martin d'Hères, France
** Ecole Nationale Supérieure de Chimie de Paris, 11 rue P. et M. Curie, 75231 Paris Cedex 05, France.

ABSTRACT

The introduction of II-VI semiconductor compounds into porous silicon layers has been investigated in order to obtain transparent and conducting contacts with the inner surface of the material. CdTe and ZnSe have been electrodeposited cathodically on n type nanoporous electrodes from acidic solutions containing the metallic cations and dissolved oxides of selenium or tellurium. CdS incorporation into p-type porous silicon has been achieved by chemical bath deposition, from solutions containing cadmium complexes and thioacetamid as a sulfur donor. Characterization of the deposits has been performed by SEM observations, X-ray analysis and RBS. Results confirm the penetration of the compounds into the porous films, with small to strong concentration gradients in thickness depending on the deposition method. After deposition and sample drying, the luminescence of CdTe embedded layers has almost disappeared, whereas those containing ZnSe and CdS show a photoluminescence efficiency which is not severely degraded.

INTRODUCTION

Since the discovery of the photoluminescent [1] and electroluminescent [2] properties of porous silicon, many efforts have been devoted to the realization of bright emitting solid state devices based on this material, but despite recent significant progress [3-4], the electroluminescence efficiency remains quite low, in the order of 0.1%. However, bright electroluminescence with an efficiency similar to that of photoluminescence (near 1%) is easily observed when the electrical contact with the porous layer is achieved by means of a liquid electrolyte [5]. In such conditions, there is an intimate contact between the internal surface of the material and the electrolyte, allowing current flow through the whole structure. In the case of a solid contact, the conducting layer does not penetrate the pore network, so that it is only a part of the porous layer which electroluminesces. Moreover, as higher polarizations must be used, this results in quite strong electric fields in the material which reduce the radiative recombination rate [6]. Then, we believe that to get more efficient emission, it is necessary to realize a contact with the inner surface of the porous film. Several attempts have been made, either by filling the pores with a conductive polymer [7-8] or by electroplating metals [9], however such materials are not transparent in the wavelength range of the emitted light. Large band-gap II-VI semiconductors seem to be better candidates to realize the contact with the porous layer: they can be chemically or electrochemically deposited, they can be chosen to be transparent in the luminescence spectral range, and they can form with silicon heterojunctions, which may lead to a good control of the charge injection into the nanocrystallites.

In this preliminary work, we show that three different compounds can be successfully deposited into the porous layer, either by electrodeposition (CdTe and ZnSe) or by chemical

Mat. Res. Soc. Symp. Proc. Vol. 452

bath deposition (CdS). The photoluminescence properties of the resulting structures are also analyzed.

EXPERIMENTS

Electrochemical deposition of CdTe and ZnSe

Binary II-VI semiconductors are formed by a metallic element (Cd or Zn) and a semi-metal (Te or Se), which have very different deposition potentials. For example, the redox potentials (referred to the SCE reference) for the couples Zn/Zn^{2+} and Se/H_2SeO_3 are -1.07 V and +0.37 V respectively . However, because of the very negative Gibbs free energy of formation of the semiconductor, the deposition potential of the less noble component is shifted to a more positive value, allowing the codeposition of the two elements [10]. By using solutions in which the metallic salt is in large excess compared to the nonmetallic oxide, codeposition is observed over a range of potentials of several hundred millivolts beyond the deposition potential of the nonmetallic element. The deposition potentials on a silicon electrode are found more cathodic when compared to deposition on a metallic electrode, because a sufficient electron flux at the silicon interface is only obtained for polarizations corresponding to near flat-band conditions. The typical current-potential curves obtained on porous silicon electrodes are shown on figure 1. The potential range found for CdTe and for ZnSe deposition are -600/-750 mV and -800/-1200 mV respectively, which is very similar to what is observed on non porous silicon electrodes.

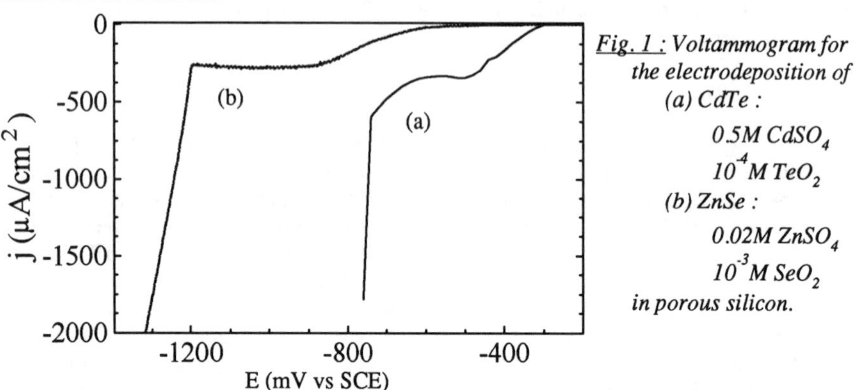

Fig. 1 : Voltammogram for the electrodeposition of (a) CdTe : 0.5M $CdSO_4$ 10^{-4}M TeO_2 (b) ZnSe : 0.02M $ZnSO_4$ 10^{-3}M SeO_2 in porous silicon.

CdTe deposition has been performed under potentiostatic and galvanostatic conditions on nanoporous silicon layers obtained by anodisation of n-type silicon (doping $N_d = 10^{17} cm^{-3}$, (100) orientation). The layer porosity was about 70% and the thickness was kept below 0.5 µm in order to prevent the formation of macropores in the substrate. Depending on the total exchanged charge, different types of samples have been prepared, which correspond either to only a few monolayers deposited on the whole porous silicon surface (with no CdTe on the outer surface), or to samples where a CdTe layer covers the outer surface. In this last case, a good adhesion of the deposit is obtained, and no layer cracking is observed, contrasting to what is observed for deposit realized on non porous silicon electrodes. The presence of CdTe is confirmed by X-ray characterization, as shown on figure 2, which shows that the growth favors

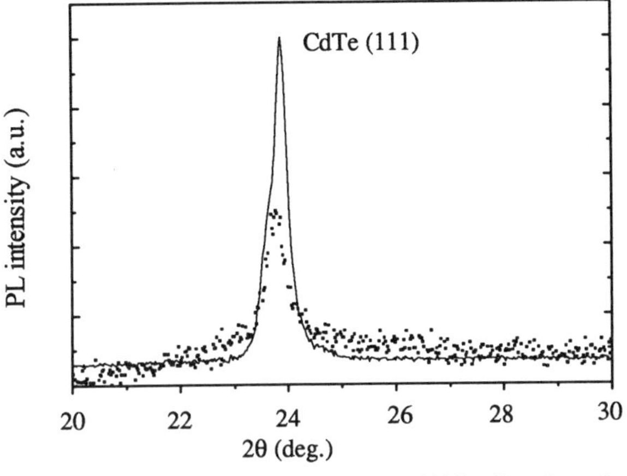

Fig. 2 : XRD pattern of CdTe electrodeposited
—— on bulk silicon
- - - - - in porous silicon

the preferential orientation along the (111) direction, for both non porous and porous electrodes. This feature might be related to a layer by layer growth.

The deposition of ZnSe has been studied on porous layers formed on heavily n-doped silicon ($N_d = 2.10^{18} cm^{-3}$). Porous silicon is formed at low current density to give a nanoporous structure over several microns, without formation of macropores. Typical samples where 2 μm thick, 70% porosity and showed efficient photoluminescence before the cathodic treatment. ZnSe deposition has been performed under potentiostatic conditions, at a potential of -900 mV. Figure 3 shows a SEM cross section picture of a cleaved sample after deposition of about 15% of ZnSe. Different contrasts appear within the porous layer, suggesting a strong concentration gradient in the thickness. Analysis by X ray fluorescence realized by focusing the beam at different locations on the cross section confirm the presence of ZnSe inside the pore network (see figure 3) and show qualitatively a large concentration difference between the top and the bottom of the porous layer. Such gradients likely result from the fact that the deposition regime is diffusion limited, which favors the electrochemical reaction in the upper part of the layer. Different deposition conditions are at present under study in order to get more homogeneous concentrations in the layer thickness.

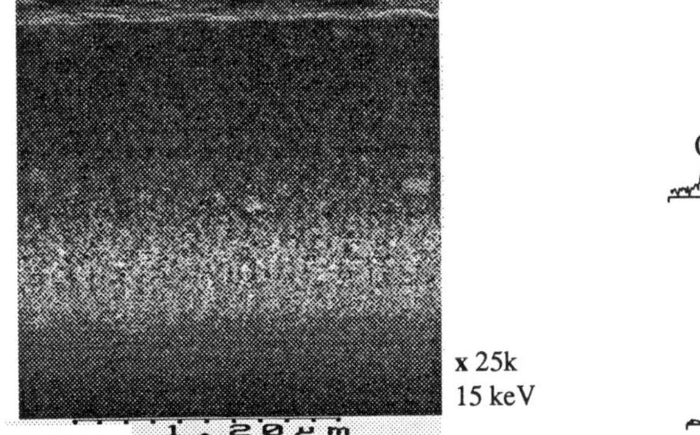

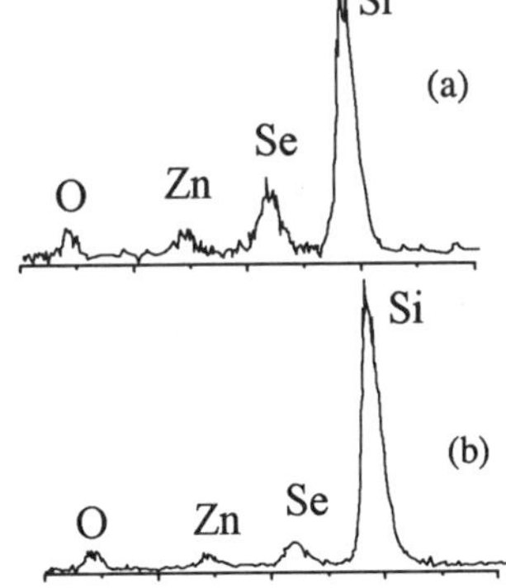

Fig. 3 : SEM image and X-ray fluorescence of a porous layer after ZnSe deposition
(a) top of the layer (b) bottom of the layer

Chemical Bath Deposition of CdS

Although CdS can also be electrodeposited, it was interesting to investigate the possibility to achieve deposition by pure chemical means which may lead to more homogeneous layers. The principle of the chemical bath deposition is in a first step to form Cd complexes in solution which precipitate on active sites at the semiconductor surface; in a second step, a sulfur donor in solution like thiourea or thioacetamid reacts with the adsorbed cadmium ions to form adsorbed molecules of CdS [11]. Cd^{++} complexation is achieved by adding ammonia to the solution, the pH being fixed between 7 and 10 by ammonium acetate, leading to the formation of two types of complexes ($Cd(OH)_i^{2-i}$ and $Cd(NH_3)_j^{2+}$) what relative concentrations depend on the pH value. It is important to use solutions of low pH in order to prevent porous silicon dissolution, which is very fast in alkaline solutions. After several negative trials, a specific procedure has been found to grow CdS in porous silicon, using a sequential deposition method: $Cd(OH)_2$ is precipitated by a contact with a cadmium acetate bath (0.1 M), then CdS formation is achieved in a thioacetamid solution (0.1M). After this treatment, the color of the porous layer changed to green, indicating the presence of CdS in the layer. SEM cross section photographs do not show any contrast variations in the layer thickness (see figure 4), however the aspect is very different to what is observed on a cross section examined before deposition : the layer appears brighter with much lower contrast after CdS deposition. The presence of CdS in the whole thickness of the layer is confirmed by X-ray fluorescence analysis. By focusing the beam on different parts of the layer, from the top to the bottom, the signature of the elements Cd and S is always found (figure 4), although a concentration gradient can be suspected from a comparison of the peak intensities obtained in the upper part and at the bottom of the layer. RBS analysis of the layer confirms that CdS is present in the whole layer, with a concentration decrease of a factor of about two between the top and the bottom. The RBS data also show a small excess of about 20% of the atomic content in sulfur compared to cadmium. These results indicate that the chemical bath deposition method can be successfully adapted to porous silicon, leading to only a small concentration gradient in the layer. It seems quite difficult to completely eliminate this gradient which is very likely due to porous nanostructure of the layer. Pores are not of regular shapes and their narrower regions can be blocked by the deposit thus preventing the complete filling of the wider parts.

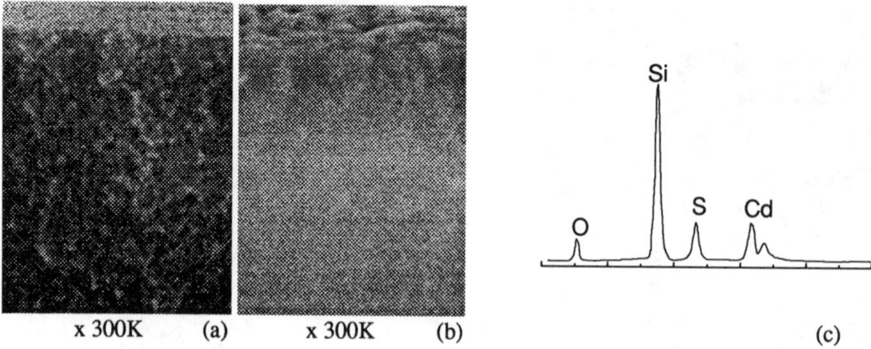

Fig. 4: SEM images of porous silicon as formed (a), of porous silicon after CdS deposition (b) and X-ray fluorescence analysis of the layer after CdS deposition (c)

RESULTS

Photoluminescence of the composite porous layers

Very different results are obtained depending on the deposited compound. After CdTe deposition, the photoluminescence spectrum which is obtained when the sample is immersed into water already shows a strongly reduced intensity, by a factor 10 to 20, depending on the samples. After drying, the emission almost disappears. This behavior might be related to the large lattice mismatch which exists between the compound and porous silicon (19%), resulting in strong stresses applied to the nanostructured material which can drastically affect the PL efficiency. In the case of ZnSe/porous silicon layers, there are no significant changes in the emitted intensity, even after sample drying, as shown on figure 5, but a blue shift of about 80 nm is observed. This spectral shift may result from different interactions due to the introduction of a new material into the pores, and similar shifts have been reported after deposition of metals, introduction of conducting polymers or porous surface methylation. It might also be partly related to some oxidation of the porous structure during the overall treatment. The very interesting result is that the emission efficiency is not strongly modified. Indeed, with ZnSe, the lattice parameter difference is much lower (4.4%) so that important stress effects are not expected. This behavior suggests that the surface passivation of the silicon nanocrystallites is not degraded by the electrochemical treatment. However, these results have to be confirmed on samples with lower concentration gradients, because in the present case, the contribution to the total spectrum of the part of the layer which has not been covered by the deposit may still be not negligible. The photoluminescence of layers containing CdS is also maintained, but with an efficiency decrease by a factor which can vary between 2 to 15 depending on the samples (figure 6). This emission intensity loss then remains tolerable, more especially as we know from RBS analysis that the whole porous structure is covered by the deposit.

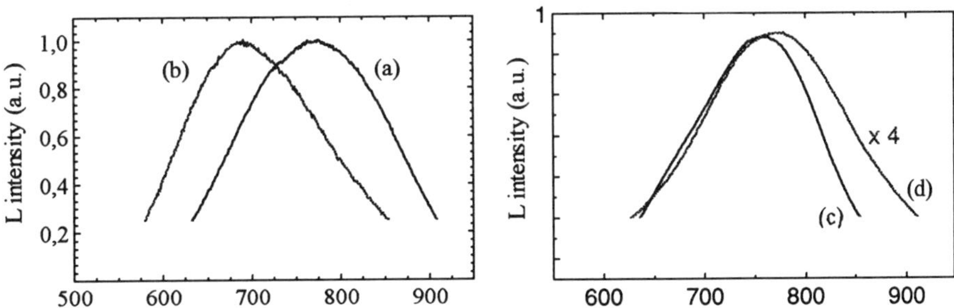

Fig 5 : Photoluminescence of a 2.5 μm thick n^+ type porous layer before (a) and after (b) ZnSe deposition. Photoluminescence of a 2 μm thick p type porous layer before (c) and after (d) CdS deposition

CONCLUSIONS

In summary, these results show that the introduction of II-VI semiconductor compounds into porous silicon layers can be successfully achieved either by electrochemical or chemical means. It seems that chemical bath deposition leads to a much better penetration of the material inside the pore network, but improvements in the electrodeposition method will be probably obtained for example by using pulsed potential techniques in order to prevent diffusion limitations. Very encouraging results have been obtained on the luminescence properties of the composite structures involving ZnSe and CdS, which show an efficient emission comparable to that of the starting material. More quantitative information about the deposits are expected from other characterization techniques like Auger profiling. The next step will be to investigate the electrical behavior of the resulting structures and to study if they can lead to improved electroluminescence characteristics.

ACKNOWLEDGMENTS

The authors are much indebted to Dr J.L. Buevoz and Dr D. Jourdain from the CNET of Meylan (France) for SEM observations and X-ray fluorescence analysis, and to Dr C. Ortega and Dr A. Grossman, from the Groupe de Physique du Solide, Université de Paris 6, France, for RBS analysis.

REFERENCES

1. L.T. Canham, Appl. Phys. Lett. **57**, 1046-1048 (1990)

2. A. Halimaoui, G. Bomchil, C. Oules, A. Bsiesy, F. Gaspard, R. Hérino, M. Ligeon, F. Muller, Appl. Phys. Lett. **59**, 304-306 (1991)

3. P. Steiner, F. Kozlowski, W. Lang, Thin Sol. Films **225**, 49-51 (1995)

4. A. Loni, A.J. Simons, T.I. Cox, P.D.J. Calcott, L.T. Canham, Electron. Lett. **31**, 1288-1289 (1995)

5. R. Hérino, J. Chim. Phys.**93**, 641-649 (1996)

6. T. Oguro, H. Koyoma, T. Ozaki, N. Koshida, J. Appl. Phys., in press (1997)

7. N. Koshida, H. Koyama, Y. Yamamoto, G.J. Collins, Appl. Phys. Lett. **63**, 2655-57 (1993)

8. A. Bsiesy, Y.F. Nicolau, A. Ermolieff, F. Muller, F. Gaspard, Thin Sol. Films **255**, 43-48 (1995)

9. M. Jeske, J.W. Schultze, M. Thönissen, H. Münder, Thin Sol. Films **255**, 63-66 (1995)

10. R.K. Pandey, S.N. Sahu, S. Chandra, Handbook of Semiconductor deposition, (Marcel Dekker, Inc., New York, 1996), pp 27-56

11. R. Ortega-Borges, D. Lincot, J. Electrochem. Soc. **140**, 3464-3473 (1993)

FABRICATION AND CHARACTERIZATION OF LIGHT EMITTING POROUS SILICON AND POLYMER NANOCOMPOSITES

S. P. Duttagupta, P. M. Fauchet
Department of Electrical Engineering, University of Rochester, Rochester NY 14627

X. L. Chen, S. A. Jenekhe
Department of Chemical Engineering, University of Rochester, Rochester NY 14627

ABSTRACT

We report the fabrication of nanocomposites by the infiltration of polymers into porous silicon. Polymers such as polyamide, polystyrene, PMMA, and PVC were chosen because they are commonly available and have been extensively studied. The pore-filling was accomplished by either diffusion of the polymer molecules into porous silicon or in-situ polymerization of the monomer. The Vickers hardness and the thermal conductivity of the samples were measured. There was a difference in the nanocomposite characteristics depending on whether the samples were as-anodized or had been annealed in oxygen. By infiltrating polyamide into an as-anodized sample, a 42% increase in hardness and a 24% increase in thermal conductivity were observed at room temperature, without any degradation of luminescence.

INTRODUCTION

Light emitting porous silicon (LEPSi) has been the focus of an intense research activity.[1,2] By annealing the as-anodized (AA) samples in a dilute oxygen environment we form silicon rich silicon oxide (SRSO) which is the active layer for our 0.1% efficient LEDs.[3] These devices have been shown to be stable over several weeks of continuous operation. Nevertheless, microhardness[4] and Young's modulus[5] data suggest that a significant improvement of the mechanical properties will be required, for further improvement in device performance. A possible means to improve the mechanical properties is to infiltrate and fill the pores by another material to form composites. Halimaoui et al. have reported the UHV deposition of Ge and Si onto PSi resulting in a progressive coverage of the inner surface and finally pore-filling.[6] We have recently demonstrated the fabrication of polymer-porous silicon nanocomposites with improved hardness.[4] The reason for choosing polymers to form composites was due to the diverse range of their properties and functionalities.[7] In this work, we report how the hardness and thermal conductivity are affected by the type of host (AA/SRSO), and the way the nanocomposite is formed.

EXPERIMENTAL

Single crystal, (100) oriented, p^+ Si (0.1 Ω-cm) substrates were selected for anodization. The porosity was varied by altering the current density.[3] The hardness of the samples was measured using the Vickers technique. This is a static indentation test, where a square based pyramidal indenter with an apex of Ø=136° is forced on the sample surface, causing a diamond shaped indent on the surface. This leads to the following expression for hardness:[8]

$$H_v = 1.8544\,(P_L.d^{-2}) \qquad (1)$$

where, d (mm) is the mean diagonal length (of the diamond shaped indent) and P_L (1Kgf = 9.8N) is the applied load. The measurements were performed by a Tukon 300BM microhardness indenter at ambient laboratory conditions. Individual Vickers hardness (H_v) values were calculated as a weighted mean over 3 identically prepared samples and ten indentations per

Mat. Res. Soc. Symp. Proc. Vol. 452 © 1997 Materials Research Society

sample. The applied load was 0.49 N and the film thicknesses were about 2 μm. The intrinsic hardness of the film was determined by eliminating the contribution of the underlying silicon substrate.[4]

The thermal conductivity (TC) of the samples was measured using the thermal comparator technique developed by Lambropolous et al.[9] This enables rapid, non-destructive TC measurement, and can be applied to samples in a conventional film-on-substrate geometry. The conductivity is determined normal to the film surface. The apparatus consists of an environmentally controlled sample chamber enclosing a sample stage, a control and readout module, and signal processing equipment. The detector is a thermocouple junction sensing tip which is maintained in contact with the sample surface, by applying a load of 5 g. The average EMF generated by the temperature difference between the sensing tip and a reference junction is proportional to the apparent conductivity (k_{app}) of the sample. The TC values were then determined based on a calibration curve, which was generated using bulk materials of known conductivity.

The apparent conductivity measured is that of the film, interface and substrate combined. A detailed modeling of the heat flow is required in order to extract the intrinsic conductivity of the film (k_f).[9] First, the effective film conductivity (k_{eff}) is determined, which also includes any contribution due to the interfaces between the film and the substrate, and the film and the probe tip:

$$(k_{eff})^{-1} = (\pi/4)\,(a/t)\,(k_{app}^{-1} - k_s^{-1}) \qquad (2)$$

where a is the heat flow radius, t is the film thickness and k_s is the substrate conductivity. The above equation is valid only if $k_{eff} << k_s$ and $t << a$. Previous TC measurement of highly porous PSi films, indicates that k_{PSi} (1.2 W/mK)[10] << k_{Si} (150 W/mK).[11] The heat flow radius was determined to be 100 μm, from thermal conductivity measurements of polymers (for example polyamide, k = 0.5 W/mK)[7] with known (and similar) conductivities. The film thickness was varied from 0.5 μm to 2.5 μm. Finally, we evaluate k_{eff} for several values of the thickness, and the intrinsic conductivity k_f is determined from the following expression:

$$t/k_{eff} = t/k_f + R_{int} \qquad (3)$$

where R_{int} denotes the thermal resistance due to any interfaces.

Following this, samples were annealed in dilute oxygen (10%) at 900 °C to form SRSO. Both AA and SRSO films (P=80%) were used to form nanocomposites. In addition, different procedures were used for infiltrating the pores with polymers. The first was to dissolve the polymers in a solvent and immerse the PSi sample in the solution for a period of several days, thereby allowing sufficient time for the polymer molecules to diffuse into the pores. The polymers used in this process were polystyrene, polymethyl methacrylate, polyvinyl chloride, polyamide, and THV 200 P (a fluoroplastic composed of tetrafluoroethylene, hexafluoropropylene, and vinylidene fluoride). Except for the last mentioned, these materials have been extensively characterized.[7] Standard organic solvents such as toluene, xylene, and acetone (for THV) were used to dissolve the polymers. The concentration was quite dilute (2-3%) for all cases (except THV, conc. ~ 6%) because of the low solubility of the polymers.

The second procedure was to infiltrate the pores with monomers such as MMA and Styrene. Subsequently, polymerization was carried out insitu to form PMMA and polystyrene. The monomers (0.8 M) were dissolved in benzene, along with an initiator ($FeCl_3$/HCl- 0.04 M). The PSi samples were immersed in a 10 mL solution, and the container was placed in a 60 °C water bath for four hours The final step before characterization for both procedures was a brief toluene rinse (acetone for THV) to remove any residual film on the surface.

Finally, the photoluminescence, microhardness, and thermal conductivity of the composites were measured and compared with the intrinsic (both as-anodized and oxidized) material. The PL exciting wavelength was the He-Cd laser line at 442 nm.

RESULTS AND DISCUSSION

Fig. 1. shows that the hardness reduces steadily with an increase in porosity. At high porosities (P>70%) the rate of reduction is higher, which may be due to morphological reasons.[3] At 80% porosity, the intrinsic hardness of both as-anodized and SRSO films is 2.2 GPa, which is about 10% of the crystalline silicon value. This agrees with the trend displayed for Young's modulus variation with porosity.[5]

The decrease in thermal conductivity (TC) as seen in Fig. 2 is even greater than that of hardness. At 80% porosity, the TC for the as-anodized sample is 1.8 W/mK (TC for c-Si: 150 W/mK). For the corresponding SRSO film, the value is 1.5 W/mK. This is of comparable magnitude as in previous reports.[10] We could not measure the thermal conductivity at lower porosities, because the thermal comparator technique is only accurate when the film conductivity is much lower (<10%) than that of the substrate.

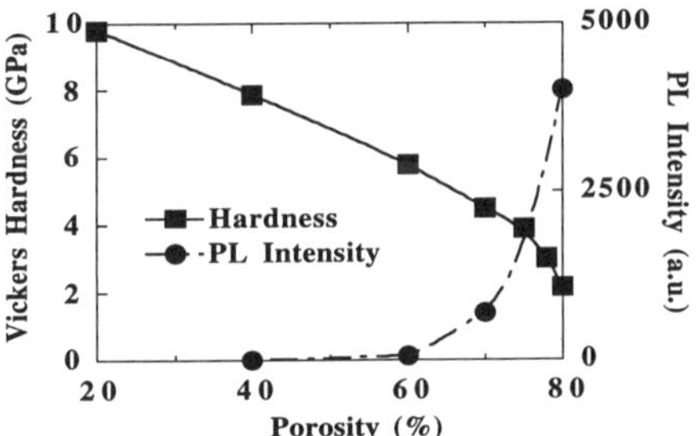

Fig. 1. Variation of hardness and PL intensity with porosity for porous silicon prepared from p+ substrates.

Fig. 2. Variation of thermal conductivity with porosity for porous silicon prepared from p+ substrates.

The photoluminescence characteristics are shown in Fig. 3 for the as-anodized case. The samples included were (a) control, and PSi-polymer composites such as (b) PSi-PMMA (infiltration by diffusion), (c) PSi-THV (diffused), and finally PSi-PMMA (insitu polymerization of MMA). The PL peak position (λ_p) was unaltered for all the cases except for THV, where we see a blueshift. This may be due to the high dielectric constant of THV (5-6), which causes a partial screening of the excitons and allow the excitonic energy levels to move closer to the bandgap.[12] When comparing the PL intensities we see that the luminescence in the composite samples is comparable to or slightly greater than the control PSi samples. Within the composites, the insitu polymerized MMA sample has the maximum intensity, followed by the THV and the PMMA-PSi (diffused) samples. The rest of the polymers (polyamide, PVC, polystyrene) which were all infiltrated by diffusion behaved similarly to the PMMA (diffused) case. The improvement in PL intensity can be attributed to better passivation, which is also linked with the effectiveness of the penetration of the polymer. From our results, it appears that the most favorable route is the insitu

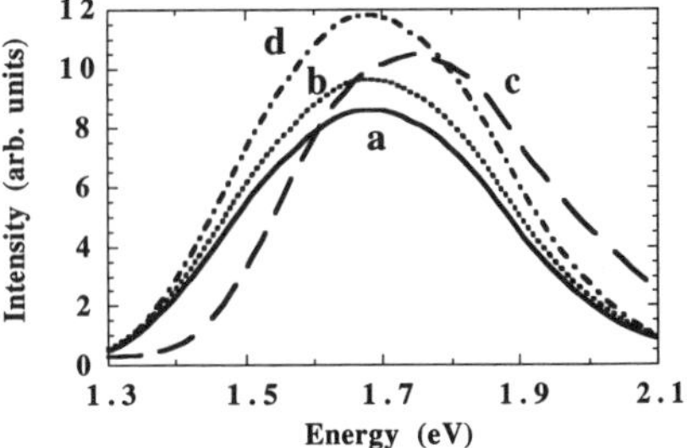

Fig. 3. PL characteristics of an (a) as-anodized PSi sample (p^- substrate, 85% porosity), and PSi-polymer composites: (b) PMMA (diffused), (c) THV (diffused), and (d) PMMA (insitu polymerized MMA).

polymerization. This is presumably because the monomer molecules could easily infiltrate the pores due to their smaller size. The higher concentration of THV probably allows for better penetration than the other polymers.

Figures 4(a) and 4(b) show that the hardness of all the nanocomposite structures is improved as compared to the 80% porous AA and SRSO films. The order of increase for the as-anodized based composites is PVC < Polystyrene < THV < insitu polymerized Styrene < PMMA < insitu polymerized MMA < Polyamide (infiltration by diffusion unless specified otherwise). The order is different for the SRSO based composites which is PVC < Polystyrene < THV < PMMA < Polyamide (all of these were diffused) < insitu polymerized Styrene < insitu polymerized MMA for the SRSO based composites. The increase in hardness for the SRSO samples is smaller than that for the AA samples. It is possible this is caused by the larger oxide structures which results in a smaller pore size as compared to the as-anodized case. This makes the penetration of large polymer molecules difficult. The relative change for the insitu polymerized samples is comparatively less, which is clearly due to easier penetration (this also explains the different order of hardness increase for AA vs. SRSO composites). Finally, we note that it was not possible to measure the intrinsic hardness of the polymer films, because the indented pattern could not be accurately measured. Instead, from Fig. 4(a) and 4(b), we observe that the hardness increases proportionately with the polymer density (with the exception of PVC).

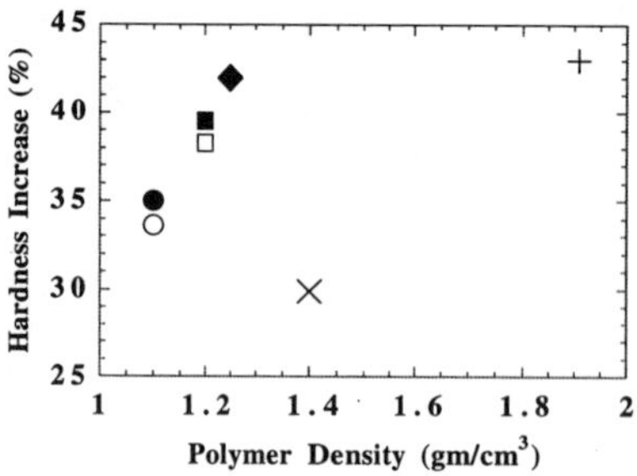

Fig. 4(a) Hardness of as-anodized (p^+ substrate, 80% porosity) - polymer nanocomposites vs. specific gravity of polymer : PSi-Polystyrene (insitu polymerized- closed circle), PSi-Polystyrene (diffused- open circle), PSi-PMMA (insitu polymerized- closed square), PSi- PMMA (diffused- open square), PVC (diffused- X), polyamide (diffused- diamond), and THV (diffused- +). The hardness of the intrinsic film was 2.2 GPa.

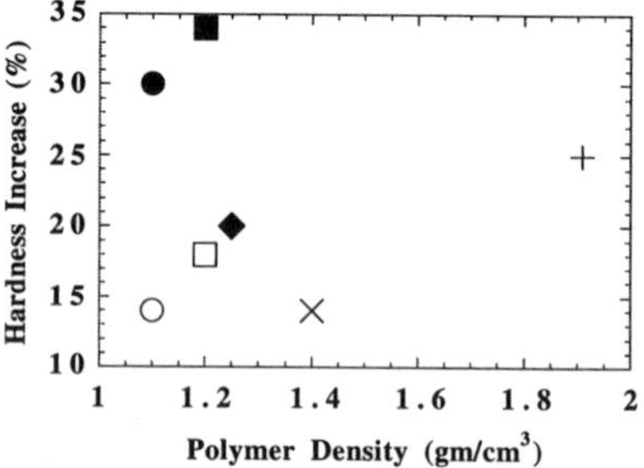

Fig. 4(b) Hardness of SRSO (p^+ substrate, 80% porosity) - polymer nanocomposites vs. specific gravity of polymer : PSi-Polystyrene (insitu polymerized- closed circle), PSi- Polystyrene (diffused- open circle), PSi-PMMA (insitu polymerized- closed square), PSi- PMMA (diffused- open square), PVC (diffused- X), polyamide (diffused- diamond), and THV (diffused- +). The annealing conditions were 90% N_2, 10% O_2, 900 °C. The hardness of the intrinsic film was 2.2 GPa.

From Fig. 5(a) and 5(b), it is seen that composite formation has also improved the thermal conductivity. The increase is in proportion to the intrinsic conductivity of the polymers. For the AA samples, the order is Polystyrene < insitu polymerized styrene < PVC < THV < PMMA < polymerized MMA < polyamide (diffused unless specified otherwise). For the SRSO samples the order is Polystyrene < THV < PVC < insitu polymerized styrene < PMMA < polymerized MMA < Polyamide. Again as in hardness, we see a smaller increase of the thermal conductivity in SRSO composites than in the AA composites. Also, the monomers are relatively less affected due to their smaller size.

CONCLUSION

The hardness of the nanoporous light emitting porous silicon is approximately an order of magnitude lower, while the thermal conductivity is about two orders of magnitude lower than that of crystalline silicon. We have fabricated LEPSi-polymer composites which resulted in the increase of both hardness and thermal conductivity. The improvement in these properties was achieved without causing any degradation in luminescence. The PL intensity was enhanced in some cases, possibly due to better passivation. Preliminary results suggest that (e.g. THV-LEPSi composite) it may be also possible to modify the PL properties by altering the effective dielectric constant of the medium. Other possibilities include the introduction of luminescent polymers. Further investigation of materials and process issues will be required in order to fabricate a composite with optimum mechanical, thermal, electrical and optical properties.

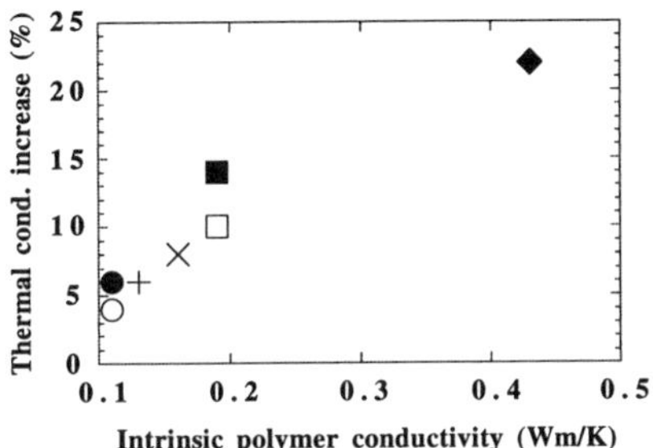

Fig. 5(a) Thermal conductivity of as-anodized PSi (p^+ substrate, 80% porosity) - polymer nanocomposites vs. specific gravity of polymer : Polystyrene (diffused - open circle), Polystyrene (insitu polymerized - closed circle), THV(diffused - +), PVC (diffused - x), PMMA (diffused - open square), PMMA (insitu polymerized - closed square), and Polyamide (diffused - diamond). The conductivity of the PSi film is 1.8 Wm/K.

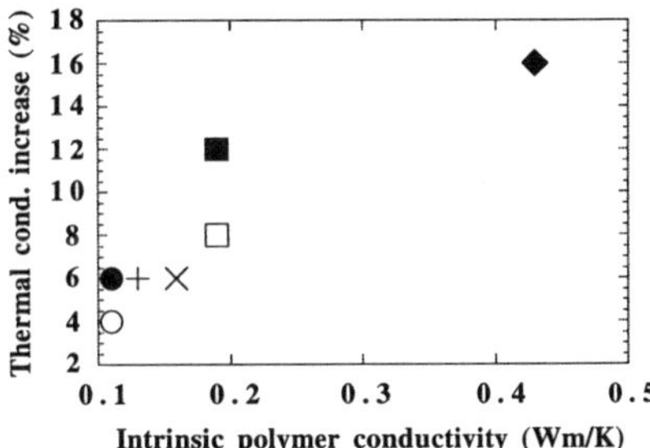

Fig. 5(b) Thermal conductivity of SRSO (p^- substrate, 85% porosity) - polymer nanocomposites vs. specific gravity of polymer : Polystyrene (diffused - open circle), Polystyrene (insitu polymerized - closed circle), THV(diffused - +), PVC (diffused - x), PMMA (diffused - open square), PMMA (insitu polymerized - closed square), and Polyamide (diffused - diamond). The conductivity of the PSi film is 1.5 Wm/K.

ACKNOWLEDGMENT

This work was supported by the NSF Science and Technology Center for Photoinduced Charge Transfer, NSF (CTS-9311741), and the Army Research Office. The authors wish to acknowledge the assistance of Birgit Puchebner of the Center for Optics Manufacturing at the University of Rochester, with the microindentation measurements. The THV polymer material was made available courtesy Dennis Hull of the 3M Corporation.

REFERENCES

[1] P.M. Fauchet, L. Tsybeskov, C. Peng, S.P. Duttagupta, J. von Behren, Y. Kostoulas, J.V. Vandyshev, and K.D. Hirschman; *IEEE Journ. Selec. Topics Quant. Elec.,* **1,** 1126 (1995).
[2] P.M. Fauchet, *Journ. Lum.,* **70,** 294 (1996).
[3] L. Tsybeskov, S.P. Duttagupta, K.D. Hirschman and, P.M. Fauchet, *Appl. Phys. Lett.,* **68,** 2058 (1996).

[4] S.P. Duttagupta, X. L. Chen, S.A. Jenekhe, and P.M. Fauchet; Microhardness of porous silicon films and composites, *Sol. St. Comm.*, In Press.
[5] D. Bellet, P. Lamagnère, A. Vincent, and Y. Bréchet; *J. Appl. Phys.,* **80**, 3772 (1996).
[6] A. Halimaoui, Y. Campidelli, P.A. Badoz, and D. Bensahel, *J. Appl. Phys.,* **78**, 3428, (1995).
[7] J. Brandrup and E.H. Immergut Eds.; *The Polymer Handbook,* John Wiley and Sons, New York, 1975.
[8] D. Chicot and J. Lesage; *Thin Solid Films* **254**, 123-130 (1995).
[9] J.C. Lamropolous, M.R. Jolly, C.A. Amsden, S.E. Gilman, M.J. Sinicropi, D. Diakomihalis, and S.D. Jacobs; *J. Appl. Phys.,* **66**, 4230 (1989).
[10] W. Lang, A. Drost, P Steiner, and H. Sandmaier; *Mat. Res. Soc. Symp. Proc.,* **358**, 561 (1995).
[11] Properties of Silicon, EMIS Datareview Series No. 4, INSPEC, The Institution of Electrical Engineers, London and New York, 1988.
[12] R. Tsu and D. Babic, Porous Silicon Science and Technology, edited by J.C. Vial and J. Derrien (Springer-Verlag, Berlin, 1995) pp. 111-119.

DEPOSITION OF POLYPYRROLE INTO POROUS SILICON

J.D. MORENO*, F. AGULLO-RUEDA*, R. GUERRERO-LEMUS*, R.J. MARTIN-PALMA*, J.M. MARTINEZ-DUART*, M.L. MARCOS**, J. GONZALEZ-VELASCO**
* Departamento de Física Aplicada C-12 and Instituto Ciencia de Materiales, CSIC, Universidad Autónoma de Madrid, 28049 Madrid, Spain.
**Departamento de Química C-9, Universidad Autónoma de Madrid, 28049 Madrid, Spain.

ABSTRACT

Polypyrrole has been electrodeposited in the interior of the pores that form the Porous Silicon structure, and a very significant increase of the electrical conductivity of the samples has been observed. The degree of filling by the polymer has been found to be highly dependent on the electropolymerization conditions. Micro-Raman Spectroscopy experiments have allowed us to measure the amount of polymer as a function to the distance from the outer PS surface.

INTRODUCTION

Porous Silicon (PS) is currently the object of much research due to its photoluminescent and electroluminescent properties [1]. These properties offer new and interesting possibilities for the use of silicon in optoelectronic devices [2], provided a good electrical contact can be achieved at the porous surface.

Different attempts have been made to obtain efficient electrical contacts to PS by deposition of metals into the porous structure [3]. The present work is aimed at getting better electrical contacts to the porous surface by means of a conducting polymer (polypyrrole) deposited into it with no prior drying of the PS sample. One important goal is to grow the polymer inside the pores, filling them completely, rather than just coating the outer surface, in which case a deficient electric contact would be stablished. Therefore, Micro-Raman Spectroscopy measurements of PS samples filled with polypyrrole have been carried out in order to obtain evidence that the polymer is actually formed inside the entire pore.

EXPERIMENT

PS layers were formed from boron-doped (*p*-type) silicon wafers of (100) orientation and a resistivity of 0.1-0.5 $\Omega\cdot$cm by electrolytic attack in a 2:1 HF (48 wt %)/ethanol (98 wt %) mixture. Wafers were etched galvanostatically under illumination at a current density of 50 mA/cm^2 for either 1 or 5 min. Once formed, samples were rinsed first in ethanol and finally in pure acetonitrile. Drying of samples was avoided in order to prevent cracking processes [4].

Deposition of the polymer was carried out galvanostatically, applying an anodic current pulse to a PS electrode immersed in a solution of the monomer, pyrrole, in acetonitrile containing 0.1 M $LiClO_4$ as supporting electrolyte. Different values for the monomer concentration in solution, C_{py}, and for the current density of the anodic pulse, i, have been employed and their influences studied.

Mat. Res. Soc. Symp. Proc. Vol. 452 © 1997 Materials Research Society

Micro-Raman spectroscopy measurements have been carried out with a Renishaw Ramascope spectrometer. Light from an argon laser was focused onto the sample surface through an optical microscope. The light scattered was collected through the same microscope and analyzed by the grating spectrometer. Spectra have been taken at different points of cross sections of the PS samples, so that each set of data corresponds to a certain depth of the original PS layer (depth resolution ca. 1 μm). A subsequent data analysis allows us to obtain the Raman intensity as a function of the distance to the outer surface.

RESULTS

The change of the potential, E (measured against a Pt quasireference electrode), with time was registered during the electrodeposition process. The general shape of the chronopotentiograms is as shown in Figure 1. Four time ranges can be distinguished. *AB* is characterized by a sharp E increase, and can be mainly associated to a partial oxidation of PS [5]. Along time interval *BC* a nearly constant E is obtained, and polymerization on the outer surface is not visually observed. E slightly increases along *CD*, where traces of external polymerization start to be detected. Finally, E stabilizes for times longer than *D*, and the formation of polymer on the outer surface is evident.

These results are similar to those reported for the electrodeposition of polybithiophene on PS [5] and seem to indicate that polymerization takes place inside the pores during time interval *BC*, reaching the external surface between *C* and *D*. After *D*, the process seems to be only superficial, leading to an external polypyrrole layer.

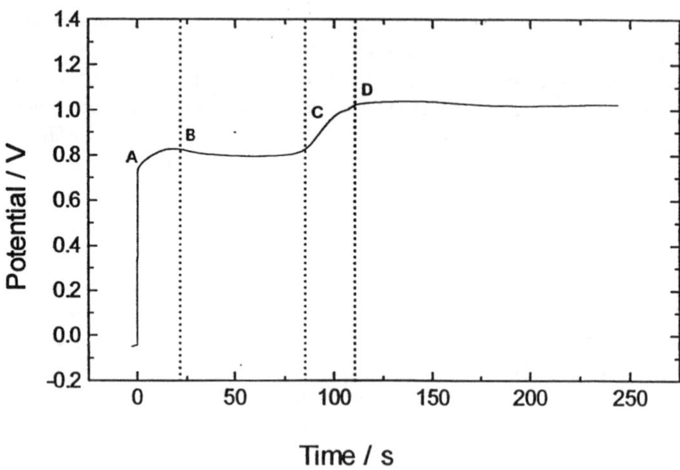

Figure 1. Chronopotentiogram for the formation of polypyrrole into a 3μm thick PS sample. C_{py} = 0.1 M; i = 4 mA/cm^2.

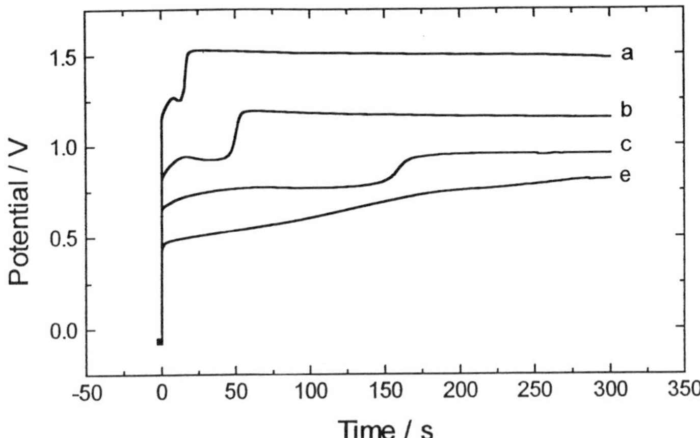

Figure 2. Chronopotentiograms for the formation of polypyrrole from a 0.05 M monomer solution into a 3μm thick PS sample under different anodic current densities: *(a)* 16 mA/cm^2; *(b)* 8 mA/cm^2; *(c)* 4 mA/cm^2; *(d)* 2 mA/cm^2.

Figure 2 shows the influence of the current density of the anodic pulse on the E-t curves obtained from a 0.05 M pyrrole solution. The time interval *CD*, which is the transition between the two constant potential regions, takes place at longer times as i decreases. At the same time, the obtained E values are smaller.

The amount of polymer formed inside the pores must be directly related to the electric charge transferred in the time interval *BC*, q_{BC}. Measurements made at different anodic current densities show that q_{BC} decreases when i increases, once a limiting value for i is surpassed. This fact seems to indicate that pores are not being completely filled by polymer, which would lead to deficient electrical contacts. For this reason, subsequent Raman experiments were carried out with samples where polypyrrole was electroformed at a relatively low i (4 mA/cm^2).

Figure 3 shows a Raman spectrum of a PS sample containing polypyrrole. The spectrum of polypyrrole deposited directly on monocrystalline silicon has been superimposed. Both spectra present the same bands for the polymer. Figure 3 also shows results for pure monocrystalline silicon, where only a Si peak at 520 cm^{-1} is obtained. It should be noticed that the latter peak is broader in the tails when it is obtained from PS. This effect is due to a relaxation of the selection rules originated by the small size of PS nanocrystals.

In Figure 4, the Raman intensity of the polypyrrole peak at 901 cm^{-1} has been plotted against the depth of the porous layer. In this way, data about the relative quantity of polymer in each section of the porous layer can be gained. It is particularly interesting to ascertain whether there is polymer at the bottom of the pores (and therefore, no void is left at the inner part of the structure), so as to study its growth along time interval *BC*.

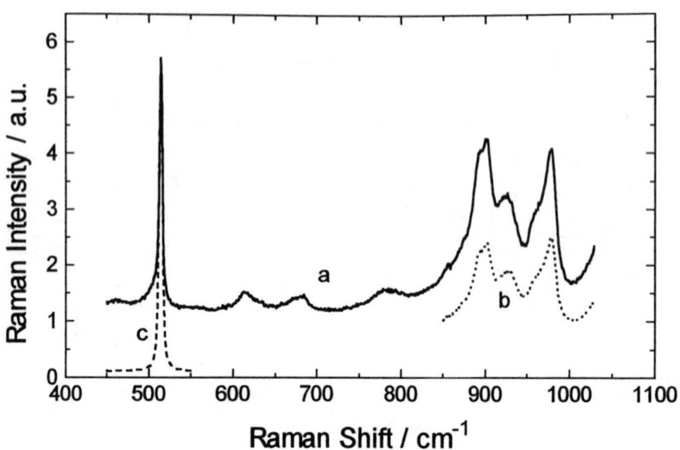

Figure 3. Raman Spectra of (*a*) polypyrrole into PS, (*b*) polypyrrole on monocrystalline Si, and (*c*) bare monocrystalline Si.

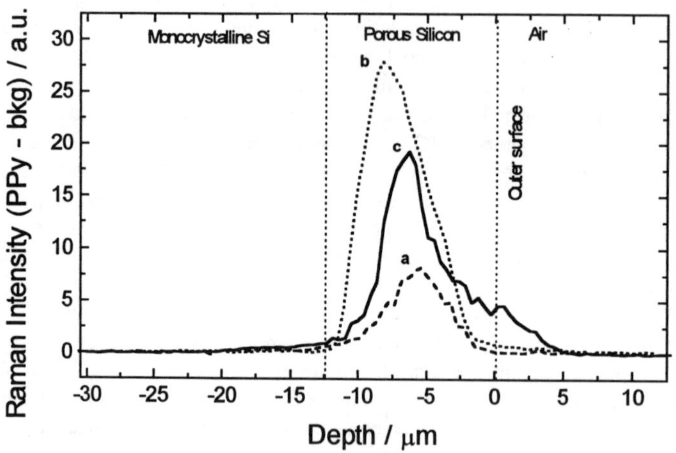

Figure 4. Raman results for polypyrrole formed into PS from a 0.1 M pyrrole solution; (*a*) at 4 mA/cm^2 stopping in the middle of the *BC* time interval; (*b*) at 4 mA/cm^2 stopping at *D*; (*c*) at 16 mA/cm^2 stopping after *D*. Depth resolution is ca. 1 μm.

Plot *a* in Figure 4 corresponds to a 12 μm thich PS sample in which polypyrrole was formed by means of an anodic pulse at 4 mA/cm^2 that was stopped just in the middle of time interval *BC*. The presence of polymer at the bottom of the pores can be noticed, although the greatest quantity is found in an intermediate region. Plot *b* corresponds to a similar sample, but one in which the anodic pulse was maintained until *D*. A much larger quantity of polymer in the pores is obtained (as compared to plot *a*), as well as a small amount on the outer surface.

Plot *c* in Figure 4 corresponds to a PS sample analogous to those of plots *a-b*, but in which a 16 mA/cm^2 anodic pulse was applied until a few seconds after point *D*. In this latter case, there is a considerable amount of polymer on the PS layer, as expected having stopped the anodic pulse at $t > D$. On the other hand, the amount of polymer in the pores is smaller than that obtained in plot *b*, where electrodeposition was carried out at a lower i. This result is in agreement with the measurements of q_{BC} as a function of i.

If these systems are to be applied for technological devices, it is necessary that the electrodeposited polymer shows appreciable electrical conductivity. Figure 5 shows conductivity measurements corresponding to the sample of Figure 4-*b*, as well as those of the same PS sample before polypyrrole was deposited. This result demonstrates the high conductivity of the electroformed contact.

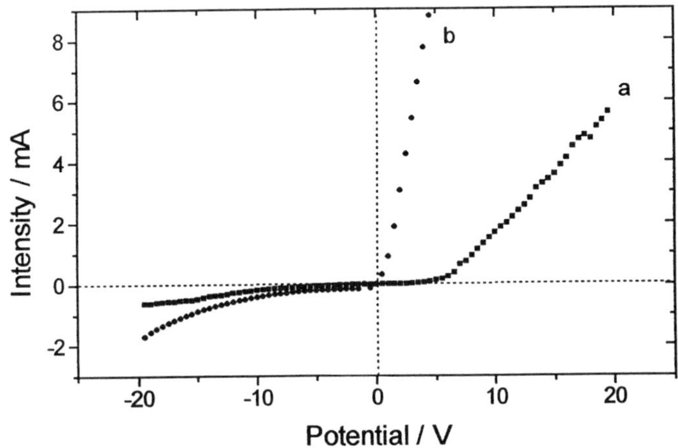

Figure 5. Conductivity measurements for the PS sample corresponding to Figure 4b, (*a*) before and (*b*) after polypyrrole was electrodeposited. Values of plot *a* are multiplied by a factor of 10^3.

CONCLUSIONS

Micro-Raman Spectroscopy measurements have allowed us to conclude that polypyrrole can be electrodeposited in the interior of the pores forming the PS structure. However, the degree of filling is highly dependent on the electropolymerization conditions, particularly the current density of the anodic pulse. When i is too high, voids are left inside the pores. Other parameters, such as monomer concentration or temperature, may also be important, and further research is currently in progress in order to optimize the process.

These preliminary results seem to indicate that, when electrodeposition conditions are good enough, the polymerization starts in the deepest porous layers and, therefore, the polymer essentially grows from the bottom to top.

The conductivity of the polypyrrole-filled PS samples greatly increases as compared to bare PS, which is a very important result from the technological point of view.

ACKNOWLEDGMENTS

We wish to thank the Fundación Ramón Areces for the financial help to carry out this research.

REFERENCES

1. L.T. Canham, Appl. Phys. Lett. **57**, 1046 (1990).

2. J. Wang, F.L. Zhang, W.C. Wang, J.B. Zheng, X.Y. Hou and X. Wang, J. Appl. Phys. **75**, 1070 (1994).

3. M. Jeske, J.W. Schultze, M. Thönissen, H. Münder, Thin Solid Films **255**, 63 (1995).

4. R. Guerrero-Lemus, J.D. Moreno, J.M. Martínez-Duart, M.L. Marcos, J. González-Velasco and P. Gómez, J. Appl. Phys. **79**, 3224 (1996).

5. J.W. Schultze and K.G. Jung, Electrochim. Acta **40**, 1369 (1995).

LOCAL STRUCTURE OF INDIUM-PLATED POROUS SILICON

Toshimichi ITO, Takashi OOIWA, Takanobu NAGAO and Akimitsu HATTA
Department of Electrical Engineering, Osaka University, Suita, Osaka 565, Japan,
ito@pwr.eng.osaka-u.ac.jp

ABSTRACT

The fine structure of porous silicon (PS) plated electro-chemically with indium at a low plating charge of 0.03 C/cm^2 has been studied mainly using Rutherford backscattering spectrometry and a transmission electron microscope. The results obtained here show (1) that the amount of plated In atoms strongly depends on the oxidation states of the pore surfaces, (2) that the average In densities, the maximum of which is usually located near the PS - Si substrate interface, are typically larger than 1% of the Si densities in the PS layers, and (3) that the plated In atoms are concentrated locally in pores larger in size. Two-step thermal oxidation of an In-plated PS specimen results in a structure of nm-sized Si crystallites surrounded mainly by Si oxide and In oxide.

INTRODUCTION

Since the discovery of strong room-temperature photoluminescence (PL) from anodically produced porous silicon (PS) in 1990 [1], many studies have been done from the viewpoints of basic understanding of the luminescence mechanism and possible applications to Si-related optical devices [2]. It turns out that a suitable passivation of as-prepared PS is required to control the strong aging properties [3] when PS-related devices could be realized. Various Si oxide passivations of PS have been reported to be useful to strongly reduce the aging phenomena due to formation of the stable Si - SiO_2 interface [4-8]. However, the electrically insulating nature of the Si oxide seems to be a significant problem for PS electroluminescence (EL) devices. Recently, metal plating of PS was reported to strikingly improve the EL properties of PS Schottky diodes [9,10]. We reported an improvement in PL from indium-plated PS [11]. However, the nanostructure of metal-plated PS has not been clarified yet. Of various metal incorporations into PS, especially indium incorporation seems to be important because the indium oxide is also electrically conductive when In-incorporated PS is oxidized for a possible efficient passivation of the PS fine structure. In the present paper, we investigate the nanostructure of In-plated PS and oxidized PS mainly using Rutherford backscattering spectrometry (RBS) and transmission electron microscope (TEM) analysis.

EXPERIMENTAL

At first, a p-type Si (100) wafer with a resistivity of 9 - 12 Ωcm was anodized in a cell filled with a HF - ethanol solution (HF : H_2O : C_2H_5OH = 1 : 1 : 2 by volume) under tungsten lamp illumination of 4.0×10^4 lx. The anodization was carried out for 120 s using a dc current source of 20 mA/cm^2. Then, the solution in the cell was changed to an indium sulfate solution of $In_2(SO_4)_3$: H_2O = 6 : 100 (by weight), and the polarity of the current source was reversed, followed by a 10-min current flow of 0.05 mA/cm^2 for In plating. During the plating, the PS specimen was illuminated (1.1×10^3 lx) by the same tungsten lamp as used in the anodization process. Rapid oxidation of the plated PS was performed at 720 ℃ for 10 s in a dry oxygen - nitrogen gas with a volume ratio (O_2/N_2) of 1/25 using a lamp furnace, followed by

Mat. Res. Soc. Symp. Proc. Vol. 452 © 1997 Materials Research Society

conventional electric furnace oxidation at 500 ℃ for 30 min.

RBS data of the In-plated specimens thus obtained were taken using a Pelletron accelerator (model: National Electrostatic Corp. 3S-R10). The beam size of 2.1-MeV He^{2+} ions employed was 2 mm in diameter. For TEM measurements using a 200-kV electron microscope (model: JEOL JEM-2010) with an energy-dispersive X-ray (EDX) analysis system (model: Noran 5500), the specimens were cut into small pieces, some of which were thinned by mechanical lapping and subsequent ion etching. The probing beam size was approximately 5 nm in diameter for TEM EDX measurements. Scanning electron microscope (SEM) images with mapping images of EDX data were taken using a field-emission SEM.

RESULTS AND DISCUSSION

As-Indium-Plated Porous Silicon

We have found that In-plating can be successfully done for freshly prepared PS specimens whereas it is difficult to plate naturally-oxidized PS specimens. Therefore, special care was required so as to restrain an unnecessary air-exposure of the PS specimens appearing below as shortly as possible when the solution in the anodization cell was changed from the HF solution to the $In_2(SO_4)_3$ solution.

The depth distribution of plated In was measured non-destructively by the conventional RBS technique. Two of RBS spectra thus obtained are typically shown in Figs.1(a) and 1(b). As schematically demonstrated by the insets in Figs.1, In metal can be plated into the PS layers. However, the In depth distribution depended on the samples, and was roughly grouped into the following two cases. For the one case (type I) as shown in Fig.1(a), the whole PS layer was plated although the amount of plated In is larger near the PS - Si substrate interface than in the subsurface PS layer. For the other case (type II), there was a depth region where almost no In atoms existed while considerable amounts of the In atoms were populated near the PS - substrate interface and near the specimen surface (Fig.1(b)). The In signal from the surface region is, however, mostly originated from an $In_2(SO_4)_3$ salt in the solution since a small amount of S was also present near the surface. The total amount of the plated In atoms ranged typically from 1×

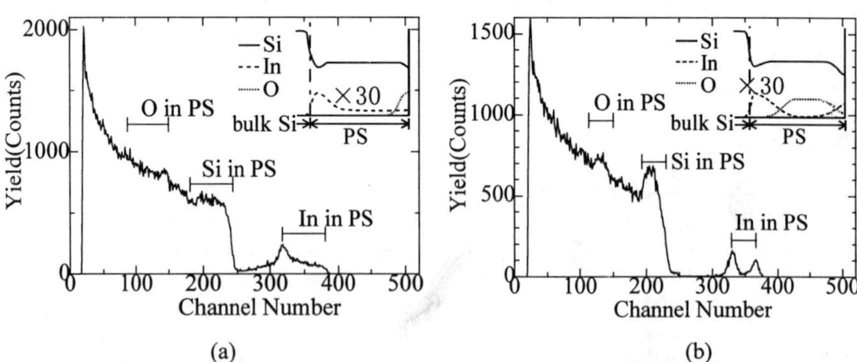

Fig.1. Rutherford backscattering spectrometry (RBS) spectra taken for a type I specimen ((a)) and a type II specimen ((b)). Insets show schematic distributions of Si, In and O atoms. A partial channeling effect was observed for the substrate Si especially in (b).

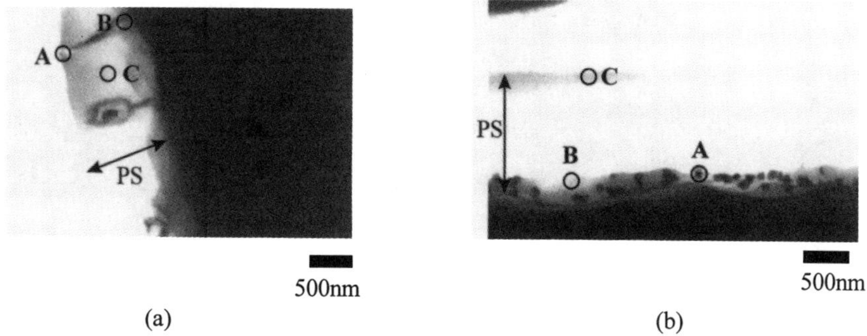

Fig.2. Transmission electron microscope (TEM) images taken for the type I specimen ((a)) and type II specimen ((b)) shown in Figs.1. Plated In atoms are concentrated more in the dark area.

10^{16} to 8×10^{16} In atoms/cm^2. These values are comparable with the areal density of 6.3×10^{16} In atoms/cm^2 which is deduced under the assumptions that all the charges fed to the PS specimen would be transferred to In^{3+} ions and that the neutralized In could be fully deposited at

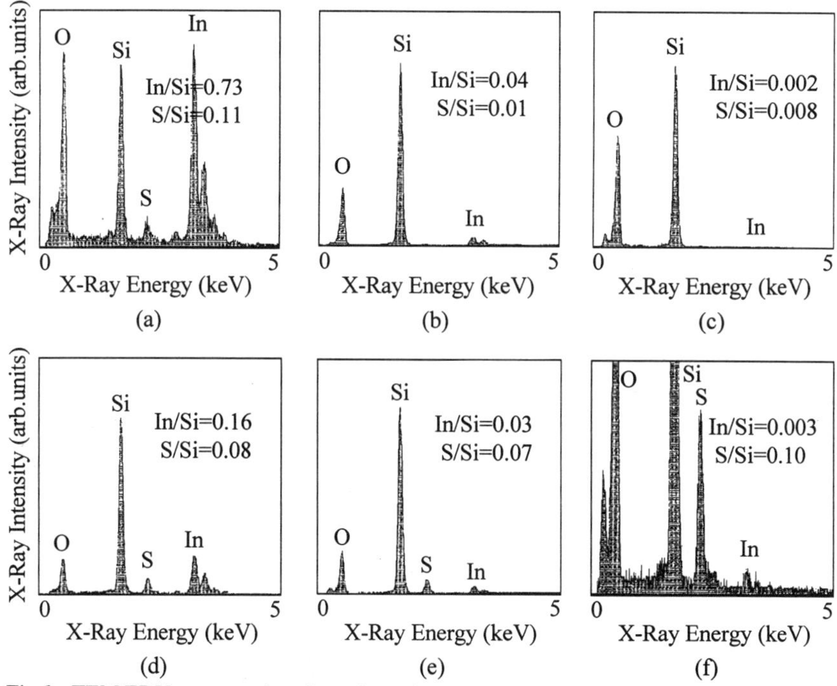

Fig.3. TEM EDX spectra taken from the points A((a)), B((b)) and C((c)) marked in Fig.2(a) (type I) and from the points A((d)), B((e)) and C((f)) marked in Fig.2(b) (type II).

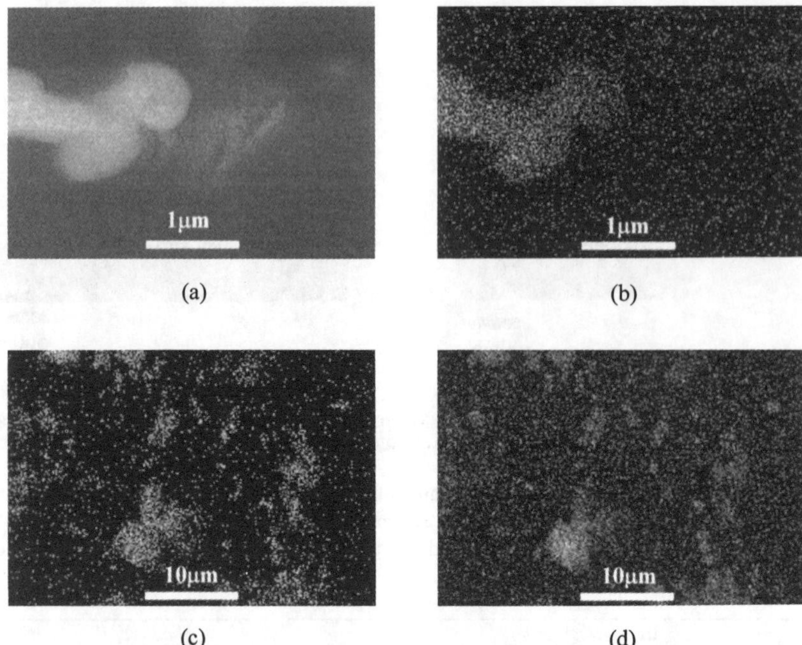

(a) (b)

(c) (d)

Fig.4. A SEM image ((a)) and a mapping image of EDX In data taken for the type I specimen (shown in Fig.1) and mapping images of EDX In ((c)) and S data ((d)) taken for the type II specimen (shown in Fig.1) using 20-kVelectrons.

the solid - solution interface. The average In density near the interface is typically more than 1% of the Si density in the PS layer. However, the scattering of the In areal densities obtained indicates an inhomogenous feature of the plating.

It should be noted that less oxygen atoms were populated in a depth region where more indium atoms existed, indicating a possible effect of the In incorporation into the PS layer on reduction of the PS layer oxidation, probably due to a capping effect of the plated In. This phenomenon differs from the difficulty in In plating observed for naturally oxidized PS since the natural oxidation occurs more preferentially at the specimen surface while the plated specimens were not significantly oxidized in the plating solution. In other words, the observed oxidation occurred naturally in air after the metal-plating process.

Since the RBS data gives only a depth information averaged over the probing beam size of 2 mm, TEM pictures were taken for these specimens in order to obtain information on the nanostructures. Figures 2(a) and 2(b) show typical TEM pictures. In the case of type I (Fig.2(a)), dark thin lines were extended from the PS - substrate interface to the top surface while most part of the specimen gave bright contrasts although a thin layer extending from the interface gave an intermediate contrast, indicating the presence of heavier element(s) in the dark region. On the other hand, in the case of type II (Fig.2(b)), many dark spots were observed near the PS - substrate interface. The typical size of these spots was about 50 nm in diameter. The thickness of the PS layer including such dark spots was below 250 nm and the other part of the PS layer was characteristic of a bright contrast.

In order to clarify the elements incorporated in the PS layer, EDX spectra were measured at three different points. The results are shown in Figs.3(a) - 3(f). It is concluded from the

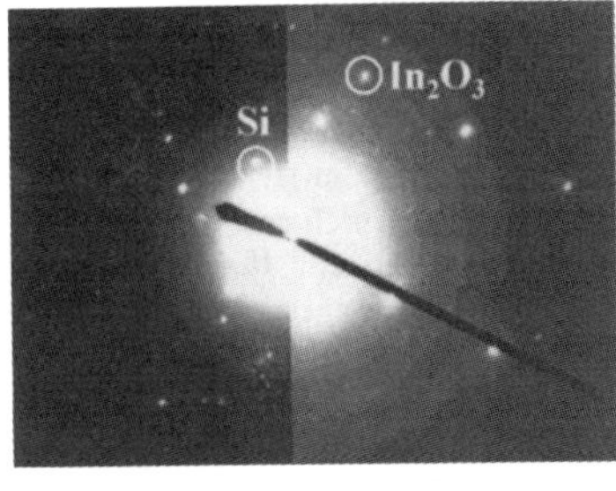

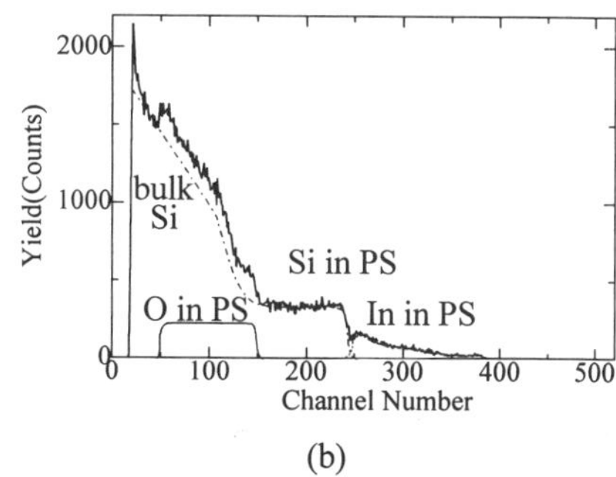

(a) (b)

Fig.5. A typical RBS spectrum ((a)) and a typical TEM diffraction image ((b)) taken from In-plated porous silicon after a two-step oxidation. In (a) diffraction spots of Si and In_2O_3 are typically indicated by circles while in (b) depth distributions of O and In are schematically shown.

comparison among the EDX data that the plated In atoms are mainly populated in the dark area of the TEM image, indicating a localized nature of the metal plating through pores in PS. The local amounts of the plated In near the PS - substrate interface, which were deduced from the EDX data, ranged from 1 to 16 % of those of the Si atoms in the same region. Mapping images of EDX data taken for the specimen surface using an SEM directly demonstrate localized distribution of the plated In metal, as shown in Fig.4(b) where the In mapping image corresponds to the SEM picture shown in Fig.4(a). In the case of type II, the In atoms near the top surface coexist with considerable amounts of S atoms (Fig.3(f)). Therefore, the surface In atoms observed are considered to be mainly originated from the plating salt left on the surface since no specimen cleaning was performed after the plating process. This can be confirmed by SEM/EDX mapping data taken for the surface of the type II specimen (Figs.4(c) and 4(d)). These facts also correspond to the RBS data shown in Fig.1(b).

Oxidation of Indium-Plated Porous Silicon

After the In-plated specimen of type I was rapidly oxidized at 720 ℃ for 10 s and subsequently oxidized at 500 ℃ for 30 min, TEM diffraction spots from the specimen were observed. These spots are found to correspond mainly to crystalline Si and In_2O_3. A typical diffraction pattern measured is shown in Fig.5(a), indicating that the crystallographical directions of the crystallites scattering the incident electron beam are aligned for the crystalline Si because of the spot pattern whereas those are more randomized for the In_2O_3 particles because of the spot and ring pattern. On the other hand, RBS spectra taken for the specimen show a uniform progress of partial oxidation in the whole PS layer (Fig.5(b)). These results indicate a possibility to realize such a structure as nm-sized Si crystallites surrounded by a Si oxide and an In oxide, although the various process parameters have not been sufficiently optimized yet so as to yield very strong PL from the oxidized In-plated PS specimens.

Possible Plating Mechanism

The following model for the In plating into PS can be proposed, based upon the above observations. The fact that the plated In is always concentrated mainly in discrete spots means the presence of the pore structures for which the metal plating occurs preferentially. Possible structures relevant to this should include pores in PS with a size distribution where relatively large pores with typical size of 50 nm in diameter are present besides the major pores with sizes of several nm. Since ions in the plating solution can move more easily through larger pores, the metal deposition by the charge transfer can take place more frequently in larger pores. If a PS layer has pipe-like or rod-like pores with relatively large diameters which run from the surface to the PS - substrate interface, then the In distribution of type I is realized. If a PS layer has no such structure, then the plated In is distributed as in type II. If the present model is the case, the pore structure in PS should be made more homogeneous in order to restrain the spot-like distribution of plated In.

Since the plating solution used was acid and contained H^+ ions beside In^{3+} ions, a considerable amount of hydrogen gas was evolved during the plating to be an insulating layer. This is one of the origins for the inhomogenous plating observed. Therefore, in order to attain a homogeneous In plating into a PS layer, a less acid plating solution should be used. Furthermore, the other parameters such as plating current, solution temperature, irradiation light intensity and plating time should also be optimized.

CONCLUSIONS

We have investigated the structure of In-plated porous silicon at a low plating charge of 0.03 C/cm^2 mainly using RBS and TEM with EDX analyses and have obtained the following results.
(1) Less oxygen atoms can populate in a depth region where more indium atoms exist, which is probably due to partial capping of the pores in porous silicon by the plated In.
(2) The In metal atoms plated into porous silicon are uniformly populated in the mm scale near the porous layer - substrate interface but are locally distributed in small regions in the nm scale. The In amount near the interface measured by the TEM EDX analysis using a 5-nm beam ranged from 1 to 16 % of the number of Si atoms in the porous layer.
(3) There were observed pipe-like or rod-like structures where the In-rich regions locally extended from the porous silicon - substrate interface to the porous silicon surface.
(4) Two-step oxidation of the In-plated PS (rapid oxidation at 720 ℃ and subsequent furnace oxidation at 500 ℃) resulted in a structure that Si crystallites were surrounded mainly by Si oxide and In oxide.

ACKNOWLEDGMENTS

This work was supported in part by a Grant-in-Aid for Scientific Research from the Ministry of Education, Science, Sports and Culture, Japan, and by a grant from Casio Science Promotion Foundation.

REFERENCES

1. L.T. Canham, Appl. Phys. Lett. **57**, 1046(1990).
2. Mater. Res. Soc. Symp. Proc. **256**, (1992); **283**, (1993); **298**, (1993); **358**, (1995); J. Lumin. **57**, (1993).
3. T. Ito and A. Hiraki, J. Lumin. **57**, 331(1993).
4. T. Ito, T. Ohta and A. Hiraki, Jpn. J. Appl. Phys. **31**, L1(1992).
5. V. Petrova-Koch, M. Muschik, A. Kuk, B.K. Meyer and F. Koch, Appl. Phys. Lett. **61**, 943(1992).
6. J.C. Vial, A. Bsiesy, F. Gaspard, R. Herino, M. Ligeon, F. Muller and R. Romestain, Phys. Rev. **B 45**, 14171(1992).
7. M. Yamada and K. Kondo, Jpn. J. Appl. Phys. **31**, L993(1992).
8. T. Ito, K. Motoi, O. Arakaki, A. Hatta and A. Hiraki, Jpn. J. Appl. Phys. **33**, L941(1994).
9. P. Steiner, F. Kozlowski and W. Lang, Jpn. J. Appl. Phys. **33**, 6075(1994).
10. P. Steiner, F. Kozlowski and W. Lang, Mater. Res. Soc. Symp. Proc. **358**, 665(1995).
11. T. Ito, T. Yoneda, K. Furuta, A. Hatta and A. Hiraki, Jpn. J. Appl. Phys. **34**, L649(1995).
12. T. Ito, K. Furuta, T. Yoneda, O. Arakaki, A. Hatta and A. Hiraki, Mater. Res. Soc. Symp. Proc. **358**, 477(1995).

EFFECTS OF NANOCRYSTALLINE STRUCTURE AND PASSIVATION ON THE PHOTOLUMINESCENT PROPERTIES OF POROUS SILICON CARBIDE

Jonathan E. Spanier[1], G. S. Cargill III, Irving P. Herman, Sangsig Kim[2]
School of Engineering and Applied Science, Columbia University, New York, NY 10027

David R. Goldstein, Anthony D. Kurtz
Kulite Semiconductor Products, Inc., Leonia, NJ 07605

Ben Z. Weiss
The Technion, Israel Institute of Technology, Department of Materials Engineering, Haifa, Israel 32000

[1]also with Kulite Semiconductor Products, Inc.
[2]current address: University of Illinois, Department of Electrical Engineering, Microelectronics Lab, Urbana, IL 61801.

ABSTRACT

We present the results of an investigation of the dependence of the photoluminescence (PL) spectra on preparation conditions, the resulting microstructure, and post-anodization treatment of porous silicon carbide films which were formed from both *p* and *n*-type 6H-SiC substrates. Porous samples were prepared by anodic dissolution under different galvanostatic conditions, resulting in different porosities and crystallite sizes. Selected-area electron diffraction patterns taken on similarly prepared porous silicon carbide (PSC) samples confirmed that the films were monocrystalline. Transmission electron microscopy of as-anodized films revealed an isotropic porous network; a dependence of porosity and nanocrystallite size on porous layer formation current density was established. Some PSC samples were passivated using a short, thermal oxidation treatment. The effects of porosity and crystallite size, and of oxide passivation in these PSC films, on PL spectra and intensity were studied using a 365 nm Kr-ion laser as excitation. Under certain conditions, the spectrally integrated PL intensity of a passivated film is more than 450x that for the original bulk SiC substrate. PL spectra are presented, and possible mechanisms are discussed.

INTRODUCTION

Studies of the optical properties of porous Si (PS) fabricated by anodic dissolution in aqueous electrolytes have attracted considerable attention and interest since the discovery of the bright visible photoluminescence from this material in 1990.[1] The possibility of its use in optoelectronic applications has been advanced by the demonstration of electroluminescence (EL)[2-5] and by improvements in the EL external quantum efficiency and longer-term device stability.[6] While there remains considerable debate as to the origin of the bright luminescence in PS, which is generally peaked at energies higher than the bulk Si band gap,[7] numerous authors have reported that the peak intensity and energy can be influenced by PS preparation conditions,[8] substrate wafer doping level and dopant type,[9] by aging[10] and also by post-preparation surface treatments, including, quenching by solvents,[11] partial oxidation,[12] passivation by rapid thermal oxidation[13,14] and by rapid thermal nitridation in NH_3.[15]

Silicon carbide possesses a much wider energy gap ($E_g \sim 3.0$ eV, 414 nm for the 6H polytype) than Si, and superior thermal conductivity, higher breakdown electric field, and higher saturation electron drift velocity. Nitrogen-aluminum donor-acceptor pair recombination in SiC has been the basis for a blue ($\lambda = 470$ nm) light-emitting pn-junction diode,[16-19] which has a small external quantum efficiency, $\sim 10^{-4}$. While these LEDs have become commercially available,[20] the quantum efficiency remains low (~0.02-0.03%) and the peak emission energy is still well below the energy gap of bulk SiC. Based on developments in PS, it is believed that porous silicon carbide (PSC) is a possible candidate for more efficient and shorter wavelength light-emitting devices *vis-a-vis* those based on bulk SiC or porous Si.

Mat. Res. Soc. Symp. Proc. Vol. 452

There have been far fewer studies of PSC than PS. The formation of PSC by anodic dissolution of both *n*-type and *p*-type substrates was first reported by Shor, *et al.*[21, 22] Researchers have reported on PL and Raman scattering,[23-26] and EL[24, 27] of PSC, electrical properties of PSC photoelectrochemically prepared from *n*-type substrates in UV light and in the dark,[28] and on the infrared reflectivity of PSC.[26, 29] We present here the first studies of the influence of the porosity and interpore spacing on the PL of PSC formed from *p*-type substrates by anodic dissolution. This also appears to be the first report on the effects of introducing a thin, passivating oxide layer onto the pore walls on the photoluminescent properties of PSC.

SAMPLE PREPARATION

Several PSC samples were prepared from aluminum-doped *p*-type 6H-SiC substrate wafer pieces (off-axis orientation of 3.5° +/- 0.5° off <0001> toward <1120>). The *p*-type SiC material is compensated with nitrogen atoms (donors) such that $N_A - N_D \sim 2\times10^{18}$ cm^{-3}. First, ohmic electrical contacts were formed on each piece by the rapid thermal annealing of sputter-deposited films of titanium and platinum. An electrical lead was attached, and each sample was encapsulated in black wax, exposing a small area, typically on the order of 1-10 mm^2. The samples were placed in a solution of dilute hydrofluoric acid with ethanol. The ethanol served as a surfactant to minimize bubble formation, which could lead to a nonuniform current distribution across the surface of the semiconductor electrode. An anodic bias, referenced to a saturated standard calomel electrode, was applied to the samples for time periods ranging from one hour to more than twenty hours, depending upon the formation current density and resulting dissolution rate, forming thick porous layers. Several PSC samples were also prepared from similarly oriented *n*-type, nitrogen-doped, 6H-SiC wafers of compensated doping concentration N_D-$N_A \sim 4\times10^{18}$ cm^{-3}. These pieces were anodized similarly, while being illuminated by a focused 1000 W Hg-Xe lamp. Free-standing PSC films with thicknesses greater than 50 μm were formed by an abrupt increase in the current density to greater than 500 mA/cm^2. Following the anodization, each sample was allowed to dry in air. Differences in the nature of the anodic etching between *p*-type and *n*-type SiC are discussed elsewhere.[31]

STRUCTURAL CHARACTERIZATION

Transmission electron microscopy (TEM) was performed on three PSC samples formed from *p*-type substrates at different anodic current densities. The samples were thinned by polishing and ion milling from the substrate side. Representative dark-field micrographs of samples prepared with J = 5 and 30 mA/cm^2 are shown in Figs. 1 and 2. Cross-section micrographs indicate that the microstructure of the PSC layer does not change with depth over the range of at least 1 μm. The porosity and average interpore spacing depend on the formation current density (as shown in Table 1) in a manner which is qualitatively similar to observations in porous Si formed from *p*-type substrates.[32] The porosity and mean crystallite diameters were tabulated directly from TEM micrographs.

Table 1
Dependence of microstructure on formation current density
for PSC formed from p-type SiC

Sample No.	*Current Density (mA/cm^2)*	*Porosity (%)*	*Mean Crystallite Size (nm)*
95P94-1	5	68	8.1
95P94-2	15	74	6.4
95P94-3	30	78	4.9

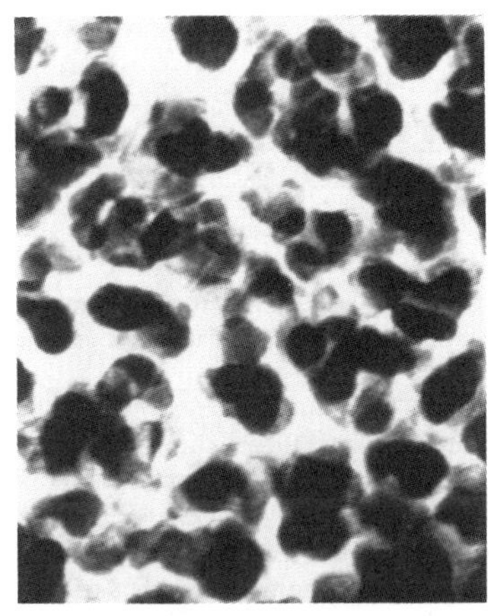

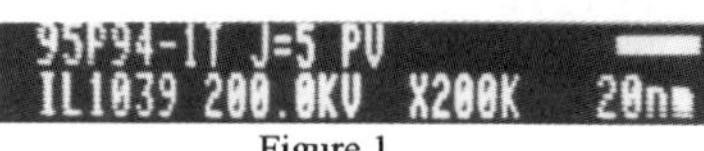

Figure 1
Planar-view TEM of *p*-PSC sample prepared at J = 5 mA/cm^2

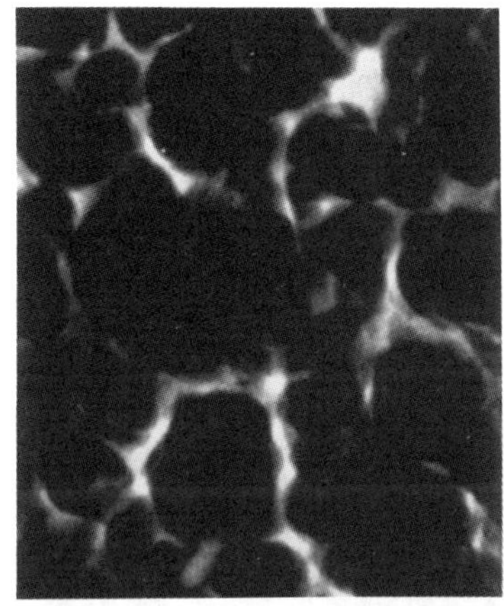

Figure 2
Planar-view TEM of *p*-PSC prepared at J = 30 mA/cm^2

PHOTOLUMINESCENCE MEASUREMENTS

The photoluminescence of a several similarly prepared PSC samples formed from *p*-type and *n*-type SiC substrates was measured at room temperature using the 365 nm line of the Kr-ion laser. The PL was dispersed by an 0.85m double spectrometer and detected by a cooled GaAs photomultiplier. The PL data were corrected for variations in spectral sensitivity of the optical apparatus by using a Xe lamp for spectral calibration. The spectral resolution was approximately 1.4nm, which is sufficient given the broad nature of the emission spectrum at room temperature. Free-standing and thick (>50μm) porous films on a bulk substrate had the same PL intensity and profile.

In Fig. 3, the room-temperature (RT) PL spectra of three as-anodized PSC films formed from a *p*-type substrate (*p*-PSC) at current densities of J = 5, 30 and 100 mA/cm^2 are shown. The integrated intensities of the samples prepared at J = 30 and J = 100 mA/cm^2 are ~2x and 8x that of the sample prepared at J = 5 mA/cm^2.

A similar behavior was observed for the as-anodized PSC films prepared from an *n*-type substrate (*n*-PSC), as shown in Fig. 4, along with the PL of bulk SiC for comparison. In this case, the integrated PL intensities of the films formed at J = 30 mA/cm^2 and at J = 60 mA/cm^2, are ~14x and 25x larger than that for the bulk SiC, respectively. A similar dependence of PL intensity on anodic current density in porous SiC films formed from *n*-type substrates has been observed by other investigators.[23, 24]

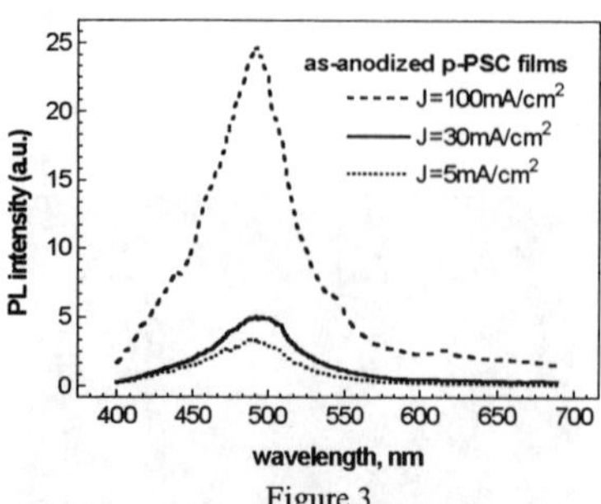

Figure 3
RT PL of as-anodized PSC films prepared from *p*-type substrates at different current densities

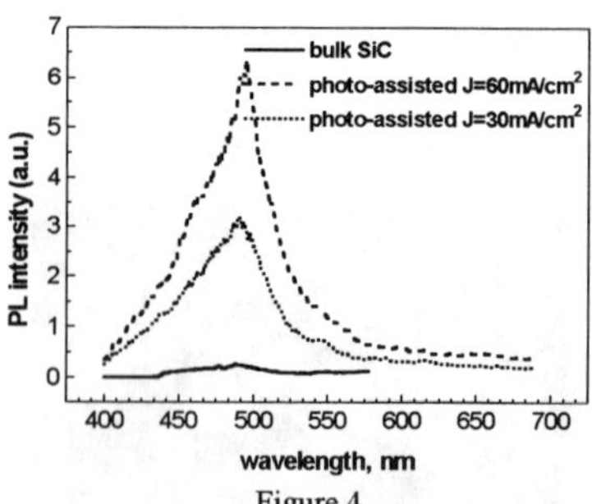

Figure 4
RT PL of as-anodized PSC films prepared from *n*-type substrates at different current densities

PASSIVATED PSC FILMS

Following anodization, several of the *p*-PSC films which were formed at different current densities were placed into an oxidation furnace at 1150°C at atmospheric pressure for five minutes and were removed and allowed to cool in ambient. From the oxidation rates of crystalline SiC,[33] it is believed that an oxide of one or two monolayers will grow on the pore walls and will passivate at least a portion of the large internal surface area by reducing the number of surface states which act as trapping centers of nonradiative or slow radiative recombination. A comparison of the PL spectrum of unpassivated and passivated films formed at J = 5 mA/cm^2 is shown in Fig. 5. The integrated intensity of the oxidized PSC film is ~8x that of the as-anodized film.

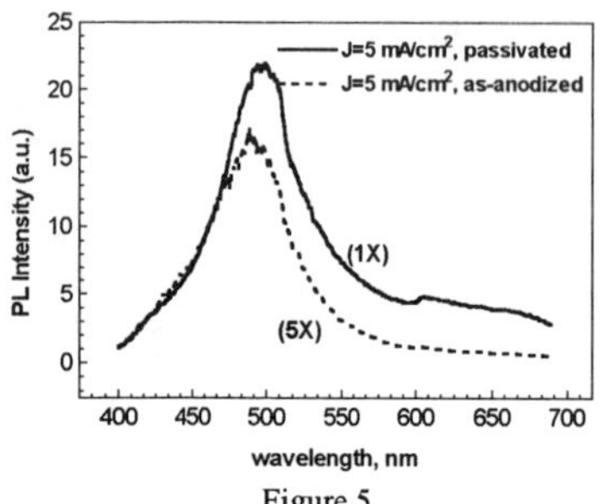

Figure 5
PL of as-anodized and passivated *p*-PSC formed at J=5 mA/cm^2

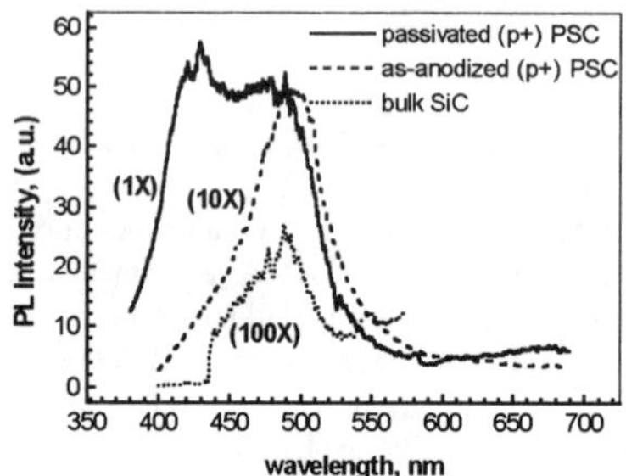

Figure 6
PL of bulk, as-anodized and passivated *p*-PSC films formed at J=30 mA/cm^2

The passivation of PSC films formed at a higher current density J=30 mA/cm^2 yielded different PL results as seen in Fig. 6. Not only is the integrated intensity of the passivated film more than 20x greater than the as-anodized film, and more than 450x that of the bulk SiC, shown for comparison, but the PL is characterized by a broad, shorter wavelength emission.

The PL response of these films was observed to be nearly linear with Kr-ion laser excitation power up to 25 mW; however, the PL measurements of the oxidized PSC films were performed at lower excitation power than measurements on the as-anodized films and the bulk SiC substrate in order to obtain good signal-to-noise ratio in the weakly luminescent bulk SiC and to minimize the reduction of the PL signal from the oxidized PSC films due to heating at higher excitation power levels.

DISCUSSION

As mentioned earlier, the observation of an increase in the room temperature PL intensity in *n*-PSC with increased formation current density is in general agreement with other investigators.[23, 24] Matsumoto, *et al*,[23] however, report that the peak of the luminescence blueshifts with increased formation current density (from 500 nm for J = 20 mA/cm^2 to 460 nm for J = 60 mA/cm^2), and suggest that this is due either to quantum confinement as originally proposed in PS,[1] or due to luminescent surface states.[30] Our findings agree with those of Petrova-Koch[24] in that no spectral shifts were observed. Given that the microstructure of the *n*-PSC formed by photo-assisted electrochemical anodization exhibits an interpore spacing of ~30 nm and that all of the luminescence peaks are well below the bulk crystal band gap of SiC, it seems highly unlikely that the luminescence is due to quantum confinement. Petrova-Koch[24] proposed that the luminescence in both bulk and PSC is due to nitrogen-aluminum donor-acceptor pair recombination, and that the enhanced efficiency in PSC may be due to a geometrical confinement effect in which a photoexcited electron-hole-pair is highly localized due to the granular nature of the material, thereby suppressing nonradiative decay processes.

A similar increase in PL intensity is observed for as-anodized *p*-PSC which is also well below the bulk energy gap. Given the smaller characteristic dimension of this material, it is not clear whether quantum confinement is the basis for the enhanced PL. The enhanced PL could be due to localization of photoexcited carriers which limits nonradiative decay as proposed above, but also it could be due to distortion of the local band structure and/or a relaxation of the momentum selection rules, so that direct transitions are permitted.

Konstantinov, *et al*[25] examined the effects of exposing *n*-PSC to oxygen ambient at 700°C for 1 hour. The result was a quenching of the blue-green PL, which was partially recoverable without the high energy component upon immersion in HF. In contrast, we observe that identically short, thermal oxidation treatments to the as-anodized *p*-PSC film further enhances the PL intensity for material of lower porosity and larger average interpore spacing (~8 nm), and there is an additional band of emission at energies higher than the original peak, extending into the bulk energy gap, in PSC of higher porosity and narrow average interpore spacing (~5 nm). The reason for this enhanced and blueshifted PL is not clear. One interpretation is that the oxidized highly porous PSC material consists of either further narrowed crystallites or local SiC clusters. A broad distribution of sizes and geometries of the remaining, passivated material could result in a virtual continuum of energy states. Another explanation, which is consistent with the first, is that a thermally grown oxide layer is passivating the large internal surface area of dangling bond sites. The striking difference could also be due to differences in the composition and properties of the surface layer of PSC following anodization at different current densities, and their associated oxidation kinetics. In contrast, the origin of the further enhancement of the PL and shorter wavelength emission could be due to the surface oxide film itself. It is clear that further work is required to characterize the structure, composition and properties adequately to explain the mechanism of PL in as-anodized and slightly oxidized *p*-PSC.

CONCLUSIONS

The room temperature PL efficiency of porous silicon carbide formed from *p*-type and *n*-type substrates is enhanced *vis-a-vis* that from the bulk substrate, and this enhancement is influenced by the preparation conditions. In addition, a further enhancement of the PL efficiency and short-wavelength emitting characteristics can be obtained by "passivating" the PSC using a short, thermal oxidation treatment. This leads to a room-temperature photoluminescence efficiency which, under certain circumstances, is more than 450x that for bulk SiC.

ACKNOWLEDGEMENTS

This work was supported at Kulite Semiconductor by a NASA Lewis Phase II SBIR Contract, No. NAS3-27247, and at Columbia University by the Joint Services Electronics Program Contract No. DAAH-4-94-G-0057. We gratefully acknowledge I. Grimberg of the Technion for performing TEM and electron diffraction studies.

REFERENCES

1. L. T. Canham, W. Y. Leong, M. J. Beale, T. J. Cox, and L. Taylor, Appl. Phys. Lett. **57**, 1046 (1990).
2. A. Richter, P. Steiner, F. Kozlowski, and W. Lang, IEEE Trans. Electron Device Lett. **EDL-12**, 691 (1991).
3. N. Koshida and H. Koyama, Appl. Phys. Lett. **60**, 347 (1992).
4. R. L. Smith and S. D. Collins, J. Appl. Phys. **71**, R1 (1992).
5. F. Namavar, H. P. Maruska, and N. M. Kalkhoran, Appl. Phys. Lett. **60**, 2514 (1992).
6. L. Tsybeskov, S. P. Duttagupta, K. D. Hirshman and P. M. Fauchet, Appl. Phys. Lett. **68**, 2058 (1996).
7. B. Hamilton, Semicond. Sci Technol. **10**, 1187 (1995).
8. S. Gardelis and B. Hamilton, J. Appl. Phys. **76**, 5328 (1994).
9. C. Pickering, M. I. J. Beale, D. J. Robbins, P. J. Pearson and R. Greef, J. Appl. Phys. C: Solid State Physics **17**, 6535 (1984).
10. K. Li, D. Diaz, J. Campbell, and C. Tsai, in Porous Silicon, edited by Z. C. Feng and R. Tsu (World Scientific, Singapore, 1994) pp. 261-274.
11. J. M. Lauerhaas, G. M. Credo, J. L. Heinrich, and M. J. Sailor, J. Am. Phys. Soc. **114**, 1911 (1992).
12. L. Tsybeskov, S. P. Duttagupta, and P. M. Fauchet, Solid State Comm. **95**, 429 (1995).
13. V. Petrova-Koch, T. Muschik, A. Kux, B. K. Meyer, F. Koch, and V. Lehmann, Appl. Phys. Lett. **61**, 943 (1992).
14. K.-H. Li, C. Tsai, J. C. Cambell, B. K. Hance, and J. M. White, Appl. Phys. Lett. **62**, 3501 (1993).
15. G. Li, X. Hou, S. Yuan, H. Chen, F. Zhang, H. Fan, X. Wang, J. Appl. Phys. **80**, 5967 (1996).
16. R. W. Brander and R. P. Sutton, Br. J. Appl. Phys. **2**, 309 (1969).
17. W. V. Muench, W. Kuerzinger, and I. Pfaffeneder, Solid State Electron. **19**, 871 (1976).
18. M. Ikeda, T. Hayakawa, S. Yamagiwa, H. Matsunami, and T. Tanaka, J. Appl. Phys. **50**, 8215 (1979).
19. S. Nishino, A. Ibaragi, H. Matsunami, and T. Tanaka, Jpn. J. Appl. Phys, **19**, L353 (1980).
20. J. A. Edmond, H.-S. Kong, C. Carter, Jr., Physica B **185**, 453 (1993).
21. J. S. Shor, I. Grimberg, B. Z. Weiss, and A. D. Kurtz, Appl. Phys. Lett. **62**, 2836 (1993).
22. J. S. Shor, L. Bemis, A. D. Kurtz, I. Grimberg, B. Z. Weiss, M. F. MacMillan, W. J. Choyke, J. Appl. Phys. **76**, 4045 (1994).
23. T. Matsumoto, J. Takahashi, T. Tamaki, T. Fugati, H. Mimura, and Y. Kanemitsu, Appl. Phys. Lett. **64**, 226 (1994).
24. V. Petrova-Koch, O. Sreseli, G. Polisski, D. Kovalev, T. Muschik, and F. Koch, Thin Solid Films **255**, 107 (1995).
25. A. Konstantinov, A. Henry, C. I. Harris and E. Janzen, Appl. Phys Lett. **66**, 2250 (1995).
26. A. M. Danishevskii, V. B. Shuman, A. Yu. Rogachev, and P. A. Ivanov, Semiconductors **29**, 1106 (1995).
27. H. Mimura, T. Matsumoto, Y. Kanemitsu, Appl. Phys. Lett. **65**, 3350 (1994).
28. A. Konstantinov, C. I. Harris, and E. Janzen, Appl. Phys. Lett. **65**, 2699 (1994).
29. M. F. MacMillan, R. P. Devaty, W. J. Choyke, D. R. Goldstein, J. E. Spanier, and A. D. Kurtz, J. Appl. Phys. **80**, 2412 (1996).
30. Y. Kanemitsu, H. Uto, Y. Masumoto, T. Matsumoto, T. Fugati, and H. Mimura, Phys. Rev. B **48**, 2827 (1993).
31. J. Shor, A. D. Kurtz, I. Grimberg, and B. Z. Weiss, R. M. Osgood, accepted for publication in J. Appl. Phys.
32. R. L. Smith and S. D. Collins, J. Appl. Phys. **71**, R1 (1992).
33. W. von Munch and I. Pfaffeneder, J. Electrochem. Soc. **122**, 642 (1990).

Part VII

Photoluminescence and Other Optical Properties of Porous Silicon

LOCALISATION OF CARRIERS IN POROUS SILICON

I. MIHALCESCU, J.C. VIAL and R. ROMESTAIN
Laboratoire de Spectrométrie Physique, Université Joseph Fourier and CNRS Boîte Postale 53 X, 38041 Grenoble cedex, France. jcvial@grenet.fr.

ABSTRACT

We analyze the intensity and decay time evolution of the porous silicon luminescence upon anodic oxidation, aging, chemical thinning and temperature variation. Strong analogies are pointed out for the photoluminescence intensity as well as for the photoluminescence decay shape evolution. They are interpreted by the variation of the extension of the carrier wavefunction induced by the modification of potential barrier efficiencies. No additional assumption such as hopping of carriers was necessary to explain the decay shapes well fitted by stretched exponential. On the contrary our observations and our simple model are in favor of a strong localization of carriers. Some experimental results are revisited within the frame of this model.

INTRODUCTION

Since the discovery of the bright visible light emission from porous silicon [1], numerous problems have been encountered. The question of wheter radiative recombination comes from localized or delocalized carriers, has not yet received convincing answer for example. Nevertheless it is an important question and the ability of porous silicon to be applied in microelectronics (electroluminescence, voltage modulation of the luminescence...) is dependent on a clear answer. Indeed the delocalization of carriers at room temperature is necessary for good transport properties, but this condition certainly affects the luminescence efficiency since carriers will find more easily non-radiative (NR) traps in the delocalized state. There is not a definite solution to this old problem, but porous silicon is an accomodant structure which can receive numerous post-formation treatments. They are expected to tailor simultaneously the carrier transport and the luminescence efficiency, therefore it is interesting to verify that luminescence can be a good tool for such characterization. The importance of the wavefunction extension from the point of view of applications is not the only motivation for this study; it has also more fundamental interest. Indeed porous silicon can be seen as a slightly disordered system compared to amorphous or alloyed semiconductors and one can expect to be in a good position to have a new approach to fundamental questions related to localization-delocalization in disordered systems. We will show in this paper that, thanks to the good optical properties of porous silicon and a good control of some post-formation treatments, an analysis of the parameters related to localization of carriers can be followed for wide ranges of their evolution. For this purpose we have chosen to focus on various experiments such as non-reversible post-formation treatments which consist of a partial anodic oxidation or in chemical thinning of the porous structure (first proposed by Canham[1]) or finally the temperature effect which is on the contrary reversible if the temperature is not higher than approximately 300°C. We will see that, despite their very different natures these experiments have very similar effects concerning carrier transport and dynamics. It is also an aim of this paper to try to understand the origins of such similarities.

The anodic oxidation process of p type porous silicon layers, now a well known technique [4], illustrates clearly the first point that we want to emphasize here, the partial oxidation of the porous layer increases simultaneously the resistivity and the luminescence efficiency. To this observation one can add the well known PL enhancement by slow dissolution of the porous layer

Mat. Res. Soc. Symp. Proc. Vol. 452

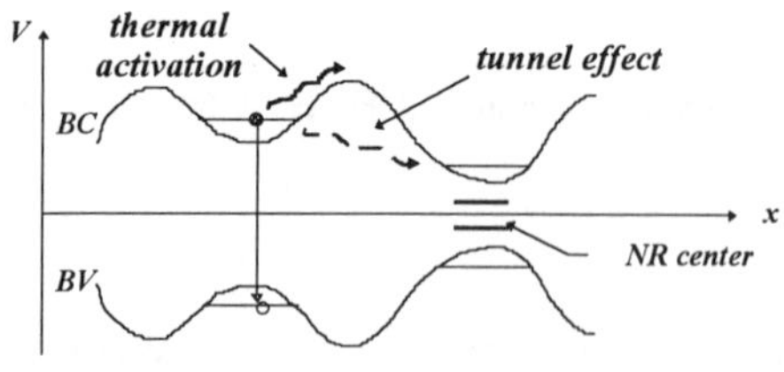

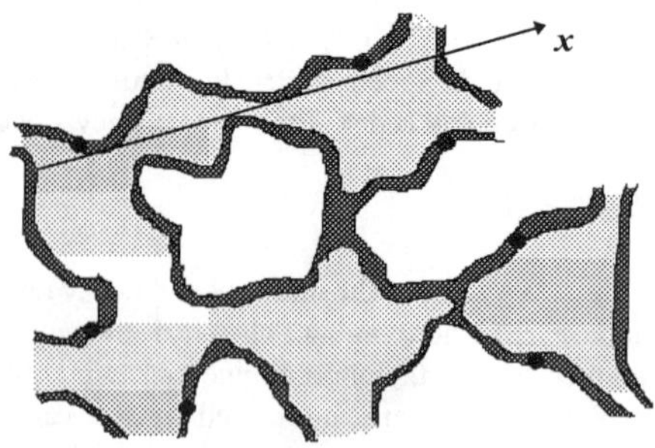

Fig 1 :
A schematic representation of porous silicon, in real space on right and in energy scale above, for a given direction as indicated by the arrow. Also shown: non-radiative centers as black spots and two examples of carrier escape by tunneling and thermal activation

in hydrofluoric acid (HF)[1] where, starting with a low porosity (and therefore a low resistivity) and a weak PL emission, the thinning leads to an increase in porosity along with a strong enhancement of the PL. Aging, has in general a positive effect on the luminescence efficiency of porous silicon. This process is not well characterized at present but in many respects can be seen as a "soft" oxidation similar to the anodic oxidation. The same correlation can be found but with the temperature as parameter. As shown by Koyama et al. [2] and Ben-Chorin et al. [3], the conductivity (in the dark) as well as the photoconductivity strongly increase for increasing temperatures while, for the same range of temperatures, the PL intensity is reduced.

These various examples point out that there exists a strong relation between the ability of carriers to reach extended states (or on the contrary to be localized) and the efficiency of the luminescence. This can be understood in an extension of the model previously presented in ref. [4] and shown in figure 1 where the porous structure is made up of slightly interconnected quantum crystallites. Electron-hole pairs are then supposed to recombine radiatively when they are present in "bright crystallites" but are dissociated and recombine non-radiatively when a carrier wavefunction is extended to "dark" crystallites . This process reduces the PL efficiency and can contribute to the free carrier population i.e. enhance the conductivity. One can imagine that a modification either via oxidation or dissolution of the interconnections between the crystallites will certainly affect this process. The temperature can also activate this extension. It is this "scenario" which will be now analyzed more carefully through the various post-formation treatments.

EXPERIMENTAL PROCEDURES

In this paper, an extensive use of porous silicon anodic oxidation and chemical dissolution will be done together with temperature variation. Carrier population decay times are monitored by the mean of time resolved optical measurements. We will focus on the « red » luminescence whose decay is in the microsecond range. Porous films formation, chemical dissolution and anodic oxidation have been presented in previous papers. Excitation of luminescence is provided by the 366 nm line of a mercury arc lamp or by a pulsed Nitrogen laser at 337 nm and the temporal resolution of the analysis chain was about 5ns. Temperature variation above and below room temperature were obtained by a « Linkam » cooling and heating stage.

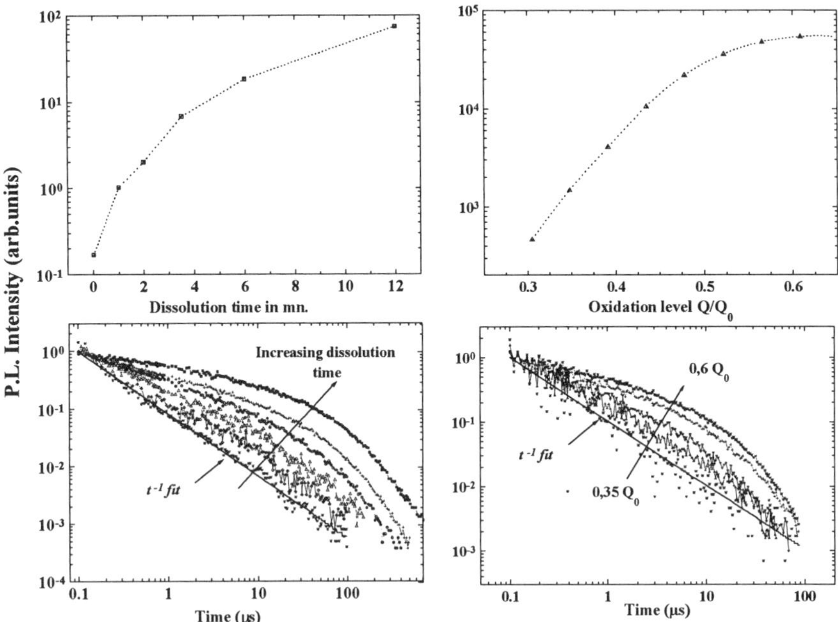

Fig 2:
CW PL intensity (upper curves) as a function of dissolution time and oxidation level Q. Q_0 represents the maximum oxidation level obtained by anodic oxidation.
Time evolution (lower curves) of the PL obtained for various dissolution time and various oxidation level.

POST FORMATION TREATMENTS AND TEMPERATURE EFFECTS

Oxidation of porous silicon can be obtained by various techniques such as the thermal oxidation, the anodic oxidation or the simple aging in ambient air. The thermal oxidation is a well known process which has been used for many years for a stable passivation of the bulk silicon. This process has been generalized for porous silicon [5]. It has been shown that the good passivation obtained only at high temperature (900°C) is accompanied by an increase in the PL efficiency and a reduction of the density of dangling bonds, explained by a substitution of the Si-H coverage by a more stable one which is mainly SiO_2. For the anodic oxidation one can be tempted to generalize this explanation since it is also accompanied by with an increase in the PL efficiency and in that sense this process has also been presented as a "passivation" of the structure. We are thinking that this word has to be avoided for the anodic oxidation, as shown by the IR transmission spectra [6] the anodic oxidation does not remove the hydrogen coverage initially installed by the electrochemical process in HF (which is the best passivation for bulk silicon).

Following our crude model we will try to give a better description for the effect of the anodic oxidation in term of localization of carriers. In previous papers anodic oxidation has already been analyzed [4] in term of reduction of the NR rates due to the reduction of the carrier tunneling through the oxide barrier which surrounds the quantum crystallites. In other words, it

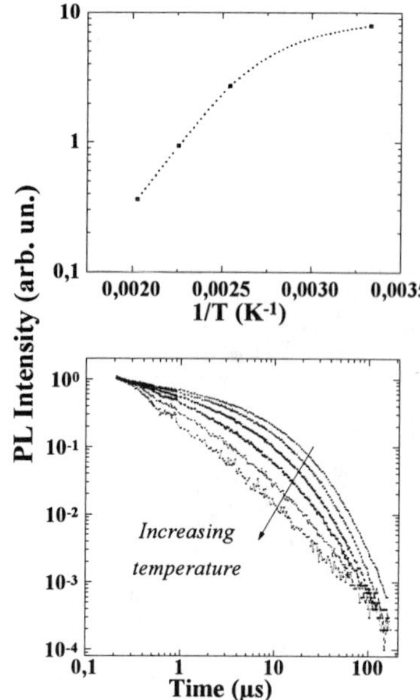

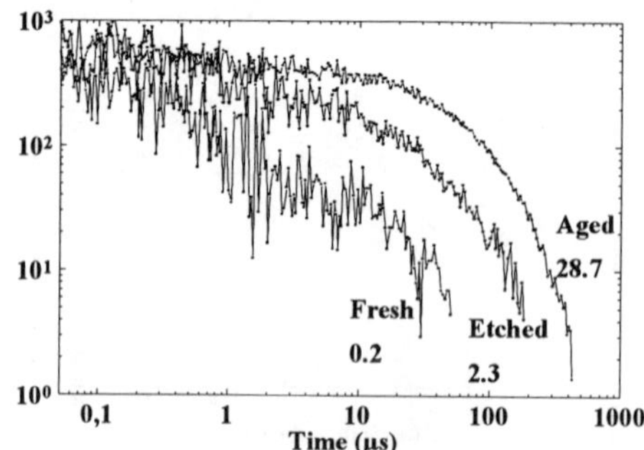

Figure 3 :
On the left: Temperature evolution of the CW PL intensity (a) and the PL decay shape (b), for a detection wavelength of 600nm. The reciprocal temperature (for the first graph) is used for the horizontal axis in order to emphasize the similarity of the evolution with the two former cases.
Above : PL decay shape dependence on aging and HF etching compare to as form sample. The CW Pl intensity is indicated for each sample.

was the growing of the SiO_2 barrier thickness which was supposed to control the carrier tunneling escape to « dark » species. In this paper, the main aspects of this interpretation will be kept but electrochemical considerations[4] show that it is not realistic to assume that the barriers are only made of SiO_2 and one have also to consider the constricted silicon bridges as an other origin for the barriers. The anodic oxidation having a thinning effect on them, their efficiency will also be enhanced by anodic oxidation. An interesting result, following the enhancement of the PL quantum efficiency, is the increase of the PL mean lifetime [4],[7]. It indicates again that the anodic oxidation controls the NR escape of carriers. In this paper, we will try to be more accurate in the description of this escape by looking at the evolution of **the shape** of the PL decays. From low oxidation level (and low PL efficiency) to high oxidation level (and high PL efficiency) the decay shape shows a continuous evolution as shown in figure 2. A stretched exponential well accounts for the shape at high oxidation level while, at low oxidation level, the shape is more accurately described by a power-law.

Aging shows the same kind of evolution with a quasi-exponential decay for aged (and efficient) samples while, fresh they are inefficient and have a power-like time dependence (see figure 3). The HF thinning reproduces the same trends (figure 2), an increase of the quantum efficiency with increasing dissolution time along with a PL decay shape evolution from a power-law (at low porosity) to a stretched exponential (at high porosity), the decay being faster for low porosity.

More surprising are the analogies of behaviour upon temperature variation in an intermediate temperature range (from room temperature to 500K). The tendency is well known, a speed up of PL decay going along with a lowering of the PL efficiency for increasing temperature. This observation has been used to prove that the dominant relaxation mechanism above room temperature is the NR escape of carriers. Concerning the decay shape evolution we record similar evolution to what has been observed upon porosity and oxidation evolution, at room temperature a stretched exponential moving to a power-law for increasing temperature (see figure 3).

DISCUSSION

In many respects the oxidation level, the porosity and the reciprocal temperature (1/T) have analogous effects on the porous silicon PL. This was already known for the PL quantum efficiency (via PL intensity measurement) or for rough measurements of the PL lifetime but we show here that this analogy stands also for more critical parameters such as the decay shape.
In order to provide a more quantitative analysis and following ref. [8] the decay shapes are fitted by the following stretched exponential function :

$$I_{PL}(t) = I_0 \cdot t^{\beta-1} \exp\left[-\left(\frac{t}{\tau}\right)^{\beta}\right].$$

This two parameter function fit (τ the characteristic lifetime and β the shape factor) gives us a simple and good description of the non-exponential decay for more than 3 orders of magnitude, it is valid for any wavelength and for all types of porous sample studied (as-prepared or aged, anodically oxidized or not), just the fitting parameters (τ and β) being changed. The particular cases of β=0 and β=1 respectively represent the t^{-1} and pure exponential law.
As shown in ref.[8] this decay shape takes its origin from a particular distribution of NR rates $D(W_{NR})$. More precisely the decay shape is proportional to the Laplace transform of $D(W_{NR})$:

$$I_{PL}(t) \propto W_R \cdot \exp[-W_R t] \cdot \int_0^{\infty} D(W_{NR}) \cdot \exp[-W_{NR} t] \cdot dW_{NR}$$

where W_R is the radiative decay rate. The problem, resolved in ref.[8] was to find the physical origin of a particular distribution whose Laplace transform gives our stretched exponential evolution and respects the general trends presented above. For this purpose the carrier wavefunction is approximated by an exponential decrease along the wire (see fig.1) therefore the NR rate (W_{NR}) will be proportional to its overlap with the trap wavefunction (here supposed to be a Dirac function).

$$W_{NR} = W_0 \cdot \exp[\, -r/R_0 \,]$$

where r is the distance to the trap and R_0 a "localization length" of the wavefunction. For randomly distributed defects in one dimension, the probability of having the first one at distance r is given by the Poisson law $f(r) = p \cdot \exp[\, -pr \,]$, where p is the linear density of traps. Using this last relation and the previous one, we obtain a power law :

$$D(W_{NR}) = f(r) \cdot \left(\frac{dW_{NR}}{dr}\right)^{-1} = pR_0 \cdot \frac{1}{W_0^{pR_0}} \cdot \left(W_{NR}\right)^{pR_0-1}$$

The inverse Laplace transform of the experimental decay (not shown here) can be fitted for a large range of recombination rates by a power-law : $(W_{NR})^{-\beta}$. So, comparing to the previous expression we deduce:

$$\beta = 1 - pR_0.$$

The shape factor of the stretched exponential decay thus gives direct information about the evolution of the localization length (R_0) or the internal surface passivation (p). This is an important result which will help us to understand what were the consequences of the post-formation treatments administrated to porous silicon samples.

Within the frame of this model it is then easy to deduce what will be the dependence of the quantum efficiency on the localization length and on the surface passivation. Ref.[8] shows that the knowledge of the parameter pR_0 and its evolution upon the various post-treatments is enough to deduce all the observed PL intensity evolution. This confirms the strong correlation between the decay shapes and the quantum efficiency. To our knowledge it is the only model which takes into account all these different observations.

CONCLUSION

This paper emphasizes the strong analogies in QE and decay shape evolution upon parameters as different as HF etching, anodic oxidation, aging and temperature. A model based on the carrier localization in porous silicon is used to explain these observations. For that we have related the quantum efficiency increase of the PL to a diminution of the carriers localization length.

The anodic oxidation, aging and the HF etching, all processes well known to increase the quantum efficiency, act not by a better surface passivation but principally by diminishing the carriers probability to reach the non-radiative centers. An interesting consequence of our model is that starting from the PL decay *shape* and their changes we can describe the whole QE evolution for more than two orders of magnitude.

We also present a set of arguments showing the necessity for the carriers to be "isolated" from the rest of the structure, inside *nanometric* "cells" in order to obtain a high QE at room temperature. This idea may be used on related materials such as the amorphous porous silicon. In our model the crystallinity of silicon is not an indispensable ingredient.

REFERENCES

1. L.T. Canham, Appl. Phys. Lett.57, 1046 (1990).

2. H. Koyama and N.Koshida, J. Lum. **57**, 293 (1993).

3. M. Ben Chorin and F. Koch, J. Lum. **57**, 159 (1993).

4. J.C. Vial, A. Bsiesy, F. Gaspard, R. Herino, M. Ligeon, F. Muller, R. Romestain, R.M Macfarlane, Phys. Rev. B **45,** 14171 (1992).

5. V. Petrova-Koch, T. Muschik, and A. Kux, Appl.Phys.Lett. **61**, 943 (1992).

6. M.A. Hory, F. Gaspard, R. Herino, M. Ligeon, F. Muller, R. Romestain, J.C. Vial, *in Porous Silicon and Related Materials* edited by R. Hérino and W. Lang (Elsevier North-Holland 1995), Thin Solid Films. **255,** p. 200. (1995).

7. I. Mihalcescu, M. Ligeon, F. Muller, R. Romestain and J.C. Vial, J. Lum. **57**, 111 (1993).

8 I. Mihalcescu, J.C. Vial and R. Romestain, J.Appl.Phys. **80**, 2404 (1996).

IN-SITU PHOTOLUMINESCENCE INVESTIGATION OF THE INITIAL POROUS SILICON FORMATION IN 0.2M NH_4F (pH 3.2)

T. DITTRICH*, V. Y. TIMOSHENKO**, J. RAPPICH***
* Technische Universität München, Physik Department E16, D-85747 Garching, Germany
** Physics Department, Moscow State University, 119899 Moscow, Russia
*** Hahn-Meitner-Institut, Abt. Photovoltaik, Rudower Chaussee 5, D-12489 Berlin, Germany

ABSTRACT

The porous silicon (por-Si) formation in 0.2M NH_4F (pH 3.2) is investigated in-situ by photoluminescence (PL). The p-type Si(100) samples are treated electrochemically in the galvanostatic regime starting from the hydrogenated surface. Single pulses of a N_2 laser are used to probe stroboscopically the radiative band-band recombination of the bulk c-Si and the PL of por-Si. The PL intensity of c-Si is correlated with the current density during the current-voltage scan and indicates changes of surface recombination by the onset of chemical reactions. The PL intensity of c-Si increases rapidly after switching off the anodic current while the PL intensity of por-Si is not influenced by the rapid current switch. This shows that the passivation of the surfaces of the Si nanostructures is not affected by the por-Si formation at the surface of the bulk c-Si.

INTRODUCTION

The in-situ control of the formation of luminescent porous Si (por-Si) at the initial state could help to get a deeper understanding about the surface processes leading to the formation of defects and porous structures. The stabilization of Si nanostructures in HF solutions can be explained with the quantum confinement leading to a strong decrease of the generation of holes which are needed for the anodization process [1]. However, the initial state of the por-Si formation should depend on the formation of specific surface defects. Such defects are created during the oxidation of Si atoms by fluorine species, for example. Breaking of Si-Si bonds is one step of the oxidation process. The created Si dangling bonds are electrically active and can influence the nonradiative surface recombination. The change of the c-Si surface recombination velocity can be detected by measuring the change of the photoluminescence. The PL is quenched with increasing surface recombination as shown by use of a cw Argon ion laser for the excitation of the PL of bulk c-Si [2]. The initial formation of por-Si can not be investigated by a cw laser due to the very high net concentration of excess carriers which increase drastically the etch rate of the developing Si nanostructures. For this reason we use a pulsed N_2 laser to probe the Si surface region stroboscopically. The duty cycle of this probe is about 10^{-5} for our experiments.

Usually por-Si is prepared in highly concentrated HF solutions at high anodic currents [3]. The in-situ control of the initial state of the por-Si formation is complicated in such solutions due to the very high rate of chemical reactions. Por-Si formation takes place also at low anodic currents in diluted aqueous fluoride solutions [4]. The structure of the formed por-Si depends sensitively on the concentration and on the pH of the solution [5,6].

This work is aimed to the in-situ investigation of the initial por-Si formation in 0.2M NH_4F solution by the use of PL of the band-band recombination of c-Si and of the characteristic PL of por-Si.

Mat. Res. Soc. Symp. Proc. Vol. 452 © 1997 Materials Research Society

EXPERIMENTAL

P-type Si(100) wafers (FZ, resistivity 5 Ωcm) are treated electrochemically in 0.2M NH_4F (pH 3.2, adjusted by addition of H_2SO_4) to control the formation of porous Si surface structures [5]. The sample is mounted into the center of a quartz tube and serves as working electrode. The diameter of the treated area amounts to 5 mm. A Pt wire and a Ag/AgCl electrode are used as counter and reference electrodes, respectively. The electrolyte is pumped continuously through the quartz tube. The equipment is described in more detail in Ref. [7].

The PL is excited with a N_2 laser (LTB-MSG200, pulse duration 0.5 ns, wavelength 337 nm, intensity 3 mJ/cm^2). The absorption length of the laser light in the c-Si is 10..20 nm. The PL signals are detected with a prism monochromator and a Si photodiode with a preamplifier in the time integrating (integration time 100 μs) regime. The PL of c-Si and por-Si are measured at the wavelengths of 1.1 and 0.6 μm, respectively.

The formation of por-Si surface layers is checked by electron microscopy. The cross section of the Si surface region is investigated by field emission scanning electron microscopy (HITACHI S4100) at the acceleration voltage of 25 kV and for a tilt angle of 30°.

RESULTS AND DISCUSSION

Fig.1 shows a current voltage scan of p-Si(100) in 0.2M NH_4F (pH 3.2). This scan is characterized by a first increase of the current at low anodic potentials up to the first maximum. Por-Si formation takes place even at these currents [4]. At higher potentials electropolishing occurs. The letters A, B, C and D mark characteristic points at the first increasing part of the current voltage curve.

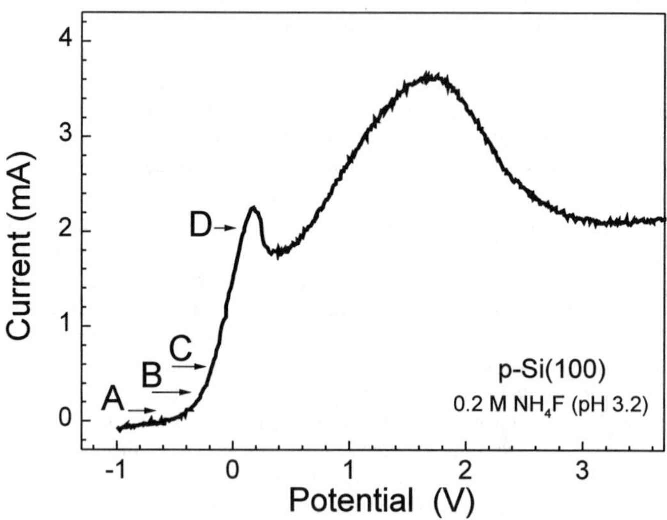

Fig.1: Current voltage scan of p-Si(100) in 0.2M NH_4F (pH 3.2).

The formation of por-Si contains the oxidation of Si backbonds by flouride species as a first step. This process is connected with a charge transfer during the formation of por-Si which characterizes the efficiency of the por-Si formation. Fig. 2 shows the cross section views of the Si surfaces after anodization at 0.03, 0.2, 0.5 and 2 mA (a-d, respectively) after passing an electric charge of 0.5 C. The currents are taken at the characteristic points A, B, C and D (see fig.1). Por-Si is formed only during anodization at the points A, B and C while treatment at point D does not lead to por-Si formation. Fig. 2b shows a double layer structure consisting of a mesoporous layer of a thickness of about 50 nm at the Si surface and a nanoporous Si top layer. The double layer structure is less pronounced in figures 2a and 2c.

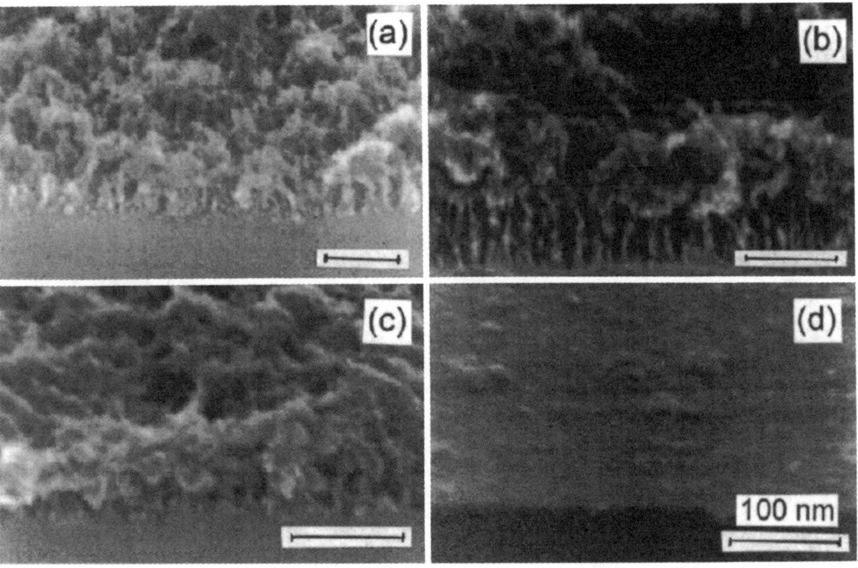

Fig.2: Cross section views of the Si surfaces after anodization at 0.03, 0.2, 0.5 and 2 mA (a-d, respectively) after passing an electric charge of 0.5 C. The currents are taken at the characteristic points A, B, C and D (see fig.1). The tilt angle of the sample is 30°. The bars mark 100 nm.

Fig.3 shows the dependence of the PL intensity of c-Si on the anodic current. In general, the higher the current the lower the PL intensity. The anodic current denotes the rate of chemical reactions at the Si surface which lead to the formation of nonradiative recombination surface defects. Further, there are also two steps of current decrease at which the PL intensity of c-Si is nearly constant with increasing anodic current. That means that not every act of charge transfer leads to the formation of a recombination active surface defect. This can be understood in terms of the dangling bond and center of reconstruction model [8] which claims that only the Si dangling bonds at Si atoms with three Si backbonds are recombination active. By other words, only the first step of oxidation of a Si atom leads to an increase of the nonradiative surface recombination. From the marked points A to B in fig.3 the formation of por-Si structures increases with increasing current (see fig.2a and b). The first plateau corresponds to the formation of the double layer porous structure (see fig.2b). At the end of this plateau (C in fig.3) the PL intensity starts to decrease again due to the onset of the competing electropolishing. As a result the thickness of the por-Si layer decreases and the active interface between the c-Si bulk and the electrolyte increases leading to an increase of the intrinsic Si dangling bond formation. Electropolishing is dominant at point D in fig.3.

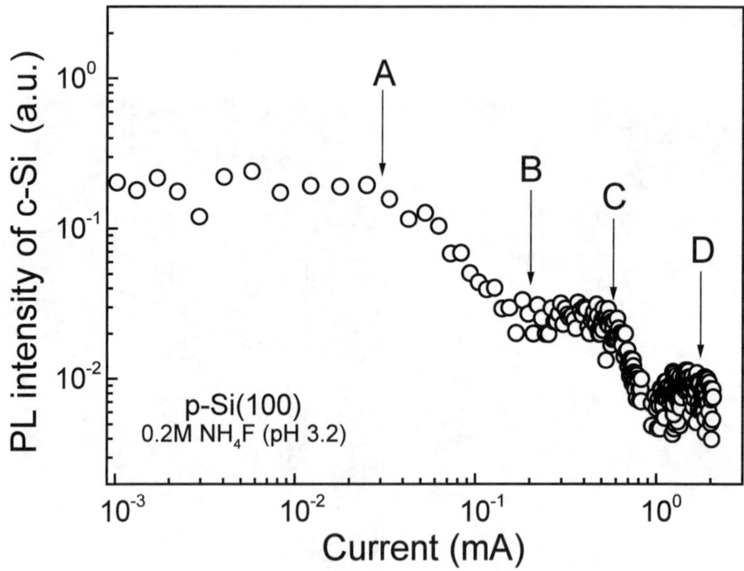

Fig.3: Dependence of the PL intensity of c-Si on the anodic current.

Fig.4 shows the time dependence of the PL intensity of por-Si and c-Si (solid triangles and open circles) during por-Si formation and after interrupting the anodization current. The current is kept between the points B and C (see fig.1 and fig.3). The PL of c-Si decreases with increasing PL of por-Si up to a saturation value. The PL of por-Si saturates for longer times. The PL of c-Si increases immediately after switching off the anodic current while the PL of the por-Si decreases slowly. The structure of the porous Si surface layer does not change during switching off the current. Consequently, the absorption of the excitation laser light remains unchanged. It should be remarked that the absorption length of the UV laser light is about 10...20 nm for c-Si and about 3...4 times larger for por-Si depending on porosity. The concentration of the excess carriers diffusing from the por-Si layer into the bulk c-Si is influenced only by nonradiative recombination at the Si surfaces. The immediate increase of the PL intensity of c-Si after switching off the current reflects the very fast passivation of the intrinsic Si dangling bonds. The increase of one order in magnitude during switching off the current is the same as for the difference between the PL intensities of the hydrogenated Si surface in solution and the level of the plateau between points B and C (fig.3). The very slow response of the PL of por-Si shows that the Si nanostructures are not involved into the chemical reactions during the anodization process. The decrease of the PL of por-Si after switching off the anodic current is induced, in our opinion, by the etching of Si nanostructures in the electrolyte. This process depends, of course, on the illumination level. The slow decrease of the PL of por-Si after switching off the anodization current gets faster if increasing the repetition rate or the intensity of the laser pulses.

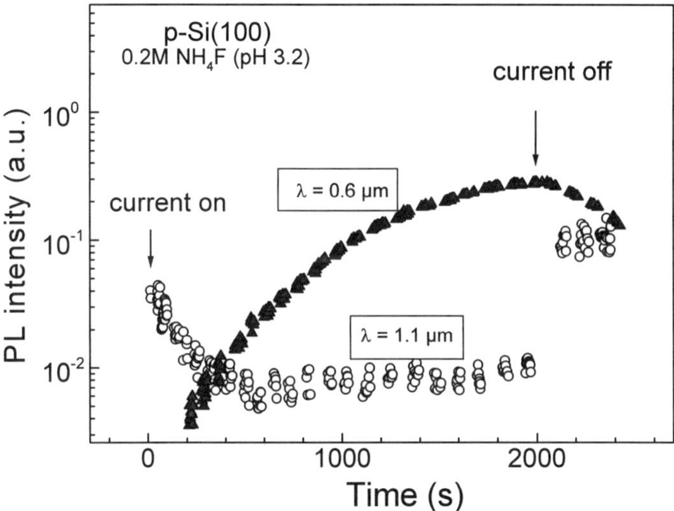

Fig.4: Time dependence of the PL intensity of por-Si and of c-Si (open and solid circles, respectively) during por-Si formation and after interrupting the anodization current. The anodization current is held at 1/3 of the current in the first maximum.

CONCLUSIONS

The PL intensity of c-Si and por-Si are measured simultaneously during the initial formation of por-Si in diluted fluoride solution. It is demonstrated that the PL can be used as a sensitive tool to investigate the changes in chemical reactions at Si surfaces which are correlated to the variation of the Si dangling bond concentration. The passivation of the Si nanostructures is independent of the por-Si formation at the c-Si/por-Si interface.

ACKNOWLEDGEMENT

We are grateful to I. Sieber for making the SEM cross section micrographs.

REFERENCES

1. V. Lehmann and U. Gösele, Mat. Res. Soc. Symp.Proc. Vol. **283**, 27 (1993).

2. T. Konishi, T. Yao, M. Tajima, H. Oshima, H. Ito, and T. Hattori, Jap. J. Appl. Phys. **31** L1216 (1992).

3. see, for example, L. T. Canham, Appl. Phys. Lett **57**, 1046 (1990) and the refs. therein.

4. D. J. Blackwood, A. Borazio, R. Greef, L. M. Peter, and J. Stumper, Electrochim. Acta **37** 882 (1992).

5. Th. Dittrich, S. Rauscher, V. Yu. Timoshenko, J. Rappich, H. Flietner, and H. J. Lewerenz, Appl. Phys. Lett. **67**, 1134 (1995).

6. Th. Dittrich, I. Sieber, S. Rauscher, and J. Rappich, Thin Solid Films **276**, 200 (1996).

7. J. Rappich, V. Yu. Timoshenko, and Th. Dittrich, J. Electrochem. Soc., accepted for publication.

8. H. Flietner in 7th Conf. Insulating Films on Semiconductors (INFOS) ed. by W. Eccleston and M. Uren (Hilger, Bristol 1991), p. 151-154.

ATOMIC LAYER ETCHING OF POROUS SILICON

I. H. LIBON, C. VOELKMANN, V. PETROVA-KOCH, F. KOCH
Technische Universität Munich, Physik-Department E16, D-85747 Garching, Germany

ABSTRACT

In this work we describe the controlled shifting of the PL peak of p^+ (10 mΩcm) porous silicon (PoSi) by means of atomic layer etching (ALEP). We hereby study the cluster-size dependence of the PL of this material. By this technique of repeated oxidation by H_2O_2 and stripping of the oxidized surface layer, we reduced the size of the crystallites layer by layer. In all previous reports the PoSi PL appeared to have a natural lower energy limit of ≈ 1.4 eV. We report for the first time a continuous shift of the PoSi PL peak between 1.01 and 1.20 eV. This observation allows us to draw conclusions for the luminescence mechanism: it proves that geometrical quantum confinement in Si crystallites is responsible for the efficient room-temperature PL in PoSi near the indirect bandgap of c-Si. Together with observations of size-independent PL peaks around 1.6 eV in thermally oxidized samples this result indicates that the PoSi PL cannot be described by one origin alone. Both the existence of molecular centers and the geometrical quantum confinement are valid in their specific range of etching and post-anodic treatment parameters.

CONCEPT AND REALIZATION

The ALEP technique is the repeated oxidation by H_2O_2 and stripping of the oxidized surface layer. The starting material for this process is as-prepared, H-terminated p^+ PoSi [1]. Immersing the sample in H_2O_2 (aqueous 30 %) for 5 minutes leads to a typical oxidation of the PoSi layer [2]. The peak at 2250 cm^{-1} observed in FTIR measurements is assigned to the O_3-Si-H_x stretching mode.
Dipping the sample afterwards in ethanoic HF (mixture 1:1 vol. of HF (aqueous 49 %) and ethanol) for 5 minutes produces the typical FTIR spectrum of H-terminated PoSi [3].
This change in surface termination can be explained by the reactivity of HF with the back-bonded oxygen. This results in the removal of the outermost Si layer bonded to the attacked oxygen layer [2]. The repeated application of the oxidation and stripping of the oxidized layer leads to a controlled size reduction, layer by layer. After a certain number of steps, ALEP was no longer a useful method, because of the internal instability of the sponge-like structure, caused by the reduced crystallite size.

PL SPECTRA

The light-emitting properties of the ALEP samples are studied as a function of the crystallite size. Fig. 1 shows the PL spectra of several H-terminated samples under 1.96 eV excitation at a temperature of 15 K collected by a 22 cm Spex monochromator and detected by a Hamamatsu PMT.

Mat. Res. Soc. Symp. Proc. Vol. 452

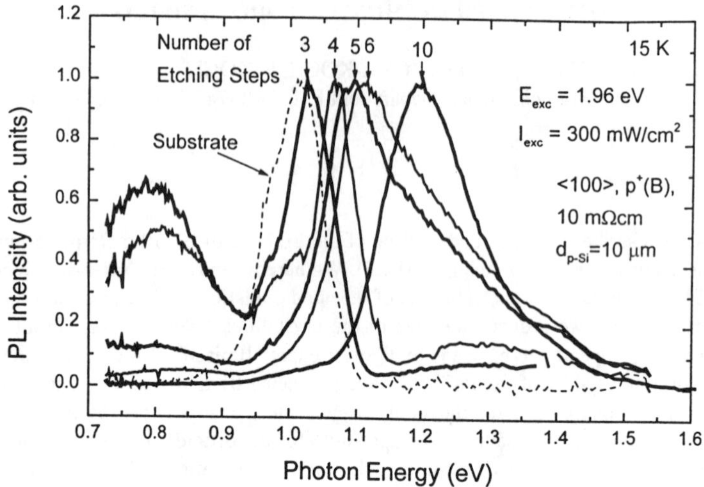

Fig. 1: PL spectra of several H-terminated PoSi samples with 1.96 eV excitation at 15 K.

The fact that the PL peak of the c-Si substrate lies at 1.02 eV is due to bangap narrowing of the highly doped material. The observed blue shift of the PL peak supports the structure-size reduction in accordance with the quantum-size model. With this ALEP method the spectral region between 1.01 eV and 1.20 eV was covered, which to our knowledge has not been verified before in the literature. The blue-shift was accompanied by a systematic broadening of the PL peak.

The observed PL peak blue shift indicates the significance of quantum confinement for the PL mechanism. For the relatively small deviations from the bulk Si bandgap it is possible to employ the effective mass theory to calculate the cluster-diameter dependence of the bandgap. Assuming a cubic infinite square-well potential the expression for the bandgap shift is

$$\Delta E_g = \frac{\hbar^2 \pi^2}{2} \cdot \frac{1}{d^2} \cdot \left(\frac{1}{m_l} + 2\frac{1}{m_t} + 3\frac{1}{m_{hh}} \right) \qquad (1)$$

where d is the square-well length, m_l and m_t denote the conduction band longitudinal and transverse masses, respectively, and m_{hh} is the valence band heavy-hole effective mass. Their values are $m_l = 0.98\ m_0$, $m_t = 0.19\ m_0$, and $m_{hh} = 0.49\ m_0$ [4]. Using square columns instead of cubes reduces the value of E_g to 2/3, since one of the 3 quantization directions of the cubic material is missing.

The observed PL peak shift, however, is bigger than this quantum confinement contribution. This is due to the decreasing number of B acceptors in the material leading to an increase in the number of crystallites containing no acceptor: In clusters with this impurity the PL is quenched due to Auger processes, which can be verified in PL intensity measurements. Therefore the PL in ALEP PoSi composed of small crystallites will be emitted mainly from crystals that do not contain B, whereas in PoSi composed of bigger crystals the PL predominantly is emitted from clusters containing B. The PL in heavily doped c-Si is red-shifted in respect to undoped material due to bandgap renormalization. The observed 60 meV PL shift in 10 mΩcm ($1 \cdot 10^{19}$ cm^{-3}) B-doped c-Si with respect to undoped c-Si [5] has an important implication for our measurements: besides the quantum confinement shift (Eq. 1) we also have to take into account the 60 meV shift between sample 0 (PL from occupied crystals) and sample 10 (PL from unoccupied crystals). The linear consideration of this bandgap narrowing shift yields the relation displayed in Fig. 2.

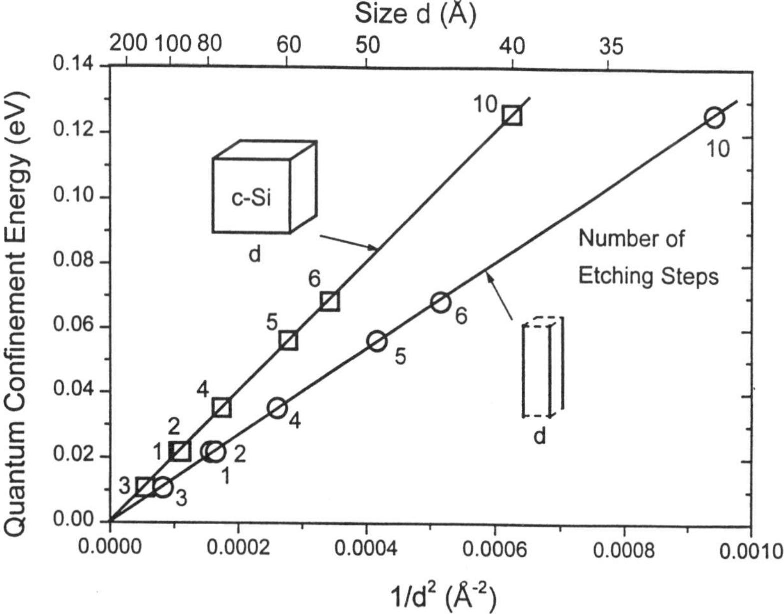

Fig.2: Si bandgap shift as a function of crystallite size calculated from effective mass theory (lines) and resulting ALEP structure sizes (squares and circles).

We calculated the cluster sizes from the 1.96 eV excited data since the absorption coefficient of typically 10^{-2} cm^{-1} for the red radiation ensures a uniform excitation of the entire 10 µm layer. Figure 2 shows the crystallite sizes of different ALEP samples resulting from this calculation. The fact that the resulting decrease in crystallite size per step of the order of 4 Å confirms the measured SAXS data [7], indicates the consistency of our procedure. With the exception of sample 3, all investigated samples lie in a consecutive order. Although the energetic position of the first 3 samples appears somewhat scattered, the size reduction between steps 5 and 10 (20 Å) is similar to the reduction between steps 1 and 5.

In all previous reports the PoSi PL appeared to have a natural lower energy limit of ≈ 1.4 eV. Conventional micro-PoSi anodized at low current density shows this PL peak energy. Meso-PoSi prepared from p^+ -type wafers and having structural features in the 10 nm range shows a weak low-temperature luminescence at ≈ 1.0 eV. After aging and weak oxidation the first PL signal that appears at RT occurs for ≈ 1.4 eV. It does so discontinuously in a way that a doublet of broad peaks at 1.0 and 1.4 eV appears. We explain this behavior as due to Auger damping: decreasing PL peak energy means an increase in particle size due to quantum confinement. A constant shallow defect concentration in the PoSi skeleton therefore implies that the number of crystallites without PL quenching defects tends to zero if the PL peak energy decreases. The critical point in PoSi is 1.4 eV. At this PL peak energy the number of particles containing no Auger quenching defects vanishes because the average crystallite size is so big that the probability of a crystal containing no defect is exceedingly small. We observe this Auger damping effect also in the PL intensity.

By use of our technique of controlled particle size reduction we here report the first continuous shift in the PoSi PL peak from the bulk Si bandgap of 1.1 eV up to 1.2 eV. These data provide evidence for the significance of geometrical quantum confinement. More importantly, the continuous shift to the bulk Si bandgap 1.1 eV proves the Si crystallites to be the origin of the PoSi PL.

PL INTENSITY

Fig. 3 shows the measured peak PL intensity as a function of the number of etching steps.

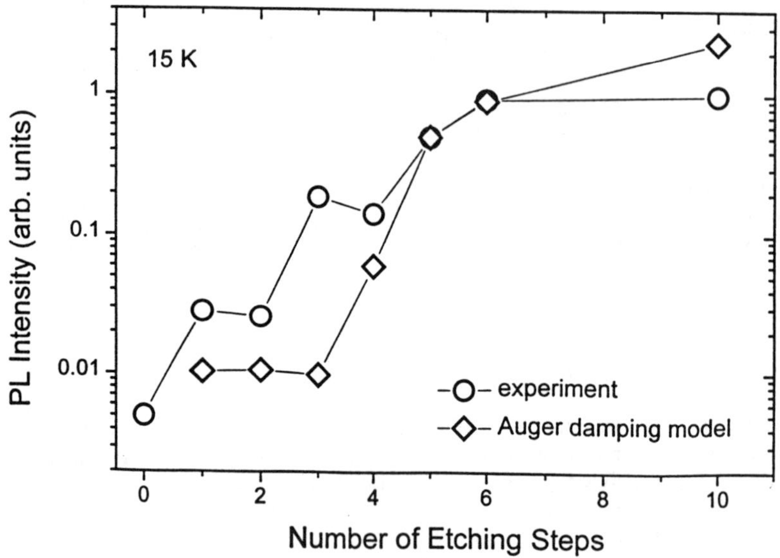

Fig.3: Experimental peak PL intensity as a function of the number of etching steps and values calculated using an Auger damping model.

The PL intensity increases exponentially with decreasing structure size until a saturation occurs in intensity beween samples 6 and 10, corresponding to a cluster size of ≈ 50 Å. We explain this behavior as due to Auger quenching; the results of this model are shown in Fig. 3. The 10 mΩcm samples exhibit a constant concentration of B acceptors of $N_A = 1 \cdot 10^{19}$ cm^{-3} [4]. As the structure size decreases with more and more etching steps, the number of acceptors in the material decreases. Thus the fraction of crystallites containing no acceptors will increase exponentially (Poisson distribution), until each crystallite contains 1 acceptor on the average. At this point the PL intensity will saturate, as observed. The cluster size for this saturation to begin is of the order of $(1/N_A)^{1/3} = 46$ Å, which correctly is the value we observe using our quantum confinement model. In the early etching steps the main part of the PL will originate from large clusters containing B showing a low, but constant radiative decay probability. Therefore in this region the PL intensity is independent of the number of etching steps.
The PL intensity I_{PL} hence consists of two contributions, one from the occupied I_{occ} and one from the unoccupied crystallites I_{unocc}: $I_{PL} = I_{occ} + I_{unocc}$ =const. $(\gamma N_{occ} + N_{unocc})$, where γ denotes the branching ratio of B acceptor bound excitons in Si to decay radiatively and nonradiatively. In applying the same constant factor for both terms we assume equal decay times for occupied and unoccupied crystals. The Poisson distribution yields the number of occupied crystals resulting in a relation between PL intensity and crystallite size d

$$I_{PL} = \text{const.} \left[\gamma \left(1 - \exp\left[\left(-\frac{d}{d_0} \right)^3 \right] \right) + \exp\left[\left(-\frac{d}{d_0} \right)^3 \right] \right]$$

with the length scale parameter $d_0 = (1/N_A)^{1/3} = 46$ Å. Using $\gamma = 1/464$ [6] this formula is used in Fig. 3.
The fact that the PL intensity can be modelled quantitatively with our Auger damping model based on structure size reduction also supports the quantum-size-effect model.

CONCLUDING REMARKS

In conclusion, by using the technique of ALEP we were able to reduce the size of PoSi crystallites in a highly controlled manner. This decrease in cluster size could be verified using small-angle X-ray scattering (SAXS) [7] : by this measurement a structure size reduction of a few monolayers per etching step is inferable. Although these preliminary X-ray measurements do not provide complete independent evidence for the quantitative evaluation of cluster sizes, the internal consistency of our model, that describes both PL peak position and intensity, is hereby strongly supported. Due to the quantum size effect the PL was blue-shifted in respect to the starting PoSi sample. We managed to produce samples with their PL peak between 1.1 and 1.2 eV, which has not been verified before. The position and shift of the PL peak is clear evidence for the quantum-size effect in Si nanocrystals being responsible for the efficient room-temperature PL in PoSi.

ACKNOWLEDGMENTS

V.P.K. thanks the Deutsche Forschungsgemeinschaft (DFG) for the Habilitationsstipendium.

REFERENCES

[1] L.T. Canham, Appl. Phys. Lett. **57**, 1046 (1990).

[2] A. Nakajima, T. Itakura, S. Watanabe, and N. Nakayama, Appl. Phys. Lett., **61**, 46 (1992).

[3] H. Ubara, T. Imura, and H. Hiraki, Solid State Commun. **50**, 673 (1984).

[4] S.M. Sze, Physics of Semiconductor Devices, (Wiley, New York, 1981).

[5] J. Wagner and J. A. del Alamo, J. Appl. Phys. **63**, 425 (1988).

[6] W. Schmid, phys. stat. solid. (b) **84**, 529 (1977).

[7] A. Misera (private communication).

ROOM TEMPERATURE BAND-EDGE LUMINESCENCE FROM SILICON GRAINS PREPARED BY THE RECRYSTALLIZATION OF MESOPOROUS SILICON

KAREN L. MOORE*, LEONID TSYBESKOV**, PHILIPPE M. FAUCHET*,** AND DENNIS G. HALL*
*University of Rochester, The Institute of Optics, Rochester, NY 14627
**University of Rochester, Department of Electrical Engineering, Rochester, NY 14627

ABSTRACT

Room-temperature photoluminescence (PL) peaking at 1.1 eV has been found in electrochemically etched mesoporous silicon annealed at 950°C. Low-temperature PL spectra clearly show a fine structure related to phonon-assisted transitions in pure crystalline silicon (c-Si) and the absence of defect-related (e.g. P-line) and impurity-related (e.g. oxygen, boron) transitions. The maximum PL external quantum efficiency (EQE) is found to be better than 0.1% with a weak temperature dependence in the region from 12K to 400K. The PL intensity is a linear function of excitation intensity up to 100 W/cm^2. The PL can be suppressed by an external electric field $\geq 10^5$ V/cm. Room temperature electroluminescence (EL) related to the c-Si band-edge is also demonstrated under an applied bias ≤ 1.2 V and with a current density ≈ 20 mA/cm^2. A model is proposed in which the radiative recombination originates from recrystallized Si grains within a non-stoichiometric Si-rich silicon oxide (SRSO) matrix.

INTRODUCTION

The light-emitting properties of bulk crystalline silicon (c-Si) have been investigated for several decades.[1, 2] Recent interest has focused on the visible photoluminescence (PL) that is observed in Si nanoclusters and in porous Si (PSi),[3] the infrared PL in silicon-germanium superlattices,[4] and the subgap PL due to impurities in c-Si.[5] Band-edge luminescence of c-Si has a low-temperature external quantum efficiency (EQE) of less than 10^{-5} due to the indirect band gap of c-Si. This EQE can be increased by better surface passivation.[6] There have been few reports of room-temperature band-edge PL in Si.[7] In this work we report room-temperature PL and EL originating from Si grains in a silicon oxide matrix.

EXPERIMENTAL RESULTS AND DISCUSSION

The samples used only for photoluminescence measurements are prepared by anodically etching boron-doped p^+ c-Si wafers with a resistivity ≈ 0.05 Ω-cm in an HF-ethanol solution (1:1) under a current density ≈ 20 mA/cm^2. After anodization, the samples are annealed in dilute oxygen (10% O_2 in N_2) for 10 minutes up to 3 hours, at temperatures ranging from 800 to 990 °C.

Mat. Res. Soc. Symp. Proc. Vol. 452 © 1997 Materials Research Society

Complete fabrication details for the Si-based light-emitting diode (LED) which is examined for PL and electroluminescence (EL) can be found in Ref. 8. The LED consists of an oxidized PSi, or SRSO, active layer formed on an n-type c-Si wafer with a poly-Si film for the top contact. Sintered aluminum provides a good ohmic contact to the poly-Si. The back contact consists of an n^+ Si layer, prepared by ion implanting phosphorous in Si followed by an anneal at 950°C. The final step in the formation of the contact is a standard aluminum metalization.

The luminescence measurements are performed using a McPherson 0.35-m grating monochromator, a North Coast liquid-nitrogen cooled Ge detector and a Stanford Research System lock-in amplifier. For low temperature data, the samples are mounted in an Air Products close-cycle He cryostat. Samples are optically excited with a chopped beam of an Ar^+ ion laser.

Typical PL spectra measured over a temperature range from 12 to 300 K for a sample annealed at 950°C for 30 minutes are pictured in Fig. 1. The inset shows a detailed PL spectrum at 12K exhibiting a weak no-phonon (NP) line at 1.158 eV, a TA-phonon line near 1.14 eV, and a TO-phonon line at 1.1 eV with a replica at 1.04 eV. The observed PL spectral lines are well known in c-Si and tabulated in Ref. 5. As the temperature is raised, the PL spectrum peak is shifted and becomes broader. The PL peak undergoes a small blueshift (to shorter wavelengths) in the temperature range from 12 to 50 K and at temperatures greater than 70 K the PL peak is redshifted (to longer wavelengths). The blueshift observed in the PL spectrum is not attributable to quantum confinement effects.

The PL spectrum broadens with increasing temperature. Examining the intensity of the PL spectrum on a semilogarithmic scale, the high energy tail of the spectrum is shown to be linear. The slope increases linearly with temperature, which is consistent with carriers following a Boltzmann distribution.[9] Defining $f(E,T)$ as the probability of energy states being occupied and $N(E)$ as the parabolic density of states, the total broadening, ΔE, of the PL line is given by the equation $\Delta E \sim f(E,T) \times N(E)$. Using this equation, it can be shown that ΔE is proportional to

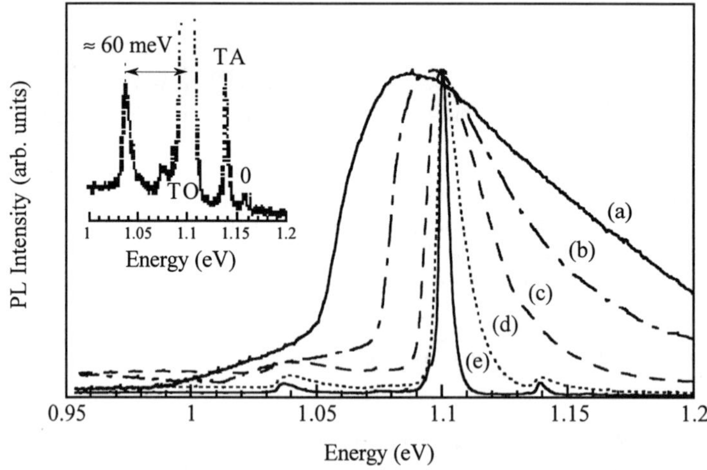

FIG. 1. PL Spectra from oxidized porous silicon measured at various temperatures: (a) 300, (b) 200, (c) 100, (d) 50, and (e) 12 K. The inset illustrates the fine structure of the 12K PL spectrum with typical phonon lines: NP line at 1.158 eV, TA line at 1.14eV, and TO line 1.1eV, with a replica at ~1.04 eV

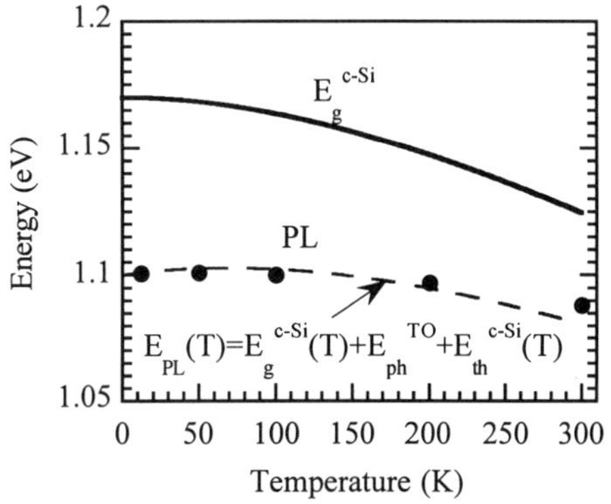

FIG. 2a. Temperature dependence of the c-Si band gap (solid line) and the PL peak position in oxidized PSi: experiment (dots) and fitting (dashed line).

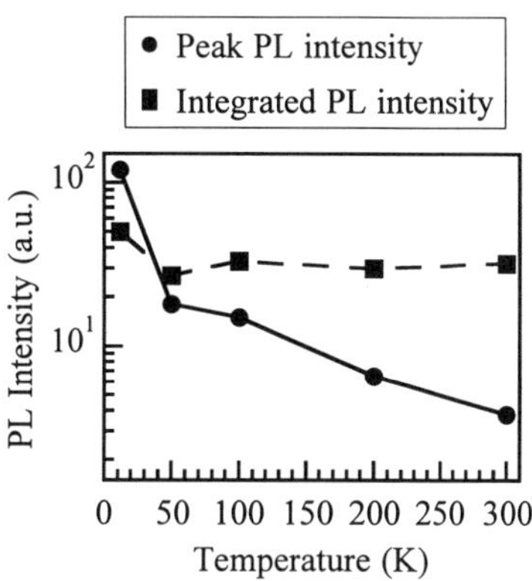

FIG. 2b. Temperature dependence of the PL peak intensity (circles) and the integrated PL intensity (squares).

2kT, which agrees well with our measurements. The temperature dependence of the PL peak follows that of the c-Si energy gap[10] with corrections for the characteristic phonon energy [TO phonon, 58 meV (Ref. 5)] and the thermal distribution of the population. Figure 2a shows good agreement between the calculated temperature dependence of the PL peak and the experimental data. The fitting formula components for the temperature dependent PL peak are defined as follows: $E_g^{c\text{-}Si}(T)$, the temperature-dependent c-Si band gap; E_{ph}^{TO}, the characteristic TO-phonon energy; and $E_{th}^{c\text{-}Si}(T)$, the most probable carrier energy for a thermal population distribution.

Exciton-related PL is thermally quenched for T≥20K in bulk c-Si.[5] For excitons bound to impurities, this quenching threshold can be shifted to higher temperatures. When measured down to 0.7 eV, the PL spectra of our samples show no impurity or defect-related transitions [including boron-related PL[5] or the oxygen-related P line[7]]. The temperature dependence of our samples is found to be entirely different from that of c-Si. The PL peak intensity decreases with increasing temperature, whereas the integrated PL intensity is essentially temperature independent from 12 to 300 K.

The PL intensity exhibits a linear dependence on the excitation intensity over the range of 1-100 W/cm^2. The PL spectrum becomes broader at the maximum excitation intensity (~130 W/cm^2); the broadening is attributed to local heating by the laser. The local temperature of the sample is calculated using the broadening of the PL spectrum and found to be ~400K. However, even at the maximum excitation level, no deviation from a linear dependence on excitation intensity is observed in the PL intensity.

The PL EQE can be estimated through a comparison of our samples with a sample that exhibits PL in approximately the sample spectral region. After ion implantation with Be and subsequent annealing, bulk c-Si has a 1% EQE at 12K.[11] Under similar experimental conditions, the PL of Si:Be is several times more efficient than that of the samples studied in this work. We estimate the EQE in our best samples to be greater than 0.1%.

Figure 3 shows spectra obtained from several samples. Spectrum (1) is PL from a

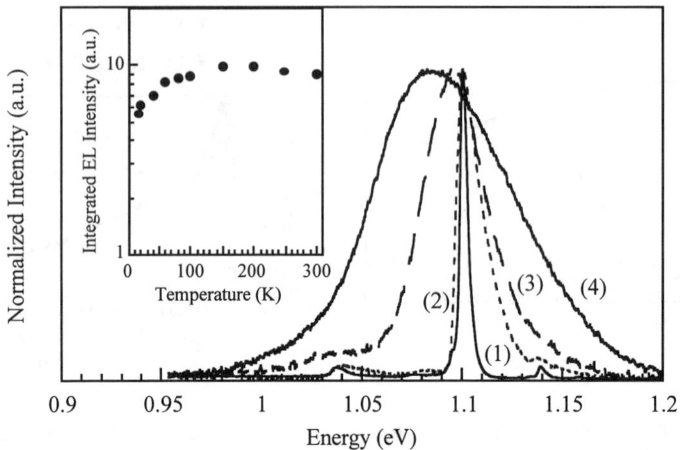

FIG. 3. PL and EL spectra: (1) the PL in SRSO active layer without poly-Si contact at 13K; (2) the PL in a device structure with top poly-Si contact at 13K; (3) the EL in a device structure at 13K; (4) the EL in a device structure at room temperature. The inset shows the integrated EL intensity as a function of temperature.

sample without a poly-Si contact (SRSO active layer on c-Si substrate) at low temperature. Spectra (2) and (3) are PL and EL measured from a complete device at low temperature, respectively. Spectrum (4) is EL from a complete device at room temperature. Once again, the PL and EL low temperature spectra show the presence of a very weak NP line at 1.158 eV, a TA-phonon line at 1.14 eV and a TO-phonon line at 1.1 eV with a replica at 1.04 eV. These PL spectral lines are broader and slightly shifted in the complete device structure, possibly resulting from the non-uniform strain produced by the preparation of the top poly-Si contact. A shift (~10 meV) of the PL peak to lower energies is observed in some devices which we attribute to a compression strain of ~20 kbar. Nevertheless, the PL spectra of the SRSO layers with and without the poly-Si top contact are very similar.

The EL spectra are broader than the PL spectrum and slightly shifted to lower photon energies. The devices have an EL threshold voltage ≥ 1 V, corresponding to a current density of less than 10 mA/cm^2. Increasing the temperature broadens the EL spectra as seen in a difference between spectrum (3) and (4) in Fig. 3. The high energy tail of the EL spectrum is an exponential, as we also observed for the PL spectrum. Under large forward biases (>5 V), the EL spectrum broadens and the high energy tail becomes non-exponential. This may be understood as a change of carrier transport mechanism from diffusion to drift. The inset in Fig. 3 shows the temperature dependence of the integrated EL intensity. From 12 to 300 K the integrated EL intensity is nearly independent of temperature.

In order to further compare the PL and EL spectra, we have measured the low-temperature PL in the presence of an external electric field (Fig. 4). The PL intensity decreases exponentially (Fig. 4a) and the PL spectrum broadens with increasing reverse bias (Fig. 4b). Under a large reverse bias, the low-temperature PL spectrum becomes similar to the EL spectrum taken from the same sample. The PL is extinguished under an external electric field ≥ 10^5 V/cm. We conclude that the PL and EL are both due to band-edge radiative recombination in

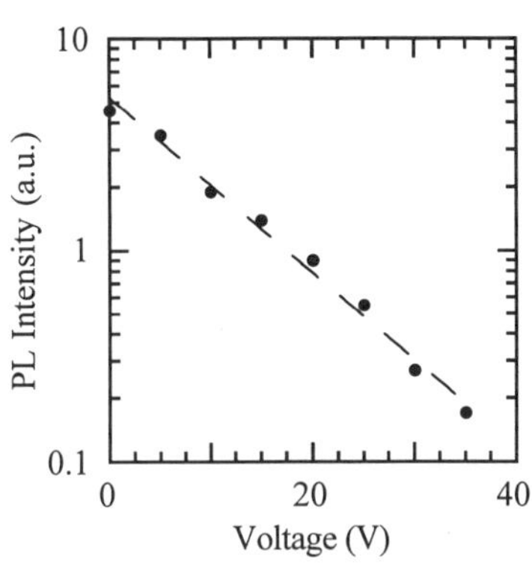

FIG. 4a. The low-temperature (T=15K) PL intensity as a function of reverse bias for a device structure.

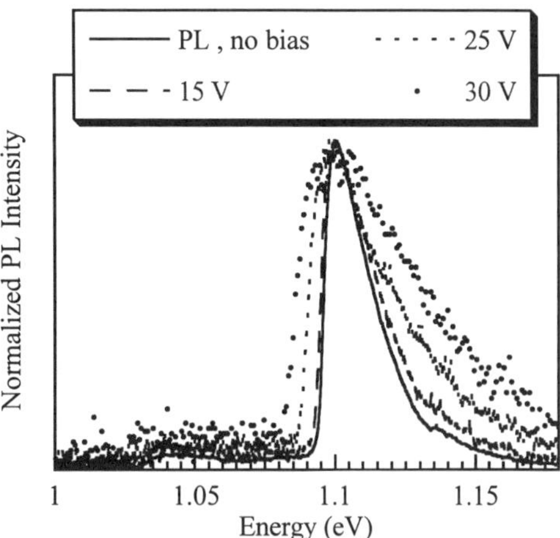

FIG. 4b. The normalized PL spectra with no bias (solid line) and with a reverse bias of 15V (large dashed line), 25V (small dashed line), and 30 V (dots).

large silicon clusters within the non-stoichiometric SRSO matrix.

SRSO is a composite material with a variation of the ratio between silicon and oxygen. The band gap of this material varies from 1.1 eV (pure Si) to ≤10 eV (SiO_{2-x}) with giant contravariant fluctuations due to different chemical compositions. These fluctuations enable a strong electron-hole pair localization. Even if the temperature is sufficient for the dissociation of excitons, the electrons and holes cannot leave the potential valleys; eventually, an exciton will reform and the electron and hole will recombine radiatively. The experimental evidence for this carrier localization is the weak PL and EL temperature dependence and the PL reduction in the presence of an external electric field E (by a factor of 10 for $E \leq 10^6$ V/cm). Typically, in c-Si the PL is quenched for an electric field $E \approx 10^4$ V/cm and the exponential dependence of the quenching is not observed.[5, 12] An exponential PL reduction with field is well known in chalcogenide glasses where a strong electron-hole pair localization takes place.[13]

CONCLUDING REMARKS

We have observed room-temperature, Si-band-edge-related PL and EL originating from a SRSO layer. The PL is estimated to have an external quantum efficiency greater than 0.1%. The PL and EL are weakly temperature dependent in the range from 12 to 300 K. From the proposed model for the SRSO structure, the PL and EL are due to radiative transitions within large Si grains immersed in SRSO.

This work was supported in part by the U. S. Army Research Office and the U. S. Air Force Office of Scientific Research.

REFERENCES

1. J. R. Haynes and H. B. Briggs, Phys. Rev. **86**, 647 (1952).
2. J. R. Haynes and W. C. Westphal, Phys. Rev. **101**, 1676 (1956).
3. L. E. Brus, P. F. Szajowski, W. L. Wilson, T. D. Harris, S. Schuppler, and P. H. Citrin, J. Am. Chem. Soc. **117**, 2915 (1995).
4. J. P. Noel, N. L. Rowell, D. C. Houghton, and D. D. Perovic, Appl. Phys. Lett. **57**, 1037 (1990).
5. G. Davies, Phys. Rep. **176**, 84 (1989).
6. E. Yablonovitch and T. Gmitter, Appl. Phys. Lett. **49**, 587 (1986).
7. O. King and D. G. Hall, Phys. Rev. B **50**, 10661 (1994).
8. L. Tsybeskov, K. L. Moore, S. P. Duttagupta, K. D. Hirschman, D. G. Hall, and P. M. Fauchet, Appl. Phys. Lett. **69**, 3411 (1996).
9. J. I. Pankove, Optical Processes in Semiconductors, Dover, New York, 1971, p. 422.
10. S. M. Sze, Physics of Semiconductor Devices, Wiley, New York, 1981, p.868.
11. T. G. Brown, P. L. Bradfield, D. G. Hall, and R. A. Soref, Appl. Phys. Lett. **12**, 753 (1987).
12. H. Weman, Q. X. Zhao, and B. Monemar, Phys. Rev. B **36**, 5054 (1987).
13. N. F. Mott and E. A. Devis, Electronic Processes in Non-Crystalline Materials, Oxford University Press, Oxford, 1979.

LIGHT EMISSION FROM INTRINSIC AND DOPED SILICON-RICH SILICON OXIDE: FROM THE VISIBLE TO 1.6 μm

L. Tsybeskov, K. L. Moore*, P. M. Fauchet*,**, and D. G. Hall*

Department of Electrical Engineering, University of Rochester, Rochester, NY 14627

*The Institute of Optics, University of Rochester, Rochester, NY 14627

**Also Laboratory for Laser Energetics and Department of Physics and Astronomy, University of Rochester, Rochester, NY 14627

ABSTRACT

Silicon-rich silicon oxide (SRSO) films were prepared by thermal oxidation (700°C-950°C) of electrochemically etched crystalline silicon (c-Si). The annealing-oxidation conditions are responsible for the chemical and structural modification of SRSO as well as for the intrinsic light-emission in the visible and near infra-red spectral regions (2.0-1.8 eV, 1.6 eV and 1.1 eV). The extrinsic photoluminescence (PL) is produced by doping (via electroplating or ion implantation) with rare-earth (R-E) ions (Nd at 1.06 μm, Er at 1.5 μm) and chalcogens (S at ~1.6 μm). The impurities can be localized within the Si grains (S), in the SiO matrix (Nd, Er) or at the Si-SiO interface (Er). The Er-related PL in SRSO was studied in detail: the maximum PL external quantum efficiency (EQE) of 0.01-0.1% was found in samples annealed at 900°C in diluted oxygen (~ 10% in N_2). The integrated PL temperature dependence is weak from 12K to 300K. Light emitting diodes (LEDs) with an active layer made of an intrinsic and doped SRSO are manufactured and studied: room temperature electroluminescence (EL) from the visible to 1.6 μm has been demonstrated.

Introduction

Optoelectronic integrated circuits may have important applications in chip-to-chip interconnects, parallel processing and the integration of photonics on silicon chips. For such applications there are specific wavelengths of interest. For example the infra-red (I-R) region near 1.5 μm is important for long distance communication due to minimum losses in silica-based waveguides. Oxidized porous Si (PSi) has already been implemented as a host for light-emitting impurities (e. g., rare earth ions), and photoluminescence (PL) in the I-R region has been reported [1, 2]. In this work we report the PL and electroluminescence (EL) in silicon-rich silicon oxide (SRSO) prepared by partial oxidation of PSi. Emission from the visible to 1.6 μm is achieved via structural modification of the SRSO and doping by rare-earth (R-E) ions and chalcogens.

Experiments and Results

In this work the samples were produced by anodic etching of boron-doped c-Si wafers with a resistivity $\rho \approx 1$ - 0.05 Ω cm in an HF-ethanol solution (1:1) under a current density $J \approx 5$ - 20 mA/cm^2. The thickness of the PSi layers varies from 0.5 to 5 μm. The intrinsic samples were annealed from 30 minutes up to 6 hours, at temperatures ranging from 400°C to 1000°C. The doping of SRSO by R-E has been performed electrochemically (according to the procedure developed in Ref. 2). Samples doped by sulfur were prepared by implantation of c-Si wafers (energy of 200 keV and dose of 10^{15}cm^{-2}), annealing at T≈1000°C (to remove crystalline damage), anodization under the conditions described above, and annealed a second time (T=950°C) for partial oxidation and additional dopant activation. In order to reduce the

Mat. Res. Soc. Symp. Proc. Vol. 452 © 1997 Materials Research Society

oxidation rate and prevent the complete conversion of PSi into SiO_2, a dilute oxygen ambient (~ 10% O_2 in N_2) was used.

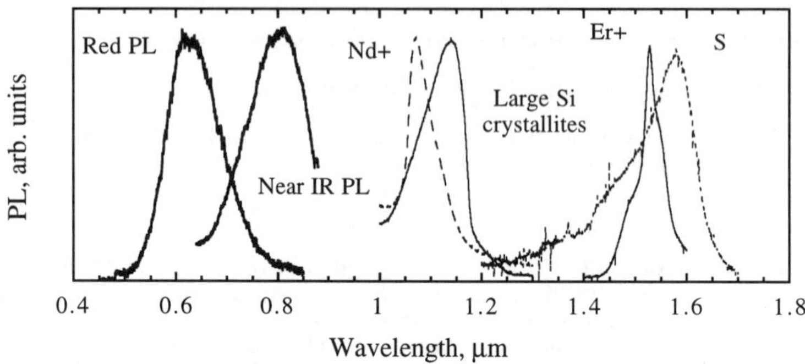

Figure 1. The PL produced by thermal-induced modification and doping of SRSO. The impurities are indicated on the figure.

Figure 1 shows the room-temperature PL spectra in an intrinsic and doped SRSO in the spectral region from the visible to 1.6 μm. We will discuss the preparation conditions and properties for each of the PL bands in detail.

The red PL band with a peak wavelength near 600-650 nm has been found in samples produced by anodization of c-Si with an initial resistivity of $\rho \approx 1\ \Omega$ cm and annealing at T < 800°C. In low-temperature resonantly excited PL spectra, the step-like structure due to phonon-assisted transitions firstly reported by P. D. J. Calcott et al. [3] remains observable [1]: therefore we concluded that the origin of the red PL is the same as in freshly anodized PSi. The room temperature PL external quantum efficiency (EQE) is estimated to be ≤ 1%. An increase of the annealing temperature shifts the PL spectrum to the near I-R with a peak at $\lambda \approx$ 750-800 nm and decreases the EQE by a factor of 3-5. The properties of the near I-R PL are similar to those reported in Ref. 4. For a comprehensive study of the red and near IR PL in these samples, see Ref. 5.

The complete activation of dopants (including a diffusion) requires a high temperature (T ≥ 1000°C) anneal. After annealing at T ≈ 950-1000°C some of our samples (low porosity PSi prepared from p+ c-Si with $\rho \approx 0.05\ \Omega$ cm) exhibit the PL attributed to phonon-assisted bandedge recombination in ultra pure c-Si [6]; therefore we conclude that large Si grains can be produced by recrystallization of PSi.

Silicon-rich silicon oxide (SRSO) made by partial oxidation and recrystallization of PSi contains a significant amount of SiO_x and the ratio of c-Si/SiO_x can be changed by the annealing conditions (i.e. temperature, time and oxygen concentration). In order to obtain luminescence in different spectral regions from this material, we have focused on several well known light-emitting impurities that can be activated in an oxygen-rich environment.

---

[1]The measurements were provided by P. D. J. Calcott and L. T. Canham, DRA, Malvern, UK.

Neodymium (Nd) doped glasses are very well characterized light-emitting materials with uses in optoelectronics and solid-state lasers [7]. The bandgap of c-Si is less than the energy of the main Nd-related radiative transition (emission at 1.06 μm). In the case of SRSO, room temperature PL due to Nd is easily observed even if the size of the Si grains remains too large for effective quantum confinement (from AFM, the grain size is ≥ 50 nm). The PL spectrum is relatively broad and almost identical to the Nd-related emission in silica (Fig. 1). In addition, the most intense PL has been found in strongly oxidized samples annealed for at least 30 minutes at T =1000°C. We conclude that the observed emission is due to Nd ions localized within a SiO matrix.

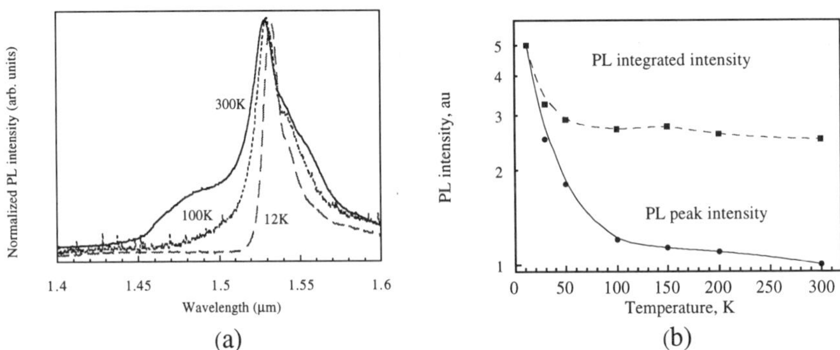

Figure 2. The PL spectra (a) and PL intensity (b) in SRSO:Er at different temperatures.

Erbium-doped silicon (Si:Er) has been the subject of extensive investigations since the 1980s when low-temperature luminescence at 1.5 μm was reported [8]. Recently, room-temperature luminescence from Si-based materials doped by Er has been demonstrated in: c-Si co-implanted by oxygen (Si:O:Er) [9]; semi-insulating polycrystalline Si (SIPOS) [10]; amorphous Si (a-Si) [11]; and porous Si (PSi) prepared by electrochemical etching of c-Si. In the case of PSi, erbium has been incorporated through ion implantation [1] or electroplating [2]. It has been suggested that the recombination of spatially confined electron-hole pairs in Si-nanograins, which is responsible for the visible luminescence in PSi , promotes efficient energy transfer and excitation of Er^{3+} ions incorporated in Si nanoclusters [1, 2]. An alternative model suggests that the Er^{3+} luminescence is mediated by photocarriers in the amorphous Si:O:H or Si:O-matrix produced by thermal oxidation of PSi [12].

In this work the PL in SRSO:Er was also observed at room temperature and, its temperature dependence is weak (Figure 2). The PL EQE in our best samples is estimated to be between 0.1% and 0.01%. The decrease in the PL intensity with increasing temperature from 12 K to 300 K is less than 50% (Fig. 2b) which is less than what has been reported in Ref. 1, 9-12. The Er-related spectra are relatively narrow (from ~ 15 meV at room temperature to ~ 5 meV at 12K), but slightly broader than the spectra in c-Si:Er (~ 10 meV at 300 K [9]). The low-temperature PL fine structure that has reported in Si:Er [8, 9] is not present. After careful examination of our data, we conclude that Er ions are not within the c-Si grains.

One of the differences between the samples studied in this work is the doping technique: in contrast to electrochemical doping (Nd, Er), sulfur has been introduced by ion implantation prior to anodization. Thus, it is possible that some S atoms will remain within the large Si grains after anodization and annealing. Room temperature PL at ~ 1.5-1.6 μm has been observed in SRSO:S (Fig. 3). The PL spectrum in SRSO:S and Si:S is quite different [13].Neververless, the PL peak which is 300 meV below the bandgap of c-Si essentially follows the temperature dependence of the c-Si bandgap (Fig. 3). The energy of ~ 300 meV is close to the ionization energy of $Si:S^{O}$ centers or B-centers [14]. The deviation between the calculated temperature dependence of c-Si bandgap and mentioned temperature dependence of the PL peak for $T < 100$ K is explained by a negative linear coefficient of thermal expansion in c-Si at $T < 80$ K [15]. Therefore, we conclude that S atoms are within the c-Si grains and not in the SiO matrix or SiO coverage.

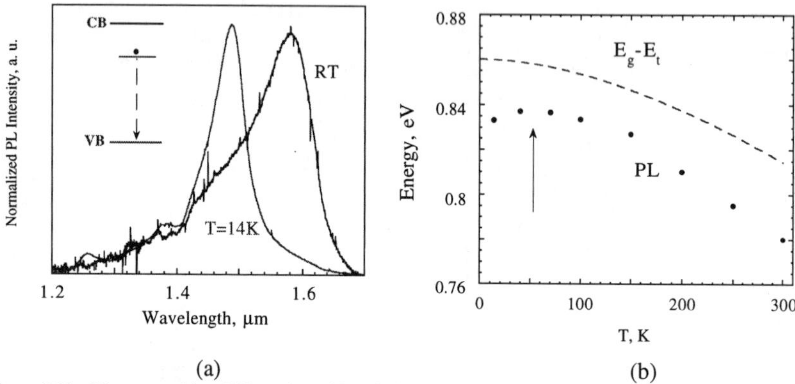

Figure 3 The PL spectra (a) and PL peak position (b) in SRSO:S compared to c-Si bandgap and ionization energy of $Si:S^{O}$ centers at different temperatures. The inset of Fig. 3a schematically shows corresponding radiative transitions. The temperature region of negative thermal expansion in c-Si is shown by arrow in Fig. 3b.

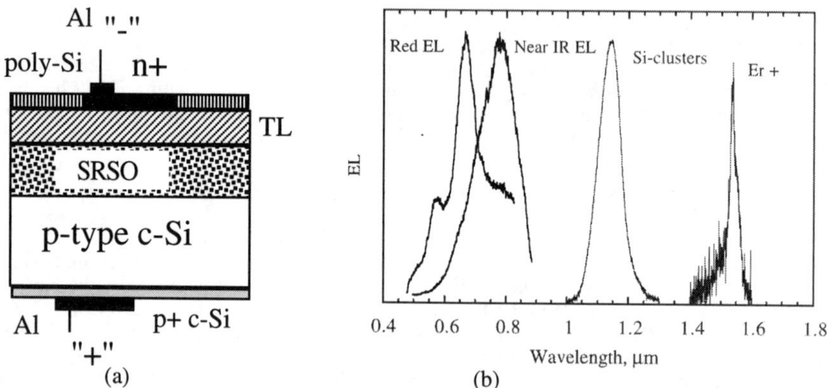

Figure 4. Schematic of SRSO-based LED (a) and room-temperature EL spectra (b). A transition layer (TL) is made from annealed PSi of very low porosity (< 30%). The active layer is made of an intrinsic or doped SRSO (dopants are indicated in Fig. 4 b). The red EL spectrum is influenced by multiple interference.

In order to demonstrate the EL in SRSO we have fabricated an LED structure similar to that reported in Ref. 16, but with either an intrinsic or doped SRSO active layer . The SRSO-based LED is a n-i-p bipolar device with a 300 nm - thick n-type poly-Si top contact, a ~ 1 μm high-resistivity SRSO active layer and a p-type c-Si substrate (Fig. 4a). The EL spectra of LEDs made from both intrinsic and doped SRSO are shown in Fig. 4b.

High-quality SRSO LEDs have an I-V rectifying ratio better than 10^3 at low bias (< 5 V). The detectable EL threshold of the LEDs is found to be: V_{thr} = 1.2 V for EL related to Si bandedge emission [17], $V_{thr} \geq 2$ V for red and near IR emission [16] and $V_{thr} \geq 10$ V for Er related emission. In LEDs made of SRSO:S and SRSO:Nd the EL has not yet been achieved. In samples made of intrinsic SRSO the PL and EL spectra are very similar (red, near IR and 1.1 μm emission [16, 17]). The dependence of EL intensity with current density is close to linear. The low voltage thresholds and similarities between the PL and EL spectra indicate that the EL is due to recombination of injected electrons and holes in the diode-like structure. In our best devices the EL EQE has been estimated to be ≤ 0.1%. The LEDs show no degradation for more than 100 hours of continuous operation.

In samples made of SRSO:Er we have achieved EL after a careful tradeoff between the electrical and light-emitting properties. The PL spectrum broader than the EL and slightly shifted to shorter wavelengths. This might be an indication of different microenvironments for the Er-ions that can be excited optically and electrically. It is possible that the Er-related PL is mostly due to Er ions located in SiO coverage of the Si nanograins and the Er-related EL is due to ions located directly at the SiO-Si interface. The high EL threshold and low efficiency (< 10^{-6}) indicates that only a small fraction of the Er-ions, possibly those located at the interface of Si-grains, can be excited by the impact of hot electrons.

Discussion and Conclusion

In this work we have demonstrated that SRSO prepared by partial oxidation and recrystallization of PSi is a flexible material from which luminescence can be obtained after high temperature induced modification or by doping with light-emitting impurities. As a result, the PL and EL have demonstrated emission from the visible to 1.6 μm. The PL and EL with peaks at 650 nm and 750 nm are intrinsic bands related to recombination in Si nanoclusters and via defects at the Si/SiO_x interface. The PL and EL at 1.1 μm is due to bandedge recombination in large recrystallized Si clusters.

Room temperature PL related to RE ions has been demonstrated at 1.06 μm (Nd) and 1.5 μm (Er). The role of oxygen in the activation of RE-ions in Si-based materials has been shown to be important [18]. In the case of SRSO:Er, the optically-active Er-ions are located in an oxygen-rich environment. The increase in local oxygen concentration in SRSO activates light-emission from the Er-ions, but also increases the thickness of SiO coverage and makes carrier transport more difficult. Since the SRSO prepared by thermal oxidation of PSi consists of Si-grains within a SiO-matrix, the Er-ions may be located within the SiO-matrix as well as at the interface of Si-grains. The SiO-coverage of the Si-grains is on the order of several monolayers, and effective carrier transport in SRSO has been observed and attributed to hot electrons [19]. Even if the Er^{3+} ions are located within the oxide layer covering the Si nanograins, they can be excited by hot electrons which have a relaxation length in order of 1-2 nanometers. In addition, the room-temperature EL in Si:O:Er has been explained by hot carriers impact excitation [20, 21]. Impact excitation is also known to be one of the pumping mechanisms of Er in InP [22].

The properties of SRSO:S are not very repeatable and are not well understood. However a correlation between the peak of the PL spectrum and value of c-Si bandgap at different temperatures is an important evidence that S is inside the Si clusters. Further work needed to develop a reliable doping technique for small Si clusters and quantum dots.

We would like thank P. H. Citrin for useful discussions. This work was supported by the U.S. Army Research Office and the U. S. Air Force Office of Scientific Research.

References

1. F. Namavar, F. Lu, C. H. Perry, A. Gremins, N. M. Kalkhoran, J. T. Daly and R. A. Soref, *in Microcrystalline and Nanocrystalline Semiconductors*, edited by R. W. Collins, C. C. Tsai, M. Hirose, F. Koch and L. Brus (Mater. Res. Soc. Symp. Proc. **358**, Pittsburgh, PA, 1995), p. 375.
2. T. Kimura, A. Yokoi, H. Horiguchi, R. Saito, T. Ikoma, and A. Sato, Appl. Phys. Lett. **65**, 983 (1994).
3. P. D. J. Calcott, K. J. Nash, L. T. Canham and M. J. Kane, *in Microcrystalline and Nanocrystalline Semiconductors*, edited by R. W. Collins, C. C. Tsai, M. Hirose, F. Koch and L. Brus (Mater. Res. Soc. Symp. Proc. **358**, Pittsburgh, PA, 1995), p. 465.
4. S. M. Prokes, W. E. Carlos, J. Appl. Phys. **78**, 2671 (1995), and S. M. Prokes, W. E. Carlos, O. J. Glembocki, Phys. Rev. B **50**, 17093 (1994).
5. L. Tsybeskov *et al.*, Preparation and characterization of an active layer for an LED based on oxidized porous silicon, this volume.
6. L. Tsybeskov, K. L. Moore, D. G. Hall and P. M. Fauchet, Phys. Rev. B 54, R8361, 1996.
7. G. E. Rindone, in *Luminescence in Inorganic Solids*, edited by P. Goldberg, Academic Press, NY & London, 1966, p. 419.
8. H. Ennen, J. Schneider, G. Pomrenke, and A. Axmann, Appl. Phys. Lett. **43**, 943 (1983).
9. J. Michel, J. L. Benton, R. F. Rerrante, D. C. Jacobson, D. J. Eaglesham, E. A. Fitzgerald, Y.-H. Xie, J. M. Poate, and L. C. Kimmerling, J. Appl. Phys. **70**, 2672 (1991).
10. S. Lombardo, S. U. Campisano, G. N. van den Hoven, A. Cacciato and A. Polman, Appl. Phys. Lett. **63**, 1942 (1993).
11. M. S. Bresler, O. B. Gusev, V. Kh. Kudoyarova, A. N. Kuznetsov, P. E. Pak, E. I. Terukov, I. N. Yassievich, B. P. Zakharchenya, W. Fuhs and A. Sturm, Appl. Phys. Lett. **67**, 3599 (1995).
12. J. H. Shin, G. N. van den Hoven, and A. Polman, Appl. Phys. Lett. **66**, 2379 (1995).
13. P. L. Bradfield, T. G. Brown and D. G. Hall, Appl. Phys. Lett. **51**, 100 (1989).
14. H. G. Grimmeiss and E. Janzen, in *Deep Centers in Semiconductors*, edited by S. T. Pantelides, Gordon & Breach Science Publishers, 2nd edition, Philadelphia, PA, 1986, p. 87.
15. G. Davis, Physics Reports **176**, 83 (1989).
16. L. Tsybeskov, S. P. Duttagupta, K. D. Hirschman and P. M. Fauchet, Appl. Phys. Lett. **68**, 2058 (1996).
17. L. Tsybeskov, K. L. Moore, S. P. Duttagupta, K. D. Hirschman, D. G. Hall and P. M. Fauchet, Appl. Phys. Lett. 69, 3411 (1996).
18. D. L. Adler, D. C. Jacobson, D. J. Eaglesham, M. A. Marcus, J. L. Benton, J. M. Poate and P. H. Citrin, Appl. Phys. Lett. **61**, 2181 (1992).
19. L. Tsybeskov, S. P. Duttagupta, K. D. Hirschman and P. M. Fauchet, in *Advanced Luminescent Materials,* edited by D. J. Lockwood, P. M. Fauchet, N. Koshida and S. R. J. Brueck (Electrochemical Society, Pennington, NJ) 1996, 34-47.
20. G. Franzo, F. Priolo, S. Coffa, A. Polman, A. Carnera, Appl. Phys. Lett. **64**, 2235 (1994).
21. B. Zheng, J. Michel, F. Y. G. Ren, L. C. Kimerling, D. C. Jacobson and J. M. Poate, Appl. Phys. Lett. **64**, 2842 (1994).
22. H. Isshiki, H. Kobayashi, S. Yugo, T. Kimura and T. Ikoma, Appl. Phys. Lett. **58**, 484 (1991).

PHOTO-IRRADIATION-INDUCED NARROWING OF PHOTOLUMINESCENCE SPECTRA FROM POROUS SILICON

M. OKAMOTO, T. NAGAO, T. OOIWA, A. HATTA and T. ITO
Department of Electrical Engineering, Osaka University, 2-1 Yamadaoka, Suita, Osaka 565, Japan, okamoto@daiyan.pwr.eng.osaka-u.ac.jp

ABSTRACT

Porous silicon (PS) has been prepared by varying photo-irradiation time in the latter period of anodization, and spectral intensity and width of photoluminescence (PL) have been measured as well as their dependences on current density during the photo-irradiation. Significant differences in PL spectra have been found when the timing of photo-irradiation is varied during the anodization. Sufficient numbers of photo-excited holes are responsible for the formation of PS layers yielding bright and sharp PL. The narrowest FWHM obtained was 0.182eV for a PL peak energy of 1.71eV while the highest peak energy of PL with FWHM $<$ 0.30 eV was 2.16 eV (575 nm).

1. INTRODUCTION

Since photoluminescence (PL) in the visible light region from porous silicon (PS) was observed[1,2], many studies on the forming process of PS and the luminescence mechanism have been performed[3]. In these studies, PL spectra are commonly measured and the broad nature of PL with full-width at half-maximum (FWHM) of 0.3 - 0.4 eV have been recognized for PS. However, an analysis based upon broader spectra becomes less clear especially when time-resolved PL spectra are discussed since the decay process strongly depends on the emitting photon energy[4]. Of course PS to be applied to light emitting devices should give narrow PL spectra. Thus, it is extremely important from both basic and application standpoints to fabricate PS samples yielding PL spectra with as narrow FWHM as possible.

It is generally known that holes are necessary for formation of PS structure and that a sufficient photo-irradiation is indispensable to create holes in an n-type Si wafer[5,6]. Even for a p-type wafer, significance of photo-irradiation during the anodization was reported[2,7]. Besides, there are many factors which change the PL spectra such as anodization current density, anodization time, concentration of solution, and oxidation after anodization[3,8,9]. It is recognized that the PL intensity increases with increasing peak energy until the peak wavelength blue-shifts to nearly 600 nm, below which the intensity usually degrade very drastically[3].

Recently we have found that PL peaks progressively exhibit a shift to the shorter wavelength side as well as a continuous increase in PL intensity and reduction in FWHM when the photo-irradiation intensity during the anodization is increased[10]. In this case, however, a further increase in the photo-irradiation intensity also results in a drastic expansion of FWHM and degradation of PL intensity although the peak can be located below 600 nm. In this study, we have investigated details of the photo-irradiation effects to prepare PS specimens which can yield strong and sharp PL with as short peak wavelength as possible. We have employed the following two-stage fabrication method : samples are (1) at first anodized without photo-irradiation for several

Mat. Res. Soc. Symp. Proc. Vol. 452

Table I Fabrication conditions used in the present study.
In the columns of current density and time, numerals before and after "/" are corresponding values in the anodization process and in the photo-irradiation process, respectively. In the anodization process, no light illumination was carried out to p-type wafers whereas a tungsten lamp of 260 lx was illuminated to n-type wafers. In the photo-irradiation process, the specimen surface was illuminated to 1.7×10^5 lx using a tungsten lamp.

condition	wafer type and resistivity (Ωcm)	anodization process / photo-irradiation process	
		current density (mA/cm^2)	time (s)
A	p(100) 14-22	2	30, 40, 50, 60
B	p(100) 14-22	2 / 2	345/15, 330/30, 315/45, 300/60
C	p(100) 14-22	2 / 0	360/20, 360/30, 360/45, 360/60
D	p(100) 14-22	2 / 20	345/15, 330/30, 315/45, 300/60
E	n(100) 1.46	2 / 2	340/20, 330/30, 315/45, 300/60
F	n(100) 1.46	2 / 20	330/30, 315/45, 308/52, 300/60
G	n(100) 1.46	2 / 40	330/30, 315/45, 300/60, 285/75

minutes, and (2) subsequently anodized under photo-irradiation for dozens of seconds with anodization current flowing.

2. EXPERIMENTS

P-type Si (100) wafers with a resistivity of 14 - 22 Ω cm were used for experiments in 3.1 and 3.2. Ohmic contacts for the p-type wafers were formed by aluminum evaporation onto their back sides and subsequent annealing at 500 °C for 10 min in a high vacuum. For experiments in 3.3, n-type Si (100) wafers with a resistivity of 1.46 Ω cm were used. Ohmic contacts of the n-type wafers were formed by annealing at 500 °C for 5 min in a high vacuum after titanium evaporation onto their back sides, and by subsequent gold evaporation onto them to prevent from oxidation. These wafers were anodized in a solution consisting of HF (47%) : C_2H_5OH = 1 : 1. A tungsten lamp was used for photo-irradiation when necessary. The fabricating conditions used are shown in Table I, where "anodization process" means the first stage for fabricating samples with sufficiently thick PS layers (thickness ≈ 1 μm) without photo-irradiation for the p-type Si wafers although a weak light of 260 lx was illuminated to the n-type Si wafers, while "photo-irradiation process" means the second stage for fabricating PS under photo-irradiation of 1.7×10^5 lx. PL measurements were performed using a spectrofluorophotometer with excitation of 300-nm ultraviolet light monochromatized from a Xe arc lamp of 150 W.

3. RESULTS AND DISCUSSION

3.1 Effect of Photo-irradiation Timing

In order to compare the effect of photo-irradiation during the PS fabrication process on the PL spectra, PS samples were fabricated from the p-type Si wafers in the following two ways. In one

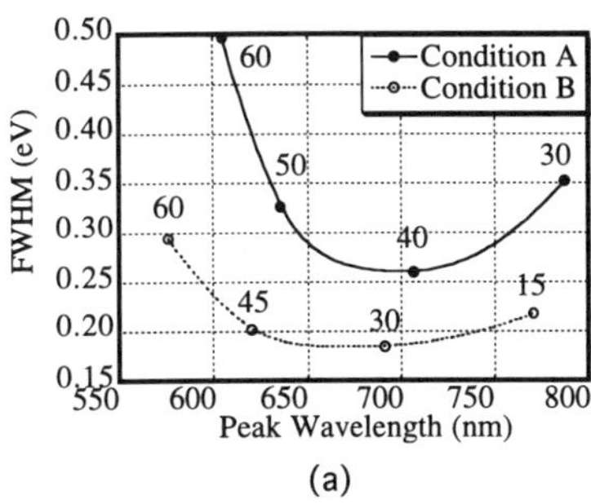

(a)

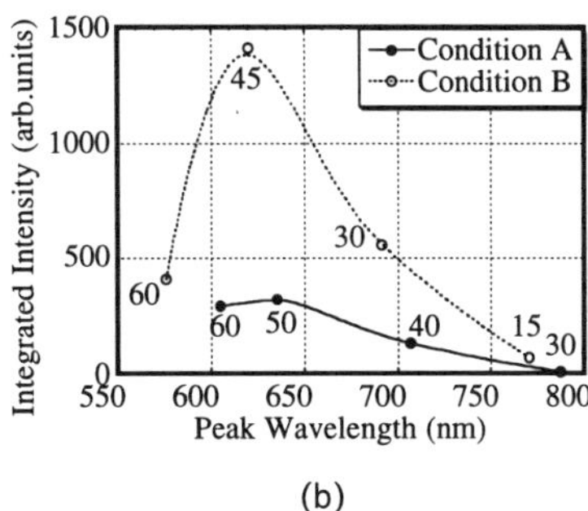

(b)

Fig.1 Dependence of full-width at half-maximum (FWHM) of photoluminescence spectra ((a)) and dependence of integrated PL intensity ((b)) on peak wavelength. Photo-irradiation time was varied under different conditions A and B. Numerals in the figure denote the photo-irradiation time used in s. Detail of conditions A and B are described in Table I.

way, samples were irradiated by a 1.7×10^5-lx light all the time when the anodization current flew (condition A). In the other way, the photo-irradiation was performed only during the last dozens of seconds in the 6-min fabrication (condition B). The anodization current density was 2 mA/cm^2 under both conditions. For condition A, four different samples were fabricated for different photo-irradiation (anodization) times of 30, 40, 50 and 60 s. For condition B, four different samples were fabricated for different photo-irradiation times of 15, 30, 45 and 60 s.

For these specimens PL measurements were carried out. Figures 1(a) and 1(b) show the FWHM and the integrated areal intensity of the PL spectra measured, respectively, as a function of the peak wavelength. The numbers denoted in these figures show the photo-irradiation times under the respective conditions. By dividing the sample fabrication process into the two processes, the first stage without photo-irradiation named "anodization process" and the second stage named "photo-irradiation process". Obvious differences in both the FWHM and the intensity of the PL spectra were observed between condition A and B. As seen in Fig.1(a) substantial narrowing of the PL spectra occurred. At most about 30% of spectral narrowing was observed for a PL spectrum with a peak wavelength of about 700 nm. Furthermore, as seen in Fig.1(b), the integrated PL intensity became 4-5 times larger by employing the photo-irradiation process separated from the anodization process.

3.2 Effect of Anodic Current in the Photo-irradiation Process

The population of photo-created holes in PS can change with that of electrons formed simultaneously. Therefore, the possible effect of current density during the photo-irradiation was examined. Samples were fabricated from p-type Si wafers in the following three ways. The first group of 4 samples were fabricated by anodizing Si at 2 mA/cm^2 for 6 min without photo-irradiation (in the first stage), and followed by white light irradiation of 1.7×10^5 lx for 20, 30, 45 and 60 s without flowing current (with open circuit) (condition C). The second group of 4 samples were fabricated under condition B. For the third group, the first stage is the same as that under condition B while at the second stage of 15, 30, 45 or 60 s the anodic current of 20 mA/cm^2 was passed under the light irradiation of 1.7×10^5 lx (condition D). The total current-flowing time was the same (6 min) for these three groups.

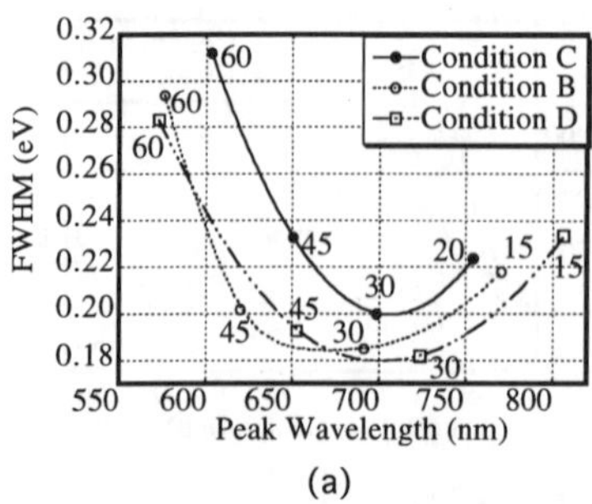

(a)

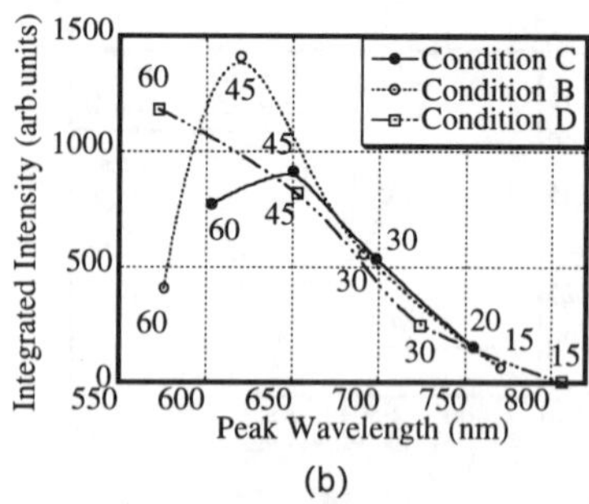

(b)

Fig.2 Dependence of spectral width (FWHM) ((a)) and dependence of integrated intensity ((b)) of photoluminescence spectra as a function of the peak wavelength. Current densities denoted by numerals in the figure were ranged from 0 to 20 mA/cm^2 during the photo-irradiation process. Three groups of PS were fabricated from p-type Si. Details of condition B,C and D are described in Table I.

Figures 2(a) and 2(b) show the FWHM and the integrated areal intensity of the PL spectra obtained in these experiments, respectively. The abscissa corresponds to the PL peak wavelength. The numbers denoted in these figures show the photo-irradiation times under respective conditions. As seen in Fig.2(a), the samples fabricated under condition B and D where the electric field is present give narrower PL spectra than those made under condition C with no current or no applied electric field in the photo-irradiation process. It is suggested from the above results that sufficient numbers of holes are responsible for the formation of the PS layers yielding the bright and sharp PL since an applied electric positive field to the sample gathers photo-excited holes to the sample surface while it extracts photo-excited electrons from the surface.

The data on the integrated intensity of PL spectra as a function of the PL peak wavelength fell on a common curve for the samples irradiated for a shorter period than 45 s. This suggests that there is a substantial relation between the PL peak energy and the intensity in the case of the sufficiently sharp PL spectra with FWHM < 0.25eV[4].

3.3 Photo-irradiation Effect for N-Type Samples

A p-type wafer itself has originally holes besides those produced by photo-irradiation. In order to investigate an effect of holes originally existing in the Si substrate on the fabrication process, similar experiments were performed for the n-type wafers. Because no sufficient anodic current can flow in an n-type wafer without photo-irradiation, the anodization was performed by irradiating white light of 260 lx, which was related to the minimum light intensity to stably maintain the anodic current of 2 mA/cm^2. The effect of the current density in the photo-irradiation process was investigated. Samples were fabricated in the three ways as follows. All the samples below were fabricated by flowing anodic currents for 6 min, and at the first anodization stage they were anodized at 2 mA/cm^2 under light illumination of 260 lx. In the photo-irradiation process under 1.7×10^5 lx illumination in the last dozens of seconds, 3 sets of 4 samples were anodized at 2 mA/cm^2 (condition E), 20 mA/cm^2 (condition F) and 40 mA/cm^2 (condition G), respectively. The light irradiation times were 20, 30, 45 and 60 s for condition E, 30, 45, 52 and 60 s for condition F, and 30, 45, 60 and 75 s for condition G.

Figures 3(a) and 3(b) show the FWHM and the integrated intensity of the PL spectra taken,

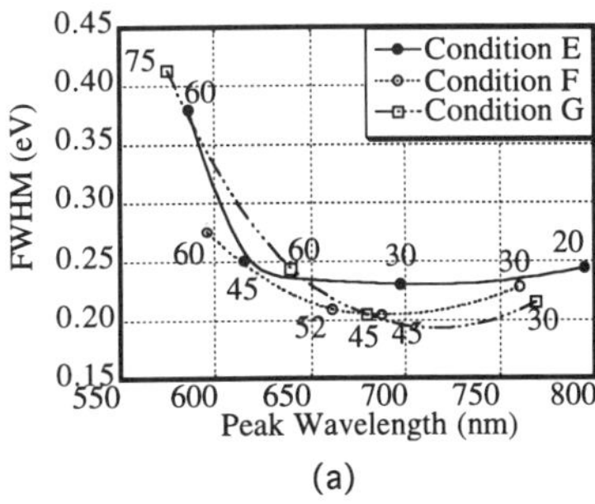

(a)

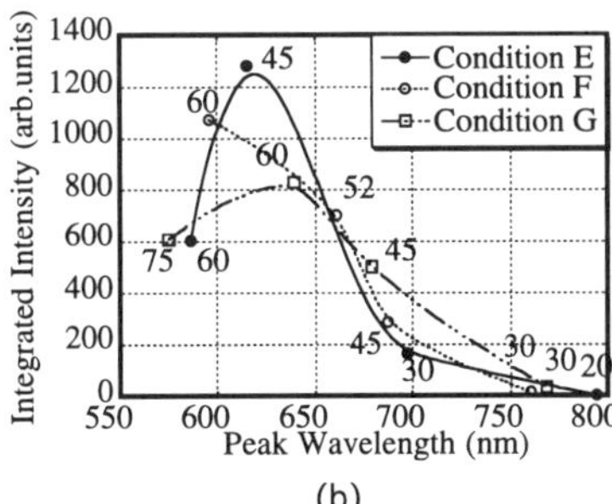

(b)

Fig.3 Dependences of spectral width (FWHM) ((a)) and integrated intensity ((b)) of photoluminescence spectra as a function of peak wavelength. Porous silicon layers were fabricated from n-type wafers under different currents during the photo-irradiation times denoted by numerals in the figures. Details of condition B,C and D are described in Table I.

respectively as a function of the observed PL peak wavelengths. The numbers shown in these figures denote the irradiation times used in the photo-irradiation process. As seen in Fig.3(a), in the case of the n-type wafers, the PL spectra with a peak wavelength ≥ 680 nm tended to become narrower for higher current densities in the photo-irradiation process, just like in the case of p-type wafers. The data plots in Fig. 3(b) demonstrate again a substantial relation between the PL peak energy and its intensity for the sufficiently sharp PL spectra (with FWHM < 0.25 eV).

3.4 Possible Model for Improvement in PL Properties

In order to explain the phenomena observed in the present study, we propose the following model for the formation of PS. Because the spectral narrowing is considered due to an increase in optical homogeneity of the specimens, it is concluded that more homogeneous PS layers than ever reported can be grown by the present two-stage fabrication method.

We consider that holes produced by the photo-irradiation can promote the hole-assisted etching reaction of Si, leading to the formation of a more uniform porous layer which yield the bright and sharp PL spectra. In other words, after a PS layer with a sufficient thickness (≤ 100 nm) for PL excitation photons (of 300 nm) to penetrate is formed in the anodization process without any light illumination, which results in relatively large and locally non-uniform pore structure (PL may not necessarily be observed at this stage), a sufficient number of photo-created holes can enhance the Si etching in nanometer scales. Since the size reduction of the Si crystallites in PS into nm scale increases their energy gap due to the confinement effect, the etching rate is faster for larger Si crystallites in size due to larger photon absorption, which can in turn increase uniformity of the crystallite size.

For PS samples conventionally prepared by the one-stage method, the intensity degradation of PL spectrum with a peak wavelength below 600nm is considered to originate from non-uniform anodization which leads to a excessive reduction of the crystallite size in special regions. This in turn results in an inhomogeneous strain distribution in the PS layer.The structurally weak regions of PS formed may be easily broken. Such structural changes may induce gap states connected to non-radiative transitions. At the contrast, for the samples fabricated by the present two-stage method, the PS layers are formed more uniformly and the creation of non-radiative gap states can be restrained effectively.

In both the p-type and n-type cases, a suitable increase of current density in the photo-irradiation process can properly gather photo-excited holes in the region of PS to be etched. Then, the PS layers with a more uniform structure can maintain the structure for smaller crystallites, which in turn may lead to the fabrication of PS giving a sharp and bright PL with a peak wavelength < 600nm.

However, because both the FWHM and the intensity of the observed PL with a peak wavelength ≤ 650 nm are scattered, the fabrication parameters should be optimized more properly to obtain PS specimens giving brighter and sharper PL spectra in this wavelength region.

4. CONCLUSIONS

(1) We have succeeded to obtain sharp PL spectra from PS by dividing the fabricating process into the anodization process and the subsequent photo-irradiation process. The narrowest FWHM obtained was 0.182eV for a PL peak energy of 1.71eV.

(2) Both spectral narrowing and high efficiency of PL can be attained by bright light irradiation to PS which is formed in the anodization process without photo-irradiation and by flowing a suitable anodic current density in the photo-irradiation process. The possible origin for this improvement is a photo-induced uniform growth of PS.

(3)PS samples having a relatively strong and sharp PL spectrum of a FWHM < 0.3 eV with a peak at approximately 575 nm were formed by the present two-stage fabrication method.

ACKNOWLEDGMENTS

This work was supported in part by a Grant-in-Aid for Scientific Research from the Ministry of Education, Science, Sports and Culture, Japan, and by a grant from Casio Science Promotion Foundation.

REFERENCES

1. L. T. Canham, Appl. Phys. Lett. **57**, 1046 (1990).
2. N. Koshida and H. Koyama, Jpn. J. Appl. Phys. **30**, L1221 (1991).
3. Mater. Res. Soc. Symp. Proc. **253**, (1992); **283**, (1993); **298**, (1993); **358**, (1995); J. Lumin. **57**, (1993).
4. K.Furuta, A. Hatta, T. Ito and A. Hiraki, in Advanced Luminescence Materials, edited by D.J.Lockwood, P.M.Fauchet, N.Koshida, and S.R.J.Brueck (The Electrochemical Society, Inc., Salem, MA, 1996), p.146-155.
5. R. L. Smith and S. D. Collins, J. Appl. Phys. **71**, R1 (1992).
6. Y. Arita and Y. Sunoda, J.Electrochem. Soc. **124**, 285 (1977).
7. T. Asano, k. Higa, S. Aoki, M. Tonouchi and T. Miyasato, Jpn. J. Appl. Phys. **31**, L373 (1992).
8. T. Ito, K. Motoi, O. Arakaki, A. Hatta and A. Hiraki, Jpn. J. Appl. Phys. **33**, L941 (1994).
9. T. Ito, K. Furuta, T. Yoneda, O. Arakaki, A. Hatta and A. Hiraki, Mater. Res. Soc. Symp. Proc. **358**, 477 (1992)
10. M. Okamoto, T. Nagao, T. Ooiwa, A. Hatta and T. Ito, to be published.

PHOTOLUMINESCENCE FROM POROUS SILICON ANODIZED WITH MONOCHROMATIC LIGHT ILLUMINATION

Takahiro Matsumoto*, Yasuaki Masumoto*, Go Arata**, and Hidenori Mimura***
*Single Quantum Dot Project, ERATO, Japan Science and Technology Corporation, 5-9-9 Tokodai, Tsukuba 300-26, Japan, tomato@sqdp.trc-net.co.jp
**Institute of Physics, University of Tsukuba, Ibaraki 305, Japan
***Research Institute of Electrical Communication, Tohoku University, Sendai 980-77, Japan

ABSTRACT

The effect of monochromatic light during the anodization of porous silicon is investigated from the points of view of structural characterization and photoluminescence. A clear correlation is observed between the photoluminescence peak energy and the bandgap energy determined by the illuminating photon energy. The difference of these two energies is analyzed by a size-dependent Stokes shift.

INTRODUCTION

Porous Si (PS) formed by electrochemical anodization has been attracting a lot of interest because it exhibits strong visible photoluminescence (PL) at room temperature [1]. Although a great deal of effort has been made to elucidate the origin of the PL, the mechanism is still a topic of discussion.

It is generally known that holes are necessary for the electrochemical dissolution [2,3], therefore in the case of anodization for n-type substrates, light illumination is necessary to supply holes into Si substrates. A consequence of this process is that the size and the bandgap of the nanostructure should be controlled by the illumination photon energy; the etching process proceeds until the illumination photon energy becomes lower than the band gap of porous Si.

In this paper, we aim to clarify the effect of monochromatic light illumination on the size of nanostructure and on the optical properties of porous Si. We have characterized the nanostructures fabricated by the various energies of monochromatic light using secondary electron microscope analysis (SEM), Raman spectroscopy, and Fourier transform infrared spectroscopy (FTIR). The correlation between the nanostructure and the PL peak energy points to the existence of size-dependent Stokes shift.

EXPERIMENTAL

Porous silicon samples were prepared by anodizing 1-5 Ωcm resistivity (100)-oriented n-type silicon wafers using HF-ethanol solution ($HF:H_2O:C_2H_5OH=1:1:2$) for 15 min. Al was

Mat. Res. Soc. Symp. Proc. Vol. 452

deposited on the back of the wafers to ensure an enough current flow to the wafer. The sample preparations were carried out at a constant current density in the range of 16 mA/cm^2 under the illumination of monochromatic light ranging from 450 nm to 700 nm (450 nm, 500 nm, 600 nm, and 700 nm). (To avoid the illumination of unnecessary light, the anodization was performed in a dark room.) These wavelengths of the monochromatic light were made by using a combination of interference filters (the transmission band width of 10 nm) and a 1 kW xenon- or a 500 W halogen lamp. The power density through the interference filters is of the order of 100 mW/cm^2. After the anodization, the samples were rinsed with ethanol and they were immediately loaded into an optical vacuum chamber to avoid oxidation. Photoluminescence spectral measurements were performed using a He-Cd laser and a specrometer with a CCD camera. The calibration of the spectral sensitivity was performed using a standard lamp. The microstructure of these samples were characterized by SEM analysis operated at 10 keV. The phonon mode study by Raman spectroscopy was performed using 488 nm laser light from Ar ion laser at room temperature. (The spectral resolution of the detection system is about 1 cm^{-1}.) Fourier transform infrared spectroscopy was also used to study the surface chemistry of these samples.

RESULTS AND DISCUSSION

Figures 1 show the SEM images of the local structure of (a) 450 nm- and (b) 700 nm-light illuminated PS films. SEM observations were performed using a Hitachi S-5000-H system operated at 10 keV. With the irradiation energy being high, both the pore size and the remaining Si structure become small; the size of the remaining Si region is of the order of 70 nm for 450 nm-light illuminated PS and of the order of 150 nm for 700 nm-light illuminated PS, however; both Si structures are too large for PL to occur in the visible range. By magnifying the observation scale, one can find many Si nanoparticles on the surface of remaining Si structures.

The nanostructure of these porous Si films was studied with Raman spectroscopy. Figure 2 shows the Raman spectra of (a) 450 nm- and (b) 700 nm-light illuminated PS films. With the increase of the illumination photon energy, the optical phonon mode of Si at 520.5 cm^{-1} broadens and downshifts compared with that of bulk Si crystal. Using a spatial correlation model [4,5], we can evaluate the average diameter of Si nanocrystals [6,7]. The fitting of the Raman signals can be performed assuming two diameters, and they are about 6 nm and 1.8 nm for the 450 nm- light illuminated PS, and about 9 nm and 3.1 nm for the 700 nm-light illuminated PS.

The above SEM and Raman structural characterizations show that there is a microstructure on the order of submicrons which consists of the backbone of porous Si and a nanostructure of the order of several nanometers which is responsible for the visible PL in porous Si. The correlation between the illumination photon energy and the local structures supports the model that the dissolution of the microstructure lasts until the nanostructure on the surface is depleted (until the bandgap of the nanostructure equals the illumination photon energy)[3].

The surface chemistry as a function of the illumination wavelength was studied with Fourier transform infrared (FTIR) measurements. Figure 3 shows the absorption spectra of the 450 nm-light illuminated PS (solid line) and the 700 nm-light illuminated PS (dashed line). In these samples, both absorption spectra are similar and three absorption peaks are characteristic in

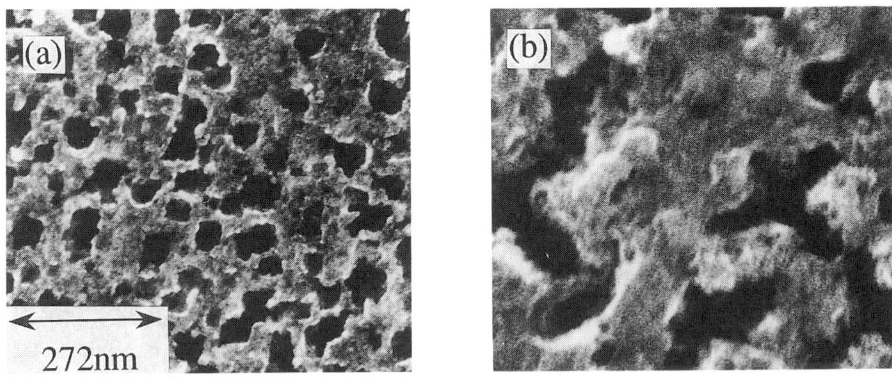

Fig. 1 SEM images of the local structure of (a) 450 nm- and (b) 700 nm-light illuminated PS films.

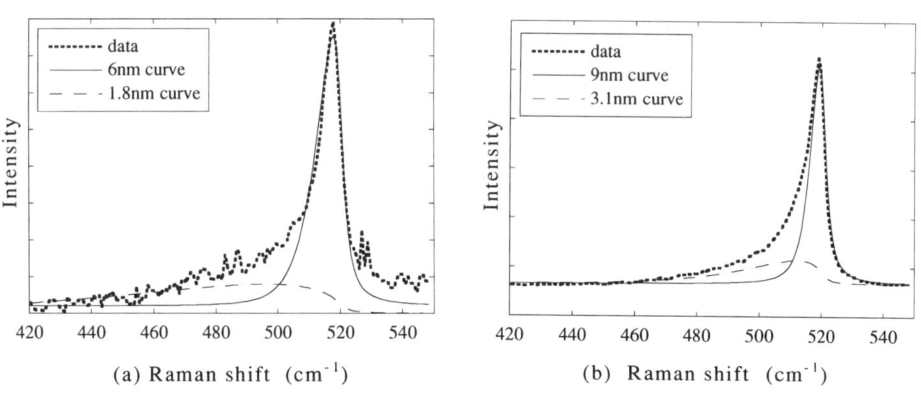

Fig. 2 Raman spectra of (a) 450 nm- and (b) 700 nm-light illuminated PS films.

these samples; absorbance due to the stretching mode of Si-H species at around 2100 cm^{-1}, the scissor modes of $Si-H_2$ at around 910 cm^{-1}, and the stretching mode of Si-O-Si at around 1100 cm^{-1} which is due to the oxidation of the samples during the FTIR measurements. Therefore, the surface of the porous Si does not change by the illumination photon energy and it can be regarded as a hydrogen terminated surface immediately after the anodization.

Figure 4 shows the PL spectra and the dependence of PL peak energy on the illumination light wavelengths (450 nm-, 500nm-, 600 nm-, and 700 nm-monochromatic light illumination) for Porous Si. As the illumination energy becomes high, the peak of the PL spectra shifts to higher energies and the full width at half maximum (FWHM) of the PL spectrum increases; 2.34 eV and 0.51 eV for 450 nm-light, 2.13 eV and 0.41 eV for 500 nm-light, 1.75 eV and 0.27 eV for 600 nm-light, and 1.49 eV and 0.21 eV for 700 nm-light as shown in the inset of Fig. 4. In spite of the use of the monochromatic light, the spectral width of 450 nm- and 500 nm-light illuminated samples is still broader than that of the samples fabricated by a non-monochromatic xenon or halogen light with the same intensity. (The peak of the PL spectrum is 1.84 eV and its FWHM is 0.34 eV.) The reason for the spectral width is not clearly understood now, however; the linear proportion of the width to the PL peak energy may indicate that the spectral broadening does not originate from the size distribution of nanocrystals but originates from the degree of the upshift energy due to the quantum confinement effect.

The clear correlation among the nanostructures, the PL peak energies and the bandgap energies is not likely to originate from the defect states. However, simple quantum confinement effects or simple Stokes shifts due to coupling with phonons also cannot explain these energy differences. Here, we assume a size dependent Stokes shift analysis to explain the discrepancy of the PL peak energy from the bandgap energy; the degree of Stokes shift becomes large as the confinement becomes strong. (This is the same problem as was observed before [8] and there is still no consensus to explain the discrepancy.) As the origin of the Stokes shift, we consider the stretching frequency (about 2100 cm^{-1}) of the surface terminated hydrogen based on the results of deuterium terminated porous Si [9]. The PL peak energies obtained here are written in both the quantum confinement term and the size dependent Stokes shift term as follows;

$$E = E_{gq} - \alpha \hbar \omega_S \quad , \tag{1a}$$

$$\alpha = \eta \cdot \left(E_{gq} - E_{gb} \right) \quad , \tag{1b}$$

where E_{gq} is the bandgap energy of confined electrons (or excitons), η is a parameter to fit the PL peak energies and E_{gb} is the bulk bandgap energy. Here, we assume the parameter α is linearly proportional to the upshift energy because the coupling constant α of the electron (or exciton)-phonon interaction becomes large as the confinement becomes strong. The above energy reduction indicates that the PL energy upshift is smaller than what is expected from the theoretical calculations, which is clearly shown by Schuppler et al. [10]. Putting Eq. (1a) into (1b), we can obtain the PL peak energy-line by selecting η equal to 1.2 as shown in the inset of Fig. 4. The parameter η used here is almost the same value as was used before [9], and this agreement strongly supports the size dependent Stokes shift model.

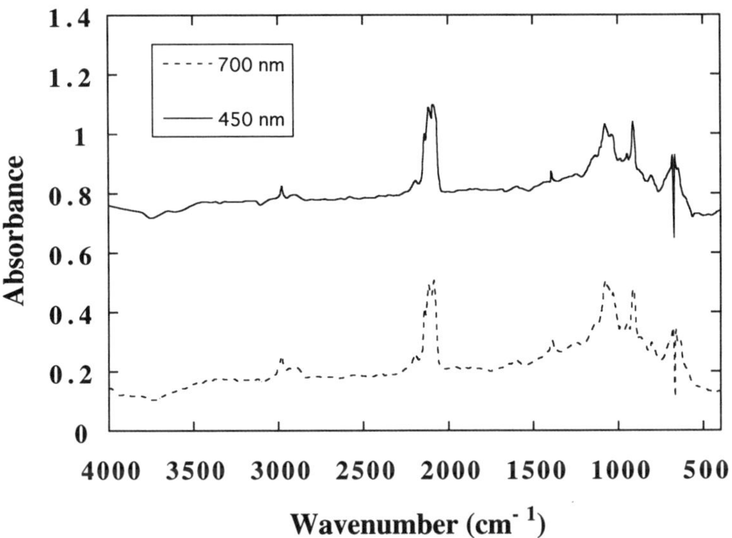

Fig. 3 FTIR absorption spectra of the 450 nm-light illuminated PS (solid line) and the 700 nm-light illuminated PS (dashed line).

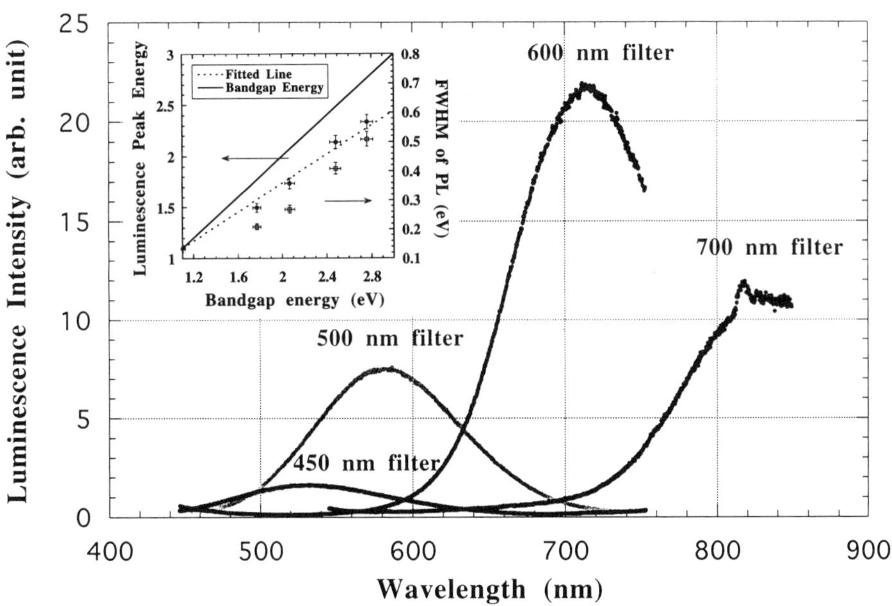

Fig. 4 PL spectra and the dependence of PL peak energy on the illumination light wavelengths for Porous Si.

CONCLUSION

We have studied the effect of monochromatic light illumination on the size of nanostructure and on the optical properties in porous Si. The bandgap of the nanostructure can be controlled by the photon energy of the monochromatic light illumination during anodization. Important problems related to the size dependence of the coupling constant in nanocrystalline system, the correlation between the spectral width and the PL peak energy, and the high PL efficiency of porous Si at room temperature still remain unresolved.

ACKNOWLEDGMENTS

The authors gratefully acknowledge S. Nair, E. Tokunaga, M. Tanaka, M. Sugisaki, H. Ren, and N. Matsura for their useful advises.

REFERENCES

[1] L.T.Canham, Appl. Phys. Lett. **57**, 1046 (1990).
[2] V. Lehmann, J. Electrochem. Soc. **140**, 2836 (1993).
[3] V. Lehmann, B. Jobst, T. Muschik, A. Kux, and V. Petrova-Koch, Jpn. J. Appl. Phys. **32**, 2095 (1993).
[4] H. Richter, Z. P. Wang, and L. Ley, Solid. State Commun. **39**, 625 (1981).
[5] I. H. Campbell and P. M. Fauchet, **58**, 739 (1984).
[6] Y. Kanemitsu, H. Uto, Y. Masumoto, T. Futagi, T. Matsumoto, and H. Mimura, Phys. Rev. B **48**, 2827 (1993).
[7] Z. Sui, P. P. Leong, I. P. Herman, G. S. Higashi, and H. Temkin, Appl. Phys. Lett. **60**, 2086 (1992).
[8] S. Schuppler, S. L. Friedman, M. A. Marcus, D. L. Adler, Y. -H. Xie, F. M. Ross, T. D. Harris, W. L. Brown, Y. J. Chabel, L. E. Brus, and P. H. Citrin, Phys. Rev. Lett. **72**, 2648 (1994).
[9] T. Matsumoto, Y. Masumoto, and N. Koshida, to be published.
[10] S. Schuppler, S. L. Friedman, M. A. Marcus, D. L. Adler, Y. -H. Xie, F. M. Ross, T. D. Harris, W. L. Brown, L. E. Brus, W. L. Brown, E. E. Chaban, P. F. Szajoski, S. B. Chrsistman, and P. H. Citrin, Phys. Rev. B **52**, 4910 (1995).

ANISOTROPY OF THE POROUS SILICON PHOTOLUMINESCENCE

G. POLISSKI, B. AVERBOUKH, D. KOVALEV, F. KOCH
Technische Universität München, Physik Department E16, Garching 85748, Germany

ABSTRACT

Polarization memory effect in the porous Si photoluminescence is studied. The anisotropy of the linear polarization degree is found in the samples etched with polarized light-assistance. The effect is explained by the anisotropic in plane distribution of the elongated Si crystallites. Under resonant optical excitation four-fold anisotropy of the photoluminescence polarization, linked to the crystalline axes of the bulk Si substrate, is observed.

INTRODUCTION

The photoluminescence (PL) from porous Si (PS) has been a subject of the intensive researches since it was observed by Canham [1]. At the present it is generally accepted that the nm-size Si crystallites with a band structure perturbed by quantum confinement are the main source of the visible light emission in PS. A lot of efforts have been exerted to find a correlation between the size distribution in the crystallite ensemble, studied by the Raman scattering, TEM measurements and other techniques, and the luminescent properties of the PS [2,3]. Some theoretical works have addressed this problem as well [4,5]. The results of these investigations might be controversial, partly because the structural methods probe the whole ensemble of particles in the porous layer, while the luminescing subsystem consists of only several percent of the nanocrystals. The qualitative result is the following: the smaller the size of the nanocrystallites, the higher the energy of the PL. The influence of the crystallite shapes on the optical properties of porous Si has gained much less attention. The structural investigations show that the crystallites are wires rather than spheres with a quite complicated shape.

The linear polarization memory effect in the PL from PS observed in [6] was explained as a consequence of the anisotropic shape of the luminescing objects in PS. Absorption preferentially occurs in those crystallites whose largest dimension coincides with the polarization of the exciting light. The light subsequently emitted by these particles is polarized along the same direction. The following works devoted to this topic [7,8] proposed the same explanation of the effect. In [9,10] the correlation between the polarization degree of PL $\rho = (I_{\parallel} - I_{\perp}) / (I_{\parallel} + I_{\perp})$, (where $I_{\parallel}(I_{\perp})$ are the components of the PL polarized parallel (perpendicular) to the polarization of the excitation), and the crystallite shapes was found for the case of the elliptical light emitting particles. It was shown that when the exciting light incidents on the cleaved edge of the (100) sample the PL is predominantly polarized normally to the surface for any direction of the excitation polarization [9]. The effect was explained by the preferential alignment of the Si crystallites along the [100] growth direction.

In a plane of the conventionally made (100) PS sample PL polarization is isotropic and does not depend on the angle between the polarization of excitation and the crystalline axes. However, in the non-linear excitation regime in the pump-probe type experiments with the two linearly polarized beams the in-plane anisotropy of the polarization degree of PL is observed [11]. The magnitude of ρ depends on the angle between the polarization vectors of the beams, and reaches the highest value for the perpendicular geometry. This anisotropy arises from the difference in population of the variously oriented crystallites by the electron-hole pairs.

Mat. Res. Soc. Symp. Proc. Vol. 452 © 1997 Materials Research Society

In the present work we show how the anisotropy in the orientation of the crystallites produces the angular dependence of the PL polarization degree in a surface plane of the PS sample. We study PL polarization of the samples made by anodization of the Si wafers of the various crystalline orientation to check whether the shape and orientation of the Si nanocrystals correlate with the crystalline axes. For (110) sample such correlation has been observed.

We found that illumination during the anodization of PS changes the polarization properties of the PL. Conventional unpolarized illumination quenches the polarization memory effect in PL, while the PS prepared with polarized light-assistance shows angular dependence of the ρ .

Finally we concern the contribution of the electronic states of the Si crystallites to the PL polarization. Under resonant excitation of (100) oriented sample ρ enhances and becomes dependent on the orientation of the polarization vector of the exciting light with respect to the crystalline axes. This effect cannot be explained in a frame of the electromagnetic theory [9,10] only and the correct description obviously requires the consideration of the electronic state symmetry of the Si nanocrystals.

EXPERIMENTAL RESULTS AND DISCUSSION

Structural investigations show that porous Si retains to some extent the initial crystalline structure of the Si wafer. The etching rate of Si is different for various crystalline directions. That causes the well-defined orientation of the pores and silicon crystallites in a porous layer correlated with the crystalline axes of the wafer [12,13]. Therefore, the crystalline orientation of the substrate determines the porous Si morphology.

The PS samples are prepared from the (100), (110), and (111) Si wafers. p-type Si with a resistivity of 1-3 Ωcm is used, the etching is performed in the dark. The details of the sample preparation and the optical setup can be found elsewhere [9]. The PL is excited by the linearly polarized light from the HeCd laser (442 nm). The sample is rotated to vary the polarization di-

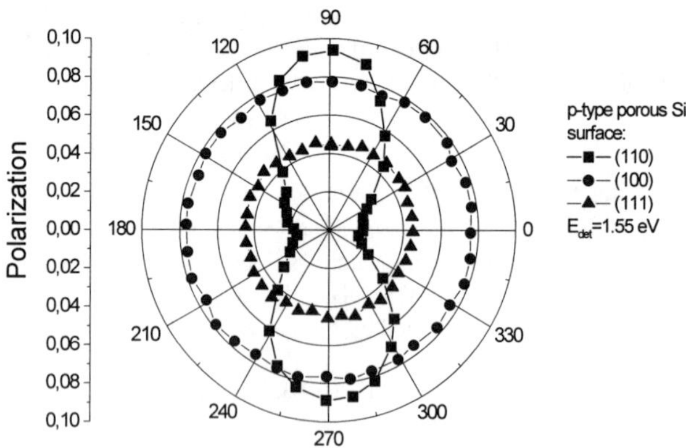

Figure 1. Polar diagram of PL polarization in porous Si prepared on the substrates of the various crystalline orientations. ρ in (100) and (111) samples is isotropic. The maximum value of ρ in (110) sample corresponds to the [100] axis.

rection of the excitation in a sample plane.

The (110) surface has a low symmetry with the [100] and [110] axes in the plane. In these samples we observed the two-fold symmetry in the angular dependence of the PL polarization degree, as depicted in Figure 1. It is assumed that the direction of the maximum value of ρ corresponds to the preferential orientation of the elongated crystallites. This result implies that the elongated crystallites or the silicon wires in the (110) porous layer are oriented preferentially along the [100] axis.

We have not observed any angular dependence of ρ in the higher symmetry (100) and (111) sample planes under non-resonant excitation. We could expect four-fold (for (100)) or six-fold (for (111)) symmetry in the crystallite orientation. We believe that the isotropy of ρ indicates the random orientation of the luminescing crystallites in these sample planes.

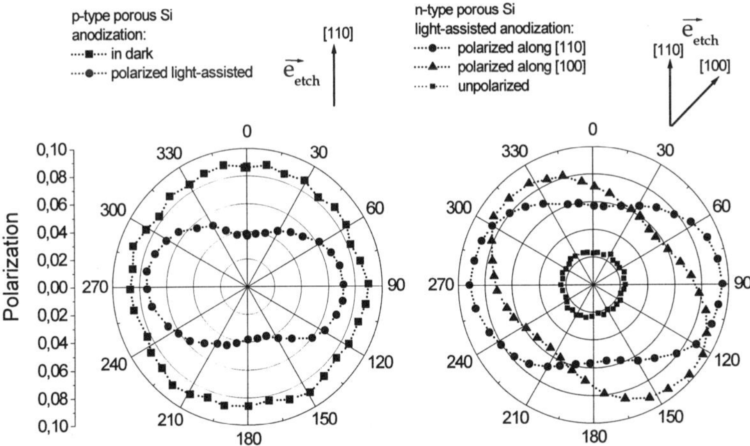

Figure 2. Angular dependence of the PL polarization in p- and n-type (100) plane porous Si. The photon energy of exciting light is 2.8 eV, the detection energy is 1.7 eV

To produce the sample with *artificially* induced and variable anisotropy of the nanocrystallite distribution, we use the linearly polarized light affecting the porous Si during the anodization. The similar technique was applied in [14]. We assume that the local anisotropy of the polarized light absorption and, therefore, of the etching rate, may lead to the anisotropy in the crystallite system. In our experiments light from the Xe-arc lamp is incident normally onto the sample surface during the etching. We optimize the intensity and the spectral content of the light to enhance the anisotropy of the PL polarization using neutral and low-pass optical filters. Figure 2 shows the angular dependence of the ρ for n- and p-type samples anodized with the presence of the 60 mW/cm^2 light passed through the 2.2 eV low-pass filter. The main experimental results for the samples produced with polarized light-assistance are the following:

1. The PL polarization degree in n-type PS increases with respect to the conventional microporous n-type sample anodized with unpolarized illumination.

2. ρ in p-type PS decreases compared with a typical sample etched in the dark.
3. The anisotropy of the ρ appears, linked to the direction of the polarization vector of the light illuminating the sample during the etching. A maximum of ρ occurs for the polarization of the PL excitation perpendicular to that of the etching illumination.

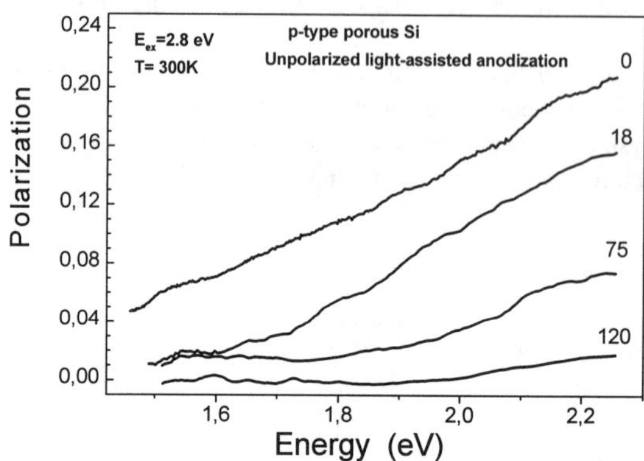

Figure 3. Spectra of PL polarization in (100) porous Si. Intensity of the light-assistance is labeled in mW/cm².

To understand the microscopic origin of these effects we studied the polarization properties of the PL from the samples anodized with unpolarized light-assistance. The result for p-type sample is shown in Figure 3. With increase of the illumination intensity along with a blue shift of the PL spectrum the quenching of ρ is observed. No anisotropy of the polarization degree in the sample plane is detected.

As the polarization memory effect is inherent for the PL from the elongated particles we conclude that etching with unpolarized illumination produces particles with a more symmetrical, "round" shape. We propose the following explanation of the light-induced changes in ρ. Asymmetric particles preferentially absorb the light with the polarization along their elongated directions. For an unpolarized illumination all of these are available. It is well established that the free hole is required to initiate the dissolution of a single Si atom from the surface of the crystallite [15]. We expect that the internal electric field in the crystallite acts to preferentially locate the both photocreated hole and F^- ion in the electrolyte near the sharply curved ends of the particle. The etching, therefore, occurs at the extremities. The result is that along with reduced size the shape of the crystallite becomes more round, the shape asymmetry is reduced and with it the value of ρ.

The reason for the light-induced anisotropy is the preferential absorption of light by the crystallites, whose longer dimension lies in the direction of polarization. These crystallites are preferentially etched and rounded off, no longer contributing to ρ. The nanocrystals oriented perpendicular absorb the light less efficiently and their shape is varied in a smaller degree. There is now a preferred axis of maximum ρ in the plane at 90^0 to the polarization of light applied during etching. The result is a uniaxial anisotropy. As only part of the particles efficiently absorbs the linearly polarized light and consequently is rounded, the value of ρ is higher in n-type porous Si anodized with polarized light with respect to the sample anodized with unpolarized illumination. On the contrary in p-type PS, prepared with linearly polarized light polarization degree is lower in comparison with conventional samples, etched in a dark (Figure 2), while it becomes higher with respect to the samples etched under unpolarized illumination. It implies, that the illumination during the anodization varies the shape of the crystallites in the similar way in n- and p-type porous Si samples.

Another way to observe the anisotropy of PL polarization in porous Si is the resonant optical excitation when the photon energy of exciting light enters the PL band. Figure 4 shows the angular dependence of the resonant ρ for several detection energies. At high energies the value of ρ is enhanced with respect to the non-resonant case. Four-fold symmetry linked to the crystalline axes is clearly seen. The direction of the maximum value of ρ changes from [110] at higher energies to [100] at the low energy edge of the PL band. None of these effects is observed under non-resonant excitation. The consideration based only on the anisotropy of the absorption and emission processes in the randomly oriented asymmetric crystallites cannot provide a sufficient explanation of the enhancement and angular dependence of the resonant ρ. We believe that the shape asymmetry of the crystallites affects the electronic state structure of the particles. In particular, the distribution of the ground state energies of excitons may become anisotropic in the ensemble of the nanocrystals, due to the valley or warped hole band splitting. The detailed description of the resonant effects will be given elsewhere [16].

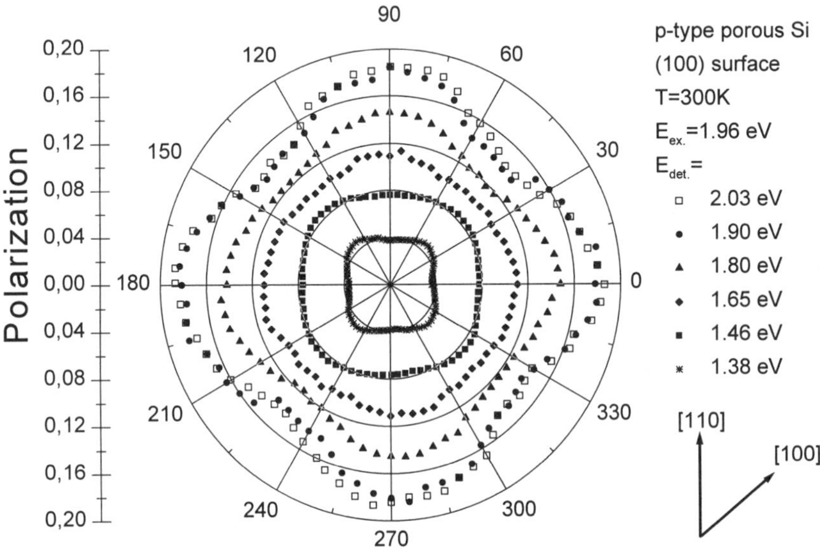

Figure 4. Polar diagram of the polarization anisotropy in the (100) plane of porous Si observed under resonant excitation condition. Note that the anisotropy in both Stokes and anti-Stokes PL regions is observed.

CONCLUSION

In summary, different cases of anisotropy of the PL polarization from porous Si are observed. The sample prepared from (110) Si substrate shows angular dependence of the polarization degree with a maximum value of ρ along [100] axis. In the (100) plane porous Si the anisotropy of PL can be induced by polarized light-assisted anodization. The locally anisotropic absorption of light during the etching induces anisotropy in the distribution of the particles. The enhancement and angular dependence of the ρ are observed under resonant excitation of PL. The effect is attributed to the contribution of the electronic state symmetry in the asymmetric Si crystallites to the polarization of the PL.

REFERENCES

1. L.T. Canham. Appl. Phys. Lett. **57**, 1046 (1990).

2. S. Schuppler, S.L. Friedman et al. Phys. Rev. B **52**, 4910 (1995).

3. Y. Monin, L. Saviot, B. Champagnon, C. Esnouf and A. Halimaoui. Thin Solid Films **255**, 188 (1995).

4. C. Delerue, M. Lannoo, G. Allan and E. Martin. Thin Solid Films, **255**, 27 (1995).

5. A.J. Read, R.J. Needs, K.J. Nash, L.T. Canham, P.D.J. Calcott and A. Qteish. Phys. Rev. Lett. **69**, 1232 (1992).

6. A.V. Andrianov, D.I. Kovalev, N.N. Zinov'ev and I.D. Yaroshetskii. JETP Lett. **58**, 427 (1993).

7. H. Koyama and N. Koshida. Phys. Rev. B **52**, 2649 (1995).

8. S.V. Gaponenko, V.K. Kononenko, E.V. Petrov, I.N. Germanenko, A.P. Stupak and Y.H. Xie. Appl. Phys. Lett. **67**, 3019 (1995).

9. D. Kovalev, M. Ben Chorin, J. Diener, F. Koch, Al.L. Efros, M. Rosen, M.A. Gippius and S.G. Tikhodeev. Appl. Phys. Lett. **67**, 1585 (1995).

10. P. Lavallard and R.A.Suris, Solid State Commun. **25**, 267 (1995).

11. D. Kovalev, B. Averboukh, M. Ben Chorin, F. Koch, Al.L. Efros and M. Rosen. Phys. Rev. Lett. **77**, 2089 (1996).

12. O. Teschke, F. Alvarez, L. Tessler and M.U. Kleinke. Appl. Phys. Lett. **63**, 1927 (1993).

13. S.-F. Chuang, S.D. Collins and R.L. Smith. Appl. Phys. Lett. **55** (1989).

14. G. Pollisski, A.V. Andrianov, D. Kovalev and F. Koch. Thin Solid Films, **255**, 235 (1996).

15. R.L. Smith and S.D. Collins. J. Appl. Phys. **71,** R1 (1992).

16. D. Kovalev, M. Ben Chorin, J. Diener, B. Averboukh, G. Polisski and F. Koch. To be published.

SYNCHROTRON RADIATION INDUCED OPTICAL LUMINESCENCE FROM POROUS SILICON: RECENT OBSERVATIONS

I. Coulthard*, T.K. Sham*,D.-T. Jiang*, K.H.Tan**
*University of Western Ontario, London, Canada, N6A 5B7, icoul@julian.uwo.ca
**C.S.R.F.,Synchrotron Radiation Centre, University of Wisconsin-Madison, Stoughton, WI.

ABSTRACT

Photoluminescence from porous silicon was examined using synchrotron radiation as an excitation source. The tunability of the excitation source permitted a wide range of excitation energies ranging from VUV to X-rays. This permitted site selective excitation where specific core levels (i.e. Si-K, O-K, Si-2p) were probed. In high porosity samples, luminescence bands of both surface and bulk origins were observed. All experiments exhibited a common luminescence maximum typically in the orange-red region of the visible spectrum. At certain specific excitation energies particularly in the VUV region additional peaks related to sites with oxygen character were also observed. The VUV excitation luminescence spectra of the porous silicon remarkably resembled that of oxygen deficient amorphous silicon dioxide glasses.

INTRODUCTION

Porous silicon has been the subject of intense study since room temperature optical luminescence was first reported [1]. The origin of the luminescence has received a great deal of scrutiny [2-7] and possible applications for porous silicon in opto-electronics have been examined [8].

Synchrotron radiation can be used as an excitation source for optical luminescence experiments to good effect. The synchrotron ring provides an intense collimated source of photons that are tunable over a wide energy range. This allows for the study of the luminescence at excitation energies near several relevant core level as well as valence binding energies such as: Si-K, Si-L, O-K, and valence. Further, by tuning the excitation energy, and using different detection schemes, the surface sensitivity of the experiment can also be adjusted.

A previous study [5] was performed at a bulk sensitive excitation energy. This study showed that in the bulk of the porous silicon layers, no species containing oxygen were involved in the luminescence. In recent studies we have used excitation energies that are more surface sensitive. These studies clearly indicate that in porous silicon samples that have been exposed to ambient atmospheric conditions, species containing oxygen are involved in luminescence from the surface of the porous silicon, albeit via a different luminescence channel. The implications of these results will be discussed.

EXPERIMENT

Porous silicon samples were prepared using a method described in detail elsewhere [9]. p^+-type (Boron doped) Si (100) wafers with a resistivity of 1-10 Ω.cm were utilized. Current densities used ranged from 1 to 500 mA/cm^2 with anodization times typically of 10-30 minutes. An electrolytic solution composed of 1:1 48% by weight HF: absolute ethanol was

Mat. Res. Soc. Symp. Proc. Vol. 452

used. The resulting porous silicon films typically appear reddish-brown in colour and exhibited orange-red to green luminescence under UV excitation. All samples were stored in ambient atmospheric conditions.

Experiments using VUV excitation were performed on the Stainless Steel (SS) Seya Beamline at the Synchrotron Radiation Centre, University of Wisconsin-Madison. X-ray excitation experiments near the Si-2p core level were performed on the Canadian Grasshopper Beamline, Canadian Synchrotron Radiation Facility, Synchrotron Radiation Centre, University of Wisconsin-Madison. The specimens were mounted in the UHV chambers of these beamlines. The optical luminescence spectra were recorded with a McPherson 234M6 spectrometer and a Hamamatsu R943-02 PMT cooled to -25 to -30°C. The system had an effective range of 200-900 nm. Optical luminescence spectra at selected excitation energies were recorded. Also recorded were luminescence yield spectra where the monochromator is set to specific wavelengths and excitation energy is scanned. Electron yield measurements relevant to this study were also carried out.

RESULTS AND DISCUSSION

The change in surface sensitivity of the experiments with a change in excitation energy can be gauged by comparing the absorption length of the photon for Si at those energies. The absorption length is defined as the depth where the incoming beam will have been reduced to $1/e$ of its original intensity as a result of passing through the sample (absorption). At 100eV and 30eV photon energies, one absorption length is 70 nm and 286 nm respectively. This is compared to an absorption length of 1.4 microns at energies near the Si-K edge utilized in a previous study [5]. The energies utilized in this study are significantly more surface sensitive. These absorption lengths compare to the attenuation of the luminescence in the visible region which is approximately 1 micron.

Figure 1 shows a range of luminescence spectra recorded for a single sample where the excitation energy was varied from 15-40 eV. One particular peak at approximately 1.5 eV is present at all energies. This peak is seen in all samples and has been shown to shift in position if any number of preparation conditions are varied. However at 30eV, indications of additional peaks begin to be emerge. At 35 and 40eV, these additional peaks are clearly evident. Of particular note is the broad peak centred at approximately 4.3 eV, which is in the UV portion of the spectrum.

The reason for this apparent "turn-on" of these additional peaks becomes evident when the partial optical yields are examined. Figure 2 shows the partial optical yields for the peak found at 1.5 eV along with the light curve of the monochromator of the SS Seya beamline. The partial yield for the 1.5 eV peak clearly resembles the light curve of beamline. This means that for this particular peak, the more excitation photons, the more intense the luminescence. This is contrasted sharply by the partial yield of the peak at 4.3 eV in Figure 3. No luminescence is produced at this energy until the excitation energy reaches above 30 eV and then the luminescence "turns on" and increases sharply. This is despite a decrease in the intensity of the excitation source as indicated by the light curve of the beamline. This is very close to the O-2s core level binding energy in silicon oxide (approx. 30 eV). Oxygen on the surface of the porous silicon must play an important role in the particular luminescence

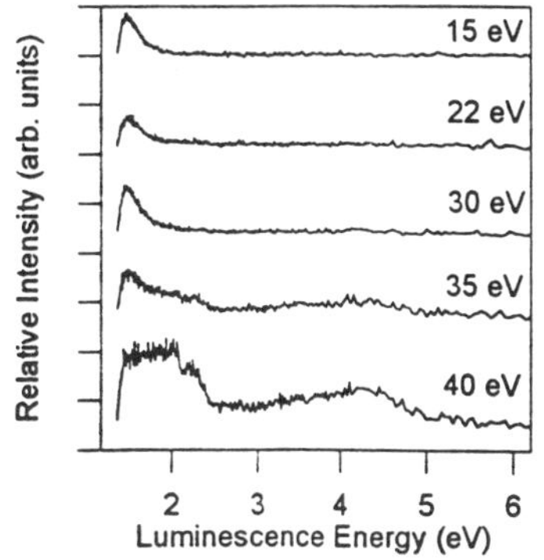

Figure 1. Luminescence spectra of porous silicon at different excitation energies. The spectra have been adjusted vertically for clarity.

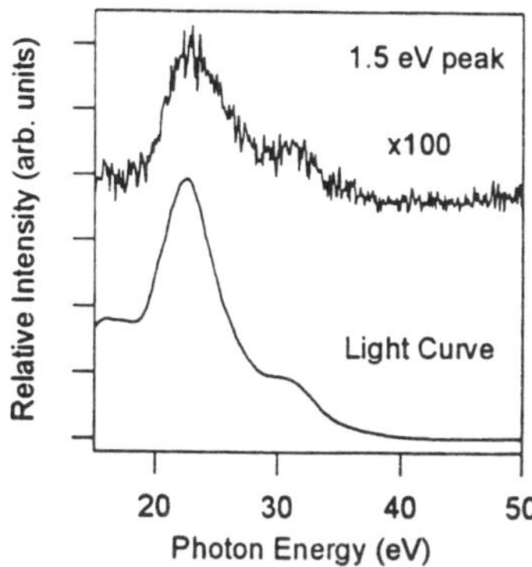

Figure 2. Partial optical yield of 1.5 eV luminescence peak compared to light curve of beamline monochromator. Vertical adjustment made for clarity.

mechanism that results in these "turn on" peaks. Figure 4 compares the energy of these peaks to the energy of luminescence bands reported for the luminescence of oxygen deficient amorphous silicon dioxide.[10] The structure of the oxygen deficient silicon dioxide is not unlike a recent model proposed for the luminescence of porous silicon [7]. It is unclear at present why a similar effect is not observed for the O-2p level (~8 eV) in the 15 eV excitation spectrum. Perhaps this level is too shallow to produce suitable conditions in the valence and conduction bands for the recombination. Further invesstigation is in progress.

The effect of surface oxygen can also be examined at excitation energies near the Si-2p core level (99.6 eV). Figure 5 shows the TEY spectra for two porous silicon samples. One has been left in ambient atmospheric conditions for upwards of a year, the other was "refreshed"

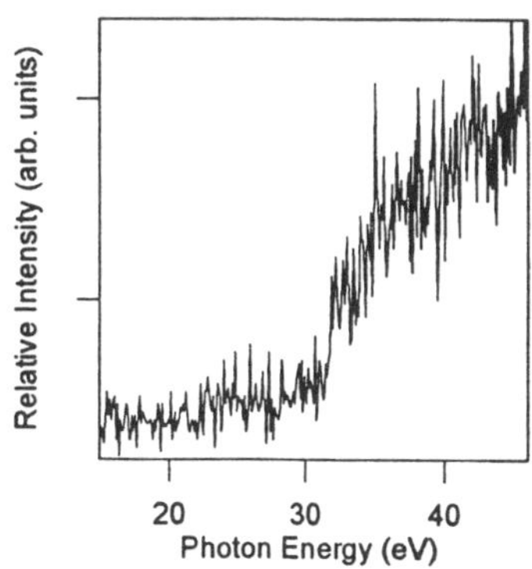

Figure 3. Partial optical yield of 4.3 eV luminescence peak.

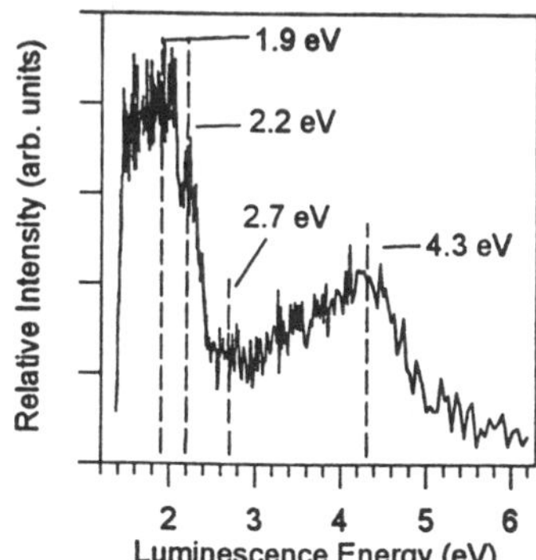

Figure 4. Luminescence spectrum of porous silicon at 40 eV. Lines show luminescence bands of oxygen deficient amorphous silicon dioxide.

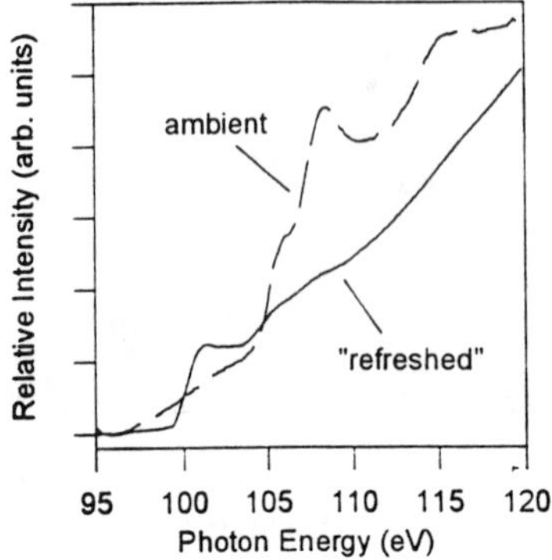

Figure 5. TEY of HF "refreshed" and ambient oxidized porous silicon.

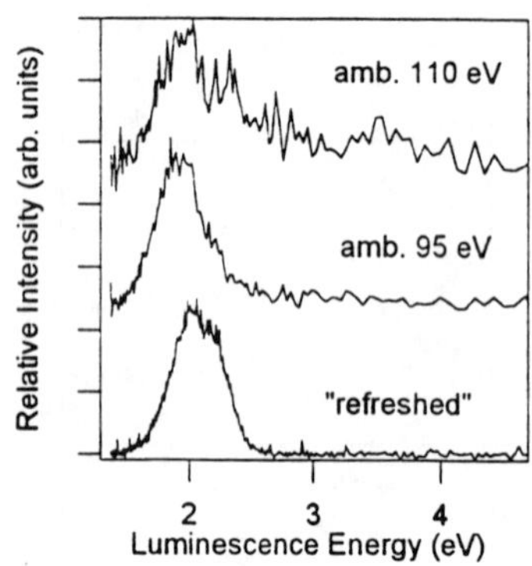

Figure 6. Comparison of luminescence spectra of HF "refreshed" and ambient oxidized porous silicon. Vertical adjustment has been made for clarity.

with HF just before introduction into the UHV chamber. The short HF treatment has the effect of removing the oxides from the surface of the porous silicon which is evident from the disappearance of the oxide peak at approximately 109 eV. The difference this makes to the luminescence spectrum can be seen in Figure 6. The spectrum of the HF treated sample (taken at 109 eV excitation energy in the figure) does not vary with excitation energy and contains only one peak. This differs from the luminescence spectra of the air-oxidized sample which do change with a change in excitation energy. At 95 eV excitation energy, only one peak is present. However, when the excitation energy is changed to 110 eV, very near the energy of the TEY oxide peak, the development of new peaks in the luminescence spectrum can be seen. The one luminescence peak that is present in all samples whether HF treatment is utilized or not, must be due to sites which locally are silicon. This observation seems to agree with quantum wire or nanocrystallite theories of porous silicon luminescence [2,6].

SUMMARY AND CONCLUSION

We have reported the luminescence spectra of porous silicon utilizing synchrotron radiation as an excitation source in the soft X-ray and VUV portions of the electromagnetic spectrum. Two types of luminescence are observed. One is attributed to sites that locally are silicon. The other is attributed to surface/interface oxides on the porous silicon as evident by the "turning on" of this luminescence at specific excitation energies related to sites containing oxygen. It would seem that recent theories of porous silicon can reasonably explain one or the other of these types of luminescence, but not both. Perhaps, for porous silicon which inevitably oxidizes in ambient conditions, the theory needs to take into account both bulk and surface/interface luminescence.

ACKNOWLEDGEMENTS

We would like to thank S.P. Frigo, and R.A. Rosenberg of SRC at the University of Wisconsin-Madison for the use of equipment and technical assistance. Financial support was provided by NSERC of Canada and the University of Western Ontario ADF Fund.

REFERENCES

[1] L.T. Canham, *Appl. Phys. Lett.* **57**, 1046 (1990).

[2] V. Lehmann, and U. Gösele, *Appl. Phys. Lett.* **58**, 856 (1991).

[3] S.M. Prokes, *J. Appl. Phys.* **73**, 407 (1993).

[4] M.S. Brandt, H.D. Fuchs, M. Stutzmann, J. Weber, and M. Cardona, *Sol. State Commun.* **81**, 307 (1992).

[5] T.K. Sham, D.-T. Jiang, I. Coulthard, J.W. Lorimer, X.H. Feng, K.H. Tan, S.P. Frigo, R. Rosenberg, D.C. Houghton, and B. Bryskiewicz, *Nature (London)* **363**, 331 (1993).

[6] L. Brus, *J. Phys. Chem.* **98**, 3573 (1994).

[7] S.M. Prokes and O.J. Glembocki, *Phys. Rev. B* **49**, 2338 (1994).

[8] MRS Symposium, vol. 283 (1993).

[9] I. Coulthard, D.-T. Jiang, J.W. Lorimer, and T.K. Sham, *Langmuir* **9**, 3441 (1993).

[10] H. Nishikara, E. Watanabe, D. Ito, and Y. Ohki, *Phys. Rev. Lett.* **72(3)**, 2101 (1994).

The recombination statistics of the visible photoluminescence of silicon nanocrystals

J. Diener[‡], D. I. Kovalev[°] *, S. D. Ganichev*, G. Polisski[°] and F. Koch[°]

[‡] *Department of Physics, University of California, Materials Sciences Division, Lawrence Berkeley Laboratory, Berkeley, CA 94720-7300, USA*
[°] *Technische Universität München, Physik-Department E16, D-85747 Garching, Germany*
* *A. F. Ioffe Physicotechnical Institute, Russian Academy of the Sciences, St. Petersburg, 194021, Russia*

ABSTRACT

A pulsed, high-power TEA CO_2 laser with lines in the region from 9.2 to 10.6 μm has been used to irradiate luminescent porous Si samples. The IR laser pulses heat the sample on a time scale much shorter than the PL decay time which is at 300 K for the PL at 1.65 eV in the order of tenth of μs. One IR pulse serves to increase the temperature of the luminescing particles in ~2 μs up to 100°C. This increase of temperature leads to a efficient reduction of the photoluminescence (PL) intensity. However, the PL decay times are almost not affected by the heating pulse. Based on this measurement a picture of the recombination statistics that takes account of the granular nature of the material is developed.

INTRODUCTION

Since the discovery of visible photoluminescence (PL) of porous silicon at room temperature 1990 by Canham [1] several models were proposed to explain this phenomenon. While the quantum-size nature of the radiative recombination processes is now generally accepted and is becoming predominant in the literature, a wide variety of models is applied to describe the time evolution of the PL under pulsed optical excitation [2,3,4,5,6]. At low temperatures, the temperature dependence of the PL intensity and the corresponding change in lifetimes has been explained by Calcott et al. [7] as a result of singlet-triplet splitting of the excitonic state. The variation of lifetimes and quantum yield versus temperature in the high temperature region has been reported by Vial et al. A model taking into account both radiative and nonradiative recombination has been proposed. We will discuss an appropriate description of the recombination statistics of the slow-red PL band at room temperature, taking into account the granular structure of the material. The general behavior of the PL lifetimes as a function of different measurement parameters is well documented. At room temperature the PL lifetimes depend strongly on the detection energy, ranging from several μs at 2.2 eV to tenth of μs at 1.6 eV. An increase of the sample temperature leads to faster luminescence lifetimes and to a decrease of the PL intensity. Several experiments were performed to clarify the physical processes responsible for the time evolution of the PL. The importance of Auger-recombination, leading to efficient damping of the PL was pointed out [8]. While this process with nonradiative lifetimes on the order of ns is important for the initial stage of the decay of the photoexcited carriers, the longtime component could only be addressed by conventional methods such as heating, oxidizing or changing the porosity of the samples.

Mat. Res. Soc. Symp. Proc. Vol. 452

In this work we show that an IR laser source is able to heat the Si nanocrystals on a time scale much shorter than the PL decay time. Due to the strong absorption of the IR radiation by the $Si\text{-}O_2$ bonds in the overlayer on the Si nanocrystals, one IR pulse serves to heat the luminescing particles on a time scale of ~2 μs up to 100°C. This gives an opportunity to investigate the temperature dependence of the slow component of the PL decay directly. Here, we focus our attention on the relation of lifetimes and intensity of the PL. The measurements show that a relation of the type:

$$\eta = \frac{\tau_{nr}}{\tau_r + \tau_{nr}}$$

(where τ_r and τ_{nr} are the radiative and nonradiative lifetimes and η is the quantum efficiency) cannot be applied at room temperatures for the red PL band. The observed decay time is argued to be to a large extent the radiative time.

SAMPLES AND EXPERIMENTAL SETUP

The samples used in this work are prepared from (100) p-type B-doped, 1 Ωcm Si wafers. Before anodisation an ohmic back contact is provided by B implantation. Anodisation is carried out in a 1:1 vol. mixture of ethanol:HF (49% in water). The etching current density is 30 mA/cm^2. The typical sample thickness is in the range of 2 μm. Following anodisation the samples are removed from the cell, washed with propanol and left to dry under ambient conditions. The freshly prepared samples have very weak luminescence. Thereafter they are aged for several weeks before the measurements in order to form a native oxide [9]. Free-standing samples are produced in a similar manner. After the etching procedure, the PS layers are detached from the substrate by an electropolishing step [10] with a current density of 500 mA/cm^2. The free-standing samples are removed from the etching cell, rinsed with ethanol and dried in CO_2 gas ambient.

A TEA CO_2 laser, tunable in the range from 9.2 μm to 10.8 μm (1087-926 cm^{-1}) is used as IR radiation source. Maximum radiation intensities used are 6 MW/cm^2 , with a pulse duration of 150 ns. The laser light is passed through a thick Ge window, in order to reject visible and near IR light. Calibrated CaF_2 attenuators are used to vary the intensity of the IR light. The intensity of the incident radiation is measured using a fast photon drag detector. The PL is excited with a pulsed N_2-laser (3.67 eV). Both lasers (UV and the IR) are synchronized using a delay generator. The luminescence is detected with a fast photomultiplier and the PL kinetics are recorded with a storage oscilloscope. Spectral resolution is achieved using a 22cm Spex monochromator. To calibrate the temperature of the sample the beam of a HeNe-laser is traversed through freestanding PS layers. The change in the transmission following one IR pulse is monitored and the temperature calibrated according to the values obtained for conventional DC heating of the same specimen. The experimental arrangement is described in details elsewhere [11].

EXPERIMENTAL RESULTS AND DISCUSSION

Fig. 1 shows the temperature of a PS sample following an IR pulse (1076 cm^{-1}, 2 MW/cm^2) over a wide time scale. Starting from room temperature the sample temperature is increased to ~ 70°C within a time scale of ~ 2 μs . Afterwards the temperature stays constant for a long time. It takes about 1 sec. until the specimen completely cools down to room temperature. The temperature increase versus the applied IR intensity has an apparent linear slope (see inset Fig. 1).

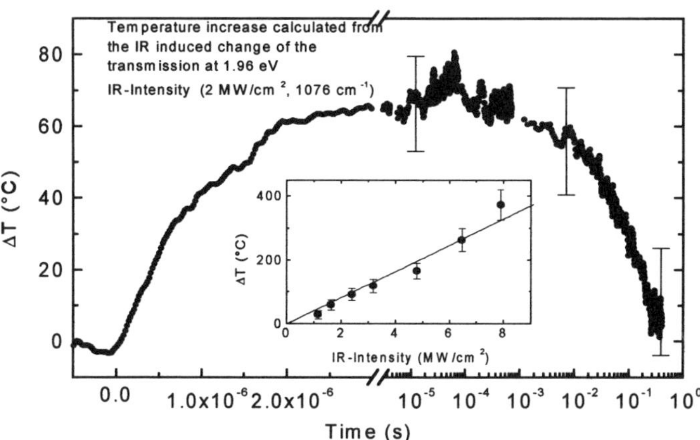

Fig. 1: The increase of the sample temperature caused by one IR pulse (2 MW/cm^2, 1076 cm^{-1}), calculated from the induced change in transmission at 1.96 eV. The inset shows the temperature increase as a function of the incident IR laser intensity.

As the IR wavelength is tuned to resonance with the Si-O absorption band, known from FTIR measurements [12], the main part of the IR energy is absorbed on the surface of the Si nanocrystallites. This gives an opportunity to heat these crystallites in a very short time scale in spite of the big heat capacity of the whole structure, obvious from the long cooling time. Comparing the temperature of the layer measured using the change in the transmission of a HeNe laser beam (mainly governed by the fraction of big crystallites) and the characteristic lifetimes of the photoexcited PL (contribution of the fraction of small crystallites) lead to similar temperatures. This indicates the homogeneity of the heating and thermal equilibrium of the sample after the IR heating step. Two points must be mentioned. First, the heating time is significantly faster than the lifetime of the PL in the red spectral region. Second, following the temperature increase, the sample remains hot in a much longer time scale than the PL decay. This is comparable to the condition achieved with conventional heating and allows to neglect cooling of the sample during the PL decay. This provides an opportunity to compare the measured lifetimes and intensities for both conditions, pulsed IR heating and conventional DC heating. The time evolution of the PL at room temperature detected at 1.65 eV and excited at 3.67 eV is shown in Fig. 2 as the dashed line. One IR pulse (1076 cm^{-1}, 3 MW/cm^2) delayed on 10 μs with respect to the UV excitation leads to a drastic change of the time behavior. Before the IR pulse the decay is identical to that of the unperturbed case (dashed line). Simultaneously with the IR pulse a fast drop of the PL is observed, with a decay time of ~ 2.5 μs. After this drop a slow time evolution is detected. The corresponding lifetime is only slightly faster than seen for the unperturbed case. Both, the lifetimes of the PL and the position of the PL maximum are temperature dependent. To clarify the point whether the observed behavior is caused by a shift of the PL spectrum to lower energies the PL kinetic was measured at different detection energies. The PL spectrum is afterwards constructed from the time evolution of the PL at different energies and normalized to the sensitivity of the detection system. The result of this measurement is shown in Fig. 3.

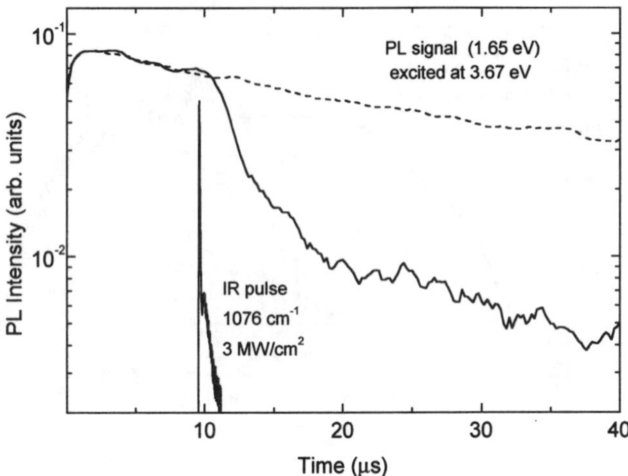

Fig. 2: The time evolution of the PL excited at 3.67 eV and detected at 1.65 eV (dashed line). The solid line shows the dynamic of the PL when an IR pulse (1076 cm^{-1}, 3 MW/cm^2) is applied. The IR pulse is shown for comparison.

Following one IR pulse the intensity of the whole PL spectrum is decreased. Only a small redshift of the PL maximum of ~ 100 meV is observed. We now turn to discuss the observed results. The mean lifetime τ_{eff} of the PL decay is in bulk semiconductors described by:

$$\frac{1}{\tau_{eff}} = \frac{1}{\tau_r} + \frac{1}{\tau_{nr}} \qquad (1)$$

where τ_r and τ_{nr} are the radiative and nonradiative lifetimes respectively. Both are temperature dependent. Within the same model the quantum efficiency η is:

$$\eta = \frac{\tau_{nr}}{\tau_r + \tau_{nr}} \qquad (2)$$

This analysis cannot explain the observed drop of the PL intensity after the heat pulse by more than one order of magnitude, while the lifetime of the PL after the IR pulse is only slightly affected (Fig. 2). We would like to mention that a similar behavior is also observed in measurements of the PL decay time for different excitation intensities [8]. At 300 K the long time component of the lifetimes is constant, in spite of the observed saturation behavior of the PL. We explain the observed response by a simple approach, which regards PS as a granular material consisting of luminescing and dark crystallites. The first are crystallites which do not contain a nonradiative center while the second part has at least one. In addition we assume that $\tau_{nr} << \tau_r$ and that the transfer of excitation between crystallites is not efficient at room temperatures.

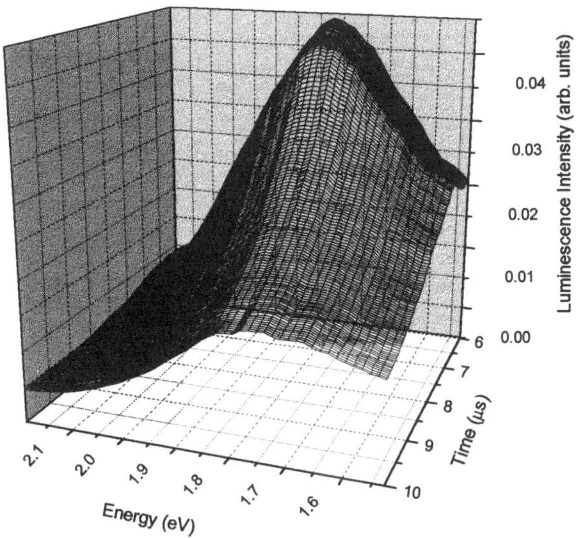

Fig. 3: The time evolution of the PL spectrum excited at 3.67 eV. An IR pulse (1076 cm^{-1}, 3 MW/cm^2) is applied within the PL decay ~ 7 µs after the UV excitation.

The probability for photoexcited electron-hole pairs to recombine radiatively in dark crystallites is nearly zero. In a similar manner electron-hole pairs recombine radiatively in luminescing crystallites. Therefore the PL intensity (or η, respectively) is determined by the ratio of these two types of crystallites. Thus the PL intensity is a function of the concentration of nonradiative defects and the observed lifetime is τ_r. In order to explain the observed drop in intensity it is straight forward to assume the creation of additional nonradiative centers by the IR pulse. This is supported by the observed, spectral independent, quenching of the whole PL spectrum (Fig. 3). On the other hand the creation of metastable defects by pulsed IR radiation has already been proposed in [13]. As a result, the IR excitation leads to an increase of the number of dark crystallites by creation of additional dangling bonds or other nonradiative centers. This leads to a decrease in the intensity of the emitted light, but only to a slight change in the radiative lifetime which corresponds now to the increased sample temperature. To be able to resolve such a fast drop, the nonradiative recombination time for the dark crystallites has to be at least as fast as the observed ~ 2.5 µs of the intensity drop. However, one of the basic assumption is that the transfer of excitation between crystallites is not efficient. This seems reasonable at room temperatures, taking into account the poor conductivity of the sample (a correlation between conductivity and PL intensity has been reported [14]). For temperatures above 370 K it was shown [2] that the temperature dependence of lifetimes and quantum efficiency behave in similar manner, following (1) and (2). This is the case when the electron-hole pairs are no longer localized in specific crystallites and have a finite probability to reach nonradiative centers. Obviously this is the limit of our assumption.

The last point to be addressed is the small redshift of the PL after the heat pulse (Fig. 3). Two effects could be accounted for this shift. On the one hand, the "bandgap" of PS is temperature dependent. It decreases ~ 24 meV from 300 K to 400 K [11]. Conventional heating results in a red shift of the PL band as well. In addition under pulsed UV excitation the same behavior could be seen with increasing delay time between excitation and detection. This is caused by the energy dependence of the lifetimes. The time resolved PL spectrum in the first few μs is usually governed by the faster decaying high energy part of the PL spectrum ~ 2 eV [4]. As a result of the relatively long delay between the UV and the heat pulse, the measurements only address the spectral region with long lifetimes. This was performed in order to ensure that the heating time is short compared to the decay of the PL. On the investigated time scale the PL maximum shifts in the unperturbed case only slightly. This proves that the observed redshift is temperature induced.

To summarize, we have shown that one IR laser pulse heats the PS sample on a time scale much shorter than the PL decay time at the low energy part of the PL. The time, temperature and spectral resolved measurements show that the PL decay time at roomtemperature is mainly the radiative one. Contrary to the recombination statistics at very high temperatures PS behaves at room temperature like a granular system of luminescing and dark crystallites. This indicates that the photoexcited carriers are still to a large extend localized in these crystallites.

Acknowledgments

J. Diener and D. I. Kovalev thank the Alexander von Humboldt Foundation for a research fellowship.

References

[1] L.T. Canham, Appl. Phys. Lett. **57**, 1046 (1990).
[2] J.C. Vial, A. Bsiesy, F. Gaspard, R. Herino, M. Ligeon, F. Muller, R. Romestain, and R. M. Macfarlane, Phys. Rev. B **45**, 1417 (1992).
[3] I. Mihalcescu, J. C. Vial, and R. Romestain, J. Appl. Phys. **80**, 2404 (1996).
[4] A.V. Andrianov, D.I. Kovalev, V.B. Shuman, I.B. Yaroshetskii, Semiconductors **27**, 71 (1993).
[5] S. Finkbeiner, J. Weber, M. Rosenbauer, and M. Stutzmann, J. Luminescence **57**, 231 (1993).
[6] S.N. Kuznetsov, V.B. Piculev, Yu. E. Gardin, I.V. Klimov, and V.A. Gurtov, Phys. Rev. B **51**, 1601 (1995).
[7] P.D.J. Calcott, K.J. Nash, L.T. Canham, M.J. Kane, and D. Brumhead, J. Luminescence **57**, 257 (1993).
[8] C. Delerue, M. Lanoo, G. Allan, E. Martin, I. Mihalcescu, J.C. Vial, R. Romestain, F. Muller, and A. Bsiesy, Phys. Rev. Lett. **75**, 2228 (1995).
[9] T. Maruyama and S. Ohtani, Appl. Phys. Lett. **65**, 1346 (1994).
[10] R.L. Smith, S.D. Collins, J. Appl. Phys. **71**, R1 (1992).
[11] D. Kovalev, G. Polisski, M. Ben-Chorin, J. Diener and F. Koch, J. Appl. Phys. **80**, 5978 (1996).
[12] H.D. Fuchs, M. Stutzmann, M.S. Brandt, M. Rosenbauer and J. Weber, Phys. Rev. B **48**, 8172 (1993).
[13] J. Diener, S. Ganichev, M. Ben-Chorin, D. Kovalev, V. Petrova-Koch, and F. Koch in Microcrystalline and nanocrystalline semiconductors, edited by Ed. L. Brus, R. W. Collins, M. Hirose und F. Koch (Mater. Res. Soc. Proc. **358**, Boston 1994) pp. 501-506.
[14] M. Ben-Chorin und A. Kux, Appl. Phys. Lett. **64**, 481 (1994).

AUGER IONIZATION OF SILICON NANOCRYSTALS

B. AVERBOUKH, D. KOVALEV, M. BEN CHORIN, F. KOCH
Technische Universität München, Physik Department E16, Garching 85748, Germany
AL.L. EFROS, M. ROSEN
Nanostructure Optics Section, Naval Research Laboratory, Washington, DC, 20375
USA

ABSTRACT

Photoluminescence saturation under intense CW optical excitation and optical degradation of photoluminescence from porous Si are studied. The anisotropy of the luminescence is observed under intense linearly polarized illumination at room temperature and after polarized light induced degradation at low temperature. The Auger process is shown to be responsible for these observations.

INTRODUCTION

The long lifetime of the photo-excited carriers in porous Si (PS) causes the Auger process to play a significant role in the recombination statistics in Si crystallites under high intensity of optical excitation. Generation of two electron-hole pairs in nm-size crystallites provides a very high effective concentration of the carriers and leads to the fast Auger non-radiative recombination. The Auger process is cited in [1] as a reason for the saturation of the photoluminescence (PL) intensity in porous Si under pulsed laser excitation. The significant decrease of the quantum yield of the PL is not accompanied by the shortening of a PL decay time. In bulk semiconductors efficiency of the light emission directly correlates with the PL lifetime. Therefore, the granular structure of porous Si, where the carriers are confined in the isolated particles, has to be taken into account to explain the observed phenomena.

In our recent work [2] we studied the PL polarization memory effect [3,4] in the regime of the nonlinear response of the PL from porous Si. Due to the local anisotropy of the light absorption in the elongated Si crystallites the probability to excite two electron-hole pairs is higher in those crystallites, whose larger dimension is directed along the polarization vector of the excitation. This leads to a reduction of the polarization degree under intense excitation.

In the present work we study two phenomena in the photoluminescence from porous Si: PL saturation under intense CW optical excitation and the low-temperature optical fatigue effect. The PL fatigue effect can be observed at room temperature as well. At room temperature the PL intensity reaches the initial value in several hours after the degradation [5], while at helium temperature no restoration of the PL is observed in a sample kept at low temperature for as long as two days [6]. The photo-darkening effect was observed in semiconductor doped glasses [7]. It was explained by the Auger ionization of semiconductor nanoparticles. We show that Auger autoionization is responsible for the optical degradation of PL from porous Si at He temperature. The non-spherical shape of the Si crystallites leads to the PL anisotropy, observed after polarized light induced degradation.

Mat. Res. Soc. Symp. Proc. Vol. 452 © 1997 Materials Research Society

EXPERIMENTAL RESULTS AND DISCUSSION

P-type Si wafers of 5 Ωcm resistivity are used for preparation of the porous Si samples. The electrochemical etching is performed in a solution of 1:1 by volume mixture of hydrofluoric acid (49% in H_2O) and ethanol. The current density is 30 mA/cm^2. The samples are fabricated in the dark. Two experimental techniques are used. In the pump-probe experiments two linearly polarized beams are incident normally onto the same point of the PS sample surface. An intense CW beam from an Ar laser (515 nm, up to 500 W/cm^2) is used as a pump. The PL is excited by the relatively weak modulated probe beam from a HeCd laser (442 nm, below 5 W/cm^2). The lock-in detection technique allows us to separate the light emission from the porous layer caused by these two beams, so that only the PL, excited by the probe one, is detected. These experiments are done at room temperature. The optical fatigue experiments are performed at liquid helium temperature. The sample is exposed to the linearly polarized CW degrading beam (515 nm, 100 W/cm^2) for 20 min. After that the PL excited by the linearly polarized probe beam (442 nm, below 1 W/cm^2) is measured. In both cases the PL is collected by a conventional two-lens condenser and focused onto the entrance slit of a monochromator. To determine the PL polarization a polarizer is placed between the lenses of condenser. A depolarizer is inserted near the entrance slit of the monochromator to avoid the polarization-dependent response of the detection system.

The lifetime of the free carriers is the essential parameter governing the efficiency of the Auger process under CW optical excitation. In the system of isolated nanocrystals a long

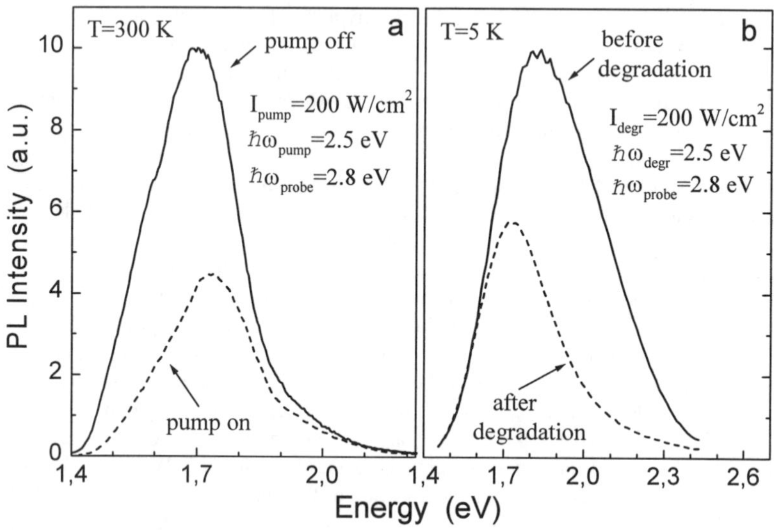

Figure 1. Spectra of the PL from porous Si with and without additional illumination by the pump light (a); before and after the optical degradation (b).

lifetime increases the probability to generate two electron-hole pairs per crystallite leading to the fast Auger recombination. The PL decay time is not a constant over the luminescence band of porous Si and becomes longer toward the red edge of the PL band. Therefore, the low energy region of the PL should be more efficiently affected by the Auger recombination process. This is confirmed by the observation shown in Fig. 1a, that the PL spectrum in the presence of the intense CW pump beam is shifted to the high energy range, the PL quenching is most pronounced at the red edge of the PL band. We would like to point out the similarity of this observation with the blue shift of the PL band of PS under intense CW optical excitation [8].

Fig. 1b exhibits the PL spectrum of porous Si before and after a 20 min exposure of the sample to the intense CW illumination at liquid helium temperature. The PL intensity is quenched mainly in the high energy range 2 - 2.2 eV and does not recover for hours. The distinct difference in these two effects is the shift of the suppressed PL band. After optical degradation the PL band shifts toward the red, while in the presence of the pump beam a blue shift is observed.

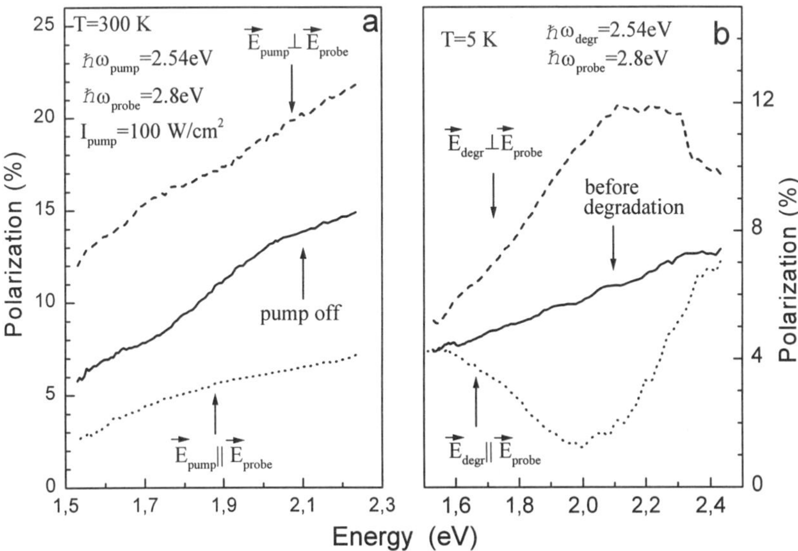

Figure 2 Spectra of the PL polarization degree of porous Si in the presence of the pump beam (a) and after the optical degradation (b) for two directions of polarization of excitation.

Fig.2 shows the spectra of the PL polarization degree $\rho = (I_\parallel - I_\perp)/(I_\parallel + I_\perp)$, where $I_\parallel(I_\perp)$ are the components of the PL, polarized along (perpendicular) to the electric field vector of the excitation. The initial values of the polarization degree are lower at low temperature. In this case the intensity of the probe beam is slightly out of the linear regime, due to the very long lifetime of the PL at liquid He temperature. The polarization degree in the presence of the pump beam is shown in Fig. 2a for the polarization direction of the pump beam parallel and perpendicular to that of the probe. For the parallel geometry ρ is reduced over the whole spectral range of the PL, while in the perpendicular case ρ increases with respect to the polarization without additional pumping light. The most significant relative changes of the ρ are found in the red edge of the PL band. The results of similar measurements performed after low temperature degradation are shown in Fig. 2b. While the polarization behavior seems similar to that for the pump-probe experiment, the polarization degree is mostly modified in the spectral region of 2-2.2 eV.

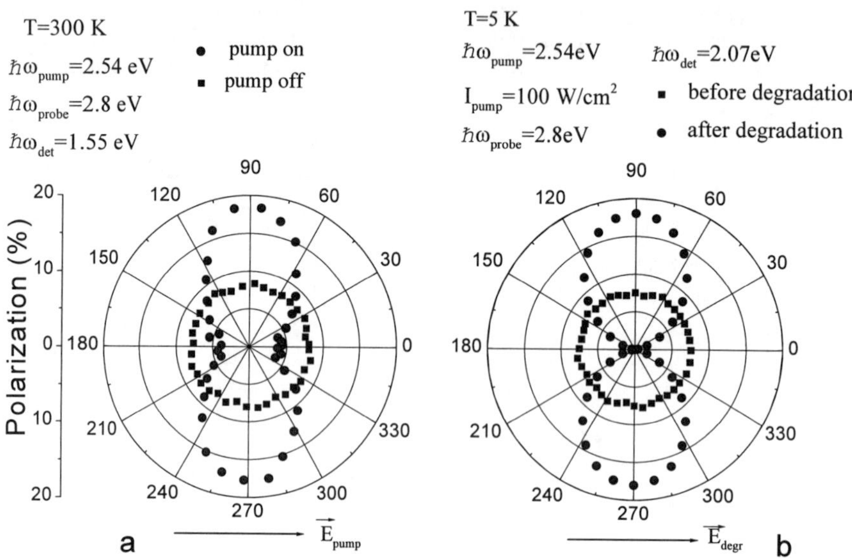

Figure 3 Polar plots of the polarization degree in the presence of the linearly polarized pump beam (100 W/cm^2) (a) and after degradation by the intense linearly polarized light (b).

Fig. 3 shows the angular dependence of ρ in the sample surface plane in the presence of the pump beam and after the low temperature optical degradation. The polarization vector of the probe beam is rotated with respect to that of the pump beam (Fig. 3a) or degrading beam (Fig. 3b). These dependencies are measured at the detection energies where the corresponding effects are well pronounced, E_{det}=1.55 eV for the pump-probe and E_{det}=2.02 eV for the fatigue experiments. Without the pump beam and before the degradation ρ is isotropic in the sample plane. Under intense CW illumination the polarization degree becomes anisotropic with the maxim value corresponding to the perpendicular directions of the pump and probe beam polarization. The degree of the anisotropy increases with the intensity of the pump beam. A similar dependence is observed after low temperature degradation (Fig. 3b). No anisotropy is found after degradation at room temperature by the linearly polarized light.

To explain these phenomena we turn to the microscopic origin of the low temperature fatigue effects and PL saturation in the presence of the intense pumping light. In the last case luminescing crystallites are occupied by electron-hole pairs generated by the pump beam. The occupation is determined by the effective concentration of the e-h pairs:

$$N = \frac{I \cdot \tau \cdot \alpha}{\hbar\omega_{pump}}, \qquad (1)$$

where I is the pump beam intensity, τ is the radiative lifetime, and α - the absorption coefficient. In our samples $\alpha \approx 2 \cdot 10^3$ cm^{-1}, τ changes from 30 μs at 2.2 eV to 100 μs at 1.45 eV. Considering the subset of the crystallites emitting light at 1.55 eV with $\tau \approx 85$ μs, we estimate that about 50% of the particles are permanently occupied by the e-h pairs already at the pump intensity I=150 W/cm^2. For the high energy part of the spectrum the occupation is lower due to the shorter lifetime. The e-h pairs generated by the probe beam in the already occupied crystallites recombine non-radiatively via the Auger process since the Auger time is much shorter than the radiative one [1]. This is observed as a quenching of the light emission from the sample. The absorption of polarized light in the elongated crystallites is anisotropic and occurs mainly in those particles whose large dimension coincides with the polarization vector of the light. So the radiative emission is suppressed mainly in the crystallites directed along the polarization vector of the intense pump beam. If the polarization vectors of the pump and probe beams are parallel, this subsystem of crystallites produces mainly the component of the PL polarized parallel with respect to the polarization of the probe beam, $I_{\parallel}$. Therefore, $I_{\parallel}$ is reduced due to the Auger process to the largest degree and with it the value of ρ. In the case of perpendicular geometry the crystallites, elongated with respect to the electric field vector of the pumping light, mainly emit $I_{\perp}$ component of the PL, excited by the probe beam. The suppression of the $I_{\perp}$ is now strongest. Therefore, the polarization degree is enhanced, the observation seen in Fig. 2a.

At low temperature Auger processes may lead to autoionization of the Si crystallites, as was proposed in [2,6]. The lifetime of carriers in Si crystallites at liquid He temperature is in the ms range. This provides a high probability of generation of two electron-hole pairs per nanocrystal. As a result of the fast Auger non-radiative recombination, the free electron can gain an energy of about $2E_g$ with respect to the top of the valence band. If this energy is high enough, the electron can escape from the crystallite to the matrix. The electron would not come back to the nanocrystal if the transport activation energy is significantly higher than kT. The hole, remaining in the crystallite, participates in further Auger damping of the radiative emission from this particle. All the e-h pairs subsequently generated in this crystallite recombine non-radiatively, so that the following PL from the sample is suppressed. Both the electron affinity of Si and the energy of the conduction band of SiO_2 with respect to the valence band of Si are in the range of 4.3 eV. Therefore, autoionization should be most efficient in crystallites with an energy gap $E_g \approx 2.15$ eV, at the high energy part of the PL band. The probability for a carrier to escape from the crystallite is the main parameter that determines the spectral dependence of the low temperature fatigue effect.

CONCLUSIONS

PL from porous Si is quenched and blue shifts under additional intense CW illumination. An anisotropy of the polarization degree of the photoluminescence is observed, linked to the polarization vector of the pump beam. After degradation of porous Si at low temperature by linearly polarized illumination the suppressed PL band shifts to the red and the same type of the polarization anisotropy is observed. These effects are explained as a consequence of the Auger process in the elongated Si nanocrystallites.

ACKNOWLEDGEMENTS

D. Kovalev acknowledges the support of the Alexander von Humbolt Foundation. This work was supported in part by the office of Naval Research.

REFERENCES

1. I. Mihalcescu, J.C. Vial, A. Bsiesy, F. Muller, R. Romenstain, E. Martin, C. Delerue, M. Lannoo and G. Allan. Phys. Rev., **B51**, p.17605 (1995)

2. D. Kovalev, B. Averboukh, M. Ben Chorin, F. Koch, Al.L. Efros and M. Rosen. Phys. Rev. Lett. **77**, p.2089 (1996)

3. A.V. Andrianov, D.I. Kovalev, N.N. Zinov'ev and I.D. Yaroshetskii. JETP Lett.,**58**, p.427 (1993)

4. H. Koyama and N. Koshida. Phys. Rev. **B52**, p.2649 (1995)

5. V. Grivickas, J. Kolenda, A. Bernussi, B. Matvienko and P. Basmaji. Brazilian J. Phys. **24**, p.349 (1993)

6. V. Grivickas, J. Linnros and J.A. Tellefsen. Thin Solid Films, **255**, p.208 (1995)

7. D.I.Chepic, Al.L. Efros, A.I. Ekimov, M.G. Ivanov, V.A. Kharchenko, I.A. Kudriavtsev and T.V. Yazeva. J. Lumin. **47**, p.113 (1990)

8. M. Koos, I. Pocsik and E. Vazsonyi. Appl. Phys. Lett., **62**, p.1797 (1993)

OPTICAL PROPERTIES OF FREE-STANDING ULTRAHIGH POROSITY SILICON FILMS PREPARED BY SUPERCRITICAL DRYING

J. VON BEHREN*, P. M. FAUCHET**, E. H. CHIMOWITZ*** and C.T. LIRA****
Department of Electrical Engineering, University of Rochester, Rochester, NY 14627
* also at the Technische Universität München, Garching, Germany
** also at the Laboratory for Laser Energetics and at the Institute of Optics
*** Department of Chemical Engineering, University of Rochester, Rochester, NY 14627
**** Department of Chemical Engineering, Michigan State University, East Lansing, MI 48824

ABSTRACT

Highly luminescent free-standing porous silicon thin films of excellent optical quality have been manufactured by using electrochemical etching and lift-off steps combined with supercritical drying. One to 50 μm thick free-standing layers made from highly (p^+) and moderately (p) Boron doped single crystal silicon (c-Si) substrates have been produced with porosities (P) up to 95 %. The Fabry-Pérot fringes observed in the transmission and photoluminescence (PL) spectra are used to determine the refractive index. At the highest P the index of refraction is below 1.2 from the IR to 2 eV. The absorption coefficients follow a nearly exponential behavior in the energy range from 1.2 eV and 4 eV. The porosity corrected absorption spectra of free-standing films made from p type c-Si substrates are blue shifted with respect to those prepared from p^+ substrates. For P > 70 % a blue shift is also observed in PL. At equal porosities the luminescence intensities of porous silicon films made from p^+ and p type c-Si are different by one order of magnitude.

INTRODUCTION

The feasibility of using light emitting porous silicon (LEPSi) [1] as a light emitter material in optoelectronics has triggered intensive research efforts over the last 6 years. Under UV excitation, the quantum efficiency of the red photoluminescence in LEPSi can be as high as 10% which is comparable to that of direct semiconductors like GaAs and dramatically higher (4-5 orders of magnitude) than that of bulk crystalline silicon. However, the highest reported external efficiencies of porous silicon LED's are in the 0.1 % regime [2][3]. The efficiency gap between the electroluminescence (EL) and PL is in large part caused by the poor electrical contacts to LEPSi and insufficient carrier transport in the film itself [4]. Increasing the porosity improves the luminescence efficiency but degrades the transport properties and the porous silicon internal network breaks down for P > 75 % after air drying. Supercritical drying of porous silicon on substrate [5] has been shown to avoid this dilemma by completely maintaining the structural integrity even at ultrahigh porosities.

As it is the highest porous silicon that has the highest quantum efficiencies we employed supercritical drying to manufacture free-standing porous silicon thin films of porosities up to 95 % and subjected them to transmission measurements from the UV to the IR. The results were compared to the PL response of identically prepared samples on substrate. These highly luminescent ultrahigh porosity films may enable new LED device structures due to direct access to the porous layer in terms of contact and light emission as well.

EXPERIMENTAL

Free-standing LEPSi layers of ultrahigh porosity were produced by electrochemical etching and subsequent lift-off as reported elsewhere [6][7][8]. Highly Boron doped and <100> polished c-Si substrates (p^+ type, 0.007-0.02 Ωcm) were anodized at current densities between 30 and 180 mA/cm^2 using ethanoic etching solutions with HF volume concentrations between 10 % and 25 %

Mat. Res. Soc. Symp. Proc. Vol. 452

to obtain porosities in the range from 50 % to 95 %. The porosity was determined gravimetrically and layer thinning during etching was accounted for by measuring the physical thickness (d) with an α-step profilometer (Tencor Instruments) and the weight of the LEPSi film. The etching time ranged between 14 sec and 30 min and resulted in thicknesses from 0.7 μm to 47.5 μm depending on the porosity. Moderately Boron doped c-Si substrates of the same orientation (p type, 5-7 Ωcm) were also in use and etched at current densities ranging from 15 to 150 mA/cm^2 in ethanoic etching solutions containing HF volume concentrations between 25 % and 40 % which resulted in porosities from 40 % to 92 %. Depending on the desired thickness (2.8 μm-54.2 μm) and porosity, the etching time was varied between 45 sec and 24 min. In all cases, the film area was 2 cm^2 and the etching solutions were stirred magnetically.

The LEPSi films were lifted off the substrate by applying a current pulse in a 10 % ethanoic HF solution. The lift-off of LEPSi made from p^+ substrates required a higher current density of 200 mA/cm^2 than that of films made from p type c-Si (50 mA/cm^2). Extreme care had to be taken during the lift-off as hydrogen bubbles often broke the fragile layers or removed them completely from the substrate. The films were soaked in four consecutive ethanol baths for 30 min before storage overnight to dilute the HF remaining in the pores [9]. Failure to perform this step resulted in the dissolution of ultrahigh porous silicon in 24 hours due to stain etching.

Liquid-vapor interfaces were avoided during drying by replacing the ethanol with liquid CO_2 and converting it into a supercritical fluid and subsequently into a gas [9]. The processing steps 1-3 are shown in Fig.1. High purity carbon dioxide (research grade 99.99 %) is pressurized to the liquid phase at room temperature and pumped into a 30 ml pressure chamber containing six porous silicon samples. When the pressure has increased to 1500 p.s.i. the ethanol is replaced by liquid CO_2 at 2 ml/min at room temperature. After purging the setup with carbon dioxide gas and three hours of flushing the vessel to guarantee that no residual ethanol was left, the chamber was completely filled with liquid CO_2 (1). The critical point of carbon dioxide (T_c = 31 °C, P_c = 1073 p.s.i.) was avoided by raising the temperature in the vessel to 40 °C. Supercritical CO_2 continued to flow for an hour at a pressure of 1500 p.s.i. and 2 ml/min (2). By releasing the CO_2 over three hours at 40 °C the supercritical fluid got converted into a gas (3). The liquid/gas phase boundary of CO_2 could also be violated by inserting the vessel into a water bath of 20 °C and flushing for an hour with liquid CO_2 at 1500 p.s.i. (1) before being vented off again for three hours (4). After each run, the bottom of the chamber had to be absolutely dry to ensure the complete exchange of the ethanol with liquid carbon dioxide.

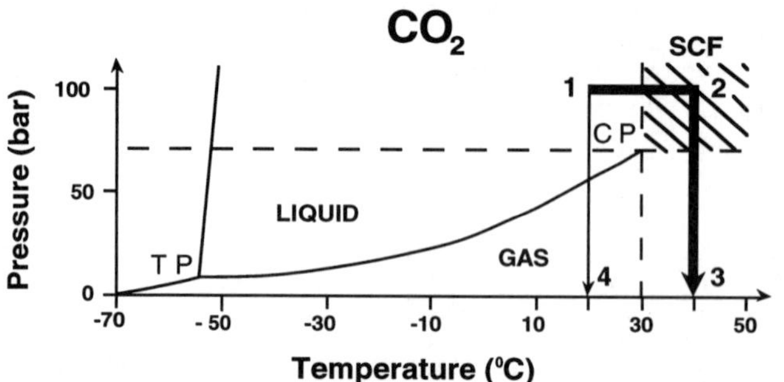

Fig.1 P-T diagram of bulk carbon dioxide. TP: Triple Point, CP: Critical Point, SCF: Supercritical Fluid. For the supercritical drying processing steps 1-3 and subcritical drying steps (1-4) see text

The ultrahigh porosity films were then carefully separated from the substrate at the border and either bonded onto transparent and highly heat conductive sapphire windows [8] for film thicknesses d < 10 μm or mounted on a washer for free-standing sampling without any background correction.

Free-standing films with 50% < P < 95% were probed using FTIR spectroscopy and optical transmission. The optical thickness of the samples mounted on sapphire was insufficient for comparing the PL intensity as a function of porosity. To avoid laser damage due to low heat conductivity during CW-PL measurements with the 442 nm line of a HeCd-laser we used optically thick films manufactured identically but still attached to the substrate. An optical microscope (magnification of 100x) connected to a CCD-camera was used to reveal the surface homogeneity and integrity of the free-standing LEPSi samples with $P \geq 90$ %. All experiments were performed under ambient conditions at room temperature. In addition to the experiments reported here, femtosecond time-resolved spectroscopy measurements have also been performed [10].

RESULTS AND DISCUSSION

1. OPTICAL MICROSCOPY: The surface quality of supercritically dried LEPSi films with $P \geq 90$ % was investigated using an optical microscope. In Fig.2 we compare the images of two identical films made a from p^+ substrate ($P \approx 95$ %, $d \approx 20$ μm), after supercritically drying (1) and

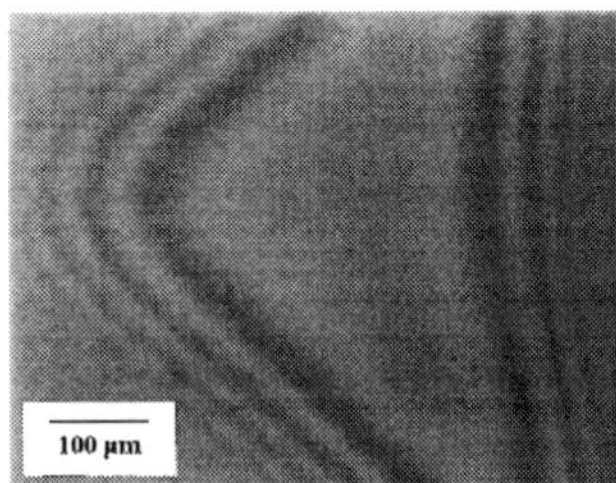

sample 1

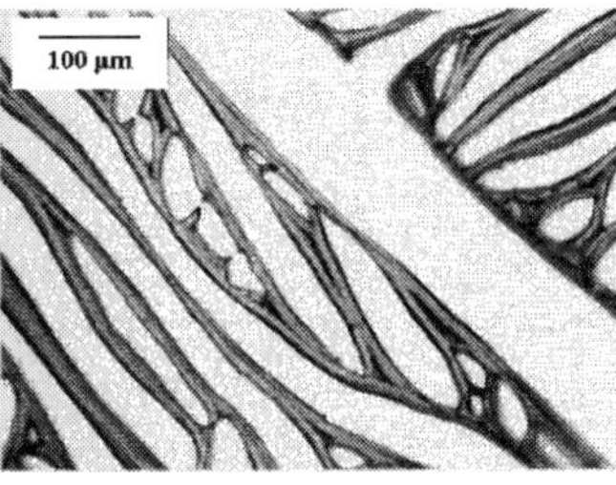

sample 2

Fig.2 Optical microscopy images of free-standing supercritically dried LEPSi film 1 and subcritically dried film 2 (p^+ c-Si substrate (0.007-0.02 Ωcm), 50 mA/cm^2, 10% ethanoic HF solution, $P \approx 95$ %, $d \approx$ 20 μm)

subcritical drying (2), respectively (see Fig.1, trajectories 1-3 and 1-4, respectively). Sample 1 is still attached to its substrate at the border and the slightly larger lattice constant of porous silicon [11] leads to a bending of the film. To improve the visibility the film in image 1 has been prepared to touch the substrate in the center whereas the outer regions to the left and right were elevated by approximately 1 mm. The Fabry-Pérot interference pattern thus induced indicates a flat and perfectly homogeneous and maintained surface on a micrometer scale. These findings are confirmed by similar images on identically prepared films on substrate (not shown). In contrast, image 2 shows a cracked and heavily disintegrated material: barely one percent of the film is left. The results for samples made from p type c-Si substrates with $P \approx 90$% are similar to those outlined in Fig.2.

2. FTIR: The surface chemistry of two free-standing LEPSi films of $P \geq 90$ % prepared from p^+ and p substrates is depicted in Fig.3. Despite identical drying and handling procedures (both samples were supercritically dried 3 weeks before the measurement) the FTIR spectra show significantly different features: the surface of sample 3 is covered by oxygen (Si-O-Si oxygen bridge mode at 1070 cm^{-1}) and hydrogen (SiH or SiH_2 mode at 622 and 664 cm^{-1}, Si-H_n scissors at 910 cm^{-1} and Si-H_x stretches at 2085 cm^{-1} and 2115 cm^{-1}). These peaks are rather sharp and resemble those detected in 40 % mesoporous silicon prepared from the same substrate [6]. There is very little backbonded oxygen which can be observed by the weak O_3-Si-H modes around 880 and 2250 cm^{-1}. We also detect a weak C-H band around 2900 cm^{-1}. Sample 4 in contrast has more backbonded oxygen and less hydrogen than sample 3 (enhanced O_3-Si-H modes around 880 and 2250 cm^{-1} and less pronounced hydrogen peaks around 664 and 2100 cm^{-1} with respect to the oxygen bridge mode around 1070 cm^{-1}). There is a significant broadening of the oxygen bridge mode around 1070 cm^{-1} indicating a high degree of structural disorder [12]. As the history of both samples is the same these obvious differences must originate from different internal morphologies. The carbon band between 2800 and 3000 cm^{-1} is experimental support for similar findings by Frohnhoff et al. [13].

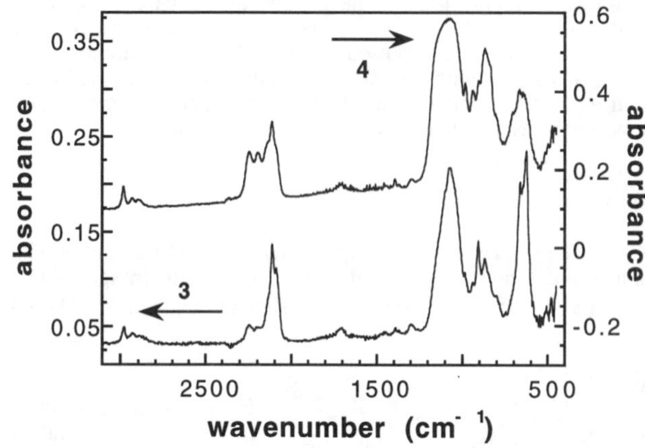

Fig.3 FTIR spectra of free-standing LEPSi films 3 weeks after preparation (sample 3: p^+ c-Si substrate (0.007-0.02 Ωcm), 50 mA/cm^2, 10% ethanoic HF solution, P ≈ 95 %, d ≈ 20 μm, sample 4: p c-Si substrate (5-7 Ωcm), 125 mA/cm^2, 25 % ethanoic HF solution, P ≈ 90 %, d ≈ 45 μm). The FTIR data of fresh samples look similar with hydrogen related peaks enhanced by a factor of approximately 2 (see text).

3. OPTICAL TRANSMISSION: The transmission spectra of four free-standing and supercritically dried films (5-8) with porosities from 85 % to 92 % made from p type substrates are shown in Fig.4. The sample thickness ranges from 14 to 20 μm and increasing the porosity

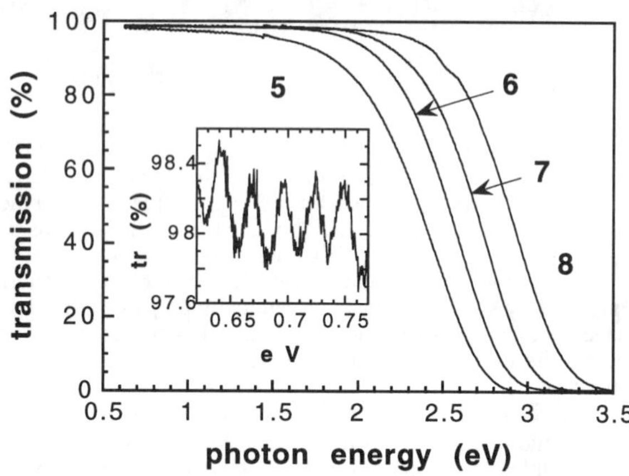

Fig.4 Transmission spectra of free-standing supercritically dried LEPSi films made from p type c-Si substrates (5-7 Ωcm). Sample 5: 20 mA/cm^2, 25 % ethanoic HF solution, P = 85 %, d = 16 μm. Sample 6: 125 mA/cm^2, 25 % ethanoic HF solution, P = 90 %, d = 20 μm. Sample 7: 125 mA/cm^2, 25 % ethanoic HF solution, P = 90 %, d = 14 μm. Sample 8: 125 mA/cm^2 + 30 min dissolution in 25 % ethanoic HF solution, P = 92 %, d = 19 μm. The inset shows the Fabry-Pérot pattern of sample 5 indicating a refractive index n = 1.4.

produces a significant blue shift of the absorption consistent with quantum confinement in smaller crystallites. Sample 6 is 4 μm thicker than sample 5 yet its absorption is blue shifted by approximately 200 meV with respect to that of sample 5 caused by an increase in porosity of 4 %. Sample 6 and 7 have equal porosities but different thicknesses. Sample 7 is 1 μm thicker than sample 8 and still there is an absorption blue shift of approximately 210 meV caused by a porosity increase of only 2 %. The optical film quality is excellent as there are neither cracks (the transmission of all samples is zero at 3.5. eV) or scattering (transmission almost 100 % in the IR). After magnification of the spectra in the IR, one can easily observe Fabry-Pérot interference fringes that are used to determine the refractive indices (n) as shown in the inset of Fig.4. Sample 5 has an index of refraction n of 1.4 and the value of the samples 6, 7 and 8 is ~ 1.2. This low index of refraction produces a low reflectivity R.

In Fig.5 the absorption coefficients of free-standing LEPSi thin films of comparable porosity made from different c-Si substrates are compared. At the porosities of interest (P = 94 % for films made from p$^+$ type substrates and P = 90 % for films made from p type substrates) the reflection is smaller than 2 % as determined from the refractive index in the near IR and is therefore not taken into account. The data of Fig. 5 is corrected for porosity [$\alpha_{PC} = \alpha/(1\text{-}P)$] [6] and compared to the absorption of crystalline silicon. Both LEPSi materials show an exponential behavior over at least 2 orders of magnitude. The absorption spectra of films made from p type Si is blue shifted by approximately 300 meV compared to that of films made from p$^+$ Si, although the porosity of the p$^+$ film is higher by 4 %. The data of Fig. 5 is not corrected for local field effects.

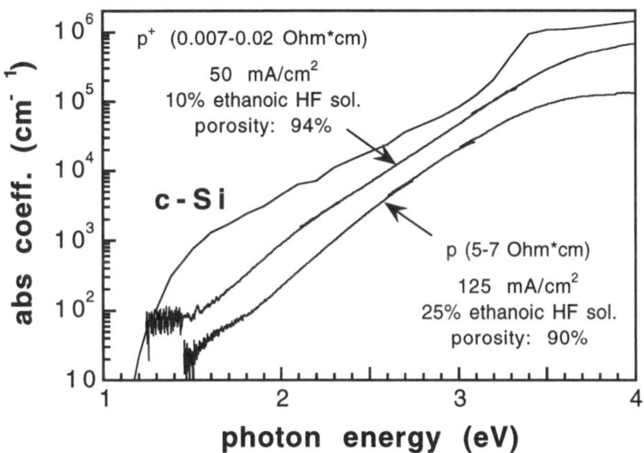

Fig.5 Optical absorption spectra of free-standing LEPSi films of different thickness and equal porosity made from p$^+$ and p type c-Si substrates. The thickness of the films made from p$^+$ c-Si was 0.7 μm, 8.1 μm and 47.5 μm. The corresponding values for the samples made from p type were 2.7 μm, 17.5 μm and 54.2 μm. The absorption coefficient is porosity corrected and compared to the absorption of bulk crystalline silicon.

4. PHOTOLUMINESCENCE: The PL response of supercritically dried LEPSi made from p$^+$ and p type Si substrates is shown in Fig.6. The PL intensity of LEPSi made from 5-7 Ωcm c-Si is about one order of magnitude larger than that of LEPSi of the same porosity prepared from

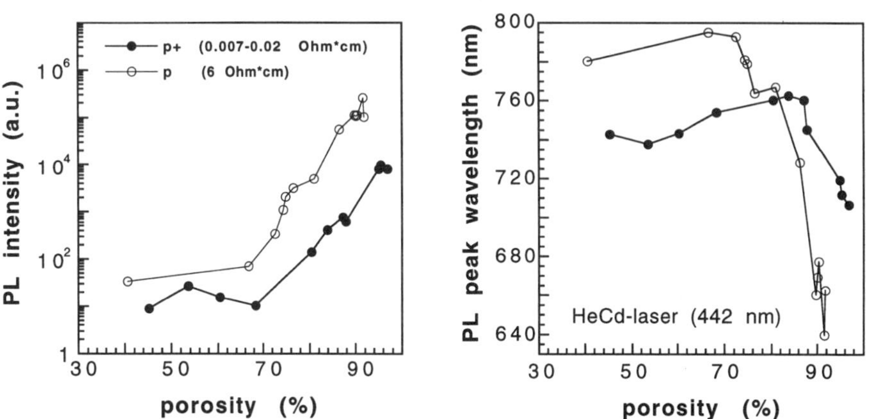

Fig.6 PL intensity and peak wavelength of supercritically dried LEPSi on substrate made from p$^+$ (0.007-0.02 Ωcm) and p (5-7 Ωcm) c-Si as a function of porosity. All samples were aged for three weeks prior to the measurements (see also Fig.3).

0.007-0.02 Ωcm c-Si wafers [13]. The PL intensity increases by three orders of magnitude and the PL peak shifts to shorter wavelengths when the porosity increases from 70 to 90 %. In the case of LEPSi made from p^+ c-Si the PL peak wavelength dependence on porosity is weak while for layers made from p c-Si the blue shift is significant. The weak PL red shift in both kinds of samples from 40 to 70 % in porosity suggests different luminescence mechanisms compared to the porosity regime of 70-90 %. Additionally, at low porosities there is no intensity dependence on porosity.

CONCLUSIONS

One to 50 μm thick free-standing porous silicon films made from p^+ and p c-Si substrates have been fabricated with porosities up to 95 % and characterized by optical microscopy, FTIR, optical transmission and photoluminescence. Optical microscopy revealed optically flat and homogeneous surfaces. These results were confirmed by Fabry-Pérot fringes in transmission and the absence of losses due to scattering. The refractive index of 90% porous films was found to be 1.2 for samples prepared from p Si. The absorption coefficients of samples with P = 90 % made from p^+ and p type Si displayed an exponential behavior over several orders of magnitude between 1.2 eV and 4 eV. The porosity corrected absorption spectra of free-standing samples made from p type c-Si substrates were blue shifted by ~ 0.3 eV with respect to those prepared from p^+ substrates. Increasing P from 70 to 90 % shifts PL peak positions to the blue and increases the PL intensities by three orders of magnitude. At equal porosity the PL intensity of p made LEPSi exceeded that of p^+ made films by approximately one order of magnitude.

ACKNOWLEDGMENTS

The authors would like to acknowledge support from the US Army Research Office, the National Science Foundation and the Science and Technology Center for Photoinduced Charge Transfer.

REFERENCES

[1] L. T. Canham, Appl. Phys. Lett. **57**, 1046 (1990)
[2] A. Loni, A. J. Simons. T. I. Cox, P. D. J. Calcott, and L. T. Canham, Electr. Lett. **31**, 1288 (1995)
[3] L. Tsybeskov, S. P. Duttagupta, K. D. Hirschman, and P. M. Fauchet, Appl. Phys. Lett. **68**, 2058 (1996)
[4] P. M. Fauchet, L. Tsybeskov, C. Peng, S. P. Duttagupta, J. von Behren, Y. Kostoulas, J. V. Vandyshev, and K. D. Hirschman, IEEE Journal of Selected Topics in Quantum Electronics **1**, 1126 (1995)
[5] L. T. Canham, A. G. Cullis, C. Pickering, O. D. Dosser, T. I. Cox, and T. P. Lynch, Nature **386**, 133 (1994)
[6] J. von Behren, L. Tsybeskov, and P. M. Fauchet, Mat. Res. Soc. Symp. Proc. **358**, 333 (1995)
[7] J. von Behren, L. Tsybeskov, and P. M. Fauchet, Appl. Phys. Lett. **66**, 1662 (1995)
[8] J. von Behren, K. B. Ucer, L. Tsybeskov, Ju. V. Vandyshev, and P. M. Fauchet, J. Vac. Sci. Technol. B **13**, 1225 (1995)
[9] B. Rangarajan and C. T. Lira, J. Supercrit. Fluids **17**, 1 (1991)
[10] K. Burak Ucer, J. von Behren, and P. M. Fauchet, in preparation
[11] O. Belmont, D. Bellet, and Y. Bréchet, J. Appl. Phys. **79**, 7586 (1996)
[12] F. Koch, V. Petrova-Koch, and T. Muschik, J. Lum. **57**, 271 (1993)
[13] St. Frohnhoff, R. Arens-Fischer, T. Heinrich, J. Fricke, M. Arntzen, and W. Theiss, Thin Solid Films **255**, 115 (1995)

SELECTION RULES IN THE RAMAN SPECTRUM OF POROUS SILICON

F. AGULLÓ-RUEDA*, J. D. MORENO**, E. MONTOYA*, R. GUERRERO-LEMUS**, R. J. MARTÍN-PALMA**, J. M. MARTÍNEZ-DUART*,**
*Instituto de Ciencia de Materiales de Madrid (CSIC), Cantoblanco, E-28049 Madrid, SPAIN
**Departamento de Física Aplicada C-12, Universidad Autónoma de Madrid, Cantoblanco, E-28049 Madrid, SPAIN

ABSTRACT

We have measured the Raman spectrum of porous silicon layers for different light polarizations. For the polarizations where the first-order Raman peak is forbidden in bulk silicon, porous silicon shows a band. This band is almost as intense as the one in the allowed polarizations but its lineshape is very different. The 'forbidden' band is usually wider and shifted to lower energies with respect to the allowed band. Both bands are analyzed in terms of sample characteristics, layer depth, and excitation wavelength.

INTRODUCTION

Raman spectroscopy has been extensively used to obtain information on the microscopic structure of porous silicon. The average size of silicon nanocrystallites can be obtained from the lineshape of the first order phonon peak around 520 cm^{-1}. When the silicon structures become very small, quantum confinement of phonons inside the particles relax the q=0 selection rule, and phonons at $q \neq 0$, which have lower frequencies, contribute to the Raman scattering. Therefore the phonon band broadens towards the low energy side and shifts slightly to lower energies. The lineshape and the shift can be described theoretically [1,2] within the standard phonon confinement model [3].

Besides the lineshape, the polarization selection rules for porous silicon are also different from the bulk. In bulk silicon, for Raman backscattering on the (001) surface, only the LO phonon is observed. For incident light polarized along a <100> direction, the LO phonon is observed only when the scattered light is collected along the perpendicular direction ('allowed' polarization configuration). No phonon is observed for the parallel case ('forbidden' polarization configuration). However, in porous silicon a peak is clearly observed in the forbidden configuration. Unfortunately, polarization data in the literature are still scarce and somewhat misleading. According to some authors [4,5] the phonon peak is mostly depolarized and it is the same for all polarization configurations. According to others [6] the peak is wider and weaker in the forbidden configuration. As we suspect that the discrepancy originates from differences in the samples, in this paper we present a number of Raman experiments on porous silicon films prepared under different conditions.

Mat. Res. Soc. Symp. Proc. Vol. 452 © 1997 Materials Research Society

EXPERIMENT

Porous silicon films were prepared on (001) boron doped p-type silicon wafers (0.1-0.5 Ω.cm). Samples were etched in a 2:1 solution of HF and ethanol. Different current densities and formation times were used. Etching was done either under illumination or in the dark. Only samples etched under illumination showed strong photoluminescence.

Raman spectra were measured with a Renishaw Ramascope spectrometer. Light from an argon laser (514.5 nm) or a helium-neon laser (632.8 nm) was focused onto the (001) surface through an optical microscope. The power density was kept smaller than 24 W/cm^2 to avoid any heating effects. Light scattered by the sample was collected through the same microscope and analyzed by a grating spectrometer. Elastically scattered light was rejected by a supernotch holographic filter. Polarization directions were along the <001> crystallographic axes.

RESULTS

Figure 1 depicts the Raman spectra measured in two samples etched for 1 minute under different conditions. One was etched in the dark (spectra on left column) and the other was etched under illumination (spectra on right column). Spectra on the upper and lower rows were

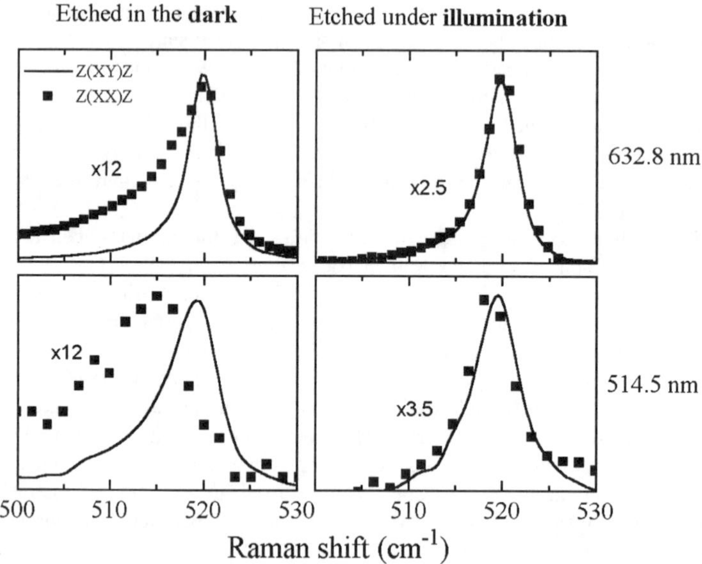

Fig. 1. Raman spectra measured for two porous silicon samples etched for 1 min at 6 mA in the dark (left column) and under illumination (right column). Upper and lower row spectra were excited, respectively, at 632.8 and 514.5 nm. Full lines stand for the allowed configuration and dots to the forbidden configuration. Latter spectra were multiplied by the indicated factor.

excited with light of 632.8 and 514.5 nm, respectively. The phonon peak is always wider for the shortest wavelengths, as has been previously observed [7]. This has been attributed to the larger penetration depth at longer wavelengths, together with a smaller pore size in the deeper layers [7], or to a resonance effect [8].

The most striking result is that for the sample etched under illumination the Raman peak is mostly depolarized, with no change in lineshape from one polarization to the other. However, for the sample etched in the dark, the peak in the forbidden configuration is one order of magnitude smaller and much broader than for the allowed configuration. Samples etched in the dark are always more homogeneous when observed under an optical microscope. Therefore we think that in the samples formed with light microscopic roughness produces a great deal of scattering and a strong depolarization of light. In the forbidden configuration one observes the peak corresponding to the allowed configuration, which is the LO phonon broadened by quantum confinement. The porous silicon film is only a few microns thick. Since the light penetration depth is of the same order of magnitude, the peak in the allowed configuration could have some contribution from the bulk silicon substrate, specially for 632.8 nm excitation.

For the samples formed in the dark, which are more homogeneous, multiple scattering is much smaller and for the forbidden configuration one observes a peak, which is mostly due to the TO phonon. This phonon, which is forbidden for scattering from the (001) surface becomes allowed in porous silicon by scattering from tilted surfaces in the porous structure [9]. Since the slope of the dispersion relations is larger for the TO than for the LO phonon, the TO band broadens more with confinement [9]. Therefore in the forbidden configuration one observes a broader peak. In the allowed configuration one observes the LO phonon peak, probably with some contribution from the substrate.

To see the effect of porosity we have also varied the current and the formation time of the porous silicon films. Fig. 2 shows the Raman spectra measured with the 514.5 nm line of an

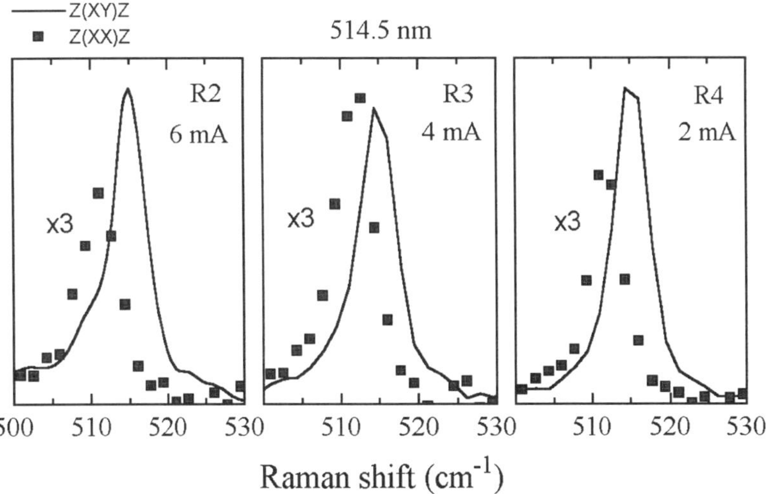

Fig. 2. Raman spectra for three porous silicon samples etched under illumination at three different currents, as indicated in the figure. Etching time was 1 minute. Spectra were excited at 514.5 nm See Fig. 1 for more details.

argon laser for samples etched during one minute under illumination at three different currents. Lineshapes are very similar for both polarizations in the three cases, independently of porosity. In the allowed configuration the peak is at higher energies, probably due to the substrate contribution, which does not have the shift due to phonon confinement. The fullwidth is larger for samples etched 4 or 6 mA in the forbidden configuration as corresponds to a larger porosity.

Finally, in Fig. 3 we compare the Raman spectra of two samples etched under illumination at 6 mA for different times. The fullwidth is larger for the sample that was etched a longer time, indicating a smaller crystallite size. Lineshapes are very similar for both polarizations, except for a shift, similar to the one in Fig. 2. Therefore, independently of the formation time or the current, samples etched under illumination have a largely depolarized Raman spectrum. How does the associated inhomogeneity correlate with the light emission ability of these samples is still unclear.

CONCLUSIONS

We have observed that polarization selection rules in the Raman spectra of porous silicon samples depend strongly on lighting conditions during formation. Samples etched in the dark, which have very poor photoluminescence emission, have a strongly polarized spectrum, with different lineshapes in the allowed and forbidden configurations. The band in the allowed configuration corresponds to the LO phonon, whereas the band in the forbidden configuration,

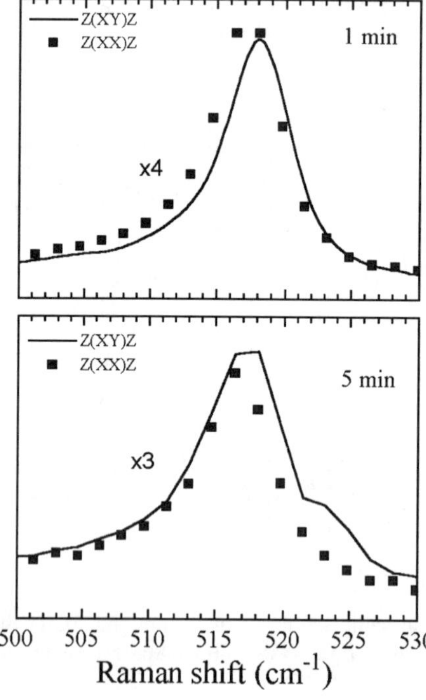

Fig. 3. Raman spectra for two porous silicon samples etched under illumination for different times, as indicated in the figure. Etching current was 6 mA. Spectra were excited at 514.5 nm. See Fig. 1 for more details.

much broader and weaker, corresponds to the TO phonon. The latter appears from scattering at pore surfaces distinct from the (001) one. On the contrary, the Raman spectra of samples etched under illumination, which have very strong luminescence, are mostly depolarized due to a much larger inhomogeneity. The LO band dominates for both configurations and probably has some contribution from the bulk silicon substrate in the allowed configuration.

ACKNOWLEDGMENTS

We acknowledge financial support from the Spanish CICyT and the Fundaciǿn Ramǿn Areces.

REFERENCES

1. R. J. Nemanich, S. A. Solin, and R. M. Martin, Phys. Rev. B **23**, 6348 (1981).

2. H. Richter, Z. P. Wang, and L. Ley, Solid State Comm. **39**, 625 (1981).

3. I. H. Campbell and P. M. Fauchet, Solid State Comm. **58**, 739 (1986).

4. R. Tsu, H. Shen, and M. Dutta, Appl. Phys. Lett. **60**, 112 (1992).

5. D. J. Lockwood, G. C. Aers, L. B. Allard, B. Bryskiewicz, S. Charbonneau, D. C. Houghton, J. P. McCaffrey, and A. Wang, Can. J. Phys. **70**,1184 (1992).

6. I. Gregora, B. Champagnon, and A. Halimaoui, J. Lumin. **57**, 73 (1993).

7. S.-L. Zhang, Y. Hou, K.-S. Ho, B. Qian, and S. Cai, J. Appl. Phys. **72**, 4469 (1992).

8. S. Guha, P. Steiner, and W. Lang, J. Appl. Phys. **79**, 8664 (1996).

9. I. Gregora, B. Champagnon, L. Saviot, and Y. Monin, Thin Solid Films **255**, 139 (1995).

Part VIII

Biochemical, Photochemical, Photoelectronic, and Optical Device Applications of Nanocrystal and Porous Semiconductors

SILICON AS AN ACTIVE BIOMATERIAL

L. T. Canham, C. L. Reeves, D. J. Wallis, J. P. Newey, M. R. Houlton, G. J. Sapsford, R. E. Godfrey, A. Loni, A. J. Simons, T. I. Cox, M. C. L. Ward.
DRA Malvern, St Andrews Road, Malvern, Worcestershire. WR14 3PS, UK.

ABSTRACT

The response of a range of porous Si and poly Si films to storage in acellular simulated body fluids is summarised and its implications discussed. It is suggested that the combination of VLSI technology, micromachining and surface microstructuring achievable with silicon, could establish this prominent semiconductor as a very useful biomaterial by the next century. The 'biocompatibility' of a variety of silicon microstructures, and even bulk silicon has received surprisingly little study, but now warrants detailed in-vitro and in-vivo assessment.

INTRODUCTION

In biology, elemental silicon is now established as one of the trace but essential elements [1], due largely to the pioneering work of Carlisle throughout the 1970's [2,3]. Its physiological role in man has been recently scrutinised by Birchall and co-workers [4,5] who emphasise the ability of silicic acid to assist biological systems in removing aluminium, a ubiquitous ecotoxicant. Despite the beneficial properties of low levels of elemental silicon, in its bulk crystalline form it has not generally been regarded or developed as a useful biomaterial. In this paper we first give examples of where Si technology is currently commercialised and at the research stage for biomedical and biotechnology applications. An overview of our recent work on rendering Si wafer surfaces 'bioactive' [6-9] with regards to calcification is then presented, and some of the implications are considered in the final section of the paper.

BIOAPPLICATIONS OF BULK SILICON TECHNOLOGY

Figure 1 illustrates some potential and already developed in-vivo applications of electronics. Commercialised products include the pacemaker and defibrillator [10], cochlear implants [11], functional electrical stimulation (FES) devices [12], and programmable drug delivery systems [13]. CMOS circuitry is generally isolated from the body by a hermetically sealed titanium package. A significant range of devices that exploit both VLSI technology and micromachining techniques are also now at the research stage. These include the 'silicon retina' [14], catheter based devices [15], glucose monitoring systems [16] and many other biomedical sensors [17-18]. In-vitro applications of Si based biotechnology span several disciplines such as clinical diagnostics, environmental monitoring and agriculture (eg disposable blood analysis chips [19], DNA chips [20], microelectrode devices for cell manipulation [21]. Once again, in most of these diverse application areas the semiconductor itself is isolated from the biological medium by either a more 'biocompatible' encapsulant or package. One motivation for the experimental work described in the next section is to explore whether the chip itself could be rendered more biocompatible. This could then enable further miniaturization of electronic implants and in general much greater flexibility in bio-electronic interfacing for both in-vitro and in-vivo applications.

Mat. Res. Soc. Symp. Proc. Vol. 452 © 1997 Materials Research Society

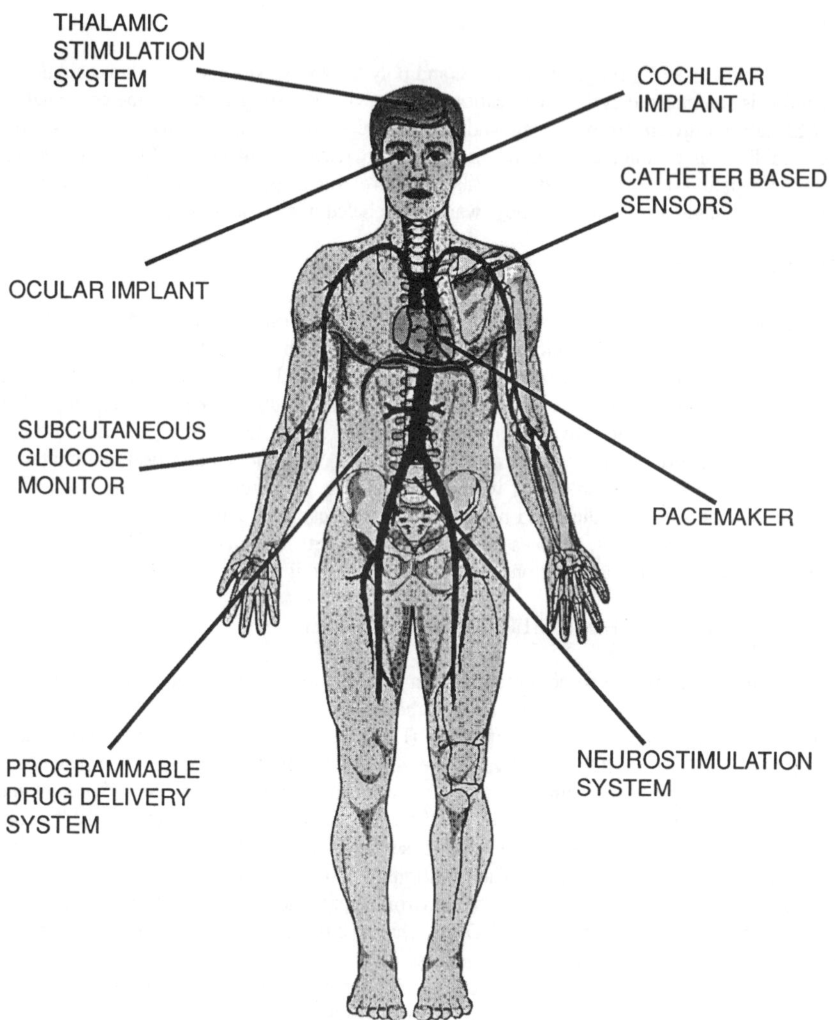

Figure 1 Examples of in-vivo applications of Si technology.

The main inorganic phase of bone, although chemically complex, closely resembles the structural state of hydroxyapatite, $Ca_{10}(PO_4)_6(OH)_2$ [22]. Many prosthetic implants are coated with synthetic hydroxyapatite films and a variety of deposition techniques have been developed [23]. The optimum type of calcium phosphate coating depends to a large extent on the biological site and function of the implant. Nevertheless, there is increasing interest in the use of coatings that mimic as closely as possible, biological apatite. In this regard there are an important class of 'bioactive' ceramics that induce calcium phosphate nucleation and growth on their surfaces in-vivo [24]. They do not therefore require in-vitro coating since they naturally bond to bone and even soft tissue in-vivo. Kokubo and co-workers have developed, and extensively utilised an in-vitro test for predicting whether a material will exhibit this type of bioactivity in-vivo [25-27]. It essentially involves storing the material at physiological pH (7.30 ± 0.05) and temperature (37 ± 1°C) in the simulated body fluid (SBF) shown in table 1, the inorganic ion content of which closely matches that of human blood plasma. Bioinert materials, such a bulk silica, alumina and titanium, have surfaces that remain largely unaffected by lengthy exposure to SBF. Bioactive materials such as Bioglass® or Ceravital® in contrast become covered by a μm thick calcium phosphate layer within days of exposure [24].

Ion	Simulated Fluid	Blood Plasma
Na^+	142.0	142.0
K^+	5.0	5.0
Mg^{2+}	1.5	1.5
Ca^{2+}	2.5	2.5
Cl^-	147.8	103.0
HCO_3^-	4.2	27.0
HPO_4^{2-}	1.0	1.0
SO_4^{2-}	0.5	0.5

Table 1 Ion concentrations (mM) of simulated body fluid used compared with those of human blood plasma. The above acellular solution was prepared by dissolving reagent grade NaCl, KCl, $NaHCO_3$, $K_3HPO_4 \cdot H_2O$, $MgCl_2 \cdot 6H_2O$, $CaCl_2 \cdot 2H_2O$ and $NaHSO_4$ in deionized water. It was pH buffered with trihydroxymethyl-aminomethane and hydrochloric acid.

We have used this in-vitro test for the first time to investigate whether semiconductor structures can exhibit such bioactivity. It has been found that whereas bulk monocrystalline Si surfaces are 'bioinert', specific porous Si [6,7] and poly Si [9] structures readily induce calcium phosphate nucleation. Figure 2 shows typical plan view images of these three types of wafer surface after various times in SBF. The bulk Si surface has a negligible level of calcification even after 4 weeks SBF exposure (figure 2(a)), whereas a high density of calcium phosphate deposits can nucleate on both microcrystalline Si ('poly-Si') and microporous Si over timescales of days to weeks (figure 2(b) and 2(c)).

Cross-sectional SEM/EDX analysis revealed that calcium phosphate could nucleate directly onto the hydrated surface of low porosity microporous silicon films which were quite stable in SBF [6,7]. However, in the case of erodible poly silicon layers nucleation occurred on an interfacial layer of hydrated silica that developed in the SBF, in analogy with Bioglass® [24]. Figure 3 shows a cross-sectional SEM and SIMS depth profile of such a structure, revealing the silica-like nature of the interfacial zone between the calcium phosphate deposit and the remaining poly silicon layer. Note also that the calcium signal drops to background levels within the latter region, suggesting that the SBF has not penetrated the grain boundaries of the residual poly silicon to any great extent.

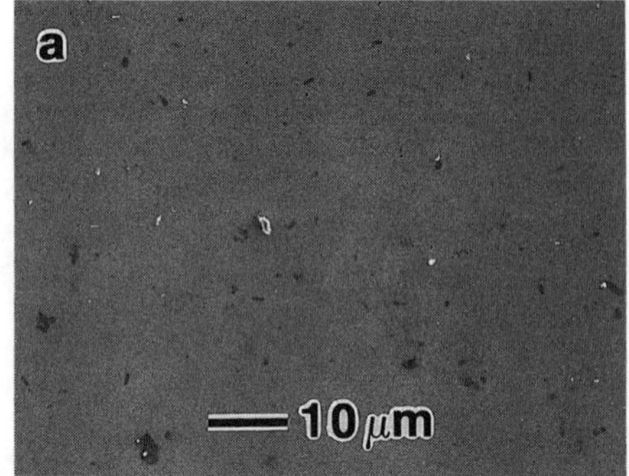

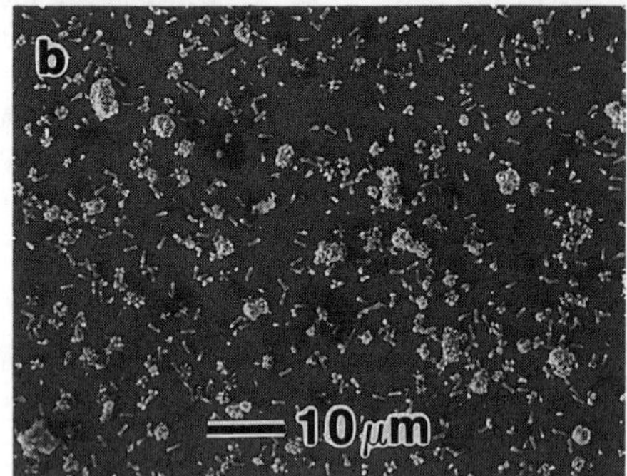

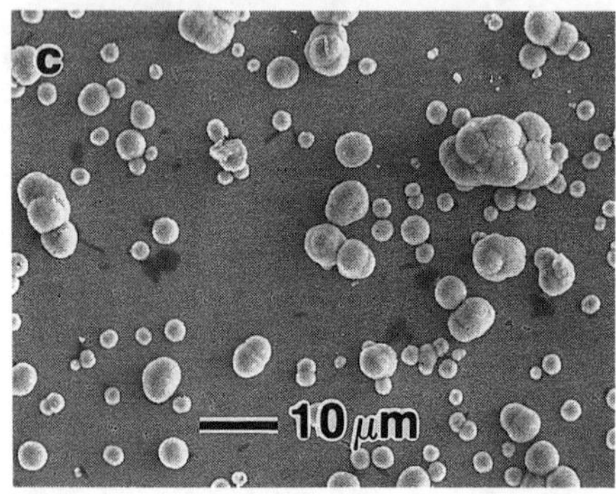

Figure 2 Scanning electron microscope images of different Si wafer surfaces after SBF exposure (a) bulk (100) surface after 4 weeks (b) 600°C poly Si layer after 3 weeks (c) microporous Si layer after 1 week. EDX analysis identified the needle shaped deposits and spherulites of (b) and (c) respectively to be calcium phosphate but not the irregular isolated surface 'debris' of (a).

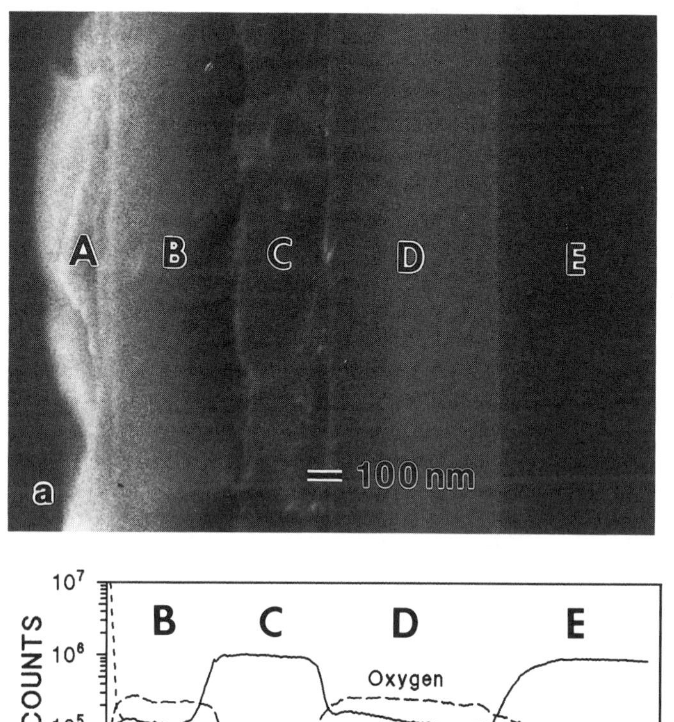

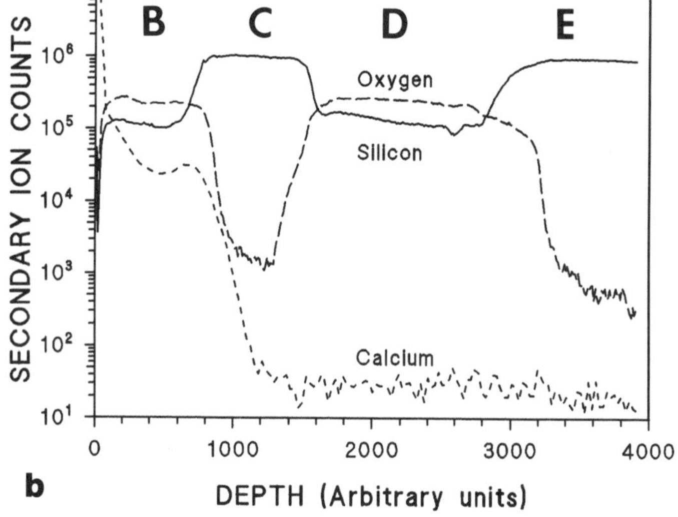

Figure 3 Poly Si layer after 2 weeks in SBF. (a) Cross-sectional SEM image showing distinguishable regions. A: Surface deposits, B: reaction zone within poly Si, C: residual poly Si, D: thermal SiO_2 layer, E: bulk silicon substrate. (b) SIMS depth profile of the same structure revealing high calcium and oxygen levels in regions A and B but low levels in C.

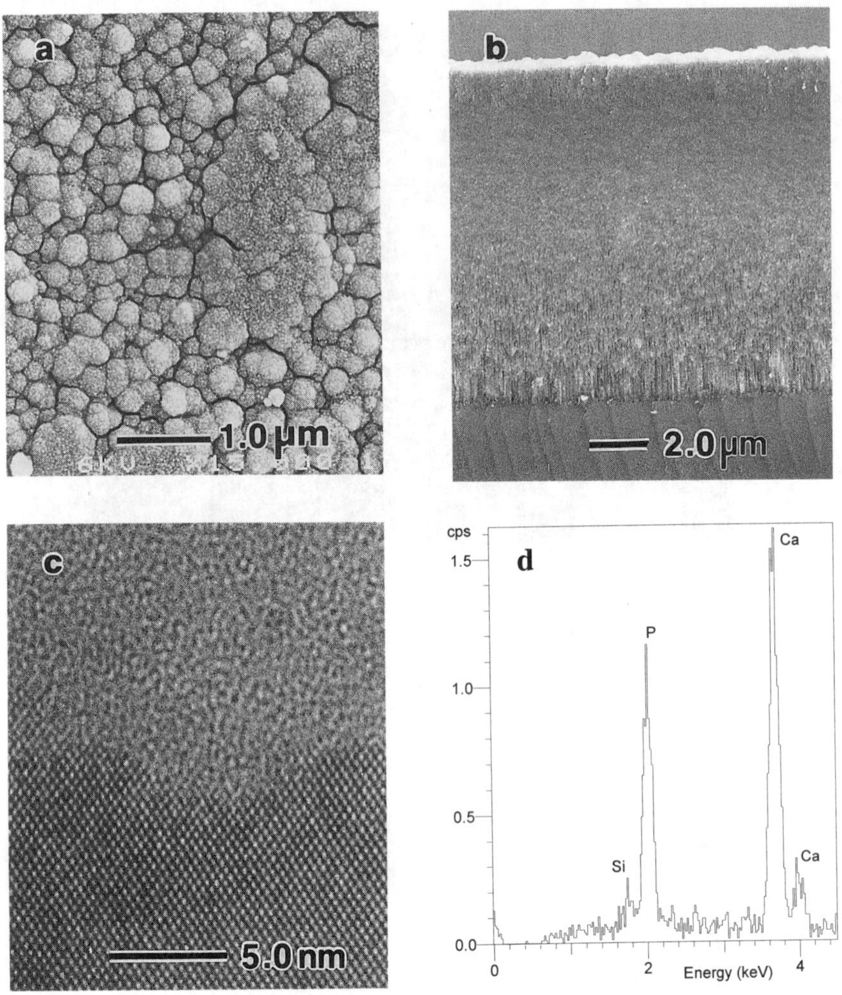

Figure 4 Electron microscopy of a microporous Si layer after cathodic bias treatment in SBF. (a) plan view SEM image of the calcium phosphate overlayer formed within 3 hours under galvanostatic conditions [8]. (b) Cross-sectional SEM image of the ~11 μm thick porous silicon layer with its thin coating of calcium phosphate. (c) HRTEM image of the interfacial region between the low porosity silicon and its overlayer. (d) HR EDX spectrum taken with a 400 KeV 10 nm spot size probe beam and through the amorphous overlayer shown in (c).

It has been found that, at least in the case of porous silicon, the kinetics of calcification are very sensitive to, and can hence be controlled by electrical bias [9]. Thin but uniform and continuous calcium phosphate coatings are achieved within hours when anodised Si wafers are cathodically biased in SBF (figure 4(a) and 4(b)). Calcium depth profiles taken by SIMS [9] also reveal that anodic bias dramatically retards the calcification process. This could be important, not only from the point of view of elucidating the mechanisms of calcification, but also with regard spatially controlling such processes electrically over a 'biochip' surface. In the case of this cathodically biased structure the calcium phosphate deposit has nucleated directly onto the porous semiconductor and not an interfacial hydrated silica layer. This is evident from figure 4(c) and (d) where HRTEM analysis of the interfacial region reveals that the crystalline Si lattice of low porosity extends right up to the calcium phosphate overlayer. X-ray analysis of this amorphous overlayer (figure 4(d)) shows negligible levels of silicon (and hence amorphous silica) from areas within 5 nm of the interface. There is hence the possibility of electrically promoting in-vivo a direct chemical bond between the semiconductor and living mineralized tissue.

FUTURE OUTLOOK

The biocompatibility of bulk silicon structures do not seem to have been investigated in depth, let alone the biological acceptability of silicon microstructures and nanostructures. There exists, for example, only isolated reports on the haemocompatibility of bulk silicon [28], and the effects of its insertion into cortical tissue [29-30].

The preliminary experiments shown here suggest that with appropriate surface microstructuring, silicon 'biochips' could be developed that have much higher levels of bioacceptability in specific sites than previously imagined. To fully exploit the potential of VLSI and MEMS (Micro Electro Mechanical Systems) technology, one of the key attributes, namely miniaturisation, could be kept, if silicon itself became the biocompatible packaging material [6]. Two examples of such a packaging concept are shown in figure 5. In figure 5(a), bulk micromachining, wafer bonding and microporous Si coatings are used to realise a miniature biochip; in figure 5(b) surface micromachining, multilayer deposition and poly Si 'bubble' techniques result in potentially even smaller devices. In both cases the bioactive coating encourages true integration with the surrounding mineralised or even soft tissue.

In a more general sense, the extremely well developed techniques for etching silicon into complex shapes over a wide range of length scales also offers exciting possibilities, provided the remaining skeleton structure is biocompatible. In analogy with the recent concept of 'living ceramics' [31] there is the potential for realising a variety of electronic-biological composites wherein living organisms are cultured, monitored and utilised in-vitro and in-vivo. With this in mind figure 6 compares the dimensionality of various organisms and biomolecules with the pore size accessible by electrochemical and lithography based etching techniques. The realisation of living semiconductor based biocomposites will require engineering of structures over a combination of very different length scales. Although for example, it has been shown that cells can enter and traverse Si membranes with micron pore size [32], the in-growth of healthy vascular tissue or bone requires > 100μm wide pores, some 10^5 times the micropore width likely to promote bonding with mineralized tissue. Figure 7 shows an example of a micromachined Si structure that has the interconnected porosity of the size needed for tissue

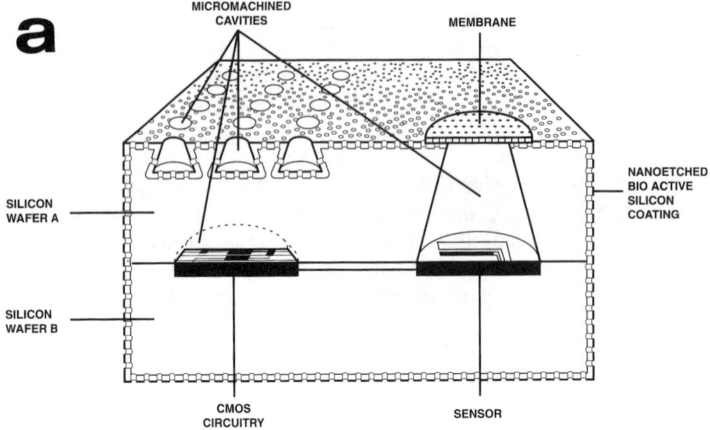

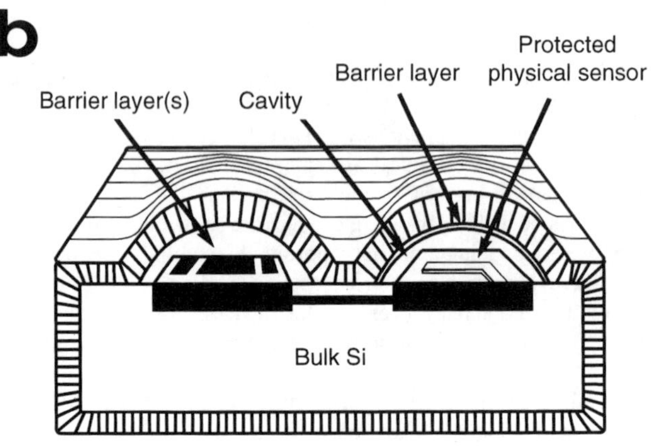

Figure 5 Schematic examples of chip packaging where Si itself is the miniaturizable package material [6,9] (a) using microporous silicon (b) using poly Si technology. In both cases the microstructured Si coating is designed to promote biological acceptance and integration. In example (a) the sensing element is electrochemical and needs to maintain contact with body fluids whereas in example (b) a physical sensor is shown which needs protection from such fluids. Clearly the type of sensor determines its in-vivo interfacing and whether or not its microenvironment needs to differ from that of the rest of the biochip surface.

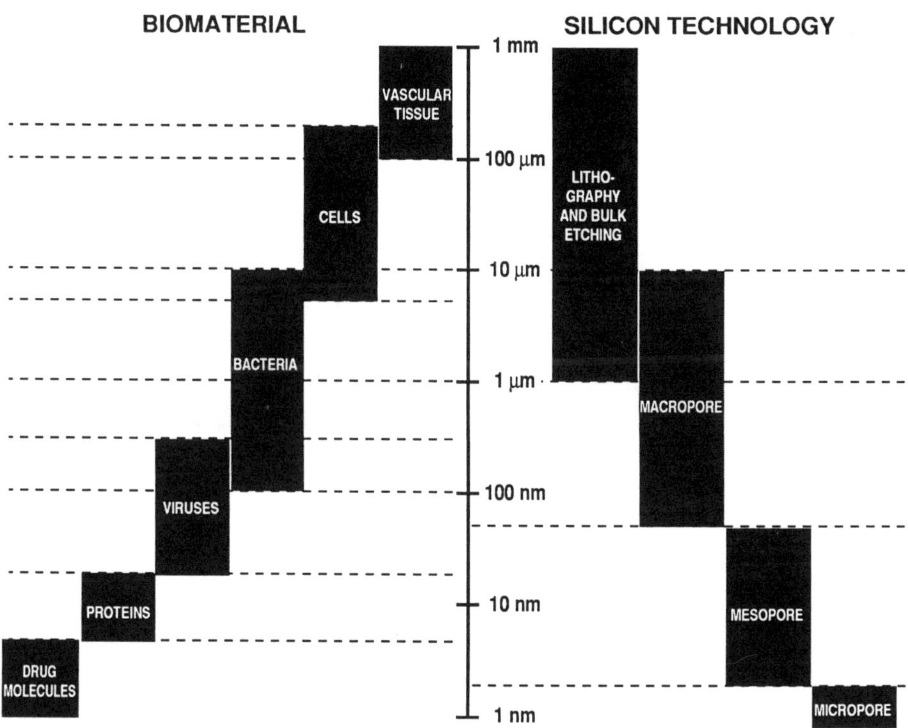

Figure 6 Comparison of the dimensionality of various biomaterials with feature sizes that can be etched into monocrystalline silicon. Whilst the building blocks of biological materials can have very small dimensionality most complex living organisms are too large to penetrate mesoporous silicon. Taking bone as an example its collagen fibrils have about 30 nm diameter and hydroxyapatite nanocrystals are only about 3 nm wide and 30 nm long. However, above the cellular level its macroscopic structural elements such as osteons are of much greater size, typically 190-230 microns width in cortical bone and 500-600 microns in cancellous bone.

Figure 7 Micromachined monocrystalline silicon cube with interconnected macroporosity suitable for tissue or bone in-growth. (a) photograph comparing cube dimensions with those of a penny coin (b) SEM image revealing interconnectivity of macropores.

in-growth studies. By subsequently coating all the faces of such a structure with bioactive poly silicon or porous silicon true integration with living complex biomaterials may be possible.

CONCLUSIONS

It has been shown in-vitro that more than one type of microstructured Si surface can induce calcium phosphate growth in a similar manner to bioactive ceramics. The 'bioactivity' and biocompatibility of a range of silicon microstructures and nanostructures needs in-vitro and in-vivo assessment. Such studies now in progress, could confirm the exciting implications for silicon based biotechnology suggested by these preliminary in-vitro experiments.

REFERENCES

[1] J. R. F. da Silva and R. J. P. Williams. The Biological Chemistry of the Elements. Clavendon Press, Oxford (1991).

[2] E. M. Carlisle. Science 178, 619 (1972).

[3] E. M. Carlisle in Silicon and Siliceous Structures in Biological Systems, edited by T. L. Simpson and B. E. Volcani. Springer Verlag (1981), Chapter 4.

[4] J. P. Bellia, K. Newton, A. Davenport, J. D. Birchall, N. B. Roberts. Eur. J. Clin. Investig. 24, 703 (1994).

[5] J. D. Birchall. Chem. Soc. Rev. 24, 351 (1995).

[6] L. T. Canham. Adv. Mater. 7, 1033 (1995).

[7] L. T. Canham, C. L. Reeves in Thin Films and Surfaces for Bioactivity and Biomedical Applications edited by C. M. Cotell, S. M. Gobatkin, G. Grobe, A. E. Meyer. MRS Proc Vol 414, 189 (1996).

[8] L. T. Canham, J. P. Newey, C. L. Reeves, M. R. Houlton, A. Loni, A. J. Simons, T. I. Cox. Adv Mater. 8, 847 (1996).

[9] L. T. Canham, C. L. Reeves, D. O. King, P. J. Branfield, J. G. Crabb, M. C. L. Ward. Adv. Mater. 8, 850 (1996).

[10] R. S. Sanders, M. T. Lee. Proc. IEEE. 84, 480 (1996).

[11] G. Wang, M. W. Vannier, M. W. Skinner, W. A. Kalender, A. Polacin, D. R. Ketten. IEEE Trans. BME. 43, 891 (1996).

[12] R. B. North, D. H. Kidd, M. Zahvrak, C. S. James, D. M. Long. Neurosurgery 32, 384 (1993).

[13] R. B. Kanoff. J.A.O.A. 94, 487 (1994).

[14] J. Wyatt, J. Rizzo. IEEE Spectrum May (1996) 47.

[15] H. Chav, K. D. Wise. IEEE Trans. Electron. Devices 35, 2355 (1988).

[16] B. D. McKeen, D. A. Gough. IEEE Trans. BME. 35, 526 (1988).

[17] W. H. Ko. IEEE Trans. BME. 33, 153 (1986).

[18] K. D. Wise, K. J. Najafi in VLSI in Medicine. Edited by N. G. Einsprich and R. D. Gold. Acad. Press Inc (1989) Chapter 10.

[19] K. A. Erickson, P. Wilding. Clin. Chem. 39, 283 (1993).

[20] S. Jacobson, J. M. Ramsey. Anal. Chem. 68, 720 (1996).

[21] M. Washizu, T. Nanba, S. Masuda. IEEE Trans. Ind. Applic. 26, 352 (1990).

[22] C. P. A. T. Klein, J. G. C. Wolfe, K. de Groot in An Introduction to Bioceramics Edited by L. L. Hench, J. Wilson. World Scientific Publ (1993) Chapter 11.

[23] W. R. Lacefield. Ibid Chapter 12.

[24] L. L. Hench. J. Am. Ceram. Soc. 74, 1487 (1992).

[25] T. Kokubo, H. Kushitani, S. Sakka, T. Kitsugi, T. Yamamuro. J. Biomed. Mater. Res. 24, 721 (1990).
[26] T. Kokubo, H. Kushitani, C. Ohtsuki, S. Sakka, T. Yamamuro. J. Mat. Sci. Mat. in Medic. 3, 79 (1992).
[27] P. Li, I Kangasniemi, K. de Groot, T. Kokubo. J. Am. Ceram. Soc. 77, 1307 (1994).
[28] Y. Kanda, R. Aoshima, A. Takada. Electron. Lett. 17, 558 (1981).
[29] D. J. Edell, V. V. Toi, V. M. McNeil, L. D. Clark. IEEE Trans. BME. 39, 635 (1992).
[30] S. Schmitt, K. Horch, R. Normann. J. Biomed. Mater. Res. 27, 1393 (1993).
[31] E. J. A. Pope. J. Sol. Gel. Sci. Techn. 4, 225 (1995).
[32] E. Richter, G. Fuhr, T. Muller, S. Shirley, S. Rogaschewski, K. Reimer, C. Dell. J. Mat. Sci.: Mat. in Medic. 7, 85 (1996).

ELECTRON MICROSCOPY OF MESOSCALE ARRAYS OF QUANTUM-CONFINED CdS NANOPARTICLES FORMED ON DNA TEMPLATES

RUSSELL F. PINIZZOTTO, YOUNG G. RHO, YANDONG CHEN, ROBERT M. PIRTLE,[*] IRMA L. PIRTLE,[*] JEFFERY L. COFFER[#] AND XIN LI[#]
Materials Science Department, University of North Texas, Denton, TX 76203
[*]Department of Biological Sciences, University of North Texas, Denton, TX 76203
[#]Department of Chemistry, Texas Christian University, Fort Worth, TX 76129

ABSTRACT

This paper describes the fabrication method and initial characterization of self-assembled mesoscale arrays of quantum-confined CdS nanoparticles using DNA as a template for the overall shape. Three DNAs were used: the circular and linear forms of the plasmid pUCLeu4, and circular ϕX174 RF II. In all three cases, the mesoscale lengths are consistent with the A-form of DNA. The structural signatures and crystallography were confirmed using conventional and high resolution transmission electron microscopy, and electron diffraction. Optical spectroscopy demonstrated that the particles display quantum-confinement effects. This research is a fundamental demonstration of the power of combining biochemical and solid-state processing techniques.

INTRODUCTION

On December 29, 1959, at an American Physical Society Meeting, Richard Feynman said, "Atoms on a small scale behave like nothing on a large scale, for they satisfy the laws of quantum mechanics. ... At the atomic level, we have new kinds of forces and new kinds of possibilities, new kinds of effects. The problems of manufacture and reproduction of materials will be quite different."[1] We now have the technology to manufacture artificial quantized structures in both two and three dimensions which exhibit new electrical and optical properties as predicted by Feynman 37 years ago. This is perhaps demonstrated best by the awarding of the 1985 Nobel Prize in Physics for the discovery of the quantized Hall effect.[2] The MRS symposium, *Advances in Microcrystalline and Nanocrystalline Semiconductors - 1996,* is a continuation of the search for improved routes for the synthesis of quantum-confined materials with unusual properties.

The fabrication method and initial characterization of self-assembled mesoscale arrays of quantum-confined CdS nanoparticles are described in this paper. A DNA template is used to control the location of nanoparticle formation, and hence the overall shape of the resulting structure. The plasmid DNAs used as templates resulted in rings less than 0.5 μm in diameter with linewidths less than 10 nm.[3,4]

Previously described methods for forming quantum-confined structures are based on the lithography of thin continuous films,[5] or on gas,[6] liquid,[7] or solid[8] phase synthesis of nanocrystalline materials from atomic precursors. The lithographic methods require expensive equipment and highly experienced personnel. Most synthetic methods result in nanoparticles that have no spatial relationship with each other: the individual nanoparticles are randomly distributed and oriented. While it is possible to fabricate thin films[9] and even three-dimensional arrays of nanoparticles[10] using volumetric restriction techniques, these methods result in unpatterned, large-scale materials with overall macroscopic dimensions that are equivalent to thin films or bulk. The difference is that they are composed of

Mat. Res. Soc. Symp. Proc. Vol. 452

nanoparticles instead of individual atoms.

The approach described here differs significantly from the recent studies of Alivisatos et al.[11] and Mirkin et al.[12] who reported the use of polynucleotides to assist in the packaging of pre-formed gold nanoparticles into clusters. We use the polynucleotide as the template for actual nanoparticle synthesis, not as a linking agent for pre-formed particles. The nanoparticles in our method are semiconductors, not metals, and they are much smaller than the gold particles used by Mirkin et al. (5 nm as compared to 13 nm). The mesoscale structures fabricated using our method are also much smaller than the clusters reported in their research (< 8 nm wide compared to 50 nm). Our methodology has already succeeded in making the large nanocrystal molecules envisioned by Alivisatos.[13]

METHODOLOGY

There are four concepts which form the basis of our methodology: (1) the synthesized material must demonstrate quantum-confinement effects, meaning that the individual nanoparticles must be less than about 6 nm in diameter for CdS; (2) the nanoparticles should be semiconductors for their unique optical and electrical properties, and for possible integration with other semiconductor technologies; (3) the array shapes should be controllable; and (4) the structures should self-assemble, that is, lithography should not be required. Previous work at TCU and UNT conclusively demonstrated that DNA can stabilize the formation of quantum-confined semiconductor nanoparticles in solution.[14,15,16] This concept was extended in the following manner for the present work: (1) DNA/metal ion complexes are formed in solution; (2) the DNA/metal complexes are adsorbed onto a solid substrate; and (3) the complexes are converted into arrays of II-VI semiconductor nanoparticles by reaction with a chalcogenide source. The resulting nanoparticle array will have the same shape and dimensions as the original DNA template. The experiments described here used three different types of DNA with known basepair sequences, shapes, and dimensions, since their structural signatures can be unambiguously identified using transmission electron microscopy.

There are numerous advantages of the approach described above. Many different types of polynucleotide templates can be used. Lines can be formed from long DNA strands, circles can be formed from plasmids, and dots can be formed from short DNA segments. Other DNA shapes such as X's and multiple branched crosses are available, so it may be possible to directly fabricate interesting and intricate device structures without using lithography.[17] At the atomic scale, different polynucleotide sequences may result in different optical and electrical properties of the nanoparticles.[18] Different substrates may be exploited, such as carbon films, mica, and silicon dioxide. Finally, different II-VI semiconductors can be prepared, such as CdS, CdSe, PbS, PbSe, ZnS, and ZnSe.[19]

RESULTS

Figure 1a is a bright field transmission electron micrograph of a typical mesoscale structure formed on a pUCLeu4 plasmid DNA template[20] using the methodology described above. The dark ring is a mesoscale assembly of CdS nanoparticles; its overall structure is similar to that of a ring of pearls or a beaded metal chain. The ring circumference is approximately 1.17 μm, as expected, since pUCLeu4 is 3455 basepairs long and each basepair contributes 0.34 nm to the length of the molecule. The individual CdS nanoparticles are approximately 5 nm in diameter. Figure 1b is a high resolution

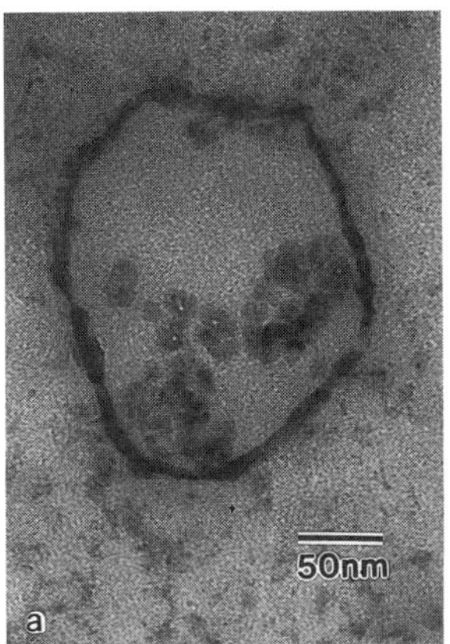

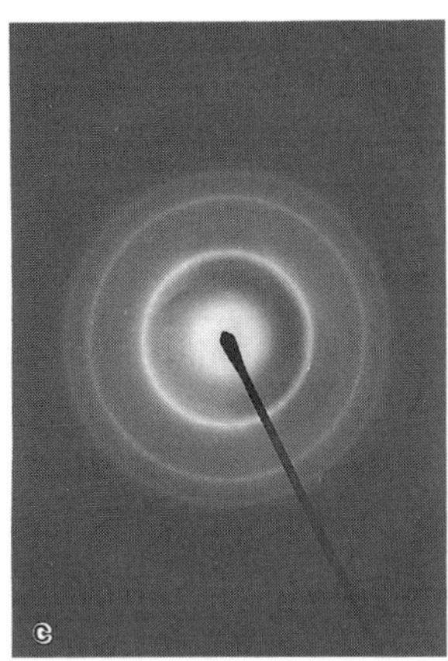

Figure 1. a: bright field transmission electron micrograph of a ring of CdS nanoparticles formed on a pUCLeu4 plasmid DNA template; b: high resolution lattice image of part of a CdS ring showing individual nanoparticles; c: selected area electron diffraction pattern of a CdS ring. The measured d-spacings are consistent with the cubic form of CdS.

transmission electron micrograph of a small section of a similar ring. The individual nanoparticles and lattice structures are visible, and the measured lattice spacings are consistent with the cubic form of CdS. There are no particular spatial relationships between the particles. Figure 1c is a selected area electron diffraction pattern from the entire mesoscale structure. The measured d-spacings match those of the cubic (zincblende or Hawleyite) phase of CdS. This is not an isolated observation; other rings have been identified in the same sample, and in other samples made using pUCLeu4 and ϕX174 RF II circular DNAs as the templates. Linear structures were observed when the linearized form of pUCLeu4 was used. We believe that Figure 1 and other published data[3,4] confirm that our methodology can be used to successfully fabricate self-assembled mesoscale arrays of quantum-confined semiconductor nanoparticles.

After repeated observations and measurements of the lengths of circular plasmid structures similar to the one shown in Figure 1a, it was discovered that the majority of the structures were significantly shorter than expected. A systematic study was undertaken to measure the lengths of three different DNA templates: circular pUCLeu4 (3455 basepairs), pUCLeu4 linearized using a restriction endonuclease (3455 bp), and circular ϕX174 RF II (5386 bp). It is well known that DNA has three basic forms, A, B and Z.[21] The B-form occurs when the DNA is hydrated, and it has a length of 0.34 nm per basepair. The A-form occurs when the DNA is not hydrated, and it is much more compact than the B-form with a length of only 0.255 nm per basepair. The Z-form is left-handed and does not apply to the current work since only right-handed DNA forms were used.

Samples of each of the templates were prepared for TEM observation using the Kleinschmidt imaging technique.[22] The lengths of approximately 25 to 30 individual molecules of each type were measured using a computerized quantitative microscopy analysis system.[23] In all three cases, the lengths of the templates match the A-form of DNA and do not match the B-form (Table 1). In future work, the DNA templates should be designed and chosen assuming that their structures are the A-form.

Table 1. Calculated and measured lengths of the DNA templates.

DNA Template	Calculated B-form Length (μm)	Calculated A-form Length (μm)	Measured Length (μm)
circular pUCLeu4	1.17	0.88	0.84
linear pUCLeu 4	1.17	0.88	0.85
circular ϕX 174 RF II	1.83	1.37	1.30

Figure 2a is a bright field TEM image of CdS/DNA structures formed using circular pUCLeu4 templates. Figures 2b and 2c are dark field TEM images of similar structures formed using linearized pUCLeu4 and circular ϕX174 RF II templates, respectively. The CdS/DNA complexes have agglomerated into web-like network structures due to the interaction between the Cd^{2+} and the DNA. Nanoparticle formation along the DNA templates is still evident, however, even in these more massive structures. The isolated rings such as the one shown in Figure 1a were found outside of these large areas of agglomeration. Network formation due to metal ion / DNA interactions has not been previously reported. This implies that future work will require the use of very dilute metal/DNA solutions to form the desired mesoscale structures.

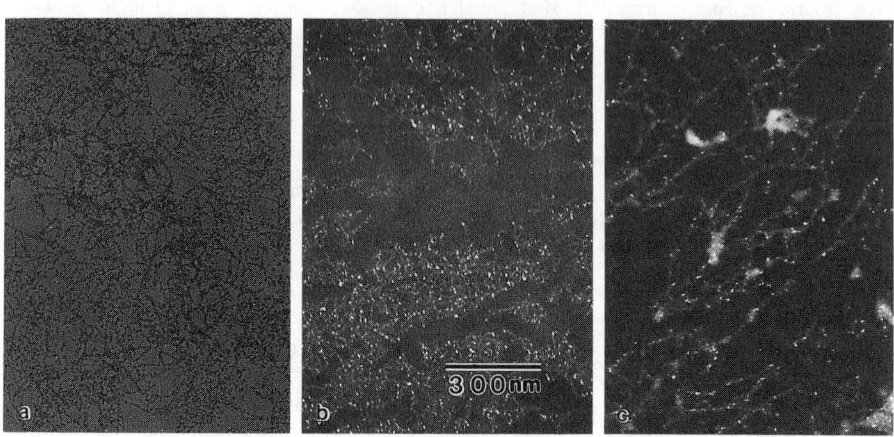

Figure 2. Web-like networks formed due to interactions between Cd^{2+} and DNA; a: bright field TEM, circular pUCLeu4; b: dark field TEM, linear pUCLeu4; c: dark field TEM, circular ϕX 174 RF II.

Figure 3 shows some of the optical characteristics of thin continuous films of CdS/DNA formed with pUCLeu4 using two different Cd/DNA concentrations. The photoluminescence peak is observed near 540 nm, which is a vivid green color. This is consistent with our previously reported results. The absorption onset is similar in both cases, near 480 nm. This is blue-shifted from the bulk value of 520 nm demonstrating that the nanoparticles display quantum-confinement effects. The results also indicate that the change in concentration did not significantly change the average particle size, but only the number of particles formed.

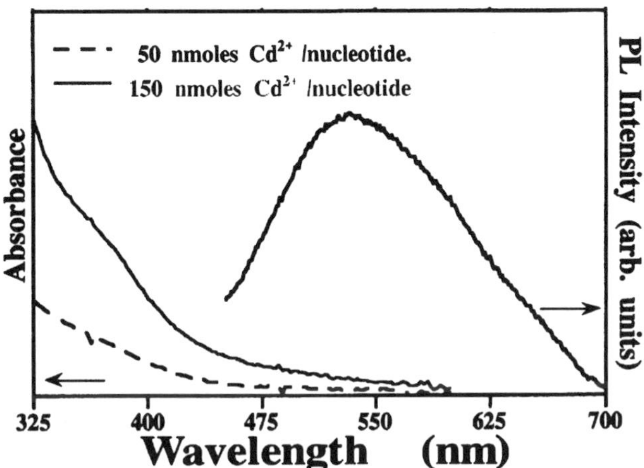

Figure 3. Room temperature UV/visible absorption and PL spectra of multilayer thin films of CdS/DNA supported on a polylysine-coated glass slide.

SUMMARY

We have demonstrated that DNA can be used as a template for the formation of mesoscale arrays of quantum-confined semiconductor nanoparticles. Three different types of DNA have been examined: the plasmid and linear forms of pUCLeu4, and circular ϕX174 RF II. In all three cases, the mesoscale lengths are consistent with the A-form of DNA. The unique structural signature of rings of nanoparticles formed on pUCLeu4 DNA was confirmed using conventional and high resolution TEM and electron diffraction. The nanoparticles are the cubic form of CdS. At higher DNA concentrations, web-like networks are formed due to interactions between the Cd and the DNA. Optical spectroscopy confirmed that the absorption edges are blue-shifted as expected for quantum-confinement, and that the complexes display photoluminescence with a peak near 540 nm.

We believe that this research has an excellent chance of being a fundamental demonstration of the power of novel nanoengineering methods. It combines biochemical specificity with the potency of sophisticated solid state processing technology suggesting exciting new possibilities in the development of quantum-confined materials and structures.

ACKNOWLEDGMENT

We appreciate the support of the National Science Foundation which partially supported this work through Grant Number DMR-9526581, L. Hess, contract monitor.

REFERENCES

1. R. P. Feynman, "There's Plenty of Room At The Bottom," American Physical Society Annual Meeting, December 29, 1959; Engineering and Science, (California Institute of Technology, Pasadena, CA, February 1960).

2. K. Von Klitzing, 1985 Nobel Prize in Physics. *The World Almanac and Book of Facts 1996,* (Funk & Wagnalls Corp., Mahwah, NJ) pg. 323 (1995).

3. N. T. Huang, S. R. Bigham, J. L. Coffer, R. F. Pinizzotto, R. M. Pirtle, and I. L. Pirtle, presented at the *Symposium on Self-Assembling Materials and Structures,* 1995 Fall Meeting of the Materials Research Society, Boston, MA, November 29, 1995, Abstract O6.8.

4. J. L. Coffer, S. R. Bigham, X. Li, R. F. Pinizzotto, Y. G. Rho, R. M. Pirtle, and I. L. Pirtle, Appl. Phys. Lett. **69,** 3851 (1996).

5. J. W. Mayer and S. S. Lau, *Electronic Materials Science: For Integrated Circuits in Si and GaAs,* (Macmillan Publishing Com., New York, 1990).

6. K. Littau, P. Szajowski, A. Muller, A. Kortan, and L. E. Brus, J. Phys. Chem. **97,** 1224 (1993).

7. L. E. Brus, Appl. Phys. A **53,** 465 (1991); H. Weller, Angew. Chem. Int. Ed. Engl. **32,** 41 (1993).

8. L. T. Canham, Appl. Phys. Lett. **57,** 1046 (1990).

9. B. O. Tabbousi, C. B. Murray, M. F. Rubner, and M. G. Bawendi, Chem. Mater. **6,** 216 (1994).

10. C. B. Murray, C. R. Kagan, and M. G. Bawendi, Science **270,** 1335 (1995).

11. A. P. Alivisatos, K. Johnsson, X. Peng, T. Wilson, C. Loweth, M. Brachez, and P. Schultz, Nature **382,** 609 (1996).

12. C. Mirkin, R. Letsinger, R. Mucic, and J. Storhoff, Nature **382,** 607 (1996).

13. R. Dagnani, "Supramolecular Assemblies: DNA used to organize gold nanoparticles," Chem. and Engr. News, August 19, 1996, pp. 6-7.

14. J. L. Coffer and R. R. Chandler, Mater. Res. Soc. Symp. Proc. **206,** 257 (1991).

15. R. R. Chandler, J. L. Coffer, C. D. Gutsche, I. Alam, H. Yang and R. F. Pinizzotto, Mater. Res. Soc. Symp. Proc. **272,** 265 (1992).

16. J. L. Coffer, S. R. Bigham, R. F. Pinizzotto and H. Yang, Nanotechnology **3,** 69 (1992).

17. R. R. Sinden, "Cruciform Structures in DNA," Chapter 4 in *DNA Structure and Function,* (Academic Press, San Diego, CA) pp. 134-178 (1994).

18. S. R. Bigham and J. L. Coffer, J. Phys. Chem. **96,** 10581 (1992).

19. R. R. Chandler, S. R. Bigham, and J. L. Coffer, J. Chem. Ed. **70,** A7 (1993).

20. Y. N. Chang, I. L. Pirtle, and R. M. Pirtle, Gene **48,** 165 (1986).

21. R. R. Sinden, *DNA Structure and Function,* (Academic Press, San Diego, CA) pp. 21-31 (1994).

22. "Preparation of nucleic acids for electron microscopy," Chapter 1 in *Electron Microscopy in Molecular Biology,* J. Sommerville and U. Scheer, eds. (IRL Press, Oxford) pp. 1-11 (1987).

23. Video-based Image Analysis System, MCID M4, Imaging Research, Inc. (St. Catherines, Ontario, Canada, 1992).

QUANTUM DOTS AS INORGANIC DNA-BINDING PROTEINS

CATHERINE J. MURPHY*, ERIC B. BRAUNS AND LATHA GEARHEART

Department of Chemistry and Biochemistry, University of South Carolina, Columbia, SC 29208
murphy@psc.sc.edu

ABSTRACT

Semiconductor quantum dots of cadmium sulfide, CdS, are approximately the size of proteins and are photoluminescent in the red, yellow, or green, depending on surface preparation. This photoluminescence is very sensitive to the nature and amount of adsorbates. We have found that DNAs with intrinsic curvature adsorb more strongly to the surface of 47 Å CdS quantum dots, as judged by concentration-dependent changes in photoluminescence. The binding constants we obtain are similar to those found for nonspecific protein-DNA interactions. The surface groups of the CdS substrate also influence DNA adsorption. Thus these protein-sized colloidal particles can be used in chemical sensing applications for curved or kinked DNA; DNA with unusual structures is thought to influence biological function such as transcription.

INTRODUCTION

One of the driving forces for research in the area of quantum dots is the hope that such materials will provide faster, smaller, and fundamentally new kinds of optoelectronic devices. A much less explored area is the use of these quantum dots as chemical sensors. We have chosen to use the photoluminescence of nanoparticles of CdS, the prototypical quantum dot, as the signal for a most unusual kind of chemical sensing: intrinsic DNA structure that is deviant from the canonical "straight" double helix. DNA that is (allegedly) intrinsically bent has direct consequences in terms of the DNA's ability to bind to proteins and hence influence gene expression [1].

The size range that makes semiconductor colloids "quantum dots" is also the size range of proteins. From the biochemical literature, it appears that certain sequences of DNA are intrinsically curved. We hypothesize that DNA that is more curved, or more likely to be curved, will adsorb stronger and/or faster to appropriately curved surfaces. We have indeed demonstrated that the photoluminescence of Cd^{2+}-rich 40 Å CdS particles and 40 Å $HOCH_2CH_2S$-capped CdS particles can distinguish between "straight," "bent," and "kinked" DNAs [2]. We use these concentration-dependent changes in photoluminescence to fit to an adsorption isotherm model to extract relative binding constants (see below). We also can perform time-resolved experiments and estimate the on-rate for different DNAs binding to the different inorganic surfaces (see below). In both sets of experiments we find that the relative binding constants and association constants are nearly identical to those observed for nonspecific protein-DNA interactions [3]. Furthermore, we find that the relative $\Delta\Delta G$ of binding (the more curved DNA binds most strongly to the curved nanoparticle surface, agreeing with our hypothesis) are very similar to the $\Delta\Delta G$'s observed for different sequences of DNA wrapping about proteins in the first step of chromosome formation [4].

In the present study we have examined aspects of our system that are more reminiscent of a biological system; we have made $H_2NCH_2CH_2S$-capped CdS particles as more "protein-like" surfaces, and have adsorbed DNAs of different sequences.

EXPERIMENTAL

Materials. Anhydrous Na_2S (Alfa), NaOH (Mallinckrodt), sodium polyphosphate (Aldrich) and $Cd(NO_3)_2{\cdot}4H_2O$ (Baker) were used as received. All reagents for buffers and electrophoresis were

Mat. Res. Soc. Symp. Proc. Vol. 452

of the highest purity available. Reverse-phase chromatographic resin (Vydac C18) was obtained from Aldrich. Oligonucleotides were synthesized by standard phosphoramidite chemistry in the USC Institute for Biological Research and Technology's Oligonucleotide Synthesis Facility: 5'–GTGTGTGTGTGTGTGTGTGTGTG–3' and complement as a "straight" duplex; 5'–AAAAAGGGCCCAAAAAAGGGCCC–3' and complement as a "bent" duplex. All oligonucleotides were purified by high pressure liquid chromatography (HPLC) on a Beckman System Gold HPLC instrument with a reverse-phase column and a triethylammonium acetate/acetonitrile gradient. Deionized and purified water (Continental Water Systems) was used in all experiments.

Instrumentation. Electronic absorption spectra were collected with a Perkin Elmer Lambda 14 UV-visible spectrophotometer. Steady-state luminescence spectra were acquired with a SLM-Aminco 8100 spectrofluorometer, with excitation at 350 nm and 4 nm resolution. Transmission electron microscopy (TEM) was performed on a Hitachi H-8000 electron microscope; samples were prepared by placing a drop of the solution onto a nitrocellulose-copper grid and drying overnight at room temperature.

Synthesis of 2-mercaptoethylamine-capped CdS. CdS particles were prepared by a modification of a literature preparation [5], with reagent weights based on a final concentration of $2x10^{-4}$ M. $Cd(NO_3)_2{\cdot}4H_2O$ (39 µL of a 0.0798 g/mL solution in water) and $(NaPO_3)_6$ (72 µL of a 0.0851g/mL solution in water) were added to 50 mL H_2O degassed with N_2, and the pH was adjusted to 8.3 with 1.0 M NaOH. Na_2S (30µL of a 0.027g/mL aqueous solution) was then added with vigorous stirring. 2-mercapto-ethylamine (2.6 µL) was then added and stirring was continued for 15 min. The pH was adjusted to 10.5, yielding a colorless solution which glowed yellow under UV light. Excess Cd^{2+} (50 µL of a 0.0798 g/mL solution) was added and the pH was again adjusted to 10.5 Particle sizing was done by TEM. Calculation of particle size from the UV-vis absorption spectrum according to the method of Brus [6] agreed well with the results from TEM (47 $\pm$ 10 Å diameter).

Luminescence Titrations. In a typical procedure, 5 µL aliquots of approximately millimolar (nucleotide) DNA solutions (5 mM phosphate buffer, pH 7.2) were added every 30 minutes to 200 µL of a 2 x 10^{-4} M colloidal CdS solution. The luminescence intensity was integrated over the wavelength range of 480-800 nm and corrected for buffer effects. The photoluminescence of the CdS solution alone was not found to change significantly over the time course of this experiment.

Kinetics of Binding by Luminescence. The amount of DNA needed to quench all of the CdS emission, as determined in the titration experiment above, was added to 200 µL of the colloidal CdS solution all at once. The emission intensity at 556 nm was recorded every 10 sec over a period of 1000 sec. Control experiments were run by adding the appropriate volume of buffer without DNA.

RESULTS

It has been previously shown that nucleic acids quench the the luminescence of quantum confined nanocrystalline semiconductors [7]. We have shown that this luminescence change is sensitive to sequence dependent DNA structure [2]. In our model, we predict that DNAs that are intrinsically "bent" will bind with higher affinity to a curved surface (the quantum dot) than would "straight" DNAs. Indeed, we find this to be the case using quantum dots having both charged and neutral surfaces [2].

In this work we have developed CdS quantum dots having an amine-rich surface, which presents a more protein-like surface to the DNA. The result of the luminescence titration of these quantum dots with "bent" DNA is shown in Figure 1. The quenching due to the straight sequence was not appreciably greater than buffer alone and is not shown here. By equating the fractional change in luminescence to fractional surface coverage [2], we have fit our data the the Frisch-Simha-Eirich adsorption isotherm for a long polymer adsorbing in short segments to a locally flat surface [8]:

$$[\theta \ \exp(2K_1\theta)] / (1-\theta) = (KC)^{1/\nu} \quad (1)$$

where θ is the fractional surface coverate (equated to fractional change in luminescence), C is the DNA concentration, K_1 is a constant that is a function of the interaction of adsorbed polymer segments and is set equal to 0.5 [2], K is the binding constant, and ν is the average number of segments attached to the surface, which has no physical meaning in our system. Using this model we find that the binding constant for "bent" DNA to an amine capped quantum dot surface is ~1100 M^{-1}. This is very similar to the 1800 M^{-1} we observed for mercaptoethanol-capped CdS quantum dots of the same size [2]. Especially noteworthy is the fact that the sequences of the DNAs used here for "straight" and "bent" are not the same as our previous studies; thus the current results reinforce the notion that DNA *shape* is the primary factor influencing adsorption to a generic curved surface.

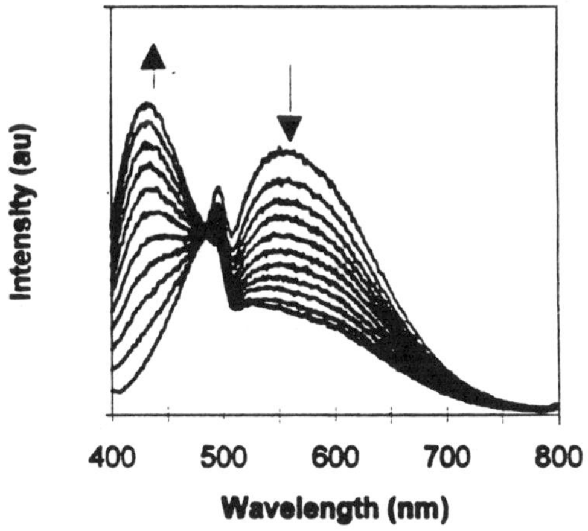

Figure 1. Luminescent titration of mercaptoethylamine capped CdS quantum dots with 10 mM (nucleotide) "bent" DNA.

Assuming a simple, second order association reaction of quantum dot with DNA (R + O ----> RO) we are able to follow the kinetics of binding by monitoring the fractional change in luminescence with time after the addition of DNA. The integrated rate equation is:

$$\left[\frac{1}{(R)-(O)}\right] \ln \left[\frac{(O)\,[(R)-(RO)]}{(R)\,[(O)-(RO)]}\right] = k_a t \quad (2)$$

where R is the concentration of free quantum dot, O is the concentration of free DNA, and RO is the concentration of DNA-quantum dot complex. This model is truly only appropriate for the kinetics of protein-DNA binding for very long DNAs [9]. We find that the second order rate constant is ~10^4 $M^{-1}s^{-1}$ which is in agreement with our previously reported results and also consistent with true nonspecific protein-DNA interactions [2].

Perhaps most intriguing about this particular system is the the new luminescence band at 425 nm that grows in as the DNA concentration increases (figure 1). We have not observed this in any of our previous experiments. It is not present in the control titration (buffer alone); however, it is present in the titration with straight DNA (data not shown) but to a much lesser extent. At the time of this writing, the origin of this new band is not clear. It may be the result of DNA damage from a photogenerated electron or hole, yielding an emissive product. It is also possible that a new DNA-colloid adduct (compared to our previous studies) has been formed with the amine surface, resulting in near band-edge luminescence. The appearance of a higher-energy photoluminescence band here is reminiscent of colloidal CdS "activation" by ions as seen by others [10].

CONCLUSIONS

We have shown that the luminescence of quantum confined CdS colloids is sensitive to sequence-dependent DNA structure. We have been able to discriminate between straight and bent DNA using CdS quantum dots capped with amines, a more biologically relevant surface. Our functional assay provides a basis for studying nonspecific DNA-protein interactions and may provide an effective way to screen for "bent" DNA.

ACKNOWLEDGMENTS

We thank the National Science Foundation (CAREER Award to C.J.M.) and the Research Corporation (C.J.M. is a Cottrell Scholar) for funding.

REFERENCES

1. For example, see (a) M. R. Gartenberg and D. M. Crothers, *J. Mol. Biol.* **219**, p. 217; (b) J. Perez-Martin and M. Espinosa, *J. Mol. Biol.* **241,** p.7 (1994).

2. (a)R. Mahtab, J. P. Rogers, J. P., and C. J. Murphy, *J. Am. Chem. Soc.* **117,** p. 9099 (1995). (b) R. Mahtab, J. P. Rogers, C. P. Singleton, and C. J. Murphy, *J. Am. Chem. Soc.* **118**, p. 7028 (1996).

3. (a) R. S. Spolar and M. T. Record, Jr., *Science* **263**, p. 777 (1994); (b) J. J. Ramsden and J. Dreier, *Biochemistry* **35**, p. 3746 (1996).

4. T. E. Shrader and D. M. Crothers, *Proc. Natl. Acad. Sci. USA* **86**, p. 7418 (1989).

5. Y. Nosaka, N. Ohata, T. Fukuyama, and N. Fuji, *J. Colloid Interface Science.* **155**, p. 23 (1993).

6. L. E. Brus, *J. Chem. Phys.* **80**, p. 4403 (1984).

7. S. R. Bigham and J. L. Coffer, *J. Phys. Chem.* **96**, p. 10581 (1992).

8. R. Simha, H. L. Frisch, and F. R. Eirich, *J. Phys. Chem.* **57**, p. 584 (1953).

9. R. B. Winter, O. G. Berg, and P. H. von Hippel, *Biochemistry* **20**, p. 6961 (1981).

10. L. Spanhel, M. Haase, H. Weller, and A. Henglein, *J. Am. Chem. Soc.* **109**, p. 5649 (1987).

STUDIES OF PHOTOREDOX REACTIONS ON NANOSIZE SEMICONDUCTORS

Jess P. Wilcoxon, Sandia National Laboratories, Org 1152, Albuquerque, NM; F. Parsapour, D.F. Kelley, Colorado State U., Dept. of Chemistry, Fort Collins, Co.

ABSTRACT

Light induced electron transfer (ET) from nanosize semiconductors of MoS_2 to organic electron acceptors such as 2,2'-bipyridine (bpy) and methyl substituted 4,4',5,5'-tetramethyl-2,2'-bipyridine (tmb) was studied by static and time resolved photoluminescence spectroscopy. The kinetics of ET were varied by changing the nanocluster size (the band gap), the electron acceptor, and the polarity of the solvent. MoS_2 is an especially interesting semiconductor material as it is an indirect semiconductor in bulk form, and has a layered covalent bonding arrangement which is highly resistant to photocorrosion.

INTRODUCTION

MoS_2 is an especially interesting semiconductor material as it is an indirect semiconductor in bulk form and has a layered covalent bonding arrangement, which resists photocorrosion. Its most common industrial applications include thermal catalysis to remove sulfur compounds from crude oil and as an excellent high temperature lubricant (e.g. axle grease). In fact, in bulk form MoS_2 in the most widely used hydrotreating catalyst and thus is vital to all chemical processing for fuels and feedstocks.

However, because MoS_2 is an indirect gap, black, IR absorbing material, it has no application as a photocatalyst. We have demonstrated that when synthesized in nanosize form, the absorption edge of MoS_2 can be significantly blue shifted.[1] This increase in the band gap energy, due to quantum confinement, is accompanied by significant shifts in both conduction and valence band energies, and implies that nanosize MoS_2 is capable of light induced electron and hole transfer (photoredox) reactions, just as are well known direct gap materials such as CdS. Experimentally, there is evidence that as MoS_2 is made smaller, the excitation becomes more like a direct transition, presumably due to the increasing importance of surface and lack of long-range translational symmetry in the lattice.

In this paper we report on static and dynamic photoluminesence measurements of nanosize MoS_2 and show that the observed increases in the bandgap energy with decreasing nanocluster size, are accompanied by increases in the conductance band potential which permit electron transfer electron to acceptor organic molecules. The rate of electron transfer is found to be dependent on both nanocluster size and solvent polarity. The fastest rates are observed in solvents of high polarity.

EXPERIMENT

Synthesis, Processing, and Physical Characterization

Nanosize MoS_2 is prepared by dissolving an anhydrous MoX_4 salt (X=Cl, Br, or I) in a water and air-free inverse micelle solution, (A typical inverse micelle solution would consist of 5-10% by weight of a quaternary ammonium surfactant in an aliphatic hydrocarbon such as octane).[2] This precursor solution is then exposed to a source of sulfide, typically H_2S gas injected through a septum in a known amount (slightly greater than 2:1 S:Mo) while rapidly stirring the solution. A brightly colored, transparent solution is formed which is then purified by extraction into an oil-immisible phase solvent, typically acetonitrile (ACN). Alternatively, we have developed high pressure liquid chromatographic (HPLC) procedures to both separate the nanoclusters from the surfactants and ionic byproducts in the solution and optically characterize

Mat. Res. Soc. Symp. Proc. Vol. 452

the purified nanoclusters on-line.[3] Gas-Chromatography/Mass Spectrometry is used to ascertain that the organic byproducts have been removed from the reaction mixture.

The purified MoS_2 clusters have been investigated using HRTEM and shown to be nanocrystalline. The larger nanoclusters have been shown to have the bulk hexagonal lattice structure by electron diffraction, but the smallest clusters (diameter, D<3 nm) have too few atoms to give unambiguous structural information based upon diffraction. However, lattice fringe images are consistent with d spacings found in the bulk material. Evidence indicates that the clusters are not spherical in shape, but are disc-like. The sandwich-like structure of MoS_2 consists of successive S-Mo-S tri-layers, with a weak Van-der-Waals interaction between layers. This graphite-like structure is responsible for the good lubricating properties of bulk MoS_2, since these planes can readily slide past one another.

Dynamic light scattering is complementary to measurements of cluster cross-sectional size by TEM since it measures the translational diffusion of the cluster in the acetonitrile solution, and via Stokes law provides an equivalent sphere hydrodynamic diameter. This diameter is comparable to the cross-sectional area for nanoclusters with D~3.0 nm, but is smaller for the D~4.5 nm clusters, implying the thickness of the latter clusters is somewhat less than the cross-sectional TEM measurement.

MoS_2 Optical Properties and Electronic Structure

Bulk MoS_2 is an indirect band-gap semiconductor, with a band gap of 1.23 eV.[4] As in the case of other indirect materials such as Si and Ge, no room temperature photoluminesence can be observed. Studies of thin crystalline films of MoS_2 reveal that the first direct absorbance in MoS_2 occurs at 1.88 eV, corresponding to ~660 nm. In bulk form, the top of the valence band is composed primarily of Mo d_{z^2} orbitals, and the bottom of the conduction band is composed of Mo $D_{x^2-y^2}$ and d_{xy} orbitals. As a result, excitation to create hole-electron pairs is metal intraband in nature and no Mo-S bonds are weakened, giving this material strong resistance to photocorrosion.[5] In contrast, II-IV semiconductors such as CdS have valence band orbitals consisting mainly of S 3p states, so that direct excitation across the gap to conduction band Cd 5s states leads to weakening of the chemical bonds holding the material together and accounts for the observed significant photocorrosion in these semiconductors.

As MoS_2 is made smaller, the momentum selection rules due to the long-range translational symmetry of the lattice are relaxed. In effect, the material becomes more like a direct band-gap material. As shown in figures 1 and 2 this results in increased room temperature photoluminesence and larger optical extinction in the visible absorbance. (The extinction coefficients can be obtained from these figures by multiplying by 1000 since the [MoS_2] ~ 1 x 10^{-3}M, and are in the 10^4 to 10^5 cm^{-1}-mol^{-1} range)

A significant portion of the observed luminesence occurs to the red of the absorbance band-edge indicating that recombination is occurring primarily from sub-band-gap trapped states at the surface of the nanocluster. Addition of an electron acceptor such as bipyridine, bpy to the MoS_2 nanocluster will decrease the lifetime of these deep trap states by funneling some of the electrons created by direct excitation to the electron acceptor bpy, from which recombination with the hole on the MoS_2 nanocluster will also occur. This allows us to follow the electron transfer kinetic indirectly. A more direct alternative approach to following the ET kinetics involves measurement of the change in the bpy absorbance due to ET as a function of time. We are currently pursuing such studies.

An obvious result of making smaller MoS_2 nanoparticles is the blue shift of the absorbance edge shown in figures 1 and 2, a corresponding blue shift of the emission, and an increased quantum efficiency for luminesence.

We can roughly estimate the conduction and valence band energies of our nanosize MoS_2 clusters by using the known potentials of the bulk material of +0.1 V and +1.33 V (vs. NHE) respectively. If we attribute most the band-gap shift observed in figure 1 and 2 to changes in the conduction band potential (because of the lower electron compared to hole mass) then we estimate a value of -.66 V and -1.46 V vs. NHE for the D=4.5 nm and D=3.0 nm samples.[6]

By photoexciting at the direct point we produce conduction band electrons with about 0.25 eV higher energies (more negative potential) than the above values. So nanoclusters will have a significant driving force for electron transfer compared to the bulk material.

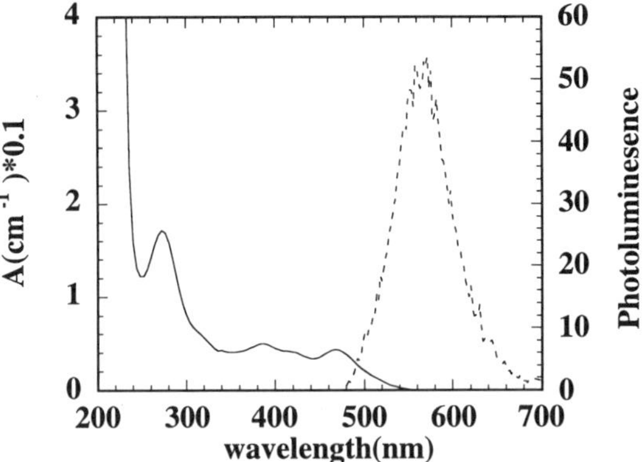

Figure 1. Absorbance (A, solid curve), and photoluminesence (excitation at 348 nm, dashed curve) from D=4.5 nm MoS_2 nanoclusters.

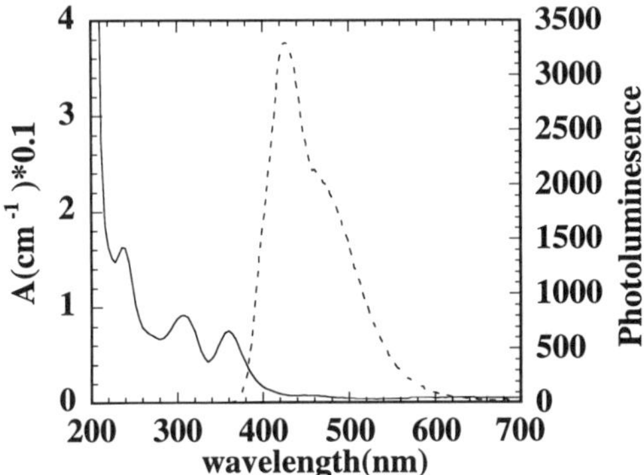

Figure 2. Absorbance (A, solid curve), and photoluminesence (excitation=348 nm, dashed curve) from D=3.0 nm MoS_2 nanoclusters.

Static PL Quantum Efficiency

Table I summarizes the results of PL integrated area, PL(area), normalized by the cluster absorbance (abs) at the excitation wavelength indicated for one size, D=3.0 nm of nanocluster. Very similar trends were observed for other sizes of cluster. Also shown in this table are the results for Coumarin 500 dye (nearly 100% Q.E. for PL) under identical instrumental conditions. We note the very significant decrease in radiative recombination (PL) as the solvent polarity decreases from ethylene glycol to o-xylene. This effect is greatest for excitation at the first direct absorbance (~340 nm, see figure 2) compared to excitation at the band edge (~400 nm). Furthermore, we observe an interesting *decrease* in PL by 5-fold when bpy is bound to the nanocluster and excitation is at 340 nm. Band edge excitation, on the other hand, actually *increases* the yield by ~15% when bpy is bound. Additionally, changing the solvent can alter the overall wavelength dependence of the PL for excitation at the first absorbance feature. (There is little effect for excitation at the band edge) Basically, solvating the clusters in very non-polar solvents such as toluene increases the amount of light emitted from trapped surface states, effectively red-shifting the PL peak by ~20 nm compared to polar solvents such as ethylene glycol.

Table I. Effect of Solvent Polarity on PL Quantum Efficiency for MoS_2 nanoclusters

Size(nm)	solvent	λ(excit)	<λ(emis)>	PL(area)/abs
Coumarin 500	methanol	350	497	1.5×10^{10}
3.0	Ethylene Glycol	350	465	1.0×10^{7}
3.0	DMF	340	481	5.8×10^{4}
3.0	acn	340	456	2.2×10^{5}
3.0	acn/bpy	340	461	4.1×10^{4}
3.0	hexanol	340	462	5.9×10^{5}
3.0	octanol	340	460	4.2×10^{5}
3.0	toluene	340	482	2.1×10^{4}
3.0	o-xylene	340	483	8.4×10^{3}
Coumarin 500	methanol	400	496	2.1×10^{10}
3.0	Ethylene Glycol	400	485	1.9×10^{7}
3.0	dmf	400	485	2.5×10^{6}
3.0	acn	400	484	1.3×10^{7}
3.0	acn/bpy	400	480	1.5×10^{7}
3.0	hexanol	400	478	7.0×10^{6}
3.0	octanol	400	479	6.1×10^{6}
3.0	toluene	400	486	8.2×10^{5}
3.0	o-xylene	400	481	2.1×10^{5}

ET Studies

The layer-like structure of both bulk and nanosize MoS_2 has important implications for its use as a photocatalyst. The basal planes of sulfur atoms are relatively inert and bifunctional electron donating ligands such as bipyridine (bpy) (used in the present study) will bind to the Mo at edge sites of the disc-like MoS_2 nanoclusters. Such binding can be demonstrated to be quite strong as a MoS_2/bpy complex elutes as a single entity during HPLC.[6] It has been further demonstrated that the molar ratio of Mo to bpy at full edge site occupation is about 2:1, and so this was the chosen concentration for bpy, and tetramethyl substituted bpy (tmb) used in our studies.

The actual reduction potentials of bpy and tmb are quite sensitive to the type of metal to which they are bound and the chemical environment (solvent polarity etc.). Never-the-less, the difference of potential of 0.29 V between the two acceptors (tmb, is more difficult to reduce), remains quite constant. So, by adding these electron acceptors to the two different sized MoS_2 nanocluster samples and observing the changes in the PL decay curves shown in figure 3 and 4, one can estimate this potential. All PL relaxations were obtained with 315 nm excitation while detection was at 420 nm for the D=3.0 nm sample and 560 nm for the D=4.5 nm sample. The decay times do depend on the detected wavelength somewhat. Addition of tmb to the MoS_2, D=4.5 nm sample results in a very small change in the PL decay curve of figure 4., and we conclude that the potentials of tmb and bpy for reduction in acetonitrile are ~-0.7 V and -.41 V vs. NHE respectively. (The decay curve of bound tmb overlaps the bare nanocluster curve of figure 4 and so is not shown).

Since the decay curves obtained are very non-exponential and represent a very wide range of decay rates, one can obtain excellent fits to these data by using a stretched exponential form, $Ae^{-(\Gamma t)^{\beta}}$ from which the decay rate Γ can be extracted. A complicating factor is that the stretched exponential exponent β changes along with Γ when bpy is bound to the cluster, as well as when different solvents are used. This implies that the various mechanisms for radiative recombination for a given size cluster are influenced by the details of the cluster interface. Complete details for such energy calculations and the underlying assumptions are given elsewhere [6].

From analysis of the blue shifts of the absorbance spectra and our ET kinetics for nanosize MoS_2, we have estimated the conduction band potential of the D=3.0 and D=4.5 nm nanoclusters to be -1.71 V and -0.91 V, respectively, in acetonitrile.[6] If we recall the estimated potential in the bulk material is +0.1 V, we can see that quantum size effects have made these nanoclusters into very strong reducing agents.

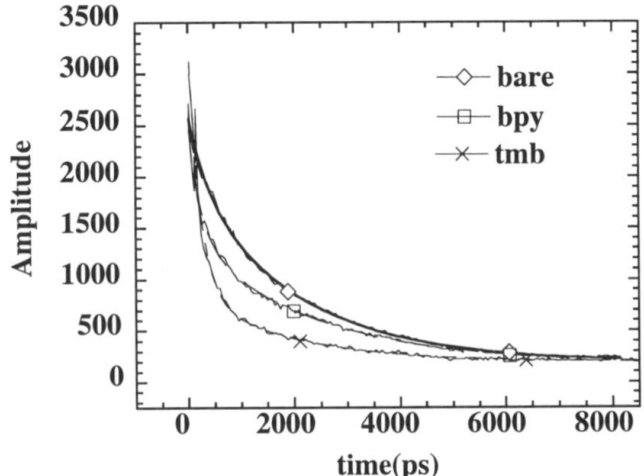

Figure 3. The decay of the photoluminesence of D=3.0 nm MoS_2 nanoclusters for various electron acceptors added.

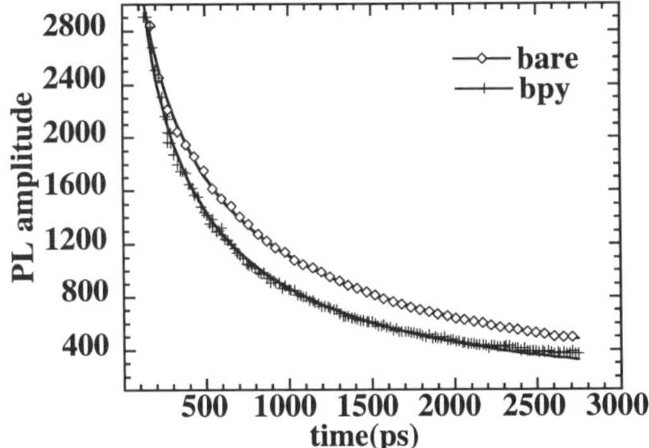

Figure 4. The decay of the photoluminesence of bare, D=4.5 nm MoS_2 nanoclusters and MoS_2 with bpy.

Size Dependence

Some of the results of our studies on ET kinetics are summarized in figure 5. We first note that for a constant solvent polarity (e.g. in acetonitrile, acn), the smallest clusters with the largest conduction band potentials are capable of driving the electron transfer to both bpy and tmb at faster rates. In fact, the larger D=4.5 nm clusters cannot effectively transfer electrons to tmb, the more difficult-to-reduce an electron acceptor. We have not shown the D=4.5 nm, tmb ET rates in figure 5 since they are so slow.

Solvent Dependence

Examination of the data of figure 5 shows that reduction of bpy to bpy radical anion is so facile for the smallest MoS_2 nanoclusters that solvent polarity has a very minor effect on ET kinetics. For the more difficult to reduce substrate, tmb, the smaller D=3.0 nm clusters show significantly slower ET once the mole fraction of the non-polar solvent, benzene is greater than 0.5. In the case of the D=4.5 nm clusters reducing the solvent polarity slows down the ET dramatically, since lower polarity solvents are less able to stabilize the charge separated state. In fact, effectively no ET to bpy occurs for solvent mixtures of greater than 20 mole% benzene.

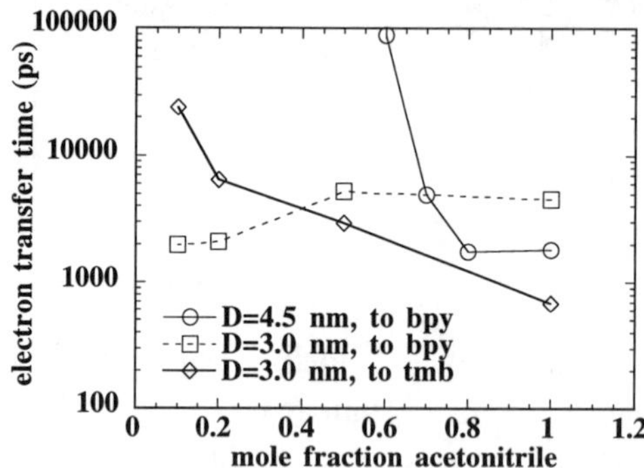

Figure 5. Effect of nanocluster size and solvent in the ET decay times (acetonitrile/benzene mixture).

CONCLUSIONS

We have shown how quantum confinement in an indirect semiconductor material can shift the conduction band potential and allow facile electron transfer to two types of substrate. Smaller clusters were demonstrated to have improved ET rates. It was also demonstrated that maximizing the solvent polarity increases the ET rate by stabilizing the charge separated state.

ACKNOWLEDGMENT

This work was supported by the U.S. dept. of Energy under contract DE-AC04-94AL85000. Sandia is a multiprogram lab operated by Sandia Corporation, a Lockheed-Martin Company, for the U.S. Dept. of Energy.

REFERENCES

[1] J.P. Wilcoxon and G.A. Samara, Phys . **51**, R7299, (1995).
[2] J.P. Wilcoxon, G. Samara, and P. Newcomer, in Nanocrystalline Semiconductors, Ed. P. Jena, (Mat Res Soc Proc, Boston, MA., 1994), pp. 277-81.
[3] J.P. Wilcoxon, P.P. Newcomer, and G.A. Samara, J. Appl. Phys., submitted, (1996).
[4] B.L. Evans in Optical and Electrical Properties (of Materials with Layered Structures), ed P.A. Lee, D. Reidel Publ, Boston, 1976, Chapt. 1, pg 1.
[5] H. Tributsch, Z. Naturforsch **32a**, 972, (1977).
[6] F. Parsapour, D.F. Kelley, S. Craft, and J.P. Wilcoxon, J. Chem. Phys., **104**, 1, (1996).

TRANSIENT PHOTOCURRENT RESPONSE OF DYE-SENSITIZED POROUS NANOCRYSTALLINE TiO_2 ELECTRODES

ALBERT GOOSSENS, G.K. BOSCHLOO, AND J. SCHOONMAN
Delft University of Technology, Laboratory for Applied Inorganic Chemistry,
Julianalaan 136, 2628 BL Delft, The Netherlands. a.goossens@stm.tudelft.nl

ABSTRACT

In order to investigate the fundamentals of electron migration in nanostructured metal-oxide semiconductors, the transient photocurrent response of dye-sensitized porous nanocrystalline TiO_2 is studied. The time-resolved photocurrent response at light steps or pulses shows a faster transient upon increasing the light intensity. Intensity-modulated photocurrent spectroscopy (IMPS) reveals that the transient photocurrent is dominated by two time constants, i.e. the geometrical one and a characteristic time related to electron trapping. A theoretical model is derived in which the occupation dynamics of a single electron trap is considered using Shockley-Read-Hall kinetics. The geometrical RC time of the electrode is also included. Excellent agreement between this model and the measured IMPS spectra is obtained.

INTRODUCTION

Porous nanocrystalline metal-oxide electrodes are the basis of new types of efficient electrochemical devices, such as photoelectrochemical solar cells, photocatalysts, Li-ion batteries, and electrochromic windows. They are prepared by low-temperature sintering of colloidal particles, resulting in a 3-dimensional porous network with a large internal surface area. The small primary particle size (5-30 nm) and the high surface-to-volume ratio give these electrodes several advantageous properties compared to dense electrodes. Investigations on this type of nanostructured electrodes have been reviewed by several authors to which the interested reader is referred [1]. Nanostructured metal-oxide electrodes are particularly useful in dye-sensitized photoelectrochemical solar cells. In particular, Grätzel and co-workers [2] developed efficient solar cells based on sol-gel derived ('fractal') TiO_2 electrodes sensitized by Ru(II) dyes. At present, efficiencies close to 10% under AM 1.5 conditions can be reached.

In porous nanocrystalline electrodes, the injected electrons travel a long distance through the porous structure before they are collected at the back contact. On their way they remain close to the semiconductor / electrolyte interface and recombination with oxidized dye molecules or electron acceptors in the electrolyte occurs to some extend. Furthermore, the probability that the electrons are trapped is high, because there is a high concentration of traps at the surface and on grain boundaries. Trapping sites are usually sources for recombination of charge carriers, leading to reduced quantum efficiencies and lower open-circuit photovoltages.

Here, in particular, intensity-modulated photocurrent spectroscopy (IMPS) is employed to study the dynamics of irradiated dye-sensitized porous nanocrystalline TiO_2

Mat. Res. Soc. Symp. Proc. Vol. 452

electrodes. IMPS is a powerful technique to measure time constants of processes that occur at illuminated semiconductor electrodes. Details about this method can be found in [3] and the references therein. In the reviews by Peter the scope and limitations of this technique is discussed in detail [4]. A mechanistic model that accurately accounts for the observed IMPS response is derived and it is shown that the use of electrical equivalent circuits is beneficial to the interpretation of IMPS data. By fitting the circuit parameters to the measured IMPS dispersion, accurate values for the electron trap time constant as well as the geometrical *RC* time constant are obtained. It appears that these values strongly correlate to the presence of background light.

RESULTS

Photocurrent transients of a dye-sensitized porous nanocrystalline TiO_2 photoelectrode behave as follows. After the light is switched on, the photocurrent responds rather slowly, typically on a time scale of 10 - 100 ms. In contrast, in dense polycrystalline or single crystalline semiconductor electrodes the photocurrent response is usually of the order of μs or faster. The steady-state photocurrent is linearly dependent on the light intensity, which is the normal behavior for a solar cell. However, the light intensity also affects the overall response time of the photoelectrochemical cell. The effect of a continuous background illumination on the photocurrent growth is shown in Fig. 1. Here, a constant illumination bias is added to a small-intensity light step. Only the response to the perturbation is shown. Addition of bias light causes a much faster response of the solar cell, but leaves the steady-state photocurrent unchanged.

In Fig. 2 the response of the photocurrent to a broad band light pulse ($\lambda > 455$ nm) is shown. The photocurrent rises quickly to a maximum after 1 ms, followed by a slow decay. A 90% decrease in the light pulse intensity results in a much slower rise of the photocurrent, i.e. it takes about 20 ms before the maximum response is observed.

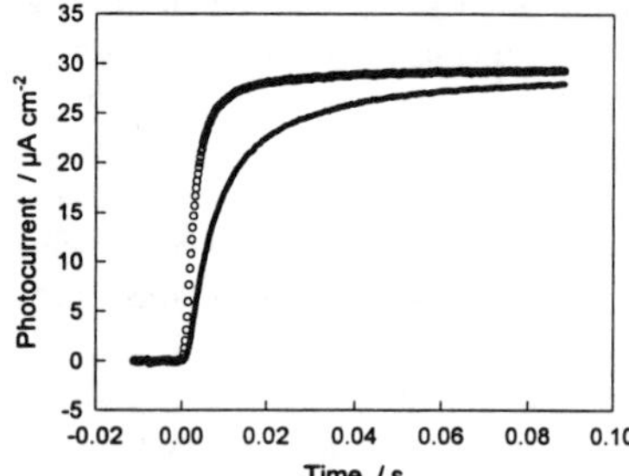

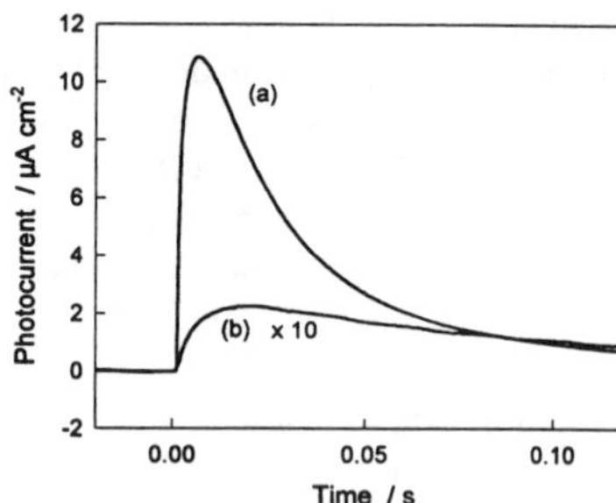

FIGURE 1 (left-hand side): Effect of a continuous white background irradiation on the photocurrent growth. Without bias light (solid line), and with bias light (open circles). The *dc* bias light gives rise to a photocurrent of 146 μA cm^{-2}, which is subtracted.

FIGURE 2 (right-hand side): Transient photocurrent response of a ZnTCPP sensitized porous nanocrystalline TiO_2 electrode to a white light flash ($\lambda > 455$ nm). The relative light intensity is 100% (a) and 10 % (b). The photocurrent of (b) has been multiplied by a factor of 10.

Intensity-modulated photocurrent spectroscopy is a relative new technique to study dynamic photocurrent responses in semiconductor electrochemistry. A small amplitude a.c. light intensity modulation is superimposed on a d.c. illumination background. The frequency response of the photocurrent is measured with a network analyzer. Both amplitude and phase of the modulated photocurrent are recorded and translated into an admittance. A complex plane representation of this admittance is commonly used.

The in-phase (real) and the out-of-phase (imaginary) components of the modulated photocurrent of a dye-sensitized porous nanocrystalline TiO_2 solar cell are presented in Fig. 3. A slightly depressed semi-circle is found, whose real component becomes negative at higher frequencies. A semi-circle with positive real and negative imaginary components is often found and can usually be ascribed to the geometrical *RC*-response of the semiconductor electrode as measured by the external circuitry. However, high-frequency IMPS response with both negative real and imaginary components is a new phenomenon that deserves further attention.

In order to separate the response of different processes, external resistors were connected in series with the solar cell. The shape of the IMPS plot remains essentially unaffected, but the characteristic frequencies change. Bias irradiation also affects the IMPS response; the shape of the IMPS semi-circle remains similar, although the time constants of the electrode processes have changed, which can be derived from the characteristic frequencies.

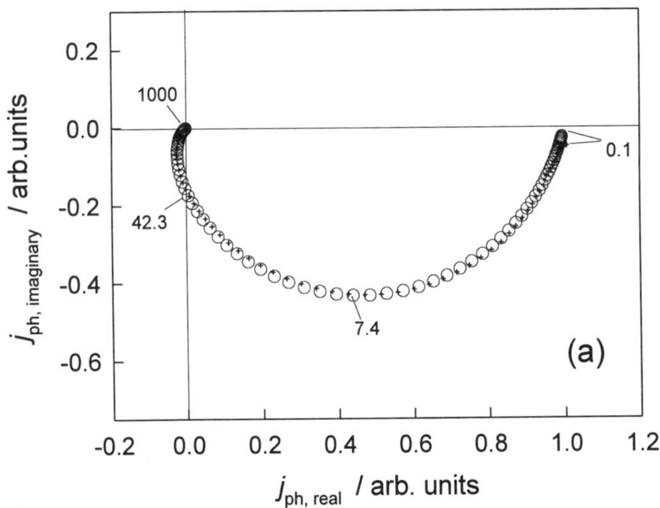

FIGURE 3: IMPS plots of a photoelectrochemical solar cell, based on ZnTCPP-sensitized porous nanostructured TiO_2. Experimental data, +, and fit results, O, are shown. The indicated frequencies are in Hz. A yellow LED (λ = 590 nm) provides a *dc* bias irradiation level of 1.5 W m^{-2} and a modulation level of 0.15 W m^{-2}.

DISCUSSION

The occurrence of slow photocurrent responses, as observed for our ZnTCPP-sensitized porous TiO_2 electrodes (Fig. 1), is typical for porous nanocrystalline photoelectrodes [5]. One reason for this behavior is the very large internal area of the porous electrodes which gives rise to a considerable interfacial capacitance. This capacitance is related to the total charge storing ability of the interface. Apart from the interfacial capacitance, the *RC*-time of the system is also governed by the resistances that are present. The sources of these resistances are manifold. The series resistance of the concentrated electrolyte and the electric leads are small, and can be neglected. Larger resistances are present in the porous TiO_2 film, the ITO back-contact, and the current measuring device. Finally, electron trapping also damps the photocurrent transients. A photo-injected electron can be trapped and released many times before it reaches the back contact, which slows down the photocurrent response. This response will be affected by the light intensity, since the time constant for trapping depends on the density of conduction band electrons.

Photodoping and other non-linear effects obscure proper determination of the *RC* and the trapping-time constant of illuminated nanostructured electrodes. Since IMPS stays within the realm of linear-response theory, it is much more powerful than transient studies where high-intensity excitation is employed. In our IMPS experiments the intensity of the modulated light is much smaller than that of the continuous background light, ensuring a linear photocurrent response. The IMPS response can be modeled accurately using Shockley-Read-Hall kinetics. Photocurrent generation is governed by an exponential law:

$$j(t) = G(1 - e^{-at})$$

in which G is the generation rate, and

$$a = k_n(n + n_1) = 1/\tau_{trap}$$

k_n being the capture constant, n the density of conduction band electrons, and n_1 a constant related to thermal emission of trapped electrons into the conduction band.

With IMPS, the generated photocurrent is convoluted with the RC time constant of the measuring network. Taking this into account an equivalent circuit can be derived given by a series connection of two RC parallel networks. The model predicts that the two capacitances involved are equal in absolute value, but opposite in sign and also the two resistances have an opposite sign. It should be noted that translation of IMPS spectra in equivalent circuits is markedly different from previously reported IMPS studies on similar porous electrodes [6].

The relative differences between the experimental data and the fit are within 5% for every applied frequency. Hence, the model accurately agrees with the found IMPS spectra. Moreover, it is found that one $(RC)_p$ circuit has positive elements while the other has negative values, in accordance with our model. An even stronger support is the fact that, apart from their sign, both capacitances (CPE elements in our fit) are equal within 10%. Since these two elements are individually optimized, this should be regarded as convincing evidence of the validity of the model.

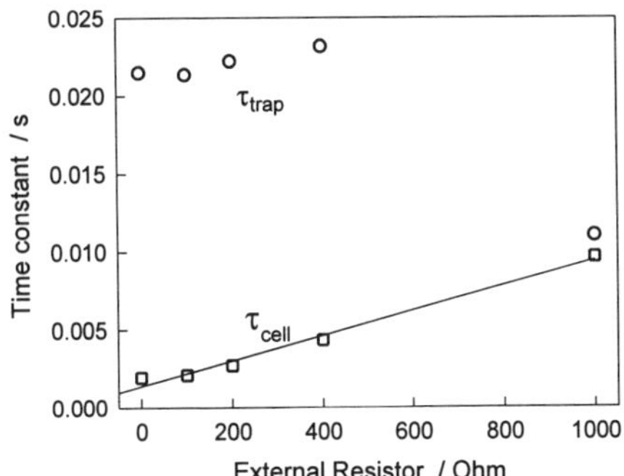

FIGURE 4 Effect of externally connected series resistors on the time constants. The circles correspond to the trapping time constant (τ_{trap}), the squares to the geometrical *RC* time constant of the cell (τ_{cell}). The drawn line is the linear regression fit to the τ_{cell} data. The time constants are calculated from the output of the non-linear least squares fit procedure using $\tau = RQ^{1/\alpha}$. The geometrical surface area of the electrode is 0.48 cm^2.

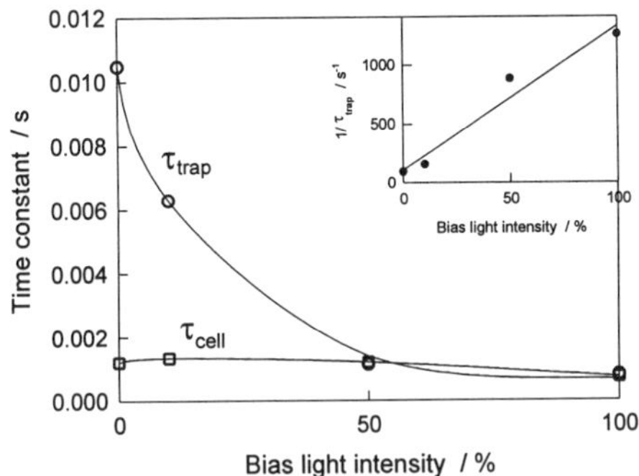

FIGURE 5 Time constants obtained from IMPS data; τ_{trap} circles, τ_{cell} squares. A bias light intensity of 100% corresponds to 25.5 Wm^{-2} of white light ($\lambda > 400$ nm). The inset shows the dependence of $1/\tau_{trap}$ on the bias light intensity.

Assignment of the two time constants using a single IMPS spectrum is impossible since both terms can be interchanged. Addition of external series resistors, however, leads to a systematic increase of the smaller of the two time constants (see Fig. 4). Hence, this time constant is the *RC*-time of the cell (τ_{cell}). The intercept with the *R*-axis of the linear regression fit of τ_{cell} shown in Fig. 4 yields an internal resistance *R* of 200 Ω. This resistance is mainly caused by the ITO support, which loses some conductivity during the heat treatment in the electrode preparation. The larger time constant remains almost unaffected when the externally connected series resistor is small (see Fig. 4). This time constant is related to the electron trapping kinetics (τ_{trap}). The decrease of τ_{trap} when a 1000 Ω external resistor is applied, is probably due to a negative shift of the Fermi level upon irradiation. This leads to trap filling, which induces a decrease of the trapping time constant.

The effect of the bias light intensity on the time constants obtained by IMPS is shown in Fig. 5. Electron trapping clearly depends on the presence of background light. With increasing light intensity, the number of electrons in the conduction band increases, resulting in a smaller time constant for trapping. In the inset of Fig. 5 the relation between $(\tau_{trap})^{-1}$ and the light intensity is shown. An almost linear relationship is found, which reveals that the electron density in the conduction band changes linearly with bias light intensity. This behavior indicates the absence of electron-hole recombination because in that case a sub-linear relationship between $(\tau_{trap})^{-1}$ and the light intensity is expected. The intensity of the bias light has little effect on τ_{cell}, although there is a small decrease of the geometrical *RC* time constant at higher intensities. This can be explained by photoconductivity effects. However, since most of the resistance is caused by the ITO support, the change in total internal resistance is small.

CONCLUSIONS

Intensity-modulated photocurrent spectroscopy (IMPS) is an excellent tool in the study of photocurrent dynamics of porous nanocrystalline electrodes. The IMPS response of dye-sensitized porous nanoscrystalline TiO_2 is determined by the geometrical *RC* time of the cell and trapping of conduction-band electrons. A single trap state in the forbidden zone, which follows Shockley-Read-Hall kinetics, in combination with the geometrical *RC* time of the cell is sufficient to describe the observed IMPS response.

ACKNOWLEDGEMENT

The authors would like to thank Novem (Netherlands Agency for Energy and the Environment) for financial support of these investigations. In addition, A.G. acknowledges the Royal Netherlands Academy of Arts and Sciences (KNAW) for his fellowship.

REFERENCES

[1] P.V. Kamat, Chem. Rev. **93** (1993) 267; M.A. Fox and M.T. Dulay, Chem. Rev. **93** (1993) 341; A. Hagfeldt and M. Grätzel, Chem. Rev. **95** (1995) 49; A.L. Linsebigler, G.Lu, and J.T. Yates Jr., Chem. Rev. **95** (1995) 735

[2] B. O'Regan and M. Grätzel, Nature **353** (1991) 737

[3] A. Goossens and D. D. Macdonald, Electrochim. Acta **38** (1993) 1965; A. Goossens and D. D. Macdonald, J. Electroanal. Chem. **352** (1993) 65; A. Goossens, M Vazquez, and D.D. Macdonald, Electrochim. Acta **41** (1996) 35 and 47, A. Goossens, Surface Science **365** (1996) 662

[4] L.M. Peter, Chem. Rev. **90** (1990) 753; E.A. Ponomarev and L.M. Peter, J. Electroanal. Chem. **396** (1995) 219

[5] K. Schwarzburg and F. Willig, Appl. Phys. Lett. **58** (1991) 2520

[6] F. Cao, G. Oskam, G.J. Meyer, and P.C. Searson, J. Phys. Chem. **100** (1996) 17021; P.E. de Jongh and D. Vanmaekelbergh, Phys. Rev. Lett. **77** (1996) 3427

ELECTRON TIME-OF-FLIGHT MEASUREMENTS IN POROUS SILICON

PRASANNA RAO,* E. A. SCHIFF,* L. TSYBESKOV,† AND P. M. FAUCHET†
*Dept. of Physics, Syracuse University, Syracuse, NY 13244-1130
†Department of Electrical Engineering, University of Rochester, Rochester NY 14627

ABSTRACT

Transient photocurrent measurements are reported in an electroluminescent porous silicon diode. Electron drift mobilities are obtained from the data as a function of temperature. Electron transport is dispersive, with a typical dispersion parameter $\alpha \approx 0.5$. The range of mobilities is $10^{-5} - 10^{-4}$ cm^2/Vs between 225 K and 400 K. This temperature-dependence is much less than expected for multiple-trapping models for dispersion, and suggests that a fractal structure causes the dispersion and the small mobilities.

INTRODUCTION

Canham's discovery that porous, crystalline silicon photoluminesces efficently near room-temperature [1] has led to enormous interest not only in the origins of this unexpected effect, but also in the possibility of preparing electroluminescent diodes from the same material. Substantial progress in this direction has been made by several groups, based on a variety of device designs [cf. 2,3,4 for early examples].

Despite these device successes, practically nothing is known conclusively about the fundamental electrical properties of porous silicons; adding to the complexity of establishing these properties is the fact that porous silicons can be made with an enormous range of structures and passivation properties, presumably with widely varying transport. In this paper, we are concerned primarily with carrier mobilities, which determine carrier drift speeds inside electrical devices and can thus dominate the speed of the device itself. The carrier mobilities of crystalline silicon, from which the porous silicon is prepared, are at best a starting point. Porous silicon's extremely fine and disordered topology would be expected to drastically reduce the drift velocity of a carrier, and the experiments published to date [5,6,7] are consistent with a qualitative change *vis a vis* the properties of the crystal. Indeed, one may speculate that it is precisely this drastic reduction in the diffusion of carriers which permits them to recombine radiatively in the porous structure.

In the present paper, we report our progress in obtaining carrier mobilities using the photocarrier time-of-flight method. In its simplest implementation, the method involves launching a sheet of photocarriers in a sample using a short laser flash. The carriers then drift under the influence of an external bias field E, and the transit time t_T required for the carriers to drift a distance L across the specimen is determined by measuring the transient current in the bias circuitry. The mobility is the ratio L/Et_T. Ref. [8] may be consulted for additional information about procedures as implemented by our laboratories.

In brief, we have observed transient photocurrents in highly porous silicon which we associate with carrier sweepout. We find mobilities in the range 10^{-5} - 10^{-4} cm^2/Vs. These are comparable to those obtained previously using electroluminescence modulation [7], although the structures involved are quite different. The *dispersion properties* of the mobilities (the dependence of mobility on the displacement of the carriers) in the two experiments are very different. Perhaps most importantly, we have measured the temperature-dependence of the drift

Mat. Res. Soc. Symp. Proc. Vol. 452

mobility. The temperature-dependence is surprisingly small, especially in comparison to other dispersive transport systems such as hydrogenated amorphous silicon (a-Si:H). This may be evidence for limitation of the mobility by true spatial disorder (fractal structure), as opposed to the temporal disorder (trapping effects) which is often involved in amorphous semiconductors.

EXPERIMENTAL DESCRIPTION

Samples

The sample for which we report measurements here was a light-emitting, porous silicon (LEPSi) diode prepared at the University of Rochester using a procedure which has yielded 0.1% efficient, stable light-emitting diodes [9]. Starting with a *p*-type Si substrate (resistivity 5-10 Ωcm), a thin layer (about 0.5 μm thick) is ion-implantated with boron. This structure is then anodized to yield a two-layer porous structure, with a low porosity layer in the ion-implanted region near the top, and a buried, high porosity layer about 0.7 μm thick. Following an oxidation step to passivate the porous layers, the structure is finished by deposition of a polycrystalline, n^+ layer (0.3 μm thick). Contact is made to this *pin* diode using evaporated Al electrodes (area about 0.1 cm^2). Bottom contact is made to the substrate. We did not etch back the layers surrounding the top Al dot, so the active area is not exactly defined.

Electrical Characterization

The transient photocurrents which we report here are measured in the electrical bias circuit following a brief laser flash. We used a nitrogen-laser pumped dye laser operating at 671 nm; the nominal pulse length for this laser is 3 ns. The actual time response of the system is

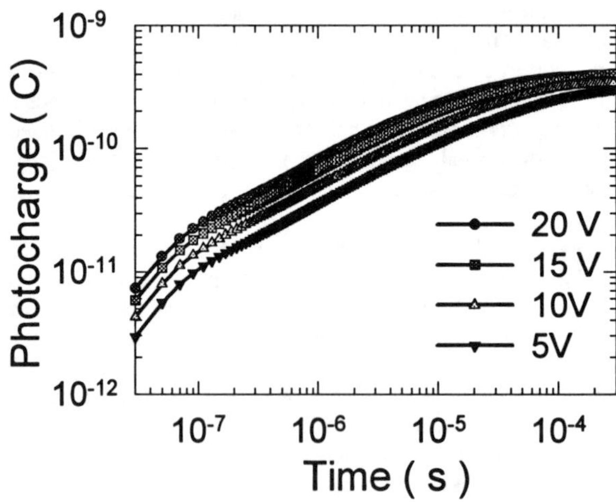

Fig. 1: Transient photocharge measured in a porous silicon *pin* diode following a 3 ns laser flash (671 nm) at 260 K; data for four bias voltages are shown. The measurements show several features associated with photocarrier drift and sweepout.

determined by the capacitance of the porous silicon diode in conjunction with the 50 Ω impedance of the external circuitry; these effects limited the time response to about 50 ns at best.

One important difficulty which we encountered was signficant reverse bias currents (ca 2 mA at -20 V bias and 300 K). This reverse bias current plays an important role in interpreting transient photocurrent measurements designed to yield mobilities through time-of-flight effects. Specifically, mobility measurements require a reasonable uniformity for the electric field inside the structure. In the conventional time-of-flight procedure, this uniformity is guaranteed by applying the reverse bias as a voltage pulse which precedes the arrival of the laser flash. The field then remains essentially uniform until charge depletion exceeds the charge CV corresponding to the geometric capacitance of the diode.

A conservative approach is to monitor the leakage current through the structure, and only to use photocurrent transient information for times t at which the total charge through the bias circuit remains about equal to CV. A short calculation reveals that a 1 mA reverse bias current and a 1 nF diode allows only 1 μs of measuring time – far too short a time to be useful. In the present measurements, we used steady-state reverse bias voltages. This creates an amibiguity regarding field uniformity which we hope to resolve subsequently; however, the measurements we report here otherwise show the features expected for photocarrier time-of-flight, and we consider it probable that a drift mobility interpretation is valid.

Transient photocurrent measurements

In Fig. 1 we illustrate the photocharge transients following the laser flash. We measured the transient current following the laser flash, subtracted the dark current, and then numerically

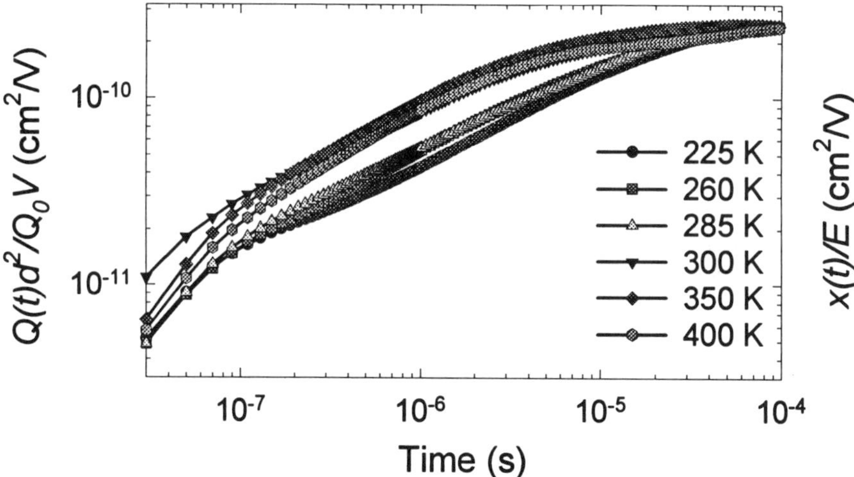

Fig. 2: Normalized transient photocharge measurements $Q(t)d^2/Q_0V$ in the porous silicon specimen for several temperatures (-20 V, 671 nm laser wavelength); normalized photocharge may be interpreted as the displacement/field ratio $x(t)/E$ for the electrons. The saturation at longer times is due to carrier sweepout at $x = d$. The times at which these curves cross $x(t)/E = 10^{-10}$ cm^2/V was used to determine the drift mobility of Fig. 3.

integrated the net current to prepare this figure. The figure shows several features we associate with sweepout of photocarriers. At larger times ($t > 10^{-5}$ s), there is a fairly well defined photocharge (about 4×10^{-10} C) which is reproduced adequately for all four voltages. This photocharge may thus be interpreted as a measure of the total photogenerated charge.

The model we favor for the photogeneration mechanism is the following. The top three porous and polycrystalline layers should be nearly transparent to the laser wavelength of 671 nm, so it is probable that photogeneration occurs mostly in the *p*-type Si substrate. Electron photocarriers generated within their diffusion length will diffuse to the porous Si/crystal Si interface. They then enter the high-field region, where their further motion is detected in the external bias circuit.

The effect of the external voltage upon the photocharge transients is also consistent with the time-of-flight interpretation. The photocharge transients for earlier times ($t < 10^{-5}$ s) are proportional to the Voltage, indicating Ohmic transport. The nominal sweepout time (the time at which charge collection reaches 50% of its ultimate value) decreases with increasing field, as anticipated for the time-of-flight interpretation.

As will be discussed subsequently, the photocharge $Q(t)$ is proportional to the mean displacement $x(t)$ of the electrons in a uniform field. Assuming this proportionality, it should be noted in Fig. 1 that $x(t)$ rises essentially as the square root of the delay time following absorption of the laser flash. This sublinearity is the phenomenon of *dispersion*, which is well known in several amorphous semiconductor systems. The dispersion parameter α corresponding to fitting $x(t)$ to the power law t^{α} is about 0.5.

DRIFT MOBILITY MEASUREMENTS

In Fig. 2 we display the transient photocharge measurements at -20 V for several temperatures. For this figure we take advantage of the time-of-flight interpretation; we have plotted $Q(t)d^2/Q_0V$, using d=700 nm (the thickness of the high porosity layer) and the total photocharge collection Q_0 (0.4 nC in Fig. 1). $Q(t)d^2/Q_0V$ may be equated to the ratio $x(t)/E$ of the electron displacement $x(t)$ to the drift field E in the high porosity region. This equation originates in an electrostatics theorem based on the fact that motion of charge between two electrodes held at constant potential induces a compensating change in the charge on the electrodes (consult ref. [8] for additional details and references). The normalization causes all transients to approach the value d^2/V for longer times, since $x(t)$ approaches d for longer times.

As is evident from the figure, reducing the temperature generally decreases the carrier drift. At 400 K, the reverse bias current had increased greatly, making subtraction of its effects less successful; we believe this effect accounts for the slight decrease in $x(t)/E$ at this temperature.

In Fig. 3 we have graphed the temperature dependence of the drift mobility μ_D corresponding to a ratio $L/E = 10^{-10}$ of the drift length L and the field E. μ_D is calculated using the *transit time* t_T, which is defined from the relation $x(t_T)/E = 10^{-10}$ cm^2/V. t_T was obtained graphically from Fig. 2. The drift mobility is defined in a conventional way as:

$$\mu_D \equiv \frac{L/E}{t_T} \quad . \tag{1}$$

It is important to note that, for dispersive transport, μ_D depends strongly upon L/E; it must not be viewed as a materials constant.

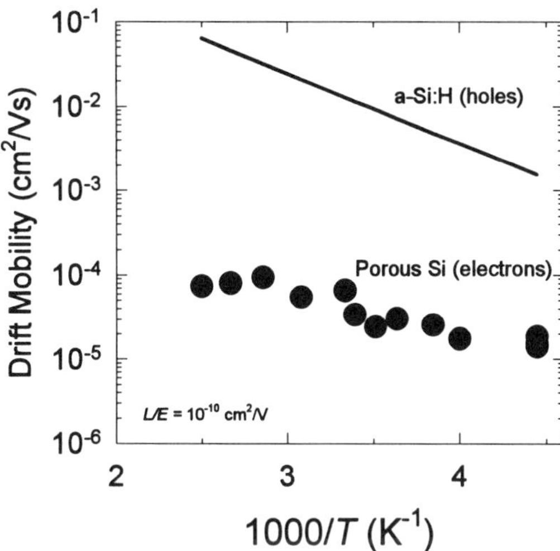

Fig. 3: Temperature-dependent mobilities in porous silicon and hydrogenated amorphous silicon (a-Si:H); both correspond to a displacement field ratio $L/E = 10^{-10}$ cm^2/V. The a-Si:H values are based on published multiple-trapping fittings consistent with measurements from several laboratories [10].

The main point of Fig. 3 is that the temperature-dependence of μ_D for porous silicon, while certainly discernible, is relatively weak compared to that of holes in a-Si:H. We chose holes in a-Si:H for comparison to porous silicon because it has similar dispersion near room-temperature ($\alpha \approx 0.5$). The line we have drawn is calculated using the standard, "exponential bandtail multiple-trapping" fitting to hole time-of-flight measurements:

$$x(t) / E = (\mu_0 / \nu)(\nu t)^{kT/E_0} \qquad (2)$$

We used the parameter values published by Gu, *et al*[10] (μ_0 = 0.27 cm^2/Vs, $\nu = 7.7\times10^{10}$ s^{-1}, E_0 = 0.048 eV), which account fairly well for results from several laboratories.

DISCUSSION

The temperature-dependences shown in Fig. 3 suggest that differing mechanisms are involved in limiting the mobilities in a-Si:H and porous Si; this conclusion is also supported by other aspects of the temperature-dependence. In particular, the dispersion parameter α is proportional to the absolute temperature for a-Si:H; this proportionality is usually considered to be strong evidence for multiple-trapping behavior. The dispersion parameters for the measurements in Fig. 2 show much weaker dispersion.

The multiple-trapping model fitted to a-Si:H is often assumed to involve an exponential distribution of bandtail states lying below a mobility edge; the latter separates localized bandtail traps from extended band transport states. However, at least two other models yield drift which can be modeled using eq. (2): hopping in bandtails [11], and long-range potential fluctuations [12]. Generally speaking, models which account for dispersion using *temporal disorder* (ie. an enormous range of trapping and release times) tend to imply large temperature-dependences, since emission from traps is a strongly temperature-dependent process.

The alternative to this model is to invoke the pure geometry of porous silicon for its mobility limitation and its dispersion. This is a property of diffusion on a *fractal* structure [13]. A complete review of this model is beyond the scope of the present paper, but we note that dispersion due to drift on a fractal structure reflects the underlying structural dimensions of the fractal (the "fractal" and "fracton" dimensions). Ben-Chorin, *et al* [6] have previously used fractal models to interpret AC conductivity measurements in porous silicon. As this type of dispersion has a structural origin, it has no intrinsic temperature-dependence, and thus appears to be a more satisfactory starting point than multiple-trapping for explaining the present measurements. These measurements reflect the properties of only a single type of porous silicon, and we hope that future work will establish mobilities for a much wider range of materials.

The authors thank Dr. Qing Gu for his help in the early stages of this research. Research at the University of Rochester was supported by the Army Research Office (ARO).

REFERENCES

[1] L. T. Canham, *Appl. Phys. Lett.* **57**, 1046 (1990).

[2] N. Koshida and H. Koyama, *Nanotechnology* **3**, 192 (1992).

[3] P. Steiner, F. Kozlowski, and W. Lang, *Appl. Phys. Lett.* **62**, 2700 (1993).

[4] F. Namavar, H. P. Maruska, and N. M. Kalkhoran, *Appl. Phys. Lett.* **60**, 2514 (1992).

[5] A. Fejfar, I. Pelant, E. Sipek, J. Kocka, G. Juska, T. Matsumoto, and Y. Kanemitsu, *Appl. Phys. Lett.* **66**, 1098 (1995).

[6] M. Ben-Chorin, F. Moller, F. Koch, W. Schirmahcer, and M. Eberhard, *Phys. Rev. B***5**, 2199 (1995).

[7] C. Peng and P. M. Fauchet, *Appl. Phys. Lett.* **67**, 2515 (1995).

[8] Q. Wang, H. Antoniadis, E. A. Schiff, and S. Guha, *Phys. Rev. B* **47**, 9435 (1993).

[9] L. Tsybeskov, S. P. Duttagupta, K. D. Hirschman, and P. M. Fauchet, *Appl. Phys. Lett.* **68**, 2058 (1996).

[10] Q. Gu, Q. Wang, E. A. Schiff, Y.-M. Li, and C. T. Malone, *J. Appl. Phys.* **76**, 2310 (1994).

[11] D. Monroe, *Phys. Rev. Lett.* **54**, 146 (1985); M. Grünewald and P. Thomas, *Phys. Status Solidi B* **94**, 125 (1979).

[12] H. Overhof, in *Amorphous Silicon Technology—1992,* edited by M. J. Thompson, *et al* (Materials Research Society, Symposia Proceedings Vol. 258, Pittsburgh, 1992), pp. 681-692.

[13] Y. Gefen, A. Aharony, and S. Alexander, *Phys. Rev. Lett.* **50**, 77 (1983); see also T. Nakayama, K. Yakubo, and R. Orbach, *Rev. Mod. Phys.* **66**, 381 (1994).

D.C. CURRENT-VOLTAGE CHARACTERISTICS AND ADMITTANCE SPECTROSCOPY OF AN Al-POROUS Si BARRIER

K. KHIROUNI*, J.C. BOURGOIN**, K. BORGI*, H. MAAREF*, D. DERESMES*** and D. STIÉVENARD***

*Laboratoire des Semiconducteurs, Faculté des Sciences de Monastir, Route de Kairouan, 5000 Monastir, Tunisia.

**Groupe de Physique des Solides, Universités Paris 6 et Paris 7, Tour 23, 2 place Jussieu, 75251 Paris Cedex 05, France

***I.E.M.N., Département I.S.E.N., C.N.R.S. (UMR 9929), Avenue Poincaré, Cité Scientifique, BP 69, 59652 Villeneuve d'Ascq Cedex, France.

ABSTRACT

We present the temperature and frequency dependence of the current-voltage (I-V) characteristics of Al barriers deposited on porous Si grown on p-type Si substrates. These barriers exhibit a rectifying behaviour when the temperature is higher than 250 K. The I-V characteristics can be understood by a conduction in porous Si taking place *via* free electrons thermally excited from P_b centers associated with the existence of a SiO_2–Si interface and *via* hopping between these P_b centers.

INTRODUCTION

The incorporation of porous Si (PSi) in an active device requires an understanding of the electrical characteristics of the structure to control its optical properties. Basically, the structure can be a metal-PSi barrier where PSi is grown on crystalline Si. The d.c. current-voltage (I-V) characteristics of such a structure have been already studied in the case of Al/PSi grown on p-type Si [1-5] and of Al and Au/PSi grown on n-type Si [6]. These studies have indicated the existence of 1) a rectifying Schottky-type barrier at the metal/PSi interface [6] with pinning of the Fermi level at an energy similar to that of the (0/–) transition of the dangling bond at the Si-SiO_2 interface, the so-called P_b center [7]; 2) a surface conduction controlled by tunneling [4]; and 3) characterized by a Poole-Frenkel effect, attributed to disorder [1].

The aim of this communication is to analyze in detail some of the above aspects by looking at the temperature dependence of the I-V characteristics. We shall also verify the existence of a tunneling regime by performing an admittance spectroscopy and correlate this regime with the nature and the density of traps at the Si-SiO_2 interface.

Mat. Res. Soc. Symp. Proc. Vol. 452 © 1997 Materials Research Society

EXPERIMENTAL

The studied structures are Al/SiP on p-type, <100>, 10^{-2} Ωcm Si. The porous layers are 9 μm thick with a porosity of 50 %, obtained by electrochemical dissolution in an HF (40 %)–ethanol (1:1) electrolyte under a current of 20 mA cm^{-2}. Aluminum is evaporated on top (area 8×10^{-3} cm^2) after the SiP surface has been cleaned by an HF treatment. The ohmic contact on the back side of the Si substrate is made prior to the growth of SiP by p^+ doping.

The measurements have been performed in the temperature range 250 – 400 K. The a.c. admittance, its real and imaginary parts (respectively the conductance and susceptance) has been measured in the range 5 Hz to 13 MHz.

RESULTS AND DISCUSSION

D.c. measurements

Typical I-V characteristics and their variation with temperature are given in Figure 1. It shows that the structure is characterized by a barrier when the temperature is high enough. The barrier is in series with a large resistance R_S. This resistance increases when the temperature decreases, which results in an ohmic characteristic below typically 250 K. Under reverse bias, the conductance varies exponentially with the inverse of the temperature T^{-1} (Fig. 2). The associated activation energy E_a is about 0.4 eV and increases slightly with the applied bias V (Fig. 3). In the forward bias regime the conductance is also characterized by an activation energy which then varies strongly with the voltage ($V - R_S I$). This regime has been analyzed in detail in ref. [3].

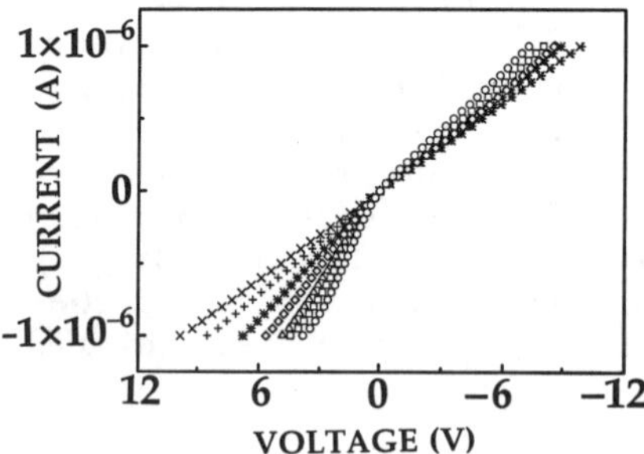

Fig. 1 : D.c. current-voltage characteristics monitored at 250 (×), 265 (+), 270 (✱), 275 (◊), 280 (Δ), 285 (□), 290 (○) K. The area of the structure is 8×10^{-3} cm^{-2}.

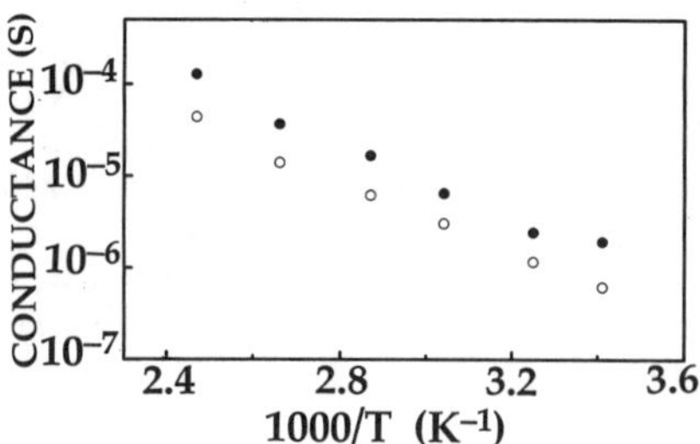

Fig. 2: Temperature dependence of the conductance measured for reverse biases of: 7 V (●) and 0 V (○).

Our results do not reproduce the preliminary study of ref. [1], in which it is reported that the current-voltage characteristics were ohmic at room temperature. Moreover, we do not detect any sign of a Poole-Frenkel effect. The variation of E_a, plotted versus $V^{1/2}$ in Figure 3 is not linear. On the contrary, E_a appears to increase slightly with V when it should decrease if a Poole-Frenkel effect was operative.

The value of the activation energy E_a is practically equal to the ionization energy of an electron from the (0/−) level of the dangling bonds (P_b centers) that are located at a $Si\text{-}SiO_2$ interface [7]. Hence, the conductance of the PSi layer appears to be limited by the concentration of electrons excited from P_b centers that are transported into Si. Consequently, there should be a direct relationship between the concentration of P_b centers and the intensity of the reverse current.

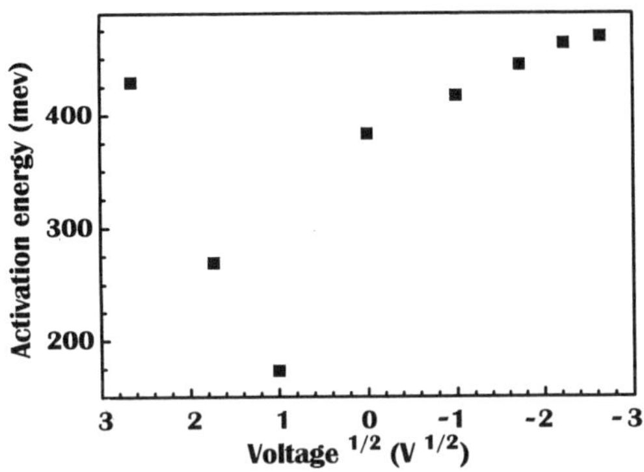

Fig. 3 : Activation energy associated with the plots of Fig. 2 versus the square root of the applied bias.

A.c. measurements

Admittance spectroscopy is a powerful technique to probe the density of states at an interface. Figures 4 and 5 show that the complex admittance of the Al/SiP structure is composed of a temperature dependent conductance which is independent of the frequency ω and of a susceptance which varies as ω^s with $s \sim 1$ and is

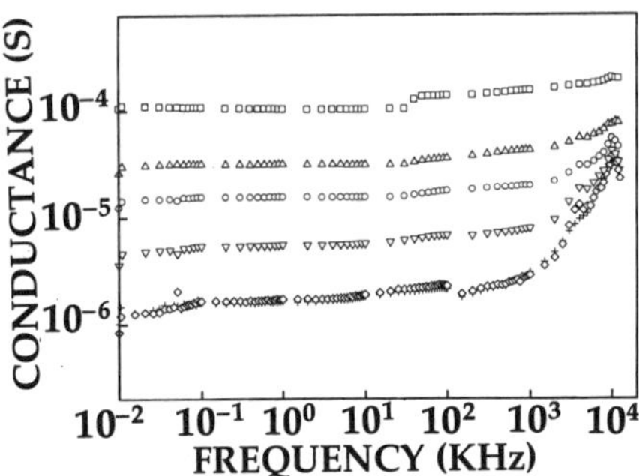

Fig. 4 : Conductance versus frequency under a reverse bias of 7 V at : 304 (◊ and +), 325 (▽), 345 (O), 370 (Δ) and 400 (□) K.

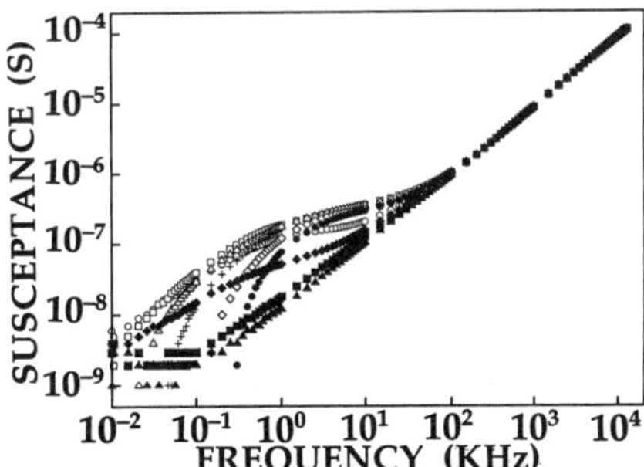

Fig. 5 : Susceptance versus frequency measured at 300 K under various reverses : 0 (O), 1 (◊), 2 (Δ), 3 (□), 4 (+), and forward 1 (●), 2 (■), 3 (▲) and 4 (◆) biases (V).

temperature independent. The frequency dependence of the susceptance is typical of a hopping process. This ω^s behaviour, which is commonly observed in insulators as well as in amorphous Si [8], is understood in terms of an electron tunneling back and forth between two adjacent localized sites having slightly different energies. When several assumptions are made (see in ref. [8] for a full treatment), the a.c. admittance is proportional to $\omega \ln(\omega_{ph}/\omega)$, where ω_{ph} is the phonon frequency, a term equivalent to ω^s with s slightly lower than 1. Thus, the admittance is composed not only of a frequency independent real part, which is thermally activated, and corresponds to band conduction induced by electrons ionized from P_b centers (see section III-1) but also of an imaginary part varying as ω^s. As a result, the admittance (modulus of the complex admittance) is constant up to a frequency ω_0 for which the susceptance becomes larger. This frequency ω_0 increases with temperature like the conductance (see Fig. 6), confirming the preliminary results of ref. [2].

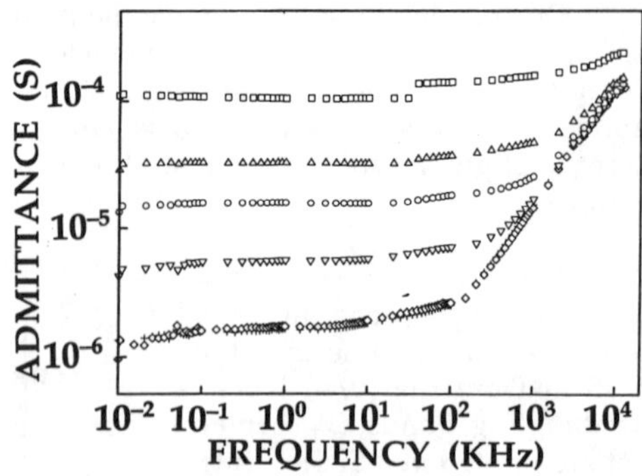

Fig. 6 : Admittance versus frequency for a reverse bias of 7 V measured at : 304 (◊ and +), 325 (▽), 345 (○), 370 (Δ) and 400 (□) K.

The hopping regime implies the existence of a large density of electronic states at the Fermi level. Since the Fermi level is pinned on the level of the P_b centers, it can be concluded that hopping occurs *via* these centers.

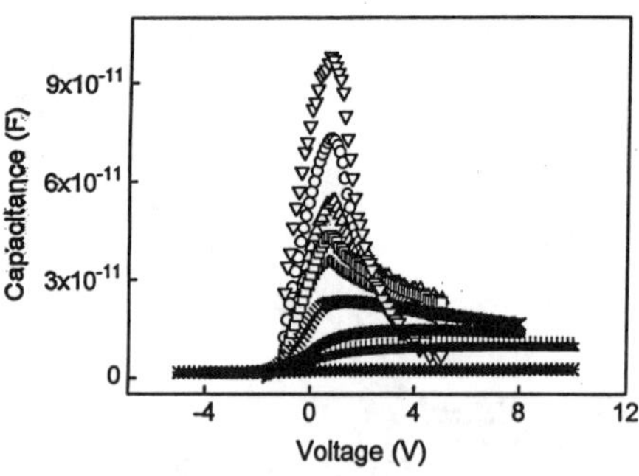

Fig. 7 : Capacitance voltage characteristics measured at 300 K for different frequencies : 0.1 (▽), 0.2 (○), 0.5 (Δ), 0.8 (□), 1 (▮), 2 (■), 5 (●), 8 (+), 10 (▲) and 50 (◆) kHz.

A density of states can be probed by capacitance or conductance measurements. Owing to the large value of the density of states involved here, the capacitance technique is best adapted. It can be shown (see for instance in ref. [9]) that the density of states is directly related to the shift of the capacitance (C) versus V between low and high frequencies. Fig. 7 demonstrates that the C-V characteristics is that of a MIS structure, with an onset of the low frequency curve around 1 KHz and corresponds to ~ 10^{11} cm^{-2} electronic charges, a result which is in agreement with observations by electron paramagnetic resonance [10] if these charges are ascribed to P_b centers.

CONCLUSION

We have demonstrated that admittance spectroscopy performed on an Al barrier deposited on porous Si is a powerful tool to understand the conductivity across such structure. The conductivity occurs *via* two channels : i) in a partially filled band of surface or interface states, by hopping; ii) in the Si conduction band by electrons thermally excited from the same states. These states appear to originate from P_b centers at a Si–SiO_2 interface. We have shown that their density, which is only estimated here, can be determined by admittance spectroscopy.

REFERENCES

1. M. Ben-Chorin, F. Möller and F. Koch, Phys. Rev. B **49**, 2981 (1994).
2. M. Ben-Chorin, F. Möller and F. Koch, J. Luminescence **57**, 159 (1993).
3. D. Deresmes, V. Marissael, D. Stiévenard and C. Ortega, Thin Sol. Film **255**, 258 (1995).
4. C. Cadet, D. Deresmes, D. Vuillaume and D. Stiévenard, Appl. Phys. Lett. **64**, 2827 (1994).
5. S. Lozarouk, P. Jaquiro, S. Katsouka, G. Masini, S. La Monica, G. Marello and A. Ferrari, Appl. Phys. Lett. **68**, 2108 (1996).
6. A.J. Simons, T.I. Cox, M.J. Uren and P.D.J. Calcott, Thin Sol. Film **255**, 12 (1995).
7. E.N. Pointdexter, Semicond. Sci. Technol. **4**, 961 (1989).
8. A.I. Yakimov, N.P. Stepina, A.V. Dvurechenskii and L.A. Scherbakova, Physica B **205**, 298 (1995).
9. S.M. Sze, *Physics of Semiconductor Devices* (J. Wiley, New York, 1981), Chap. **7**.
10. H.J. von Bardeleben, D. Stiévenard, A. Grosman, C. Ortega and J. Siejka, Phys. Rev. B **47**, 10899 (1993).

POROUS MICROCRYSTALLINE SILICON SOLAR CELLS

S. P. Duttagupta, S.K. Kurinec,* and P.M. Fauchet

Department of Electrical Engineering, University of Rochester, Rochester NY 14627

*Department of Microelectronic Engineering, R.I.T., Rochester, NY 14623

ABSTRACT

We report the fabrication of photovoltaic devices by the anodization of microcrystalline silicon films on single crystal silicon substrates. The porosity of the films was varied from 20% to 60% by changing the anodization conditions. An unetched μc-Si based device was used for reference. The influence of the porosity on the series resistance (R_S), the reflectance, and the spectral response of the devices was studied in detail. In order to determine R_s, the current-voltage characteristics were analyzed, both in the dark and under illumination. We observed that the value of R_s increased from 3.1 Ω to 97 Ω and the value of the reflectance decreased from 24% to 7% when the porosity increased from 20% to 60%. Initially, an optimum device performance (fill factor of 0.53 and efficiency of 7.2%) was achieved for a porosity of 40%, which was about a 40% improvement as compared to the reference (unetched) μc-Si based device. Due to a further reduction in R_s by using an intermediate ITO layer and a superior grid-contact architecture, a device efficiency of 10% has been recently achieved.

INTRODUCTION

Porous silicon (PSi) has been the subject of an intensive investigation following the report of efficient photoluminescence (PL) by Canham in 1990.[1] Since then it has been shown that porous polycrystalline (PPSi) and microcrystalline silicon (PMSi) films may also be formed, with very similar optical properties.[2] While the current research on porous silicon has focused on the fabrication of LEDs, the use of PSi in photovoltaic devices has been reported as early as 1982.[3] This involved the anodization of p^+/n c-Si solar cells to form a thin porous layer, which was oxidized at 500-600 °C. As a result, the average reflectance ($\Re$) was reduced from 37% to 8% in the 300 to 800 nm region, which increased the short circuit current density (J_{sc}) by 25%. Since then there have been several reports concerning PSi solar cells and photodetectors.[4-7] The potential advantages of light emitting porous silicon (LEPSi) based photovoltaic devices include: (a) Its highly textured surface increases light trapping and reduces loss due to reflection.[4] (b) The bandgap of PSi may be optimized for the absorption of sunlight, since the maximum theoretical solar cell efficiency vs. bandgap curve peaks at around 1.5 eV.[5] (c) The efficient photoluminescence of LEPSi may be used to down convert higher energy solar radiation (blue/UV) into longer wavelength light (red/IR) which is absorbed more efficiently into bulk silicon.[5] This might explain the improved spectral response of LEPSi/c-Si solar cells as compared to a c-Si device at incident photon energies greater than 2 eV.[6]

However, in spite of these advantages, only a few applications have been reported so far. Zheng et al.[4] have fabricated Al/LEPSi/Si photodetectors, and the lateral photovoltaic effect in PSi was utilized in fabricating a position sensor.[7] The efficiency of LEPSi solar cells remains poor (<3%),[6] as compared to c-Si devices. The main problems were that the LEPSi film is highly resistive, which results in a lower current density, and the effective bandgap may have been too large for efficient absorption in the long wavelength part of the solar spectrum. The PL down conversion effect is also not significant (~ 5%).[8] Finally, the efficiency may suffer from a large

Mat. Res. Soc. Symp. Proc. Vol. 452 © 1997 Materials Research Society

defect density which pins the Fermi level and reduces V_{oc}. Thus even for point contact cells (backside contacts for both polarities), anodization resulted in a loss of device performance.[5]

The situation is different for microcrystalline Si films (grain size < 100 nm), which are characterized by a high intrinsic defect density (~$10^{17}/cm^3$).[9] The degradation in the transport properties following anodization is accordingly expected to be less drastic here than in the c-Si case. Another unexplored aspect is the influence of porosity on the device performance. Since some optical properties improve with increasing porosity, while the electrical properties worsen, it is possible that an optimum porosity value exists for which the device performance is maximized.

EXPERIMENTAL

The μc-Si films were deposited on n-type (100) c-Si substrates (1 Ω-cm) at 580 °C using low pressure chemical vapor deposition. The films were doped in-situ with phosphorus (SiH_4/PH_3) and boron (SiH_4/B_2H_6). A 4 μm thick n-type (1 Ω-cm) film was deposited first, followed by a 1 μm thick p-type (0.1 Ω-cm) film. The p^+-n films were subsequently anodized (with the exception of a control sample), and the average porosity ($\langle P \rangle$) was determined gravimetrically. During anodization, the HF concentration was kept at 25% per weight and the applied current density (J) was 6 mA/cm^2 ($\langle P \rangle$=20%), 9 mA/cm^2 ($\langle P \rangle$=30%), 14 mA/cm^2 ($\langle P \rangle$=40%), 22 mA/cm^2 ($\langle P \rangle$=50%), and 33 mA/cm^2 ($\langle P \rangle$=60%). The film thickness was determined by ellipsometry to be approximately 3 μm. X-ray diffraction analysis showed that the as-deposited films had a (110) preferred orientation with a particle size of 25 nm. Following anodization, we observed the presence of 10 nm crystallites, which is consistent with our results obtained with porous films prepared from p^+ c-Si under the same anodization conditions. The final step was contact metallization. First, a 200 nm thick Al film was evaporated. Next, a grid pattern was generated using photolithography and an Al etch step ($H_3PO_4/HNO_3/CH_3COOH$, 40 °C, 2 min.). The samples were then sintered in forming gas (95% N_2, 5% H_2) at 425 °C for 15 min. in order to reduce the Al/PMSi interfacial resistance. The device area was 2 cm^2.

RESULTS AND DISCUSSION

From transmission measurements of a 40% porous free-standing porous microcrystalline (PMSi) film, the approximate penetration depth was calculated to be 2.8 μm at a wavelength of 620 nm. This indicates that a significant amount of the incident radiation is absorbed in the 3 μm porous layer. In addition, the highly absorbing 2 μm unetched μc-Si layer has a penetration depth of less than 0.1 μm at 514.5 nm.[10] As a result, very few photons reach the c-Si substrate.

The I-V characteristics of the devices were measured at room temperature, both in the dark and under AM1 illumination (P_{in}=100 mW/cm^2). The dark current-voltage relationship of a diode may be expressed as:[11]

$$I = I_o \exp [q/mkT (V-IR_s)] \quad (1)$$

where m is the ideality factor and R_s is the lumped series resistor element which includes the contribution of the Al/PMSi contact resistance, the inter-grid resistance (lateral) and the resistance of the PMSi film (vertical). The dynamic resistance of the diode is derived from (1):

$$r = dV/dI = mkT/qI + R_s \quad (2)$$

From Eqn. 2 we see that the influence of R_s increases at higher current levels resulting in a deviation of the voltage (ΔV) from the ideal behavior (i.e., R_s = 0). We have determined the series resistance by plotting ΔV with current in the high current regime ($\Delta V = IR_s$) in Fig. 2.[12]

The dark I-V response of the devices with an average film porosity of 20% (A), 40% (B), and 60% (C) is shown in Figure 1. At low injection levels (V< 0.2 V), the space charge current dominates.[13] The ideality factor m is 2, 2.08, and 2.1; and the reverse saturation current I_o (determined by extrapolation to zero voltage), is 1 nA, 0.5 nA, and 0.1 nA for devices A , B , and C, respectively. At V = 0.2V, the onset of the diffusion regime occurs for devices A and B and their respective ideality factors become m=1.1 and m=1.26, respectively. In contrast, the behavior of device C remains unchanged. For V > 0.3 V, we observe the influence of R_s on the three I-V characteristics. From Figure 2, the series resistance of device B (<P>=20%) is calculated to be 5.47 Ω from the slope of the ΔV-I plot. Similarly, R_s is determined to be 3.06 Ω (<P>=20%), 4.1 Ω (<P>=30%), 26 Ω (<P>=50%), and 97 Ω (<P>=60%). Such a highly nonlinear increase in series resistance with porosity is consistent with the reported R_s of 500 Ω for LEPSi (P~80%) devices.[11] The response for the control device was almost identical to that of device A, with a slightly smaller series resistance (2.8 Ω).

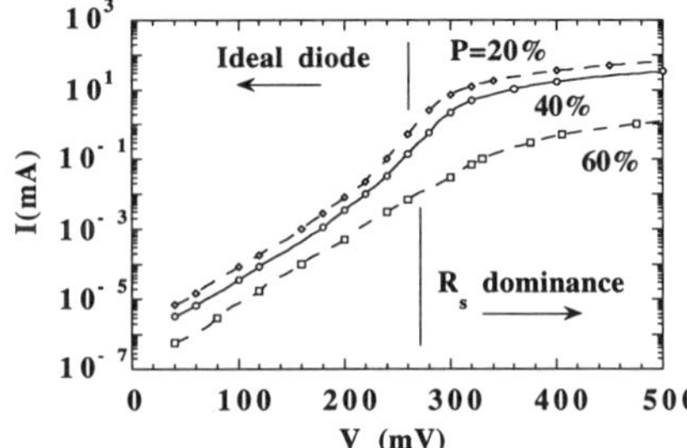

Fig. 1. The dark I-V response of photovoltaic devices based on p+/n porous µc-Si films with (A) 20%, (B) 40%, and (C) 60% porosity. The influence of the series resistance results in nonideal behavior for V>300 mV.

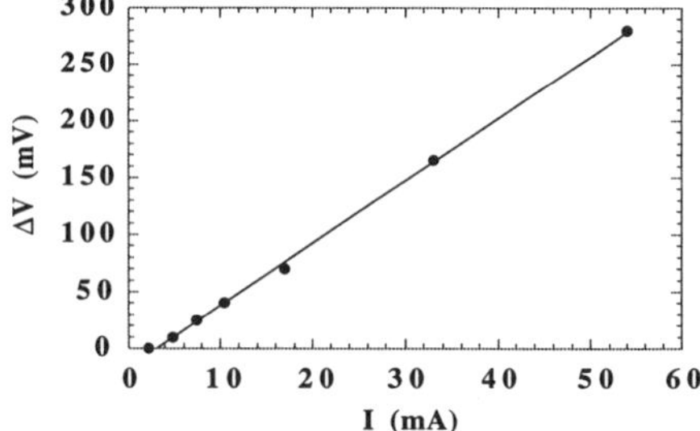

Fig.2. Plot of the deviation of the voltage from the ideal behavior (zero series resistance) vs. current. The slope of the curve gives series resistance. Ideal diode behavior (ΔV=0) is observed at low injection levels. The PMSi film is 40% porous.

Under illumination, an ideal solar cell is equivalent to a constant current generator in parallel to a p-n junction diode. Including the R_s element, the current-voltage relationship is as follows:[14]

$$I_L = I_R - I_o [\exp(q/mkT (V_L + I_L R_s))] \qquad (3)$$

where I_L is the load current, and I_R is the photocurrent (short-circuit current). Figure 3 shows the fourth quadrant response of the three devices under simulated AM1 illumination. The values of V_{oc} are 0.57 V, 0.55 V, and 0.5 V and those of I_{sc} are 40 mA, 50 mA, and 20 mA for devices A, B, and C, respectively. The voltage and current corresponding to the maximum power points were determined by applying constant power contours to Fig. 5. Their values are found to be{V_m, I_m} = {0.4 V, 28 mA} for device A; {0.37 V, 39 mA} for device B; and {0.2 V, 13 mA} for device C. Again, the I-V response of the control device was similar to device A, with a slightly smaller I_{sc} value (37 mA).

Using (3), R_s can be determined independently by the variable illumination intensity technique proposed by Wolf and Rauschenbach :[15]

$$R_s = (V_{L1} - V_{L2})/(I_{L2} - I_{L1}) \qquad (4)$$

where V_{L1}, V_{L2} (I_{L1}, I_{L2}) are the load voltage (current) under two different intensities, such that (I_{R1} - I_{L1}) = (I_{R2} - I_{L2}); i.e. the diode currents are the same in both cases. In Figure 4, we have plotted the I-V response of device B at 100% and 50% intensities. The difference in V_{oc} is negligible, while I_{sc} drops from 50 mA to 28 mA. To calculate R_s, we chose $V_{L1} = V_m$, and $I_{L1} = I_m$. The value of R_s is then found to be [(490-370)/(39-17)] = 5.4 Ω for device B. Similarly, R_s is determined to be 2.8 Ω (control device), 3.1 Ω (<P>=20%), and 4 Ω (<P>=30%). The agreement with the values of R_s determined earlier validates our analysis of the dark I-V response. The same technique could not be applied for high porosity film (P=50%, 60%) based devices, because of the high series resistance.

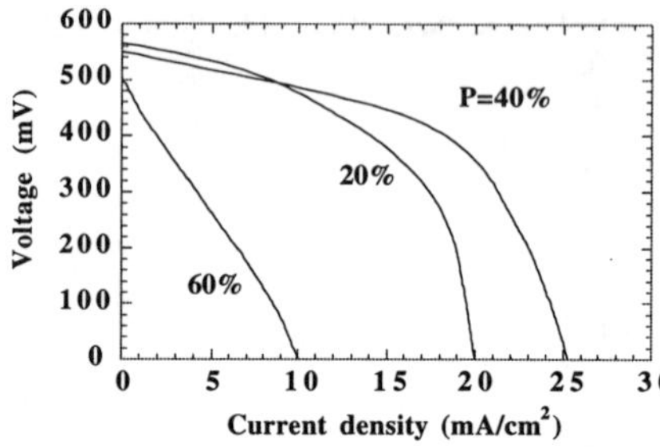

Fig. 3. The I-V response (fourth quadrant) of devices (A) (<P>=20%), (B) (<P>=40%), and (C) (<P>=60%) under simulated AM1 illumination. The values of V_{oc} (I_{sc}) are 0.57 V (40 mA), 0.55 V (50 mA), and 0.5 V (13 mA) for devices A, B, and C, respectively. The maximum power points are given by {J_m ,V_m} = {0.4 V, 28 mA}, {0.37 V, 39 mA}, and {0.2 V, 13 mA} for devices A, B, and C.

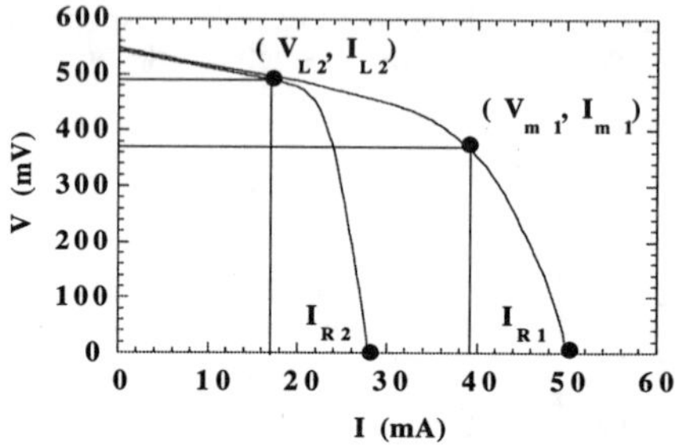

Fig. 4. The I-V response (fourth quadrant) of device B (<P>=40%) under (a) 100% and (b) 50% intense AM1 illumination for the determination of series resistance. The data points chosen (a) (0.37 V, 39 mA), and (b) (0.49 V, 17 mA), are such that the diode current is the same for the two cases.

The integrated reflectance (ℜ) of the devices was measured using a spectrophotometer with an integrating sphere attachment. At 633 nm, the values for ℜ were determined to be 43% (control device), 24% (<P>=20%), 16% (<P>=30%), 11% (<P>=40%), 8.2% (<P>=50%), and 6.8% (<P>=60%) respectively, indicating a rapid decrease in reflectance with increasing porosity. Due to the lower ℜ, device C is able to absorb short-wavelength photons more efficiently than device A.

The spectral response (SR) was determined by measuring the ratio of J_{sc} to the incident photon flux at different energies and normalizing the result to the known response of a commercial silicon solar cell, in order to eliminate any uncertainties associated with the determination of the incident photon flux over a wide spectral range. Figure 5 compares the normalized quantum efficiency of the control device with devices A, B, and C, in the range of 1-3.1 eV.

From Figure 3, the fill factor [FF = (J_m V_m)/(J_{sc} V_{oc})] for the devices are calculated as 0.49 (<P>=20%), 0.53 (<P>=40%), and 0.26 (<P>=60%), respectively. The series resistance and reflectance values are plotted in Fig. 6(a). The power efficiency for all the devices [η=(FF. J_{sc} V_{oc})/P_{in}] is plotted in Fig. 6(b). The low efficiency at high porosities is due to the large R_s and the dominant role of the space-charge limited current, a result similar to what has been obtained with LEPSi films.[6] When comparing devices A and B, we observe that while the diode I-V characteristics are similar, the optical properties ($\Re$, SR) of device B are superior, leading to higher values of FF and η. Still, the value of V_{oc} for device B is slightly lower (by 20 mV) than that of device A, presumably because of a higher density of interface states. Comparing the control and A devices, we observe a difference in spectral response only at high energies (E>2 eV).

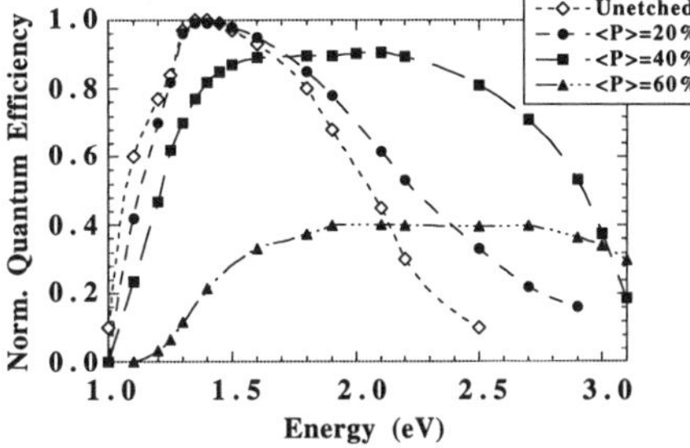

Fig. 5. The normalized quantum efficiency of the control device (unetched) and devices A (P=20%), B (P=40%), and C (P=60%). For device B, the small drop in quantum efficiency compared to device A is accompanied by a large improvement in the high energy response. The difference between control and A devices is observed only for E > 2 eV.

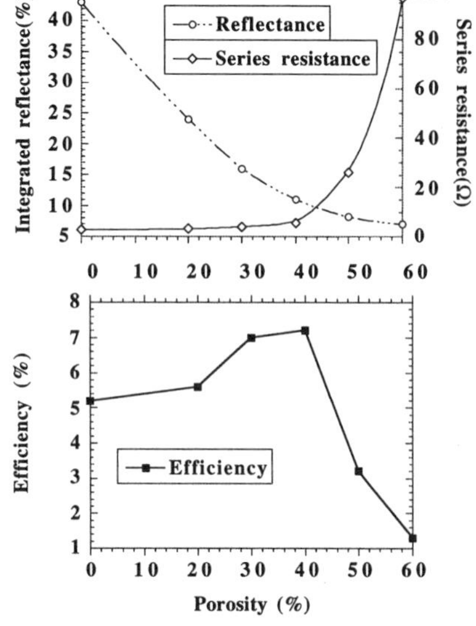

Fig. 6(a). Upper plot showing the variation of R_s and reflectance with porosity.
Fig. 6(b) Lower plot shows how the maximum device efficiency is achieved because of optimum R_s and reflectance. The respective efficiency values are as follows: 5.2% (control device), 5.6% (<P>=20%), 7% (<P>=30%), 7.2% (<P>=40%), 3.2% (<P>=50%), and 1.3% (<P>=60%).

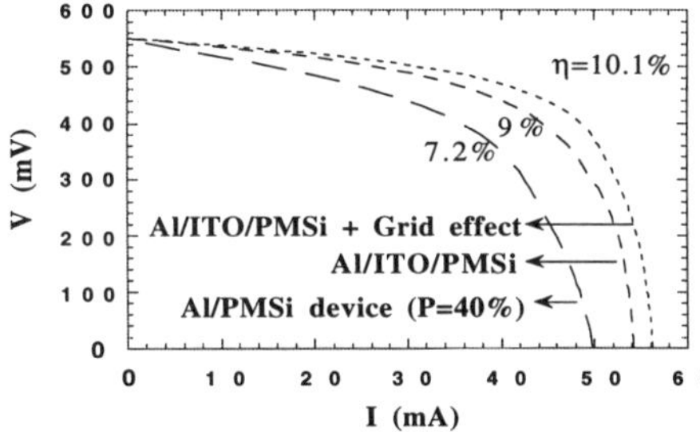

Fig. 7. Enhancement of device efficiency with an intermediate ITO layer and improved grid architecture, with thinner and more frequent gridlines.

The R_s of 5.4 Ω for device B is still rather high. Since porous silicon (as a rough approximation) mainly comprises of vertical columns, the resistance against the lateral motion of carriers (intra-grid) may be significant. We have reduced this lateral resistance by depositing a 200 nm thick layer of transparent, conducting indium tin oxide (ITO) film between the Al contact and PMSi. Also, the original Al grid contact used to cover about 20%-30% of the total area. A new architecture, with thinner and more frequent gridlines created a greater percentage of exposed area. The effect of these process steps caused a reduction in the overall R_s value from 5.4 Ω to 3Ω. Figure 7 shows that as a result both the FF and the I_{sc} have improved (V_{oc} unchanged). The efficiency of the ITO/PMSi solar cell is 10%, a 39% improvement over device B.

Finally, we have considered the possible influence of the c-Si substrate on the device performance, even though carrier penetration is negligible. Preliminary measurements of the spectral response of a free-standing film based device (Al/PMSi/ITO back contact) shows that the response is almost identical to the on-substrate case, except at E~1 eV. This indicates (even though the device areas were different- 2 cm^2 vs. 0.5 cm^2), that the contribution of the c-Si substrate is not significant.

CONCLUSIONS

We have reported photovoltaic devices based on porous microcrystalline Si. In order to optimize the device performance, the influence of the porosity on parameters such as the series resistance, the reflectance and the spectral response has been analyzed. The optical properties improve with increasing porosity. However, the series resistance increases with increasing porosity. The optimum performance was observed for a device based on a 40% porous PMSi film. By using an intermediate ITO layer, and changing the grid structure we have succeeded in achieving a 10% efficient device. Further optimization of the ITO layer and grid design is expected to improve the device efficiency further. Another goal would be to use different substrates.

ACKNOWLEDGMENT

This work was supported in part by the New York State Energy Research and Development Authority with additional support from the NSF Science and Technology Center for Photoinduced Charge Transfer. The Cornell Nanofabrication Facility was used for performing some of the measurements. The authors wish to acknowledge Dr. T.K. Hatwar and Dr. C. Brucker at the Eastman Kodak Company, for assistance.

REFERENCES

1] L. T. Canham, *Appl. Phys. Lett.*, **57**, 1046 (1990).
2] P. Guyader, P. Joubert, M. Guendouz, C. Beau, and M. Sarret, *Appl. Phys. Lett.* **65**, 1787 (1994).
3] A. Prasad, S. Balakrishnan, S.K. Jain, and G.C. Jain, *J. Electrochem. Soc.*, **129**, 596 (1982).
4] J.P. Zheng, K.L. Jiao, W.P. Shen, W.A. Anderson, and H.S. Kwok, *Appl. Phys. Lett.* **61**, 459 (1992).
5] Y.S. Tsuo, M.J. Heben, X. Wu, Y. Xiao, C.A. Moore, P. Verlinden, and S.K. Deb, *Mater. Res. Soc. Symp. Proc.*, **283**, 405 (1993).
6] C. Peng, PhD Thesis, University of Rochester, 1995.
7] D. W. Boeringer and R. Tsu, *Appl. Phys. Lett.*, **65**, 2332 (1994).
8] V.A. Skryshevsky, A. Laugier, V.A. Vikulov, *Mat. Sci. & Engg. B*, **40**, 54 (1996)
9] Y. Hamakawa and H. Okamoto, Amorphous Semiconductors Technology and Devices, edited by Y. Hamakawa, Vol. 16, (North Holland, Tokyo, 1985) p. 271 .
10] G. Harbeke, L. Krausbauer, E.F. Steigmeier, A.E. Widmer, H.F. Kappert, and G. Neugebauer, *J. Electrochem. Soc.*, **131**, 675 (1984).
11] H.P. Maruska, F. Namavar, and N.M. Kalkhoran, *Appl. Phys. Lett.*, **61**, 1338 (1992).
12] G.W. Neudeck, The PN Junction Diode, (Addison-Wesley, Reading, MA, 1989).
13] D. K. Schroder, Semiconductor Material and Device Characterization, (John Wiley and Sons, Inc., New York, NY, 1990).
14] R.K. Kotnala and N.P. Singh, Essentials of Solar Cells, (Allied Publishers Pvt. Ltd., New Delhi, India, 1986).
15] M. Wolf and H. Rauschenbach, *Adv. Energy Conver.*, **3**, 455 (1963).
16] S.P. Duttagupta, P.M. Fauchet, A.C. Ribes, H.F. Tiedje, T.E. Dixon, D.E. Brodie, and S.K. Kurinec, Submitted to Solar Energy Materials and Solar Cells.

«Holographic grating in Porous Silicon»

G. Lérondel*, M. Thönissen**, S. Setzu*, R. Romestain* and J.C. Vial*
* Laboratoire de Spectrométrie Physique, Université J. Fourier - CNRS (UMR 5588), B.P. 87, 38402 St. Martin d'Hères cedex, France, lerondel@spectro.grenet.fr
** Institut für Schicht-Ionentechnik (ISI), KFA Jülich, 52425 Jülich, Germany

ABSTRACT

We have produced lateral porosity modulation in porous silicon layers on a micron scale. Using the photosensitivity of the etching process and optical interferences, stripes are formed periodically on the top but also in the depth of the layer depending on the illumination wavelength. The periodicity of the structure is easily modified by changing the wavelength or the incident angles of the two laser beams used to create the modulated illumination. By rotating the sample, two dimensional structures have also been obtained. The samples formed by this procedure are characterised by light diffraction and photoluminescence both of which show the alternance of high and low porosity ranges. With this kind of in-depth lithography, easy and very cheap fabrication of porous silicon gratings promises a large potential range of applications of this material to integrated optics and photonics.

INTRODUCTION

An example of the photosensitivity of silicon was given in 1992 by Nogushi, who simply synthesized porous silicon (PS) by illuminating a Si substrate in hydrofluoric acid solution [1]. Photodissolution is expected to be amplified when applied to PS, on account of the extraordinarily large specific surface area of this material, and was used to increase the room temperature photoluminescence [2]. Moreover, by changing the porosity of the porous silicon it is possible to cover a broad range of refractive indices [3], which enables to produce waveguiding structures [4], as well as optical superstructures, Bragg reflectors and Fabry Perot filters [5]. In these previous examples the porosity along the direction of dissolution is modulated through the current density of formation. In our case, we endeavoured to produce PS structures in the layer plane. The principle is based on the photosensitivity of the material and an interference technique. The holographic technique is already used in industry to obtain diffraction gratings as a substitute for standard mechanical techniques. Basically in this case a mask and a photosensitive coating are needed to define the ranges of the material, which will be chemically dissolved. The result is surface lithography. In the case of porous silicon the material is itself photosensitive in its depth and therefore allows a new type of lithography.

METHOD AND EXPERIMENTAL DETAILS

As shown in the figure 1, the interferences were obtained by two laser beams of wavevector, $\mathbf{k_1}$ and $\mathbf{k_2}$ falling on the sample with the same angle of incidence, θ. The lattice vector amplitude $G = 2\pi/a$, where a is the periodicity of the pattern, is simply given by $\mathbf{G} = \mathbf{k_1} - \mathbf{k_2}$. If we project this quantity on the layer plane, a is directly derived from the following relation:

$$a = \frac{\lambda}{2\sin\theta} \qquad (1)$$

Mat. Res. Soc. Symp. Proc. Vol. 452

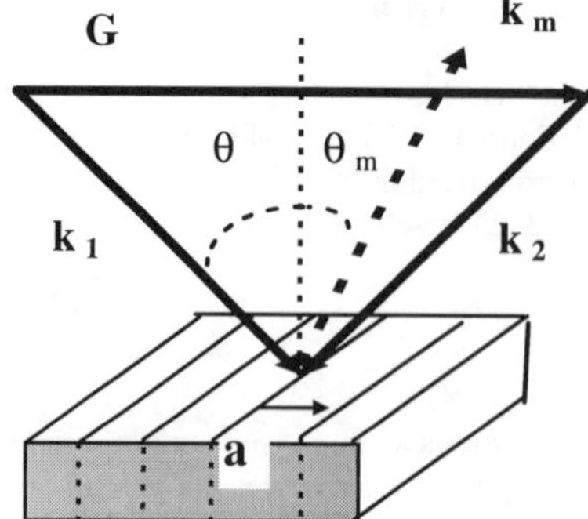

where λ is the wavelength of the illumination. This relation is independent of the refractive index of the porous silicon layer so that the periodicity remains the same throughout the depth of the layer. It should be mentioned that the lowest limit obtained for $\theta = \pi/2$ is given by $\lambda/2$.

For illumination, we used the 457 and 515 nm Ar laser lines. Due to the lower absorption of the PS in the red a HeNe laser was also used. p type PS layers were structured either during formation (electro-photodissolution) or after (electroless photodissolution).

Fig. 1: Schematic representation of the interference pattern resulting from the two incident wavevectors k_1 and k_2 and illustration of the diffraction that results from the interaction between k_1 and the lattice vector G.

RESULTS AND DISCUSSION

Figure 2 shows the surface of a porous silicon layer after illumination by interferences in the electrolyte. We used p type substrate (4 - 6 ohm.cm) and standard anodysation process in an electrolyte with 35% (in volume) of HF and 30% of ethanol. The layer is 5 μm thick and the porosity of the order of 60%. For the photodissolution we used the same electrolyte as for the formation. The Ar laser illumination time was 15 min, and the power was about 10 mW/cm^2. These parameters were optimised in order to avoid mechanical destruction of the material at high laser power. As we can see, well defined stripes are formed, leading to a pattern of periodicity 2.71 μm, which can be estimated from the photograph. In order to clarify the effect of the photodissolution, scanning electron microscope photograph of the same sample was performed. Figure 3 shows a side view of the structured layer slightly tilted so that the surface of the sample (upper part of the photograph) is also visible. The porous silicon layer can be divided in two regions. The upper part corresponds to a succession of

Fig. 2: Optical micrograph of a p-type PS lateral superlattice. Stripes are clearly visible with a regular step of 5 μm. They correspond to the image of the interference pattern.

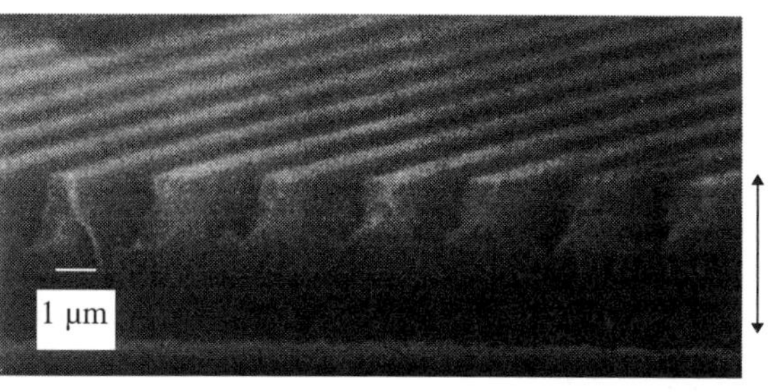

Fig. 3: Tilted SEM photograph of the PS structured layer (same as fig. 1). The upper part of the layer display clearly grooves, corresponding to the illuminated part of the sample. About 2.5 μm of the total layer thickness (5μm) have been structured.

walls and holes periodically distributed in the layer plane. The bottom half remains unchanged after the photodissolution. The shape of the structure will be discussed later in more detail in the paragraph devoted to luminescence. The interference pattern is well reproduced in the PS. The cosine profile resulting from the interferences gives rise to different density ranges. The limit in the depth is given by the penetration depth of light in the material, here between 2.5 and 3 μm at 515 nm. To partially summarise, we observed that the photodissolution is limited to the illuminated region and occurs on the surface but as well as in the depth of the layer, which seems to confirm previous results PS photodissolution obtained in our group [7]. Modulation of the refractive index in the direction parallel to the surface has been obtained (lateral superlattices).

It is possible to add one direction of modulation by successively performing two photodissolutions in two different directions. This was simply obtained by rotating the cell after the first dissolution (for example 90°). Figure 4 is a tilted side view of a double structured PS layer. The PS formation and the parameters for one photodissolution were the same as those in the previous paragraph. The photograph shows clearly a 2D array of dots, due to the two modulations of illumination.

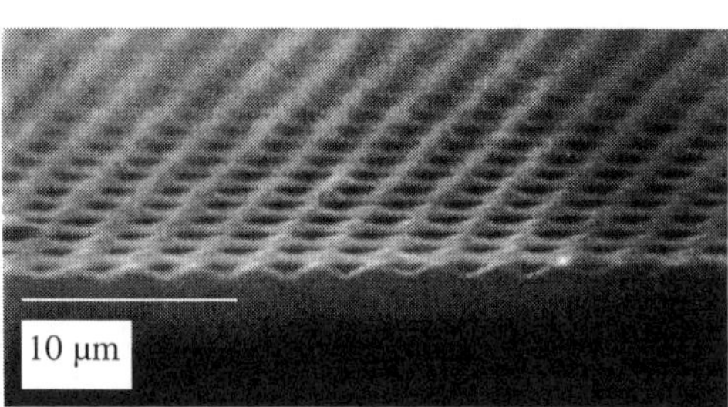

Fig. 4: SEM photograph of a 2D PS superlattice obtained by two successive photodissolutions.

In order to determine the spatial resolution of the photodissolution process, samples were prepared by increasing the incidence angle of the laser beams to reach submicron periodicity. The other electrochemical and light illumination parameters were the same as presented in the first paragraph. The smallest size that we have obtained is 0.67 μm. This value is to be compared with the diffusion length given in the literature for the carriers in the PS. Chazalviel et al. found a value of 1 μm [9]. It probably depends on the porosity and the morphology of the material but it is interesting to note that photodissolution assisted by holographic techniques can give fundamental information on the carrier diffusion in PS.

As we have previously seen porous silicon can be structured periodically at the micrometric scale by modulated photodissolution. To complete the characterisation we used light diffraction, expected from such structures. Figure 5 shows a diffraction pattern of the porous silicon lateral superlattice, described above. Since the layer is doubly structured, the brightest point at the centre of the figure, corresponding to the zero order (specular reflection) is surrounded in several directions by a series of sharp points corresponding to the different orders of diffraction of both structures. The two highest diffraction intensities are obtained in the vertical (up-down)

Fig. 5: 2D Diffraction pattern of the PS lateral superlattice presented in the figure 4. The most intense spot, at the centre of the photograph, corresponds to the zero order of diffraction. It is surrounded in several directions by spots corresponding to multiple orders of diffraction.

and horizontal (left-right) directions, respectively, corresponding to the two perpendicular grating directions, defined by the reciprocal lattice vectors $\mathbf{G_1}$ ($G_1=2\pi/a_1$) and $\mathbf{G_2}$ ($G_2=2\pi/a_2$). Integer linear combinations of G_1 and G_2 give rise to the secondary orders of diffraction observed in the oblique directions. The difference observed in the diffraction intensities in the two principal directions is attributed to the standard chemical dissolution occurring in the first grating. To avoid such effects, 2 D gratings could be made simultaneously by using interferences of four laser beams. In this case the beams should be coupled in pairs with their orthogonal polarisations. The angular mode positions are given by the generalised formula for the grating:

$$a = m\frac{\lambda}{(\sin\theta_m + \sin\theta_i)} \qquad (2)$$

θ_m is the angular position of the mth diffraction order and θ_i the incident angle of the beam. Since this formula has to be verified for each order a precise determination of a is then possible. We found for the grating corresponding to the figure 3, 2.67 +/- 0.02 μm, the error being given by the square deviation to the mean value.

To detect the effect of the photodissolution in the bulk, we investigate the luminescence of such structured layers. Figure 6 shows the photograph of the light emitted from the sample under UV excitation. Low porosity samples of about 60% (i.e. non luminescent porous silicon) exibit after photodissolution a very strong luminescence pattern which follows the superlattice periodicity, as it is shown on the figure. This confirms clearly that porosity is modified by the illumination. The shape of the luminescence, figure 7, is well described by the modulated porosity due to the interferences. At the position (x,y) in the layer plane if we assume that the luminescence is proportionnal to the modulated intensity, we obtained the following relation:

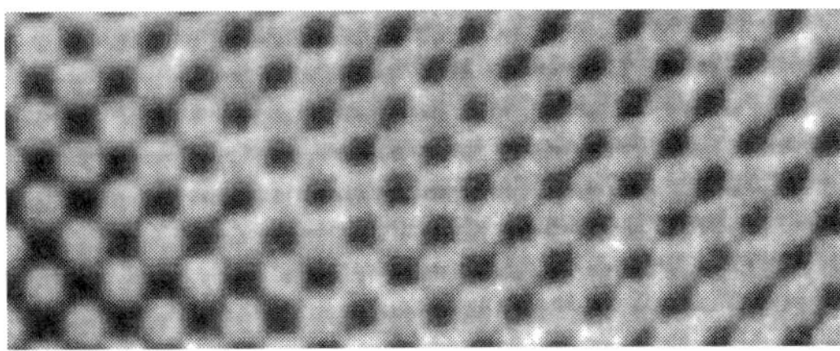

Fig. 6: Luminescence pattern of a 2D structured layer. The spatially resolved photodissolution enhances the initial porosity of the layer, giving rise to highly efficient localised luminescence. The characteristic size of the luminescent dots is 2 μm.

$$I \propto \sin^2(G/2\,x) + \sin^2(G/2\,y) \qquad (3)$$

G is the amplitude of $\mathbf{G_1}$ and $\mathbf{G_2}$, the reciprocal vectors of the 2D lattice Studies of the reflectivity show a change in the mean value of the refractive index after the photoprocess, that confirms the porosity modulation. Since holes mean 100% porosity, we explained the shape of the figure 3 by a drying effect. Just after the photodissolution, when the sample is kept in the HF solution, the mirror aspect of the PS surface remains, only a change in the colour is observed. After drying the sample has a milky aspect. This could be explained by the kind of rough lattice observed on the figure 3. The porosity of the illuminated region was so high that during the drying this part of the PS layer, not mechanically stable, is lost in the solution [8]. To increase the highest stable porosity we used for long time dissolution the pentane drying procedure [8]. In such conditions very high efficient, micrometric localised luminescent structures have been obtained.

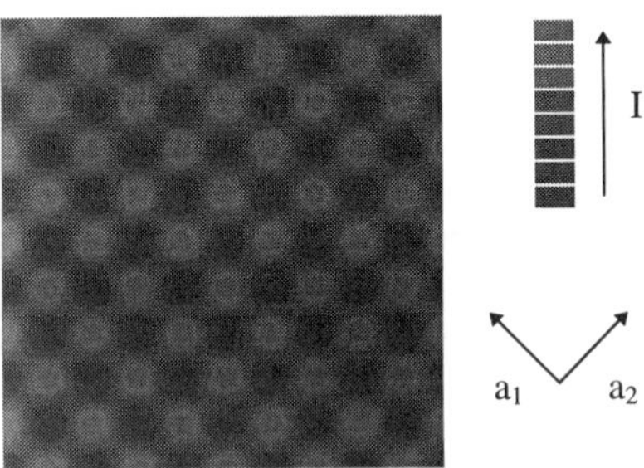

Fig. 7: Simulation of the modulated density in the layer plane resulting from the optical interferences. This figure compares well with the previous figure representing the luminescence pattern. The shape of the luminescence is due to the local porosity changes which are directly correlated to the illumination.

Until now we have shown that a modulated photodissolution can be applied successfully to porous silicon formed with low doped Si substrate. We have also obtained similar PS structures in p+ type (0.01 ohm.cm). In fact the photodissolution process occurs whatever the doping level of the Si substrate. However for the electro-photodissolution, which means the illumination during the formation, similar structures have only be obtained for substrate resistivities larger than 0.1ohm.cm. This can be explained by the fact that the electro-photodissolution is, at least at the beginning of the process, strongly correlated to the ratio between the number of photo-generated and extrinsic carriers in the material, which is negligible in highly doped substrates. Higher laser power will increase the number of photo-generated carriers but also destroy the porous silicon structure.

CONCLUSION

To conclude, the first PS structures using a holographic technique and the photosensitivity of the material have been obtained. The interference pattern can be reproduced very precisely in the PS layer even down to the submicron range. It is then possible to modulate the refractive index in the layer plane. Thanks to the diffraction property of such structures, this will give rise to applications in integrated optics and photonics.

ACKNOWLEDGEMENTS

We gratefully acknowledge S. Barret for S.E.M. photographs.

REFERENCES

[1] N. Noguchi and I. Suemune, Appl Phys Lett. **62** 1429 (1992).

[2] A. Bsiesy, J. C. Vial, F. Gaspard, R Herino, M. Ligeon, F. Muller, R. Romestain, A. Wasiela and A. Halimaoui and G. Bomchil, Surf. Sci. **254** 195 (1991).

[3] C. Pickering, M.I.J. Beale, D.J. Robbins, P.J. Pearson, R. Greef, J. Phys. C (Solid State Physics) **17** 6335 (1984).

[4] M. G. Berger, S. Frohnhoff, W. Theiss, U. Rossow, H. Muender, *Porous silicon science and technology*, ISBN 3-540-58936-8, Springer (1994), p. 345

[5] A. Loni and L. T. Canham, M. G. Berger, R. Arens-Fischer, H. Munder and H. Luth, H. Arrand and T. M. Benson, Thin Solid Films **276**, 143 (1996).

[6] M. Ligeon, F. Muller, R. Herino, F. Gaspard, J.C. Vial, R. Romestain, S. Billat and A. Bsiesy, J. Appl. Phys. **74** 1265 (1993).

[7] S. Letant and J.C. Vial, to be published in J. Appl. Phys. (15th December 1996).

[8] O. Belmont, D. Bellet and Y. Bréchet, J. Appl. Phys. **79**, 7586 (1996).

[9] V. M. Dubin, F. Ozanam and J.-N. Chazalviel, Phys. Rev. B **50**, 14867 (1994).

OPTICAL SENSORS BASED ON POROUS SILICON MULTILAYERS: A PROTOTYPE

W.Theiß*, R.Arens-Fischer*, S.Hilbrich*, D.Scheyen*, M.G.Berger**, M.Krüger**, M.Thönissen**,
*I. Phys. Inst., Aachen University of Technology (RWTH), D-52056 Aachen, Germany
**Institut für Schicht- und Ionentechnik, Forschungszentrum Jülich, D-52425 Jülich, Germany

ABSTRACT

Porous silicon multilayers are easily produced by a variation of the current density during the electrochemical etching process. The obtained porosity profile in depth is equivalent to a refractive index profile. Preparing appropriate multilayer systems one can design optical filters with sharp reflectivity peaks from the infrared through the whole visible spectral range (dielectric mirrors). Stability in time is achieved applying a suitable oxidation process.
In many cases of optical process and material control the required spectral decomposition of the probing light can be done with sufficient accuracy by a set of porous silicon optical filters whose spectral characteristics can be adapted easily to the given problem. In this work we present the prototype of an optical sensor that is a low-cost alternative to 'real' (and expensive) spectrometers. Application examples in the infrared and visible are given to demonstrate the sensor performance.

INTRODUCTION

Porous silicon has been investigated mainly with respect to its extraordinary luminescence properties which have been discovered a few years ago [1]. Meanwhile it is clear that fast optoelectronic devices will be difficult to make using porous silicon. This is due to the fact that typical decay times of porous silicon luminescence are in the µs range.
On the other hand, display applications of porous silicon layers are still under discussion. The easy fabrication of complex porous layer stacks (such as porosity superlattices) by current variation during the electrochemical etching process [2] favours the application of porous silicon in optical thin film systems. This is because the produced porosity profiles are also refractive index profiles which can be used to design quite freely the optical properties of such a system.

In this work we apply porous silicon layer stacks as optical reflectance filters. They are used as selective elements in optical sensors following

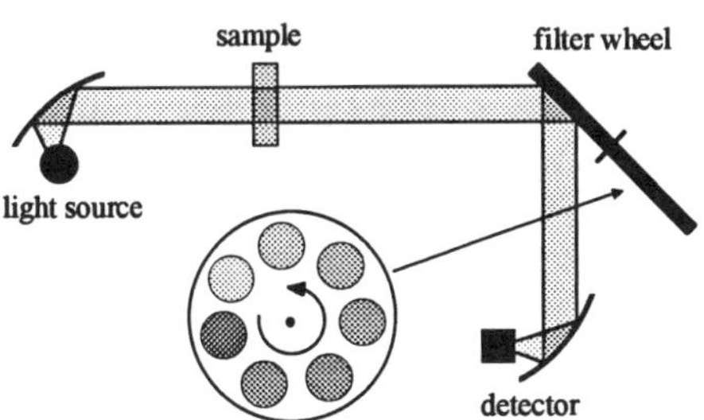

Fig.1: Basic scheme of the optical sensor. A reflection design has been realized since – up to now –no transmission filters can be produced easily.

Mat. Res. Soc. Symp. Proc. Vol. 452

the scheme shown in fig.1. A wide variety of filter characteristics, i.e. the frequency-dependence of the reflectivity, can be achieved. The most intuitive case is used in this work: so-called Bragg filters are applied which exhibit a large reflectivity in a narrow frequency range only and a low reflectivity everywhere else. These dielectric mirrors can be tuned from the mid infrared spectral range up to the near UV by proper choice of porosity and thickness of the individual sublayers. Of course, the frequency position and the width of these band pass filters must be adapted to the given analytic problem.

After the description of some experimental details concerning the sensor design we shortly discuss data analysis algorithms appropriate for fast online process control. Finally some examples are given demonstrating the performance of the sensor prototype in the visible and IR spectral ranges.

EXPERIMENTAL DETAILS

The electrochemical way to dielectric mirrors

The production of porous silicon reflectance filters has been described in detail previously [2][3]. A short summary is given here for completeness.

It is well know that silicon dissolution during the preparation of porous silicon takes place at the pore tips only. The 'etch front' propagates uniformly into the silicon wafer – the porous silicon layer thickness increases linearly with time. The porosity of the layer obtained this way is constant in depth and depends on the applied current density. Switching to a new current density means etching a new layer with different porosity whose thickness is once again a linear function of time. A current density variation periodic in time leads to a refractive index variation periodic in depth – a porosity superlattice with remarkable optical properties is produced. It has been shown that stability in time can be achieved by appropriate thermal oxidation of the filter [3].

The so-called Bragg filters used in this work are prepared by a step-like switching between two distinct current densities. Examples are given in fig.2.

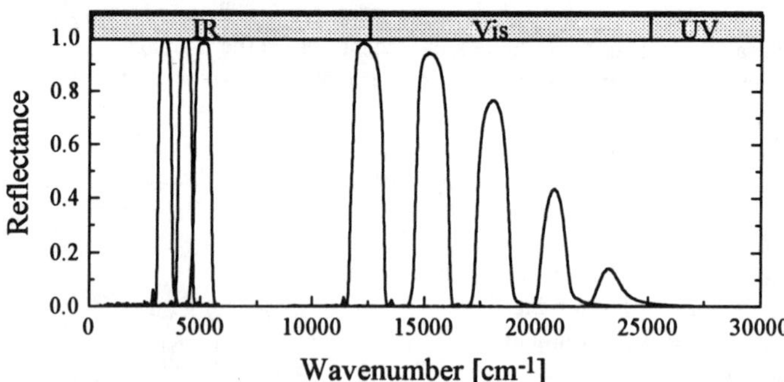

Fig.2: Characteristics of the optical filters used in this work. Four reflections have been used which is necessary to suppress the filter sidelobes.

Sensor setups

Porosity superlattices could be removed from the silicon substrate by a 'current shock' (i.e. increasing the current density into the electro-polishing regime) but their mechanical stability is poor. Thus it is advisable to leave them on the Si substrate which is opaque in the visible and – due to a metal backside contact – also in the infrared. Therefore a reflection setup has to be employed.

The sensor prototype follows directly the scheme shown in fig.1. To allow for multiple reflections (which improves the filter selectivity) the filter wheel was designed as shown in fig.3. The filters on the front and the back side of the wheel must not neccessarily be the same. Instead it can be useful in some cases to combine different individual filter characteristics to a new total one. The present version of the filter wheel allows to use up to 8 filter combinations which is certainly sufficient for the majority of online analysis problems.

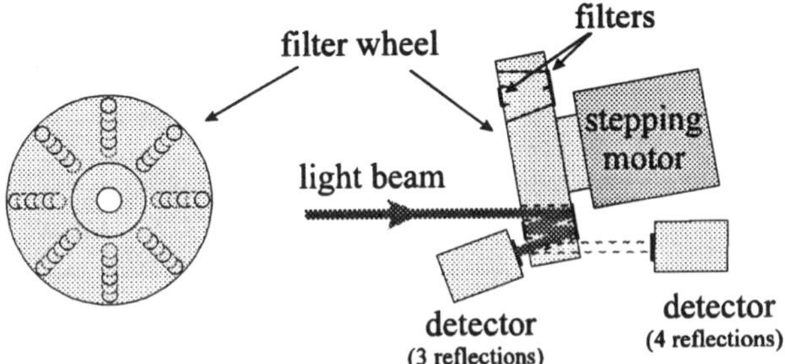

Fig.3: Sensor setup used to obtain the results shown below. Options for three or four reflections are shown, but a larger number of reflections is also possible.

The step motor of the filter wheel is controlled by a computer which also reads the detector signal and does the data analysis and output. For the investigations in the visible a halogen light source and a Si-photodiode detector have been used. In the infrared a SiC light source and a MCT (Mercury-Cadmium-Telluride)-detector were applied. With the exception of the MCT-detector and the PC only low-cost components are involved.

Data analysis

The sensor output for a sample to be analyzed is the set of detector readings for each filter combination. Looking at the filter characteristics given in fig.2 the sensor can be considered as an optical 'spectrometer' of very low resolution.

The analysis of the measured data must be fast, reliable and simple in order to fulfill the requirements of online analysis. We have chosen to use a chemometric method (PLS, i.e. Partial Least Squares, see [5]) which is established in mid and near infrared spectroscopy. The basic procedure is the following: In a training session the software analyzes the relation between the parameters to be determined (e.g. layer thicknesses or concentrations of components) and the measured spectra. Obviously, samples must be used in the training for which all parameters are known, which means that they must have been obtained by alternative methods. Typically 10 to 40 training spectra are required for the learning phase. In the case of PLS the training knowledge is then used to setup a calibration matrix which is then applied to compute for each online sample very rapidly the desired parameters from the measured spectrum.

Ideally the parameters that are to be determined should depend linearly on some spectral features, or at least in a monotonuous way. This puts some restriction on the type of problems that can be treated with PLS. However, the examples shown below are all well behaving with respect to the applicability of PLS.

EXAMPLES

A color sensor

The filters in the right half of fig.2 cover the whole visible spectral range. Hence a sensor equipped with these filters should be able to recognize colors. This was verified and results are given here as a first example.

Table 1: Typical results of color recognition achieved with the optical sensor in the visible.

Sample	prepared (r,g,b)	measured (r,g,b)
1	(150,50,255)	(145,50,214)
2	(150,50,150)	(160,49,178)
3	(255,200,150)	(240,196,144)
4	(255,200,100)	(239,198,123)
5	(0,0,200)	(0,15,182)
6	(200,200,200)	(198,210,192)

Samples were prepared by printing colored rectangles with an inkjet printer on overhead transparencies. The colors were defined by the specification of their rgb values (red, green, blue) which are conventional 'color coordinates' in computer graphics. After the training of the PLS software using 25 samples of various sets of rgb values the sensor was able to determine the color of unknown samples. Examples are given in table 1. The largest deviations are found for the blue component which is due to the fact that in this spectral range the light source intensity as well as the detector sensitivity are quite low.

Thickness monitoring

The optical characterization of thin films during or immediately after deposition becomes more and more important since subsequent production steps may critically depend on correct

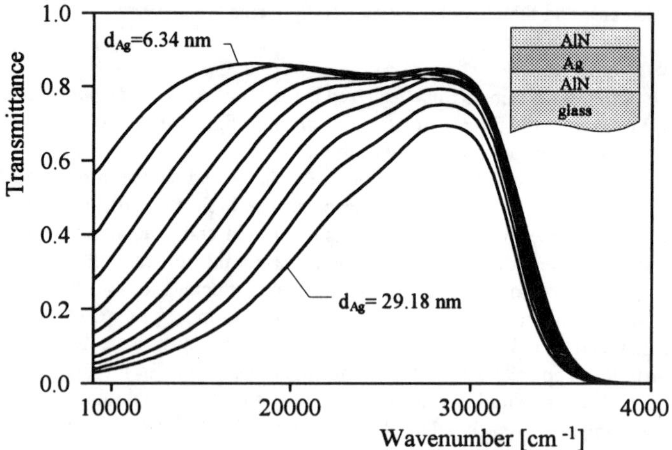

Fig.4: Transmittance spectra of thin film samples (produced by sputtering) with varying thickness of silver embedded between two aluminum-nitride (AlN) layers (25 nm each). As substrates conventional microscope slides have been used.

thicknesses in a layer stack. Also lateral homogeneity is required in many cases and must be monitored nondestructively. The desired characterization of thin film systems can be achieved by optical spectroscopy in most cases.

If the optical spectra exhibit broad spectral features depending monotonuously on one of the thicknesses an optical sensor as presented in this work can be a low-cost alternative to conventional spectrometers. For the system presented in fig.4 the thickness of the silver layer can be measured with the sensor very accurately. This is shown in fig.5 where the true layer thickness (obtained from the known sputtering rate and verified by a full simulation of the complete spectra shown in fig.4) is compared to the value determined by the sensor.

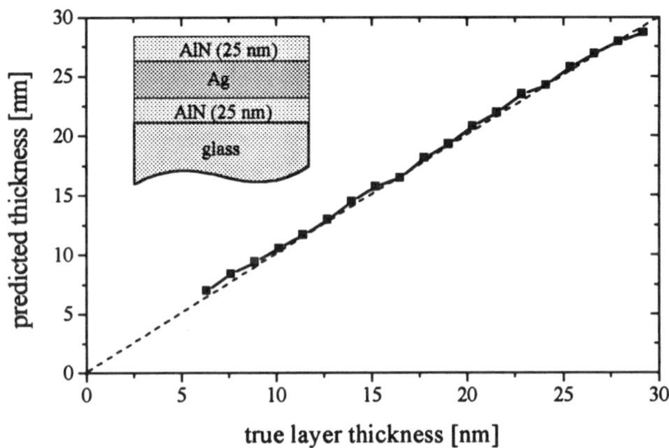

Fig.5: comparison of the true Ag layer thicknesses to the ones predicted by the sensor.

Analysis of infrared absorption bands

Finally we give an infrared example to demonstrate the wide spectral variation of sensor applicability. It was verified by conventional mid infrared spectroscopy that the transmission spectrum of adhesive tape (based on a poly-ethylene substrate) shows broad absorption bands around 3350 and 4250 cm^{-1}, almost no absorption is found above 4500 cm^{-1}. Typical spectra of stacks of adhesive tape are shown in fig.6.

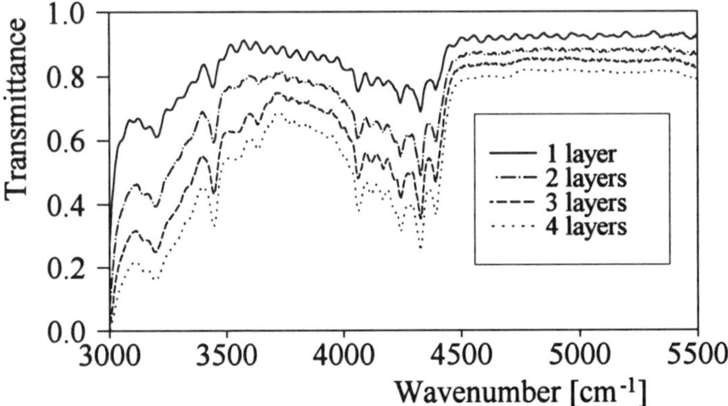

Fig.6: IR spectra of various stacks of adhesive tape differing in the number of layers.

Three filter combinations (see fig.2) were used to analyze stacks of adhesive tapes which differed from each other by the number of layers. The filter sets were prepared to have maximum reflectivity at 3400, 4250 and 5200 cm^{-1}, respectively. Hence the sensor can distinguish between the absorption bands (depending strongly on the total thickness of the stack) and the background above 4500 cm^{-1} which is almost independent of the number of layers. The slight transmission decrease with layer number found in this spectral range is due to diffuse light scattering at the tape interfaces.

As fig.7 shows the sensor can determine from the three detector signals successfully the number of stacks used in individual samples. More details on this example can be found in [4].

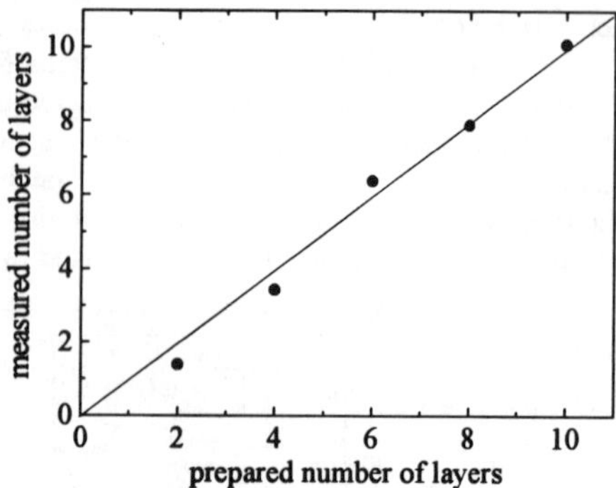

Fig.7: Results of infrared absorption band analysis done by the optical sensor. The comparison of the prepared number of tape layers and the prediction for this quantity shows satisfying agreement.

CONCLUSIONS AND ACKNOWLEDGMENT

It was shown in this article that optical filters based on porous silicon multilayer stacks can be applied as frequency selective components in optical sensors. The examples show a wide variety of possible applications from the infrared to the visible spectral range.

We thank T.Müggenburg for the preparation of the AlN-Ag-AlN thin films.

REFERENCES

[1] L.T. Canham, Appl. Phys. Lett. **57**, p. 1046 (1990)

[2] M.G.Berger, C.Dieker, M.Thönissen, L.Vescan, H.Lüth, H.Münder, W.Theiß, M.Wernke, P.Grosse, J. Phys. D: Appl. Phys. **27**, p. 1333 (1994)

[3] M.G.Berger, M.Thönnissen, W.Theiß, H.Münder, 'Microoptical devices based on porous Si', in Structural and Optical Properties of Porous Silicon Nanostructures, edited by G.Amato, C.Delerue and H.J. von Bardeleben, Gordon & Breach, in press

[4] S.Hilbrich, R.Arens-Fischer, L.Küpper, W.Theiß, M.G.Berger, M.Krüger, M.Thönissen, 'The application of porous silicon interference filters in optical sensors', submitted to Thin Solid Films

[5] H.Martens, T.Naes, Multivariate Calibration, (John Wiley & Sons, New York 1989)

IMPROVED INTERFERENCE FILTER STRUCTURES MADE OF POROUS SILICON

M. Thönissen, M.G. Berger, M. Krüger, W. Theiß *, S. Hilbrich *, R. Arens-Fischer, S. Billat, H. Lüth
KFA Jülich, ISI, D-52425 Jülich, Germany
*RWTH Aachen, I. Physik. Inst., D-52056 Aachen, Germany

ABSTRACT

Recently passive optical devices like filter structures or waveguides based on porous silicon have attracted high interest due to their easy and cheap fabrication. We have formed interference filters using porous silicon by changing the current density during formation. For the specific design of these filter structures a calibration of the etch rate and the refractive indices is required. Therefore we have determined the effective dielectric function for different substrate doping levels and anodization current densities by fitting reflectance spectra. Based on these results different kinds of reflectance filters consisting of discrete layers (Bragg reflectors, Fabry-Perot filters) as well as filters with a continuous change of the refractive indices with depth (rugate-filters) can be realised. Furthermore we present applications of these filter structures such as anti-reflectance coatings and high quality mirrors.

INTRODUCTION

The formation process of porous silicon superlattices is known very well and the resulting multilayer structures have been intensively studied [1-7]. Possible applications of these structures are for example interference filters [8], waveguides [9] and color sensitive photodiodes [10]. For some of these applications well defined reflectance/transmittance characteristics are required. This makes high demands on the control of the optical thickness. For filters with improved and more complex reflectance/transmittance characteristics, rugate filters are the best choice. These kinds of filters are characterized by a continuous change of the refractive index with depth. They are well known in optics to allow the realization of fascinating filter properties [11,12]. We will demonstrate that PS is very well suited for the formation of rugate filters, which show indeed better filter characteristics as compared to conventional PS multilayer stacks. Moreover this technique allows the formation of anti-reflectance coatings and filters with multiple reflectance peaks in the VIS and IR.

EXPERIMENTAL

PS layers were formed by anodization of Czochralski grown boron doped silicon (100) wafers. PS layers on substrates with resistivities of 200 mΩcm (p) were investigated. A metal contact was evaporated on the back side of the wafer to allow a homogeneous current flow. Immediately before anodization the substrates were cleaned in propanol in an ultrasonic bath and rinsed in deionized water. Anodization was performed under standard galvanostatic conditions using a mixture of H_2O:HF:C_2H_5OH = 1:1:2. The anodization current was supplied by a Keithley 238 high precision constant current source which is controlled by a computer to allow the formation of PS multilayers. To prevent the photogeneration of carriers, the anodization is performed in the dark. After formation the samples are rinsed with pure ethanol and dried with nitrogen gas. Reflectance spectra of the samples were taken with a Perkin Elmer λ2 photospectrometer in the range of 9000-50000 cm^{-1} (1111-200 nm). The reflectance was

Mat. Res. Soc. Symp. Proc. Vol. 452 © 1997 Materials Research Society

measured under an incidence angle of 8° using s-polarized light. From the reflectance spectra of single porous layers the effective dielectric function of PS was deduced by a fitting procedure (for details see [13]). The etch rate as a function of the anodization current density was determined by a thickness measurement of the PS layer using a scanning electron microscope (SEM).

RESULTS AND DISCUSSION

For the formation of interference filters the optical thickness $d_0 = n \cdot d$ (n refractive index, d geometrical thickness) must be known. We have investigated both parameters n and d as a function of the anodization current density for single layers on p and p^+ doped substrates. The refractive indices n were determined by a fit of the reflectance spectra in the range from 9000 up to 25000 cm^{-1} using effective medium theory. A typical fit is shown in Fig. 1, where the high accuracy of the fit can be seen. For simplicity in Fig. 2a the effective refractive indices are shown only for 12000 cm^{-1} for both doping levels. The etch rate of these samples determined by SEM is shown in Fig. 2b.

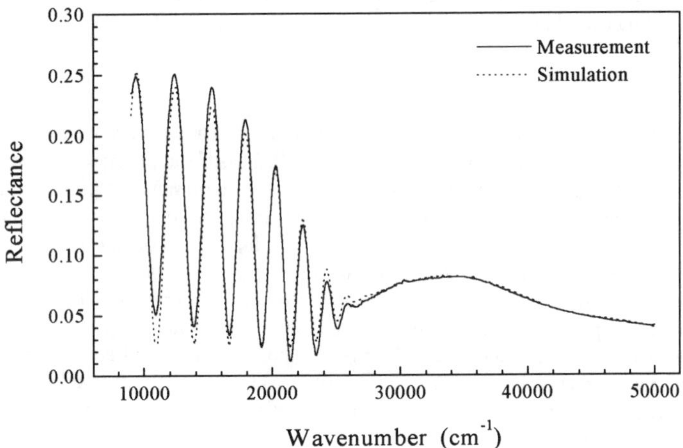

Fig. 1: Fit and measurement of a reflectance spectrum. The porosity of the porous layer on the p-doped substrate was about 71%, the thickness 1μm. For the fit effective medium theory was used.

Using these results Bragg reflectors can be formed very easily: Each layer in the Bragg reflector must be of the optical thickness of a quarter of the wavelength that should be reflected. The high refractive layer H corresponds to the lower porosity, the low refractive layer to the high porosity. The computer controlled etch process makes possible the formation of different Bragg reflectors in the visible and infrared range (Fig. 3). For these filters the stack of [HL] layers was repeated 20 times. Disadvantages of the presented Bragg filters are the sidelobes of the Bragg-reflectors and additional second order peaks. These can be avoided by using more complex filter structures: In contrast to other dielectric filter structures gradual variations of the refractive index with depth can be realized very easily with porous silicon: Knowing the relation of anodization current density and refractive index (see Fig. 2) the continuous variation of the refractive index with depth could be realized by a continuous variation of the current density. These types of filters are the so called rugate filters.

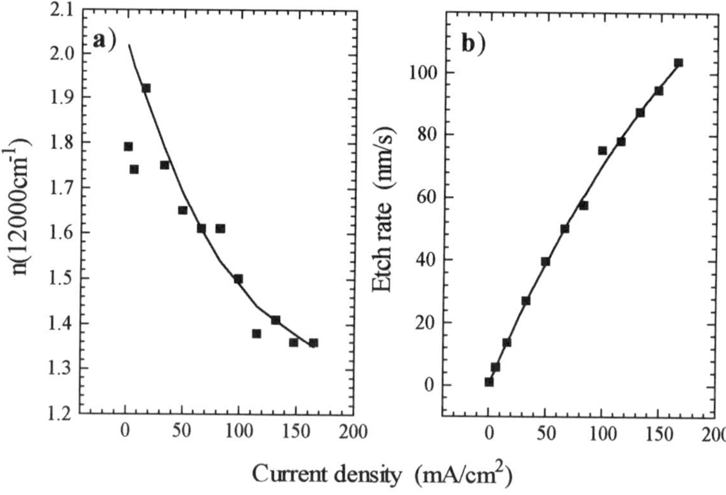

Fig. 2: Effective refractive index for 12000cm^{-1} (a) and etch rate (b) as a function of the anodization current density.

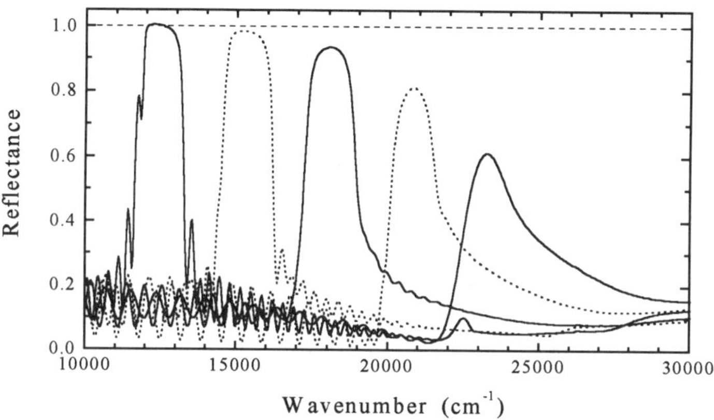

Fig. 3: Reflectance spectra of different Bragg reflectors $[HL]^{20}$ exhibiting reflectance peaks over the whole visible spectral range. The samples were formed on p-doped substrate using current densities of 30/120mA/cm² for the H/L layer, respectively. The decrease of the peak maximum is due to light absorption in the filter structure.

Typically the refractive index is changed sinusoidal as shown in Fig. 4, for an improvement of the reflectance an exponential index matching to the refractive indices of the substrate and the one of air and an additional apodization by a gaussian function was introduced. For special demands an almost free design of the reflectance becomes possible: By the Fourier transformation of the refractive index as a function of the layer thickness the reflectance (as a function of the wavenumber) can be calculated.

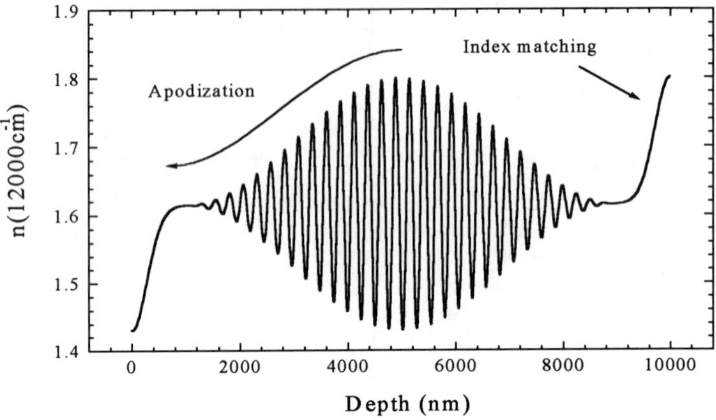

Fig. 4: Depth profile of the refractive index of a rugate filter. For the improvement of the filter characteristics an apodization and index matching was added.

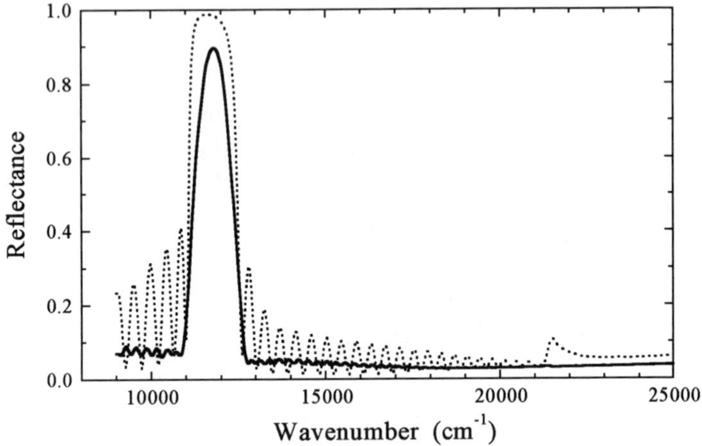

Fig. 5: Reflectance spectrum of a rugate filter formed on p-doped substrate. For comparison the reflectance spectrum of a Bragg reflector $[LH]^{24}$ is shown. In contrast to the Bragg reflector the rugate filter is characterized by the absence of sidelobes around the reflectance peak.

For a sinusoidal change of *n* with depth a delta-shaped peak of the reflectance is expected from theoretical calculations [11]. This means clean reflectance spectra without strong sidelobes and second order peaks. A rugate filter with the data from Fig. 2 was formed on a p-doped substrate.

The resulting reflectance spectrum is shown in Fig. 5. For comparison a standard Bragg reflector $[LH]^{24}$ is shown. The rugate filter is characterized by the almost complete absence of sidelobes and a very low reflectance value close to the reflectance peak.

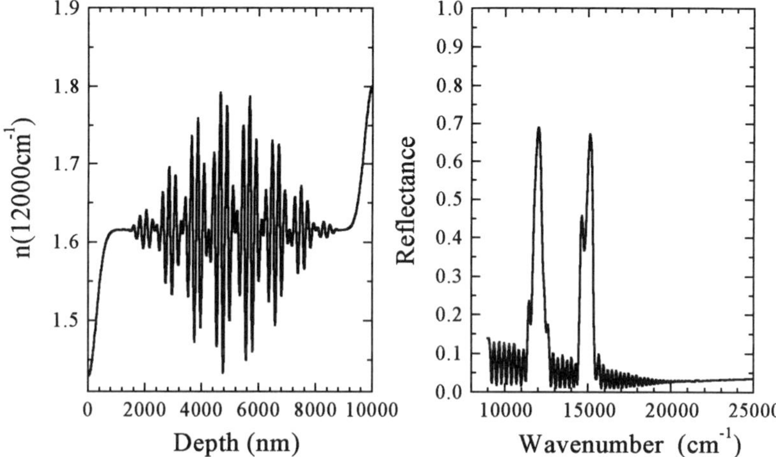

Fig. 6: Depth profile of the refractive index (a) and reflectance spectrum (b) of a rugate filter formed on p-doped substrate. The index profile is a superposition of two sine functions with different periods resulting in two reflectance peaks at different wavelengths.

Using the technique of a gradual variation of the refractive index double peak structures and antireflectance coatings easily can be realized: For the double peak filter the refractive index was varied using a superposition of two sine functions with different periodicities (Fig. 6). Also the formation of anti-reflectance layers becomes possible by a gaussian modulation of n with depth (Fig. 7). All these filter-structures show the advantages of using porous silicon for these complex filter structures. The quality of the presented filters made of porous silicon and their thermal stability is high enough to use them as high quality mirrors: We have formed a Bragg reflector ($[LH]^{20}$) for a wavelength of 647 nm and used it as a back side mirror in a Krypton Ion Laser (Innova 200) operating at this wavelength. The laser has worked with a reduced output power (about 30% in comparison to the commercial mirror). The laser has worked without interruptions for more than two days without loosing performance.

CONCLUSION

By the determination of the parameters refractive index n and geometrical thickness d for two doping levels we are able to form Bragg reflectors very precisely over the whole visible range of wavenumbers. Disadvantages like sidelobes and second order peaks can be reduced by using more complicated filter structures like the so called rugate filters with a continuous variation of n with depth. By these technique also double peak structures and anti reflectance coatings can be realized very easily. Further applications of these filter structures are high quality mirrors that can be used as mirrors in lasers.

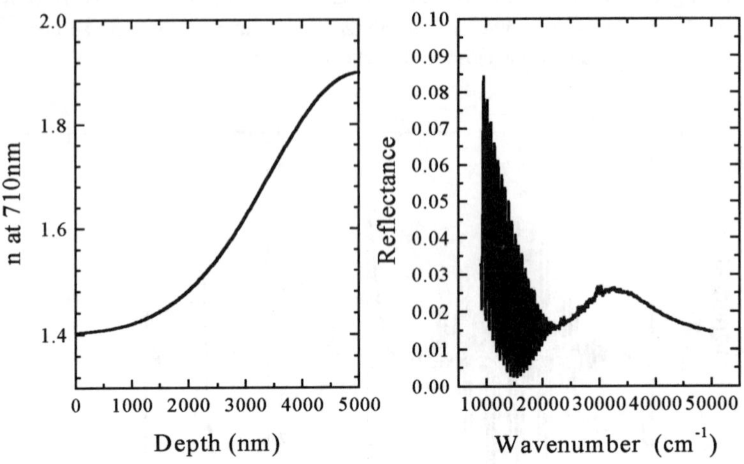

Fig. 7: Depth profile of the refractive index (a) and the reflectance spectrum (b) of an anti-reflectance coating formed on p-doped substrate. The index profile is a gaussian function. The resulting reflectance spetrum shows an decrease of the reflectance over the whole observed range of wavenumbers.

REFERENCES

1. M.G. Berger et al., J. Phys. D: Appl. Phys. **27**, 1333 (1994)
2. M.G. Berger et al., Mat. Res. Soc. Symp. Proc. **358**, 327 (1995)
3. G. Vincent, Appl. Phys. Lett. **64**, 2367 (1994)
4. D. Buttard et al., Thin Solid Films **276,** 69-72 (1996)
5. C. Mazzoleni, L. Pavesi, Appl. Phys. Lett. **67**, 2983 (1995)
6. Xing-Long Wu et al., Appl. Phys. Lett. **68**, 611 (1996)
7. M. Araki et al., Jap. J. Appl. Phys. **35** (2B), 1041 (1996)
8. M.G. Berger et al., in "Optical interference coatings", Florin Abeles ed., Proc. SPIE **2253**, 865 (1994)
9. A. Loni et al., Proceedings of the IEE colloquium "Microengineering Applications in Optoelectronics", 27th February, 1996, London, (Digest No.96/39)
10. M. Krüger et. al., Proceedings of the 26th ESSDERC, ISBN 2-86332-196-X
11. B.G. Bovard, Applied Optics **32** (28), 5427 (1993)
12. W.H. Southwell, Applied Optics **28** (23), 5091 (1989)
13. W. Theiß et al., Mat. Res. Soc. Symp. Proc. **358**, 435 (1995)

FABRICATION OF 10-NM GRATINGS USING STM CHEMICAL VAPOR DEPOSITION

A. ARCHER, J.M. HETRICK, M.H. NAYFEH
Department of Physics, University of Illinois at Urbana-Champaign, Urbana, IL, USA

ABSTRACT

Using STM assisted chemical vapor deposition (STM-CVD), we have successfully fabricated what we believe are the finest gratings to date. Our best grating consists of half-micron long evenly spaced lines which have a width of 10 nm and a 10 nm interspacing. The gratings were fabricated on passivated p-type silicon <111> using trimethylaluminum (TMA) as a precursor gas. Although the composition of the nanostructures is not exactly known at this time, we believe that the lines are partially metallic. We made some systematic investigations to optimize the resolution, including studies of the effect of tunneling current, tip speed, tip-sample biasing voltage, and the effect of formation of very fine TMA whiskers on the tip.

INTRODUCTION

Recently, advances in scanning probe technology have made it possible to manipulate matter on the submicron scale, opening new possibilities for research into the behavior of matter in this regime. Of particular importance is the fact that these new fabrication techniques might make it possible for the first time to observe quantum behavior in transport under conditions in which it has remained previously unobserved. Both the potential for new technologies based on quantum transport phenomena and the inherent scientific interest in this new field have motivated research groups in recent years to develop effective and consistent techniques to fabricate structures on the submicron scale.

In this paper, we present the results of our initial attempts to fabricate structures on the nanometer scale. Using a technique known as STM chemical vapor deposition (STM-CVD), we were able to fabricate what we consider to be the smallest gratings to date, consisting of a series of lines each 10 nm wide with a 10 nm interspacing. We will describe the experimental procedure which yielded such results as well as our initial attempts to study the effect of three of the fabrication parameters on the process. We also discuss our initial attempts to study the electrical properties of these structures by looking at the tip-sample IV characteristics. Finally, we propose a mechanism which might explain our ability to fabricate these extremely fine features.

EXPERIMENT

Sample Preparation and Initial Set-Up

The substrate used for our experiment was a <111>-oriented p-type boron-doped silicon sample with a resistivity between 0.01 and 0.02 Ω-cm. The sample was part of a batch originally purchased from Virginia Semiconductors and had been RCA cleaned prior to shipping. In order to remove any organic contaminants and to passivate the sample, we followed a procedure first proposed by Ishizaka and Shiraki [1]. To remove contaminants, the sample was ultrasonically cleaned in methanol, boiled in trichloroethylene, ultrasonically cleaned in methanol for a second time and finally rinsed with deionized water. Passivation was accomplished by soaking the silicon sample first in a 30% aqueous solution of H_2O_2 and then in a 50% aqueous solution of HF.

Mat. Res. Soc. Symp. Proc. Vol. 452 © 1997 Materials Research Society

The tip used in our experiment was electrochemically etched from a tungsten wire in a 1 M KOH solution. Special care was taken to avoid any vibrations during the etching process in order to obtain as sharp a tip as possible.

Both the tip and the sample were introduced into the STM chamber which was then pumped down to a pressure as low as 10^{-8} torr prior to introducing the organometallic gas. Trymethylaluminum was then leaked into the chamber until the pressure increased to between 10^{-4} and 10^{-2} torr.

Imaging and Fabrication

The silicon substrate was imaged prior to introduction of the TMA by using a 4 V tip-sample bias (with the sample positively biased with respect to a grounded tip), a current of 1 nA and a tip speed of 1220 nm/s. Images of the sample prior to fabrication show a smooth sample with a peak-to-valley height between 6 and 10 nm, approximately.

Successful fabrication took place when we increased the voltage to 6 V and the current to 10 nA and reduced the scanning speed to 1.2 nm/s. The lines were fabricated by setting the tip to perform a single pass from one end to the other of our imaging window with the abovementioned parameter set. The sample was reimaged after each fabrication attempt. Our best results are shown in Figure 1a, with a cross section of the fabricated lines shown in Figure 1b. This figure shows a variety of attempts at fabrication. The smallest lines shown in the cross section are 10 nm wide and can have approximately an interspacing as small as 10 nm.

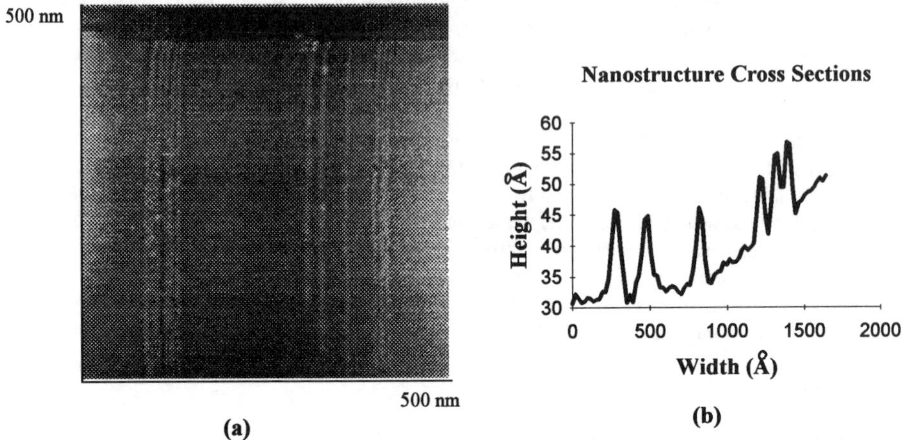

Figure 1a. Nanometer-scale lines formed by STM-CVD of TMA on silicon. The lines are approximately 10 nm wide and 2 nm high. They were fabricated using a tip-sample bias of 6V, a current of 10 nA and a tip speed of 1.2 nm/s.

Figure 1b. Cross section of the group of nanofabricated lines on the left. The cross section was obtained by averaging height information for every single line of data over the entire length of the scan.

We investigated the effect of the fabrication parameters on our ability to write with the STM. Our results indicate that it is possible to nanofabricate over a wide range of currents and tip speeds, but that the ability to fabricate was extremely sensitive to the tip-sample bias. We shall discuss the results of this experiment in greater detail later.

Electrical Characterization

The ultimate goal in our research project is to use STM-CVD to create structures which can be used to study transport phenomena on the nanometer scale. In order to do so, it will be necessary to try to determine the composition of these structures, specifically their metallic content. Determining the metallic content of our structures will allow us to further develop the technique so as to improve their conductivity without compromising their low dimensionality. Ultimately, we would like to deposit our nanometer-scale lines between larger electrical contacts in order to perform current-voltage (IV) characterization. Although we are currently pursuing this goal by implementing our STM-CVD technique on a microscope with a larger scanning range, we have taken a preliminary step in this direction by performing IV characterizations using the STM as an electrical probe. Specifically, a deposit is made by using a large tip-sample voltage in the presence of TMA. The large voltage is necessary in order to allow for a large tip-sample separation and therefore a larger area of fabrication. The tip is then positioned above the deposit and the STM control unit is set to perform an IV measurement. The results of such measurements are shown in Figures 2a and 2b. Figure 2a is an IV curve of the tip and the sample prior to fabrication showing the expected rectifying behavior. Figure 2b is an IV curve done under similar conditions after the fabrication procedure outlined above. It appears from Figure 2b that the electrical characteristics of the tip-substrate system have changed as a consequence of the fabrication. The IV curve shows what appears to be ohmic, rather than rectifying behavior.

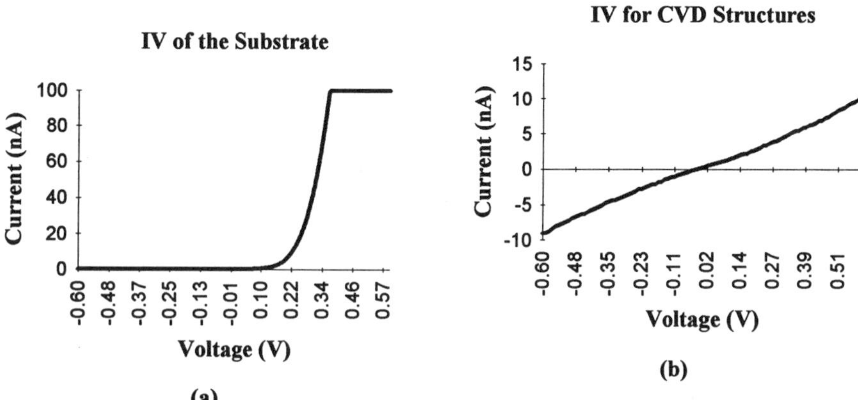

Figure 2a. IV curve between the tip and the sample prior to fabrication. The tip sample bias was set initially to a value of 0.1 V in order to establish the tip-sample spacing. The tip-sample distance was then held fixed while performing the IV measurement.

Figure 2b. IV curve between the tip and a large structure created after pulsing the tip with a large voltage. The procedure for performing the IV measurement was similar to the one outlined in Figure 2a. The initial tip-sample bias was again initially set to 0.1 V.

RESULTS

Influence of fabrication parameters

Successful fabrication was possible for a range of tip speeds (1.2 to 610 nm/s) and a range of currents (.01 to 10 nA). It was only when the highest tip speeds and the lowest currents were used that the fabricated structures began to deteriorate significantly, becoming spottier as the currents decreased and the tip speeds increased. It should be noted here that as of this writing we have not explored all the possible combinations of current, voltage, and tip speed for this system, but only varied one parameter keeping the other parameters constant.

The strongest determining factor in our ability to fabricate was the tip-sample bias. Although the exact value of the bias varied from one experimental run to the next, within the same experimental run there was a very definite bias voltage below which no fabrication took place. The fact that fabrication seems to depend so strongly on the bias and so weakly on the current and the tip speed might provide strong evidence in favor of a fabrication model which is field or energy dependent rather than one which is thermally dependent. Specifically, we conjecture that TMA is at least partially dissociated by the action of the electron beam and that the resulting fragments are deposited on the surface.

Further evidence of this dissociation process is given by the change in the local electrical properties of the sample when a deposit is formed. The IV shows a change from rectifying to ohmic behavior which might indicate that the metallic content of the substrate has increased.

Tip whisker formation

It is possible that our ability to fabricate nanostructures might be greatly improved by the formation of whiskers on the tip during the fabrication process itself. Other researchers [2] have noted the formation of filaments as long as 200 nm on their tips when a positive voltage is applied to the tip in the presence of an organometallic gas. Here, we wish to offer some qualitative observations which provide indirect evidence of the accumulation of material on the tip. (As of this writing, we have not examined the tips under an SEM). We have noticed a drastic extension of the tip when the polarity of our tip-sample substrate is switched and the current is increased. Specifically, when we switched from our scanning parameters (4 V, 1 nA) to a voltage of -4 V and increased the current to 10 nA, the z-piezo extended to its maximum range, approximately 200 nm according to our digital readout. In contrast, when we kept the tip-sample bias at 4 V and increased the current we did not notice such a radical change in the z-piezo reading until we got to very high currents, 70-90 nA.

We can offer at this time two possible explanations for the z-piezo extension: either the tip itself was damaged (bent or curled) due to the change in parameters, or if it was not damaged then something which had accumulated on the tip was damaged or destroyed by the change in parameters. In the case of the damaged tips the fact that we were still able to fabricate nanometer-scale lines (as small as 10 nm) even after we performed several of these polarity and current switches opens up the possibility that some sort of structure accumulated on these blunted tips, structures which might have been sharp enough to allow for the fabrication of the nanostructures mentioned above. On the other hand, the presence of material accumulated on the tip might strongly depend on the polarity of the tip-sample bias and the current. By switching the bias and increasing the current, we might have destroyed these whiskers. This dissolution would also manifest itself as a z-piezo extension.

CONCLUSIONS

In this paper we reported the initial results of fabrication using STM-CVD. Using this technique, we have been able to fabricate what we believe are the smallest gratings to date. Since our ultimate goal is to make our nanostructures as sharp as possible without losing electrical continuity, we made initial investigations regarding the influence of tip-sample bias, current, and tip speed on the quality of the fabricated nanostructures. We also made initial attempts to determine the metallic content of our fabricates by using the STM tip to perform IV measurements on the fabricated structures and to compare them to similar IV's done on the surface. Our initial results show that the parameter which dominates our ability to fabricate is the tip-sample voltage, and that fabrication seems to modify locally the electrical properties of the substrate.

We intend to continue our study of STM-CVD by determining the metallic content of our nanostructures using Auger analysis and by testing for electrical conductivity by depositing nanometer-scale lines between electrical contacts which have been patterned using standard lithographic techniques. These techniques have been used by other researchers [3,4,5] in investigating other similar STM-CVD processes.

REFERENCES

1. A. Ishizaka and Y. Shiraki, J. Electrochem. Soc. **133,** 666 (1986).
2. A. Kent, T.M. Shaw, S. von Molnar, and D. D. Awschalom, Science **262,** 1249 (1993).
3. M. H. McCord, D.P. Kern, and T.H.P. Chang, J. Vac. Sci. Technol. B **6**, 1877 (1988).
4. E.E. Ehrichs, W.F. Smith, and A.L. deLozanne, Ultramicroscopy **42-44**, 1438 (1992).
5. S. Rubel, M. Trochet, E.E Ehrichs, W.F. Smith, and A.L. deLozanne, J. Vac. Sci. Technol. B **12**, 1894 (1994).

Part IX

Electroluminescence and Light-Emitting Devices of Nanocrystal and Porous Silicon

ELECTROLUMINESCENCE DEVICE PERSPECTIVES OF SI^+-IMPLANTED SIO_2

F. KOZLOWSKI, H. E. PORTEANU, V. PETROVA-KOCH, F. KOCH
Technische Universität München, Physik-Department E 16, D-85747 Garching, Germany

ABSTRACT

Thermal oxide layers on Si which have been implanted with Si conduct electrical current and emit light. The electroluminescence effect (EL) has an efficiency which is comparable to the best values that have been reported for the porous Si based devices. Generally, the EL spectrum differs substantially from the photoexcited luminescence. It is linked with that excited by external high energy electrons in cathodoluminescence (CL) experiments. We suggest to consider the effect as internal CL. Based on studies of transport and EL characteristic we evaluate the possibilities of an electroluminescent device based on such layers.

INTRODUCTION

The newest addition to the various schemes for preparing a system of photoluminescing nanoparticles is Si implantation into a thermal SiO_2 layer [1-4]. In contrast to electrochemical etching (porous Si) and gas-phase deposition (CVD particles), this yields a compact and closed layer. The particle size is controlled in a thermal precipitation step by varying the length of time and the temperature. The implantation technique affords precise dose and depth control. It is a processing technology that is an accepted tool in the manufacturing of electronic devices. The photoluminescence is tunable in the visible range [4-6] and has an efficiency comparable to that of porous silicon (PoSi) in its red-orange range of color.

The implanted layers have an electrical conductivity that depends on the implantation dose and annealing treatments. The SiO_x material that forms the layer is related to the CVD-System studied by DiMaria *et al.* [7] for its electroluminescence properties.

All of these considerations suggest that one explore the prospect of an electroluminescent device based on electrical current passing through the thin nonstoichiometric SiO_x covering the surface of the wafer. We report here on our experience with electroluminescence and discuss the problems and prospects encountered in trying to make light-emitting devices from such implanted oxide layers.

SAMPLE PREPARATION

Thermally grown Si-oxide, 100 nm thick on a (100) Si-wafer was implanted with Si^+ ions with an energy of 50 keV, and a dose of 5×10^{16} cm^{-2}. The Si-substrate was either of p- or n-type. The implantation energy was chosen so that the highest Si density lies in the middle of the oxide layers. Afterwards the samples were annealed in a range of temperatures (300 - 1150 °C) for a time between 2 and 40 minutes in the nonoxidizing atmosphere of forming gas (H_2 / N_2). The implanted side was covered with several 4×4 mm pads of thin semitransparent Au films (~ 100 Å).

Mat. Res. Soc. Symp. Proc. Vol. 452 © 1997 Materials Research Society

PHOTOLUMINESCENCE INVESTIGATIONS

The samples are studied using a prism spectrometer with a Si-diode detector or photomultiplier. The excitation source is a HeCd laser operating at 325 nm. The results depend sensitively on the annealing history and implantation dose of the sample. In Fig. 1 we summarize our observation for a few representative examples.

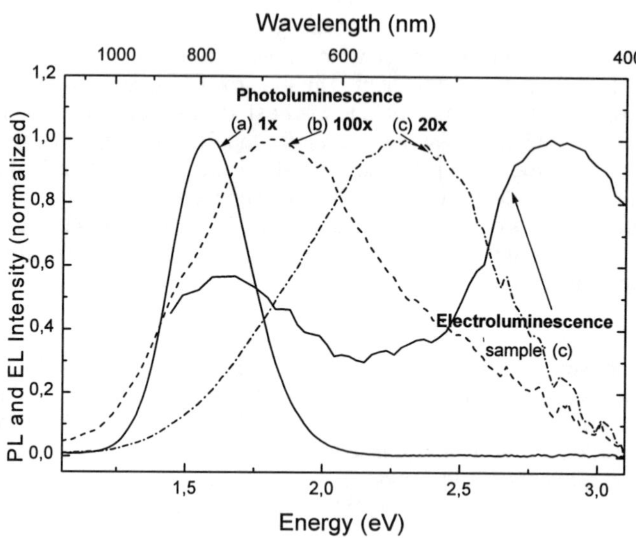

Fig. 1 The normalized photoluminescence of different samples *a*, *b* and *c*. (annealing temperature / time: *a*) 1150 °C / 15 min., *b*) 900 °C / 15 min. *c*) 900 °C / 3 min.). For comparison the EL spectrum of the sample *c* also appears in the figure.

Sample *a* (Fig. 1) is based on a p-type wafer ($\rho \approx 1\ \Omega$cm) and was annealed at 1150 °C for 15 min. The samples *b* and *c* are prepared on n-type substrates and were annealed at 900 °C for 15 min. and 3 min. respectively. The photoexcited luminescence of the sample *a* is two orders of magnitude higher than for sample *b*. It is comparable from the point of view of the intensity with the very best PoSi but is considerably more stable than PoSi luminescence. The response time of the red luminescence (sample *a*) is few tens of microseconds while for the blue region (sample *c*) it is in the nanoseconds range. This is consistent with the usual observations for PoSi. The dramatic improvement of luminescence intensity is not found in the electroexcitation case. Moreover, the spectral distribution does not at all follow the photoluminescence. For sample *c* we record in Fig. 1 a two-peak curve with the dominant emission in the blue. More discussion of this will be found in a later section. We note that the emission occurs in red and blue bands, as noted previously for the electroluminescence of SiO_x CVD-layers [7].

TRANSPORT PROPERTIES

There are two contradictory aspects which concern the electroluminescence of the Si nanoparticle system. The smaller and more isolated are the particles, the more efficient is the photoluminescence. On the other hand an efficient, electrically excited luminescence requires good transport properties (luminescence at low voltage and current). Most of the other work on electroluminescent Si-devices is based on PoSi structures [8,9,10,11,12] which in the active layer contain a higher fraction of Si than the oxidic material discussed here. The current paths in PoSi are probably better confined to the Si component than in the case of the randomly distributed precipitated Si particles in SiO_2.

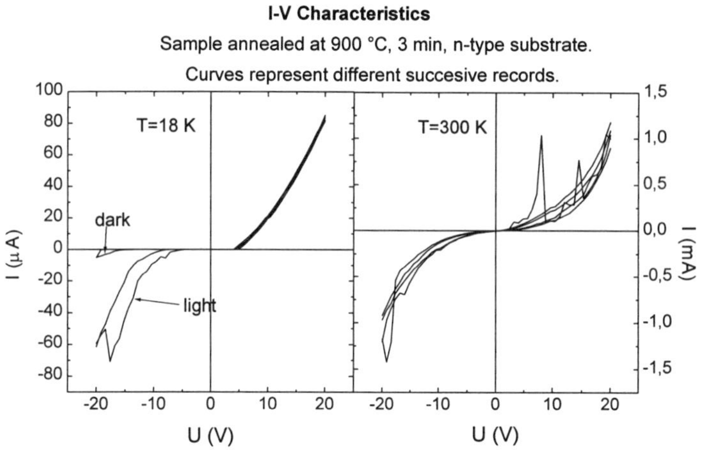

I-V Characteristics

Sample annealed at 900 °C, 3 min, n-type substrate.

Curves represent different succesive records.

Fig. 2 Current-voltage relation for an implanted layer (sample c) at high and low T. The influence of light shows the existence of depleted interfacial layer between the substrate and the oxide.

Electrical excitation of the luminescence in the implanted SiO_2 layers involves hot electrons tunneling between the nanoparticles. The current-voltage characteristic of the layer (Fig. 2) has a low voltage regime in which the current varies continuously and smoothly. The curves, except for the sudden steps, are reproducible without much hysteresis. One notices an asymmetric low temperature curve with the reverse bias for negative gate voltage. The influence of illumination is explained by the existence of a depletion layer at the substrate-oxide interface.

Electroluminescence with an efficiency comparable to that previously observed [9,10,11,12] for PoSi-based structures is found above the 20 V of the low voltage regime. From the I-V characteristic this appears as breakthrough with highly localized avalanche current. In a high resolution optical microscope we observe sharply defined local intensity fluctuations for the emission. In order to prevent a catastrophic breakthrough discharge and shorting of the oxide layer it is necessary to use a very thin gate layer contacted by a sharp probing needle and a series resistance of about 2 kΩ. It proved impossible to operate electroluminescence in a stable way with massive contacts. All of this suggests that the luminescence is excited by local and controlled avalanche breakdown involving hot electrons.

ELECTRO- AND CATHODOLUMINESCENCE MEASUREMENTS

The implanted layers were studied with regard to electrically excited luminescence in two different modes. We distinguish between electroluminescence (EL) with ~60 V applied to the tip and current 1-5 mA, and cathodoluminescence (CL) using a beam of 25 keV electrons. Representative results for two samples are given in Fig. 3.

Both samples are on n-type substrates. The implantation dose is 5×10^{16} cm^{-2}, the thermal oxide thickness 100 nm. The annealing is done at 900 °C for times of 30 s and 30 min. respectively.

The EL spectra are predominantly blue with a broad peak at 2.7 eV. The longer annealing time has caused the red part of the spectrum to develop a sharp peak. Comparison with the "3-minute" sample in Fig. 1 shows how this peak has evolved. Under EL and CL excitation the spectra age and change somewhat with time.

Fig. 3 shows that there is a close relation of EL with the case of external electron excitation using a 25 keV beam source (CL). EL appears as internal cathodoluminescence. A very striking result is that the cathodic bombardment of the thermal SiO_2, material which never was subject to implantation with Si, leads after some time (~15 min.) to a very similar CL spectrum. The luminescence intensity of the CL-damaged oxide is much less.

The comparison with the photoexcited luminescence (PL) in the lower part of Fig. 3 shows that the electrical excitation spectra EL and CL have very little in common with PL. This same behavior have previously been noted in Fig. 1.

CONCLUSIONS.

The data on transport and luminescence suggest that different mechanisms apply for the two types of radiative emissions. The PL requires a photoabsorption step and predominantly involves the precipitated nanoparticles of the Si in the oxide. The electrical excitation EL makes use of the particles as a current and field concentrators that define the local discharge conditions. The light however stems from excitations of luminescing defect centers in the SiO_2 matrix. The particle luminescence as excited in PL occurs mainly in the red between 1.4-2.0 eV. The dominant luminescing defect of the matrix is in the blue at 2.7 eV. Nevertheless, there is a significant overlap in the red part of the spectrum where both EL and PL give significant emissions. As seen in both figures 1 and 3 the spectral shapes in the red are not identical. It appears that red luminescence can be generated by distinct points defects such as for the electron damaged thermal oxide as well as from Si nanoparticles and their surface layers where they form an interface to the matrix.

The present work points to the possibility that in some of the previously studied devices involving oxidized PoSi, the EL originates more from red luminescing point defects than from nanoparticles.

For the EL from the implanted SiO_2 layers we judge that the nanoparticles themselves do not luminesce. The predominant light is that of the matrix in the blue (2.7 eV) band. Even the red portion of the EL spectrum involves point defects in the quartz.

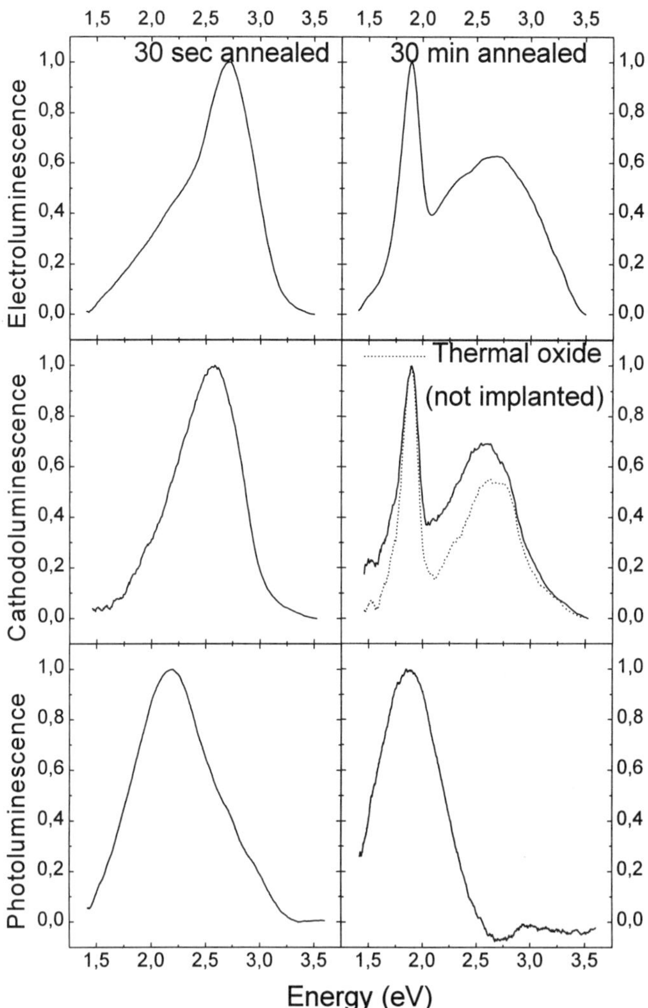

Fig. 3. The electro- cathodo- and photoluminescence of two implanted oxide layers which differ only in their annealing treatments.

The blue EL in the devices is reasonably efficient and has a fast response (~1 ns). The technical difficulties of making a good device lie in the electrical excitation. From our observations we conclude that a local avalanche breakthrough is required in order to generate light. This explains the significant aging effect and the general instability.

The very encouraging developments in the PL of the implanted layers, in which the emissions are increased by two orders of magnitude after annealing above 1100 °C [13], have not yet brought an improvement of the EL. It remains to look for EL from layers which have a higher fraction of Si than the present samples (~10% Si in SiO_2).

REFERENCES

1. T. Shimizu-Iwayama, S. Nakao, K. Saitoh, Appl. Phys. Lett. **65**, 1814 (1994).

2. P. Mutti, G. Ghislotti, S. Bertoni, L. Bonoldi, G. F. Cerofolini, L. Meda, E. Grilli, M. Guzzi, Appl. Phys. Lett. **66**, 851 (1995).

3. H. A. Atwater, K. V. Sheglov, K. J. Vahala, R. C. Flayan, M. L. Brougersma, A. Polman, MRS Proc. **316**, 409 (1994).

4. T. Fischer, V. Petrova-Koch, K. Sheglov, M. S. Brandt and F. Koch, Thin Solid Films, **276**, 100 (1996).

5. V. Petrova-Koch, T. Fischer, K. Sheglov, M. S. Brandt and F. Koch, Int. Symp. Advanced Lum. Mat., Electrochem. Soc., Chicago 1995.

6. V. Petrova-Koch, T. Muschik, T. Fischer, and A. Fojtik, in: Optics of Small Particles, Interfaces and Surfaces, R. Hummel (editor), 1995.

7. D. J. DiMaria, J. R. Kirtley, E. J. Pakulis, D. W. Dong, T. S. Kuan, F. L. Pesavento, T. N. Theis, J. A. Cutro, S. D. Brorson, J. Appl. Phys. **56**, 401 (1984).

8. P. Steiner, F. Kozlowski, W. Lang, Appl. Phys. Lett. **62**, 2700 (1993).

9. A. Loni, A. J. Simons, T. I. Cox, P. D. J. Calcott, L. T. Canham, Electronics Letter, **31**, 1288 (1995).

10. L. Tsybeskov, K. L. Moore, S. P. Duttagupta, K. D. Hirschman, D. G. Hall, P. M. Fauchet, Appl. Phys. Lett. **69**, 3411 (1996).

11. N. Lalic, J. Linnros, E- MRS Spring Meeting, Strassbourg, France, May 22-26, 1995.

12. P. Steiner, F. Kozlowski, M. Wielunski, W. Lang, Jpn. J. Appl. Phys. **33**, 6675 (1994).

13. T. Schuster, T. Dittrich, H. E. Porteanu, T. Fischer, E. Hechtl, V. Petrova-Koch, F. Koch, "Improvement of the luminescing behavior of Si^+-implanted SiO_2 films", this volume.

ELECTROLUMINESCENT DEVICES BASED ON ZERO- AND ONE- DIMENSIONAL SILICON STRUCTURES

A. G. NASSIOPOULOS*, P. PHOTOPOULOS*, V. IOANNOU-SOUGLERIDIS*,
S. GRIGOROPOULOS*, D. PAPADIMITRIOU**.
*Institute of Microelectronics, NCSR Demokritos,
P.O. Box 60228, 15310 Aghia Paraskevi Attikis, Athens-Greece.
** National Technical University of Athens, Physics Department, Zografou Campus, GR 157 73 Athens-Greece

ABSTRACT

Stable solid state electroluminescent devices, based on one- and two- dimensional silicon structures were fabricated. One- dimensional structures were in the form of silicon nanopillars on a silicon substrate, fabricated by using lithography and etching techniques. Isolation and planarization of nanopillars was made by using polymethyl methacrylate (PMMA), which is a non conducting polymer, totally transparent in the visible region. Gold (Au) was used as top contact metal and the back ohmic contact was based on aluminum. Zero-dimensional structures were in the form of silicon nanocrystallites, deposited by low pressure chemical vapour deposition (LPCVD) on a thin SiO_2 layer, thermally grown on silicon. Gold and aluminum were also used as contact metals. In both cases the devices showed stable electroluminescence at room temperature, visible with the naked eye. Devices based on nanocrystallites showed more uniform emission. The onset of light emission was between 5 and 7 Volts. The devices were checked for several hours of continuous operation and no degradation was observed.

INTRODUCTION

Following the demonstration of efficient, visible, room temperature photoluminescence from porous silicon [1], there has been considerable effort to produce other low dimensional silicon structures in order to study their luminescence properties with the overall aim to fully understand the light emission mechanism from these systems, as well as from porous silicon. The other motivation was to create new materials with size tunable optical and electronic properties, based on Si. It was demonstrated that in the nanometer range the band gap of low dimensional silicon systems increases with decreasing size [2,3]. Light emission in the visible range is also observed and it is attributed to quantum confinement within nanometer size structures. The interest as it concerns the applications is focussed on the fabrication of efficient, stable, solid state electroluminescent devices. 0-D, 1-D and 2-D silicon structures were produced. 0-D structures were for example produced in the form of silicon nanocrystallites in an SiO_2 matrix by silicon ion implantation [4], by plasma decomposition of SiH_4 and H_2 gas mixture in a resonant cavity [5] and by molecular beam epitaxy of CaF_2/Si superlattices [6]. 1-D structures in the form of silicon nanowires were produced either by using lithography and etching techniques (monocrystalline nanowires) [7-11] or by plasma enhanced chemical vapour deposition of α-Si:H through the pores of alumina (amorphous nanowires) [12]. 2-D systems were also produced by several methods. By lithography and etching techniques 2-D silicon walls were fabricated [10,13]. SiO_2/2-D Si/SiO_2 structures were also fabricated by oxidation of a SIMOX wafer [14,15]. Si/SiO_2 superlattices were made by using MBE deposition [16]. In this paper we report on the fabrication of electroluminescent devices, based on silicon nanowires and silicon

Mat. Res. Soc. Symp. Proc. Vol. 452 © 1997 Materials Research Society

nanocrystallites. Nanowires were produced by deep UV lithography and highly anisotropic silicon etching, followed by further thinning using high temperature thermal oxidation. Nanocrystallites were produced by depositing by low pressure chemical vapour deposition (LPCVD) a very thin silicon layer on silicon dioxide, thermally grown on silicon.

EXPERIMENTAL RESULTS

Silicon nanowires in the form of silicon nanopillars on an n-type (100) silicon substrate were fabricated by using deep UV lithography, based on a surface imaging technique (silylation process) and highly anisotropic silicon etching based on fluorine chemistry. The pillars were further thinned by high temperature thermal oxidation [9]. Device fabrication has been described in detail elsewhere [17]. Isolation and planarization has been achieved by using a non conducting transparent polymer (PMMA) to fill-in the whole area, containning the pillars. An oxygen plasma was then used for a few seconds in order to remove the resist from the tops of the pillars and the silicon dioxide was removed from them by diluted HF. Contact was made by using a thin gold layer, sputtered on the device area through a mechanical mask. A back side Aluminum ohmic contact was also formed.

The current-voltage characteristics (I-V curves) of this device show rectifying behavior. Electroluminescence was observed at forward bias with an onset at around 5 Volts. The reverse current is important, probably due to leakages to larger silicon structures which co-exist with the nanopillars within the pattern. The diode area was of the order of a few mm^2 but nanopillars were covering a much smaller area. An example of EL spectra from the above device is shown in fig.1.

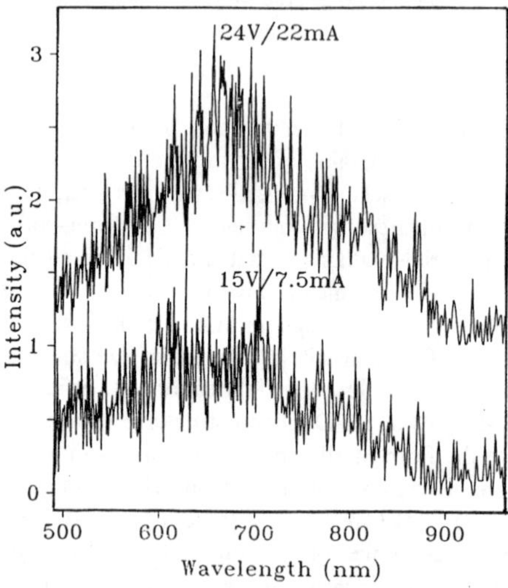

Fig.1 Electroluminescence from a device based on silicon nanowires.

EL, visible with the naked eye, was only emitted from areas containing the thin pillars. The PL spectrum of the same device area is similar to the EL spectrum with a slight blue shift of the signal maximum. The signal is broad in both cases. The EL signal is weak since the emitting area is small. It is smaller than the photoluminescenent area because a part of the nanowires is not contacted, due to their smaller height.

Analogous devices were formed by using silicon nanocrystallites as the light emitting material, in the place of the nanowires (fig. 2). The nanocrystallites were produced on a thin silicon dioxide layer, 20 nm thick, thermally grown on silicon, by low pressure chemical vapour deposition at 610°C. The deposited silicon layer was 18 nm thick and it was polycrystalline. Transmission Electron Microscopy characterization revealed the existence within the layer of silicon nanocrystallites of sizes between 1.5 and 15 nm. By electron diffraction it was verified that they were randomly oriented. Photoluminescence experiments showed that the above crystallites were luminescent at room temperature. The signal was in general broad with a peak in the red. An example of the obtained spectra is shown in fig. 3. The above nanocrystallites were used to produce light emitting devices. UV lithography and reactive ion etching were used to define the diode area as shown schematically in fig.2. A mesa structure was formed by selectively etching the layer containning the crystallites in areas outside the diode area. The top contact metal was a thin gold layer, deposited through a mechanical mask. An aluminum ohmic contact was formed on the back side of the wafer.

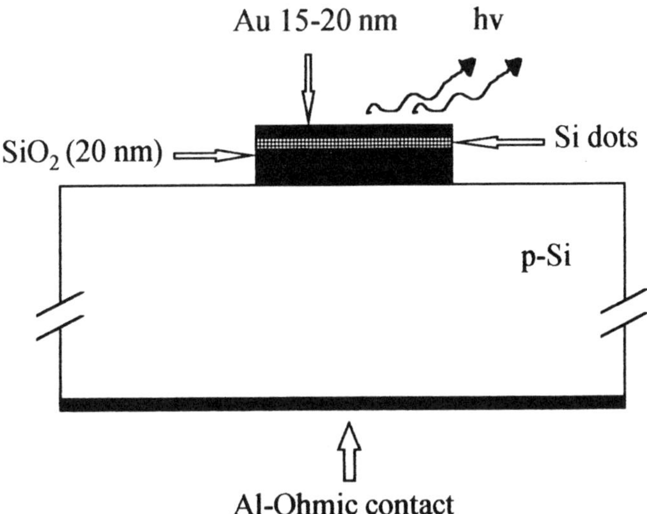

Fig.2 Schematic representation of a device based on silicon nanocrystallites.

Electrical measurements were performed also on those devices. Fig.4 shows an example of I-V curves obtained at room temperature. The devices show rectifying behaviour and the reverse leakage current is low. The leakage current is in this case much lower than that of the device based on silicon nanowires. This is attributed to the better diode isolation by the mesa structure, defined by lithography and etching. The onset of luminescence was also at around 5-7 Volts.

An example of EL spectra from the devices based on nanocrystallites is shown in fig. 5 for different bias voltages. The signal is broad with more than one maxima, probably due to an inhomogeneous distribution of crystallite sizes. Device stability was checked over more than 30 hours of operation without obvious device degradation. Only a small current drift was observed at constant bias voltage. The same devices were tested periodically during a several months period and they did not show significant aging.

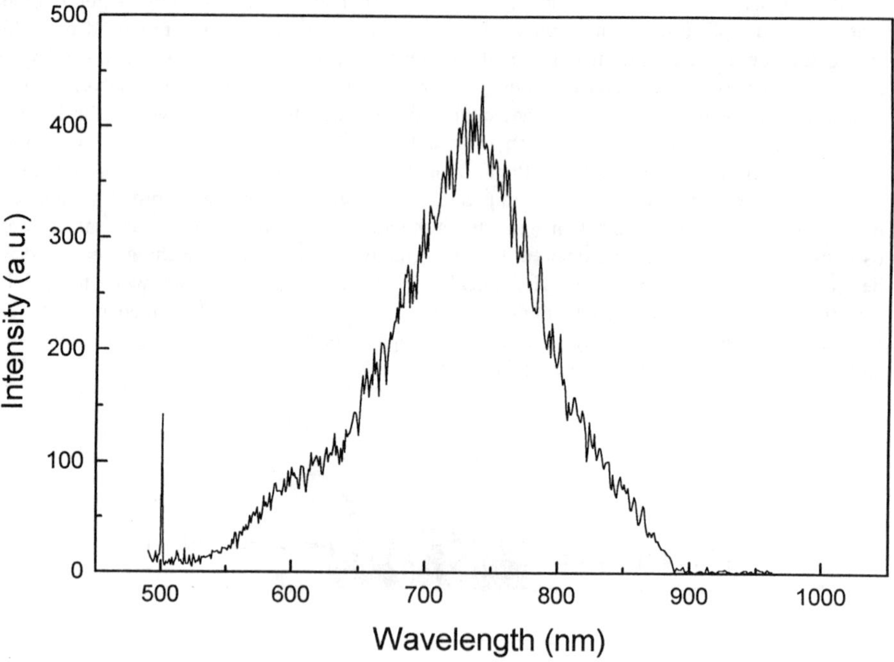

Fig.3 Photoluminescence spectrum from silicon nanocrystallites on a thin silicon dioxide layer (20 nm thick).

DISCUSSION

Both zero-dimensional and one-dimensional silicon structures fabricated by different methods were found to emit efficiently light at room temperature. In the case of silicon nanowires the structures emit either with hydrogen surface passivation or with an SiO_2 passivation. Devices from nanowires emit light at forward bias with an onset of EL as low as a few Volts (5-7 Volts). The emission spectrum is approximately the same as that obtained by light excitation. Since in this case the bottom silicon surface was isolated and only the tops of the silicon pillars were contacted, the observed light emission is clearly coming from these structures. An ionization mechanism by electron injection from the n-type substrate under the influence of the high electric field is proposed.

In the case of zero-dimensional structures the nanocrystalline material is separated from the silicon substrate by a silicon dioxide layer, 18 nm thick. The substrate was p-type. EL was observed when a negative bias voltage was applied on the device. Under the corresponding electric field, holes are accumulated at the Si/SiO_2 interface and electrons are injected from the top metal through the isolator, thus ionizing the crystallites in their way to the interface. The EL signal from the crystallites was much broader than that from the nanowires, attributed to crystallites of different sizes. The devices based on the crystallites were very stable and efficient (the emission was visible with the naked eye under day light) and their characteristics were easily reproducible.

No aging effects were observed in both cases for several hours of operation of the devices.

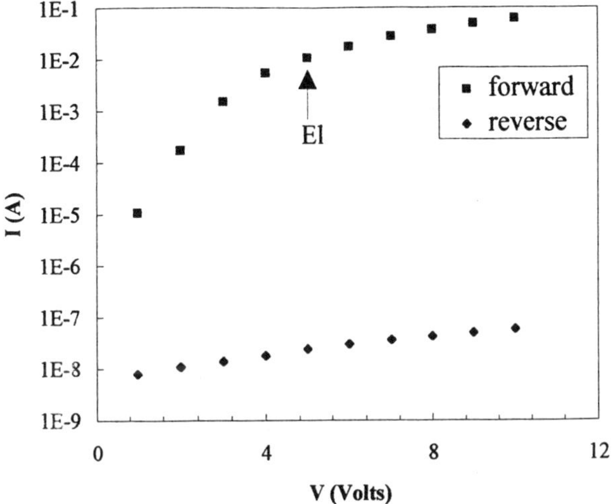

Fig.4 Current-Voltage characteristic of the device of fig.2. The onset of electroluminescence is ≈ 5 Volts.

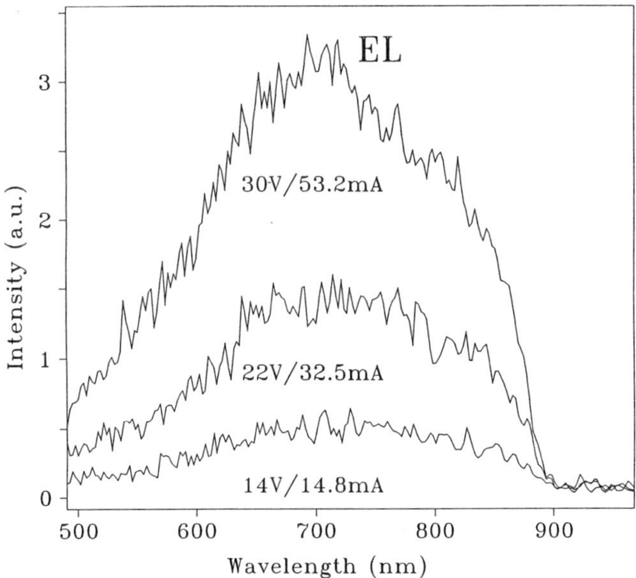

Fig.5 EL spectra of the device of fig 2 obtained at different bias voltage.

CONCLUSION

The results presented above open new important possibilities to the use of silicon nanostructures in silicon integrated optoelectronics. Efficient and stable light emitting devices may be produced. Ordered silicon nanowires of well defined dimensions may be fabricated by using the process developed for anisotropic silicon etching, provided that a high resolution electron beam machine is used for the initial pattern definition. Very clean structures are obtained after etching and their diameter is controlled by high temperature thermal oxidation. Nanocrystalline silicon produced
by low pressure chemical vapor eposition on thin silicon dioxide layers is also a challenging issue for the fabrication of very stable and efficient light emitting devices. These devices do not suffer from stability problems and their fabrication and operation is fully C-MOS compatible.

REFERENCES

1. L.T.Canham, Appl. Phys. Lett. 57, 1046, (1990)
2. G. D.Sanders and Y.C.Chang, Phys. Rev. B, 45(16), 9202, (1992)
3. S.Ossicini, A.Fasolino, F.Bernardini, Phys. Rev. Lett. 72, 1044, (1994)
4. Tsutomu Shimizu-Iwayama, Setsuo Nakao and Kazuo Saitoh, Appl. Phys. Lett., 65 (14), 1814, (1994)
5. H.Takagi, H.Ogawa, Y.Yamajaki, A.Ishijaki and T.Nakagiri, Appl. Phys. Lett. 56(24), 2379, (1990)
6. F.Arnaud d'Avitaya, L.Vervoort, F. Bassani, S.Ossicini, A.Fasolino, F.Bernardini, Europhysics Letters, 31, 25, 1995.
7. H.I.Liu, D.K.Biegelsen, F.A.Ponce, N.M.Johnson and R.F.N.Pease, Appl. Phys. Lett. 64(11), 1383, (1994)
8. P.B.Fischer,K.Dai, E.Chen and S.Y.Chen, J.Vac.Sci.Thechnol. B11, 2524, (1993)
9. A.G.Nassiopoulos, S.Grigoropoulos, E.Gogolides and D.Papadimitriou Appl. Phys. Lett., 66, 1115, (1995)
10. A. G.Nassiopoulos, S.Grigoropoulos, D.Papadimitriou, E.Gogolides, Phys. Stat. Sol., (b) 190, 91, (1995)
11. A.G.Nassiopoulos, S.Grigoropoulos and D.Papadimitriou, Electrochemical Soc. Proceed. 95-25, 297, (1995)
12. S.Katsuba, N.Kazuchits, G.De Cesare , S.La Monica, G.Maiello, E.Proverbio and A.Ferrari, Mat. Res. Soc. Symp. Proceed., 358 93, (1995)
13. An-Shyang Chu, Saleem H.Zaidi and S.R.J.Brueck, Appl. Phys. Lett. 63(7), 905, (1993)
14. Y.Takahashi, T.Furuta, Y.Ono, T.Ishiyama, Jpn. J. Appl. Phys., 34, 950, (1995)
15. P.N.Saeta and A.C.Gallagher, J. Appl. Phys. 77(9), 4639, (1995)
16. D.J.Lockwood, J.M.Baribeau and Z.H.Lu, Advanced Luminescent Materials edited by D.J. Lockwood, P.M. Fauchet, N.Koshida and S.R.J.Brueck, Electrochem. Soc. Proceed. 95-25, 339 (1995)
17. A.G.Nassiopoulos, S.Grigoropoulos and D.Papadimitriou, Appl. Phys. Lett. , 69(15), (1996).

ELECTROLUMINESCENT DEVICES MADE FROM SILICON NANOCRYSTALS EMBEDDED IN VARIOUS HOST MATRICES

Gildardo R. Delgado, Howard W. H. Lee, and Khashayar Pakbaz
Lawrence Livermore National Laboratory, P.O. Box 808, L-174, Livermore, CA 94551

ABSTRACT

Si nanocrystals produced by ultrasonic fracturing of porous silicon (PSi) were used to fabricate electroluminescent (EL) devices. The active EL material consists of Si nanocrystals embedded in various host matrices such as polyvinylcarbazole (PVK), polymethylmethacrylate (PMMA), and silica sol-gels. Several device configurations were used to induce EL processes that rely on radiative electron-hole recombination via carrier injection or impact excitation of the nanocrystals. We report on the optical and electrical properties of these devices. We discuss relevant physics pertaining to the Si nanocrystals/host and discuss advantages and disadvantages among the different host matrices.

INTRODUCTION

Light emission from nanometer size Si structures have been well documented. The several orders of magnitude enhancement in visible light emission of these structures promote tremendous interest due to its potential commercial applications. Silicon nanocrystals may provide the means for advancement of optoelectronic devices such as photodetectors, optical switches, semiconductor lasers, and flat panel displays. Our objectives are to characterize Si nanocrystals to better understand the physics behind the mechanism for visible emission and secondly realize devices based on Si nanocrystals. Our optical studies revealed that visible emission is due to quantum confinement effects as a direct result of their nanometer size. More importantly, our optical results revealed that the emission wavelength is tunable thus making Si nanocrystals ideal for optoelectronic devices. We proceeded to fabricate low cost Si nanocrystal electroluminescent devices. The active EL material consists of Si nanocrystals embedded in various host matrices such as PVK, PMMA, and silica sol-gels. Major advantages of this composite material system are the ease of producing high quality, thin, conformal EL films. Several device configurations were used that rely on radiative electron-hole recombination within the nanocrystals for light emission. We report on the optical and electrical properties of these devices including PL and current-voltage (I-V) behavior. These EL devices have applications as highly efficient light emitting devices. The cost and ease of processing of these materials make them ideal for optoelectronic and flat panel display technologies.

EXPERIMENT

Si nanocrystals were fabricated by the ultrasonic fracturing of porous Si (PSi) [1]. Porous silicon is made by anodic etching of silicon wafers in hydrofluoric acid [1-8]. Polished, single crystal p-type silicon (100) wafers with resistivities of 15-25 Ω-cm were used. Porous silicon was produced by etching in a current density of 20 mA/cm^2 in a 49 wt% hydrofluoric acid (HF) solution for 20 minutes (figure 1). Fabrication of PSi forms nanometer silicon structure resulting in quantum confinement effects as depicted in Figure 2. The structure of porous Si has been well characterized [4-8]. Silicon nanocrystals in a toluene colloidal suspension were prepared by scraping the nanostructured surface of the PSi and ultrasonicating the particles. Upon removal from the ultrasonic bath, the colloid was centrifuged and filtered through 0.45 μm filters. Spectroscopic grade toluene was used since it is an inherent solvent for PVK as well as PMMA.

Mat. Res. Soc. Symp. Proc. Vol. 452 © 1997 Materials Research Society

Nanocrystal colloidal suspensions in methanol and isopropyl were prepared for fabrication of the silica sol-gel films.

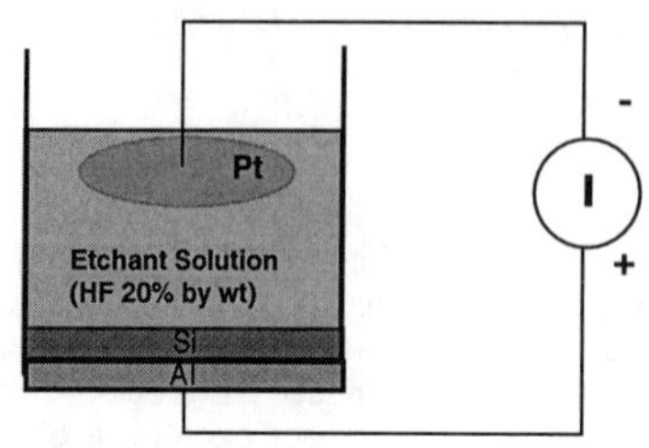

Fig. 1 Setup used for fabrication of PSi

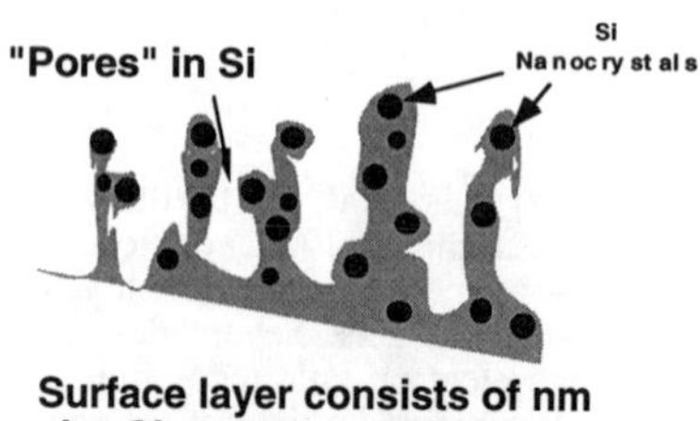

Fig. 2 Cross section diagram depicting Si nanocrystals in the PSi structure

RESULTS

Optical Characterization

Our optical characterization included absorption, photoluminescence (PL) and PL excitation (PLE) spectroscopy. Si nanocrystals of the appropriate size and shape yield efficient PL that spectrally shift with excitation wavelength. Fig. 3 shows PL spectra of Si nanocrystals in a toluene colloidal suspension that illustrates the spectral shifts. Peak wavelengths varied from about 380 nm to 430 nm when excited from 300 nm to 360 nm. Our optical analysis revealed that quantum confinement effects are responsible for the blue emission and the red peak is most likely due to traps arising from quantum confinement effects[9]. The spectral shifts in the PL are due to quantum confinement effects that result from the different size and shape distribution in the samples. The spectral shifts are crucial since this indicates that the emission wavelength can be controlled by controlling the size and shape of the nanocrystals. Colloidal suspension in various solvents, such as hexane, methanol, and toluene, all showed identical behavior thus indicating that the solvent does not interfere with the mechanism for light emission. Our experimental analysis and comparison with theoretical models [10,11] indicate that our samples range in size approximately between 1.9-2.6 nm diameters.

Device Structure and Processing/Fabrication

A thin single-layered electroluminescent (EL) device using Si nanocrystals embedded in a host matrix as the emitter material was achieved. Glass slides with patterned, thin film Indium-Tin-Oxide (ITO) were used as the anode. Thin (~ 2000 Å) EL films were easily deposited on the ITO substrate by spin coating techniques, resulting in a highly transparent layer. The active media were comprised of Si nanocrystals embedded in thin films of various matrices including polyvinylcarbazole (PVK), polymethylmethacrylate (PMMA) and silica sol-gels. The cathode consisted of evaporated thin (1000 - 1500 Å) films of Al or Ca.

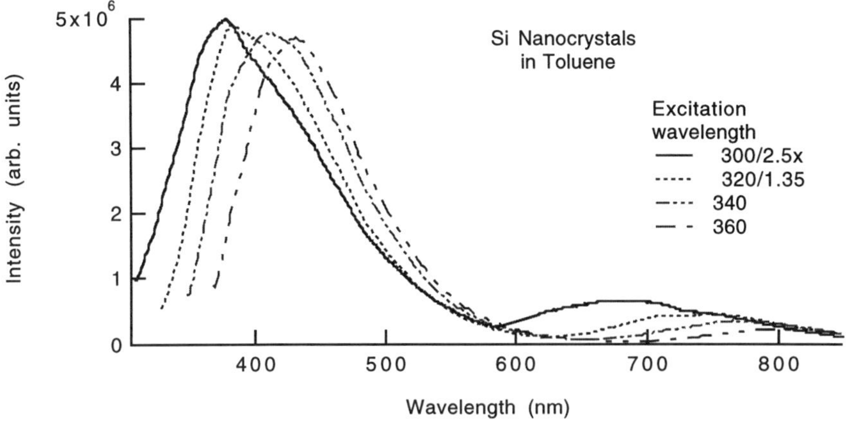

Figure 3 Photoluminescence of Si nanocrystals in a toluene colloidal suspension. This figure illustrates the shifting of the PL peaks with excitation wavelength.

Device Characterization

The device was designed to operate via an electron-hole (e-h) recombination process. For the device to work, a transparent host matrix of larger energy gap than the Si nanocrystal 's energy gap is needed. The matrices mentioned above were chosen for this property, along with their capacity to form high quality thin films. The mechanism for light emission from the active medium involves using the Si nanocrystals as recombination centers in the host matrix as illustrated in figure 4. Thus, for example, e-h injection into the matrix will allow for carrier trapping followed by radiative e-h recombination within the matrix. Bright emission in the blue and red spectral regions was achieved at room temperature in a nitrogen atmosphere. Devices had a purple emission. PL from thin films of Si nanocrystals embedded in PVK and PMMA revealed the same blue and red emission peaks observed from the colloidal suspension (Fig. 1-2). This explains the purple emission from the EL device and indicates that the nanocrystals are responsible for the observed electroluminescence (EL).

Current–voltage measurements are shown in Fig. 5 for the PVK devices. The turn-on voltage for the devices ranged from 30-40 volts for the PVK film and 100 volts for the sol-gel film. We attribute the large turn-on voltages to the relatively thick EL films (~2000 Å) and the effectiveness of carrier injection with the metals used. Figure 6 shows a photograph of an operating device based on Si nanocrystals in PVK.

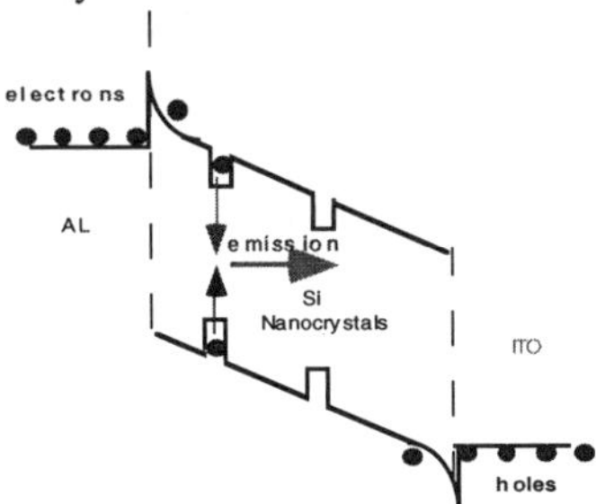

Fig. 4 This figure depicts the e-h injection mechanism for the Si nanocrystal/host media

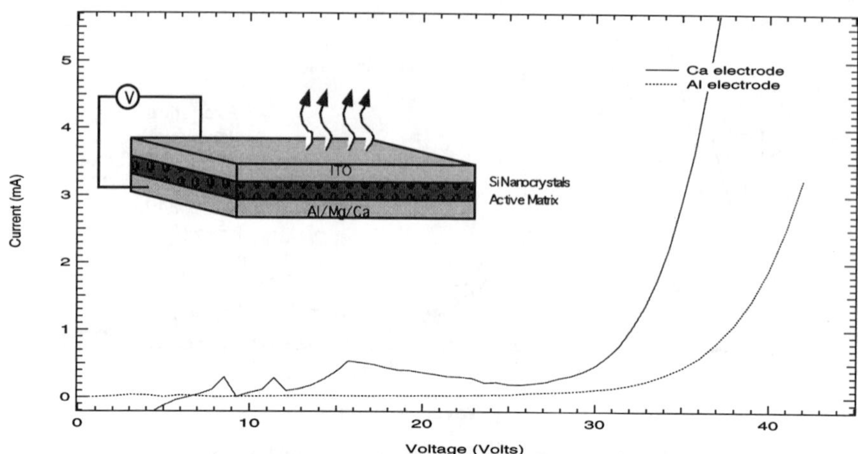

Figure 5 Current-Voltage characteristics for Si/PVK device with Al and Ca electrodes. Device schematic is shown in the inset

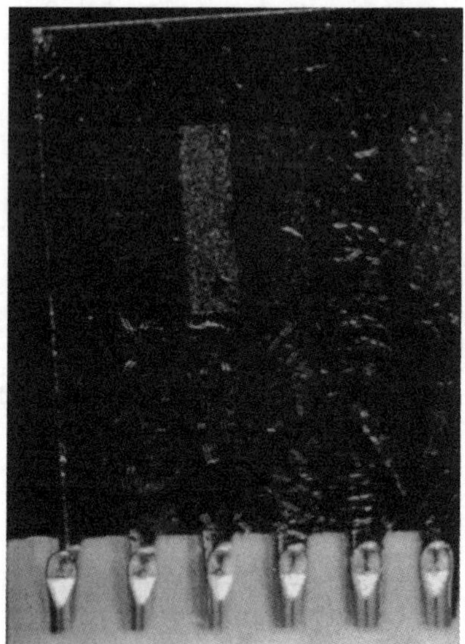

Figure 6 Photograph of an operating device based on Si nanocrystals in PVK

CONCLUSION

Novel thin EL films comprised of nanocrystals embedded in appropriate host matrices show significant potential for EL device applications. Presently, research is being conducted on other semiconductor nanocrystals and their use in optoelectronic devices. The possibility of embedding different nanocrystals into a matrix will enable tunability throughout the visible spectrum. Concurrently we are investigating other transport materials to facilitate electron and hole injection, along with various device configurations and host matrices to reduce turn-on voltages and increase efficiency. Many advantages can be expected from visible light-emitting semiconductor nanocrystalline materials. Among them are low cost, easy processing, high efficiency, conformability, and most importantly, the capability for tuning the output wavelength with a single material system.

Work at LLNL performed under the auspices of the USDoE under contract number W-7405-ENG-48.

REFERENCES:

1. J.L. Heinrich, , C.L. Curtis, G.M. Credo, K.L. Kavanagh, and M. J. Sailor, Science **255**, 66 (1992).
2. L. T. Canham, Appl. Phys. Lett. **57**, 1046 (1990)
3. Porous Silicon Science and Technology, edited by J.C. Vial and J. Derrien, Springer-Verlag, New York, (1995).
4. S.F. Chuang, S.D. Collins and R.L. Smith, Appl. Phys. Lett. **55**, 675(1989)
5. G. Bomchil, R. Herino, K. Barla and J.C. Pfister, J. Electrochem. Soc. **130**, 1611 (1983)
6. Porous Silicon, edited by Z. C. Feng and R. Tsu, World Scientific Publishing Company, River Edge, New Jersey, 1994.
7. See, for example, Materials Research Society Symposium Proceedings **256**, (1994).
8. See, for example, Materials Research Society Symposium Proceedings **283**, (1992), Chapters I - V.
9. G.R. Delgado and H.W.H. Lee, Phys. Rev. Lett, submitted.
10. C. Dellerue, G. Allan, M. Lannoo, Phys. Rev. B. **48**,11024 (1993).
11. L.W. Wang, A. Zunger, Phys. Rev. Lett. **73**,1039 (1994).

VISIBLE LIGHT EMITTING DIODE EMPLOYING ELECTROCHEMICALLY ANODIZED NANOCRYSTALLINE SILICON THIN FILM

T. TOYAMA, T. YAMAMOTO, T. MATSUI, and H. OKAMOTO
Department of Electrical Engineering, Faculty of Engineering Science, Osaka University, Toyonaka, Osaka 560, Japan

ABSTRACT

Visible electroluminescence (EL) has been achieved on the entirely solid state thin film light emitting diode (TFLED) employing electrochemically anodized nanocrystalline Si (*nc*-Si) as a light emitting active layer. The TFLED consisting of *p*-type *nc*-Si, and *intrinsic* and *n*-type amorphous layers was fabricated on a SnO_2-coated glass substrate. The *nc*-Si was formed in HF aqueous solution from boron doped microcrystalline Si (μc-Si) deposited by rf plasma chemical vapor deposition (CVD). The TFLED exhibits clear rectification with a forward threshold voltage of about 1.5 V, whereas visible EL emission is observed upon applying reverse bias voltages. The diode ideality factor is more than 2, and the light output increases with the square of the diode current. The EL emission color is orange-red and the spectral peak energy is 1.8 eV.

INTRODUCTION

Extensive efforts have been carried out to develop a novel light emitting diode (LED) based on nanocrystalline silicon (*nc*-Si) formed by electrochemically anodized Si (porous Si) which shows strong visible photoluminescence (PL) [1,2]. Since the first observation of electroluminescence (EL) in Schottky barrier diodes [3] and p-*n* junction diodes [4], the emission color extends over most of the visible spectral region (including blue [5]) and the external quantum efficiency can reach about 0.2% [6]. Therefore it would be an attractive approach to fabricate light emitting *nc*-Si thin films due to combining both fabrication techniques of microcrystalline (μc-Si) thin films and porous Si for development of large- area light emitting devices. Little work has been done to obtain EL emission from electrochemically formed nc-Si thin films produced from plasma-chemical-vapor-deposition (CVD)-prepared μc-Si [7], as well as from polycrystalline Si fabricated by low pressure CVD [8], however EL emission has been obtained using an electrolyte junction in both cases.

Microcrystalline Si deposited by plasma CVD have been successfully applied to optoelectronic devices such as solar cells, image sensors and thin film light emitting diodes (TFLEDs) due to the higher dark conductivity than that of amorphous Si (*a*-Si) by several orders of magnitude [9,10]. The device fabrication utilizes fully the benefit of the plasma CVD method (i.e., large area deposition at low temperature on several substrates involving glass, steel, and polymer films). In this article, we present a solid state *p-i-n* junction TFLED which incorporates *p*-type *nc*-Si electrochemically anodized from μc-Si. Device performance of the nc-Si TFLED including EL and current–voltage characteristics is discussed. Raman and PL studies are also presented on the *nc*-Si thin films [10].

EXPERIMENTAL

The structure of the TFLED is illustrated in Fig. 1. The *p*-type *nc*-Si thin film was electro-

Mat. Res. Soc. Symp. Proc. Vol. 452

chemically formed on a SnO_2-coated glass substrate from boron-doped *p*-type μc-Si. The μc-Si was deposited by the rf plasma CVD method with a low deposition temperature of 180°C. The rf power was roughly 2.5 times higher than that for an *a*-Si film. The thickness of the μc-Si was 200 nm as deposited. A mixture of SiH_4, B_2H_6 and H_2 gases was used as a source gas. The B_2H_6/SiH_4 flow rate ratio was varied 0.1 to 2.5 vol.% to change the crystalline volume fraction in the film. The SiH_4/H_2 flow rate ratio was set to be less than ~3 vol.%. Electrochemical anodization was performed in an HF aqueous solution. A 50% HF solution was further diluted by H_2O into 1/20–1/50. The anodizing current density and time were changed in the range of 1.4–9.7 mA/cm^2 and 0.5–5 min, respectively. In this experiment, illumination was not employed during the anodization. A 50-nm-thick *intrinsic a*-Si, and a 50-nm-thick *n*-type *a*-SiC were sequentially deposited on the *p*-type *nc*-Si layer by the plasma CVD method at the deposition temperature of 180°C. The *n*-type a-SiC layer has used in a-SiC *p-i-n* junction TFLED as a wide-gap injection layer [11]. Finally, evaporation of Al back contact with an area of 0.033 cm^2 completed the construction of the TFLED.

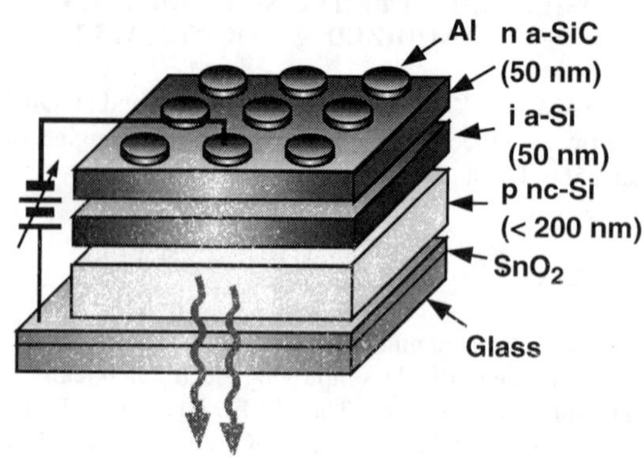

FIG. 1. Structure of *p*-type nc-Si/ *intrinsic* a-Si/ *n*-type a-SiC heterostructure TFLED. The *nc*-Si layer was electrochemically anodized from plasma-CVD-deposited μc-Si. The thickness of the *p* layer must decrease from the μc-Si thickness of 200 nm as deposited.

The PL and EL measurements were carried out at room temperature. The PL sample was excited by the He-Cd laser light with a wavelength of 325 nm. The PL and EL emission light was dispersed by a 20-cm monochromator with a 600-nm grooved grating and detected by a cooled GaAs(Cs) photomultiplier (Hamamatsu Photonics R943-02). The spectra were corrected for the spectral response of the optical system. Raman scattering spectrum was measured at room temperature with the use of a microprobe system including a 514.5-nm Ar^+ laser (JEOL JRS-SYSTEM1000).

RESULTS AND DISCUSSION

Properties of *nc*-Si thin films

Figures 2 and 3 show the Raman scattering and PL spectra for samples formed with the various anodizing current densities ranging from 1.4 to 9.7 mA/cm^2. The Raman spectrum for 1.4 mA/cm^2 is almost unchanged from the spectrum of as-deposited μc-Si. The Raman spectra are found to be twofold lying at 510 cm^{-1} and around 480 cm^{-1} corresponding to the TO phonon mode scattering due to the crystalline component and the amorphous-like component in the films, respectively [12]. With an increase in the anodizing current density, the fraction of amorphous-like component in the spectrum becomes larger comparing with the crystalline component. The greater amorphous/crystal ratio for the higher anodizing current density is likely due to; (1) a reduction in the grain size of

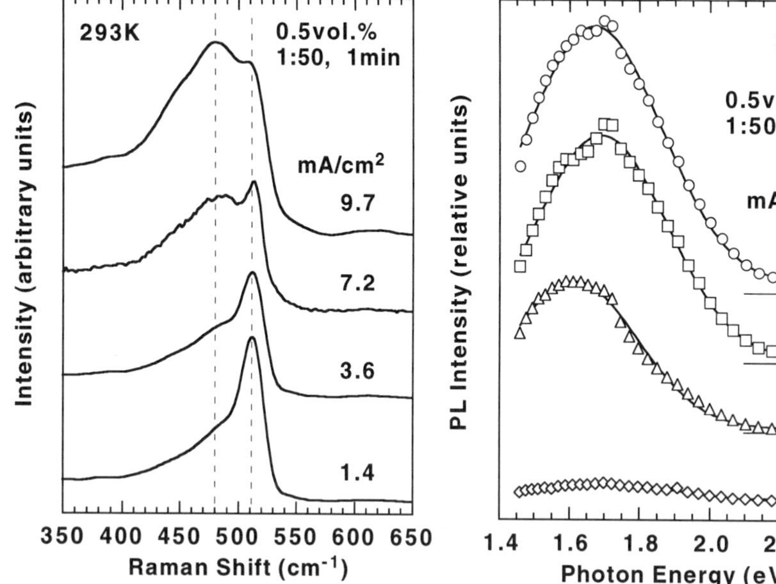

FIG. 2. Raman scattering spectra for samples with changing the anodizing current densities ranging in 1.4–9.7 mA/cm^2. The spectra consist of two components to be ascribed to a crystal component peaking at about 510 cm^{-1} and amorphous-like broad component at about 480 cm^{-1}.

FIG. 3. PL spectra from nc-Si excited by a He-Cd laser (λ_{exc} = 325 nm) at room temperature. The samples were anodized for 1 min at different anodizing current densities [9.7 mA/cm^2, ○ ; 7.2 mA/cm^2, □ ; 3.6 mA/cm^2, △; 1.4 mA/cm^2, ◇]. The lines are fitting results to the Gaussian form.

the crystallites in the film; (2) an increase in the surface states on the crystallites would induced due to the reduction in the grain size; (3) an increase in volume fraction of amorphous-like tissue in the film. These three issues have been proposed for the origin of PL in nc-Si [2].

As shown in Fig. 3, the broadband PL spectra are observed from the anodized *nc*-Si thin films except for the sample anodized at 1.4 mA/cm^2 of which the Raman spectrum is almost unchanged from the μc-Si. Gaussian spectral shape with the peak energies of 1.61–1.69 eV and the full widths at half maximum (FWHMs) of 0.52 eV is similar to the 'S'-band PL spectrum from porous Si [2]. A comparison of the PL result in Fig. 3 with the Raman results in Fig. 2 yields that the PL spectra shift slightly toward higher energies and the PL intensities increase gradually with an increase in the amorphous/crystal ratio in the Raman spectrum. This correlation alone can not identify whether the luminescence is induced by the decrease in the size of the crystallites or the increase in the amorphous part.

In contrast, the correlation between the crystalline volume fraction in the μc-Si and the PL intensity is indicative of the light emission induced by finite size crystallites. As we have already mentioned in the previous report [10], a comparison between both Raman and PL results with changing the crystalline fraction in μc-Si due to varying gas phase boron doping concentration indicates that stronger PL emission tends to occur on the samples of which the crystalline fraction is larger before the anodization. In the case of anodized a-Si thin films, PL emission was not detected after the anodization at any anodizing current. This implies that the luminescence is mainly ascribed to the crystallite-related origins, i.e., the quantum confinement effects or surface states on the crystallites [2]. Visible PL emission in anodized *a*-Si thin films, however, has been reported by Bustarret *et al.* [13] and Wehrspohn *et al.* [14]. Since the occurrence of the quantum confinement effects in *a*-Si have been already evidenced in *a*-Si-based multilayer by optical [15] and electrical measurements [16], the PL emission in the anodized *a*-Si is expected basically due to the quantum or the spatial confinement of the photocreated carries [14]. Nevertheless, our Raman and PL results indicate that the luminescent efficiency of the *nc*-Si produced from the *μc*-Si is superior to that of the anodized *a*-Si.

Device performance of TFLED

As shown in Fig. 4, the entirely solid state TFLED consisting of *p*-type *nc*-Si, and *intrinsic* and *n*-type amorphous layers exhibits clear rectification in the current – voltage (*J–V*) characteristics with a threshold voltage of about 1.5 V in the forward direction (applying positive bias on SnO_2) and the soft breakdown voltage of about 4–5 V for the reverse direction. The rectification ratio at 3 V is 33 and the series resistance is found to be almost negligible. According to the conventional diode *J–V* characteristics [17], the diode ideality factor *n* is estimated to be quite large (more than 10). Furthermore, as shown in the inset of Fig. 4, the exponential approximation can apply to the "reverse" current characteristics with the huge ideality factor *n* of 50. The diode ideality factor more than 2 is also found in *n*-type indium tin oxide (ITO)/ *p*-type porous Si heterojunction diode in which the large factor (n = 8–10) is explained as due to the interface layer at the Si crystallite boundary [18,19].

The EL emission was detected only under the reverse bias voltage within the measurable spectral range of the photomultiplier (300–850 nm). In Fig. 5, the unmonochromated light output *L* is plotted against the current density *J* ranging in 0.1–2 A/cm^2. The light output increases as the square of the current density. The square-law in the *L–J* characteristics is also observed in the *n-type* ITO/ *p*-type porous Si heterojunction at current densities less than about 2 A/cm^2 in which it is likely to be ascribed to recombination at defects in the space charge region [18,19]. Since both *J–V* and *L–J* characteristics in the TFLED behaves as those in the *n*-type ITO/ *p*-type porous Si heterojunction diode, it maybe expected that the EL emission in the TFLED occurs at the *n*-type SnO_2 / *p*-type *nc*-Si junction region rather than at the *p-i-n* heterojunction.

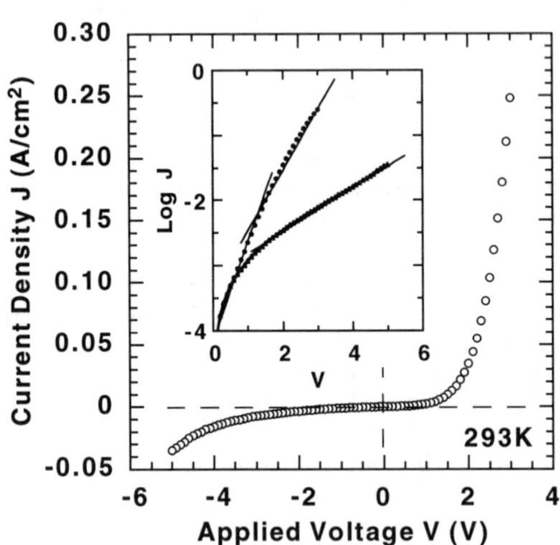

FIG. 4. *J–V* characteristics of the nc-Si TFLED at room temperature showing clear rectification. The diode ideality factor is quite large (more than 2). Inset shows the logarithmic plot of the *J–V* characteristics.

Figure 6 shows the EL spectrum of the *nc*-Si TFLED under a reverse bias voltage of 4 V. The light emis-

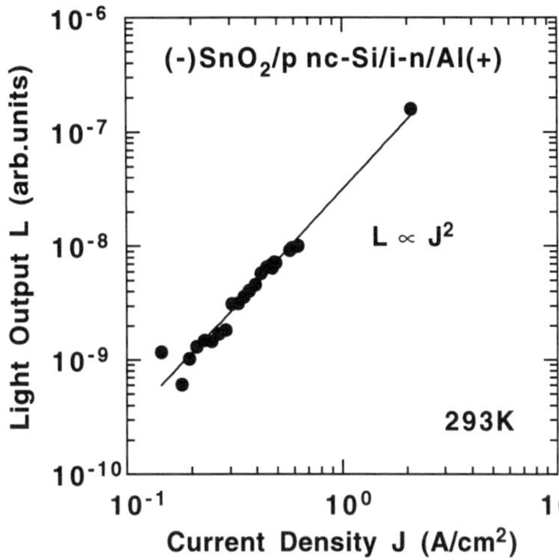

FIG. 5. *L–J* characteristics of the *nc*-Si TFLED at room temperature. The light output increases with the square of the current density.

sion from the TFLED can be detected by the naked eye and the emission color is orange-red. Also shown in Fig. 6 is the PL spectrum of the *p*-type *nc*-Si layer fabricated by the same conditions including the μc-Si deposition and the electrochemical treatments. PL and EL measurements were carried out at room temperature employing the same detection system but with the observation directions from the *nc*-Si film side for PL and from the SnO_2 / glass substrate side for EL. If PL is measured from the glass side, the visible PL emission from the commercial glass substrate merges

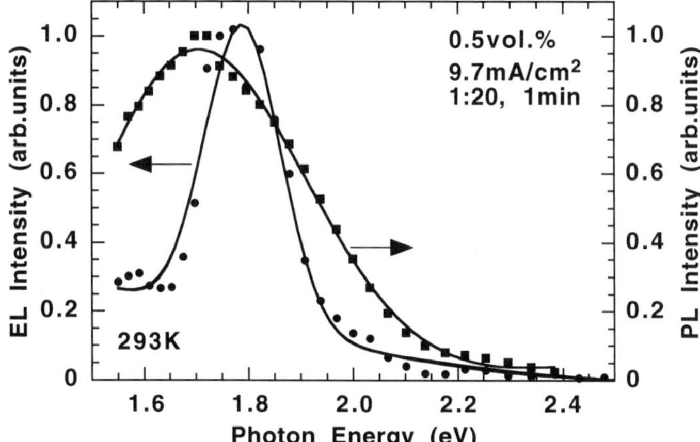

FIG. 6. EL (●) and PL (■) spectra from nc-Si thin film fabricated under the same conditions. EL spectrum was measured applying a reverse bias voltage of 4 V on the *p-i-n* structure TFLED. The lines are the fitting results to the Gaussian form, and the EL peak energy lies in 1.8 eV with FWHM of 0.2 eV.

into the PL emission from *nc*-Si. The EL spectrum from the TFLED shows the peak energy of about 1.8 eV and FWHM of 0.2 eV. The EL spectrum shifts to the higher energy and becomes narrower with a pronounced reduction of its lower energy emission in comparison with the PL spectrum. The spectral difference between the EL and the PL spectra maybe due to Fabry-Perot interference which weakly superposed on the EL spectrum.

CONCLUSIONS

We have demonstrated the possibility of an entirely solid state TFLED on an SnO_2 / glass substrate using electrochemically anodized *p*-type *nc*-Si as a light active layer. Judging from the Raman scattering and PL measurements of the *nc*-Si, the presence of the crystallites is likely to play a crucial role in determining the light emitting performances. The TFLED consisting a *p-i-n* structure of the *nc*-Si and amorphous layers exhibits clear rectifying *J–V* characteristics, while the visible EL emission is observed with applying the negative bias voltage on the SnO_2 electrode. The diode ideality factor more than 2 and the quadratic increase in the light output against the current density are found. The EL emission peak energy locates at 1.8 eV and observed emission color is orange-red.

REFERENCES

1. as recent studies, Microcrystalline and Nanocrystalline Semiconductors, edited by R.W. Collins, C.C. Tsai, M. Hirose, F. Koch and L. Brus (Mater. Res. Soc. Proc. **358**, Pittsburgh, PA, 1995).
2. L.T. Canham, Phys. Stat. Sol. (b) **190**, 9 (1995) and references there in.
3. A. Richter, P. Steiner, F. Kozlowski, and W. Lang, IEEE Elect. Device Lett. **12**, 691 (1991).
4. Z. Chen, G. Bosman, and R. Ochoa, Appl. Phys. Lett. **62**, 708 (1993).
5. P. Steiner, F. Kozlowski, M. Wielunski, and W. Lang, Jpn. J. Appl. Phys. **33**, 6075 (1994).
6. J. Linnros and N. Lalic, Appl. Phys. Lett. **66**, 3048 (1995).
7. E. Bustarret, M. Ligeon, J. C. Bruyère, F. Muller, R. Hèrino, F. Gaspard, L. Ortega, and M. Stutzmann, Appl. Phys. Lett. **61**, 1552 (1992).
8. N.M. Kalkhoran, F. Namavar, and H.P. Maruska, J. SID, **3/1**, 3 (1995).
9. J. Kanicki, Amorphous and Microcrystalline Semiconductor Devices: Optoelectronic Devices (Artech House, London, 1991) Vol. 1.
10. T. Toyama, T. Matsui, Y. Kurokawa, H. Okamoto, and Y. Hamakawa, Appl. Phys. Lett. **69**, 1261 (1996).
11. Y. Hamakawa, D. Kruangam, T. Toyama, M. Yoshimi, S.M. Paasche, and H. Okamoto, Optoelectron. Devices and Technol. **4**, 281(1989).
12. Z. Iqbal and S. Veprek, J. Phys. C **15**, 377 (1982).
13. E. Bustarret, M. Ligeon, and M. Rosenbauer, Phys. Stat. Sol. (b) **190**, 111 (1995).
14. R.B. Wehrspohn, J.-N. Chazalviel, F. Ozanam, and I. Solomon, Phys. Rev. Lett. **77**, 1885 (1996).
15. K. Hattori, T. Mori, H. Okamoto, and Y. Hamakawa, Phys. Rev. Lett. **60**, 825 (1988).
16. S. Miyazaki, Y. Ihara, and M. Hirose, Phys. Rev. Lett. **59**, 125 (1987).
17. S.M. Sze, Physics of Semiconductor Devices, 2nd ed. (Wiley, New York, 1981).
18. F. Namavar, H.P. Maruska, and N.M. Kalkhoran, Appl. Phys. Lett. **60**, 2514 (1992).
19. H.P. Maruska, F. Namavar, and N.M. Kalkhoran, Appl. Phys. Lett. **61**, 1338 (1992).

ELECTROLUMINESCENCE AND CARRIER TRANSPORT IN LEDS BASED ON SILICON-RICH SILICON OXIDE

L. Tsybeskov, K. D. Hirschman*, S. P. Duttagupta, and P. M. Fauchet**

Department of Electrical Engineering, University of Rochester, Rochester, NY 14627
* also Department of Microelectronic Engineering, Rochester Institute of Technology, Rochester NY 14623
** also Laboratory for Laser Energetics, Department of Physics & Astronomy, and The Institute of Optics, University of Rochester, Rochester NY 14627

ABSTRACT

LEDs based on silicon-rich silicon oxide (SRSO) have been fabricated utilizing thermal oxidation of electrochemically etched c-Si and standard microelectronic processes including LPCVD, photolithography, ion implantation and metallization. The bipolar devices consist of heavily-doped polycrystalline Si, a transition layer made of oxidized mesoporous Si, an active SRSO layer (doped or intrinsic) and a crystalline Si substrate. The LED's electrical properties exhibit a low interface state density which is explained by the partial filling of the micropores within the transition layer by polycrystalline Si. The dominant carrier transport in SRSO is due to thermally-assisted tunneling at low fields and electric field-assisted (Fowler-Nordheim) tunneling at high fields. The electroluminescence (EL) is stable and the EL modulation is limited by the carrier transition time (not by the carrier lifetime) which explains the high modulation speed ($\geq$ 10 MHz). We have fabricated a prototype alphanumeric 7-segment display with an isolated version of the LEDs and standard microelectronic processing techniques.

INTRODUCTION

The electroluminescence (EL) in porous Si (PSi) based light-emitting diodes (LEDs) has been a subject for investigation for the last 5 years [1]. Currently, the EL efficiency ~0.2% is achieved under CW operation with relatively low operating bias [2]. However, the lifetime of PSi-based LEDs is less than one hour. The EL degrades drastically due to unstable surface passivation (see Ref. 1). We have recently reported the fabrication of LEDs from partially oxidized porous silicon (PSi) which demonstrates electroluminescence (EL) without degradation over 250 hours of pulsed mode operation [3]. The oxidation procedure is responsible for altering the electrical characteristics and electroluminescent properties of the device [4]. Si grains with high-quality passivation produces primarily red-band EL (~ 1.8-2.0 eV), where Fowler-Nordheim (field-assisted) tunneling is the dominant mechanism of carrier transport. On the other hand samples with poor quality surface passivation primarily exhibit near infrared (IR) EL with a spectral peak at ~ 1.6 eV, where carrier transport is controlled by a thermally-assisted trap and escape mechanism. The EL behavior is explained by two different mechanisms of radiative recombination of injected carriers. This is responsible for the coexistence of red and near IR EL bands. At low temperatures (< 200K) the difference between the two types of samples is clear, however at room temperature the two mechanisms of carrier transport and recombination compete. Similar two-band PL spectra have been observed. The quality of surface passivation has been found to be responsible for several device parameters including reverse-bias leakage current, device charging, transient characteristics and the EL efficiency [4]. In this work we will focus on accelerated aging and a new device design which enhances EL output.

Mat. Res. Soc. Symp. Proc. Vol. 452

EXPERIMENTAL

Pre-anodization Processing

The initial substrates are (100) p-type crystalline Si with a resistivity of ~ 10 Ω cm. A high dose BF_2 implant is done on the backside of the substrate to provide a low resistance contact for anodization. A 1500Å Si_3N_4 film is deposited via low pressure chemical vapor deposition (LPCVD). The frontside of the wafers is then patterned with the first masking level (active), which defines the areas of local anodization. The Si_3N_4 film is removed via reactive ion etching (RIE) to open windows to the silicon substrate. The film is also completely removed from the backside of the wafer using the same method. A second BF_2 implant (dose ≈ $10^{15}cm^{-2}$) is done in the active regions. After photoresist removal and a cleaning procedure, the wafers are annealed at 1000°C for 15 min in N_2 to activate the impurities. The wafers are then ready for anodization.

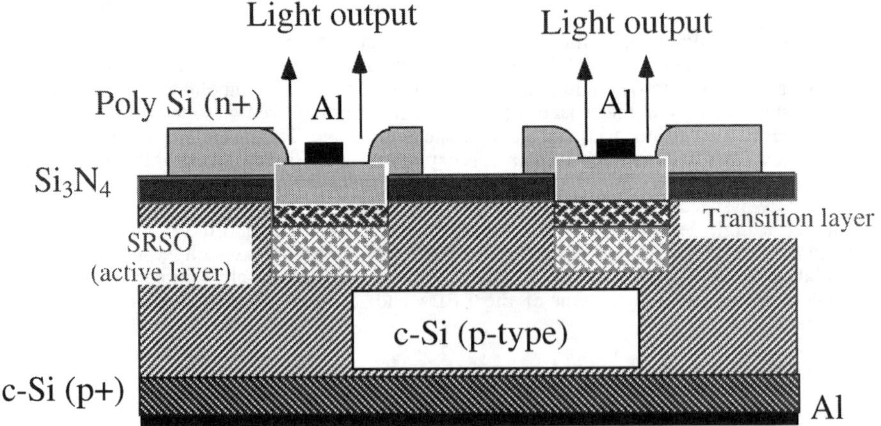

Figure 1 Complete device structure (metal/heavily doped /c-Si substrate/high porosity well (active layer)/low porosity well (transition layer) /insulating layer/ poly Si/ Metal)

Anodization and Annealing

A 10 Ω cm p-type crystalline silicon wafer with a heavily doped p+ surface layer is anodized in an $HF:H_2O:C_2H_5OH$ (1:1:2) solution under a constant current density J ≈ 15 mA/cm^2. The p+ region is transformed into a mesoporous layer (porosity ~ 40%) whereas the underlying p-type silicon is transformed into a nanoporous silicon layer (porosity ~ 75-80%). Annealing at 800-900°C is performed in a dilute oxygen ambient (10% O_2 in N_2). Hydrogen on the surface of the Si nanoclusters is replaced by oxygen, changing the surface coverage of Si grains from Si-H to Si-O bonds. The material produced by controllable thermal oxidation of PSi can be described as silicon-rich silicon oxide (SRSO).

Cathode and Contact Formation:

Poly-Si is chosen as the top contact material (Fig. 3). A 0.3 μm poly-Si film, which is transparent in the red-IR part of the spectrum, is produced via LPCVD performed at 610°C. This process does not damage samples which were previously annealed at 800-900°C. The poly-Si layer is patterned photolithographically to define the device contact region. To provide efficient electron injection during device operation, the poly-Si layer is heavily doped (n+) by phosphorus ion implantation with a dose of 10^{15} cm^{-2} at an energy of 50 KeV, followed by an activation anneal which produces a low resistivity polysilicon contact ($\rho \leq 0.1$ Ω cm). Aluminum contacts are then made to the n+ poly-Si film and the bulk substrate. The gap between top Al contact and the border of the active region is ~ 10 μm. The complete device structure can be seen in fig. 1.

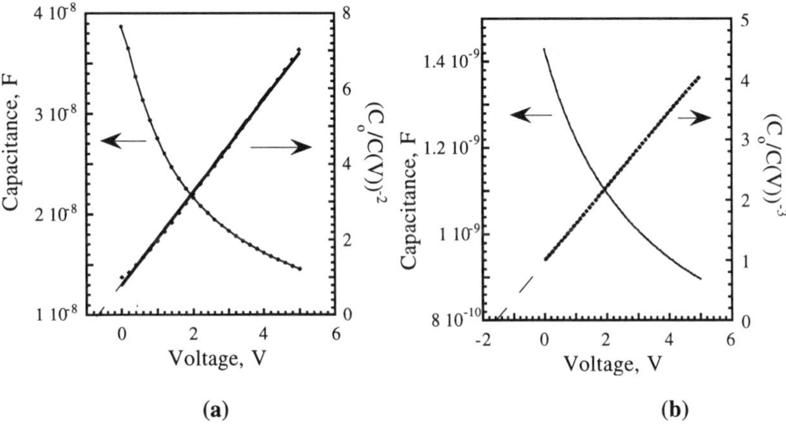

Figure 2. C-V data for sample (a) which has a "step" junction ($C \sim V^2$), and sample (b) which has a "quasi-linearly" graded junction ($C \sim V^3$) [5] correlates well with Supreme-3 1-D process simulation. The results indicate a relatively low interface state density (less than 10^{12} cm^{-2}) between poly-Si and the mesoporous transition layer.

ANALYSIS AND DISCUSSION

The role of mesoporous (≤ 30 % porosity) transition layer is to provide good electrical contact and improve thermal conductivity in the devices [3, 4]. Fig. 2 shows the C-V analysis results which indicate a relatively low interface state density (less than 10^{12} cm^{-2}) between poly-Si and the mesoporous transition layer. The partial filling of pores is one of possible reason for a smooth transition: in mesoporous Si the size of pores are larger than in nanoporous Si [6]. SEM picture of device cross-section has confirmed the smooth transition between poly Si and mesoporous Si. The refractive index of SRSO ($n \leq 2$) is significantly lower then that of c-Si (n=3.5). Therefore, the big part of the EL emission can be reflected back and absorbed by the c-Si substrate due to total internal reflection. In our latest design using local anodization we have introduced an additional scattering near the border of SRSO well, and an improvement of EL intensity has been found. The detectable EL threshold in our best devices has been found to be near 2 V [3, 4]. Therefore, we conclude that the major part of the voltage drop is over the p-i-n junction. In addition, the EL is relatively fast. In our test structure the risetime was less than 100 ns and decay time was near 2 μs (Fig. 3). The exponential decay of EL (Fig. 4) may be governed by an RC constant, where R and C are the resistance and capacitance of the device.

This is not a contradiction with long microseconds PL lifetime [8], because in the submicron-thick active layer the EL is limited by a transition time and not a carrier lifetime.

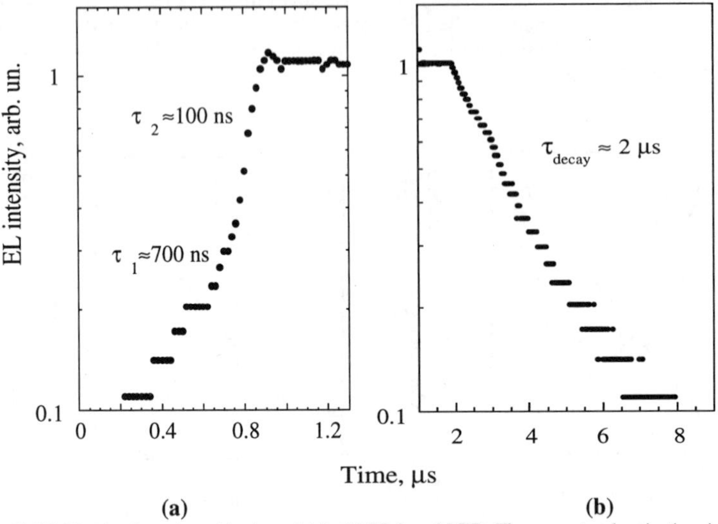

Figure 3. EL kinetics (on a logarithmic scale) in SRSO-based LED. The current pulse risetime is shorter than 20 ns. The EL risetimes (a) and decay time (b) extracted from the EL kinetics are indicated. The bandwidth of the detector preamplifier is near 10 MHz and may be responsible for the EL risetime. The exponential shape of the EL decay is most likely due to the RC time constant, where R and C are the resistance and capacitance of the device

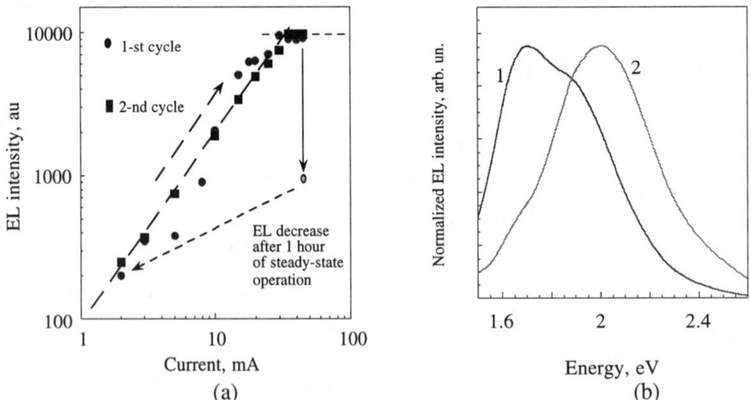

Figure 4. Accelerated aging test shows a reversible EL decrease(a) and normalized EL spectra (b) in linear regime (1) and after 3 hours of operation in saturated regime (2). The near-IR EL band is suppressed and the red EL band is unchanged.

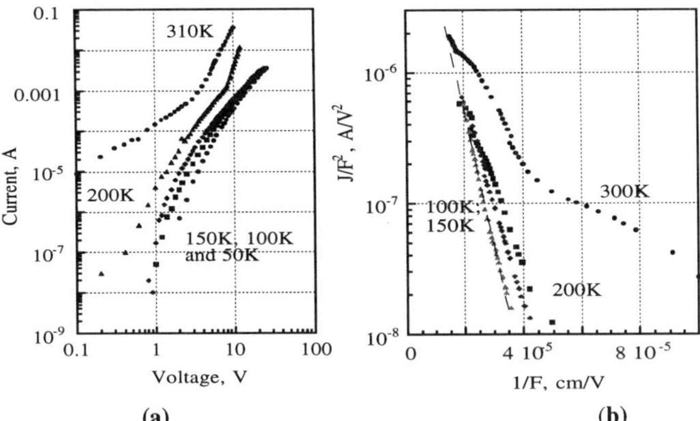

Figure 5. I-V curves at different temperatures in samples with thermally-assisted carrier transport (a) and field-assisted tunneling (b). In Fig. 5b the ratio of the current density J over the square of the average electric field F is plotted versus 1/F (Fowler-Nordheim plot). The average field is calculated as the ratio of the applied voltage and the average thickness of the PSi layer.

The EL stability has been tested under pulsed current in the regime of low power output; no degradation has been found during weeks of continuos operation [3, 4]. In CW operation the EL intensity is linear with respect to current (linear regime) and saturates at high current density (saturated regime) (Fig. 5 a). The accelerated aging test has been performed under steady-state current in the saturated regime. Figure 5a shows that under these conditions the EL intensity drops an order of magnitude after one hour. However, the EL is totally recovered after several minutes of zero bias. The EL spectrum in linear regime is similar to the PL spectrum with a peak near 1.65 eV and a shoulder at ~1.9 eV (Fig. 5b) [7]. However, after several hours of operation in saturated regime, the near-IR EL is suppressed and the EL peak shifts to ~ 2 eV. Our resent study has shown, that the red PL band in SRSO is similar to the PL in as anodized PSi, and the near infra-red (near-IR) PL is defect related [7]. The properties and origin of the PL at ~ 1.6 eV have been studied and discussed in Ref. 9. The reversible EL decrease we explained as charging of defect states due to localization of injected carriers according to mechanism proposed in Ref. [4]. Therefore, we conclude that the near I-R EL also related to recombination via defect states. The red EL band is not affected by charging; currently the origin of red EL remains unclear.

The current-voltage (I-V) relationship in our devices exhibits rectifying behavior with the ratio between direct and reverse current ~ 100-1000 at low bias ($V \leq 5$ V): A leakage current seems to be a serious problem for such an LED. Recent measurements show that the leakage current can be suppressed by a surrounding guard-ring that can improve the rectification ratio [10] and increase the power efficiency which currently remains to be ~ 0.1% for our best devices. The carrier transport is found to be thermally-assisted or close to the Fowler-Nordheim law with an additional component which possibly corresponds to thermally-assisted carrier transport via defect states (Fig. 5) [4]. At low temperatures a defect-related component is quenched (Fig. 5), however the calculation of a potential barrier yields an unreasonably low value for the standard field-assisted mechanism of tunneling (less than 100 meV). The details of carrier transport is not well understood.

An application of the improved devise structure is demonstrated by the fabrication of a prototype alphanumeric 7-segment display (Fig. 6). This type of display can find an application in virtual reality (VR) computer simulation. Currently, a submicron resolution is photolithography limited, and transient characteristics are better than 10 MHz. The VR display operates at low-light level for a dark-accommodated eye. Figure 6 shows the display design with several glowing segments. The emission is bright and can be clearly seen with unaded eye.

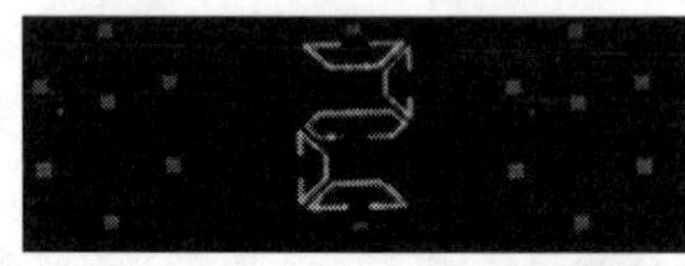

Figure 6. A prototype alphanumeric 7-segment display with several glowing elements.

CONCLUSION

The design, fabrication and light-emitting properties of an improved SRSO-based LED have been reported. The transient and steady-state characteristics of an LED and the device stability have been tested. The EL intensity decrease during the test of accelerated aging is found to be reversible, and a charging of defect states seems to be responsible for EL steady-state and transient responses. The EL spectra in SRSO-based LED is similar to the PL spectra, and the EL kinetics are relatively fast.. We conclude that the observed EL is due to recombination of injected carriers in a diode-like structure. The complete model of EL requires additional investigation of carrier recombination and transport. The device efficiency and quality of surface passivation must be improved in order to achieve performance level approaching that of commercial LEDs.

This work was supported by the U.S. Army Research Office.

REFERENCES

1. P. M. Fauchet *et al.*, IEEE J. Selected Topics in Quantum Electronics **1**, 126 (1995).
2. A. Loni *et al.*, Electron. Lett. **31**, 1288 (1995).
3. L. Tsybeskov *et al.*, Appl. Phys. Lett. **68**, 2058 (1996).
4. L. Tsybeskov *et al.*, in *Advanced Luminescent Materials,* edited by D. J. Lockwood, P. M. Fauchet, N. Koshida and S. R. J. Brueck (Electrochemical Society, Pennington, NJ) 1996, 34-47.
5. S. M. Sze, Physics of Semiconductor Devices (J. Wiley & Sons, New York, NY), 1969, pp. 812.
6. R. L. Smith and S. D. Collins, J. Appl. Phys. **71**, R1 (1992).
7. L. Tsybeskov *et al.*, Preparation and characterization of the active layer for an LED based on oxidized porous silicon, this volume.
8. J. C. Vial, in *Porous Silicon Science and Technology,* edited by J. C. Vial and J. Derrien, Springer-Verlag, Berlin, 1995, p. 137.
9. S. M. Prokes and W. E. Carlos, J. Appl. Phys. **78**, 2671 (1995).
10. K. D. Hirschman *et. al.*, Nature **384**, 338 (1996).

PREPARATION AND CHARACTERIZATION OF THE ACTIVE LAYER FOR AN LED BASED ON OXIDIZED POROUS SILICON

L. Tsybeskov, K. D. Hirschman*, L. F. Moore**, P. M. Fauchet*** and P. D. J. Calcott****

Department of Electrical Engineering, University of Rochester, Rochester, NY 14627

* Also Department of Microelectronic Engineering, Rochester Institute of Technology, Rochester NY 14623

** Department of Physics & Astronomy, University of Rochester, Rochester NY 14627

*** Also Laboratory for Laser Energetics, Department of Physics & Astronomy and The Institute of Optics, University of Rochester, Rochester NY 14627

**** DRA Malvern, St Andrews Rd, Malvern, Worcs WR14 3PS, UK

Abstract

We have studied the photoluminescence (PL) in oxidized porous silicon (PSi), prepared from anodized crystalline Si followed by annealing at temperatures ranging from 700 to 1000°C. It has been found that two PL bands with spectral peaks at 1.6 eV (near-IR band) and near 2 eV (red band) exist with a strong dependence on preparation (annealing) conditions. Recent experimental results show a correlation between the intensity of the near-IR band and the level of leakage current in the diode-like structure. The suppression of the near-IR emission results in improved carrier transport, and the enhancement of the red band emission maximizes the electroluminescence (EL) efficiency. The PL study of thermally oxidized PSi indicates different recombination mechanisms. The red PL band is associated with a mechanism similar to band-tail-recombination within the quasi-bandgap of Si nanograins, whereas the near infra-red PL is associated with recombination via defect centers. These mechanisms will be discussed.

Introduction

Crystalline silicon (c-Si) is the dominant semiconductor material for microelectronics, and there has been an effort over several decades to find a place for silicon in optoelectronics. Major approaches are band-structure engineering in SiGe superlattices [1] and doping by isoelectronic impurities [2] or rare-earth metals [3]; however, none have been found to provide efficient luminescence at room temperature. In contrast, electrochemically prepared porous Si (PSi) exhibits strong visible room temperature photoluminescence (PL) with ≥ 1 % external quantum efficiency (EQE) [4]. However, the PL in PSi is not stable and can degrade after several hours of storage at the room temperature in a laboratory environment. The degradation of the PL and the electroluminescence (EL) has been found and studied intensively [5, 6, 7]. The proposed mechanism is the effusion of hydrogen from the as-anodized PSi surface [7]. The rate of hydrogen effusion increases drastically at high temperatures ($T > 400°C$). The irreversible thermal quenching of the PL and EL makes it impossible to use standard microelectronic thermal processes ($T \sim 1000°C$) in the fabrication of PSi-based devices.

The EL in PSi was reported shortly after the discovery of PL [8]. A typical PSi-based LED consists of a transparent or semitransparent contact (Au, ITO or conducting polymers) and a PSi layer (thickness between 1-10 μm) on a crystalline silicon (c-Si) substrate (p- or n-type) [8-10]. The characteristics of these structures, which can be modeled as a type of Schottky

Mat. Res. Soc. Symp. Proc. Vol. 452

junction, are not promising due to the lack of an ohmic contact to the PSi layer and a large number of defects at PSi/metal interface. These devices have exhibited low EL external quantum efficiency ($\leq 10^{-4}$) and high EL threshold conditions ($V \geq 10$ V and $J \geq 10$ mA/cm^2). Improvements in the external quantum efficiency have recently been reported, however the stability of such LEDs continues to be a serious problem for practical applications because the efficiency decreases drastically during the first hour of operation [5, 6].

A strategy where hydrogen coverage is replaced by oxygen passivation has been explored. A controllable, low-rate oxidation in a diluted oxygen ambient is used to convert PSi to silicon-rich silicon oxide (SRSO) [11]. Ideally this material consists of Si nanoclusters covered by a thin SiO film, and it may combine the attractive light-emitting properties of PSi with a stable structure of strongly non-stoichiometric SiO. Carrier transport in SRSO has been studied and attributed to Fowler-Nordheim tunneling [12]. Unfortunately, the process of PSi thermal oxidation can not be compared with conventional oxidation of c-Si; the reaction starts at significantly lower temperature (~ 400°C) during hydrogen desorption; the process is fast and depends on initial surface morphology which is produced by anodic etching [13]. The PL in SRSO has been found in the spectral region from 1.5 μm to 470 nm [14]. A blue PL band has been reported in strongly-oxidized PSi and explained as radiative recombination within SiO [15]. The PL properties of red and near IR bands as well as the PL origin is not yet clear [14].

In this work we report comprehensive studies of light-emitting properties of SRSO, which will provide essential information for device engineering.

Experiments and Results

The samples for PL measurements were prepared by anodization of boron-doped single-crystal Si wafers with a resistivity ~ 5-10 Ω cm in an $HF:H_2O:C_2H_5OH$ (1:1:2) solution under a constant current density $J \approx 5$-15 mA/cm^2. The thicknesses of PSi layers varies from 0.5 μm to 2 μm. The PL measurements were done in an Air Products close-cycle He cryostat at temperatures between 300 K and 12 K. Visible/near IR PL was excited by He-Cd laser (440 nm) and detected by an Optical Multichanel Analyzer (OMA).

The strongest red PL band has been found in relatively high porosity samples from an initial substrate resistivity of ~ 10 Ω cm after annealing in oxygen diluted in N_2 (~ 10 %) during 20-25 minutes at T= 700°C-800°C. The PL spectrum has a peak at 1.8-1.9 eV with a full width of half maximum (FWHM) near 300 meV (Fig. 1). The maximum EQE has been estimated to be close to 1% at room temperature. The PL was stable under unfocused 440 nm He-Cd laser radiation with a beam diameter of 2 mm and power of 12 mW. An increase of the excitation intensity leads to PL degradation, possibly due to direct heating by the laser beam and the poor thermal conductivity of the high-porosity surface. The PL temperature dependence was found to be weak with a maximum PL intensity near 200 K (Fig. 1). A decrease in sample temperature leads to a 150-200 meV blue shift of the PL spectrum and to a FWHM increase up to 400 meV (Fig. 1). The increase in the annealing temperature $T_{ann} > 850^oC$ provides a suppression of the red PL. However, another PL band rises with a peak energy near 1.6 eV (Figure 2). This near-IR PL band is narrower with a FWHM of 150-200 meV. The PL spectrum of the near-IR band was found to be unchanged when the sample temperature decreases. In most of our samples a coexistence of the red (1.8 eV) and near IR (1.6 eV) PL bands is observed (Figure 2). The single near IR PL band with peak at ~ 1.6 eV also has been found in samples annealed at the same temperature (800°C-850°C) but for longer time (~ 2 hours). The PL intensity dependence as a function of excitation intensity is linear for near IR PL band and sublinear for red PL band.

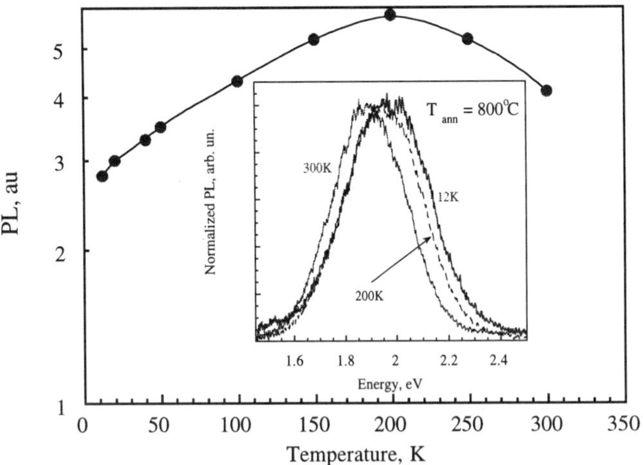

Figure 1. Normalized PL spectra in PSi annealed at T=800°C in diluted O_2 (~10% in N_2) measured at different temperatures and PL intensity temperature dependence.

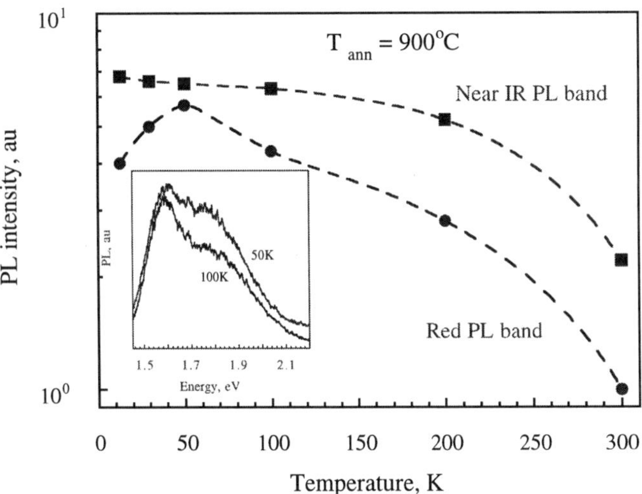

Figure 2. Temperature dependence of the red and near-IR PL bands in PSi annealed at T=900°C in diluted O_2 (~10% in N_2).

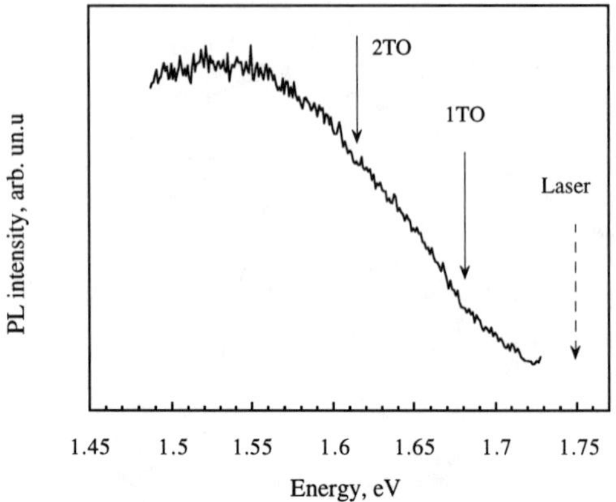

Figure 3. Resonantly excited PL spectra measured at T = 4.2K in oxidized PSi

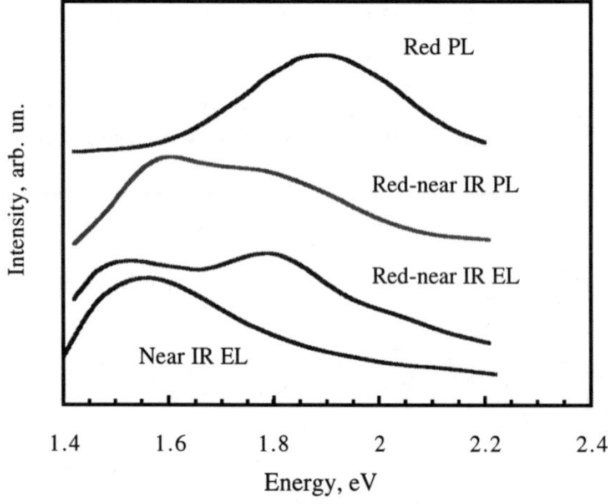

Figure 4. The PL and EL spectra in different samples of oxidized PSi: Double-band luminescence.

Resonantly excited PL spectroscopy has shown to be informative in as anodized PSi. The step-like structure of the PL spectrum indicates phonon-assisted transitions and, the energy of 57 meV corresponds to TO phonon in c-Si [16]. The resonantly excited PL spectrum in our samples exhibits very weak reminiscence of the step-like structure (Fig. 3). Possibly, the PL in our samples only partially originates from Si nanoclusters.

Discussion

Recently we demonstrated the EL in an LED based on SRSO and concluded that the origin of the EL is radiative recombination involving both defect states and silicon nanocrystals [11, 12]. In addition to reasonably good efficiency ($\leq 0.1\%$), our device has low EL threshold ($\geq 2V$).

This indicates that the EL might be due to carrier injection in the diode-like structure. Since a correlation between the PL and EL spectra is important in order to draw conclusions about the EL mechanism, let us compare the PL and EL spectra in our samples. Figure 4 shows the PL and EL spectra in several samples of oxidized PSi. In general, the EL as well as the PL shows coexistence of two bands located near 1.6 eV and 1.8 eV.

Two observed PL bands are similar. However, the major difference is the PL temperature dependence. The observed blue shift of the red PL peak with decrease in sample temperature is near 100 meV, which is much greater than the variation of c-Si bandgap in the temperature region of 10 K - 300 K ($\Delta E_g \sim 50$ meV). Therefore, the blue PL shift can not be explained by c-Si linear thermal expansion. Another possible explanation is that size distribution of Si nanocrystals leads to fluctuation in the effective bandgap; thus, it can be modeled as bandtails. The recombination of photocarriers will take place in such bandtail states, and the energy of the highest occupied states will correspond to the highest PL energy. In the case of CW excitation and steady-state recombination, the energy of the highest occupied states will be equal to the mobility edge. However, the energy of the mobility edge increases and thermally-assisted escape of carriers from the bandtail states will be suppressed when the temperature decreases (Fig. 5). Therefore, the blue shift of the PL peak with temperature decrease can be explain due to an increase of the mobility edge energy. The PL intensity temperature dependence shows a maximum at $T \approx 200$K that is similar to that reported for as anodized PSi and can be explained within a model suggested by Vial et al. [17]. Proposed explanations and observations reminiscent of phonon-assisted transitions in the resonantly excited PL spectra lead us to the conclusion that the origin of the red PL in thermally oxidized PSi is the radiative recombination in the bandtail states within Si nanocrystals quasi-bandgap.

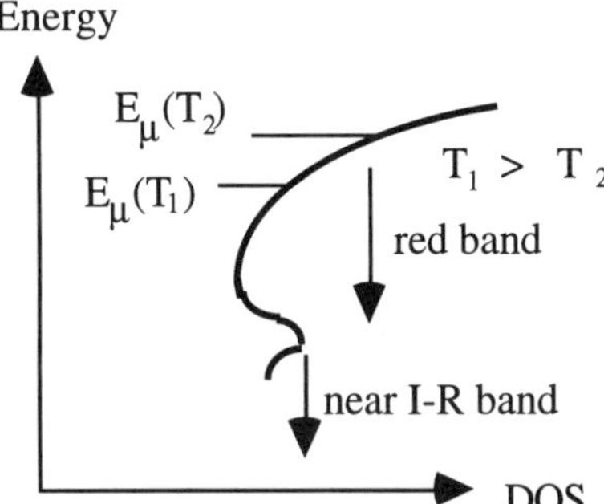

Figure 5. Schematic of proposed model for recombination via bandtail states and deep defect-related states in quasi-bandgap of Si nanocrystals. The energy of mobility edge $E_\mu(T)$ is shown for two temperatures.

In contrast to the red PL, the near-IR PL spectrum is not sensitive to temperature and the PL intensity temperature dependence shows monotonic increase with a temperature decrease (Fig. 2). The PL at ~ 1.6 eV has been intensively studied by Proks et al, and has been explained as defect related radiative recombination [18]. In our model, the 1.6 eV PL band corresponds to recombination via relatively deep localized states. The energy levels of these states are not monolevels, but in case of steady-state recombination the deep states are completely occupied. Thus, the PL spectra are not temperature sensitive. In the proposed model (Fig. 5) the ionization energy of such states is ~ 0.2-0.4 eV and thermally-assisted escape of carriers at room temperature is totally suppressed. However, in samples with presumably high number of such defects, where the near-IR PL band dominates, carrier transport and the EL are temperature dependent. Assuming that the major carrier transport mechanism in these samples is thermally-assisted hopping the contradiction is explained [12]. The hopping via defect states does affect the rectification ratio of our diode-like structure as well as efficiency of the EL. These defects are also responsible for reversible EL degradation which is discussed in Ref. 19. Further work needs to be done to minimize the number of the defects which will improve the performance of the device .

Conclusion

We have studied the luminescence in thermally oxidized PSi which has been used as an active layer for an LED. Two PL bands with peaks at 1.6 eV and near 2 eV have been found: the red PL is explained as radiative transitions via bandtails and the near I-R attributed to defect-related transitions in quasi-bandgap of oxidized PSi. These studies has been used for optimization of an LED (see Ref. 19). This work was supported by the U.S. Army Research Office.

References

1. R. Zachai *et al.*, Phys. Rev. Lett. **64**, 1055 (1990).
2. G. S.Mitchcard *et al.*, Solid State Commun. **29** (1979).
3. H. Ennen *et al.*, Appl. Phys. Lett. **43**, 943 (1983).
4. L. T. Canham, Appl. Phys. Lett. **57**, 1046 (1990).
5. J. Linnros and N. Lilic, Appl. Phys. Lett. **66**, 3048 (1995)
6. A. Loni *et al.*, Electronics Letters **31**, 1288 (1995).
7. M. A.Tischler *et al.*, Appl. Phys. Lett. 60, 639 (1992).
8. A. Richter *et al.*, IEEE Electron Device Letters **12**, 691 (1991).
9. N. Koshida and H. Koyama, Appl. Phys. Lett. **60**, 347 (1992).
10. F. Namavar, H. P. Maruska, and N. M. Kalkhoran, Appl. Phys. Lett **60**, 2514 (1992).
11. L. Tsybeskov *et al.*, Appl. Phys. Lett. **68**, 2058 (1996).
12. L. Tsybeskov *et al.*, in *Advanced Luminescent Materials,* edited by D. J. Lockwood, P. M. Fauchet, N. Koshida and S. R. J. Brueck (Electrochemical Society, Pennington, NJ) 1996, 34-47.
13. See for example, *Microcrystalline and Nanocrystalline Semiconductors*, edited by R. W. Collins, C. C. Tsai, M. Hirose, F. Koch and L. Brus (Mater. Res. Soc. Symp. Proc. **358**, Pittsburgh, PA, 1995).
14. P. M. Fauchet *et al.*, IEEE J. Selected Topics in Quantum Electronics **1**, 126 (1995).
15. L. Tsybeskov, J. V. Vandyshev and P. M. Fauchet, Phys. Rev. B **49**, 7821 (1994).
16. P. D. J. Calcott *et al., in Microcrystalline and Nanocrystalline Semiconductors*, edited by R. W. Collins, C. C. Tsai, M. Hirose, F. Koch and L. Brus (Mater. Res. Soc. Symp. Proc. **358**, Pittsburgh, PA, 1995), p. 465.
17. J. C. Vial in *Porous Silicon Science and Technology,* edited by J. C. Vial and J. Derrien, Springer-Verlag, Berlin, 1995, p. 137
18. S. M. Prokes and W. E. Carlos, J. Appl. Phys. **78**, 2671 (1995).
19. L.Tsybeskov *et al.*, Electroluminescence and carrier transport in an LED based on silicon-rich silicon oxide, in this volume.

A STUDY OF THE FACTORS WHICH DETERMINE THE MODULATION SPEED OF A SHALLOW PN JUNCTION POROUS SILICON LED

A.J. SIMONS, T.I. COX, A. LONI, P.D.J. CALCOTT, M. J. UREN and L.T. CANHAM
Defence Research Agency, St. Andrew's Road, Great Malvern, Worcestershire, WR14 3PS, UK.

ABSTRACT

The effect of the chemical thinning of the porous silicon structure on the speed and efficiency of electroluminescent devices, produced by the anodisation of a pn junction in bulk silicon is investigated. Thinning of the silicon wires results in an increase in the efficiency but at the expense of a reduction in operating speed. It is demonstrated that the operating speed is limited by the photoluminescence lifetime for small signal excitation. However, for large signals, the electroluminescence can be turned off more than 5 times faster than the photoluminescence lifetime, indicating that this need not necessarily limit device operating speed.

INTRODUCTION

The demonstration [1] that it is possible to obtain efficient visible photoluminescence (PL) at room temperature from porous silicon (PS) opened up the possibility of all silicon optoelectronics. Applications envisaged include PS light emitting devices (LEDs) for optical interconnection and emissive displays integrated with silicon circuitry. For optical interconnection, it is important to be able to modulate the electroluminescence at frequencies greater than 100 MHz and with an external power efficiency greater than 1%. EL efficiencies greater than 0.1% have been reported in PS LEDs, produced by porosifying pn junctions made in bulk silicon, under both CW operation [2] and pulsed operation [3]. Several groups have made measurements which indicate that modulation of PS LEDs at frequencies greater than 1 MHz may be possible, albeit from devices with lower than 0.1% power efficiency [4-6]. The maximum -3dB frequency reported for small signal modulation of the EL of a PS LED is greater than 1 MHz [4]. For excitation of the EL with a square pulse, rise times as small as 100 nsec have been reported by several groups [4-6]. Fall times as short as 30 nanoseconds have also been reported [6].

In this paper, we investigate the factors limiting the modulation speed of PS devices which are similar to those we previously reported [2] to have a CW external power efficiency of 0.1%. The fabrication of this device has been described in detail [7]. In summary, the device has an indium tin oxide (ITO) contact (100 nm thick) on top of a porous silicon region of thickness ~400 nm on an n-type substrate of resistivity in the range 10-20 Ωcm. The PS region is formed by the anodisation of a pn junction prepared by implanting the substrate with boron at an energy of 35 keV to a dose of 10^{16} cm^{-2}. The light assisted anodisation is performed at a current density of 3 $mAcm^{-2}$ in 40 wt% aqueous hydrofluoric acid for four minutes. During anodisation, the silicon skeleton formed by the electrochemical process is also reduced in size by photochemical etching by the electrolyte.

Here, we examine two devices which are similar to that reported in reference [2]. Device A is chosen to have a shallower junction than the prior more efficient device [2] and thereby to reduce the chemical etching component. This is achieved by reducing both the implantation energy and the anodisation time. The PS layer for device B is formed by chemically thinning the silicon wires within a device A type PS layer. Thus, devices A and B offer a vehicle for the investigation of the effect of chemical etching on the efficiency and dynamics of the device operation.

Mat. Res. Soc. Symp. Proc. Vol. 452 © 1997 Materials Research Society

EXPERIMENTAL

The fabrication of devices A and B is identical to that previously described [7] except that:
Device A: the pn junction is made by using a boron implant at an energy of 15 keV. The anodising time is 2 minutes which produces a porous silicon layer of thickness 200±30 nm.
Device B: the porous silicon layer is formed by leaving the PS layer of device A in the electrolyte under illumination under zero bias for three minutes.

The area of the ITO contacts is 0.013 cm^2 and the devices were operated under a vacuum of 0.1 torr. Devices A and B are both rectifying such that the ratio of the currents under a forward and reverse bias of 3 V is greater than 100. The threshold voltage and the corresponding current which must be applied for the appearance of EL are 1.3 V and 15 μA for device A and 1.7 V and 100 nA for device B. For device A, the maximum external quantum efficiency (EQE) of the EL is 0.0004% and the peak wavelength of the EL is at 780 nm. The corresponding values for device B are an EQE of 0.04% and a peak wavelength of 730 nm. For both devices, the differential resistance, as determined from the slope of the static current voltage curve, decreases approximately exponentially in the voltage range 1.5 to 3 V. The resistances at 1.5 and 2.5 V for device A are 1.4×10^4 and 3×10^3 Ω respectively. The corresponding values for device B are 3×10^7 and 3×10^5 Ω. Thus the resistance of B is at least a factor of 100 greater than that of A in the range 1.5 to 2.5 V.

IMPEDANCE MEASUREMENTS

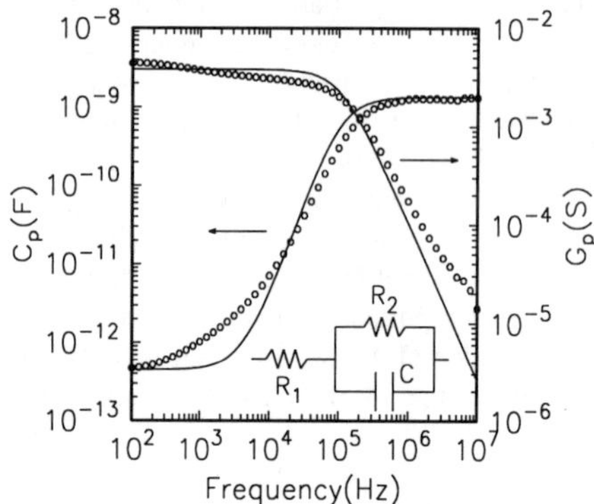

Figure 1 : Parallel capacitance (C_p) and conductance (G_p) for device B for a forward bias of 3V. data (O) and fit (—) (see text).

Small signal impedance measurements were made as a function of frequency for a range of applied forward biases. The behaviour of the impedance of this type of device under reverse bias has already been discussed [7]. A plot of the parallel capacitance and conductance as a function of frequency is shown in figure 1 for device B under a forward bias of 3.0 V. In general, the data could be approximated by the equivalent circuit shown in the inset in figure 1. The solid line shows a fit to the data for R_1=500 Ω , R_2=3×10^5 Ω , C=3 nF. R1 was found to be 450±50Ω for both devices A and B and also to be independent of applied bias. R_1 will include contributions from contact resistances, the lateral resistance of the ITO (~100Ω$\square^{-1}$) and the substrate resistance. To a first approximation, the parallel combination of R_2 and C may therefore be regarded as representing the active region of the device. This is also a reasonable first order description of a pn junction at a given bias. For both devices, the capacitance C is similar in value and, as predicted for a pn junction, increases in value from about 1.5 to 3 nF as the bias is increased from 1.5 to 3.0 V. It is surprising that C is similar for both devices as it implies the same diffusion capacitance for two structures with widely different resistances. R_1+R_2 has the same bias dependence and within a factor of two the same values as those described for dV/dI, extracted from the DC current voltage curve, in the experimental section. Thus, R_1+R_2 is much greater for device B than device A at a given bias. For example, for a bias of 2 V, R_1+R_2 is 6500 Ω and 10^6 Ω for devices A and B respectively. The fit of the active region of the device to a single RC element is only an approximation and there is clearly considerable dispersion in the data as compared to the simple lumped model.

PHOTOLUMINESCENCE (PL) MEASUREMENTS

The relative integrated CW PL efficiencies, excited by λ=442 nm, of the PS films in device A to device B is 1:10. The peak of the PL for device A is at 720 nm whereas that for device B is at 680 nm. The chemical thinning process will have produced wires with smaller dimensions which therefore explains the blue shift in the PL and EL spectrum of B relative to A [1]. In turn, the smaller dimensions give rise to a greater degree of localisation of carriers within the silicon nanostructures. This greatly reduces the non-radiative recombination processes and hence leads to the greater PL efficiency observed for B relative to A. The PL was also excited for device B using laser pulses of duration 200 μs at a wavelength of 457.9 nm. The rising and falling edges of the PL transient were recorded for different emission wavelengths. The time (t) dependence of the rising edge was found to be well described by the function $(1-\exp(-t/\tau_r^{PL}))$. The rise time, τ_r^{PL}, was 2.7 μs and was independent of the wavelength of detection. The falling edge is shown in figure 2 on a logarithmic scale for an emission wavelength of 700 nm. The PL signal falls by about 30% in the first two microseconds. For times longer than about three microseconds, the decay is approximately exponential with a decay time of 12±2 μs. The initial fast component is similar for all emission wavelengths in the range 600 - 800 nm. The decay time for the long time region increases with increasing wavelength from 6±1 μs at 600nm to 20±3 μs at 800 nm. The differing dynamics of the rise and fall transients is dissimilar to that previously reported in porous silicon [8]. The increase with wavelength in decay time for the long time tail is similar to that previously reported [9]. When the slow component of the PL is excited at 2 K, in similar structures using the resonant excitation method [10], then phonon onsets are observed which shows that the slow component, at least, is due to luminescence in single crystal silicon. The signal was too weak to record PL transients for device A.

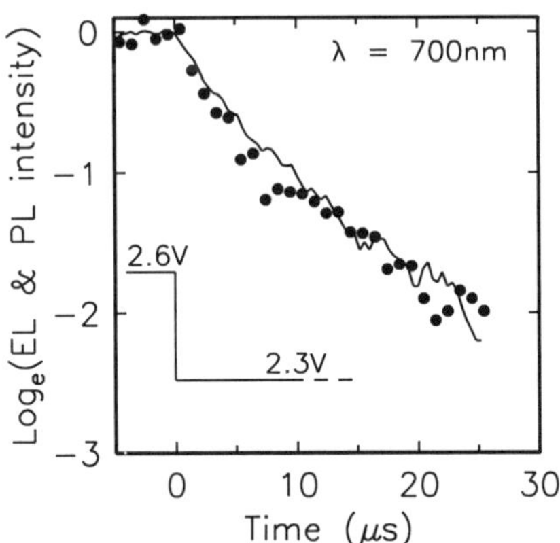

Figure 2 : Comparison for device B of the falling edge of the PL (●) and EL (—) on changing bias from 2.6 to 2.3V.

ELECTROLUMINESCENCE MEASUREMENTS

The small signal response of the EL to a 200 mV p-p sinusoidal waveform superimposed on a DC level was determined as a function of the DC level and wavelength of emission. The EL was detected using a photomultiplier coupled to a spectrum analyser. Figure 3 shows the response of device A as a function of frequency for three different DC bias levels. As shown in figure 3, the fall-off in response may be characterised by the -3dB frequency. However, it is apparent that there is dispersion in this time constant which increases for lower applied biases. Reference to table I shows that, for both devices, the -3dB frequency increases as the DC bias is increased. This suggests that the dynamics are at least partially controlled by the field across the device and therefore the transport of carriers to the active luminescent regions in the device. The penalty for the increased efficiency of device B is a reduction in the -3dB frequency at a given bias by a factor of at least five.

Table I. The -3dB frequency (Hz) as a function of the DC bias level for devices A and B

Bias (Volts)	1.5V	2.0V	2.5V	3.0V
Device A	$3.6x10^4$ Hz	$7.9x10^4$ Hz	-	$2.1x10^5$ Hz
Device B	-	$1.5x10^4$ Hz	$2.5x10^4$ Hz	$3.2x10^4$ Hz

EL was also excited using a 300 mV p-p square pulse superimposed on a DC level. The current showed transients on both the rising and falling edges with a time constant of about one microsecond. This time constant is, as expected, of the same order as the product $R_1.C$ which is ~10^{-6} s for both devices in the bias range 1.5 to 3 V. The time (t) dependence of the rising edge of the EL is well described by the function $(1-\exp(-t/\tau_r^{EL}))$. The rise time (τ_r^{EL}) of the EL signal is plotted in figure 4 as a function of the DC bias level.

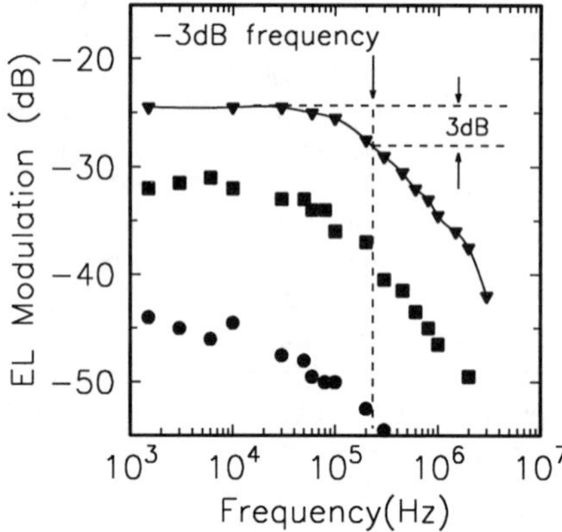

Figure 3 : EL small signal response for device A for different DC biases. ▼ : 3.2V, ■ : 2.0V and ● : 1.6V.

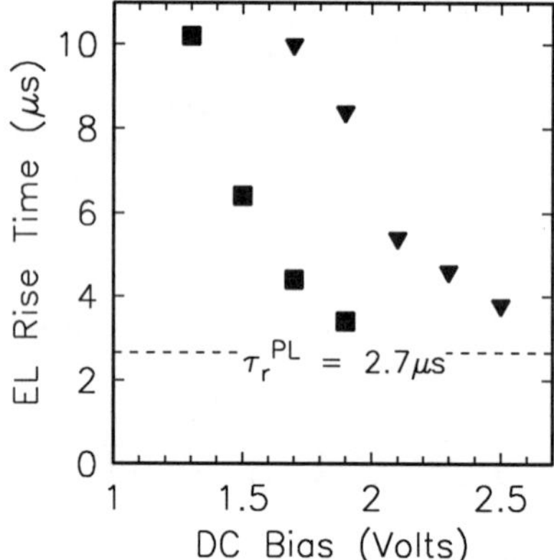

Figure 4 : EL rise time following excitation with a 300 mV square pulse as a function of the DC level; ■ : device A; ▼ : device B.

As suggested by the bias dependence of τ_r^{EL}, the dynamics of the rising edge will be controlled by the transport of carriers to the luminescent centres and their subsequent radiative recombination with a time constant which we envisage should be equal to τ_r^{PL}. In a simple mobility picture, the time for carriers to reach the luminescent regions will be inversely proportional to the voltage across the transport region. Thus, as the bias is increased, τ_r^{EL} should fall and tend towards the PL rise time at high bias. Reference to figure 4, where the dotted line represents the PL rise time for device B, shows the expected asymptotic behaviour. The fact that, at a given bias, τ_r^{EL} is shorter for device A than device B reflects the fact that transport to the luminescent centres is faster in device A at a given bias.

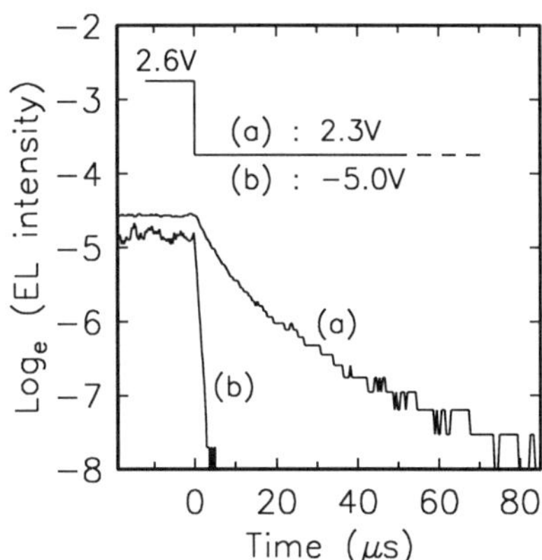

Figure 5 : Falling edge of EL following a change in bias from 2.6V to (a) 2.3 V and (b) -5.0V.

The falling edge of the EL is shown in figure 2 for a change in bias from 2.6 to 2.3 V for an EL emission wavelength of 700 nm. Reference to figure 2 shows that the EL and PL falling edges are very similar. The time for the signal to fall by 1/e did not vary with DC level over the range 1.5-3.0 V and was 5.5±1.0 μs and 12±2 μs for A and B respectively.

In an attempt to speed up the falling edge of the EL transient, the effect of applying a large negative going pulse of -5 V onto the falling edge of the square pulse was investigated. This was achieved by applying the waveform shown in the inset in figure 5. Reference to figure 5 shows that the time to fall by 1/e decreases from 14 to 1.2 microseconds as the value of the lower voltage is decreased from 2.3 to -5.0 V. The value of 1.2 microseconds is similar to the timescale of the initial fast component of the PL falling edge and is much faster than the long time tail. The bias of -5 V may be sufficient to remove electron hole pairs from the luminescent regions and to prevent any carriers already in the structure from reaching the luminescent regions.

DISCUSSION

It has been shown [11] that red emitting porous silicon is composed of ~2 nm wide silicon quantum wires of undulating diameter. It has been proposed [11] that there are corresponding undulations in the conduction and valence band edges. When carriers are photoexcited, the electron hole pairs so created relax almost instantaneously to the local minima in the band edges. One of the principal reasons that the PL can be so efficient is that the undulating band edges inhibit the movement of excitons along the wires to non-radiative sites. It therefore follows that electrical transport will probably be difficult through this type of material which displays efficient PL. In the electroluminescence process, electrons and holes must be injected electrically into the luminescent undulations from the contacts. Thus the EL response involves a convolution of the dynamics of carrier injection and the radiative recombination process.

Taken together, devices A and B illustrate the trade-off between carrier injection and radiative efficiency when considering the electroluminescence process. Thus A is a structure with silicon wires of larger diameter and hence smaller undulations in the conduction and valence band edges. Therefore electrical transport through device A is easier than through device B. Thus, for a given bias, A operates at a higher speed than B. At the same time the density of undulations, sufficiently quantum confined to give efficient PL, in A is smaller than for B. Thus, it is much easier for carriers to find paths in A which shunt out the luminescent regions. This is reflected in the much higher threshold currents for the appearance of EL in A (15 μA) than B (100 nA), and also the lower EL efficiency of A (0.0004%) than B (0.04%). It also explains why the ratio of EL efficiencies (100:1) is a factor of ten greater than the relative PL efficiencies (10:1).

Transport of the carriers to the luminescent regions, as reflected in the EL rise time, can be speeded up by the application of an increasingly positive going pulse. In the high bias limit, the rising edge of the EL is similar to that of the rising edge of the PL. With a small negative going pulse, e.g. the bias is changed from 2.6 to 2.3 V, the EL decay is dominated by the recombination of carriers. With the application of a large negative going pulse, e.g. the bias is changed from 2.6 to -5 V, it was found to be possible to quench the EL on a timescale considerably shorter than the PL decay time. Thus the highest speed operation would seem to be obtained with a pulse with both high positive and negative going components. This is compatible with integration into a circuit where the EL device could be driven by, for example, plus and minus five volts. Another possible route to higher speed operation is to use a PS film which has a reduced PL lifetime while maintaining a high PL efficiency. The quantum confinement model implies that such material will be a more highly quantum confined structure and that carrier transport and therefore high frequency operation may be difficult, unless thinner silicon wires without an undulating width can be produced.

CONCLUSIONS

The prospects for realising PS LEDs with an efficiency greater than 1% which can also be modulated at greater than 100 MHz, as required for optical interconnection, are currently not very encouraging. One possible route lies in the use of optical cavities to reduce the PL lifetime without degrading the electrical transport properties of the material. It is, however, much more likely that EL devices can be produced which have the performance required for emissive displays. Once produced, whether these devices can find a place in the market remains to be seen. When judging this issue, sight should not be lost of fact that these devices are made from silicon, the material dominating the electronics industry today.

REFERENCES

1. L. T. Canham, Appl. Phys. Lett., **57**, 1046 (1990).
2. A. Loni, A. J. Simons, T. I. Cox, P. D. J. Calcott and L. T. Canham, Electronics Letters, **31**, 1288 (1995).
3. J. Linnros and N. Lalic, Appl. Phys. Lett., **66**, 3048 (1995).
4. L. Tsybeskov, S. P. Duttagupta, K. D. Hirschman and P. M. Fauchet, Appl. Phys. Lett., **68**, 2058 (1996).
5 S. Lazarouk, P. Jaguiro, S. Katsouba, G. Masini, S. L. Monica, G. Maiello and A. Ferrari, Appl. Phys. Lett., **68**, 2108 (1996).
6. J. Wang, F. L. Zhang, W. C. Wang, J. B. Zheng, X. Y. Hou and X. Wang, J. Appl. Phys., **75**, 1070 (1994).
7. A. J. Simons, T. I. Cox, A. Loni, L. T. Canham, M. J. Uren, C. Reeves, A. G. Cullis, P. D. J. Calcott, M. R. Houlton and J. P. Newey, in Proceedings of the International Symposium on Advanced Luminescent Materials, edited by D.J. Lockwood, P.M. Fauchet, N. Koshida and S.R.J. Brueck (Electrochemical Society, PV92-25, 1996) pp 73-86.
8. P. D. J. Calcott, K. J. Nash, L. T. Canham, M. J. Kane and D. Brumhead, J. Phys.: Condensed Matter, **5**, L91 (1993).
9. Y. H. Xie, W. L. Wilson, F. M. Ross, J. A. Mucha, E. A. Fitzgerald, J. M. Macaulay and T. D. Harris, J. Appl. Phys., **71**, 2403 (1992).
10. P. D. J. Calcott, K. J. Nash, L. T. Canham, M. J. Kane and D. Brumhead, J. Lumin., **57**, 257 (1993).
11. A. G. Cullis and L. T. Canham, Nature, **353**, 335 (1991).

NONLINEAR ELECTRICAL FUNCTIONS OF POROUS SILICON LIGHT-EMITTING DIODES

K. UENO, T. OZAKI, H. KOYAMA, and N. KOSHIDA
Division of Electronic and Information Engineering, Faculty of Technology,
Tokyo University of Agriculture and Technology, Koganei, Tokyo 184, Japan

ABSTRACT

Some nonlinear electrical characteristics in electroluminescent porous silicon (PS) diodes with a relatively thin PS layer (0.5–5 μm thick) are described. The experimental PS diodes were composed of a semitransparent Au film, a PS layer, p- or n-type Si substrate, and an ohmic back contact. The PS layers were prepared by anodizing Si wafers in an ethanoic HF solution. In some cases, the PS layers were treated by rapid thermal oxidization (RTO) process. When the bias voltage is applied, the PS diodes show the electrical behavior like the metal-insulator-semiconductor (MIS) diodes. The negative-resistance characteristics and memory effect are also observed. These results indicate that the quantum-structured nature of the PS layer appears not only in the optical properties but also in the electrical properties.

INTRODUCTION

Since room-temperature visible photoluminescence (PL) [1] and electroluminescence (EL) [2] from porous silicon (PS) were reported, many studies have been conducted toward PS-based optoelectronic integration. Because PS consists of a great number of Si nanocrystallites, visible light-emission from PS has been discussed in relation to quantum confinement effects. The quantum structure of PS should have effects not only on the optical properties but also on the electrical properties. Recently-reported cold electron emission from PS light-emitting diodes [3] reflects the characteristic property of electrical conduction in PS. To apply the properties of PS to optoelectronic devices, more detailed information about the carrier injection and subsequent transport is required.

The experimental PS light-emitting diodes were fabricated in the form of Au electrode/PS/c-Si/ohmic contact. The experimental analyses of EL [4], cold electron emission, and electrical PL quenching [5] effects in PS suggest that there is a significantly high electric field in the region of the outer surface near the Au electrode under the biased condition. The region of about 0.5 μm below the outer surface is an important determining factor in the optoelectronic properties of PS. To clarify the carrier transport in this region in PS layer, we studied the electrical properties of EL-emissive PS diodes with a relatively thin PS layer (0.5–5 μm thick). It is reported here that PS diodes show nonlinear electrical functions. Their mechanisms are also discussed.

EXPERIMENTAL

Sample Preparation

The experimental PS diodes were composed of a top contact (semitransparent thin Au film), a PS layer, a Si substrate, and an ohmic back contact. The following three types of PS layer were prepared.

Sample A: The PS layer was formed on nondegenerate p-type Si wafers by anodization in the dark. The thickness was about 1.5 μm.

Sample B: The PS layer was formed on degenerate n-type Si wafers by photo-anodization under illumination by a 500 W tungsten lamp from a distance of 20 cm. The thickness was about 5 μm.

Sample C: The PS layer was formed on nondegenerate p-type Si wafers by anodization in the dark. The thickness was 0.5 μm. Immediately after anodization, this sample was treated by rapid thermal oxidization (RTO).

In all cases, a solution of 55% HF : ethanol = 1 : 1 was used for anodization. The anodized

Mat. Res. Soc. Symp. Proc. Vol. 452 © 1997 Materials Research Society

samples were transferred into a vacuum chamber, and semitransparent Au films were deposited onto the PS layers. The active area of the PS diodes was 5 mm in diameter.

Measurements

The current–voltage (I–V) and capacitance–voltage (C–V) characteristics of PS diodes were measured under the forward and reverse bias conditions at room temperature and low temperatures. The EL intensities are also measured at the same time. For the PS diodes fabricated on p-type Si substrates, the forward bias condition corresponds to the case in which a negative voltage is applied to the Au electrode with respect to the substrate, and vice versa in the case of n-type substrates.

RESULTS

Electrical Properties as a MIS Diode

Typical I–V curves of the PS diode with sample A at room temperature and 50 K are shown in Fig. 1 by the solid and dashed curves, respectively. The experimental PS diode under the forward bias condition is also schematically shown as the inset of Fig. 1. The diodes exhibit a definite rectifying behavior at room temperature. At low temperatures of 50 K, a significant decrease in the diode current is clearly observed at low reverse bias voltages. The reverse diode current, however, shows a rapid increase at a critical reverse bias voltage in a similar way in conventional MIS tunnel diodes.

To investigate these peculiar characteristics of the experimental PS diodes, a band model is schematically shown in Fig. 2. The PS layer includes a great number of quantum-sized silicon crystallites surrounded by some electronic barriers. Under the threshold reverse bias voltage V_{th}, the inversion effect occurs at the PS/p-Si interface, and consequently a number of electrons are generated there. The generated electrons are injected into PS by tunneling. The observed value of V_{th} is close to the estimated value in this system.

The calculated curve of the diode current in accordance with this model is shown in Fig. 3 by the solid curve. The inset shows the potential barrier diagram, where Δd and Φ_0 are the barrier width and the barrier height, respectively. We assumed that the generation-recombination current is dominant below V_{th}. The calculated result at $\Delta d = 6.0$ nm and $\Phi_0 = 0.3$ eV explains well the experimental behavior indicated by open circles. These values of Δd and Φ_0 used for the fit in Fig. 1 are reasonable in view of the structural property of PS sample A and the previously observed activation energy ($\cong$0.3 eV) [6] in the PS conductivity near room temperature.

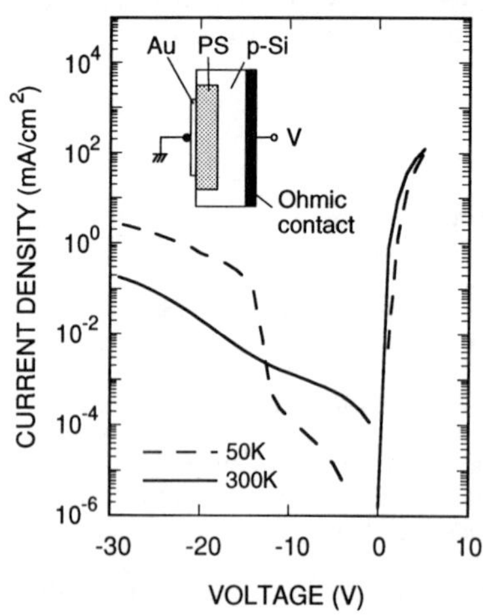

Fig. 1. The I–V curves of a PS diode with sample A at room temperature and 50 K. The inset is a schematic illustration of the sample under the reverse bias condition.

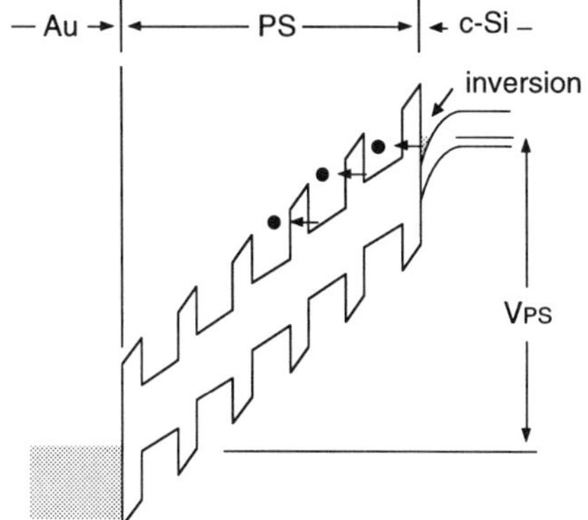

Fig. 2. Schematic model of a PS diode under the condition that a sufficiently positive bias voltage is applied to the Au electrode with respect to the p-type silicon substrate.

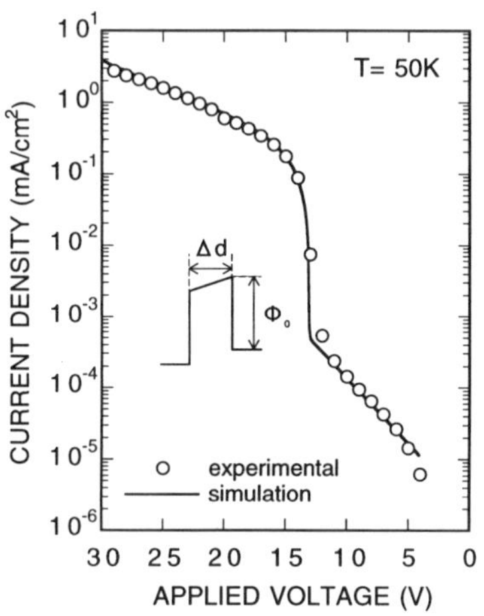

Fig. 3. The simulation result the tunneling model and the experimental results obtained from Fig. 1. The inset shows a potential barrier used in the simulation. The fitting parameters used in this case are $\Delta d = 6.0$ nm and $\Phi_0 = 0.3$ eV.

The Negative-Resistance Characteristics

The I–V curve and corresponding EL intensity of the PS diode with sample B are shown in Fig. 4 by the closed circles and the open circles, respectively. The forward bias condition corresponds to the case in which a positive voltage is applied to the Au electrode with respect to the substrate. Beyond a critical forward bias voltage, the diode current shows a significant decrease. The peak-to-valley current ratio is about 2. This negative-resistance characteristic was reversible, and observed even at low temperatures of about 10 K. It should be noted that the EL intensity begins to increase at the onset of the negative-resistance behavior. To clarify this relationship, the diode current density dependence of EL intensity is plotted in Fig. 5. Obviously, the EL emission is switched on with a change in the carrier transport mode.

The photo-anodized PS layer includes a great number of electrically-isolated Si crystallites. Under the forward bias condition, the electrons are injected from the Si substrate into the PS layer through the narrow potential barrier, and then drifted toward the Au electrode by the electric field. Beyond the critical bias voltage, electrons are injected into the electrically-isolated luminescent Si crystallites by tunneling. When this electron injection extends to the entire region in PS, further injection should be suppressed because of a significant change in the potential distribution along the thickness direction of the PS layer. Under this situation, the electric field at the PS/Au interface becomes higher. It makes possible radiative recombination of carriers in Si crystallites through field-promoted hole generation mechanism [4]. Thus the EL emission is switched on despite a decrease in the transmittance of carriers through the PS layer.

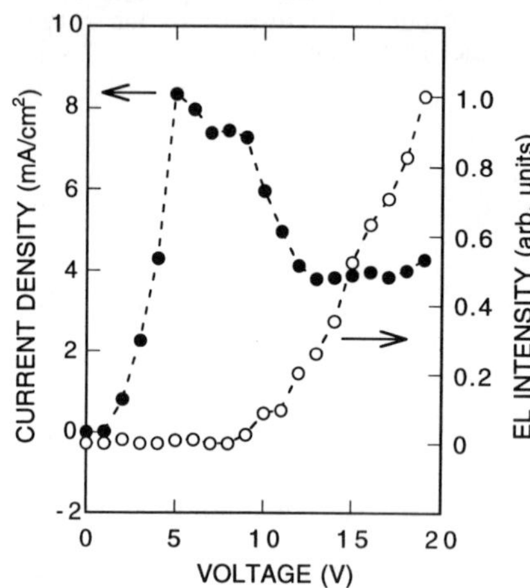

Fig. 4. The I–V curve of a PS diode with sample B and the corresponding EL intensity at room temperature as a function of forward bias voltage. The diode begins to emit visible light at the onset of the negative-resistance effect.

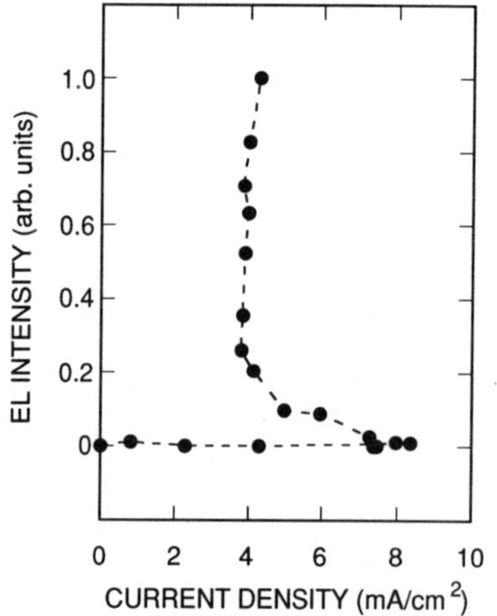

Fig. 5. The current density dependence of the EL intensity obtained from the result of Fig. 4. The voltage-controllable EL emission is switched on at a critical carrier injection.

Memory Effects

The I–V curves of the experimental PS diode with sample C are shown in Fig. 6. The diode current shows a significant hysteresis. Though the hysteresis is observed in as-prepared PS diodes, the effect become more significant when the sample is treated by RTO processing. As the bias voltage is swept from 0 V toward the reverse direction (negative), the diode current density at first remains extremely low (~0.2 μA) as shown in Fig. 6 by the closed circles (off-mode). At a threshold voltage of about -25 V, however, the diode operation changes from the off-mode to the on-mode, and the high diode current begins to flow as shown by the open circles. Once the diode turns on, the low resistivity is kept on even at low bias voltages.

The on-mode can be turned to the off-mode when a sufficient forward bias voltage (18 V in this case) is applied. At this point, the high-resistivity state appears again. These results shows that the RTO-PS diode operates as a memory device, and the reverse and forward bias voltages correspond to the writing and erasing voltages, respectively. The difference in the diode current between the on and off states reaches three to five orders of magnitude. This behavior was also observed at low temperatures (10 K). In addition, the EL intensity behaves in a similar way to the diode current.

The RTO-PS layer consists of a great number of Si crystallites which are isolated electrically by interfacial thin oxide. When the writing voltage is applied to the diode, we can assume that electrons in the valence band of crystallites tunnel to the conduction band of neighboring ones through thin oxide layers. The C–V characteristics at the on-mode suggest that significant positive charges are stored in the PS layer. When the erasing voltage is applied to the PS diode, injected and stored carriers are swept away from PS, and then the distorted potential distribution returns to the initial off-state such that the high-electric field region promoting tunnel effect disappears near the PS/Si interface.

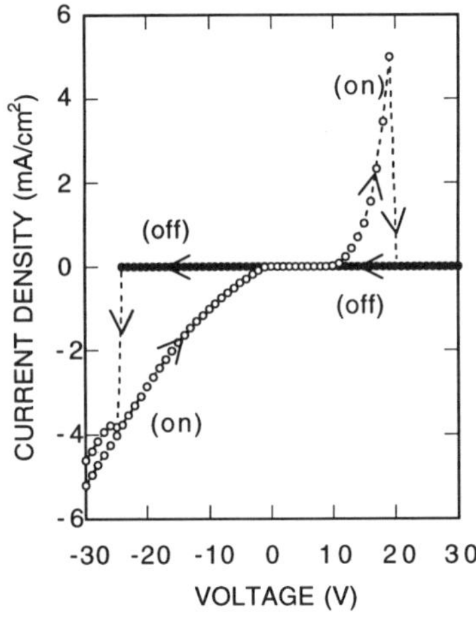

Fig. 6. A typical example of the hysteresis I–V curve observed in PS diodes with sample C. The two different electrical states (on and off) can be controlled simply by an external bias voltage.

CONCLUSIONS

The following three nonlinear electrical functions have been observed in the PS light-emitting diodes.

(a) The PS diodes formed on the p-type Si wafer show the step-like behavior in the I–V curve when the reverse bias voltage is applied. This can be explained by the MIS diode model.

(b) The negative-resistance effect appears in the I–V curve of the PS diodes formed on n-type Si substrates. The EL emission becomes available correspondingly to the onset of field-induced holes generation.

(c) The PS diodes treated by RTO process exhibit a significant hysteresis in the I–V characteristics. The devices have two distinct stable states (on and off). The two states can be simply controlled by the external bias voltage.

These functions are related to the electrical properties of PS as a quantum-sized crystalline system.

ACKNOWLEDGEMENTS

This work was partially supported by the Nissan Science Foundation, the Research Foundation Opto-Sience and technology and a Grant-in-Aid from the Ministry of Education, Science, Sports and Culture of Japan.

REFERENCES

1. L. T. Canham, Appl. Phys. Lett. **57**, 1046 (1990).
2. N. Koshida and H. Koyama, Appl. Phys. Lett. **60**, 347 (1992).
3. N. Koshida, T. Ozaki, X. Sheng, and H. Koyama, Jpn. J. Appl. Phys. **34**, L705 (1995).
4. T. Oguro, H. Koyama, T. Ozaki, and N. Koshida, J. Appl. Phys. **81** (1997) (in press).
5. H. Koyama, T. Oguro, and N. Koshida, Appl. Phys. Lett. **62**, 3177 (1993).
6. T. Ozaki, M. Araki, S. Yoshimura, H. Koyama, and N. Koshida, J. Appl. Phys. **76**, 1986 (1994).

INTEGRATING BIPOLAR JUNCTION TRANSISTORS WITH SILICON-BASED LIGHT-EMITTING DEVICES

K.D. HIRSCHMAN*, L. TSYBESKOV, S.P. DUTTAGUPTA AND P.M. FAUCHET†
Department of Electrical Engineering, University of Rochester, Rochester, NY 14627

ABSTRACT

Silicon-based optoelectronic devices enable the realization of optoelectronic systems that are compatible with integrated circuit manufacturing technology. This work reports on silicon-based visible light-emitting devices (LEDs) that have been successfully integrated into a standard bipolar fabrication sequence. The basic LED structure consists of a 0.5-1.0μm thick silicon-rich silicon oxide (SRSO) active light-emitting layer formed on a p-type silicon wafer by partial oxidation of porous silicon (PSi), with an n+ doped polysilicon cathode. The LEDs exhibit bright electroluminescence (EL) with a spectral peak between 1.75 and 1.90eV. The LEDs are connected in a common-emitter configuration to integrated vertical pnp bipolar driver transistors. This is the first demonstration of an all-silicon visible light emitter / bipolar transistor optoelectronic integrated circuit. The LED device fabrication, process integration, and optoelectronic device characteristics are discussed.

INTRODUCTION

The need for silicon-based optoelectronic components that can be integrated into microelectronic technology has stimulated a significant development effort toward advancing silicon-based optoelectronics. Photodetectors, waveguides, wavelength demultiplexers and modulators have all been fabricated in silicon-based technology [1]. The device for which conventional silicon technology falls short is the light emitter, since crystalline silicon, an indirect bandgap semiconductor, is an inefficient light-emitting material. In contrast, porous silicon (PSi) prepared by anodic etching has demonstrated high efficiency room temperature visible photoluminescence [2,3]. Reasonably efficient PSi-based light-emitting devices (LEDs) have been fabricated (external power efficiency > 0.1% [4]), however the stability of such LEDs continues to be a serious problem for practical applications because the efficiency decreases drastically during the first hour of operation [4,5]. In addition, the as-anodized material cannot be integrated into conventional silicon process technology because of its extreme reactivity and inherently fragile structure. Partial oxidation of porous silicon in a dilute oxygen ambient produces silicon nanoclusters within an oxide matrix, or silicon-rich silicon oxide (SRSO). This material exhibits appropriate light-emitting and carrier transport properties and is compatible with conventional processing techniques.

---

* also Department of Microelectronic Engineering, Rochester Institute of Technology, Rochester, NY 14623
† also Laboratory for Laser Energetics, Department of Physics & Astronomy, and The Institute of Optics, University of Rochester, Rochester NY 14627

Mat. Res. Soc. Symp. Proc. Vol. 452 © 1997 Materials Research Society

We have previously reported [6,7] on a surface-emitting SRSO-based LED which has characteristics among the best reported in silicon-based technology [4,8] with respect to electroluminescence (EL) efficiency, operating threshold conditions and frequency response, and demonstrates a significant improvement over porous silicon based device structures with respect to device stability. In contrast to porous silicon, the active SRSO layer satisfies several critical microelectronic material processing requirements including tolerance to thermal processing (T~1000°C) and chemical resistance. A simple optoelectronic circuit which integrates the SRSO-based LED with a bipolar driver transistor has now been designed and fabricated, and test results verify its compatibility with the IC process.

EXPERIMENTAL

LED Fabrication

The fabrication sequence for the SRSO-based LED is now briefly discussed (for a more detailed description of the specific processing conditions refer to [6]). A 10Ωcm p-type crystalline silicon wafer with a heavily doped p+ surface layer is anodized in an HF/ethanol solution. The p+ region is transformed into a mesoporous layer (porosity ~ 40%), whereas the underlying p-type silicon is transformed into a nanoporous silicon layer (porosity ~ 75-80%) which extends 0.5-1.0μm into the substrate. A dilute oxygen anneal (10% O_2 in N_2) at 800-900°C is then performed to transform the as-anodized hydrogen passivated PSi into SRSO. A polysilicon film is deposited via LPCVD, and subsequently doped n+ to form the cathode. Aluminum contacts are then made to the n+ polysilicon film and the bulk substrate. Fig. 1 shows the cross-section of the LED.

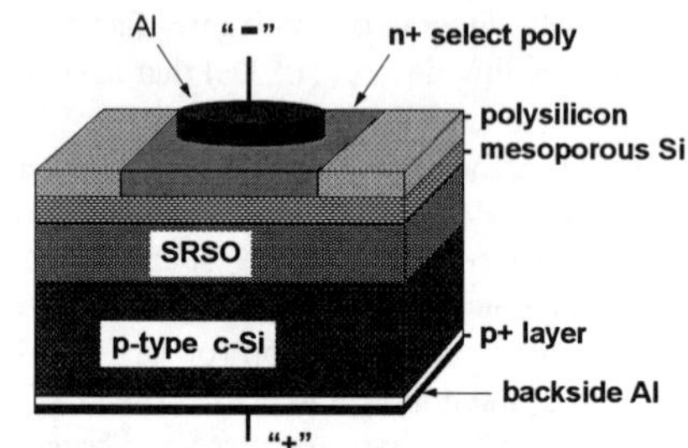

Fig. 1. Cross-section of the SRSO-based LED multilayer structure including the aluminum contact, n+ polysilicon cathode, mesoporous Si transition layer, and SRSO active layer. Also included is a p+ layer and aluminum contact to the backside of the crystalline Si substrate.

Carrier Transport and Electroluminescence

The resulting multilayer structure provides efficient bipolar carrier injection into the semi-insulating SRSO layer. The thin n+ polysilicon layer provides low contact resistance with minimal light absorption. The density of interface states between the polysilicon and the mesoporous "transition" layer has been measured to be less than $10^{12}cm^{-2}$ [7], thus providing effective carrier transport from the n+ polysilicon cathode to the active light-emitting SRSO layer. The I-V characteristics taken at temperatures ranging from 100 to 300K are shown in fig. 2. The current through the device is basically limited by the carrier transport within the SRSO layer. At room temperature there are competing mechanisms, including field-assisted tunneling and thermally-assisted escape and trap controlled transport [7,9]. The low temperature curves (<200K) exhibit Fowler-Nordheim behavior ($I \sim V^2 exp[-1/V]$), which is typically used to describe tunneling current through SiO_2 under high electric fields. Although the average electric field in our experiments is ~ 10^4 V/cm, it is possible that an enhanced electric field of the required order of

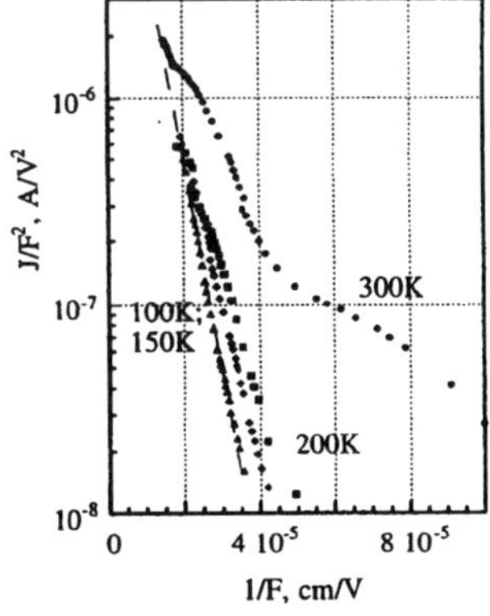

Fig. 2. I-V Characteristics of an SRSO-based LED taken at different temperatures. The characteristics are plotted as the ratio of the current density over the square of the average electric field (J/F^2) versus 1/F (Fowler-Nordheim coordinates). The average field is calculated as applied voltage divided by the average thickness of the PSi layer.

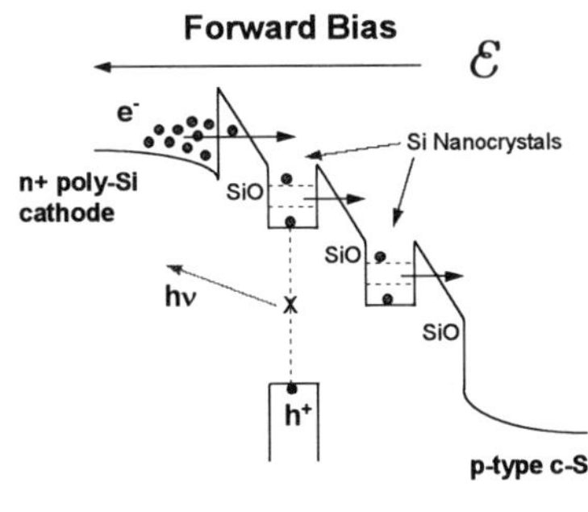

Fig. 3. Schematic of carrier (electron) transport in SRSO. Field-assisted tunneling through thin oxide barriers enables carriers to reach the silicon nanocrystals. The localized electron-hole pairs can recombine resulting in the emission of a photon.

magnitude (10^6-10^7V/cm) exists due to local electric field distortions near the Si nanocrystals [10], resulting in high-field injection across the thin passivating oxide film. The effectiveness of tunneling increases if quantum confinement creates resonant states [11]. Fig. 3 shows a representative schematic of carrier transport and photon emission. Adequate nanocrystallite surface passivation and improved carrier injection have resulted in an increase in the quantum efficiency and significant improvements in the EL. The SRSO-based devices have the following specifications at room temperature: EL peak from 1.7 to 2.0eV; detectable light emission at an applied voltage of ~ 2V and a current density of ~ 10mA/cm^2 ; maximum light intensity of ~ 1mW/cm^2 ; highest external power efficiency ~ 0.1%; modulation bandwidth exceeding 1MHz; and stability without degradation for several weeks of continuous operation.

Device Integration

We have most recently reported [12] a significant milestone for silicon-based optoelectronics; the integration of SRSO-based LEDs into a microelectronic circuit. A simple optoelectronic circuit which integrates the SRSO-based LED with a surrounding bipolar driver transistor has been designed and fabricated (see fig. 4). The driver transistor is connected in a common-emitter configuration and modulates light emission by amplifying a small base input signal and controlling current flow through the LED. Fig. 4 shows a micrograph of an integrated structure (a) along with its cross section (b) and equivalent circuit (c). The circular design is area-efficient and scalable, provides effective electrical isolation and demonstrates a truly integrated structure. Structures of various sizes were fabricated, with the active area ranging from 0.005 to 2mm^2.

The standard bipolar process was modified to include five additional lithography levels and several processing steps in order to accommodate bipolar transistor fabrication, isolated SRSO-based LED fabrication, and formation of the transistor-LED local interconnects. The

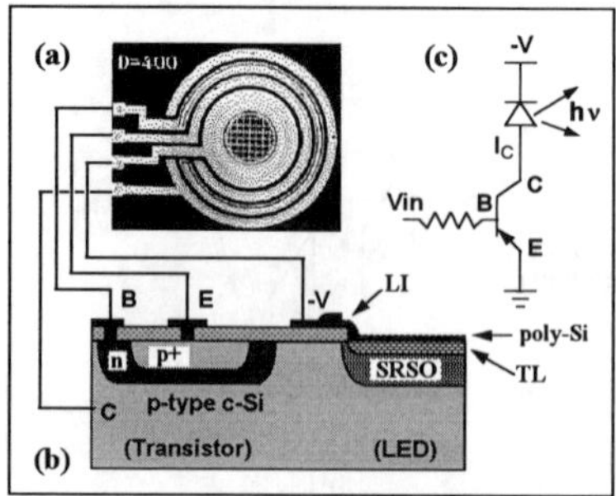

Fig. 4. Micrograph of an integrated LED-bipolar transistor structure (a) along with the cross section (b) and equivalent circuit (c). The LED is in the center of the structure, and has a 400μm diameter active light-emitting area. The surrounding bipolar transistor is identified by the concentric emitter (E), base (B) and collector (C) terminals. The cross section is taken through the center of the integrated structure, and can be mirrored on the right edge because of the symmetric design. The right half of the cross section shows the multilayer structure of the LED, including the polysilicon local interconnect (LI) and cathode, the transition layer (TL) and the SRSO layer.

bipolar process utilized implanted base and emitter regions inside an implanted p-well collector (or a bulk p-type wafer for non-isolated structures). Thermal SiO_2 was grown during the emitter and base drive-in steps for electrical isolation. Following the standard bipolar device fabrication sequence the oxide was patterned and etched, opening windows to bare silicon for construction of the LED. A Si_3N_4 protective layer was deposited via LPCVD, and then patterned to enable formation of selectively anodized regions [13]. The LED fabrication sequence, as previously discussed, continued with a high-dose BF_2^+ implant and anneal (for p+ surface formation), anodization, and SRSO transformation. An n+ polysilicon layer was then produced and patterned to form the LED cathode, and serve as a local interconnect as shown in fig. 4. Contact windows were patterned and etched through the Si_3N_4 (which remained for electrical isolation) and the underlying SiO_2 layer to the bipolar driver terminals. Finally, an aluminum layer was sputtered, patterned and sintered to form the device contacts and interconnect lines.

RESULTS AND DISCUSSION

Electrical Device Characteristics

Electrical and EL testing of the integrated structure have verified that the performance of both the LED and the bipolar device is consistent with the performance of the individual devices. Fig. 5 shows the pnp bipolar transistor electrical characteristics. The device exhibits a forward active current gain $\beta \sim 100$, which is typical of the standard bipolar process. The LED I-V characteristic in fig. 6 demonstrates efficient carrier injection under forward bias. The p+ collector contact (not shown in fig. 4) completely surrounds the integrated structure and functions as a guard ring, providing improved electrical isolation by blocking field surface inversion. The LED exhibits a rectifying ratio (J_F/J_R) of $\sim 10^5$ under light-emitting conditions. Applying a current density of $\sim$ 100A/cm^2 (corresponding to a total dissipated power of $\sim$ 1KW/cm^2) for several minutes of continuous operation was nondestructive which indicates effective heat dissipation in the improved device structures.

Electroluminescence Response

The integrated structures demonstrated functionality as driver/LED pairs. Fig. 7 shows the LED EL response in both DC and pulsed modes of operation. During the pulsed mode test a function generator was used to apply a current pulse at 500Hz (10% duty) to the base of the

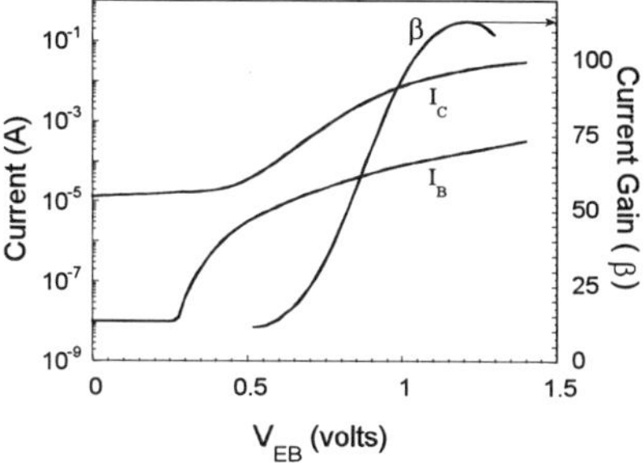

Fig. 5. PNP transistor Gummel-plot which shows the collector current (I_C) and base current (I_B) as the emitter-base voltage (V_{EB}) sweeps the device into the forward active region of operation. The current gain $\beta = I_C/I_B \sim 100$, which is typical of the standard bipolar process.

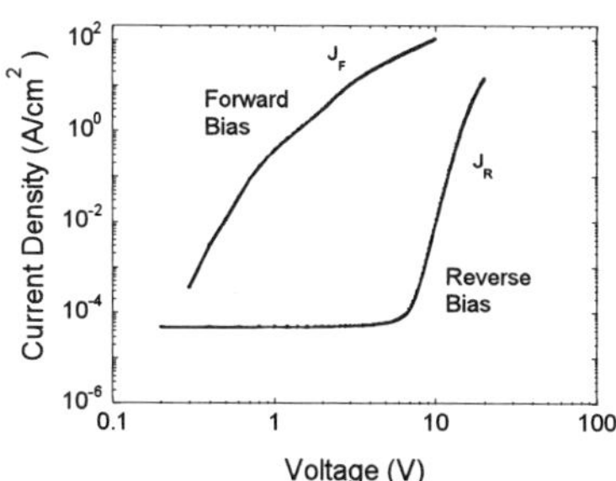

Fig. 6. LED I-V characteristics under forward bias (light emitting) and reverse bias. J_F and J_R are the forward and reverse bias current densities, respectively. The rectifying ratio (J_F/J_R) is $\sim 10^5$ at 5-10V bias, the region in which bright EL is observed.

integrated driver transistor, which induced an amplified current pulse through the LED. The resulting light signal was detected by a photomultiplier tube, which sent a corresponding voltage signal to an oscilloscope that was triggered from the function generator. The behavior under DC and pulsed operation is almost identical, which confirms that the light is not due to a thermal emission process. The lower right inset in fig. 7 shows the EL spectrum at different levels of applied current. The peak position and shape of the spectrum are essentially independent of the applied current. The drivers were able to amplify low-level base input signals and modulate the LEDs at frequencies in the kilohertz range. This does not represent the upper limit of the LED switching speed, as confirmed by the EL waveform shown in fig. 8.

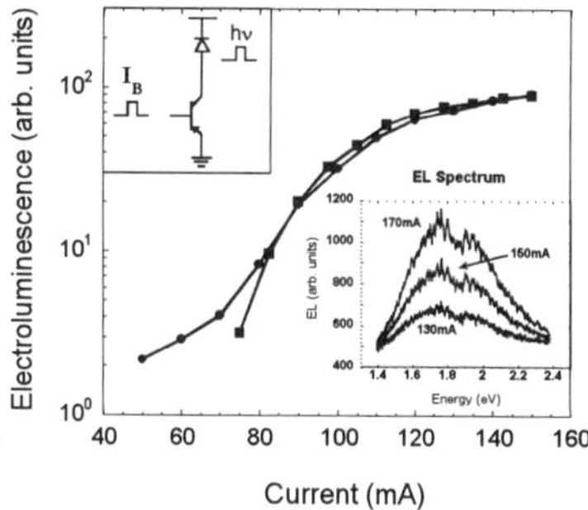

Fig. 7. EL response of the LED. The plot shows the EL signal as a function of the drive current in both the DC mode (circles) and pulsed mode (squares) of operation. The inset shows that the peak position (near 1.8eV) and the shape of the EL spectrum are independent of the appliied current.

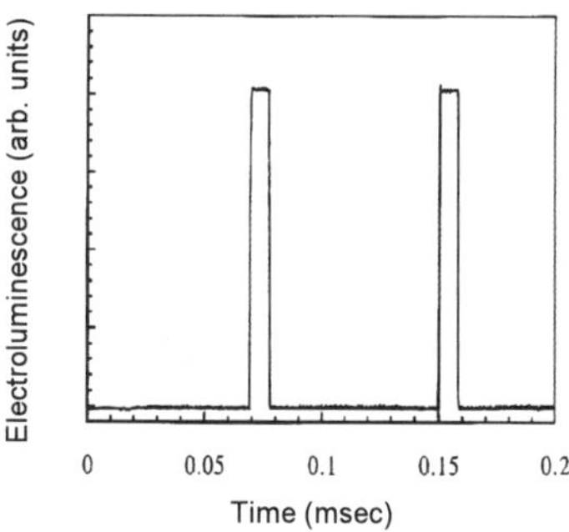

Fig. 8. EL response of the LED under direct current pulse modulation. The EL follows the current exactly on the msec time scale. The shape of the EL waveform demonstrates that modulation exceeding 1MHz is possible.

CONCLUSION

Successful integration of silicon-based LEDs with bipolar driver transistors has been reported. Improvements made in the active SRSO light-emitting material and the device design have advanced the LED performance levels toward that of commercial devices, and allowed processing in a standard IC fabrication sequence without degradation of either the LED or driver transistor characteristics. There are still significant challenges which must be overcome before silicon-based LEDs become practical light emitters. Improvements in efficiency and power dissipation are necessary for display applications, while increased modulation speed is critical for high speed optical interconnects. In addition, the device structures must be integrated with other optoelectronic components. The development of silicon-based integrated optoelectronic components will enable the incorporation of such structures into advanced systems in order to overcome existing performance limitations. Fig. 9 shows an addressable LED array utilizing the integrated local drivers. The capability of combining the new device structures with other silicon-based optoelectronic components and complex microelectronic circuits will offer increases in system integration that cannot yet be realized.

Fig. 9. Micrograph of an addressable LED Array utilizing the integrated driver transistors. The LEDs are placed at a pitch of 0.35mm, and addressed by the intersecting row (emitter) and column (base) select lines.

ACKNOWLEDGMENTS

The authors wish to acknowledge Dr. Lynn F. Fuller at the Center for Microelectronic Engineering, Rochester Institute of Technology, for use of the integrated circuit fabrication facilities. This work was supported by the US Army Research Office.

REFERENCES

1. D.G. Hall, *Mat. Res. Soc. Symp. Proc.* **298**, 367 (1993)
2. L.T. Canham, *Appl. Phys. Lett.* **57**, 1046 (1990)
3. P.M. Fauchet, *J. Lumin.* **70**, 294 (1996)
4. A. Loni *et al.,Electronics Lett.* **31**, 1288 (1995)
5. L. Zhang *et al., Mat. Res. Soc. Symp. Proc.* **358**, 671 (1995)
6. L. Tsybeskov, S.P. Duttagupta, K.D. Hirschman and P.M. Fauchet, *Appl. Phys. Lett.* **68**, 2058 (1996)
7. L. Tsybeskov, S.P. Duttagupta, K.D. Hirschman and P.M. Fauchet, in *Advanced Luminescent Materials*, edited by D.J. Lockwood, P.M. Fauchet, N. Koshida and S.R.J. Brueck, (The Electrochemical Society, Pennington, NJ, 1996) p. 34.
8. S. Lazarouk *et al., Appl. Phys. Lett.* **68**, 1646 (1996)
9. L. Tsybeskov, K.D. Hirschman, S.P. Duttagupta and P.M. Fauchet, "Electroluminescence and Carrier Transport in LEDs Based on Silicon-rich Silicon Oxide", *Mat. Res. Soc. Symp. Proc.* **452** (1997), this volume.
10. D.J. DiMaria and D.W. Dong, *J. Appl. Phys.* **51**, 2722 (1980)
11. D.J. DiMaria *et al., J. Appl. Phys.* **56**, 401 (1984)
12. K.D. Hirschman, L. Tsybeskov, S.P. Duttagupta and P.M. Fauchet, *Nature* **384**, 338 (1996)
13. S.P. Duttagupta *et al., Mat. Res. Soc. Symp. Proc.* **358**, 647, (1995)

Elaboration and light emission properties of low doped p-type porous silicon microcavities

G. Lérondel, P. Ferrand and R. Romestain
Laboratoire de Spectrométrie Physique (UMR 5588), BP 87, 38402 Saint Martin d'Hères Cedex, France Gilles.LERONDEL@ujf-grenoble.fr

ABSTRACT

We account for the elaboration of Bragg reflectors and microcavities based on efficiently luminescent porous silicon. A characterisation of very thin porous silicon layers obtained with current densities of formation varying from 1.5 mA to 300 mA is presented. The resulting refractive index variation (typically from 1.37 to 1.86 at 700 nm) enables the elaboration of high quality Bragg reflectors and Fabry-Perot filters from the yellow to the near infrared. Although low doped p-type porous silicon develops rougher interfaces than highly doped p-type porous silicon, its better luminescence efficiency has enabled us to elaborate microcavities with a strong emission in a narrow band.

INTRODUCTION

A few years after the discovery of its luminescence at room temperature, porous silicon (PS) appears more and more as a new optical material [1]. The possibility of modulating the porosity by simply modulating the formation current density enables to elaborate well defined thin layers in a large range of refractive indices. This leads to fabrication of optical structures such as Bragg reflectors, filters, optical waveguides [2,3] and more recently emitting microcavities [4]. Since highly doped p-type substrates permit smoother interfaces and a wider range of indices, theses substrates were usually preferred. However it is well known that a substrate with low doping level leads to a better efficiency of the luminescence. In this paper we will present results showing that this type of substrate can be used to obtain passive and bright devices.

EXPERIMENTAL DETAILS

The multilayer structures were obtained by modulating the current density of formation (j) in substrate of 5.3-6.7 ohm.cm resistivity. The anodisation parameters were chosen to cover the wider porosity range: the HF concentration was 35% and the current densities were varied between 1.66 and 333 mA/cm^2. The dissolution times for a $\lambda/4$ layer of Bragg structures are typically from 60 s to 0.5 s. Porosities were determined by gravimetry on 3 inch wafers. The PS multilayer have been characterised by Scanning Electron Microscopy using a JSM35C JEOL instrument. Reflectivity spectra were measured with a double beam Lambda 9 UV/VIS/NIR Perkin Elmer spectrometer. The violet line (457 nm) of an Ar laser was used for the photoluminescence excitation. It allows to excite the porous silicon down to 5 μm for usual porosities.

RESULTS AND DISCUSSION

Figure 1 shows the variations of the refractive index, n and the formation velocity, v as a

Mat. Res. Soc. Symp. Proc. Vol. 452 © 1997 Materials Research Society

function of the current density, Fig. a) and b), respectively. The values for n have been obtained by a fit of the experimental reflectivity spectrum obtained for single layer. The values of the

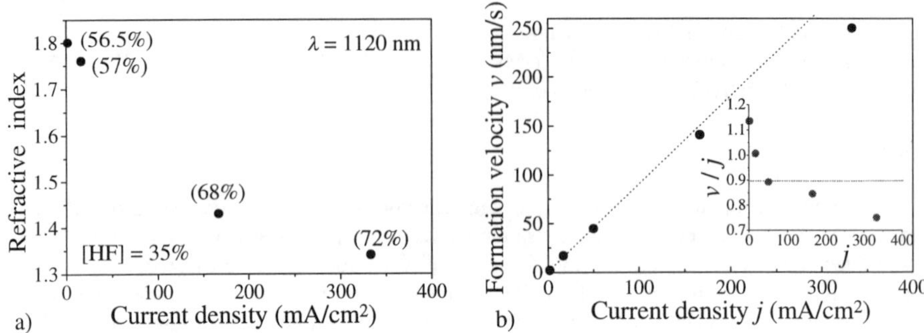

Fig 1: Calibration curves of the refractive index (a) and the formation velocity (b) as a function of the anodisation current density.

refractive indices are given at a wavelength of 1120 nm. The same layer was later dissolved to perform thickness measurements in order to determine the formation velocity. The values of j (1.66 to 333 mA) were chosen so that well defined porosities and thicknesses could be obtained. As it is well known the porosity increases with j. After optimization, the minimum and maximum current densities enable us to obtain 56.5% and 72%, respectively, which leads to a variation of the refractive index by 0.47. On figure 1b) we first observed a linear dependence of the formation velocity as a function of the current density. Then, a saturation begins to appear for high value of j. The inset illustrate this effect showing the dependence of the ratio v/j, (i.e. the valence of the anodisation) as a function of j.

These calibration curves were used to control the vertical modulation of the porosity during silicon anodisation. Figure 2 shows a SEM photograph of a multilayer structure. A periodical change of the contrast is clearly observed

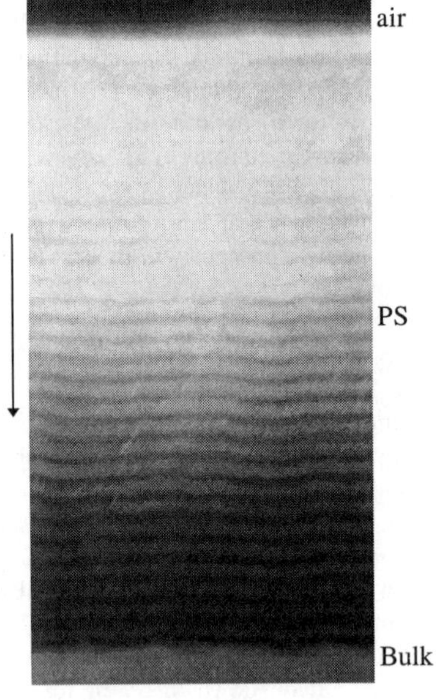

Fig. 2: SEM photograph of a PS multilayer. The structure was obtained repeating 30 times the current sequence of 150 mA/cm^2 - 1.25 s and 15 mA/cm^2 - 10.25 s. The periodicity and the total layer thickness are 0.39 and 11.7 μm, respectively. The overexposure near the top is due to surface charge effect. The arrow indicates the dissolution (i.e. formation) direction.

throughout the entire thickness of the PS film. Due to the low conductivity of the porous silicon, surface charge effects can not be avoided in the microscope and give rise to the overexposure observed at the top of the layer. Since this level of brightness is much higher than the bulk one (bottom part of the photograph) it can not be attributed to a porosity gradient in the depth. The periodicity of this structure (bright and dark stripes) is 0.39 μm and remains unchanged in the depth, attesting for the control of the layer thickness during the formation. However, the photograph also shows irregular undulations at the interfaces. This is more pronounced near the bulk (bottom) interface. This effect is due to fluctuations in the formation velocity [5]. A direct consequence of this on the light propagation is the increase of the losses in the structure due to scattering [6]. However, comparing two successive interfaces one see that these fluctuations are correlated which explains why the optical interferences are not destroyed.

Since a porosity (i.e. refractive index) and a thickness can be determined it is then possible to derive the periodic optical thickness ($n_1d_1 + n_2d_2$) of such structure. n_1, d_1 and n_2, d_2 are respectively the refractive index and the thickness of the two elementary layers of the periodic structure. When the constructive interference condition in a multilayer system is obtained it is well known that maxima in the reflectivity spectrum appear for different wavelengths λ_k given by:

$$k\lambda_k/2 = (n_1d_1 + n_2d_2) \quad (1)$$

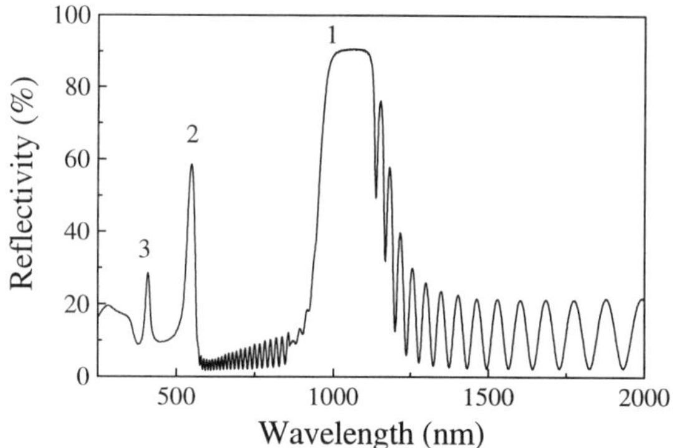

Fig. 3: Typical reflectance of a low doped p-type PS Bragg Reflector, centred in the NIR. The numbers correspond to the three orders of the Bragg mirror. A stop band and maximum of reflectivity for the first order of 230 nm and 92% were measured.

Figure 3 shows a typical reflectivity spectrum of a low doped p-type Bragg structure. Moving from the IR towards the UV, we observed first regular fringes due to the total optical thickness of the structure and secondly three peaks, which correspond to the three orders (k=1,2 and 3) of the Bragg condition. The shape of the three peaks is partially due to the absorption in the material which increases for the smaller wavelengths. The maximum of reflectivity is smaller than the expected value for the Δn and the number of periods. This is due to the losses in the multilayer. In our case they have two origins, scattering at the successive interfaces and absorption, as previously discussed.

The periodicity in the multilayer can be broken by adding a layer with an optical thickness n_cd_c, in the middle of the structure. If the relation (1) and the conditions $n_cd_c = \lambda/2$ are verified for the same wavelength, a Fabry-Perot cavity is formed. Figure 4 shows a typical reflectivity spectrum of a low doped p-type PS Fabry-Perot (FP) filter. We observed a broad stop band due to the Bragg mirrors and at the centre the sharp notch of the Fabry-Perot mode. The FWHM of the mode and the reflectivity stop band are respectively 16 nm and 235 nm. The first and second Bragg numbers of period have been chosen as 7 and 13, respectively. This dissymmetry improves

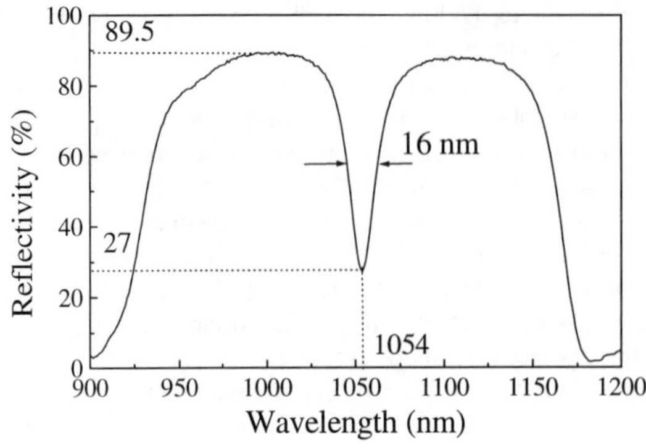

Fig. 4: low doped p-type PS microcavity.

the outcoupling (minimise the minimum of reflection) of the FP mode by compensating for the absorption in the material. The curvature of the stop band also indicates that losses occur in the structure as discussed in the previous paragraph. By decreasing d_1 and d_2, Fabry-Perot microcavities centred in the visible region (i.e. at the position of the PS luminescence) have been obtained. These microcavities have comparable quality so that we can conclude that passive optical structure could be equally obtained from the IR to the visible range with low doped p-type PS as for substrate presenting higher doping level [7].

Since porosity higher than 70% can be be reached by varying the current density such structures are expected to present luminescent properties. Figure 5 shows the luminescence of a microcavity centred in the visible range at 747 nm. The effect of the FP cavity is clearly obtained,

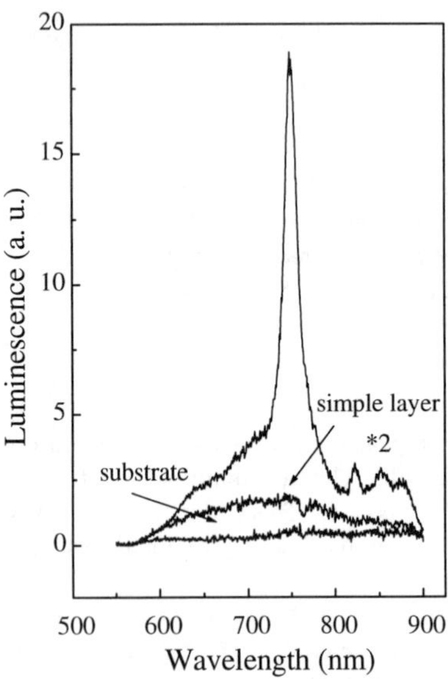

Fig. 5: Low doped p-type PS luminescent microcavity. The position and the FWHM of the luminescence peak are 747 and 14 nm, respectively. The 457 nm Ar laser line was used as excitation. No correction due to the absorption was done.

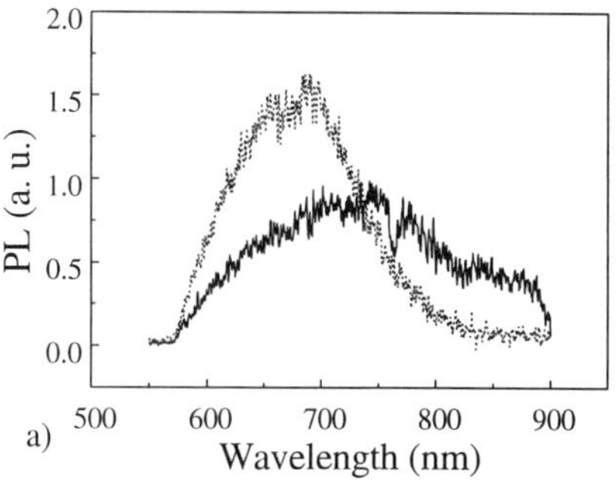

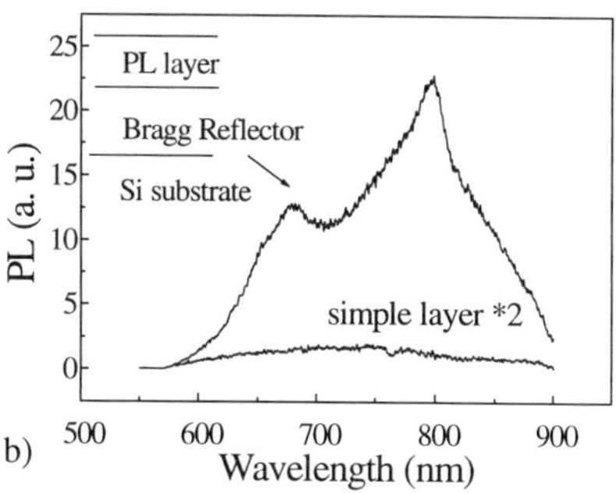

Fig 6: Effect of the second Bragg formation on the microcavity layer. a) PS half microcavity;b) The layer corresponding to the luminescence curve representing with dots was kept in the HF solution for 862s corresponding to the second Bragg formation time (Fig. a). The scheme of the structure is given in the inset (Fig. b).

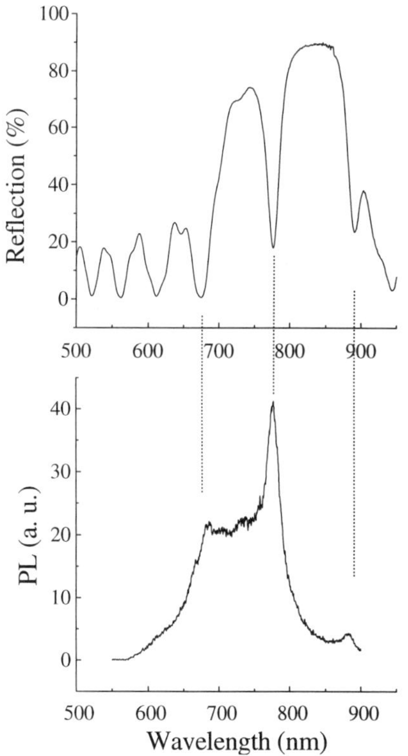

Fig. 7: Correlation between reflection and photoluminescence. Maxima of PL correspond to minima of reflection.

the luminescence presents a narrow peak, whose position (747 nm) and FWHM are closed to the reflectivity one. Comparing to the single layer spectra we observed an enhancement of the luminescence intensity. An estimation of the quantum efficiency of this structure give an amount of 15%. This value results from the combination of the highly efficient luminescence observed for the low doped p-type PS and the cavity effect. However, one observes a relatively large luminescence background in addition to the cavity peak. This seems to indicate that the Bragg mirrors, 68% and 56.5% of porosity, which are initially not luminescent, became weakly luminescent during the formation due to the dissolution. This effect is illustrated in figure 6 a) for the case of a single layer equivalent to the microcavity layer. After a time corresponding to the second Bragg formation, the luminescence is enhanced by a factor of 2. A blue-shift is also observed. It has to be taken into account in order to overlap the position of the cavity mode and the luminescence maximum. To precise the effect of the light confinement, we compared the luminescence of the single layer (i.e. luminescent layer) and a structure described in the figure 6 b), which consist in a layer on top of a Bragg reflector. The two samples exhibit very different intensities of luminescence. The presence of a good quality mirror behind the layer would only increase the luminescence by a factor of two. In fact it appears that interferences in the structure have to be invoked to explain luminescence of such a layer. As a matter of fact a strong correlation between luminescence and reflectivity is shown in the figure 7, which represents the reflection and luminescence spectra of a microcavity centred in the visible region. Maxima of luminescence corresponds to minima of reflection both for the FP cavity mode and fringes.

CONCLUSION

To conclude, high quality structures, like Bragg Reflector and Fabry-Perot filters have been elaborated using low doped p-type PS. Strong luminescence and light confinement give rise to highly efficiency PS microcavities. Up to now the mechanism of the porous silicon luminescence is seen as an essentially non radiative process, eventual increase of the radiative recombination rate due to the cavity will probably generate new interest on this material.

REFERENCES

[1] L. T. Canham, Appl. Phys Lett. **57**, 1046 (1990).

[2] M. G. Berger, S. Frohnhoff, W. Theiss, U. Rossow, H. Muender, *Porous silicon science and technology*, ISBN 3-540-58936-8, Springer (1994), p. 345

[3] A. Loni and L. T. Canham, M. G. Berger, R. Arens-Fischer, H. Munder and H. Luth, H. Arrand and T. M. Benson, Thin Solid Films **276**, 143 (1996)

[4] V. Pellegrini, A. Treducucci, C. Mazzoleni and L. Pavesi, Phys. Rev. B **52** (20), R14328 (1996)

[5] G. Lérondel, R Romestain and S. Barret J. Appl. Phys. to be published

[6] G. Lérondel and R. Romestain, Thin Solid Films in press

[7] M. Araki, H. Koyama and N. Koshida, Jpn. J. Appl. Phys. Vol. **35**, 557 (1996)

ENHANCEMENT OF THE SPONTANEOUS EMISSION RATES IN ALL POROUS SILICON OPTICAL MICROCAVITIES

L. PAVESI, M. CAZZANELLI, AND O. BISI
INFM and Dipartimento di Fisica, Università di Trento,
via Sommarive 14,I-38050 Povo (Italy) [pavesi@science.unitn.it]

ABSTRACT

The emission properties of a porous silicon layer placed in an optical microcavity is investigated by photoluminescence and time resolved photoluminescence measurements. The microcavity is formed by an all porous silicon Fabry-Perot filter made by two distributed Bragg reflectors separated by a λ or $\lambda/2$ porous silicon layer. Our main findings are that the spontaneous emission spectrum is drastically modified: the linewidth is narrowed, the time decay of the emission is shortened by a factor of about 2/3 at room temperature and the peak emission intensity is increased by a factor of more than 10. These facts are caused by the redistribution of the optical modes in the microcavity due to the presence of the optical resonator and to the variation of the dielectric environment.

INTRODUCTION

The observation of room temperature visible photoluminescence from porous silicon (PS) has renewed the interest of the scientific community in the field of all-silicon based photonics. [1] Soon light emitting devices based on PS appeared. [2] However, some problems related to the PS physics emerged: low external quantum efficiency, wide emission band and long recombination lifetimes.

In 1995, we realized the first all PS optical planar microcavity (PSM) with the aim to address these draw-backs. [3] In fact, the PSM could be used to improve PS based devices by the use of Resonant Cavity Light Emitting Diodes. [4] A better spectral purity, an increase in the emission efficiency and a large directionality of the emitted light could be envisaged and have been demonstrated in the photoluminescence properties of PSM. [3]

All PSM consist of an active central layer embedded between a pair of distributed Bragg Reflectors (DBR's). [3] The mirrors act as a "1-D confiner" of the light spontaneously emitted from the central layer.[5] In a simple scheme, the spontaneous emission rate of a semiconductor (Γ) at T=0 K is given by the Fermi golden rule

$$\Gamma(\hbar\omega) = \frac{2\pi}{\hbar} |\langle M_{eh} \rangle|^2 \rho(\hbar\omega) \qquad (1)$$

where $\langle M_{eh} \rangle$ is the interaction hamiltonian matrix element between the initial and final electronic state and $\rho(\hbar\omega)$ is the photon mode density. If one modifies the photon mode density, e. g., by placing the active medium in an optical microcavity, the spontaneous emission rate is modified accordingly. [5] Indeed, in the PSM a narrowing in the emission lineshape of PS down to 3.7 nm, [6] an increase in the peak emission efficiency up to 22 times and a directionality of the emission about a cone of 30° is actually achieved. [3] These facts are attributed to the spatial redistribution of the optical modes due to the planar microcavity. [5]

An interest exists in verifying whether the time decay of the luminescence of PSM is changed with respect to that of a simple PS layer. This is the subject of the present paper.

Mat. Res. Soc. Symp. Proc. Vol. 452 © 1997 Materials Research Society

TABLE I. Summary of the time resolved luminescence measurements. The column labeled n_C reports the measured refractive index of the layers at the central wavelength λ_c which is shown in the third column. I_{integ} refers to the normalized spectrally integrated luminescence intensity, $I(\lambda_c)$ to the normalized peak emission intensity at the wavelength of the microcavity peak (λ_c), and τ_{λ_c} and β_{λ_c} to the parameters extracted from a stretched exponential fit of the luminescence decay at λ_c (for the reference DBR they are reported for the different λ_c of the various PSM). The PS 45% sample had a luminescence too weak to be detected by our system.

	n_C	λ_c (nm)	I_{integ}	$I(\lambda_c)$	τ_{λ_c} (μs)	β_{λ_c}
PS 45 %	2.1		-	-	-	-
PS 62 %	1.52		0.13	0.03	22±2	0.50±0.03
PS 75 %	1.30		0.26	0.08	21±0.5	0.72±0.01
DBR (45%)	1.52/2.2	720	0.84	0.23	7.5±0.3	0.47±0.01
(62%)				0.20	9.2±0.4	0.49±0.01
(75%)				0.24	7.5±0.3	0.47±0.01
PSM 45 %	2.1	692	1	0.81	8.7±0.1	0.62±0.01
PSM 62 %	1.52	741	0.95	0.66	13.3±0.1	0.62±0.01
PSM 75 %	1.30	692	0.92	1	15.2±0.1	0.62±0.01

MICROCAVITY FABRICATION AND CHARACTERIZATION

An all PSM consists of a Fabry-Perot structure (two mirrors separated by a thin central layer), where the porosity of the different porous silicon layers are changed to obtain the desired refractive index profile. This is realized by an appropriate modulation of the anodization time and the current density during the electrochemical etch. [7,8] A detailed discussion of the procedure to obtain the PSM has been already presented in [8]. In this work, we present data on various PSMs which differ only in the porosities and thicknesses of the central layer in order to obtain different refractive indexes n_C in the central layer. All the PSM were obtained by electrochemical anodization of 0.01 Ωcm p-type doped Si-wafers with a 15% HF solution. They were formed by two distributed Bragg reflectors as mirrors (constituted by 12 periods of 62%-45% porosity layers, with optical thicknesses of $\lambda_c/4$) which sandwiched a central layer of optical thickness $\lambda_c/2$. λ_c is the wavelength of the Fabry-Perot resonance. In the following we label the PSM by adding to "PSM" the porosity of the central layer, i. e. "PSM 45%" means a microcavity sample with a central layer of 45% porosity. Porous silicon samples obtained with the same etching parameters (same porosity) as those used for the microcavity central layers constitute the reference PS. A distributed Bragg reflector was obtained by using the same etching parameters as the microcavity mirrors. Some characteristics of the samples are presented in Table I.

The CW photoluminescence (PL) measurements were performed using the 488-nm line of an Argon ion laser, with an incident intensity of about 10 W/cm^2. The PL was collected along the normal to the sample surface and dispersed with a 0.75-m double monochromator equipped with a GaAs photomultiplier. Reflectivity measurements were carried out with a monochromatized 1000 W Xe lamp and were calibrated with respect to the known reflectance of an Al mirror.

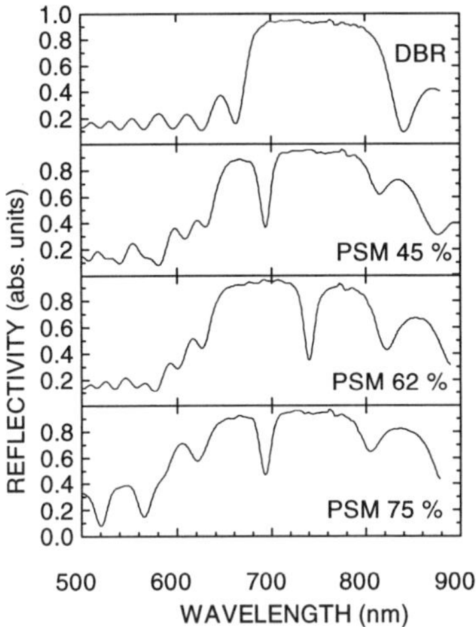

Figure 1. Room temperature reflectance spectra of the porous silicon microcavities. For reference the reflectance of a distributed Bragg reflector obtained by using the same etching parameters as those for the PSM mirrors is reported in the top panel.

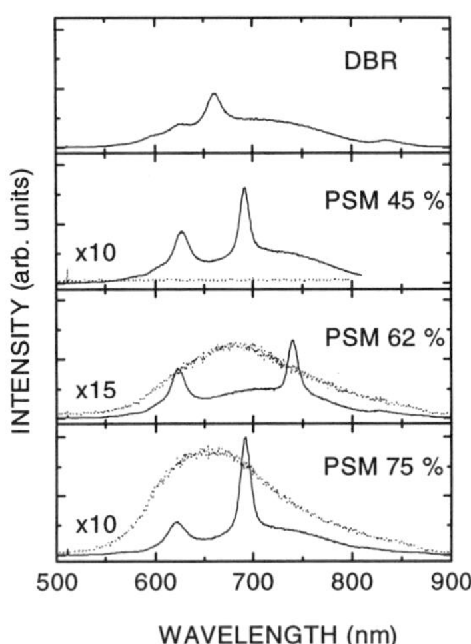

Figure 2. Room temperature luminescence spectra (lines) of the porous silicon microcavities (PSM). For each PSM the corresponding spectrum of the reference porous silicon layer is reported (dotted lines) after multiplication by the factor indicated on the left side to allow a better view. The spectra are in relative units.

The time resolved PL measurements were accomplished by using a frequency-doubled mode-locked Ti:Al_2O_3 laser operated at a repetition rate of 4 KHz with an excitation wavelength of 410 nm, a peak power density of 500 W/cm^2 and a pulse duration of 1.5 ps. A 0.2 m monochromator coupled to a streak camera was used to record the time resolved PL. Only the PL inside a cone of about 0.03 sterad about the sample normal was collected.

CW AND TIME RESOLVED LUMINESCENCE

The reflectivity spectrum of the reference DBR is shown in Fig. 1 top panel. A region of high reflectivity (stop band) is present in the spectral range 680-820 nm about the central wavelength λ_c =720 nm. The insertion of a central layer $\lambda_c/2$ thick in the middle of two symmetric DBR, as for the PSM, causes the appearance in the reflectivity spectrum of a dip at λ_c, which corresponds to the microcavity optical resonance. The increased transmittance for λ_c is a manifestation of the propagating optical mode through the microcavity. The optical modes for $\lambda \neq \lambda_c$ are not propagating and thus are reflected. This represents the two-dimensional confinement of the photons inside the microcavity. [5] Let us note that these samples are not fully optimized with respect to the etching parameters and, hence, the linewidth of the λ_c resonance is only reduced to 10 nm. Optimized samples yield linewidths as narrow as 3.7 nm. [6] The PL spectra of these samples are shown in Fig. 2 and compared with those of the reference PS samples. A summary of the PL results is also reported in Table I.

One can notice that the presence of the microcavity modifies the emission properties of PS both with respect to the emission intensity and the lineshape. A sharp emission peak instead of a wide emission band is present in the PL of the PSM. This peak is caused by the resonant coupling at λ_C of the excitons, photo-excited in the central layer, with the two-dimensional confined photons. [5] In the case of perfect photon confinement, i. e. ideal mirrors, the photon emission will be only allowed at λ_C. However, dielectric porous silicon microcavities are characterized by the fact that: 1) the photon mode penetrates into the DBR, 2) the emission spectrum is wider than the photon resonance, 3) the reflectivity in the stop band is not 100%, and 4) the mirrors have a luminescence emission which is overlapping the emission of the central layer (see top panel in Fig 2). All these facts cause the appearance of a weak emission for $\lambda \neq \lambda_C$ and of side bands (e. g. the peak at 620 nm). [5] It is interesting to note that the enhancement in the luminescence is dependent on the refractive index value n_C of the central PSM layer. By increasing n_C, an increase in the ratio of the peak intensity at λ_C of PSM with respect to reference PS is observed.

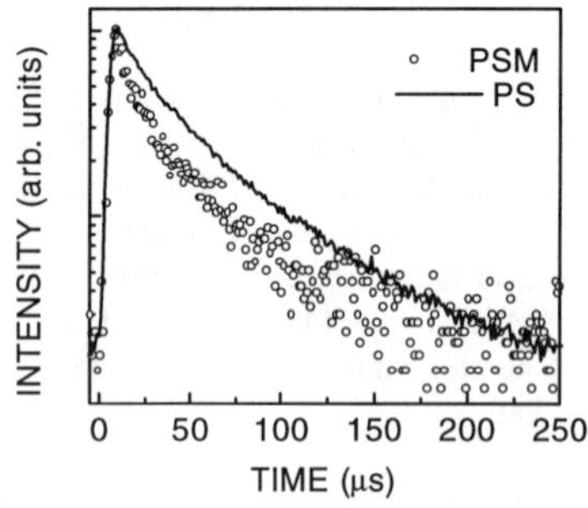

Figure 3. Room temperature time decay of the luminescence for a λ/2 microcavity (PSM and indicated by the dots) and the reference sample (PS and indicated by the line) at the PSM peak wavelength.

Time resolved measurements have been also performed. An example of the luminescence decay at λ_C is shown in Fig. 3 for a PSM and reference PS. The luminescence decays faster for the PSM than for the reference PS. As is already known, [9] the observed decay is not exponential but it has a stretched exponential lineshape:

$$I(t)=I_o \exp(-t/\tau)^{\beta}. \quad (2)$$

A least-square fit of the luminescence decay with the stretched exponential lineshape yields the τ and β parameters reported in Table I. For simplicity, in the following we will concentrate only on the τ values. In general, the luminescence of the PSM decays faster than that of the PS for $\lambda=\lambda_C$. The ratio between the τ of the PSM and that of the reference PS increases as n_C increases. A more detailed study of the time resolved luminescence of PSM samples will be presented elsewhere.

DISCUSSION

The theoretical treatment of the coupling of the excitonic transition with the optical mode inside a microcavity has been reviewed in [5]. It has been shown that dielectric microcavities

TABLE II Effective refractive index (n_{DBR}) of the DBR, penetration length of the optical mode on the Si side (ℓ_{Si}) and on the air side (ℓ_{air}), and effective refractive index (n_{eff}) (see Eq. 5) for the optical mode at λ_c for the three microcavities studied in this work.

	n_{DBR}	ℓ_{Si} (nm)	ℓ_{air} (nm)	n_{eff}
PSM 45%	1.835	459	559	1.872
PSM 62%	1.867	327	338	1.774
PSM 75%	1.835	469	511	1.721

differ from ideal or metallic microcavities for the finite reflection angle and the penetration of the optical mode into the dielectric mirrors. As a consequence in a planar dielectric microcavity the spontaneous emission rate is slightly modified and what one observes is the redistribution of the emission in space. This explains the modified CW luminescence lineshapes. In addition, it is also observed that the variation in the emission lifetime will be of the order of unity in a dielectric microcavity and mostly of the variations are due to the change of the local dielectric environment.

In fact, the spontaneous emission rate for a bulk material $\Gamma(\hbar\omega)$ is proportional to the photon mode density $\rho(\hbar\omega)$, where

$$\rho(\hbar\omega) = \frac{n^3(\hbar\omega)^2}{\pi^2 c^3 \hbar^3}, \tag{3}$$

and n is the refractive index of the material. In a PSM, the optical mode coupled to the excitonic transition extends into the DBR. Consequently, the refractive index entering into Eq. 3 is not the one of the central porous layer reported in Table I, but an effective index (n_{eff}) averaged over the volume where the optical mode extends, i. e. over the central layer and part of the DBR.

In [10], the penetration length (ℓ) of the optical mode in a dielectric mirror has been calculated:

$$\ell = \frac{\lambda_c}{4n_c}\frac{q}{1-p}\frac{\left(1+a^2p^{m-1}\right)\left(1-p^m\right)}{\left(1+q^2a^2p^{2m-2}\right)} \tag{4}$$

where $q=n_c/n_L$, $p=n_L/n_H$, $a=n_H/n_{air}$ or n_H/n_{Si} depending on whether one considers the top or the bottom DBR, n_L and n_H are the refractive indexes of the high and low porosity layers of the DBR and m is their number. Hence, n_{eff} is

$$n_{eff} = \frac{n_{DBR}\left(\ell_{air}+\ell_{Si}\right)+\lambda_c/2}{\ell_{air}+\ell_{Si}+\lambda_c/(2n_c)} \tag{5}$$

where the effective refractive index n_{DBR} of the DBR is the spatial average between n_L and n_H. Results of the calculation of n_{eff} are given in Table II.

According to Eq. (3), the different lifetimes measured for the PSM and for the reference PS should fit a cubic law dependence on the refractive index. This trend is shown in Fig. 4 where the inverse of the measured lifetimes are reported as a function of the refractive index.

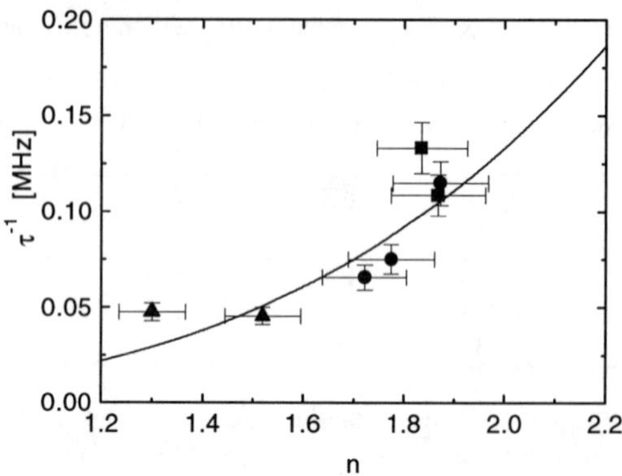

Figure 4. Luminescence decay rates as a function of the refractive index for the PS reference (triangles), the DBR (squares) and the PSM (discs).

A power law fit of the experimental data with $1/\tau \propto n^{\alpha}$ gives α=3.5±0.8. In view of the simplicity of the model, the agreement is satisfactory.

CONCLUSIONS

The use of microcavities has been shown to modify both the luminescence lineshape and the luminescence decay rates of porous silicon. In view of the novelty of the PSM, many aspects of their physics need to be clarified still, but some interesting features have emerged in this study. In particular, the role of the local dielectric environment in determining the actual recombination rates of excitons have been established in porous silicon microcavities.

ACKNOWLEDGEMENTS

The experimental set-up used in this work has been financed by the University of Trento through the "Progetto Speciale 1995: Spettroscopia veloce nel Silicio Poroso" and by grants from I.N.F.M. and C. N. R. Fruitful discussion with G.Bjork are acknowledged.

REFERENCES

[1] L. T. Canham, Appl. Phys. Lett. **57**, 1046 (1990).
[2] A. Loni et al., Electronics Lett. **15**, 1288 (1995).
[3] L. Pavesi et al., Appl. Phys. Lett. **67**, 3280 (1995) and Phys. Rev. B **52**, R14328 (1995).
[4] L. Pavesi, R. Guardini and C. Mazzoleni, Solid St. Commun. **97**, 1051 (1996).
[5] G. Bjork and Y. Yamamoto, in Spontaneous Emission and Laser Oscillation in Microcavities, ed. H. Yokoyama and K. Ujihara (CRC Press, London 1995) 189.
[6] L. Pavesi and P. Dubos, to be published.
[7] M. G. Berger et al., J. Phys. D: Appl. Phys. **27**, 1333 (1994); M. G. Berger et al., Thin Solid Films **255**, 313 (1995).
[8] C. Mazzoleni and L. Pavesi, Appl. Phys. Lett. **67**, 2983 (1995).
[9] L. Pavesi, J. Appl. Phys. **80**, 216 (1996); E. H. Roman and L. Pavesi, J Phys: Condens. Matter **8**, 5161 (1996); P. Maly et al., Phys. Rev. B **54**, 7929 (1996).
[10] D. I. Babic and S. W. Corzine, IEEE J. Quantum Electr. **28**, 514 (1992).

Part X

Preparation and Properties of Nanocrystalline, Microcrystalline, and Microstructured Thin Films

GROWTH AND STRUCTURE OF MICROCRYSTALLINE SILICON PREPARED WITH GLOW DISCHARGE AT VARIOUS PLASMA EXCITATION FREQUENCIES

F. FINGER*, R. CARIUS*, P. HAPKE*, L. HOUBEN*, M. LUYSBERG**, M. TZOLOV***
* Forschungszentrum Jülich, ISI, 52425 Jülich, Germany,
** University of California - Berkeley, Dept. MS &ME, Berkeley, CA,
*** Bulgarian Academy of Sciences, Solar Energy &New Energy Sources, 1784 Sofia, Bulgaria

ABSTRACT

Microcrystalline silicon was prepared with glow discharge deposition from silane/hydrogen mixtures at plasma excitation frequencies in the range 13.56 MHz - 116 MHz. The influence of the plasma excitation frequency on the growth and the structural properties of the material is investigated. At high excitation frequencies, higher growth and etching rates, larger grain sizes with less disorder within the grains, higher crystalline volume fractions, a reduced amorphous but more porous interface layer on glass and quartz substrates, and faster nucleation on amorphous silicon substrates are obtained. The results are discussed within a schematical growth model.

INTRODUCTION

Microcrystalline silicon (μc-Si:H) for application in thin film devices like flat panel displays and thin film solar cells is easily prepared with plasma enhanced chemical vapour deposition (PECVD) by only little adaptation from the well established amorphous silicon (a-Si:H) technology. Compared with a-Si:H the microcrystalline material has higher conductivities and mobilities and a higher absorption in the near infrared all of which are of importance for the several technological applications. The use of microcrystalline silicon for contact layers in thin film solar cells is well established [1]. Recently the material gained more interest as active layer in a multi-gap cell structure [2, 3] and there is an upcoming research activity in thin crystalline (polycrystalline) silicon for photovoltaic applications where low temperature processes are requested [4]. For the various applications a couple of technical issues of the PECVD of μc-Si:H need further investigation and optimization like (I) the deposition rate and (II) the nucleation behaviour of μc-Si:H on various substrate materials.

(I) The deposition rate of μc-Si:H in standard PECVD processes with SiH_4 and H_2 is much too low for deposition of active photovoltaic layers of 2-3 μm, not to speak of "thin" crystalline materials of 30 μm. Generally a correlation between crystallinity and suppressed growth rate is found [5]. Techniques providing high growth rates while maintaining good material quality are needed. A remarkable improvement in growth rate of quality μc-Si:H is achieved in PECVD using fluorinated gases [6]. Also with hot wire CVD, much increase in deposition rate is achieved [7]. While with the latter technique a further post-deposition hydrogen passivation of grain boundary defects could be necessary, with fluorinated gases some additional experimental precautions might be needed. It is therefore of much interest that also PECVD processes with plasma excitation frequencies (ν_{ex}) higher than the standard 13.56 MHz lead to higher growth rates of good quality μc-Si:H [8-12].

(II) Related to the maintenance of good material quality at high growth rate is the control of the nucleation of μc-Si:H on various kinds of substrates as the resulting interfaces will strongly affect the performance of thin film devices. The substrate dependence of the nucleation behaviour of μc-Si:H has in the past been studied on a variety of substrate materials such as glass, metal, transparent conductive oxides, crystalline and amorphous silicon [13-19]. Generally nucleation depends on type and morphology of substrate. The differences are in terms of crystallinity, porosity, columnar structure, inhibition of crystallization especially on amorphous silicon as a substrate or an amorphous or porous interface layer. Thus nucleation has to be optimized for each substrate.

We present results on μc-Si:H material prepared with PECVD in the frequency range 13.56 MHz - 116 MHz. The material is investigated with transmission electron microscopy (TEM), X-

Mat. Res. Soc. Symp. Proc. Vol. 452 © 1997 Materials Research Society

ray diffraction at grazing incidence (XRD), Raman spectroscopy, optical band-to-band and infrared absorption (IR), photoluminescence (PL), electron spin resonance (ESR) and electrical conductivity. It is shown that the plasma excitation frequency ν_{ex} strongly affects the growth and nucleation process of μc-Si:H and consequently the structural and electronic properties. The influences of ν_{ex} on the process plasma and on the μc-Si:H film growth, which are related to a reduction of the plasma sheath potentials and sheath thickness, a higher dissociation of the process gases and a higher ion current density at the substrate [20-24], will be discussed.

EXPERIMENTAL DETAILS

The samples were prepared by PECVD with plasma excitation frequencies between 13.56 MHz and 116 MHz at 200 °C in a commercial deposition system with base pressure $\leq 10^{-9}$ mbar [25]. The reaction chamber is of standard diode type with 10*10 cm^2 substrates. We use gas mixtures of SiH_4 in H_2 with PH_3 as doping gas. For plasma excitation a broad band generator (10 - 125 MHz) and a T-network matchbox with a stepwise variable inductance is used. No special precautions are taken to optimize network connections at the various frequencies (like cable length and low pass filters) which might affect power coupling and power measurements. The forward and reflected power measured with a power meter in the 50 Ohm line between generator and matchbox was 5W and < 0.5W, respectively. To get a stable and laterally homogeneous plasma, the total gas pressure is reduced (p=500...300 mTorr) as one increases the plasma excitation frequency. Samples are grown on borosilicate glass (Corning 7059), polished Si wafer for IR and TEM, Chromium (e-beam evaporated on glass), intrinsic a-Si:H on glass, roughened quartz for photoluminescence and on aluminium foils for preparation of powder material for ESR. Deposition and etch rates are determined by measuring film thicknesses with a step profiler. For the etch rate an a-Si:H film is exposed to a pure hydrogen plasma with otherwise the same conditions like in the H_2/SiH_4-plasma for deposition. The average grain size is determined from X-ray diffraction (XRD) at grazing incidence using the Scherrer formula. The problems with evaluation of XRD spectra of μc-Si:H with columnar structure as compared with the spherical shape assumed in the Scherrer formula are discussed elsewhere [26]. Raman back scattering and transmission electron microscopy (TEM) are used to investigate the crystalline volume fraction and to obtain detailed information on the film structure [25-29]. Optical absorption in the infrared and near the fundamental absorption edge and electronic conductivity is measured as a function of film thickness to study the initial growth of thin layers and to get information on the H-bonding in the material. Photoluminescence is used to study carrier recombination [30] and electron spin resonance is used to measure the defect density [31, 32].

RESULTS

Growth and etch rate, grain size and volume fractions

With VHF-GD one can simultaneously increase the deposition rate, the average grain size and the crystalline volume fraction of μc-Si:H [12, 27].These results are summarized in Figs. 1a-c for samples with thicknesses of about 3000Å. First in Fig. 1a the net deposition rate of μc-Si:H and the etch rate for amorphous material as a function of frequency are shown. Both etching of amorphous material and the total deposition rate for μc-Si:H increase. Next the average grain size as determined from XRD is shown in Fig. 1b. Note that the evaluation of XRD with the Scherrer formula only gives an average number and can under-estimate the maximum grain size [28, 33]. Maximum grain sizes as seen in high resolution TEM are much larger with dimensions of up to 200 nm in growth direction at 94 MHz (see below).

As a qualitative measure of the crystalline volume fraction the intensity ratio of the crystalline (518 - 520 cm^{-1} and 500 cm^{-1}) and amorphous (480 cm^{-1}) contributions in Raman spectra of μc-Si:H is shown in Fig. 1c. While this is not an absolute measure of the crystalline volume fraction (the amorphous content is overestimated and comparison with TEM measurements show that the true crystalline volume fractions are much larger), the increase in crystallinity as a function of ν_{ex} can be considered as proof and is confirmed by XRD and TEM [27-29].

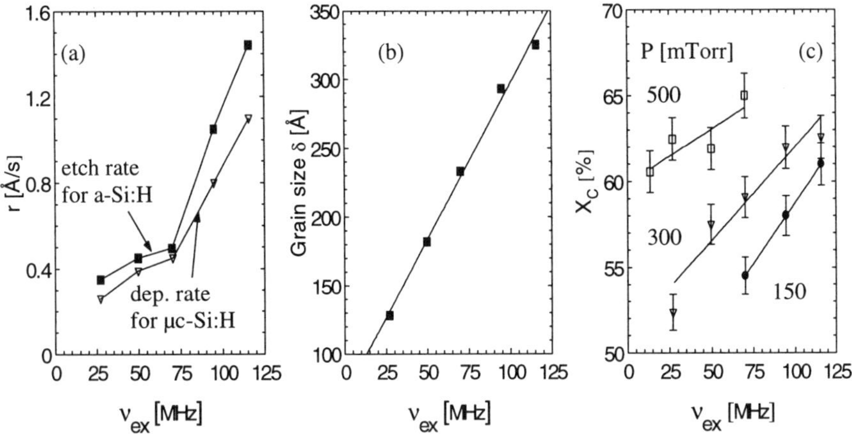

Fig. 1 (a) Growth and etch rates (b) average grain size and (c) Raman intensity ratio vs. ν_{ex}.

Transmission electron microscopy

The structural details of µc-Si:H were studied with TEM [27]. In Fig. 2 (a) dark, (b) high resolution and (c) bright field TEM micrographs of samples prepared at 27 MHz and 94.7 MHz, representative of material prepared at low and high ν_{ex} resp., are shown. The samples are deposited on crystalline Si wafer with a native oxide as for TEM sample preparation this type of substrate is easier to handle than glass. Concerning the nucleation process, the SiO_2 cover layer is considered to be similar to the borosilicate glass. At both frequencies we find the typical columnar structure (Fig. 2a). In the sample prepared at 94.7 MHz many columns consisting of crystallites of the same orientation (corresponding to the bright contrasts) extend over the total film thickness of 440 nm while for the sample prepared at 27 MHz most columns are restricted in length. Some of the columns exhibit a cone-like shape. The average size of individual grains and the lateral extent is smaller at lower frequencies. The dark regions here do not correspond to amorphous phase but rather crystalline regions with different orientation which is confirmed by tilting the samples. In both samples irregular fringe contrast points to the presence of twins. The region close to the interface between substrate and film is characterized by the occurence of small columns and small crystallites which are not attached to columns. These small features (more disorder) are more pronounced in the sample prepared at 27 MHz. The high resolution images (Fig. 2b) nicely visualize the crystalline structure of individual grains. Little amorphous phase is seen for both frequencies between the grains and at the films substrate interface. An amorphous layer (2 nm) is probably the oxide layer of the crystalline silicon wafer. The crystalline volume fraction can be estimated to be larger than 90% in both films. Note that the contrast of a partially amorphous region is dominated by the crystalline contrast. Therefore the amorphous volume fraction can be underestimated in the TEM images. The crystalline phase itself is more disordered at low frequencies in a sense that there are more twins and stacking faults and changes in the crystallographic orientation within one grain. In the bright field images (Fig. 2c), where crystalline regions are seen as dark contrasts a large number of spherical and crack-like voids can be indentified. The size and the density of the crack like voids increases towards the substrate film interface and the density of the voids is higher in the sample prepared at high plasma excitation frequency. As an additional feature the surface roughness (12 nm) can be seen for the sample prepared at low frequency. By comparison with dark field images the roughness is found to correlate with the column structure. A similar value for the surface roughness is found in the other sample where only the lower part is imaged in Fig. 2c.

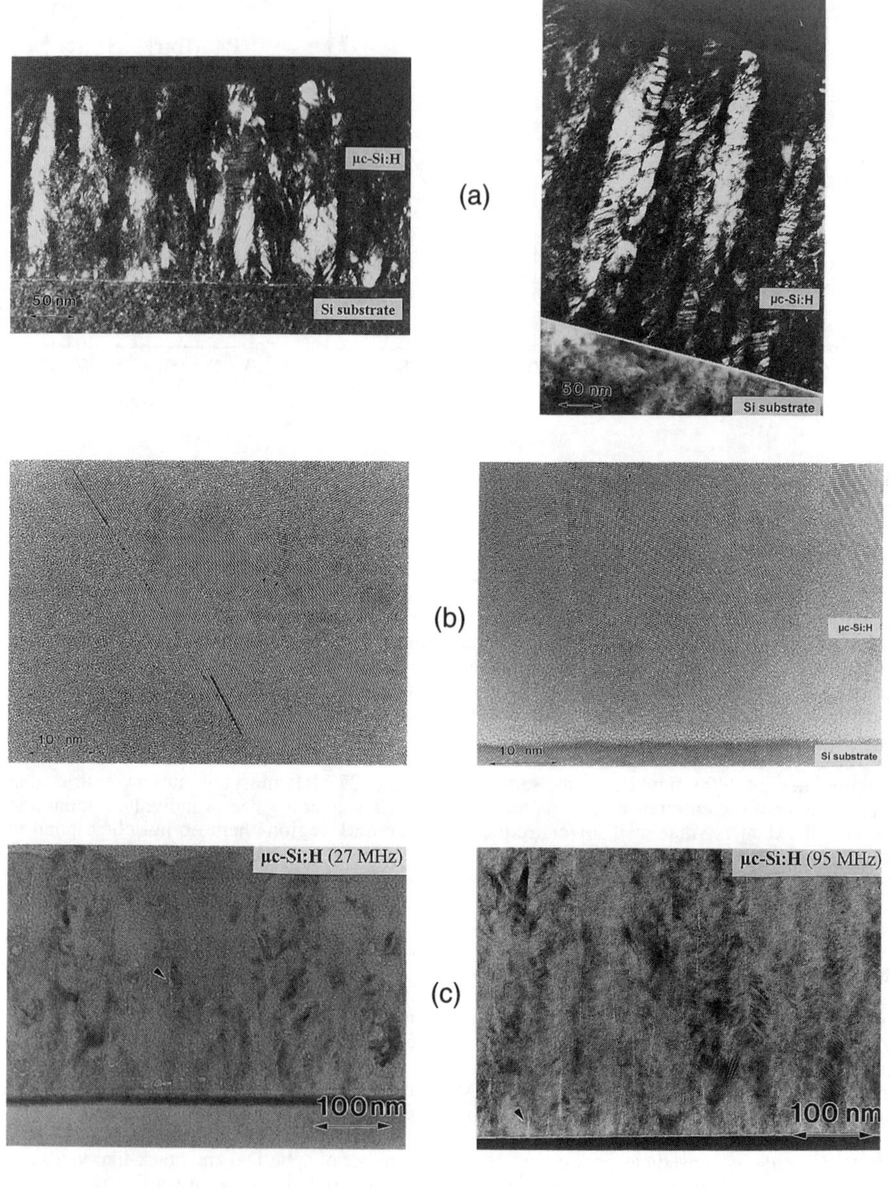

Fig. 2 (a) Dark field, (b) high resolution and (c) bright field images of μc-Si:H samples prepared at 27 MHz (left) and 94.7 MHz (right), respectively. In (b, left - 27MHz) the arrows indicate twins and the line a tilting of the crystalline plane. In (b, right - 94.7 MHz) the arrow indicates a stacking fault. In (c) the arrows indicate spherical (left) and "crack-like" (right) voids.

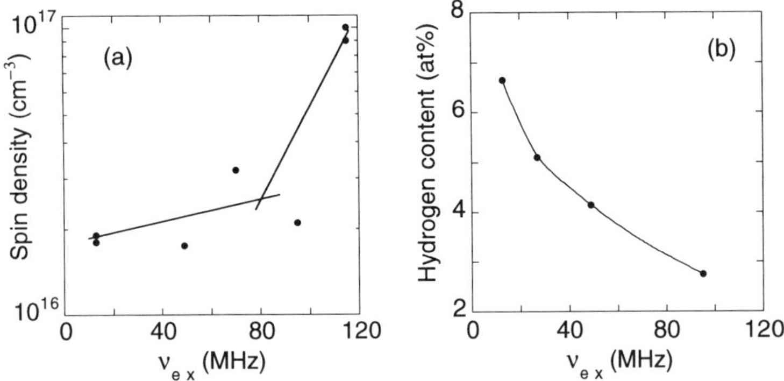

Fig. 3 (a) Spin density of the dangling bond resonances at g=2.0046 and g=2.0052 (measured at room temperature) and (b) hydrogen content determined from IR absorption spectra in µc-Si:H vs. ν_{ex}. The lines are guides to the eyes.

Spin density and hydrogen content

In the next figure (Fig. 3) the spin density measured at room temperature and the total hydrogen content estimated from IR measurements both for undoped material as a function of the plasma excitation frequency ν_{ex} are shown. The spin density is calculated from a superposition of resonances at g=2.0046 and g=2.0052 which are attributed to dangling bonds at grain boundaries, in the amorphous phase and in the crystalline grains [31, 32]. We find spin densities in the range 2...4*10^{16} cm^{-3} for $\nu_{ex} \leq 95$ MHz but a strongly increased spin density at the highest $\nu_{ex} = 116$ MHz. The hydrogen content is calculated from the 640 cm^{-1} absorption mode according to the procedure used for a-Si:H [34]. The hydrogen content decreases from 6.5 to 2.5 at% with ν_{ex}. This is to be expected for a decreasing amorphous phase as very little hydrogen can be accomodated in the crystalline phase of undoped material.

The increase of the spin density and the decrease of the hydrogen content is in accordance with a higher etch rate at high ν_{ex} (Fig 1a) and the appearance of cracks and voids in the TEM images Fig. 2c). Further evidence for such porosity is found when looking at films of a few ten nanometers. While TEM and HRTEM seem ideal techniques for these investigations one is confronted with a complicated sample preparation and especially cross-sectional images might lack resolution for the initial growth region if a too large surface roughness is present. Therefore application of easily accessible, non-destructive techniques (IR and band-to-band optical absorption, Raman, PL, conductivity) is favoured. Data from these techniques on thin films with thicknesses 20-100 nm will be presented in the following:

Infrared absorption

In Fig 4 the IR spectra of thin (20 nm) µc-Si:H films grown with low and high ν_{ex} on SiO_2 covered Si-wafer are shown in the initial stage and after an HF dip. The spectra of these thin highly doped n-type films have been corrected for interferences and free carrier absorption. We find besides the Si-H absorption bands at 640 cm^{-1}, 850-890 cm^{-1}, and 2000-2100 cm^{-1} strong absorption at 1100 cm^{-1} which is indicative of indiffusion of oxygen into a porous material. The samples deposited at low ν_{ex} show much lower intensity of the Si-O band near 1100 cm^{-1} than

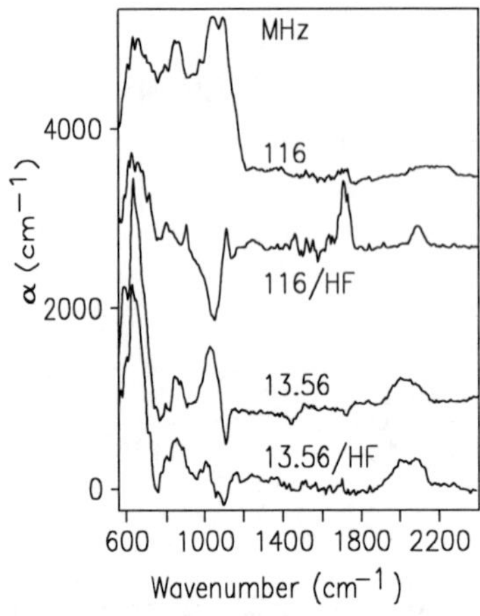

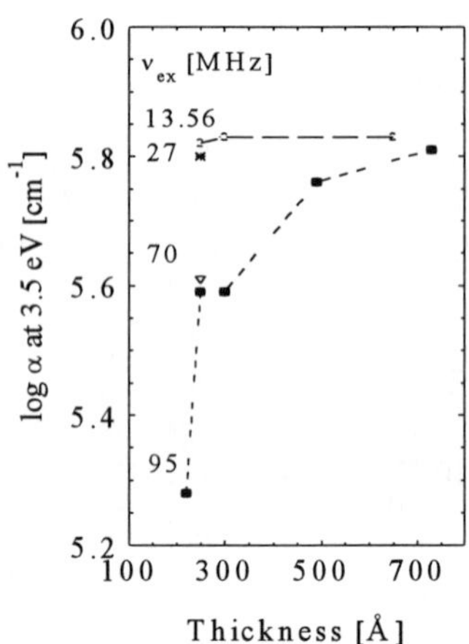

Fig. 4 Infrared absorption for 20 nm thin μc-Si:H films prepared at low and high ν_{ex} before and after HF dip.

Fig. 5 Optical aborption coefficient of μc-Si:H prepared at different ν_{ex} vs. film thickness.

the samples deposited at higher ν_{ex}. After an HF dip the band at 1100 cm^{-1} is completely removed. The negative peak is due to the removal of the native oxide from the back of the wafer substrate. The additional dip of the sample deposited at 116 MHz in ethanol leads to strong adsorption of carbon related radicals at 1700 cm^{-1}. In contrast, thick μc-Si:H samples (above 150 nm) do not show the Si-O related band at 1100 cm^{-1}. Thus thin microcrystalline layers, especially those deposited at high ν_{ex} undergo strong and fast oxidation by air exposure whereas at larger thicknesses this is not the case.

Optical absorption at 3.5 eV

The porous interface for material prepared at high ν_{ex} is also seen in optical absorption. In Fig. 5 the absorption coefficient α at 3.5 eV (where α can be easily evaluated for these thin films) for layers grown on glass at 13.56 MHz and at 95 MHz as a function of the film thickness, and in addition for layers of d=25 nm deposited at different frequencies are shown. We find a clear difference between low frequency (13 & 27 MHz) and high frequency (≥70 MHz) material. For very thin films (25 nm) the absorption is higher for material prepared at low frequency indicating a more dense material. Material prepared at high frequency has a considerable increase in absorption with layer thickness, while low frequency material has not. In agreement with the TEM and IR results we assume that at the very beginning of the deposition the high frequencies material has more and larger voids (but high crystallinity) while low frequency material starts as a more dense structure with amorphous phase, which fills out the void regions.

Electrical conductivity

The presence of an initial amorphous nucleation layer and incomplete coalescence of crystallites leads to a strong thickness dependent conductivity of microcrystalline silicon films with a steep decrease of the conductivity below a certain thickness [1, 35]. We also observed this phenomenon for the films grown at different frequencies on glass and a-Si:H. In Fig. 6 the thickness dependence of the conductivity of μc-Si:H films on glass and a-Si:H is shown for material prepared at different plasma excitation frequencies. On glass (Fig. 6a), μc-Si:H material

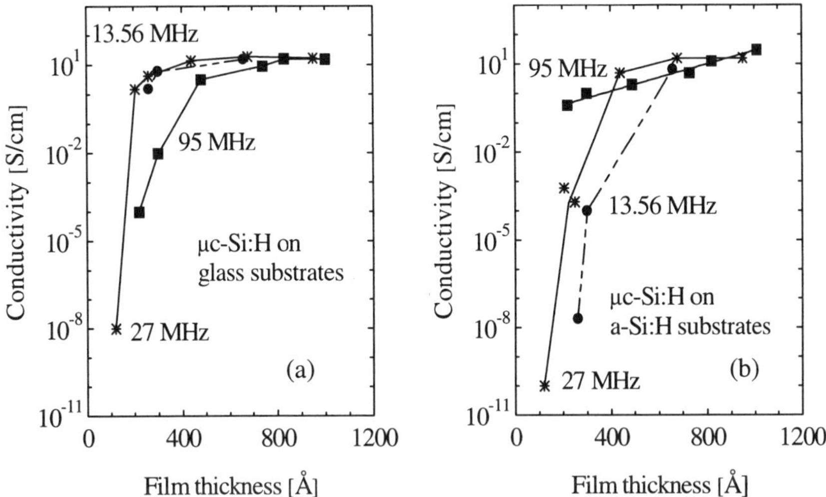

Fig. 6 Electrical conductivity at room temperature of highly doped (2% PH_3) µc-Si:H grown on (a) glass and (b) a-Si:H at different ν_{ex} as a function of the film thickness.

prepared at low excitation frequency maintains a high conductivity down to about 20 nm. The conductivity in high frequency material instead already decreases around 60 nm. Surprisingly, on a-Si:H substrates (Fig. 6b) the opposite behaviour is found: material prepared at high frequencies remains highly conductive to a much lower thickness.

Raman spectroscopy

The improved crystallite formation on a-Si:H at high ν_{ex} - assumed from above conductivity results - is also seen in Raman spectroscopy. In Fig. 7 Raman spectra of thin films (22 - 100nm) grown on a-Si:H at 13.56 MHz and 95 MHz are shown. For samples grown at 13.56 MHz (Fig 7a) the crystalline components are much weaker than in the material prepared at 95 MHz (Fig 7b) and not observable below 30 nm while at 95 MHz at 22 nm there is still a clear crystalline peak.

On glass (not shown) a Raman signal from crystalline components is observed for both materials already at the lowest thicknesses studied here (22-25 nm). To estimate the amorphous contribution one has to consider that the Raman signal from the glass substrate is similar to that of a-Si:H, i.e. for small thicknesses significant contributions from the substrate are expected and a quantitative analysis is difficult. For a comparison we deposited material on chromium (not shown) where no background scatter from the substrate is expected. Neglecting differences in the nucleation on chromium and glass, it is found that low and high ν_{ex} material with thicknesses of the order of 10-30 nm have strong Raman signals at 480 cm^{-1} correlated to amorphous phase. This is in contrast to HRTEM pictures where on SiO_2 no a-Si:H interface was observed.

Photoluminescence

Finally the high content of amorphous phase at the film substrate interface for material prepared at low ν_{ex} on glass or quartz is also seen in photoluminescence. In Fig. 8 PL spectra at 40 K for material prepared at (a) 13.56 MHz and (b) 49 MHz with excitation light (2.34 eV) from the front surface and through the substrate are shown. At low ν_{ex} a clear difference between both excitation directions are seen. From the film surface side a single PL band centred at 0.85 eV with FWHM of 0.15 eV is seen which is known from previous studies on µc-Si:H [36]. While

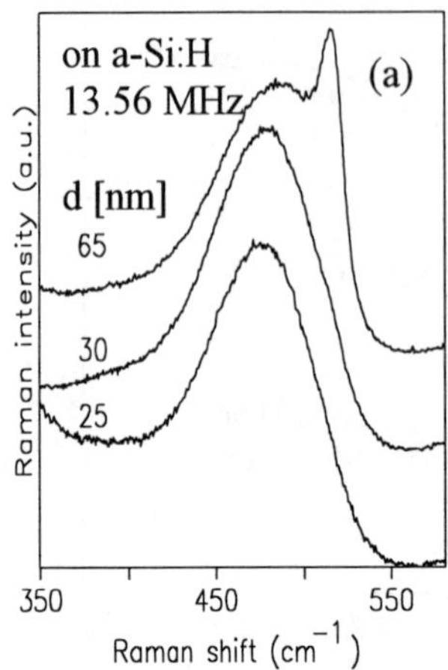

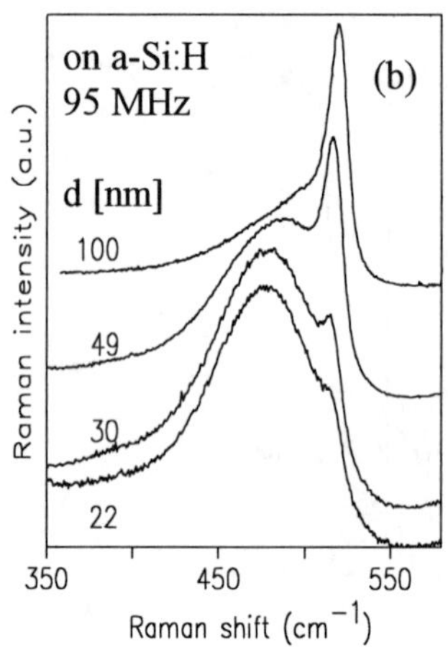

Fig. 7 Raman spectra of material grown under microcrystalline growth conditions on a-Si:H at (a) 13.56 MHz and (b) 95 MHz with various thicknesses.

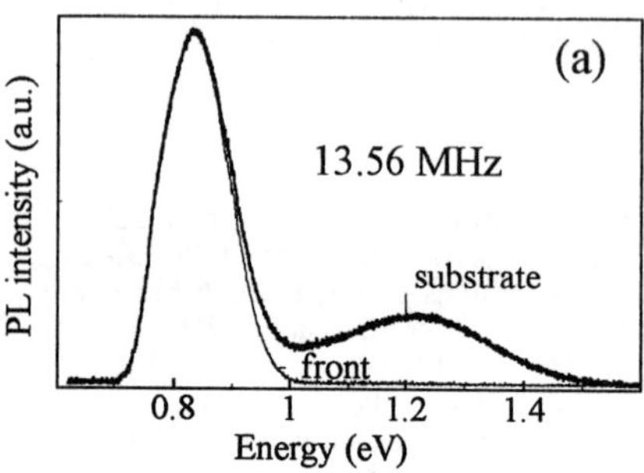

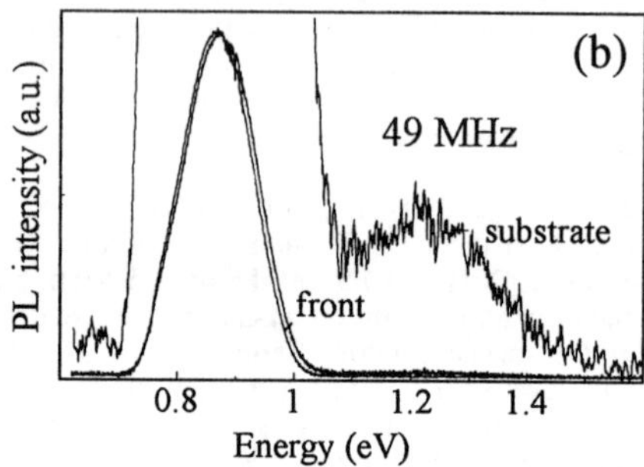

Fig. 8 Photoluminescence spectra of μc-Si:H prepared at (a) 13.56 and (b) 49 MHz for excitation on the film surface and through the substrate

this band has a similar energetic position like the defect luminescence in a-Si:H, it is much narrower and the origin is unclear. From the substrate side an additional band centered at about 1.2 eV (FWHM = 0.27 eV) is observed which is known from the tail-tail recombination in a-Si:H. At high ν_{ex} much less difference is found between both excitation directions and only after strong expansion of the spectra with excitation through the substrate the amorphous phase PL signal can be distinguished. So again evidence is found for a higher content of amorphous phase at the film substrate interface for material prepared at low plasma excitation frequency ν_{ex}.

DISCUSSION

We first briefly summarize above results. The increase of ν_{ex} leads to:

- an increase in the deposition rate of μc-Si:H films,
- an increase in the etching rate of disordered material,
- an increase in grain size and crystalline volume fraction,
- a decrease of disorder between and within the grains,
- an increase in the void fraction in the initial stages of growth on glass and SiO_2,
- a decrease in the amorphous fraction in the initial stages of growth on glass and SiO_2,
- an increase in spin density,
- a decrease in hydrogen content and
- an improved and faster nucleation on a-Si:H.

These results are used to discuss the growth process of μc-Si:H and the implications of a variation of the plasma excitation frequency based on concepts proposed in the literature. Although considerable work has been devoted to the study of growth of μc-Si:H there is still much controversy about the detailed mechanisms involved. Important topics - which will be used for our discussion - are: (i) high H coverage of the film surface to promote surface diffusion of precursors [37], (ii) etching - possibly selective for disordered material - as cause for structural relaxation [33, 38-40], and (iii) a growth zone process as proposed for the layer-by-layer (LbL) techniques [40, 41]. We argue that all (i - iii) above mentioned effects have to be considered for an explanation of the results presented here. However, we think that the hydrogen coverage mainly influences the deposition rate and is not required for optimum crystalline growth.

Also concerning the influence of changes of the plasma excitation frequency on the process plasma there is still uncertainty. However, some experimental effects are obvious. The increase of ν_{ex} leads to a slight increase in the radical density in the bulk plasma [24], a decrease in the sheath thickness whereby more radicals can diffuse without loss reactions to the substrate [21, 22], a decrease in the plasma sheath voltages, that is a decrease in the maximum ion energies [20, 21, 23] and an increased ion flux at the substrate [24].

Both radicals and ions will lead to a higher surface reactivity. In particular it was found that the ion flux is closely related to the deposition rate of μc-Si:H [42] and thus straightforwardly explains the observed increase in deposition rate with ν_{ex} as a result of the ion induced desorption of hydrogen from the surface. But this does not explain the simultaneous increase in grain size and crystalline volume fraction. At low temperatures grain sizes are limited by the fact that crystallites grow from many nucleation centers only until they touch an adjacent grain. In the picture of a balance between etching and deposition for the growth of μc-Si:H the simultaneous increase of deposition rate and grain size requires that from the very beginning of the deposition, regions between grains (amorphous or grain boundaries) and even some of the crystalline nucleation seeds are etched away to allow subsequent crystalline growth and development of large grains. It further requires sufficient supply of favourable gas phase species and/or changes in the surface reaction processes. This is provided by the above described changes in the plasma properties at VHF.

For a graphical illustration of the initial growth stage we recall a model for the structural development of μc-Si:H grown by PECVD proposed by Collins and Yang [13] (Fig. 9) from their in-situ ellipsometry studies. Let us consider the two situations on an "etch-resistant" substrate like glass or SiO_2 (as natural oxide on c-Si) and on a-Si:H which is more susceptible to be affected by the plasma process.

Growth on glass and SiO_2

Small crystallites are formed at the beginning of the deposition which develop in size with film thickness. The space between them is filled with disordered material and some voids. At a certain film thickness the separate crystallites merge and the film consists of silicon crystallites and grain boundaries (Fig. 9a). The disordered material remains in low frequency material to a

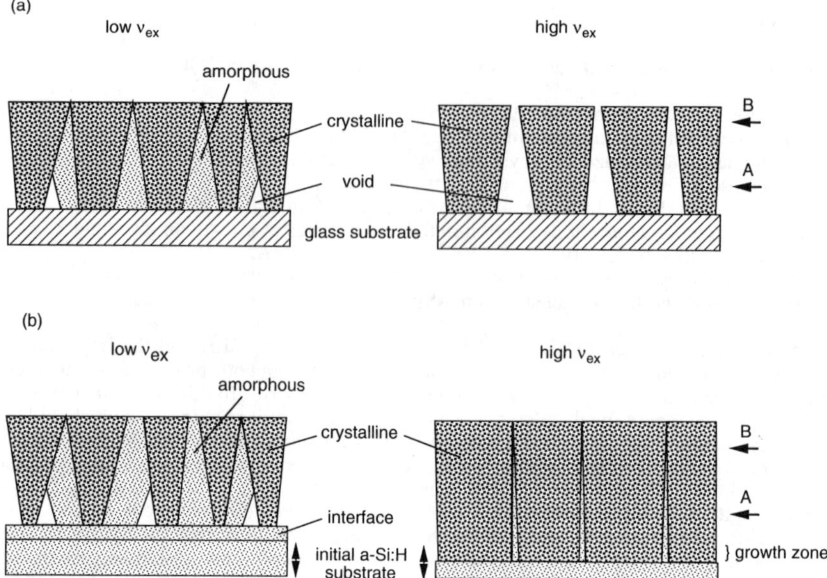

Fig. 9 Schematical illustration of growth of μc-Si:H on (a) glass and (b) a-Si:H at low and high plasma excitation frequencies ν_{ex} .

large extent but is etched away in the high frequency material thus increasing the void fraction at the interface (compare TEM results). The large void fraction in the thin (<25 nm) high frequency samples leads to lower values of the absorption coefficient. The fraction of voids decreases with increasing film thickness and the value of the absorption coefficient increases with the film thickness reaching the bulk value (Fig. 5). If the inter-crystalline space is filled with amorphous material (preparation at low frequency) the absorption coefficient will have no thickness dependence as the difference between the absorption coefficient of crystalline and amorphous silicon at 3.5 eV is small.

The enhanced etching at high frequency leads to a reduction of the nucleation site density. Crystallites will have more space for lateral growth as confirmed by XRD and TEM measurements. If still they start at the substrate with a cone-like shape, crystallites will touch each other and form highly conductive transport paths at a later stage (=thickness) as is shown in Fig. 6a. If the deposition is stopped at point B in Fig. 9a the high frequency material will be strongly oxidized by air exposure because of the large inner surface whereas the low frequency material will undergo oxidation mainly on the front surface. The removal of the oxide by HF dip confirms that the oxidation takes place only on the surfaces (Fig. 4). The appearance of the narrow band, typical for micro- or polycrystalline silicon, at 2100 cm^{-1} in the high frequency samples (Fig. 4) is due to the saturation of the Si surface bonds with hydrogen atoms, an effect typical for HF etching.

Growth on a-Si:H

On a-Si:H there is much improved microcrystalline growth at higher ν_{ex}. This can be explained if the growth process is not restricted to the plasma/film interface but extends into a growth zone (Fig. 9b), similar as in the hydrogen treatment step in the LbL technique [40,41]. The hydrogen (more effective at high ν_{ex}) penetrates into the a-Si:H substrate forming a spongy [43] or high-surface-roughness (imagine a porous silicon structure) material from which crystalline nucleation can proceed. It seems obvious that only such creation of more degree of freedom can lead to phase transformation while a local bond re-arrangement in a bulk material or at the surface is not plausible. Another approach proposes an intermediate liquid stage in this transformation process [44]. Apparently this transformation process is much more effective at VHF and thus fast nucleation of μc-Si:H on a-Si:H is obtained at high frequencies whereas at conventional conditions the plasma is not reactive enough to form the porous layer and only after a considerable additional amorphous growth (interface layer in Fig 9b), crystallites start to form. Strong evidence for the existence of a deep growth zone below the plasma-film interface comes from TEM bright field images where spherical voids of dimensions of up to 10 nm [27] are observed which are difficult to reconcile with a surface process while one can imagine the formation of burried-in voids from a porous material or material with a large surface roughness. Finally we point out that formation of voids as a result of an amorphous-to-crystalline phase transformation is to be expected due to the densitiy deficiency between the two phases if no material (Si) transport is provided from the plasma/film interface into the growth zone or if the so-called "free volume", present in the disordered state, is not allowed to migrate out of the film during the crystallization process [45]. It remains open if generation of porosity can be avoided in the layer-by-layer process via fine tuning of the LbL technique with ν_{ex} as optimization parameter [46].

CONCLUSIONS

Very high frequency plasmas allow an increase of crystallinity and growth rate of μc-Si:H at low deposition temperatures. This is a result of the high "reactivity" of the VHF plasma, due to modification of the growth process through higher ion fluxes and radical densities at high plasma excitation frequencies ν_{ex}.

While the initial layers of low frequency material on glass and SiO_2 still have amorphous phase left, the highly effective etching of amorphous phase at high frequency leads to a less dense interface with cracks and voids between the crystallites as all amorphous tissue is etched away. This points to the need of a controlled, substrate adapted nucleation process.

On a-Si:H the nucleation of microcrystalline material is promoted at high frequencies. Through interaction of the more "reactive" VHF plasma with the a-Si:H substrate material, the growth process is taking place not only at the film-plasma interface but is extended to a growth zone of more than 10 nm. By etching of disordered phase and penetration of hydrogen into the growth zone a phase transformation of the a-Si:H substrate from the amorphous into a microcrystalline structure is mediated. The initial stage of growth on a-Si:H substrates at high ν_{ex} therefore corresponds to the hydrogen treatment step in the layer-by-layer technique.

ACKNOWLEDGEMENTS

We thank Kshem Prasad for his pioneering work in this field. The work at the Forschungszentrum Jülich is supported by the BMBF.

REFERENCES

1 L. Yang, L. Chen, S. Wiedemann and A. Catalano, Mat. Res. Soc. Proc. **283**, 463 (1992)
2 C. Wang and G. Lucovsky, Proc. 21st IEEE PVSC (IEEE, New York) 1614 (1990)
3 J. Meier, P. Torres, R. Platz, S. Dubail, U. Kroll, J.A. Anna Selvan, N. Pellaton Vaucher, C. Hof, D. Fischer, H. Keppner, A. Shah, K.D. Ufert, P. Giannoules and J. Koehler, Mat. Res. Soc. Symp. Proc **420**, 1996
4 J.H. Werner, R. Bergmann and R. Brendel, Festkörperprobleme, **34**, 115 (1995)
5 C. C. Tsai, in Amorphous Silicon and Related Materials ed. H. Fritzsche, World Scientific Publishing Company, Singapore, 123 (1988)
6 M. Nakata, A. Sakai, T. Uematsu, T. Namikawa, H. Shirai, J.-I. Hanna and I. Shimizu; Phil. Mag. **B 63**, 87 (1991)

7 A.R. Middya, A. Lloret, J. Perrin, J. Huc, J.L. Moncel, J.Y. Parey and G. Rose, Mat. Res. Soc. Symp. Proc. **377**, 119 (1995)
8 Y. Hattori, D. Kruangam, K. Katoh, Y. Nitta, H. Okamota and Y. Hamakawa, Proc. 19th IEEE PVSC (IEEE, New York) 689 (1987)
9 S. Oda, J. Noda and M. Matsumura, Mat. Res. Soc. Symp. Proc. **118**, 117 (1988)
10 K. Prasad, F. Finger, H. Curtins, A. Shah and J. Baumann, Mat. Res. Soc. Symp. Proc. **164**, 27 (1990)
11 R. E. Hollingsworth and P. K. Bhat, Appl. Phys. Lett. **64**, 616 (1994)
12 F. Finger, P. Hapke, M. Luysberg, R. Carius, H. Wagner and M. Scheib, Appl. Phys. Lett. **65** 2588 (1994)
13 R. Collins and B.Yang, J.Vac. Sci. Technol. **B7**, 155 (1989)
14 G. N. Parsons, J. J. Boland and J.C. Tsang, Jpn. J. Appl. Phys. **31**, 1943 (1992)
15 B. Drevillon, Prog. Crystal Growth and Charact. **27**, 1 (1993)
16 P. Roca i Cabarrocas, N. Layadi, T. Heitz, B. Drevillon, Appl. Phys. Lett. **66**, 3609 (1995)
17 W. Westlake and M. Heintze, J. Appl. Phys. **77**, 879 (1995)
18 H. Shirai, Jpn. J. Appl. Phys. **34**, 450 (1995)
19 C.C. Tsai, G.B. Anderson and R. Thompson, J. Non-Cryst. Solids **137&138**, 673 (1991)
20 S. Oda, Proc. Jpn. Symp. Plasma Chem. **5**, 7 (1992)
21 C. Beneking, F. Finger and H. Wagner, Proc. 11th EC PVSEC (Harwood Academic Press, Chur), 586 (1993)
22 U. Kroll, Y. Ziegler, J. Meier, H. Keppner, A. Shah, Mat. Res. Soc. Symp. Proc. **336**, 115 (1994)
23 W. Schwarzenbach, A.A. Howling, M. Fivaz, S. Brunner and Ch. Hollenstein, J. Vac. Sci. Tech. **A 14**, 132 (1996)
24 M. Heintze, Solid State Phenom., **B 44-46**, 181 (1995)
25 P. Hapke, Thesis, Technical University Aachen (1995)
26 L. Houben, Thesis, University Düsseldorf (1995)
27 M. Luysberg, P. Hapke, R. Carius and F. Finger, Phil. Mag. A, in press (1996)
28 L. Houben, M. Luysberg, P. Hapke, R. Carius, F. Finger and H. Wagner, unpublished
29 F. Finger, R. Carius, P. Hapke, L. Houben, M. Luysberg and M. Tzolov, 9th International School On Condensed Matter Physics, Varna, 1996, to be published by World Scientific,
30 R. Carius, F. Finger, M. Luysberg, P. Hapke and U. Backhausen, in ref. 29
31 C. Malten, F. Finger, P. Hapke, T. Kulessa, C. Walker, R. Carius, H. Wagner, R. Flückiger, Mat. Res. Soc. Symp. Proc. **358**, 757 (1995)
32 P. Hata, P.C. Taylor and F. Finger, Mat. Res. Soc. Symp. Proc. **420** (1996)
33 S. Veprek , Mat. Res. Soc. Symp. Proc. **164**, 39 (1990)
34 H. Shanks, C. J. Fang, L. Ley, M. Cardona, F. J. Demond and S. Kalbitzer, Phys. Stat. Sol. **B110**, 43 (1980)
35 K. Prasad, U. Kroll, F. Finger, A. Shah, J.-L. Dorier, A. Howling, J. Baumann and M. Schubert, Mat. Res. Soc. Symp. Proc. **219**, 383 (1991)
36 P.K. Bhat, G. Diprose, T.M. Searle, I.G. Austin, P.G. LeComber and W.E. Spear, Physica **B+C 117&118**, 917 (1983)
37 A. Matsuda, J. Non-Cryst. Solids **59-60**, 767 (1983)
38 B.Drevillon, I. Solomon and M. Fang Mater. Res. Soc. Proc. **283**, 455 (1992)
39 M. Heintze, W. Westlake and P.V. Santos, J. Non-Cryst. Solids **164-166**, 985 (1993)
40 H. Shirai, B. Drevillon and I. Shimizu, Jap. J. Appl. Phys. **33**, 5590 (1994)
41 N. Layadi, P. Roca i Cabarrocas, B. Drevillon and I. Solomon, Phys. Rev. **B 25**, 5136 (1995)
42 S. Veprek, M. Heintze, F.-A. Sarott, M. Jurik-Rajman and P. Willmot, Mat. Res. Soc. Symp. Proc. **118**, 117 (1988)
43 This follows also from hydrogen diffusion studies on plasma hydrogenated material. W. Beyer, private communication and J. Non-Cryst. Solids **198-200**, 40 (1996),
44 C. M. Fortmann and I. Shimizu, J. Non-Cryst. Solids **198-200**, 1146 (1996)
45 V. Dimitrov and L. Anestiev, to be published in Mat. Sci. and Engineering A; and M. Baleva, private communication (1996). We thank M. Baleva for drawing our attention to this work.
46 P. Hapke, R. Carius, F. Finger, A. Lambertz, O. Vetterl, H. Wagner, this conference

PREPARATION OF MICROCRYSTALLINE SILICON WITH THE LAYER-BY-LAYER TECHNIQUE AT VARIOUS PLASMA EXCITATION FREQUENCIES

P. HAPKE, R. CARIUS, F. FINGER, A. LAMBERTZ, O. VETTERL, H. WAGNER
Forschungszentrum Juelich, ISI-PV, 52425 Juelich, Germany, p.hapke@kfa-juelich.de

ABSTRACT

For application as nucleation layer in thin film devices, microcrystalline silicon was deposited with the layer-by-layer technique using plasma excitation frequencies between 27 and 95 MHz, various hydrogen treatment times and various film thicknesses per layer. An optimum phase transformation is found at an intermediate plasma excitation frequency, i.e. at this frequency the shortest hydrogen annealing time is necessary for an effective amorphous-to-crystalline phase transformation.

INTRODUCTION

The preparation of microcrystalline silicon (μc-Si:H) for use in thin film devices requires a controlled nucleation of crystalline seeds which in general is critically influenced by the type and morphology of the substrate and which is usually different from growth conditions for bulk material. For the preparation of μc-Si:H with plasma enhanced chemical vapour deposition (PECVD) it was found that the increase of the plasma excitation frequency ν_{ex} beyond the standard frequency of 13.56 MHz has a beneficial effect on the growth and structural properties of "bulk" μc-Si:H material in terms of increased deposition rate, increased crystalline volume fraction and grain sizes and less disorder between and within individual grains [1-3]. On the other hand an increased density of voids at the film-substrate interface on glass and SiO_2 was observed at high ν_{ex} [4]. For an optimum device performance - which is strongly affected by the interface properties - and also for an optimum bulk material growth, it is desirable to have a large area, compact and homogeneous crystalline layer in the beginning of the growth. One possibility to achieve this could be to deposit microcrystalline silicon with the *layer-by-layer* technique (LbL), where deposited a-Si:H is treated by a hydrogen plasma and under certain conditions transformed into crystalline material [5-9]. As it was also observed that at high ν_{ex} there is an improved growth of the crystalline phase on a-Si:H substrate material - which was attributed to a phase transformation similar to that in the LbL process [10] - it is of interest to study the influence of ν_{ex} in the LbL process in some more detail. This is the topic of the present paper.

EXPERIMENTAL

The films are prepared in a conventional diode-type reactor with Very High Frequency Plasma Enhanced CVD processes using excitation frequencies between 27 and 95 MHz. As process gases we used silane and hydrogen at a 1:1 ratio during the deposition. For the doping of the a-Si:H layers a gas mixture of 10 ppm PH_3 diluted in H_2 was used. For the H-treatment a pure hydrogen plasma with the same ν_{ex} as for the deposition process was applied. The plasma input power was 100 mW/cm^2 for all depositions and hydrogen treatments, the reflected power was below 10%. The entire process is computer controlled and runs automatically. The change between the deposition and the H-treatment step is done by switching the gas flows without cutting the plasma. Typical response times of the mass flow controllers are 3 s. All samples are deposited by repeating the deposition and H-treatment sequence 15 times. The deposition temperature was

Mat. Res. Soc. Symp. Proc. Vol. 452 © 1997 Materials Research Society

290 °C and the pressure was kept at 300 mTorr. As substrates we used glass (Corning 7059), glass covered with e-beam evaporated Cr (with a thickness of about 1000 Å) and glass/Cr/a-Si:H. The thickness of the a-Si:H substrate layer is about 130 Å.

For a first series of samples the deposition time t_{Dep} of the amorphous layer was kept constant at 12 s and the hydrogen plasma treatment time t_H was varied between 90 s and 330 s at different ν_{ex}. For a second series the deposition time was adjusted to obtain similar individual film thicknesses, again at different ν_{ex} but at a constant H-treatment time t_H of 180 s.

The films are characterized by Raman spectroscopy. The Raman measurements have been performed using the 488 nm line of an Ar laser, with an excitation power of 100 mW. The spectra were recorded with a charge-coupled device attached to a double monochromator with a spectral resolution less than 1 cm^{-1}. Because the films are very thin, the use of Cr as substrate has the advantage that signals from the glass substrate are suppressed.

The thickness of the films has been determined with a step profiler (Sloan Dektak 3030). The steps were made by using KOH. For every substrate the thickness has been measured ten times on various places of the samples and the average value is given.

RESULTS AND DISCUSSION

Since it is known that changes in the plasma excitation frequency ν_{ex} - keeping all other deposition parameters fixed - lead to changes of the deposition and etch rates of amorphous and microcrystalline silicon [1,2,11], the deposition and the etch rates for a-Si:H films deposited under continuous growth conditions, which are the same as in the deposition step of the layer-by-layer technique (SiH_4/H_2=1:1), are determined for a comparison between the continuous and the LbL-process. To determine the etch rates, the reduction in film thickness after a hydrogen plasma treatment (30 min. H_2-plasma) was measured. An overview of the deposition and etch rates as a function of the plasma excitation frequency is given in Tab. 1. We observe an increase of the deposition and etch rates with ν_{ex}. Compared to the deposition rate of a-Si:H the etch rate is small but also increases with ν_{ex}.

Frequency [MHz]	growth rate [Å/s]	etch rate [Å/s]
27	2.1	0.2
49.7	2.6	0.3
95	6.25	0.5

Tab. 1 Overview of the deposition and etch rates as a function of the plasma excitation frequency for continuously grown a-Si:H films on Cr-substrates

In Fig. 1 the thickness per layer for films prepared by the layer-by-layer technique at different ν_{ex} is plotted as a function of the hydrogen plasma treatment time. The thickness per layer increases with the plasma excitation frequency. With increasing hydrogen plasma treatment time the thickness per layer is reduced; i.e. the films are getting etched. Raman measurements show that the films deposited at 27 and 49 MHz are microcrystalline for t_H=120 and 150 s (see also Fig. 2). For t_H=180 s the films are etched away. The Raman spectra of films deposited with 95 MHz show no crystalline component until an annealing time t_H < 150 s. For $t_H \geq 150$ s a weak crystalline signal at 520 cm^{-1} can be observed (Fig. 2) which increases further with t_H and at t_H=300 s the film is completely etched away (see Fig. 3b).

If we compare the values of the deposition and etch rates from the continuous process (Tab. 1) with the results of Fig. 1 we note that the film thickness per layer can not be determined by using

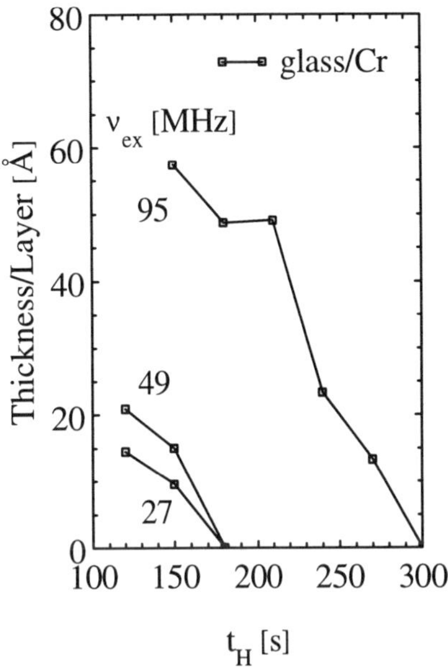

Fig. 1 Thickness per layer as a function of t_H for films deposited with various plasma excitation frequencies.

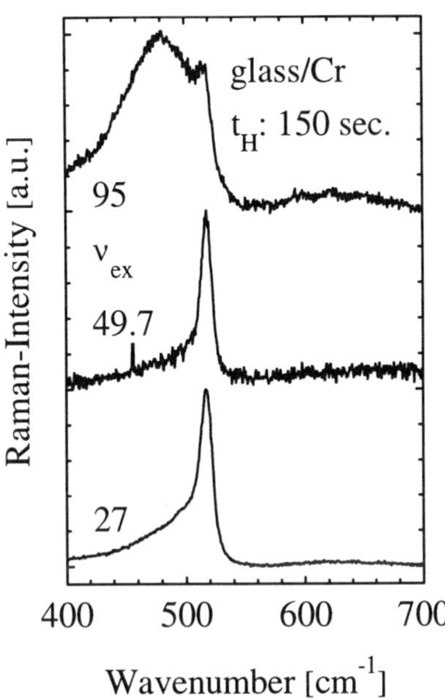

Fig. 2 Raman spectra of three Layer-by-Layer films deposited at various excitation frequencies. The hydrogen treatment time t_H is fixed at 150s.

the values of the continuous process; e.g. a film deposited on glass/Cr at an excitation frequency of 95 MHz (r=6.25 Å/s, $t_{Dep=}$12 s) should have a film thickness of about 75 Å but an etch rate of about 0.5 Å/s (t_H=150 s) should result in a complete removal of the film. Instead, we observe an film thickness per layer of about 57 Å.

The Raman spectra on films treated by hydrogen for 150 s are shown in Fig. 2. For an excitation frequency of 27 and 49.7 MHz the Raman spectra are dominated by the intensity at ≈520 cm^{-1} which is attributed to the crystalline phase. In contrast, the spectrum of a film deposited at 95 MHz is dominated by the intensity at 480 cm^{-1} and the intensity at 520 cm^{-1} is only weakly marked, i.e. this film has a considerable amount of amorphous phase. The ratio of the integrated intensities R ($R=I_c/(I_c+I_a)$) as a semiquantitative measure for the crystalline volume fraction X_c shows that the film deposited at 49.7 MHz has the highest crystalline content. This could mean that at an intermediate frequency a range is found where the phase transformation is beneficial influenced by the frequency [6]. The relatively weak crystalline signal in the 95 MHz sample could be partly due to the increased individual layer thickness deposited at high ν_{ex}.

To study the influence of the film thickness on the phase transformation first the t_H time was further increased at the highest excitation frequency of 95 MHz. In Fig. 3 the Raman spectra of material prepared at ν_{ex} = 95 MHz on a-Si:H (Fig. 3a) and chromium (Fig. 3b) at various t_H times are shown. On the a-Si:H substrate the crystalline component at 520 cm^{-1} first decreases for t_H>210 s and than increases again at t_H≥270 s. This means the individual layer gets completely etched away at t_H=240 s and only the a-Si:H substrate layer remains. Further increase of t_H is getting the a-Si:H substrate also transformed into μc-Si:H before eventually it should be also

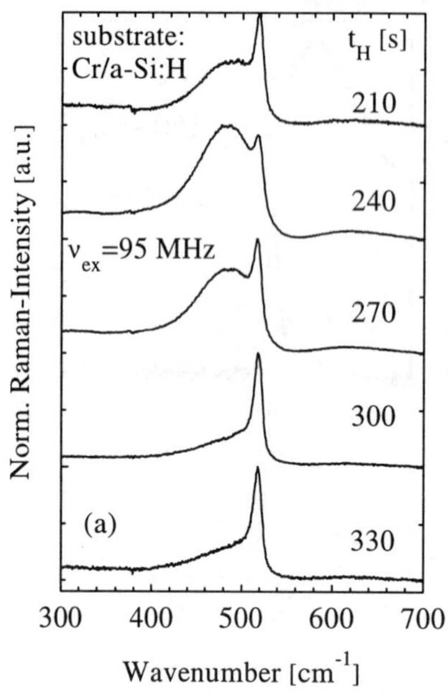

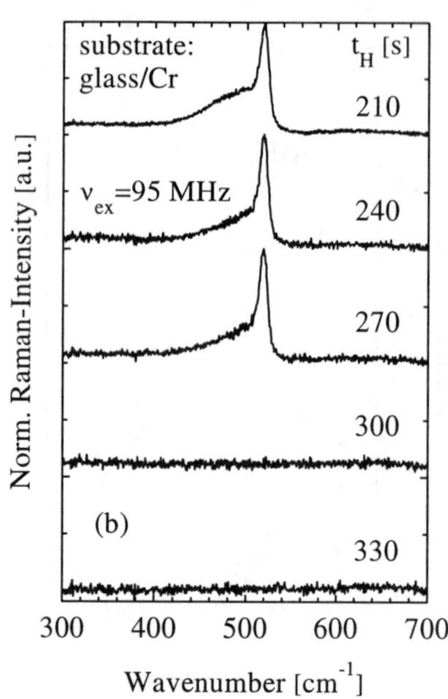

Fig. 3 Raman spectra of samples prepared by LbL at ν_{ex}=95 MHz on (a) glass/Cr/a-Si:H and (b) glass/Cr at various hydrogen treatment times t_H.

completely removed. On the Cr substrate instead, the increase in t_H results in a continuous increase of the crystalline component and then a rather abrupt removal of the film.

To separate the effects occurring from the film thickness and the phase transformation caused by the hydrogen treatment, we deposited films with approximately the same thickness. Because the minimum deposition time for our deposition system is at t_{Dep}=12 s we increased t_{Dep} for the films deposited at 27 and 49 MHz to 40 s and 30 s, respectively. In Tab. 2 an overview of the deposition, the H-treatment times and the film thicknesses of the films are given.

Frequency [MHz]	t_{Dep} [s]	t_H [s]	thickness [Å]
27	40	180	710
49.7	30	180	840
95	12	180	730

Tab. 2 Deposition, H-treatment times and thicknesses of LbL films deposited at various frequencies.

In Fig. 4 the Raman spectra of the three films are shown. Again, similar to Fig. 2 the highest Raman intensity at 520 cm^{-1} can be observed for the sample deposited at 49.7 MHz. The film deposited at 95 MHz also exhibits a strong Raman intensity at 520 cm^{-1} but the ratio of the crystalline to amorphous components is smaller as compared with the 49.7 MHz sample. For the film deposited at 27 MHz only an amorphous signal is observable.

To discuss this result in more detail one has to keep in mind that i) the film thickness increases

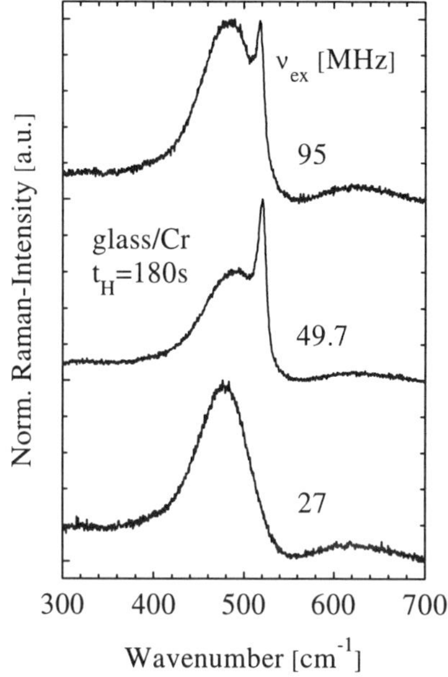

Fig. 4 Raman spectra of films deposited with similar thickness for various ν_{ex}.

with frequency ii) probably the structure of the deposited amorphous film changes with frequency and iii) the plasma excitation frequency influences the hydrogen plasma and thereby the phase transformation. The film thicknesses as noted in Tab. 2 are very similar, i.e. changes of the phase transformation using various ν_{ex} are due to changes in the hydrogen plasma and/or changes in the structure of the deposited amorphous film. The strong crystalline signal observed for the 49.7 MHz sample gives further evidence that this plasma excitation frequency is favorable for the phase transformation amorphous-crystalline.

As mentioned, not only the plasma excitation frequency and the film thickness but also the ″quality″ of the deposited amorphous material should be considered. Although it is known that the material properties of a-Si:H do not deteriorate with ν_{ex} [11], the transfer of these results from the continuous to the LbL process is difficult. For example, the deposition and etch rates of the continuous process for the determination of the thickness of films with the layer-by-layer technique can not be applied. Furthermore, it has been shown [9] that with the number of sequences the structural composition of the ″growth zone″ changes continuously, which implicate that the substrate for every deposition changes.

CONCLUSION

Microcrystalline has been deposited with the layer-by-layer technique using various plasma excitation frequencies. The increase of ν_{ex} results in an increase of the thickness per layer and in considerable differences of the structural properties as determined with Raman spectroscopy: the film grown with ν_{ex}=49.7 MHz has the largest crystalline content. To separate the influence of the film thickness from the influence of the hydrogen treatment, films with approximately the same thickness are prepared. Again, the film grown with an intermediate frequency exhibits a Raman spectrum with the strongest peak at 520 cm^{-1}, indicating the largest crystalline content.

ACKNOWLEDGMENTS

We thank C. Fortmann for helpful discussions. This work is supported by the Bundesministerium für Bildung und Forschung BMBF, Germany.

REFERENCES

1. F. Finger, R. Carius, P. Hapke, L. Houben, M. Luysberg and M. Tzolov, this conference.
2. M. Luysberg, P. Hapke, R. Carius and F. Finger, Phil. Mag. **A** (1996), in press.

3. F. Finger, P. Hapke, M. Luysberg, R. Carius, H. Wagner and M. Scheib, Appl. Phys. Lett. **65**, 2588 (1994).

4. P. Hapke, M. Luysberg, R. Carius, M. Tzolov, F. Finger and H. Wagner, J. Non-Crystalline Solids **198-200**, 927 (1996).

5. A. Asano, Appl. Phys. Lett. **56**, 533 (1990).

6. H.V. Nguyen, I. An, R. Collins, Y. Lu, M. Wagaki and C. Wronski, Appl. Phys. Lett. **65**, 3335 (1994).

7. C. Fortmann, P. Hapke, A. Lambertz and F. Finger in Amorphous Silicon Technology-1996, edited by M. Hack, E.A. Schiff, S. Wagner, A. Matsuda, R. Schropp (Mater. Res. Soc. Proc. **420**, Pittsburgh, PA, 1996), in press.

8. C. Fortmann and I. Shimizu, J. Non-Crystalline Solids **198-200**, 1146 (1996).

9. N. Layadi, P. Roca i Cabarrocas, B. Drevillon and I. Solomon, Phys. Rev. B **52**, 5136 (1995).

10. P. Hapke, M. Luysberg, R. Carius, M. Tzolov, F. Finger and H. Wagner, Solid State Phenomena **51-52**, 161 (1996).

11. H. Curtins, N. Wyrsch, M. Favre and A. Shah, Plasma Chem. and Plasma Processing **7**, 267 (1987).

SUBSTRATE-SURFACE EFFECT ON INITIAL GROWTH PROCESS OF MICROCRYSTALLINE SILICON FILMS

K. IKUTA*, J. W. PARK**, L. H. KUO**, T. YASUDA*, S. YAMASAKI*, and K. TANAKA*
*Joint Research Center for Atom Technology (JRCAT) - National Institute for Advanced Interdisciplinary Research (NAIR), 1-1-4, Higashi, Tsukuba, Ibaraki 305, Japan
**Joint Research Center for Atom Technology (JRCAT) - Angstrom Technology Partnership (ATP), 1-1-4, Higashi, Tsukuba, Tsukuba, Ibaraki 305, Japan

ABSTRACT

Initial growth processes of hydrogenated microcrystalline silicon (μc-Si:H) films have been investigated by scanning tunneling microscopy (STM), high-resolution transmission electron microscopy (HRTEM), and reflection high energy electron diffraction (RHEED). The μc-Si:H films were prepared by plasma enhanced chemical vapor deposition (PECVD) on H-terminated Si(111) and plasma-oxidized SiO_2/Si(111) surfaces that were made atomically-flat by a careful wet processing. On H-terminated Si(111) the initial growth was epitaxial as evidenced by HRTEM and RHEED, while on SiO_2/Si(111) the initial process was nucleation of amorphous Si followed by formation of randomly oriented μc-Si:H structure. STM observation revealed that, on both H-terminated and SiO_2-terminated surfaces, initial growth processes proceed through the nucleation-and-coalescence mechanism.

INTRODUCTION

Microcrystalline silicon (μc-Si:H) films have been intensively studied for device applications such as solar cells and thin-film transistors. Being a mixed-phase material, the microscopic structure of μc-Si:H films depends not only on the deposition methods [1-5] but also on the nature of the substrate surface where the interface between the μc-Si:H film and the substrate is formed. The initial stage of hydrogenated silicon (Si-H) growth has been studied using spectroscopic ellipsometry [6-9], infrared absorption spectroscopy [10,11], x-ray photoelectron spectroscopy [12] and scanning probe microscopy [13-16]. Among these approaches, scanning probe methods are powerful to obtain information about the local structures, while other methods mainly provide average information. Previously we have demonstrated results of STM observation on the initial stage of Si-H growth on the graphite substrate [16].

In this work, we report the results of STM observation of initial growth process of μc-Si:H on two different atomically-controlled substrate surfaces: hydrogen terminated Si(111) and ultra-thin SiO_2 layers. We show that on both surfaces the initial growth processes proceed through the nucleation-and-coalescence mode. We discuss the initial growth processes combining the results of HRTEM and RHEED.

EXPERIMENT

Atomically-flat Si(111) surfaces were prepared by wet-chemical processing [17] as summarized below. P-type Si(111) substrates were first degreased in acetone and ethanol. After an RCA-SC1 clean (boiling in a solution of NH_4OH: H_2O_2:H_2O = 0.05:1:5), the samples were dipped in an HF solution (1 %) to remove an oxide layer. The sample was then treated in NH_4F to prepare atomically-flat H-terminated surface by anisotropic etching. Following a rinse in pure water the sample was inserted into the multi-chamber CVD system through an oil-free loadlock chamber. We will call this surface H-Si(111) in the following text.

The multi-chamber CVD system we used consists of a loadlock chamber in which plasma oxidation was made, a PECVD room, Auger electron spectroscopy (AES) room, and STM room in which STM and RHEED observations were made. These rooms are connected through gate valves through which the sample was transferred in vacuum.

Mat. Res. Soc. Symp. Proc. Vol. 452

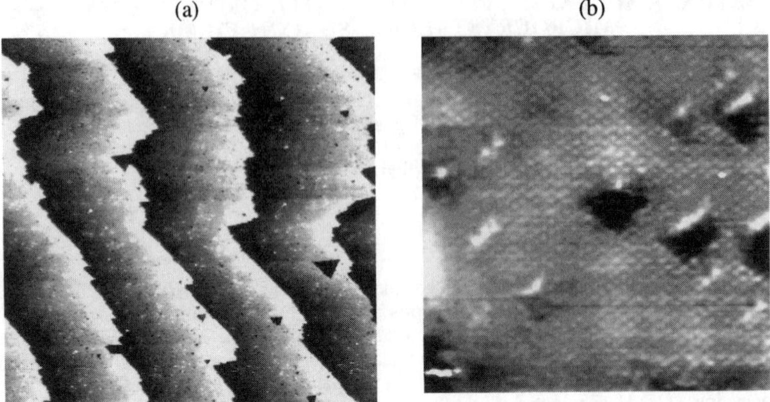

Fig. 1 Typical STM images of the surface of H-terminated Si(111) observed just after a wet-chemical treatment. In Fig. 1(a), 400 x 400 nm^2 topography, monolayer-height steps can be seen. In Fig. 1(b), 14 nm x 14 nm^2 topography, H-terminated Si(111)-1 x 1 structure is observed.

SiO_2-terminated surface was prepared by exposing the H-Si(111) surface to O_2 plasma in the loadlock chamber [18]. Oxidation conditions were as following: a source gas was O_2 (0.1 sccm) diluted by He (50 sccm), rf power and total gas pressure were 20 W and 100 mTorr, respectively. O_2 plasma was exposed to the sample at room temperature for 15 s. The formation of silicon oxide layers was confirmed by AES. Thickness of the silicon oxide layer was estimated to be 0.5 nm from the lineshape of Si_{LVV} (92 eV) Auger features. The plasma-oxidized H-Si(111) surface will be called SiO_2/Si(111) in the following text.

Deposition of hydrogenated silicon films was carried out on the H-Si(111) and SiO_2/Si(111) surfaces by plasma-enhanced CVD using 13.56 MHz rf excitation of SiH_4 (1 sccm) diluted by H_2 (99 sccm). The rf power density was 0.42 W/cm^2. The triode-type configuration was used, where a negative bias of - 50 V was applied to the third electrode placed at 3 cm below of the power electrode and 2 cm above the grounded electrode (the sample holder). The substrate temperature was 350 °C. In all experiments, uncontrolled deposition during the plasma ignition and stabilization was prevented by a mechanical shutter located above the substrate.

After depositions, the samples were transferred through a gate valve to an ultra-high vacuum chamber (less than 10^{-10} Torr), where STM observation was performed. The conditions of STM measurements are as following: the tunneling current is 200 pA, the sample voltages are 2.0 V before oxidation and larger than 3.5 V after oxidation.

RESULTS AND DISCUSSIONS

Initial Surfaces

Figure 1 shows STM topographies of the surface of H-Si(111) observed just after a wet-chemical treatment. In Fig. 1(a), a 400 x 400 nm^2 image, atomically flat terraces with about 100 nm in width can be seen along with several monolayer-height steps. There are triangular etch pits with a lateral dimension up to a few hundreds nm. The flat terraces are terminated by silicon mono-hydride bondings, as shown in Fig. 1(b). Bright spots are arranged in a separation of 0.39 nm. This is known as a H-terminated Si(111)-1 x 1 structure, which reproduced the results of the previous reports on the STM observation of similarly-prepared Si(111) surfaces [19]. Holes with a lateral size of around several nms can be seen among a mono-hydride regular lattice.

Fig. 2 Typical STM images of the surface of an ultra-thin SiO_2 layer prepared on H-terminated Si(111) by O_2 remote plasma. The triangular holes and monolayer-height steps observed in Fig. 1(a) are conserved even after the oxide formation.

Figure 2 shows a typical, 400 x 400 nm^2, STM topography of SiO_2/Si(111). The triangular holes and monolayer-height steps observed in Fig. 1(a) are conserved even after the oxide formation.

Taking into account the wide band gap of SiO_2 (9.3 eV[20]) which is significantly larger than the sample bias in our observation as mentioned above, STM topography in the Fig. 2 represents not the surface morphology of SiO_2 layer but the morphology of the interface between SiO_2 layer and c-Si substrate. Namely, electrons emitted from the tungsten tip directly tunnel through both the vacuum gap and the thin SiO_2 layer into the conduction band of silicon. The tunneling probability in the SiO_2 layer, however, should be different from that in the vacuum barrier. Actually, a relatively high sample bias (> 3.5 V) was necessary for stable image acquisition on the surface of SiO_2 layer comparing to on that of H-Si(111). This possibly implies that tunneling through the SiO_2 layer needs a higher bias than that through the vacuum barrier with the same gap width as a thickness of the SiO_2 layer to keep the same tunneling current.

The structure of Si-H films

Before discussing the STM images of the Si-H films, we will briefly discuss the structure of Si-H films prepared under the deposition condition employed (identified above). On the H-Si(111) substrates, the RHEED patterns were streaky up to a nominal thickness of 2.5 nm although week spotty pattern was superposed. As the film thickness was further increased they changed to a superposition of spotty and halo-ring patterns. This indicates that the initial growth is epitaxial to the Si substrate, and that the further growth induced roughening and formation of μc-Si:H structure.

The structure of the Si-H thin films on H-Si(111) was characterized in detail using HRTEM as shown in Fig. 3. In this figure, the interface region of Si-H/H-Si(111) is marked by two arrows indicating a smooth epitaxial growth of the Si-H thin film within a thickness of 2-3 nm. A high density of stacking errors (labeled by arrowheads) near Si-H/H-Si(111) interface, however, initiated the columnar growth and caused formation of amorphous tissues as clearly seen in the image. These stacking errors may be nucleated from the defects existing on the initial surface such as holes shown in Fig. 1(b).

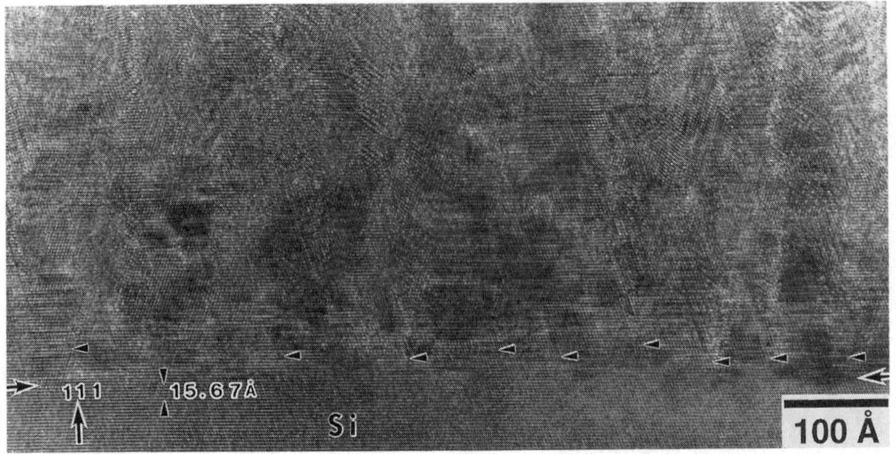

Fig. 3 HRTEM image of the Si-H film deposited on H-terminated Si(111) substrate. The interface region of Si-H/H-Si(111) is marked by two arrows. Stacking errors are labeled by arrowheads near the interface.

On the SiO_2/Si(111) substrates, halo RHEED patterns were obtained from the beginning of the growth, indicating deposition of an amorphous film. For 50 nm thick film, a ring pattern characteristic of randomly oriented microcrystallites was obtained.

STM observations on initial stage of Si-H films

Figure 4 shows the STM topographies of the surface of Si-H films with a nominal thickness of 0.3 nm deposited on H-Si(111). In Fig. 4(a), 400 x400 nm^2 topography, monolayer steps and triangular holes are clearly observed even after a deposition, which implies that Si-H deposition took place homogeneously on the surface. A magnified image shown in Fig. 4(b) reveals that the first step in Si-H growth on H-Si(111) is island formation, each terrace being decorated.

Figure 5 shows the STM topographies of the Si-H films on SiO_2/Si, where the nominal film thickness was 0.3 nm, the same as in Fig. 4. Similarly to Fig. 4, monolayer steps and triangle holes exist after a deposition, and the surface is covered homogeneously with islands of amorphous Si as evidenced by the RHEED observation above. By analyzing the images in Figs. 4 and 5, we have found a quantitative difference in the average island dimensions between the two cases. On H-Si(111) the average lateral size was about 2 nm and the height was about 0.3 nm, while on SiO_2/Si(111) the average lateral size and height were 3 nm and 0.5 nm, respectively. The cause of the differences seen in the dimensions of nuclei is unclear at the moment.

The results of STM observations are qualitatively consistent with the results of spectroscopic ellipsometry [21], which indicated nucleation and coalescence at an initial stage of deposition of a-Si:H and μc-Si:H.

The key point to understand the nucleation process is to identify the nucleation sites on H-Si(111) and SiO_2/Si(111) surfaces. In either case, defective sites such as surface Si dangling bonds are likely to act as nucleation centers. Atomic H, which exist at a significant density in PECVD environment, can generate such defects, for example, Si dangling bonds by H-abstraction reaction on H-Si(111) and E' center formation on SiO_2. A puzzling observation,

(a) (b)

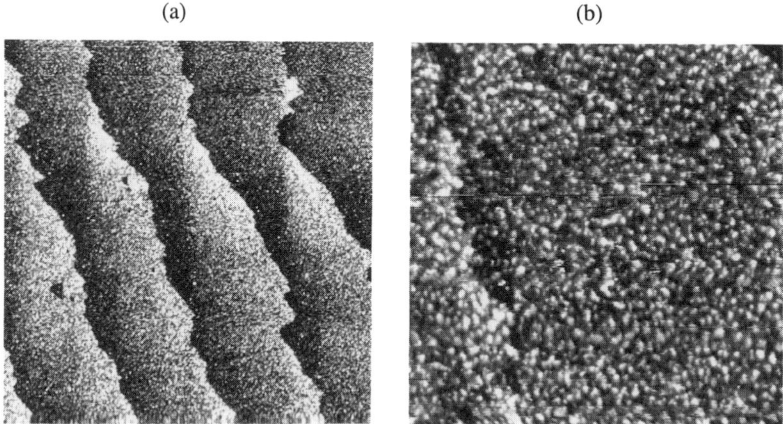

Fig. 4 Typical STM images of the surface of Si-H film deposited H-terminated Si(111) surface. In Fig. 4(a), 400 x 400 nm^2 topography, monolayer-height steps can be seen. In Fig. 4(b), 100 x 100 nm^2 topography, islands-formation is observed.

(a) (b)

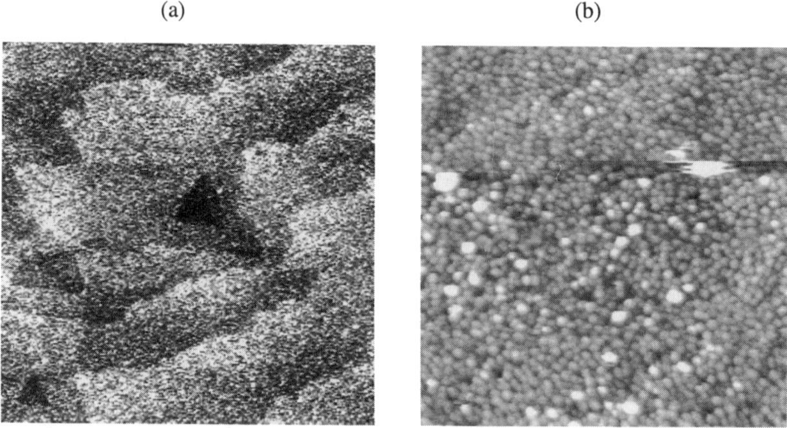

Fig. 5 Typical STM images of the surface of Si-H film deposited on an ultra-thin SiO_2 surface. In Fig. 5(a), 400 x 400 nm^2 topography, monolayer-height steps are also seen. In Fig. 5(b), 100 x 100 nm^2 topography, nucleated islands cover the surface, which is similar to that in Fig. 4(b).

however, is that the nucleation density is very similar on the two surfaces, which are quite different in the chemical nature. It is unlikely that the densities of defects or dangling bonds are so close on these two surfaces. Apparently further investigation, such as a measurement of the temperature dependence of the nucleation density, is needed to clarify this issue.

CONCLUSIONS

Initial stage of μc-Si:H deposition on both hydrogen-terminated Si(111) and plasma-oxidized Si(111) substrates were investigated using STM, RHEED, and HRTEM. STM images show that island structures, with an average lateral size of about 2 nm and with about 0.3 nm in height, are formed at very initial stage of deposition on H-Si(111). From RHEED and HRTEM measurements, it is found that these nuclei coalesce to grow epitaxially although some stacking errors and amorphous tissues are included during the film growth. On SiO_2/Si(111) substrates, STM images also indicate the formation of islands with dimensions of 3 nm in width and 0.5 nm in height. However, from RHEED observations, the initial structure of Si-H is amorphous, and becomes randomly oriented μc-Si:H with a thickness larger than 50 nm.

ACKNOWLEDGMENTS

This work, partly supported by NEDO, was performed in the Joint Research Center for Atom Technology (JRCAT) under the joint research agreement between the National Institute for Advanced Interdisciplinary Research (NAIR) and the Angstrom Technology Partnership (ATP).

REFERENCES

1. A. Matsuda, J. Non-Cryst. Solids **59&60**, 767 (1983).
2. J. J. Boland and G. N. Parsons, Science **256**, 1304 (1992).
3. S. Koynov, S. Grebner, P. Radojkovic, E. Hartmann, R. Schwarz, L. Vasilev, R. Krankenhagen, I. Sieber, W. Henrion, and M. Schmidt, J. Non-Cryst. Solids **198-200**, 1012 (1996).
4. N. Imajo, J. Non-Cryst. Solids **198-200**, 935 (1996).
5. M. Scheib, B. Schröder, and H. Oechsner, J. Non-Cryst. Solids **198-200**, 895 (1996).
6. A. M. Antoine and B. Drevillon, J. Non-Cryst. Solids **97&98**, 1403 (1987).
7. I. An, H. V. Nguyen, N. V. Nguyen, and R. W. Collins, Phys. Rev. Lett. **65**, 2274 (1990).
8. T. Akasaka, Y. Araki, M. Nakata, and I. Shimizu, Jpn. J. Appl. Phys. **32**, 2607 (1993).
9. H. Shirai, J. Non-Cryst. Solids **198-200**, 931 (1996).
10. Y. Toyoshima, K. Arai, A. Matsuda, and K. Tanaka, J. Non-Cryst. Solids **137&138**, 765 (1991).
11. S. Miyazaki, H. Shin, Y. Miyoshi, and M. Hirose, Jpn. J. Appl. Phys. **34**, 787 (1995).
12. M. Kawasaki, and H. Suzuki, J. Appl. Phys. **75**, 3456 (1994).
13. M. Matsuse, M. Kawasaki, and H. Koinuma, Proc. 1994 IEEE 1st World Conf. on Photovoltaic Energy Conversion, p. 425 (IEEE, USA, 1994).
14. D. M. Tanenbaum, A. Laracuente, and A. C. Gallagher, Proc. Mat. Res. Soc. Symp. **377**, 143 (1995).
15. H. Deki, M. Fukuda, S. Miyazaki, and M. Hirose, Jpn. J. Appl. Phys. **34**, L1027 (1995).
16. K. Ikuta, Y. Toyoshima, S. Yamasaki, A. Matsuda, and K. Tanaka, Spring Meeting of Mat. Res. Soc. (San Francisco, California, 1996).
17. H. Tokumoto, Y. Morita, and K. Miki, Mat. Res. Soc. Symp. Proc. **259**, 409 (1992).
18. T. Yasuda, Y. Ma, Y. L. Chen, G. Lucovsky, and D. Maher, J. Vac. Sci. Technol. **A 11**, 945 (1993).
19. R. S. Becker, Y. J. Chabal, G. S. Higashi, and A. J. Becker, Phys. Rev. Lett. **65**, 1917 (1990).
20. Z. A. Weinberg, G. W. Rubloff, and E. Bassous, Phys. Rev. **B 19**, 3107 (1979).
21. M. Fang and B. Drevillon, J. Appl. Phys. **70**, 4894 (1991).

GRAIN-SIZE CONTROL OF NANOCRYSTALLINE SILICON BY PULSED GAS PLASMA PROCESS

A. ITOH, T. IFUKU, M. OTOBE, and S. ODA,
Research Center for Quantum Effect Electronics and Department of Physical Electronics, Tokyo Institute of Technology, Meguro-ku, Tokyo 152, Japan.

ABSTRACT

A new method for the formation of nanocrystalline Si (nc-Si) in the SiH_4 plasma using pulsed-H_2 supply with very-high-frequency (VHF;144 MHz) excitation is proposed to control the size of nc-Si. Nanocrystalline Si is formed in the gas phase of SiH_4 plasma cell by coalescence of radicals. The principle of size control is based on the separation of nucleation and growth process. Supplying H_2 into a SiH_4 plasma enhances nucleation of nc-Si and suppresses growth rate of nc-Si. The nucleated nc-Si grows larger in a SiH_4 plasma during the off state of the H_2 supply. As the newly supplied H_2 forces nc-Si grown in the previous cycle out of the plasma cell into the deposition chamber, the next nucleation of nc-Si is enhanced simultaneously. Using this method, we fabricated 8 nm-diameter nc-Si with small dispersion ($\pm$1 nm) successfully.

INTRODUCTION

Nanocrystalline silicon (nc-Si) has received a great deal of attention for application to quantum-effect optoelectronic devices and single-electron tunneling (SET) transistors for the next generation ultra large scale integrated circuits (ULSI's). Recently, SET devices have been proposed and fabricated using compound semiconductors, metal/insulator, and Si/SiO_2 nanostructure systems [1-6]. Although room-temperature operation has been demonstrated [5,6], realization of stable characteristics has been still difficult at high temperature because of the difficulty in fabrication of nanometer-scale structures. For realization of SET devices operating at room temperature, reproducible fine structures less than 10 nm-scale are strongly required.

We have fabricated nc-Si particles with 3–30 nm diameter by a SiH_4 plasma process using very-high-frequency (VHF;144 MHz) excitation [7,8]. The merits of VHF plasma are higher efficiency of radical formation [9] and lower self-bias of plasma [10,11] compared with radio-frequency (RF;13.56 MHz) plasma. However, it has been difficult to control the size uniformity of nc-Si, causing a problem of reproducibility in current-voltage characteristics of SET devices. In order to apply nc-Si to future electron devices, it is crucially required to fabricate as small particles as possible (less than 10 nm) and to control the size distribution precisely.

In this paper, the effects of H_2 supply into a SiH_4 plasma on the fabrication of nc-Si is described. Based on the effects, a method of pulsed-H_2 supply into a SiH_4 plasma is proposed to control the size of nc-Si. Size and distribution of nc-Si were evaluated by transmission electron microscopy (TEM) observation. Successful fabrication of nc-Si less than 10 nm with a small spread of size was accomplished by means of the pulsed-H_2 supply method.

EXPERIMENTAL

A schematic diagram of the nc-Si deposition system, which is a modified Si molecular beam epitaxy equipment, is shown in Fig. 1. A plasma cell is attached instead of Knudsen cells. The electrodes of the plasma cell are capacitively coupled. The stainless steel plate with an orifice of 6 mm diameter and 2 mm length, separating the UHV chamber from the plasma cell, is used for grounded electrode. The volume of the plasma cell is 230 cm^3. The deposition rate of nc-Si is monitored by a quartz crystal sensor. The pressure in the cell is monitored by a capacitance manometer and is controlled by the flow rate of source gas.

Nanocrystalline Si is formed by coalescence of radicals produced from a SiH_4 plasma in the plasma cell and is extracted out of the plasma cell through the orifice to the UHV chamber. Deposition of nc-Si is carried out on an amorphous carbon micro-grid substrate, which is for TEM observation, at room temperature.

Mat. Res. Soc. Symp. Proc. Vol. 452 © 1997 Materials Research Society

The pressure in the cell was 0.42 Torr for SiH_4 before plasma activation, and 0.40 Torr during activation, where the flow rate of SiH_4 was set $8\,cm^3/min$. The flow rate of H_2 was varied from 0 to $25\,cm^3/min$, while the flow rate of SiH_4 was kept constant at $8\,cm^3/min$. For a method of pulsed-H_2 supply into a SiH_4 plasma, fabrication of nc-Si was carried out with a H_2 flow rate of $24\,cm^3/min$. Both on-time and off-time of H_2 were set at 0.5 sec. The VHF power density was varied from 0.42 to $2.8\,W/cm^2$.

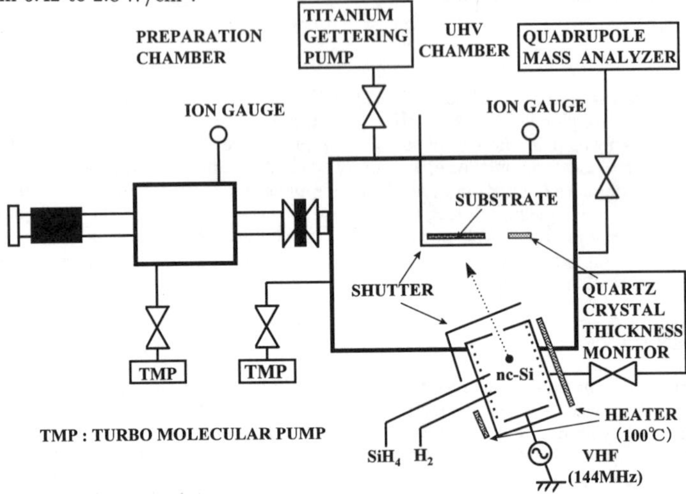

Figure 1: Schematic diagram of nc-Si deposition apparatus.

RESULTS AND DISCUSSION

Figure 2 shows a typical high-resolution TEM image of nc-Si with 15 nm diameter, which was fabricated using a pure SiH_4 plasma at the VHF power density of $2.8\,W/cm^2$. The Si(111) lattice image indicates that nc-Si grown by the process is single domain crystal. The surface amorphous

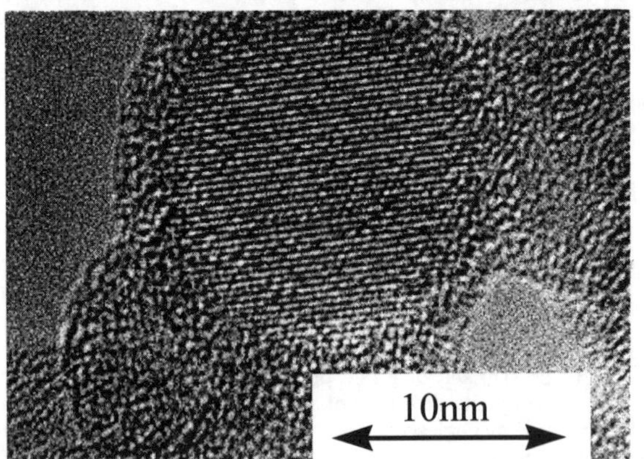

Figure 2: High-resolution TEM image of nc-Si fabricated by pure SiH_4 plasma at VHF power density of $2.8\,W/cm^2$.

region of nc-Si is a naturally-formed oxide layer with a thickness of 1.5 nm, which is observed for all nc-Si with different size. Thus, the 1.5 nm-thickness oxide layer was unintentionally formed at the surface in air at room temperature. In the case of using a pure SiH_4 plasma, the size of nc-Si was random ranging from 4 nm to 40 nm at VHF power density of 2.8 W/cm^2. Although the average size of nc-Si could be reduced with decreasing VHF power, the size-distribution of nc-Si shows large dispersion. Hence, it is difficult to fabricate nc-Si with small spread of size by control of VHF power.

From the analysis of infrared absorption spectra, the surface of nc-Si just shortly after taken out of the UHV chamber was covered with hydrogen atoms [12]. Since the surface hydrogen which terminates dangling bonds of Si should suppress the growth rate of nc-Si in the cell, H_2 supply is effective to reduce the size of nc-Si. Besides, the conductance for the evacuation is proportional

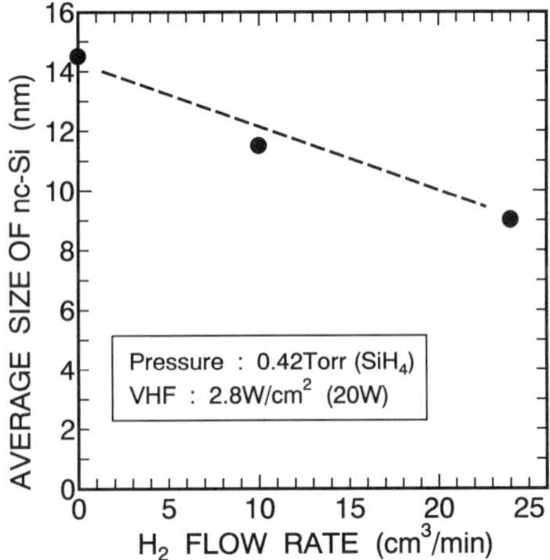

Figure 3: Dependence of average size of nc-Si on H_2 flow rate.

to the reciprocal square-root of molecular weight [13], and H_2 supply into a SiH_4 plasma should reduce the residence time of source gas in the cell owing to small molecular weight of H_2, which causes the decrease in growth time of nc-Si. As a result, the size of nc-Si should be small with increasing H_2 flow rate due to the reduction of growth time. Hence, the effects of H_2 dilution for a SiH_4 plasma on the size of nc-Si was investigated in detail as follows.

Figure 3 shows the dependence of average size of nc-Si on H_2 flow rate. The VHF power density was set at 2.8 W/cm^2. The average size of nc-Si decreased with the increase of H_2 flow rate. This result indicates that the average size of nc-Si can be reduced feeding H_2 into a SiH_4 plasma.

According to the theory of diffusion-controlled growth [14], the radius of nc-Si is given by

$$r = \sqrt{2Dn(C-s)t}, \tag{1}$$

where r is the radius of nc-Si, D the diffusion coefficient of SiH_x (x=0,1,2,3), n the reciprocal number of Si atom per unit volume in crystal, C the concentration of SiH_x at the position far from nc-Si surface, s the concentration of SiH_x at the surface of nc-Si, and t the growth time. From eq.(1), the size of nc-Si should be proportional to the square-root of growth time, which is determined by the residence time in the plasma cell.

Figure 4 shows the average size of nc-Si versus square-root of residence time of source gas in the cell. The residence time (t_r) was calculated from the volume of the plasma cell (V), the pressure in the cell (P), and the gas throughput (Q) obtained from the flow rate of gas as follows [15];

$$t_r = \frac{PV}{Q}. \tag{2}$$

To obtain longer residence time in the cell, an orifice of 3 mm diameter and 0.3 mm length was used. The average size of nc-Si is proportional to the square-root of residence time, indicating that the growth mechanism of nc-Si is well explained by a diffusion-controlled growth model.

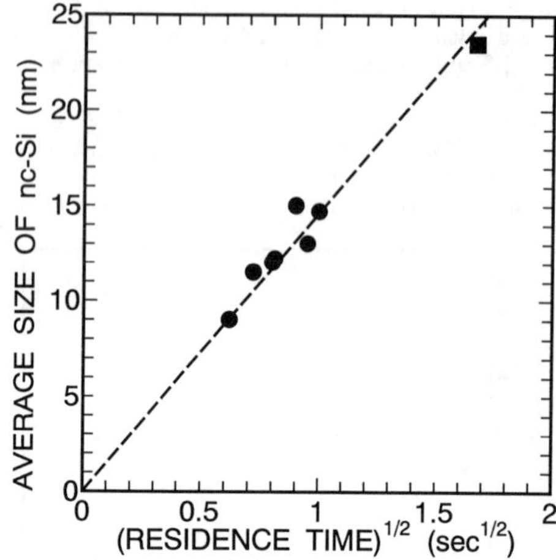

Figure 4: Average size of nc-Si as a function of the square-root of residence time. Deposition performed using an orifice of 6 mm diameter is denoted by circles and a square denotes deposition using an orifice of 3 mm diameter.

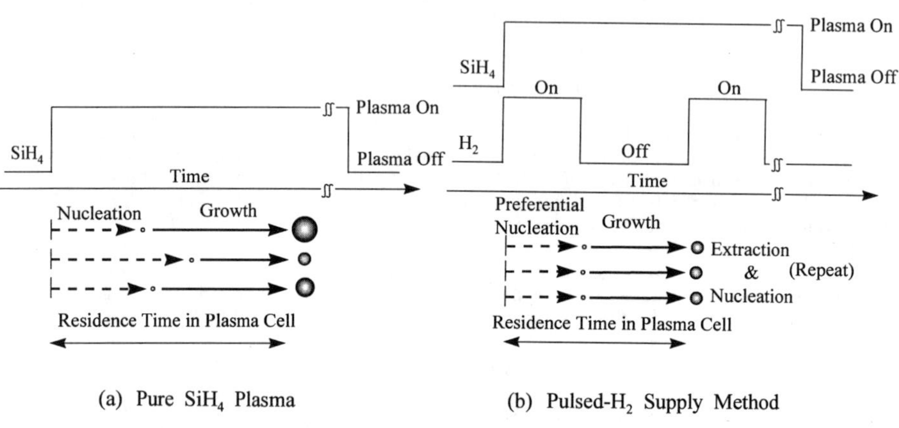

Figure 5: Schematic diagram of gas flow sequence and its relation to nucleation and nc-Si growth for (a) pure SiH_4 plasma and (b) pulsed-H_2 supply method.

At the low VHF power density of 0.71 W/cm^2, the density of nc-Si on the substrate increased with the increase in H_2 flow rate. Since the nc-Si rarely combine together during growth in the cell, the density of nc-Si on the substrate is proportional to the nucleation rate in the cell. The increase in the density of nc-Si on the substrate suggests the enhancement of nucleation rate for nc-Si in the plasma cell. At the high VHF power condition, since the number of SiH_x radicals is larger than that of hydrogen radicals, the growth of nc-Si progresses. However, the proportion of hydrogen radicals in the plasma cell increases against nuclei of nc-Si with decreasing VHF power. Excess hydrogen radicals may terminate the surface of nc-Si and the number of dangling-bonds at the surface of nc-Si decreases. As a result, the surface energy of nc-Si decreases, and growth of nc-Si may be suppressed. Therefore, in presence of a large amount of hydrogen radicals, SiH_x radicals contribute to nucleation rather than growth.

Based on the above results, we consider the formation mechanism of nc-Si in the plasma cell, and propose a novel method in order to obtain the nc-Si with a small diameter and a small spread of size. As shown schematically in Fig. 5 (a), nucleation and growth takes place randomly in the case of using a pure SiH_4 plasma, and the size of nc-Si is not uniform. To control the size of nc-Si, separation of nucleation and growth is required. Taking advantage of preferential nucleation of H_2 dilution, we employed a pulsed-H_2 supply method as shown in Fig. 5 (b). When H_2 gas is supplied into a SiH_4 plasma, nucleation of nc-Si is enhanced. When H_2 gas supply is off, nucleated nc-Si grows larger in a SiH_4 plasma. In the next cycle of H_2 supply, nc-Si grown in the previous cycle are pushed out of the cell into the deposition chamber, and nucleation of nc-Si for the next cycle takes place simultaneously.

Figure 6 shows the size distribution and TEM images for nc-Si fabricated (a) by pure SiH_4 plasma, and (b) by a pulsed-H_2 supply method. The VHF power density is set as low as $0.42\,W/cm^2$ to get large effect of H_2 dilution on the size of nc-Si as possible. As expected, the size distribution prepared by a pulsed-H_2 supply method is more uniform than that by a pure SiH_4 plasma. Half-width of size distribution of nc-Si fabricated by a pulsed-H_2 supply method was 1 nm, and successful fabrication of nc-Si with a diameter around 8 nm was realized.

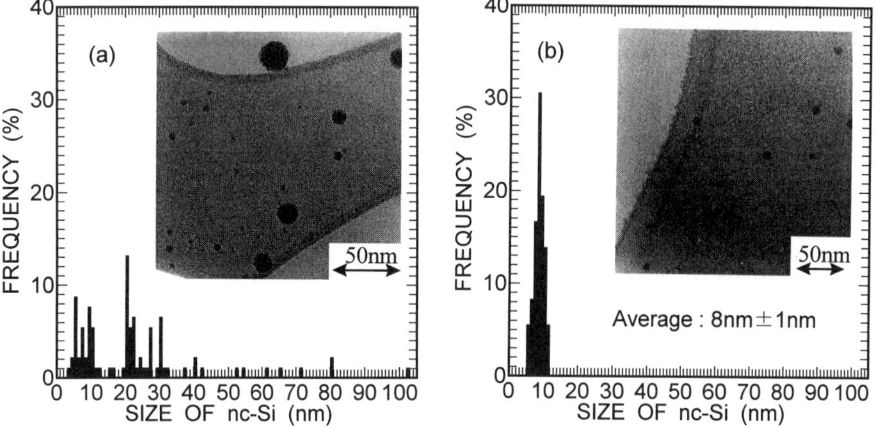

Figure 6: Size distribution and TEM images for nc-Si fabricated by (a) pure SiH_4 plasma and (b) pulsed-H_2 supply method.

CONCLUSIONS

We have investigated nc-Si formation in the plasma cell with VHF (144 MHz) excitation. When a SiH_4 plasma was diluted by H_2, the average size of nc-Si decreased with increasing dilution ratio. Since the average size of nc-Si is proportional to the square-root of residence time of the gas in the plasma cell, growth mechanism of nc-Si is explained by diffusion-controlled growth. The density of nc-Si increases with the increase in H_2 flow rate.

Based on the effects of H_2 dilution on the size of nc-Si, we proposed a method of pulsed-H_2 supply into a SiH_4 plasma to control the nucleation and growth time. Using this method, we fabricated nc-Si with a diameter around 8 nm±1 nm successfully. This result serves a promising step toward implementation of integrated circuits based on Si SET devices.

ACKNOWLEDGMENTS

The authors would like to thank Professor K. Yagi and Professor N. Yamamoto for use of the TEM apparatus and their useful suggestions. This work was partially supported by a Grant-in-Aid for Scientific Research from Ministry of Education, Science, Sports, and Culture, Japan.

REFERENCES

1. K. K. Likharev, IBM J. Res. & Dev. **32**, 144 (1988).
2. M. A. Kastner, Rev. Modern Physics **64**, 849 (1992).
3. E. H. Visscher, J. Lindeman, S. M. Verbrugh, P. Hadley, J. E. Mooij, and W. van der Vleuten, Appl. Phys. Lett. **68**, 2014 (1996).
4. K. Matsumoto, M. Ishii, and K. Segawa, J. Vac. Sci. Technol. B **14**, 1331 (1996).
5. K. Yano, T. Ishii, T. Hashimoto, T. Kobayashi, F. Murai and K. Seki, IEEE Trans. Electron Devices **41**, 1628 (1994).
6. Y. Takahashi, H. Namatsu, K. Kurihara, K. Iwadate, M. Nagase and K. Murase, IEEE Trans. Electron Devices **43**, 1213 (1996).
7. M. Otobe and S. Oda,in Amorphous Silicon Technology–1995, edited by M. Hack, E. A. Schiff, A. Madan, M. Powell, and A. Matsuda (Mater. Res. Soc. Proc. **377**, San Francisco, CA, 1995) pp.51-56.
8. M. Otobe, T. Kanai, T. Ifuku, H. Yajima, and S. Oda, J. Non-Crystalline Solids **198–200**, 875 (1996).
9. S. Oda, J. Noda and M. Matsumura, Jpn. J. Appl. Phys. **29**, 1889 (1990).
10. S. Oda and M. Yasukawa, J. Non-Cryst. Solids. **137&138**, 677 (1991).
11. S. Oda, Plasma Sources Sci. Technol. **2**, 26 (1993).
12. S. Oda and M. Otobe, in Microcrystalline and Nanocrystalline Semiconductors, edited by R. W. Collins, C. C. Tsai, M. Hirose, F. Kock, and L. Brus, (Mater. Res. Soc. Proc. **358**, Boston, MA, 1995) pp.721-731.
13. M. Knudsen, The Kinetic Theory of Gas, (Methuen, 1950).
14. A. E. Nielsen, Kinetics of Precipitation, (Pergamon, Oxford, 1964).
15. S. Dushman and J. M. Lafferty, Scientific Foundation of Vacuum Technique, 2nd ed. (John Wileys & Sons, Inc., New York, 1962).

REAL-TIME MONITORING OF HYDROGEN ELIMINATION PROCESSES IN PULSED-GAS PECVD USING *IN SITU* MASS SPECTROSCOPY

Easwar Srinivasan, Jeremy S. Bordeaux and Gregory N. Parsons
Dept. of Chemical Engineering, North Carolina State University, Raleigh, NC 27695.

ABSTRACT

In situ mass spectroscopy is used to monitor and analyze the hydrogen elimination reaction products during cyclical exposure of thin films of amorphous silicon to a flux of atomic deuterium. Mass spectroscopy results that atomic deuterium etches deposited silicon forming SiD_4 and abstracts hydrogen bonded to silicon in the film to form HD. The relative signal intensities show that abstraction is the primary hydrogen elimination mechanism. The energy of activation for the abstraction reaction is obtained from the mass spectroscopy signals through a first order kinetic analysis and is found to be approximately zero, indicating that abstraction is not thermally activated.

INTRODUCTION

Amorphous and polycrystalline silicon formation at temperatures less than 400°C is important for application in photovoltaics and thin film transistors. Hydrogen dilution of silane or intermittent exposure of amorphous silicon surface to hydrogen plasma has been used to cause a transition from amorphous to microcrystalline silicon [1-7], but have generally been unsuccessful in making large crystallite polysilicon on glass. Understanding the role of hydrogen in the amorphous to microcrystalline transition will provide the insight necessary to use this process to form polysilicon on glass. Microcrystalline silicon films deposited by varying experimental parameters have been used to propose mechanisms for the structural transition [1-7]. However, the vastly varied experimental conditions that have been used have not yielded a universal mechanism. *In situ* mass spectroscopy has been used to monitor hydrogen elimination mechanisms in post-deposition deuteration of sputtered hydrogenated amorphous silicon (a-Si:H) [8]. Similar post-deposition deuteration experiments have been performed on plasma or hot-wire assisted CVD a-Si:H [9,10]. In this article, we will discuss results obtained using *in situ* mass spectroscopy to monitor the hydrogen elimination mechanisms during atomic hydrogen exposure in a cyclic process to form microcrystalline silicon. These results will be used to obtain kinetics of critical hydrogen elimination mechanisms; abstraction and etching.

EXPERIMENT

The rf plasma deposition system [11] and the mass spectroscopic chamber [12] have been discussed in detail elsewhere. Films have been formed by varying the extent of atomic hydrogen (deuterium) exposure during pulsed-gas deposition. In this process, a thin layer of amorphous silicon is deposited during t_{ON} and then the growth surface is exposed to hydrogen (deuterium) plasma during t_{OFF}. The expected flow of gases in the chamber during pulsing (without plasma) is shown in

Mat. Res. Soc. Symp. Proc. Vol. 452 © 1997 Materials Research Society

Fig. 1. Trace amounts of HD contamination is present in the D_2 bottle supplied and will be seen during t_{OFF}. The flow of gases after shutting off the valves decrease slowly to zero due to the residence time in the chamber. This cycle is repeated to obtain thick films. The mass spectrometer has been set up to probe the products and unused reactants exiting the chamber. The mass spectrometer is connected to the deposition system using a 75 μm orifice that enables the mass spectroscopic chamber to be maintained at $4x10^{-6}$ torr during deposition at typical pressures of 0.6 torr. Simple flow analysis using fundamentals of mass transport show that flow in the deposition chamber is laminar, diffusive and well-mixed, ensuring that the species probed using the mass spectrometer is representative of those in the deposition chamber. For mass spectroscopic studies, deuterium is used in place of hydrogen. Products from surface reactions during atomic deuterium exposure are monitored using mass spectroscopy. Mass spectroscopic data used to determine reaction kinetics have been obtained using t_{ON}=60 s and t_{OFF}=90 s. These long cycle times have been used to minimize signal fluctuations due to pulsing of gases. The temperature was varied between room temperature (25°C) and 150°C and rf powers of 10 W and 50 W were used. Other conditions for deposition were, a pressure of 0.6 torr in the deposition chamber, 2% silane diluted in helium with a flow rate of 100 sccm and a hydrogen (deuterium) flow of 60 sccm.

RESULTS

Fig. 2 shows infrared spectra of films deposited at a substrate temperature of

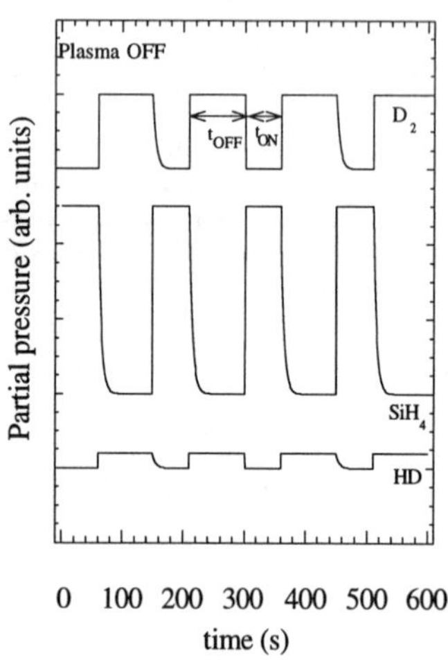

Fig. 1: Expected flow of gases in the chamber, taking into account gas residence time in the chamber.

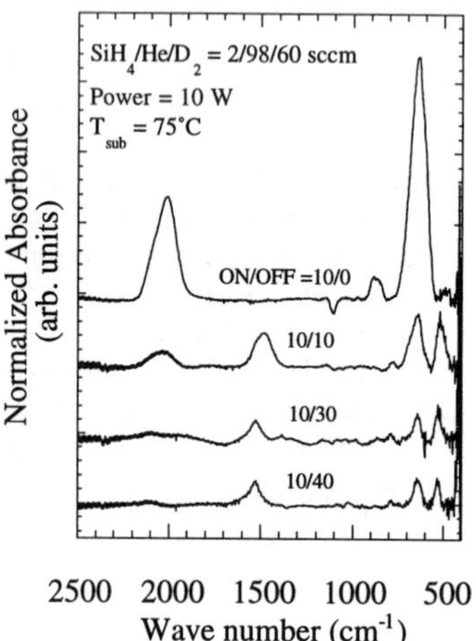

Fig. 2: Infrared spectra of films deposited at 75°C, with varying degree of atomic D exposure. Absorbance normalized to thickness of films.

75°C and rf power of 10 W with atomic deuterium exposure varying between 0 s and 40 s. The film deposited without deuterium exposure shows Si-H related features at 630, 870 and 2000 cm^{-1}, corresponding to Si-H bending, scissors-wagging and stretching vibrations respectively. As t_{OFF} is increased, Si-H features decrease while Si-D features at 1480 (SiD stretch) and 1520 cm^{-1} (SiD_2 stretch) increase. Degeneracy in Si-H bending modes split with the appearance of a feature at 540 cm^{-1}. For t_{OFF}= 30-40 s, the Si-H modes are near the detection limit of the ir transmission, and the spectra are dominated by Si-D modes. The deuterium incorporation is seen more clearly for a deposition temperature of 200°C. It is evident that with increasing t_{OFF}, SiD_2 stretch mode increases, consistent with observations in pulsing with hydrogen [12]. Similar trends have been obtained at substrate temperatures of 25 and 250°C, though the transition from Si-H to Si-H_2 and SiD to SiD_2 modes occurs at lower atomic D exposure times (t_{OFF}) for higher temperatures and higher t_{OFF} for lower temperatures.

Mass spectroscopy signals obtained at amu 3 (HD^+) and amu 34 (SiD_3^+) at a substrate temperature of 150°C under typical pulsed deposition conditions (t_{ON}/t_{OFF} = 10/40 s) are shown in Fig. 3. Silicon related species can be expected to have signals between amu 29 and amu 34. However, mixed hydride-deuteride species with signals at amu 31 ($SiHD^+$), and 33 ($SiHD_2^+$) have not been observed under all conditions investigated. Signals at amu 32 (SiD_2+ or SiH_2D^+) and 34 (SiD_3^+) are proportional for all experiments, confirming the absence of mixed hydrides. Fig. 3 shows that signals at amu 3 is an order of magnitude larger than the signal at amu 34. The signal at amu 3 originates from abstraction reaction 1 and amu 34 originates from etching reaction, 2.

$$(Si)_3 - Si - H + D \xrightarrow{k_a} (Si)_3 - Si\bullet + HD(g) \tag{1}$$

$$(Si)_3 - Si - Si - D_3 + D \xrightarrow{k_e} (Si)_3 - Si\bullet + SiD_4(g) \tag{2}$$

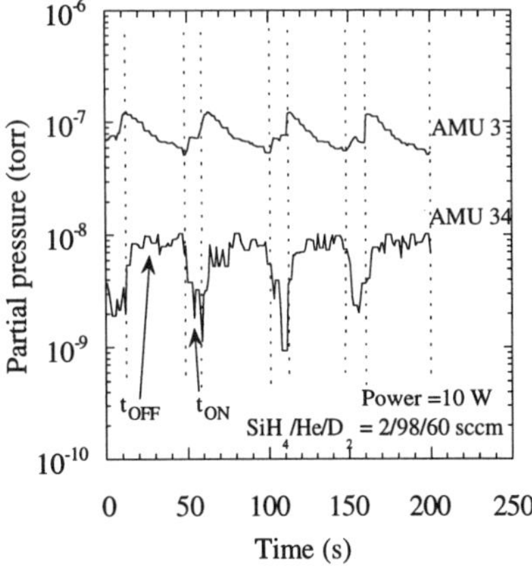

Fig. 3: Signals at amu 3 and 34 obtained at 150°C with t_{ON}/t_{OFF} = 10/40 s.

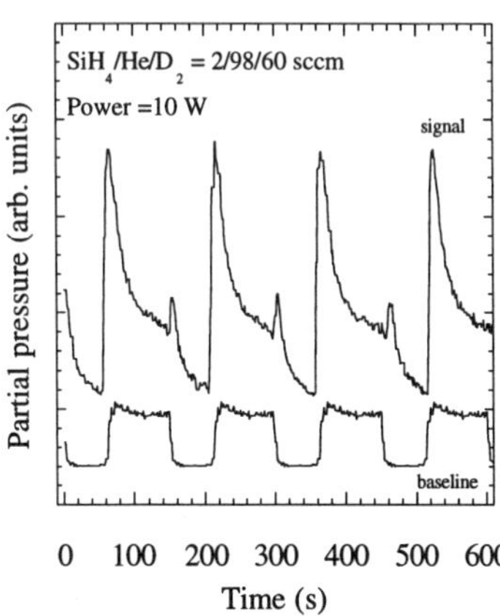

Fig. 4: Signal and baseline for amu 3, at 150°C with t_{ON}/t_{OFF} = 60/90 s.

Significant changes in both signals are observed during t_{OFF} (deuterium exposure). The signal at amu 34 is nearly constant during t_{OFF}, whereas the signal at amu 3 reaches a maximum value when deuterium plasma is turned ON and then slowly decays. The mass spectroscopic signals can be used to obtain the kinetics of the abstraction and etch reactions. Longer deuterium exposure time of 90 s is used for studying the kinetics. Fig. 4 shows the amu 3 signal (plasma ON) and the baseline (plasma OFF) over four cycles. The signal is offset for clarity. The signal is collected over four cycles while the baseline is obtained once and reproduced four times. The HD signal decay can be modeled as the sum of an exponential arising from a first order abstraction of the surface hydride and a constant rate of abstraction from the bulk hydrides. The exponential decay and the eventual leveling is clearly seen in Fig. 5, where the abstraction rate is plotted as a function of time. Abstraction rates are calculated from the partial pressure data (Fig. 4) obtained from the mass spectrometer using equation 3.

$$r_{HD} = \left[\left(\frac{p_{HD}}{P}\right)_{ms} \times \dot{n}\right] + \left[\left(\frac{dp_{HD}}{dt}\right)_{ms} \times \frac{N}{(P)_{ms}}\right] \tag{3}$$

In this equation subscripts 'ms' refer to the mass spectrometer, and p_{HD} refers to the partial pressure of HD in torr, P, the total pressure in torr, $\dot{n}$, the total molar flow rate in the deposition chamber in moles/s, and N, the total number of moles of gas in the chamber. Assuming an elementary reaction, the total rate of abstraction can be expressed as

$$r_a = [k_a C_{SiH}]_{surf} + [k_a C_{SiH}]_{bulk} \tag{4}$$

where the two terms separately yield the surface and bulk abstraction. In Fig. 5, the surface and bulk abstraction rates at any time t during a representative atomic D exposure cycle is shown. Using an Arrhenius type expression for the rate constant,

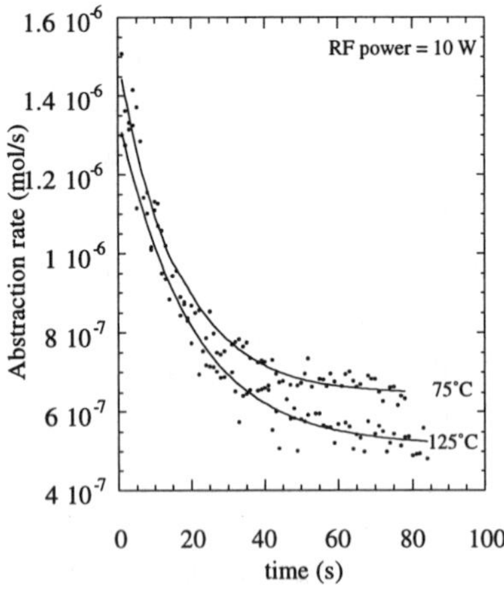

Fig. 5: Abstraction rates as a function of atomic deuterium exposure time. For times less than 40 s, surface abstraction dominates.

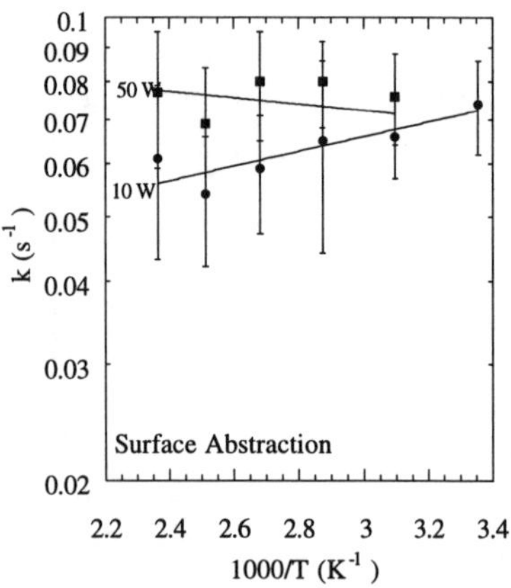

Fig. 6: Activation energy determination for surface abstraction at 10 and 50 W from rate constants at different temperatures.

surface abstraction activation energies are calculated by plotting the logarithm of rate constant, k against the inverse temperature and fitting the data to a straight line, as shown in Fig. 6. Similarly, bulk abstraction and etching rates at different temperatures are used to obtain the activation energy of these reactions. Table 1 shows the energies of activation obtained in this method for the two different rf powers used.

Table 1: Activation energy estimated from rate constant data.

Reaction	Energy of Activation (kcal/mol)	
	10 W	50 W
Surface Abstraction	-0.3±0.3	0.1±0.5
Bulk Abstraction	-0.4±0.3	-0.3±0.5
Etching	-0.5±0.3	-0.4±0.2

DISCUSSION

The infrared spectra in Fig. 2 show that atomic deuterium exposure of the surface results in a rapid substitution of hydrogen with deuterium, consistent with previous post-deposition deuteration experiments [9]. This substitution is accompanied by a decrease in the total hydrogen content and the films become more di-deuteride bound, due to preferential elimination of the mono-deuteride bonded units, consistent with results obtained on pulsing hydrogen [12] and *ab initio* quantum chemistry calculations [14]. The mass spectroscopy signals in Fig. 3 show that the important hydrogen elimination mechanism is abstraction. Depending on the conditions used, the abstraction signals at amu 3 (HD) are 2-10 times larger than the etch signals at amu 34 (SiD_3^+ fragment in the cracking of SiD_4). In the temperature regime studied, abstraction is clearly dominant. For small times of deuterium exposure, surface abstraction rates are much higher than bulk abstraction rates, consistent with excess surface hydrogen concentration observed in sputter deposited amorphous silicon using mass spectroscopy [8] and *in situ* infrared spectroscopy [13]. Etching reaction proceeds through repeated insertion into strained silicon-silicon bonds till the trihydrides are formed, which can then diffuse on the surface and form relaxed Si-Si bonds or be etched (reaction 2). Etching rates will be determined by the slowest step in the series of reactions leading to etching. The electronegativity of hydrogen is larger than that of silicon, and consequently, insertion into the monohydride bond is likely to have the largest barrier and slowest rates. Rate equations will be useful in understanding the reaction kinetics and will indicate the method to enhance hydrogen elimination leading to larger crystallite formation. The insertion rate equation will depend on the monohydride concentration, as does the abstraction rate. Lower etching rates can be explained only by a larger barrier for insertion than abstraction. Dangling bonds formed during these hydrogen elimination processes may be saturated with atomic deuterium or serve as a site for silicon hydride diffusion.

The abstraction rate obtained in Fig. 5 is treated as a sum of two components: (i) surface abstraction that decays exponentially, and (ii) bulk abstraction that is a constant. The exponential decay in rate follows the surface Si-H concentration, which is depleted in a simple first order, elementary reaction. The activation energies of abstraction (both surface and bulk) and etching in the investigated temperature regime are near zero, indicating that both reactions are not thermally activated. These results are consistent with the microcrystalline silicon formation model that includes dangling bond formation by hydrogen elimination, with subsequent diffusion of silicon hydrides on the surface [3, 4, 15]. The data presented here clearly shows that hydrogen elimination proceeds predominantly through hydrogen abstraction.

CONCLUSIONS

In this article, real time mass spectroscopy has been used to determine that hydrogen abstraction is the important hydrogen elimination mechanisms during atomic hydrogen exposure. Kinetics of abstraction and etch reactions have been performed to determine that these reactions are not thermally activated between 25°C and 150°C. This implies that hydrogen elimination rates are not going to be sufficiently enhanced by increasing temperature. However, a higher deposition temperature will yield a film with a lower hydrogen content. Hydrogen elimination is important in creating dangling bond sites for a relaxed network formation.

ACKNOWLEDGMENTS

The authors would like to thank funding from DARPA High Definition Systems Program (Contract # DABT 63-94-C-0004)

REFERENCES

1. A. Matsuda, J. Non-Cryst. Solids **59/60**, 767 (1983).
2. C.C. Tsai, R. Thompson, C. Doland, F.A. Ponce and B. Wacker, Proc. Mat. Res. Soc. Symp. **118**, 49 (1988).
3. M. Otobe and S. Oda, Jpn. J. Appl. Phys. **31**, 1948 (1992).
4. A. Asano, Appl. Phys. Lett. **56**, 533 (1990).
5. S. Koynov, R. Schwarz, T. Fisher, S. Grebner and H. Munder, Jpn. J. Appl. Phys. **33**, 4534 (1994).
6. H. Shirai, B. Drevillon and I. Shimizu, Jpn. J. Appl. Phys. **33**, 5590 (1994).
7. G.N. Parsons, Appl. Phys. Lett. **59**, 2546 (1991).
8. J.R. Abelson, L. Mandrell and J.R. Doyle, J. Appl. Phys. **76**, 1856 (1994).
9. S. Miyazaki, Y. Kiriki, Y. Inoue and M. Hirose, Jpn. J. Appl. Phys. **30**, 1539 (1991).
10. C.M. Chiang, S.M. Gates, S.S. Lee, M. Kong and S.F. Bent, submitted to J. Chem. Phys.
11. E. Srinivasan, D.A. Lloyd and G.N. Parsons, J. Vac. Sci. Technol. A, Jan/Feb (1997).
12. E. Srinivasan and G.N. Parsons, J. Appl. Phys, March (1997).
13. Y. Toyoshima, K. Arai, A. Matsuda and K. Tanaka, J. .Non-Cryst. Solids **137/138**, 765 (1991).
14. E. Srinivasan, H.Yang and G.N. Parsons, J. Chem. Phys. **105**, 5467 (1996).
15. J.R. Abelson, Applied Physics A 56, 493 (1993).

OPTICAL AND ELECTRICAL PROPERTIES OF UNDOPED MICROCRYSTALLINE SILICON DEPOSITED BY THE VHF-GD WITH DIFFERENT DILUTIONS OF SILANE IN HYDROGEN

N. BECK*, P. TORRES*, J. FRIC**, Z. REMEŠ**, A. PORUBA**, HA STUCHLÍKOVÁ**, A. FEJFAR**, N. WYRSCH*, M. VANĚČEK**, J. KOČKA**, A. SHAH*

* Institut de Microtechnique, Université de Neuchâtel, Rue A. L. Breguet 2, 2000 Neuchâtel, Switzerland
** Institute of Physics, Academy of Sciences of the Czech Republic, Cukrovarnická 10, 162 00 Praha 6, Czech Republic

ABSTRACT

We show that the optical and electrical properties of microcrystalline silicon (μc-Si:H) deposited by the VHF-GD technique at 110 MHz can considerably be tuned by changing the dilution ratio of silane to hydrogen.

With increasing silane dilution we observe enhanced optical absorption for energies below 2 eV due to the transition of the material from amorphous / microcrystalline mixture to a pure microcrystalline phase. Simultaneously, the light scattering and the defect absorption increases. Strong dilution also promotes the incorporation of impurities into the material, leading to a pronounced extrinsic behaviour as seen from the decrease of the activiation energy of the electrical conductivity.

The electrical properties were investigated in the dark by the Time of Flight technique. We measured drift mobilities at room temperature which slightly increase with dilution, reaching values of 3 cm^2/Vs for electrons and 1.2 cm^2/Vs for holes. The ratio between electron and hole drift mobilities is found to be around 2 for all samples studied, similar to that of crystalline silicon.

Furthermore, post-transient Time of Flight measurements revealed detrimental electron deep traps in low dilution material.

INTRODUCTION

With an enhanced optical absorption in the infrared region together with a low sub-bandgap absorption, the microcrystalline silicon produced by the VHF-GD deposition technique is a promising material for solar cell applications [1]. The absorption properties of such material is almost "crystalline-like" [2], but can also be tuned by changing the deposition parameters [3].

On the other hand, only very little is known, so far, about electrical transport in μc-Si:H; the collection mechanism for carriers in corresponding p-i-n cells is therefore still unclear. Entirely microcrystalline p-i-n solar cells with thicknesses of up to 3.6 μm [4] have been shown to collect almost the totality of photo-generated carriers even though coplanar transport (photoconductivity and ambipolar diffusion length) is only slightly higher in such μc-Si:H layers compared to amorphous silicon a-Si:H [5].

With the aim to gain more insight into absorption and transport mechanisms of microcrystalline silicon, we produced a series of μc-Si:H layers using different dilution ratios of silane in hydrogen for deposition. This enabled us to create materials with very different qualities, ranging from purely amorphous to highly microcrystalline material. The optical absorption measurements as well as the transport properties parallel to the growth direction reveal important changes within this series, mainly with respect to the density and activity of deep defects, which will be presented and discussed in the present paper.

Mat. Res. Soc. Symp. Proc. Vol. 452 © 1997 Materials Research Society

EXPERIMENTAL

The samples were deposited by the VHF-GD deposition technique at a plasma excitation frequency of 110 MHz using a gas purifier [6]. Thereby the ratio of silane flow over total gas flow $[SiH_4]/[SiH_4 + H_2]$ was varied between 1.25% and 7.5%. The sample deposited at 7.5% dilution is amorphous, all the others are microcrystalline. With deposition rates ranging from less than 0.7 Å/s (1.25% dilution) to 3.5 Å/s (7.5% dilution) layer thicknesses between 1.6 and 1.9 μm were obtained. All films were deposited either on uncoated glass (Schott AF 45) or on chromium-coated glass (for Time of Flight). Semitransparent, small size (~2 mm^2 area) chromium top contacts were deposited for TOF samples, coplanar aluminium contacts with a gap of 0.5 mm were used for dark conductivity and CPM measurements.

The optical absorption of the materials was determined by Transmission/Reflection spectroscopy, by Constant Photocurrent Method (CPM) and by Photothermal Deflection Spectroscopy (PDS). The transport properties were studied by dark conductivity measurements at temperatures ranging from 25°C to 180°C. Time of Flight (TOF) measurements in both short-time and post transient mode were performed respecting the guidelines for TOF measurements on μc-Si:H presented in [7].

X-ray diffraction measurements on all microcrystalline samples presented here show a preferential growth in the <220> direction. A medium grain size of around 200 Å and columnar shape of the grains has already been found in the past for other samples produced using similar deposition conditions [1].

RESULTS

Optical properties

It has already been shown [3] that the transition from microcrystalline to amorphous material is abrupt and typically takes place around 7.5% dilution for the deposition conditions used here. The optical absorption spectra measured on the present series (represented in figure 1) confirms the drastic change of optical absorption with a variation of the dilution ratio. Thereby, three distinct effects can be observed: the

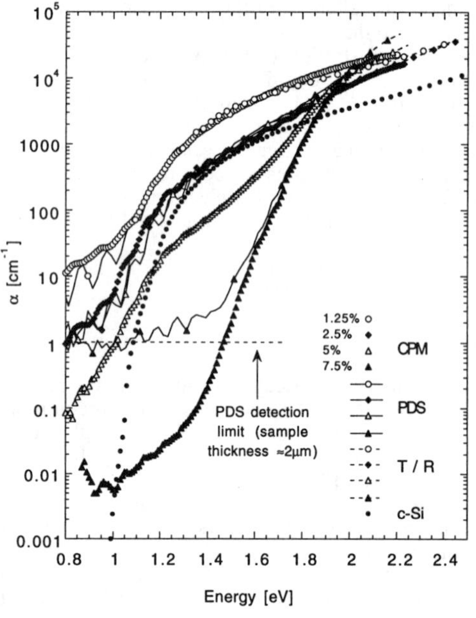

Fig. 1 Optical absorption coefficient for the 110 MHz dilution series determined by PDS, CPM and Transmission / Reflection measurements: the transition from amorphous to microcrystalline silicon takes place at around 7.5% dilution. Crystalline silicon data [8] is also plotted for comparison. Note the difference between the CPM and PDS spectra for the 5% sample, discussed in the text.

transition from a-Si:H to μc-Si:H, scattering and the effect of mixed-phase material.

First, the main phenomenon observed is of course the transition from non-direct absorption in amorphous material with a Tauc gap around 1.7 eV to the optical absorption in the crystalline material with an indirect gap of about 1.1 eV. However, there are also significant changes in the absorption curve within the microcrystalline range of deposition conditions. As a second effect, for the sample with the highest dilution, further enhancement of optical absorption by scattering is apparent. We measured the scattered part of the total reflected light (and thus also of the penetrating light) and observed an increase from about 8% at 650 nm (5% dilution sample) to more than 50% for the highly diluted sample (1.25% dilution). As the scattering events result in a longer effective optical path of the photons, an enhanced apparent absorption is observed [9].

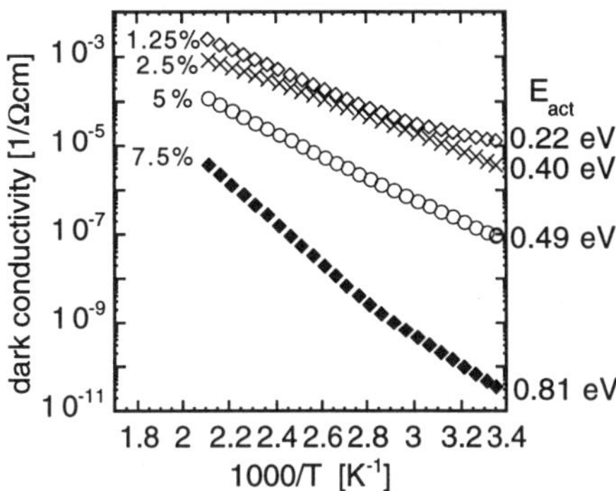

Fig. 2 Measured dark conductivities as a function of temperature for different dilutions. The sample with 7.5% dilution is amorphous, all the others are microcrystalline. The activation energies were determined at room temperature.

Third, the difference between CPM and PDS deduced optical absorption (as clearly seen in the sample with 5% dilution) is due to the mixture of amorphous and crystalline phase prevalent in this particular case. PDS sees the true optical absoprtion, whereas CPM curve can be deconvoluted into two contributions: now, above approx. 1.7 eV the transport in amorphous tissue with higher photosensitivity gives an additional contribution to the "apparent optical absorption coefficient". Because of the calibration procedure used this results in an artificial shift of "α" to lower values for photon energies below 1.7 eV.

When looking at the subbandgap optical absorption (below 1 eV), PDS cannot see the differences between the samples (due to its "detection limit"). On the other hand, CPM absorption, which is not masked by the substrate, interface and surface absorption, can detect the changes in the defect-connected absorption: the subbandgap absorption connected to defects is found to increase with increasing dilution ratios. This can indicate an increased number of dangling bonds, unsaturated by hydrogen, at the grain boundaries [3]. The possible role of oxygen impurities on the optical spectrum has not yet been studied, experiments are under preparation. However, post-transient TOF measurements performed on the same samples show a different trend. This apparent inconsistency will be addressed in more detail in the following section.

Electrical properties

Even though all layers were produced using a gas purifier in order to prevent the incorporation of impurities [6], the microcrystalline materials deposited with high dilutions show a pronounced extrinsic behaviour (Fig. 2) as is usually obtained in the presence of [O] contaminants. The measured dark conductivities vary from 1 x 10^{-7} 1/Ωcm (typical for "truly" intrinsic μc-Si:H) to 1.3 x 10^{-5} 1/Ωcm with activation energies ranging from 0.49 eV to 0.22 eV. In fact, the low deposition rates of the highly diluted μc-Si:H (0.67 Å/s for 1.25% compared to 2.2 Å/s for 5% dilution) can possibly promote the incorporation of oxygen as a donor impurity in the material and cause the observed shift of the Fermi level towards the conduction band edge. The layer deposited with 7.5% dilution shows values of dark conductivity and activation energy that are typical for intrinsic amorphous material (Fermi level close to midgap).

The characterisation of the carrier transport in the dark was made by the Time of Flight technique (TOF). Thereby, mainly two features designated this method as appropriate to study the carrier transport in our layers: first, the transport of electrons **and** holes can be measured independently and secondly the transport can be monitored in the growth direction of the layer, in a configuration close to that of a p-i-n cell. The results obtained for drift mobilities (μ^D) at room temperature are presented in figure 3. Even though the transport properties vary less drastically than the optical properties, some of the observed features are remarkable. First of all, the drift mobilities clearly increase from 1 to 3 cm^2/Vs (electrons) and from 0.4 to 1.2 cm^2/Vs (holes) with increasing dilution at room temperature. Furthermore, in all measured μc-Si:H layers the drift mobilities for electrons are about twice the hole drift mobilities; this ratio is in clear contrast with the much higher ratio observed for a-Si:H ($\mu^D_e/\mu^D_h \approx 100$). This observation strongly indicates that the drift mobility of carriers in μc-Si:H is **not** governed mainly by the amorphous tissue, as here band tail trapping would give rise to the typical strong mobility asymmetry. Thus, other mechanisms related to grain boundaries and/or scattering must be taken into account.

The measurements of drift mobilities of electrons and holes as a function of temperature (300 - 380 K) are plotted in figure 3. Whereas the drift mobility μ^D of electrons significantly decreases with increasing temperature for all microcrystalline samples, the opposite is observed for holes (except for the 1.25% dilution sample). The activation energies for hole mobilities are low, 59 meV (5% dilution) and 21 meV (2.5% dilution), respectively. Thereby, numerous processes can influence the drift mobility with changing temperature: thermoionic emission of carriers over barriers [10], scattering effects (which can reduce the mobility at high temperatures), but also deep trapping and tunneling may be involved. Thus, at this stage it is not possible to discriminate the dominant transport process in our μc-Si:H material and more extensive measurements will be necessary here.

Whereas short-time TOF monitors the transit of the photogenerated carrier package through the layer, the extension of the TOF experiment to a much longer time scale (post-transient TOF [11], [12]) gives information about the density of states (DOS) in the gap. Thereby, electrons and holes which were deep-trapped during transit are thermally reemitted at longer times and their collection becomes a direct fingerprint of the DOS. In our microcrystalline material the post transient currents from holes are featureless and no particular deep trap states for holes can be observed. On the other side, for electron transients, very pronounced detrapping at times around 10^{-3} - 10^{-4} s appears for low-dilution samples (see Fig. 4b and c), indicating thus the presence of efficient electron traps in the middle of the gap (about 0.5 eV

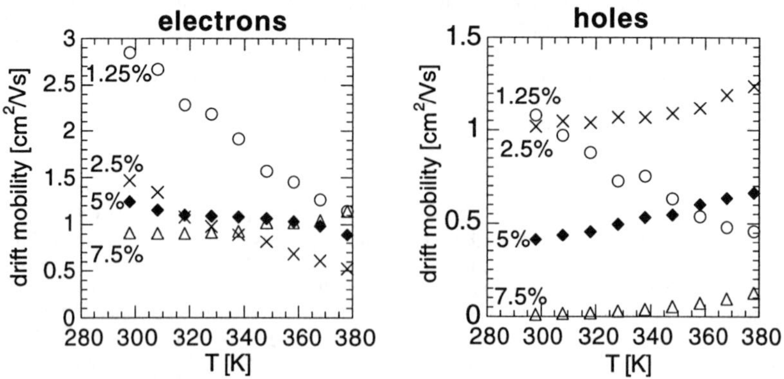

Fig. 3 Temperature dependency of electron and hole drift mobilities $\mu^D{}_e$ and $\mu^D{}_h$ as determined by Time of Flight.

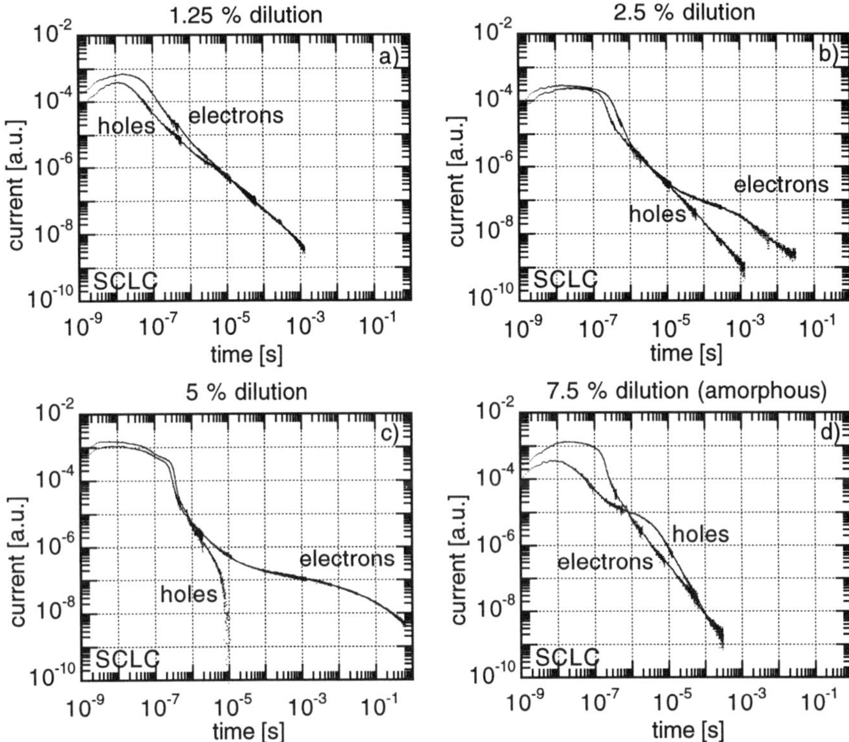

Fig. 4 Time of Flight post transient currents (in space charge limited mode SCLC) for electrons and holes: the low dilution microcrystalline samples (fig.4c, but also b) show a pronounced detrapping current for electrons, indicating the presence of a detrimental type of electron deep trap in this material. Note that the characteristic time in the SCLC mode is the extraction time which is different from the transit time used for drift mobility determination.

deep). As these traps are only active for electrons, we attribute them to positively charged defects at the grain boundaries. However, it is not clear at this stage, if they are due to dangling bonds or to impurities (e.g. metal impurities or oxygen complexes). In addition, the observed Fermi level shift (with increasing dilution) may also fill up defect states with electrons, making them inactive for electron trapping (and invisible for TOF) even though they remain present. Thus, we cannot exclude the presence of such states in highly diluted materials on the basis of our TOF measurements. This also means, that our observation of important trap state density in low-dilution material is not necessarily contradictory to the optical measurements by CPM, which indicate a higher density of deep defect states for high-dilution material. CPM data reflects the optical transitions from gap states below the Fermi level (occupied by electrons), hence, CPM sees different defect states than post-transient TOF.

CONCLUSIONS

We have shown that the optical and electrical properties of μc-Si:H prepared by glow discharge can strongly be influenced by changing the dilution ratio of silane to hydrogen.

Thereby, not only structural changes are involved, but also an enhanced incorporation of impurities at low deposition rates (typical for high dilution materials), leading to deposition of material with an extrinsic character.

Spectral dependence of the optical absorption coefficient reflects the material changes that occur due to an increased dilution: it directly shows the transition from amorphous to microcrystalline material and also reflects residual amorphous fractions. In the subbandgap region absorption measurements are sensitive to defects and impurities. To extract more quantitative data, more has to be known about the influence of oxygen and on the distortion of absorption spectra due to light scattering.

Time of Flight measurements on the μc-Si:H samples show a strongly increased drift mobility of holes when compared to typical a-Si:H. The measured values increase with higher dilution ratios from 1 to 3 cm^2/Vs for electrons and from 0.4 to 1.2 cm^2/Vs for holes. The ratio between electron and hole mobilities observed at room temperature is about 2 and is close to that of crystalline silicon; this clearly indicates a transport mechanism which is different from the multiple trapping transport found in a-Si:H.

By using the post-transient TOF technique we have also found in low-dilution material an efficient type of deep electron trap in the middle of the gap whose origin is not yet clear. Further studies are necessary in order to clarify if these traps are due to dangling bonds or to impurities in the material. Furthermore, the Fermi level shift which accompanies the dilution effect, can render "invisible" certain defect states for TOF by filling them up with electrons; thus a direct interpretation of the results becomes difficult.

ACKNOWLEDGMENTS

This work was supported by the Swiss National Science Foundation under grants FN-39377 and FN-45696.95 and by the Swiss Federal Office of Energy under Research grant EF-REN (93)032.

REFERENCES

[1] J. Meier, S. Dubail, R. Flückiger, D. Fischer, H. Keppner and A. Shah, Proc. of the 1st WCPVEC, Hawaii, 405 (1994).

[2] N. Beck, J. Meier, J. Fric, Z. Remes, A. Poruba, R. Flückiger, J. Pohl, A. Shah and M. Vaněček, J. Non-Cryst. Sol. **198-200**, 903 (1996).

[3] U. Kroll, J. Meier, A. Shah, S. Mikhailov and J. Weber, J. Appl. Phys. **80,** 4971 (1996).

[4] J. Meier, P. Torres, R. Platz, S. Dubail, U. Kroll, J.A. Anna Selvan, N. Pellaton Vaucher, C. Hof, D. Fischer, H. Keppner, A. Shah, K.-D. Ufert, P. Giannoulès, J. Koehler, to be published in Mater. Res. Soc. Symp. Proc. (San Francicso, 1996).

[5] M. Goerlitzer, N. Beck, P. Torres, J. Meier, N. Wyrsch and A. Shah, J. Appl. Phys. **80,** 5111 (1996).

[6] P. Torres, J. Meier, R. Flückiger, U. Kroll, J.A. Anna Selvan, H. Keppner, A. Shah, S.D. Littelwood, I.E. Kelly and P. Giannoulès, Appl. Phys. Lett. **69**, 1373 (1996).

[7] N. Wyrsch, M. Goerlitzer, N. Beck, J. Meier and A. Shah, to be published in Mater. Res. Soc. Symp. Proc. (San Francicso, 1996).

[8] M. A. Green, M. J. Keevers, Progress in Photovoltaics: Research and Applications **3**, 189 (1995).

[9] M. Vaněček, D. Cervinka, M. Favre, H. Curtins and A. Shah, Mater. Res. Soc. Symp. Proc. **192**, 639 (1990).

[10] J. Y. W. Seto, J. Appl. Phys. **46**, 5247 (1975).

[11] G. F. Seynhave, R.P. Barclay, G.J. Adriaenssens, J.M. Marshall, Phys. Rev. **B39**, 10196 (1989).

[12] J. Kočka, M. Vaněček, P. Macháček, A. Fejfar, E. Šípek, Ho-The-Ha, I. Pelant, J. Fric, J. Rosa, Z. Remeš, A. Poruba, M. Konagai and W. Kusian, Proc. of the 1st WCPVEC, Hawaii, 437 (1994).

ENLARGED PARAMETER SPACE BY USE OF VHF-GD FOR DEPOSITION OF THIN <p> TYPE μc-Si:H FILMS

P. TORRES, J. MEIER, R. FLÜCKIGER, H. KEPPNER, AND A. SHAH

Institut de Microtechnique, Rue A.-L. Breguet 2, CH-2000 Neuchâtel, Switzerland

ABSTRACT

In this paper new results on thin p - type μc-Si:H films deposited at low temperatures of 170 °C by the Very High Frequency - Glow Discharge technique (VHF-GD) are presented. The "tolerated" amount of diborane added in the gas phase ratio as well as the influence of three different plasma excitation frequencies (70, 100 and 130 MHz) in obtaining high electrical conductivity are investigated. The goal is to optimise very thin (< 400 Å) and hence optically transparent films by maintaining high conductivities for the application as window layers of solar cells.

INTRODUCTION

Hydrogenated microcrystalline silicon deposited at low temperatures by plasma methods is in general an interesting photovoltaic material [1, 2], and it is indeed also a promising material for optoelectronic applications. Doped μc-Si:H films in particular, result in excellent contact layers for photovoltaic applications due to their higher conductivity compared to amorphous silicon [3].

However, the major drawback to grow very thin boron doped μc-Si:H layers is that it is technologically quite a delicate task to find appropriate parameters. One has to ensure that the layer is indeed microcrystalline and thus avoid a phase transition to amorphous silicon. This morphological requirement is mainly achieved by highly diluting silane in hydrogen. Standard 13.56 MHz PE-CVD requires, in general, dilution ratios higher than 1% SiH_4/total gas flow; on the other hand, the VHF-GD technique allows for lower dilution ratios, i.e. of gas flows up to 5 % SiH_4/total for μc-Si:H growth [4]. As our earlier studies on p - type μc-Si:H have shown [5] doping has to be carefully adjusted: The optimal diborane doping ratio (B_2H_6/SiH_4) for highly conductive films has to be determined empirically: On one hand, enough doping is needed to push the Fermi level close to the valence band, on the other hand, a too high doping will result in the amorphisation of the layers.

Earlier studies [4, 6] have shown that VHF - conditions are favourable for microcrystalline growth. The better growth of μc-Si:H at higher plasma excitation frequencies motivated us to further increase the excitation frequency in order to try to obtain even higher conductivities. So far, comparative excitation frequency studies for μc-Si:H film growth have been carried out only on n-type μc-Si:H [6]; however, growth of excellent p-type μc-Si:H has recently been demonstrated at the single high excitation frequency of 170 MHz [7].

Mat. Res. Soc. Symp. Proc. Vol. 452 © 1997 Materials Research Society

EXPERIMENTAL

All films in this study were deposited on AF45 glass from Schott (alkali free) in a single-chamber parallel-plate reactor by the VHF-GD technique at plasma excitation frequencies of 70, 100 and 130 MHz. To avoid cross-contamination from outgasing, the chamber walls were heated overnight, thus leading to a base pressure of 5×10^{-8} mbar at room temperature. To achieve good reproducibility, all layers were deposited with the same chamber history. The electrodes have a surface of 139 cm^2 and are separated by a gap of 15 mm. The substrate is positioned upside down on the upper electrode, where the effective substrate temperature is kept at 170 °C for all the films. The radio frequency is capacitively coupled to the lower electrode providing impedance matching for the entire frequency range of 55 to 200 MHz.

The employed gases were silane (SiH_4), hydrogen (H_2), and diborane (B_2H_6) diluted at 500 ppm in H_2. The gas pressure was kept at 0.4 mbar and the dilution level of silane was 1.5 % over a total gas flow of 100 sccm.

The effective power, determined by the subtractive method [8, 9] was kept approximately at 3,5 W. This corresponds to setpoint values of 4, 5 and 9 W for 70, 100 and 130 MHz respectively. One can observe from this measurements that the power coupling into the reactor becomes worse at higher frequencies. Therefore it is important for a comparative study like this one, where the frequency is changed, to actually check the effective power in the plasma.

We selected as a monitor to determine the optimal doping, the value of the room temperature dark conductivity σ_d of μc-Si:H layers as measured on a coplanar electrode configuration under an N_2 atmosphere at 10 mbar after a standard temperature annealing step up to 150 °C.

RESULTS AND DISCUSSION

The layer thickness was kept constant at around 400 Å for optimisation of the boron doping experiment. This was done in order to avoid errors due to the strong thickness effect on σ_d, as observed for very thin layers [7]. In order to optimise σ_d, a doping-study was carried out for the different plasma excitation frequencies: 70, 100 and 130 MHz. From Fig. 1 it is obvious that the doping range for highly conductive p - type μc-Si:H films is broadened when the plasma excitation frequency is increased. Thus, we can deduce, that higher excitation frequencies will favour the growth of p - type μc-Si:H layers. This effect is also interesting from the point of view of plasma physics, since it provides further experimental evidence that the VHF-GD technique at high plasma excitation frequencies is indeed an appropriate tool for the growth of μc-Si:H. Here, we observe that even higher "doses" of diborane in the gas phase ratio do not hinder crystalline nucleation if one moves to higher plasma excitation frequencies.

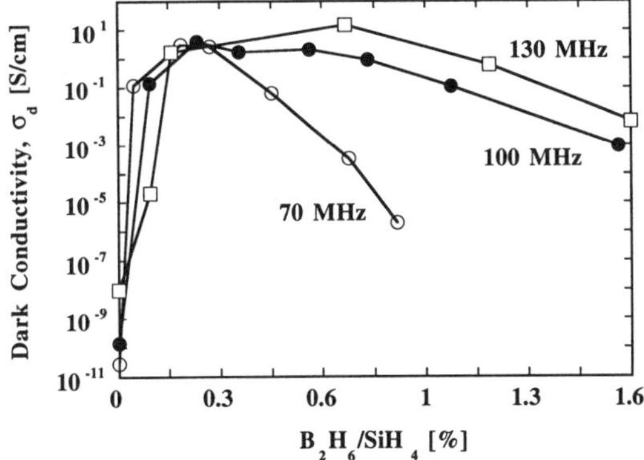

Fig. 1: Room temperature dark conductivity as a function of diborane doping for three different rf-plasma excitation frequencies. Film thickness is nominally 400 Å.

From Fig. 1 we deduce that the optimal diborane doping ratio (B_2H_6/SiH_4) for the 70 MHz series is about 0.2 % and that this optimum is shifted to a higher value of about 0.7 % for the 130 MHz series. To compare this 130 MHz p - type films with an earlier study done at 70 MHz [5], we deposited a thickness series at 0.7 % B_2H_6 doping as assumed to be optimal for 130 MHz (fig. 1). It is well known that for thicknesses under a critical value, the dark conductivity σ_d will drop substantially.

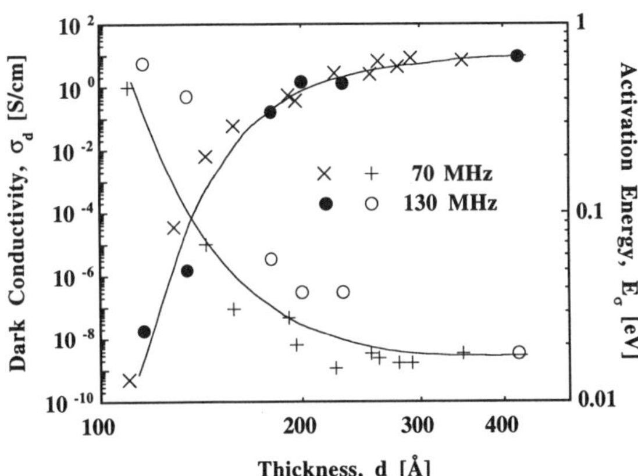

Fig. 2: Room temperature dark conductivity and dark conductivity activation energy, as a function of film thickness. Note that the 70 MHz series are taken from a previous work [5] and are deposited as the present series, in an equivalent reactor under similar conditions.

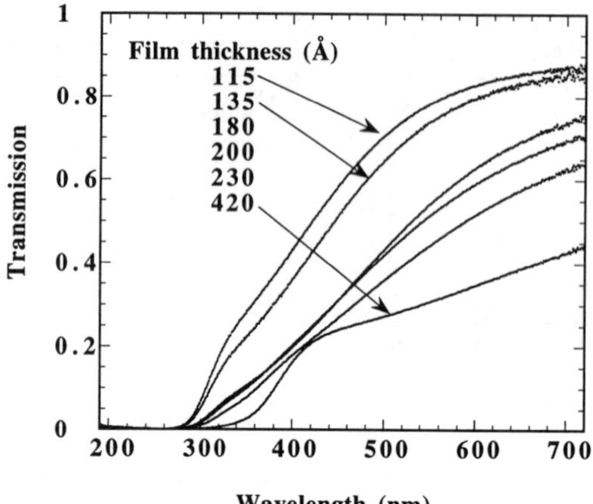

Fig. 3: Optical transmission of p - type layers deposited at a plasma excitation frequency of 130 MHz for a set of layer thicknesses as shown above in Fig. 2.

In Fig. 2 we compare the 130 MHz series with a 70 MHz series already published [5] which was deposited in an equivalent reactor and under similar conditions. It is surprising that even under such different growth conditions the films have same properties. In Fig. 3 we show the optical transmission of the 130 MHz thickness series. The highly conductive films (d > 180Å) show considerable absorption in the visible range, but the appropriate thickness e.g. for a window layer in a solar cell has to be worked out in the device itself. Just to give an example, the nucleation behaviour on substrates other than glass (as used for this work) can in certain cases result even in epitaxial growth [10] (this is observed on crystalline silicon for identical deposition parameters as those used in this work). For growth on underlying amorphous silicon layers a careful pretreatment is required, otherwise no nucleation occurs and hence the film remains amorphous [11, 12].

The deposition rate of boron-doped films depends strongly on the B_2H_6/SiH_4 gas phase ratio. To compare the deposition rates of highly conductive films obtained by different plasma excitation frequencies, we kept the gas phase doping ratio at the constant values of 0.2 % B_2H_6 / SiH_4. Fig. 4 shows the deposition rate in function of the frequency. For comparison, the 130 MHz film with optimal doping of 0.7 % is also shown in the same figure. We see that an increased in excitation frequency results in higher deposition rates, which is even further increased by enhancing the diborane doping ratio. However, this increased deposition rate does not affect the films quality in terms of electrical conductivity as seen in Fig. 1 provided that the excitation frequency is high enough.

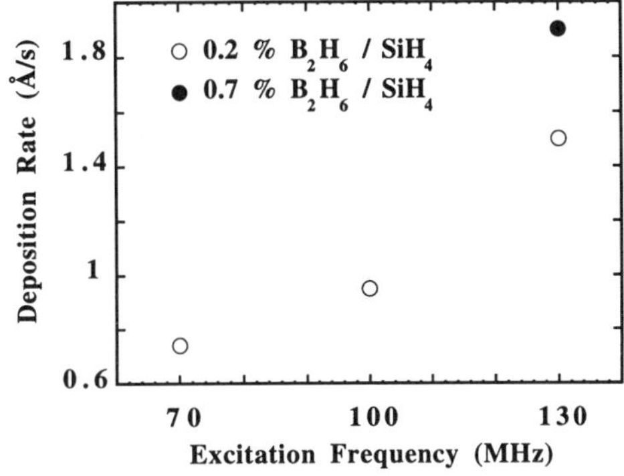

Fig. 4. Deposition rates of diborane doped microcrystalline silicon films deposited at several plasma excitation frequencies on glass. Further doping even more increases the deposition rate.

P_{eff} = 3.5 W for all films

CONCLUSIONS AND OUTLOOK

Starting from previously published results on p - type μc-Si:H, deposited at an excitation frequency of 70 MHz, where the VHF-GD method had already been shown to give excellent layers , the authors have extended in this paper the plasma excitation frequencies to even higher frequencies of 100 and 130 MHz. Thereby, they found that the parameter range for the diborane gas phase doping ratio is substantially enlarged. This effect supports earlier findings that higher plasma excitation frequencies lead to higher crystallinity and higher deposition rates of the films. The larger parameter space at the higher plasma excitation frequency of 130 MHz should probably also facilitate the growth of boron-doped μc:Si,C:H layers, i.e. of silicon - carbon alloy. The methane used there in the gas phase is also known to hinder nucleation [13]. Higher plasma excitation frequencies may possibly favour the growth of SiC - crystallites as was found by Hamakawa et al. when using the ECR - process [14]: The enlarged parameter space demonstrated in this work at higher frequencies may possibly constitute a tool to overcome this limitation.

ACKNOWLEDGEMENTS

The technical assistance of S. Dubail is gratefully acknowledged. This work was supported by the Swiss Federal Office of Energy under Research Grants EF-REN (90)045 and (93) 032.

REFERENCES

[1] J. Meier, P. Torres, R. Platz, S. Dubail, U. Kroll, J.A. Anna Selvan, N. Pellaton Vaucher, Ch. Hof, D. Fischer, H. Keppner, A. Shah, K.-D. Ufert, P. Giannoulès, and J. Koehler, Mat. Res. Soc. San Francisco (1996), to be published.

[2] P. Torres, J. Meier, R. Flückiger, U. Kroll, J.A. Anna Selvan, H. Keppner, A. Shah, S.D. Littelwood, I. E. Kelly, and P. Giannoulès, Appl. Phys. Lett. **69** (10), 1373, (1996).

[3] L. Yang, L. Chen, S. Wiedeman, and A. Catalano, Mat. Res. Soc. Symp. Proc. **283**, 463, (1993).

[4] R. Flückiger, J. Meier, G. Crovini, F. Demichelis, F. Giorgis, C. F. Prirri, E. Tresso, J. Pohl, V. Rigato, S. Zandolin, and F. Caccavale, Mat. Res. Soc. Symp. Proc. **358**, 751 (1994).

[5] R. Flückiger, J. Meier, H. Keppner, M. Götz, and A. Shah, Proc. of 23rd IEEE PVSEC, 839 (IEEE, New York, 1993).

[6] F. Finger, P. Hapke, M. Luysberg, R. Carius, H. Wagner, and M. Scheib, Appl. Phys. Lett. **65** (20), 2588, (1994).

[7] M. Heintze, M. Schmitt, Mat. Res. Soc. San Fancisco (1996), to be published.

[8] V. A. Godyack and R. B. Piejack, J. Vac. Sci. Technol. A **8** (5), 3833 (1990).

[9] A. A. Howling, J. L. Dorier, Ch. Hollenstein, U. Kroll, and F. Finger, J. Vac. Sci. Technol. **A 10**, 1080 (1992).

[10] P. Torres, R. Flückiger, J. Meier, H. Keppner, U. Kroll, V. Shklover, and A. Shah, Proceedings of 13th EC PVSEC, edited by H. S. Stephens &Associates (Bedford, UK, 1995), p. 1638.

[11] N. Pellaton Vaucher, B. Rech, D. Fischer, S. Dubail, M. Goetz, H. Keppner, C. Beneking, O. Hadjadj, V. Shklover, A. Shah (1996), to be published in Techn. Dig. PVSEC-9 (Japan).

[12] P. Pernet, M. Goetz, H. Keppner, and A. Shah, this conference (Mat. Res. Soc. Boston 1996).

[13] R. Flückiger, J. Meier, A. Shah, A. Catana, M. Brunel, H.V. Nguyen, R.W. Collins, and R. Carius, Mat. Res. Soc. Symp. Proc., **336**, 511 (1994).

[14] Y. Hamakawa, Y. Matsumoto, G. Hirata, and H. Okamoto, Mat. Res. Soc. Symp. Proc., Vol. **164**, 291 (1989).

GROWTH MECHANISM OF MICROCRYSTALLINE SILICON DEPOSITED BY ECRCVD

I. BECKERS, E. CONRAD, P. MÜLLER, N. H. NICKEL, I. SIEBER, AND W. FUHS
Hahn-Meitner-Institut, Rudower Chaussee 5, 12489 Berlin, F. R. Germany

ABSTRACT

Microcrystalline silicon (μc-Si) films were prepared by electron cyclotron resonance assisted chemical vapor deposition (ECRCVD) using helium, argon and hydrogen dilution. The crystalline fraction was estimated from Raman backscattering spectra and scanning electron-microscopy (SEM) was used to obtain information on roughness and homogeneity of the films. For hydrogen dilution the highest crystallinity (X_C = 85 %) occurs at a ratio of $\Delta_H = [H_2]/([H_2]+[SiH_4]) = 0.98$. At the same time the deposition rate decreases continuously with increasing H_2 dilution. These results are consistent with the idea that H etching promotes the growth of μc-Si. At $\Delta_H > 0.98$ a X_C decreases due to a H mediated transition of small crystallites into amorphous tissue. The implications of these results for the growth mechanisms are discussed.

1. INTRODUCTION

Microcrystalline silicon (μc-Si) is a material of growing interest for a variety of applications like solar cells or thin-film transistors. For better control of the deposition process it is important to understand the growth process. Previously, it was shown that hydrogen dilution of silane causes the transition from the amorphous to the microcrystalline phase [1]. Comparing hydrogen dilution to argon dilution Tsai [2] demonstrated that hydrogen effectively etches silicon during the deposition of μc-Si. By preferentially eliminating energetically unfavorable configurations during the film growth, hydrogen controls the structure, film growth mechanism, defect concentration, grain size, electrical properties, and H concentration of the resulting films.

In this paper we compare the structural properties of samples deposited in a remote electron cyclotron resonance CVD (R-ECRCVD) to samples prepared by conventional PECVD. The influence of the excitation gases (argon, helium, and hydrogen) on the structure, crystallinity, and deposition rate is investigated as a function of the silane dilution.

2. EXPERIMENTS

Electron cyclotron resonance chemical vapor deposition (ECRCVD) has been used to prepare the samples for the present study. The potential advantage of this technique is the fact that silane decomposition occurs downstream from the plasma by impact with excited and ionized excitation gases. Moreover a remote ECRCVD provides a high-density plasma at low pressure. Three sets of samples were deposited at a temperature of 325 °C and a pressure of 5 mTorr using He, Ar, and H_2 as excitation gas, respectively. The excitation gases are primarily used to achieve the desired silane dilution. The plasma was ignited using a microwave frequency of 2.45 GHz, a

Mat. Res. Soc. Symp. Proc. Vol. 452 © 1997 Materials Research Society

microwave power of 1000 W, and a magnetic field of 875 G. In order to obtain stable plasma conditions when using He or H_2 it was necessary to add 10 % argon. In addition to the silane dilution with the excitation gases it was possible to directly dilute silane with hydrogen before entering the process chamber.

The specimens were characterized by scanning electron microscopy (SEM) and by step profiling with a mechanical stylus to determine the deposition rate. The crystallinity was obtained from Raman measurements.

3. INFLUENCE OF THE EXCITATION GASES

Deposition Rate

Helium or argon dilution results in small deposition rates (circles and squares in Fig. 1). The values are comparable to data obtained by conventional PECVD. Extra-polating the data r_d becomes zero at $\Delta_X = [X]/([X]+[SiH_4]) = 0$. This is a surprising result considering that our specimens were deposited at 5 mTorr and 1000 W. However, since we limited the dilution range to high He or Ar concentrations we could not observe the monotonic increase of r_d as reported previously [2].

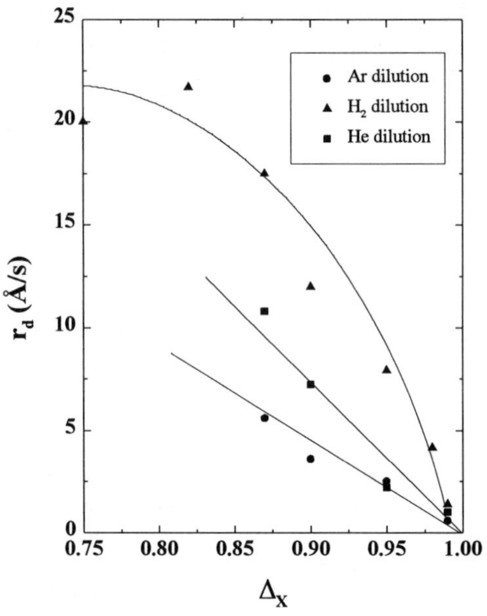

Upon adding hydrogen to the reactant gas the formation of silicon films is determined by a balance between deposition and etching. Although plasma deposition of silicon films is a complex process involving complicated radical-surface reactions [3] film deposition and etching are the main processes and can be expressed by the following reaction [4].

$$SiH_{X,Plasma} \Leftrightarrow Si_{Solid} + xH_{Plasma} \quad (1)$$

Fig.1: Deposition rate, r_d, vs dilution $\Delta_X = [X] / ([X] + [SiH_4])$ for argon (circles), helium (squares) and hydrogen (triangles).

Increasing the H dilution results in an increase of $SiH_{X,Plasma}$ and hence reduces the deposition rate. The deposition rate, r_d, for H dilution is represented by the solid triangles in Fig. 1. The dependency of r_d on H dilution is consistent with data reported previously, however, compared to conventional PECVD [2] the deposition rate is enhanced by a factor of 13. The equilibrium between deposition and etching occurs at a hydrogen dilution of $[H_2]/([H_2]+[SiH_4]) < 1$. Extrapolating the curve, r_d becomes zero at about $\Delta_H \approx 0.991$. The higher deposition rate and the lower H dilution necessary to reach chemical equilibrium (Fig. 1) are most likely due to difference in power and pressure. While our samples were deposited at 5 mTorr and 1000 W Tsai used 400 mTorr and 10 W in a conventional PECVD system.

Crystalline Fraction and Surface Roughness

The crystalline fraction of our samples was estimated from Raman backscattering data. Typical Raman data are plotted in Fig. 2 as a function of Ar dilution and at a constant dilution with different dilution gases. With increasing Ar dilution the TO mode at 520 cm^{-1} increases while the broad peak at 480 cm^{-1} decreases (Fig. 2(a)). The TO mode at 520 cm^{-1} and the broad gaussian peak at 480 cm^{-1} are characteristic of the crystalline and amorphous phase, respectively. Thus, with increasing Ar dilution the crystalline phase increases at the cost of the amorphous phase and decreases again. Fig. 2 (b) shows Raman spectra of samples that were prepared with a He, Ar, and H_2 dilution of 0.99, respectively. It is obvious that H dilution results in microcrystalline silicon films with the high crystallinity while the specimen prepared with He dilution reveals only the broad amorphous peak.

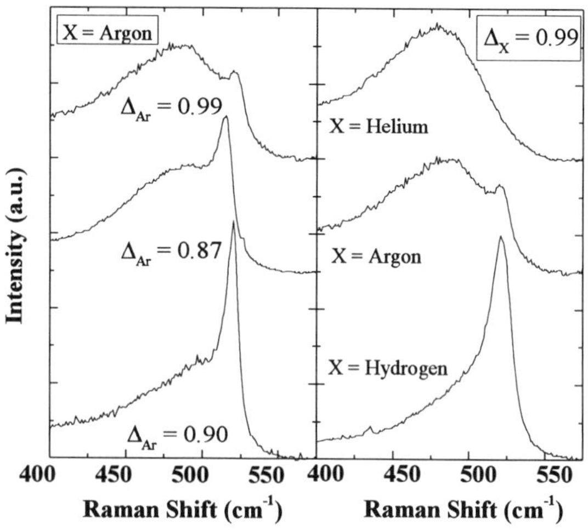

Fig.2: Raman spectra of samples prepared with (a) various argon dilutions and (b) at constant dilution (Δ_X = 0.99) of the indicated gases.

To obtain the crystalline fraction, X_C, of our samples the Raman spectra were deconvoluted into the amorphous and crystalline peaks. Then the crystalline fraction was derived from the integrated intensities of these peaks [5].

However, in order to obtain good fits to the data it was necessary to include an additional peak between 500 and 512 cm^{-1}. This additional peak is commonly attributed to grain boundaries [6] or to wurtzite structure [7, 8] and hence has to be taken into account when estimating the crystallinity of μc-Si films. To obtain a quantitative value for X_C the data has to be corrected since Raman measurements indicate the presence of an amorphous phase (≈ 20 %) even in μc-Si that is completely crystallized [7]. Consequently, the intensity of the amorphous phase was corrected by subtracting 25 % of the corresponding crystalline fraction.

The crystalline fraction as a function of the silane dilution with He, Ar, and H_2 is plotted in Fig. 3. Microcrystalline growth was achieved independent of the dilution gas. The lowest maximum crystalline fraction of 31 % was observed in He diluted samples at a dilution of 0.95. Argon dilution yielded a higher crystalline fraction over the entire dilution range compared to He with maximum values of X_C = 51 %. This is accompanied by a shift of the dilution towards smaller values for the maximum position.

On the other hand, hydrogen diluted samples are amorphous at low dilution. Significant crystallinity is observed for H dilution values larger than 0.90. X_C increases drastically and reaches values of up to 85 %. This maximum value for X_C occurs at a H_2 dilution of 0.98. With further increasing hydrogen dilution X_C decreases. Similar results were observed by Wang et al.

They found the largest grain size at a dilution of 0.98 [9]. Another really interesting aspect should be mentioned here. When increasing the hydrogen dilution by adding hydrogen directly to silane, instead of dilution via the excitation gas, the crystalline fraction of the films increases further to a new maximum ($X_{C,max}$ = 94%) and then decreases (crosses in Fig. 3). X_C first exceeds and then falls short of the value expected for the corresponding H_2 dilution when using hydrogen as excitation gas only. This behavior was reproduced for a second dilution range. Since H is used as excitation gas and premixed with the reactant gas the deposition pressure increases. In order to elucidate this effect a set of samples was prepared by keeping the H dilution at a fixed value of 0.95 and increasing the deposition pressure. In the inset of Fig. 3 X_C is plotted vs. the deposition pressure. With increasing pressure X_C increases to a maximum value and then decreases at $p > 15$ mTorr. This result suggests that the increase of the pressure causes the observed increase of X_C (crosses in Fig. 3).

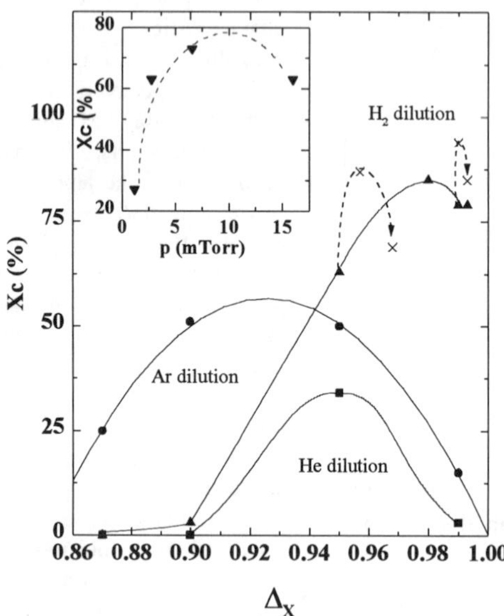

Fig.3: Crystalline fraction as a function of the dilution $\Delta_X = [X] / ([X] + [SiH_4])$. The circles, squares, and triangles were obtained on samples diluted with X = argon, helium, and hydrogen, respectively. The crosses represent films deposited with hydrogen also added to the reactant gas. The Inset shows the dependence of deposition rate and crystallinity on the pressure. The dilution is $\Delta_{H2} = 0.95$.

μc-Si films deposited with H dilution via the excitation gas have a rough surface. However, premixing SiH_4 with H_2 has a remarkable impact on the surface roughness as can be seen in the SEM micrographs in Fig. 4. On the other hand, helium seems to prevent a distinct roughness and samples prepared by argon dilution are all smooth.

4. DISCUSSION

The main processes that influence film growth are plasma processes and surface reactions. In the following discussion we are focusing primarily on the latter process since there is no direct link between the quality of the plasma and the properties of the resulting μc-Si samples.

In ECRCVD the most probable precursors are believed to be SiH_X (x= 0-2) [10]. These precursors have a higher sticking coefficient than SiH_3 radicals and are more reactive. The activation energy for surface diffusion of SiH_X (x= 0-2) is larger and hence, the diffusion length is shorter. However, for the growth of high quality material the SiH_3 precursor is considered to be important [11]. On the other hand, chemical reactions on the surface can still lead to the formation of SiH_3.

In case of noble gas diluted silane plasmas, the plasma processes have to be taken into account, since the amount of hydrogen coming from the dissociation of SiH_4 is neglectable at high dilution. The higher deposition rate for He dilution could be due to a higher production of SiH_X (x= 0-2) radicals which would be consistent with the lower crystal fraction found in these samples. A second possibility to account for the differences observed in Ar and He diluted samples could be the higher mass of Ar. Argon ions can deposit more energy on the growing surface than He ions either promoting crystalline growth or damaging the growing sample. This is consistent with the observed shift of the maximum of X_C to lower dilution (Fig. 3).

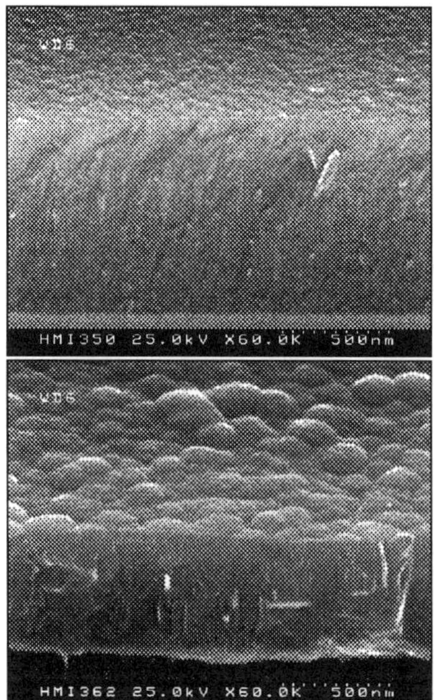

Fig.4: Cross-sectional scanning electron-microscopy (SEM) pictures showing two films deposited at similar Δ_H. When hydrogen is added to the reactant gas, the surface is smoother (top: rms < 50 nm), than using hydrogen as excitiation gas only (bottom: rms ≈ 100 nm).

In case of H dilution the most important processes for film growth occur on the growing sample surface. The increase in crystallinity X_C with increasing H_2 dilution is commonly attributed to the etching effect of hydrogen [2]. However, at $[H_2]/([H_2]+[SiH_4])>0.98$ the crystalline fraction decreases. A similar behavior was observed when measuring the crystalline fraction as a function of the deposition pressure. Moreover, the chemical equilibrium of hydrogen is an important factor in determining the structure of thin films deposited from SiH_4/H_2 plasmas. During growth the H distribution can be described by a hydrogen chemical-potential, μ_H [12]. Raising μ_H by adding hydrogen to the reactant gas causes a reduction of the disorder of amorphous silicon films and eventually leads to the growth of microcrystalline silicon. This could be due to a decrease of the H concentration in the resulting film and hence, a decrease of μ_H with increasing crystallinity. H concentrations were measured by SIMS and determined from FTIR measurements. For H dilution $\Delta_H > 0.95$ a constant H concentration of $\approx 2\times10^{21}$ cm^{-3} was measured indicating that μ_H did not decrease. Also ion bombardment cannot account for the decrease of X_C solely. With increasing pressure the average ion and electron energies decrease because of increasing impact probabilities. Hence, with increasing pressure the crystalline fraction of a μc-Si film should increase monotonically. This, however, is not the case (inset in Fig. 3). A variation in grain size between 1000nm and 500nm just leads to a change in the crystallinity of about 2 % [13]. So surface kinetics also can expected to be responsible here. Finally, the observed decrease in X_C must be due to hydrogen reactions on surface. The most prominent reaction is etching of weakly bound silane radicals. However, H is also known to break strong Si–Si bonds [14]. Under high H dilution of silane etching of strong bonds may be enhanced leading to vacancies, weak Si–Si bonds, Si–H, and Si dangling-bonds in the growing lattice of a Si grain (Fig. 5).

5. SUMMARY

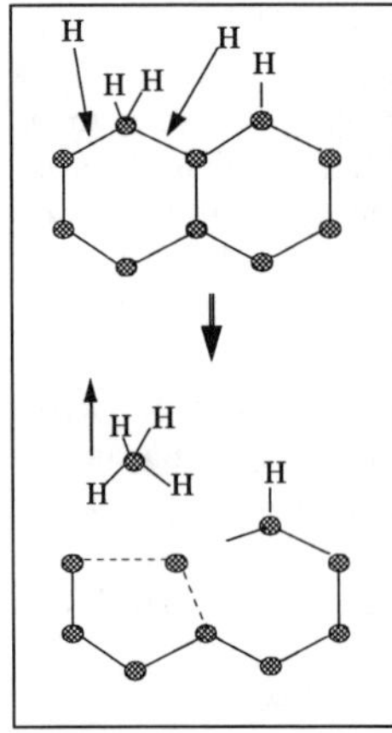

In summary, microcrystalline silicon films were prepared by remote ECRCVD technique using Ar, He, and H_2 as excitation gases. Maximum crystalline fractions of 34 %, 51 %, and 94 % were obtained for He, Ar, and hydrogen dilution, respectively. For hydrogen dilution the crystalline fraction revealed a maximum at Δ_H = 0.98. With increasing deposition pressure an increase of the crystalline fraction was observed. Premixing of hydrogen and silane results in samples with a smoother surface. The decrease of the crystalline fraction with increasing Δ_H cannot be explained with ion bombardment solely. We suggest that reactive H species transform already crystalline fractions of the growing film into an amorphous phase.

Fig. 5: Sketch of a part of a grain in a silicon lattice (top). Under high H dilution etching of strong bonds may be enhanced leading to an amorphous tissue (bottom).

ACKNOWLEDGMENTS

The authors are grateful to W. Pilz of the Humboldt University in Berlin for Raman measurements, and to M. Poschenrieder of the Hahn-Meitner-Institut for the measurement and interpretation of FTIR spectra. We also like to acknowledge technical support from J. Krause.

REFERENCES

1. A. Matsuda, J. Non-Cryst. Solids. **59&60**, 767 (1983).
2. C. C. Tsai, in Amorphous Silicon and Related Materials, ed. H. Fritzsche (Advances in Disordered Semiconductors 1, Chicago, 1988), 123-147.
3. A. Gallagher, in Material Issues in Amorphous-Semiconductor Technology, edited by D. Adler, Y. Hamakawa, and A. Madan (Mater. Res. Soc. Proc. 70, Pittsburgh, PA 1986), 3.
4. W. E. Spear, and P. G. LeComber, in The Physics of Hydrogenated Silicon I, edited by J. D. Joannopoulos, and G. Lucovsky, New York, 1984, 63.
5. E. Bustarret, M.A.Hachida, Appl. Phys. Lett. **40**, 1675 (1988).
6. H. Xia, Y. L. He, L. C. Wang, W. Zhang, X. N. Liu, X. K. Zhang, D. Feng, J. Appl. Phys. **78**, 6705 (1995).
7. P. Hapke, Dissertation, 15-20 (1995).
8. R. Kobliska, S. Solin, Phys. Rev. B. **8**, 3799 (1973).
9. K. C. Wang, H. L.Hwang, P. T. Leong, T. R. Yew, J. Appl. Phys. 77, 6542 (1995).
10. M. Zhang, Y. Nakayama, Solid State Phenomena **44-46**, 135-180 (1995).
11. D. Das, Solid State Phenomena **44-46**, 227-258 (1995).
12. R.A. Street, Phys. Rev. B **43**, 2454 (1991).
13. N. H. Nickel, unpublished
14. N. H. Nickel, W. B. Jackson, Phys. Rev. B **51**, 4872 (1995).

POST-DEPOSITION ANNEALING AND HYDROGENATION OF HOT-WIRE AMORPHOUS AND MICROCRYSTALLINE SILICON FILMS

J.P. CONDE[a)], P. BROGUEIRA[b)], V. CHU[c)]

[a)] Department of Materials Engineering, Instituto Superior Técnico, Av. Rovisco Pais, 1096 Lisboa Codex, Portugal, pcfconde@alfa.ist.utl.pt

[b)] Department of Physics, Instituto Superior Técnico, Av. Rovisco Pais, 1096 Lisboa Codex, Portugal

[c)] Instituto de Engenharia de Sistemas e Computadores, Rua Alves Redol 9, 1000 Lisboa, Portugal

ABSTRACT

Amorphous and microcrystalline silicon films deposited by hot-wire chemical vapor deposition were submitted to thermal annealing and to RF and electron-cyclotron resonance (ECR) hydrogen plasmas. Although the transport properties of the films did not change after these post-deposition treatments, the power density of a Ar^+ laser required to crystallize the amorphous silicon films was significantly lowered by the exposure of the films to a hydrogen plasma. This decrease was dependent on the type of hydrogen plasma used, being the strongest for an ECR plasma with the substrate held at a negative bias, followed by an ECR hydrogen plasma with the substrate electrode grounded, and finally by an RF hydrogen plasma.

INTRODUCTION

Hot-wire chemical vapor deposition (HW-CVD) has gained recent interest as a technique which can be used to deposit high-quality amorphous silicon (a-Si:H) and microcrystalline silicon (μc-Si:H) at high growth rates. a-Si:H and μc-Si:H have been deposited by HW-CVD from hydrogen and silane decomposed on the surface of a heated tungsten wire.[1-4] High deposition rates (r_d >10 Å/s) have been obtained for both a-Si:H and μc-Si:H.[4-7] In addition, the use of high substrate temperatures (T_{sub}~400-500°C) resulted in a-Si:H with improved stability against light-induced degradation.[8] The use of high substrate temperatures (T_{sub}>300°C) with high filament temperatures (T_{fil}~2500°C) in the absence of hydrogen dilution resulted in films with high growth rates (r_d>15 Å/s), characteristic of a-Si:H deposited by HW, and high conductivities characteristic of μc-Si:H.[9] The objective of this paper is to study the effects of low-temperature thermal annealing (T_{ann}<400°C) and post-deposition hydrogen plasma treatments on these films, as well as on HW a-Si:H and μc-Si:H deposited at 200 °C with varying hydrogen dilutions.

EXPERIMENT

Sample Preparation

A group of six samples was prepared by hot-wire deposition (HW) using a single tungsten filament (7 cm long and 0.5 mm diameter) placed approximately 3 cm from the substrate. Table 1 shows the deposition conditions used, the growth rates obtained and the structure of the films determined by Raman spectroscopy. One a-Si:H and one μc-Si:H sample prepared by capacitively coupled RF glow discharge are also included as reference. Table 2 lists some of the properties of these samples in the as-deposited state and after 6 months of storage in the dark (aged state). Samples S467, S488 and S489 show typical a-Si:H properties, while S490 shows

Mat. Res. Soc. Symp. Proc. Vol. 452 © 1997 Materials Research Society

typical μc-Si:H behavior. Samples S492 and S493 are unusual in that they combine the very large growth rates (>15 Å/s) typical of a-Si:H deposited by HW as well as an a-Si:H-like Raman spectrum, with the high dark conductivities and low activation energies characteristic of μc-Si:H.

Table 1. Deposition conditions and growth rates

Name	Dep. type	F_{SiH4} (sccm)	F_{H2} (sccm)	$\%H_2$	T_{sub} (°C)	T_{fil} (°C)	P_{rf} (W)	p_{dep} (Torr)	r_d ($Ås^{-1}$)	Raman
S492	HW	5	0	0	300	2500	--	0.02	15.5	a
S493	HW	5	0	0	400	2500	--	0.02	25.0	a
S467	HW	5	0	0	220	2500	--	0.02	27.7	a
S488	HW	4	6	60	220	2000	--	0.02	22.2	a
S489	HW	4	16	80	220	2500	--	0.02	13.3	a
S490	HW	5	45	90	220	2500	--	0.02	2.67	μc
MC217	RF	10	0	0	200	--	5	.1	1.25	a
S259	RF	1	50	98	200	--	57	.1	0.53	μc

Post-deposition Annealing and Hydrogenation

The six HW-deposited were subjected to the following post-deposition treatments: (i) thermal annealing at 300 °C in a vacuum chamber in an atmosphere of 0.1 Torr of Ar; (ii) thermal annealing at 400 °C in a vacuum chamber in an atmosphere of 0.1 Torr of Ar; (iii) RF glow discharge hydrogen plasma using a power of 10 W, a pressure of 50 mTorr, a substrate temperature of 300°C and a flux of 50 sccm of hydrogen; (iv) ECR (electron cyclotron resonance) hydrogen plasma using a power of 150 W, a pressure of 10 mTorr, a substrate temperature of 300°C and a flux of 28.5 sccm of hydrogen. Each sample was divided into pieces and were independently submitted to these different treatments. The RF electrodes were placed 3 cm apart and had an area of ~100 cm^2. The ECR source was an Astex Compact ECR Source and the substrate was placed 6 cm downstream from the source. In all the above cases, the substrate was grounded. An additional experiment was performed in which the substrate exposed to an ECR hydrogen plasma was biased with -200 V DC.

Table 2. Dark conductivity and activation energy

Name	as deposited		aged		RF H_2		ECR H_2		400 °C anneal	
	σ_d $\Omega^{-1}cm^{-1}$	E_A eV	σ_d $\Omega^{-1}cm^{-1}$	E_A eV	σ_d $\Omega^{-1}cm^{-1}$	E_A eV	σ_d $\Omega^{-1}cm^{-1}$	E_A eV	σ_d $\Omega^{-1}cm^{-1}$	E_A eV
S492	$3.1x10^{-5}$	0.33	$5.2x10^{-5}$	0.31	$7.4x10^{-5}$	0.49	$1.5x10^{-4}$	0.23	$3.7x10^{-6}$	0.45
S493	$3.7x10^{-4}$	0.23	$3.9x10^{-4}$	0.22	$1.5x10^{-3}$	0.14	$3.0x10^{-4}$	0.18	$1.2x10^{-4}$	0.22
S467	$6.5x10^{-11}$	0.89	$8.8x10^{-11}$	0.90	$9.9x10^{-11}$	0.84	$4.3x10^{-10}$	0.84	$1.6x10^{-8}$	0.53
S488	$3.2x10^{-10}$	0.85	$5.2x10^{-10}$	0.85	$1.5x10^{-8}$	0.73	$4.7x10^{-10}$	0.73	$2.8x10^{-10}$	0.84
S489	$3.1x10^{-10}$	0.85	$2.2x10^{-10}$	0.79	$4.7x10^{-10}$	0.84	$4.0x10^{-10}$	0.73	$2.3x10^{-11}$	0.87
S490	$3.4x10^{-5}$	0.47	$2.9x10^{-2}$	0.10	$5.2x10^{-5}$	0.08	$6.8x10^{-3}$	0.10	$3.7x10^{-3}$	0.15
MC217	$1.9x10^{-10}$	0.89	$9.4x10^{-10}$	0.81	--	--	--	--	--	--
S490	$1.9x10^{-6}$	0.48	$4.4x10^{-5}$	0.34	--	--	--	--	--	--

Film characterization

The samples were characterized in their as-deposited, aged, annealed and post-hydrogenated states. The dark conductivity (σ_d) and photoconductivity (σ_{ph}) were measured on parallel contacts with 6 mm length and 1 mm spacing. The activation energy of σ_d ,E_A, was extracted from the temperature dependence of σ_d between room-temperature and 100 °C. σ_{ph} was measured using uniformly absorbed light and is quoted for a generation rate (G) of 10^{21} $cm^{-3}s^{-1}$. The gamma factor of the photoconductivity was extracted from the dependence of the σ_{ph} on G ($\sigma_{ph} \propto G^{\gamma}$) for $10^{15}<G<10^{21}$ $cm^{-3}s^{-1}$. In selected samples the subgap absorption was measured using the constant photocurrent method (CPM).

Raman microprobe spectroscopy was used to determine whether the structure of the film was amorphous or microcrystalline and, in the latter case, to estimate the grain size and crystal fraction.[10-12] The 514.5 nm line of an Ar^+ laser was focused on the sample through a microscope. By increasing the power of the laser it was observed that, beyond a certain

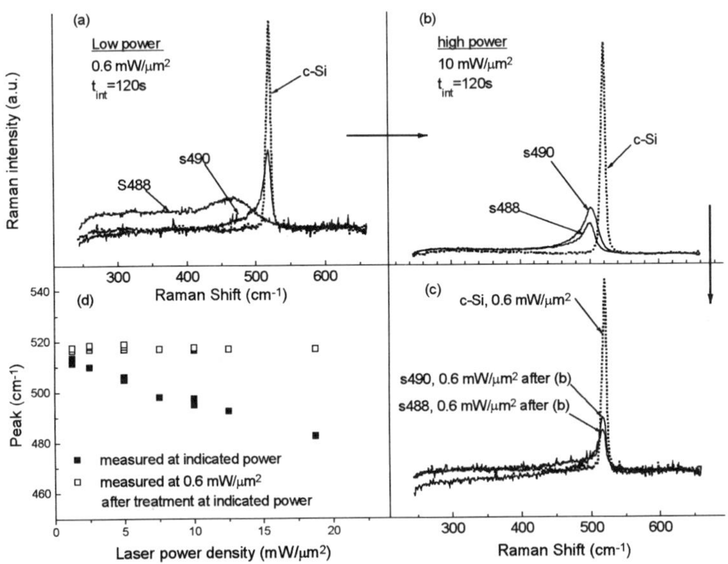

Figure 1. (a) Raman spectra of a-Si:H sample (S488) and μc-Si:H sample (S490) in the as-deposited state and of c-Si. (b) Raman spectra of the same samples measured using 10 mW/μm^2. (c) Raman spectra remeasured in the same spot using lower laser power (~0.6 mW/μm^2). (d) Raman peak position of μc-Si:H as a function of laser power (solid symbols). The open symbols indicate the position of the peak when the measurement was taken afterwards using low laser power (<0.6 mW/μm^2) in the same spot.

threshold, it was possible to crystallize the a-Si:H films during the integration time of the measurement. The spot size was 1 μm^2 and the maximum power used was ~10^6 W/cm^2. This observation suggests an interesting application of a Raman microprobe spectrometer. By keeping a constant integration time (120s) and ensuring that the film thickness is larger than the penetration depth of the light (in this case, approximately 1000 Å) one can measure the threshold power necessary to crystallize an amorphous sample. A comparison of this threshold power, either between samples or on the same sample after different treatments, can give an estimate of the relative temperature of crystallization. Fig. 1 shows that the laser used in the Raman measurement can induce crystallization of the amorphous film (Fig. 1(b) and (c)). This figure also indicates that, due to localized heating during the recrystallization step (which coincides with the measurement at high laser power), the frequency of the optical phonon shifts to a lower wavenumber (Fig. 1(b)).

This shift is approximately proportional to the power used (Fig. 1(d)). This shift does not occur for c-Si at the power levels used because of its higher thermal conductivity. In order to obtain power-independent spectra, it is necessary to remeasure the crystallized film on the same spot at a low power level (<0.6 mW/μm^2), at which no significant heating effects occur.

RESULTS AND DISCUSSION

Sample Aging

Table 2 compares the dark conductivity and activation energy of the samples in their aged state with those in the as-deposited state. Only the μc-Si:H samples S490 (HW) and S259 (RF) showed a major change: a 3 and 1.5 orders of magnitude increase in σ_d , respectively. This effect is generally attributed to the diffusion of oxygen through pores in the structure or along the grain boundaries, followed by unintentional doping of the crystallites.[7] These results suggest that the HW μc-Si:H sample is more permeable to oxygen than the RF μc-Si:H sample. The typical a-Si:H samples, both HW (S467, S488 and S489) and RF (MC217) a-Si:H samples showed no aging effects. It is worth mentioning that the samples S492 and S493, which show σ_d characteristic of μc-Si:H but which were deposited at high growth rates (>15 Å/s) and are Raman-amorphous, did not show any aging effects in σ_d (to within the error of the measurement) in marked contrast with the behavior of the microcrystalline sample S490.

The Effect of Thermal Annealing and Plasma Hydrogenation on the Transport Properties

No significant change in σ_d, its activation energy, σ_{ph}, the γ factor of σ_{ph}, and the subgap absorption measured by CPM, was observed after as long as 15 h of thermal annealing at 300 °C, 7 h of thermal annealing at 400 °C, 7h of exposure to RF hydrogen plasma at 300°C, or 7 h of exposure to ECR hydrogen plasma at 300°C. Table 2 illustrates this by showing values of σ_d and its activation energy after these post-deposition treatments. The Raman spectra of these samples did not show significant changes after aging, thermal annealing or exposure to an RF or ECR hydrogen plasma.

The Effect of Plasma Hydrogenation on the Laser Recrystallization Properties

Although it does not significantly change the transport properties of the HW a-Si:H films, plasma hydrogenation treatments have an important effect on the temperature needed to crystallize these films (as revealed by the power of the laser necessary to induce a crystalline-like phonon peak). Fig. 2 summarizes the results obtained. In the as-deposited state, most of the

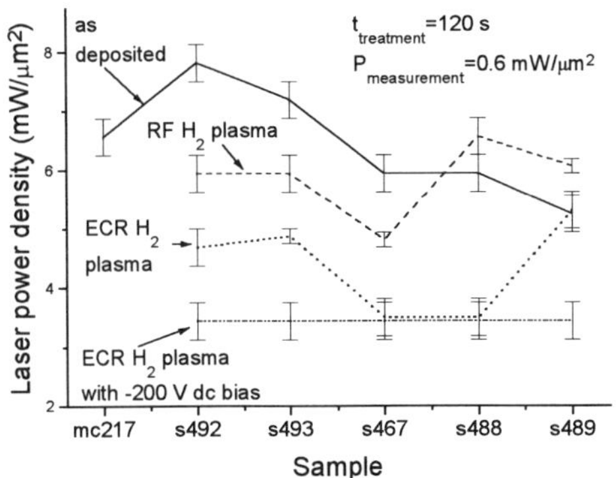

FIG. 2. The effect of the different types of hydrogen plasma treatments on the threshold power of the laser needed to induce crystallization. The bottom of each vertical bar indicates the highest power measured for which the sample showed an amorphous Raman spectrum. The top of each vertical bar indicates the lowest power measured for which the sample showed a microcrystalline optical phonon peak in the Raman spectrum

samples require a power of 5.6-6.25 mW/μm^2 to crystallize. The notable exceptions are S492 and S493, which require 6.9-7.5 mW/μm^2. Upon hydrogenation with an RF plasma, a significant drop in the threshold power for crystallization was observed for the samples deposited with no hydrogen dilution (S492, S493 and S467). When the samples are submitted to an ECR hydrogen plasma, a strong decrease of the threshold power for crystallization is observed for all the samples except the one prepared with 80% hydrogen dilution (S489). The decrease is even more dramatic when the samples are submitted to an ECR hydrogen plasma with a negative bias applied to the substrate. In this case, all the samples require only ~3.75 mW/μm^2 for crystallization. Fig. 3 highlights the evolution of the Raman spectra for the different hydrogen plasma treatments in the case of one sample (S492). These results suggest that the plasma density (and, as a consequence, the activated hydrogen concentration) is the dominant factor in the decrease of crystallization temperature, since ECR plasmas have plasma densities 2-3 orders of magnitude higher than capacitively-coupled RF discharges. Ion bombardment plays a role in

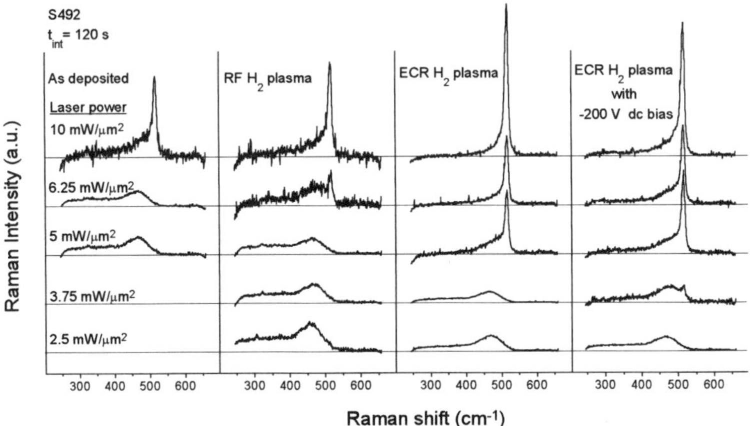

FIG. 3. Raman spectra for sample S492 after several different hydrogen plasma treatments. For each treatment, Raman spectra are shown for different laser powers. After the appearance of the crystalline peak, the spectra are taken at low power following exposure at the stated power.

decreasing the crystallization temperature, since the application of a negative bias increases the energy of the impinging positive ions.

The position of the Raman peak is related to the crystallite size.[12] For all samples shown in fig. 2, the peak was located at ~517 cm^{-1}, indicating a grain size of approximately 50 Å. Increasing the power of the laser did not result in an increase in the grain size, although it did result in the crystalline fraction increasing to near unity.

CONCLUSIONS

A hydrogen plasma treatment at low temperature (T_{sub}=300 °C) induces structural changes in HW a-Si:H, significantly lowering the laser power required to induce crystallization, strongly suggesting a parallel decrease in the temperature necessary for thermal crystallization of HW a-Si:H. These structural changes did not change the transport properties. Of the plasma treatments studied in this work, an ECR hydrogen plasma coupled with a negative d.c. bias on the substrate produced the most profound effect in lowering the thermal crystallization temperature. For the laser power levels used, although the crystalline fraction increased with the power, reaching values close to unity, there was no observable change in the crystallite grain size.

ACKNOWLEDGEMENTS

The authors gratefully acknowledge H. Silva for help with the measurements, and R. Almeida for the use of the Raman spectrometer. This work was supported by JNICT through Financiamentos Plurianuais to ICEMS and INESC and through the project PRAXIS/3/3.1/TIT/11/94.

REFERENCES

1. H. Wiesmann, A.K. Ghosh, T. McMahon, and M. Strongin, J. Appl. Phys. **50**, 3752 (1979).
2. H. Matsumura, Jpn. J. Appl. Phys. **25**, L949 (1986); J. Appl. Phys. **65**, 4396 (1989); Jpn. J. Appl. Phys. **30**, L1522 (1991).
3. J. Doyle, R. Robertson, G.H. Lin, M.Z. He, and A. Gallagher, J. Appl. Phys. **64**, 3215 (1988).
4. A.H. Mahan, J. Carapella, B.P. Nelson, R.S. Crandall, and I. Balberg, J. Appl. Phys. **69**, 6728 (1991).
5. R. Zedlitz, F. Kessler, and M. Heintze, J. Non-Cryst. Solids **164-166**, 83 (1993).
6. J. Cifre, J. Bertomeu, J. Puigdollers, M.C. Polo, J. Andreu, and A. Lloret, Appl. Phys. A **59**, 645 (1994).
7. A.R. Middya, A. Lloret, J. Perrin, J. Huc, J.L. Moncel, J.Y. Parey, and G. Rose, Mater. Res. Soc. Symp. Proc. **377**, 119 (1995).
8. A.H. Mahan, M. Vanacek, AIP Conf. Proc. **234**, 195 (1991).
9. J.P. Conde, P. Brogueira, R. Castanha, V. Chu, Mat. Res. Soc. Symp. Proc. **420**, in press (1996).
10. T. Kaneko, M. Wagaki, K. Onisawa, and T. Minemura, Appl. Phys. Lett. **64**, 1865 (1994).
11. S. Veprek, F.A. Sarott, and Z. Iqbal, Phys. Rev. B**36**, 3344 (1987).
12. Y. He, C. Yin, G. Cheng, L. Wang, X. Liu, and G.Y. Hu, J. Appl. Phys. **75**, 797 (1994).
13. D.L. Smith, Thin-Film Deposition: Principles and Practice, McGraw-Hill, New York, 1995, pp. 453-555.

STRUCTURAL ANALYSIS OF NANOCRYSTALS EMBEDDED IN AMORPHOUS Si FILMS

B. GARRIDO *, A. ACHIQ **, J. MACÍA *, J.R. MORANTE *, A. PÉREZ-RODRÍGUEZ *, P. RUTERANA **, R. RIZK **

* EME, Departament de Física Aplicada i Electrònica, Universitat de Barcelona, Diagonal 645-647, 08028 Barcelona, Spain, blas@iris1.fae.ub.es

** LERMAT-ISMRA, URA-CNRS 1317, 6, Boulevard du Maréchal Juin, 14050 CEDEX, Caen, France

ABSTRACT

Silicon nanocrystals embedded in amorphous silicon films have been obtained by RF sputtering using two approaches : i) with a pure H_2 plasma at different substrate temperatures (between 50° and 250°C) and ii) with the substrate at 250°C, from a mixture of Ar and a variable partial pressure of H_2, and annealing at 750°C. In the first case (i) Raman spectra show strong phonon confinement effects and an increase in the crystalline fraction as temperature increases. In the second case (ii), a higher compressive stress is estimated and a close correlation is found between the optical band gap E_0 and the hydrogen content in the basic amorphous layer a-Si:H (i.e. before annealing). These results correlate with the TEM data, which corroborate an enhancement of the crystallinity of the layers and an increase in the particle size when the substrate temperature increases. Furthermore, the crystalline fraction remains nearly constant with the H_2 partial pressure. The role of hydrogen and voids in the structure of the films is presented and discussed.

INTRODUCTION

Nanocrystalline silicon films (nc-Si) are of great interest due to a wide range of potencial applications. When crystallite size decreases below 10 nm and becomes comparable with the diameter of the bulk exciton (4-5 nm), quantum size effects occur leading to a widening of the band gap and a concomitant increase of the probability for optical transitions and these new effects can be useful for optoelectronic applications [1,2]. Moreover, hydrogenated nc-Si films show much higher conductivities than amorphous silicon (a-Si:H) [2]. This makes suitable the application of nc-Si for thin film devices. Nc-Si films can be obtained either directly, i.e. during deposition, or through recrystallization of a-Si:H. While the former approach is achieved by the use of a pure hydrogen plasma at relatively low temperatures that prevent softening of the glass substrate and also offers a saving in post-processing, the latter consists in a subsequent annealing of layers deposited with a mixture of $Ar+H_2$ plasma. In all the techniques used so far hydrogen is a major component of the gas atmosphere. We present in this paper the results of the analyses of nanocrystallized Si layers grown by sputtering using both approaches, as a function of substrate temperature for the direct deposition (pure hydrogen plasma) or as a function of hydrogen partial pressure for the thermally annealed samples (hydrogen-argon mixtures). We present the results of the structure with emphasis in the grain size and crystallinity and their relationship with the substrate temperature and composition of the atmosphere during deposition, and in the light of previous results about measurements of optical band gap and hydrogen concentration [2].

Mat. Res. Soc. Symp. Proc. Vol. 452

EXPERIMENT

The films investigated in this study were deposited by sputtering on glass (Corning 7059) and single-crystal silicon substrates. Two set of samples were prepared following the approach of direct and indirect growing of silicon nanocrystallites. The first set of samples (labelled hereafter as A) were deposited in pure hydrogen gas (99.9999%) at a pressure of 0.4 Torr and by varying the substrate temperature (T_S) between 50 °C and 250 °C. The thicknesses were compressed between 0.5 and 1.0 μm. The second set (labelled hereafter as B) were prepared by fixing the substrate temperature at 250 °C and by varying the hydrogen content (r_H) of the mixed Ar-H_2 atmosphere from 0% to 75% of hydrogen; an annealing step for thermal crystallization was performed afterwards at 720 °C under N_2 atmosphere and with a duration of 90 min. The thicknesses were between 0.9 and 1.4 μm in this case.

The size of the crystallites, stress and the crystalline fraction of the layers were determined from the analysis of the Raman spectra. The measurements were performed in a Jobin Yvon T64000 instrument by exciting the sample with an Ar laser working at a wavelength of 514.5 nm. The sample was mounted in the sample holder of a microscope and the light was collected in the backscattering configuration. The diameter of the spot on the sample was of about 0.7 μm and the power on the sample was of about 1 mW. X-Ray Diffraction (XRD) experiments and Transmission Electron Microscopy (TEM) observations were also performed as an independent way of measuring crystallite size and preferential orientations. Furthermore, the results are discussed in the light of Optical Transmittance (OT) measurements of the samples from 0.4 to 4.1 eV, performed to obtain the refractive index and the absorption coefficient and by applying a suitable model we determined the fraction of voids relative to the total volume. The hydrogen content of the layers was measured from the InfraRed (IR) absorption spectra recorded for the films deposited on silicon substrates.

RESULTS AND DISCUSSION

The Raman, XRD and TEM measurements of the samples investigated have shown evidence of variable proportion of crystalline fraction and/or crystalline size with T_s (set A) or r_H (set B). The Raman experimental spectra of all the samples presented both silicon amorphous and silicon crystalline bands. The Raman spectrum of the amorphous silicon is characterized by four main bands which were fitted with gaussian shapes centered at frequencies given approximately by 480 cm^{-1}, 380 cm^{-1}, 310 cm^{-1} and 150 cm^{-1} [3]. The spectrum of bulk crystalline silicon is characterized by a lorentzian shaped band centered at 520 cm^{-1} and having a width at half maximum of approximately 3 cm^{-1}. Nevertheless, we found the crystalline band of the Raman spectra of our samples to be highly asymmetric in the low frequency side. This is a clear indication of important confinement effects due to the small size of the crystallites as has been reported before [3,4]. Therefore, we fitted the crystalline bands of the Raman spectra with the Spatial Correlation Model (SCM), in which the spatial part of the wave function of the vibrational modes is multiplied by a weighting function that localizes the mode inside a finite space [3,5]. Spherical crystallite shapes were assumed as an approximation for all the weighting functions. The outputs of the fitting procedure were the correlation length of the phonons in the crystallites (L) and the remaining shift of the band relative to the bulk silicon reference was assumed to correlate linearly with the stress (Σ) in the layers as given by the factor Σ=-0.31 GPa/cm^{-1} [5]. We have to remark that confinement can be due not only to the spatial limitation of the grain boundary, but also to a high density of defects which can give rise to strong phonon confinement. Fig. 1 shows some Raman spectra for samples with different crystalline fractions. We also calculated the crystalline fraction (ρ_C) of the layers from the Raman experiments by integrating the intensities of the crystalline (I_C) and amorphous (I_A) bands, as determined from the fittings. The ratio of the Raman efficiencies of crystalline and amorphous silicon (Y) is a function of the correlation length and is given by Y=0.1+exp(-L/250) [3]. Hence, the crystalline fraction is calculated from the expression $\rho_C=I_C/(I_C+YI_A)$.

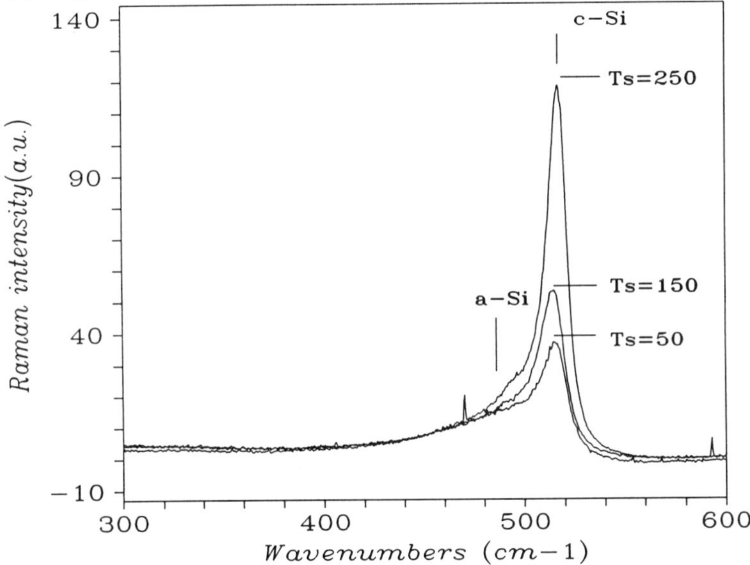

Fig. 1. Raman spectra of the nanocrystalline films for set A, showing the increasing amount of the crystalline component when increasing T_S.

The results of L and ρ_C for the set of samples A are plotted in Fig. 2. Both correlation length and crystalline fraction increase with T_s. L is increasing from 50 Å for T_S=50 °C to 62 Å for T_S=250 °C and there is also a clear correlation between the substrate temperature and ρ_C, being for T_S=250 °C the crystalline fraction as high as 74%. Negligible stresses were found for all these samples.

For the set B of samples, the crystalline fraction seems to be independent on r_H, being all the values between 70-75%. Correlation length for set B is 66 Å for r_H=0 and increases slightly for higher r_H given a value between 70-75 Å which seems to be almost independent of the r_H concentration of the plasma. Stresses are compressive for all the samples of set B, showing a non-monotonic behaviour with r_H with a minimum of -681.3 MPa for r_H=20% and near -1 GPa for lower and higher r_H concentrations.

From the OT measurements we deduced the optical band gap E_0, according to the Tauc plots, $(\alpha h\nu)^{1/2}$ versus $(h\nu-E_0)$ where $h\nu$ is the photon energy. For the set of samples A the variation of E_0 versus T_S was previously reported in [2] and found to decrease from a value as high as 2.40 eV for T_S=50 °C to approximately 1.95 eV for T_S=250 °C. The decrease of the grain size for low T_S in these samples is consistent with the increase of E_0 due to quantum size effects as reported by B. Delly et al. [6]. The growth of crystallites is described as resulting from complex interactions of silicon hydrides (SiH_x) and the growing surface layer, through continual breaking or etching of strained near surface Si-Si bonds, leading to an enhanced surface mobility of the deposited species, and hence allowing the relaxation of the network towards the crystalline state [7,8]. This is accompanied by desorption of hydrogen by forming volatile species such as H_2, SiH_4 and Si_2H_6 [1, 8]. A concomitant reduction of hydrogen content with increasing T_S was revealed from the IR experiments as desorption of molecular hydrogen and/or silicon hydrides is expected when the substrate temperature increases. The quasi-absence of the Si-H IR stretching mode at 2000 cm^{-1} attributed to isolated bonds in a-Si and the presence of an strong absorption at 2090 cm^{-1} usually assigned to Si-H_2 and Si-H bonds at the crystalline surface suggest that formation of crystallites is accomplished as due mainly to the high surface diffusivities of silicon hydrides. The almost negligible stresses found from Raman in these samples can be explained by the presence of an important fraction of voids (f_V) as determined from the low refractive indexes measured. For this set of samples (A), large f_V values were obtained, ranging from 37.5% for the lowest T_S=50 °C and decreasing down to 28.5% for the highest T_S=250 °C substrate temperature.

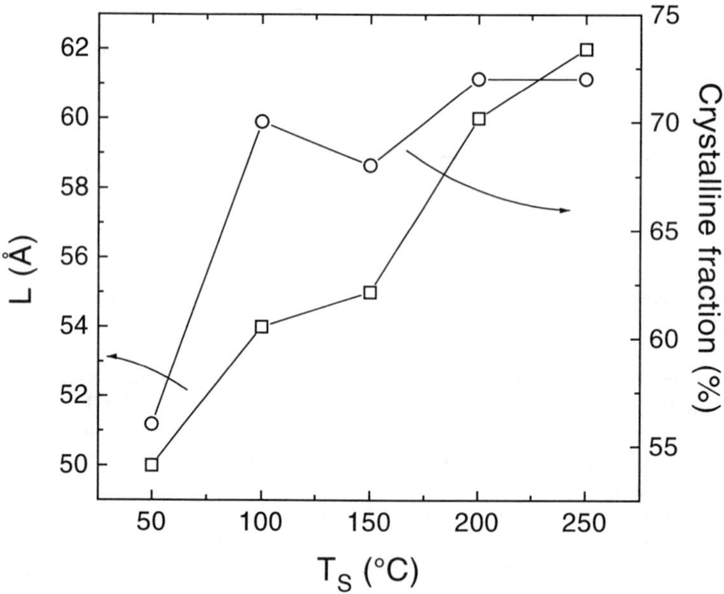

Fig. 2. Correlation length (L) and crystalline fraction (ρ_C) as a function of T_S (set A).

For the set of samples B, L was between 70-75 Å which corresponds to a band gap of 1.6-1.7 eV [6]. Hence, the experimental results showed that E_0 dropped below 1.8 eV in the whole range of r_H concentrations, showing a non-monotonic behaviour which shows a striking resemblance to that of hydrogen content (c_H) of the pre-annealed samples (Fig. 3). The variation of E_0 could be originated from the variation of the extent of the band tails induced by the disorder at the grain boundaries. The higher c_H corresponds to the larger E_0 for $r_H=20\%$. This means that hydrogen contributes to the relaxation of the structure at the growth and during thermal annealing and then to the narrowing of the band tail width and consequently to the widening of the band gap E_0. This explanation is supported by the minimum stress noticed for the higher c_H that corresponds to $r_H=20\%$. Compressive stress can be understood within a model of growth with a mechanism of competitiveness between neighbouring grains and maybe there is also a competition with other processes that makes the stress to vary non-monotonically. On the other hand, the fraction of voids (f_V) is much more lower for the set B, decreasing its value from 19.4% for $r_H=0\%$ to 5.9% for $r_H=75\%$ The behaviour of f_V for the set B is a clear indication of the increasing compactness of the samples when more hydrogen is incorporated in the basic material even though the hydrogen is desorbed after annealing.

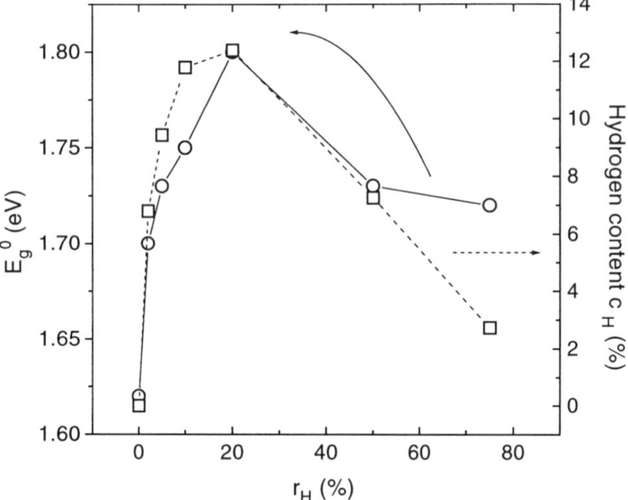

Fig. 3. Optical band gap (E_0) and hydrogen concentration (c_H) as a function of r_H.

The TEM images showed a clear evidence that a large fraction of silicon is crystallized in both sets of samples. For the samples of set A grown at the lowest temperatures ($T_S=50$ °C and $T_S=100$ °C) we have noticed the growth of an amorphous buffer layer (130 nm thick) before the crystallization takes place (cone-shaped). The existence of this amorphous layer was reported previously [2,9] and explained by the limited nucleation rate at low temperatures. The crystalline phase is characterized by a distribution of nanocrystalline grains with predominance of 50 Å particles for $T_S=50$ °C and larger particles (near 100 Å) for higher temperatures. The set B also showed the existence of an amorphous buffer layer but only for the unhydrogenated layer, i.e. $r_H=0\%$. However, it must be admitted that Raman results are only indicative of the superficial region of the layers due to the limited penetration depth of the light in silicon. Particle size measured by TEM is found to agree qualitatively with the values obtained from Raman. These features suggest the important role of hydrogen in the enhancement of the nucleation rate an growth of the crystallites.

Further independent data concerning the size of the silicon nanocrystals was obtained from the XRD experiments. Moreover, the intensities of the diffraction lines relative to a silicon powder oriented at random can give also information about the preferential orientations in the growth of the nanocrystals (texture). The average size of the crystallites was determined from the width of the diffraction peaks by a standard procedure and the results obtained were systematically larger by a 30-40% -sizes between 80-120 Å- compared to that obtained from the correlation length in the Raman measurements (65-75 Å). In fact, the eventual existence of extended defects within the grains (microtwins, dislocations,...) can confine the phonons and then give results of correlation lenght systematically smaller than grain size. The analysis of the relative XRD intensities showed that the crystalline particles are far from being spherical, specially for samples with a high crystalline fraction where there seems to be a preferential orientation for high <hkl> indices.

CONCLUSIONS

We have shown that control of the grain size and hence of the optical band gap is possible by varying the substrate temperature for the case of direct growth of the nanocristalline layer. However, hydrogen incorporation was found to induce void formation and this allows relaxation though not significant stresses could be measured. A more compact tissue and almost constant crystalline fraction and grain size was obtained for the amorphous layers deposited in a mixture of H_2-Ar and crystallized after thermal annealing. To some extent, optical band gap can be monitored by the hydrogen content in the pre-annealed material, which correlates well with the softening of the stress at the boundaries and hence the decrease of the density of localized states at the band tails.

REFERENCES

1. S. Veprek in Polycrystalline Semiconductors IV, edited by S. Pizzini, H.P. Strunk, J.H. Werner (Proceedings of the 4th International Conference, Scitec Publications, vol. 51-52, 1996), p. 225-235.

2. A. Achiq, R. Rizk, R. Madelon, F. Gourbilleau, P. Voivenel, Thin Solid Films (In press).

3. E. Bustarret, M.A. Hachicha and M. Brunel, Appl. Phys. Lett., **52**, p. 1676 (1988).

4. Y. He, C. Yin, G. Chen, L. Wan, X. Liu and G.Y. Hu, J. Appl. Phys., **75**, p. 797 (1994).

5. J. Macía, Doctoral thesis, Publications of the University of Barcelona (1996).

6. B. Delly and E.F Steigmeier, Phys. Rev. B, **47**, p. 1397 (1993).

7. N.V. Nguyen, I. An, R.W. Collins, Y. Lu, M. Wakagi and C.R. Wronski, Appl. Phys. Lett., **65**, p. 3335 (1994).

8. T. Akasaka and I. Shimizu, Appl. Phys. Lett., **66**, p. 3441 (1995).

9. Y.H. Yang and J.R. Abelson, Appl. Phys. Lett., **67**, p. 3623 (1995).

ROLE OF SiH_2 IN ^1H NMR OF μc-Si:H DEPOSITED WITH DIFFERENT PLASMA EXCITATION FREQUENCIES AND SILANE CONCENTRATIONS

P. HARI*, P.C. TAYLOR*, AND F. FINGER**
*Department of Physics, University of Utah, Salt Lake City UT 84112
**ISI-PV, Forschungszentrum Julich, D-5170 Julich, Germany

ABSTRACT

Previous ^1H NMR and IR studies of six samples of μc-Si:H prepared under plasma excitation frequencies ranging from 13 MHz to 95 MHz and silane concentrations ranging from 3% to 8% revealed three important results: (1) for a fixed plasma excitation frequency (95 MHz) the hydrogen content increases with silane concentration; (2) for fixed silane concentration the hydrogen content is roughly constant over a wide range of plasma excitation frequencies; and (3) the ^1H NMR free induction decay exhibits beat frequencies which correspond to the calculated frequencies due to SiH_2 in microcrystalline silicon. In this study we investigate the role of SiH_2 in the μc-Si:H structure using ^1H NMR measurements. We studied two samples prepared at two different plasma excitation frequencies (13 MHz and 95 MHz). The ^1H NMR lineshapes of these samples were measured at 300 K and 77 K. The motionally narrowed component of the ^1H NMR is not as rapid at room temperature as that observed previously, but the linewidth of this component increases significantly at 77 K. Because of this difference between our results and those reported previously, it is possible that H_2 molecules are not responsible for the motional narrowing in our samples. We present plausible arguments for the hindered motion of SiH_2 groups in our μc-Si:H samples.

INTRODUCTION

Hydrogenated microcrystalline silicon has been studied for many years. The earliest ^1H NMR measurements by Hayashi et al. [1], Kumeda et al. [2] and Boyce et al. [3,4] showed clearly the presence of both broad and narrow NMR lines as well as the presence of significant hydrogen motion contributing to the narrow NMR line even at 300 K. The films of μc-Si:H are inhomogeneous mixtures of crystalline grains and amorphous regions. Presumably, the hydrogen resides predominantly on the external surfaces of the crystalline grains and in the amorphous regions. There is still no agreement on which ^1H NMR lines occur in which regions. Also, although hydrogen motion is universally observed for a fraction of the hydrogen atoms, the species responsible for this motion are still not well identified. In the earlier studies on samples that usually contained more than 10 at. % hydrogen, the moving hydrogen was attributed to molecular hydrogen (H_2). In the samples studied here we suggest that the component of hydrogen that is moving is probably SiH_2 groups that are undergoing hindered rotation . Our previous study [5] tentatively identified a beat pattern in the ^1H NMR free induction decays as due to SiH_2 groups. Since calculations [3] suggest that the second moment for the SiH_2 doublet is about 9 kHz and our observations yield a value of about 9.6 kHz, we have attributed the doublet in our samples to this bonding configuration. In this study we report the temperature dependence of ^1H NMR measurements on μc-Si:H films prepared by the plasma-enhanced chemical vapor deposition (PECVD) technique using variable radio-frequency (rf) excitation.

From previous studies it is well known that NMR is a very useful tool to study the hydrogen environment in both hydrogenated amorphous silicon (a-Si:H) [6] and μc-Si:H [1–5,7]. In the present study we use ^1H NMR to investigate the broad and narrow free-induction-decay (FID)

Mat. Res. Soc. Symp. Proc. Vol. 452 © 1997 Materials Research Society

components as functions of the temperature in samples of μc-Si:H grown at different plasma excitation frequencies. We have previously measured the hydrogen concentration and the degree of crystallinity as functions of the PECVD frequency in μc-Si:H samples [5,8–10]. In the samples we studied the degree of crystallinity increases with the frequency of the PECVD for constant SiH_4 partial pressure (constant H incorporation). This situation is shown quantitatively in Fig. 1. As seen in this figure the hydrogen concentration is essentially constant up to about 80 MHz after which it appears to decrease slightly. Such estimates have been compared to those obtained by infrared absorption (IR) spectroscopy [4] and the results indicate that the hydrogen concentrations estimated by NMR and IR are essentially the same. We note that there is no *a priori* guarantee that these two experimental measurements will give the same result.

In addition to the total hydrogen concentration, measurements of the broad and narrow components of the ^1H free induction decay give detained information concerning the hydrogen incorporation in the μc-Si:H samples as a function of the PECVD frequency. As mentioned above previous measurements have shown that motional narrowing of the narrow ^1H NMR line occurs at 300 K in essentially all samples of μc-Si:H [1-5,7]. In the earlier studies this motionally narrowed component was attributed to molecular hydrogen (H_2) [1-4]. Because of the different narrowing characteristics shown by the narrow line in our previous work [5], we suggested that in our samples the narrowing might be due to the hindered rotation of SiH_2 groups. Our previous work also suggested that the hydrogen concentration increases as the volume fraction of μc-Si:H decreases. At a fixed silane concentration, the crystalline volume fraction increases from about 0.6 to essentially one as the frequency increases from 13.56 to 95 MHz. Since it is much more difficult to incorporate hydrogen in the crystalline phase of Si, the decrease in hydrogen concentration at the highest plasma frequency (Fig. 1) is attributable to the increased crystalline volume fraction.

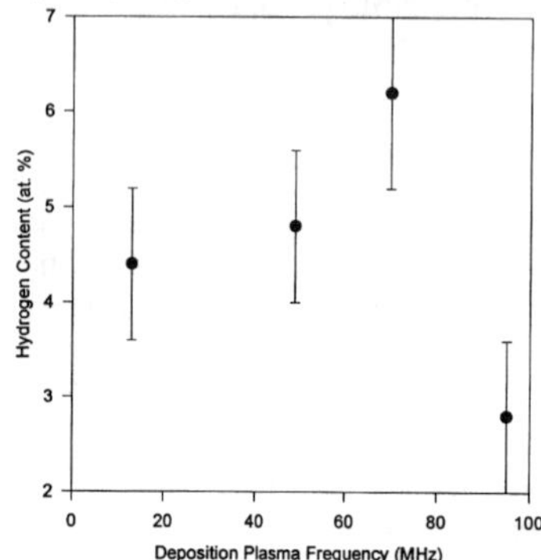

Figure 1 Hydrogen content in four μc-Si:H samples (3 at. % silane concentration) measured from ^1H NMR as a function of plasma deposition frequency.

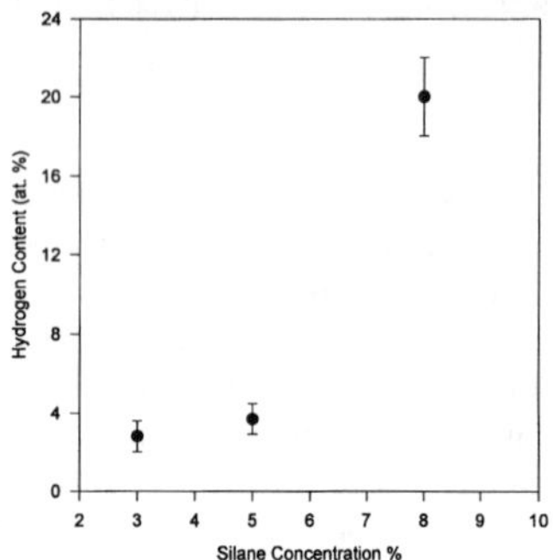

Fig. 2. The hydrogen content of three samples of μc-Si:H at 95 MHz plasma deposition frequency as a function of silane concentration. Hydrogen concentrations were estimated using ^1H NMR as described in the text.

This trend is also supported by the results shown in Fig. 2. In this figure are shown the hydrogen concentrations measured by ^1H NMR (using the $t = 0$ extrapolations of the free induction decays, FIDs) in samples made at a plasma deposition frequency of 95 MHz at different silane concentrations. From Fig. 2 it is apparent that the hydrogen concentration dramatically increases as the silane concentration increases. This result is consistent with the results of TEM measurements on the same samples [8] which show that the volume fraction of crystallinity decreases dramatically with increasing silane concentration. In the first samples studied, those with high hydrogen concentrations, the doublet contribution to the ^{1}H NMR lineshape was attributed to H_2 or to SiH_2 It is well known that the doublet due to molecular hydrogen in a-Si:H appears only at low temperatures because at higher temperatures (> 20 K) the H_2 molecules rotate fast enough that doublet is motionally narrowed. Because of this fact and because of the magnitude of the splitting, it appears that SiH_2 groups are the most likely candidates for the resolved doublets in μc-Si:H. The temperature dependent data to be presented below support this interpretation.

SAMPLE PREPARATION

Samples were prepared by plasma enhanced chemical vapor deposition (PECVD) at 200°C using plasma excitation frequencies from 13.56 to 95 MHz. For the NMR studies the samples were deposited on aluminum foil and etched off using dilute HCl to form powdered samples. Details of the sample preparation procedures are available elsewhere [5,9,10]. The a-Si:H sample was prepared by PECVD at 230°C.

EXPERIMENTAL DETAILS

The ^{1}H NMR experiments were performed on a Matec 5100 system with an Oxford superconducting magnet at 2 Tesla. Details are available elsewhere [5]. Variable temperatures were obtained using a helium gas flow system. Over 50,000 traces were accumulated to compute the NMR free induction decay (FID) lineshapes. A typical 90° pulse for the experiments was approximately 1.4 μs.

RESULTS

As in most a-Si:H samples, the FIDs in all μc-Si:H samples exhibit both broad and narrow lines. However, in the μc-Si:H samples the narrow line (in the frequency domain) is narrower than the corresponding line in a-Si:H by a factor of about three. The broad and narrow components were estimated by fitting the FID with Gaussian and Lorentzian lineshapes, respectively. For the narrow Lorentzian lineshapes, the FID is given by

$$I(t) = Ae^{-t/T_2^*} \quad \text{(Lorentzian)} \tag{1}$$

where I(t) is the free induction decay intensity and A is a constant. Table I shows the various narrow (Lorentzian) components for the μc-Si:H samples. These components are compared to a typical a-Si:H sample. (A 10^{-3} P-doped a-Si:H sample with 10 at. % of hydrogen was used as the reference in this study.) The values for T_2^* listed in Table I differ from those shown in ref. 5 because of a calculational error in the previous work.

The free induction decays of four μc-Si:H samples (2095, 2395, 2595 and 3095 in Table I) each exhibit a beat pattern. The beat frequency in all these samples was previously estimated [5] to be about 9.5 ± 0.5 kHz. There is no obvious trend in beat frequency with plasma excitation frequency even through these samples vary somewhat in hydrogen content as shown in Fig. 1.

From the free induction decay of μc-Si:H the hydrogen content for each sample was estimated in comparison with the free induction decay of the 10^{-3} P-doped a-Si:H sample. The first four samples (2095, 2395, 2595 and 3095) shown in Table I were prepared at a fixed SiH_4/H_2 gas ratio of 0.03. NMR data indicate that as the silane concentration increases (at a fixed plasma excitation frequency), the hydrogen content is essentially constant and decreases slightly at the highest

Table I. Narrow ^{1}H NMR lines in the free induction decay of μc-Si:H samples at different plasma frequencies.

Sample	Plasma Frequency (MHz)	Silane Concentration (%)	Narrow Line (T_2^*) (μS)	
			300 K	77K
2095	13	3	290 ± 20	200 ± 20
2395	49	3	295 ± 20	
2595	95	3	255 ± 15	200 ± 20
3095	70	3	330 ± 15	
3595	95	5	200 ± 15	
3695	95	8	195 ± 15	
a-Si:H	13	~10	140 ± 10	140 ± 10

frequency. When similar measurements are made on samples deposited at 95 MHz and different silane concentrations (Fig. 2), the hydrogen concentration is seen to increase significantly from ~ 4 at. % to ~ 20 at. %.

The FIDs for a typical sample of μc-Si:H and for a typical sample of (phosphorus-doped) a-Si:H are shown in Fig. 3 on a semilog scale. Two essential features are apparent from the data shown in Fig. 3. First, the narrow line (long, exponential time decay) is much narrower (longer in time) for the μc-Si:H sample than it is for the a-Si:H sample. Second, the oscillatory behavior (beating) in the FID of the μc-Si:H sample is also apparent, especially around 200 μs. The beating is attributed to SiH_2 groups as mentioned above.

Table II shows the temperature dependence of the narrow ^{1}H NMR FID component in two samples of μc-Si:H, one deposited at 13 GHz and one at 95 GHz. The broadening of the narrow component of the NMR line is similar for both samples. As will be discussed below, this motional narrowing is much less pronounced than that observed previously on samples containing much greater concentrations of hydrogen (up to 20 at. %). Typical values at 300 K for the samples previously studied with greater hydrogen concentrations are given in Table II for comparison.

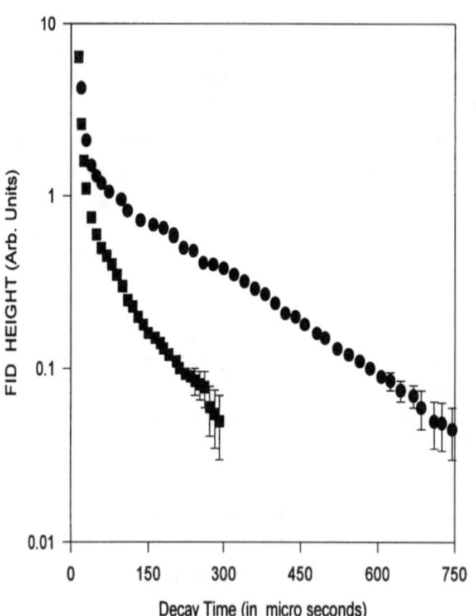

Fig. 3. FIDs for a-Si:H (squares) and μc-Si:H (3595, circles). See Text for details.

DISCUSSION

The width in the frequency domain of the narrow component of the ^1H NMR (see Table II) suggests that, in all the μc-Si:H samples studied in this work, there exists significant motional narrowing of this component. This conclusion is based on the fact that the linewidths at room temperature are a factor of approximately two narrower than that commonly observed in a-Si:H and the fact that in two μc-Si:H samples that we have checked (Table II) the linewidth of this component increases at 77 K. Previous studies of μc-Si:H [1–4,7] have also observed motional narrowing of the narrow line, although the narrowing at room temperature is more pronounced (a factor of approximately four), and the line remains this narrow at least down to 50 K. In a-Si:H the linewidths of the broad and narrow components are on the order of 20 to 30 kHz and 2 to 5 kHz, respectively, independent of the temperature. If we assume that the narrow line in μc-Si:H is Lorentzian, then the linewidth (full width at half maximum) is given by $1/\pi T_2$, which from Table II is about 1.5 to 2.0 kHz for the samples containing about 5 at. % hydrogen which were made at essentially all of the plasma frequencies. At the highest deposition frequency (95 MHz), the linewidth increases. When the hydrogen concentration increases at fixed deposition frequency, the linewidth of the narrow Lorentzian line increases and approaches that observed in a-Si:H. The latter trend is consistent with the fact that the degree of micro-crystallinity decreases with increasing hydrogen concentration.

Since the motional narrowing we observe is not as rapid at room temperature as that observed previously, and since the linewidth increases significantly at 77 K, it is possible that H_2 molecules may not be responsible in our case. A second possibility for the motionally-narrowed line in our samples of μc-Si:H is the motion of SiH_2 groups. Although further experiments are necessary to be certain which attribution is correct, the attribution to SiH_2 groups is supported by two pieces of

Table II. Narrow ^1H NMR lines in free induction decay of various μc-Si:H samples.

Sample	Plasma Frequency (MHz)	Hydrogen Content (at.%)	Narrow Line FWHM (kHz)	Temperature
2095	13	4	1.1 1.6	300 77
2595	95	4	1.3 1.6	300 77
Ref. 7	13	6-12	0.4 - 0.5	300
Ref. 1	13	9-12	0.5 - 0.55	300

experimental evidence. First, the magnitude of the splitting (about 9.6 kHz) is consistent with that expected for SiH_2 groups but not with that expected for H_2 molecules. Second, the motional narrowing of the line is less pronounced in our samples than it is in the samples previously studied where the motion was attributed to H_2 molecules.

With regard to the Pake doublet component of the ^{1}H NMR in the μc-Si:H samples, it is clear that there must be a distribution of hindered rotation rates since this doublet is observed both at 300 and 77 K. That is, for our interpretation to be correct, some of the SiH_2 groups are motionally narrowed and some "stationary," as far as the ^{1}H NMR is concerned, at both 300 and 77 K. Experiments at lower temperatures are necessary to confirm this interpretation.

ACKNOWLEDGMENTS

The work at the University of Utah was supported by the National Renewable Energy Laboratory under subcontract number XAD3121142.

REFERENCES

1. S. Hayashi, S. Yamasaki, A. Matsuda and K. Tanaka, J. Non-Cryst. Solids **59**, 779 (1983).

2. M. Kumeda, Y. Yonezawa, A. Morimoto, S. Ueda and T. Shimizu, J. Non-Cryst. Solids **59**, 775 (1983).

3. J.B. Boyce, M. Stutzmann and S.E. Ready, J. Non-Cryst. Solids **77&78**, 265 (1985).

4. J.B. Boyce in *Hydrogen in Disordered and Amorphous Silicon*, edited by G. Bambakidis and R.C. Bowman (NATO Study Inst. Proc., Rhodes, Greece, 1985), p. 101.

5. P. Hari, P. C. Taylor and F. Finger, MRS Symp. Proc. **420**, 491 (1996).

6. J.A. Reimer, R.W. Vaughan and J.C. Knights, Phys. Rev. B **24**, 3360 (1981).

7. S.E. Ready, J.B. Boyce and C.C. Tsai, Mat. Res. Soc. Symp. Proc. **118**, 103 (1988).

8. F. Finger, M. Tzolvov, P. Hapke, M. Luysberg, L. Houben, R. Carius and H. Wagner, 8th Annual Sunshine Workshop (Tokyo, 1995), p. 89.

9. F. Finger, C. Malten, P. Hapke, R. Carius, R. Fluckiger and H. Wagner, Philo. Mag. Lett. **70**, 247 (1994).

10. C. Malten, F. Finger, P. Hapke, T. Kulessa, C. Walker, R. Carious, R. Fluckiger and H. Wagner, Mat. Res. Soc. Symp. **358**, 757 (1995).

Influence of hydrogen incorporation into Silicon on the room-temperature photoluminescence

J. Rappich[a], Th. Dittrich[b], Y. Timoshenko[c], I. Beckers[a] and W. Fuhs[a]

[a] Hahn-Meitner Institut, Abteilung Photovoltaik, Rudower Chaussee 5, D-12489 Berlin, Germany
[b] TU-München, Physik-Department E-16, D-85747 Garching, Germany
[c] Faculty of Physics, M.V. Lomonosov University, 119899 Moscow, Russia

Abstract

We investigated the possibility to use the photoluminescence (PL) emission at room temperature for the characterization of grain boundary and surface passivation of µc-Si and single crystalline silicon. The PL-spectra of c-Si and µc-Si taken at 300 K consist of an emission band around 1.1 and 1 eV, respectively. Measuring the PL intensity we study the influence of electrochemical and plasma passivation by hydrogen. We compare the behavior of the films to that of Si crystals. In this case the PL intensity is high for a well H-terminated c-Si surface and decreases during cathodic polarization (hydrogen evolution) which points to the formation of additional non-radiative recombination centers. Whereas the PL intensity of the as prepared µc-Si films is unchanged, that of the annealed films increases during hydrogen evolution indicating the electrochemical passivation of defects on grain boundaries.

Introduction

Recently we have shown that pulsed photoluminescence (PL) at room-temperature can be used as an in-situ method to investigate the formation and passivation of defects during electrochemical treatments at crystalline silicon surfaces [1]. The exciton emission at about 1.1 eV decreases strongly when the hydrogenated surface is anodically oxidized in fluoride solutions and increases again after hydrogen passivation. This behavior points to a better passivation of c-Si by hydrogen than by wet and disordered anodic oxides. The passivation of bulk defects or grain boundaries in µc-Si or poly-Si films can be achieved by use of hydrogen in order to optimize the electronic properties [2]. In this study we attempt low temperature passivation by hydrogen using both H-plasma and cathodically induced H-evolution. This bases on the assumption that the electrochemical behavior of µc-Si is similar to that of c-Si. The aim of this work is not to optimize the µc-Si electronically but to find out as to whether it is possible to use PL to monitor the passivation in situ.

Experimental

The investigations were performed on p-type single crystalline wafers (c-Si, 1 Ωcm resistivity) and microcrystaline Si films (µc-Si) which had been deposited by ECR-

Mat. Res. Soc. Symp. Proc. Vol. 452

PECVD at a substrate temperature of 325 °C onto n-Si(111) (60-70 Ωcm resistivity). The c-Si samples were cleaned by electropolishing in 0.1 M NH_4F (pH 4.2) by applying the oscillation treatment to get an optimized hydrogen passivated surface [3]. The oxidized surface of the μc-Si films, which have a thickness of about 800 nm, were etched back by diluted HF. The μc-Si samples were treated cathodically after the deposition and after annealing at 450 °C for 12 h in 0.2 M K_2SO_4 (pH 3) under galvanostatic condition (galvanostat/potentiostat Jaissle IMP 88 PC). The used electrolyte (0.2 M K_2SO_4, pH 3) consists of p.a. grade chemicals, highly purified water and was purged with nitrogen. In addition, some films from the same run were exposed to a hydrogen plasma in the ECR or RF system at 325 °C (MW:1500 W, 100mTorr H_2 or RF:25W, 100mTorr H_2).

The PL is excited by a N_2-laser (LTB-MSG200, wavelength 337 nm, pulse width 0.5 ns). The light intensity is about 0.5 mJ/cm^2. The PL transients of the c-Si and μc-Si samples are detected using a prism monochromator, an InGaAs-photodiode with a high impedance preamplifier (EMM) and a digital oscilloscope (HP 54510 A) at a 1.13 eV or 1.0 eV, respectively. The experimental setup is described in more detail elsewhere [1].

Results and Discussion

Fig. 1 shows the PL spectra of p-Si(100) before (solid line) and after (dotted line) electrochemical hydrogen evolution.

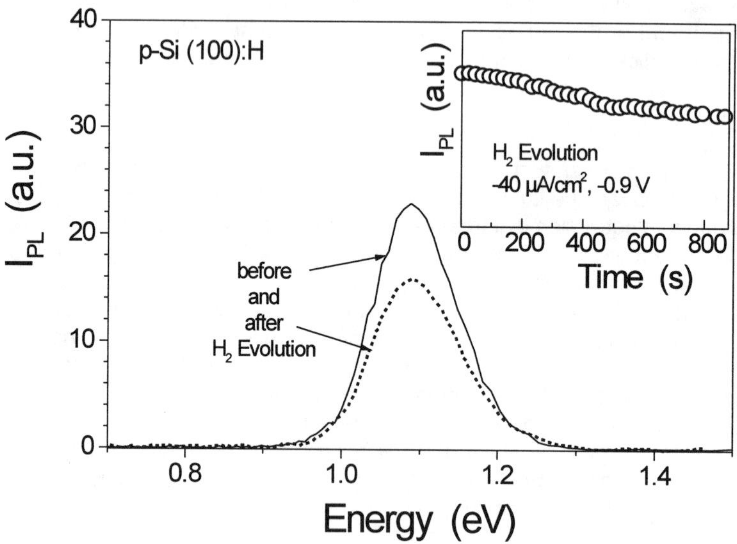

Fig.1 PL spectra of p-Si(100) before (solid line) and after (dotted line) electrochemical hydrogen evolution (i=-40 μA/cm^2, 0.2 M K_2SO_4, pH 3). Inset: time dependence of the PL intensity at 1.13 eV during cathodic current flow.

The solid line represents the behavior of the well hydrogenated c-Si surface after the oscillation treatment [3,4]. It exhibits a single broad structure which has a peak at about 1.1. eV. This emission my be assigned to the exciton emission. This surface is then cathodically treated in the acidic aqueous solution at a current density of i=-40 μA/cm^2 which leads to potential of -0.9 V at the c-Si electrode. The in situ measured PL intensity tends to decrease with increasing time as seen from the inset in fig.1. The increase of time is correlated to an increase of electric charge which passes the electrode and leads to the formation of H_2 at the surface where adsorbed H-atoms occurs as an intermediate. These H-species can diffuse into the Si bulk, break Si-Si bonds by forming Si-H [5] and thus create defects such as dangling bonds. Those act as non-radiative recombination centers and therefore the PL intensity decreases. This, however, is not connected with a spectral change. The dotted curve in fig.1 which was measured after t=800 s has the same shape as the original curve.

In μc-Si the PL emission at 300 K is considerably lower than in c-Si. Nevertheless there is a single emission band which peaks around 1 eV (fig.2). It is tempting, to assign the emission to excitons. However, the band can be due as well to defect luminescence [6,7].

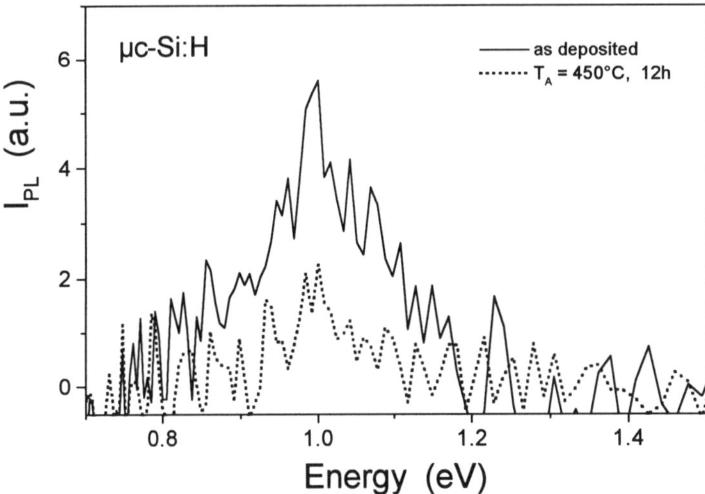

Fig.2 PL spectra of the as deposited (solid line) and the annealed (450°C, 12 h) μc-Si thin film (dotted line).

At low temperature the intensity is considerably higher and the spectra have a more complicated structure. In fig.2 we compare the spectra of the as deposited sample with that after a 450 °C anneal. Annealing at this temperature causes hydrogen effusion and an enhancement of the defect density. ESR measurements of samples deposited under similar conditions indeed showed a continuous increase of the dangling bond density from 10^{16} cm^{-3} at 200 °C to 10^{18} cm^{-3} at 450 °C. The low temperature at which this H-effusion takes place suggests that the hydrogen evolves from internal surfaces. This is connected with a decrease of the PL-emission as can be seen from figure 2. This result suggests that the defects on the grain boundaries act as non-radiative recombination centers. Obviously it is difficult to passivate µc-Si films by electrochemical H-treatment. Fig.3 shows a plot of the PL intensity at 1 eV as a function of time during the cathodically induced H-evolution.

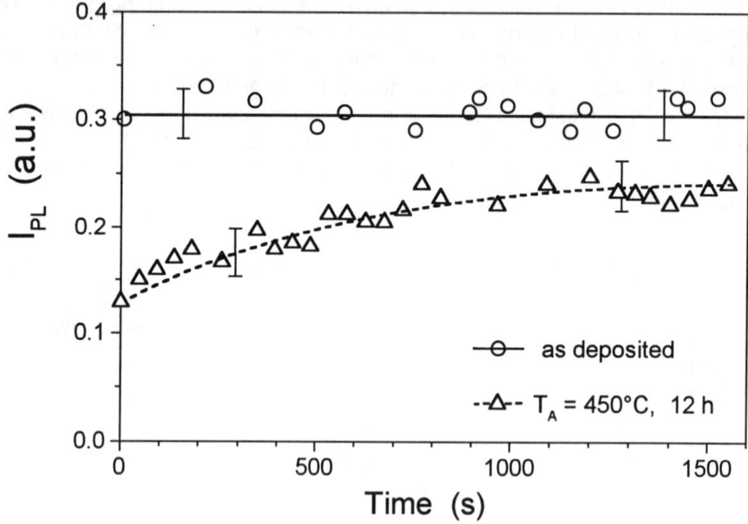

Fig.3 Time dependence of the PL intensity at 1 eV during the cathodic hydrogen evolution in 0.2 M K_2SO_4 (pH 3) of the as deposited (open circles) and annealed (up triangles) µc-Si films.

The PL intensity of the sample in the as deposited state does not change at all though hydrogen is developed and supposed to diffuse into the bulk of the film. This is different from the behavior after the annealing process where the intensity increases as we expect if hydrogen passivates defects. Apparently only the defects created during the effusion can be passivated by incorporation of hydrogen. The PL intensity clearly increases by about 60 - 80 %. Neither in the as deposited state nor after the annealing process did we observe any change in the PL intensity and spectra after the plasma treatments.

The above results show that the electrochemical treatment has only little influence on the defect passivation. It also has no detectable influence on the structure of µc-Si. Fig.4 shows SEM images of the µc-Si surfaces after deposition (a) and after annealing and electrochemical treatment (b). There is no visible change in the surface morphology of the outer spheres (diameter about 50 to 100 nm) of the columnar-grown crystallites which are seen in the cross sectional view of the layer (30° tilt angle).

Fig.4 SEM photos of the μc-Si thin film (a: as deposited, b: after annealing and electrochemical hydrogen evolution and c: cross sectional view of the layer, 30° tilt angle).

Thus this investigation led to the surprising conclusion that the electrochemical treatment and plasma passivation have no beneficial effect on the radiative recombination in case of the μc-Si films. This may be due to the fact that the radiative lifetime in these heterogeneous materials is not determined by the grain surfaces. On the other hand the origin of the PL emission is not quite clear in μc-Si. It might be due to recombination at defects or defect impurity complexes and thus be less affected by grain boundary passivation. Of course the insensitivity of the PL emission to the electrochemical treatment does not mean that there is no improvement of other electronic properties such as transport. These interesting questions need further study.

Conclusion

We have shown that hydrogen evolution at well passivated c-Si leads to the formation of defects which decreases the PL intensity. As deposited μc-Si films show no change of the PL emission after the electrochemical treatment. The annealing at 450 °C leads to H-effusion which is accompanied by a decrease of the PL intensity which can be reduced again by cathodically induced hydrogen evolution in aqueous solution at roomtemperature. Surprisingly H-plasma treatments do not affect the PL emission.

Acknowledgment

The authors thank Ina Sieber (HMI) for recording the SEM images and Dr. Mell (Uni Marburg) for preparing some µc-Si films. The ESR measurements have been performed by Dagmar Will. This work has been supported by the Bundesministerium für Bildung und Forschung BMBF under the contract number 0329513A.

References

1. J. Rappich, Th. Dittrich and V. Yu. Timoshenko; J. Electrochem. Soc. in press
2. N.H. Nickel, N.M. Johnson and W.B. Jackson; Appl. Phys. Lett. **62**, 3285 (1993)
3. S. Rauscher, Th. Dittrich, M. Aggour, J. Rappich, H. Flietner and H.J. Lewerenz; Appl. Phys. Lett. **66**, 3018 (1995)
4. J. Rappich and H. J. Lewerenz; J. Electrochem. Soc.**142**, 1234 (1995)
5. K.C. Mandal, F. Ozanam and J.-N. Chazaviel; Appl. Phys. Lett. **57**, 2788 (1990)
6. P.K. Bhat, G. Disprose, T.M. Searle, I.G. Austin P.G. Le Comber and W.E. Spear; Physica **117B&118B**, 917 (1983)
7. M. Yamaguchi and K. Morigaki; J. Phys. Soc. Jap. **62**, 2915 (1993)

MICROSTRUCTURES OF LUMINESCENT nc-Si BY EXCIMER LASER ANNEALING OF a-Si:H

Xinfan Huang, Wei Wu, Honghui Shen, Wei Li, Xiaoyuan Chen, Jun Xu, and Kunji Chen

Department of Physics and State Key Laboratory of Solid State Microstructures, Nanjing University, Nanjing 210093, CHINA, kjchen@netra.nju.edu.cn
Center for Advanced Studies in Science and Technology of Microstructures, Nanjing 210093, CHINA

ABSTRACT

We have reported for the first time on visible photoluminescence (PL) in crystallized a-Si:H/a-SiN_X:H multilayer structures by CW Ar ion laser annealing treatments. In this paper we present new results on visible PL from crystallized a-Si:H by using KrF excimer pulse laser (wavelength 248 nm) irradiating treatments. The transmission electron microscopy and Raman scattering studies reveal the microstructures of crystallized Si films, which depend on the pulse number and the pulse energy density of KrF laser. When the laser pulse energy density is higher than 520 mJ/cm^2, the nanosized Si crystallites (nc-Si) can be formed from a-Si:H layers with a thickness of 100 nm and strong PL with a peak wavelength of 610 nm has been observed at room temperature.

1. INTRODUCTION

To prepare and study luminescent silicon based materials is one of the active subjects in the field of semiconductor optoelectronics, especially after the discovery of strong room temperature photoluminescence (PL) from porous silicon [1]. Recently many improvements of the electrochemical dissolution method for fabrication of porous silicon have been reported [2] but all these approaches were based on chemical solution (wet) synthesis techniques. A dry process for fabricating light emitting silicon films would be ideal for integrating this material into the current silicon based microelectronics and very large scale integrated circuits technology and consequently would provide new possibilities in the realization for monolithic integration of silicon with optical signal processing [3-5].

We have reported on room temperature visible PL in crystallized Si:H/SiN_X:H multiquantum well structures by a CW Ar ion laser annealing technique [6]. This new approach for fabricating luminescent silicon is a fully dry process which is different from the electrochemical dissolution method to obtain luminescent porous silicon. This is also a good way to avoid the surface contamination which plays a significant role in strong visible luminescence in porous silicon layers [7,8]. In this paper, we present new results on room temperature visible PL from crystallized a-Si:H films by using KrF excimer pulse laser irradiating technique. The main purpose of this work is to investigate the influence of the pulse

Mat. Res. Soc. Symp. Proc. Vol. 452 © 1997 Materials Research Society

number and pulse energy density on the microstructures of luminescent nc-Si films. The distinct microstructural changes were observed for energy densities from 160 to 560 mJ/cm^2, and the changing tendency is similar to the observations reported in reference 9.

2. EXPERIMENTAL

2.1 Fabrication of a-Si:H films

The a-Si:H thin films were prepared in a capacitively coupled single reaction chamber by a rf plasma enhanced chemical vapor deposition (PECVD) method with a reactant gas of pure silane (SiH_4). The films were deposited on fused quartz and SiO_2/Si substrates which were heated to 250 °C. The SiO_2/Si substrates were prepared by means of thermal oxidation of c-Si wafers and the thickness of the SiO_2 layer was about 700 nm. In addition, before the deposition of a-Si:H, an a-SiN_x:H layer of 300 nm was grown on the substrate as a buffer layer. The deposition rate was 0.1 nm/s under a total reaction pressure of 20 Pa.

2.2 Laser crystallization of samples

Laser crystallization of samples was performed by using a KrF (λ= 248 nm) excimer laser with a 30 ns pulse length. During irradiation the samples were held in a stainless-steel vacuum chamber at room temperature. The KrF excimer laser pulse irradiated normal to the sample surface through a quartz window. The beam energy density was controlled by changing the voltage of the laser discharge and the distance between the lens and the sample surface. The size of the laser beam was 4×11 mm^2. In our experiments an energy density of laser in the range of 0.1-1.0 J/cm^2 was used.

2.3 Microstructure characterization and PL measurements of crystallized samples

The crystallinity and microstructures of laser crystallized samples were characterized by Raman scattering spectroscopy and transmission electron microscopy (TEM). The PL measurements were carried out at room temperature by excitation with the 488 nm Ar ion laser line.

3. RESULTS AND DISCUSSION

Figure 1 shows Raman scattering spectra of crystallized samples irradiated by single pulse laser with various energy densities. It is clear that when the laser energy density $\geq$ 160 mJ/cm^2, the samples show asymmetric broad Raman peaks, which indicates that the Si films have been crystallized, but some amorphous phases remain in the film. On the other hand, in Fig. 1 it can be also seen that, with increasing laser energy density the Raman peak shifted from 514 cm^{-1} to 517 cm^{-1} , and the shape of Raman peak became sharper and more symmetrical, which means that the crystalline component in the films increases and the average grain size increases from

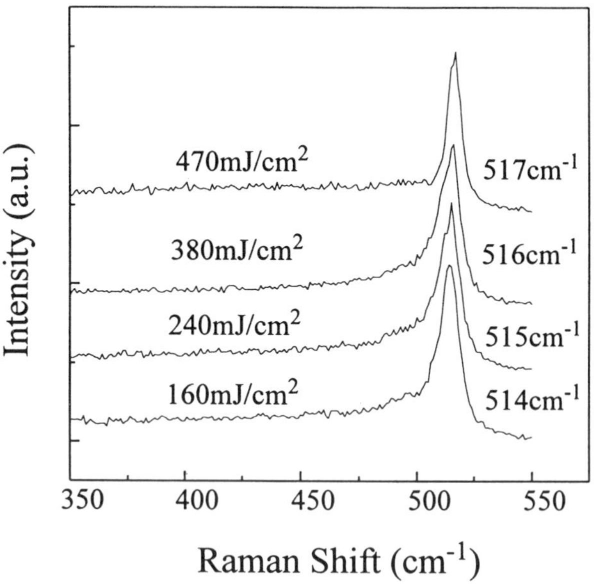

Fig.1 Raman spectra of crystallized a-Si:H samples irradiated by various laser energy density.

3.5 nm to 5.5 nm, as estimated by an empirical model of phonon confinement in crystallized a-Si:H structures [10].

The energy density dependence of the grain size in crystallized samples was also discovered by bright-field TEM micrographs which are shown in Fig. 2. Figure 2 (a)-(c) correspond to films, which were same as in Fig. 1, irradiated by 160 mJ/cm^2, 240 mJ/cm^2 and 380 mJ/cm^2 , respectively. In this low energy density regime (< 500 mJ/cm^2), the behavior of the increase in grain size with increases in the energy density was also observed, which corresponds to the results of Raman spectra shown in Fig. 1. But in the high energy density regime ($\geq$ 560 mJ/cm^2), as shown in Fig. 2(e), a reversal in the microstructural trend was observed in that fine-grained nanosize Si crystallites were obtained. It is interesting that, there is a very narrow regime between the high and low energy density regimes, where an extremely sharp increase in the grain size occurred and the average grain radius was approximately equal to the film thickness which is shown in Fig. 2(d). The changing tendency in the microstructures with energy density is similar to the observations made by Im et al [9].

The PL spectrum of the crystallized sample at room temperature is shown in Fig. 3. The sample was deposited on SiO_2/Si substrate and irradiated by a single pulse with 560 mJ/cm^2. The PL spectrum peaked at around 610 nm (2.03 eV) with full width at half maximum (FWHM) of 0.3 eV. A planar view TEM photograph of this crystallized sample is shown in Fig. 2(e), which

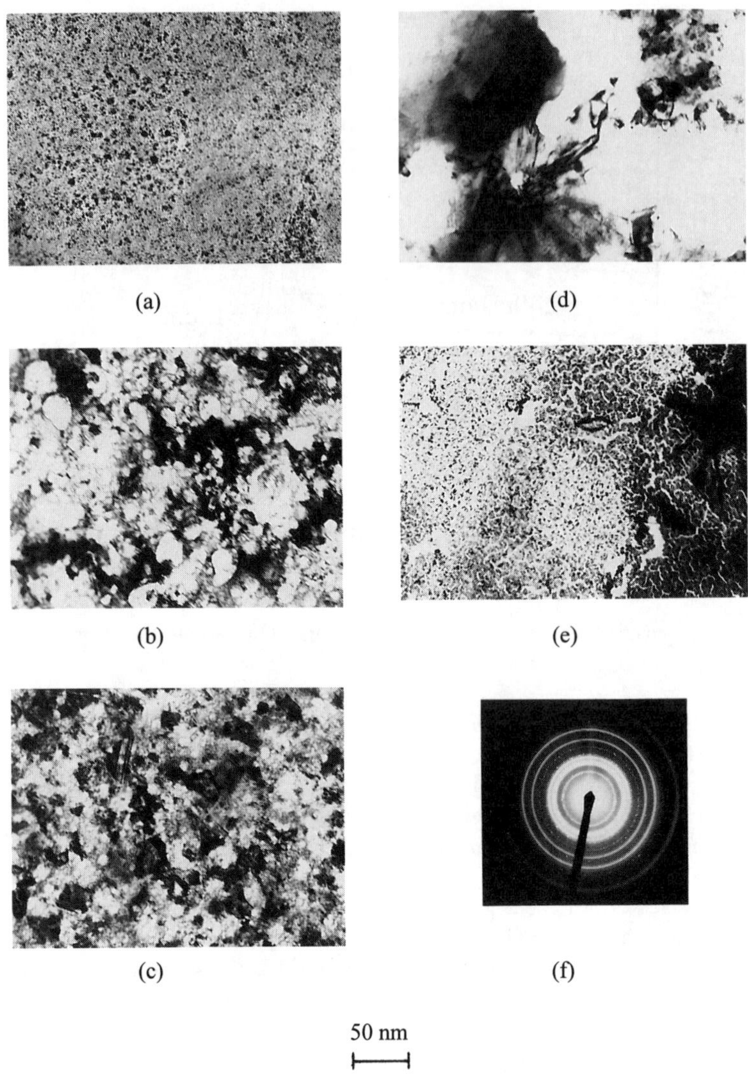

Fig.2 Planar TEM Micrographs of single-pulse laser crystallized Si thin films at various energy densities: (a) 160 mJ/cm, (b) 240 mJ/cm^2, (c) 380 mJ/cm^2, (d) 520 mJ/cm^2, (e) 560 mJ/cm^2 ; (f) Electron diffraction pattern of sample irradiated at 560 mJ/cm^2.

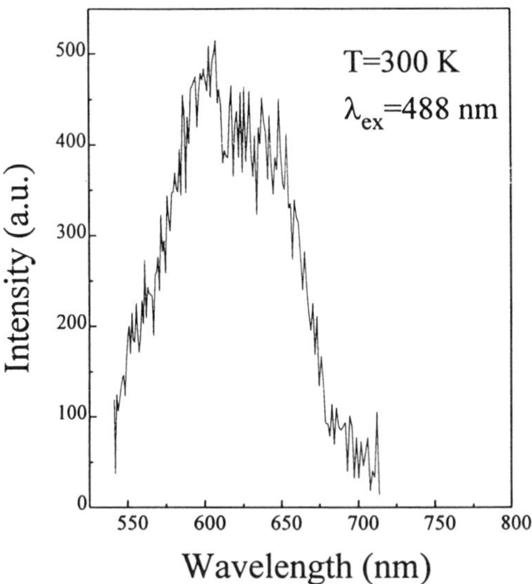

Fig.3 PL spectrum of crystallized a-Si:H samples irradiated by single pulse with 560 mJ/cm^2.

reveals very fine Si crystallites. From the electron diffraction pattern of Fig. 2(f) we estimated the average grain size of Si crystallites is about a few nanometer. We have not observed PL from crystallized samples irradiated in low energy density regime. The study of differences in the optical properties between two kinds of crystallized Si films is under way.

4. CONCLUSION

We have developed a new method for fabricating luminescent nano-crystal Si films by KrF excimer pulse laser crystallization from a-Si:H films prepared by PECVD. The visible PL has been obtained from crystallized a-Si:H at room temperature. The crystallinity and uniformity have been improved comparing with that in Ar ion laser crystallized samples. This method would provide new possibility in the realization for monolithic optoelectronic integration of silicon through low-temperature and dry processes.

5. ACKNOWLEDGMENTS

This work is supported by the Climbing Program --National Key Project for Fundamental Research and the National Science Foundation of China, and is partially supported by the National Integrated Optoelectronics Laboratory, Institute of Semiconductors Academia Sinica.

REFERENCES

1. L. T. Canham, Appl. Phys. Lett., 57, 1046 (1990)

2. for review see MRS Symp. Proc., No. 358, Boston (1994)

3. S. Furukawa and T. Miyasato, Jpn. J. Appl. Phys., 27, L2207 (1988)

4. H. Takagi, H. Ogawa, Y. Yamazaki, A. Ishizaki and T. Nakagiri, Appl. Phys. Lett., 56, 2379 (1990)

5. E. Edelberg, S. Bergh, R. Naone, M. Hall and E.S. Aydil, Appl. Phys. Lett., 68, 1415 (1996)

6. K. J. Chen, X.F. Huang, J. Xu and D. Feng, Appl. Phys. Lett., 61, 2069 (1992)

7. M. Stutzman, M.S. Brandt, E. Bustarret, H.D. Fuchs, M. Rosenbauer, A. Hopner and J. Weber, J. Non-Cryst. Solids, 164-166, 931 (1993)

8. P. Deak, M. Rosenbauer, M. Stutzman, J. Weber and M.S. Brandt, Phys. Rev. Lett., 69, 2531 (1992)

9. J. S. Im and H.J. Kim, Appl. Phys. Lett., 63, 1969 (1993)

10. X. G. Cheng, H. Xia and K.J. Chen, Phys. Stat. Sol. A, 118, K51 (1990)

DEPOSITION OF PHOTOLUMINESCENT NANOCRYSTALLINE SILICON FILMS BY SiF_4-SiH_4-H_2 PLASMAS

G. Cicala*, G. Bruno*, P. Capezzuto*, L. Schiavulli**, V. Capozzi** and G. Perna**
*Centro di Studio per la Chimica dei Plasmi C.N.R, c/o Dipartimento di Chimica, Università di Bari, Via Orabona, 4 - 70126 Bari, Italy, cscpgc07@area.ba.cnr.it
**Dipartimento di Fisica, Università di Bari, and Istituto Nazionale di Fisica della Materia, Via Orabona, 4 - 70126 Bari Italy.

ABSTRACT

Visible photoluminescence at 1.62 eV has been observed at room temperature from fluorinated and hydrogenated nanocrystalline silicon (nc-Si:H,F) produced in a typical plasma enhanced chemical vapor deposition system. The use of SiF_4-SiH_4-H_2 mixture, because of the H_2 dilution and the presence of SiF_4, favours the amorphous - crystalline transition through the etching process of the amorphous phase. The x - ray diffraction measurements give an average grain size of about 100 Å. The presence of these nanocrystals shifts the absorption edge of the films towards higher energy. An energy gap of 2.12 eV is estimated, although the hydrogen content in the material is only 4.5 at. %. The temperature dependence of the photoluminescence behaves similarly to that of porous silicon.

INTRODUCTION

Photoluminescent silicon based materials are widely investigated for their optoelectronic applications. Nowadays, several growth processes based on plasma enhanced chemical vapor deposition (PECVD) technique are being developed for the production of nanocrystalline silicon (nc-Si), which presents photoluminescence (PL) even at room temperature (RT). Veprek et al. [1] prepared nc-Si:H films by hydrogen diluted silane plasmas or by chemical transport of silicon in hydrogen plasmas. However, similarly to porous silicon (PS) [2,3], these films exhibit a PL improvement only after a post processing treatment, like oxidation and annealing in forming gas (H_2 - N_2). Very recently, Toyama et al. [4], have obtained nc-Si by using the PECVD technique for the deposition of boron doped microcrystalline silicon (μc-Si) and subsequently by treating them electrochemically in HF aqueous solution. The material, produced by this hybrid technique, exhibits a visible photoluminescence at 1.7 eV and a visible electroluminescence at 1.8 eV in a p-i-n heterostructured diode. The method utilizing hydrogen diluted silane, commonly employed for the production of micro- or polycrystalline silicon [5, 6], has also been used by Liu et al. [7], who report on nanocrystalline silicon grown at temperatures lower than 200 °C and

Mat. Res. Soc. Symp. Proc. Vol. 452 © 1997 Materials Research Society

exhibiting RT visible PL at 1.82 eV. Room temperature and visible PL, detected *in situ* and in real time during the formation of nanopowders in silane r.f. plasmas, has been reported by Courteille et al. [8]. These authors have found that the oxygen addition does not affect the intensity of 1.6 eV PL peak, in contrast to what commonly found for PS [3] or post processed nc-Si [1]. Here, it is important to emphasize that Veprek et al. [1] have reported the appearance of a new PL peak at 2.7 eV after the oxidation of the nc-Si:H films.

As for the growth process, the deposition of nanocrystalline silicon is related to the important selective etchant role of hydrogen atoms towards the amorphous phase with respect to the crystalline structure. This etching process promotes the nucleation and the crystallization of a-Si:H at low temperature with desired grain sizes [5, 6]. Thus, the use of silicon tetrafluoride as silicon precursor can be of some interest, since the presence of fluorine atoms in the growth environment can favour the amorphous-crystalline transition, due to their etchant activity. Tachibana et al. [9] have studied the effect of SiF_4 addition to the conventional H_2 diluted silane mixture, and they have confirmed, with an *in situ* ellipsometric monitoring, that the crystalline fraction increases at high values of SiF_4/SiH_4 ratio. However, to our knowledge no evidence of the visible PL at room temperature from plasma deposited fluorinated materials is reported without a post processing treatment [10].

In the present contribute, we report on preliminary results of fluorinated nanocrystalline silicon (nc-Si:H,F) films, deposited by PECVD technique starting from SiF_4-SiH_4-H_2 mixtures. Emphasis is given on the material characterization and on the most relevant results like PL peak position, electrical activation energy E_a, energy gap E_g, hydrogen content and size of nanocrystals.

EXPERIMENTAL

nc-Si:H,F films have been prepared by means of PECVD technique in a parallel plate ultra high vacuum (UHV) reactor from SiF_4-SiH_4-H_2 mixtures. The growth parameters were: nominal deposition temperature 450 °C, pressure 1 Torr, gas flows of the mixture components $SiF_4/SiH_4/H_2$ 10/0.2/82 sccm, and the r.f. (13.56 MHz) power was varied in the range 10 - 50 W. The film growth is carried out on the grounded electrode and simultaneously on different substrates (c-Si, molybdenum and Corning 7059 glass). The deposition rates, r_D, evaluated *in situ* by laser interferometry technique, have been found to be strongly affected by r.f. power. In fact, r_D increases from 0.1 to 0.5 Å/sec when the power increases from 10 to 50 W.

The infrared absorption measurementss of samples deposited on double polished c-Si substrates have been performed by using a FTIR spectrometer (Bomem Michelson 102), which covers the range 400 - 4000 cm^{-1} in order to detect the presence of O and H by the correspondent absorption of Si-H and Si-O bonds and to evaluate the H content.

The x-ray diffraction measurements were performed (Philips PW 1800 diffractometer) on films deposited on various substrates materials such as Corning 7059 glass, crystalline silicon and foils of polycrystalline molybdenum. The x-ray peaks have been found to be not affected by the material type.

The absorption coefficient and the energy gap of the nc-Si films deposited on glass were determined at room temperature by a UV-Vis spectrophotometer (Varian Cary 219) in the region 200 - 900 nm.

Electrical dark conductivity as a function of temperature and photoconductivity under AM1 illumination were carried out in a small stainless steel chamber, working at a pressure of 10^{-5} mbar, on 0.5 μm thick films deposited on glass. The electrical silver contacts were evaporated at a gap size of about 0.2 cm. The measurements were automatized with a digital data acquisition system, with a programmable temperature scan controlled by computer.

As for the PL measurements, the samples have been put in a liquid helium-closed-cycle cryostat, whose temperature could be varied from 10 to 300 K, within 0.5 K of uncertainty. The films have been photoexcited using the 514 nm line of an Ar-ion laser whose light beam has been focused on a spot of about 100 μm diameter, at $\pi/4$ angle with respect to the normal to the film surface. The emission light has been collected from the film surface and has been analyzed at $\pi/2$ from the incident laser beam direction by a double-grating spectrometer (linear dispersion of 1 meV/mm). The signal has been detected by means of a cooled GaAs photomultiplier and has been recorded using photocounting techniques.

RESULTS AND DISCUSSION

Typical x-ray diffraction patterns of the nc-Si:H,F films grown on glass substrate are reported in Fig. 1. The Bragg peaks are principally oriented along the (111) and (220) planes. Thus, the nanocrystalline materials consist of polycrystals with crystallite size of about 100 Å as evaluated by the approximate Sherrer's equation from the FWHM (full width at half maximum) of Bragg peak. At the highest power value (50 W) the (110) texture of the nanocrystalline film seems to be preferred with respect to (111) texture. From the SEM (scanning electron microscopy) images, the surface results to be smooth without any evident morphology.

The infrared data analysis evidences that nc-Si:H,F films incorporate H amount up to 4.5 at %, while in typical fluorinated amorphous silicon the H content is much higher (about 20 at %). There is also difference in the H atom incorporation in the material, being prevalent as Si-H bonds (2000 cm^{-1}) in amorphous and as SiH_2 configuration (2100 cm^{-1}) in nanocrystalline films. In addition, the absorption peak of the Si-O bonds (1100 cm^{-1}) is also present in nc-Si:H,F.

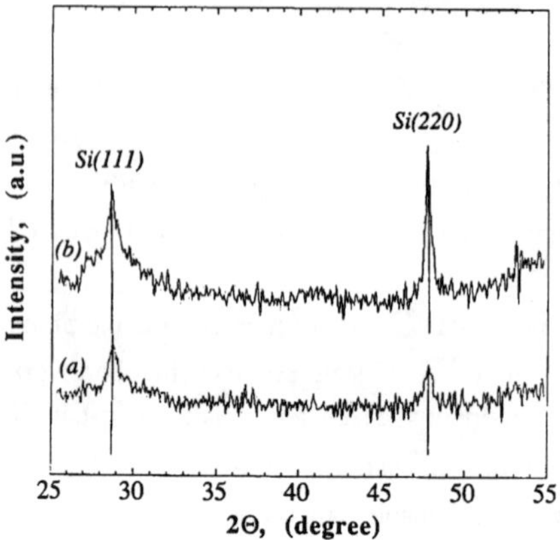

Fig. 1. X-ray Bragg diffraction patterns from nc-Si:H,F films deposited on Corning 7059 glass at two different values of r.f. power: (a) 10 and (b) 50 W.

The optical absorption results in UV - Vis region show that the nc-Si:H,F samples exhibit a lower absorption coefficient, as expected, with respect to the amorphous material. Figure 2 shows the absorption data for a-Si:H and nc-Si:H samples in terms of Tauc's plot, which gives optical energy gap values of 1.8 and 2.12 eV, respectively.

The nanocrystalline nature of the films is evidenced by the temperature dependence of

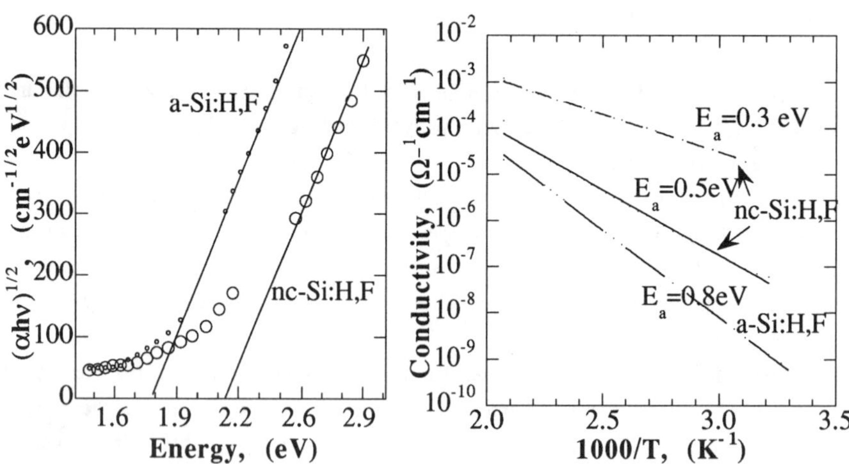

Fig. 2. $(\alpha h\nu)^{1/2}$ values of a-Si:H,F and nc-Si:H,F films as a function of energy.

Fig. 3. Temperature dependence of dark conductivity for nc-Si:H,F and a-Si:H,F films.

the electrical dark conductivity, as reported in Fig. 3. In fact, nc-Si:H,F films exhibit at room temperature conductivity values 2-4 orders of magnitude higher than those of amorphous, and comparatively lower values of activation energy. The values of 0.3 and 0.5 eV refer to two nc-Si:H,F films grown at r.f. power values of 10 and 50 W, respectively. This last result is probably due to the increased presence of various defect levels within the energy gap of nc-Si:H,F material. Accordingly, the photo-to-dark conductivity ratio decreases from values as high as 10^6 for photovoltaic grade a-Si:H,F films to values of 2-10 for nc-Si:H,F samples.

Figure 4 shows the PL spectra of a nc-Si:H,F film measured at laser intensity of $2KW/cm^2$ and at different temperatures. Since the Corning glass substrate exhibits a PL band centered at 1.75 eV, the spectra of Fig. 4 have been corrected by taking into account the PL

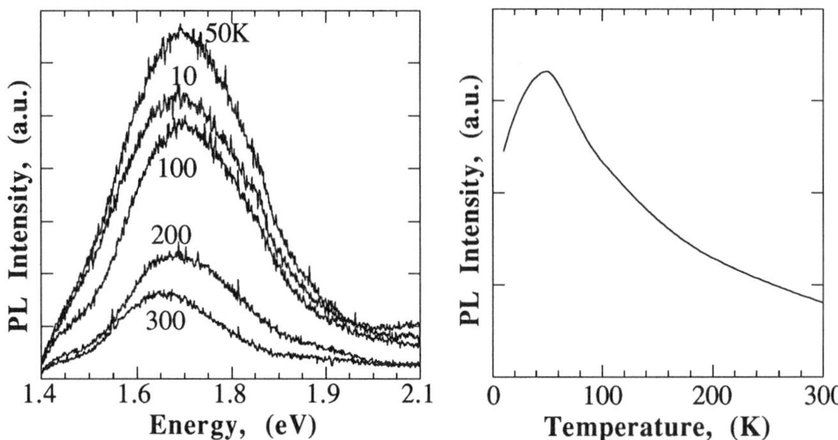

Fig. 4. The photoluminescence spectra of the same nc-Si:H,F sample at different temperature values.

Fig. 5. Temperature dependence of the PL intensity (detected at 1.62 eV) for a nc-Si:H,F film.

spectrum of the substrate, the penetration depth of the laser line, and its self-absorption through the film. The PL band of spectra of Fig. 4 is centered at 1.62 eV and its spectral position remains almost constant by increasing the temperature from 10 K up to 300 K.

Figure 5 shows that, by decreasing the temperature, the PL intensity exhibits the expected increase. The anomalous decrease observed at temperatures below 50 K is similar to that observed for porous silicon films below 200 K [11]. The linewidth of this PL band is about 0.2 eV, remarkably narrower than that reported (0.35 eV) for nc-Si:H films [7]. This smaller width could be related to a narrower energy distribution of defect levels localized in the energy gap.

Moreover, it is worth remarking that the spectral position of the PL band (1.62 eV) is 0.5 eV away from the energy value of the optical gap (2.12 eV). This difference is in good agreement with the activation energy (0.5 eV) obtained by the temperature dependence of the dark conductivity (see Fig.3).

CONCLUSIONS

Nanocrystalline films obtained by PECVD technique starting from SiF_4-SiH_4-H_2 mixture evidence room temperature photoluminescence with a narrow linewidth (0.2 eV). The presence of nanocrystals induces a widening of the energy gap up to 2.12 eV. The position of the photoluminescence peak with respect to the energy gap value can be related to the activation energy of the electrical transport mechanism.

ACKNOWLEDGMENTS

This work has been partially supported by "Progetto Strategico Materiali Innovativi-C.N.R." and by a grant from "Ministero della Ricerca Scientifica e Tecnologica".

REFERENCES

1. S. Veprek, T. Wirschem, M. Ruckschloβ, H. Tamura and J. Oswald, in Microcrystalline and Nanocrystalline Semiconductors, edited by R. W. Collins, C. C. Tsai, M. Hirose, F. Kock and L. Brus (Mater. Res. Soc. Symp. Proc. **358**, Pittsburgh, PA 1995), pp. 99 - 110.

2. L. T. Canham, Appl. Phys. Lett. **57**, 1046 (1990).

3. V. Lehmann and H. Foll, J. Electrochem. Soc. **37**, 653 (1990).

4. T. Toyama, T. Matsui, Y. Kurokawa, H. Okamoto and Y. Hamakawa, Appl. Phys. Lett. **69**, 1261 (1996).

5. I. Solomon, B. Drévillon, H. Shirai and N. Layadi, J. Non-Cryst. Solids **164-166**, 989 (1993).

6. S. Oda and M. Otobe, in Microcrystalline and Nanocrystalline Semiconductors, edited by R. W. Collins, C. C. Tsai, M. Hirose, F. Kock and L. Brus (Mater. Res. Soc. Symp. Proc. **358**, Pittsburgh, PA 1995), pp. 721 - 731.

7. X. Liu, X. Wu, X. Bao and Y. He, Appl. Phys. Lett. **64**, 220 (1994).

8. C. Courteille, J.-L. Dorier, J. Dutta, Ch. Hollenstein, A. A. Howling and T. Stoto, J. Appl. Phys. **78**, 61 (1995).

9. K. Tachibana, T. Shirafuji, Y. Hayashi, S. Maekawa and T. Morita, Jpn. J. Appl. Phys. **33**, 4191 (1994).

10. G. Cicala, P. Capezzuto, G. Bruno, L. Schiavulli, G. Perna and V. Capozzi, J. Appl. Phys. **80**, (11) xx (1996).

11. X. L. Zheng, W. Wang and H. C. Chen, Appl. Phys. Lett. **60**, 986 (1992).

CHARGE CARRIER ABSORPTION IN DOPED MICROCRYSTALLINE SILICON FILMS

M. HEINTZE*, E. LOTTER**, C.-D. ABEL*, AND M.B. SCHUBERT*
* IPE, University of Stuttgart,Pfaffenwaldring 47, 70569 Stuttgart, Germany
** Center for Solar Energy and Hydrogen Research Baden-Wuerttemberg, Heßbruehlstraße 21c, 70569 Stuttgart, Germany

ABSTRACT

The infrared absorption in doped microcrystalline silicon thin films is analyzed by modelling the complex permittivity as the sum of the contributions resulting from interband transitions and from absorption by charge carriers. Their density and the drift mobility within grains is described by free carrier motion according to the Drude theory. However, for a good fit to experimental data a small trapping energy, reflecting the effect of grain boundaries, is included in the model. By comparing results obtained from the analysis of infrared data with conductivity and Hall measurements in films grown in a very high frequency (VHF) plasma and by hot-wire chemical vapor deposition (CVD), we show that infrared transmission measurements provide a simple access to transport parameters in these films.

INTRODUCTION

Microcrystalline silicon (μc-Si) is gaining considerable interest as a thin film semiconductor. Possible applications include thin film solar cells, where highly conductive doped layers may be combined with a thin amorphous absorber or alternatively with a microcrystalline absorber, if the thickness is raised to a few micrometers. For the growth of these films at relatively high rates from the gas phase and at temperatures between 200 and 400°C two methods are currently being investigated, plasma enhanced chemical vapor deposition (PECVD) in a capacitively coupled reactor using very high frequencies (VHF) [1] or hot-wire CVD [2].

Highly conductive doped microcrystalline silicon films, as envisaged for contact layers in solar cells, show significant absorption due to free charge carriers [3]. In crystalline silicon this phenomenon has been frequently employed for the material characterization, deducing the carrier density and mobility, if the effective mass of the charge carriers is known. This analysis has also been extended to doped poly- or microcrystalline thin films. In an early study [3], as well as in later work on phosphorous doped polycrystalline films prepared by CVD [4], the carrier density and mobility was evaluated using the Drude theory. Under the assumption of a wavelength independent refractive index n, the inverse of the absorption coefficient α^{-1} is proportional to λ^{-2} and this was shown to lead to satisfactory results for wavelengths smaller than 4 - 5 μm [3][4]. However, typical highly doped microcrystalline silicon films exhibit a broad minimum in the IR-transmission around λ = 10 μm, clearly in contradiction with $\alpha^{-1} \propto \lambda^{-2}$. This observation has been tentatively attributed to grain size effects [5]. In order to overcome the simplification of assuming a constant refractive index, both transmittance and reflectance data may be analyzed thus yielding the wavelength dependence of both n and α [6].

In this contribution we show that an evaluation of transmittance spectra is possible without additional reflectance measurements by modelling the dielectric function of the films and fitting the transmission predicted by the model to the experimental results. Transport

Mat. Res. Soc. Symp. Proc. Vol. 452 © 1997 Materials Research Society

properties evaluated from infrared optical absorption are compared with electrical measurements for samples prepared by VHF-PECVD and by hot-wire deposition.

EXPERIMENT

The preparation of microcrystalline silicon films by VHF-PECVD was carried out in a parallel plate reactor at an excitation frequency of 170 MHz. Two series of boron doped samples were investigated, in one of which the conductivity was controlled by varying the gas phase doping ratio B_2H_6/SiH_4 up to 0.5% and a second series, in which the gas pressure was changed at a constant doping ratio of 0.5%. Increasing the gas pressure from 0.3 to 1 mbar leads to an increase in conductivity by a factor of 5 as the result of an increased doping efficiency [7]. The hot-wire deposition of μc-Si:H was carried out in a high vacuum system using a tungsten filament, 4 cm in length and 250 μm in diameter and centered above the substrate [8]. These films were doped with phosphorous and boron by adding up to 2% phosphine or diborane, respectively, to the silane. Films deposited on glass substrates were characterized electrically by DC conductivity measurements and for some VHF-films by Hall effect measurements. Fourier transform IR spectra were measured on samples grown on high-resistivity crystalline (100) oriented wafers.

MODELLING

The carrier absorption in μc-Si:H thin films was analyzed by modelling the complex permittivity. The contribution of the interband absorption is regarded separately and independently of the effect of carrier absorption. Light absorption due to interband transitions leads to a nearly constant value for ε_r and a refractive index of 3.2 - 3.3 in undoped microcrystalline silicon films in the infrared range and to negligible values for α. As a first step, the refractive index $n(\lambda)$ and film thickness d_f are determined by analyzing the width and the amplitude of interference fringes at energies below the optical gap [9]. If needed, additional information on the thickness may be obtained with a step profilometer.

Upon including free carrier absorption in the dielectric function according to the Drude theory, a broad maximum in the absorption coefficient is obtained [10]. However, the experimental transmission data in μc-Si:H are not well reproduced by the simple free carrier model. This is not surprising, since charge transport in μc-Si:H is subjected to considerable trapping in grain boundaries or dislocations [11]. Therefore we introduce a small trapping energy, thus describing the dispersion relation by a single oscillator model in which the damping is expressed by the electrical conductivity. The real part of the permittivity ε_1 is the sum of ε_r resulting from interband transitions and the contribution of the "quasi-free" carriers:

$$\epsilon_1(\omega) = \epsilon_r + \frac{e\,\sigma}{\epsilon_0\, m^*\, \mu} \frac{\omega_0^2-\omega^2}{(\omega_0^2-\omega^2)^2 + \frac{e^2}{\mu^2 m^{*2}}\omega^2} \tag{1}$$

For the imaginary part ε_2 the following is obtained:

$$\epsilon_2(\omega) = \frac{e^2\,\sigma}{\epsilon_0\, m^{*2}\, \mu^2} \frac{\omega}{(\omega_0^2-\omega^2)^2 + \frac{e^2}{\mu^2 m^{*2}}\omega^2} \tag{2}$$

Here ω, ε_0, m^*, are the angular frequency, free space permittivity and effective mass respectively, and the trapping energy is expressed by $\hbar\omega_0$. Using these equations, the complex permittivity (ε_1 and ε_2) due to interband and carrier absorption is computed. The film transmittance predicted by the model is then fitted to the experimental data by varying the three parameters N, μ and ω_0. The absorbance α of the films can be calculated either in a coarse approximation using n only from the interference analysis or, more refined, with the wavelength dependent n from eqs. 1 and 2.

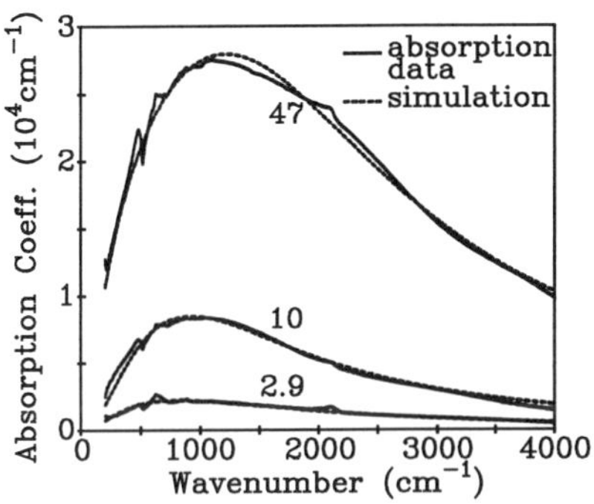

Fig. 1: Comparison of experimental IR-absorption spectra of boron doped μc-Si:H with modelled carrier absorption. The numbers indicate the carrier densities in units of 10^{19} cm^{-3}.

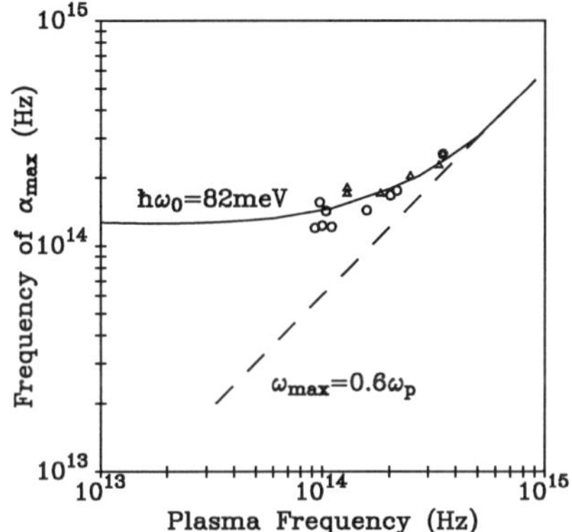

Fig. 2: Frequency of the absorption maximum vs. plasma frequency in VHF-samples, o: doping series, Δ: pressure series.

RESULTS

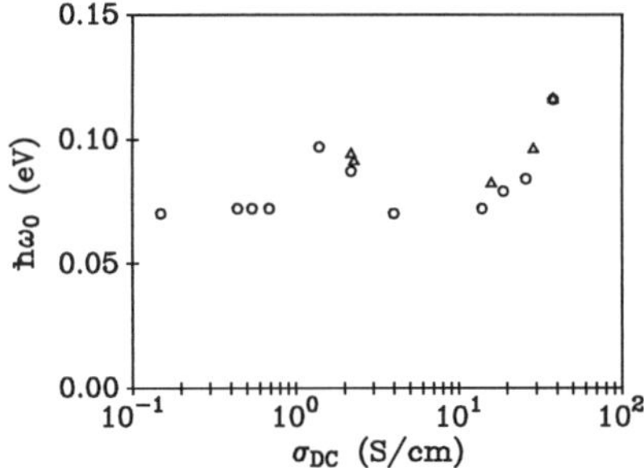

Fig. 3: Trapping energy $\hbar\omega_0$ as a function of the DC-conductivity in VHF-samples, o: doping series, Δ: pressure series.

The quality of the fit obtained to the experimental data is illustrated in Fig. 1. The absorption coefficient α, calculated from transmission data using $n(\lambda)$ as obtained from the complex permittivity, is shown as a function of the wavenumber for three differently doped samples. It can be seen that the maximum in α is closely reproduced by the model calculation.

The position of the maximum in α is determined by the trapping energy and the plasma frequency $\omega_p = (e^2N/\varepsilon_0\varepsilon_r m^*)^{1/2}$. This is shown Fig. 2. For $\omega_0 = 0$ the frequency of the absorption maximum closely follows the relation $\omega_{max} = 0.6\omega_p$, indicated by a dashed line. If the trapping energy $\hbar\omega_0$ is included, it determines the lower limit of α_{max}, shown by the solid line. The data points included in the graph are values for VHF samples and show that $\omega_{max} \geq \hbar\omega_0$ in typical films. In Fig. 3 the trapping energy is plotted as a function of the DC conductivity for the p-type VHF-samples. In these films the DC conductivity is determined by the charge carrier density, whereas the effect of preparation conditions on the Hall mobility is negligible [7]. The trapping energy lies between 70 and 80 meV, somewhat higher values of up to 110 meV are

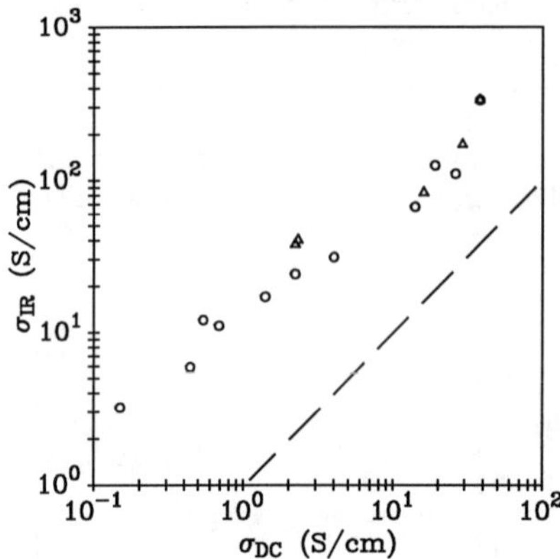

Fig. 4: The electrical conductivity obtained from IR-data as a function of the DC conductivity in VHF-samples,
○: doping series, △: pressure series.

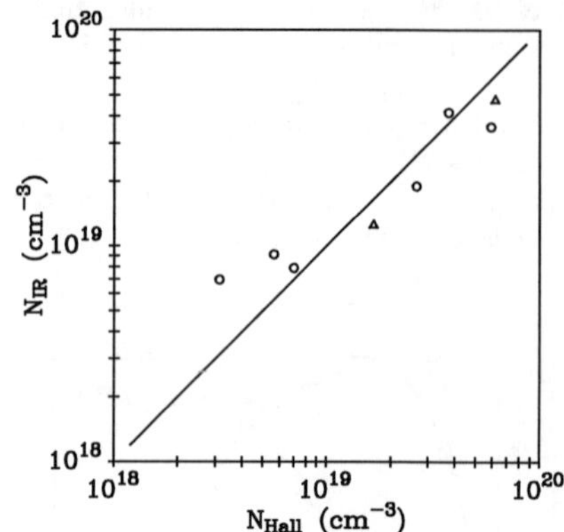

Fig. 5: Carrier density from IR-data plotted against results from Hall measurements in VHF-samples,
○: doping series, △: pressure series.

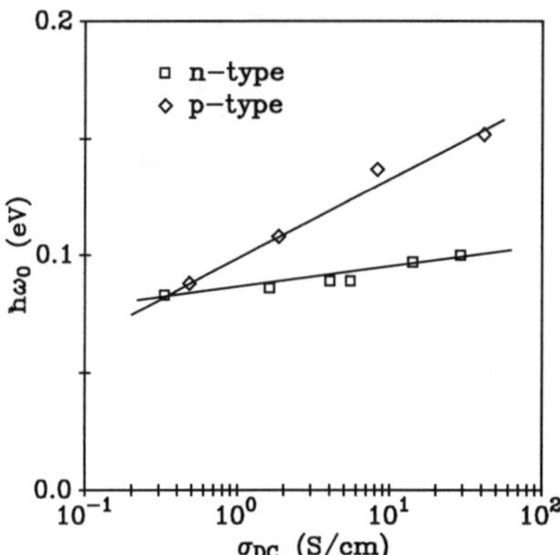

Fig. 6: Trapping energy in hot-wire samples as a function of the DC conductivity.

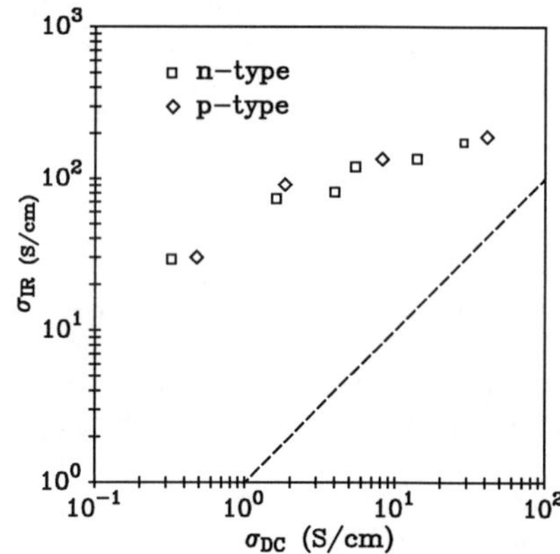

Fig. 7: The electrical conductivity obtained from IR-data as a function of the DC conductivity in hot wire samples.

found in heavily doped films.

In Fig. 4 the conductivity σ obtained in VHF samples from the IR-absorption is plotted versus data from DC conductivity measurements at room temperature. It can be seen that over a range of almost three orders of magnitude σ_{IR} is nearly proportional to σ_{DC}, with $\sigma_{IR} \approx 10\ \sigma_{DC}$. The corresponding carrier density obtained from Hall measurements is compared with IR results in Fig. 5, using the effective mass for holes in silicon $m/m^* = 2.1$ [12]. A close agreement of the infrared and Hall data is observed, with the scatter of the data only reflecting experimental uncertainties.

The trapping energy in n- and p-type hot wire samples is shown in Fig. 6. $\hbar\omega_0$ is nearly independent of the doping in phosphorous doped films, whereas a pronounced increase is observed for boron doped films. The more pronounced increase observed in the p-type hot wire films as compared to the plasma-CVD samples is most likely due to the higher boron concentration of 2% employed as opposed to the maximum concentration of 0.5% used in the VHF deposition. The comparison of the conductivities is shown in Fig. 7. Similar to the VHF material $\sigma_{IR} > \sigma_{DC}$, however with a more pronounced dependence on the doping.

DISCUSSION

The optical model presented here allows a good fit to experimental IR absorption data up to 40 μm. A benefit of the present method lies in the ability to describe the carrier absorption in the range of its maximum and not only at wavelengths below 5 μm. In addition, structural information may be deduced from the trapping energy employed in the model. The "Drude mobility" is considerably higher than the Hall mobility. This discrepancy has been observed previously and was attributed to the grain boundaries [3] or to different averaging over the carrier distribution [6]. The data on the VHF samples show that the carrier density may be evaluated from the IR-absorption and independent from Hall measurements.

The observation of a trapping energy of around 70 meV and somewhat higher in heavily boron doped films underlines the role of the amorphous grain boundary tissue in controlling the transport properties of μc-Si:H: in amorphous silicon a similar widening of tail states upon boron doping has been found by electron spin resonance [13] and by optical absorption in the near infrared [14].

The ratio of the electrical conductivity from IR-absorption to the DC conductivity is much larger in lightly doped hot-wire films (Fig. 7) than in VHF-PECVD material (Fig. 4). This is likely to be due to structural differences resulting from the different preparation methods: whereas the hot wire films currently investigated are subject to oxidation [8], the films prepared in a VHF-plasma show no oxygen incorporation. In a VHF plasma the growth surface is subject to a considerable flux of positive ions [15]. Energetic ions have been shown to lead to reduction in grain size, but to compressive film stress, which prevents oxygen incorporation at the grain boundaries [16][17]. Apart from the very large grain size reported in hot wire material [2], it is remarkable that the films could be doped with up to 2% boron without inducing a transition to amorphous material [8]. Possibly the large increase of $\sigma_{IR} / \sigma_{DC}$ in lightly doped hot-wire films is the result of the different grain boundary structure: whereas the DC conductivity has been reported to depend significantly on the oxygen incorporation [16], we have not observed an effect on the IR absorption.

CONCLUSIONS

We have shown that the broad absorption band of doped microcrystalline silicon films in the infrared can be understood quantitatively in terms of carrier absorption, if the complex permittivity of the films is modelled accounting for carrier trapping in grain boundaries. The carriers are treated as "quasi" free by incorporating a small trapping energy into the Drude model. The charge carrier absorption in the infrared is well suited for the investigation of structural and electronic properties resulting from the preparation method.

ACKNOWLEDGMENTS

We would like to thank S. Schöninger for Hall measurements. Funding by the BMBF under contract 0329617 is acknowledged.

REFERENCES

1. S. Oda, J. Noda, and M. Matsumura in Amorphous Silicon Technology, ed. by A. Madan, M.J. Thompson, P.C. Taylor, P.G. LeComber, and Y. Hamakawa, (Mater. Res. Soc. Symp. Proc. **118**, Pittsburgh PA 1988) p. 117
2. A.R. Middya, J. Guillet, J. Perrin, A. Lloret, and J.E. Bourée, Proc. 13th European Photovoltaic Solar Energy Conference (1995), ed. by W. Freiersleben, W. Palz, H.A. Ossenbrink, and P. Helm, (H.S. Stephens & Ass., Bedford UK) p. 1704
3. S. Usui and M. Kiguchi, J. Non-Cryst. Sol., **34**, 1 (1979)
4. Y. Mishima, M. Hirose, and Y. Osaka, J. Appl. Phys. **51**, 1157 (1980)
5. F. Finger, K. Prasad, S. Dubail, A. Shah, X.-M. Tang, J. Weber, and W. Beyer in Amorphous Silicon Technology - 1991 edited by A. Madan, Y. Hamakawa, M.J. Thompson, P.C. Taylor and P.G. LeComber (Mat. Res. Soc. Symp. Proc. **219**, Pittsburgh, PA 1991), p. 383
6. K. Peter, G. Willeke, K. Prasad, A. Shah, and E. Bucher, Phil. Mag. **B 69**, 197 (1994)
7. M. Heintze and M. Schmitt in Amorphous Silicon Technology - 1996, ed. by M. Hack, E.A. Schiff, S. Wagner, A. Matsuda, and R. Schropp (Mat. Res. Soc. Symp. Proc **420**, Pittsburgh PA), to be published
8. H.N. Wanka, R. Zedlitz, M. Heintze, and M.B. Schubert in Proc. 13th European Photovoltaic Solar Energy Conference, ed. by W. Freiesleben, W. Palz, H.A. Ossenbrink, and P. Helm, (H.S. Stephens and Ass., UK 1995), p. 1753
9. R. Swanepoel, J. Phys. E: Sci. Instrum. **16**, 1214 (1983)
10. M. Heintze, C.-D. Abel, E. Lotter, and M.B. Schubert, to be published.
11. G. Willeke, in Amorphous and Microcrystalline Semiconductor Devices, Vol. II edited by J. Kanicki (Artech House, Boston 1992) p. 55
12. "Semiconductors" by R.A. Smith, (Cambridge at the University Press, 1959)
13. M. Stutzmann, D.K. Biegelsen, and R.A. Street, Phys. Rev. **B 35**, 5666 (1987)
14. K. Pierz, W. Fuhs, and H. Mell, Phil. Mag. **B 63**, 133 (1991)
15. M. Heintze and R. Zedlitz, J. Non-Cryst. Sol. **164-166**, 55 (1993)
16. H. Curtins and S. Vepřek, Sol. State Commun., **57**, 215 (1986)
17. M. Komuna, H. Curtins, F. -A. Sarott, and A. Vepřek, Phil. Mag. **B 55**, 377 (1987)

PULSED ESR STUDY OF THE CONDUCTION ELECTRON SPIN CENTER IN μc-Si:H

JIANG-HUAI ZHOU [a], SATOSHI YAMASAKI [b], JUNICHI ISOYA [b,c]
KAZUYUKI IKUTA [b], MICHIO KONDO [d], AKIHISA MATSUDA [d]
KAZUNOBU TANAKA [b]
[a] Joint Research Center for Atom Technology-Angstrom Technology Partnership
1-1-4 Higashi, Tsukuba, Ibaraki 305, Japan, zhou@jrcat.or.jp
[b] Joint Research Center for Atom Technology-National Institute for Advanced Interdisciplinary Research, 1-1-4 Higashi, Tsukuba, Ibaraki 305, Japan
[c] University of Library and Information Science, 1-2 Kasuga, Tsukuba
Ibaraki 305, Japan
[d] Electrotechnical Laboratory, 1-1-4 Umezono, Tsukuba, Ibaraki 305, Japan

ABSTRACT

The spin-lattice relaxation times (T_1) of conduction electron (CE) and dangling bond (DB) centers in μc-Si:H have been directly measured using the 3-pulse inversion recovery method. For both CE and DB, the inversion recovery curve follows a stretched exponential form. T_1 of DB is about twice the T_1 of CE, however the temperature dependence of T_1 seems to be the same for both CE and DB and can be approximated by T^{-4}. While the DB echo decay is modulated by both ^{29}Si and ^1H nuclei, we found no modulation of the CE echo decay by the H nucleus, indicating that CE centers are located in H-depleted phases in μc-Si:H. The modulation results are direct evidence that CE centers are located in the crystalline grains and DB centers in the amorphous phases.

INTRODUCTION

Hydrogenated microcrystalline silicon (μc-Si:H) is attracting increasing attention because of the potential to use it for realizing high-stability and high-efficiency low-cost solar cells [1,2]. μc-Si:H is a multi-phase material; it consists of crystalline grains and amorphous tissue regions. As with amorphous silicon, electron spin resonance (ESR) is a powerful tool for studying the electronic and defect structures of μc-Si:H. Two ESR centers have been observed in μc-Si:H; one is the dangling bond (DB) center at g~2.005 and the other is the so-called conduction electron (CE) at g~1.998. Both centers are believed to play an important role in carrier transport in μc-Si:H. In the dark, the CE signal may or may not appear, depending on the sample. The exact reason for this is not clear, however the CE signal is readily observed at low temperatures if the sample is illuminated with light.

It has been generally assumed that the CE resonance signal comes from free electrons in the crystalline grains. This assumption is based mainly on the similarity between the CE signal in μc-Si:H and that observed in n-type crystalline silicon [3], and there has been no direct evidence for this assignment. Furthermore, there has been little work on the dynamic properties (i.e. spin relaxation) of the two spin centers in μc-Si:H. The nature of the spin relaxation processes contains important information on the structure of the lattice in the vicinity of the spin center. In this paper, we report on the pulsed-ESR study of the spin relaxation processes in μc-Si:H and discuss the spatial distribution of the spin centers and the microscopic structure of μc-Si:H.

Mat. Res. Soc. Symp. Proc. Vol. 452 © 1997 Materials Research Society

Table I. Main characteristics of the samples studied in this work.

sample No	thickness (μm)	vol. fraction (%)	dark DB density (cm^{-3})	RT σ_d (S/cm)	RT E_a (eV)	CE signal in the dark
#33002	0.42	90	$1.4x10^{16}$	$2x10^{-5}$	0.25	Yes
#33003	0.75	60	$1.2x10^{16}$	$3x10^{-5}$	0.22	No

EXPERIMENT

The thin film μc-Si:H samples studied in this work were prepared from highly H_2-diluted SiH_4 using the conventional PECVD method [4]. The film was deposited on Al foil held at 250 °C. The chamber pressure during deposition was 50 mTorr. Flakes of the film were collected by etching off the Al foil. For the ESR measurements, the flakes were first annealed at 190 °C in vacuum for 2 hrs and then sealed into a quartz tube filled with 1 atmosphere He gas. Two samples were studied and the main characteristics of the samples are given in Table I.

In a pulsed ESR experiment, one measures the echo signal intensity of the spins following a particular sequence of pulsed excitations. To measure the spin-lattice relaxation time (T_1), we used the 3-pulse inversion recovery pulse sequence [5]

$$P(180^\circ) - t - P(90^\circ) - \tau - P(180^\circ) - \tau - \text{echo},$$

where t is scanned and τ is fixed. Here the degree value in the parentheses is the turning angle of that pulse. The phase memory time was measured using the classical 2-pulse Hahn echo pulse sequence

$$P(90^\circ) - \tau - P(180^\circ) - \tau - \text{echo},$$

where τ is scanned.

In order to generate a sufficient signal intensity for the CE resonance, a red laser light (710 nm, 50 mW/cm^2) was used to illuminate the sample.

RESULTS

Echo-Detected ESR Spectrum

Fig. 1 shows the ESR spectrum of one (#33003) of the μc-Si:H samples measured using our pulsed-ESR spectrometer. One can see that in the dark, only the DB signal at g=2.005 appears. Under illumination, however, the CE signal at g=1.998 appears in addition to the DB signal, the separation between the two signals being about 10 G. For sample #33002, the spectra are similar to those shown in Fig. 1, except that this sample shows the CE signal, although weaker than the DB signal, even in the dark. Compared with the spectra measured with the conventional cw ESR, the echo-detected ESR spectra have the same line characteristics.

Spin-Lattice Relaxation

Fig. 2 shows the typical 3-pulse inversion recovery curves for CE and DB centers in our μc-Si:H samples. We found that the inversion recovery curve cannot be described by a simple exponential form for either CE or DB; instead, it exhibits a stretched exponential behavior

$$I(t) = I_0\{1 - 2\exp[-(t/T_1)^{\beta}]\}, \quad (1)$$

where I(t) is the signal intensity at time t, I_0 is a constant, T_1 is the spin-lattice relaxation time, and β (< 1) is the stretch parameter. The solid lines in Fig. 2 are fits to the

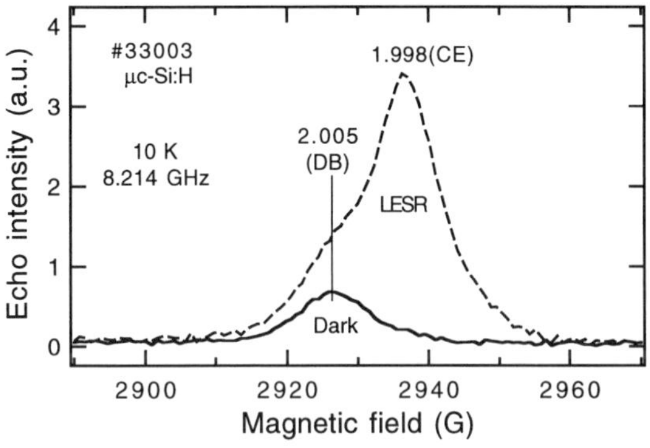

Fig. 1. Echo-detected ESR spectra of μc-Si:H in the dark and under illumination.

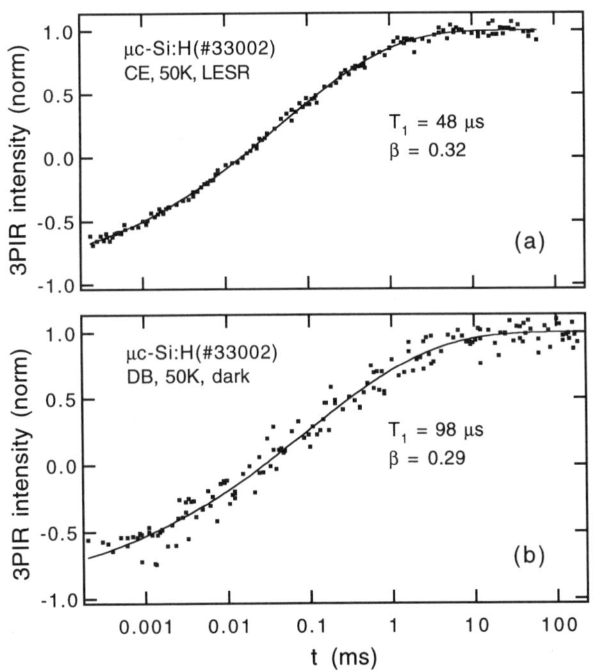

Fig. 2. Typical 3PIR curves for conduction electrons (a) and dangling bonds (b) in μc-Si:H. the solid lines are fits to the experimental data using eq. (1).

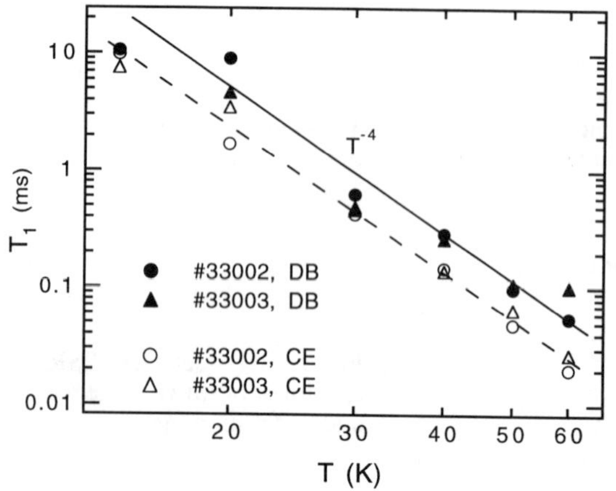

Fig. 3. Temperature dependence of spin-lattice relaxation time (T_1) of CE and DB centers in μc-Si:H.

experimental data using eq. (1).

The temperature dependence of T_1 is shown in Fig. 3. One can see that, despite the different volume fractions of the two samples, there is no apparent difference between the spin-lattice relaxation times in these samples, and this is true for both CE and DB centers. However, T_1 of DB is about twice that of CE. For both CE and DB centers, the time dependence of T_1 can be described roughly by T^{-4} over the temperature range studied in this work.

Two-Pulse ESEEM

Two-pulse ESEEM here stands for two-pulse electron-spin-echo envelope modulation [6,7]. It can be shown that, during the spin echo decay, hyperfine interactions between the unpaired electron and its surrounding nuclear spins (S-I interactions) can produce a regular time variation of the magnetic field at the electron spin. This results in a modulation of the echo signal as the time scale of the pulse sequences is changed [7]. The modulation frequency depends on the nuclear type, and the amplitude of the modulation depends on the distance between the electron and nuclear spins and on the number of equivalent nuclei at a given electron-nuclear distance. An analysis of the echo modulation pattern can give information on the identity, number, and distance of weakly interacting nuclei, and thus on the structure of the material.

The nuclear modulation frequencies can be revealed by Fourier transform of the time-domain two-pulse ESEEM curve. Fig. 4 shows the Fourier transforms of the two-pulse ESEEM curves of CE and DB centers in μc-Si:H. We find that in the CE two-pulse ESEEM, only the double frequency of ^{29}Si nucleus appears, whereas in the DB two-pulse ESEEM, both the double frequency of ^{29}Si nucleus and the fundamental frequency of ^{1}H nucleus appear. The fact that the CE echo decay is modulated by only Si nucleus and the DB echo decay is modulated by both Si and H nuclei indicates clearly that the CE centers are located in H-depleted regions and the DB centers in regions with a relatively high content of H. We believe that this is the first direct evidence that conduction electrons and dangling bonds in μc-Si:H are located in different phases of the material.

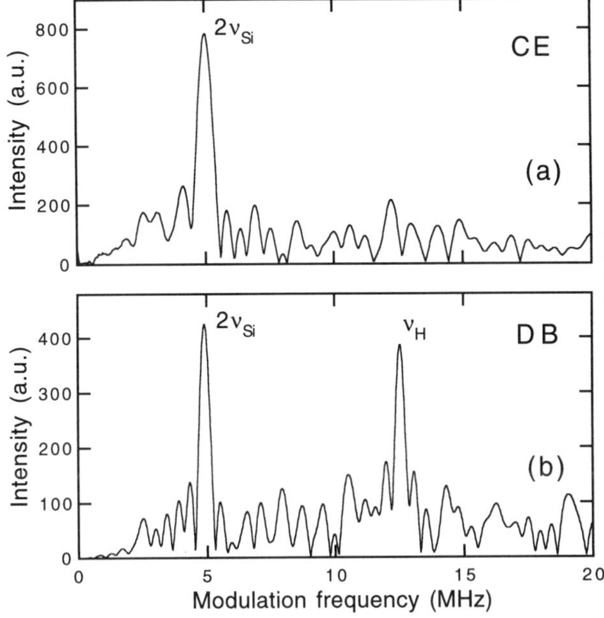

Fig. 4. Fourier transform ESEEM spectra for CE and DB centers in μc-Si:H.

DISCUSSION

The T^{-4} temperature dependence of T_1 (Fig. 3) shows that the spin-lattice relaxation process is neither a simple two-level systems (TLS) process [8,9] nor a simple two-phonon Raman process [10]. A similar temperature dependence of T_1 has been observed for the dangling bond in annealed a-Si:H [11], although T_1 of the DB in a-Si:H is some 10 times longer than T_1 of the DB in μc-Si:H. The T^{-4} temperature dependence of T_1 is difficult to understand, however the fact that CE and DB have the same temperature dependence of T_1 seems to suggest that CE and DB spin-lattice relaxations have a common origin. One candidate for this is the boundary regions between the grains and amorphous tissue regions. It has been suggested that the boundaries may contain high density of surface H and/or SiH clusters [4, 12]. It is possible that the H atoms in the boundaries may be very effective in causing spin-lattice relaxation for both CE and DB in μc-Si:H. Obviously, deuteration will help clarify whether or not H is involved in the spin-lattice relaxations. Another possibility is that certain mobile electron spin species may exist in μc-Si:H. These mobile spins serve as a channel for spin-lattice relaxation of both CE and DB, yet do not show up in the ESR spectrum due, for example, to a broad line width.

Another point worth discussing is the value of T_1 of CE. Our results show that T_1 of CE is only a factor of 2 shorter than that of DB (Fig. 3). However, recently Muller and coworkers mentioned in a paper that T_1 of CE in their μc-Si:H sample can be three orders of magnitude smaller than that of DB [13]. In our experiment, the pulse separation τ was set to be 240 ns, thus spins with T_1 shorter than 240 ns could not be detected. If T_1 of CE was three orders of magnitude smaller than that of DB, then the CE T_1 would be 200 ns at 40 K and still shorter at higher temperatures (see Fig. 3), and thus the CE signal would not show above 40 K. Clearly this is not the case, at least for our samples. There is of course a possibility that some of the CE centers have T_1 shorter than, say, 240 ns and were not picked up by our machine. However, since the CE signal clearly appears in our echo-detected ESR spectrum, we believe that the number of such short-T_1 CE centers, if any, is very small.

CONCLUSIONS

Using pulsed ESR, we found that, for both CE and DB centers in μc-Si:H, the spin-lattice relaxation curve exhibits a stretched exponential behavior (eq. (1)). T_1 of DB is about twice the T_1 of CE, however the temperature dependence of T_1 seems to be the same for both CE and DB and can be approximated by T^{-4}. The ESEEM measurements show that CE and DB centers are spatially separated in μc-Si:H; the former are located in H-depleted regions and the latter in regions with a relatively high content of H.

ACKNOWLEDGMENTS

The authors would like to thank Drs C. Malten and J Muller of Julich, Germany for their kind comments on the CE signal in μc-Si:H. This work, partly supported by NEDO, was performed at the Joint Research Center for Atom Technology (JRCAT) under the joint research agreement between the National Institute for Advanced Interdisciplinary Research (NAIR) and the Angstrom Technology Partnership (ATP).

REFERENCES

1. J. Meier, P. Torres, R. Platz et al. in Amorphous Silicon Technology (MRS Spring Meeting, San Francisco, 1996), to be published.
2. S. Guha, J. Yang, P. Nath, and M. Hack, Appl. Phys. Lett. **49**, 218(1986).
3. F. Finger, C. Malten, P. Hapke, R. Carius, R. Fluckiger, and H. Wagner, Philo. Mag. Lett. **70**, 247(1994), and references therein.
4. A. Matsuda, S. Yamasaki, K. Nakagawa, H. Okushi, K. Tanaka, S. Iizuma, M. Matsumura, and H. Yamamoto, Jpn J. Appl. Phys. 19, L305(1980).
5. M. K. Bowman and L. Kevan, in Time Domain Electron Spin Resonance, edited by L. Kevan and R. N. Schwartz (John Willey & Sons, New York, 1979), p. 67.
6. J. Isoya, S. Yamasaki, H. Okushi, A. Matsuda, and K. Tanaka, Phys. Rev. B **47**, 7013(1993).
7. L. Kevan, in Time Domain Electron Spin Resonance, edited by L. Kevan and R. N. Schwartz (John Willey & Sons, New York, 1979), p. 279.
8. W. A. Phillips, Proc. Roy. Soc. Lond. A **319**, 565(1970).
9. P. W. Anderson, B. I. Halperin, and C. M. Varma, Phil. Mag. **25**, 1(1972).
10. P. G. Klemens, Phys. Rev. 125, 1795(1962).
11. R. Durny, S. Yamasaki, J. Isoya, A. Matsuda, and K. Tanaka, J. Non-Cryst Solids **164-166**, 223(1993).
12. K. Tanaka and S. Hayashi, in Glow-Discharge Hydrogenated Amorphous Silicon, edited by K. Tanaka (KTK Scientific, Tokyo, 1989), p. 39.
13. J. Muller, C. Malten, F. Finger, and H. Wagner, Proc. of the 23rd Intern. Conf. on the Phys. of Semicond., Berlin, July, 1996, to be published.

PHOTOCARRIER RECOMBINATION IN MICROCRYSTALLINE SILICON STUDIED BY LIGHT INDUCED ELECTRON SPIN RESONANCE TRANSIENTS

J. MÜLLER, F. FINGER, C. MALTEN, H. WAGNER
Forschungszentrum Jülich, ISI, D-52425 Jülich, Germany

ABSTRACT

To get information on the density of states distribution and the photocarrier recombination in microcrystalline silicon (μc-Si:H), samples with various amounts of n- and p-type doping are studied with electron spin resonance (ESR) and stationary and time-resolved light induced ESR. The intensity of the dark ESR signals from dangling bonds (DB) and conduction electrons (CESR) is investigated as a function of the doping level. The DB signal has a flat distribution over a wide doping range while the CESR signal strongly increases with n-type doping. Upon illumination with white or infrared light both resonances are enhanced with an intensity that depends on the doping level. The decay of the light induced signal and the dependence in time and intensity of the residual signal on different initial excitation energies and dark/light - sequences is studied. The results are discussed with a schematic band diagram for μc-Si:H. The existence of a potential barrier is proposed which spatially separates photogenerated carriers. A large band-offset between crystalline and disordered regions is further suggested.

INTRODUCTION

Microcrystalline silicon (μc-Si:H) is a phase mixture of amorphous and crystalline regions where the position of the electronic states of both phases and their charge state influence the carrier recombination mechanisms and the electronic transport. It was shown [1,2], that the electron spin resonance (ESR) signal of undoped and slightly doped μc-Si:H material contains contributions from resonances at g=2.0046 and g=2.0052 attributed to dangling bonds (DB) at grain boundaries, in the amorphous phase and in the crystalline grains and a resonance at g=1.997 attributed to conduction electrons (CESR). The CESR resonance was found to correlate with the n-type character of the material while the DB signal maintains a constant intensity over a fairly wide range of Fermi level positions. At low temperatures (T<60K) both resonances are enhanced upon steady state white light illumination. After switching off the illumination the resonances decay with a wide distribution of decay times. In the previous work [1,2] the shift of the Fermi level was obtained only by residual gas doping and thus not well defined. Furthermore samples for conductivity measurements were not available in all cases. In the present paper we used controlled small amounts of doping gases and always prepared samples in the same run on quartz glass for conductivity measurements. The dependence of the dark signal on conductivity can therefore be monitored and the corresponding Fermi level shift can be estimated. In addition, stationary and time resolved light induced ESR (LESR) with various illumination sequences with white light and infrared (IR) light is employed to study the photocarrier recombination mechanisms. The results are discussed in a schematic band model.

EXPERIMENT

Samples were prepared by plasma enhanced chemical vapour deposition at a silane to hydrogen ratio of 3 : 97 and plasma excitation frequencies between 13.56 and 115 MHz. The substrate temperature was 200 °C, the total gas pressure p = 300 mTorr and the discharge power P = 5 - 10 W. These preparation conditions result in material with a high crystalline volume fraction (>90%) and average grain sizes of 20 - 30 nm [3,4]. Phosphine or diborane were used as doping gases. Samples with a thickness of 2-7 μm were deposited on roughened quartz glass substrates for conductivity measurements and on Al foils, which were etched away by HCl after deposition yielding 30 - 120 mg of powder material for ESR measurements. The powder is sealed under He atmosphere in quartz tubes (inner diameter = 4 mm). ESR was measured with an X-band spectrometer and a cylindrical cavity (transverse magnetic mode) with optical

Mat. Res. Soc. Symp. Proc. Vol. 452

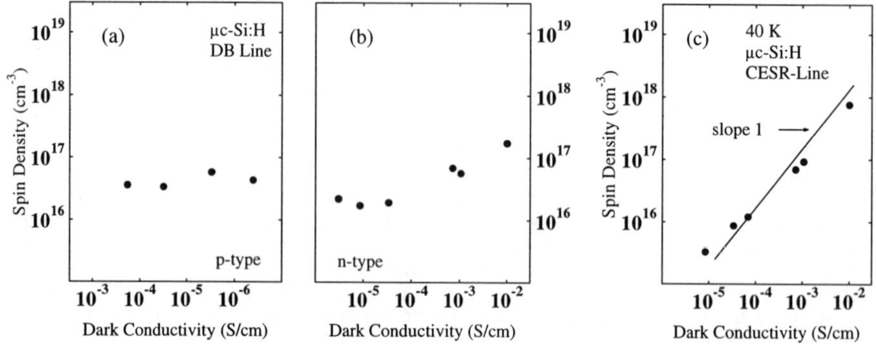

Fig. 1 Dependence of the dangling bond (a,b) and the conduction electron (c) signal on the dark conductivity at room temperature.

viewport in the temperature range 20 - 300 K with 100 kHz modulation frequency and 2-4 G modulation amplitude. For LESR we used a 100 W halogen lamp with a variety of band-pass and cut-off filters. The light source can be switched off with a mechanical shutter and the subsequent decay of the LESR lines is measured at constant magnetic field at the position of the respective resonances with a modulation amplitude of 10 G.

RESULTS

In Fig. 1 the dependence of the DB (a,b) and the CESR (c) signals on the room temperature dark conductivity are shown. The DB signal has a flat distribution over a wide range of conductivities and thus Fermi level positions. The spin densities are in the range $2\text{-}6*10^{16}$ cm^{-3} but increase for the highest n-type samples. If the DBs are considered as defects in an amorphous silicon (a-Si:H) like region, the energy dependence of the DB intensity is much different from what is observed in a-Si:H, where a defect peak with a FWHM of 0.2 eV is found near the center of the mobility gap of 1.75 eV [5]. The CESR line shows a clear correlation (close to 1:1) with the n-type conductivity (Fig. 1c). Note, however, that the CESR line can still be observed in material which is nominally undoped with conductivities in the range 10^{-5} S/cm.

Upon illumination with heat filtered white light from a halogen lamp, generally both resonances are enhanced. Only the sample with the highest n-type conductivity does not show an additional LESR signal. Also the spectra of p-type samples, with no CESR line observable in the dark, contain an LESR-CESR signal. In Fig. 2 the LESR spectra of n- and p-type samples with dark conductivities of $\sigma = 6.9*10^{-5}$ S/cm and $\sigma = 3.0*10^{-6}$ S/cm, respectively, are shown together with the dark signals. For comparison also dark and LESR spectra of an a-Si:H sample are shown. The LESR-CESR line is assigned to photoexcited conduction electrons, while holes are trapped in DB-states making a transition in the charge state of the dangling bonds from D^- to D^0 and thus enhancing the DB signal. This is similar to the LESR of n-type a-Si:H where the holes also get primarily trapped in charged D^- states and thus no holes in deep tail states can be observed. Note that generally conduction holes in crystalline silicon are not detected by ESR as the six-fold degenerate valence band splits up under internal strain which causes a strong lifetime reduction (line broadening). Only under additional strong external strain or with extremely pure material hole resonances can be observed [6,7].

Upon switch off of the illumination the LESR signals decay with a wide distribution of decay times (recombination times) leaving a residual signal at temperatures below 40K for several hours. Such a behaviour is well known from amorphous silicon and related alloys and is attributed to "deep" trapped carriers and a spatial separation of recombination partners. Similar

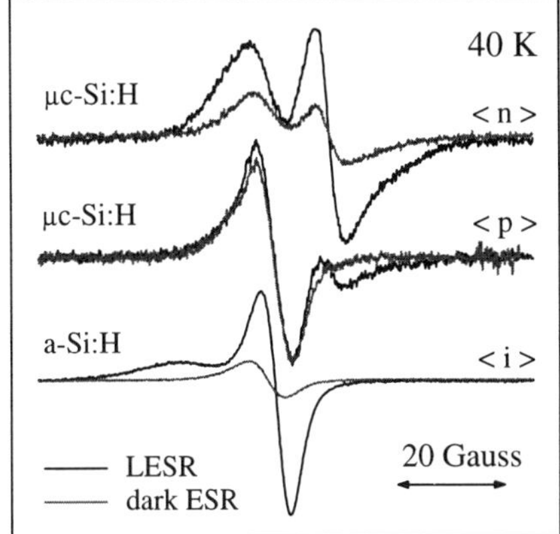

Fig. 2 ESR and steady state LESR signals of n- and p-type μc-Si:H and a-Si:H.

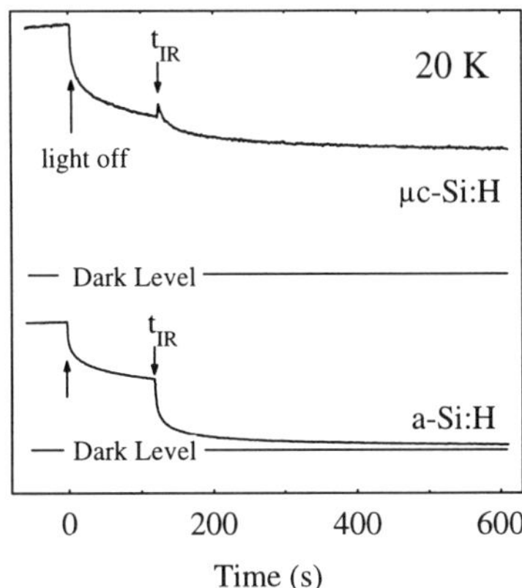

Fig. 3 Decay of the LESR signal in μc-Si:H and response of the residual signal to infrared light.

findings were reported for porous silicon [8] and also in early studies on crystalline silicon [9]. In a-Si:H this residual LESR signal can be quenched optically (by IR illumination) and thermally by heating the sample to 150K through re-excitation of deep trapped carriers into the band from where they can recombine more easily [10]. In Fig. 3 LESR transients of a-Si:H and μc-Si:H samples are shown. For a-Si:H the resonance of electrons in conduction band tail states, for the μc-Si:H the CESR line is monitored. After steady state illumination with white light ($h\nu$>1.75eV) the light is switched off at t=0. The resonance decays with a broad distribution of decay times (very similar for both types of samples). At time t_{IR} the samples are re-illuminated with IR light through a Ge-filter ($h\nu$<0.67eV). In a-Si:H this causes the residual signal to be quenched down to the dark level. In some cases - depending on the dark Fermi level position - a small residual signal remains, which can be quenched only thermally. In μc-Si:H the IR light causes an initial transient enhancement of the LESR signal before a quenching effect sets in. Still the remaining signal is considerably above the dark level. These results are similar for the CESR and the DB signal.

In Fig. 4 the behaviour of this LESR-IR transient in μc-Si:H at two different illumination times t_{IR} is demonstrated. For each trace the dark signal intensity is indicated. The traces are shown with the same vertical resolution so that the signal intensities are directly comparable. With increasing t_{IR} the quenching contribution of the IR light decreases. The IR-LESR signal reaches the same level in both cases. For long enough waiting times, when the residual signal has dropped below the IR-LESR signal, the IR illumination only causes an enhancement. The initial transient enhancement of the CESR line upon illumination is observed for all times t_{IR}, but the decay of the transient becomes slower as t_{IR} increases. For the DB-line a similar behaviour is observed. However, the time constants for the initial decay of the residual signal are faster and for long waiting times t_{IR} no transient is observed anymore.

Also the signal intensity upon IR illumination without preceding white light is shown. The intensity reached is the same as after a white light / dark / IR-light sequence for undoped and weakly doped p-type samples. For weakly doped n-type samples (as the one shown in Fig. 4) the steady state IR-LESR level without preceding white light remains below the level obtained after a sequence white light / dark / IR-light in the time frame investigated here.

In the next figure the LESR transient of the CESR line with repeated IR illumination and dark sequences after preceding white light is shown (Fig. 5). Again the first IR illumination (after 2 minutes) causes a transient enhancement and a leveling out on a steady state level.

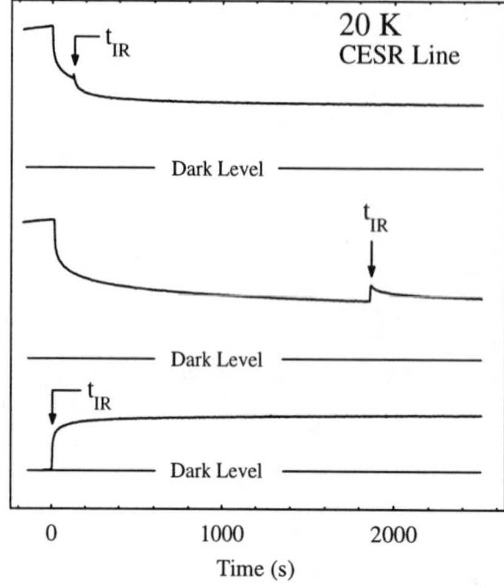

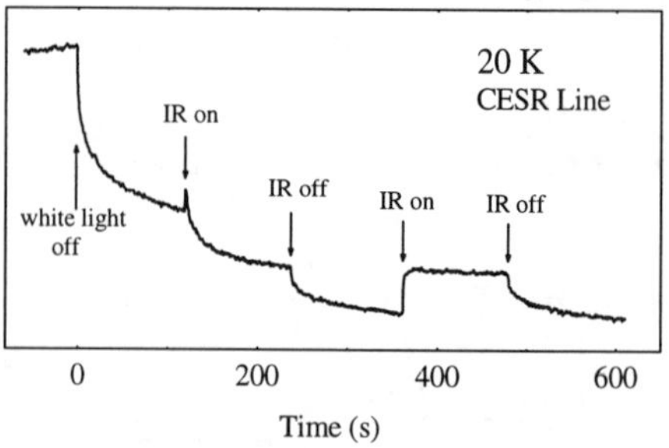

Fig. 5 Repeated IR and dark sequences within a residual CESR-LESR signal in μc-Si:H.

Fig. 4 Response of the residual CESR-LESR signal in μc-Si:H upon IR illumination after two different waiting times t_{IR} (upper two curves) and IR-LESR signal without preceding white light (lower curve).

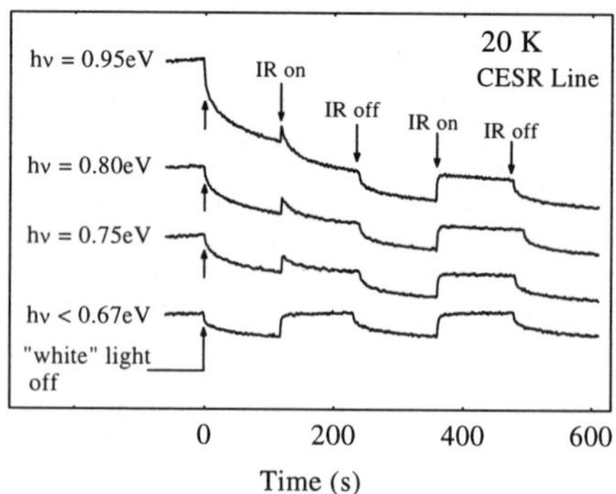

Fig. 6 Repeated IR (hν<0.67eV) and dark sequences within LESR transients in μc-Si:H at different initial excitation energies.

Turning off the IR light causes a decay of the signal below the level which would have been reached without the IR light induced quenching effect. Later IR illumination only causes an enhancement (no fast transient) of the resonance to the steady state IR illumination level. Again a similar behaviour is found for the DB resonance.

In a further experiment the excitation energy of the "white" light was varied. In Fig. 6 the response of the CESR line upon different excitation energies (0.95 eV, 0.80 eV, 0.75 eV, <0.67 eV), a dark sequence and IR quenching with E < 0.67 eV is shown. The excitation bandwidth is about 0.8% FWHM (11 nm). For the Ge-filter (E<0.67eV) the low energy cut-off is determined by the IR-transmission of the optical system (i.e. 0.56 eV). As can be seen, the behaviour upon the IR light after the dark sequence changes considerably with a decrease in the excitation energy. The first transient becomes slower and disappears completely at 0.67 eV. In addition the quenching contribution decreases and for 0.67 eV only enhancement upon IR light is found. Intensity effects can be excluded as a reason for the observed behaviour as optical filters, sample tubes and the quartz glassware all have the same transmission for the different energies used. Also the spectral characteristic of the halogen lamp is very flat in that energy range. Therefore the initial light intensity was almost the same for the upper 3 curves in Fig. 6.

DISCUSSION

To obtain a better overview, in the following the observations will be illustrated with transition diagrams in a schematic band model of μc-Si:H for the case of undoped and weakly doped n-type material.

Two regions of the material are considered: (A) the crystalline grains and (B) a disordered/amorphous/grain boundary region. For region (A) the band gap of c-Si (1.1 eV) is assumed with additional band-tail and midgap states (DB) due to crystalline imperfections and surface effects. For region (B) a simple density-of-states picture for a-Si:H with a mobility gap of 1.75 eV, tail states and deep defects is assumed [5]. For the band-offset between the two regions no conclusive data is available. While several contradictory reports on the band-offset between μc-Si:H or c-Si and a-Si:H exist, the situation for crystalline grains in an a-Si:H-like matrix could again be different. We put a large offset in the conduction band. Together with a Fermi level position close to the conduction band of region (A), this takes into account that in undoped and weakly doped material resonances from CESR and DB are observed simultaneously. Finally we propose a potential barrier between the two regions (A) and (B). This potential barrier could be caused by charge accumulation.

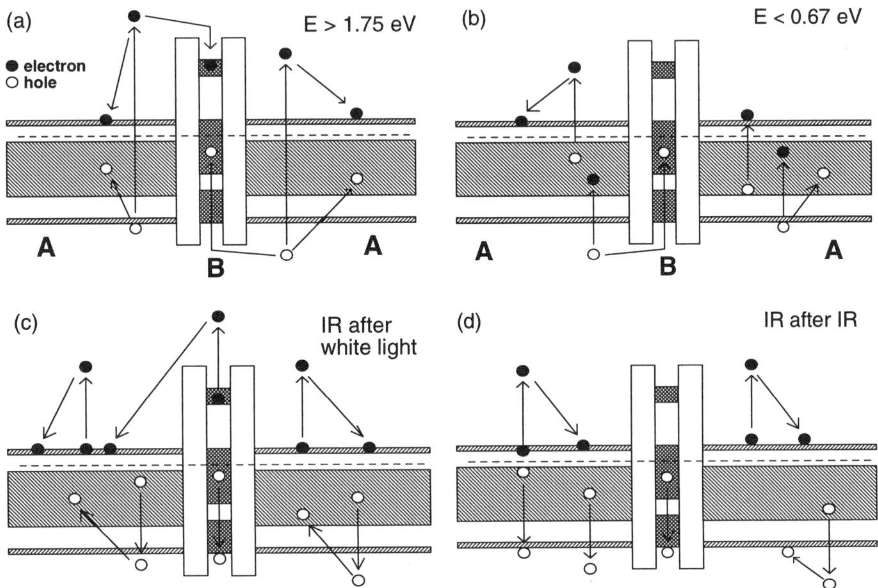

Fig. 7 Schematic band diagram of μc-Si:H and optical transitions for white and infrared light excitation.

The optical excitations mainly involve region (A), as the disordered regions only have a small volume fraction. After excitation the carriers thermalize: electrons to the conduction band edge or shallow tail states, holes into DB states making the transition $D^- \rightarrow D^0$. In addition, with white light illumination (Fig. 7a), carriers have enough excess energy to cross the potential barrier and get trapped in region (B). With IR light (<0.67 eV) this is not possible (Fig. 7b). The threshold energy is between 0.75 and 0.67 eV (see Fig. 6). The carriers which are transferred to region (B) are thus spatially separated from the carriers in region (A). This results in long residual LESR relaxation times. From the residual state, which is different after white- or IR-light illumination, additional IR light will have different effects. After white light, IR light can re-emit electrons from region (B) (Fig. 7c). This will give the transient enhancement signal. It will also re-emit holes from regions (A) and (B) and this will increase the recombination resulting in a quenching signal. Finally, a steady state under IR illumination is obtained. For repeated IR/dark sequences, only the first IR illumination will give a transient enhancement. Once the electrons are re-emitted from their traps in region (B) these can not be filled again with IR.

At present it is not clear if the two regions may be identified with the crystalline grains and the amorphous or grain boundary regions, respectively. It is evident that there must exist some kind of potential barrier which separates states that can be only populated by white light from the remainder of the material. Also there must be some charge separation which results in very long recombination times.

The existence of deep (DB) defect states with different energetic positions in both regions is in good agreement with the dark ESR data, where indications for defect states with different g-values (i.e. different environments) and also a very broad distribution in energy of these defect states are found. We have to stress that a general high potential barrier between adjacent grains is not compatible with the high conductivities and the low activation energies of μc-Si:H at low temperatures [11], i.e. if such structures exist, they can only be present in part of the material.

CONCLUSIONS

Results from ESR and LESR on μc-Si:H with various degrees of n- and p-type doping are used to discuss a simple band diagram and the carrier recombination in this compound material. Dangling bond states have a distribution over a wide energy range involving both the disordered regions and the crystalline grains. The resonance attributed to conduction electrons correlates nicely with the n-type conductivity. The existence of states is proposed, which are separated from the remainder material by a potential barrier. These states can be filled with white light illumination and act as traps. From here carriers can be re-emitted by IR light. Possibly these states are located in the conduction band tail of the disordered/amorphous parts. Further, a considerable conduction band-offset between grain regions and disordered regions is proposed.

ACKNOWLEDGMENTS

We thank M. Stutzmann for pointing out the references on ESR of holes in crystalline silicon. This work is supported by the BMBF.

REFERENCES

1. F. Finger, C. Malten, P. Hapke, R. Carius, R. Flückiger and H. Wagner, Phil. Mag. Lett. **70**, 247 (1994)
2. C. Malten, F. Finger, P. Hapke, T. Kulessa, C. Walker, R. Carius, H. Wagner and R. Flückiger, Mat. Res. Soc. Symp. Proc. **358**, 757 (1995)
3. M. Luysberg, P. Hapke, R. Carius, F. Finger, Phil. Mag. A, in press (1996)
4. F. Finger, R. Carius, P. Hapke, L. Houben, M. Luysberg, M. Tzolov, this conference
5. M. Stutzmann, D.K. Biegelsen, R.A. Street, Phys. Rev. B **35**, 5666 (1987)
6. G.L. Bir, E.I. Butikov, G.E. Pikus, J. Phys. Chem. Solids **24**, 1467 (1963)
7. H. Neubrand, Phys. Stat. Sol. (B) **86**, 269 (1978)
8. H.J. v. Bardeleben, C. Ortega, C. Grosman, V. Morazzani, J. Siejka, D. Stieveenard, J. Lumin. **57**, 301 (1993)
9. A. Honig, Bull. Am. Phys. Soc. **6**, 118 (1961)
10. R. Carius, W. Fuhs, AIP Conference Proc. **120**, 125 (1984)
11. U. Backhausen, P. Hapke, R. Carius, F. Finger, U. Zastrow, H. Wagner, this conference

HALL-EFFECT STUDIES ON MICROCRYSTALLINE SILICON WITH DIFFERENT STRUCTURAL COMPOSITION AND DOPING

U. BACKHAUSEN, R. CARIUS, F. FINGER, P. HAPKE, U. ZASTROW, H. WAGNER
Forschungszentrum Jülich, ISI-PV, 52425-Jülich, Germany, p.hapke@kfa-juelich.de

ABSTRACT

Hall-effect experiments on <n>-type microcrystalline silicon samples with a wide range of structural composition and doping have been performed. For highly doped samples the conductivity σ and the mobility μ show a non-singly activated behaviour while the carrier density is almost temperature independent. The comparison of the carrier density with the phosphorous concentration in conjunction with the conductivity gives strong evidence that the Hall-effect data have to be corrected with the crystalline volume fraction X_c. Furthermore, the increase of the mobility with X_c, which is linked in our case to the grain size, can be explained when the length of the transport paths is taken into account. Our results will be discussed in the framework of different transport models. It is concluded that transport in μc-Si:H can not be explained in terms of thermionic emission over barriers with a well defined barrier height; instead a distribution of barrier heights have to be considered. A transport model is suggested where μc-Si:H is viewed as an interconnected network.

INTRODUCTION

The understanding of the electronic properties of μc-Si:H is an essential requirement for the use of this material in device applications. One of the crucial experiments to get more insight into the electronic properties is the Hall-effect which yields the carrier density and in conjunction with the measurement of the conductivity, the mobility. Previously we reported about an experimental set-up which allows the measurement of the Hall-effect on μc-Si:H down to 10 K [1]. Highly doped samples with a wide range of crystalline volume fractions and grain sizes were investigated. One of the most interesting results was an increase of the mobility with the grain size δ and the crystalline volume fraction X_c. Furthermore, the conductivity and mobility exhibit a non-singly activated behaviour as a function of temperature whereas the carrier density is almost temperature independent. This result can not be explained with the usually applied transport model developed for polycrystalline Si [2].

In this paper we compare at first some results from the Hall-effect measurements published in [1] with SIMS investigations. In the second part, a series of samples with various doping levels is investigated. The results are discussed in the framework of transport models, where thermionic emission over barriers, tunneling through the barriers and percolation is viewed.

EXPERIMENTAL

The Hall-effect measurements were performed with a modulated current technique in a temperature range between 10K-350K which is described in detail in [1]. All samples investigated were deposited with plasma enhanced CVD in the very high frequency range from SiH_4 highly diluted in H_2. The samples are grown at a substrate temperature of 200°C, a total gas pressure of 300 mTorr and a discharge power of 50mW/cm^2. The film thicknesses of the samples is about 500 nm. The first set of samples (series I) is deposited at a plasma excitation frequency of 94.7 MHz at various silane concentrations between 2% and 6.1% in the gas phase. For this series the doping

Mat. Res. Soc. Symp. Proc. Vol. 452 © 1997 Materials Research Society

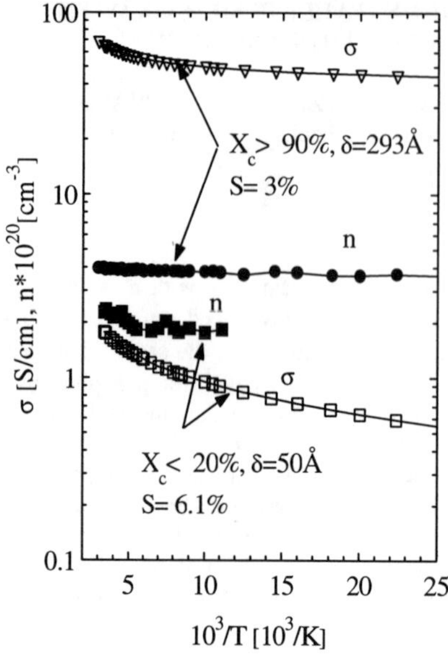

Fig. 1 Conductivity and carrier density for two samples with high and low X_c and grain size δ.

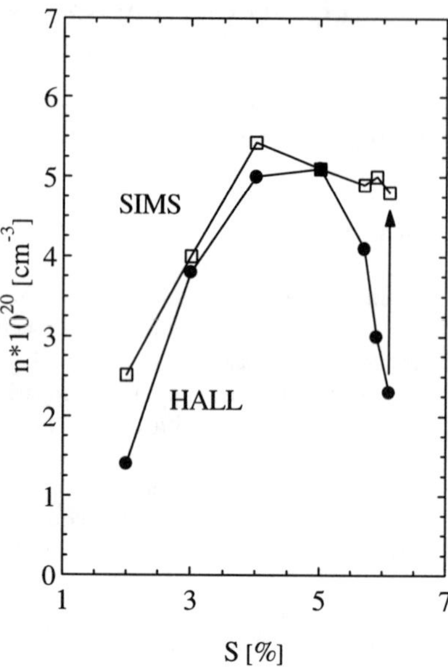

Fig. 2 Carrier density and phosphorous concentration as a function of the silane concentration in the gas phase.

concentration in the gas phase was 2% PH_3 in SiH_4. The SiH_4 concentration S ($S=SiH_4/(SiH_4+H_2)$) is one of the crucial parameters for the structural composition of μc-Si:H. With increasing S the grain size decreases (here: S=2%: δ=350Å→S=6.1%: δ=50Å) and the structure changes from crystalline to amorphous (here: S=2%: X_c>90%→S=6.1%: X_c<20%). The second series (series II) consists of samples with a high crystalline volume fraction (S=3%, X_c>90%) and various phosphorous doping levels between 2% and 20 ppm.

For the structural characterization we performed Raman scattering (excitation wavelength: 488 nm of an Ar-Laser) and X-ray diffraction (grazing incidence). For the determination of the phosphorous concentration we carried out secondary ion mass spectroscopy (SIMS) using near-normal oxygen bombardment at an energy of 6kV and detecting positive secondary ions. The phosphorous concentrations were calibrated by comparing the integrated SIMS ion signals of a phosphorous implanted sample with the implanted ion dose.

RESULTS

In Fig. 1 the conductivity and the carrier density for two highly doped samples (series I) with different crystalline volume fractions X_c and grain sizes δ are plotted as a function of temperature. The conductivity of the sample with a high crystalline volume fraction above 90% (see also Ref. [3], sample A) exhibits the non-singly activated behaviour which is typical for highly doped μc-Si:H. Decreasing the crystalline volume fraction and the grain size (here: X_c<20%, δ=50Å,S=6.1%) leads to a significant change in the conductivity: i) the conductivity decreases and ii) the temperature dependence (slope) is more pronounced. This sample shows also the non-singly activated behaviour of the conductivity. The Hall-effect measurements show that the carrier density is almost temperature independent for both samples, i.e. the temperature dependence of the conductivity is dominated by the mobility (assuming single carrier dominated transport in the crystalline phase, i.e. $\sigma = \mu ne$).

To determine the phosphorous concentration we performed SIMS measurements on this set of samples (series I). In Fig. 2 the results of the SIMS investigations and the Hall-effect data are shown for comparison. The data from the SIMS measurements show an initial increase up to $\approx 5*10^{20}$ cm^{-3} (S=4%) and saturates for larger S. For samples with a high crystalline volume fraction X_c (S < 5%) the SIMS results agree well with the Hall-data. For low X_c (S > 5%, X_c < 50%) differences between the SIMS data and the carrier density appear. The saturation of the phosphorous concentration for S≥4% means that the build-in coefficient of the doping atoms does not change very much. Instead the carrier density changes strongly for S>5%, i.e. in the range where the strongest changes of X_c can be observed.

We recall that the conductivity of the sample with the smallest crystalline volume fraction (Fig. 1, X_c < 20%) is at least two orders of magnitude higher as for amorphous material with the same doping level. Therefore it is proposed that the conductivity is determined by the crystalline phase. This means further that the data obtained from the Hall-effect measurements have to be corrected with the crystalline volume fraction (see arrow in Fig. 2). In this picture the entire crystalline phase contributes to transport, dead ends and not connected grains are neglected. Despite the uncertainty of the exact crystalline volume fraction (depending on the method X_c differs by a factor of three for samples with low X_c [6]) a correction of the data seems not appropriate. But we can conclude from the results of the SIMS and the Hall-effect investigation that the doping efficiency in our highly doped μc-Si:H films is close to unity. We can further conclude that the carrier density is only weakly influenced by the structure (X_c and δ) and the latter dominantly influence the mobility [1,4,5]. TEM investigations on the S=6.1% sample shows that the crystallites are statistically distributed over the film cross-section, i.e. the highly conducting microcrystalline phase embedded in the low conducting amorphous phase can be treated by a percolation network. This implies further that the transport paths are no straight connection between one contact to the other, i.e. the length of a transport path has to be taken into account, which influences the mobility (Fig. 5).

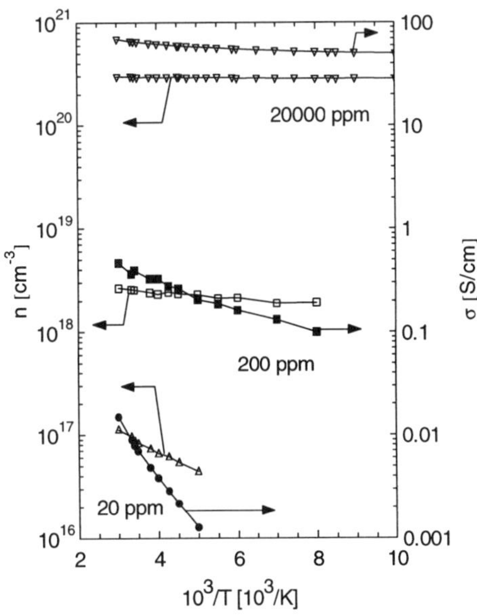

Fig. 3 Carrier density and conductivity for series II as a function of temperature.

Next we turn to series II, which consists of samples of different doping. The Raman spectra of this samples are strongly dominated by the Γ_{25} transition located at 520 cm^{-1} and the ratio of the integrated crystalline and amorphous Raman intensities (as a measure for the crystalline volume fraction, here: X_c >90%) does not change with doping. Fig. 3 shows the temperature dependence of the carrier density and the conductivity. Down to a doping level of 200ppm the carrier density is almost temperature

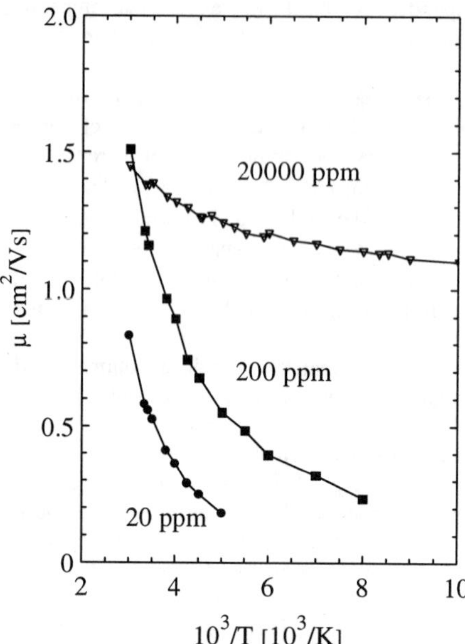

Fig. 4 Mobility for the samples of series II as a function of temperature.

independent. At a doping level of 20ppm a decrease of the carrier density with decreasing temperature can be observed. The conductivity exhibits again the non-singly activated behaviour and decreases with doping. The y-axis range of both quantities is the same, thus n and σ can be compared directly.

In Fig. 4 the mobility of the samples of series II is plotted. Again, we observe a non-singly activated behaviour as a function of temperature, the mobility decreases slightly and the temperature dependence is more pronounced with decreasing doping. The comparison of Fig. 3 and 4 reveals that for this samples with high X_c and high doping levels (2% and 200 ppm) the absolute values of the conductivity is dominantly determined by the carrier density and only weakly dependent on the mobility (as expected, because of the constant crystalline volume fraction). Instead, the temperature dependence is dominantly determined by the mobility, because the carrier density is only weakly temperature dependent (similar to series I). For the sample with the lowest doping level (20ppm) the temperature dependence of both the carrier density and the mobility determines the temperature dependence of the conductivity.

DISCUSSION

At first we summarize the results given above. For highly doped samples the conductivity and the mobility changes strongly with the crystalline volume fraction and the grain size while the carrier density remains high. The comparison of Hall-effect and SIMS measurements shows that the doping efficiency is close to unity. Varying the doping level of samples with a high crystalline fraction leads to a strong dependence of the conductivity on the carrier density, while the mobility is only weakly dependent on the doping. The non-singly activated behaviour of the conductivity gives evidence that the transport in μc-Si:H is controlled by a distribution of barrier heights and/or additional transport processes like tunneling through the barriers have to be considered.

The results described above will be discussed now in the framework of transport models. A popular transport model which is often used to explain the electronic properties of μc-Si:H is the transport over grain boundary barriers with well defined barrier heights [2]. One of the characteristics of this model is that for highly doped poly-Si the conductivity and the mobility (the carrier density is temperature independent) shows a singly activated behaviour as a function of temperature [2,7]. Trivial enough, if the transport data of μc-Si:H are evaluated over a limited temperature range [5,8,9] a linear dependence of $\ln\sigma$ (and μ) over 1/T is received, which is in general not the case if a wider temperature range is investigated [10-12]. To describe the non-singly activated behaviour of the conductivity as a function of temperature (see Fig. 1 and 3) we assume in a first approximation parallel transport paths which are characterized by different barrier heights

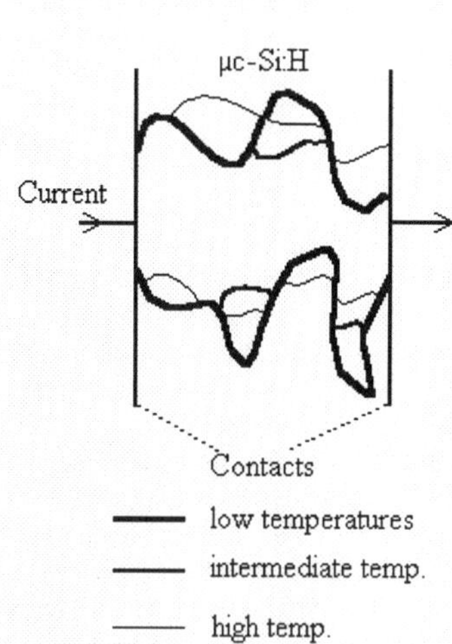

Fig. 5 Model considering parallel and serial transport paths within individual paths.

(E_{B1}<E_{B2}...<E_{Bj}). At low temperatures the path with the lowest activation energy (E_{B1}) is used for the transport. While the temperature increases, more and more parallel transport paths (E_{B2}...E_{Bj}) are added. Indeed, the temperature dependence of the conductivity (mobility) can be described very well but the physical meaning of the exponential prefactor (which is direct proportional to the grain size) is questionable. Even in the low temperature regime grain sizes in the order of 1Å are necessary to achieve a reasonable description of the temperature dependence, which is of cause nonsense. Another experimental fact, which can not be explained with this model, is the increase of the mobility with the crystalline volume fraction (density of parallel transport paths).

In the next step not only the thermionic transport over the barrier but also tunneling through the barrier is considered. Simulations of the transport properties considering thermionic emission and tunneling [12] show that it is possible to model the experimental data in the temperature range 300-70 K. But, one of the characteristics of tunneling processes is the temperature independence of the tunnel current at very low temperatures, i.e. the conductivity and the mobility should be constant at very low temperatures which is contrary to our results, even for low temperatures (<70K) the conductivity decreases further [1]. Because of these discrepancies, the transport over grain boundaries with well defined barriers [2] and the thermionic emmission together with tunneling [12] has been rejected for a description of the transport properties in μc-Si:H.

From the experimental data described above the following qualitative transport model is suggested [13]. Similar to poly-Si the transport properties are strongly dominated by potential barriers created through interface states at the grain boundaries. Because of a wide range of grain sizes, amorphous interface layers and voids it is reasonable to assume a (broad) distribution of barrier heights. But in contrast to our considerations above (parallel paths) one should consider parallel and serial transport paths, i.e. an interconnected network (Fig. 5). At low temperatures the carriers have to path a long way from one contact to the other, i.e. the carriers are following the path with the lowest resistance (small activation). Increasing the temperature enables the carriers to overcome barriers with higher activation energies within an individual transport path. This means that the average barrier height (activation) of an individual transport path will increase, which explains the non-singly activated behaviour of the mobility and conductivity as a function of temperature. From the qualitative picture drawn in Fig. 5 follows also, that with increasing temperature the length of an individual transport path is shortened. At high crystalline volume fractions the path length is the shortest because the probability to find a short low resistance path is high, which also means the mobility is high. Instead, decreasing the crystalline volume fraction results in longer transport paths (percolation) and therefore in a lower mobility. If we consider samples with high crystalline volume fractions but different doping we expect from our model that the mobility remains high compared to the strong changes of the carrier density (Fig. 3 & 4) because the path length is short. The doping mainly influences the carrier density and the height of the grain boundary potential barriers and thereby the temperature dependence of the mobility which is more pronounced at low doping levels. Therefore, for an explanation of the decrease and the temperature dependence of the mobility with doping, one has to consider the length of transport paths and the barrier height.

CONCLUSIONS

Hall-effect measurements on μc-Si:H combined with SIMS investigations revealed that the doping efficiency in highly doped μc-Si:H is close to unity and the carrier density in μc-Si:H has to be corrected by the crystalline volume fraction. Conductivity measurements on a sample with a small crystalline volume fraction (X_c <20%) give evidence that transport in highly doped μc-Si:H should be considered as a percolation problem, which implicates that the crystalline volume fraction and the length of a transport path has to be taken into account to explain the electronic properties. From this point of view the increase of the mobility with the crystalline volume fraction can be easily explained, because with increasing X_c the length of a path decreases.

For an explanation of the non-singly activated behaviour of the conductivity, μc-Si:H is viewed as an interconnected network. Starting at low temperatures the transport path is long. With increasing temperature more and more additional transport path with rising activation energies within one path are opened up and shorten the path length. This implies that the average activation energy continuously increases which leads to an increase of the conductivity and mobility with increasing temperature with the typical non-singly activated behaviour.

ACKNOWLEDGMENTS

The authors thank R. Otto for the development of the electronics concerning the Hall-effect apparatus and L. Houben for performing X-ray diffraction and TEM measurements. This work is supported by the Bundesministerium für Bildung und Forschung (BMBF), Germany.

REFERENCES

1. P. Hapke, U. Backhausen, R. Carius, F. Finger and S. Ray, in Amorphous Silicon Technology-1996, edited by M. Hack, E.A. Schiff, S. Wagner, A. Matsuda, R. Schropp (Mater. Res. Soc. Proc. **420**, Pittsburgh, PA, 1996), in press.
2. J.W. Seto, J. Appl. Phys. **46**, 5247 (1975).
3. M. Luysberg, P. Hapke, R. Carius and F. Finger, Phil. Mag. A (1996), in press.
4. A. Matsuda, J. Non-Crystalline Solids **59&60**, 767 (1983).
5. W.E. Spear, G. Willeke, P.G. LeChomber and A.G. Fitzgerald, J. de Phys. **42**, C4-257 (1981).
6. F. Finger, R. Carius, P. Hapke, L. Houben, M. Luysberg and M. Tzolov, 9th International School on Condensed Matter Physics, (Varna, Bulgaria, 1996), to be published by World Scientific.
7. J.W. Orton and M.J. Powell, Rep. Prog. Phys. **43**, 1263 (1980).
8. G. Willeke in Amorphous & Microcrystalline Semiconductors Devices, Vol. 2, edited by J. Kanicki (Artech House, Boston, London, 1992), 55.
9. G. Lucovsky and C. Wang, in Microcrystalline Semiconductors: Material Science & Devices, edited by P. Fauchet, C.C. Tsai, L. Canham, I. Shimizu, Y. Aoyagi (Mater. Res. Soc. Proc. **283**, Pittsburgh, PA, 1992), pp. 443-454.
10. M. Nakata and I. Shimizu, see book reference 9, pp. 591-596.
11. F. Finger, C. Malten, P. Hapke, R. Carius, R. Flückiger and H. Wagner, Phil. Mag. Lett. **70**, 247 (1994).
12. A. Di Nocera, A. Mittiga and A. Rubino, J. Appl. Phys. **78**, 3955 (1995).
13. R. Carius, F. Finger, M. Luysberg, P. Hapke and U. Backhausen, see book reference 6.

NANOCRYSTALLINE Ge SYNTHESIS BY PICOSECOND PULSED LASER INDUCED MELTING AND RAPID SOLIDIFICATION

J. SOLIS*, J. SIEGEL*, C. GARCIA**, J. JIMENEZ**, R. SERNA*
*Instituto de Optica, CSIC, Serrano 121, 28006-Madrid (SPAIN)
Tel: +34-1-5616800, Fax: +34-1-5645557
e-mail: iodjs37@pinar1.csic.es
**Departamento de Física de la Materia Condensada, Cristalografía y Mineralogía, Universidad de Valladolid, 47011-Valladolid, (SPAIN).

ABSTRACT

Melting and rapid solidification has been induced in DC-sputtered amorphous Ge (a-Ge) films on glass substrates by irradiation with picosecond laser pulses at 583 nm. The melting-solidification kinetics has been followed by means of real time reflectivity measurements with ns resolution and the structure of the rapidly solidified material has been analyzed by means of Raman spectroscopy in micro-Raman configuration. The results obtained show that for laser pulse fluences above a certain threshold recalescence occurs during solidification leading to the formation of nanocrystalline Ge embedded in an amorphous matrix. The phonon correlation length (L) obtained from the Raman spectra of the irradiated regions has been used to analyze the evolution of the crystallite size with the laser fluence. For low fluences above the recalescence threshold, the values of L are in the 8-10 nm range. However, if the fluence is sufficiently increased, the crystallite size shows a clear linear dependence on the fluence with phonon correlation lengths scaling from 6 to 13 nm.

INTRODUCTION

Optical and electronic properties of semiconductor nanocrystals embedded in a solid matrix have attracted much interest during the last few years [1] because they exhibit quantum confinement effects with a high potential for the development of new photonic devices [2]. This interest has been accompanied by a large experimental effort devoted to the development of new methods for synthesizing semiconductor nanocrystals with homogeneous and controllable sizes. In the case of Ge, nanocrystals have been successfully produced both in germanium or silicon dioxide matrices by several procedures such as the annealing of GeO films in He [3], the cosputtering of Ge and SiO_2 [4] followed by thermal annealing, ion implantation of Ge in SiO_2 accompanied by thermal annealing [5], the deposition of Si_xGe_{1-x} films followed by oxidation and annealing under H_2 [6] or, more recently, the use of atmospheric pressure chemical vapor deposition combined with thermal annealing [2]. In this context, the use of pulsed laser induced rapid solidification might provide a new approach for the synthesis of nanocrystalline material since many studies have already shown the strong interdependence between the conditions upon which rapid solidification occurs and the nature and quality of phases finally produced. In particular, it has been shown that rapid solidification in Si under severe supercooling conditions may give rise to the presence of bulk solidification and recalescence phenomena with extremely large crystalline nucleation rates [7 ,8]. The same kind of phenomena has been recently reported in amorphous Ge (a-Ge) films upon irradiation with ps laser pulses, leading to the formation of a mixture of amorphous and crystalline material [9 ,10]. This work reports on the synthesis of nanocrystalline material by picosecond pulsed laser induced melting and rapid solidification.

Mat. Res. Soc. Symp. Proc. Vol. 452 © 1997 Materials Research Society

EXPERIMENT

The samples used for the present study are 50 nm-thick a-Ge films grown at room temperature on glass substrates. The films are deposited by DC-sputtering from a Ge (99.999%) target in a vacuum system with a residual pressure of $3x10^{-6}$ Torr and with an Ar operating pressure of $4x10^{-3}$ Torr. Details about the characterization of the as-deposited films by means of transmission electron microscopy and ellipsometry are given elsewhere [11].

The as-deposited samples were irradiated by laser pulses of about 10 ps at 583 nm in order to induce melting and rapid solidification. The irradiation pulses are delivered by a synchronously pumped dye laser whose output is amplified by a pulsed dye amplifier pumped by a frequency doubled Nd-YAG laser. The amplified beam is focused on the sample surface to a size of about 1 mm while the reflectivity of the surface at 633 nm is monitored in real time with ns resolution by means of a HeNe laser beam focused at the center of the irradiated spot to a size of approximately 50 μm. After each exposure the sample is moved to a fresh region. The fluence of the laser pulses at the sample site is varied in the 0-90 mJ/cm^2 range. Further details about the experimental setup for ps laser pulse irradiation and real time reflectivity measurements are given in Ref.[9].

After irradiation the structure of the irradiated regions was analyzed by means of Raman spectroscopy [12 ,13] in micro-Raman configuration [14]. The measurements were performed with a DILOR XY Raman spectrometer attached to a metallographic microscope. The excitation beam is delivered by an Ar^+ laser operating at 514 nm. The laser beam is focused on the sample surface to a size of about 10 μm, and it is sufficiently attenuated (a few mW) in order to prevent both the presence of spectral broadening or local annealing.

RESULTS AND DISCUSSION

Figure 1 shows two representative real time reflectivity transients obtained upon melting and rapid solidification of the as-deposited material induced by irradiation with ps laser pulses. The sharp reflectivity increase observed following the absorption of the irradiation pulse

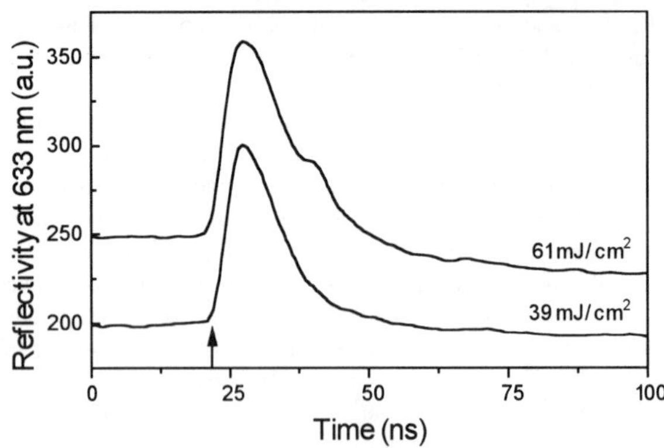

Figure 1: Real time reflectivity transients obtained upon irradiation of as-deposited a-Ge films with ps laser pulses. The arrow indicates the temporal position of the irradiation pulse and the laser pulse fluences are given in each transient.

indicates the formation of a molten metallic layer on the surface. For melt depths below approximately ≈ 25 nm the maximum transient reflectivity level is a function of the molten layer thickness. For larger melt depths, the liquid layer becomes optically thick at the probe beam wavelength leading to a constant reflectivity value. This transient reflectivity behavior has been widely reported upon laser induced melting of semiconductors like Si, Ge or GaAs [15]. In our case the minimum pulse fluence required for inducing surface melting is about 20 mJ/cm^2. Upon solidification and for fluences below ≈55 mJ/cm^2, the reflectivity decreases abruptly after the maximum and the process is completed in about 200 ns (not shown in the plot). In this case, the final reflectivity level of the surface is nearly the same as the initial, thus suggesting that the rapidly solidified material is amorphous. For fluences above ≈ 55 mJ/cm^2, the decrease after the reflectivity maximum shows the presence of a shoulder where the reflectivity remains nearly constant for a few ns. After the shoulder, the reflectivity decreases again but at a slower rate and the final reflectivity value observed is clearly smaller than the initial one. This latter type of optical reflectivity transients have been earlier reported and are related to recalescence and crystalline phase formation during solidification [7,9]. During recalescence, the release of the solidification enthalpy (at the initial nucleation stages) gives raise to an increase in the liquid temperature (supercooling decrease) thus reducing the rate at which liquid is consumed. This generates the shoulder in the transients and promotes the formation of crystalline material.

Figure 2 shows the evolution of the reflectivity after irradiation as a function of the laser pulse fluence. The reflectivity values have been normalized to the initial reflectivity of the as-deposited material (R_0) as $R=(R_f-R_0)/R_0$ where R_f is the reflectivity value after irradiation. Three different regimes can be observed in the plot. At low fluences, above the melting threshold (Regime I: $20 < E < 48$ mJ/cm^2), the final reflectivity of the surface is a slightly decreasing function of the pulse fluence which reaches normalized values around -0.05. If the fluence is further increased, (Regime II, $48 < E < 72$ mJ/cm^2), the final reflectivity shows a marked decreasing behavior. The discontinuous line in the plot denotes the fluence threshold above which recalescence effects are clearly observed in the real time reflectivity transients. For even

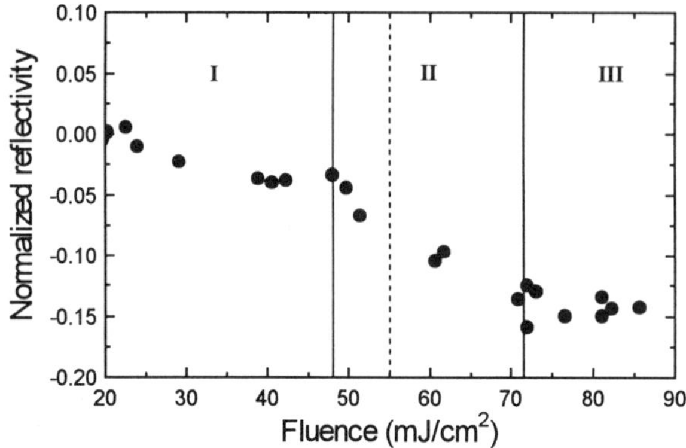

Figure 2: Normalized reflectivity of the surface after irradiation versus laser pulse fluence. The reflectivity values have been normalized as $R=(R_f-R_0)/R_0$ where R_0 and R_f denote respectively the reflectivities before and after irradiation. The solid lines indicate the separation between the observed regimes in the evolution of the reflectivity and the discontinuous line indicates the fluence threshold above which recalescence effects are clearly observed in the reflectivity transients.

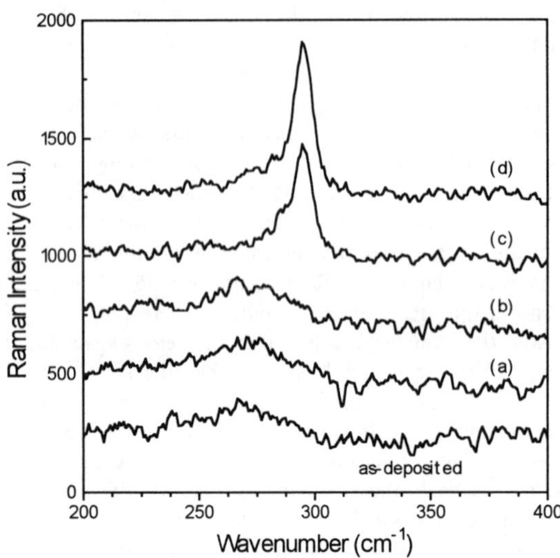

Figure 3: Raman spectra corresponding to the as-deposited material and to regions irradiated by ps laser pulses at fluences of (a) 39, (b) 51, (c) 61 and (d) 81 mJ/cm^2.

larger fluences, (Regime III, E > 72 mJ/cm^2) the reflectivity evolution shows a plateau with normalized values close to -0.15, clearly indicating the formation of crystalline material [9].

Figure 3 shows some representative Raman spectra obtained from regions irradiated with pulse fluences within each of the three regimes above mentioned. Within Regime I, the Raman spectra of the irradiated surfaces show the characteristic broad transverse optic (TO) phonon band of a-Ge at 270 cm^{-1}, thus showing that the rapidly solidified material re-amorphizes, (spectrum (a)). The same behavior is observed for low fluences whithin regime II, (spectrum (b)). However, as soon as the fluence is increased above the recalescence threshold, the spectra show clearly the development of the narrow optic phonon band of crystalline Ge at 298 cm^{-1} (spectrum (c)). The same applies for fluences within Regime III, as can be seen in spectrum (d).

The phonon correlation length of the crystalline material has been used to investigate the influence of the pulse fluence on the average size of the crystalline material induced upon rapid solidification. Several studies have shown that Raman lineshape analysis allows us to estimate the average crystallite size with a reasonably good agreement with the values obtained by other techniques such as transmission electron microscopy [2,12,13]. The phonon correlation length (L) in the crystalline material was determined from the experimental Raman spectra using the ratio Γ_a/Γ_b, where Γ_a and Γ_b are respectively the half widths of the optical phonon band of c-Ge at the low and the high frequency sides of the spectrum. The relation between L and the Raman line shape was determined by J. Gonzalez-Hernandez et al. assuming a gaussian spatial correlation model [12]. The ratio Γ_a/Γ_b is independent on the apparatus correction function and therefore an accurate correlation length can obtained in a simple way.

Figure 4 shows the evolution of L as function of the laser pulse fluence in the irradiated regions. It can be seen that for fluences within Regime II the phonon correlation lengths observed seem to increase slightly with fluence reaching values in the 8-10 nm interval. When

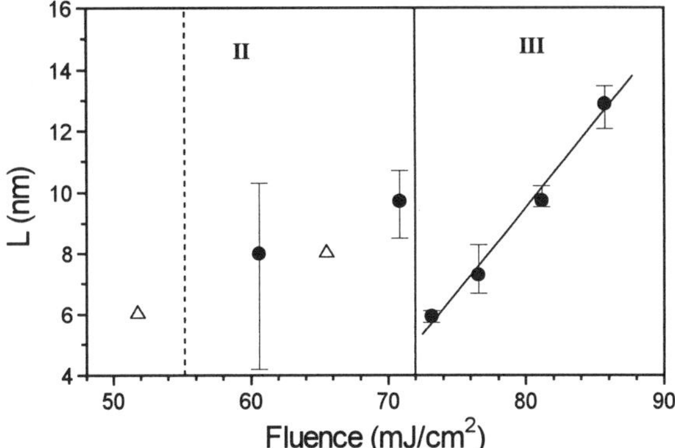

Figure 4: Phonon correlation length (L) (●) of the crystalline material versus laser pulse fluence. Results from Ref.[10] (Δ) have been also included for comparison. The solid line is a guide to the eye.

the fluence is increased just above 72 mJ/cm^2 (Regime III), the observed L values become smaller, reaching a minimum close to 6 nm. For larger fluences the evolution of the phonon correlation length shows a clear linear dependence on the pulse fluence with L values scaling from ≈6 nm to ≈13 nm in the 72-86 mJ/cm^2 fluence interval. The figure includes also preliminary results earlier reported by C. García et al. [10] in similar experimental conditions but in a shorter fluence interval. The comparison of both sets of data might suggest the existence of a linear dependence between the crystallite size and the fluence also in Regime II, but with a smaller slope. Because of the large data dispersion within in this regime and an error in the comparison of the absolute fluence of both sets of data amounting to about 10%, a definitive conclusion can not be obtained. The different behavior in the evolution of L in Regimes II and III is in any case not unexpected. At this point one must consider the fact that the supercooling prior to nucleation is related to the pulse fluence [9], and therefore it is expected that the solidification conditions, especially the nucleation rate, and the maximum crystallite size achievable [16] should be modified as the fluence is increased from values very close to the recalescence threshold to values clearly above it. The supercooling decrease in the larger fluences regime could account for the observed linear increase of the crystallite size in Regime III.

CONCLUSION

The present results demonstrate that ps pulsed laser induced melting and rapid solidification may be applied for producing nanocrystalline Ge. It has been also shown that for fluences above a certain threshold the average size of the crystallites formed depends linearly on the pulse fluence, which can be therefore used for controlling the crystallite size. Even when the present results have been obtained in pure a-Ge films, the feasibility of using this experimental approach in structures formed by a-Ge and GeO_2 or SiO_2 opens the possibility for a novel method for synthesizing Ge nanocrystals in insulator matrices without the need of long time post-process annealing.

ACKNOWLEDGMENTS

This work has been partially supported by CICYT (Spain), (Project TIC93-0125). Jan Siegel acknowledges the funding of the European Community through a grant (ERB40001GT954352) within the Training and Mobility of Researchers Programme.

REFERENCES

1. See for instance: L. Brus, Appl. Phys. **A 53**, 465 (1991), A.G. Cullis and L.T. Canham, Nature **353**, 335 (1991) or Microcrystalline and Nanocrystalline Semiconductors edited by L. Brus, M. Hirose, R.W. Collins, F. Koch and C.C. Tsai, (Mater. Res. Soc. Proc. **358**, Pittsburgh, PA, 1995).
2. A.K. Dutta, Appl. Phys. Lett. **68**, 1189 (1996) and references cited therein.
3. N.N. Ovsyuk, E.B. Gorokhov, V.V. Grishchenko, and A.P. Shebanin, JETP Lett. **47**, 298, (1988).
4. Y. Maeda, N. Tsukamoto, Y. Yzawa, Y. Kanemitsu and Y. Matsumoto, Appl. Phys. Lett. **59**, 3168 (1991).
5. K.S. Min, K.V. Shcheglov, C.M. Yang, H.A. Atwater, M.L. Brongersma and A. Polman, Appl. Phys. Lett. **68**, 2511 (1996).
6. D.C. Paine, C. Caragianis, T.Y. Kim, T. Shigesato, T. Isikawa, Appl. Phys. Lett. **62**, 2841 (1993).
7. S.R. Stiffler and M.O. Thompson, Phys. Rev. Lett. **60**, 2519 (1988).
8. T. Sameshima and S. Usui, J. Appl. Phys. **74**, 6592 (1993).
9. J. Siegel, J. Solis, C.N. Afonso, C. García, J. Appl. Phys. **80**, (15 December 1996).
10. C. García, A.C. Prieto, J. Jimenez, J. Siegel, J. Solis, C.N. Afonso, M. Chafai in Advanced Laser Processing of Materials: Fundamentals and Applications edited by R. Singh, D. Norton, L. Laude, J. Narayan, J. Cheung (Mater. Res. Soc. Proc. **397**, Pittsburgh, PA, 1996) pp. 435-440.
11. J.C.G. de Sande, C.N. Afonso, J.L. Escudero, R. Serna, F. Vega and E. Bernabeu, Appl. Opt. **31**, 6133 (1992).
12. J. Gonzalez-Hernandez, G.H. Azarbayejani, R. Tsu and F.H. Pollak, Appl. Phys. Lett. **47**, 1350 (1985).
13. M. Fujii, S. Hayashi and K. Yamamoto, Jap. J. Appl. Phys. **30**, 687 (1991) and references cited therein.
14. J. Jimenez, E. Martín, A. Torres, B. Martín, F. Rull and F. Sobrón, J. Mater. Sci. **4**, 217 (1993).
15. G.E. Jellison Jr., D.H. Lowndes, D.N. Mashburn, and R.F. Wood, Phys. Rev. **B 34**, 2497 (1986).
16. S.R. Stiffler and M.O. Thompson, Phys. Rev. **B 43**, 9851 (1991).

ELECTRICAL PROPERTIES OF THE MULTILAYER STRUCTURES BASED ON ULTRATHIN DIAMOND-LIKE CARBON FILMS

V.I. POLYAKOV *, P.I. PEROV *, N.M. ROSSUKANYI *, A.I. RUKOVISHNIKOV *, A.V. KHOMICH *, and A.M. BARANOV**

*Institute of Radio Eng. & Electronics, RAS, 11 Mohovaya str., Moscow 103907, Russia, vip197@ire216.msk.su
** NPO Vacuummashpribor, Nagornyi pr.7, Moscow 113105, Russia

ABSTRACT

The electrical characteristics of multilayer structures based on amorphous ultrathin diamond-like carbon films were investigated including dynamic and quasi-static current-voltage characteristics, capacitance-voltage characteristics, deep level transient spectra. The effect of illumination and temperature on these characteristics was also investigated. For the multilayer structures composed of lower band gap amorphous carbon layers separated with higher band gap ones, there were observed well-defined regions of negative differential resistance and sharp 20-fold changes in capacitance at definite voltages. Activation energies, capture cross sections, and locations of trapping centers were defined. The effects observed are discussed in terms of trap-assisted tunneling and, also, in terms of resonant tunneling between energy levels in superlattices and charge filling of the quantum wells and trapping centers.

INTRODUCTION

Abeles and Tiedje (1983) were the first to produce periodic multilayers (superlattices) of amorphous material (Si and related alloys) [1]. The observed peculiarities in the optical and electrical characteristics of these amorphous superlattices were explained in terms of quantum size effects. In 1987 Miyazaki, Ihara and Hirose observed negative-differential-resistance (NDR) in amorphous double-barrier structures, which could be explained in terms of resonant tunneling [2]. Later, it was shown that the phase compositions and, consequently, the optical and mobility band gap of diamond-like carbon (DLC) films can be varied over a wide range by changing the conditions under which they are deposited [3-7]. This enables one to made superlattice structures using only amorphous DLC films with different band gaps. The NDR effect for the amorphous superlattices using DLC multilayers were described for the first time in [4, 9]. The quantum size effect was considered as one of the possible models to describe the bumps in current-voltage characteristics of the structures investigated.

However, some authors disputed resonant tunneling as the possible mechanism for the samples based on amorphous silicon and ascribed the observed features to a combination of alternative effects such as charge limited currents, a multiple-hopping transport mechanism, and trap-assisted tunneling processes [10, 11]. It seems that there is the same ambiguity in the explanation of the real conduction mechanism for amorphous DLC superlattices. Therefore, the peculiarities in carrier transport through of the amorphous DLC superlattices are questionable and need additional studies.

In this work, we report new data on the electrical characteristics of amorphous carbon superlattices including dynamic and quasi-static current-voltage characteristics, capacitance-voltage characteristics, deep level transient spectra, and the effect of the illumination and

Mat. Res. Soc. Symp. Proc. Vol. 452 © 1997 Materials Research Society

temperature on these characteristics. Activation energies, capture cross sections, and locations of trap states were defined. Different mechanisms of the effects observed are discussed including trap-assisted tunneling, resonant tunneling between energy levels in superlattices and charge trapping with bulk and interface traps.

EXPERIMENTAL DETAILS

The multilayer structures of hydrogenated amorphous carbon (a-C:H) films with quantum size were fabricated using ion-plasma deposition onto the n-type silicon substrates. Changing the deposition regimes, the structures of alternating a-C:H films with two different energy (mobility) gaps were prepared. The multilayer structures contained more than 20 layers of a-C:H with thicknesses 2 - 5 nm. Several semitransparent 1 mm^2 area Ni electrodes were deposited on top of the structures using thermal evaporation. Indium electrical contacts to conductive Si substrates were made using a current pulse annealing technique. The proposed energy band diagram for the DLC superlattice of the structure consisting of alternating amorphous carbon layers with two different effective band gaps E_{g1} and E_{g2} is presented in Fig. 1.

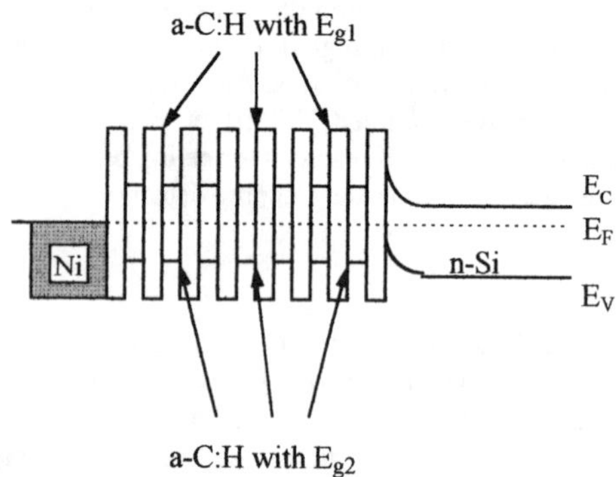

Fig.1. The band diagram of a DLC multilayer structure.

Current-voltage, capacitance-voltage characteristics, deep level transient spectroscopy, and the effects of the illumination and annealing on these characteristics were used to study the electrical properties of the structures.

In order to obtain the trapping center parameters, we used Charge-based Deep Level Transient Spectroscopy (Q-DLTS). Unlike the widely used capacitance DLTS method [12], our technique allows one to study insulating materials [13, 14]. The cycling DLTS algorithm used in the present work is different from that of Lang [12]. In Lang's algorithm the rate window $\tau_m = (t_2 - t_1)/\ln(t_2/t_1)$, where t_1 and t_2 are the times from the beginning of discharge, is kept fixed while the sample temperature is scanned to obtain the DLTS spectrum. In the alternative algorithm used in the present work, we obtain the spectrum by scanning the rate window τ_m while keeping the temperature of the sample fixed. If we keep the ratio $t_2 / t_1 = \alpha$ constant and vary τ_m, then a maximum, ΔQ_{max}, in the DLTS spectra, $\Delta Q(\tau_m)$, occurs at the rate window equal to the emission rate of the traps at temperature T. In the present measurements, $\alpha = 2$ and trapping center densities can be obtained as $N_t = 4\,\Delta Q_{max} / qA$, where q is the electron charge and A is the contact area. The trapping center concentrations were found from Q-DLTS spectra

$\Delta Q(\tau_m)$ taken at room temperature, and the activation energy E_a and capture cross section σ were found from the Arrhenius dependence after measurements of Q-DLTS at different temperatures (up to 500 K). All the Q-DLTS measurements were made with the computerizing system ASEC-3.

RESULTS AND DISCUSSION

The current-voltage (I-V) characteristics were measured under dark and light conditions at room and other temperatures. Some measurements were made after sample annealing at about 250 C. The most interesting I-V characteristics for amorphous DLC multilayer structures (the well width = 3 nm) are shown in Fig. 2. As seen, the curves display the pronounced bump at negative applied voltages. Note that negative-differential-resistance features appear at room temperature and are still observable even if the temperature was increasing above 100 K (Fig. 2a). The position and shape of NDR feature were strongly dependent on illumination (Fig. 2b) and, especially, from rate of the applied voltage sweeps (Fig. 2c). In dynamic C-V characteristics the more pronounced NDR feature was observed.

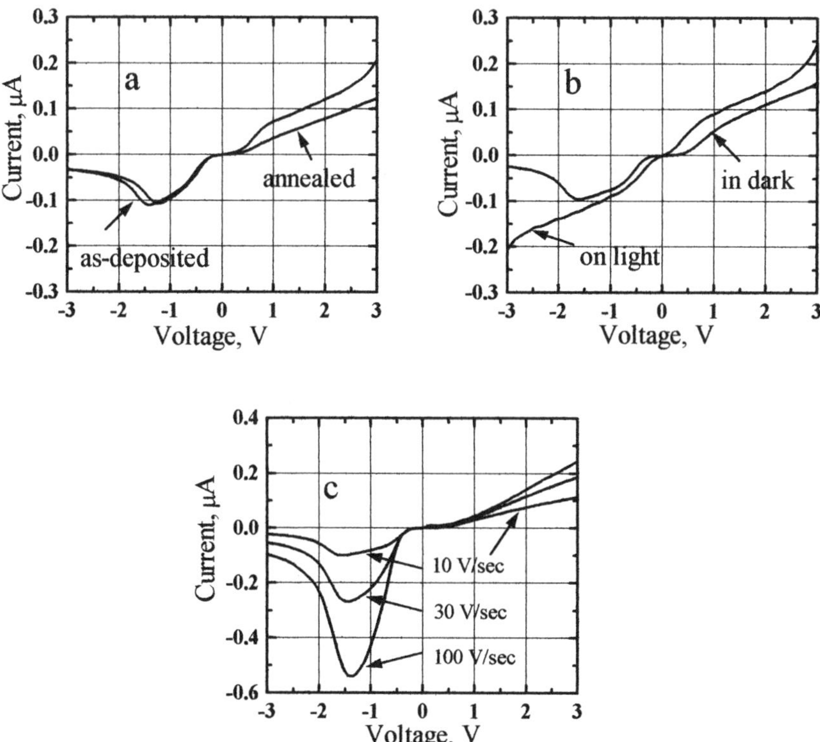

Fig. 2. Current-voltage characteristics of the DLC multilayer structure with well width = 3 nm.

In approximately the same NDR range, a sharp change in the capacitance of the structure was observed in the capacitance-voltage (C-V) characteristics, with both the threshold voltage and the capacitance dependent on test signal frequency (Fig. 3). Such a strong frequency dependence of the capacitance is evidence of trap-defined effects as the main mechanism of the experimentally observed C-V and possibly, I-V characteristics.

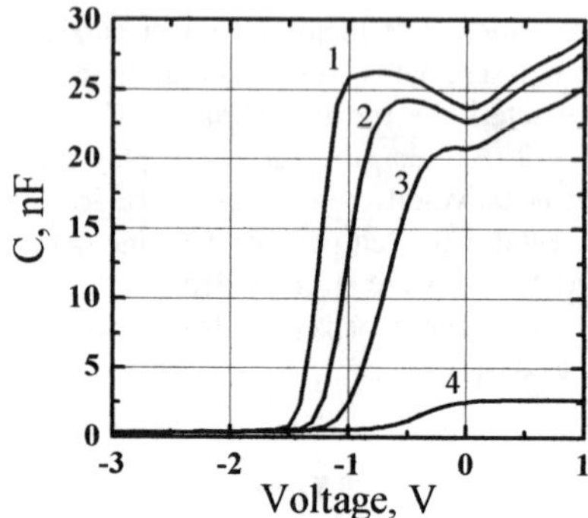

Fig. 3. The capacitance-voltage dependence at different test signal frequencies, kHz: 1- 0.01; 2 - 0.1; 3 - 1; 4 - 100.

For determining the trapping center concentrations, activation energies and capture cross section Q-DLTS spectra, $\Delta Q(\tau_m)$, were measured. Fig. 4 shows the DLTS spectra as a function of charging pulse amplitude before (a) and after (b) annealing at 250 C (20 min.). As can be seen, all filling of the trapping centers takes place after applied voltages of about -2 to -3V. Note that the peak A in the DLTS spectra appeared after applied voltages in the same NDR range. The calculated integrated trap-center concentrations N_t were about $(1\text{-}5)\cdot 10^{12}$ cm^{-2} and not strongly affected by annealing, while the capture cross section could be changing over large range, from 10^{-14} to 10^{-21} cm^2 and was not related to the other parameters. The observed range of the activation energy was 0.3 - 0.7 eV for all samples prepared here. The increase in the activation energy was observed for the structures with smaller well width. The activation energy and capture cross section for the trap-centers which take part in the formation of peak A amounted to 0.4 eV and about 10^{-20} cm^2 respectively.

From the analysis of the experimental data we can see the essential role of the trapping centers in the carrier transport mechanism for DLC multilayer structures prepared here. The resonant tunneling mechanism and large density of trapping centers can explain the bumps and negative differential resistance regions in I-V characteristics. At definite voltages, the energy level in one "quantum well" coincides with the next level in the adjacent well resulting in an increase in current through the structure. However, due to the trapping of the charge carriers within deep levels or, possibly, within the wells themselves, the potential distribution across the structure is changing with time, displacing the energy levels and decreasing the current through the structure. In such a case, the dependence of the current bump in the I-V characteristics on the voltage sweep rate could be understood as a result of the charge trapping within the deep levels.

However, macroscopically visible quantum size effects (resonant tunneling) should only be present if the mean free path is at least on the order of magnitude of the barrier-well system

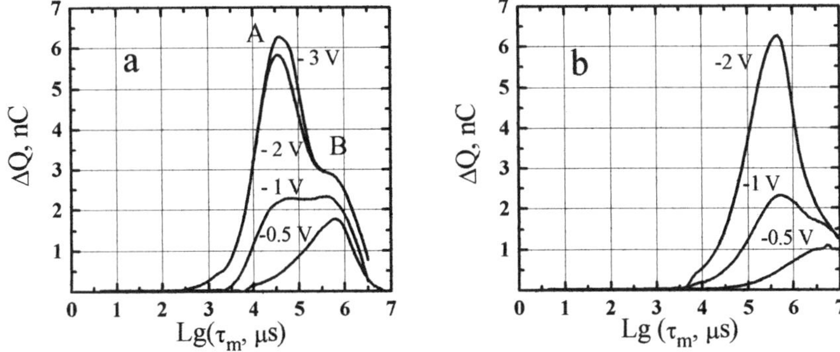

Fig. 4. The Q-DLTS spectra of the DLC multilayer structure measured at different charging pulse amplitudes before (a) and after (b) annealing.

spatial extent. The mean free path for amorphous materials is very small at room temperature, less than 0.5 nm, and the mobility is very small also. Furthermore, the calculated NDR features could be pronounced in amorphous superlattices only at low temperatures (see, for example, a result of numerical calculations for amorphous superlattices [11]). These calculations are based on a very simplified model; thus, reports of pronounced NDR peculiarities at room temperature [4, 9, 15, 16] can be questioned in terms of their interpretation as resonant tunneling phenomena. The reproducible anomalies in the I-V curves reported by many workers might be due to fortuitous coincidences with bumps predicted from a resonant tunneling picture. It seems, in our case that the NDR features in the I-V characteristics and all other the experimentally observed phenomena may qualitatively be explained if conduction through amorphous multilayer structures is considered as a superposition of area conduction and conduction in small channels (trap-assisted tunneling). The trapping centers and potential wells (synthetic trapping centers) in the prepared amorphous superlattices play the role of the seeds for channel formation. Since most of the applied voltage is dropped across the insulating (barrier) parts, only moderate voltages (in our case 2-4 V) could be sufficient to cause very high electrical fields and consequently extreme physical conditions. These might result in carrier injection, leading to hopping conduction of the carriers through the trapping centers (extreme filling of traps as in the case of space-charge-limited currents), and the formation of channel transport paths. In this case the resistance is changing only in a section of the channel path that leads to a redistribution of voltage drop, and this might yield a feedback of the process and a NDR region forming [2, 11]. When the resistances of some parts (of the barrier layers) are decreasing, whereas those of other parts (of the silicon depletion region) simultaneously are increasing, a sharp change in capacitance results (see Fig. 3). In this picture, all the experimental phenomena depend on trapping center parameters, have strong frequency dispersion and can be observed at room temperature.

ACKNOWLEDGMENTS

The authors are indebted to A.I.Krikunov for manufacture of contacts. This work was partly supported by grants No 96-02-18373 of the Russian Fund of Fundamental Research.

REFERENCES

1. B. Abeles, and T. Tiedje, Phys. Rev. Lett., **51**, 2003 (1983).

2. S. Miyazaki, Y. Ihara, and M. Hirose, Phys. Rev. Lett., **59**, 125 (1987).

3. V.V. Sleptsov, V.M. Elinson, O.N. Ermakova, M.G. Ermakov, V.I. Polyakov, P.I. Perov, G.F. Ivanovsky, Pis'ma v Zh.Techn.Fiz., **16**, 15 (1990).

4. V.I. Polyakov, P.I. Perov, M.G. Ermakov, O.N. Ermakova, V.M. Elinson, V.V. Sleptsov, Thin Solid Films, **212**, 226 (1992); V.I. Polyakov, M.G. Ermakov, O.N. Ermakova, P.I. Perov, V.V. Sleptsov, V.M. Elinson, A.M. Baranov, Diamond and Related Materials, **1**, 626 (1992).

5. V.V. Sleptsov, A.A. Kuzin, A.M.Baranov, V.M. Elinson, Diamond and Related Materials, **1**, 570 (1992).

6. K.K. Chan, S.R.P. Silva, and G.A.J. Amaratunga, Thin Solid Films, **212**, 232 (1992).

7. V. Polyakov, M. Ermakov, O. Ermakova, P. Perov, V. Sleptsov, V. Elinson, A. Baranov, 2nd International Conference on the Applications of Diamond Films and Related Materials, edited by M. Yoshikawa, Y. Tzeng, and W.A. Yarbrough (The 2nd International Conference - ADFRM Proc., Tokyo, Japan 1993), p.381-386.

8. S.R.P. Silva, G.A.J. Amaratunga, Rusli, S. Haq, E. Salje. Thin Solid Films, **253**, 20 (1994).

9. S. Ravi P. Silva, Gehan A.J. Amaratunga, Charles N. Woodburn, Mark E. Welland, Sajad Haq, Japan. Journ. Appl. Phys., **33**, 6458 (1994).

10. C.J. Arsenault, M. Meunier, M. Beaudoin, B. Movaghar, Phys. Rev. **B 44**, 11521 (1991).

11. N. Bernard, B. Frank, B. Movaghar, G.H. Bauer, Philosoph. Magazine, **B 70**, 1139 (1994).

12. D.V. Lang, J.Appl. Phys. **45**, 3023 (1974).

13. S. Nath, J.I.B. Wilson, Diamond and Related Materials, **5**, 65 (1995).

14. V.I. Polyakov, P.I. Perov, A.I. Rukovishnikov, O.N. Ermakova, A.L. Aleksandrov, B.G. Ignatov, Mikroelektronika, **16**, 326 (1987); B.M. Arora, S. Chakravarty, S. Subramanian, V.I. Polyakov, M.G. Ermakov, O.N. Ermakova, P.I. Perov., J.Appl.Phys., **73**, 1802 (1993); V.I. Polyakov, A.I. Rukovishnikov, P.I. Perov, A.V. Khomich, A.A. Sukhanov, B.F. Dorfman, B.N. Pypkin, M.G. Abraizov, B. Druz, Thin Solid Films (1996), to be published.

15. I. Pereira, M.N.P. Carreno, R.K. Onmori, C.A. Sassaki, A.M. Andrare, F. Alvarez, Journal Non-Crystalline Solids, **97&98**, 871 (1987).

16. Y.L. Jiang, and H.L. Hwang, IEEE Trans. Electron. Devices, **36**, 2816 (1989).

CRYSTALLINE AND ELECTRICAL PROPERTIES OF ITO SEMICONDUCTIVE FILMS DEPOSITED BY ATMOSPHERIC RF PLASMA TECHNIQUE

R.W. Moss, S. Misture, D.H. Lee, R.A. Condrate, Sr., X.W. Wang
Alfred University, Alfred, New York, 14802

ABSTRACT

Indium tin oxide (ITO) semiconductive films were deposited by an atmospheric RF plasma technique. Indium-to-tin (In:Sn) ratios varied from 10:0 to 0:10. A small amount of antimony was doped into some ITO samples for comparative studies. Substrate materials were soda-lime-silicate (SLS) float glass and fused silica glass. Structural, electrical, and optical properties were dependent on the In:Sn ratio, precursor material feeding rate, oxygen feeding rate, and other deposition conditions.

INTRODUCTION

An atmospheric RF plasma deposition technique was developed a few years ago,[1] which may have potential for large scale applications. Our earlier study indicated that semiconductive indium tin oxide (ITO) films were conductive and transparent.[2] However, crystalline and electrical properties of those films were not well understood. To further investigate this newly developed process, a series of experiments were performed.

EXPERIMENT

An RF plasma set up consisted of a starting material assembly, a plasma reactor, and a substrate assembly, as shown in Figure 1. The plasma reactor was powered by an RF generator (35 kW, 4 MHz). Argon was the primary plasma gas. The starting solution contained indium nitrate pentahydrate (In = 0.1919 Molar), and tin chloride pentahydrate (Sn = 0.2139 Molar); the indium to tin (In:Sn) ratio varied from 10:0 to 0:10, in intervals of one. A standard antimony solution (Sb = 0.08213 Molar) was used as the Sb source, with Sb/(In+Sn) ratio of approximately 8%.

After stirring, a precursor solution was poured into a plastic container housed in an ultrasonic nebulizer. Excited by the ultrasonic energy, mist was produced from the solution, fed into the plasma reactor, where it was then completely evaporated. The resultant vapor was deposited onto the substrate outside of the reactor. The precursor material feeding rate, or mist rate, was varied between 0.10 to 0.13 ml/min. Oxygen gas flow was introduced through either the tangential plasma gas inlet or the carrier gas inlet. Most films reported in the following section were fabricated with the tangential oxygen flow rate of 5 l/min., and the oxygen carrier gas flow rate of 4 l/min. The substrate temperature was 500 °C, and the deposition time was 60 min. Substrate materials were soda-lime-silicate (SLS) float glass and fused silica glass. The as-deposited films coated on the substrates were adhesive, and subsequently analyzed.

RESULTS

The films were nearly 100% transparent in the visible region. Elemental analysis showed that the initial stoichiometric ratios of In:Sn in the precursor solution are close to that of the films.

Mat. Res. Soc. Symp. Proc. Vol. 452 © 1997 Materials Research Society

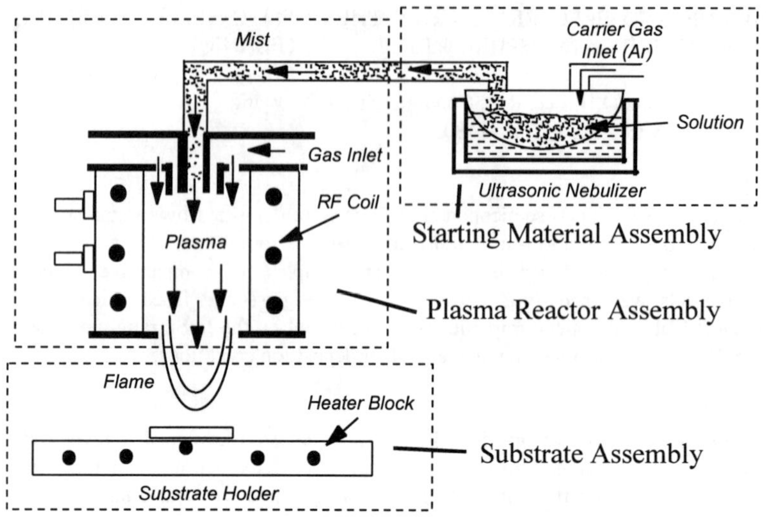

Figure 1. Experimental setup for atmospheric RF plasma mist deposition technique.

To study the crystal formation and lattice parameter, X-ray diffraction (XRD) was utilized to analyze films deposited on SLS glass. When the In:Sn ratio was between 10:0 and 6:4, cubic indium oxide (In_2O_3, ICDD No. 6-416, a = 10.118 Å) was formed. When the In:Sn ratio was between 4:6 and 0:10, tetragonal tin oxide (SnO_2, ICDD No. 21-1250, a = 4.738 Å, c = 3.188 Å) was formed. Thus, the phase transition from the In_2O_3 to the SnO_2 phase took place between In:Sn = 6:4 and 4:6 in the film samples. As tin was doped into cubic In_2O_3, the location of a characteristic diffraction peak shifted to lower angles, indicating a lattice parameter expansion. The lattice parameter, a, increased as the tin concentration increased, see Figure 2. The total relative shift was approximately 0.78%. On the other hand, when indium was doped into tetragonal SnO_2, lattice parameters, a and c, changed slightly, see Figure 3 (a) and (b) respectively. Such trends in those cubic In_2O_3 and tetragonal SnO_2 films are consistent with that of previously determined powder samples[3].

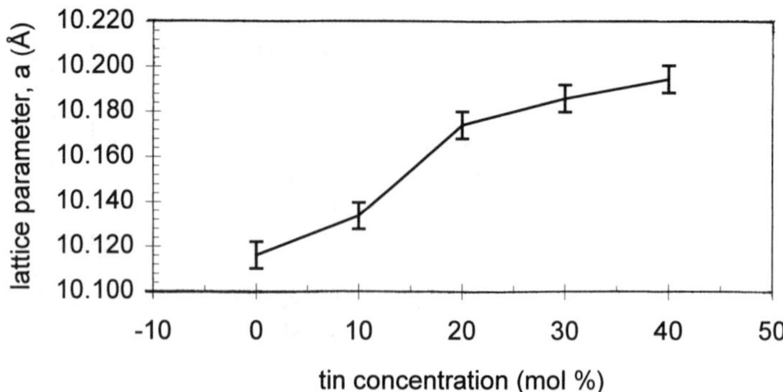

Figure 2. Lattice parameter, a, for cubic indium oxide films with increasing tin concentration.

(a)

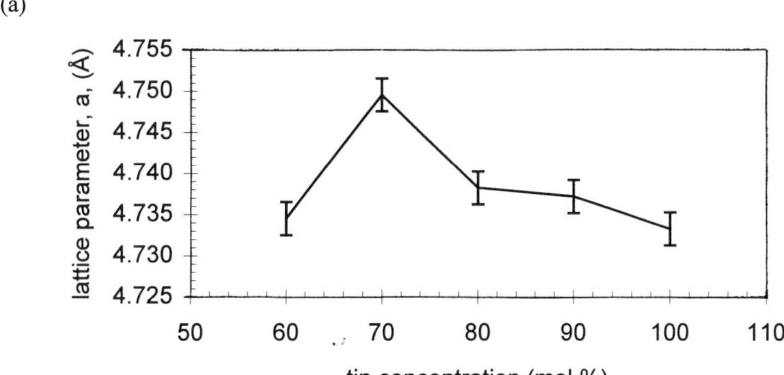

(b)

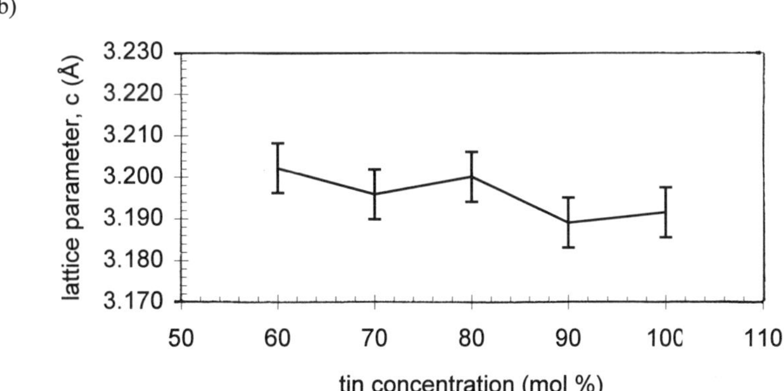

Figure 3. Lattice parameters, a (top) and c (bottom), for tetragonal tin oxide.

The crystallite size is inversely proportional to the full width at half maximum (FWHM) of diffraction peak(s), as described by the Scherrer relation. In Table A, the reciprocal values of FWHM, given as D, are shown, with various In:Sn ratios and oxygen feeding rates. If the oxygen flow rate was 5 l/min. and tin was initially doped into indium oxide, the values of D decreased correspondingly. Further addition of tin did not reduce the D values substantially. Coincidentally, this trend was similar to the trend for the lattice parameter, a, illustrated in Figure 2. On the other hand, as indium was doped into tin oxide, D values did not change much.

The misting rate affected optical transparency, uniformity, particle size, density, and resistivity, see Table B. When In:Sn ratio was 9:1 and the misting rate was 0.127 ml/min., the resistivity was the lowest among all samples examined. Correspondingly, the surface of this film contained round grains, see Figure 4 (a). Thickness of this film was 600 nm, and the film growth appeared to be columnized, see Figure 4 (b). As the sample temperature decreased, its resistivity increased, resembling a semiconductor's characteristic as illustrated in Figure 5. The activation energy (E_a) was derived from the Arrhenius Equation: $\sigma = \sigma_o \exp[-E_a / k_b T]$; where T is the sample temperature in

Kelvin, σ_o is the electrical conductivity at 300 K, σ is the electrical conductivity at temperature T, and k_b is Boltzman's constant. The activation energy of this sample (In:Sn=9:1) is approximately 3 meV. In comparison, the activation energy of pure indium sample (In:Sn=10:0) is 10 meV.

	O_2 flow rate of 5 l/min (tangential inlet)	**O_2 flow rate of 4 l/min (carrier gas inlet**	**Sb dopant (8 mol%)**
In:Sn = 10:0	2.370	2.770	
In:Sn = 9:1	2.612	2.817	3.394
In:Sn = 8:2	1.340	2.301	
In:Sn = 7:3	1.202		
In:Sn = 6:4	1.100		
In:Sn = 4:6	0.679		
In:Sn = 3:7	0.728		
In:Sn = 2:8	0.592		
In:Sn = 1:9	0.663		
In:Sn = 0:10	0.658		1.137

Table A. D values versus the In:Sn ratio and the oxygen flow rate.

Figure 4. SEM micrograph of (a) the coating surface, and (b) a cross-sectional image for an In:Sn=9:1 sample.

CONCLUSIONS

Indium tin oxide (ITO) semiconductive films were deposited by an atmospheric RF plasma technique. When In:Sn ratio was between 10:0 and 6:4, cubic indium oxide was formed. The lattice parameter of the cubic structure, a, increased as tin concentration increased. When In:Sn ratio was between 4:6 and 0:10, hexagonal tin oxide was formed. Electrical and optical properties were also dependent on the In:Sn ratio, precursor material feeding rate, oxygen feeding rate, and other deposition conditions.

In:Sn Ratio	Misting Rate (ml/min)	Thickness (SLS glass)	Particle Size (nm)	Film Density	Resistivity (ohm-cm)
10:0	0.100	800 nm	50-75	~ 100%	0.1700
9:1	0.100	500 nm	50-75	95 +	0.1750
9:1	0.117	850 nm	75-100	95 +	0.0900
9:1	0.127	700 nm	100-150	95 +	0.0067
9:1	0.167		150-200	85 +	----
8:2	0.100		100-150	90 +	----
9:1 ; 8% Sb	0.117	700 nm	75-100	95 +	0.0740
9:1 ; 8% Sb	0.133	800 nm	100-150	90 +	0.1000
0:10 ; 8% Sb	0.133		300-500	75 +	----

Table B. Film density, particle size, resistivity, and thickness for various In:Sn ratios and misting rates.

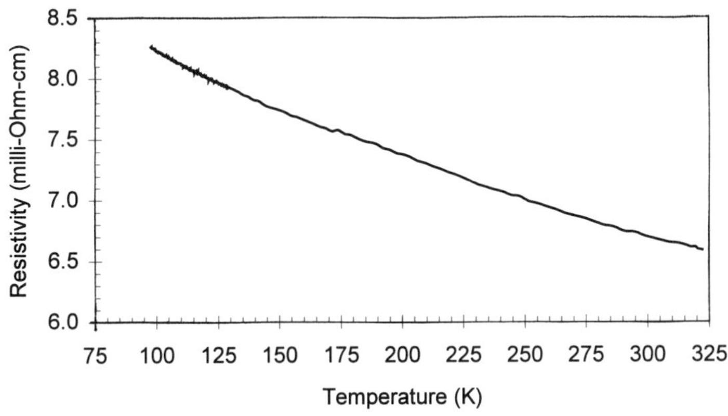

Figure 5. Electrical resistivity for In:Sn = 9:1 sample deposited at 500 C.

ACKNOWLEDGMENT

This project was partially supported by the NSF Industry-University Center for Glass Research, and the New York State Science and Technology Center for Advance Ceramic Technology.

REFERENCES

1. X. W. Wang, H. M. Zhong, and R. L. Snyder, Appl. Phys. Lett. 57 1581 (1990).
2. D. H. Lee, K. D. Vuong, J. A. A. Williams, J. Fagan, R. A. Condrate, Sr., and X. W. Wang, J. Mater. Res. 11 895 (1996).
3. D.H. Lee, R.W. Moss, K.D. Vuong, R. A. Condrate, Sr., and X. W. Wang, to be published in Thin Solid Films.

MORPHOTROPIC PHASE BOUNDARY IN THE NANOCRYSTALLINE $PbZr_xTi_{1-x}O_3$ SYSTEM

R. S. KATIYAR, J. F. MENG *
Department of Physics, University of Puerto Rico, San Juan, Puerto Rico 00931-3343

ABSTRACT

Micro Raman spectra for $PbZr_xTi_{1-x}O_3$ (PZT) with a grain size of 60 nm have been recorded. The results show that the lowest E(TO) phonon mode of $PbTiO_3$ (soft mode) displays a decrease in frequency and increase in linewidth with increasing Zr concentration. A discontinuous behavior of the phonon energy of the soft mode occurs at x=0.4 and it can be attributed to a phase transformation from the tetragonal ferroelectric to rhombohedral ferroelectric phase in PZT. The coupling between the soft mode and an additional impurity mode, has been observed, which plays an important role in the phase transformation. A lowest phonon mode at 10 cm^{-1} is detected and it is found to exhibit an increase in intensity for x above 0.2. The grain size dependence of the mode seems to be associated with the effect of grain size on the microstructure in PZT. The mechanism of grain size - induced new morphotropic phase boundary, which is lower than that of the corresponding bulk materials, is discussed.

INTRODUCTION

PZT ($PbZr_{1-x}Ti_{1-x}O_3$) - type materials are of commercial interest on account of their excellent piezoelectric, pyroelectric, ferroelectric, and dielectric properties. These characteristics facilitate applications in many areas, including hydrophones, thermal imaging, nonvolatile memories, surface acoustic wave generators, and gas sensors [1-3]. It is well known that enhanced piezoelectric effects exist, such as a maximum electromechanical coupling coefficient and dielectric susceptibility, for compositions near the morphotropic phase boundary (MPB) between the tetragonal and rhombohedral ferroelectric phases in the PZT solid solution system [4-6]. This makes PZT the most useful piezoelectric ceramic for preparing some advanced devices. For these compositions, there are fourteen possible poling directions over a very wide temperature range, explaining why the piezoelectric coefficients are largest near the MPB [6]. Most of the effort has been concentrated on lead titanate and PZT solid-solution compositions near the morphotropic phase boundary. That is, the determination of the MPB as well as the mechanism of the phase transformation and dielectric enhancement have received increasing interest and attention. Once the phase boundary was considered as the composition where these two phases were present in equal quantity [7], Benguigui and coworker have suggested the coexistence of the two phases in MPB over the region of 15 mol % [8], but Isupov has reported a possible existence of an extended region of the MPB composition [9]. Subsequent investigations by Kakegawa *et al.* [10] and Multani *et al.* [11] have proved that no coexistence of the two phases occurs in any range of composition without compositional fluctuations. Nakamura had also pointed out that the MPB is the phase boundary of the first order transition between tetragonal and rhombohedral phases [12]. However, even in the case of bulk PZT, the exact composition of the morphotropic phase boundary (MPB) has never been precisely defined; it ranges from 45 to 50 mol % of $PbTiO_3$ [13-15]. There is a coexistence region of the tetragonal and rhombohedral phases whose width is also not well defined [14-16], ranging from 2 to 15 mol % of $PbTiO_3$ around the composition Ti/Zr=48/52. On the other hand, one can find many theoretical

Mat. Res. Soc. Symp. Proc. Vol. 452 © 1997 Materials Research Society

investigations on MPB. For example, the ionic polarizabilities in PZT compositions were calculated by the Clausius-Mossotti-Lorentz-Lorentz equation and the resultant gap of the polarizability appearing at the MPB composition suggested that the phase transition boundary is to be due to the electrostatic long-range dipole moment [17]; A statistical model was proposed to address the problem of two-phase coexistence near the MPB [18]. In this paper, our investigations will be focused on these interesting problems using Micro Raman scattering on nanocrystalline PZT materials.

EXPERIMENTAL

Nanocrystalline $PbZr_xTi_{1-x}O_3$ (with x = 0.0, 0.1, 0.2, 0.3, ... ,1.0) materials were separately prepared, essentially using a sol-gel in which lead acetate, titanium butoxide and zirconate butoxide were used as the precursor materials. The processes involved dissolving the metal-containing compounds in the solvent, hydrolyzing and drying the resulting solution into various gels, and finally heat treating these gels at different temperatures and for different time intervals to form nanocrystalline PZT powder. Chemical phase analysis was done from power XRD measurements, using a computer controlled x-ray diffractometer model Siemen D-5000 with a typical scan speed of 0.3° min^{-1}.

The average grain size (d_{XRD}) was calculated from the full width at half maximum(FWHM) of the (111) diffraction peak using the Scherrer equation $d_{XRD} = K\lambda/\beta(\theta)\text{Cos}\theta$. Where λ is the x-ray wavelength, $\beta(\theta)$ is the FWHM of the diffraction line, θ is the angle of diffraction, and the constant $K \approx 1$. This equation assumes the small crystallite size to be the only cause of line broadening.

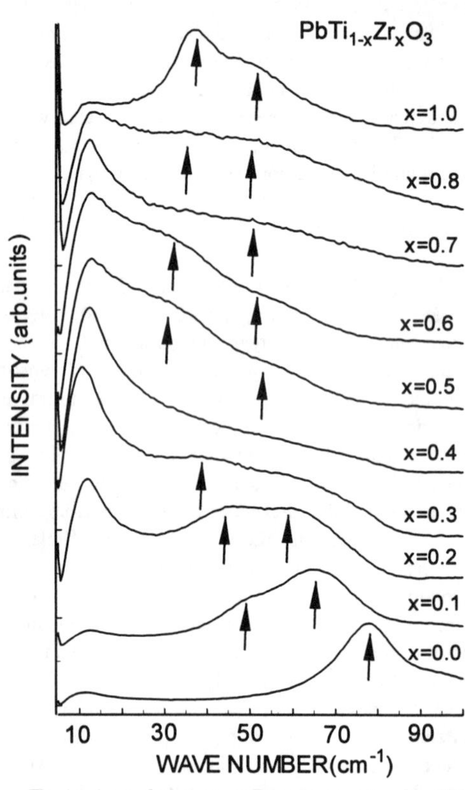

Fig.1a Low frequency Raman spectra for PZT with grain size of 60 nm

The Raman spectra were excited using the 514.5 nm radiation line from a Coherent Innova 99 Ar-ion laser. The power impinging the sample was kept at 10 mW. Jobin-Yvon T64000 Spectrophotometer with subtractive pre-monochromators coupled to the third spectrograph monochromator with 1800 grooves/mm grating was used for recording the spectra. The liquid nitrogen cooled charge coupled device (CCD at 140 K) detector was employed to record the Raman signals. All of the spectra were taken in the back-scattering geometry with a 20 × microscope objective that allows 10 mm of working distance and ≈ 5×10^3 μm^2 of illuminated area. A resolution of 1.9 cm^{-1} was used as measured from the width of Ar-ion plasma lines.

RESULTS AND DISCUSSION

Low and high frequency Raman spectra for PZT with grain size of 60 nm are respectively shown in Fig. 1a and 1b. Fig. 2 displays the dependence of Raman phonon modes with varying Zr concentration. The lowest Raman spectrum, exhibiting all features of the tetragonal phase as observed in bulk $PbTiO_3$(PT) [19], corresponds to that of nanocrystalline PT. In fact, the critical grain size from the tetragonal phase to the cubic structure in PT is determined to be about 12-20 nm [20]. From Fig.1, it is found that the lowest E(TO) phonon mode (soft mode) decreases its frequency and appears as a widening in linewidth with increasing Zr concentration. As x reaches 0.4, the soft mode displays a very weak intensity and an unmeasureable frequency position. On increasing x further, its intensity begins to increase again up to x=1.0. Another low frequency mode, occurring at 50 cm^{-1} for x=0.1, is considered an impurity ty mode, as observed in bulk PZT ceramics. It is interesting to find that the impurity mode has a similar Zr dependence to the soft mode, in which a discontinuous behavior about its phonon energy is also observed to appear at x=0.4. On the other hand, the modes $E(TO_2)$ at 200 cm^{-1}, B_1+E at 290 cm^{-1} and $A_1(TO_3)$ at 580 cm^{-1}, in the high frequency region exhibit an abrupt decrease in intensity at x=0.4. These experimental facts suggest that a phase transformation from the ferroelectric(FE) tetragonal to ferroelectric rhombohedral phase occurs x=0.4 in nanocrystalline materials with grain size of 60 nm.

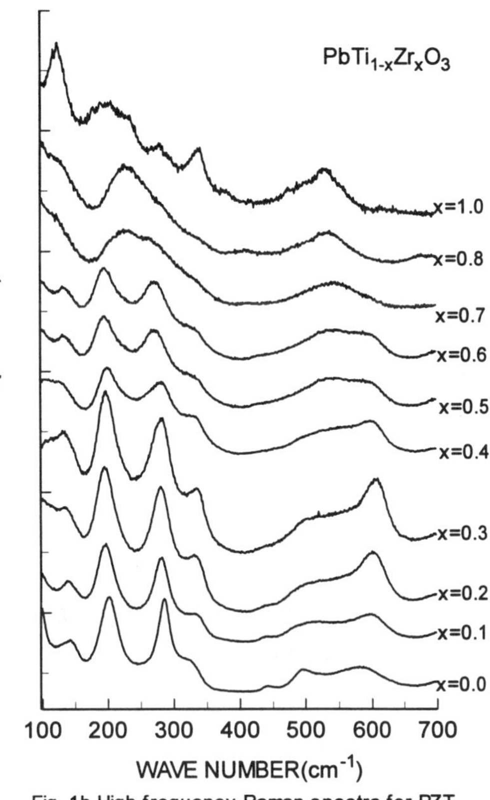

Fig. 1b High frequency Raman spectra for PZT with grain size of 60 nm

The solid solution of bulk PZT shows a FE tetragonal to FE rhombohedral phase transition at x ≈ 0.52 at T = 295 K. Raman scattering experiments in the case of bulk PZT showed that the substitution of Ti by Zr leads to a softening of the lowest E(TO) mode in PT and the softening of the soft mode may play an important role in the tetragonal-rhombohedral phase transition of

PZT [21]. Therefore, it is reasonably suggested that the soft mode for nanocrystalline PZT still plays an key role for the phase transition at x=0.4.

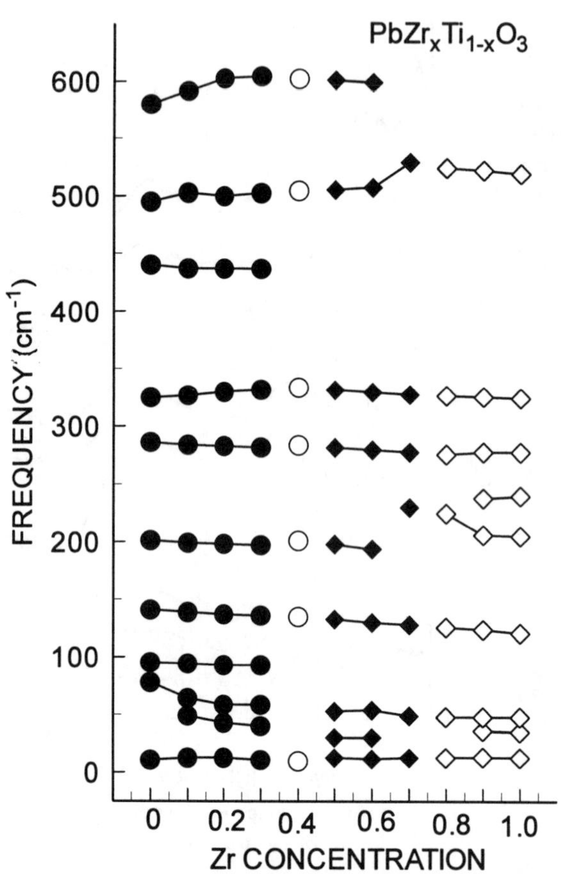

Fig.2 The Zr dependence of Phonon frequencies

In Fig. 3 we have shown the dependence of the squared frequency of the soft mode and other two low frequency modes on Zr concentration. It should be noted that the soft mode does not decrease its frequency to zero at x=0.4, which means that this phase transition should be of first order. This is in agreement with the measured results in the bulk PZT. However, in previous experiments, it was considered that the square of the E(TO) is linear in the concentration x, and goes to zero as the concentration approaches the critical concentration x_c. This sugges tion would imply that the tetragonal-rhombohedral phase transition in PZT is of second order. According to our measurements, the non zero soft mode frequency at the phase boundary in nanocrystalline PZT can attribute to a coupling between the soft mode and the impurity mode as described in bulk PZT, in which both high temperature Raman spectra of $PbTi_{0.75}Zr_{0.25}O_3$ [22] as well as room Raman spectra of PZT with different Zr concentrations [21] indicate a coupled mode behavior.

In the case of bulk PZT, it is known that there are two rhombohedral ferroelectric phase for $0.52 < x < 0.64$ and $0.64 < x < 0.94$, and an antiferroelectric region for $x > 0.94$. According to Fig. 1, the Raman spectrum at x=0.5 differs obviously from that at x=0.4, especially for the low frequency region. This may be explained as the region of the first rhombohedral ferroeletric phase, i. e. $0.4 < x < 0.5$. As x is increased to 0.7, careful observations find that the Raman spectrum seems to undergo another discontinuous change in phonon energy and intensity, and therefore the second rhombohedral ferroelectric phase is considered to occur in $0.5 \leq x \leq 0.7$.

The fourth region, x > 0.7, indicating a continuous spectrum, is considered as the antiferroelectric one.

Finally, let's go back to discuss another interesting phonon mode shown in Fig. 1a, i.e. a lowest frequency phonon mode at 10 cm^{-1}. For pure $PbTiO_3$ and $PbZrO_3$, the mode is found to have a weak intensity. For x to be 0.2, the mode increases its intensity up to x=0.4. Then, it begins to decrease in intensity for Zr the concentration above 0.7. Obviously, the Zr concentration dependence of the mode is closely correlated with the mentioned several phase regions above. It should be pointed out that the mode for bulk PT displays a much weaker intensity than nanocrystalline PT. Thus, the change of the mode in intensity can be attributed to effects of Zr concentration and grain size. In previous work, the polarization of the Raman spectrum of $BaTiO_3$ showed the existence of a lattice mode(propagating along the c or a axis) in the vicinity of 5 cm^{-1} with a definite E_1 symmetry [23]. It was concluded that BT has a " low frequency" transverse E_1 mode at room temperature. However, the origin of the mode is not clear.

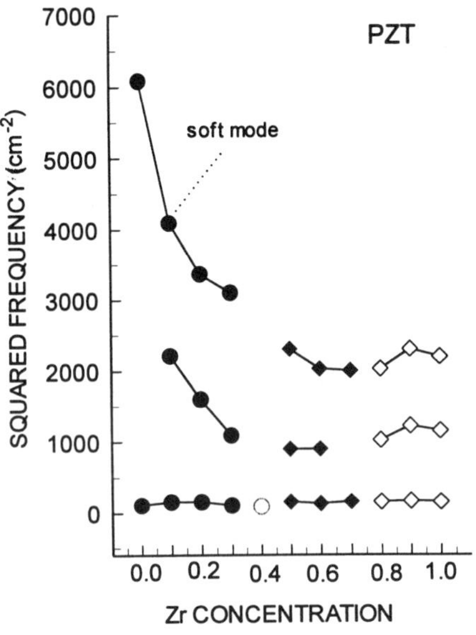

Fig. 3 The Zr dependence of the squared frequency

CONCLUSION

Raman spectra for nanocrystalline PZT have been obtained. The grain size has an obvious effect on Raman phonon modes, especially on the low frequency modes. The Zr concentration dependence of the soft mode plays a key role in the phase transformations in PZT. The first phase transformation at x=0.4 from ferroelectric tetragonal phase to rhombohedral phase is driven by " softening" of the soft mode with Zr concentration. Non-zero frequency of the soft mode at the phase boundary, responsible for first order phase transition, results from a coupling between the soft mode and an additional impurity mode. The decrease of grain size lead to every phase region in PZT towards a lower Zr concentration, i.e. there are two rhombohedral ferroelectric phase for $0.4 < x < 0.5$ and $0.5 \leq x \leq 0.7$, and an antiferroelectric region for $x > 0.7$. The Zr dependence of the observed lowest frequency mode in nanocrystalline PZT is closely associated with the changes of the phase regions.

* On leave from Department of Physics, Henan University, Kaifeng, P. R. China

ACKNOWLEDGMENTS

The work was supported in part by the DOE Grant DE-FG02-94ER75764, NASA-NCCW-0088 and NSF-OSR-9452893

REFERENCE

1. C. A. Paz de Araujo and G. W. Taylor, Ferroelectrics, **116**, 215 (1991).
2. J. F. Scott, C. A. Pazde Araujo, L. D. McMillan, H. Yoshimori, H. Wanatabe, T. Mihara, M. Azuma, T. Ueda, D. Ueda and G. Kano, Ferroelectrics, **133**, 47 (1992).
3. J. F. Scott and C. A. Paz de Araujo, Science, **246**, 1400 (1989).
4. B. Jaffe, R. S. Roth and S. Marzullo, J. Appl. Phys. **25,** 809 (1954).
5. B. Jaffe, R. S. Roth and S. Marzullo, J. Res. Nat. Bur. Standards, **55**, 239 (1955).
6. R. E. Newnham, Rep. Progr. Phys. **52,** 123 (1989).
7. B. Jaffe, W. R. Cook and H. Jaffe, Piezoelectric Ceramics, Academic Press, London , 1971, pp.136.
8. L .Benguigui, Solid State Commun. **19,** 979 (1976).
9. K. Kakfgawa, J. Mohri, K. Takahashi, H. Yamamura and S. Shirasaki, Bull. Chem. Soc. Japan, **5**, 717 (1976).
10. M. S. Multani, S. G. Gokarn, R. Vijyaraghavan and V. R. Palkar, Ferroelectrics, **37,** 652 (1981).
11. T. Nakamura, Bull. Ceram. Soc. Japan, **14,** 894 (1979).
12. E. Sawguchi, J. Phys. Soc. Jpn. **8**, 615 (1953).
13. A. V. Turik, m. F. Kupriyamov, E. N. Sidorenk and S. M. Zaitsev, ZH. Tekh. Fiz. **50,** 2146(1980) [Sov. Phys. Tech. Phys. **25**, 1251 (1980)].
14. P. Gr Lucuta, F. L. Constantinesan and D. Barb, J. Am. Ceram. Soc. **68**, 533 (1985).
15. P. Ari-Gur and L. Benguigui, Solid State Commun. **15,** 1077 (1978).
16. Lu Hanh, K. Uchino and S. Nomura, Jpn. J. Appl. Phys. **17**, 637 (1978).
17. M. Fukuhara, A. S. Bhalla and R. E. Newnham, Phys. Stat. Sol.(a), **122,** 677 (1990).
18. Wenwu Cao and L. Eric Cross, **47,** 4825 (1993).
19. G. Burns and B. A. Scott, Phys. Rev. Lett. **25,** 1191 (1970).
20 J. F. Meng, G. T. Zou, Q. L. Cui, Y. N. Zhao, Z. Q. Zhu, J. Phys.: condense matter **6,** 6543 (1994).
21. D. Bauerle, Y. Yacoby and W. Richter, Solid State Commun. **14,** 1137 (1974).
22. R. Merlin, J. A. Sanjurjo and A. Pinczuk, Solid State Commun. **16,** 931 (1975).
23. A. Pinczuk, W. Taylor, E. Burstein and L. Lefkowitz, Solid State Commun. **5**, 429 (1967).

Part XI

Applications of Nanocrystalline and Microcrystalline Thin Films

THE "MICROMORPH" CELL: A NEW WAY TO HIGH-EFFICIENCY-LOW-TEMPERATURE CRYSTALLINE SILICON THIN-FILM CELL MANUFACTURING?

H. Keppner, P. Torres, J. Meier, R. Platz, D. Fischer, U. Kroll, S. Dubail, J. A. Anna Selvan, N. Pellaton Vaucher, Y. Ziegler, R. Tscharner, Ch. Hof, N. Beck, M. Goetz, P. Pernet, M. Goerlitzer, N. Wyrsch, J. Veuille, J. Cuperus, A. Shah, J. Pohl*

Institut de Microtechnique, A.-L. Breguet 2, Université de Neuchâtel, CH-2000 Neuchâtel, Switzerland

*University of Konstanz, D-78434 Konstanz, Germany

ABSTRACT

In the past, microcrystalline silicon (μc-Si:H) has been successfully used as active semiconductor in entirely μc-Si:H p-i-n solar cells and a new type of tandem solar cell, called the "micromorph" cell, was introduced [1]. Micromorph cells consist of an amorphous silicon top cell and a microcrystalline bottom cell. In the paper a micromorph cell with a stable efficiency of 10.7 % (confirmed by ISE Freiburg) is reported.

Among sofar existing crystalline silicon-based solar cell manufacturing techniques, the application of microcrystalline silicon is a new promising way towards implementing thin-film silicon solar cells with a low temperature deposition. Microcrystalline silicon can, indeed, be deposited at temperatures as low as 220°C; hence, the way is here open to use cheap substrates as, e.g. plastic or glass. In the present paper, the development of single and tandem cells containing microcrystalline silicon is reviewed. As stated in previous publications, microcrystalline silicon technique has at present a severe drawback that has yet to be overcome: Its deposition rate for solar-grade material is about 2Å/s; in a more recent case 4.3 Å/s [2] could be obtained. In the present paper, using suitable mixtures of silane, hydrogen and argon, deposition rates of 9.4 Å/s are presented. Thereby the dominating plasma mechanism and the basic properties of resulting layers are described in detail. A first entirely microcrystalline cell deposited at 8.7 Å/s has an efficiency of 3.15%.

1. INTRODUCTION.

Microcrystalline Silicon (μc-Si:H) is generally obtained by using a strongly hydrogen-diluted silane discharge [3]. Apart from some isolated approaches [4, 5] presented in the past, μc-Si:H was never seriously taken into account as being able to play a role as active semiconductor in a solar cell; this was basically due to its generally high density of gap-states. On the other hand, microcrystalline silicon has since some time already been successful in providing highly doped p and n contacts, as needed for such devices as solar cells and thin film transistors (TFT). It should be emphasised that the Very High Frequency Glow-Discharge (VHF-GD) deposition technique can be considered to play an important role in obtaining "solar grade" μc-Si:H material [1, 6-10]

The deposition of "mid gap" material is not straightforward; as-deposited intrinsic μc-Si:H turns out to be slightly n-type. According to Veprek [11] this is basically due to oxygen impurities. Two techniques have in the past been tried out to overcome this problem:

1. The compensation technique: Adding small traces of diborane to the feedgas compensate the n-type behaviour [6].
2. The purifying technique: a gas purifier (oxygen getter) is installed close to the reactor in order to remove oxygen-containing impurities in the feedgas [12]. It can be concluded at the present

Mat. Res. Soc. Symp. Proc. Vol. 452

state, that the purifying technique has the capacity to completely replace the compensation technique.

2. SOLAR CELLS BASED ON MICROCRYSTALLINE SILICON: REVIEW OF PAST WORK

Striking features of the entirely microcrystalline solar cells deposited by VHF-GD using either the compensation method or the purifier method, are the following:

1. complete stability of the efficiency under intense long term light soaking; a property that sofar could never be achieved using amorphous silicon as base material (see Fig. 1 taken from Meier et al. [1]).
2. strongly enhanced extension of the spectral response towards infrared wavelengths, as compared to amorphous silicon (see Fig. 2, taken from Meier et al. [1]).

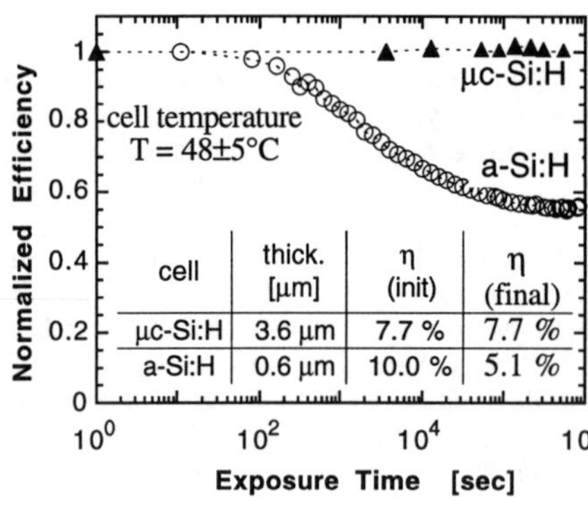

Fig. 1 Light-soaking of an amorphous and of a μc-Si:H cell under 6 and 10 suns respectively (According to Meier et al. [9]).

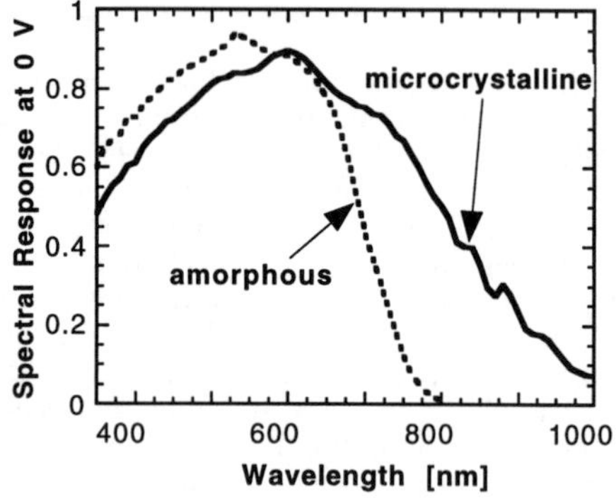

Fig. 2 Comparison of the spectral response of an a-Si:H cell and of an entirely microcrystalline cell (Meier et al. [1]).

The combination of an entirely microcrystalline solar cell used as bottom-cell with a "classical" amorphous silicon solar cell forms a "true" tandem cell; this combination is called by the authors a "micromorph" cell. Thereby, the top cell is rendered more stable using a relatively thin intrinsic amorphous layer deposited under strong hydrogen dilution. As shown in previous publications [7], the top cell alone gives rise to the full light-induced degradation of the total micromorph tandem cell. Table 1 gives an overview of the stable efficiencies sofar obtained for entirely μc-Si:H cells and micromorph tandem cells.

cell type	η [%]	Isc	Voc	FF	area [cm2]	remark
μc-p-i-n	7.7	25.3	0.448	67.9	0.33	10 suns, 48±5°C, 400 h
Micromorph	10.7*	11.0	1.359	63.0	0.33	1000 h Am 1.5; 50°C

* Confirmed by the ISE Freiburg (D).

Tab. 1 Solar cell parameters of stable entirely μc-Si:H cells and of micromorph cells.

Limits and potential for further improvements.

From Tab. 1. one can conclude that the top cell produces about two thirds of the total output of the tandem cell [10]; this is based on the strong difference in Voc between top and bottom cell. From this fact, one has to deduce that the stability of the top cell remains a key feature, as long as the open circuit voltage of the bottom cell remains limited. This also means that a further increase of the V_{oc} of the bottom cell alone contains the highest potential for a further improvement of the stable efficiency of the micromorph cell. The strongest limitation in the future industrial potential of all types of thin-film crystalline silicon solar cells is, at present, the too long manufacturing times (recrystallization and/or deposition rates). For the deposition of μc-Si:H, the VHF-GD process in itself brings sofar at least some improvement, as reported by Finger et al. [13] Attempts at obtaining yet higher deposition rates are described by Torres et al. [2]. In this paper we present an alternative approach using argon plus hydrogen dilution of silane. Argon dilution and mixtures of Ar, H_2 and silane have often been reported in literature for a-Si:H deposition with the attempt to obtain more stable material [14]. As a beneficial effect Kroll et al. [15] report on significant silane savings using argon dilution. In another work, an increase of the deposition rate of a-Si:H due to Ar-dilution is reported by Hautala et al. [16]. For μc-Si:H, DC plasma gun experiments for high-rate deposition of μc-Si:H are reported by Imajyo [17]; in the latter case, a mixture of Ar, H_2 and SiH_4 was used. In other publications an enhanced dissociation of SiH_4 is reported due to molecular quenching via metastables [15, 18, 19]. In the latter case, a pure Ar / SiH_4 system was studied.

3. EXPERIMENTAL: PROCEDURES USED FOR ARGON /HYDROGEN DILUTION

All the experiments presented here were undertaken using the VHF-GD technique at 70 MHz in a parallel plate reactor. Three series were deposited; Tab. 2. shows the strategy used.
For the M-series (**M**ixing-series), the mixture of the diluting gases Ar and H_2 was varied at constant silane flow. (6 sccm). The sum of flows of the diluting gases (Ar + H_2) was 150 sccm.
For the D-series (**D**ilution-series), the mixture of the two diluting gases was kept constant, at 90 sccm H_2 and 60 sccm Ar, whereas the silane flow was varied.
For the P-series the **P**ower of the plasma was varied for a fixed silane flow of 8.4 sccm in 90 sccm hydrogen and 60 sccm Argon.

Serie	Φ(SiH_4) [sccm]	Φ(H_2) [sccm]	Φ(Ar) [sccm]	Power [W]	Pressure [mbar]
M	6	**140 - 0**	**10 - 150**	30	0.8 mbar
D	**6-12**	90	60	40	0.9 mbar
P	8.4	90	60	**10 - 70**	0.9 mbar

Tab. 2. Strategy of H_2 / Ar dilution experiments for deposition of μc-Si:H.

Within the study of these series M, D, and P we will now lay the emphasis on plasma properties, deposition rate, structural properties of the μc-Si:H films (X-Ray measurements), infrared measurements, and sub-bandgap absorption (PDS).
In order to correlate the properties of the layers of series M, D and P with the corresponding properties of the plasma, optical emission spectroscopy was carried out. For this purpose, a monochromator with a detector was mounted at the VHF-reactor. Before every deposition, a new glass plate had to be inserted into the reactor in order to protect the window from being coated. The substrate and the powered electrode were heated up to 220°C. For this study, no gas purifier was applied.

4. RESULTS AND DISCUSSION

Results on the deposition rate.

In several publications a link between the intensity of the SiH* 412 nm emission line and the deposition rate of a-Si:H was shown (Perrin et al. [20]). Hamasaki et al. [21] reported on a criterion for the deposition of microcrystalline silicon if the relationship of the emission intensities H(656nm)/Si(412nm) exceeds unity. As in our present study three gases ($SiH_4/H_2/Ar$) are involved, a criterion for the growth rate of μc-Si:H from the emission spectrum of these gases was looked for. Fig. 3 shows a typical OES (optical emission spectrum) of a SiH_4 / H_2 / Ar plasma. The lines of interest are therein particularly mentioned:

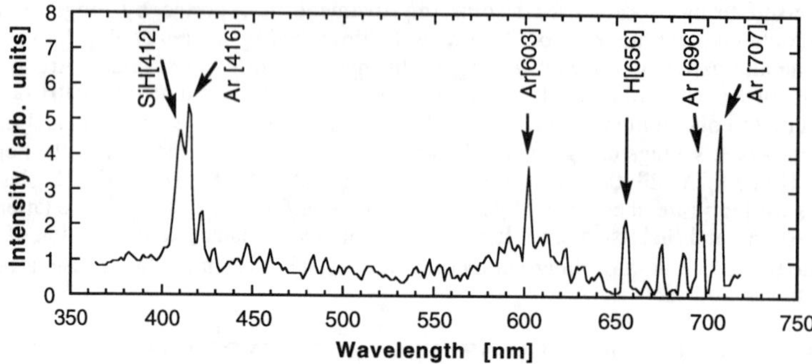

Fig. 3: Typical OES of a SiH_4 / H_2 / Ar plasma with the lines of interest for this study.

Result of M-series.

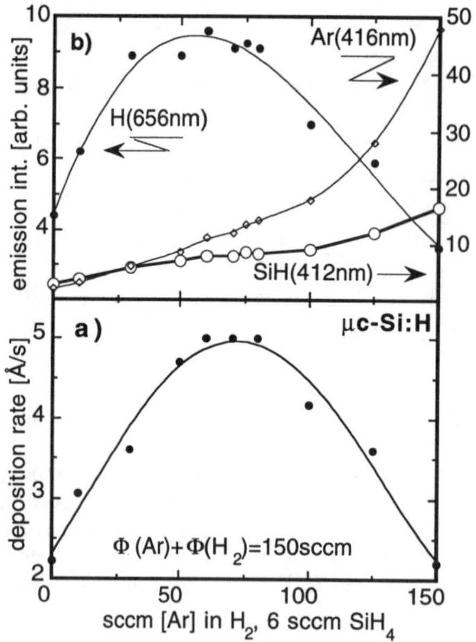

Fig. 4 M-series: a: deposition rate of μc-Si:H in function of the Ar/H_2 mixture at constant silane flow; b: emission lines from the plasma. Note that the Hα (656 nm) emission intensity correlates well with the deposition rate as observed in 4a.

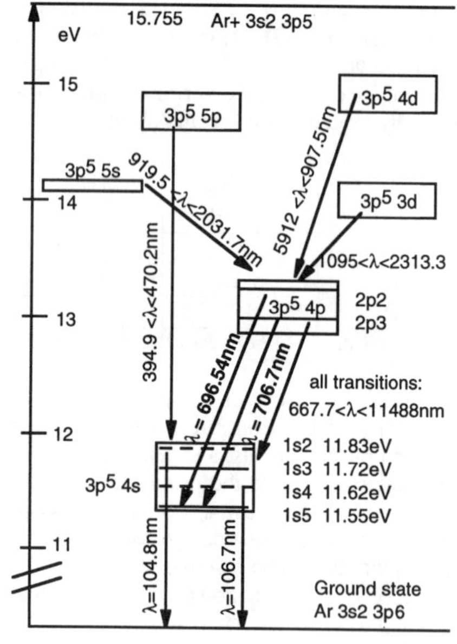

Fig. 5. Schematic energy levels of the argon atom, according to Ferreira et al [23]. The transients that are marked give are related to the rate of populating of the metastable state $1s_5$

As soon as argon partly replaces hydrogen in the diluting gas at constant silane concentration, the deposition rate of μc-Si:H increases. Note that all layers presented in the M series are microcrystalline.

Looking at plasma chemistry, the symmetrical shape of the plot in Fig 4a might be helpful for understanding the significant increase in deposition rate. As is well known from Hot-Wire experiments [22], the rate-limiting factor for μc-Si:H deposition is not related to the depletion of silane but rather to its dissociation.

Let us first consider the extreme cases at the boundaries of Fig. 4a phenomenologically: On one side (100% hydrogen dilution), a supporting contribution to the further radicalisation of SiH_4 due to the presence of argon is lacking; on the other side (100% argon dilution), a supporting contribution due to the presence of hydrogen is lacking. In both extreme cases, the lowest rate of about 2.2 Å/s is observed. In the intermediate range, both gases contribute equally to enhanced dissociation of silane. In order to explain this, we have to consider the different reaction channels with their respective reaction rates: The supporting contribution of argon in the silane plasma can be attributed to the creation of metastable states, as was already explained before in detail by several publications [18, 19, 23]. Hereby, the following fact should be mentioned: The creation of argon metastables (e.g. the Ar* ($^3P_2$) state) from the argon ground state, is due to electron impact excitation. Note that electron impact does not necessarily directly lead to the creation of metastable states (called here Ar*), but first mostly to the population of states of higher argon levels (e.g. Ar $3p^54p$), see Fig. 5. This first step of excitation requires a threshold energy of 13.27 eV, i.e. it requires hot electrons. In a second step the emission of light at 696 and 706 nm (see in Figs. 3, 5) due to the rate of depopulation of the states $2p_2$ or $2p_3$ gives information on the rate of populating of the metastable state $1s_5$. On the other hand, collision of silane molecules with an argon atom in the metastable state can lead to the so-called molecular quenching reactions of metastables due to silane (2):

$$SiH_4 \xrightarrow{Ar^*} SiH_3 + H \qquad \text{or} \qquad SiH_4 \xrightarrow{Ar^*} SiH_2 + (2)H_{(2)} \tag{1}$$

These reactions become the dominant loss mechanism for the occupation of metastable states **only** in strongly diluted silane plasmas (4% according to [18]), because comparatively more "hot" electrons are than available for the creation of metastables. If more silane is in the feedgas, more hot electrons of the energy distribution function are consumed for exciting silane via the different excitation channels instead of being available for the creation of metastables. We may assume, hence, that the dissociation of silane via metastable state quenching can indeed play a more important role in the deposition of μc-Si:H, as compared to the deposition of a-Si:H, because the conditions related to highly diluted silane are indeed prevalent in the case of μc-Si:H. For an explanation of the results given in Fig. 4a we propose the following model:

$$C(\text{total}) = \underbrace{C(SiH_4)}_{\text{const}} + \underbrace{C(H_2)}_{=0 \text{ for } 0\%\ H_2} + \underbrace{C(Ar)}_{=0 \text{ for } 0\%\ Ar} \tag{2}$$

where C(total) is the concentration of all radicals that contribute to the formation of μc-Si:H; $C(SiH_4)$, $C(H_2)$ and C(Ar) are the contributions of radicals formed from SiH_4, H_2 and Ar, respectively. The reactions and their rate constants are given in the columns of Tab. 3.

Looking at Tab. 3, the respective reaction rates can be calculated from their rate constants and their concentration of prime components, using equations similar to the following equation (4). In equation (4) we will take, as an example, the Ar* quenching due to silane:

$$\frac{dSiH_4}{dt} = k_{Ar^*}[SiH_4]\cdot[Ar^*] \tag{3}$$

C (SiH_4)	C (H_2)	C(Ar)	
$SiH_4 \xrightarrow{e} SiH_3 + H$ k=1.59 x 10^{-10} $SiH_4 \xrightarrow{e} SiH_2 + (2)H_{(2)}$ k=1.87 x 10^{-11}	$H_2 \xrightarrow{e} 2\,H$ k=4.49 x 10^{-12}	$Ar \xrightarrow{e} Ar^*$ k=1.21 x 10^{-12}	electron impact
$SiH_4 \xrightarrow{H} SiH_3 + H_2$ k=2.68 x 10^{-11}	$SiH_4 \xrightarrow{H} SiH_3 + H_2$ k=2.68 x 10^{-11}		hydrogen abstraction
	$H_2 \xrightarrow{Ar^*} 2\,H$ k=7 x 10^{-11}	$SiH_4 \xrightarrow{Ar^*} SiH_3 + H$ k=1.4 x 10^{-10} $SiH_4 \xrightarrow{Ar^*} SiH_2 + (2)H_{(2)}$k =2.6 x 10^{-10}	molecular quenching
3.8 % SiH_4 in total	0-92.3% H_2 in total	0-92.3% Ar in total	

Tab. 3. Most important reaction channels for silane dissociation due to electron impact, hydrogen abstraction and molecular quenching. The rate constants are given in [cm^3/s]; the data are taken from Kushner [24] with references therein.

On the left side of Fig. 5a, the ensemble of all radical contributions given in the columns C(SiH_4) and C(H_2) of Tab. 3 lead to the poor rate of 2.2Å/s. On the right side of Fig. 4a, all equations of the columns C(SiH_4) and C(Ar) lead to about the same low deposition rate. One might be led to assume, now, looking at Tab. 3 that a combination of all three columns at equal contributions of C(H_2) and C(Ar) could perhaps explain the increase of the deposition rate as observed in Fig. 4a. However, a more detailed analysis is called for and will lead to a different conclusion. In our simple model only the most important reaction channels are taken into account; Furthermore, no steady-state concentrations can be determined due to the lack of data regarding the loss mechanisms. However, we know that the C(Ar) column in Tab. 3 should contribute, in the case of 100% Ar dilution much more than in the corresponding case of 100 % H_2, the C(H_2) column; this results from the high production rate of metastables and its high rate constants for the molecular quenching channels. The only common reaction between argon and hydrogen that is characteristic for equal dilution with hydrogen and argon, is the quenching reaction of argon metastables via molecular hydrogen. However, both the rate constant and the concentration of the components of this reaction can be estimated to be much smaller than the contribution of C(Ar). Hence, we may indeed conclude that our simple plasma chemistry model given in Tab. 3 can in fact not explain the increase of the deposition rate of more than the factor of two as we observe in the M-series.

Hence, there must be an additional, much stronger contribution due to hydrogen that competes with the enhanced production of radicals as given by the column C(Ar). As a possible further explanation, we may assume that the chemisorbed hydrogen at the growing surface is etched away by atomic hydrogen and allows, by that, better sticking of radicals. This effect was postulated [25, 26] to be the rate-limiting mechanism for the deposition of amorphous and microcrystalline silicon. The rate limitation at 100% hydrogen dilution is due to the lack of radicals (left side of Fig. 5a); at 100% argon dilution, however, sufficient radicals are created but, according to this model, they all can not stick and are repelled into the plasma. Hence, at 100 % argon dilution, the deposition rate limiting factor is, as we now propose, due to sticking. Fig. 4b shows a good correlation between the emission intensity of the Hα line at 656 nm and the deposition rate, whereas the intensities of all the other lines have the tendency to increase for increasing argon dilution. We propose that for the M-series, the Hα line should be considered to be, indeed, a good monitor for the deposition rate.

Result on D-series.

Looking at the **D series** in Fig. 6a and Fig. 6b, the situation changes w.r.t. Fig. 4a and Fig. 4b: The increase in silane concentration in the feedgas from 3.2% silane to 7.4% in the mixture leads to an increase in deposition rate up to values as high as 9.4 Å/s the highest deposition rate we have observed in the present study. As found also for the M-series, all samples in the D-series are microcrystalline.

The evaluation of the spectral lines given in Fig. 6b show an increase of the intensity of the SiH* line at 412nm that correlates with the deposition rate. The line intensities of Ar and H, however, remain unchanged for lower concentrations of SiH_4 in the feedgas; for higher concentrations all intensities have a tendency to decrease. From Fig. 6 a,b one has to conclude for the D series that the emission intensity of the Hα line at 656 nm can here no more be used as a monitor of the deposition rate, whereas the SiH* line at 412 nm correlates better, in this specific case.

The **P-series** represented in Figs. 7a,b shows a surprising behaviour: The samples at 10W and 20W are amorphous silicon. At 30 W, μc-Si:H silicon is grown, at the maximum deposition rate within the P-series. Going to higher power, a pronounced reduction in deposition rate can be observed. All the emission lines in Fig. 7b increase with the power until a plateau is reached. Comparing Fig. 7a and 7b, no correlation between the deposition rate and any spectral line can be deduced.

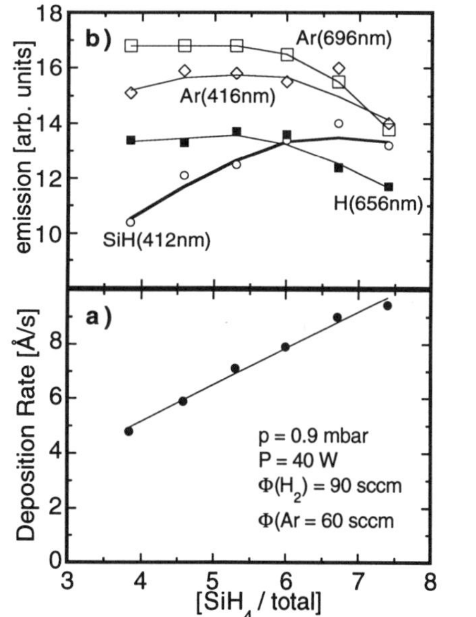

Fig. 6 a, b: D-series: Deposition rate of μc-Si:H in function of silane dilution and its comparison with the intensity of certain OES lines.

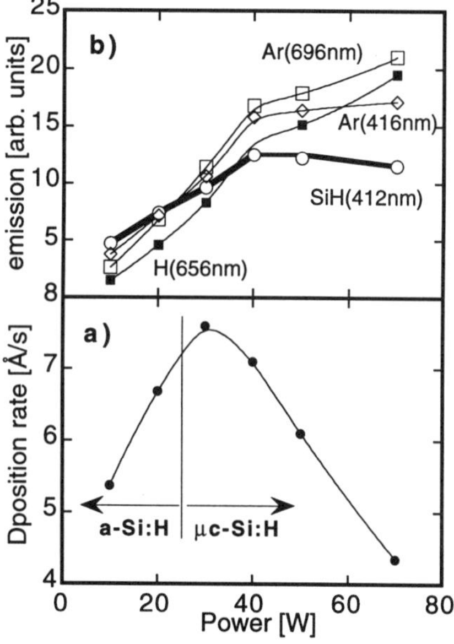

Fig. 7 a, b: P-series: Deposition rate of μc-Si:H in function of plasma power and its comparison with the intensity of OES lines.

From this behaviour of the P-series one can conclude, that the deposition rate of μc-Si:H can not be increased only by simply increasing the VHF power; for each set of H_2/Ar dilution an optimum set of pressure, SiH_4 flow and plasma power has to be individually determined.

Structural Properties

In the following, the structural consideration are mainly presented for the D-series; in fact this it is within this series that the highest deposition rate could be obtained in accordance with the main emphasis of this work.

X-Ray diffraction.

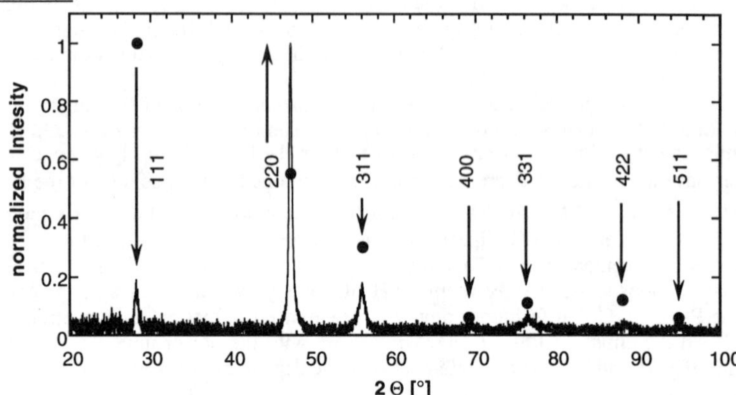

Fig. 8: Typical X-Ray diffraction pattern of a μc-Si:H thin film grown in an argon /hydrogen mixture. In contrast with powder samples, the 220 peak is very pronounced. The comparison with the round filled symbols (Si powder sample) is an indication of the strongly oriented growth of these μc-Si:H samples.

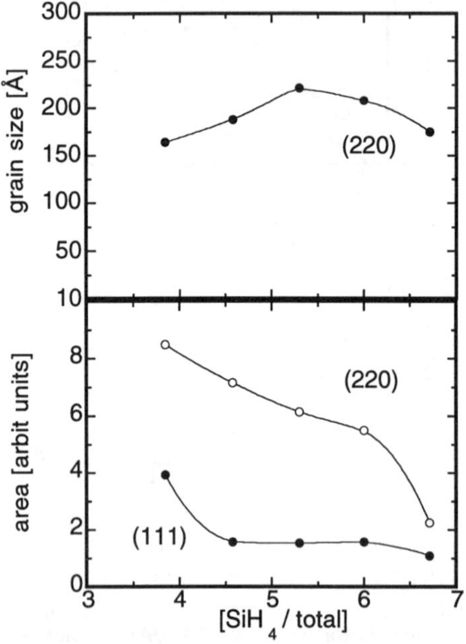

Fig. 9 Peak areas of the (111) and (220) X-Ray diffraction of the D-series; as well as apparent grain size of the D-series taken from the 220 signal and calculated by Scherrer's equation.

The X-Ray diffraction patterns of the samples of series M, D, P are all similar to the one shown in Fig. 8. We have strongly anisotropical columnar growth in the 220 direction. The specific pattern shown in Fig. 10 belongs exceptionally to the M series (75 sccm H_2; 75 sccm Ar; 6 sccm SiH_4). From Fig. 8 we may conclude that the crystalline growth in our μc-Si:H samples is strongly oriented. We will now try to calculate the gain-size by applying Scherrer's formula, based on the (220) diffraction. Thereby, we will not determine the true grain-size, because our crystallites are not spherical, rather, we will find an apparent grain-size.

From SEM and TEM images the grains were found to be of columnar shape, see work of Meier et al. [9]. Still, a large apparent grain-size can be assumed to give, together with a strong anisotropy (as indicated by an anomalously large difference in the (111) and (220) X-ray diffraction value), means that there are probably no or only few grain-boundaries perpendicular to the direction of the photocurrent in a cell. Details of such a study are given elsewhere [27].

Infrared measurements.

The infrared measurements reveal two features: first, the Si-H_x stretching mode does not show a pronounced shift from 2000 cm^{-1} to 2100 cm^{-1} as usually seen for the a-Si:H-μc-Si:H transition for the case of low concentrations of silane in an Ar/H_2 mixture (Fig. 10a). This is in agreement with results recently reported from Das et al. [19]; these authors diluted silane in pure argon. Presumably, due to the presence of argon in the plasma, hydrogen is differently bonded in the resulting μc-Si:H layers. As reported by Kroll at al. [28] (using VHF-GD and only H_2-dilution). The phase transition from amorphous to microcrystalline silicon is usually related to a shift of the peak from 2000 to 2100 wavenumbers.
For μc-Si:H the 2100 cm^{-1} peak is attributed to bonded hydrogen on crystallite surfaces.

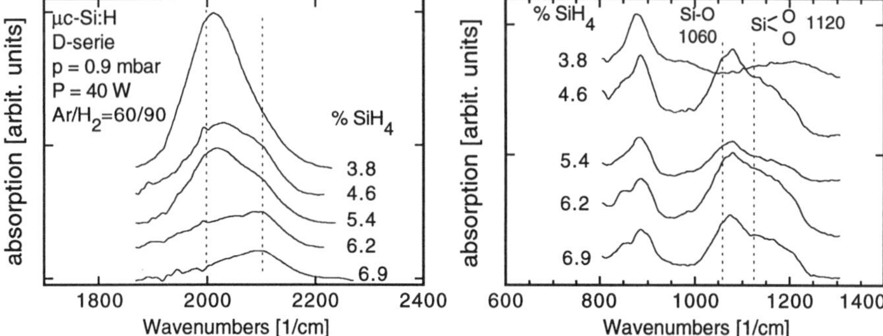

Fig. 10 a. Si-H_x stretching mode of the samples of D-series. Note, usually the peak at 2100 1/cm is typical for microcrystalline silicon.

Fig. 10 b. In some of the samples of D-series, fingerprints for incorporated oxygen are found. The amount can not be correlated neither to the deposition conditions, nor to exposure at air.

Secondly, significant quantities of bonded oxygen in the layers are observed from the 1060 cm^{-1} and 1120 cm^{-1} modes (see Fig. 10b). The relative quantity of oxygen can not be correlated w.r.t. either the control parameters of the plasma (power, gas flow, pressure, dilution) nor with the post-oxidation after air-exposure. Note that in some cases, oxygen was observed to diffuse within two weeks into the layers [11]. This, however, does not seem to be the case for our samples.

As an explanation for the origin of oxygen in these layers, we postulate that the metastable states of the argon atoms play an important role in connection with water vapour in the feed gas. Enhanced oxygen incorporation was in fact previously already found by Kroll et al. [15] by SIMS measurements performed on a-Si:H layers that were deposited using argon-diluted silane. Using the gas-purifying technique, these authors could successfully suppress oxygen in the a-Si:H layers.

As all the experiment on individual layers we present here (series M, D, P) were undertaken without using the purifying technique, we may assume, that the following reaction gave rise to oxygen incorporation:

$$Ar^* + H_2O \xrightarrow{4.8\times10^{-10}\,cm^3/s} OH^* + H + Ar \qquad (4)$$

Velazco et al [29] found, that water vapour is an even more efficient "quencher" for Ar* metastables than silane. We deduce therefore, that in the case of Ar/H_2-dilution high deposition rates of μc-Si:H rate deposition becomes possible; however, water-vapour quenching of Ar* metastables has to be suppressed here, e.g. using a purifier in the gas lines.

Subbandgap absorption

Fig. 11 shows a comparison between the infrared absorption of our sofar best μc-Si:H samples according to previous work of Meier et al. [9] (called here "μc-Si:H Ref"), crystalline silicon, amorphous silicon, and our present high-rate deposited μc-Si:H, obtained using Ar/H_2 dilution.

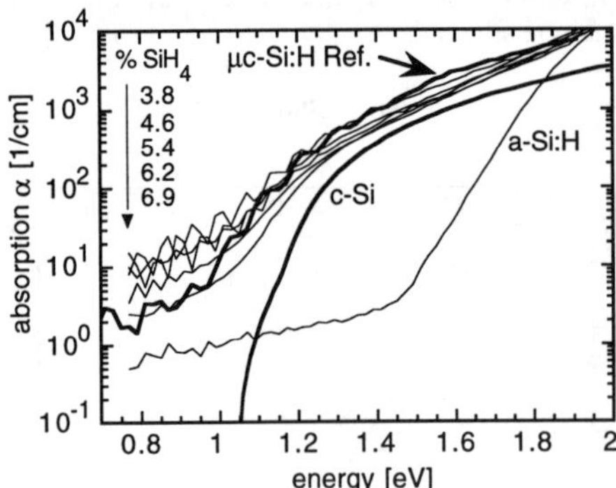

Fig. 11. Comparison of our sofar best μc-Si:H ("μc-Si:H Ref."), c-Si, a-Si:H and high-rate deposited μc-Si:H from the present series D.

It can be seen from Fig. 11 that the absorption around 1.2 eV for the new samples turns out to be lower than the previous values obtained with the previous sample indicated as "μc-Si:H Ref." This is an undesired effect for future solar cell applications, because the gain in deposition rate is partly offset by a loss in absorption around 1.2 eV; a fact which indeed requires thicker cells.

However, at the present state of the work, and looking at the strong oxygen incorporation of most of the samples in our present of the D-series, it should not be concluded in any way that high-rate deposited material has, as a general feature, a lower infrared absorption.

It was furthermore observed in this study that the material with the highest deposition rate (9.4 Å/s) showed the lowest subbandgap absorption within the present series; furthermore, all layers have an overall higher absorption than c-Si.

Results on solar cells.

A first attempt in fabricating an entirely μc-Si:H n-i-p solar cell with an i-layer deposited at 8.7 Å/s using Ar/H_2 diluted silane resulted in a cell efficiency of 3.15 %. The thickness of the i-layer was about 5.5 μm [27]. Here, a gas purifier was used while depositing this cell.

5. CONCLUSIONS

A silicon-based tandem solar cell called the "micromorph" solar cell containing a microcrystalline silicon bottom cell and an amorphous silicon top cell has obtained a stable efficiency of 10.7 %. Because of the relatively low absorption coefficient in the near infrared, a thick microcrystalline i-layer had to be applied: hence, this results in rather long deposition times result; thus this remains a major problem to overcome. Argon / hydrogen dilution of silane, as the method that was used here, in combination with the VHF-GD process, basically provides the possibility to achieve deposition rates up to 10 Å/s. Argon/hydrogen dilution experiments show, furthermore, that the columnar growth structure and the average grain-size of the crystallites can,

indeed, be controlled within a certain range. Thanks to the addition of argon as dilution gas, the parameter space that allows one to "tune" μc-Si:H w.r.t solar cell requirements is strongly expanded. Special care must thereby be taken in looking at all gases involved in the plasma: it was e.g. found that water vapour in the discharge leads to a pronounced incorporation of oxygen impurities into the growing film. It is concluded, therefore, that the use of a gas purifier becomes imperative for this particular method. The infrared absorption around 1.2 eV was lower in the present high-rate deposited microcrystalline silicon layers than in previous layers deposited by the VHF-GD in our group. This again, may be due to the enhanced oxygen incorporation that results from the use of argon as one of the dilution gas.
A first trial check with such high-rate material already result e.g. in a cell with 3.15 % efficiency. Thereby, an oxygen purifier was used, but the cell parameters were sofar not optimised.

ACKNOWLEDGEMENTS

This work was supported by the Swiss Federal Energy Department under Research Grant EF-REN (93)032

REFERENCES

1. J. Meier, R. Flückiger, H. Keppner, A. Shah, Appl. Phys. Lett., Vol. 65 (7), pp. 860-862, 1994.
2. P. Torres, J. Meier, M. Goetz, N. Beck, U. Kroll, H. Keppner, A. Shah, this conference.
3. S. Veprek and V. Marecek, Solid State Electronics Nr. 11, p. 683, 1968.
4. C. Wang and G. Lucowsky, Proc. 21st IEEE Photovoltaic Specialists Conference, Orlando 1990, Vol. 2, pp 1614-1618.
5. M. Faraj, S. Gokhale, S. M. Choudhari, and M. G. Takwale, Appl. Phys. Lett. 60, p. 3289, 1992.
6. R. Flückiger, J. Meier, H. Keppner, U. Kroll, A. Shah, O. Greim, M. Morris, J. Pohl, P. Hapke, R. Carius, Proceedings of the 11th EC Photovoltaic Solar Energy Conference, Montreux, 1992, p. 617.
7. J. Meier, S. Dubail, D. Fischer, J. A. Anna Selvan, N. Pellaton Vaucher, R. Platz, C. Hof, R. Flückiger, U. Kroll, N. Wyrsch, P. Torres, H. Keppner, A. Shah, K.-D. Ufert, Proceedings of the 13th EC Photovoltaic Solar Energy Conference, Nice, 1995, p. 1445.
8. J. Meier, S. Dubail, R. Flückiger, D. Fischer, H. Keppner, A. Shah, Proceedings of the 1st World Conference on Photovoltaic Energy Conversion, Hawaii, 1994, Vol. 1, pp. 409-412.
9. J. Meier, P. Torres, R. Platz, S. Dubail, U. Kroll, J.A. Anna Selvan, N. Pellaton-Vaucher, Ch. Hof, D. Fischer, H. Keppner, A. Shah, K.-D. Ufert, P. Giannoulès, J. Koehler, to be published in the Proc. MRS 1996 Spring Meeting San Francisco.
10. D. Fischer, S. Dubail, J.A. Anna Selvan, N. Pellaton-Vaucher, R. Platz, Ch. Hof, U. Kroll, J. Meier, P.Torres, H. Keppner, N. Wyrsch, M. Goetz, A. Shah, K.-D. Ufert, Proc. 25th IEEE Photovoltaic Specialists Conference , Washington , 1996, Vol. 2 pp. 1053-1056.
11. S. Veprek, Z. Iqbal, R. O. Kühne, P. Capezzuto, F-A Sarott and J. K. Gimzewski, J. Phys. C: Solid State Physics, 16, pp. 6241-6262, 1983.
12. U. Kroll, J. Meier, H. Keppner, and A. Shah, J. Vac. Sci. Technol. A 13(6) p. 2742, 1995.
13. F. Finger, P.Hapke, M. Lysberg, R. Carius, H. Wagner, Appl. Phys. Lett. 65(20), p. 247,1994.
14. A. Matsuda, S. Mashima, K. Hasezaki, A. Suzuki, S. Yamasaki and P.J. McElhenny, Appl. Phys. Lett., 58, p. 2494, 1991.

15. U. Kroll, PhD thesis University of Neuchâtel, Hartung & Gorre Verlag, Konstanz, ISBN 3-89191-905-0, 1995.
16. J. Hautala, Z. Saleh, J.F.M. Westendorp, H. Meiling, S. Sherman, and S. Wagner, to be published in the Proceedings of the MRS Spring Meeting, San Francisco, Vol. 420, 1996.
17. N. Imajyo, J. of Non-Cryst. Solids, 198-200, pp. 935-939, 1995.
18. L. Sansonnens, A.A. Howling, Ch. Hollenstein, J-L. Dorier, and U. Kroll, J. Phys. D: Appl. Phys. 27, pp 1406-1411, 1994.
19. U. K. Das and P. Chaudhuri, S.T. Kshirsagar, J. Appl. Phys. 80(9) pp. 5389-5397, 1996.
20. J. Perrin, J. Schmitt, Chem. Phys. 67, p. 167, 1982.
21. T. Hamasaki, H. Kurata, M. Hirose, and Y. Osaka, Appl. Phys. Lett. 37, p. 1084, 1980.
22. A. R. Middya, J. Guillet, J. Perrin, A. Lloret, and J. E. Bourrée, Proceedings of the 13th EC PV Conference, Nice, 1995, p. 1704.
23. C. M. Ferreira and J. Loureiro, J. Appl. Phys. 57(1), p. 82, 1985.
24. M. J. Kushner, J. Appl. Phys. 63(8), p. 2532, 1988.
25. S. Veprek, F. -A. Sarrott and S. Rambert, E. Taglauer, J. Vac. Sci. Technol. A 7(4) p. 2614, 1989.
26. M. Heintze and R. Zedlitz, Progress in Photovoltaics: Research and applications, 1, p. 213, 1993.
27. H. Keppner, U. Kroll, P. Torres, J. Meier, R. Platz, D. Fischer, N. Beck, S. Dubail, J.A. Anna Selvan, N. Pellaton Vaucher, M. Goerlitzer, Y. Ziegler, R. Tscharner, Ch. Hof, M. Goetz, P. Pernet, N. Wyrsch, J. Vuille, J. Cuperus, and A. Shah, to be published at the NREL/SNL Photovoltaics Program review Meeting, Lakewood Co, 1996.
28 U. Kroll, J. Meier, A. Shah, S. Mikahaiov, J. Weber, J. Appl. Phys. 80, p. 4971, 1996.
29. J. E. Velazco, J. H. Kolts, D. W. Setser, J. Chem. Phys., 69(10), p. 4357, 1978.

N-I-P MICROMORPH SOLAR CELLS ON ALUMINIUM SUBSTRATES

M.GOETZ, P.TORRES, P.PERNET, J.MEIER, D.FISCHER, H.KEPPNER, A.SHAH
Institut de Microtechnique (IMT) de l'Université de Neuchâtel, Rue Breguet 2, CH-2000 Neuchâtel, Switzerland, goetz@imt.unine.ch

ABSTRACT

The first successful deposition of 'micromorph' silicon tandem solar cells of the n-i-p-n-i-p configuration is reported. In order to implement the 'micromorph' solar cell concept, four key elements had to be prepared: First, the deposition of mid-gap, intrinsic microcrystalline silicon (μc-Si:H) by the 'gas purifier method', second, the amorphous silicon (a-Si:H) n-i-p single junction solar cell, third, the microcrystalline silicon n-i-p single junction solar cell and fourth, the ability of depositing on aluminium sheet substrates.

All the solar cells presented have been deposited on flat aluminium sheets, using a single layer antireflection coating to couple the light into the cell. It is shown, that this antireflection concept- together with a flat substrate- holds for amorphous single junction solar cells, but it reaches its limit with the extended range of spectral response of the 'micromorph' cell.

The best initial efficiencies for each category of n-i-p cells on flat substrates were: 8.7% for the amorphous silicon single junction cell, 4.9% for the microcrystalline silicon single junction cell and 9.25% for the 'micromorph' tandem cell.

INTRODUCTION

The mixed amorphous / microcrystalline tandem solar cell concept was introduced in 1994 [1] by our group (IMT). These so-called 'micromorph' cells are 'real' tandem cells, employing two materials of a different optical gap for the top and the bottom cell (1.7eV and 1eV, respectively [2]). The enhanced infrared absorption and the lack of light-induced degradation [3] of the microcrystalline silicon are successfully combined with the high V_{OC} of the amorphous silicon solar cell. This lead sofar to solar cells of the p-i-n-p-i-n superstrate structure on classical TCO on glass of a confirmed stabilised efficiency of 10.5% [4].

Conceptionally, cells of the p-i-n deposition order are limited to transparent substrates like glass or some (rare!) highly transparent and temperature resistant polymers, as the light always enters the cell *through* the substrate in this configuration. To be able to use a wider choice of substrates, including opaque metal sheets, the inverse n-i-p deposition order has to be applied. In this case, the light enters the cell from the 'top' side, i.e. the side opposite to the substrate. One advantage of the n-i-p configuration is the better control of the delicate <p>/<i>-interface, as it is deposited last, on the opposite side of the nucleation base [5]. Microcrystalline window layers are used more successfully in the n-i-p structure than in p-i-n cells, leading to high V_{OC}-values even without any buffer layer. In this study aluminium sheets are used as substrates. The long term goal is the direct integration of silicon thin film solar cells into a type of façade construction elements, which consists of an aluminium-polymer compound. This type of façade construction elements- but without the incorporated solar cell- is a standard commercial product of which millions of square meters have been sold [6]. Aluminium has also been chosen for its ductility which offers the (not yet employed) possibility of controlled geometrical texturing for

Mat. Res. Soc. Symp. Proc. Vol. 452 © 1997 Materials Research Society

improved light-trapping, allowing to deposit thinner and therefore more stable a-Si:H solar cells.

EXPERIMENT AND RESULTS

Amorphous silicon n-i-p single junction solar cells

To be able to deposit a-Si:H solar cells on *commercial* aluminium sheets, as was our aim, several substrate-related problems had to be solved. Metal diffusion from the substrate or from the back contact had to be suppressed by an antidiffusion layer, as is reported elsewhere [7]. Surface topography of the aluminium sheets turned out to be a hard problem. The surface of the metal sheet has to be treated in an appropriate way to avoid 'pinholes' in the solar cells or in general to get high values of parallel resistance.

In the n-i-p configuration the top contact is deposited last. A thin film of Indium Tin Oxide (ITO) can serve as electrical contact *and* antireflection (AR) coating at the same time. An antireflection coating is always optimised for a certain wavelength and has a certain 'spectral width', i.e. a range of wavelength, where the reflectivity is actually reduced. In our first case, the flat cell of amorphous silicon, the minimum of reflection is set to 550nm and the important range (with respect to the spectral response curve of the cells) from 400 to 700nm is more or less covered. The excellent coupling of the most important green light (reflectivity near 0%) more than compensates the losses at the borders of the spectral range of the AR coating, where some of the blue and some of the red or infrared light is reflected. A flat cell of a thickness of 0.4 microns with a good back reflector and the presented AR coating, but without any further light-trapping, furnishes about 15mA/cm^2 at I_{sc} conditions. It has an efficiency of more than 8%.

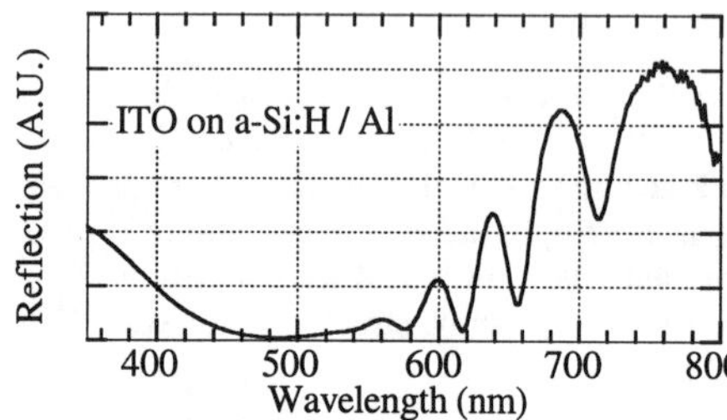

Figure 1. Reflection of an ITO antireflection coating on a-Si:H on a flat aluminium sheet vs. wavelength. The optimum in this example is set to about 500nm.

Note: A solar cell with this AR coating has a violet aspect; the colour violet is the sum of red and blue (reflected) light.

Microcrystalline silicon n-i-p single junction solar cells

Microcrystalline silicon, deposited without special precaution, has an n-type character. For a long time μc-Si:H was therefore not considered to be a potential candidate for an absorbing <i>-layer in a solar cell. By the 'microdoping' approach [8], a first microcrystalline 'mid-gap' material could be realised and first entirely microcrystalline silicon solar cells were deposited. In the meantime, it could be shown that the use of a gas purifier can replace the quite unreliable technique of 'microdoping' [9, 10]. In the p-i-n configuration on textured glass / TCO substrates, efficiencies of 7.7% have been achieved by our group [9].

First entirely microcrystalline n-i-p cells on aluminium substrates have been deposited. Microcrystalline cells are very sensitive to contamination of the <i>-layer; as the pronounced effect of the gas purifier prooves. It is therefore surprising, that the same antidiffusion scheme, as has been introduced for the (less sensitive) amorphous silicon cells, also holds for the μc-Si:H cells. The low contamination level allows one to deposit thick cells of up to 4 microns to obtain high short circuit currents. (Microcrystalline material deposited by the 'microdoping' technique limited the thickness of the cell to 2 microns [9].)

The absorption range of the microcrystalline cell is extended towards higher wavelengths compared to amorphous cells. For μc-Si:H solar cells, the thickness of the ITO antireflection layer should therefore be optimised to have its minimum of reflection at a higher wavelength than for a-Si:H cells. For the flat single junction microcrystalline cell of 4 microns thickness, with a good back reflector and an ITO AR layer, but without any further light trapping, a current density of 24 mA/cm^2 has been measured. The best efficiency of a μc-Si:H n-i-p cell on an aluminium substrate was sofar 4.9%.

The microcrystalline silicon deposited by Very High Frequency -Glow Discharge (VHF-GD) has an enhanced absorption coefficient. Its absorption over the whole important wavelength range is higher than the absorption of crystalline silicon. The reason for the high absorption is still under discussion: strain-induced change of the absorption, the high relative number of silicon atoms at grain boundaries and light scattering within the film are probable explications [11,4].

The films have a columnar structure [9] and- under certain deposition conditions- a 'milky' aspect. Further research should show, which factor of the deposition conditions is responsible for a higher or for a lower scattering coefficient of the films. For the solar cells presented, a microcrystalline material with high light scattering has been chosen. The comparison of the spectral response curve of an amorphous and a microcrystalline single-junction cell (figure 3, circles and squares) shows much more important interference fringes for the a-Si:H cell than for the μc-Si:H cell; the substrate is a flat aluminium sheet in both cases. The light seems to be scattered within the microcrystalline film, thereby reducing the effect of interference. (The rougher surface of the microcrystalline film compared to an amorphous film, however, could also be responsible for reduced interference). Scattering could act as 'internal' light trapping, enhancing the optical thickness of the cell, and this might be an important advantage of microcrystalline films. Deposition conditions could be optimised to get the highest scattering of the infrared light. (Note, however, that light scattering can only work on the fraction of the light, which has actually *entered* the solar cell. It has no effect on the *surface* of the cell. An effective AR coating still remains important.)

One drawback of the microcrystalline silicon <i>-layers is the rather low deposition rate (typically 1.2 Å/sec for VHF-GD), combined with the fact, that the cells are about 10 times thicker than their amorphous counterparts. Economic production needs short deposition times. For this reason, a big effort was undertaken to raise the deposition rate. The cells, which are presented here, were deposited at a rate of 4.3 Å/sec [12] and ongoing research promises even higher deposition rates (9.4 Å/sec are reported in [13]).

'Micromorph' silicon n-i-p-n-i-p heterojunction solar cells

All the individual elements (a-Si:H n-i-p cell, μc-Si:H n-i-p cell and a tested substrate) were first tested and the whole 'micromorph' cell had then to be 'put together'. The two cells were deposited in different deposition chambers and the samples were transferred at the air. Some samples were even stored for one or two weeks between the two depositions. The inevitable native oxide layer on the bottom cell (μc-Si:H) was not removed before deposition of the top cell (a-Si:H). Still, the tunnel junction between the two cells of the tandem worked well, and reasonable values of series resistance were measured. Both of the two adjacent layers of the tunnel junction (<p> of the bottom cell and <n> of the top cell) are microcrystalline layers, a combination which has been reported to be favourable for tunnel junctions [14].

First attempts to deposit the 'micromorph' solar cells were promising, leading to an initial efficiency of 9.25% on a flat substrate.

The 'micromorph' cell is a real tandem cell, employing materials of a different gap for the two cells. The advantage is, that a wider wavelength range of the sun's spectrum can be used by the cell. The 'micromorph' cell is sensitive from 350nm up to more than 1000nm. The wider absorption range, however, forms a more stringent requirement for the antireflection concept. Our structure (a flat cell with a single layer AR coating) reached its limits. The wavelength range, which is covered by the ITO single layer AR coating is too small for the 'micromorph' cell. Possibilities to resolve this problem would be (expensive?) multilayer AR coatings with a wider range or combinations of the single layer AR coating with other light trapping features. Rather wide V-grooves in the substrate with an angle of 90° at the bottom, conformally covered with the solar cell [15,16], for instance, would give every reflected photon a second chance to be coupled into the cell, leading to higher short circuit currents. The reflectivity R for each wavelength would be the square of its value without light trapping. (For example: if R at a certain wavelength without V-groove is 20%, then R with V-grooves of 90° is 20% x 20% = 4%).

Typically, the optimisation of a tandem cell (p-i-n-p-i-n on glass / TCO) is a problem of two variables: the thickness of the top and the thickness of the bottom cell. The two thicknesses have to be adjusted to have equivalent currents in both cells; the top cell has to be thin enough to let enough light pass for the bottom cell. In our case, we can play on *three* variables: the thickness of the top cell, the thickness of the bottom cell and the thickness of the ITO layer. The last parameter moves the 'window' for best coupling of the light into the cell more towards one or more towards the other cell.

The first depositions showed that, for the current matching, we can aim on $24mA/cm^2$ of total current at I_{sc} conditions, i.e. $12mA/cm^2$ for the top cell and $12mA/cm^2$ for the bottom cell. This total current is limited by the AR range of the ITO layer.

Sofar, the I_{sc}-values of bottom and top cell were not yet balanced ($11.1 / 13.1mA/cm^2$). With a V_{oc} of 1.24V and a fill factor of 67.3% this still counts up to 9.25% initial efficiency. (With equal currents of 12.1 mA/cm^2 an initial efficiency of 10% would have been reached).

No degradation experiments have been done with these cells as yet. Still, it is obvious that we confront the same problem as has been described for the p-i-n-p-i-n cells in [3]: The microcrystalline bottom cell can reach very high current levels. Therefore the amorphous top cell, which does not profit of the back reflector as it does in a single junction device, must be rather thick to furnish the same current as the bottom cell in the tandem device. In our case, the top cell must have a thickness of about 3500Å to deliver 12 mA/cm^2 at I_{sc} conditions. Cells of this thickness still suffer from pronounced degradation. (Note: only the top cell has to be considered for degradation, microcrystalline cells have shown to be stable even under very strong illumination [9]). In order to reduce degradation of the device, thinner amorphous cells would be preferable. Solutions like an interior mirror layer between the two cells to reduce the thickness of the top cell are discussed in [3]. The same concepts could be applied in the n-i-p-n-i-p configuration.

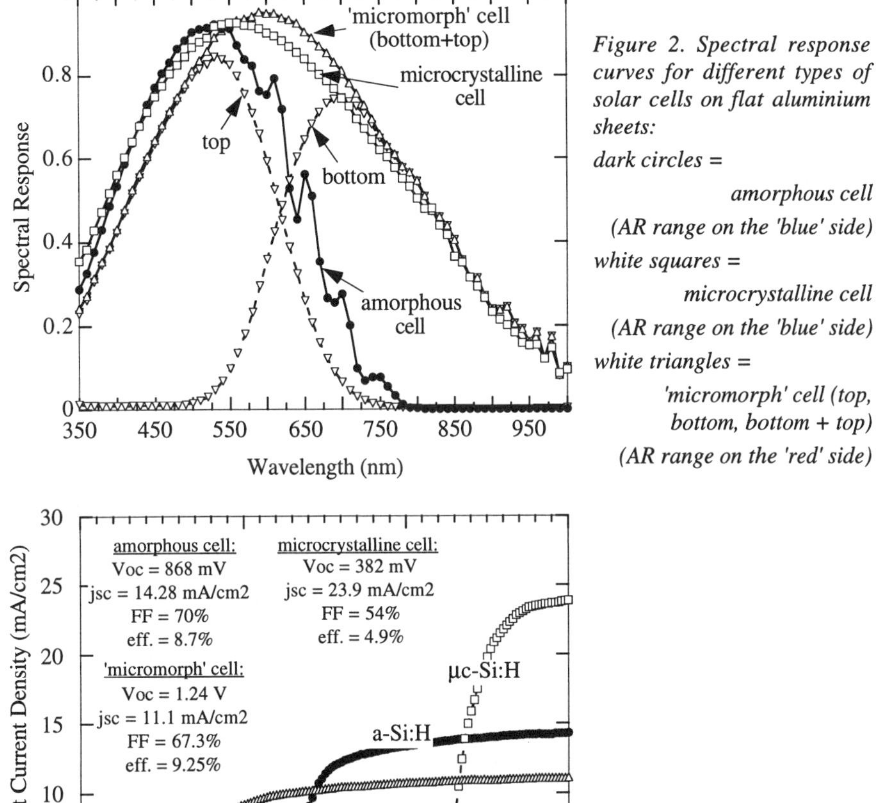

Figure 2. Spectral response curves for different types of solar cells on flat aluminium sheets:
dark circles = amorphous cell (AR range on the 'blue' side)
white squares = microcrystalline cell (AR range on the 'blue' side)
white triangles = 'micromorph' cell (top, bottom, bottom + top) (AR range on the 'red' side)

Figure 3. I(V)-characteristics for three types of solar cells on flat aluminium sheets: an amorphous cell, a microcrystalline cell and a 'micromorph' cell.

CONCLUSIONS

Several elements were needed for the successful deposition of the first 'micromorph' solar cells of the n-i-p-n-i-p configuration on flat aluminium sheets: First, the deposition of mid-gap, intrinsic microcrystalline silicon by the 'VHF-GD/gas purifier method', second, the amorphous n-i-p single junction solar cell, third, the microcrystalline n-i-p single junction solar cell and fourth, the experience of depositing on aluminium sheets. 'Micromorph' solar cells of an efficiency of 9.25% could be realised, and this even without matched currents. It was found, that the tunnel junction between the two cells works well without any optimisation. From an optical point of view, it seems that light scattering within the microcrystalline bottom cell acts as 'internal light-trapping', enhancing the optical thickness of the flat solar cells. On the other hand, the applied single-layer antireflection coating has too small a spectral range for this type of solar cells, as these cells are sensitive to a wide spectral range of light from 350nm to more than 1000nm. Therefore, a multilayer AR coating or supplementary light trapping schemes would be needed to take advantage of the full range of light, to which the 'micromorph' cell is sensitive.

ACKNOWLEDGEMENTS

This work has been supported by the Swiss National Energy Research Foundation under grant NEFF 682 and by the Swiss Federal Office of Energy (OFEN) under contract EF-REN (93) 032. We would like to thank Alusuisse Technology & Management Ltd. for supplying the aluminium substrates.

REFERENCES

1. J. Meier, S.Dubail, R.Flückiger, D.Fischer, H.Keppner, A.Shah, 1st WCPEC, Waikoloa, Hawaii, Dec. 5-9, 1994

2. D.Fischer, S.Dubail, J.A.Anna Selvan, N.Pellaton Vaucher, R.Platz, Ch. Hof, U.Kroll, J.Meier, P.Torres, H.Keppner, N.Wyrsch, M.Goetz, A.Shah, K.-D.Ufert, 25th IEEE PV Spec. Conf., Washington, 1996

3. J.Meier, S.Dubail, D.Fischer, J.A.Anna Selvan, N. Pellaton Vaucher, R.Platz, Ch.Hof, R.Flückiger, U.Kroll, N.Wyrsch, P.Torres, H.Keppner, A.Shah, K.-D.Ufert, 13th European Photovoltaic Solar Energy Conference, Nice, 1995, p.1445-1450

4. J.Meier, P.Torres, R.Platz, S.Dubail, U.Kroll, J.A.Anna Selvan, N. Pellaton Vaucher, Ch.Hof, D.Fischer, H.Keppner, A.Shah, K.-D.Ufert, PVSEC-9, Miyazaki, 1996

5. P.Pernet, M.Goetz, H.Keppner, A.Shah, this conference

6. M.Wuest, P.Toggweiler, 1st WCPEC, Waikoloa, Hawaii, Dec. 5-9, 1994

7. M.Goetz, H.Keppner, P.Pernet, W.Hotz, A.Shah, MRS 1996 spring meeting, San Francisco

8. R.Flückiger, J.Meier, M.Goetz, A.Shah, J.Appl.Phys. **77** (2), 15 January 1995, p.712-716

9. J.Meier, P.Torres, R.Platz, S.Dubail, U.Kroll, J.A.Anna Selvan, N.Pellaton Vaucher, Ch.Hof, D.Fischer, H.Keppner, A.Shah, K.-D.Ufert, P.Giannoulès, J.Koehler, MRS 1996 spring meeting, San Francisco

10. P.Torres, J.Meier, R.Flückiger, U.Kroll, J.A. Anna Selvan, H.Keppner, A.Shah, S.D. Littlewood, I.E. Kelly, P. Giannoulès, Appl. Phys. Lett. **69** (10), 2 September 1996

11. N.Beck, J.Meier, J.Fric, Z.Remes, A.Poruba, R.Flückiger, J.Pohl, A.Shah, M.Vanecek, J. of non-cryst. solids, 198-200 (1996), p.903-906

12. P.Torres, J.Meier, M.Goetz, U.Kroll, H.Keppner, A.Shah, this conference

13. H.Keppner, U.Kroll, P.Torres, J.Meier, R.Platz, D.Fischer, S.Dubail, J.A.Anna Selvan, N.Pellaton Vaucher, Y.Ziegler, R.Tscharner, Ch.Hof, N.Beck, M.Goetz, P.Pernet, M.Goerlitzer, N.Wyrsch, A.Shah, J.Pohl, E.Bucher, this conference

14. T.Yoshida, K.Maruyama, O.Nabeta, V.Ichikawa, H.Sakei, Y.Uchida, 19th IEEE PV Spec. Conf., 1987, p.1095-1100

15. G.Schumm, H.-D.Mohring, G.H.Bauer, 11th Photovoltaic Solar Energy Conference, Montreux, Oct. 12-16,1992, p.207-210

16. D.Thorp, P.Campbell, S.Wenham, 25th IEEE PV Spec. Conf., Washington, 1996

MICROCRYSTALLINE SILICON SOLAR CELLS AT HIGHER DEPOSITION RATES BY THE VHF-GD

P. TORRES, J. MEIER, M. GOETZ, N. BECK, U. KROLL, H. KEPPNER, AND A. SHAH

Institut de Microtechnique, Rue A.-L. Breguet 2, CH-2000 Neuchâtel, Switzerland

ABSTRACT

A 7.7 % single junction cell efficiency for an entirely microcrystalline silicon (μc-Si:H) device has recently been reported by our group [1]. This was achieved by applying the purifier technique, a technique which is indeed easier to handle than the earlier used "microdoping" approach. The purpose of the present paper is twofold: First to show in detail the impact on device performance when a gas purifier is used; and second to illustrate that the deposition rate of the active, absorbing i-layer can be increased from the former 1.55 Å/s up to 4.3 Å/s while still maintaining reasonable device performances. In the latter case a first n-i-p solar cell structure on an aluminium sheet could be fabricated with an efficiency of 4.9 %.

INTRODUCTION

As-grown undoped μc-Si:H has a strong n–type character which often hinders its application in devices. Two possible explanations for this non-intrinsic behaviour of undoped μc-Si:H have been given in the literature: structural native defects within the material or extrinsic oxygen impurities [2, 3]. Conceptually, both effects can be overcome by "microdoping", i.e. low-level boron doping. This method has been applied to standard GD [4], remote PECVD [3], reactive magnetron sputtering [5] and VHF-GD [6]. Outstanding early work by Vepřek et al. [e.g. 7, 8, 9] already showed the strong influence of incorporated oxygen impurities in μc-Si:H; however, film growth was performed here by means of chemical transport in a hydrogen discharge: this specific deposition method may allow to grow relatively pure films and hence give the possibility to study fundamental aspects of the material, but it is definitely not suited for large-area applications as required for the deposition of photovoltaic devices. Indeed, the ultimate goal is to deposit "device-grade" μc-Si:H by means of a technique, as e. g. VHF-GD, with large-area capability.

The detrimental n-type character of undoped μc-Si:H is overcome in the work described in the present paper by the purifier technique [10, 11]; thereby, the oxygen content in μc-Si:H can be reduced from the usual value of 10^{19} - 10^{21} atoms/cm^3 down to 2.5×10^{18} atoms/cm^3. This leads to "device-grade" μc-Si:H with a "midgap" character (Fermi level near the middle of the gap); consequently, "microdoping" can be omitted [12]. For comparison, note that in Czochralsky-grown silicon crystals, oxygen from the quartz crucible is generally present in concentrations of the order of 10^{18} atoms/cm^3 [13].

A second key issue addressed in this paper is to increase even further the deposition rates of the active, absorbing i-layer. This is a major concern, since μc-Si:H cells are about 10 times thicker than a-Si:H cells; note that this situation which applies to intrinsic μc-Si:H layers is in contrast with the situation prevailing for the very thin doped μc-Si:H layers that are used as contact layers in amorphous and other cells. In the present work deposition rates up to 4.3 Å/s for

Mat. Res. Soc. Symp. Proc. Vol. 452

"device-grade" μc-Si:H have been achieved and a first solar cell was fabricated with such a material.

EXPERIMENTAL

All layers and cells in this study were deposited in a single-chamber parallel-plate reactor by the VHF-GD technique. Doped contact layers were the same as those optimised in previous work [14, 15]. The substrate is positioned upside down on the upper electrode, where the effective substrate temperature is kept at 170 °C - 220 °C. The radio frequency is capacitively coupled to the lower electrode providing impedance matching for the entire frequency range of 55 to 200 MHz. The gases employed were silane (SiH_4), hydrogen (H_2), phosphine (PH_3) and diborane (B_2H_6) diluted at 500 ppm in H_2. Gas pressure was kept in the range of 0.4 - 0.8 mbar and the total gas flow was always 100 sccm. Dilutions are given in the following as % SiH4/total gas flow.

The approach followed for the purifying technique applied here, is to try to control all possible oxygen contamination sources: this implied both achieving a sufficiently low reactor outgasing rate of less than 1.5 x 10^{-6} mbar l /sec as well as using a gas purifier (the latter reduces the oxygen contamination down to the sub-ppb range in the feedgas). The combination of these two measures leads to a successful reduction of the oxygen content in a-Si:H as well as in μc-Si:H layers [10, 11].

Two p-i-n photovoltaic devices have been deposited, one with and one without applying the gas purifier to directly compare the effect on the active i-layer. The "standard" parameters employed here (110 MHz plasma excitation frequency, 6 W of applied rf-power, 2.5 % dilution) result in μc-Si:H deposition rates of 1.55 Å/s, as is typical for VHF-GD. This is already an order of magnitude higher when compared to deposition rates achieved by standard GD at 13.56 MHz excitation frequency. Deposition is performed on glass coated with textured SnO_2 (Asahi type U). Back contacts for these "standard" cells are made of highly textured ZnO and Ag [16].

To increase the deposition rate, we even further increased the plasma excitation frequency to 130 MHz and also increased the silane concentration to a dilution of 5%. Two layers of a thickness of 2-3 μm were deposited on AF45 glass from Schott: one at a low rf-power of 7 W (at 0.4 mbar) and one at a high rf-power of 16 W (at 0.8 mbar), with deposition rates of 2.2 and 4.3 Å/s respectively. The deposition pressure had thereby to be increased to keep the plasma confined between the electrodes. To determine the optical absorption of these layers, we performed photothermal deflection spectroscopy (PDS) and Transmission/Reflexion measurements.

The layer with the highest deposition rate (130 MHz, 16 W) was thereafter implemented in a device: An inverted n-i-p photovoltaic cell was deposited on an Al sheet. The front contact was a merely sputtered ITO film, which optically performs quite well for a-Si:H based solar cells, but which is not an optimal antireflecting coating for μc-Si:H solar cells with their extended spectral response. This particular optical problem results in reduced short circuit current densities and is discussed elsewhere [17]. Back contacts are flat, sputtered Ag and ZnO to obtain a reflecting mirror.

RESULTS AND DISCUSSION

Impact of the purifier technique on the solar cell performance

It is well known that the quality of the i-layer has a strong influence on the performance and, specifically, on the spectral response (SR) of a μc-Si:H solar cell device [18]. Fig. 1 shows the SR of an integral μc-Si:H p-i-n photovoltaic device ("standard" deposition parameters at 110 MHz, deposition rate of 1.55 Å/s) with the i-layers prepared with and without the feedgas purifier. The effect of reducing the oxygen content by means of using the gas purifier is remarkable and leads to a substantial increase in SR.

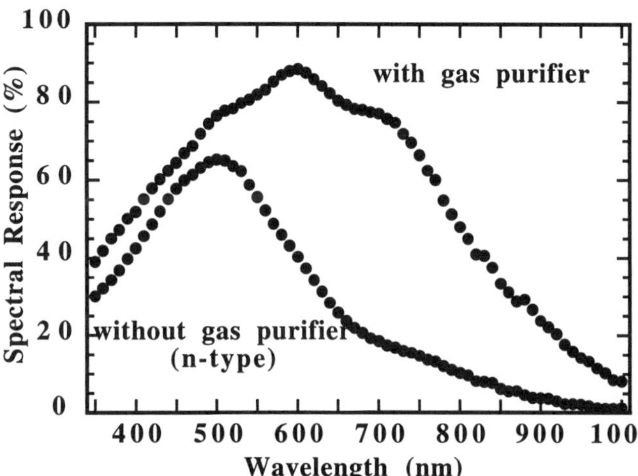

Fig. 1: Comparison of two 2.8 μm thick μc-Si:H p-i-n solar cells. One was produced with and one without the feedgas purifier. "Standard" deposition conditions of 110 MHz, 6 W result in a deposition rate of 1.55 Å/s for both cells.

degradation	Voc (mV)	Jsc (mA/cm^2)	FF	eff. (%)
before	407	22.6	57.8	5.3
after	418	22.4	57.2	5.3

Table 1: Performance of the cell deposited with purifier (Fig. 1) before and after light soaking during 264 h at 50 °C with light from a high-pressure sodium lamp (light intensity equivalent to 8 sun).

For the cell shown here (deposited with the purifier technique) all photogenerated carriers are separated and collected at an i-layer thickness of 2.8 μm . Indeed, the SR measured at a bias tension of -4 V remains, within measurement errors, identical to the one measured at 0V. Full characterisation under AM 1.5 conditions shows a cell efficiency of 5.3 %. Degradation experiments on this μc-Si:H solar cell support our earlier finding that μc-Si:H can be considered as a photovoltaically stable material: Exposure to light from a high-pressure sodium lamp at an intensity of about 8 suns, for a time period of 264 h and at a temperature of 50 °C did not show any light-induced degradation effect of this cell, as represented in Table 1.

Layers at high deposition rates

To get an answer to the question whether it is possible or not to further increase the deposition rates of μc-Si:H within the frame work of VHF-GD, several measures have simultaneously been applied: The silane concentration in the gas phase ratio was set to a dilution of 5%, the rf-excitation frequency was increased to 130 MHz and furthermore significantly higher plasma powers of up to 16 W were applied. Deposition rates increased thereby up to 4.3 Å/s. The gas purifier was used, here also, during deposition to avoid any additional ambiguity.

In Fig. 2 PDS absorption measurements are shown for two layers deposited under these new deposition conditions. It is surprising, that increasing the plasma power from 7 to 16 W does not lead to a higher defect density of the layer as evidenced by the low subbandgap absorption (Fig. 2), or to be more precise, any increase in the defect density of the layer - that may indeed be present - is not "seen" by PDS.

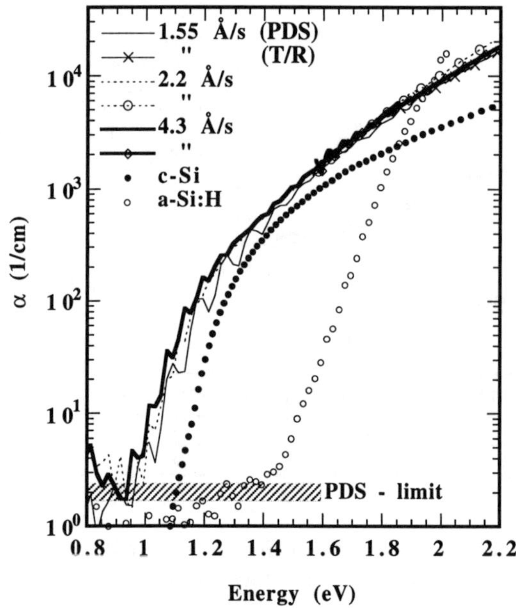

Fig. 2. PDS spectra of μc-Si:H layers deposited at different deposition rates: 1.55 Å/s with the "standard" conditions (110 MHz), 2.2 and 4.3 Å/s with a plasma excitation frequency of 130 MHz, at 7 and 16 W, respectively.
The enhanced absorption as compared to crystalline silicon persists on the whole spectral range and the subbandgap absorption is close to the PDS detection limit, even for the layer deposited with the highest plasma power. For comparison purposes typical values of the absorption coefficient α are given for c-Si [19] and a-Si:H.

Special care has to be taken, when the dilution is changed, as was done here, from 2.5 to 5 % [20], since a substantial (and detrimental) change in the "useful part" of the optical absorption could possibly result while doing this. Therefore, we also show in Fig. 2, the PDS curve of a layer deposited under "standard" deposition conditions (110 MHz, 2.5% dilution, 6 W): one can thereby clearly see that all layers represented show (when compared with the curve for c-Si [19]) the enhanced optical absorption that is typical for most μc-Si:H layers produced by VHF-GD [21] (Note, thereby, that the use of a combined Ar/H-dilution in VHF-GD leads to μc-Si:H layers that sofar do not (fully) show the enhanced optical absorption [22]).

Application of layers deposited at high rates to devices

The layer introduced above, which was grown at 130 MHz and 16 W, was thereafter implemented in a photovoltaic device to actually check its solar cell potential. The purifier technique was applied and a 3.85 μm thick device was grown at a deposition rate of 4.3 Å/s. The device was deposited in the n-i-p configuration on an aluminium sheet. This option was chosen because such supporting substrates have many interesting application aspects, as e.g. omitting expensive textured TCO on glass as well as better potential for building integration [23]. Furthermore, the ductile Al sheet allows for a better stress relaxation. The spectral response (SR) of this device is shown in Fig.3 . Relatively high short circuit current densities of 23.89 mA/cm^2 could be achieved. This is surprising because a non-optimal front contact is used and no texturing is applied in this cell. Note that for comparison, usual p-i-n cells are deposited on highly textured TCO. Reverse biasing of the cell at -3V during SR results in an even higher Isc of 26.0 mA/cm^2. Full characterisation under AM 1.5 conditions shows a cell efficiency of 4.9 % with a Voc of 382 mV and a FF of 54 %. By integrating an identical device in a micromorph tandem structure (μc-Si:H / a-Si:H) a double-junction cell with 9.25 % efficiency was obtained [17].

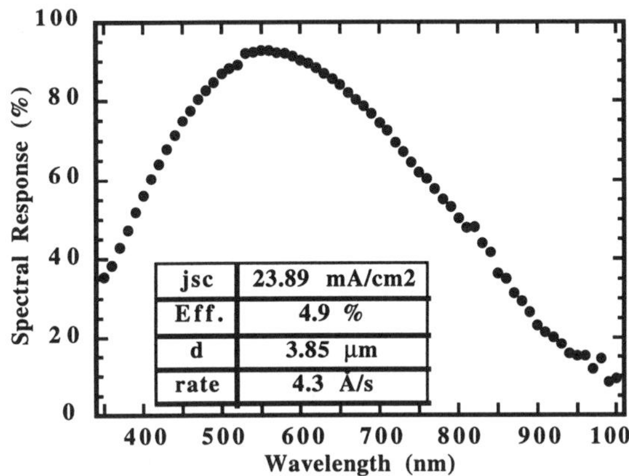

Fig. 3: μc-Si:H n-i-p solar cell on aluminium sheet deposited at an increased deposition rate of 4.3 Å/s. The back contact is flat sputtered ZnO and Ag. The front contact is flat sputtered ITO which also acts as a single layer antireflecting coating.

CONCLUSIONS AND OUTLOOK

The purifier technique used in this work results in substantially enhanced device performance and even higher solar cell efficiencies than the earlier "microdoping" approach used by us and by other groups. Higher deposition rates up to 4.3 Å/s are reported here for the VHF-GD deposition method at a plasma excitation frequency of 130 MHz and first experiments with such layers result in a single-junction cell efficiency of 4.9%. Higher deposition rates as demonstrated in this paper may also be helpful in further reducing oxygen impurities in the layers.

However, not only the improvement of device performances but also the problem of deposition rates has to be further investigated in the future, in order to obtain entirely microcrystalline silicon solar cells that can be industrially relevant. The key issue is thereby not just to show that very fast deposition rates can be achieved for individual layers, but also to show that these very same layers actually perform efficiently in devices.

ACKNOWLEDGEMENTS

The technical assistance of S. Dubail is gratefully acknowledged. This work was supported by the Swiss Federal Office of Energy under Research Grant EF-REN (93) 032. We would like to thank Alusuisse Management Ltd. for supplying the aluminium substrates.

REFERENCES

[1] J. Meier, P. Torres, R. Platz, S. Dubail, U. Kroll, J.A. Anna Selvan, N. Pellaton Vaucher, Ch. Hof, D. Fischer, H. Keppner, A. Shah, K.-D. Ufert, P. Giannoulès, and J. Koehler, Mat. Res. Soc. Symposium, San Francisco (1996), to be published.
[2] C. Wang and G. Lucovsky, Proc. of 21st IEEE Spec. Conf., p. 1614, (1990).
[3] G. Lucovsky, C. Wang, M.J. Williams, Y.L. Chen, and D.M. Maher, Mat. Res. Soc. Symp. Proc. **283**, 443 (1993).
[4] W. E. Spear, G. Willeke, and P. G. LeComber, Physica **117B &118B**, 908 (1983).
[5] W. A. Turner, M.J. Williams, Y.L .Chen, D.M. Maher, G. Lucovsky, Mat. Res. Soc. Symp. Proc. **283**, 567 (1993).
[6] R. Flückiger, J. Meier, M. Götz, and A. Shah, J. Appl. Phys. **77**, 712 (1995).
[7] S. Veprek, Z. Iqbal, R. O. Kühne, P. Capezzuto, F.-A. Sarott, and J. K. Gimzewski, J. Phys. C: Solid State Phys., **16** , 6341 (1983).
[8] F. Mattenberger and S. Veprek, CHEMTRONICS, **1**, 107 (1986).
[9] M. Konuma, H. Curtins, F.-A Sarott and S. Veprek, Phil. Mag B, **55 - 3**, 377 (1987).
[10] U. Kroll, J. Meier, H. Keppner, S.D. Littlewood, I.E. Kelly, P. Giannoulès, and A. Shah, Mat. Res. Soc. Symp. Proc. **377**, 39 (1995).
[11] U. Kroll, J. Meier, H. Keppner, A. Shah, S.D. Littlewood, I.E. Kelly, and P. Giannoulès, J. Vac. Sci. Technol. A **13**, 2724 (1995).
[12] P. Torres, J. Meier, R. Flückiger, U. Kroll, J. A. Anna Selvan, H. Keppner, A. Shah, S. D. Littlewood, I. E Kelly, and P. Giannoulès, Appl. Phys. Lett. **69**, 1373 (1996).
[13] M.A. Green, High Efiiciency Silicon Solar Cells, Trans Tech Publications, ISBN 0-87849-537-1, 126 (Aedermannsdorf, 1987).
[14] K. Prasad, Ph. D. thesis, Institute of Microtechnology, University of Neuchâtel (1991).
[15] R. Flückiger, Ph. D. thesis, Institute of Microtechnology, University of Neuchâtel, ISBN 3-89191-965-4 (1995).
[16] J.A. Anna Selvan, H. Keppner, and A. Shah, Mat. Res. Soc. Symp., San Francisco (1996), to be published.
[17] M. Goetz, P. Torres, P. Pernet, Meier, D. Fischer, H. Keppner, and A. Shah, this conference.
[18] J. Meier, S. Dubail, R. Flückiger, D. Fischer, H. Keppner, and A. Shah, Proc. 24th IEEE, 409 (1994).
[19] M. A. Green, M. J. Keevers, Progress in Photovoltaics: Research and Applications **3**, 189 (1995).
[20] N. Beck, P. Torres, J. Fric, Z. Remeš, A. Poruba, Ha Stuchlíková, A. Fejfar, N. Wyrsch, M. Vaněček, J.Kočka, A. Shah, this conference.
[21] N. Beck, J. Meier, J. Fric, Z. Remeš, A. Poruba, R. Flückiger, J. Pohl, A. Shah and M. Vaněček, J. Non-Cryst. Solids **198 - 200**, 903 (1996).
[22] H. Keppner, U. Kroll, P. Torres, J. Meier, R. Platz, D. Fischer, N. Beck, S. Dubail, J.A. Anna Selvan, N. Pellaton Vaucher, M. Goerlitzer, Y. Ziegler, R. Tscharner, Ch. Hof, M. Goetz, P. Pernet, N. Wyrsch, J. Vuille, J. Cuperus, A. Shah, J. Pohl, and E. Bucher, this conference.
[23] M. Goetz, H. Keppner, P. Pernet, W. Hotz, and A. Shah, Mat. Res. Soc. San Francisco (1996), to be published.

GROWTH OF THIN <p> µc-Si:H ON INTRINSIC a-Si:H FOR <nip> SOLAR CELLS APPLICATION

P. Pernet, M. Goetz, H. Keppner, A. Shah.

IMT, Université de Neuchâtel, A.-L. Breguet 2, 2000 Neuchâtel, Switzerland.

ABSTRACT

The <p> µc-SiC:H / <i> a-Si:H junction can be considered to be a sub-system of a n/i/p solar cell. Optimised performance of this junction can be assumed to be a key feature for obtaining high efficiency solar cells.

In this paper the authors present results on the conductivity of boron doped microcrystalline hydrogenated silicon (<p> µc-Si:H) thin films deposited on amorphous substrates (e.g. glass or glass/<i> a-Si:H). It is shown that, without any treatment of the substrate or of the underlying surface, the <p> layers showed a strongly reduced conductivity. This indicates either a bad nucleation or a poor microcrystalline behaviour. By using an appropriate surface treatment of the substrate, a gain in photoconductivity of about three orders of magnitude could be obtained (σ > 3 S/cm at a layer thickness of 400Å). We conclude from this, that for thin <p> type µc-Si:H layers the nucleation conditions are essential for obtaining best electric properties of the film w.r.t. solar cell performance.

Based on these results, interface treatment was successfully implemented in n/i/p solar cells deposited on TCO coated glass and stainless steel. The results of these experiments are also presented.

INTRODUCTION

Amorphous hydrogenated silicon (a-Si:H) solar cell technology is one of the leading techniques for low-cost and large-area photovoltaic cell manufacturing. In contrast to the well established p/i/n junction deposited on a glass/TCO, the less studied "inverted" structure, i.e. the n/i/p structure, offers the possibility to use substrates such as stainless steel, aluminium or plastic. For obtaining the best performance of n/i/p structures with respect to the <p> type window layer and the a-Si:H junction, the following factors have to be taken into account:

a) the doped layers in the device play a key role in building up a strong internal electric field across the <i> layer,

b) the <p> doped window layer should have high electric conductivity, a wide optical bandgap and a high transmission within the spectral range of the sun,

c) the compatibility in terms of electrical contact with both the <i> layer and the top contact (TCO) is required to guarantee a high fill factor (FF). A good solution consists of using a <p> type µc-SiC:H layer [11]; nevertheless, the microcrystalline character of such a layer strongly depends on its nucleation condition on the substrate [4-8], i.e. on its immediate microcrystalline growth [6-8].

In this paper, we focus our attention, in a first part, on the understanding of H_2 and CO_2 plasma treatments which allow the growth of a <p> µc-Si:H thin layer (400Å) by the nucleation or oxidation of the <i> layer. In a second part, we show how these plasma treatments can be applied to n/i/p cells deposited on glass/TCO and stainless steel substrates.

Mat. Res. Soc. Symp. Proc. Vol. 452 © 1997 Materials Research Society

EXPERIMENTAL

All films and solar cells were deposited with the VHF-GD technique at an excitation frequency of 70 MHz in a single chamber deposition system [1]. The two coplanar electrodes have a surface of 133 cm^2, the substrate is fixed face down on the upper electrode. The RF is applied on the lower electrode. The gap between the electrodes is fixed at 14 mm during the deposition of the films and solar cells.

The effective temperature of the electrodes is kept between 170°C and 240°C (depending on the layer), the walls of the chamber are kept close to ambient temperature.

The gases used for the film and cell deposition were silane (SiH_4), diborane (B_2H_6), phosphine (PH_3), hydrogen (H_2) and methane (CH_4).

The film thicknesses were measured with a step profiler on steps etched out by a KOH solution and averaged over 5 measurements. The deposition rate was determinated separately for the <i> and <p> layers deposited on glass of type Schott AF45 (4 x 4 cm^2).

The plasma treatment was performed using either a CO_2 or a H_2 plasma. Gas pressure, VHF - power and time of treatment were varied, respectively.

a) Test structure: A cross-section of the test structure used is given in Fig. 1: The thickness of all <i> a-Si:H layers was 1500 Å, and 400 Å for the <p> µc-Si:H layers. All samples were exposed to an annealing step close to the deposition temperature of the <p> layer under N_2 atmosphere. The (lateral) conductivity of the p-layer was measured using a four-point probe measurement system. In order to check homogeneity, the measurement was carried out at 9 sites distributed on the surface. Only homogenous layers were taken into consideration for the plots presented here.

When necessary, the conductivity was measured, after the same annealing step, by using two coplanar aluminium ohmic contacts (1000 Å) under N_2 atmosphere at a pressure of 10 mbar. As shown in Fig. 1, the test structure was used to study the conductivity of the <p> doped layer in function of the plasma treatment.

For the deposition from undiluted SiH_4 of the undoped layer, a pressure of p = 0.4 mbar and a power of 4W was applied.

The <p> µc-Si:H layers were deposited at T_{dep} = 170°C; power input P = 4W; total gas flow =100 sccm; SiH_4/tot = 1%; B_2H_6/SiH_4 = 0.4%.

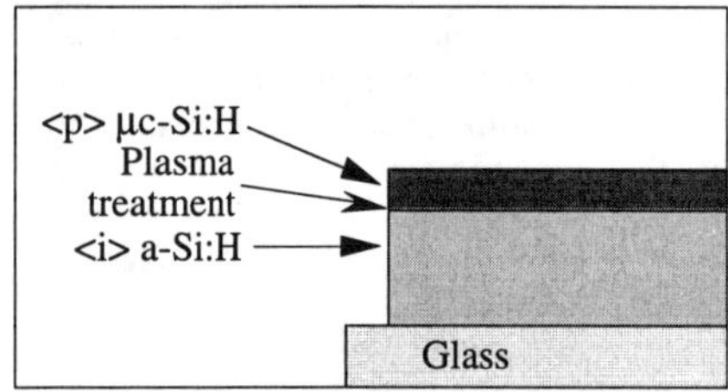

Fig. 1: Test structure of a <i> a-Si:H / <p> µc-Si:H on glass.

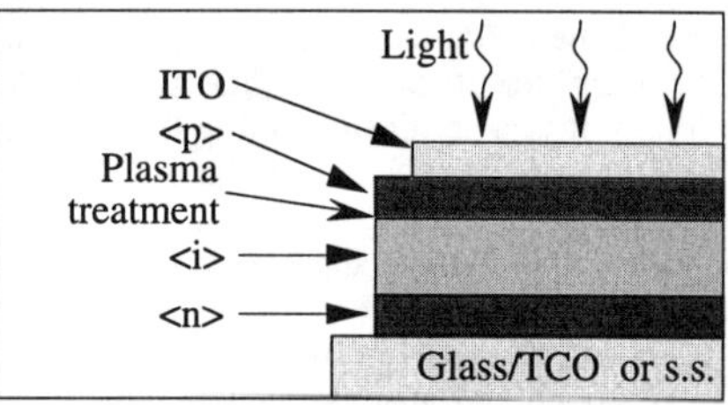

Fig. 2: Structure of a <n/i/p> cell deposited on glass / TCO or stainless steel.

b) Cell structure:
The complete n/i/p cells were developed on glass / TCO (Asahi type U) and stainless steel in the same deposition system as mentioned before. The thicknesses of the layers were: <n> µc-Si:H: 250Å; <i> a-Si:H: 4500Å; <p> µc-SiC:H: 250Å.

The only difference between all cell structures measured consisted in a different plasma treatment at the i/p interface. The front contact ITO was sputtered through a mask defining 15 cells of 0.2 cm^2.

Like in test structure a) the cells were also annealed at the same temperature in an N_2 atmosphere. The cells were characterised by the illuminated j-V curve, thereby a two source solar simulator was used. From the j-V characteristics, the open circuit voltage V_{oc} and the fill factor FF were determined. The short circuit current (Isc) was determinated by integration of the spectral response curve [3] of a selected cell.

RESULTS AND DISCUSSION

Conductivities as high as 3 S/cm could be obtained for a <p> doped µc-Si:H (400Å) layer deposited on glass treated with an Ar – H_2 plasma [2]. If no plasma treatment was applied, the deposition of <p> µc-Si:H, on glass or in the test-structure a), yielded very poor lateral conductivities of the doped layer of around 10^{-3} – 10^{-4} S/cm. This clearly shows – and is already well known [4-8, 10]– that the microcrystalline nature of a <p> doped layer is highly dependant on the treatment of the underlying surface.

With a H_2 or a CO_2 plasma treatment at the i/p interface, it is possible to obtain high conductivities, however. Figs 3 and 4 show that, for a set of parameters of pressure and power, the conductivity of the <p> layer depends on the exposure time of the substrate to the H_2 or CO_2 plasma .

For a long H_2 plasma treatment (not shown on the graphic), the conductivities are already reduced, furthermore the layer became inhomogeneous. It seems that, on one hand, the surface is saturated by the hydrogen and, on the other hand, the amorphous layer is etched away by the prolonged treatment. A conductivity of 3 S/cm of the <p> doped layer was achieved with a H_2 plasma treatment of 3 min., corresponding to the conductivity of the same <p> doped layer deposited directly on treated glass [2]. Thus, for a fixed set of parameters (pressure and power), an optimum in surface treatment can be found.

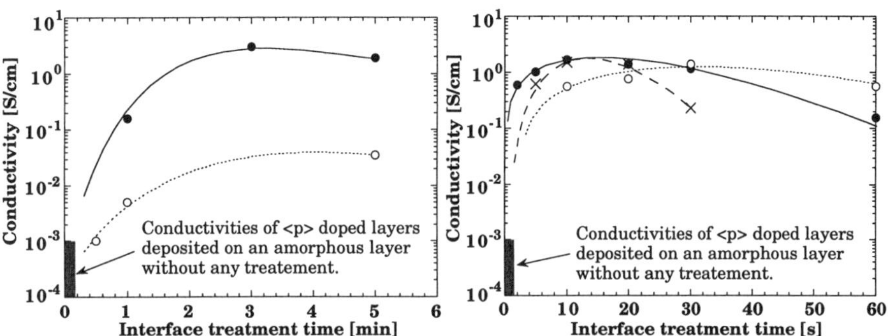

Fig. 3: Conductivity of 400Å thick <p> doped layers on an amorphous intrinsic layer of 1500 Å in function of the exposure time of the interface by a H_2 plasma. Parameters: ●: 0.035 mbar, 10W; ○: 0.1 mbar, 8W. The variable is the duration of the treatment.

Fig. 4: Conductivity of 400Å thick <p> doped layers on an amorphous intrinsic layer of 1500 Å in function of the exposure time of the interface by a CO_2 plasma. Parameters: ●: 0.1 mbar, 3W; ○: 0.2 mbar, 2W; ×: 0.03 mbar, 2W. The variable is the duration of the treatment.

Using a CO_2 plasma, it was found that the conductivity of the <p> doped layer changes drastically within the first 15 sec. The maximum-value strongly depends on the pressure and power. For too high power and/or too long exposure time, the CO_2 plasma forms species (similar as water vapour) that increase strongly the base pressure of the vacuum system and that are difficult to be removed by downpumping. Apart from that, it seems that there is an oxidation on the top of the amorphous surface (formation of a-SiO:H) [9] which favours the nucleation of the microcrystalline growth of the <p> layer. Using the CO_2 plasma treatment, conductivities in excess of 1 S/cm were achieved.

Assuming that, in a first step, the optimum growth condition w.r.t. H_2 or CO_2 plasma treatment are the same for the <p> doped μc-SiC:H layers incorporated in the cells as for the unalloyed individual μc-Si:H layers mentioned before, we applied these treatments in n/i/p cells. In the cells, thinner <p> layers were incorporated; hence, it can be assumed, that the influence of the nucleation properties of the plasma-treated substrate (or, of the underlying <i> layer) should be even more pronounced.

Table 1 summarises the characteristics of these cells deposited on glass/TCO.

a-Si:H / <p>μc-SiC:H interface treatment:	V_{oc} [mV]	j_{sc} [mA/cm^2]	FF [%]	Eff [%]
a: none	667	15.95	53.4	5.7
b: H_2 plasma	823	15.7	59.1	7.6
c: CO_2 plasma	831	15.8	63.4	8.3

Table 1: Characteristics of n/i/p cells deposited on glass/TCO. The different layers have the same parameters of deposition, the only difference is the treatment of the interface i/p.

Fig. 5 shows the j-V curves of cells deposited on glass/TCO. The H_2 and CO_2 plasma treatments correspond to the optimum of the curves ● Fig. 3 and × Fig. 4, respectively. As expected, there is a spectacular increase of the V_{oc} value between the cells *a* (without treatment) and *b* and *c* (with treatment), which is due to a better microcrystallinity of the p-doped layer. The increase in the fill factor is probably due to a better contact of the <p>/ITO interface (tunnel junction ?) as deduced from the evaluation of the series resistance.

In a second step, by further adjustment of the p-layer (<p> μc-SiC:H; 250 Å) deposition sequence and the plasma treatment condition, a further increase of the V_{oc} of over 40 mV could be achieved for both H_2 and CO_2 plasma treatment (Fig. 6).

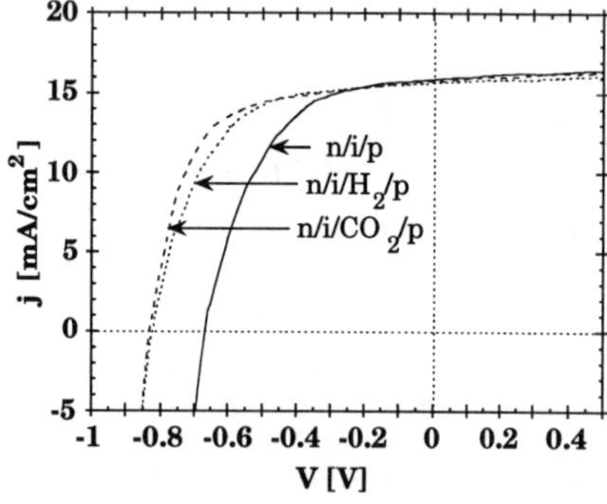

Fig. 5: j-V curves of n/i/p cells deposited on glass/TCO. The <p> μc-Si:H layers have been optimized as individual layers only. The only difference is the plasma treatment at the interface.

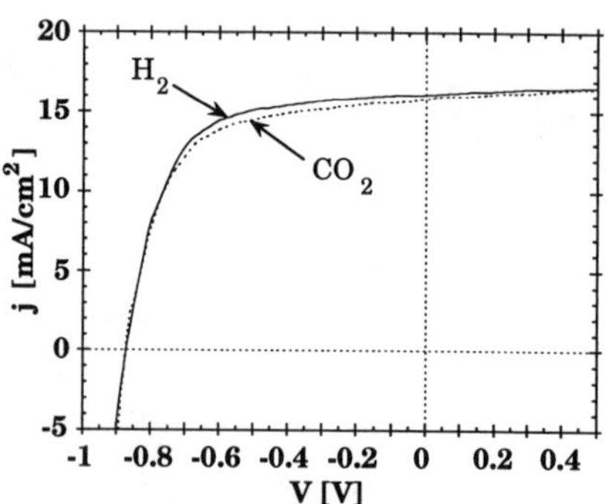

Fig. 6: j-V curves of n/i/p cells deposited on glass/TCO. The <p> μc-SiC:H layer is further optimised in the cell, w.r.t. the plasma treatment.

Table 2 summarises the characteristics of these cells:

<p>μc-SiC:H optimised w.r.t. plasma treatment.	V_{oc} [mV]	j_{sc} [mA/cm^2]	FF [%]	Eff [%]
d: H_2 plasma	869	16.2	64.1	9.0
e: CO_2 plasma	874	15.9	63	8.7

Table 2: Characteristics of the n/i/p cells deposited on glass/TCO. The cells d *and* e *are, respectively, the same as the cells* b *and* c *(Table 1), but with improved deposition conditions for the <p> μc-SiC:H.*

Table 3 resumes the characteristic of cells deposited on stainless steel.

H_2 plasma treatment:	V_{oc} [mV]	j_{sc} [mA/cm^2]	FF [%]	Eff [%]
f: no	687	14.1	55.0	5.3
g: yes	890	13.3	59.7	7.2

Table 3: Characteristics of the n/i/p cells deposited directly on stainless steel with the H_2 plasma treatment mentioned; the cells f *and* g *are, respectively, the same as the cells* a *and* d *deposited on glass/TCO. The difference in efficiency is probaly due to the lack of pronounced light trapping in the case of stainless steel substrate.*

For the ideal growth condition of <p> doped μc-Si:H films on amorphous substrates a plasma treatment prior to <p>-layer deposition was found to have a beneficial impact. For the example of optimised hydrogen plasma exposure the best overall cell performance could be achieved. In a previous publication [12], similar results were published regarding an effect called "chemical annealing". In this case exposure to atomic hydrogen (or to a hydrogen plasma) resulted in a post-recrystallization of the already grown amorphous layers. However a relatively complicated procedure was used. Surprisingly, CO_2 exposure (no hydrogen involved) shows similar results. We assume, because of this, that plasma exposure generally leads to more favourable nucleation conditions for doped μc-Si:H, somewhat independently of the gas and the substrate / underlying layer involved.

CONCLUSIONS

Starting from the deposition parameters of good-quality, thick good microcrystalline <p> doped layers deposited on glass (glass/<p> μc-Si:H) with a conductivity of 3 S/cm, we have demonstrated that the microcrystalline structure is completely lost if such layers are deposited as thin layer directly on an underlying amorphous layer (glass/a-Si:H/<p> "μc"-Si:H) without any plasma treatment at the <i/p> interface. By using, first, a H_2 or a CO_2 plasma treatment, high conductivities of the <p> doped layers can again be found. The introduction of such a plasma treatment was successfully implemented in solar cells with an inverted structure (<n/i/p>) deposited on glass/TCO and stainless steel. A significant improvement of the FF (53.4 –> 64.1%), V_{oc} (667 –> 874mV) and efficiency (5.4 –> 9.0%) could thereby be achieved.

ACKNOWLEDGMENTS

This work was supported by the Swiss Federal Office for Energy grant EF(REN) 93 (032) and by the Swiss Projekt und Studiernfonds der Elektrizitäts-Wirtschaft (PSEL) grant No. 88.

REFERENCES

1 H. Curtins, N. Wyrsch and A. Shah, Electronics Lett. **23**, 228 (1987).

2 R. Flückiger, Ph.D. thesis, Institute of Microtechnology, University of Neuchâtel, ISBN 3-89191-965-4 (1995).

3 D. Fischer, Ph.D. thesis, Institute of Microtechnology, University of Neuchâtel (1994).

4 P. Torres, R. Flückiger, J. Meier, H. Keppner, U. Kroll, V. Shklover, A. Shah, Proc. of 13th EC PVSEC, 1638 (1995).

5 N. Pellaton Vaucher et al., to be presented at PVSEC-9, Miyasaki.

6 B. Goldstein, C. R. Dickson, I. H. Campbell and P. M. Fauchet, Appl. Phys. Lett. **53**, 2672 (1988).

7 L. Yang, L. Chen and A. Catalano, MRS Symp. Proc. **283**, 463 (1993).

8 A. Catalano et al., Solar Cells, 27, p 25, 1989.

9 S. Fujikake, H. Ohta, A. Asano, Y. Ichikawa and H. Sakai, MRS Symp. Proc. **258**, 875 (1992).

10 Y.-M. Li, F. Jackson, L. Yang, B.F. Fieselmann and L. Russell, MRS Symp. Proc. **336**, 663 (1994).

11 R. Flückiger, J. Meier, A. Shah, J. Pohl, M. Tzolov and R. Carius, MRS Symp. Proc. **358**, 793 (1995).

12 K. Nakamura, K. Yoshino and I. Shimizu, 1st WCPEC, **1**, 480 (1994).

NUCLEATION AND GROWTH OF μc-Si:H n- AND p-TYPE LAYERS IN a-Si:H p-i-n AND n-i-p SOLAR CELLS: REAL TIME SPECTROELLIPSOMETRY STUDIES

Joohyun Koh, H. Fujiwara, C.R. Wronski(1), and R.W. Collins(2)
Materials Research Laboratory and Departments of (1) Electrical Engineering and (2) Physics
The Pennsylvania State University, University Park, PA 16802.

ABSTRACT

We have applied real time spectroellipsometry (RTSE) to study the growth of microcrystalline silicon n- and p-layers [μc-Si:H:(P,B)] incorporated into amorphous silicon (a-Si:H) p-i-n and n-i-p solar cells, respectively. In previous research, we have applied RTSE to characterize a-Si:H solar cells having only amorphous component layers. The μc-Si:H:(P,B) component layers, however, pose a more difficult RTSE analysis problem for two reasons. First, the near-surface of the underlying i-layer is modified in the μc-Si:H:(P,B) growth process, and second, the microstructural evolution near the i/(n,p) interfaces is very complicated. From RTSE spectra ($1.5 \leq h\nu \leq 4$ eV) collected every ~4-15 s during growth, we have extracted the time evolution of the μc-Si:H:(P,B) layer microstructure, thicknesses, and optical properties along with the modifications that the near-surface i-layer properties undergo in the formation of the i/(n,p) interfaces. We suggest that the beneficial optical properties of the microcrystalline layers may be due to size effects in the crystallites that make up the films.

INTRODUCTION

In previous reports, we have described results obtained in analyses of real time spectro-ellipsometry (RTSE) data collected during hydrogenated amorphous silicon (a-Si:H)-based solar cell fabrication [1,2]. For the separate a-$Si_{1-x}C_x$:H:B, a-Si:H, and a-Si:H:P materials in both p-i-n and n-i-p cell structures, the time evolution of the bulk and surface roughness layer thicknesses, along with the thicknesses of the roughness layers at each of the interfaces have been measured. In addition, the dielectric functions and optical gaps have been determined for each of the bulk layers in the same analyses. In the previous studies, these capabilities have been restricted to the amorphous layers of the cell since these layers exhibit a homogeneous growth pattern and thickness-independent bulk layer optical properties.

In the following, we have successfully expanded the RTSE analysis approach to include the more complex microcrystalline silicon n- and p-layers [μc-Si:H:(P,B)] that are incorporated into the highest performance solar cells [3]. In this case, RTSE provides information on the modification that the underlying i-layer undergoes upon exposure either to the H_2-rich plasma from which the μc-Si:H:P layer is formed or to the pure H_2 plasma used to pretreat the i-layer prior to μc-Si:H:B deposition. In addition, we can determine the thickness at which isolated crystalline nuclei make contact to form a coalesced doped layer. This information is important because, when the nucleation density is too low, coalescence will not occur within the desired layer thickness of 200-300 Å, and the resulting discontinuous doped layer will lead to poor device performance. The dielectric function ε obtained in the same RTSE analysis can provide information on the optical gap and microstructure of the μc-Si:H:(P,B) layers in the actual device configuration. Because the optical properties at different stages of the μc-Si:H:(P,B) growth process can be determined, the uniformity of the bulk microcrystalline material can be assessed as a function of thickness.

EXPERIMENTAL DETAILS

The solar cells studied here were deposited on different substrates held at 200°C. For the p-i-n cell configuration, codeposition was performed on SnO_2:F and ZnO for devices and c-Si for RTSE measurements. For the n-i-p configuration, the same Cr-coated c-Si substrate was used for the device and for RTSE measurements. The intended thicknesses of all n- and p-layers were 300 and 200 Å, respectively. The a-Si:H i-layers which served as the substrates for the μc-Si:H:P n-layer in the p-i-n and for the μc-Si:H:B p-layer in the n-i-p were prepared to 0.5 μm using pure SiH_4, an

Mat. Res. Soc. Symp. Proc. Vol. 452

rf plasma power flux density of 70 mW/cm^2, and a pressure of 0.18 Torr.

The a-$Si_{1-x}C_x$:H:B (x~0.05) p-type window layer for the p-i-n cell was prepared using flow ratios of 6:4:0.02:1 SiH_4:CH_4:$B(CH_3)_3$:He, a power of 70 mW/cm^2, and a pressure of 0.13 Torr, whereas the n-type microcrystalline layer for the p-i-n was deposited using flow ratios of 2:0.012:100 SiH_4:PH_3:H_2, a power of 700 mW/cm^2, and a pressure of 0.50 Torr. The latter gas mixture leads to microcrystallinity in a ~1.8 μm thick layer, as determined from features in the dielectric function ε at the E_1 and E_2 transitions of bulk c-Si. The conductivity activation energy in the thick μc-Si:H:P layer was ~0.03 eV, as compared with 0.25 eV for 0.8 μm thick a-Si:H:P prepared under the same conditions as the n-layer in the n-i-p cell.

The a-Si:H:P n-layer for the n-i-p cell was prepared using a flow ratio of 10:0.2 SiH_4:PH_3, a power of 70 mW/cm^2, and a pressure of 0.12 Torr, whereas the p-type microcrystalline film was deposited using flow ratios of 0.6:0.4:200:0.2:10 SiH_4:CH_4:H_2:$B(CH_3)_3$:He, a power of 700 mW/cm^2, and a pressure of 0.90 Torr. Although we have been identifying the p-type material here as μc-Si:H:B for simplicity, a ~1.1 μm thick layer prepared under these conditions is in fact a mixed-phase p-type μc-Si:H/a-$Si_{1-x}C_x$:H composite material, as determined from an effective medium theory analysis of ε. The conductivity activation energy in the thick mixed-phase layer was ~0.52 eV, as compared with 0.55 eV for 0.8 μm thick a-$Si_{1-x}C_x$:H:B prepared under the same conditions as the p-layer in the p-i-n cell. Prior to the deposition of the p-layer in the n-i-p, the i-layer surface was exposed to a pure H_2 plasma for 133 s, using the same power and partial pressure as for the p-layer. This treatment led to an improvement in open-circuit voltage (V_{oc}) of the cell from ~0.4 to 0.85 V, attributed to an increase in microcrystal nucleation density.

RTSE was performed using a multichannel spectroscopic ellipsometer with an acquisition time of 160 ms and a repetition time that ranged from 4 s during nucleation of the microcrystalline layers to 15 s in the later stages of growth. Typical RTSE data analyses of microcrystalline (n,p)-layer growth were confined to the spectral range where the underlying i-layer is opaque and the (p,n)-layer/substrate structure need not be included in the optical model.

RESULTS AND DISCUSSION

Structural Evolution

The results of the structural analysis of RTSE data collected during μc-Si:H:P n-layer growth are summarized in Figs. 1(a)-(d). The basic approach is similar to that described elsewhere, but with the addition of part (a) [1,2]. Four structural parameters are shown in Fig. 1: (a) the volume fraction f_{Si-H} associated with new Si-H bonds that form from Si-Si bonds in the top ~300 Å of the underlying i-layer as a result of its exposure to the H_2-rich n-layer plasma; (b) the volume fraction f_i of the n-layer material that penetrates the i-layer surface roughness, ultimately yielding i/n interface roughness; (c) the surface roughness thickness d_s on the n-layer; and (d) the bulk thickness d_b of the n-layer. In the analysis of Fig. 1, we assume that, once the bulk n-layer begins to grow [i.e., after the interface is completely filled in (f_i=0.5), denoted by the horizontal dashed line in Fig. 1(b)], the underlying i-layer is no longer modified [denoted by the dashed line in Fig. 1(a)]. This approach is reasonable since it leads to the same degree of sub-surface i-layer modification upon saturation as when the i-layer is exposed to a pure H_2 plasma under the same conditions of power and H_2 partial pressure [corresponding to f_{Si-H}=0.14; see Fig. 2(a)].

Figure 1 shows that in the initial 30 s, the i-layer is modified such that ~0.14 volume fraction of additional Si-H bonds develops. Previous correlations have suggested that this effect corresponds to an increase in the bonded H content of ~3.5±1 at.% [4]. In addition, the 25 Å thick surface roughness on the i-layer fills with n-layer material, and the n-layer surface roughness also develops to a thickness of ~25 Å. With the start of bulk n-layer formation at 30 s, the underlying structure is assumed to be stable, and the bulk layer grows linearly as a function of time at a rate of 0.77 Å/s. Concurrently, roughness thickness on the n-layer gradually increases to 54 Å. The final effective thickness of the layer is given by substituting the final values into $d_{eff}=0.5(d_i+d_s)+d_b$. This yields 272 Å, somewhat less than the intended value of 300 Å.

The corresponding results of the structural analysis during a 133 s H_2 plasma exposure of the i-layer, followed by mixed-phase μc-Si:H:B/a-$Si_{1-x}C_x$:H:B p-layer growth, are summarized in Fig. 2. Parts (a) and (b) show that during the H_2 plasma exposure, (i) the i-layer is modified such that ~0.14 volume fraction of additional Si-H bonds develops and (ii) the 28 Å thick surface

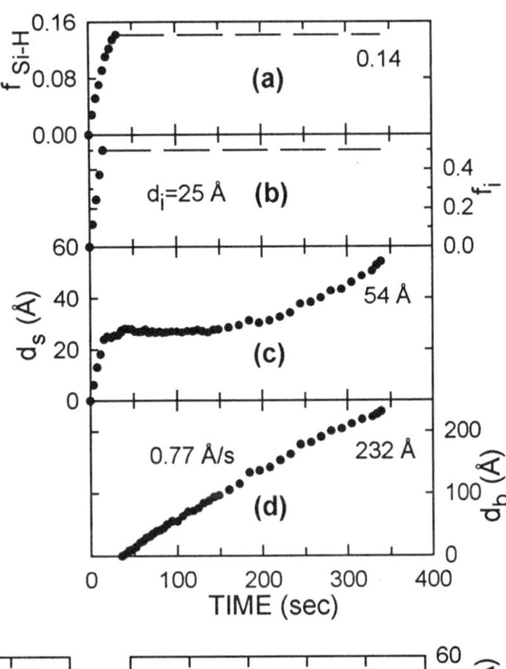

FIGURE 1. RTSE analysis results for μc-Si:H:P n-layer growth on the a-Si:H i-layer of a p-i-n solar cell, including (a) modification of the top 300 Å of the i-layer by incorporated H, expressed in terms of f_{Si-H}, the volume fraction of new Si-H bonds that are formed from Si-Si bonds in the near-surface of the i-layer; (b) filling of the i-layer surface roughness, expressed in terms of f_i, the volume fraction of n-layer material within the surface modulations of the i-layer; (c) development of n-layer nuclei (t<30 s) and surface roughness (t>30 s) thicknesses, d_s; and (d) evolution of the n-layer bulk film thickness, d_b.

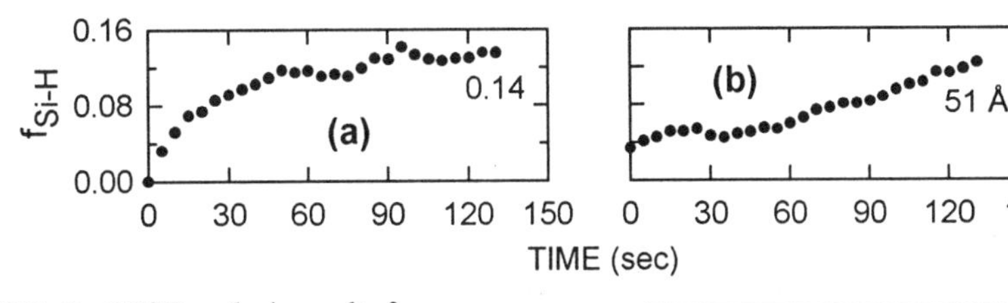

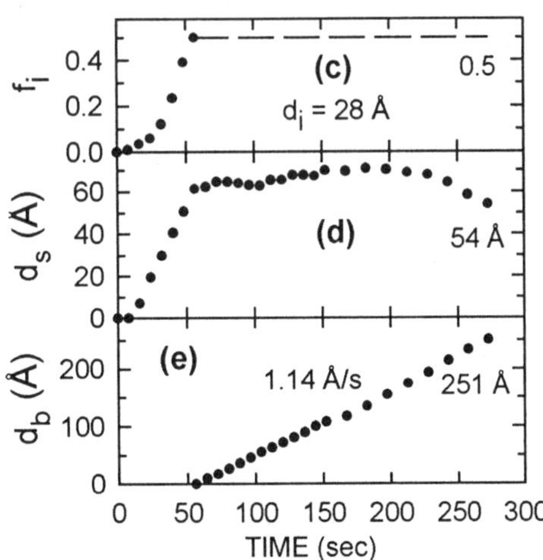

FIGURE 2. RTSE analysis results for (a-b) H_2 plasma exposure of the a-Si:H i-layer of an n-i-p solar cell, followed by (c-e) mixed-phase μc-Si:H:B/ a-$Si_{1-x}C_x$:H:B p-layer growth on the i-layer. Processes depicted here include: (a) modification of the top 300 Å of the i-layer by incorporated H, expressed in terms of f_{Si-H}, the volume fraction of Si-H bonds that are formed from Si-Si bonds in the near-surface of the i-layer; (b) evolution of the surface roughness thickness, $d_{s,sub}$, on the i-layer; (c) roughness filling at the i-layer surface expressed in terms of f_i, the volume fraction of p-layer material within the surface modulations of the i-layer; (d) development of the p-layer nuclei (t<60 s) and surface roughness (t>60 s) thicknesses, d_s; and (e) evolution of the p-layer bulk thickness, d_b.

roughness on the i-layer increases by 23 Å to 51 Å. A study correlating atomic force microscopy (AFM) and RTSE measurements has shown that the additional 23 Å of roughness that develops during H_2 plasma exposure is in the form of atomic-scale, surface-connected nanovoids that are trapped at the i/p interface (i.e., not filled in) upon overdeposition by the p-layer. In contrast, the initial 28 Å of roughness on the i-layer is in the form of surface modulations with an in-plane scale of 300-400 Å, which are filled in during the first 60 s of p-layer growth as shown in Fig. 2(c). In the p-layer nucleation process, nuclei increase in thickness to 62 Å before making contact at t=60 s, the onset of bulk film growth [see Fig. 2(d)]. With the start of bulk microcrystalline p-layer formation at 60 s, this layer grows linearly as a function of time at a rate of 1.14 Å/s as shown in Fig. 2(e). Concurrently, the roughness thickness on the p-layer gradually decreases to 54 Å. The final effective thickness of the layer is 292 Å, much greater than the intended value of 200 Å.

The p-layer smoothening in Fig. 2(d) versus n-layer roughening in Fig. 1(c) may result from the fact that the p-layer develops a two-phase μc-Si:H:B/a-$Si_{1-x}C_x$:H:B structure in which the amorphous component fills in the regions between coalescing crystallites. In contrast, the n-layer appears to remain single-phase μc-Si:H. These conclusions are also consistent with the measured optical and electrical properties of the thick microcrystalline n- and p-layers. Because of the complexity of the optical model for the i/(n,p) structure, we have correlated the results for the roughness layer thickness by RTSE at the end of the depositions of Figs. 1 and 2 with ex situ AFM on the final doped-layer surfaces. We find that the RTSE-deduced result for the final roughness thickness on the microcrystalline n-layer lies 8 Å above the trend observed for a series of a-Si:H-based films, given by d_s(RTSE) = 1.5d_{rms}(AFM) + 4 Å, where 'rms' represents the root-mean-square roughness from ex situ AFM. In contrast, the RTSE-deduced p-layer roughness is ~25 Å larger than would be expected based on this trend. We attribute this observation to a smaller in-plane roughness scale on the p-layer, leading to roughness components below the AFM resolution (~50 Å, controlled by the tip), possibly associated with the amorphous phase.

Optical Properties

In addition to the structural evolution shown in Figs. 1 and 2, the optical properties of the different components of the structure can be extracted in the same RTSE analyses. Although the increasing roughness with time in Fig. 1(c) and the low nucleation density in Fig. 2(d) (nuclei contact at 62 Å) reveal inhomogeneous microstructures that offer indirect evidence of crystallite formation, more direct evidence of the presence of crystallites must come from the optical properties. The fact that the optical properties of 1-2 μm thick n- and p-layers prepared under the same conditions as those in Figs. 1 and 2, respectively, exhibit clear evidence of crystallinity, however, does not imply that the 200-300 Å layers are microcrystalline, owing to the possible development of the crystalline structure with increasing thickness. As a result, we must consider in detail the optical properties in the actual device configurations.

The dielectric function in the neighborhood of the absorption onset is shown in Fig. 3 for the bulk microcrystalline n-layer after 105 Å of growth and at the end of growth when the bulk layer is 232 Å. In spite of the fact that the Fig. 3 results are obtained at different stages in the evolutionary pattern of the growth process, the results are nearly identical. This suggests that the bulk film is relatively uniform and that the use of a single layer to represent the bulk n-layer is valid. It can be noted in Fig. 3 that the μc-Si:H:P n-layer in the p-i-n configuration exhibits a much higher energy onset in ε_2 than the a-Si:H:P used in the n-i-p configuration. (The latter results are given in Fig. 3 as the solid lines.) Applying the $\varepsilon_2^{1/2}$ extrapolation for the optical gaps (assuming relaxed k-conservation, parabolic bands, and a constant dipole matrix element [5]), values of 1.59 eV and 2.28 eV would be obtained for the a-Si:H:P and the μc-Si:H:P films, respectively. However, the $\varepsilon_2^{1/2}$ plot for the μc-Si:H:P shows poor linearity, and this observation, coupled with the very high optical gap provides our strongest evidence of a microcrystalline film structure.

Corresponding results are shown in Fig. 4 for the mixed-phase μc-Si:H:B/a-$Si_{1-x}C_x$:H:B after 55 and 174 Å of bulk layer growth. The differences in this case are greater than those observed for the n-layer, possibly due in part to the development of the two-phase structure. In addition, the much lower magnitudes of ε in comparison to the μc-Si:H:P n-layer suggest that even though these optical properties are representative of a "bulk" p-layer, there remains a high density of voids in the mixed-phase structure that may also evolve with thickness. This would also be consistent with the higher nuclei contact thickness for the microcrystalline p-layer (62 Å) in comparison with the

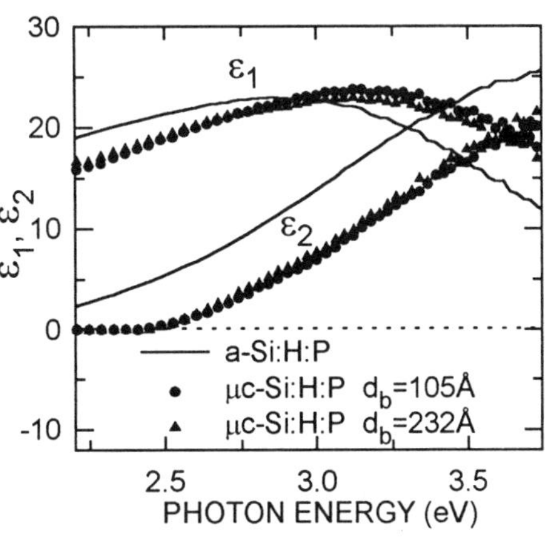

FIGURE 3. Near-band-edge dielectric function for the bulk μc-Si:H:P n-layer of the p-i-n cell of Fig. 1 deduced at two different stages in the growth process, at 105 Å (circles) and at 232 Å (triangles) bulk layer thickness. The corresponding results for an a-Si:H:P n-layer in the n-i-p cell configuration are given for comparison (lines). All data were obtained at a temperature of 200°C.

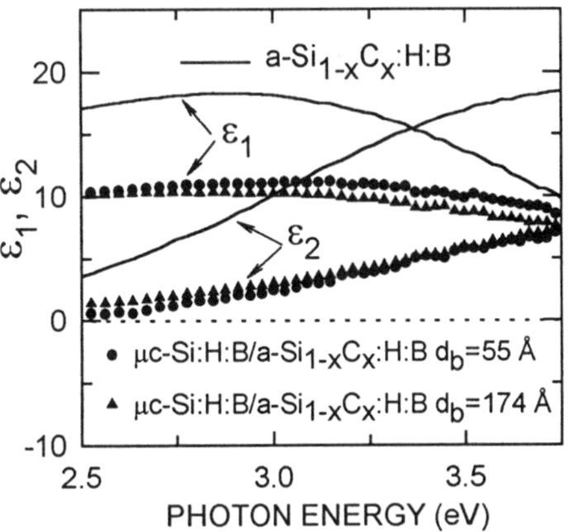

FIGURE 4. Near-band-edge dielectric function for the mixed-phase μc-Si:H:B /a-$Si_{1-x}C_x$:H:B p-layer of the n-i-p cell of Fig. 2 deduced at two different stages in the growth process, at 55 Å (circles) and at 174 Å (triangles) bulk layer thickness. The results for an a-$Si_{1-x}C_x$:H:B p-layer in the p-i-n cell configuration are given for comparison (lines). All data were obtained at a temperature of 200°C.

n-layer (25 Å). Figure 4 also shows that the mixed-phase μc-Si:H:B/a-$Si_{1-x}C_x$:H:B p-layer in the n-i-p configuration exhibits a much higher energy onset in ε_2 than that of the a-$Si_{1-x}C_x$:H:B used in the p-i-n configuration. (The latter results are given in Fig. 4 as the solid lines.) Using the $\varepsilon_2^{1/2}$ extrapolation for the optical gaps, values of 1.74 and 2.04 eV would be obtained for the a-$Si_{1-x}C_x$:H:B and the μc-Si:H:B/a-$Si_{1-x}C_x$:H:B, respectively. Here again, the much higher optical gap provides evidence for crystallites in the latter material.

The optical gap behavior for the μc-Si:H:P n-layer is characteristic of confinement effects in Si nanocrystals [6]. An estimated room temperature gap of ~2.35 eV would lead to a nanocrystal size of ~20-25 Å, based on plane-wave pseudopotential calculations of the relationship between size and bandgap [7]. This size scale is consistent with the nuclei contact thickness of the film, suggesting that nuclei contact may limit the in-plane grain size. The narrower gap for the microcrystalline p-layer, along with the enhanced absorption below the abrupt absorption onset which increases with bulk layer thickness, may be attributed to the development of the amorphous component of the two-phase structure. In fact, this amorphous component appears to dominate the electrical properties of the layer, leading to a V_{oc} in the n-i-p cell (0.85 V) similar to that obtained for an a-$Si_{1-x}C_x$:H:B p-layer (0.83 V). In spite of their wider optical gaps, the dielectric functions of the microcrystalline n- and p-layers in Figs. 3 and 4 show no sharp features near 3.3 eV due to the E_1 transition along the Λ-point in the band structure of Si, as might be expected from a crystalline phase [8]. This is most likely due to a combination of broadening effects including (i) the elevated substrate temperature (200°C) at which all RTSE measurements are made, (ii) electron scattering at defects within the crystallites or at boundaries between crystallites, or (iii) differing degrees of stress in the coalesced crystallites. In general, we conclude that the beneficial optical properties of the microcrystalline doped layers in the device configuration are attributable at least in part to confinement effects in small Si nanocrystals unique to the structure of thin layers.

SUMMARY

We have extended real time spectroellipsometry (RTSE) capabilities to the analysis of a-Si:H-based solar cells that incorporate doped microcrystalline layers. This has required additional complexity in the optical analysis, namely the inclusion of the modification of the underlying i-layer during the initial nucleation of the microcrystalline layer on its surface or during an initial H_2 plasma pretreatment of the i-layer. The importance of the RTSE approach lies in the ability to establish the structure, thickness, and optical properties of 200-300 Å microcrystalline layers in the actual device configuration, rather than from thick layers often used for property and growth rate evaluations. For a 272 Å thick μc-Si:H n-layer incorporated into the p-i-n cell of this study, growth is quite uniform with nuclei beginning to coalesce after a cluster thickness of ~25 Å. The resulting n-layer optical properties provide evidence for confinement effects in 20-25 Å nanocrystals, since an abrupt absorption onset and optical gap are observed near 2.4 eV. For a ~290 Å thick intended μc-Si:H:B p-layer incorporated into the n-i-p cell of this study, growth is less uniform with nuclei beginning to coalesce after a cluster thickness of ~60 Å. In this case, the optical properties are less clearly characteristic of nanocrystallites owing to the development of an a-$Si_{1-x}C_x$:H phase. In future research, we will study the structural and optical properties of microcrystalline p-type layers prepared using different gas mixtures in order to relate the resulting information to the device performance.

ACKNOWLEDGMENT

This research was supported by NREL under Subcontract No. XAN-4-13318-03 and by NSF under Grant Nos. DMR-9217169 and DMR-9622774.

REFERENCES

1. R.W. Collins, J.Koh, Y. Lu, S. Kim, J.S. Burnham, and C.R. Wronski, *Proceedings of the 25th IEEE Photovoltaics Specialists Conference*, May 1996, Washington DC, USA, (IEEE, 1996), p. 1035.
2. J. Koh, Y. Lu, S. Kim, J.S. Burnham, C.R. Wronski, and R.W. Collins, Appl. Phys. Lett. **67**, 2669 (1995).
3. S. Guha, J. Yang, P. Nath, and M. Hack, Appl. Phys. Lett. **49**, 218 (1986).
4. I. An, Y.M. Li, C.R. Wronski, and R.W. Collins, Phys. Rev. B **48**, 4464 (1993).
5. G.D. Cody, B.G. Brooks, and B. Abeles, Solar Energy Mater. **8**, 231 (1982).
6. L. Brus, J. Phys. Chem. **90**, 2555 (1986).
7. L.-W. Wang and A. Zunger, J. Phys. Chem. **98**, 2158 (1994), and references therein.
8. P. Lautenschlager, M. Garriga, L. Vina, and M. Cardona, Phys. Rev. B **36**, 4821 (1987).

Thin Microcrystalline Silicon by a Microwave Remote Plasma Deposition Scheme for Heterojunction Solar Cells and Thin Film Transistors

B. Jagannathan, R. L. Wallace* and W.A. Anderson, Center for Electronic and Electro-optic Materials, Dept. of Electrical and Computer Eng, State University of New York at Buffalo, Amherst, NY 14260 USA
*Present Address: Evergreen Solar, Inc., 211 Second Ave., Waltham, MA 02154

Abstract

A microwave remote plasma system, operating in an electron cyclotron resonance (ECR) condition, has been designed to deposit microcrystalline silicon (μ c-Si). The plasma properties have been studied by Langmuir probe measurements in the deposition region. Microcrystalline silicon films of thickness < 0.3 μm have been grown by the decomposition of 2% SiH_4/Ar diluted in H_2. The films were analyzed for dark and photoconductivity, crystallinity and total H_2 content. Raman spectroscopy indicates >50% crystallinity for films deposited between 300-400 °C at pressures of 10 mTorr with an input power in the range of 200-600 W for a H_2/SiH_4 ratio of 60. An increase in deposition temperature from 300 to 400 °C resulted in an increase in conductivity along with a decrease in the activation energy. H_2 evolution studies indicate that the purely μ c-Si films show a total 2.5% H_2 content compared with the 16.5% seen for the totally amorphous films. Better structural and electrical properties have more recently been obtained using 2% SiH_4/He mixtures.

Introduction

In the last few years there has been a great thrust to develop microcrystalline silicon for applications in thin film transistors (TFT) and solar cells. Some research groups are reporting success in applying μ c-Si in solar cell structures [1], and there has also been some reports of use in TFT's [2]. Normally, such a growth has been achieved by an rf glow discharge decomposition of pure SiH_4 excessively diluted in H_2 and He [3] at low temperatures of 300 °C. Models for μ c-Si growth in excess H_2, postulate that the weak Si-Si bonds of the growing a-Si film are etched away by the increased surface diffusion of H_2, causing an equilibrium condition between growth and etching that results in the crystallinity of the film. Grain sizes of 0.03-0.1 μm have been reported [4][5] for such films. Although good structural property of the films, and controllability of the process have been demonstrated, devices with this material have been limited by the rather high conductivity exhibited even by the undoped films. Recently, it has been argued to be an effect of oxygen impurities in the film [6]. At any rate, the higher defect density translates to lower carrier generation in solar cells, and a higher leakage in the TFT's.

To address some of the issues stated above, here the μ c-Si growth was studied using 2% SiH_4 in Ar or He with further dilution in H_2 in a microwave discharge. Microwave ECR plasma deposited a-Si has been reported to be more stable compared with the rf glow discharge deposited material [7]. This technique is particularly useful because it is electrodeless, and it is possible to obtain higher electron densities compared with the rf glow discharge technique. A low cost system, employing a commercial magnetron and a modified power supply is being used for the purpose. Raman spectroscopy has been used to monitor the crystallinity of the films as the growth conditions were varied. This was supplemented with hydrogen effusion

Mat. Res. Soc. Symp. Proc. Vol. 452

experiments to relate the H_2 content and binding to some of the observed electrical properties in these films.

Experimental

The schematic of the deposition system is shown in Fig.1. The system comprises of a Pyrex cross, mated to a quartz tube within a resonant cavity. The magnetron tube (of a commercial microwave oven) is mounted on the cavity, which is a rectangular Al waveguide. The modified power supply is controlled by a variac so that the input power can be continuously varied up to 1000 W. The plasma is excited in a background gas in the quartz tube and the process gases are fed at the front end of the quartz tube. The two permanent magnets (885 Gauss) provide the ECR condition. A 2" stainless steel block with heater and thermocouple serves as the substrate holder. The chamber is pumped by a diffusion pump, and has a base pressure of < 10^{-6} Torr. A cold trap is used to reduce backstreaming.

The probe measurements were conducted using a 1 mm diameter wire with 1 cm of exposed length with the rest being insulated. The collected current by the probe was measured for H_2, He, and Ar discharges for normal operating conditions of pressure and power. Microcrystalline silicon was deposited on Corning glass and oxidized wafer at 10-15 mTorr pressures for various growth conditions. Microstructural changes in the film were studied by Raman measurements at the Ar 5145 Å line on the films deposited on glass substrates. Al contacts 1mm square spaced 40 μm apart were evaporated for conductivity measurements and the photoconductivity was evaluated at a illumination of 100 mW/cm^2. Hydrogen content and evolution profile in the samples was studied by heating the film deposited on oxidized wafer in a vacuum at a constant rate of 20 °C/s while measuring the partial pressure of H_2^+ ions with a quadrupole mass spectrometer.

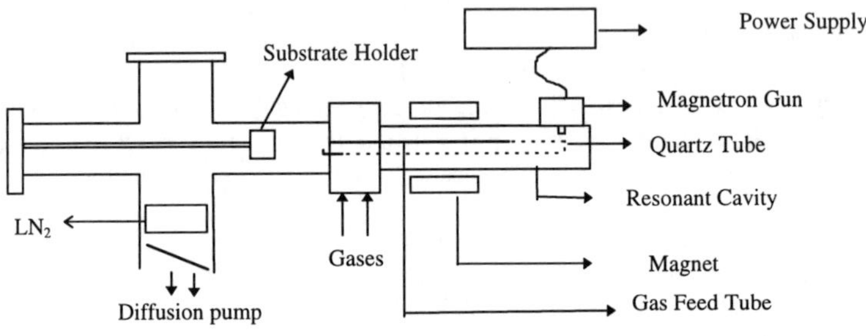

Fig.1 Schematic of the plasma deposition system

Results

The probes to evaluate the plasma were placed about 14 cm away from the magnets, downstream in the plasma, to avoid the magnetic field effects on the collected current. Fig.2 shows the electron current for a H_2 discharge operating at two conditions, one for 400 W at a pressure of 10 mTorr, and another for a condition of higher power and lower pressure. Electron density (n_e), electron temperatures (T_e), and plasma potential (V_P) are some of the parameters that are extracted from this curve. Simple probe theory to estimate these parameters assumes that the electrons in the plasma are Maxwellian distributions [8]. The collected current is defined as $I=I_e^* \exp(-qV/kT_e)$, where I_e^* is the electron saturation current, also $I_e^*=Aqn_e\sqrt{T_e / 2\pi m_e}$, where m_e is the mass of an electron and A is the probe area. For the 10 mTorr discharge, T_e was 12.5 eV, and the electron density estimated from this was about $2\times10^8 cm^{-3}$ with a V_P of 18 volts. The collected current is seen to increase with decreasing pressure and higher powers consistent with the fact that this is expected to create more electrons with a larger mean free path, whereas T_e is insensitive to such variation (slope for the two curves are similar).

Microcrystalline Si was deposited with 2% SiH_4/Ar as the source gas for various levels of H_2 dilution (H_2/SiH_4 ratio) with a maximum dilution of 60 at 10-15 mTorr pressures. Fig.3 shows the Raman spectra following the morphology of the deposited Si as a function of dilution for deposition temperatures between 250-300 °C, and input powers of 200-400 W. For films formed with dilution ratio of 35 and 60, the film is predominantly crystalline as seen from the strong T0 phonon peak near 520 cm^{-1} and no amorphous component is discernible at 480 cm^{-1}. For a ratio of 60, the FWHM of the crystalline peak is about 9.48 cm^{-1} and increases to 12.9

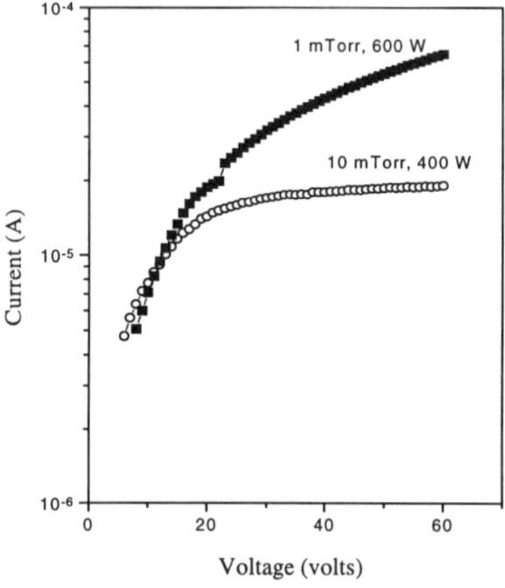

Fig.2 Electron current collected by a probe for H_2 discharge

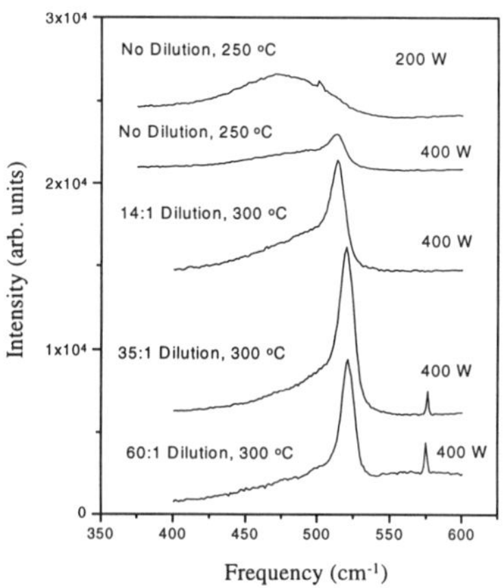

Fig.3 Raman spectra for Si films deposited by H_2 dilution of 2%SiH_4/Ar

cm^{-1} at a dilution of 14, implying a smaller grain size for a lower dilution ratio. As the amorphous component develops, the integrated intensity ratio from the crystalline and amorphous peaks is used for computing the crystalline content. For a dilution of 60, the crystalline content is >>50%, thus it is likely that all of the film is crystallized, since no amorphous band is seen.

The surprising observation comes for deposition performed without any H_2, at 400 W power. Even for this situation, a strong peak is observed at 513 cm^{-1} with a FWHM of 16.5 cm^{-1} attributed to the crystalline component, along with an amorphous shoulder. Using the integrated intensity for amorphous and crystalline components it is seen that 45% of the film still crystallizes. As such, in Fig.3, the crystallinity does not seem to be a strong function of dilution in H_2. Tsai et al [4] have observed that adding H_2 to a SiH_4 discharge favors the "etching" process that leads to a crystalline network in the growing film. But, the μ c-Si growth has also been observed in SiH_4 discharges in excess He at temperatures near 330 °C [3]. In this case again, etching and network relaxation have been ascribed to energetic He species on the surface of the growing film. In the present experiment, species present comprise only those from SiH_4 and Ar dissociation, but it is not clear if the "etching" of weak Si-Si bonds was being caused by Ar radicals or if the H radical from the dissociation of SiH_4 was causing the observed effect. Reducing the input power to 200 W leads to the formation of a totally amorphous phase as seen in Fig.3. The crystallinity of the deposited Si was also seen to be insensitive to the substrate temperatures between 300-400 °C. The FWHM of the crystalline peak for depositions done at 400 °C was 10.13 cm^{-1}, which would indicate similar grain sizes for depositions done at 300 and 400 °C.

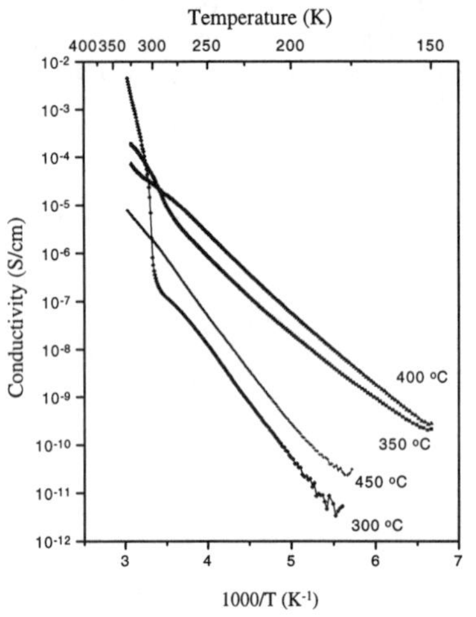

Fig.4 Dark conductivity of μ c-Si films prepared using 2% SiH_4/Ar diluted in H_2

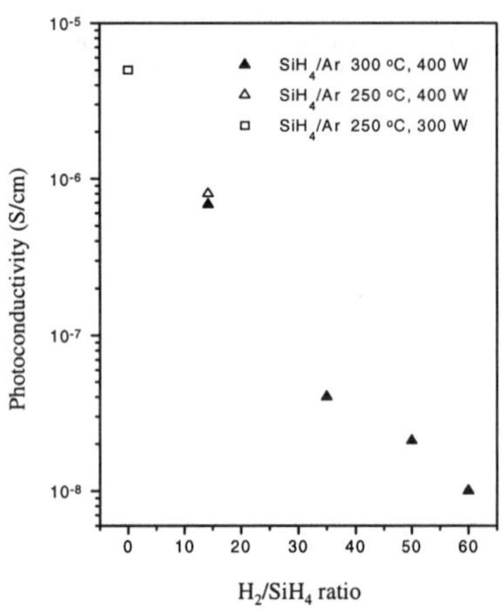

Fig.5 Photoresponse of a-Si and μ c-Si films prepared by H_2 dilution of 2% SiH_4/Ar

Fig.4. shows the dark conductivity of the films as a function of deposition temperature for a fixed H_2 dilution ratio of 60. The films seem to show two conduction regions, especially evident for the growth carried out at 300 °C. In one, region for temperatures lower than 285 K, the conductivity is thermally activated agreeing with the model of the form $\sigma=\sigma_o\exp(-\Delta E_{act}/kT)$, where ΔE_{act} is the activation energy. ΔE_{act} for conduction decreases from a value of 0.45 eV for depositions at 300 °C to 0.33 eV for deposition at 400 °C. It increases to 0.48 eV for the growth at 450 °C. In this region, the overall conductivity seems to increase as the deposition temperature is increased from 300 to 400 °C. A second conduction region is observed for measurement temperatures greater than 300 K where there is a rapid increase in conductivity, the extent of which seems to decrease with increasing deposition temperature. A hopping carrier transport in the amorphous regions of the film would not explain such high conduction at temperatures of 300 K. Since this region reduces with growth temperatures, it is likely that the lower growth temperature leads to more film stress related defects which cause the sudden change in conductivity, whereas for the higher temperature growth, a more relaxed network with fewer such defects would be expected.

Fig.5 shows the photoconductivity of the Si films as a function of H_2 dilution for growth near 300 °C. The photoresponse is seen to increase as the H_2 content decreases in the plasma, coinciding with an increasing amorphous content in the films. Thus, the films prepared at dilution of 60 show a lower photoresponse compared to the films prepared at a dilution ratio of 14, which show a photoconductivity of 7×10^{-7} S/cm. Completely amorphous films are formed with no H_2 in the discharge by decreasing the input power and increasing the deposition pressure [40 mTorr]. The best amorphous films deposited had a photoconductivity of 5×10^{-6} S/cm and this is shown in Fig.5 for comparison. However, the dark transport in these amorphous films shows an activation energy of 0.56 eV, much lower than the device quality a-Si prepared using 100% SiH_4.

The improved photoresponse as the dilution is reduced may be attributed to the increased H_2 content in films, passivating the defects in the film. To investigate this, H_2 content and binding in the films was studied by the effusion profiles. Fig.6. shows the normalized H_2 pressure as the film is heated from 200 to 800 °C. Since no significant H^+ contribution was seen, the area under the curve was used to determine the total H_2 content (c_H) in the film. For films prepared with a dilution of 60 and substrate temperature of 300 °C, the H_2 evolution starts at 400 °C and is completed by 530 °C. As the dilution decreases, and the amorphous content in the films increases, the evolution starts at 350 °C, and for a totally amorphous film, this process starts at 300 °C and a peak at 350 °C is also observed in addition to the peak at 425 °C (which has also lowered). Two conclusions are drawn from this: firstly, that there is more included H_2 in the film as the temperature and power of the deposition is reduced; secondly, the H_2 content in the films is seen to increase from 2.5% when the film is microcrystalline to 16.5 % when the film is amorphous. Previous H_2 evolution studies [9] indicate that the low temperature peak is associated with molecular H_2 trapped in the voids, and the high temperature (450-500 °C) peak is associated with the bonded H_2. Thus, for deposition with Ar, the void formation in the amorphous film is high, whereas the material is obviously more dense in the microcrystalline form.

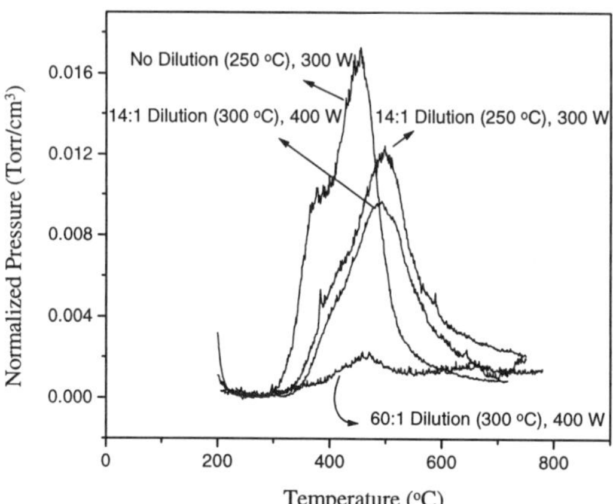

Fig.6 H_2 evolution in a-Si and μ c-Si films

Much improved structural and electrical properties have been observed in μ c-Si depositions performed with 2% SiH_4/He diluted in H_2. For films deposited within 300-400 °C no amorphous shoulder was observed in Raman spectra for a dilution ratio of 30. FWHM in the

Raman spectra for films deposited at 300 and 400 °C were 7.4 and 8.8 cm^{-1} respectively, implying a better grain size compared to the films deposited with 2% SiH_4/Ar. Dark conductivity of the μ c-Si films deposited at 400 °C with a dilution of 30 was $8x10^{-6}$ S/cm, with an activation energy of 0.49 eV which was also better than that observed for the Ar mixture.

Conclusions

Microcrystalline Si was grown in a microwave remote plasma system using 2% SiH_4/Ar diluted in H_2. Strong crystallinity was observed in the silicon deposited at dilution ratios between 30-60. Crystallinity was even observed in films prepared without any H_2 for 400 W of input power. The microstructure changed to amorphous for subsequent reduction in power. Dark conductivity in the films increases with increasing substrate temperatures in the 300-400 °C range with a decrease in activation energy for conduction from 0.47 to 0.3 eV. Microcrystalline films prepared at 300 °C also have a low photoresponse, which increases as the amorphous content in the film increases, and is maximized at $5x10^{-6}$ S/cm using a combination of lower deposition temperature and input power. The hydrogen evolution studies indicate that the H_2 content in the films prepared at 300 °C increases from 2.5% for the μ c-Si case to 16.5% for the purely amorphous case. Films prepared with 2% SiH_4 in He show better crystallinity and exhibit higher conduction activation energy compared to the films prepared with Ar.

Acknowledgment

This research was initiated with financial support from National Renewable Energy Laboratory (NREL) and New York State Energy Research and Development Authority (NYSERDA). More recently, funding has been provided by NYSERDA through a local industry, Advanced Refractories Technology, Inc.

References

1. D. Fisher, S. Dubail, J. A. Anna Selvan, N . Pellaton Vaucher, R. Pltz, Ch. Hof, U. Kroll, J. Meier, P. Torrese, H. Keppner, N. Wyrsch, M. Goetz, A. Shah, K. D. Ufert, Proc. 25th IEEE PVSC, p 1053 (1996)
2. J. I. Woo, H. J. Lim, and J. Jang, Appl. Phys. Lett., **65**, p 1644 (1994)
3. J. Jang, S. C. Kim, K. C. Park, and Sung Ki Kim, J. Appl. Phys., **75**, p 3184 (1994)
4. C. C. Tsai, R. Thompson, C. Donald, F. A. Ponce, G.B. Anderson, and B. Wacker, Mat. Res. Soc. Proc., Vol. **118**, p 49 (1988)
5. K. C. Wang, H. L. Hwang, P. T. Leong, and T. R. Yew, J. Appl. Phys., **77**, p 6543 (1995)
6. P. Torres, J, Meier, R. Fluckiger, U. Kroll, J. A. Anna Selvan, H. Keppner, and A. Shah, Appl. Phys. Lett., **69**, p 1373 (1996)
7. V. Dalal, S. Kaushal, R. Girvan, S. Hariasra, and L. Sipahi, Proc. 25th IEEE PVSC, p 1069 (1996)
8. N. Hershkowitz, Plasma Dignostics, Vol 1 (Discharge Parameters and Chemistry), edited by O. Auciello, and D. L. Flamm, Academic Press, p 113 (1989)
9. W. Beyer, and H. Wagner, J. Appl. Phys., **53**, p 8745 (1982)

PROPERTIES OF AMORPHOUS SILICON THIN FILM TRANSISTORS WITH PHOSPHOROUS-DOPED HYDROGENATED MICROCRYSTALLINE SILICON

J. H. CHOI, C. W. KIM, H. G. YANG, J. H. SOUK
LCD R&D Group II, AMLCD Division, Semiconductor Business, SAMSUNG Electronics Co., Ltd., 449-900, San #24 Nongseo-Ri, Kiheung-Eup, Yongin-City, Kyungki-Do, KOREA

ABSTRACT

Phosphorous (P) doped hydrogenated microcrystalline silicon ($n^+\mu$c-Si:H) films have been prepared by using the hydrogen-diluted plasma enhanced chemical vapor deposition (PECVD) method. The crystallinity of films deposited over the range of SiH_4/H_2 flow ratios and RF-power is studied by Raman spectroscopy. For a 900 Å thick film deposited at 250°C, a conductivity of $71\Omega^{-1}cm^{-1}$ and an average crystallinity of 49% is obtained. $n^+\mu$c-Si:H films as well as n^+a-Si:H films are used for both etch stopper and back channel etch type TFTs and the I_d-V_g characteristics are compared. For the etch stopper type TFT, the field effect mobility of 0.85 cm^2/V.sec, threshold voltages of 2 - 3 V and I_{on}/I_{off} ratio of $\sim 10^7$ are obtained.

INTRODUCTION

In recent years, Active-Matrix Liquid Crystal Displays (AMLCDs) have been widely used for notebook PCs, desk-top monitors, TVs, and other displays [1-2]. One of the most important characteristics required for a TFT in LCD panel is the low photo-induced leakage current. TFTs with high off current will degrade the contrast ratio of the display because of the loss of its video information before the frame is being refreshed. Minimization of the photo leakage-current in a-Si:H TFTs is important to obtain flicker-free, high display quality TFT-LCD. By decreasing the thickness of the undoped a-Si:H layer, the photo-leakage current of the a-Si:H TFT can be reduced. [3] Microcrystalline silicon is another good candidate for low photo-leakage current because of its high electrical conductivity and low optical absorption coefficients compared to a-Si:H. Especially, heavily phosphorous doped μc -Si:H is widely used for ohmic contact layers in a-Si:H TFTs. [4] The contact resistance between source/drain metal and n^+ layer as well as the resistance of the n^+ layer can be reduced remarkably by using the $n^+\mu$c-Si:H instead of using the conventional n^+a-Si:H as an ohmic contact layer in the a-Si:H TFT.

In the present work we have developed etch-stopper type 15.1" diagonal XGA color TFT-LCD. We also investigate the behavior of the microcrystalline silicon (μc-Si:H) layer with respect to the photo & thermally induced leakage currents.

EXPERIMENT

The single layer $n^+\mu$c-Si:H films are deposited on the bare substrates and the film properties are examined. In order to get an $n^+\mu$c-Si:H film, the hydrogen dilution method is used where the $H_2/(H_2+SiH_4)$ ratio is higher than 98%. Raman spectroscopy and SEM are used to check the crystallinity and microstructure of the films. Resistivity of $n^+\mu$c-Si:H is calculated from the film thickness and sheet resistance measured by four-point probe (RT-8A made by Napson). Two types of bottom gate a-Si:H TFTs with $n^+\mu$c-Si:H layers are fabricated.

Figure 1 illustrates two types of a-Si:H TFTs. Type A (Etch-Stopper) is fabricated by depositing a SiN_x gate dielectric, a-Si:H active layer, and SiN_x for the etch-stopper (E/S) of the TFT channel on top of the gate line metal. To reduce the parasitic capacitance (C_{gs}) between gate electrode and source, the channel passivation layer is patterned by the self-alignment technique. Data line metal and active layer are patterned after the $n^+\mu$c-Si:H is

Mat. Res. Soc. Symp. Proc. Vol. 452

deposited. Type B (Etch-Back, E/B) is formed by depositing SiN_x gate dielectric, a-Si:H active layer, and $n^+ \mu$c-Si:H on top of the gate line metal. After active etching, the S/D metal is patterned. Finally, the TFT channel is defined by removing the n^+layer.

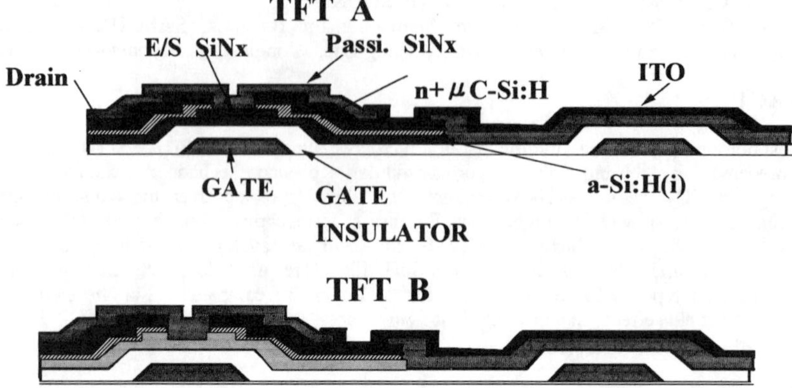

Fig. 1 Cross Section View of Two types of a-Si:H TFTs. Type A is the etch stopper type and the type B is the back channel etch type TFT.

RESULTS

1. Properties of $n^+ \mu$c-Si:H

Fig. 2(A) and (B) show the RF-power dependence of Raman spectra for $SiH_4/(H_2+SiH_4)$ ratios of 2.0% and 1.4% each. Typical Raman peak of amorphous state is $480cm^{-1}$ and crystalline state is $520cm^{-1}$.

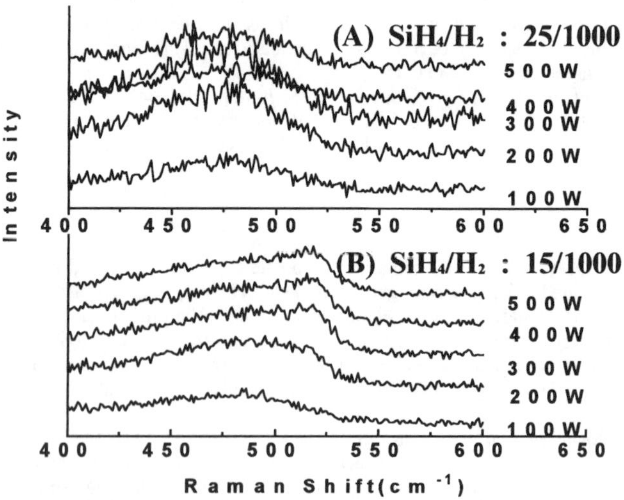

Fig. 2 Raman scattering spectra for various deposition RF-power.

Fig. 2(A) shows all amorphous states regardless of RF power where the hydrogen ratio is relatively small. But for higher hydrogen ratio and for the RF-power of higher than 300 W, the Raman peaks shift toward 520cm^{-1} which shows the formation of crystalline state as in Fig. 2(B). This suggests that the formation of μc-Si:H film requires high RF-power as well as high hydrogen dilution. In Fig. 2(B) the 520cm^{-1} peak is very broad. This suggests that the film is formed by the mixture of microcrystalline and amorphous states.

Figure 3 shows the deposition time dependence of film thickness and resistivity. The deposition thickness is linear to deposition time, but the resistivity which should be the intrinsic property of $n^+\mu$c-Si:H film is not constant with respect to deposition time (film thickness). For very thin film the resistivity is much higher than thick film which is almost constant. This suggests the existence of the transition layer at the beginning of the film formation. [5] Since the interface between a-Si:H and source/drain metal is critical for good ohmic contact formation the transition layer should be removed. More work needs to be done to solve this problem. Theoretically, large contact resistance between data line and a-Si:H reduces the on-current of TFT. By replacing n^+a-Si:H to $n^+\mu$c-Si:H, contact resistance is reduced and the TFT on-current is increased effectively. [6]

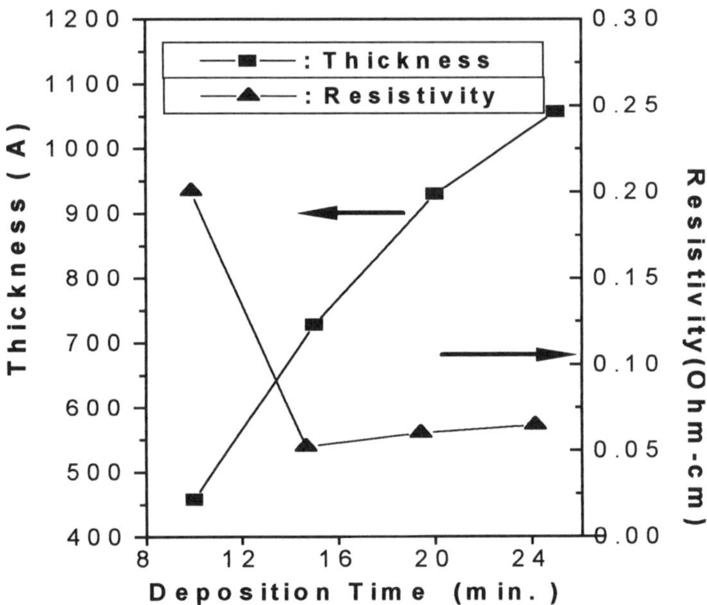

Fig. 3 Thickness & resistivity vs Depo-Time of the film deposited at 275°C

Properties of the present $n^+\mu$c-Si:H and the conventional n^+a-Si:H films are shown in Table I. The resistivity of $n^+\mu$c-Si:H film is about three orders of magnitude lower than that of n^+a-Si:H as shown in Table I. It suggests that the fraction of electrically active donors in $n^+\mu$c-Si:H film is larger than that in n^+a-Si:H film. Therefore $n^+\mu$c-Si:H is better for ohmic contact layer than n^+a-Si.

Table I. Properties of n⁺μc-Si:H and n⁺a-Si:H.

	$n^{+}\mu$c-Si:H	n^{+}a-Si:H
Thickness (Å)	900	500
Sheet Resistance (kΩ/□)	1.6	60000
Resistivity (Ω-cm)	0.014	300

2. TFT characteristics

The Id-Vg characteristics of E/S and E/B TFT with $n^{+}\mu$c-Si:H are shown in Fig. 4. The field effect mobility (μ) and the threshold voltage (Vth) are determined by plotting the saturation region of $I_d^{1/2}$ versus Vg. For E/S TFT μ is about 0.85 cm^2/Vs, Vth is about 2 - 3 V and on/off current ratio is ~ 10^7. The light of 4000 cd/m^2 is illuminated from the back side of the TFT to measure the photo-leakage current of E/S and E/B TFT. Under back side illumination, the leakage current of E/S TFT with $n^{+}\mu$c-Si:H is found to be one order of magnitude lower than that of E/B TFT. The light-induced leakage current in the off state is due to the generation of photo-induced electron-hole pairs. [7] The smaller off-current under light illumination for E/S TFT is due to the thin undoped a-Si:H layer in a-Si:H TFT. According to A. Sugahara et. al, [8] the photo-leakage-current mainly flows through side area of the channel, named side-channel. The reason is that the photo carriers generated in the active channel are more easily recombined underneath the the top etch-stopper insulating layer than around the side-channel. The interface between intrinsic a-Si:H and etch -stopper SiNx is the effective electron-hole recombination site for E/S TFT.

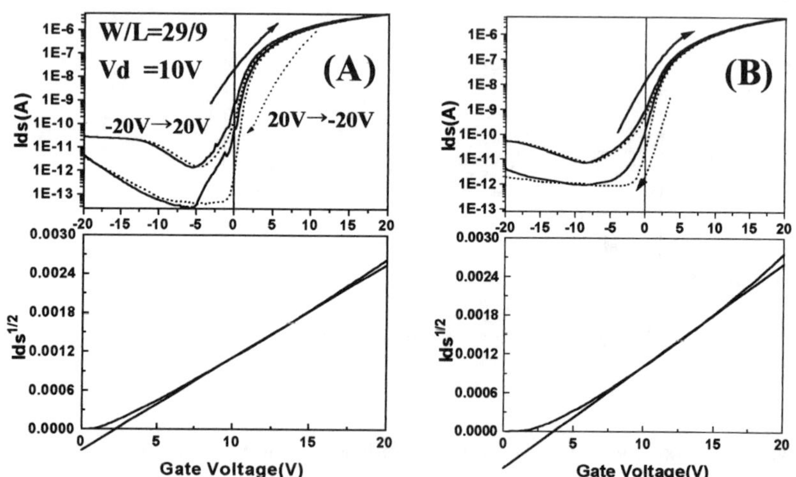

Fig. 4 Transfer curves of the a-Si:H TFTs with $n^{+}\mu$c-Si:H under dark and back-light illumination. (A) E/S TFT, (B) E/B TFT.

Practically, AMLCDs are operated higher than the room temperature, e.g. around 50°C. The off current issue is even more important for high temperature operation because the off current is exponentially increased with temperature. Several techniques have been proposed to decrease the off current, such as the control of the Fermi level [9], and the optimization of the passivation deposition conditions. [10,11] In Fig. 5, the transfer curves of E/S TFT with n^+ μc-Si:H for various operation temperatures are shown.

The hydrogenation effect for thermally induced off current of the E/S TFT is investigated. The hydrogenation is done by exposing a-Si:H surface to H_2 plasma before n^+ μc-Si:H deposition. The hydrogenation can decrease the thermally-induced off-current by more than one order of magnitude. It is reported that there exists a current path through which the leakage-current mainly flows. Since no change of the on current is observed, the decrease of the off current is related to the change of side channel characteristics by hydrogenation. It is assumed that the number of interface states at the side-channel is increased through hydrogenation. These increased interface states can prevent the formation of the accumulation layer of electrons in a-Si:H films at the side-channel interface. The increased density of states at the side-channel interface is also assumed to pin Fermi level at a relatively deep level because the increase of off current activation energy is observed.

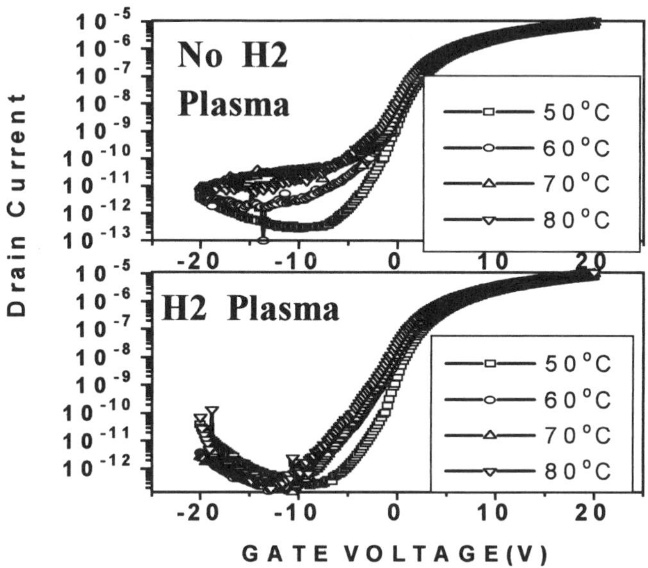

Fig. 5. Dependence of Id-Vg characteristics under the operating temperature. (A) without hydrogenation, (B) with the hydrogenation

Conclusion

It is found that the resistivity of the phosphorous doped microcrystalline film deposited under 98% hydrogen dilution is several thousand times lower than that of normal n+a-Si:H film. Using μc-Si:H as n+ layer, 15.1"-diagonal XGA E/S TFT LCD is fabricated. The photo and thermal-leakage current is one of the major problem to be solved. H_2 plasma treatment before n+μc-Si:H deposition for E/S TFT can decrease the thermally induced off-current of TFTs by more than one order of magnitude.

References

1. A. Ban, Y.Nishioka, T.Shimada, M. Okamoto, M.Katayama, SID 96 Digest, 93 (1996)
2. K.Kawai, T.Sakurai, M. Katayama, T.Nagayasu, N.Kondo, Y.Nakata, S.Mizushima, K.Yano, M.Hijikigawa, SID 93 Digest, 743 (1993)
3. N.Hirano, N.Ikeda, H.Yamaguchi, S.Nishida, Y.Hirai and S.Kaneko, IDRC 94, 369 (1994)
4. J.Kanichi, E.Hasan, J.Griffith, T.Takamori and J.C.Tsang,Mat. Res. Soc. Symp. Proc. 149, 239 (1989)
5. K.C.Hsu, B.Y.Chen, H.T.Hsu, K.C.Wang, T.R. Yew and H.L.Hwang, Jpn. J. Appl. Phys. Vol. 33 (1994) pp. 639-642
6. J. Kanichi, Appl. Phys. Lett., 53, 1943 (1988)
7. K. Suzuki, "Flat panel displays using amorphous and microcrystalline semiconductor devices" in Amorphous and Microcrystalline Semiconductor Devices. Ed. J.Kanichi (Artech House, Boston, 1991) p.111
8. A. Sugahara, M.Seiki, Y.Miura and M. Shibusawa, AMLCD 94 Japan Chapter 184 (1994)
9. N. Ibaraki, K.Furuda and H.Takata, Conf. Rec. Int. Top. Conf. on Hydrogenated Amorphous Silicon Devices and Technology, 182 (1988)
10. M.J.Powell, Mat. Res. Soc. Sympo. Proc., 33, 259(1984)
11. M.J.Powell, Insulating Films on Semiconductors Eds. J.F.Verweij and D.R.Wolters (North-Holland), 245 (1983)

INFLUENCE OF NANOCRYSTALLINITY ON PROPERTIES OF PHOTODIODE AND TFT IMAGE SENSOR STRUCTURE

A. KOLODZIEJ, S. NOWAK, P. KREWNIAK
Department of Electronics, University of Mining and Metallurgy, al.Mickiewicza 30, 30-059 Krakow, Poland, kolodzie@uci.agh.edu.pl

ABSTRACT

This paper reports technological experiments which have strongly affected the properties of the multilayer image sensor structure. The high deposition rate and large hydrogen content during the reactive magnetron deposition process causes nanocrystallization of prepared silicon films. We also report experiments with laser crystallization. The effect of microstructure in a-Si:H films on TFT and pin diode characteristics has been investigated. Crystallinity was confirmed by Raman scattering, small angle X-ray diffraction (SAXD) and transmission electron microscopy (TEM). The parameters of the photodiodes such as quantum efficiency, dark and light currents, etc. have been tested using the measurements of samples in the configuration of a linear image sensor consisting of two rows, 30 to 160 pixels per row, with dimensions $0.1mm^2$-$1mm^2$.

INTRODUCTION

Large area image sensors based on a-Si:H have been proposed for a variety of applications. Among these are line sensors for document scanning or X-ray computer tomography, as well as two-dimensional matrix sensors for radiation or contact imaging [1,2]. It has been shown that high detection efficiency and stable performance may be limited by long term changes in properties of a-Si:H and the entire structure, by the dynamic response of photodiodes and the read out transistors especially under illumination.

The principles of operation of the image sensors are as follows: the incoming radiation is detected by the a-Si:H photodiode and the photogenerated charge is stored as the diode capacitance C_p. After one frame time, the transistor is switched on and the signal charge is transferred to the readout amplifier. The dynamic performance of the image sensor is limited by the readout of the pixel capacitance C_p, through the transistor on-resistance R_{on}, and by thermal emission of trapped charge within the photodiode. With an optimized transistor technology, readout can be accomplished in about 10µs. The influence of inconvenient phenomena such as charge trapping can be lessened when using microcrystalline silicon instead of amorphous silicon.

Throughout this paper we consider a few new technological experiments used during the process of creating integrated multilayer structures of a-Si:H image sensors consisting of switching TFT and pin photodiodes. Our proposal is to use the reactive magnetron sputtering technique to create aforementioned multilayer structures, particularly nc-Si:H layers [3]. The premises of creation of a photodiode based on a nc-Si:H layer lie in its ability to retain high optical absorption and photo-response, characteristic of the a-Si:H layer, and electronic stability as well as thermal trapped charge on the other hand. The grain size in these films is a relevant parameter closely related to the electronic properties. In case of the switching TFT, the needs of increased R_{on} and channel mobility μ can be achieved by means of transistor channel recrystallization [4]. In this paper results are described of channel direct recrystallization using a YAG laser, mainly because of the simplicity and low cost of this method. Because of the degrading electronic features, influenced by the interface states between the nc-Si:H and SiN_x films as well as multilayer diffusion

Mat. Res. Soc. Symp. Proc. Vol. 452

between n+, p+ and i a-Si:H layers, additional nitridation by strong flux on N ions has been used.

The aforementioned experimental results actually are mean values collected over 160 samples.

EXPERIMENTAL DETAILS

The TFT and diode structures in applications as LCDs, ISFETs and linear and 2D image sensors (visible range) have been studied over the last few years. For these applications we used to make test samples of the form: two rows and 30 to 160 pixels per row with dimensions $0.1mm^2$ to $1mm^2$.

During our research, the DC reactive magnetron sputtering technique was used to obtain appropriate films and multilayer structures, and a considerable number of preparatory parameters had to be optimized to produce low defect n+, p+, i a-Si:H, a-SiNx structures as well as their interfaces. It can be observed that the properties of the mentioned films are functions of substrate temperature, T_S, applied power, P_r, argon partial pressure, p_{Ar}, hydrogen partial pressure, p_{H_2}, ammonia partial pressure, p_{NH_3}, phosphine partial pressure p_{PH_3}, diborane partial pressure, $p_{B_2H_6}$, total pressure, p_2, shield potentials, surface quality and degree of consumption of silicon target, geometry of magnetron-screen-shield-substrate system, base pressure and others [3]. Highly nanocrystalline material was obtained by modification of the magnetron process to high power ($15W/cm^2$), high hydrogen content in the chamber (about 75%), low pressure (about $8 \cdot 10^{-2}$ Torr) and higher substrate temperature (350°C) than in the case of a-Si:H. The schematic cross-sections of various fabricated structures of image sensor are shown in Fig.1.

There are a few forms of thin film silicon materials with respect to the grain size: amorphous (a-Si:H), nanocrystalline (n-Si:H), microcrystalline (μc-Si:H) and polycrystalline (poly-Si:H). One alternative method for growing μc-Si is to start from an a-Si:H film and crystallize it [5,6,7]. We report a study of laser induced crystallization of nc-Si:H films as well as of nc-Si:H bottom gate TFT structures deposited on glass substrates. The laser was a Nd:YAG pulsed at 1064nm. Each substrate was laser crystallized by scanning the laser beam over the sur-

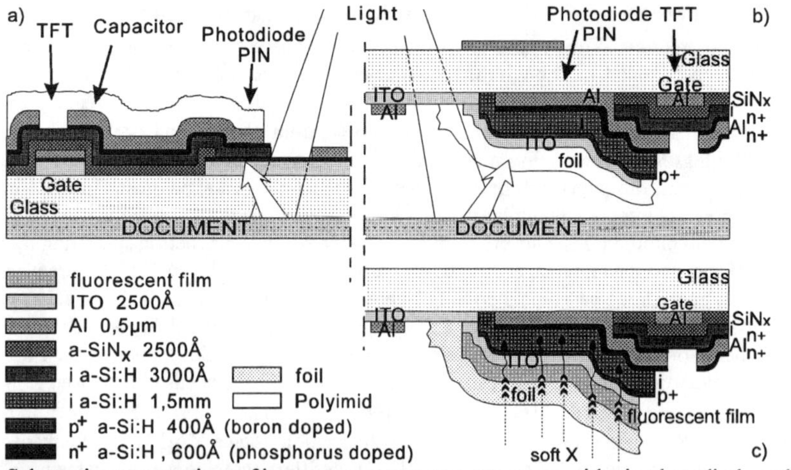

Fig.1. Schematic cross-sections of image sensor two-row structures with pin photodiode and switching TFT. Three tested configurations are shown.

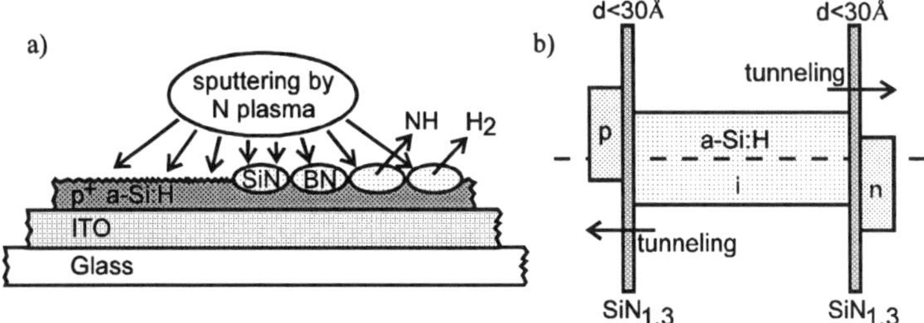

Fig.2. (a) Idea of surface nitridation of p[+] film and (b) adequate junction scheme.

face of the sample in a raster pattern normal to the surface in dry nitrogen ambient.

There are other technological experiments stabilizing and creating new properties of the examined structure. Additional thin and very thin films which play a role of diffusion blocking barriers, selective controllers of ions, elastic layers with respect to mechanical stress, electrical isolation and tunneling layers, are used. The specialized films may be very well defined (stoichiometric, homogenous, etc.). To improve properties of the intersurface layers, structures were treated by a pure nitrogen plasma flux for a 1 to 4 minutes time span. The surface was partially sputtered and partially nitrided leading to thin layers of 20-50A $SiN_{1.3}$. The layers and their roles have been schematically presented in Fig.2. It should be stated that all those experiments were performed in an additional vacuum chamber.

RESULTS AND DISCUSSION

Raman spectra of deposited amorphous and nanocrystalline films are presented in Fig.3. Raman scattering spectroscopy is a useful technique in investigation of microcrystalline materials [5]. The Raman spectra of a-Si:H and nc-Si:H can be decomposed into three peaks by least square fitting, Fig.3b. The peak at 480cm^{-1} is identified as arising from the amorphous phase, the peak at 520cm^{-1} is identified as arising from the crystalline phase and the peak at 500cm^{-1} is considered to be caused by the distribution of crystalline phase sizes. The crystallinity X(R) determined by the Raman scattering method can be calculated from the following equations:

$$X(R)=(I_{500}+I_{520})/(\delta I_{480}+I_{500}+I_{520}) \quad (1)$$

where: I_{480}, I_{500}, I_{520} are integrated intensities of the peaks at 480, 500 and 520 cm^{-1}, respectively. In the case "b" we obtain X(R)=0.58. Instead of fitting the Raman spectra by a distribution

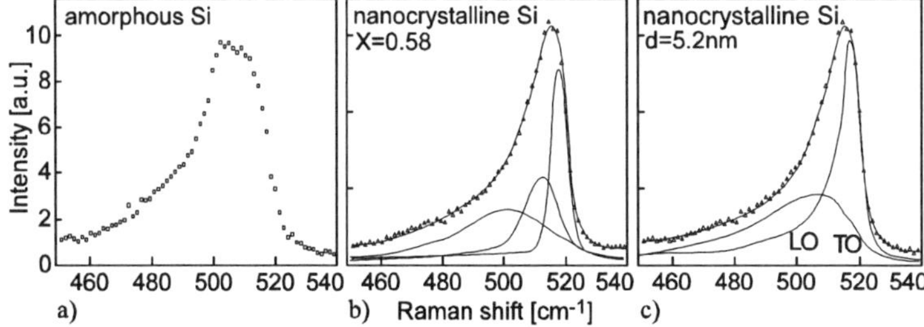

Fig.3. Raman spectra of Si films: (a) amorphous, (b) nanocrystalline - three phase deconvoluted, (c) nanocrystalline - combination of 70% LO and 30% TO.

of particle size (only LO mode), the measured spectra can be perfectly fitted with one particle size and a mixture of LO and TO modes [8]. In this method the diameter of the particles was d=5nm. This result leads one to question the size distribution for the nanocrystalline materials.

Laser crystallized 1μm thick nanocrystalline structure samples, processed at room temperature with pulse rates of 10 Hz with various laser energies and shot densities, show very strong changes of photoresistivity and mobility as is seen in Fig.4. The laser power ranges from $400mJ/cm^2$ to $800mJ/cm^2$.

The properties of the films produced in various ranges of laser power are as follows:

Range A - nanocrystalline silicon nc-Si:H, $\rho_{dark} \approx 10^{12}\Omega/\square$, photo-response $\rho_{dark}/\rho_{light} \approx 10^5$, mobility $\mu \approx 1 cm^2V^{-1}s^{-1}$, optical gap $E_g \approx 1.79eV$.

Range B - highly defective nc-Si:H, $\rho_{dark} \approx 10^{13}\Omega/\square$, $\rho_{dark}/\rho_{light} \approx 5$, $\mu \approx 0.1\ cm^2V^{-1}s^{-1}$, $E_g \approx 1.6$ eV.

Range C - microcrystallized μc-Si, $\rho_{dark} \approx 5\cdot 10^6\Omega/\square$, $\rho_{dark}/\rho_{light} \approx 10^2$, $\mu \approx 50\ cm^2V^{-1}s^{-1}$, $E_g \approx 1.6eV$.

Range D - μc-Si strongly destroyed by explosively evolved hydrogen from the film, $\rho_{dark} \approx 10^{10}\ \Omega/\square$, $\rho_{dark}/\rho_{light} \approx 10$, $\mu \approx 15\ cm^2V^{-1}s^{-1}$, $E_g \approx 1.2eV$.

We have studied I_{DS}-V_{GS} characteristics of laser crystallized TFTs as it is seen in Fig.5. For laser energy parameters from Range C, the results shown in Fig.4 were obtained.

The initial hydrogen contents in the films were 25% and 10%, as determined from infrared absorption measurements on film codeposited on Al/glass substrates [3].

Hydrogen contents changed explosively during laser annealing. Transmission electron microscopy (TEM) and X ray diffraction of the multilayer's cross-section were employed to study the crystallinity of the films and quality of their surfaces [3,4].

Some aspects of the physics of the process do not appear to be fully understood. Explosive hydrogen evolution is believed to cause surface roughness. It was observed that nanocrystalline films containing 25% hydrogen deposited on silicon nitride, crystallized without bubbling. The study of the transient phenomena and materials properties shows that the hydrogen out-diffusion strongly depends on the laser energy density and the initial film structure [3,4,6,7]. In order to improve linearity and stability of the field and time dependent characteristics of the pin photodiode as well as to increase the short wavelength QE some modifications of interfaces relative to the conventional pin structure were developed. The idea of surface nitridation of various interfaces is shown in Fig.2. The basic role of this process is the creation of a very thin $SiN_{1.3}$ film which is resistant to migration of boron, phosphorus and other ions. The process leads to improved homo-

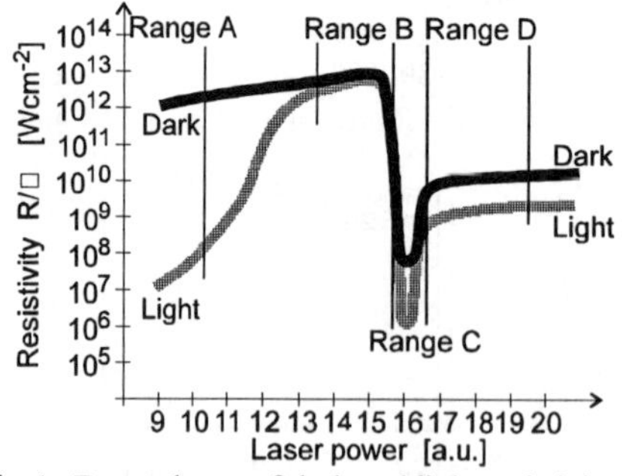

Fig.4. Dependence of dark and light resistivity of a 1μm thick a-Si:H film YAG laser crystallized versus laser power.

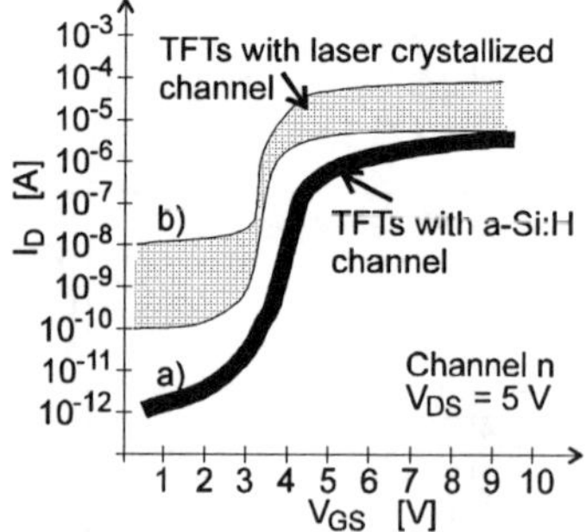

Fig.5. I_{DS} vs. V_{GS} for an n-channel transistor. Thickness of the characteristics signifies the statistical spread of all TFTs in the sensor.

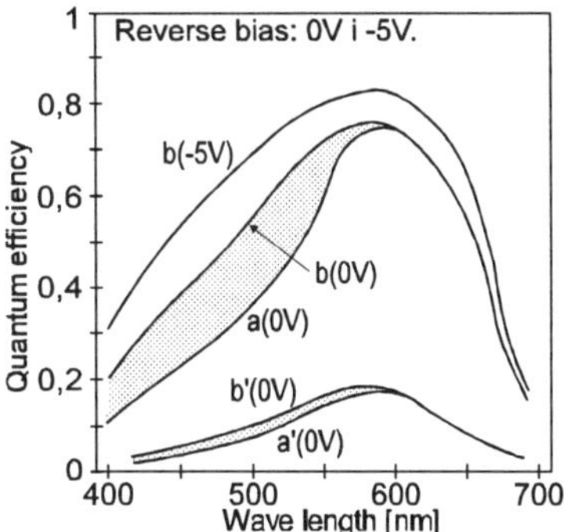

Fig.6. Influence of polarization and p^+ film surface nitridation on the quantum efficiency QE of pin diode. Curves a and b like in Fig.7; curves a', b' - under "back" light and reflection from a white paper.

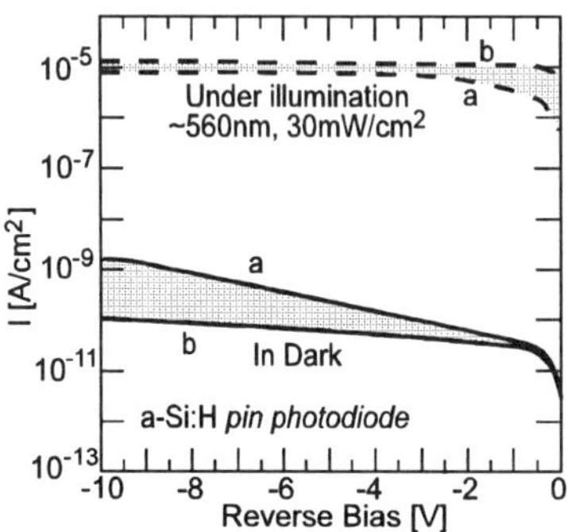

Fig.7. Typical reverse a-Si:H PIN diode I(V) characteristics. Curve a - pin diode with 150A p^+ a-Si:H film; curve b - pin diode with 150A p^+ nc-Si:H film and additional nitridation of intersurfaces.

geneity and lower thickness of p^+ and n^+ films. These factors have an impact on the limitation of surface recombination processes and, in turn, increase the QE. Fig.6 shows the QE, measured with normal incidence light and back illumination from white paper. The QE decreases for increasing λ into the red, due to incomplete absorption in the i-layer; so increasing the i-layer thickness increases the QE. The QE decreases for decreasing λ into the blue, for several reasons. There is a wavelength dependence of the reflection loss and small percentage of absorption in the ITO layer. In addition, there is a finite absorption of carriers in the p-layer and possibly some back diffusion of carriers generated near the top of the i-layer. The latter effects result in electron-hole pairs generated in the a-Si that are not collected. In Fig.6 the bias dependence of the QE is shown. At zero bias, there is incomplete charge collection, but we find a nearly saturated QE for reverse biases greater than 4V. The effect of nitridation influences an increasing of QE in the short part of the wavelength range. It leads to lower effective thickness of p^+ film and halts back diffusion of carriers. The forward characteristics of the diode do not generate significant changes because of the tunneling effect of the carriers by those layers. However, the barrier characteristics are more advantageous as seen in Fig.7. Dark reverse current is visibly smaller after nitridation. The quantity of the effect is not understood by us. Fig.8 shows typical a-Si:H TFT $I(V_G)$ curves. The architecture of our line sensor is shown in Fig.9. Parameters of image sensors with amorphous films and in the case of nanocrystallized films and nitrided interfaces are, respectively, as follows: $I_{on}/I_{off}=10^6\text{-}5\cdot10^7$, $\mu=0.5\text{-}15cm^2V^{-1}s^{-1}$, $R_{on}=0.4\text{-}1.7M\Omega$, threshold voltage $V_{th}=0.6\text{-}2.5V$, responsivity 110-140Vcm2/μJ at 550nm. For 1μm thick diodes and with 0.1mm2 surface areas the RC time reached about 12μs. We have built readout electronics which operated the sensor at high speed to access the imaging performance in real time, i.e. a line time t_l of 120μs, a clock frequency of 16MHz, and a frame time of 40ms.

CONCLUSIONS

This paper presents a series of successful technological experiments (nanocrystallization, nitridation). The main conclusions, leading to the video sensor parameter optimization, have been stated. The nc-Si:H layers with properties similar to those obtained with CVD methods have been

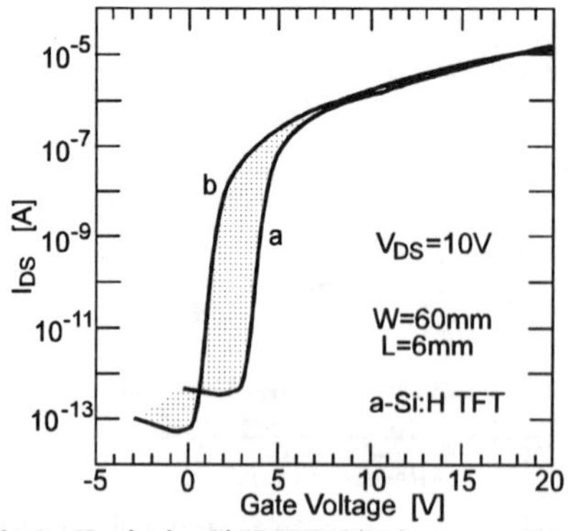

Fig.8. Typical a-Si:H TFT I(V_G) curves. Curve a - TFT with amorphous a-Si:H, curve b - TFT with nanocrystalline nc-Si:H film and with additionally nitrided channel interface.

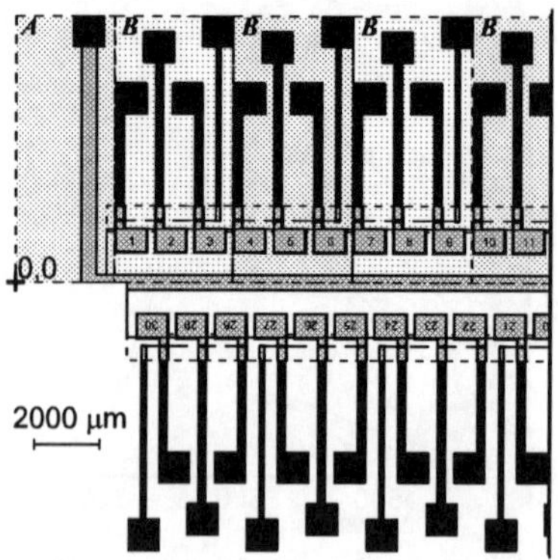

Fig.9. Fragment of the line sensor architecture.

obtained here. Range C in Fig.4 represents optimal parameter set for the laser crystallization process. The range C is very narrow and it is difficult for repetition of the laser parameters and crystallization results. The process of crystallization is easier in the case of a-Si:H films with 25at.% hydrogen content than for films of 10at.% hydrogen content. Hydrogen rich silicon films are partially nanocrystallized and therefore crystallization goes smoothly in contrast to films with low hydrogen content. In the case of the transistor, the diameter of the laser beam was too wide and the influence of the laser energy on source and drain electrodes destroyed them. Therefore the yield of acceptable transistor after crystallization was low, about 20%. The sensitivity threshold of the diode has been moved towards the UV and the the RC constant of the transistor-diode circuit has been significantly diminished. The observed results provide the possibility of making a 2D image sensor with a resolution of 100μm in the format A4. Those parameters are comparable to data in [1,9].

REFERENCES

1. I.Fujieda, S.Nelson, P.Nylen, R.A.Street, R.L.Weisfield, J. Non-Cryst. Solids **137&138**, 1321 (1991).
2. M.J.Powell, I.D.French, J.R.Hughes, N.C.Bird, O.S.Davies, C.Glasse, J.E.Curran, Mat. Res. Soc. Symp. Proc. **258**, 1127 (1992).
3. A.Kolodziej, J. Non-Cryst. Solids **185**, 168 (1995).
4. A.Kolodziej, S.Nowak, P.Krewniak, in Proceedings of the 19-th Conference of ISHM Poland (International Society for Hybrid Microelectronics Poland Chapter, Wroclaw, 1996), p. 113.
5. J.Kanicki, in Amorphous and Microcrystalline Semiconductor Devices: Optoelectronic devices (Artech House, Boston, 1991).
6. M.Hack, P.Mei, R.Lujan, A.G.Lewis, J. Non-Cryst. Solids **164-166**, 727 (1993).
7. R.I.Johnson, G.B.Anderson, J.B.Boyce, D.K.Fork, P.Mei, S.E.Ready, S.Chen in Amorphous Silicon Technology (Mater. Res. Soc. Proc. **297**, 1993) pp. 533-538.
8. I.Gregora, B.Champagnon, L.Saviot, Y.Monin, Thin Solid Films **255**, 139 (1995).
9. D.K.Biegelsen, W.B.Jackson, R.Lujan, D.Jared, R.Weisfield in Amorphous Silicon Technology (Mater. Res. Soc. Proc. **377**, 1995) pp. 833-838.

INJECTION ELECTROLUMINESCENCE FROM THIN FILM p-i-n STRUCTURES MADE FROM NANOCRYSTALLINE HYDROGENATED SILICON

A.A. ANDREEV*, B.Y. AVERBOUCH*, P. MAVLYANOV*, S.B. ALDABERGENOVA**, M. ALBRECHT**, D. STENKAMP**, and H.P. STRUNK**
* A.F.Ioffe Physical-Technical Institute, 194021, St. Petersburg, Russia
**Universität Erlangen-Nürnberg, Institut für Werkstoffwissenschaften, Mikrocharakterisierung, D-91058 Erlangen, Germany

ABSTRACT

Nanocrystalline silicon films are prepared by plasma enhanced chemical vapour deposition of silane under the conditions of high hydrogen dilution (3:100). The film structure consists of nanoclusters 0.8 to 5 nm in size (volume fraction 30%) embedded in an amorphous matrix. The Tauc gap of the amorphous matrix is 1.95 to 2.05 eV depending on deposition parameter. These films are characterized as regards photoluminescence (PL) and, prepared to p-i-n structures, electroluminescence (EL). The PL and EL agree in (i) luminescence peak at 1.9 eV, i.e. small Stokes shift, (ii) almost no temperature dependence between 77 K and 293 K, (iii) fast kinetics with time constant of a few 10^{-8} s. These data can be understood in terms of quantum confinement in Si nanocrystallites smaller than around 2 nm. The EL in addition exhibits a luminescence band extending up to 3 eV, which can be interpreted by interband transition due the hot carriers.

INTRODUCTION

Because of the indirect fundamental bandgap monocrystalline Si shows only weak band-to-band emission in the near infrared. The recent developments in the field of microscopically structured silicon in the form of porous silicon [1,2] have opened a new way to obtain bright interband emission. This way is based on the quantum confinement effect for small silicon clusters, which leads to change in nature of the optical transitions. Therefore, it seems to be quite attractive to study other possible approximations to produce microstructured silicon. One of the most promising technological direction is the preparation of microcrystalline or nanocrystalline silicon (diameter in the order of nanometer) in a matrix of amorphous hydrogenated silicon.

The main experimental effect of the quantum confinement is the blue-shift of the absorption edge. Theoretical models based on the effective mass approximation have established that for crystallites with of 3-5 nm phonon assisted transitions are dominant and for smaller crystallites direct optical transitions become allowed [3]. According to [4], emission in the energy range of 1.8 - 2.8 eV could be possible only in very small two or three-dimensional structures and fast optical properties in the nanosecond range are only possible for crystallites or wires with diameters smaller than around 1 nm. Thus, the size of 1 nm seems to be a critical value for nanocrystalline Si, where one would expect unique properties.

In this paper, we present experimental results on microstructure, optical absorption and luminescent properties of Si:H films prepared under conditions of high hydrogen dilution of silane. This material exhibits an optical gap up to 2.05 eV as is also discussed in the literature [5,6]. This new modification of amorphous wide gap Si:H is not well-studied. We investigate further electroluminescence (EL) of p-i-n structures made from wide gap Si:H. We explain the specific features of the EL and PL by the effect of confinement in Si nanocrystallites. Especially, we emphasize the strong effect of the smallest Si clusters with sizes of less than 2 nm, which are clearly distinguished in our HRTEM experiments.

EXPERIMENTAL PROCEDURES

Hydrogenated silicon (Si:H) films are prepared by chemical vapour deposition in a silane-hydrogen mixture (3% SiH_4 in H_2) in a reactor of the type used in systems with plasma excitation by cyclotron resonance. We use a two-mode excitation with frequencies of 13.56 and 75 MHz with a system of quadrupole radio-frequency vibrators. An axial magnetic field of about 150 Gs is applied

Mat. Res. Soc. Symp. Proc. Vol. 452 © 1997 Materials Research Society

in the excitation region. Substrate and substrate-holder are isolated from excitation electrodes as well as from the chamber walls (floating potential). RF power dissipation is measured to range around (2 - 5)$\cdot 10^{-2}$ W/cm^3. The substrate temperature is varied between 60 and 180°C and the flow rate between 2.5 - 15 cm^3/min (normal conditions). The growth rate is controlled in situ and is kept less than 0.7 Å/sec. We use two types of Mo glass substrates: covered with transparent oxides (indium-tin oxide (ITO) and SnO_2) and uncovered. For electroluminescence (EL) measurements we prepared p-i-n structures: glass/ITO/p-Si:H(100Å)/i-Si:H(700-1000Å)/n-Si:H(250Å)/Al. To obtain doped layers we use boron (B_2H_6) and phosphin (PH_3). For transmission electron microscopy analysis the samples are prepared by mechanical thinning followed by chemical etching in HF/H_2O (1:10) to strip them from their SiO_2 substrates. The PL is excited by pulse N_2 laser with wavelength of 337 nm. The PL is investigated at 78 and 300 K.

EXPERIMENTAL RESULTS

Structure

It is well-established that high hydrogen dilution of silane strongly stimulates chemical etching by the hydrogen plasma and leads to the growth of layers consisting of nanocrystalline (nc-Si:H) and µc-Si:H + amorphous (a-Si:H). To observe directly the microstructure at the atomic level we use high resolution electron microscopy (HRTEM, point resolution 1.75 Å). HRTEM experiments show that the deposited Si:H films contain nanocrystalline and amorphous phases. The sizes of the nanocrystallites can be controlled by choosing appropriate growth parameters such as gas flow rate and substrate temperature. Figures 1 and 2 show TEM micrographs of a Si:H film. Individual microcrystallites appear as ordered areas consisting of parallel sets of fringes. The orientation of the fringes is random which indicates the spontaneous nucleation. The volume fraction of crystalline phase is 25...33%. From the high resolution micrographs we measured directly the size of the crystallites. In Fig.1 the size is 4 - 5 nm in diameter. We find in other areas in between these (Fig.2) a high density of nanocrystallites with diameters of around 1 nm. There seems to exist a tendency to a bimodal distribution of the diameter of the nanocrystallites peaking at 4-5 and at around 1 nm.

Optical absorption

The optical band gap E_g of the layers is estimated from measurements of the optical absorption. The value of E_g was determined by fitting the absorption data to the Tauc expression that is used for amorphous materials: $\alpha h\nu = B\cdot(h\nu - E_g)^2$, where B is a constant, hν is a photon energy. The experimental curves in the Tauc coordinates define the absorption edge rather well. The extrapolated values of E_g vary between 1.95 and 2.05 eV depending on the growth parameters and are significantly wider than those of a-Si:H (1.75 eV).

Photoluminescence

The PL spectra in Si:H samples show several specific features, which are very different from those of ordinary a-Si:H. Fig.3 shows PL spectra taken at room temperature (curve 1) and at 77K (curve 2): (i) The PL spectrum peaks at 1.9 eV (room temperature) and 1.8 eV at (77 K). (The high frequency band with a peak at 2.8 eV originates from the emission of the glass substrate.) (ii) The PL peak at room temperature is shifted from the Tauc gap energy towards lower energies by only 0.1 eV, in a-Si:H, however, such a shift exceeds 0.4 eV. (iii) The PL depends weakly on temperature whereby in a-Si:H the intensity of the PL peak at T=77 K is more than two orders of magnitude higher than at T=300K. (iv) The decay after pulse excitation is very fast and characterized by a time less than 30 nsec (as opposed to 10 µs of a-Si:H).

Electroluminescence

The properties of the fabricated p-i-n structures will now be described. We have observed two different ranges of current-voltage (I-V) characteristics, an initial diode type behaviour at low currents and voltages, and a secondary characteristic above a critical applied voltage, where visible (bright white-yellow) light emission sets in.

The I-V characteristics at low voltages can be described by a relation: $J = J_s \cdot \{\exp(eV/nkT) - 1\}$, with a diode saturation current density $J_s = 10^{-9} A \cdot cm^{-2}$. The fit yields ideality factors n in between 3 and 10; the forward/reverse current ratio is larger than 10^5. This I-V behaviour is typical of p-i-n structures based on a-Si:H.

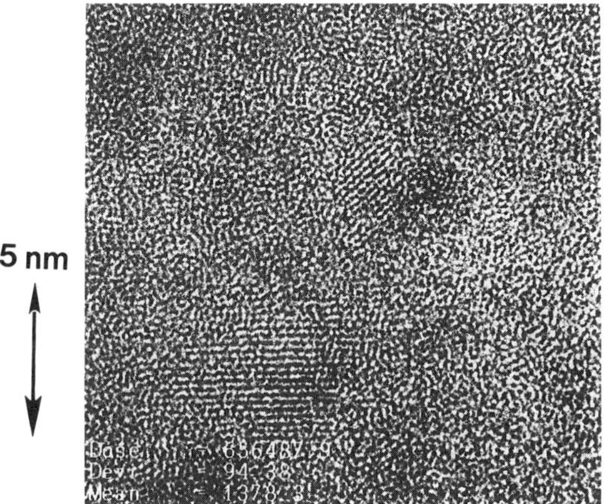

Fig.1. High resolution image of the Si:H film. Nanocrystallites appearing as stacks of parallel fringes are embedded in an amorphous matrix.

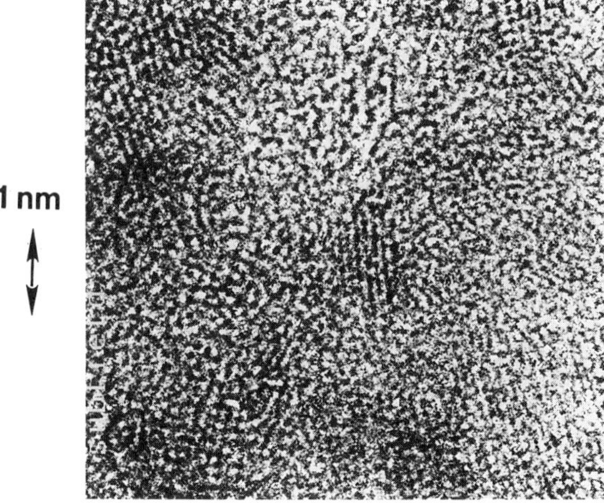

Fig.2. High resolution image of amorphous network between comparatively large (4-5 nm) crystallites. Stacks of three/four short fringes indicate crystallites with a mean size of about 1 nm.

The secondary characteristics, eg. Fig.4, appears after the forming, ie after exceeding a critical current. The measured I-V curves in Fig.4 can be fitted by the Fowler-Nordheim relationship, that describes tunneling processes: $J \sim V^2 \cdot \exp\{-const(\alpha\sqrt{\varphi})\}$, where φ and $\alpha=\varphi/eV$ are the barrier height and width, respectively. Consistent with a tunneling mechanism of conduction these I-V characteristics are practically independent of temperature, at least in the temperature range of the measurements, 100 - 279 K.

Fig.5 shows the EL spectrum. Its peak is centered at 1.9 eV as is the peak of the PL spectrum (to within 0.05 eV). Also, the EL decay time as determined with pulse current experiments is short, 10 ns, and independent of pulse duration. These facts strongly suggest that both EL and PL originate from the same mechanism.

One noteworthy difference exists in the EL and PL spectra: a broad intensity band is visible region (see insert in Fig.5) extending up to 3 eV. We will briefly consider this in the discussion.

We have measured the integral intensity of this spectrum as a function of temperature. This intensity increases exponentially with falling temperature below ~ 200K and we obtain an activation energy of E_a=0.028 eV (eg Fig.6). If we associate this value with an excitonic ground state, then it is higher than the corresponding value of ~15 meV [7] for the bulk Si.

DISCUSSION

We will focus on two topics in the following and discuss possible mechanisms for the secondary characteristics observed in the p-i-n structure, and for the EL and PL luminescence in view of the microstructure observed in our a-Si:H films. This microstructure consists of an amorphous matrix that contains nanocrystals with diameters below 5 nm with a high number density at around 1 to 2 nm. The crystalline volume fraction amounts to one third and, thus, the nanocrystallites can be considered to be separated from each other by a thin amorphous tissue.

The most interesting result of the p-i-n measurements is that obviously a change in the conduction mechanism occurs above a certain applied voltage/current density concomitant with light emission. The I-V curve changes from the diode behavior typical of completely amorphous Si:H materials to a behavior that can best be described by a tunnel controlled carrier transport according to the Fowler-Nordheim model. It is obvious that the diode behavior of our films can be attributed to the amorphous phase that is the only continuous phase in our films. In comparison the secondary characteristics with its large increase of current and different transport mechanism, should be interpreted with an additional efficient current path. Such a path is from the amorphous phase through the nanocrystallites. The necessary tunnel barrier can be ascribed to the amorphous/crystalline and crystalline/amorphous boundaries. Our results are not sufficient to elaborate on the physical nature of the barrier, however, several possibilities are conceivable: (i) the band bending effect, (ii) strain shift of the band edge at the crystalline/amorphous boundary, (iii) polarization effect due to the strong polarity of Si:H bond. H atoms bonded to Si are negatively charged with roughly 0.1 of an electron charge. This effect may be considerable because the large number of hydrogen atoms passivate the dangling bonds on the surfaces of the crystallites.

An important consequence of this notion is that the EL, which indeed sets in with onset of the second characteristics, is due to the presence of the nanocrystallites. We will discuss this aspect further in the context of the PL results.

We have found two agreeing properties of EL and PL: the luminescence peaks are situated at 1.9 eV and the kinetics is fast as characterized by time constants 1x(EL) and 3×10^{-8}(PL) sec. This clear correspondence indicates the same origin for EL and PL. Therefore, we will discuss the mechanism of luminescence using our absorption and PL measurements. We will find that the whole body of results can be described by interpreting the dominant role of the nanocrystallites in luminescence, as found above from EL, within the framework of a quantum confinement model. We first consider the large optical gap of ~ 2 eV found from our absorption studies and show that it can be shown to be ultimately caused by the rather high hydrogen content in the amorphous silicon phase. Because of the quite strong SiH bond (3.3 eV) as compared to the Si-Si bond (2.4 eV) the hydrogen forms new bonding states deep within the Si valence band. As a consequence a recession of the valence band top edge occurs [8,9] and the valence band maximum shifts to lower energies with increasing hydrogen content. Further, the antibonding states associated with the Si-H

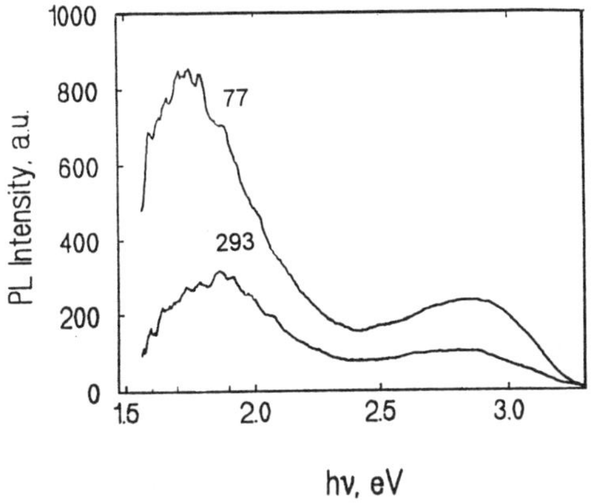

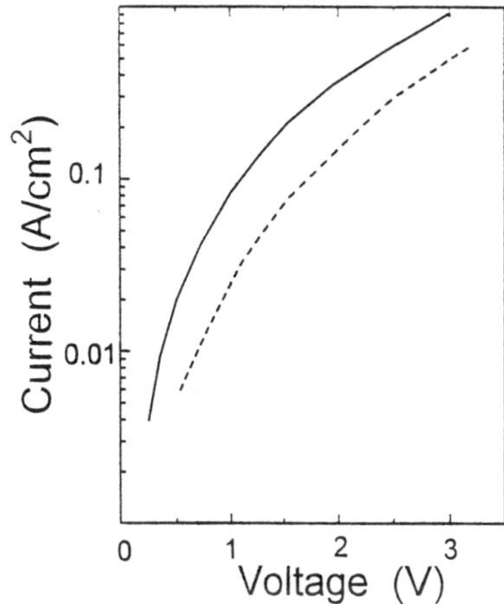

Fig.3. Photoluminescence spectra from Si:H film at room temperature and 77 K. Photoluminescence peak centered at 2.9 eV originates from the emission of substrate.

Fig.4. Second I-V characteristics of p-i-n structures based on nano-Si:H, solid lines - direct bias, dashed lines - reversed bias.

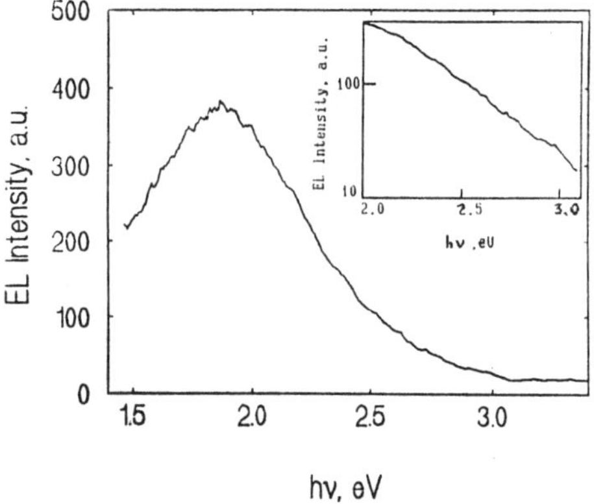

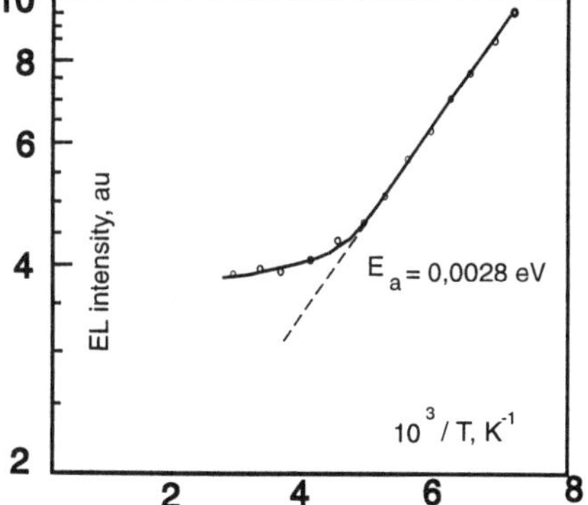

Fig.5. Electroluminescence from p-i-n structures based on nano-Si:H. Insert shows high energy wing of electroluminescence spectrum in semi-logarithmic dependence.

Fig.6. Temperature dependence of integral electroluminescence from p-i-n structure.

bonds lie one or very few tenths above the conduction band edge. Both effects contribute to the opening of the band gap and, according to Ashida and Fukuda [5], E_g reaches 2.0 eV (as observed here) at a hydrogen content of 30%. Other authors [6] find a hydrogen content of 20%.

Discussing now the photoluminescence we have to note the two very prominent aspects of it: (1) a fast kinetics (less than 30 nsec) as compared to that usually observed in a-Si:H (10 μsec) and (ii) the small Stokes shift of 0.1 to 0.2 eV. Further aspects are the weak temperature dependence of the PL intensity. All these aspects point to a direct recombination which cannot occur in the amorphous phase. We can explain these features, however, by considering the nanocrystals in terms of the quantum dot model, which is applicable to structures of sizes below around 2 nm.

The critical step in the application of the quantum confinement model is in our case to show that a bound state exists although there are no physical reasons to expect a large potential barrier between crystalline and amorphous Si regions. The barrier thus might amount essentially to the difference Δ between the respective energy gaps of a-Si:H and c-Si: $\Delta=E_g$(a-Si:H)-E_g(c-Si)=2.0 eV-1.1 eV=0.9 eV. Modifications of this value, for example due to the polarity of the Si-H bonds at the surface of the crystallites, are conceivable. The quantum mechanical treatment [10] shows, that decreasing the well size at a given well depth pushes the levels of the bound states up and out of the well. The ground state reaches the top of the well (and thus the level of the band edge) at the minimum well depth: $U_O=\pi^2\hbar^2/8ma^2$ ($\hbar$ - Planck's constant, m-electron mass, a-size of quantum dot, eg. of crystallite). We take a crystallite size a=1 nm, for m the electron rest mass, and find U_O=0.084 eV. For well depths, U, slightly larger than U_o the ground states lie within the wells but still near the band edge. This fact can explain the small Stokes shift observed in this work.

The proposed model fails in explaining the broad high energy band up to 3 eV observed in electroluminescence. These energies are far beyond the discussed band structure model. This luminescence might be due to recombination of hot carriers that are injected from the electrodes and accelerated in the high bias field. In fact, at a bias of $3xE_g/e$ and a p-i-n structure width of 100 nm we obtain a field of $5\cdot10^5 V\cdot cm^{-1}$. In addition, the intensity of this wide band, when plotted semi-logarithmically yields a linear dependency on the photon energy as expected [11,12] for such a type of luminescence.

SUMMARY

We have reported on the electro- and photoluminescence properties of highly hydrogenated amorphous silicon films that contain in a volume fraction of 0.3 crystalline Si particles in the diameter range up to 5 nm as confirmed by electron microscopy. Absorption measurements indicate a band gap of 2 eV for the amorphous matrix. The main (electro- and photo-) luminescence at around 1.9 eV, its small Stokes shift and the fast decay time point to a direct transition due to quantum wells that are represented by silicon nanocrystals with diameter of around 1 nm.

REFERENCES

[1] L.T.Canham, Appl.Phys.Lett.,**57**, 1046 (1990).
[2] P.D.Calcott, K.J.Nash, L.T.Canham, M.J.Kane and D.Brumhead, J. Phys.: Condens.Matter **5**, L91 (1993).
[3] M.S.Hybertsen, Phys.Rev.Letters, **72**, 1514 (1993).
[4] G.Allan, C.Delerue, and M.Lannoo, Phys.Rev.B, **48**, 7951 (1993).
[5] Y.Ashida, N.Fukuda, Solar Energy Materials and Solar Cells,**34**, 291(1994).
[6] K.Nakamura, K.Yoshino, I.Shimizu in Proceedings of the First World Photovoltaic Energy Conversion Conference, p.480, December 1994, Hawaii.
[7] Ch.Kittel, Einführung in die Festkörperphysik, 6 ed., (Oldenbourg, München, 1983), p.361.
[8] D.A.Papaconstantopoulos, E.NEconomou, Phys.Rev.B, **24**, 7233 (1981)
[9] L.Ley in The Physics of Hydrogenated Silicon II, edited by J.D.Joannopoulos and Lucovsky (Springer-Verlag, 1984), p.130.
[10] L.D. Landau and E.N.Lifshitz, Quantum Mechanics, (Nauka, Moscow, 1974, third edition), p.139.
[11] W.Haecker, Phys.Stat.Sol.(a), **25**, 301 (1974).
[12] P.J. Wolff, Phys.Chem. Soc. **16**, 184 (1960).

MICROSTRUCTURE OF THIN FILM PHOTOCONDUCTORS AND ITS CORRELATION WITH OPTICAL AND ELECTRONIC PROPERTIES

U. KLEMENT*, D. HORST**, F. ERNST*
*Max - Planck Institut für Metallforschung, Seestrasse 92, 70174 Stuttgart, Germany
**Institut für Netzwerk- und Systemtheorie, Universität Stuttgart, Pfaffenwaldring 47, 70550 Stuttgart, Germany

ABSTRACT

The objective of this work is to find a material to replace amorphous hydrogenated silicon used as photosensitive part in the "retina" of an "electronic eye". For that reason, ZnS, ZnSe, CdS and CdSe were chosen for investigations. Thin films, prepared by chemical vapour deposition, were characterized by transmission electron microscopy. The observed microstructures were correlated with the optoelectronic properties of these materials. CdSe was found to be the most promising material for our application. Hence, the influence of a dielectric interlayer and the effects of additional annealing treatments were analyzed for CdSe and will be discussed with respect to the optimization of the material.

INTRODUCTION

Unlike a CCD-camera the "electronic eye" will be based on thin film technique. Optical signal processing and neuronal nets will allow a fast and easy object recognition by massive parallel image processing. The main part of the electronic eye is the "retina", which is supposed to collect light intensities and colors. Similar to the human eye retina, an array of sensors will be constructed, in which each sensor contains a photosensitive resistor (PR) to collect light informations and a thin film transistor (TFT) to read out the information (Fig. 1). Large-area production is feasible, as thin-film technology can handle large-area microelectronics on substrates larger than the A4 format.

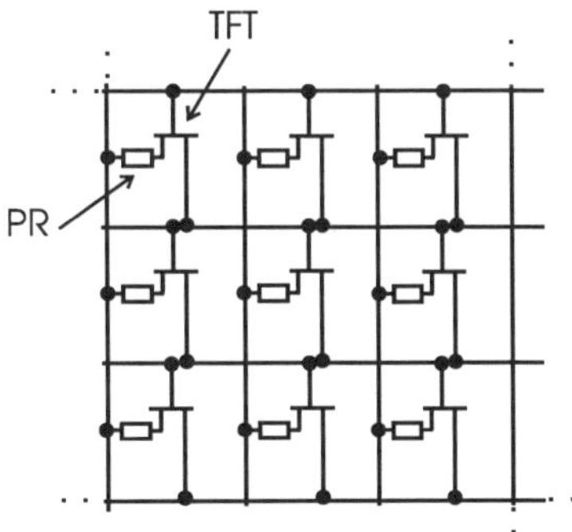

Fig. 1: Section of the retina - sensor array with photoresistor (PR) as light sensitive part and thin film transistors (TFT) for information read-out.

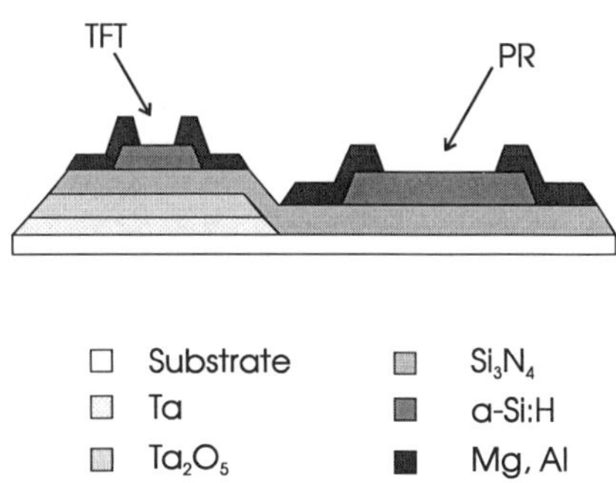

Fig. 2: Outline of an amorphous silicon thin film transistor (TFT) combined with photoresistor (PR).

Mat. Res. Soc. Symp. Proc. Vol. 452

Up to now amorphous hydrogenated silicon (a-Si:H) has been favored for application in the retina. The outline of an amorphous silicon TFT combined with photoresistor is shown in Fig. 2. Such a sensor can be processed in four steps [1]. However, photoconductivity of a-Si:H degrades when exposed to light (Staebler-Wronski effect) [2], which can have disadvantages for technical application. Hence, polycrystalline photosensitive materials are investigated with respect to their feasibility to replace amorphous hydrogenated silicon. Suitable materials have to operate at room temperature and must possess a spectral response in the visible range of the spectrum. Furthermore, long term stabilization has to be guaranteed. The properties of those new materials will determine the efficiency of the camera.

EXPERIMENT

Thin films of ZnS, ZnSe, CdS and CdSe, about 50 to 120 nm in thickness, were prepared on glass substrates by chemical vapour deposition (CVD). Furthermore, CdSe-films were deposited on glass with an additional dielectric SiO_2 interlayer. The microstructure was characterized with a JEOL 2000 FX transmission electron microscope (TEM). Cross sectional samples were prepared by ion milling. For optoelectronic investigations halogen light illumination is used. Postdepositional annealing treatments were performed i) for 1 h at 380°C under vacuum followed by ii) a two-step anneal for 30 min at 300°C in air and 250°C under vacuum, respectively.

RESULTS

Table I shows the photoconductive properties known from literature [3-6], like band gap energy, maximal spectral response and cut off wavelength of the chosen semiconductors. ZnS has a very narrow range of sensitivity, which is located at the ultraviolett end of the visible spectrum. ZnSe, in contrast, provides a wide range of sensitivity, but like CdS the maximal spectral response is found to be in the blue and green region, respectively. CdSe, is quite similar to a-Si:H, which has a band gap energy of 1.6 eV and a maximal spectral response around 550 nm. Furthermore, polycrystalline CdSe possesses a much higher charge carrier mobility than ZnSe or CdS.

Table I

Band gap energy E_g, maximal spectral response λ_{max}, cut off-wavelength $\lambda_{cut\ off}$ and charge carrier mobility μ_n in polycrystalline material for different semiconductors.

	E_g [eV]	λ_{max} [nm]	$\lambda_{cutt\ off}$ [nm]	μ_n [cm^2/Vs]
ZnS	3.65	330-370	398	-
ZnSe	2.60	470	1300	0.01-0.1
CdS	2.40	520	850	5
CdSe	1.70	750	1250	20-380

The dynamic range of a sensor array is limited by the current which can be conducted by the TFT ("ON-current"). This current can be raised by increasing the charge carrier mobility. In CdSe, charge carrier mobility is higher by a factor of 100-200 than in a-Si:H. Thus, the combination of CdSe-TFT and a-Si:H-photoconductor - which possess a high sensibility and a relatively low conductivity - leads to a high dynamic range of the sensor. But in this case photoconductor and TFT are not compatible in processing. Additional fabrication steps, like photolithographic structuring, would be required. Since, less fabrication steps are of technological interest, TFT and photoconductor need to be realized from the same material. Light sensitive resistors with CdSe can be fabricated with the same process as the TFTs [7-9]. Used as photoconductor, CdSe leads to an increase of the conductivity by a factor of 100. However, the homogeneity of the photoconductive properties in CdSe is not yet satisfying. But still, CdSe is the most promising material to replace a-Si:H in the electronic eye. Since many of the optoelelectronic properties of the materials are determined by the corresponding properties of the grain boundaries, we expect to achieve better optoelectronic properties after optimizing the microstructure. Since the coexistence of both phases - especially in CdSe -is frequently observed [10-12], there seems to be only a small energy difference between the cubic zincblende structure and the hexagonal wurzite structure in ZnS, ZnSe, CdS and CdSe. The microstructures of the investigated thin films is in each case formed by columnar growth and the grains, about 20-25 nm in width, always contain multiple twins. Fig. 3a and b show the microstructure of CdS and ZnSe thin films.

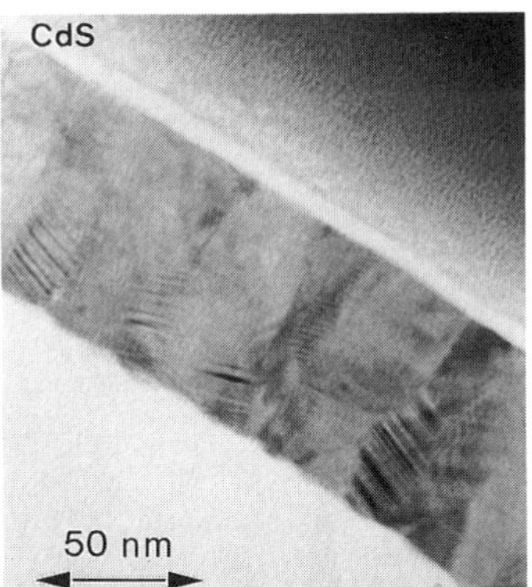

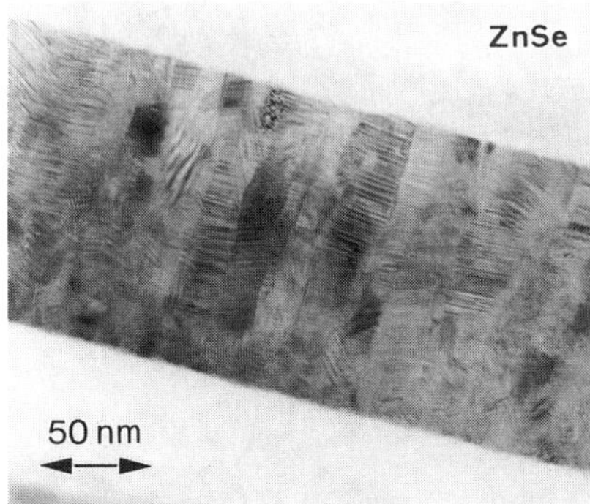

Fig. 3a and b: TEM micrograph of a CdS- and a ZnSe thin film.

Even though the grains have a columnar shape, they do not necessarily extend from back (substrate) to front (surface) of the thin film, i.e. there are also grain boundaries parallel to the film surface. This type of grain boundary is known to be disadvantageous for carrier transport. Being defects in the crystal structure, grain boundaries introduce allowed energy levels in the forbidden band gap of the semiconducting material and act as effective recombination centers [13]. Consequently, the lateral dimension of the grains within the film must be large compared to minority carrier diffusion lengths to avoid significant loss in current output. In other words, the distance from front to back side of the film should be shorter than the distance to the lateral grain boundaries. According to investigations in silicon the character of the grain boundary is as important: random, large-angle grain boundaries are strong recombination centers, but their electrical activity can vary considerably. Small-angle grain boundaries ($\Theta < 10°$) which consist of an array of dislocations still show an efficient current degradation of up to 20 to 30%. However, coherent twins and other low-energy (near-) CSL boundaries are usually not active or are only weakly active [14,15]. Hence, a more desirable microstructure of a polycrystalline thin film consists of large columnar grains extending across the film thickness. Existing grain boundaries are preferentially CSL-boundaries of low energy.

Optimizing the microstructure by increasing the grain size and the photoconductive response [6], CdSe thin films were annealed for 1 hour at 380 °C under vacuum (R). In some cases a two-step heat treatment for 30 min at 300°C (air) and 250°C (vacuum) (T), respectively, followed the previous one. Shown in Fig. 4a and b, grain sizes in the range of the film thickness (70 - 100 nm) are achieved. Similar results were obtained by Norian [16] investigating the effect of annealing on CdSe-TFTs. Texture analysis revealed a weak <111>-fibre texture after both heat treatments (Fig. 5).

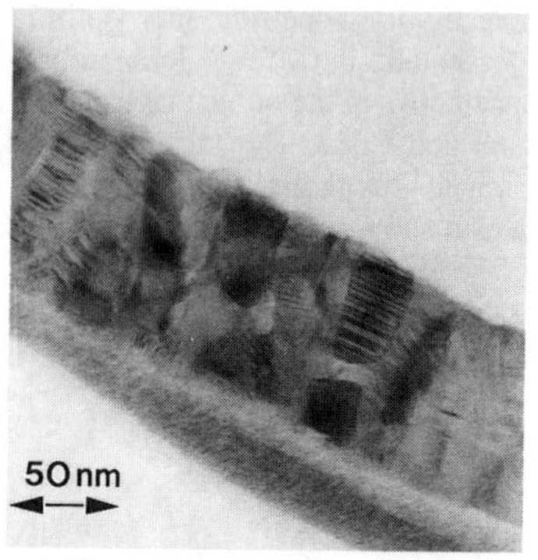

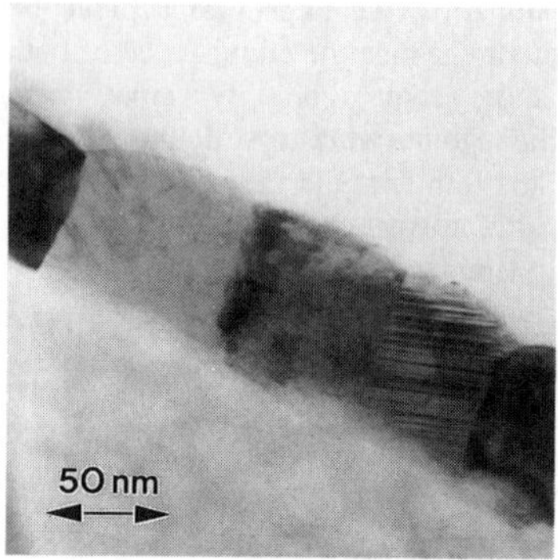

Fig.4a and b: TEM micrographs of CdSe thin a) as prepared and b) film after annealing for 30 min at 380°C (vacuum)

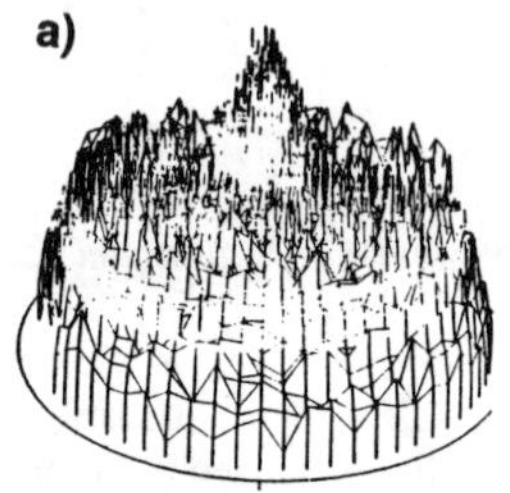

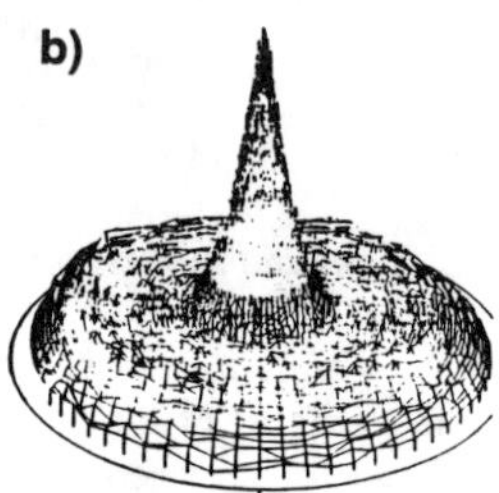

Fig. 5a and b: Texture analysis of the CdSe thin films: a) as prepared b) after annealing for 1 hour at 380°C under vacuum

Neither deposition of a dielectric SiO_2-interlayer nor the second annealing treatment change the grain boundary configuration, but the optoelectronic properties are influenced. Fig. 6 shows conductivity versus illumination for CdSe, as prepared and annealed like described above (R and R+T). As can be seen, the first heat treatment already raises conductivity. We attribute this effect to occurring grain growth. The second two-step treatment leads to further even more pronounced increase in conductivity. Since no visible change in microstructure is observed, the longer annealing time and recovery of point defects are possible explanations but a strong influence of the annealing in air (surface of the crystallites are oxidized along grain boundaries [8]) is actually expected. Depth profile analyses (XPS) of the annealed samples are in progress.

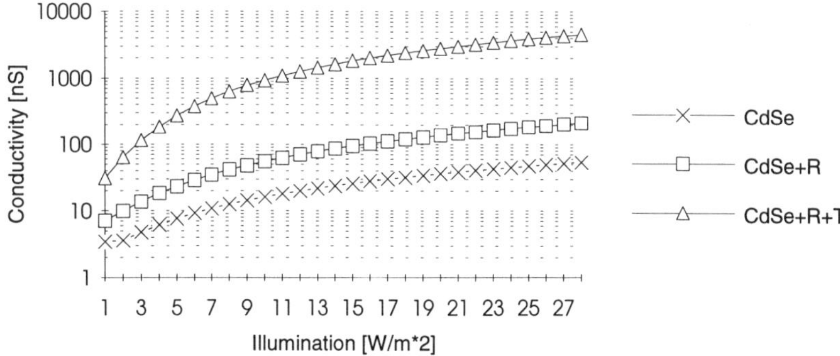

Fig. 6: Conductivity versus illumination for CdSe after different annealing treatments

Photoconductive properties [17] described by

$$G = a \cdot E^b + c \qquad (1)$$

of CdSe samples are listed in Table II in comparison to a-Si:H. *G* stands for the photoconductivity, *E* for the normalized illumination and *c* represents the dark current value. The conductivity coefficient *a* is supposed to be between 10 and 100 nS. The relative sensibility *b*, which is a measure of linearity, has to be maximal or at least greater than 0.5.

The photoconductive properties of the material are depending on the annealing treatment. Table III shows the parameters *a* and *b* in dependence of the annealing treatment (C: depositing contacts, A: annealing in air, V: annealing under vacuum).

Table II
Parameters a and b to characterize photoconductive properties of a-Si:H and CdSe after different annealing treatments

	a *[nS]*	*b*
a-Si:H	11.2	0.89
CdSe	10.4	0.48
CdSe+R	35.9	0.52
CdSe+R+T	562.0	0.67

Table III
Parameter a and b in dependence of the processing for different CdSe-samples

	annealing treatment	*a* *[nS]*	*b*
sample 1	- - C - -	111	0.52
sample 2	A - C - -	37	0.92
sample 3	V - C - -	282	0.53
sample 4	AVC - -	249	0.87
sample 5	AVCA -	116	0.89
sample 6	AVCAV	194	0.87

Already after a first attempt to optimize the microstructure, the photoconductive properties of annealed CdSe thin films are comparable to a-Si:H. With CdSe, we see high potential to achieve properties superior to a-Si:H. Further experiments will aim to achieve a stronger texturing of the material by use of crystalline substrates like Si-wafers and the application of higher annealing temperatures. Another option for improvement is the use of other dielectric interlayers. The influence of annealing in air will be determined by XPS-analyses.

CONCLUSION

We have shown that CdSe has the potential to replace amorphous silicon as the light sensor in the retina of an electronic eye. Prerequisites like the application as both, photoconductor and thin film transistor can be accomplished with this material, too. Already at the present state of microstructural optimization optoelectronic properties are comparable to a-Si:H. Further optimization of the microstructure, i.e. grain growth and texturing by use of crystalline substrate seem to be promising.

ACKNOWLEDGMENTS

We acknowledge the financial support of this project (contract number 01 M 3025 D) by the Bundesminesterium für Bildung, Forschung und Technologie.

REFERENCES

1. D. Horst, E. Lüder, M. Habibi, T. Kallfass and J. Siegorder, *Proceedings of the 2nd Symposium on Thin Film Transistors Technologies*, Miami Beach 1994, Electrochemical Society, Inc., **Vol. 94-35**, pp. 381-391.
2. C. Wronski and D. Staebler, Appl. Phys. Lett. **31**, 292 (1977).
3. Landolt-Börnstein, *Zahlenwerte und Funktionen aus Naturwissenschaft und Technik, Neue Serie, Gruppe III: Kristall- und Festkörperphysik. Band 17d*, (Springer Verlag, Heidelberg, New York, Tokyo 1984), p. 162.
4. H.-J. Lewerenz and H. Jungblut, *Photovoltaik, Grundlagen und Anwendungen*, (Springer Verlag, Berlin, Heidelberg, New York, Tokyo, 1995), p. 210 and 330.
5. H.J. Möller, *Semiconductor for Solar Cells*, (Artec House, Boston, London 1993), p. 285.
6. R.A. Mickelsen in *Polycrystalline and Amorphous Thin-Films and Devices, Materials Science Series* ed.: L.L. Kazmerski (Academic Press, New York, 1980), pp. 209-227.
7. A. Van Calster, A. Vervaet, I. De Rycke, J. De Baets and J. Vanfleteren, Journal of Crystal Growth **86**, 924 (1988).
8. J. Spachmann, E. Lüder, T. Kallfaß and W. Otterbach, in *Polycrystalline Semiconductors, Springer Proceedings in Physics, Vol. 35*, eds.: J.H. Werner, H.J. Möller and H.P. Strunk (Springer Verlag, Berlin, Heidelberg, 1989), pp. 262-267.
9. H.F. Heek, Solid State Electronics, **Vol. 11**, 459 (1968).
10. F. Raoult, B. Fortin and Y. Colin, Thin Solid Films **182**, 1 (1989).
11. L. Däweritz and M. Dornics, Phys. Status Solidi A **20**, K37 (1973).
12. H.M. Naguib, H, Nentwich and W.D. Westwood, J. Vac.Sci. Technol. **16(2)**, 217 (1979).
13. M.A. Green, *Solar Cells: Operating Principles, Technology and System Application, Series in Solid State Electronics*, (Prentice Hall Inc., Englewood Cliffs, New York 1982), p. 188.
14. H.J. Möller, *Semiconductor for Solar Cells*, (Artec House, Boston, London 1993), p. 254.
15. S. Martinuzzi, in *Polycrystalline Semiconductors, Springer Proceedings in Physics, Vol. 35*, eds.: J.H. Werner, H.J. Möller and H.P. Strunk (Springer Verlag, Berlin, Heidelberg 1989), pp. 148-157.
16. K.H. Norian, Thin Solid Films **47**, 195 (1977).
17. D. Horst, Doctoral thesis, Universität Stuttgart, 1996.

Part XII

Preparation, Characterization, and Applications of Polycrystalline Silicon and Other Crystalline Systems

CRYSTALLIZATION AND EVALUATION OF NON-STOICHIOMETRIC SILICON-CARBON FILMS FOR HETERO THIN-FILM TRANSISTORS

Kwangsoo CHOI and Masakiyo MATSUMURA
Department of Physical Electronics, Faculty of Engineering, Tokyo Institute of Technology, 2-12-2 O-okayama, Meguro-Ku , Tokyo 152, JAPAN, matumura@pe.titech.ac.jp

ABSTRACT

Experimental works have been reviewed on poly-Si/poly-SiC_X hetero TFTs aiming at extremely low off-current even under intense light illumination conditions with reasonable field-effect mobility. The results indicated that the hetero TFT having the stacked poly-Si/poly-SiC_X layers is promising as a switching device in matrices

INTRODUCTION

It is well known that the conversion efficiency of amorphous-silicon (a-Si) based solar-cells has been improved dramatically by introduction of a-Si/a-SiC_X hetero structures. One role of the newly-introduced wide-bandgap a-SiC_X films in the cell is to suppress effectively the leakage current to a lower level in the dark, resulting in an enlarged open-circuit voltage, since minority carrier density in the contact regions is inversely proportional to the exponent of the bandgap, E_g. The wide-E_g material will be also useful in reducing the leakage current of the other a-Si based devices, such as thin-film-transistors (TFTs). The a-Si/a-SiC_X hetero TFTs and also the a-SiC_X TFTs, however, are not interesting because the most serious problem for the present a-Si TFTs is not in the high leakage (off-) current in the dark, but in the low field-effect mobility together with the photo-enhanced off-current. Furthermore, there will be no room for mobility enhancement by using an a-SiC_X film with lower mobility than the a-Si film. The hetero structures and wide-E_g materials, however, are very interesting for the poly-Si based TFTs, where their mobility, i.e., the on-current is sufficiently high but their off-current is extremely high for matrix applications.

This paper reviews our preliminary works on crystallization and evaluation of the non-stoichiometric SiC_X film, and on its application to hetero TFTs.

FILM PROPERTIES AND DESIRABLE HETERO STRUCTURE

Cross sections of the proposed poly-Si/poly-SiC_X hetero TFTs [1,2,3] are shown schematically in Fig.1, and desirable distributions of localized states in the poly-SiC_X film are shown in Fig.2 for the hetero TFTs shown in Fig.1. There are two ways of applying the poly-Si/poly-SiC_X hetero structure into TFTs. Although the operation principles are different between these two devices, the research target of both devices is the same, i.e. the low off-current even under intense visible light illumination conditions. Thus the distribution of the localized states in the poly-SiC_x film is the most important parameter for selecting the device structure.

The first TFT has hetero junctions across the current flow similar to hetero bipolar transistors [4]. The source and drain are of poly-SiC_X with wide E_g, and the channel is of poly-Si with high mobility. Dense electrons induced in the poly-Si channel by the field-effect of the positive gate voltage can move along it with high mobility, and the n-doped poly-SiC_X film is

Mat. Res. Soc. Symp. Proc. Vol. 452 © 1997 Materials Research Society

highly conductive. Thus the on-current is as high as that of the conventional poly-Si TFTs. Since hole density in the source and drain of n-doped poly-SiC_X is very low by the desirable effects of wide E_g, the off-current is reduced to much less than that of the poly-Si TFTs. Furthermore, by the help of small light absorption coefficient of the poly-Si film compared with that of the a-Si film, the off-current is kept low enough even under illumination conditions.

There should be a lot of discussions on the optimum E_g value for the poly-SiC_X film, since semiconductor quality of the SiC_X films depends strongly on the C content, x, i.e., E_g and also on the crystallization method. The ideal distribution of the localized states in the poly-SiC_X film is that only the energy of the states moves, as shown in Fig.2(b), upwards with the conduction band edge for those above the midgap and downwards with the valence band edge for those below the midgap, compared with the state distribution in the poly-Si films shown in Fig.2(a) for reference. Since the minority carrier density is proportional to $exp(-E_g/kT)$, where T is ambient temperature, a 120meV ($=(2kT/q)log(10)$) increase of E_g from that (E_g=1.1eV) of the poly-Si film seems sufficient for reducing the off-current level of the hetero TFTs by two orders in magnitude, i.e., to as low as that of the conventional a-Si TFTs. However, the generation-recombination current of holes born in the high-field region near the drain is much more than the diffusion current of holes flowing out from the n-type drain for the poly-Si TFTs. And the generation rate of holes is proportional to intrinsic carrier density. In order to decrease the field-enhanced generation rate by two orders in magnitude, i.e., the tunneling probability of carriers trapped at the localized states [5,6], the high-field region should not be in the poly-Si channel but in the poly-SiC_X drain region having E_g of 250-300meV wider than that of the poly-Si film. This means that the gate electrode should extend into the drain in order that the n-doped poly-SiC_X layer near the insulator interface is converted to p-type as shown schematically in Fig.1(a), under the large drain voltage, V_d, and large negative gate voltage, V_g, conditions. Then, the localized state distribution shown in Fig.2(c), where the states are dense below the midgap, is not preferable since the poly-SiC_X interface is difficult to be converted to p-type by the effects of negative V_g. The state distribution should be as shown in Fig.2(b) for the high performance hetero TFT having the cross section shown in Fig.1(a).

The second device has the hetero junction along the current flow similar to high electron mobility transistors [7]. Since the poly-SiC_X upper layer should have no serious influences for electron accumulation and transport in the channel formed by the poly-Si bottom layer, there should be a few localized states above the midgap in the poly-SiC_X film. For low off-current, the poly-SiC_X film should act to pin the energy band of the poly-Si layer near the intrinsic condition. This requirement can be satisfied if the poly-Si film is very thin and if there are dense localized states below the midgap in the poly-SiC_X film. Thus the state distribution shown in Fig.2(c) is preferable. The poly-Si channel is not converted to accumulate holes even under the negative V_g conditions, and V_d is not localized near the drain but spreads uniformly over the channel. Since the field is uniform and weak along the channel, the carrier generation is not enhanced by the field. The generation rate of carriers is thus kept very low, resulting in the low off-current. There is another advantage of this structure. Namely, the poly-Si film is very thin and the poly-SiC_X film has a short carrier life-time. Thus the photo-generated off-current will be extremely low.

CRYSTALLIZATION AND EVALUATION OF SiC_X FILM

Neither conventional solid-phase crystallization (SPC) method nor chemical-vapor-deposition (CVD) method seems suitable for forming the poly-SiC_X film since there is only one

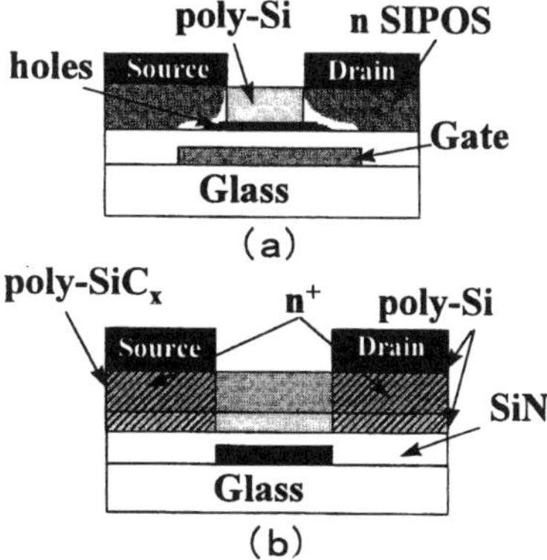

Fig.1 Schematic cross-sectional views of the proposed poly-Si/poly-SiC$_X$ hetero TFTs.

(a) A wide-E_g source and drain TFT. Holes should be accumulated at the wide-E_g drain/insulator interface, and the depletion region should be formed inside the drain in order to suppress field-enhanced thermal-generation of carriers.

(b) A wide-E_g overlayer TFT. Band of the ultra-thin poly-Si channel is pinned by dense localized states below the midgap of the poly-SiC$_X$ overlayer.

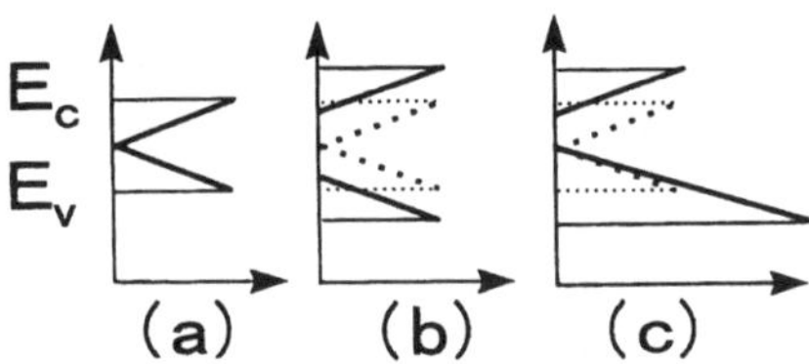

Fig.2 Possible localized states distribution for the poly-SiC$_X$ films.

(a) The distribution of the poly-Si (x=0) film for reference.

(b) The shifted distribution of the localized states from that of the pure poly-Si film.

(c) The distribution having dense localized states below the midgap.

thermally stable phase in the Si/C mixture, i.e., silicon-carbide with equal numbers of Si and C atoms. There are several structures in the SiC crystal but their E_g are more than 2.2eV. Since we need much narrower E_g, C content, x, should be much less than unity. Introduction of a small amount of C into the Si film by the SPC and CVD methods, however, will cause precipitation of small SiC grains in the poly-Si network, resulting in inferior electronic properties without widening total E_g. In order to obtain the non-stoichiometric, uniform but crystallized SiC_X film with E_g of about 300meV greater than that of the pure Si film, a highly non-isothermal process is preferable. Thus we had studied the excimer-laser crystallization method for the SiC_X film since the several 10nm-thick film can be crystallized within several 10ns by this method.

The a-SiC_X films were deposited on glass substrates as a starting material by the low-temperature CVD method using a mixture of disilane (Si_2H_6) and allene (C_3H_4). The deposition temperature, pressure and disilane flow rate were fixed at 460°C, 1Torr and 2sccm, respectively. The growth rate was increased only slowly with adding allene with its flow rate R as shown in Fig.3. x increased linearly with R, and was about 0.2 at R=0.5sccm and about 0.4 at R=1sccm. Increment, ΔE_g, of E_g for the a-SiC_X film with respect to that of the pure a-Si film is also shown in the figure. ΔE_g increased steadily with R, i.e., with x, and was about 300meV at R=0.5sccm (x=0.2) and about 500meV at R=1sccm (x=0.4). Since an excess increase in R is expected to deteriorate seriously semiconductor properties of the poly-SiC_X film, we selected R=0.5sccm (x=0.2) as the standard flow rate.

A 50-nm thick a-SiC_X film on glass substrate was dehydrogenated at 550°C for 1 hr in a high vacuum furnace in order to prevent the H eruption during the excimer-laser crystallization process. It was then crystallized using single shot ArF excimer-laser light (λ=193nm) pulse with various energy densities, E. The crystallized film was post-hydrogenated in atomic H ambient at 400°C for 20min to terminate dangling bonds before measuring its resistivity, ρ. Figure 4 shows ρ as a function of E for several values of R. ρ decreased abruptly at E=100~150mJ/cm^2 and then was saturated at low values. These low ρ films were confirmed to be crystallized by observing the ring RHEED patterns. The threshold E value at which a rapid decrease in ρ occurred increased with R, and the saturated ρ value also increased with R. ρ was about 3 orders in magnitude higher for the x=0.2 film than that for the x=0 film. The activation energy of ρ was 520meV for the x=0 film and 620meV for the x=0.2 film. These values seem correspond with a 200-300meV increase of ΔE_g. The film had very low photo-sensitivity, as expected.

Raman spectra and X-ray diffraction spectra were measured for the poly-SiC_X films, but their crystal structure and uniformity of composition were not clarified since the films were too thin. Thus spatial variation of x on the film surface was measured by μ-AES for the x=0.4 film, but there was no clear variation. Since the beam diameter was as large as 100nm, it can be concluded that there were no Si or SiC grains having a size as large as 100nm. In order to make clear the fact that the Si or SiC grains, even if they exist, are extremely small, that is, the film is nearly uniform, the film was pre-etched using a HF+HNO_3 solution which can remove selectively Si grains but not SiC grains, and depth profile of x was evaluated by sputtering AES. A schematic cross-sectional view of the etched film is shown in Fig.5(a). Since average volumes of hypothetical SiC and Si grains are approximately the same for the x=0.4 film, if they are, the etched surface will be occupied by dense SiC grains and the film surface will have an x value very near to unity, i.e., much higher than the average value. After non-selective sputtering, however, these small surface-precipitated SiC grains will disappear as shown in Fig.5(b), and x will be reduced to the average value of 0.4. The characteristic distance for the x saturation should be equal to an average size of the hypothetical SiC grain. The measured depth profile showed,

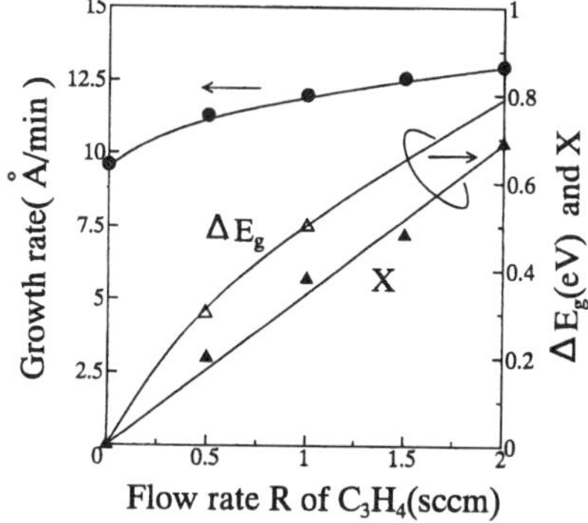

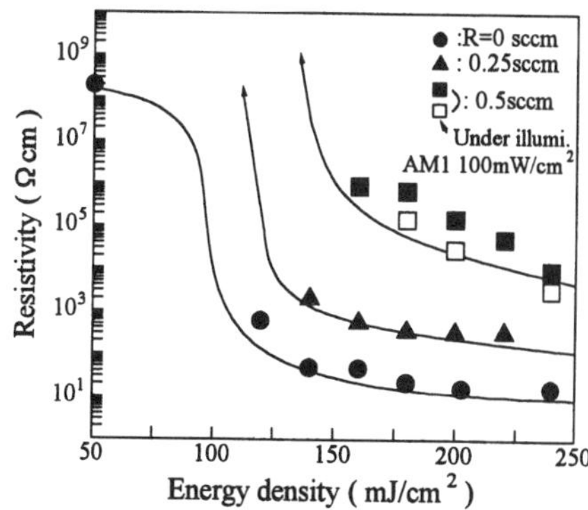

Fig.3 Deposition rate, C content, x, and incremental E_g of the a-SiC_X film with respect to those of the a-Si film as a function of allene flow rate.

Fig.4 Resistivity of the SiC_X film as a function of energy density of ArF excimer-laser light.

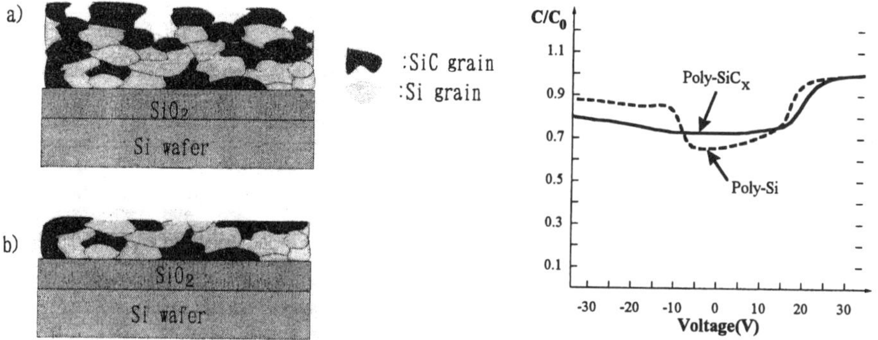

Fig.5 Schematic view of the poly-$SiC_{0.4}$ film etched selectively by a $HF+HNO_3$ mixture.
(a) Before sputtering
(b) After sputtering

Fig.6 MOS C-V characteristics for the poly-Si and poly-$SiC_{0.2}$ films.

however, only a small bump of x within 3nm from the surface, where O was also observed. Even if we do not take the surface effect into account, this result indicates that the x fluctuation, that is, the hypothetical SiC grain size was less than 3nm. Therefore, the excimer-laser crystallization method seems able to form the non-stoichiometric crystallized SiC_x film with uniform C concentration.

MOS capacitance-voltage characteristics are shown in Fig.6 for the poly-Si film and the poly-$SiC_{0.2}$ film. When large positive V_g was applied to the substrate (gate), dense electrons were accumulated near the SiO_2/semiconductor interface, resulting in a large capacitance. The capacitance took a small value around V_g=0 since both electrons and holes were not induced near the interface. When negative V_g was applied, the capacitance took a large value in the case of the poly-Si film since dense holes were accumulated near the interface. In the case of the poly-$SiC_{0.2}$ film, however, the capacitance did not increase but remained at a small value. These results can be interpreted that there are dense localized states near the midgap or below the midgap in the poly-$SiC_{0.2}$ film, and that they keep the $SiC_{0.2}$ film to n-type. The hetero TFTs having the cross section shown in Fig.1(a) are difficult to be operated with good performances, and the poly-Si/poly-$SiC_{0.2}$ stacked hetero TFT shown in Fig.1(b) is preferable.

Poly-Si/Poly-SiC_x STACKED HETERO TFT

The hetero TFT shown in Fig.1(b) was fabricated by using the process steps shown in Fig.7. After the Cr gate was formed on glass, the 200nm-thick SiN_x film, the 20nm-thick a-Si film and the 30nm-thick a-$SiC_{0.2}$ film were deposited by the CVD method. After dehydrogenation at 550°C for 1hr, 3 shots of ArF excimer-laser light pulse were irradiated from the top to crystallize the stacked films. And then, the 10nm-thick phosphorus-doped a-Si film was deposited. XeF excimer-laser light (which can transmit glass substrate) was then irradiated from the back side of the glass by using the Cr gate pattern as a mask in order to dope the Si and $SiC_{0.2}$ layers outside the channel region to n type. After the post-hydrogenation at 400°C for 20min, pattering the active layers and opening the contact holes, Al was evaporated to form the source and drain electrodes. Finally, the n^+ a-Si film remaining on the channel region was etched away selectively by using largely different etching rates between the n-type a-Si and undoped poly-$SiC_{0.2}$ films.

The depth profile is shown in Fig.8 for the stacked Si/$SiC_{0.2}$ film on the thermal oxide. A dotted curve is for the as-deposited film. x was changed sharply at the a-Si/a-$SiC_{0.2}$ interface. The profile is also shown in the figure for the laser-crystallized film by a solid curve. There was no clear mixture or inter-diffusion, as desired, between the Si film and the $SiC_{0.2}$ film by 3 shots of excimer-laser light irradiation.

Semi-logarithmic I_d-V_g characteristics are shown in Fig.9 for the fabricated hetero TFT. Channel width and length were 90μm and 22μm, respectively. The TFT showed the relatively high on-current and the low off-current. Current increase under negative V_g conditions seems come from the fact that the poly-Si film was too thick to be pinned its energy band by the poly-$SiC_{0.2}$ film. The on/off current ratio was more than 10^7. Current was increased slightly even by direct illumination of intense sun-light with 10k lux to the channel as shown by a dotted curve in the figure. The electron field-effect mobility calculated from the slope in the linear I_d-V_d region was about 2.2cm^2/Vs.

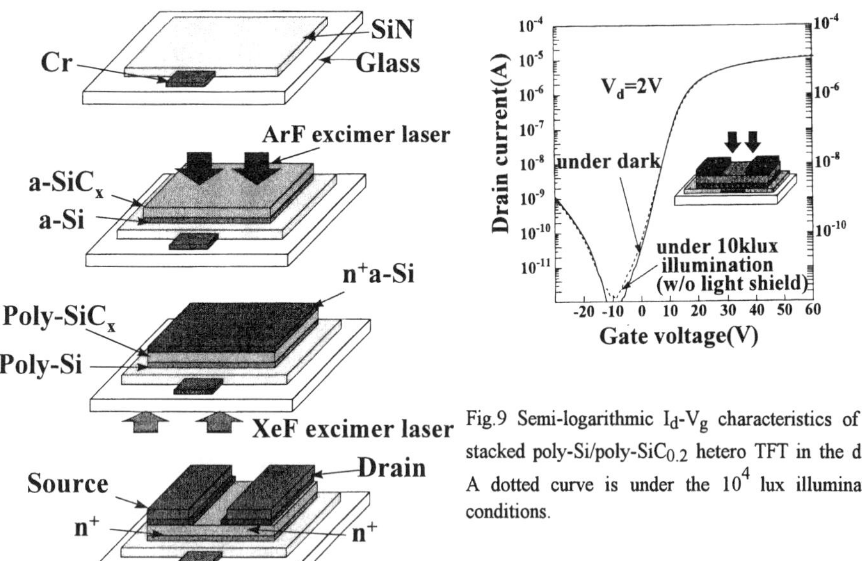

Fig.9 Semi-logarithmic I_d-V_g characteristics of the stacked poly-Si/poly-$SiC_{0.2}$ hetero TFT in the dark. A dotted curve is under the 10^4 lux illumination conditions.

Fig.7 Stacked Hetero TFT fabrication process.

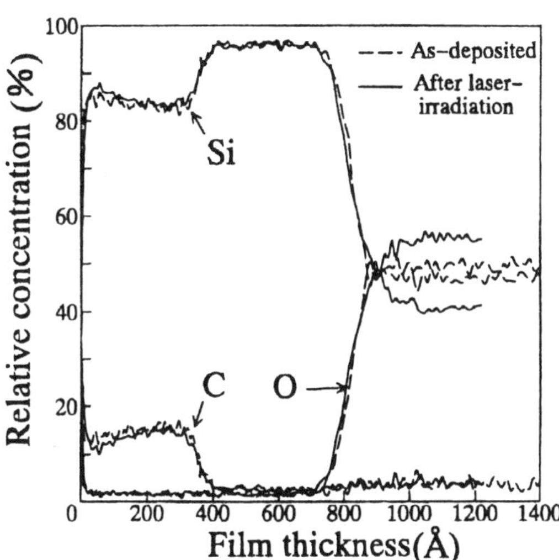

Fig.8 Depth profile of the stacked Si/$SiC_{0.2}$ films on thermal oxide before and after the excimer-laser crystallization. The Si film and the $SiC_{0.2}$ film are 20nm-thick and 30nm-thick, respectively.

CONCLUSIONS

We have proposed two types of hetero TFTs having the non-stoichiometric poly-SiC_X film. The poly-SiC_X film has been crystallized by the excimer-laser annealing method to avoid the SiC micro-crystal formation. Since the film has dense states below the midgap, the hetero TFT having the stacked poly-Si/poly-$SiC_{0.2}$ film has been concluded preferable. This type of hetero TFT has been fabricated and its performances were evaluated. The results showed that the hetero TFT is promising for matrices since it has the reasonable on-current and the satisfactorily low off-current even under strong light illumination conditions.

REFERENCES

1. K.S.Choi, Y.Uchida and M.Matsumura: Jpn. J. Appl. Phys. 34, 349 (1995).
2. K.S.Choi, Y.Uchida and M.Matsumura: Jpn. J. Appl. Phys. 35,1648 (1996).
3. K.S.Choi, Y.Uchida and M.Matsumura: 1996 Int'l Display Research Conf., 17 (1996).
4. H.Kroemer, Proc. IRE, 45, 1535 (1967).
5. J.G.Fossum, A.Ortiz-Conde, H.Shichijo and S.K.Banerjee: IEEE T-ED, 32, 1878 (1985) .
6. S.K.Madan and D.A.Antoniadis: IEEE T-ED, 332, 1518 (1986) .
7. R.Dingle, H.L.Stormer, A.C.Grossard and W.Wiegmann, Appl. Phys. Lett., 33, 665 (1978).

EXCIMER-LASER CRYSTALLIZATION OF SILICON FILMS: NUMERICAL SIMULATION OF LATERAL SOLIDIFICATION

Vikas V. Gupta, H. Jin Song, and James S. Im

Department of Chemical Engineering, Materials Science, and Mining Engineering, Columbia University, New York, New York 10027

ABSTRACT

We have utilized a recently developed transient two-dimensional model for simulating localized beam-induced melting and solidification of thin silicon films on SiO_2. Specifically, by tailoring the lateral beam profile, we simulate those situations that are encountered in the artificially-controlled superlateral growth (ACSLG) method, in which various techniques are utilized to irradiate the sample in preselected regions of a silicon film. The spatially and temporally localized character of heating is simulated by introducing a time-dependent two-dimensional heat-source function. The evolution of melt-creation and ensuing solidification is studied as a function of incident energy density and film thickness. The results show two distinct types of behavior as a function of incident energy density: at low energy densities, partial melting and predominantly vertical solidification occur; while at high energy densities, complete melting of the irradiated portion of the film is followed by rapid lateral solidification.

INTRODUCTION

Excimer-laser crystallization of thin Si films on glass substrates is being acitvely investigated for thin-film transistor applications. Much of the previously conducted investigations consist of experimentally inclined effort, and proportionally less work has been devoted to a quantitative analysis of the process. Previous experimental and theoretical studies on the relevant phase-transformation scenarios have definitively revealed that far-from-equilibrium conditions can readily prevail and that a two-dimensional character—in terms of heat flow and transformations—is integral to the process [1,2]. In part, such scenarios are introduced as a consequence of the fact that the films are very thin and that the films rest on top of an SiO_2 surface; the chemical stability and amorphous nature of SiO_2 make the surface a non-participating entity as far as the interface-led transformation is concerned.

Although there exist a number of numerical models that can simulate various types of energy-beam-induced melting and solidification of surfaces and thin films [3,4,5,6,7], close examination of these existing models reveals that they lack various capabilities and features that are needed in order to effectively accommodate and properly simulate the situations that are encountered in excimer-laser-induced melting and solidification of thin Si films on SiO_2. In this paper, we utilize the recently developed two-dimensional transient numerical model [8] to simulate the melting and solidification encountered in ACSLG techniques: Localized complete melting of the film via irradiation with spatially confined beam, and the ensuing lateral solidification, which is initiated from the unmelted regions of the film and proceeds into the completely molten area. We also extract technologically and scientifically relevant details involved in the phase transformation induced by the ACSLG technique.

NUMERICAL MODEL

In addition to being two-dimensional and properly handling the localized heat consumption or generation at the solid-liquid interface, the model includes (1) the interface response function, (2) the proper treatment of the nonequilibrium conditions that occur (i.e., undercooling of the interface and supercooling of molten Si), and (3) the tracking of the liquid-solid interface, including treatment of the separation and merger of solid regions, and (4) the nonparticipating nature of the interface between liquid Si and SiO_2 (i.e., the oxide interface does not initiate regrowth nor does it act to effectively catalyze nucleation of solids).

The spatial and temporal discretization and iterative computation of the following heat-transport equation in a manner consistent with the finite-difference method, and utilizing the Alternate-Direction Explicit (ADE) scheme [9] form a basis of our model. The equation contains two source terms: one that deals with thermalization of the deposited laser energy, and another that treats the localized consumption (during melting) and generation (during solidification) of the latent heat at the solid-liquid interface—i.e.,

Mat. Res. Soc. Symp. Proc. Vol. 452

$$C_p(\phi,T)\frac{\partial T}{\partial t} = \frac{\partial\left(K(\phi,T)\frac{\partial T}{\partial x}\right)}{\partial x} + \frac{\partial\left(K(\phi,T)\frac{\partial T}{\partial y}\right)}{\partial y} + S_1(t) + S_2(t) \quad (1)$$

with the *initial condition* T(0) = preheat temperature, and insulating boundary conditions at the top and the sides, where C_p is the heat capacity, *K* the thermal conductivity, S_1 the absorbed laser energy heat source function, *f* the phase-identifier function, and S_2 the latent heat function.

We utilize the temperature, phase, and vertical- and lateral-position-dependent heat source function in order to model the position-dependent energy deposition process that occurs when a spatially tailored beam, whose energy density may be high enough to initiate melting of the irradiated portion of the film, is incident on the film surface.

The model further incorporates the interface response function (IRF) — i.e., systematic variation of the interface velocity as a function of the interfacial undercooling and overheating — at the phase boundary. (A linearized IRF with the proportionality constant of 6.7 cm/sK was used.) A similar approach was taken previously in the one-dimensional model that was developed for analyzing pulsed laser melting and solidification of Si surfaces [6,10]. Further details of the model can be found in reference [8].

SIMULATED SAMPLE CONFIGURATION

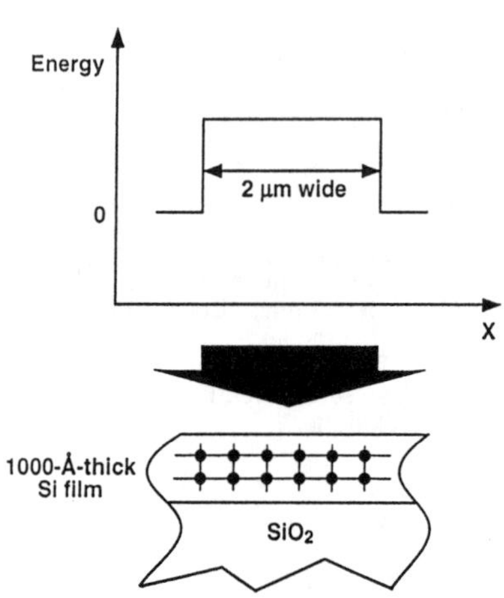

Figure 1: A schematic of the sample configuration and the lateral beam profile. The pulse duration of the excimer laser 36ns.

We analyze the thermal response of a thin Si film on SiO_2 (Fig. 1) when it is irradiated with a laterally tailored incident beam so as to induce spatially confined melting and solidification of the film. Such a situation, in which complete melting of the Si film is induced at — *and only at* — predesignated regions, forms the basis of a new category of ELC processes collectively referred as the artificially-controlled superlateral growth method (ACSLG) [11].

In order to simplify the situation, we have carried out simulations corresponding to melting and solidification of thin crystal Si films, at various energy densities, using a square beam profile whose beam width is fixed at 2 μm. Specifically, the following values are used in the calculation: a node size of 20 nm was used for the Si film and energy densities of 500 mJ/cm^2 and 800 mJ/cm^2. As to the specific details of the simulation, we have used the grid of 250 by 5 nodes for representing the Si film, 250 by 55 nodes for representing the substrate. A variable node scheme was used in order to reduce the total computation time without sacrificing the accuracy of the results. The sample configuration that was utilized for the simulation consisted of a 1,000-Å-thick Si film on oxide.

RESULTS

The simulation results reveal that there are, as shown in Figs. 2 and 3, two distinct types of solidification behavior that are observed depending on the energy density of the incident beam.

Figures 2a and 2b show the evolution of the solid-liquid interface during melting and solidification, respectively, when the film is irradiated at 500 mJ/cm^2. The results are typical of low-energy-density beam-induced melting and solidification and can be characterized as essentially being vertical and one-dimensional. It is only near the edges of the irradiated portion (i.e., within the thermal diffusion distance) where the two-dimensional aspect of heat flow and phase transformation are manifested. This scenario is obtained for all irradiation conditions that lead to partial melting of the film. Typical vertical regrowth velocities correspond to approximately 2m/s.

Figures 2c and 2d on the other hand, show the evolution of the solid-liquid interface during melting and solidification when the film is irradiated at the higher energy density of 800 mJ/cm^2. Here, the evolution of the interface can be described as being mainly vertical during the melt-in period but is clearly *lateral* during solidification. The same scenario is obtained for those irradiation conditions that lead to complete melting of the irradiated portion of the film. (The exact width of the completely melted region (1) is always greater at the top surface of the film than at the oxide interface and (2) can be actually smaller or larger — by the corresponding thermal diffusion distance — than the width of the beam.) Typical lateral regrowth velocities of 10 m/s are obtained.

Figure 3 reveals the temperature profile of the nodes corresponding to the bottom portion of the Si film during lateral solidification. It is clear that an interface temperature that is typically well below the melting point of Si, corresponds mostly to the hottest point of the film within this layer; the temperature decreases in both directions as a function of increasing distance from the interface.

Figure 4 shows the maximum molten widths at the top and the bottom of the Si film as a function of the energy density. As the bottom width comes closer to the top width the bottom width variation with energy density also becomes similar to the top width. From 700 mJ/cm^2 onwards until 900 mJ/cm^2 the top and bottom widths are nearly, but not exactly, coincidental.

Figure 5 shows the time evolution of temperature vs. depth profile. It reveals that the temperature rise due to the thermal diffusion of the absorbed excimer laser energy is confined mostly to a thin surface region of the substrate (within 10,000 Å), i.e., the substrate underneath this portion is well insulated from the heating caused by excimer laser irradiation. Thus, our results indicate that with a buffer SiO_2 layer of 1 μm or so, a glass substrate would essentially be thermally insulated from the heating by the laser.

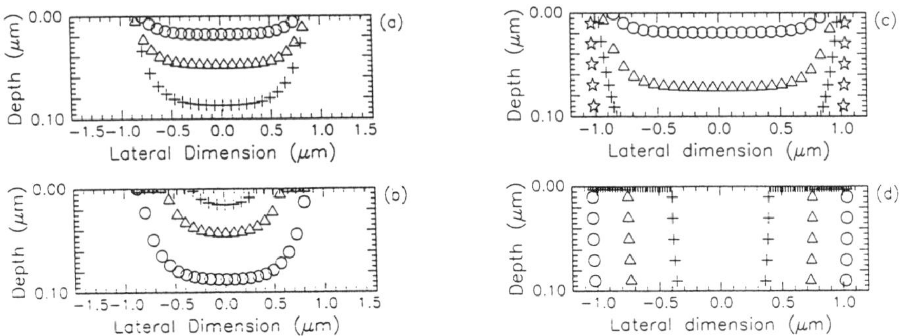

Figure 2: The location of the solid-liquid interface, under laser irradiation of 500 mJ/cm^2, at 15ns(o), 20ns(Δ), 30ns(+), for melting is shown in (a) and at 35ns(o), 50ns(Δ), 60ns(+), for solidification in (b); under laser irradiation of 800 mJ/cm^2, at 10ns(o), 15ns(Δ), 20ns(+), 30ns(★), for melting in (c) and at 35ns(o), 65ns(Δ), and 95ns(+) for solidification in (d). Note that the x and y axes have different scales.

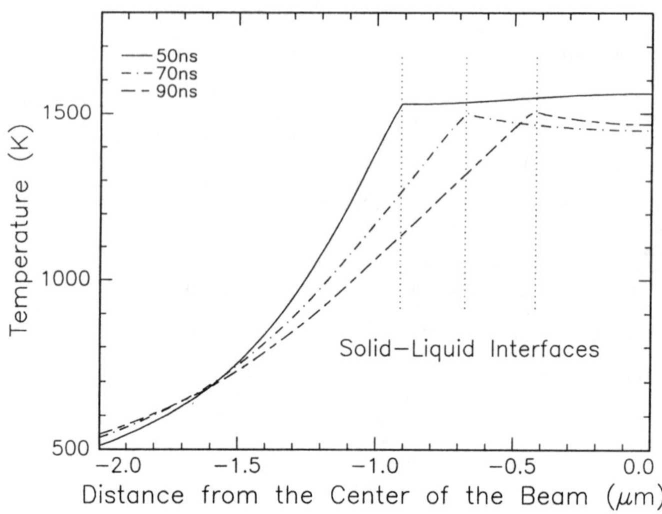

Figure 3: The temperature profiles at 50ns, 70ns, and 90ns, of the nodes corresponding to the bottom portion of the Si film during lateral solidification is shown.

DISCUSSION

The most prominent aspect of the lateral temperature profile shown in Fig. 4 pertains to its unusual shape at the interface. Its general shape is qualitatively identical to the lateral temperature profile that was predicted previously in the course of analyzing the superlateral growth phenomenon [2]. The origin of this thermal contour was identified and attributed to (1) the localized release of the heat of fusion that occurs at the interface during the growth process, and (2) the significant lateral and vertical heat loss that occurs on *both* sides of the interface. Note that the three physical variables that were previously identified as being transient and important in determining the extent to which the lateral growth can proceed — the interface temperature as a function of time, the bulk liquid temperature as a function of time, and the characteristic length associated with interfacial recalescence as a function of time — can all be extracted from the simulation results.

It is also observed in Figure 3 that as time evolves, the interfacial temperature as well as the bulk liquid temperature initially decreases and then rises again. As two growth fronts come closer, the heat of fusion released from each front starts to overlap and leads to this rise of temperature. By keeping track of the center node (x=0) temperature, which is typically at a local minimum during solidification, it may be possible to eventually estimate whether or not nucleation would occur in the liquid.

One of the definitive conclusions that can be drawn from the results presented in this paper is that the solidification velocities that are observed during lateral growth are significantly higher (approximately 10 m/s) than those observed during vertical regrowth (approximately 2 m/s). This outcome plainly reflects the deeper interfacial undercooling attained during lateral solidification of thin films that occurs as a simple consequence of the far more efficient two-dimensional removal of the smaller total amount of heat of transformation generated per solidification distance — due to its thin film configuration — at the solidifying interface.

It is important to note that this observation points out the fallacy associated with an often-invoked argument which contends that a slower solidification rate leads to a larger grain sized

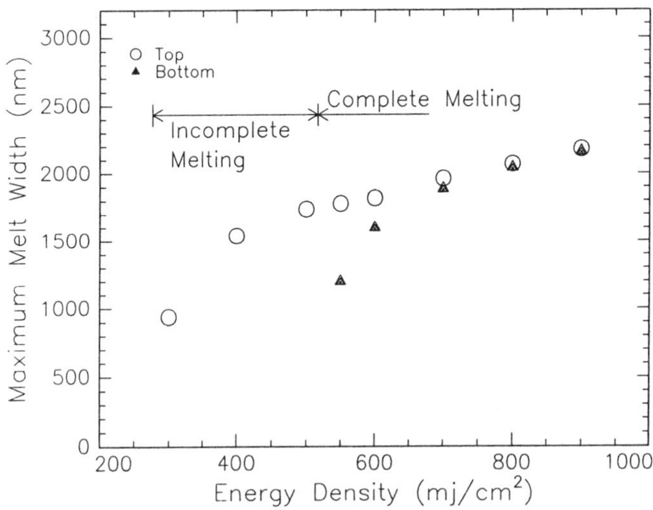

Figure 4: The maximum melt widths at the top and bottom of Si film as a function of energy densities.

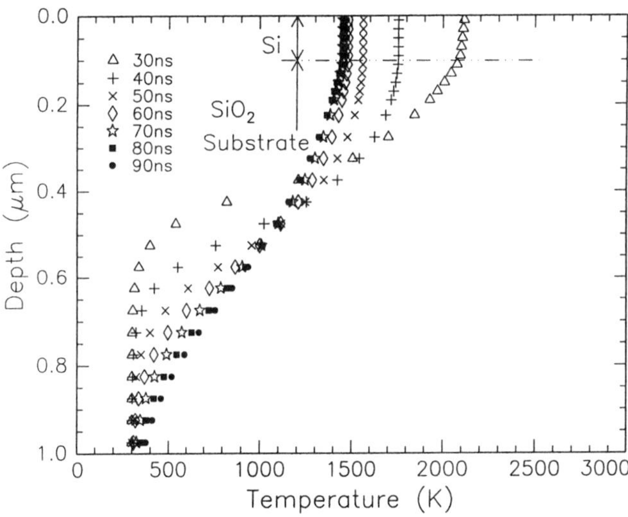

Figure 5: The time evolution of temperature vs. depth.

polycrystalline material [12,13,14]. If this were in fact the case, one would expect to obtain polycrystalline Si films with smaller grains for those conditions that lead to lateral solidification; experimental results clearly indicate otherwise [11]. In the absence of nucleation of solids in a supercooled liquid, the evolution of the microstructure during epitaxial regrowth is determined mainly by (1) the microstructure of the unmelted portion of the material in contact with the liquid at the start of solidification and (2) the occlusion among growing grains that occurs during the ensuing solidification. (Fracturing of dendrites [15] and breakdown of epitaxial growth are two other means through which the microstructure can be affected). These principles, when viewed together with the current results (which indicate that significantly greater solidification distances are attained during lateral growth) explains — and is consistent with — the experimentally obtained microstructures wherein significantly larger grains are observed for all those conditions that induce substantial lateral solidification of Si films.

CONCLUSIONS

We have used a recently-developed, transient, two-dimensional model that properly accounts for essential physical features of the ELC process in order to simulate localized excimer-laser beam-induced melting and solidification of thin silicon films on SiO_2. The evolution of melt-formation and ensuing solidification is studied as a function of incident energy density. The entire time-resolved evolution of the transformation sequence is obtained from the simulation. The results reveal — and demonstrate in quantitative detail — two distinct types of solidification behavior as a function of incident energy density: at low energy densities, partial melting and predominantly vertical solidification occurs within the irradiated area, while at high energy densities, complete melting of the portion of the film is followed by rapid lateral solidification.

REFERENCES

1. J.S. Im, H.J. Kim and Mike O. Thompson, Appl. Phys. Lett. **63**, 1969 (1993).
2. J.S. Im and H.J. Kim, Appl. Phys. Lett. **64**, 2303 (1994).
3. J.S. Im, J. D. Lipman, I. N. Miaoulis, C. K. Chen, and C. V. Thompson, Mat. Res. Soc. Symp. Proc. **157**, 455 (1990).
4. R.F. Wood and G.E. Giles, Phys. Rev. B **23**, 2923 (1981).
5. P. Baeri and S.U. Campisano, *Laser Annealing of Semiconductors,* edited by J.M. Poate and J.W. Mayer (Academic, New York, 1982), 75.
6. M.J. Uttormark and M.O. Thompson, Unpublished document.
7. R.K. Singh and J. Narayan, Mater. Sci. and Eng. B **3**, 217 (1989).
8. V.V. Gupta, H.J. Song, and J.S. Im, Mat. Res. Soc. Proc. **397**, 465 (1996)
9. H.Z. Barakat and J.A. Clark, J. Heat Trans. **88,** 421 (1966).
10. P.H. Bucksbaum and J. Bokor, Phys. Rev. Lett. **53**, 182 (1984).
11. H.J. Kim and J.S. Im, Appl. Phys. Lett. **68** 1513 (1996); H.J. Song and J.S. Im, Appl. Phys. Lett. **68** 3165 (1996); J.S. Im and R.S. Sposili, MRS Bulletin, Volume **xxi**, No. 3 (1996); R.S. Sposili and J.S. Im, Appl. Phys. Lett. **69**, 2864 (1996).
12. K. Shimizu, O. Sugiura, and M. Matsumura, IEEE Trans. Electron Devices **40**, 112 (1993).
13. H. Kuriyama, S. Kiyama, S. Noguchi, T. Kuwahara, S. Ishida, T. Nohda, K. Sano, H. Iwata, H. Kawata, M. Osumi, S. Tsuda, S. Nakano, and Y. Kuwano, Jpn. JAP. **47**, 3700 (1991).
14. T. Sameshima and S. Usui, J. Appl. Phys. **74**, 6592 (1993).
15. Chalmers. B. Chalmers, *Principles of Solidification* (Wiley, New York, 1964).

THE EFFECT OF FILM THICKNESS AND PULSE DURATION VARIATION IN EXCIMER LASER CRYSTALLIZATION OF THIN Si FILMS

J.P. LEONARD, M.A. BESSETTE, V.V. GUPTA, and JAMES S. IM
Department of Chemical Engineering, Materials Science, and Mining Engineering, Columbia University, New York, NY 10027

ABSTRACT

Recognizing that the processing window in conventional excimer laser crystallization corresponds mainly to the partial melting regime, and that this can be properly simulated using a one-dimensional model, we investigate numerically the melting and solidification of thin silicon films on SiO_2. Here a portion of the silicon film is melted and subsequent vertical solidification is initiated from the lower interface bounding the unmelted region. Upper and lower energy density limits for this regime are calculated for crystal silicon films of thickness 10 to 300 nm, and for pulse duration ranging from 10 to 200 ns. These calculations show that increasing pulse duration requires proportionally more incident energy density to partially melt the film, while decreasing film thickness reduces the range of energy densities over which partial melting can occur. The results are explained in terms of characteristic thermal diffusion distances and the enthalpy change associated with melting. In view of the results we discuss optimization of the conventional excimer laser crystallization and the avoidance of complete melting during the process.

INTRODUCTION

Excimer laser crystallization of amorphous silicon thin films on glass [1,2] is emerging as one of the first widely accepted applications of lasers as a production tool in the electronics industry. In typical conventional excimer laser crystallization processes [3] (e.g. multiple pulse based scanning or single pulse based stationary irradiation methods), a silicon film is partially melted by a homogenized excimer laser beam and rapidly resolidified. Results from previous investigations [4] indicate that it is very important to avoid complete melting of the film over a large area. Such melting can, for example, arise from pulse-to-pulse energy fluctuations, spatial inhomogeneities in the beam, and microstructural or phase non-uniformities in the silicon film. With complete melting, the absence of a local liquid-crystal interface allows undercooling and nucleation [2,5] in the melt to occur during the quench, resulting in a very fine-grain polycrystalline material. This in turn can severely degrade the uniformity of the resulting devices.

In this paper, the conditions for partial melting are investigated. Currently, various combinations of film thickness (~50–200 nm) and laser pulse duration (~30–200 ns) are utilized in conventional excimer laser crystallization processes. A proper assessment of the effect of the variation of such key parameters on the processing window constitutes a rudimentary, yet essential investigation of the excimer laser crystallization process. The relevant quantities associated with partial melting can be readily and efficiently obtained using a well-developed one-dimensional numerical program [6,7]. Specifically, these include the threshold energy densities needed for surface melting and complete melting, as well as vertical solidification velocities. Our work is in contrast to previous studies [8,9,10] that used a one-dimensional numerical model to study melting and vertical regrowth, but were incorrectly applied for these conditions corresponding to complete melting of the film. Such treatments of melting and vertical regrowth are valid only when the film is partially melted.

BACKGROUND

Processing Window

When viewed in terms of a maximum primary melt depth, excimer laser crystallization of silicon on SiO_2 can be divided into several distinct regimes [11].

Mat. Res. Soc. Symp. Proc. Vol. 452 © 1997 Materials Research Society

Irradiation with low energy densities below the threshold for surface melting will have a negligible effect on microstructural evolution. At the other extreme, irradiation with energy densities above the threshold required to completely melt the film down to the substrate leaves only a chemically inert liquid-silica interface. In the absence of an underlying liquid-crystalline interface, solidification must occur through either the nucleation of crystal embryos in the undercooled liquid, or, in the case of near-complete melting— through lateral growth from pre-existing seeds. Such transformations cannot be rigorously modeled by one-dimensional numerical simulation, but requires a multi-dimensional model. In typical sample configurations the quench, nucleation, and growth rates will result in a very fine to small grain polycrystalline microstructure [2,5].

In the partial melting regime, the maximum extent of the melt lies within the silicon film, as determined by the total absorbed energy, pulse duration, enthalpy associated with melting, and rate of heat conduction into the substrate. In this case, residual unmelted material provides a solid-liquid interface from which rapid regrowth can occur upon cooling. This can be considered the regime in which conventional excimer laser crystallization processes [11] typically operate, and is accurately characterized by a one-dimensional numerical model.

Numerical Model

Although analytical expressions for temperature evolution exist for cases of pulsed laser melting of some bulk materials, complications that include a temperature dependent thermal conductivity, heat capacity and refractive index, non-equilibrium melting and solidification, as well as multilayer films of varied composition— can only be treated through numerical simulation. Finite differences methods have been applied with much success in pulsed laser annealing of silicon substrates [12]. The model used in this investigation is derived from the one-dimensional numerical code originally developed [7] for studying pulsed laser annealing of silicon surfaces. Specific features of the model include the following: (1) Temperature dependent absorption and reflection are determined from published optical properties for crystal silicon at a wavelength of 308 nm. (2) Interfacial motion is based on a linearized interface response function of 25.0 and 6.7 cm/(sec °K) respectively, for melting and solidification. (3) The temporal profile of the pulse is gaussian, with a full width at half maximum duration ranging from 10 to 200 ns. (4) The silicon film is modeled as crystalline. In an actual crystallization process, this is encountered after the as-deposited amorphous layer has been crystallized by an initial pulse. This simplification avoids a number of complicated melt-mediated transformation scenarios, which may in turn depend on the microstructural and compositional details of the as-deposited amorphous film.

Processing window thresholds were determined by repeated simulations at varying pulse energies while monitoring the maximum melt depth. An algorithm runs successive simulations until a target maximum melt depth is obtained within a prescribed tolerance, and records the energy. The incident energy density corresponding to a maximum melt depth 5±4 Å was taken to be lower bound of the partial melting window, while the upper bound was defined as a melt depth to within 5±4 Å of the bottom of the film. In practice the algorithm found energies providing maximum melt depths within the tolerance in fewer than 20 iterations. In this manner, the processing window can be determined for any combination of film thickness or pulse characteristics. As well, details of temperature evolution in elements at various depths, interfacial temperature, velocity, and position as a function of time were recorded. Node sizes were chosen to provide convergent solutions where further reductions in size did not change the results. The substrate thickness was increased well beyond the maximum thermal diffusion distance for the longest pulse duration.

RESULTS

The upper and lower threshold energy densities were calculated for varying pulse duration and silicon film thickness. In fig. 1, the variation of threshold energy density as a function of film thickness is shown for 30 and 200 ns. The lower threshold, which is the incident energy required to initiate melting, is largely invariant as a function of film thickness. For a silicon film thickness greater than 50 nm, the upper threshold scales approximately linearly with increasing thickness. Below 50 nm, the energy density required to melt actually increases with decreasing film thickness. The effects of varying pulse duration are shown in fig. 2 for several film thicknesses common in conventional excimer laser crystallization. The upper and lower threshold energy densities tend to increase smoothly with increasing pulse duration. Typically an increase by a factor of 2 in energy density is observed from 10 ns to 200 ns pulses. The lower threshold energy is again observed to be relatively independent of film thickness, while the upper threshold energy density (and also the width of the partial melting window) scales with increasing film thickness.

The relative width of the partial melting window is important when considering the effect pulse-to-pulse variation has on the crystallization process. The normalized width is plotted in fig. 3 (width of partial melting window divided by the lower threshold energy density) as a function of pulse duration. Smaller relative widths require greater laser stability for the crystallization process to remain in the partial melting regime and avoid complete melting. Relative widths scale with film thickness and also decrease sharply around 30 ns pulse duration, then more gradually at longer pulse widths. Typically, the relative window is seen to decrease by 30-40 percent when pulse duration is increased from 30 ns to 200 ns (e.g. 0.52 to 0.32 for 50 nm films).

The temperature evolution of the underlying SiO_2 substrate was also investigated as a function of varying pulse duration for irradiation of a 100 nm silicon film. Figure 4 plots the peak temperature recorded at various depths in the SiO_2 as a function of the pulse duration. The upper threshold energy density for partial melting was used in each case. The plot shows that the peak temperature approaches the silicon melting temperature at depths less than 0.1µm. For pulses 30 ns and below, the peak temperature falls off sharply beyond 0.5 µm depth. However for longer pulses, peak temperatures are well above 800 °K even at depths of 1 µm.

Vertical regrowth velocities were recorded for conditions corresponding to the upper threshold energy at varying pulse durations, allowing solidification initiated from the thin unmelted crystal layer to propagate the full thickness of the film. Velocities were relatively constant during solidification, with a maximum velocity of approximately 2.0 m/sec in the case of a 10 ns pulse, to 1.0 m/sec for a 200 ns pulse. Such velocities are consistent with the one dimensional interfacial growth model used, and are well below interfacial amorphization velocities [13].

DISCUSSION

Given accurate temperature dependent physical properties, the simulation adequately models simple heat flow and interfacial melting and regrowth. Interpretation of these numerical results relative to actual excimer laser crystallization processes requires a consideration of a number of complications: The melt dynamics in real systems are complicated by explosive crystallization [14] in the case of as-deposited amorphous silicon. As well, the presence of a capping oxide or other thin dielectric layers may affect the optical coupling. The dependence of melt dynamics on the microstructural configuration can also be significant when the irradiation conditions correspond to near-complete melting—and must also be considered [11]. Beam-related causes of non-uniformity include spatial variation and pulse-to-pulse fluctuation. While the former can be dramatically improved through beam homogenization techniques (e.g. a multi-lenslet homogenizer), pulse-to-pulse variation cannot be as effectively handled. This is an important factor to consider with regard to avoiding complete melting, especially when operating the laser in the near-complete melting regime [2] where the largest crystals are typically obtained. All these complications will affect the absolute values of the threshold energies in actual experiments. However, in terms of relative threshold energies, the trends observed as a function of film thickness and pulse duration can be considered valid and provide an important guideline for optimizing excimer laser crystallization processes.

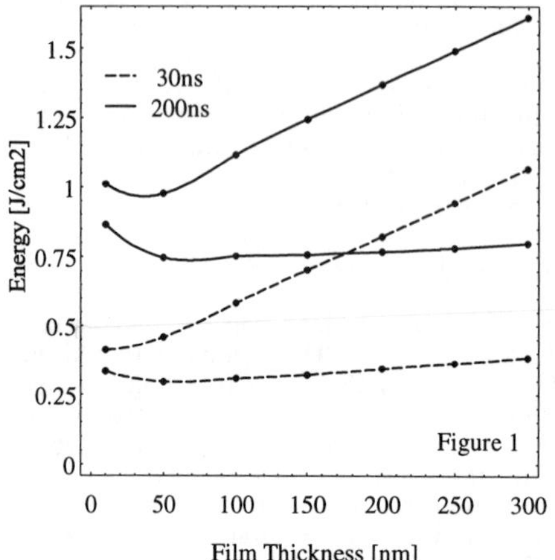

Fig 1. Threshold energy densities for partial melting calculated as a function of silicon film thickness irradiated with 30 and 200 ns pulse duration. Lower threshold is energy density required to melt the top 1-9 Å of the silicon film. Upper threshold is energy density required to melt within 1-9 Å of bottom of film.

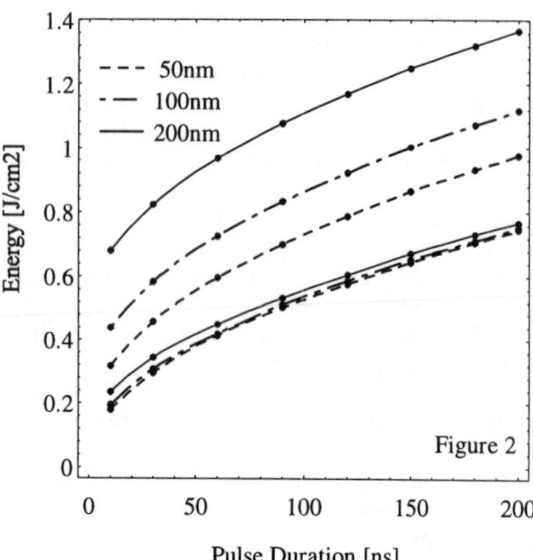

Fig. 2. Threshold energy densities for partial melting calculated as a function of pulse duration for 50,100, and 200 nm silicon films. Lower threshold is energy density required to melt the top 1-9 Å of the silicon film. Upper threshold is energy density required to melt within 1-9 Å of bottom of film.

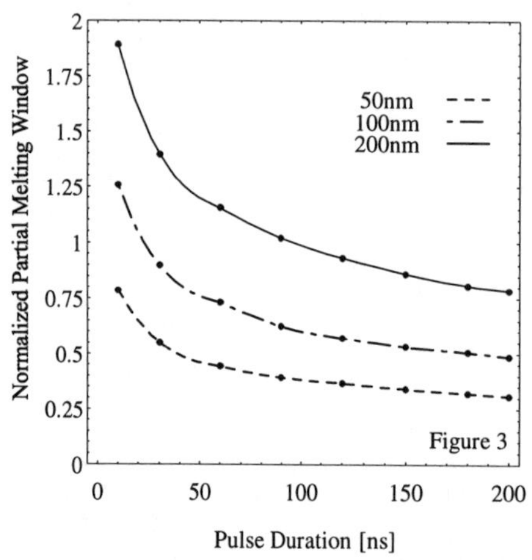

Fig. 3. Normalized partial melting window (partial melting window width as a fraction of lower threshold energy density) for 50, 100, and 200 nm films as a function of pulse duration.

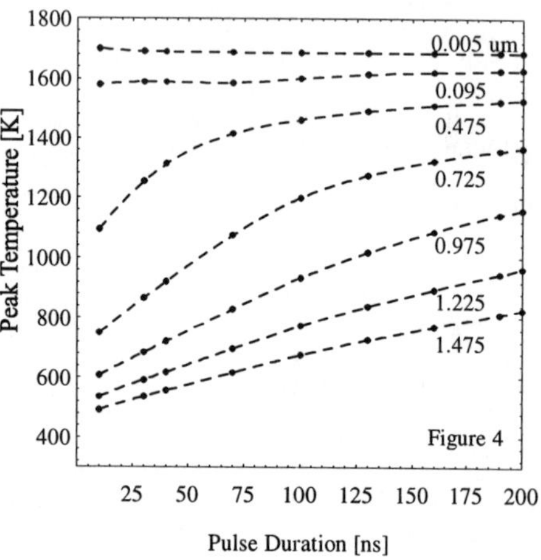

Fig. 4. Peak temperature recorded as a function of excimer pulse width for various depth in SiO_2 substrate. Simulation was run for 1200 ns in each case, and temperature recorded. For depths beyond approximately 1.2 μm, the temperature was still rising after 1200 ns.

The lower energy density threshold for partial melting is determined primarily by heat conduction in the solid, for which a simple one-dimensional thermal diffusion model [15] can provide a conceptual understanding and confirms the validity of the numerical simulation results: Assuming a constant thermal diffusivity (D) and no phase change, the pulse energy is absorbed as heat in a shallow surface layer (typically 10 to 20 nm for a 308 nm excimer pulse incident on silicon). Conduction of heat into the sample during the duration (t_p) of the pulse results in an approximately gaussian temperature profile of characteristic width $(2Dt_p)^{1/2}$. By this argument, it is expected that longer pulses will produce flatter temperature profiles extending deeper into the substrate, thus requiring more incident energy to reach the silicon melting temperature at the surface. As well, optical coupling effects are evident in the rise in threshold energies for pulse durations below 50 nm. Although thin film interference effects are not incorporated into the model, decreasing the film thickness below the characteristic absorption depth does allow increasingly more optical energy to pass through the film— where it is lost. This increases the amount of total incident energy needed to melt to a comparable depth relative to thicker films.

The upper energy density threshold is determined by both the enthalpy associated with melting and heat conduction into the substrate. Although conduction into the substrate occurs during the melting of the film, the enthalpy required to melt silicon is much larger. Thus, the width of the partial melting energy regime is determined primarily by the film thickness, increasing linearly as shown in fig. 1., while changing only slightly as a function of pulse duration. Understanding the excimer laser crystallization process in terms of a partial melting window bounded by well defined threshold energy densities allows insight into selecting various film and laser configurations. It is important to stress that the upper threshold energy density, above which the film is completely melted, represents a critical parameter that must not be exceeded in the conventional excimer laser crystallization process. Whenever complete melting of isolated areas larger than that which can be recovered by lateral growth from the existing interface [16] occurs, it will lead to subsequent undercooling in those regions, and result in nucleation of very fine-grain polycrystalline material— or amorphization in the case of very thin films [10,17].

It should be noted that there exist laser crystallization techniques that rely on inducing complete melting of the film. With the artificially controlled super-lateral growth (ACSLG) method [3], for example, the film is completely melted only in pre-determined regions. Solidification then takes place by lateral growth inward from unmelted solid at the edges of the completely melted region. A proper analysis of such solidification behavior requires a two-dimensional treatment [18] of melting and regrowth, and is not addressed in the context of this one-dimensional numerical investigation. Also, in view of the difficulties associated with operation in the partial melting regime for excimer laser irradiation of ultra-thin films or with long pulses, processes that completely melt large areas of the film should be considered as an alternative crystallization method. Such processes could yield a small, but uniform, polycrystalline material for those devices in which uniformity is critical but performance requirements are lower (e.g. thin film transistors for controlling pixels in active matrix liquid-crystal displays). The solidification in this case depends on nucleation of solids in the liquid at deep undercooling, and cannot be treated with an interfacial regrowth model.

Nevertheless, the partial melting results presented here are directly applicable to large-area crystallization processes using new high power excimer systems [19] with a pulse duration up to 200 ns. In particular, the higher energy densities required to operate in the partial melting regime at a longer pulse duration, as well as reduced (relative) processing windows at longer pulse durations or thinner films must be considered in the development and optimization of these systems.

CONCLUSION

We define the partial melting regime as the processing window for conventional excimer laser crystallization, and use a one-dimensional numerical simulation to calculate threshold energy densities that are the boundaries of this regime. These are the critical processing parameters for conventional excimer laser crystallization, and this information constitutes a basic but important conceptual framework for the development of conventional excimer laser crystallization processes. Systematic and unambiguous trends with ramifications for optimization of the process are revealed. As expected from simple thermal diffusion considerations, the combination of a thinner film and a longer pulse duration is shown to reduce the partial melting processing window. In particular, increases in the pulse duration lead to progressive delocalization of the thermally affected area, which in turn increases the absolute energy densities needed to effect a given melting condition, and more significantly, decreases the normalized width of the energy-density processing window. Together, these trends point to more difficulty in obtaining a spatially uniform microstructure in laser crystallization processes. Avoidance of complete melting is of utmost importance, as regions that are melted completely to the substrate may form an undesirable fine-grain polycrystalline structure. Therefore, spatial and pulse-to-pulse variations must be considered when operating within the partial melting window. Knowledge of the trends outlined here with respect to the partial melting window give insight into the performance of specific laser crystallization systems, specifically the difficulties that may be encountered with increasing pulse duration and decreasing film thickness.

This work was supported by DARPA under project N61331-94-K-0033.

REFERENCES

1. T. Sameshima, S. Usui, Mat. Res. Soc. Symp. Proc. **71**, 435 (1986).
2. J.S. Im, H.J. Kim, M.O. Thompson, Appl. Phys. Lett. **63**, 1969 (1993). H.J. Kim, J.S. Im, M.O. Thompson, Mat. Res. Soc. Symp. Proc. **283**, 703 (1993).
3. J.S. Im, R.S. Sposili, MRS Bulletin, XXI, **3**, 35 (1996).
4. R.I. Johnson, G.B. Anderson, J.B. Boyce, D.K. Fork, P. Mei, S.E. Ready, S. Chen, Mat. Res. Soc. Symp. Proc. 297, 533 (1993).
5. S.R. Stiffler, M.O. Thompson, Phys. Rev. Lett. **60**, 2519 (1988).
6. P.H. Bucksbaum, J. Bokor, Phys. Rev. Lett. **53**, 182 (1984).
7. M.J. Uttomark, M.O. Thompson, Unpublished document.
8. K. Shimizu, O. Sugiura, M. Matsumura, IEEE Trans. Elect. Dev. **40**, 112 (1993).
9. H. Kuriyama, S. Kiyama, S. Noguchi, T. Kuwahara, S. Ishida, T. Nohda, K. Sano, H. Iwata, H. Kawata, M. Osumi, S. Tsuda, S. Nakano, Y. Kuwano, Jpn. J. Appl. Phys. **47**, 3700 (1991).
10 T. Sameshima, S. Usui, J. Appl. Phys. **74**, 6592 (1993).
11. H.J. Kim, J.S. Im, Mat. Res. Soc. Symp. Proc. **321**, 665 (1994).
12. Laser Annealing of Semiconductors, edited by J.M. Poate, J.W. Mayer, Academic, NY, (1982).
13. M.O. Thompson, J.W. Mayer, A.G. Cullis, H.C. Webber, N.G. Chew, Phys. Rev. Lett. **50**, 896, (1983).
14. M.O. Thompson, G.J. Galvin, J.W. Mayer, P.S. Peercy, J.M. Poate, D.C.Jacobson, A.G. Cullis, N.G. Chew, Phys. Rev. Lett. **52**, 2360 (1984).
15. N. Bloembergen, Laser Solid Interactions and Laser Processing, AIP conf. proc. **50**, 1 (1978).
16. J.S. Im, H.J. Kim, Appl. Phys. Lett. 64, 2303 (1994).
17. T. Eiumchotchawalit, J.S. Im, Mat. Res. Soc. Symp. Proc. **321**, 725 (1994).
18. V.V. Gupta, H.J. Song, J.S. Im, Mat. Res. Soc. Symp. Proc. **397**, 465(1995).
19. M. Stehle, Laser Focus World, 101 (May 1996).

SINGLE-CRYSTAL Si FILMS VIA A LOW-SUBSTRATE-TEMPERATURE EXCIMER-LASER CRYSTALLIZATION METHOD

Robert S. Sposili, M. A. Crowder, and James S. Im
Columbia University, Department of Chemical Engineering, Materials Science, and Mining Engineering, New York, NY 10027

ABSTRACT

This paper describes an extension of the sequential lateral solidification (SLS) method for producing single-crystal Si regions in predetermined locations on thin Si films on SiO_2. This is accomplished by manipulating the shape of the solidification front in order to exploit the tendency of grain boundaries to propagate nearly perpendicular to the melt interface. Specifically, we employ a chevron-shaped beamlet to select and grow the grain at the chevron's apex for propagation, resulting in a well-defined single-crystal region. Many such chevrons are processed concurrently—each one leading to a single-crystal region at a precisely determined location, and each one being large enough for complete inclusion of an entire thin-film transistor (TFT) device. Microstructural examination of defect-etched SLS material using optical microscopy reveals unambiguously that single-crystal regions free of high-angle grain boundaries are produced.

INTRODUCTION

It has long been recognized that a thin single-crystal Si film is the best material on which to fabricate thin-film transistors (TFTs), as the performance of devices fabricated on grain-boundary-free material can approach that of monolithic-wafer-based devices [1]. Such a material can potentially enable a number of applications—such as integrated active-matrix liquid-crystal displays (AMLCDs), static random-access memory (SRAM), and three-dimensional integrated circuits—to be optimally addressed. Presently attainable TFT devices, using amorphous Si (a-Si) or polycrystalline Si materials, have a number of limitations. Amorphous Si is inadequate due to its low carrier mobility, which prevents a-Si-based TFT devices from achieving the requisite level of performance. The carrier mobility of polycrystalline Si is higher than that of a-Si, but still falls short of single-crystal material. Although the mobility of polycrystalline Si can be improved somewhat by increasing the grain size, the random location of grain boundaries can degrade device-to-device uniformity [2].

While some high-cost and technically challenging silicon-on-insulator (SOI) methods do exist for producing single-crystal thin Si films, the nature of the applications listed above require that, in some cases, the substrates be transparent, and in all cases, that the substrates not be subjected to excessively high temperatures during the crystallization process. These requirements, along with economic considerations, preclude consideration of the SOI or SOI-derived methods as a practical means of achieving single-crystal films for these applications.

SEQUENTIAL LATERAL SOLIDIFICATION

Previously, we reported on the development of a new excimer-laser crystallization method, called sequential lateral solidification (SLS), that enables the conversion of as-deposited amorphous (a-Si) or polycrystalline Si films on SiO_2 to a directionally solidified microstructure consisting of long, columnar grains [3, 4]. The SLS process, which is based on the controlled application of the super-lateral growth phenomenon [5, 6], involves complete melting of preselected regions of a Si film via projection irradiation through a patterned mask, and controlled translation of the film relative to the mask between pulses.

Here, we describe how the SLS process can be used to create single-crystal regions in initially amorphous or polycrystalline Si films. We exploit the well known phenomenon that the grain boundaries in directionally solidified materials tend to form so as to always be approximately perpendicular to the melt interface [7]. We accomplished this by carrying out SLS using a spatially tailored beam contour.

Mat. Res. Soc. Symp. Proc. Vol. 452 © 1997 Materials Research Society

EXPERIMENTAL METHOD

The samples consisted of Si substrates with a 1.9-μm-thick SiO_2 isolation layer, on top of which 2,000-Å-thick a-Si films were deposited via LPCVD. The experimental apparatus consisted of an excimer laser operating at 308 nm (XeCl), a UV projection system, and a sub-micrometer-precision translation stage. The projection system included a variable-energy attenuator for fine control of the beam energy, and a two-element imaging lens for imaging the mask features onto the sample. The UV mask used in these experiments was fabricated out of molybdenum silicide on a quartz substrate, and the pattern consisted of an array of chevron-shaped apertures on an opaque background such that only the laser radiation that passed through the narrow slits was transmitted. The relative positions of the mask, imaging lens, and sample were adjusted to yield a pattern demagnification ratio of approximately 5X. At this demagnification, the image of each chevron had a slit width of 5 μm at the sample plane, and was 50 μm across. The samples were translated for 50 μm.

Processing was carried out by iteratively (1) irradiating the sample at an energy density sufficient to induce complete melting of the Si film in the exposed areas (approximately 900 mJ/cm^2 for the 2,000-Å thickness), and (2) translating it relative to the beam over a distance (typically 0.75 μm) approximately one-half of the single-pulse lateral growth distance. This process was repeated over a total translation distance of 50 μm. Subsequent to processing, the majority of the samples were defect-etched (Secco etchant), and examined via optical microscopy. Some samples were left unetched so that their surface morphology could be examined using a stylus-type profilometer.

RESULTS

Figure 1 shows a low-magnification optical micrograph of a defect-etched SLS-processed sample. Each individual crystallized region represents the area processed by a single chevron feature on the mask, which contained a regular array of many such chevrons. All of these regions were crystallized simultaneously; the total area that can be processed is limited only by the available laser energy. The number and locations of the individual regions can be controlled through design of the mask.

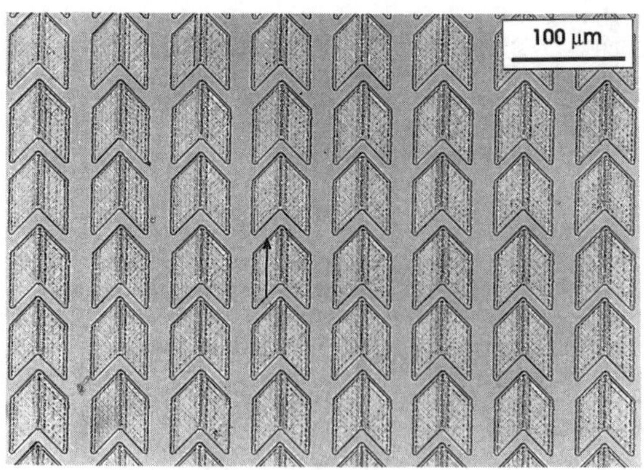

FIG. 1. Low-magnification optical micrograph of an SLS-processed film. The arrow shows the solidification direction, and the magnitude of translation. Each crystallized region was processed by an individual beamlet, and the entire area was processed simultaneously.

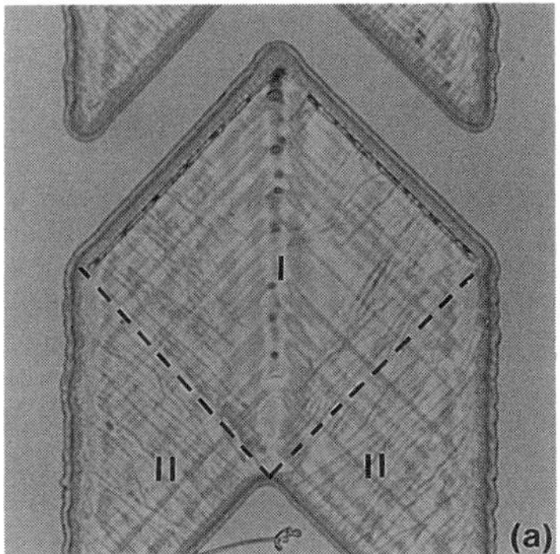

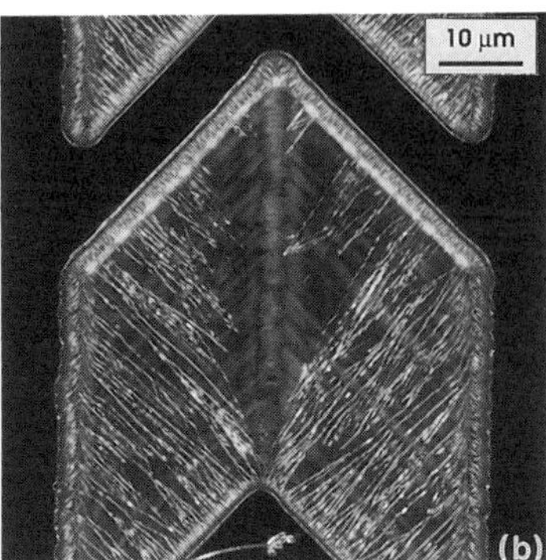

FIG. 2. High-magnification optical micrographs of SLS-processed films; (a) brightfield, with dotted lines demarcating the boundaries between the single-crystal and columnar crystal regions; (b) darkfield.

Figure 2(a) is a higher magnification optical micrograph (also defect-etched), showing the region crystallized by a single chevron. Figure 2(b) is the corresponding darkfield image, which gives better contrast. The dashed lines in Fig. 2(a) delineate the boundaries between the single-crystal region (I) and regions II, which have a lateral columnar microstructure identical to that observed in previous experiments that used a mask consisting of straight slits [5]. Within region I, there is a nearly perfect, defect-free central region, bordered on each side by regions with a number of linear defects. Careful examination of these linear defects shows that they terminate in the interior of the single-crystal region, and therefore cannot be high-angle grain boundaries, but rather are low-angle subboundaries. A narrow polycrystalline transition border lies between the interior of the crystallized area and the outer amorphous matrix area.

Figure 3 shows surface profile scans of unetched SLS-processed samples near the center of the chevrons, where the scan direction is parallel to the translation direction. Figure 3(a), which represents a typical section in the interior of the crystallized region, shows that the surface is relatively smooth, with only some gentle, long-spatial-period thickness undulations on the order of

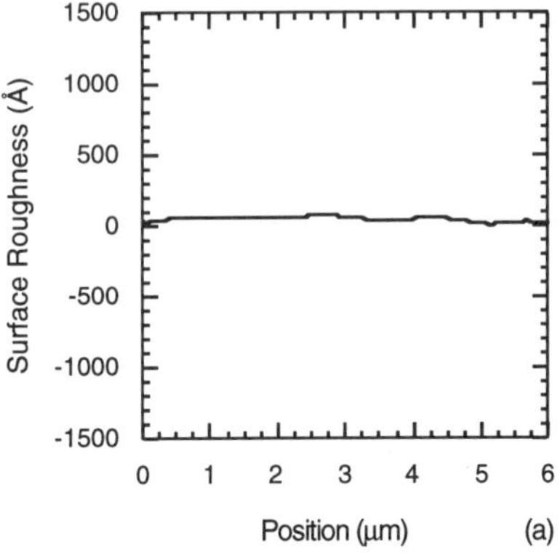

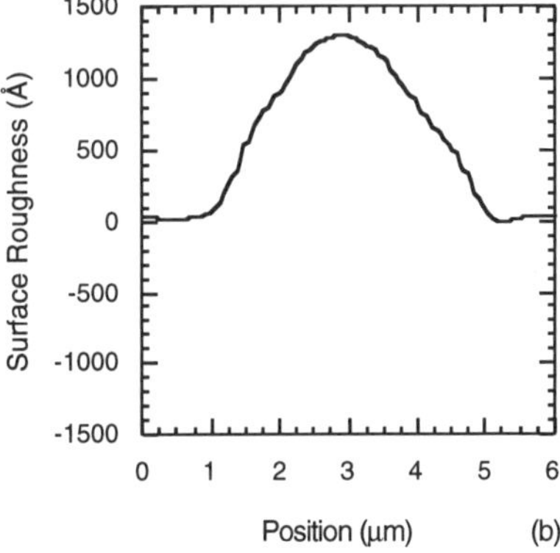

FIG. 3. Surface profiles of SLS-processed material, scanned along crystallization direction; (a) in interior of single-crystal region; (b) across the last position to be irradiated.

50 Å or less in amplitude. Figure 3(b) shows the profile of the prominent protrusion located at the end of the scanned area, corresponding to the last point to be irradiated.

DISCUSSION

The distinguishing features of the SLS process—especially in comparison to other previously investigated and developed low-temperature methods for producing crystalline Si films on SiO_2 [3]—are: it produces single-crystal regions large enough for complete inclusion of an entire TFT device, and it allows for precise control of their location on the film. Without this control, the ability to produce large grains—but with randomly located grain boundaries—is of limited practical utility in TFT fabrication.

Previously, we described the essential features of the SLS process, and explained the mechanism through which it creates directionally solidified crystalline films of arbitrary length [3, 4]. Figure 4 is a schematic diagram showing the development of the microstructure during the initial steps of the SLS process. In Figure 4(a), a narrow region of the film, bounded by the dashed lines, is irradiated at an energy density sufficient to induce complete melting. Subsequently, lateral grain growth proceeds from the unmelted regions adjacent to the narrow strip. Depending on the width of the molten region, lateral growth will cease when either (1) the two opposing growth fronts collide at the center, or (2) the molten region becomes sufficiently supercooled so that bulk nucleation of solids occurs—whichever occurs first. Due to such considerations, the maximum lateral growth distance that can be achieved with a single pulse is limited to less than 2 μm, depending on the film thickness and the incident energy density. Subsequently, the film is translated relative to the beam image over a distance less than the single-pulse lateral growth distance, as shown in Figure 4(b), and irradiated again. As lateral growth recommences from the edges of the completely molten region, one of which is located within the grains grown during the previous irradiation step, the length of the grains is increased beyond the single-pulse lateral growth distance. This process can be repeated indefinitely leading to grains of any desired length; the final microstructure obtained in this fashion is shown in Figure 5(a).

In this work, we have redesigned the basic mask feature as a chevron, or elbow-shaped slit rather than a straight slit [see Figure 4(c)]. The effect is that the grain formed at the apex of the chevron experiences lateral growth not only in the translation direction, but also transverse to it, due to the well known and commonly exploited fact that the grain boundaries form roughly perpendicular to the melt interface [5]. Thus, the negative curvature of the molten zone at the apex of the chevron leads to widening of the grain, as shown in Figure 4(d), such that it eventually

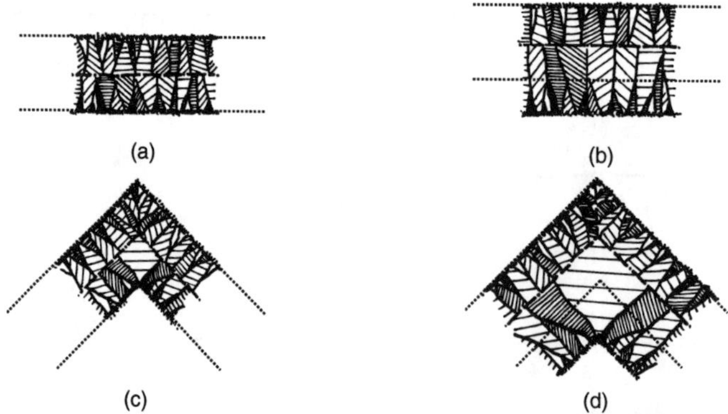

FIG. 4. Schematic views of the SLS process; (a) first irradiation pulse through a straight slit mask, as indicated by the dashed outline; (b) second pulse, also as indicated by the dashed outline; (c, d) analogous sequence using a chevron-shaped mask pattern.

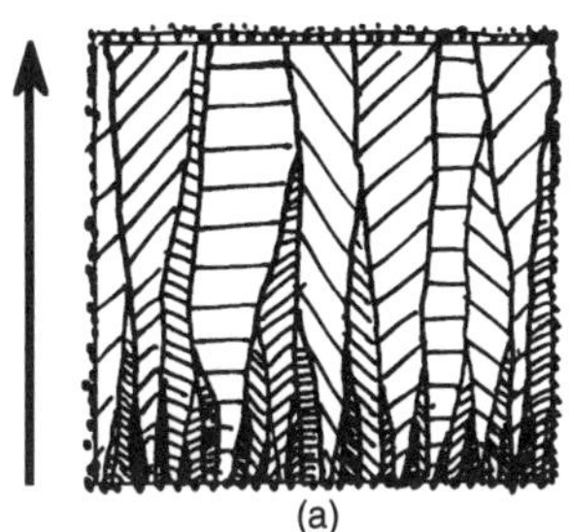
(a)

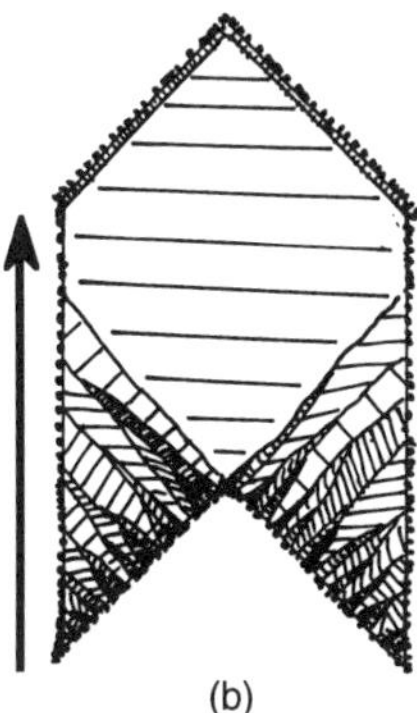
(b)

FIG. 5. Schematic showing the SLS microstructure for (a) a straight-slit and (b) a chevron-shaped beamlet. The arrow shows the solidification direction.

encompasses all of the irradiated zone, while all other grains run out at the edges. Figure 5(b) shows how this leads to the creation of a single-crystal region. Note that (1) the size of the single-crystal region is determined by, and can be controlled by varying the width of the beamlet incident on the sample surface and the total translation distance, and (2) the locations of these regions can be precisely controlled by using an appropriately patterned mask.

The small- and fine-grained polycrystalline transition region surrounding Regions I and II originates from vertical and lateral explosive crystallization, as well as from solidification following partial melting and near-complete melting of the film [5].

The low-angle subboundaries present within the single-crystal region have been reported previously as a major defect in SLS-processed material, and have also been observed in zone-melting recrystallization (ZMR) of Si films—a high-temperature SOI method [8]. In general, subboundary formation can be traced to a number of sources [7]. In SLS, the most likely cause is the thermal-gradient-induced stress present at the melt interface. A higher gradient is expected to lead to higher thermal stresses in the solidifying front, which in turn increases the likelihood of defect formation. This, along with other factors such as the details of generation and coalescence of dislocations, may account for the observed absence of defects and subboundaries in the central region of the crystallized area. Based on heat-flow considerations, the temperature gradient at the melt interface is expected to be lower at the apex of the chevron than elsewhere. Further investigation must be carried out in order to clearly identify the exact cause of subboundary formation in the SLS process. In any event, *the fact that the subboundary-free region observed in the center of the crystallized area is large enough (typically on the order of 10-μm wide) to accommodate the active-channel region of a TFT can be systematically exploited in fabrication of TFT devices.*

In contrast to the microscopically rough surface sometimes seen in polycrystalline Si films [9], the SLS-crystallized films have a smooth surface. This is remarkable in that a protrusion of the type seen in Fig. 3(b) is produced at the conclusion of growth during each individual pulse [10]. The fact that such protrusions are not observed—except, as anticipated, at the center of the last irradiated zone—indicates that they are re-melted and planarized with subsequent processing.

Preliminary examination of the texture of the single-crystal regions using TEM electron diffraction shows that many of them have an orientation close to (111), but orientations close to other low-index planes such as (100) are also observed. The distribution of grain orientations among the single-crystal regions, as well as the possible ways through which one could control the texture, will be considered in more detail in a future publication.

There are several factors that, if not properly addressed, can have a negative effect on the SLS process. These include (1) placement of the incident beam image, which can be affected by vibration and imprecise translation, and which can lead to discontinuity in crystal growth; (2) the incident energy density seen by the film, which if outside the processing window can lead to discontinuity in lateral growth or agglomeration of the films; and (3) interference-caused intensity variations within the patterned beam, which can effectively reduce the energy-density processing window.

In conclusion, we have developed the first technically viable method for producing single-crystal silicon films on glass substrates. Our excimer-laser-based method leads to formation of location- and dimension-controlled single-crystal regions on an SiO_2 surface. We obtain such a material via sequential lateral solidification of a thin Si film in which we employ contoured excimer-laser beamlets tailored such that the shape of the beamlets dictates the development and evolution of the microstructure during lateral solidification. The method provides a means to realize in a practical manner the goal of fabricating low-temperature single-crystal Si TFT devices on glass substrates.

ACKNOWLEDGMENTS

This work was funded by the Defense Advanced Research Projects Agency (DARPA) under project N61331-94-K-0033.

REFERENCES

1. L. L. Kazmerski, *Polysilicon and Amorphous Thin Film Devices* (Academic, New York, 1980).
2. R. I. Johnson, G. B. Anderson, J. B. Boyce, D. K. Fork, P. Mei, S. E. Ready, and S. Chen, Mater. Res. Soc. Symp. Proc. **297**, 657 (1993).
3. James S. Im and Robert S. Sposili, Mater. Res. Soc. Bull. **21** (3)39 (1996).
4. Robert S. Sposili and James S. Im, Appl. Phys. Lett. **69**, 2864 (1996).
5. James S. Im, H. J. Kim, and Michael O. Thompson, Appl. Phys. Lett. **63**, 1969 (1993).
6. James S. Im and H. J. Kim, Appl. Phys. Lett. **64**, 2303 (1994).
7. Bruce Chalmers, *Principles of Solidification* (Wiley, New York, 1964).
8. E. I. Givargizov, *Oriented Crystallization on Amorphous Substrates* (Plenum, New York, 1991).
9. T. I. Kamins, *Polycrystalline Silicon for Intergrated Circuit Applications* (Kluwer Academic, Boston, 1988).
10. H. J. Kim and James S. Im, Mater. Res. Soc. Symp. Proc. **397**, 401 (1995).

Crystallization of $Si_{(1-y)}C_y$ films by excimer laser annealing: characterization of the microstructure of the films

P. BOHER, M. STEHLE and J.L. STEHLE,
SOPRA S.A., 26 rue Pierre Joigneaux, 92270 Bois-Colombes (France)

E. FOGARASSY, J.J. GROB, A. GROB, D. MULLER,
Laboratoire PHASE, UPR du CNRS N° 292, BP20, 67037 Strasbourg (France)

ABSTRACT

Epitaxial $Si_{(1-y)}C_y$ substitutional alloy layers are prepared on monocrystalline silicon substrates by carbon multiple energy ion implantation followed by XeCl excimer laser annealing on large surfaces. Structural analysis of the films before and after laser annealing are made very precisely using spectroscopic ellipsometry (SE), x-ray diffraction (XRD) and Rutherford backscattering (RBS) techniques. We show that annealing energy densities higher than $2J/cm^2$ result in monocrystalline epitaxial layers with low quantity of defects. The lattice contraction due to the carbon inclusion increases with the implanted C concentration up to about 1.1%. For higher values a more complex behaviour is observed with partial (or total) relaxation of the layer and/or carbide formation..

INTRODUCTION

The $Si_{(1-x)}Ge_x$ system has received considerable attention for applications in the field of optoelectronics and high speed heterojunction bipolar transistors. These applications require a very good control of composition and stress of the epitaxial layers. To reduce the layer stress, inclusion of carbon has been studied by different techniques like molecular beam epitaxy, chemical vapor deposition (1), implantation (2) or laser annealing after implantation (3-4). However, due to the formation of carbide precipitates, only two ways of preparation can be applied; the growth can be made at low temperature (<850°C) to prevent carbide formation, or the process can be faster than the carbide formation using for example rapid thermal annealing (5) or liquid phase epitaxy induced by pulsed laser heating (3-4,6). In this work, the latter method is used to recrystallize silicon layers amorphized by implantation of carbon in view of futher application to $Si_{(1-x-y)}Ge_x C_y$ structures. Very high energy excimer laser system (VEL) developped by SOPRA which allows homogeneous annealing on large surfaces is used. Non destructive structural characterization after implantation and after laser annealing are made using spectroscopic ellipsometry, x-ray diffraction and Rutherford backscattering techniques.

EXPERIMENTAL TECHNIQUES.

Sample implantation

Float zone <100> monocrystalline silicon wafers are used. Different samples have been implanted with variable carbon content and four successive energies (15, 25, 40 and 50keV) to achieve flat carbon profiles over about 150nm. Three wafers have been implanted with different doses to get 0.67, 1.1 and 1.32% for the mean carbon content.

Mat. Res. Soc. Symp. Proc. Vol. 452

Laser annealing

Annealing has been performed using a SOPRA VEL excimer laser with maximum total energy on the sample surface of 45J/pulse. This is a XeCl excimer laser working at 308nm with x-ray preionization to initiate the discharge (7-8). Three different laser cavities have been associated to provide 45J/pulse on the sample surface. The pulse duration is around 200ns. All the treatments of the study have been made with only one laser shot in air at room temperature.

Characterization techniques

SE measurements were made with a commercially available SOPRA ES4G rotating polarizer instrument with a high resolution double monochromator and photon counting detector. The angle of incidence was fixed in this study to 75° close to the pseudo Brewster angle of silicon for optimum sensitivity. A 3mm diameter beam with <0.5mrad divergence is produced by a mirror system. The polarisation state of the light is modulated by a rotating quartz Rochon polarizer and analyzed by another fixed Rochon after reflection from the sample. The wavelength range is selected from 0.25 to 0.85µm. XRD measurements were made with a conventional Philips HR1 diffractometer especially adapted to the grazing measurements. A cobalt x-ray tube with fine focus is setup with a monochromator consisting of four Ge crystals to select the Co K-α line. After reflection the beam is detected by a proportional counter. Details on the setup can be found elsewhere (9). Depth resolution of Rutherford backscattering was enhanced using 2MeV He+ and a glancing angle detection geometry. The tetragonal expansion (or contraction) of the films was estimated from the measurement of the angular shift between oblique incidence channeling dips respectively recorded on distorted layers and substrate as reported in reference (10).

EXPERIMENTAL RESULTS

SE analysis of $Si_{(1-y)}C_y$ layers

SE has been applied systematically on all the samples before and after laser annealing. As reported in Figure 1 in the case of the 1.32% C implanted sample, SE measurements show a great variation with the energy of the laser beam. The as-implanted sample shows a smooth shape both for the tan Ψ and cos Δ (Cf. Figure 1) ellipsometric parameters. On the contrary the sample annealed at 1.2J/cm2 shows very well defined interference fringes. In the UV range the measurement changes drastically compared to the as-implanted sample and the shape of the curves are also very similar for all the annealing energies. Since in the UV range, the light absorption allows to see only the top surface of the sample (penetration depth < 200Å), we can conclude that the samples are crystallized almost at the top surface even for the lower laser energy. For the higher laser energies (2.1 and 2.3J/cm2), there are no more interference fringes and the sample seems completely crystallized.

The SE measurements reported in Figure 1 have been simulated using graded multilayer structures including Gaussian distribution of amorphous silicon inside crystalline silicon. Optical indices of both materials have been extracted from our database (crystalline silicon indices have been extracted from measurement on monocrystalline silicon wafers, amorphous silicon has been extracted from measurement on LPCVD hydrogen free amorphous silicon layers).

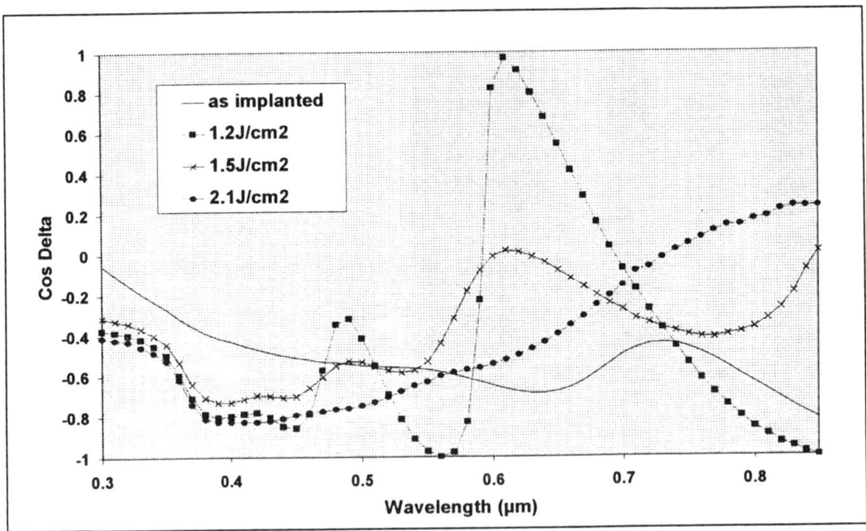

Figure 1: CosΔ parameter measured by SE on the 1.32% C implanted silicon sample before and after laser annealing.

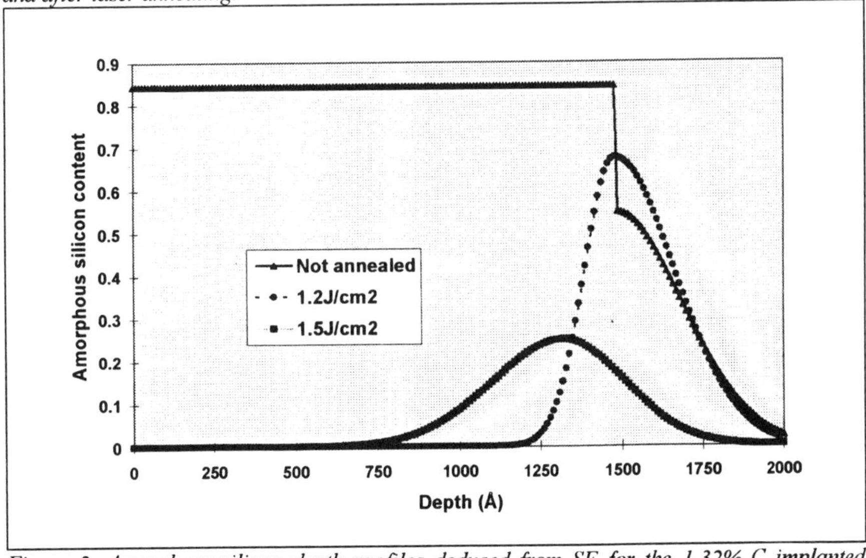

Figure 2: Amorphous silicon depth profiles deduced from SE for the 1.32% C implanted silicon sample. No more amorphous layer is detected after annealing at 2.1J/cm2.

More precisely, the sample structure is split into 30 layers with the same thickness and the total thickness is fitted during the regression with the amorphous content of each layer. In each case, an equivalent SiO_2 top layer of 40-60Å in thickness, is added to the structure to take into account the non stoechiometric oxide layer detected by RBS at the surface of the layers after laser annealing. We have used a fixed composition on a top layer and a Gaussian shape for the amorphous content decrease at the bottom interface to account for the measurement on the as-implanted samples. For the samples annealed at low density of energy we have used a double asymetric Gaussian profile for the remaining amorphous region located in depth of the sample surface. The different amorphous profiles used for this simulation of the measurement of Figure1 are reported in Figure 2 versus the depth inside the silicon sample. We found that a 500Å thick zone at the bottom of the implanted region, stay amorphous after annealing at 1.2J/cm2. After irradiation at higher density of energy (1.5J/cm2), the amount of amorphous material is reduced in the perturbed region (Cf. Figure 2). After irradiation at 2.1J/cm2 or more, SE cannot detect any remaining amorphous region and the sample seems completely crystallized in the entire depth but with a slightly different optical index.

RBS and XRD analysis of $Si_{(1-y)}C_y$ layers

As shown in Figure 3, the SE model is confirmed by RBS analysis of the same samples. Indeed, an amorphous region is clearly detected in depth of the sample whereas the top surface is crystallized. The depth of the perturbated region can be evaluated around 1500Å from the RBS measurement in agreement with the SE measurements (Cf. Figure 2). After annealing at 2.1J/cm2 the RBS spectrum is very similar to bare silicon showing that the layer is really epitaxial and monocrystalline with a low quantity of defects (Cf. Figure 2).

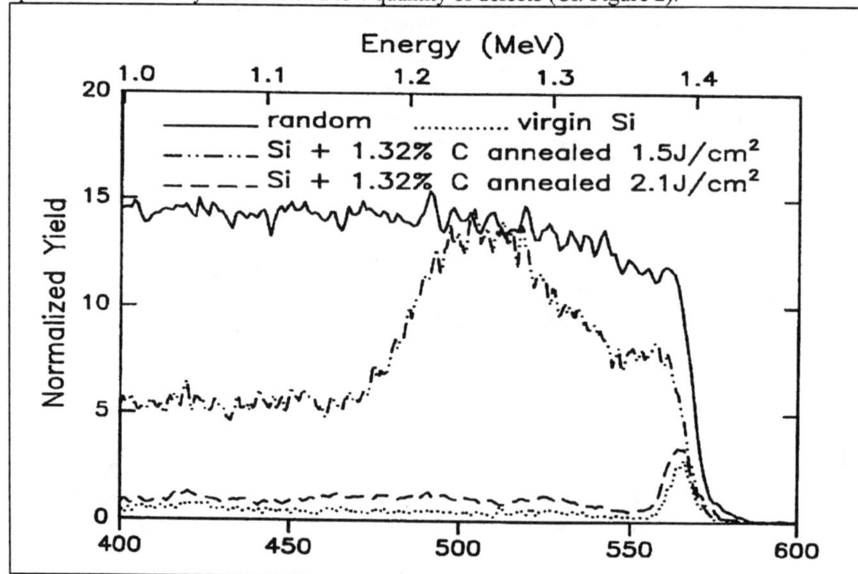

Figure 3: Experimental RBS spectrum of the 1.32% C implanted silicon sample annealed at 1.5J/cm2 and 2.1J/cm2 compared to random and channeling Si spectra.

Examples of XRD spectra obtained after annealing of the 1.1% C implanted silicon sample are reported in Figure 4. In addition to the Bragg peak of Si some additional contributions appear at higher angles. These contributions are more or less well defined depending on the annealing conditions. For energy densities not sufficient to get complete recrystallization of the implanted zone (1.2 and 1.5J/cm2 from SE measurements), the contribution is very broad. On the contrary for energy densities sufficient to melt the entire layer (2.1 and 2.4J/cm2), the contribution becomes very well defined. In the case of 1.1% C implanted sample, at 2.1J/cm2 an additional well defined Bragg peak can be detected (Cf. figure 4). The additional Keissig fringes can be used to determine the thickness of the $Si_{(1-y)}C_y$ epitaxial layer in this case (~1500Å in agreement with SE measurements). The position of the peak can be used to extract the deviation of the lattice parameter from silicon. Results obtained on the different epitaxial layers are summarized in Table I. As expected, the strain of the layer increases with the carbon content from 0.67% to 1.10%. For the highest carbon content, the stress is reduced due to either some silicon carbide precipitation in agreement with other studies (4-11), or partial relaxation of the layer. For the highest density of energy (2.4J/cm2) used in the study, the structural quality seems also slightly reduced compared to 2.1J/cm2. The occurance of two contributions for the 1.1% C implanted sample at this energy (Cf. figure 4) confirms the partial relaxation of the layer and the occurance of dislocations. Further studies are needed to clear this point. Oblique incidence channeling dips around the <111> plane have been also performed to provide the lattice deviation of the $Si_{(1-y)}C_y$ layer. Similar results as those deduced from XRD are found except for the highest carbon concentration (1.32%), for which the abscence of any angular shift indicates that the layer is fully relaxed.

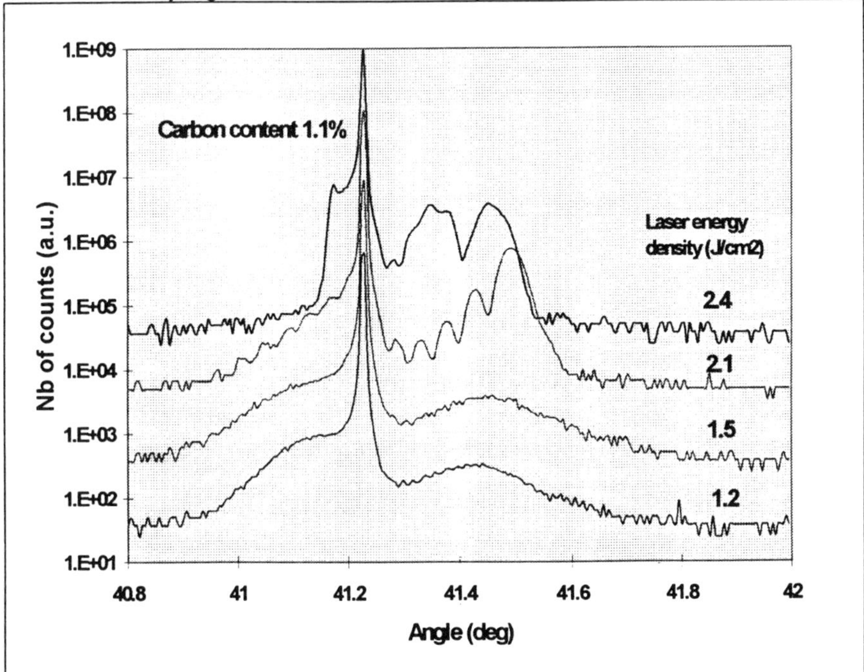

Figure 4: Experimental XRD spectra of the 1.1% C implanted silicon samples after annealing conditions. For figure clarity the curves have been shifted.

CONCLUSION

In this study we have shown that excimer laser annealing of carbon implanted silicon can provide high quality epitaxial layers when the implantation profile is flat and when the energy density is optimized accurately. The crystalline quality is very good as observed by RBS and

XRD techniques. Moreover, the lattice mismatch observed in these films increases with the carbon content showing that the carbon inclusion is probably free of carbide formation up to 1.1%. Further studies will be necessary to get $Si_{(1-x-y)}Ge_xC_y$ epilayers in the same way using implantation of carbon and germanium in silicon and laser annealing. Nevertheless, the possibility up to now to anneal large surfaces (up to 20 cm² in one laser shot) is a key point for further industrial applications.

Carbon concentration (%)	**Energy density (J/cm2)**	**Lattice deviationfrom XRD (%)**
0.67	2.1	-0.34
0.67	2.4	-0.32
1.10	2.1	-0.52
1.10	2.4	-0.46
1.32	2.1	-0.45
1.32	2.4	-0.40

Table I: Lattice deviation by XRD technique for the different epitaxial $Si_{(1-y)}C_y$ *layers*

ACKNOWLEDGMENTS

Mr. **L. Hennet** from University Henry Poincaré (Nancy, France) is kindly acknowledged for the XRD measurements.

REFERENCES

1) S.C. Jain, H.J. Osten, B. Dietrich, H. Richter, Semicond. Sci. Technol., 10, p. 1289 (1995)

2) J.W. Strane, H.J. Stein, S.R. Lee, B.L. Doyle, S.T. Picraux, J.W. Mayer, Appl. Phys. Lett., 63, N°20, p. 2786 (1993)

3) E. Fogarassy, D. Dentel, J.J. Grob, B. Prevot, J.P. Stoquert, R. Stuck, MRS Symp. Proceed., V 354, p. 585 (1995)

4) Z. Kantor, E. Fogarassy, A. Grob, J.J. Grob, D. Muller, B. Prevot, R. Stuck, Appl. Phys. Lett., 69, N°7, p. 969 (1996)

5) J.W. Strane, S.R. Lee, H.J. Stein, S.T. Picraux, J.K. Watanabe, J.W. Mayer, J. Appl. Phys., 79, N°2, p. 637 (1996)

6) A.G. Cullis, R. Series, H.C. Weber, N.G. Chew, Semiconductor silicon 1981, Edited by R.F. Huff (Electrochemical Society, Pennington, N.J.), p. 518 (1981)

7) B. Godard, P.Murer, M. Stehle, J. Bonnet, D. Pigache, GLC 92; Heraklion, Greece, September (1992)

8) M. Stehle, Laser Focus World, june, p. 135 (1993)

9) J.Bobo, B.Baylac, L.Hennet, O.Lenoche, M.Piecuch, B.Raqet, J.Oussel, V.Viel, E.Snoeck, J. of Magn. And Magn.Mat., 121, 291 (1993)

10) A. Grob, J.J. Grob, D. Muller, B. Prevot, R. Stuck, E. Fogarassy, EMRS Spring Meeting, Symposium D, paper D-P4 (1996)

FABRICATION OF POLYCRYSTALLINE SI THIN FILM FOR SOLAR CELLS

M. TANAKA, S. TSUGE, S. KIYAMA, S. TSUDA AND S. NAKANO
New Materials Research Center, Sanyo Electric Co., Ltd.
1-18-13 Hashiridani, Hirakata, Osaka 573, Japan

ABSTRACT

The a-Si/poly-Si thin film tandem solar cell is a promising candidate for low-cost solar cells. We have conducted R&D on poly-Si thin film using the Solid Phase Crystallization (SPC) method from amorphous silicon (a-Si). To improve the film quality of SPC poly-Si, we have developed a new SPC method called the partial doping method. This method features two stacked starting a-Si layers, a P-doped layer and a non-doped layer. Nucleation occurs in the P-doped layer, and the non-doped layer is the crystal growth layer. For the nucleation layer, we developed a Si film with a unique structure, which features relatively large crystallites (~1000A) embedded in a matrix of amorphous tissue. By combining these technologies, a conversion efficiency of 9.2% was obtained for poly-Si thin-film solar cells. For further improvement in the conversion efficiency, based on the concept of "independent control of nucleation and crystal growth", it is necessary to combine the best fabrication methods for each layer. A high conversion efficiency of more than 12% was found possible by using the CVD method and a new back surface reflection structure.

INTRODUCTION

Recently, solar cells have been gathering much attention as a clean energy source which may help to solve many problems of environmental pollution. Sanyo Electric has been conducting R&D on amorphous silicon (a-Si) solar cells. We have developed various original technologies, such as the integrated type structure[1], the separated reaction chamber method[2], the laser patterning method[3], and the super chamber method[4]. In 1980, Sanyo started the mass production of a-Si solar cells for the first time[5]. Our advanced technologies have enabled us to achieve a stabilized conversion efficiency of 10.6% for an a-Si/a-SiGe tandem solar cell, and 8.9% for a single junction a-Si solar cell[6]. In the field of poly-Si thin-film, a high field effect mobility of more than $400cm^2/Vs$ was obtained by using the laser recrystallization method[7].

Based on the above technologies, we have continued to conduct R&D for solar cells with high conversion efficiency and low cost. Although the a-Si solar cell is one of the candidates, its efficiency is not sufficiently high because of its limited photosensitivity. To achieve solar cells with a much higher conversion efficiency, it is necessary to develop materials with high photosensitivity in the long-wavelength region where a-Si films have no photosensitivity. Polycrystalline silicon (poly-Si) thin-film is an effective material for this purpose and a-Si/poly-Si tandem solar cells are the most feasible[8], considering the cost of solar cells as shown in Fig.1.

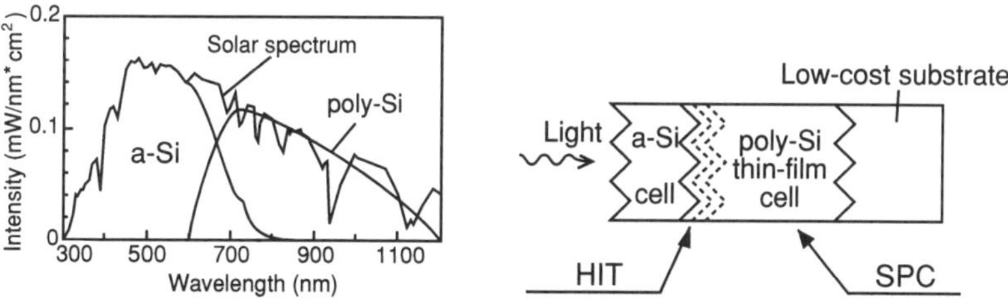

Fig. 1 Solar spectrum and target structure of high-efficiency, low-cost solar cells

Mat. Res. Soc. Symp. Proc. Vol. 452 © 1997 Materials Research Society

Based on our policy for this development, we investigated a device structure for a heterojunction between a-Si and crystalline Si and the material properties of thin-film poly-Si. For the device structure, we have developed a new-structure solar cell, called HIT[9] based on a p-type a-Si/n-type crystalline silicon heterojunction solar cell. The HIT solar cell offers high performance with low cost. For materials, we have been investigating poly-Si thin films fabricated by the solid phase crystallization (SPC) method[10] as a material for solar cells with photosensitivity in the long-wavelength region. This SPC method has many features suited to the fabrication process of solar cell materials, such as a low process temperature.

In this paper, we review our R&D on poly-Si thin-film solar cells using the HIT structure and the SPC method. We also discuss topics for further improvement.

HIT STRUCTURE SOLAR CELL

Concept of the New HIT Structure Solar Cell

Figure 2 shows the structures and doping concentrations of the conventional p/n homojunction (a), p-type a-Si/n-type crystalline Si heterojunction (b), and new HIT (c). As the conventional homojunction needs a high temperature around 800℃, this structure cannot be used in our low-cost solar cell process. In addition, it is difficult to fabricate an abrupt junction because of cross-contamination of the dopants in the homojunction. The p-type a-Si/n-type c-Si heterojunction can solve these problems because this structure has the following attractive characteristics:

(1) neither a high temperature nor much energy is necessary for processing.
(2) a top cell (a-Si cell) can be fabricated in the same process for tandem use.
(3) a shallow, sharply-profiled junction can be easily obtained through a deposition process.

But it was found that a slight cross-contamination of dopants remains in the heterojunction.

To improve the properties of the p/n heterojunction, we have developed a new "HIT" structure, in which a non-doped a-Si thin layer is inserted between the p/n heterojunction. We call it HIT, which stands for Heterojunction with Intrinsic Thin-layer.

In order to investigate the HIT structure, single crystalline Si (c-Si) was used as a substrate instead of poly-Si thin film.

Comparison between HIT and p/n Heterojunction

The dark I-V characteristics for the HIT cell and the p/n heterojunction solar cell were measured. Using the HIT structure, the backward current density was reduced from 10^{-6} to about 10^{-8} A/cm^2, in spite of a slight change in the forward current. The non-doped a-Si layer

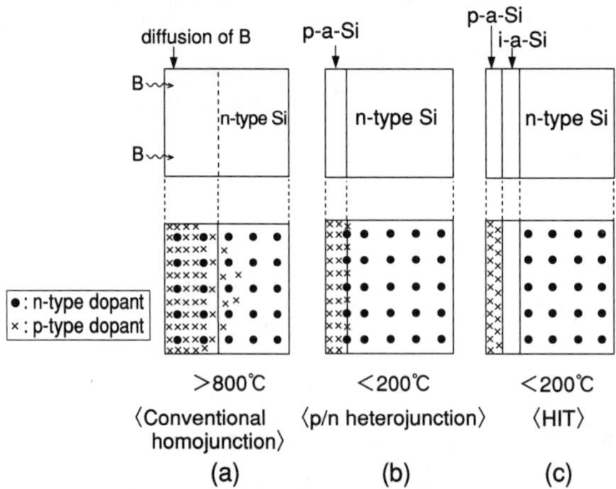

Fig. 2 Structures and dopant concentrations of various structure solar cells

seems to reduce the recombination of carriers near the interface.

Figure 3 shows the output characteristics of the HIT-structure solar cell as a function of the non-doped a-Si layer thickness. In this figure, the data at a film thickness of zero are those of p/n heterojunction solar cells. Apparently, with the HIT structure , both the V_{OC} and the F.F. were improved. The V_{OC} was especially improved, by about 30mV, and a F.F. of more than 0.8 was obtained. The I_{SC} became lower as the thickness of the non-doped a-Si layer increased. Although the total thickness of the a-Si layers (p-layer in the p/n heterojunction, p- and i-layers in the HIT structure) was the same, the I_{SC} was higher in the HIT cells than in the p/n heterojunction solar cells. This was due to higher collection efficiency in the short wavelength region. In the HIT structure , therefore, the non-doped a-Si layer seems to be a slightly "active" layer.

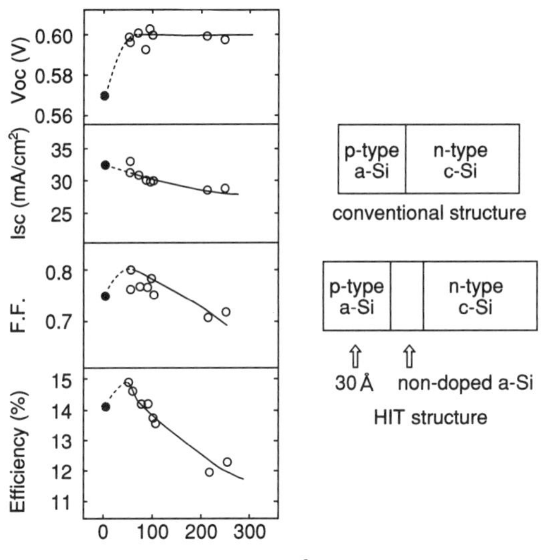

Fig. 3 Output characteristics of HIT dtructure solar cells as a function of the thickness of non-doped a-Si layer

Improvement in Conversion Efficiency in HIT-Structure Solar Cells

To achieve higher efficiency, various technologies were applied to the HIT structure. Using a c-Si substrate with a textured surface and HIT structure on both sides, a conversion efficiency of 20.0% was achieved with a size of $1cm^2$ cell as shown in Fig. 4. This is the highest value ever reported for a solar cell with an a-Si/c-Si heterojunction structure or a junction that is prepared at a low temperature (~120℃).

Stability

a-Si based solar cells exhibit light-induced degradation, called the Staebler-Wronski effect[11]. This might also be the case in the HIT solar cell. Figure 5 (a) shows the result of a light-soaking test for the 20%-efficient HIT solar cell. After 5 hours' soaking at 5 suns, no degradation was observed in the HIT solar cell. This is probably due to the fact that the a-Si layers are very thin and contribute little to the power generation. Figure 5 (b) shows the result of a high-temperature exposure test for the above cell. Although the junction is fabricated at a temperature below 150℃, no degradation was observed. Thus it can be concluded that the HIT solar cell offers good stability in practical applications.

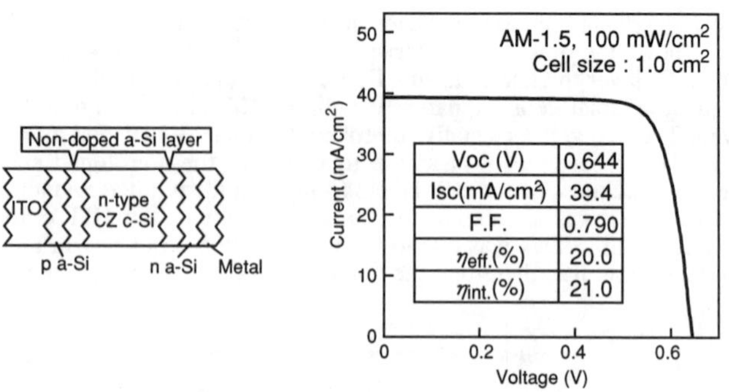

Fig. 4 Illuminated I-V characteristics of a HIT solar cell confirmed at JQA. 1-5 Ωcm, ~400 μ m-thick CZ water is used

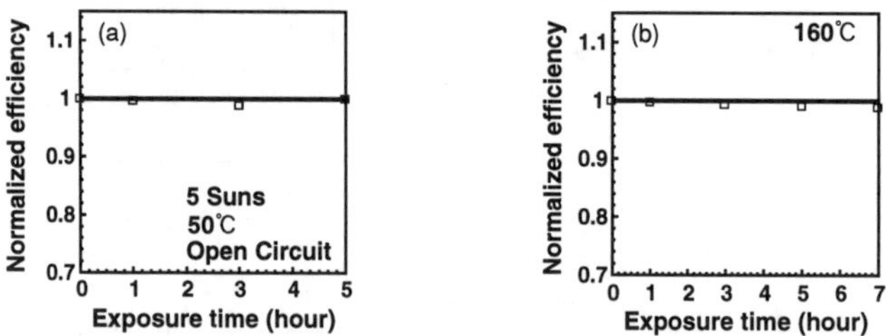

Fig. 5 Results of (a) alight-soaking test and (b) a high-temperature exposure test for the 20% efficient HIT solar cell

THIN-FILM POLY-SI BY SPC METHOD

Solid Phase Crystallization (SPC) Method

The SPC method which we have developed is a way to prepare poly-Si thin film from a-Si film made using the plasma-CVD method by thermal annealing. This SPC method has following features:

(1) The process temperature is low at around 600℃.
(2) The process is very simple.
(3) Poly-Si thin film with a large area can be prepared.

Figure 6 shows a diagram of the SPC method, which consists of two processes. In the first process, phosphorus (P) doped a-Si films were deposited by the plasma-CVD method onto a substrate. In the second process, a-Si films were transferred to the poly-Si by thermal annealing in a vacuum at a low temperature.

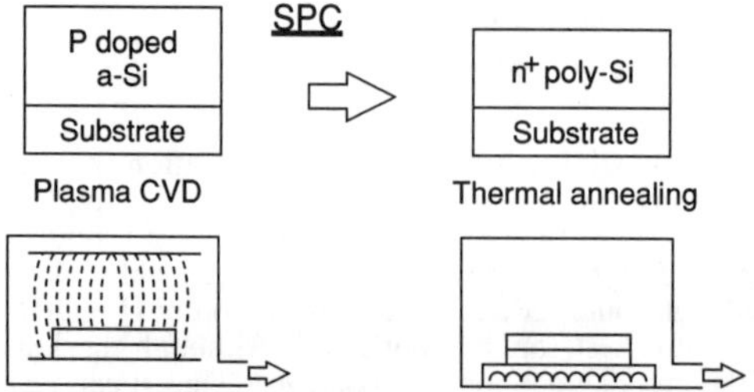

Fig. 6 The diagram of the SPC method which consists of two processes

Properties of Poly-Si Thin-Film Prepared by the SPC Method

P-doped a-Si films were prepared by the plasma-CVD method on quartz substrates. Phosphorus was in-situ doped by mixing PH_3 with SiH_4 as a source gas for the SPC process. The substrate temperature was in the range from 450℃ to 650℃.Thermal annealing was performed for 3 ~ 100 hours at temperatures around 600℃ in a vacuum. First, the effect of phosphorus atoms on SPC was investigated. The SPC of non-doped a-Si film did not occur and remained a-Si. On the other hand, the SPC of P-doped a-Si film did occur. So, it was found that SPC occurred easily when phosphorus doping was used.

Partial Doping Method

In order to improve the photovoltaic ability of poly-Si films in the long wavelength region, it is essential to fabricate poly-Si films with a columnar structure (no horizontal grain boundary). We found it effective to separate the crystal nuclei generation layer and crystal growth layer in order to control the direction of crystal growth and obtain films with a columnar structure as shown in Fig. 7. The most effective of these approaches is a partial doping method[12], which features two stacked starting a-Si layers (a P-doped layer and an undoped layer). The P-doped layer serves as a nucleation layer since nucleation preferentially occurs in the P-doped layer in the initial stage of the SPC process.

Figure 8 shows cross-sectional SEM photographs of n-type poly-Si prepared by bulk doping and partial doping, where the phosphorus concentration is $1.2\times10^{20}cm^{-3}$ and the SPC condition is 600℃, over 10 hours. In bulk doping and partial doping, n-type poly-Si after SPC have almost the same cross section and grain size. But the phosphorus concentration in n-type poly-Si prepared by partial doping is lower than that prepared by bulk doping, and its concentration is about $2\times10^{18}cm^{-3}$. So, by using the partial doping method, n-type poly-Si with large grain was obtained in the region of low phosphorus concentration.

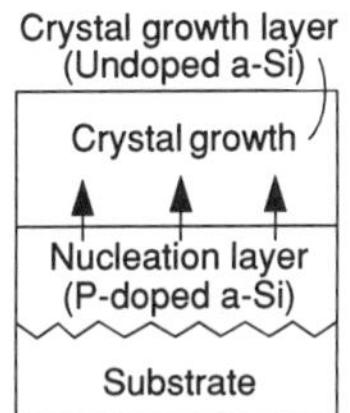

Fig. 7 Conceptual figure of our SPC method

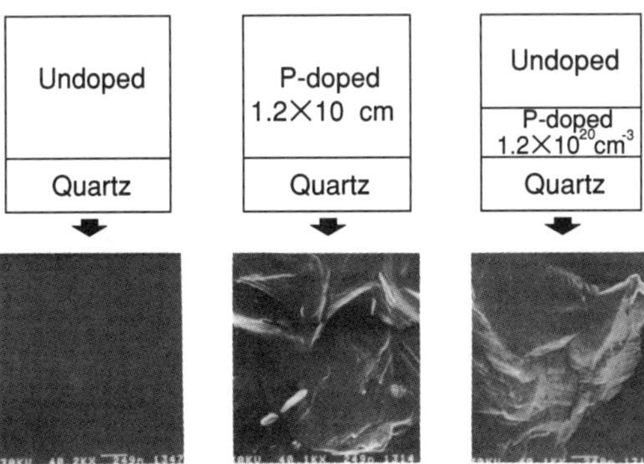

Fig. 8 Cross-sectional SEM photographs of non-doped and n-type poly-Si prepared by bulk doping and partial doping

Next, the minority carrier trap density[13] of n-type poly-Si prepared by partial doping was investigated. Figure 9 shows a comparison of the minority carrier trap density between partial doping and bulk doping. The minority carrier trap density of n-type poly-Si prepared by partial doping decreased further, compared to that prepared by bulk doping. A minimun value of $4.6\times10^{11}cm^{-2}$ was obtained.

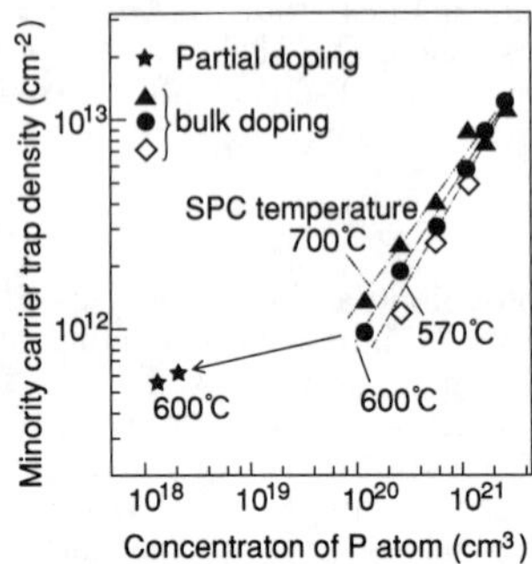

Fig. 9 Comparison of minority carrier trap density in partial doping and bulk doping as functions of phosphorus concentration and SPC temperature

Improvement in the Crystal Growth Layer

The structural disorder of the a-Si films before the SPC process was analyzed to investigate the a-Si film's suitability to solid phase crystallization. Figure 10 shows the typical Raman spectrum of an a-Si film. Two prominent peaks are observed at about $480cm^{-1}$ and $150cm^{-1}$. These peaks are attributed to the transverse optical (TO) like phonon mode and transverse acoustic (TA) like phonon mode, respectively. It has been reported that the TA peak height divided by the TO peak height is a sensitive probe of the structural disorder of a-Si[14]. So, TA/TO was used as a parameter for structural disorder. The peak heights were measured from the broken line in Fig. 10. The correlation between the grain size of poly-Si and the structural disorder of the a-Si films was investigated. Figure 11 shows the relationship between the average grain size of poly-Si and the TA/TO value of the a-Si films. The average grain size was defined as the average of all grains visible in surface and cross-sectional SEM images. The average grain size of poly-Si increased with increases in the TA/TO value of the a-Si films, and a maximum average grain size of 2.5 μm was obtained at a TA/TO value of 0.47. This relationship was clarified through this study for the first time. It is remarkable that the suitable TA/TO value is small in device-quality a-Si films, but large in a-Si films for solid phase crystallization.

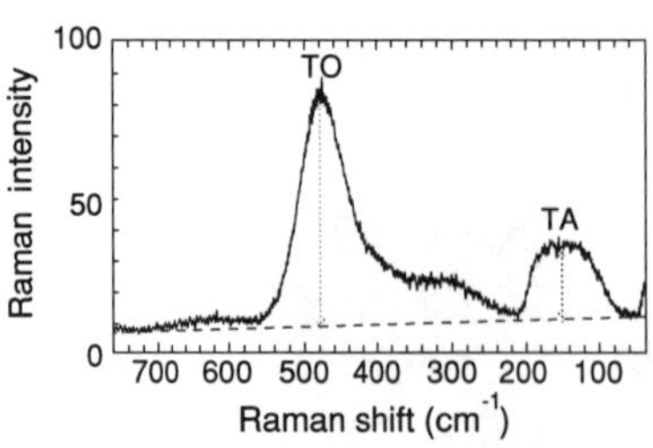

Fig. 10 Typical Raman spectrum of an a-Si film

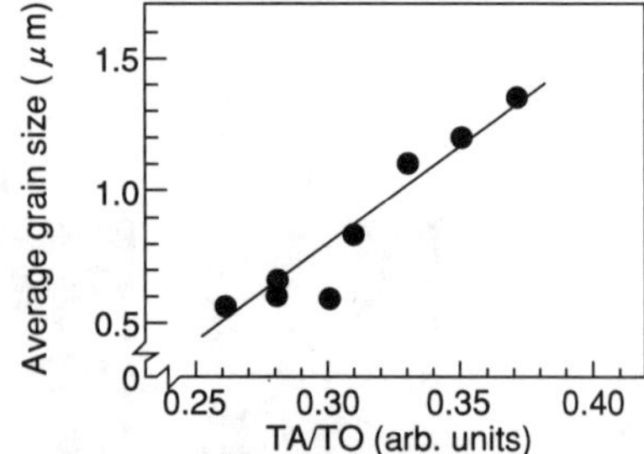

Fig. 11 Relationship between the average grain size of poly-Si after the SPC process and the TA/TO value of a-Si film before the SPC process

Improvement in the Nucleation Layer

To further improve the film quality, we have focused on the film structure of the nucleation layer and obtained a silicon film with unique structure.

Nucleation layers were deposited by plasma-CVD from silane (SiH_4) gas diluted with hydrogen on flat or textured quartz substrate. The typical deposition conditions are summarized in Table 1. The degree of roughness for the textured substrate is about 10 μm. Crystal growth layers with a thickness of 5 μm were subsequently deposited by plasma-CVD from SiH_4 gas onto the nucleation layers.

Table 1 Deposition conditions of Si films

	Flow rate (sccm)			RF power (W)	Pressure (Pa)	Temperature (°C)
	100%SiH_4	0.1%PH_3/H_2	H_2			
Nucleation layer	3–10	1–10	40–300	20–50	10–100	550–650
Crystal growth layer	10–50	—	—	80	50–100	550–600

Figure 12 shows the deposition temperature dependence on the volume fraction of crystallites for microcrystalline silicon (μc-Si) films fabricated by plasma-CVD from SiH_4 diluted with hydrogen. The volume fraction is estimated by X-ray diffraction (XD) spectroscopy. The open symbols represent the work done by Matsuda[15], which shows that a crystalline-to-amorphous transition region exists at ~500℃, and that the deposited films are amorphous above that temperature. From the viewpoint of obtaining poly-Si with a larger grain size, a film with a smaller proportion of crystallites to amorphous tissue is preferable to a nucleation layer, because the density of the nuclei can be smaller. Thus, we experimented at the temperature region above 500℃. Our experimental results are indicated by the solid symbols in Fig. 12. As can be seen, it seems that no crystallite exists in the obtained films. However, the results of the SEM photographs are different. In Fig. 13, SEM photographs of a conventional μc-Si film and a silicon film deposited at 550℃ are shown. Our film features relatively large crystallites (~1000 Å) embedded in a matrix of amorphous tissue.

This film has a favorable structure for the nucleation layer since it has many crystallites which act as seeds for the subsequent crystallization. The film was applied to the nucleation layer. 5 μm-thick undoped a-Si was deposited by plasma-CVD onto the layer. After that, the film was crystallized by thermal annealing. Figure 14 shows a cross sectional Transmission Electron Microscopy (TEM) image of the obtained poly-Si. As can be seen, a large grain with no horizontal boundary from the bottom to the top surface is obtained.

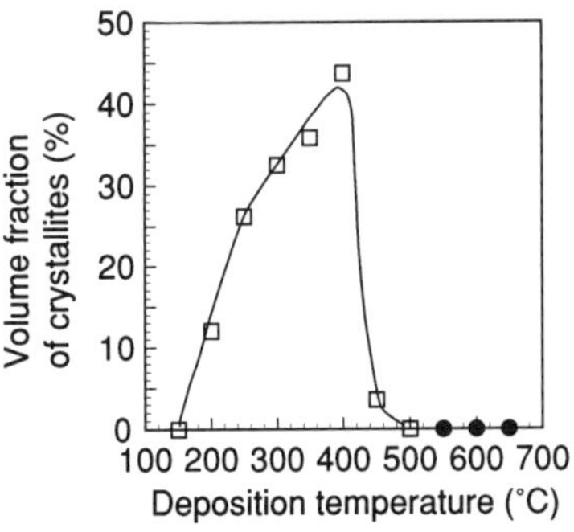

Fig. 12 Deposition temperature dependence on the volume fraction of crystallites. Open symbols are quoted from ref.[15]

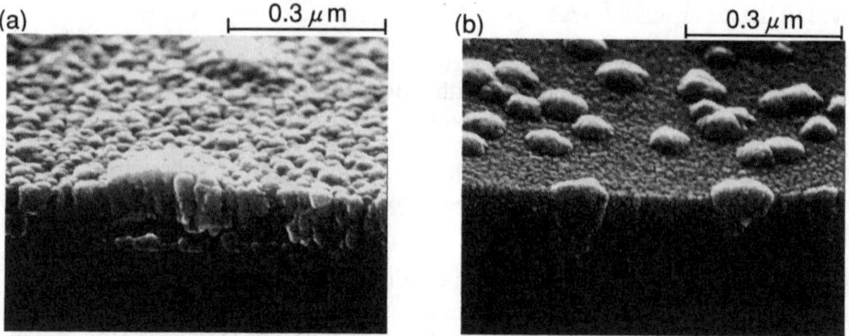

Fig. 13 SEM photographs of (a) a conventional μ c-Si and (b) the newly developed silicon film

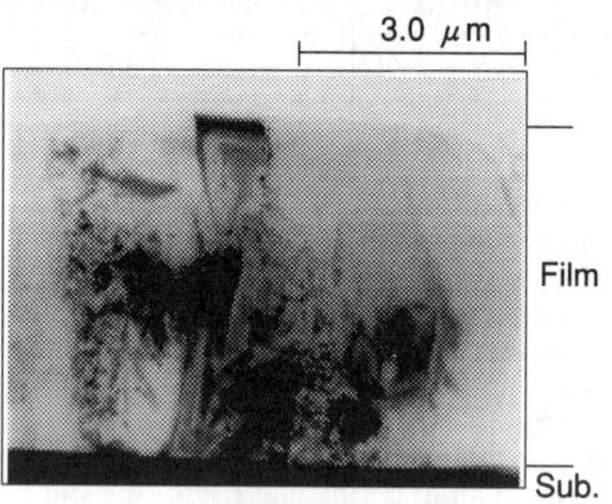

Fig. 14 Cross-sectional TEM image of an SPC poly-Si film
The SPC temperature is 600℃

Poly-Si Thin-film Solar Cell

To test the photovoltaic ability of the poly-Si films, we fabricated solar cells with the structure of ITO/p-type a-Si/i-type a-Si/n-type poly-Si (10 μ m)/n^+-type poly-Si/metal. The i-type and p-type a-Si:H layers were deposited by the plasma-CVD method. The n^+-type poly-Si layer functions as a Back Surface Field (BSF) layer and provides good ohmic contact to metal. The illuminated I-V characteristics were measured under AM 1.5 (100mW/cm^2) condition, and a total area conversion efficiency of 9.2% was achieved as shown in Fig. 15. This high conversion efficiency for a thin-film poly-Si solar cell indicates the considerable potential for the active layer (only 10 μ m thick) of thin-film solar cells.

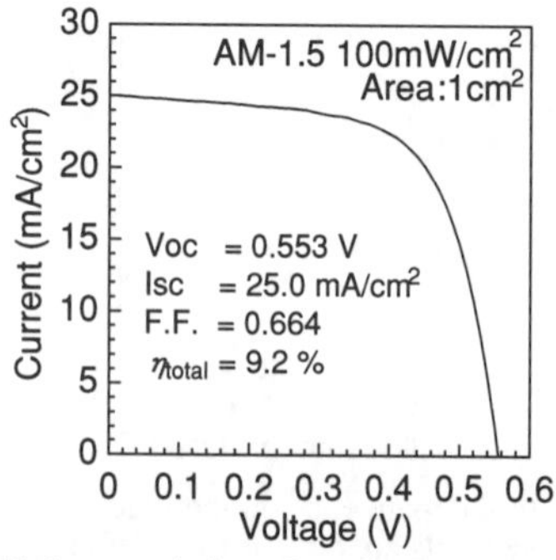

Fig. 15 Illuminated I-V characteristics of a poly-Si cell prepared by the SPC method

NEW FABRICATION METHOD FOR FURTHER IMPROVEMENT

Important Issues of SPC Method

For further improvement in poly-Si thin-film solar cells, there are two categories of important issues: One is for high performance, the other is for industrialization. For high performance, we will develop technologies based on the independent control of the nucleation layer and crystal growth layer, which is the concept of the partial doping method. There are important issues which include the passivation of the grain boundary, improvement of the interface properties between the silicon layer and the substrate, and so on. For industrialization, the SPC method requires an improvement in reliability, enlargement of the device area, and so on. However, the most important issue is the achievement of higher through-put.

New Fabrication Method with a High Through-put

For a higher through-put, we have tried a new fabrication method using a CVD-Si layer on the poly-Si seed layer fabricated by the SPC method, as shown in Fig. 16. This method is based on the independent control of nucleation and crystal growth, and the CVD method offers the possibility of a high deposition rate. Our target for the CVD-Si is a high deposition rate of more than 1 μ m per minute, high quality, and a low process temperature .

The Preparation conditions of our CVD are as follows: Si_2H_6 gas was used as the material gas, because a high deposition rate was obtained at a low temperature. The substrate temperature was around 750℃. The deposition rate was around 0.1~0.2 μ m/minute in this experiment.

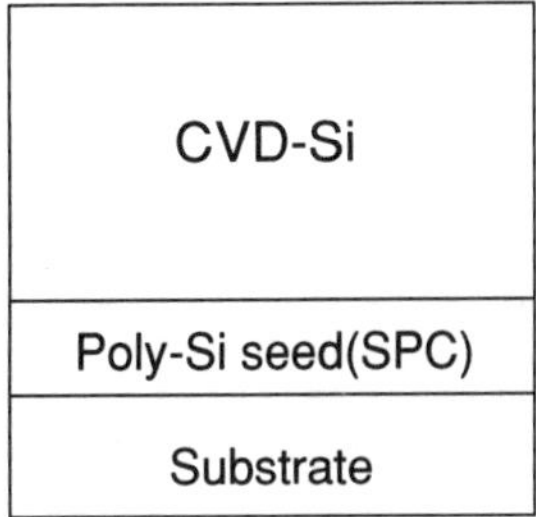

Fig. 16 New fabrication method using CVD

Film Quality of CVD-Si

In order to investigate the quality of the CVD-Si film, we deposited CVD-Si film onto a single crystalline silicon wafer instead of the SPC seed layer. Figure 17 shows a TEM image of epitaxial Si on (111) single crystalline silicon deposited by the CVD method. From this figure, it is found that interference patterns are continuous from c-Si to epi-Si and there are no defects.

We also fabricated solar cells onto a c-Si wafer using the CVD method. The substrate was a low-resistivity n-type c-Si wafer. The junction was a HIT structure. The thickness of poly-Si layer was only 4 μ m. We obtained a conversion efficiency of about 14%. From these results, the CVD-Si layer is found to be high in quality.

As a next step, we fabricated silicon films onto metal. SPC silicon film was fabricated on W film and a subsequent silicon layer was deposited by the CVD method. Figure 18 shows the TEM image of the sample. The right part of the figure shows the electron diffraction patterns of three different positions. From these patterns, it is found that the poly-Si thin-film has good crystallinity.

However, solar cells using this method showed relatively low performance. The conversion efficiency of these cells is less than that by the SPC method.

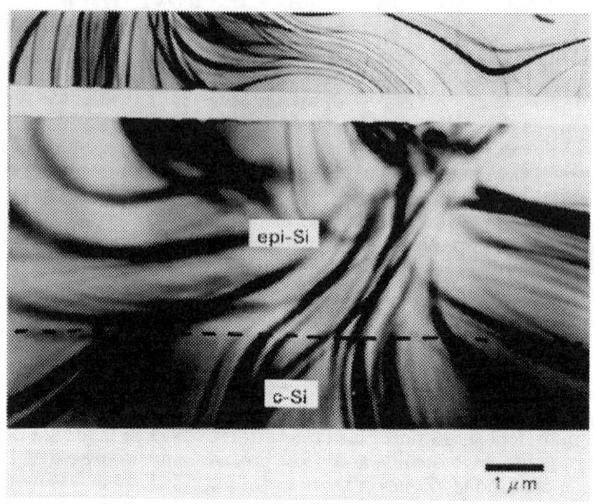

Fig. 17 TEM image of epitaxial Si / (111) c-Si

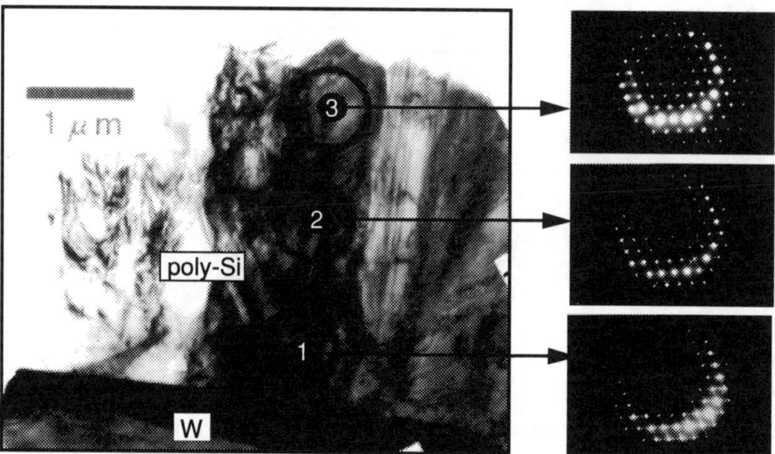

Fig. 18 TEM image and diffraction patterns of poly-Si on W

Improvement in Back Reflectance

One of the main reasons why the performance of the solar cells is relatively low is the low optical reflectance at the back surface. In order to solve this problem, we investigated various kinds of back surface structures.

Figure 19 shows the calculation results of optical absorption in silicon films with various kinds of back electrodes. By using a silicon dioxide layer or ITO layer, optical absorption can be improved.

Based on another calculation, using PC-1D, it was found that a high conversion efficiency of more than 12% is possible with a thickness of 10 μm by combining the new fabrication method using the CVD method with the new back reflectance structure.

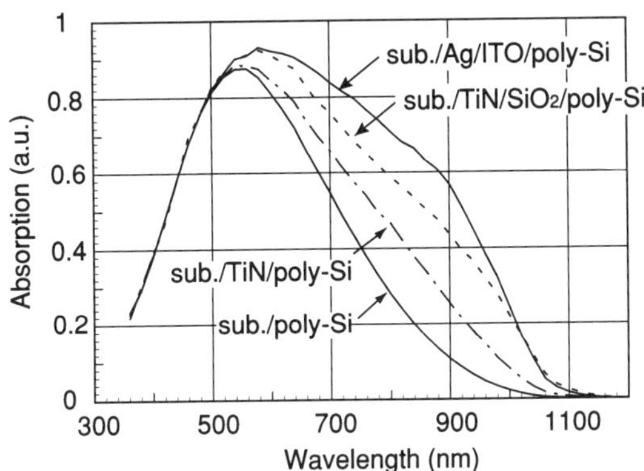

Fig. 19 Calculated results of optical confinement performance by three-dimensional simulator. The thickness of the poly-Si layer is assumed to be 4 μ m

CONCLUSION

In order to improve the conversion efficiency of thin-film poly-Si solar cells, we have developed the HIT structure as a device technology and a novel SPC method as a material technology.

The highest conversion efficiency of 20% was obtained for a HIT solar cell using a c-Si wafer instead of thin-film poly-Si.

For the material technology, we have developed a new SPC method called the partial doping method. We also developed a Si film with a unique structure as a nucleation layer, which features relatively large crystallites (~1000A) embedded in a matrix of amorphous tissue. By combining the above technologies, a conversion efficiency of 9.2% was obtained for poly-Si thin film solar cells.

To further improve the conversion efficiency, based on the concept of "independent control of nucleation and crystal growth", it is necessary to combine the best fabrication methods for each layer. A high conversion efficiency of more than 12% was found possible by using the CVD method and a new back surface reflection structure.

ACKNOWLEDGEMENT

This work was supported by NEDO (New Energy and Industrial Technology Development Organization) as a part of the New Sunshine Project under the Ministry of International Trade and Industry.

REFERENCES

1. Y. Kuwano, T. Imai, M. Ohnishi and S. Nakano, Proc. of 14th IEEE Photovol. Spec. Conf., San Diego (1980) 1408.

2. Y. Kuwano, M. Ohnishi, S. Tsuda, Y. Nakashima and N. Nakamura, Jpn. J. Appl. Phys., 53 (1982) 5273.

3. S. Nakano, T. Matsuoka, S. Kiyama, H.Kawata, N. Nakamura, Y. Nakashima, S. Tsuda, H. Nishiwaki, M. Ohnishi, I. Nagaoka and Y. Kuwano, Jpn. J. Appl. Phys., 25(1986) 1936.

4. S. Tsuda, T. Takahama, M. Isomura, H. Tarui, Y. Nakashima, Y. Hishikawa, N. Nakamura, T. Matsuoka, H. Nishiwaki, S. Nakano, M. Ohnishi and Y. Kuwano, Jpn. J. Appl. Phys., 26 (1987) 33.

5. Y. Kuwano and M. Ohnishi, Proc. 9th Int. Conf. Amorphous & Liquid Semicond., Grenoble (1981) C4-1155.

6. K. Wakisaka, M. Tanaka, M. Isomura, H. Haku, S. Kiyama and S. Tsuda, Technical Digest International PVSEC-9, Miyazaki (1996) 583.

7. H. Kuriyama, T. Nohda, Y. Aya, T. Kuwahara, K. Wakisaka, S. Kiyama and S. Tsuda, Jpn. J. Appl. Phys., 33 (1994) 5657.

8. H. Takakura, K. Miyagi, T. Kanata, H. Okamoto and Y. Hamakawa, Proc. of 4th Int. Photovoltaic Science & Engineering Conf. (1989) 403.

9. M. Taguchi, M. Tanaka, T. Matsuyama, T. Matsuoka, S. Tsuda, S. Nakano, Y. Kishi and Y. Kuwano, Technical Digest of 5th International PVSEC (1990) 689.

10 T. Matsuyama, K. Wakisaka, M. Kameda, M. Tanaka, T. Matsuoka, S. Tsuda, S. Nakano, Y. Kishi and Y. Kuwano, Jpn. J. Appl. Phys., 29 (1990) 2327.

11. D.L. Staebler and C.R. Wronski, Appl. Phys. Lett., 31 (1977) 292.

12. T. Matsuyama, M. Tanaka, S. Tsuda, S. Nakano and Y. Kuwano, Jpn. J. Appl. Phys. 32, (1993) 3720.

13. John Y. W. Seto, J. Appl. Phys., 46 (1975) 5247.

14. T. Shimada, Y. Katayama, K. Nakagawa, H. Matsubara, M. Migitaka and E. Maruyama, J. Non-Cryst. Solids., 59&60 (1983) 783.

15. A. Matsuda, J. Non-Cryst. Solids., 59&60 (1983) 767.

PURELY INTRINSIC POLY-SILICON FILMS BY HOT WIRE CHEMICAL VAPOR DEPOSITION

J.K. RATH, K.F.FEENSTRA, D. RUFF*, H. MEILING AND R.E.I. SCHROPP
Utrecht University, Debye Institute, P.O.Box 80000, 3508 TA Utrecht, The Netherlands
*Philipps-Universitat Marburg, D-35032 Marburg, Germany.

ABSTRACT

Poly-silicon films have been prepared by hot-wire chemical vapor deposition (HWCVD) from hydrogen diluted silane gas at a low temperature (430 °C). The optical gap of the poly-silicon films is 1.1 eV, though with a higher optical absorption than c-Si. The grains have a preferential orientation (220) perpendicular to the substrate with an average crystallite size of 70 nm. The crystalline volume fraction is 95% with complete coalescence of grains. Large structures up to 0.5 μm could be observed in the AFM micrograph. The activation energy (0.54 eV) and the low carrier concentration (10^{11} cm^{-3}) indicate a fully intrinsic nature of the films. The $\mu\tau$ product of carriers is 7.1×10^{-7} cm^2V^{-1} whereas the ambipolar diffusion length (L_D) is 334 nm. The excellent photo-conductive properties are attributed to the low ($\sim10^{17}$ cm^{-3}) defect density. The HWCVD poly-silicon films showed a very small temperature dependence of mobility, indicating negligible trapping of carriers at the grain boundaries. Preliminary n-i-p cells incorporating poly-silicon i-layer yielded 3.15 % efficiency.

INTRODUCTION

The successful utilisation of microcrystalline silicon as the active layer of the solar cell [1] has opened up the possibility of achieving stable cells made by large-area deposition in a one step process. Recent advances are not only the increased efficiency of single junction cells but also the demonstration of a good efficiency in the "micromorph" tandem cell structure [2]. However, to overcome the obvious problem of deposition rate in the PECVD process, a fast growth process is highly desirable. Hot-wire deposition of poly-Si films [3] seems to be a promising technique to achieve this target. The materials structure depends on the deposition conditions. A detailed knowledge of the structure i.e., the size and the orientation of grains, the nature of the grain boundaries and the H bonding configuration is necessary to evaluate the trapping and recombination kinetics of the photogenerated carriers in the material. However the assignments of the vibrational spectra to various hydrogen bonds are in dispute because of the obvious shift of the bands from their respective positions in a-Si:H [2]. An attempt has been made in the present study to understand the structure and correlate the structure with the opto-electronic properties to evaluate the quality of the material. At the end we will present the results of a preliminary solar cell in an n-i-p configuration.

EXPERIMENT

Poly-silicon films were deposited on 10 cm x 10 cm Corning 7059 glass and c-Si wafer substrates by HWCVD in one of the chambers of an ultra high vacuum multichamber system (PASTA). The thickness of the films was measured by Dektak profilometer and reflection/transmission measurement. Samples were characterised by x-ray diffraction (XRD), Raman spectroscopy, Fourier transform infrared (FTIR) spectroscopy, photothermal deflection spectroscopy (PDS), electron spin resonance (ESR), steady state photo-carrier grating technique (SSPG), Hall mobility measurement and electrical conductivity in dark and white light. The $\eta\mu\tau$ of the carriers was estimated by 700 nm monochromatic light with a flux of 10^{15} cm^{-2} s^{-1}. The

Mat. Res. Soc. Symp. Proc. Vol. 452 © 1997 Materials Research Society

crystallite size, x, was estimated from the XRD by Scherrer formula [4] $x=k\lambda/(\Delta\theta)\cos\theta$, where $k=0.9$, λ is the wavelength of Cu K_{α} x-ray radiation, $(\Delta\theta)$ is the FWHM of the peaks (in units of 2θ) and θ is the angular position of the peaks. The crystalline volume fraction (V_f) was estimated from the Raman spectrum by the equation, $V_f=I_c/(I_c+mI_a)$, where I_c and I_a are the deconvoluted intensities of the Raman spectrum at 520 cm^{-1}, corresponding to the crystalline part, and at 480-500 cm^{-1} corresponding to the amorphous-grain boundaries respectively [5]. An n-i-p cell was made in the configuration n^+-c-Si/poly-Si:H (HWCVD)/p-μc-Si:H (PECVD)/ITO. The cells were characterised by I-V and spectral response measurements.

RESULTS

To obtain polycrystalline films, the source gas SiH_4 was diluted with hydrogen. The crystallinity depends strongly on the silane to hydrogen ratio ($\Phi(SiH_4)/\Phi(H_2)$) r [6]. At a deposition pressure (P_r) of 0.1 mbar and wire temperature (T_w) of 1900 °C the films obtained at a low r of 1% exhibit random orientation of the crystallites whereas at a higher r of 10% the grains are preferentially oriented along (220) direction perpendicular to substrates. Estimation of the grain size by XRD revealed that at a low r of 1% the grain sizes along the (220) direction and the (111) directions are equivalent, suggesting isotropic growth of the grains. At higher r the grains grow anisotropically and the grain size along the (220) direction is longer than that along the (111) direction. This structural transformation with change of gas ratio can also be seen from the AFM micrographs. Fig. 1 shows the AFM pictures of the poly-silicon films at different deposition conditions. At low r values of 1% (Fig 1a), the structures are round and ball-like. However, at a higher r of 10%, they are bigger but needle type with complete coalescence of grains (Fig.1b). Structures of 0.5 μm can be observed in this micrograph. However, at a higher r value of 15%, the crystalline structures are isolated from each other and embedded in an amorphous matrix (Fig.1c). Here the amorphous volume is large (25%). The structures are also needle type but much bigger than those of sample b because of the smaller nucleation density.

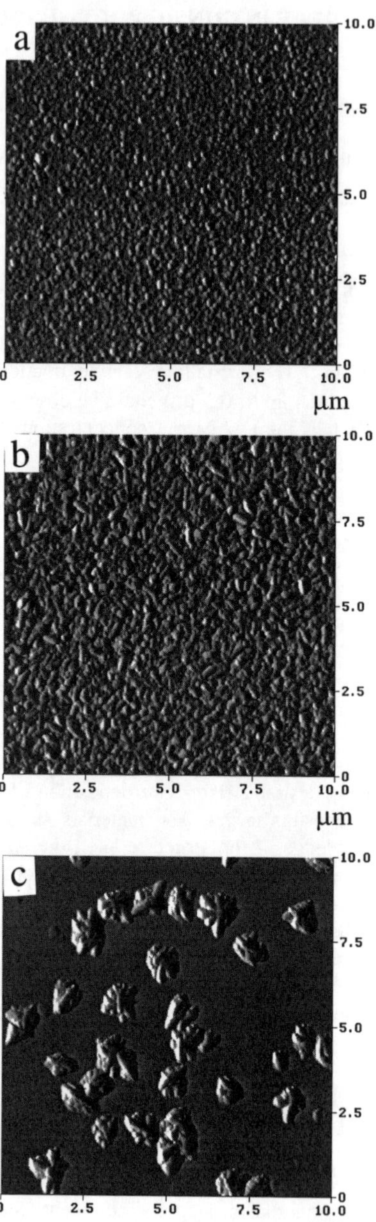

Fig. 1 AFM micrograph of poly-Si:H films deposited at SiH_4/H_2 of a) 1%, b) 10%, c)15%

The AFM pictures also reveal the roughness of the surface of the films. Table 1 lists the root mean square peak height (ΔZ_{RMS}) and the maximum height of the peaks (ΔZ). It is observed that the ΔZ_{RMS} increases with dilution from 8.9 nm at r=1% to 46.0 nm at r=15% whereas the ΔZ increases from 66nm to 283 nm. It should be noted that for the r=15% film the ΔZ_{RMS} in the amorphous region is only 1.1 nm whereas in the polysilicon part it is 69.3 nm. With such roughness we would expect scattering of light . We will discuss later the beneficial aspect of this in the solar cell structure.

The role of hydrogen in the passivation of grain boundary defects is a crucial one to improve the photosensitivity of the material. To that end we have looked into the bonding configuration of the hydrogen in the network and its behavior at various deposition conditions. Fig. 2 shows the IR spectrum of the film prepared at 10% SiH_4/H_2 and P_r of 0.1 mbar. In Fig. 2a, at the wire temperature (T_w) of 1900 °C the stretching mode has two main bands at 2000 cm^{-1} and ~2100 cm^{-1} . The 2000 cm^{-1} mode can be attributed to the amorphous like Si-H mode . The absence of any vibrational mode at 800-900 cm^{-1} supports the fact that the hydrogen bonds at 2100 cm^{-1} mode do not originate from $(SiH_2)_n$ polyhydrides. At a deposition temperature of 430 °C, generally, hydrogen evolves from the SiH_2 and SiH_3 bonds. Hence, we do not expect SiH_2 or SiH_3 modes in the IR spectrum [7]. The Raman spectrum of the sample does not show any band at 480 cm^{-1}, indicating a negligible amorphous volume fraction in the structure. Only a weak band at 495 cm^{-1} corresponding to grain boundaries has been identified [5]. This result suggests that the 2100 cm^{-1} mode can not be related to any voids in the amorphous region. Analogous with the interpretation of the IR spectra of hydrogen passivated c-Si [8], we attribute this band to the hydrogen at the grain boundaries. The splitting of the 2100 cm^{-1} mode into narrow peaks can be attributed to the hydrogen at (111) and (110) surfaces in the grain boundaries [9]. The samples deposited at a lower wire temperature (1850 °C) do not show any absorption at 2100 cm^{-1} ; only the 2000 cm^{-1} mode is observed. This suggests that the grain boundary defects are minimised. In Fig. 2b, the rocking mode vibration at 600-700 cm^{-1} also shows considerable changes with variation of crystallinity. The broad absorption can be deconvoluted into two gaussian bands centered around 620 cm^{-1} and 640 cm^{-1}. The 620 cm^{-1} becomes stronger with increasing crystallinity and in the best crystalline material 640 cm^{-1} is almost absent. An examination of the IR spectrum (Fig. 3) shows no absorption at 1050 cm^{-1} corresponding to the Si-O mode even after several days of exposure to air . This demonstrates the compactness of the structure which prevents the indiffusion of oxygen into the material.

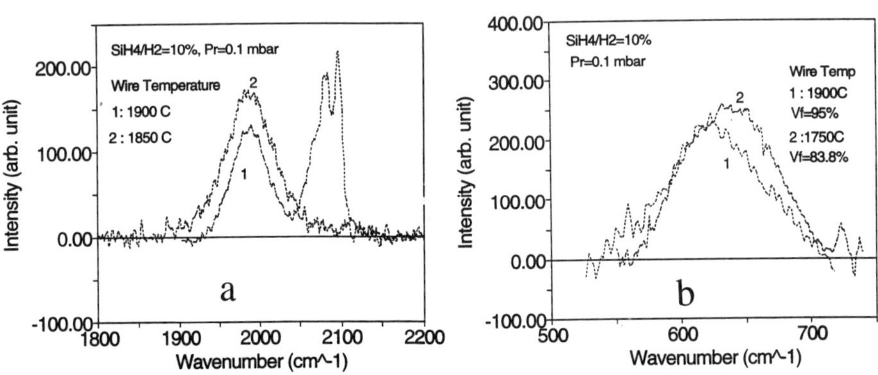

Fig.2 : IR spectra of poly-silicon films. a) stretching modes, b) rocking modes.

Table 1. Surface texture of the Poly-Si:H films derived from AFM micrographs.

Sample	T_w (C)	SiH_4/H_2	P_r(mbar)	ΔZ (nm)	ΔZ_{RMS}(nm)
p1028	1900	1%	0.1	66.0	8.9
p1063	1900	5%	0.1	89.0	10.3
p1073	1900	10%	0.1	201.0	20.1
p1067	1900	15%	0.1	283	46.0
p1211	1850	10%	0.1	180.0	19.6
p1210	1800	10%	0.1	150.0	19.8
p1209	1750	10%	0.1	120.0	11.2
p1191	1900	10%	0.125	- -	- -
p1186	1900	10%	0.15	206.0	25.0
p1117	1900	10%	0.3	300.0	37.0

Fig.4 shows the temperature dependence of the dark conductivity of an optimized polycrystalline film . Conductivity was measured with coplanar Ag contacts which showed ohmic characteristics. It is observed that there is single activation energy of 0.54 eV. The band gap of this film has been estimated from $\sqrt{\alpha}$ versus E plot as 1.1eV, and the same as for c-Si. Correlation of the band gap with the activation energy implies that the Fermi level is exactly at the middle of the gap. This demonstrates that purely intrinsic film has been made even without employing any gas purifier or boron compensation. This is possible due to mainly (i) a high deposition rate (>0.5 nm/sec) and (ii) the compactness of the structure due to the large crystalline volume fraction.

Fig. 5a shows the temperature dependence of the Hall mobility (μ_H) of the polycrystalline films. The sign of the Hall voltage suggests that all the samples are n-type. It is observed that at P_r of 0.1 mbar with increase of *r* value, μ_H increases and becomes less temperature dependent. The mobility further improves for materials made at a slightly higher deposition pressure. The activation energy of the Hall mobility, estimated from the slope of the temperature dependence plot is very small (0.012 eV). This suggests negligible trapping of the carriers at the grain boundaries or small barrier to the electron transport. However at very high deposition pressures the mobility decreases and activation energy increases to 0.035 eV at 0.3 mbar. For comparison, we studied the temperature dependence of the mobility of a poly-Si film made by solid phase crystallization (SPC), i.e., by annealing an a-Si:H film at 700 °C for one hour. The material is mostly crystallized at these conditions as confirmed by Raman spectroscopy. This film has defect

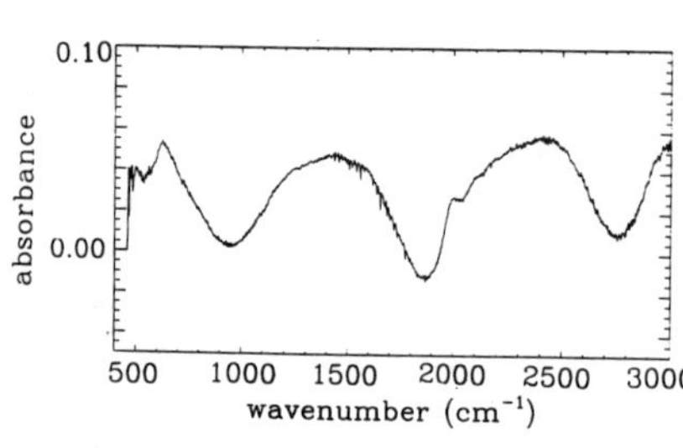

Fig. 3. IR spectra of a Poly-Si film (SiH_4/H_2=10%) after long exposure to air.

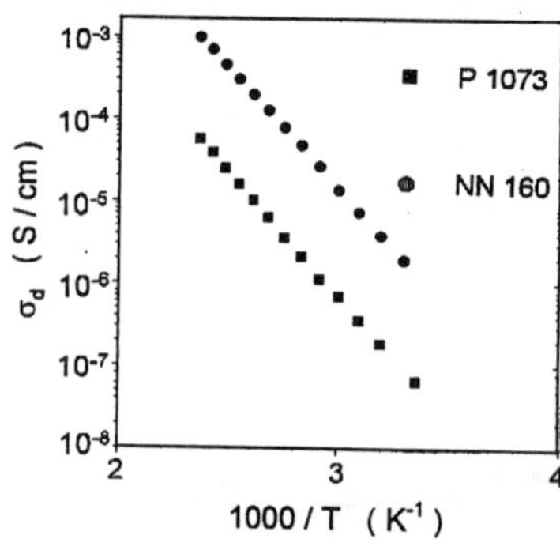

Fig.4 : The temperature dependence of the dark conductivity of Poly-Si and an SPC film (NN160)

density (ESR) of 5×10^{17} cm^{-3}, and white light photoconductivity of 8.4×10^{-6} $\Omega^{-1}cm^{-1}$ at room temperature. The SPC material shows higher mobility than the HWCVD film, though with a slightly higher activation energy. Fig 5b shows the carrier concentration of the poly-silicon films, calculated from the mobility and the conductivity data. It is observed that the carrier concentration in the HWCVD poly-Si films is much lower than that of the SPC film. This confirms the intrinsic nature of the HWCVD films. The comparatively low mobility of the HWCVD films can also be attributed to the intrinsic nature of the films [10].

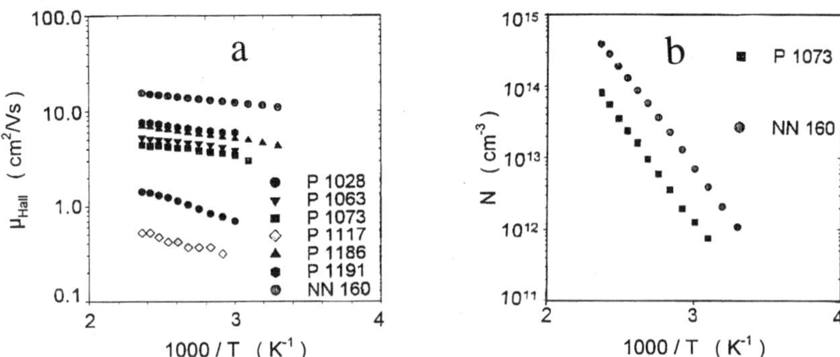

Fig.5: Temperature dependence of Hall mobility (μ_H) and carrier concentration (N). The sample numbers are described in Table 1.

The optoelectronic quality of the films were evaluated by the photoconductivity and SSPG measurements. It is observed that the minority carrier diffusion length L_D (obtained from SSPG) increases with increasing silane to hydrogen ratio and reached a value of 220 nm at r of 10% . At higher r, the L_D value drops. However, the L_D value reaches 334 nm at a reduced wire temperature of 1800 °C. At wire temperatures less than $\leq$ 1750 °C the material tends to become amorphous and the carrier diffusion length decreases. The $\eta\mu\tau$ product of the optimized film is 7.1×10^{-7} cm^2 V^{-1}. The material showed photoconductivity of 1.9×10^{-5} Ω^{-1} cm^{-1} and a photo sensitivity of 1.4×10^{2}. The above results confirm the device quality of our HWCVD poly-Si materials. The defect density (dangling bond) estimated from the ESR spectrum is ~10^{17} cm^{-3}. The improvement of the carrier transport in our samples can be attributed to the small defect density at the grain boundaries.

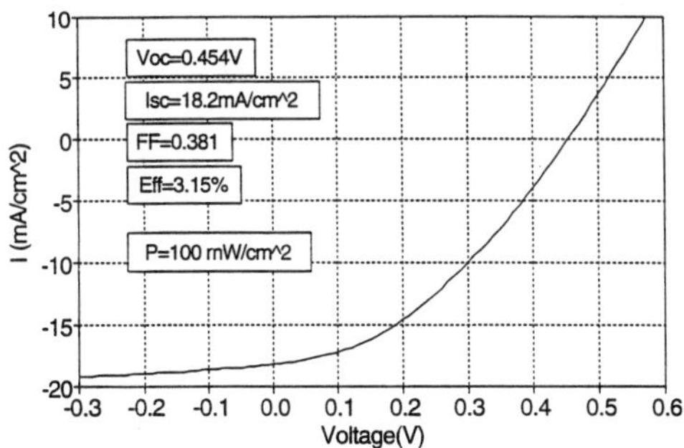

Fig.6 : I-V characteristics of an n-i-p cell in the configuration n^+-c-Si/poly-Si:H/p-μc-Si:H/ITO

We fabricated an n-i-p cell in the configuration on n^+-c-Si/poly-Si:H/p-μc-Si:H/ ITO where highly doped (0.01Ω cm) n^+-c-Si was used both as the substrate as well as the n-layer of the cell. The p-layer (p-μc-Si:H) was deposited by plasma CVD in another chamber of the multichamber system PASTA. The p-μc-Si:H layer has already shown excellent window characteristics in a-Si:H based solar cell device [11]. The thickness of the i-layer is 1.5 μm. The I-V characteristics of the cell is shown in Fig. 6. The current of 18.2 mA/cm^2 is possible due to (i) higher absorption of the poly-Si:H films compared to c-Si and (ii) better light trapping due to scattering at the textured surface. Optical modelling predicts such currents for this thickness[12]. In an n-i-p cell we expect to collect all the photogenerated carriers due to a drift field. We do not expect any appreciable contribution from the n^+ wafer because of the high defect density and small carrier mobility. A reasonably good V_{oc} of 0.45 V has been achieved. It is possible due to the low defect density in the film and the columnar orientation of the grains. As a result a high field is sustained in the bulk. However, the FF needs improvement. Optimization of the interfaces is in progress to further improve the performance.

CONCLUSIONS

Device quality poly-Si films have been prepared by HWCVD method at a relatively fast growth rate of 0.55 nm/sec. Changes in the growth kinetics with deposition conditions have been identified by AFM, XRD and Raman Spectra. The poly-Si films surface shows surface texture which is beneficial for light trapping. The materials are truly intrinsic as proved by activation energy and Hall measurements. This has been achieved without any gas purifier or compensation mechanism. Carrier transport shows negligible trapping at the grain boundaries. A crystalline volume fraction of 95%, an average grain size of 70 nm, diffusion length of minority carriers of 334 nm and a $\mu\tau$ product of 7.1×10^{-7} cm^2 V^{-1} demonstrated device quality. An n-i-p cell made using this poly-Si:H film as the window layer showed 3.15 % efficiency.

ACKNOWLEDGEMENT

The authors thank Dr. H. Mell , Phillips Univ. Marburg for PDS and Hall measurements, M.W.M. van Cleef for ITO deposition and C.H.M.van der Werf for film deposition. This research was partially funded by NOVEM (Netherlands Organisation for Energy and Environment) and is part of a film-Si project co-ordinated by ECN, Petten.

REFERENCES

[1] M.Faraji, Sunil Gokhale, S.M.Choudhari, M.G.Takwale and S.V.Ghaisas, Appl. Phys. Lett., **60**, 3289 (1992).
[2] J.Meier, Proc. MRS Spring Meeting, **420**, San Francisco (1996) (in print).
[3] R.O.Dusane et al, Appl. Phys. Lett, **63**, 2201 (1993).
[4] H.P.Klug and L.E.Alexander, X-ray Diffraction Procedure, John Wiley & Sons, New York (1974).
[5] E.Bustarret, M.A.Hachichia and M.Brunel, Appl. Phys. Lett., **52**, 1675 (1988).
[6] J.K.Rath et al, Tech. Digest Int. PVSEC-9, Miyazaki. Japan (1996) p227.
[7] S.Veprek, Z.Iqbal, H.R.Oswald and A.P.Webb, J.Phys C: Solid State Phys., **14**, 295 (1981).
[8] G.E.Becker and G.W.Gobeli, J.Chem. Phys., **38**, 2942 (1963).
[9] Takahashi Satoh and Akio Hiraki, Jpn. J. Appl. Phys., **24**, L491 (1985).
[10] E.H.Putley, The Hall effect and related phenomena, Butterworth & Co. (1960).
[11] J.K.Rath et al, Proc. 25th IEEE PVSC, Washington D.C., (1996) p1101.
[12] R.Krankenhagen et al , Proc. 13th EPVSEC, Nice (1995) p1700.

LOW-TEMPERATURE FORMATION OF DEVICE-QUALITY POLYSILICON FILMS BY CAT-CVD METHOD

Hideki Matsumura, Akira Heya, Ritsuko Iizuka, Akira Izumi, An-Qiang He and Nobuo Otsuka
JAIST (Japan Advanced Institute of Science and Technology),
Tatsunokuchi, Ishikawa-ken 923-12, Japan

ABSTRACT

Polycrystalline silicon (poly-Si) films are deposited at temperatures lower than 300-400℃ by the cat-CVD method. In the method, a SiH_4 and H_2 gas-mixture is decomposed by catalytic cracking reactions with a heated tungsten catalyzer placed near substrates. Carrier transport, optical and structural properties are investigated for this cat-CVD poly-Si. The films show both large carrier mobility and large optical absorption for particular deposition conditions. The cat-CVD poly-Si films are found to be one of the useful materials for thin film transistors and thin film solar cells.

INTRODUCTION

Polycrystalline silicon (poly-Si) has been widely applied to various electronic devices such as thin film transistors (TFTs) for liquid crystal displays, thin film solar cells and large scale integrated circuits. However, for use of low cost glass substrates, poly-Si has to be prepared at low temperatures as in amorphous silicon (a-Si) deposition.

Excimer laser annealing is one of the methods to obtain poly-Si films at low temperatures, and such poly-Si exhibits a high field effect mobility up to 350 cm^2/Vs [1] . However, this method does not appear practical because of low throughput due to the laser scanning problem.

On the other hand, one of the authors has developed a new low-temperature deposition method, named "catalytic chemical vapor deposition (cat-CVD) method"[2-6]. In the method, deposition gases are decomposed by the catalytic cracking reactions with a heated catalyzer placed near the substrates. For this cat-CVD method, recently, some people like to say "hot-wire CVD" probably because of influence of the paper from SERI (NREL) group [7]. However, the shape of the catalyzer is very important particularly for industrial application of this method and it is not limited in "wire". Additionally, it has been already verified by our previous papers that deposition gases are not decomposed simply by heat but decomposed by catalytic cracking[6,8]. The name of "hot-wire" dose not appear suitable when we take account of industrial application of this method.

This cat-CVD method is also useful for deposition of many kinds of films such as silicon nitride[9], and it is also known that poly-Si films can be obtained at temperatures lower than 400 ℃ by this method using a SiH_4 and H_2 gas mixture[10]. The films can be deposited on a large-area substrate simply by spreading the catalyzer widely. In addition, poly-Si films of various properties are obtained by adjusting deposition conditions.

In this paper, carrier transport, optical and structural properties of cat-CVD poly-Si are systematically studied in consideration of their application to TFTs and solar cells. It is found that the cat-CVD poly-Si films are promising materials for wide-area devices such as TFT.

EXPERIMENTAL PROCEDURES

The cat-CVD apparatus used in this study is schematically illustrated in Fig.1. SiH_4 and H_2 gases are introduced into a stainless-steel chamber through nozzles at the bottom. At the upper part of the chamber, a substrate holder with a heater is installed. Substrate temperatures are measured by a thermocouple (TC) attached on the substrate holder. The distance from the nozzles to the substrate holder is about 8cm. A catalyzer wire is set at the middle between the nozzles and substrate holder. A coiled tungsten (W) wire with a diameter of 0.75 mm is used as the catalyzer. The catalyzer is heated at about 1600 ℃ and its temperature is measured by an infrared thermometer through a quartz window. Poly-Si films are deposited mainly on fused quartz substrates at a pressure of about

Mat. Res. Soc. Symp. Proc. Vol. 452

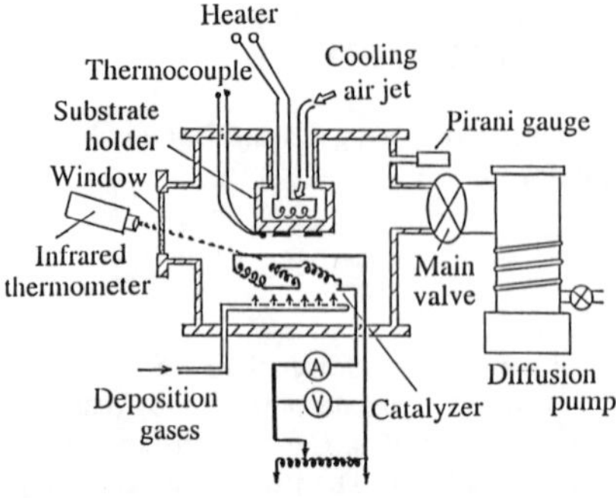

Fig1. Schematic diagram of cat-CVD apparatus.

Table Ⅰ. The deposition conditions of poly-Si films.

FR(SiH_4)	0.5, 1.0, 1.5 sccm
FR(H_2)	30 sccm
Substrate temperature	300 ℃
Catalyzer power	1050 W
Catalyzer temperature	1600 ℃
Gas pressure	1.2 mTorr

1.2 mTorr. The flow rate of SiH_4, FR(SiH_4), is varied using values of 0.5, 1.0 and 1.5 sccm, while the flow rate of H_2, FR(H_2), is fixed at 30 sccm. Deposition conditions of poly-Si films are summarized in Table Ⅰ. The properties of another poly-Si film, which is deposited with FR(SiH_4) = 1.0 sccm and FR(H_2) = 30 sccm and annealed at 900 ℃ in N_2 for 1h is studied for comparison.

Structural properties of poly-Si films are studied by the X-ray diffraction (XRD) technique using Cu K α X-rays and by the Raman scattering spectroscopy using the excited light of 514.5 nm wavelength of an argon ion laser. The cat-CVD poly-Si films are preferentially oriented along (220) direction. The grain size of poly-Si films is estimated using Scherrer's formula [11] from the FWHM of (220) peak. The fraction of crystalline to amrphous phase is evaluated from the TO phonon-made signal of the Raman spectra. Crystalline fraction is determined from the areal ratio of the signal due to the crystalline phase at 520 cm^{-1} to the sum of the signals due to both the crystalline phase and the amorphous phase at 480 cm^{-1}. The transmission electron microscopy (TEM) is also used to observe the crystalline structure.

Optical properties of poly-Si films are characterized by the absorption coefficient α, which is measured by optical transmittance measurement.

Electrical properties of poly-Si films are characterized by the carrier mobility μ and the barrier height ϕ_B at grain boundary. The mobility μ of poly-Si films is obtained using the Van der Pauw method. The barrier height ϕ_B is derived from the temperature dependence of μ. Additionally, to confirm the quality of cat-CVD poly-Si, a TFT is finally fabricated by using cat-CVD films.

RESULTS AND DISCUSSION

Structural properties of poly-Si films

Figure 2 shows the XRD patterns for cat-CVD poly-Si films. For films deposited with FR(SiH_4) = 0.5 and 1.0 sccm and annealed poly-Si films, diffraction signals of (220) peak are clearly observed although for the films deposited with FR(SiH_4) = 1.5 sccm no diffraction signal is observed. The grain size is estimated to be about 370 Å for FR(SiH_4) = 0.5 sccm, about 270 Å for FR(SiH_4) = 1.0 sccm and about 740 Å for annealed sample, respectively.

The Raman spectra for cat-CVD poly-Si films are shown in Fig.3. For the film deposited with FR(SiH_4) = 0.5 sccm and annealed sample, TO photon signal at 520 cm^{-1} is clearly observed and that at 480 cm^{-1} is scarcely observed. However, with an increase in FR(SiH_4) from 0.5 sccm to 1.5 sccm, the signal at 520 cm^{-1} becomes small and that at 480 cm^{-1} becomes large. These films consist of the mixture of crystalline and amorphous phases. Crystalline fraction estimated is about 70 % for FR(SiH_4) = 0.5 sccm, 50% for FR(SiH_4) = 1.0 sccm, only several % for FR(SiH_4) = 1.5 sccm. The crystalline fraction of annealed sample is over 90 %. From these results, it is found that the crystallization is enhanced and the grain size becomes large with an increase of the flow rate ratio of H_2 to SiH_4. The results imply that the volume fraction of amorphous phase in poly-Si films can be controlled by changing the flow-rate ratio of SiH_4 and H_2.

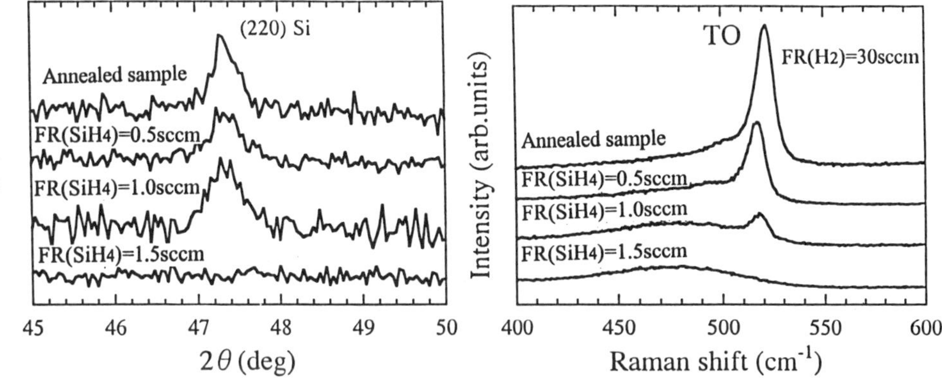

Fig.2. X-ray diffraction patterns for poly-Si films deposited with FR(SiH_4) = 0.5, 1.0 and 1.5 sccm and annealed sample.

Fig.3. Raman spectra for poly-Si films deposited with FR(SiH_4) = 0.5, 1.0 and 1.5 sccm and an annealed sample.

A cross sectional TEM bright field image of a poly-Si film deposited with FR(SiH_4) = 0.5 sccm is shown in Fig.4. The TEM image confirms the coexistence of crystalline and amorphous phases. The TEM image also shows that crystalline grains are columnarly grown immediately from the surface of substrates.

Fig.4. A cross sectional TEM image of a poly-Si film deposited with FR(SiH_4) = 0.5 sccm.

Optical properties of poly-Si films

Figure 5 shows the absorption coefficient α vs. photon energy for cat-CVD poly-Si films deposited with FR(SiH_4) = 0.5 and 1.0 sccm and for annealed sample. It is found that the α for as deposited films with FR(SiH_4) = 0.5 and 1.0 sccm is higher than that for hydrogenated a-Si (a-Si:H) films, single crystalline Si and conventional low-pressure CVD (LPCVD) poly-Si films in the photon energy range between 1.0 and 2.0 eV. It should be noted that the α for as-deposited cat-CVD films with FR(SiH_4) of 0.5 and 1.0 sccm in low photon energy region from about 1.2 to 1.6 eV is apparently larger than that of the conventional LPCVD poly-Si. The reason is not clear at this moment. However, it can be said that relatively large α is one of features for cat-CVD poly-Si films.

Electrical properties of poly-Si films

It may be worried that the degradation of electrical properties will occur if this large α originates simply from defect absorption. Thus, next, the electrical properties of cat-CVD poly-Si films are investigated. The temperature dependence of the Hall mobility is shown in Fig.6. It is found that the mobility is about 20 cm^2/Vs in spite of low deposition temperature. From the gradient of $\mu\sqrt{T}$ vs. 1/T plots ϕ_B is evaluated. The ϕ_B is about 0.007eV and usually lower than the thermal activation energy at room temperature. The results demonstrate that the carriers can move smoothly in cat-CVD poly-Si films.

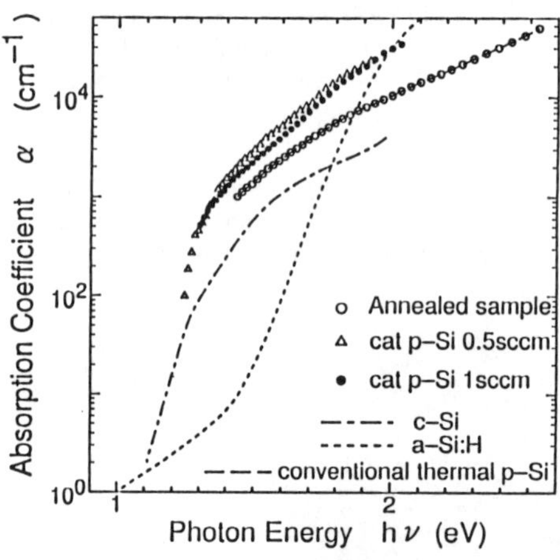

Fig.5. Absorption coefficient of poly-Si films deposited with FR(SiH_4) = 0.5 and 1.0 sccm and an annealed sample along with a-Si:H films, single-crystalline Si and LPCVD poly-Si films.

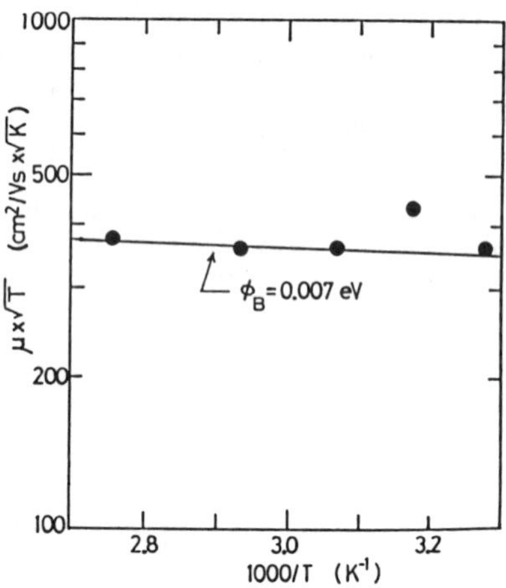

Fig.6. Temperature dependence of Hall mobility.

cat-CVD poly-Si TFT

To confirm the quality of cat-CVD poly-Si films, finally, TFT is fabricated by using this cat-CVD films. The structure of TFT is very simple. A 200-nm-thick silicon-dioxide film grown thermally on low-resistivity crystalline Si is used as the gate insulator and the crystalline Si itself is used as a gate electrode. Poly-Si islands with a size of 50 μm × 300 μm are formed on such silicon-dioxides by cat-CVD deposition and subsequent photolithographic process. Source and drain electrodes are formed by aluminum evaporation on the poly-Si islands. The channel length L and width W are 15 μm and 300 μm, respectively.

The source-drain current I_{DS} vs. gate voltage V_{GS} characteristics is demonstrated in Fig.7. The steep increase in I_{DS} is found with an increase in V_{GS}. This implies the low density of defect level in the band gap. The field effect mobility estimated from $\sqrt{I_{DS}}$ vs. V_{GS} relation is about 70 cm^2/Vs. This also confirms the smooth carrier transport in cat-CVD poly-Si films.

SUMMARY

It is possible to prepare poly-Si films with both crystalline and amorphous phases by cat-CVD method. These cat-CVD poly-Si films show high absorption coefficient, large mobility and smooth carrier transport at the same time. In cat-CVD poly-Si, crystalline grains appear columnarly grown from the surface of substrates. The TFT fabricated by using cat-CVD poly-Si films can operate well. The cat-CVD poly-Si films can be expected as new useful material for TFT and solar cell application.

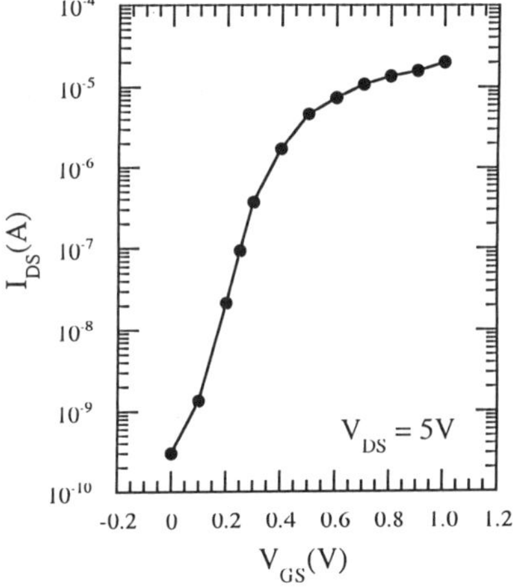

Fig.7. I_{DS} - V_{GS} characteristics of cat-CVD poly-Si TFT.

ACKNOWLEDGEMENTS

The authors are very grateful to Dr. A. Masuda at JAIST for his discussion and suggestions to this work. They are also grateful to Prof. S. Horita at JAIST for their helpful support. This work is partially supported by 1994 - 1996 Research Foundation for Electrotechnology of Chubu and by 1996 Grant-in-Aid for Scientific Research from the Ministry of Education, Science, Sports and Culture of Japan.

REFERENCES

1. S. Usui, Optoelectronics **4**, 235 (1989) .
2. H. Matsumura, H. Ihara and H. Tachibana, Proceedings of the 18th IEEE Photovoltaic Specialists Conference, (1985, Las Vegas) p.1277.
3. H. Matsumura, Jpn. J. Appl. Phys. **25**, L949 (1986).
4. H. Matsumura, Appl. Phys. Lett. **51**, 804 (1987)
5. H. Matsumura and H. Ihara, J. Appl. Phys. **64**, 6505 (1988).
6. H. Matsumura, J. Appl. Phys. **65**, 4396 (1989).
7. A. Mahan, J. Carapella, B. P. Nelson, R. S. Crandall and I. Balberg, J. Appl. Phys. **69**, 6728 (1991).
8. H. Matsumura, Proceeding of 19th International Conference on the Physics of Semiconductors Vol. 2,(1988,Warsaw) p.1685.
9. H. Matsumura, J. Appl. Phys. **66**, 3612 (1989).
10. H. Matsumura, Jpn. J. Appl. Phys. **30**, L1522 (1991).
11. T. Kamins, Polycrystalline Silicon for Integrated Circuits Applications (Kluwer Academic Publication, Boston, 1988) Chap.5.

LARGE CRYSTALLITE POLYSILICON DEPOSITED USING PULSED-GAS PECVD AT TEMPERATURES LESS THAN 250°C

E. Srinivasan, S.J. Ellis, R.J. Nemanich*, and G.N. Parsons
Dept. of Chemical Engineering, N.C. State University, Raleigh, NC 27695
* Dept. of Physics, N.C. State University, Raleigh, NC 27695

ABSTRACT

A pulsed-gas intermittent deposition technique is used to deposit high crystallinity hydrogenated micro- or poly-crystalline silicon using silane and hydrogen. This method has been used to deposit crystallites that are comparable to those obtained using PECVD of fluorinated silanes. RHEED and TEM have been used to understand the nucleation process. The pulsed-gas method is promising for depositing polycrystalline silicon and subsequent use in thin film transistor applications.

INTRODUCTION

Polysilicon is conventionally deposited using low pressure chemical vapor deposition at temperatures greater than 600°C [1]. Corning 7059 glass has a strain point temperature of 600°C [2]. Processing temperature of polysilicon for thin film transistor application has to be below 600°C for compatibility with glass. Rapid thermal annealing and excimer laser assisted annealing techniques have been developed to keep the processing temperature below 600°C [3, 4]. A low temperature, direct deposition process for polycrystalline silicon on glass could lead to high throughput and improved performance of large thin film transistor arrays. Microcrystalline silicon (μc-Si:H), consisting of small crystallites between 30-500 Å in an amorphous matrix, has been formed by PECVD using silane/hydrogen mixtures at temperatures between 200 and 350°C [5]. Larger crystallite sizes have been reported recently due to improved removal of hydrogen in the films [6,7]. Real-time mass spectroscopy measurements in a pulsed-gas deposition process shows that intermittent exposure of the growth surface to atomic deuterium (H) yielded HD (abstraction product) and SiD_4 (etch product) [6]. These two reactions reduce the hydrogen content and provide dangling bond sites for further deposition. During the atomic D(H) exposure, trideuterides (SiD_3) or trihydrides (SiH_3) are formed by D(H) insertion into strained Si-Si bonds. These can diffuse on the surface and form relaxed Si-Si bonds, improving the crystallinity [8]. Recently, the use of fluorinated silanes has yielded large crystallites at temperatures of 250-350°C [9,10]. The enhanced crystallite formation process is not well understood, though it is speculated that the high electronegativity of fluorine is more efficient in hydrogen removal, thus enhancing crystallite formation. However, the highly toxic nature of silicon tetrafluoride and the possibility of formation of hazardous HF makes the ability to form large crystallites with SiH_4/H_2 mixtures attractive.

Mat. Res. Soc. Symp. Proc. Vol. 452 © 1997 Materials Research Society

EXPERIMENT

Rf parallel plate plasma enhanced chemical vapor deposition (PECVD) system described elsewhere has been used in this study [11]. The pulsed-gas deposition scheme used has also been discussed in detail in another report [6]. The films are deposited at temperatures between room temperature and 275°C. The deposition pressure has been maintained at 0.6 torr in all the experiments. 2 % silane diluted in helium (flow rate = 100 sccm) and hydrogen (flow rate = 60 sccm) have been used for all results reported in this study. The duty cycle, which is defined as t_{ON}/t_{OFF}, has been varied between 10/0 (no atomic hydrogen exposure) and 10/40 s (40 s of H exposure per 10 s of deposition), where t_{ON} is the time of deposition when silane flows into the chamber and hydrogen flows into the pump and vice versa during t_{OFF}. Experiments have been performed with rf powers of 10 W and 50 W (electrode area = 625 cm^2). Typical thicknesses of deposited films have been between 1000 and 3000 Å. These films have been analyzed using infrared spectroscopy to observe changes in Si-H bonding, Raman spectroscopy to understand film crystallinity, cross-sectional transmission electron microscopy (XTEM) to observe the crystallite structure, atomic force microscopy (AFM) to monitor the crystallinity and surface roughness, and reflection high energy electron diffraction (RHEED) to observe surface structure. RHEED has also been used to understand the nucleation process by depositing a few thin films between 30-300 Å thick. In this article the consistency between the results obtained from the different techniques used will be shown.

RESULTS

Fig. 1a and 1b show the Raman spectra of films deposited on glass obtained at

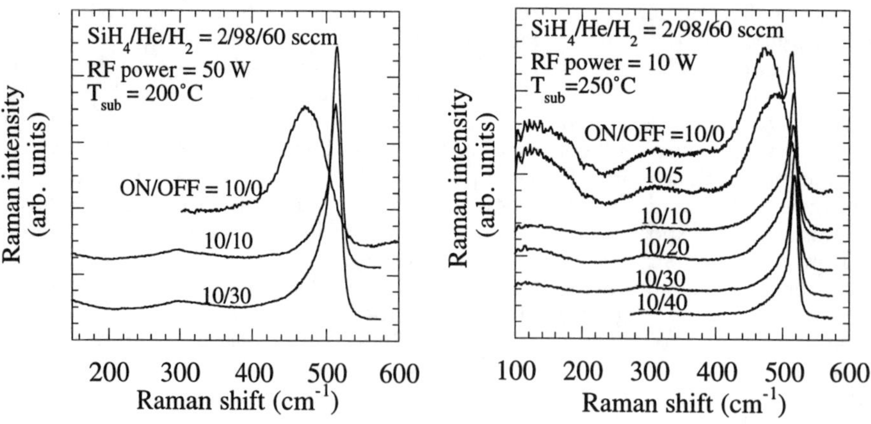

Fig. 1: Raman spectra of films deposited with varying duty cycles at (a) 200°C, 50 W rf power and (b) 250°C, 10 W rf power

varying the duty cycle between 10/0 (no atomic hydrogen exposure) and 10/40 s. Films deposited without any H exposure are amorphous characterized by the broad feature at 480 cm^{-1}. As H exposure is increased, the crystalline fraction increases as evidenced by the sharp feature near 520 cm^{-1}. For any duty cycle, the film deposited using rf power of 50 W is more crystalline than the corresponding film deposited using 10 W. These films show that the crystallinity increases with increasing exposure of atomic hydrogen. In Figs. 2a, 2b and 2c, cross-sectional bright field transmission electron micrographs of films on crystalline silicon (c-Si) substrates with varying structural morphology are shown. Fig. 2a shows the cross-section of a thick amorphous silicon (a-Si:H) film grown at 250°C and with no atomic hydrogen exposure. The cross-section of a film which is diphasic (distinct 480 and 520 cm^{-1} features in Raman spectra), deposited at 75°C and with a duty cycle of 10/15 s is shown in Fig. 2b. Fig. 2c shows a highly crystalline μc-Si:H film obtained at 250°C using 40 s of atomic hydrogen exposure per 10 s of deposition is shown. For the diphasic and microcrystalline silicon film, there is significant substrate damage as compared to the amorphous film. The substrate-film interface is not distinct and is indicated in the figures.

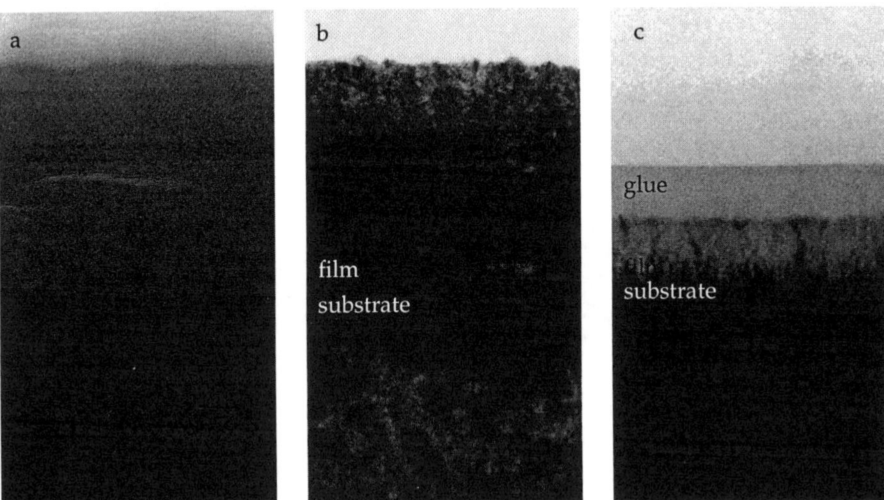

Fig. 2: Films showing distinctly different morphology (a) amorphous, (b) diphasic, and (c) microcrystalline.

Film morphology depends on the substrate and this can be seen in the atomic force micrographs shown in Fig. 3a and 3b. Fig. 3a shows the surface of a large crystallite μc-Si:H film obtained with a duty cycle of 10/40 s deposited on a c-Si substrate at 250°C. Fig. 3b shows the surface of the film on a glass substrate from the same run. Roughness analysis shows that the rms roughness of the film on c-Si is significantly larger than that for film on glass. Co-planar aluminum electrodes are evaporated onto films deposited on glass using a shadow mask and these are used to measure the photo and dark conductivity. Photoconductivity is performed using

AM 1 white light illumination. The dark conductivity of all the μc-Si:H films (10^{-3} S/cm) were high compared to those for a-Si:H films (10^{-9} S/cm). The films were annealed at 200°C in a nitrogen ambient for 2 hours, and the dark conductivities

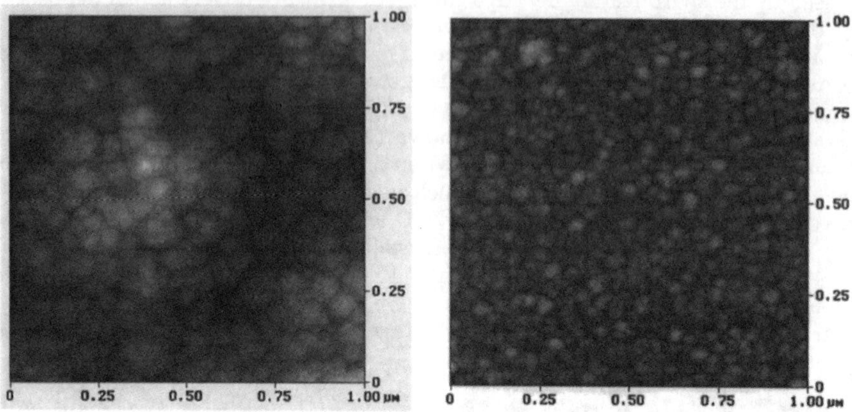

Fig. 3: Atomic Force Micrographs of microcrystalline silicon films (a) on silicon and (b) on glass.

were unaltered. Crystallite nucleation has been studied using hydrogen dilution rather than an intermittent deposition process. Films were deposited at 150°C using an rf power of 50 W. The flow rates used were the same as that for intermittent deposition, indicated in the experimental section. These films are characterized using RHEED to understand crystallite nucleation. Figs. 4a and 4b shows RHEED patterns of films approximately 300 Å thick on c-Si and glass, respectively. The distinct rings that can be seen in Fig. 4a indicates that the film is microcrystalline,

Fig. 4: RHEED images of thin hydrogenated silicon films (a) on c-Si substrates, and (b) on glass (about 300 Å) and (c) 30 Å film on c-Si.

whereas the haze in Fig. 4b is characteristic of an amorphous film. Fig. 4c shows the RHEED pattern of a 20 Å thick film deposited on c-Si. RHEED pattern for the corresponding film on glass could not be obtained because the electron beam impinging the surface charged the substrate very fast. However, it is clear that these thin films on c-Si substrates are amorphous, as evidenced by the haze in the RHEED pattern. RHEED patterns of thick microcrystalline silicon films (~ 1000 Å) deposited on a nitride and c-Si using a duty cycle of 10/30 s at 150°C are shown in Fig. 5a and 5b, respectively. Fig. 5b shows that the film deposited on c-Si substrates are near polycrystalline, due to the distinct spots in the RHEED pattern, and the film on nitride is microcrystalline. Infrared spectroscopy of these films and the link between crystallinity and bonding nature in these films have been established recently [6].

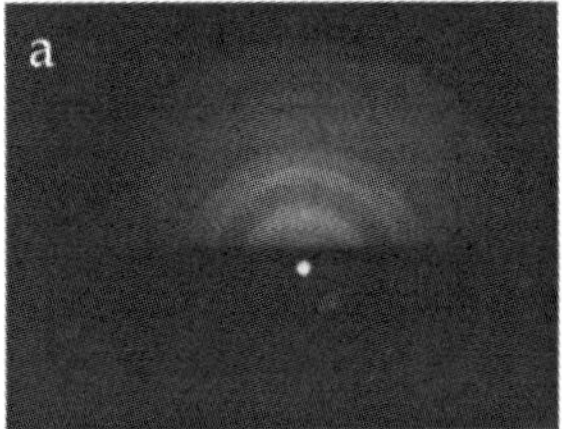

Fig. 5: RHEED patterns showing substrate dependence (a) on nitride and (b) on c-Si substrates.

DISCUSSION

Raman spectra (Fig. 1) show that films deposited at higher rf power and lower temperature (50 W, 200°C) are crystallized for the same atomic hydrogen exposure time as films deposited at lower rf power and higher temperature (10 W, 250°C). Infrared spectroscopy analysis shows that films deposited at 50 W have a larger dihydride fraction ($\alpha_{2090}/\alpha_{2000}$), where α_{2090} and α_{2000} are the absorption coefficients obtained without any spectral deconvolution. The deposition rate is also higher at 50 W. The micrographs in Fig. 2 clearly shows that the crystallinity increases with increasing extent of atomic hydrogen exposure. The crystallite sizes are comparable to those obtained in films using hot wire CVD [12] and from fluorinated silanes [9, 10]. Atomic force microscopy results showing lower surface roughness in films deposited on glass is consistent with previous results comparing μc-Si:H deposited on chromium and glass [13]. These micrographs show that the crystallites and roughness are larger in the film deposited on c-Si substrates. The RHEED patterns in Fig. 4 and 5 suggest that crystallite nuclei formation is faster on c-Si compared to glass or nitride surfaces. In addition, Fig. 4 shows no crystallites in thin films (< 50 Å) deposited on c-Si substrates and that in films as thick as 300 Å, crystallites are not observed on glass substrates. The RHEED and AFM results clearly show that nucleation processes are substrate dependent. These results can be used to propose a nucleation mechanism for crystallite formation. c-Si provides a surface favorable to large crystallite formation and hence a rough surface. Amorphous substrates like glass or nitride offers a large number of nucleation sites which results in a large number of crystallites with small sizes and consequently, lower roughness.

Dark conductivities in the deposited micro or polycrystalline films are relatively high. It has been suggested that the high dark conductivities may be related to a high oxygen content in the films. Reducing the oxygen content to near 10^{19} at/cm^3 can potentially lower the dark currents [14]. Annealing these films at 200°C in a nitrogen ambient did not change the dark conductivity levels, consistent with the theory that high dark conductivities may be related to oxygen contamination.

The RHEED patterns in Fig. 4 show that crystallites form after an incubation time sufficient to deposit approximately 100 Å. This time is slightly larger for deposition on glass as evidenced by the distinct difference in the RHEED patterns obtained from 300 Å thick films on glass and c-Si substrates. The first 100 Å is amorphous in nature and is consistent with TEM results in Fig. 2.

CONCLUSIONS

Pulsed gas deposition process has been used to obtain large grain crystallite hydrogenated silicon materials. The roughness of the film is related to the crystallinity of the film, as evidenced by TEM and AFM measurements. Film deposition is substrate dependent and crystallite nucleation occurs after a certain incubation time which depends, not only on the deposition parameters, but on the substrate also. The films have a high dark conductivity which may be related to the oxygen contamination in the films. This is expected to be lowered in the next generation deposition system in our laboratory equipped with a load-lock and with lower base pressures of 10^{-8} torr.

ACKNOWLEDGMENTS

The authors gratefully acknowledge funding for this research from DARPA (Contract #DABT 63-94-C0004). We also thank Prof. D.M. Maher and Dr. K.N. Christensen for the TEM analysis.

REFERENCES

1. I.W. Wu, A.G. Lewis, T.Y. Huang and A. Chiang, IEEE Electron Device Lett. **10**, 123 (1989).
2. J.C. Lapp, D.M. Mofatt, W.H. Dumbaugh, P.L. Bocko and M. Anma, SID 94 Technical Digest, 851 (1994).
3. G. Liu and S.J. Fonash, Jpn. J. Appl. Phys. **30**, L269 (1991).
4. D.H. Choi, S. Imai and M. Matsumura, Jpn. J. Appl. Phys. **34**, 459 (1995).
5. J.R. Abelson, Applied Physics A **56**, 493 (1993).
6. E. Srinivasan and G.N. Parsons, submitted to Journal of Applied Physics.
7. I. Sieber, I. Urban,I. Dorfel, S. Koynov, R. Schwarz and M. Schmidt, Thin Solid Films **276**, 314 (1996).
8. A. Matsuda, J. Non-Cryst. Solids **59/60**, 767 (1983).
9. K.Y. Choi, C.Y. Lee and C. Lee, Jpn. J. Appl. Phys. **34**, 4673 (1995).
10. H. Kakinuma, M. Mohri and T. Tsuruoka, J. Appl. Phys. **77**, 646 (1995).
11. E. Srinivasan, D.A. Lloyd and G.N. Parsons, J. Vac. Sci. Technol. A, Jan/Feb (1997).
12. P. Brogueira, J.P. Conde, S. Arekat and V. Chu, J. Appl. Phys. **79**, 8748 (1996).
13. H. Shirai, Jpn. J. Appl. Phys. **34**, 450 (1995).
14. P. Torres, J. Meier, R. Fluckiger, U. Kroll, J.A. Anna Selvan, H. Keppner, A. Shah, S.D. Littlewood, I.E. Kelly and P. Giannoules, Appl. Phys. Lett. **69**, 1373 (1996).

EPITAXY-LIKE GROWTH OF POLYCRYSTALLINE SILICON ON THE SEED CRYSTALLITES GROWN ON GLASS

Y. Miyamoto, A. Miida and I. Shimizu, The Graduate School, 4259 Nagatsuta, Midori-ku, Yokohama, Japan 226

ABSTRACT

Polycrystalline silicon thin films were grown on glass by two-steps, i.e., deposition of seeds on glass (1) and growth epitaxy-like on the seeds (2). For the growth of seeds, the surface reaction was intentionally enhanced by impingment of atomic hydrogen at rather high temperature (450 °C). Strongly textured polycrystalline Si exhibiting (220) preferential orientation was grown epitaxy-like on the seeds.

INTRODUCTION

In recent years, the thin film solar cells made of microcrystalline (μc-) silicon have become a current topic because of the excellent stability for light soaking and rather high energy conversion efficiency as high as 8 % or more. [1], [2] High quality polycrystalline(poly-) silicon thin films have been fabricated as well by Plasma Enhanced (PE-) CVD or Hot Wire techniques in an analogous manner to the fabrication of μc-Si. [3], [4] However, these thin films grown at rather low temperature are characterized by a structure that depends strongly on the thickness since the chemical reactions on the growing surface are ruled greatly by the topology of the surface. [5] In particular, it is quite difficult to make a crystalline film in the vicinity of the glass substrate.

In this study, we attempted to grow poly-Si thin films from fluorinated precursors (SiF_mH_n: n+m=3) on glass substrates by the two steps, i.e., the deposition of seeds on glass (1) and the growth epitaxy-like on the seeds (1).

EXPERIMENT

A remote type PE-CVD apparatus was used to grow the poly-Si from a gaseous mixture of SiF_4 and H_2 as described elsewhere. [3] A spectroscopic ellipsometer (SE) (Jovin Yvon) was equipped on the reactor for in situ real time monitoring of the growing surface. For the growth of seeds on glass, the deposition of a very thin layer (4 nm thick) and the treatment with atomic hydrogen were alternately repeated to enhance the crystallization (Layer-by-layer (LBL-) technique. [3]

The structures of the films were evaluated from the measurements of SE, XRD, Raman and RHEED. The measurements of the depth profiles of impurities

Mat. Res. Soc. Symp. Proc. Vol. 452 © 1997 Materials Research Society

such as F, H and O were made by SIMS, which provide information on the structures depending on the thickness.

Schottky diodes were fabricated by stacking N-type seed crystals, undoped poly-Si and Pt top electrode deposited in this order on ZnO coated glass.

RESULTS AND DISCUSSION

Growth of seeds on glass substrate

High quality poly-Si thin film consisting of grains (200 nm dia) was made on glass at T_s=300 °C by LBL from fluorinated precursors. [3] However, amorphous-rich layer was grown at the vicinity of glass substrate. Some device performances were limited by the properties of these thin layers. Generally speaking, it is quite difficult to make μc-Si or poly-Si exhibiting an abrupt interface on glass substrate when the films are grown at rather low temperature. A fairly large amount of amorphous phase remains in the thin layer at the vicinity of the glass substrate due to the incubation period for generation of crystallites on glass. It implies that the surface reactions for construction of the Si-network are affected greatly by the surface topologies. [5]

We verified from SE measurements that amorphous phase remaining in the layer grown in the vicinity of the glass was efficiently altered into crystals by treating with atomic hydrogen. [6] Figure 1 shows the changes in the pseudo-dielectric constant (ε_2) measured at hν=4.2 eV corresponding to the critical point (E_2) of c-Si monitored during growth by LBL at Ts=300 °C and 450 °C, respectively. As shown in the inset of this figure, (ε_2) values change periodically synchronized with the deposition and the H-treatment. The (ε_2) spectra for the films grown to 50 nm thick are illustrated in figure 2. The (ε_2) spectrum (at T_s=450 °C) is characterized by sharp critical points, E_1 (3.2 eV) and E_2 (4.2 eV) due to its highly ordered structure, whereas the one (at Ts=300 °C) is attributed to that of the amorphous phase. According to the analysis of these spectra with the effective medium approximation, the volume fraction of crystals are 100 % and <30 % for the thin films made at Ts=450 C and 300 C, respectively. These remarkable differences observed in the (ε_2) spectra give a strong support to an idea that the crystalline seeds were made at the vicinity of glass by enhancing intentionally the surface reaction with atomic hydrogen at rather high T_s as high as 450 °C.

Figure 3 (left) are Raman spectra for the films100 nm thick grown on glass in various flow rates of SiF_4 at Ts=380 °C. Highest crystallinity is obtained for the flow rate of 30 sccm. On the right of the figure, the curves of

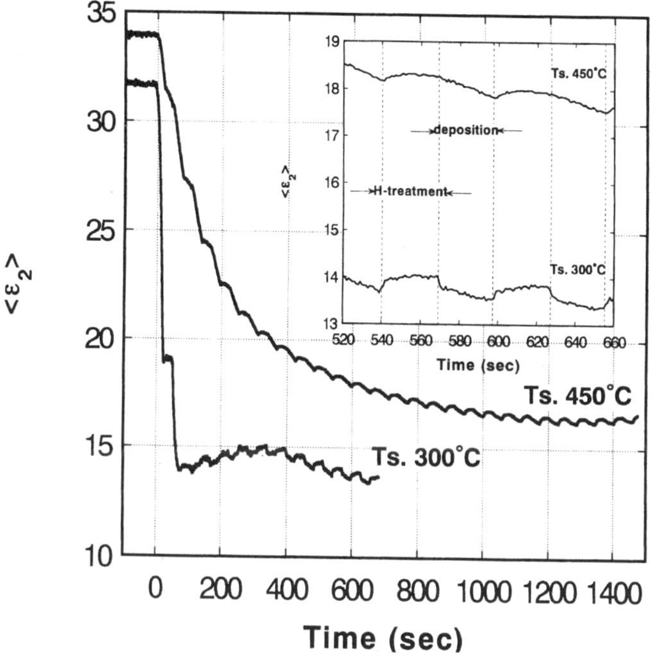

Fig.1 Real time monitoring of the growth of seeds on SiO_2 coated c-Si substrate by LBL
The numbers are the substrate temperatures.

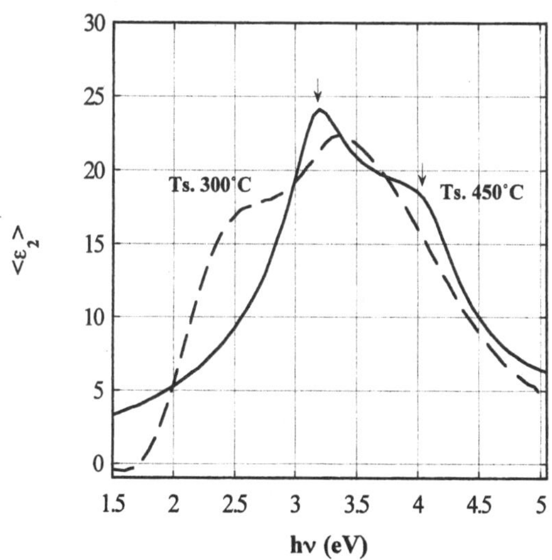

Fig.2 Spectra of pseudo-dielectric function (ε_2) for the films grown at Ts=300 °C and 450 °C

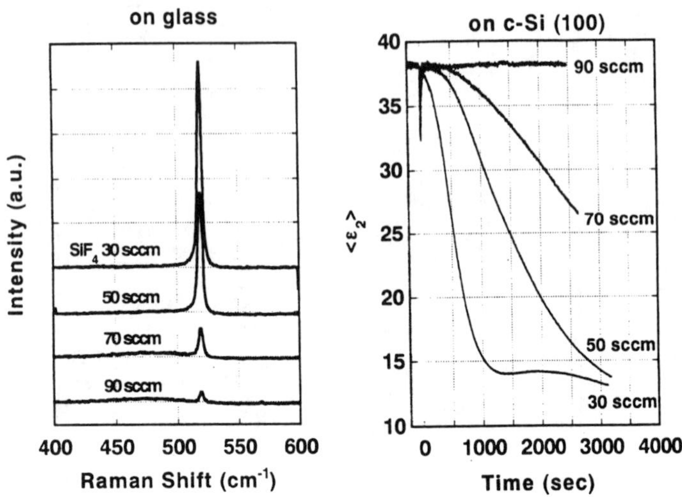

Fig.3 Effects of substrates on the structure of Si-thin films grown by varying the mixing ratio of source gases
The numbers indicate the flow rate of SiF_4 (sccm).

$\langle\varepsilon_2\rangle$ monitored during the growth of films on c-Si(100) with various flow rates are shown as well. Single crystal Si with an atomically flat surface is epitaxially grown at the flow rate of 90 sccm, while the crystallinities are deteriorated obviously with increasing the flow rates. These results lead us to a conclusion that the surface topologies affect markedly the chemical reactions to the construction of ordered structure of silicon.

Growth of poly-Si on the seeds

Poly-Si thin films were grown on the Si-seeds made on glass by LBL using varying conditions such as T_s and the flow rate of SiF_4. Figure 4 shows the $\langle\varepsilon_2\rangle$-spectra (top) and XRD spectrum(bottom) for the films of 100 nm thickness grown in various flow rates of SiF_4 at Ts=350 °C. Poly-Si films exhibiting a texture of (220) preferential orientation were grown on the seeds at flow rates lower than 50 sccm, while amorphous-rich films are obtained at flow rates over 70 sccm. According to the real time monitoring of $\langle\varepsilon_2\rangle$ during the growth, the $\langle\varepsilon_2\rangle$ of the seeds is smoothly connected to that of the layer grown on them. We, therefore, conclude from this evidence that the texture of the seed crystals is printed epitaxy-like onto the crystals grown by selecting the condition for the surface chemical reactions, i.e., chemical species and substrate temperature.

Schottky diodes and their performances

Schottky diodes were fabricated by growing an undoped poly-Si layer (about 1 μm thick) on N-type seed crystals [P-doded poly-Si made by LBL]. A semi-transparent Pt layer was deposited on the top. The current vs voltage curves were measured in the dark and under light illumination (100 mW/cm^2) for the Schottky diodes consisting of poly-Si grown by the Two-Step-Growth. Excellent diode performances, i.e., a rapid increase in the current under the forward bias and a well-saturated current under the reverse bias, are observed in the devices fabricated at T_s higher than 300 °C. In addition, high photocurrent is obtained in the device under light illumination. At T_s lower than 300 °C, on the other hand, amorphous-rich film was grown on the seeds, which results in the obvious deterioration in the device performances. The leakage current under the reverse bias was reduced efficiently by two orders of magnitude by falling T_s from 380 °C to 300 °C without accompanying obvious changes in the photocurrent. Inter-diffusion of phosphorous at high Ts is responsible for the origin of an increase in the leak current. These evidences lead us to a conclusion that the Two-Step-Growth is a promising method to grow high quality poly-Si layers on glass.

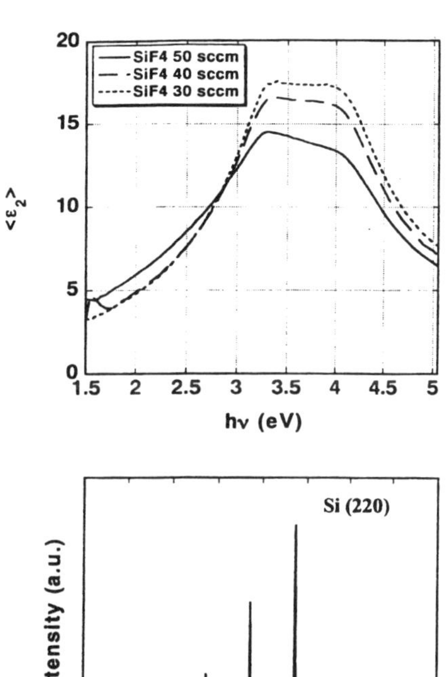

Fig.4 (ε_2) spectra (top) and XRD spectra for the films grown on the seeds made by LBL
The numbers indicate the flow rate of SiF_4.

CONCLUSION

1: Two-Step-Growth consisting of the growth of seed crystals on glass (1) and growth of poly-Si on the seeds (2) was proposed with the aim of fabricating poly-Si thin film on glass substrate.
2: The surface reactions are intentionally enhanced to promote crystallization by raising the substrate temperature and supplying excessive flow of atomic hydrogen.
3: Poly-Si is grown epitaxy-like on the seeds by printing the texture at T_s of 300 °C or higher by enhancing the surface chemical reactions.

ACKNOWLEDGEMENT

This work was supported in part by NEDO under the "New Sun Shine Project".

REFERENCES

1. J.Meier, R.Fluckiger, H.Keppner, A.Shah, Appl. Phys. Lett.,65,860 (1994)
2. K.Yamamoto, T.Suzuki, M.Yoshimi and A.Nakajima, 10th "Sunshine" workshop on thin film solar cells, Technical Digest, (Tokyo 1996) pp19
3. S.Ishihara, D.He, M.Nakata and I.Shimizu, Jpn. J. Appl. Phys.,32, 1539 (1993)
4. H.Matsumura, Y.Tashiro, K.Sasaki, and S.Furukawa, Jpn. J. Appl. Phys.,33, L1209 (1994)
5. D.Meakin, J.Stoemenos, P.Migliorato, N.A.Economou, J. Appl. Phys.,61,5031 (1987)
6. T.Akasaka and I.Shimizu, J.Non-Cryst.Solids,198-200,883 (1996)

Growth of High Quality Poly-SiGe on Glass Substrates

Kunihiro Shiota*, Daisuke Inoue, Kouichirou Minami, Masaji Yamamoto, and Jun-ichi Hanna
*Electronic Component Development Div., NEC Corp., Kawasaki 221, Japan
Imaging Science & Engineering Lab.,Tokyo Institute of Technology, Yokohama 226, Japan

ABSTRACT

The composition variation and strutural properties of poly-SiGe thin films were investigated by Reactive Thermal CVD with Si_2H_6 and GeF_4 . Deposition of the films was carried out at a low temperature of 450°C on oxidized silicon substrates using different growth parameters, i.e., the source gas flow ratio (Si_2H_6/ GeF_4) and thegas flow rate. The structural profiles of as-deposited films were characterized by X-ray diffraction (XRD) and Raman scattering spectroscopies, scanning electron microscopy (SEM) and transmission electron microscopy (TEM).

All these films show (220) preferential orientation. The mole fractions of Si in poly-Si_xGe_{1-x} films were estimated to be from 0.95 to 0.05 for x by using Vegard's law for the XRD peaks. TEM observation revealed that high crystallinity was well established even in poly-$Si_{0.95}Ge_{0.05}$ films owing to the direct nucleation on the substrate surface.

INTRODUCTION

Polycrystalline Si thin films have many applications in integrated circuit technology. Examples span a wide range of applications, from metal oxide semiconductors (MOS) to bipolar silicon technologies. In recent years, a new area has emerged where the application of poly-Si films to large-area electronic devices such as thin film transistors (TFTs)[1] for manufacturing active matrix liquid crystal displays (AM-LCDs) and solar cells[3] has been addressed. For these devices, low temperature processing (lower than 450°C) , which is required so that low cost glass substrates can be used, and establishing the film uniformity over large areas are desirable.

High quality poly-Si films have been obtained so far by deposition of amorphous silicon thin films and subsequent crystallization upon a low-temperature anneal (thermal annealing)[2] or laser anneal[3]. The films crystallized are well-suited for TFT application and such TFTs exhibit high field effect mobility (>100 cm^2/Vs)[4]. For thermal annealing, the associated processing time for the crystallization, however, was shown to increase significantly as the annealing temperature was lowered.[2] Recent efforts have been made in metal-induced crystallization in order to solve this problem. In fact, considerable reduction of both annealing temperature and time, e.g., at 450°C for 5 hours, was achieved in nickel-induced crystallization[5].

On the other hand, for laser annealing, there is no serious problem in terms of process time and temperature. Thus, its practical application to TFTs is in progress, although there remains the technical problem of the difficulty in establishing the large-area uniformity of the films because of the stepwise laser illuminations of limited area and the beam homogeneity.

Alternatively, low temperature CVD is a potential technique for cost-effective and productive deposition of large-area poly-Si thin films because of the one-step process and the feasibility of depositing large-area thin films. Plasma CVD has been most extensively studied for the low temperature growth of poly-Si films and has achieved it at considerably high growth rates of sub-

Mat. Res. Soc. Symp. Proc. Vol. 452 © 1997 Materials Research Society

nm/sec[6],[7],[8]. However, there remains a serious problem to be solved before the practical application of these films to the TFTs, i.e., crystalline inhomogeneity along with film growth[6],[9]. This is characterized typically by formation of an amorphous layer prior to polycrystalline growth on the substrate. Consequently this inhomogeneity makes it difficult to establish high crystallinity in a very thin poly-Si film of 50 to 100 nm in thickness which is required to reduce the off-current in the poly-Si TFT.

In our previous works, a new concept of low temperature CVD for large-area thin films, i.e., Reactive Thermal CVD, was proposed and its effectiveness was demonstrated in the low temperature deposition of poly-Ge and heteroepitaxial Ge with Si_2H_6 and GeF_4 as source gasses[10,11]. This CVD process features chemical reactions between Si_2H_6 and GeF_4 in order to promote their decomposition in the vicinity of heated substrate.

Here composition and crystallinity of as-deposited poly-SiGe films were studied by by X-ray diffraction (XRD) and Raman scattering spectroscopies, scanning electron microscopy (SEM) and transmission electron microscopy (TEM). Deposition of the films was performed under different growth condotions, i.e., the source gas ratio (Si_2H_6 /GeF_4) and the gas flow rate. It is found that homogeneous crystal growth takes place just on the substrate surface, leading to high crystallinity even in Si-rich poly-Si_xGe_{1-x} films up to x=0.95. This is the first reported case of high crystallinity poly-Si_xGe_{1-x} (x=0.95) prepared by thermal CVD at a low temperature of 450℃.

EXPERIMENT

The reactor for the film growth consists of a main chamber connected to a turbo-molecular pump through a gate valve, a fore-chamber for loading substrates without exposing the main chamber to the air, and a transfer rod equipped with a substrate platform and a heater, which is backed by a mechanical booster pump and a rotary pump.

GeF_4 (>99%, Central Glass Inc.) and Si_2H_6 (99.99%, Mitsui Toatsu Chemicals Inc.) are used as source materials with He. GeF_4 and Si_2H_6 flow rates were 0.1-2.7 sccm and 2.2-20 sccm, respectively, while He flow rate was kept constant at 300 sccm.

The total pressure was kept constant at 0.45 Torr for all the experiments by controlling the pumping rate of the mechanical-booster pump. The growth temperature was fixed at 450℃, which is a typical condition for polycrystalline films in the present CVD process. The substrates used were Corning 7059 glass plates for Raman and XRD studies and silicon (100) wafers having a 100 nm thick SiO_2 layer thermally grown by dry oxidation at 1100℃ for TEM study . The growth time was fixed at 20 min for all the samples except a sample prepared at a rate of 3 nm/min for 60 min. The film thickness was estimated from a cross sectional micrograph obtained by the SEM observation. The structural profiles of poly-SiGe films were characterized by XRD and Raman scattering spectroscopies, and SEM and TEM studies. The composition fractions of these films were estimated by using Vegard's law from the XRD peaks.

RESULTS

Under fixed growth conditions, i.e., a Si_2H_6 flow rate of 20 sccm, a pressure of 0.45 Torr and a growth temperature of 450℃, growth rates were roughly promotional to the GeF_4 flow rate over the entire range studied. Thus, the resulting film thicknesses were varied from 720 nm to 250 nm. Fig. 1 shows Raman spectra of these films deposited at different GeF_4 flow rates under Ar laser excitation. All the spectra exhibit peaks around 510 cm^{-1} for a TO mode of crystalline Si and around 410 cm^{-1} for a TO mode of crystalline Si-Ge and/or around 300 cm^{-1} for a TO mode of crystalline Ge, indicating deposition of polycrystalline films at 450℃. It is a good indication of increasing Si incorporation in the films with decreasing GeF_4 flow rate that new peaks around 410 cm^{-1} and 510 cm^{-1} take clear shape with a decrease in the peak intensity around 300 cm^{-1}. The full width at the half maximum (FWHM) of the peak at 510 cm^{-1} for the most Si-rich film is 12 cm^{-1} and that at 300 cm^{-1} for the most Ge-rich film is 4 cm^{-1}. These full widths indicate high crystallinity of the films near the surface.

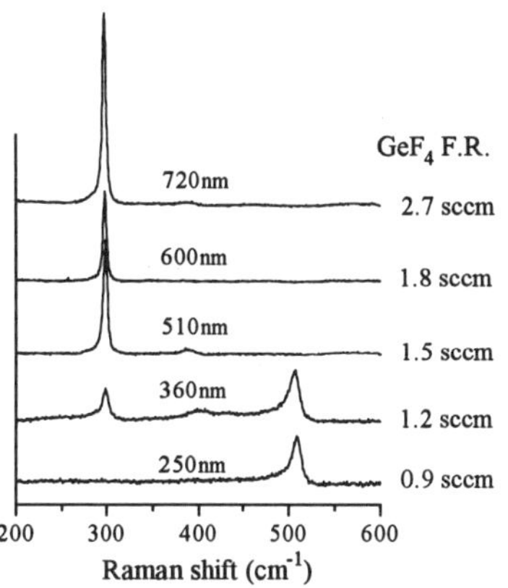

Fig.1 Raman spectra of poly-SiGe thin films deposited at different GeF_4 flow rates.

In order to evaluate bulk crystallinity of these films, a XRD study was carried out. Fig. 2 shows XRD spectra for these films. All the films show preferential orientation of the (220) plane in spite of film composition.A clear shift of peaks for the (220) plane was observed from 45.3 to 47.3 degree when the GeF4 flow rate was decreased down to 0.9 sccm as is expected due to Si incorporation in the films. Si mole fractions of these films were estimated by using Vegard law for the peak of the (220) plane.These results are plotted as a function of GeF4 flow rate in Fig. 3. The Si mole fraction was increased from 0.05 at 2.7 sccm to 0.95 at 0.9 sccm as is indicated by open circles. An abrupt increase of Si fraction occurred at around 1.2 sccm. There is good agreement with the abrupt change of Raman spectra at the same flow rate as shown Fig. 1. The solid circles in Fig. 3 indicate a relative deposition yield of GeF4 calculated from mole fractions of Ge and deposition rates of the films. A similar abrupt change in the gas yield was seen at the same GeF4 flow rate of 1.2 sccm in spite of linear change in the growth rate. These results suggest two different growth mechanisms are associated with the present film growth process, with the dividing line at around 1.2 sccm. The FWHM of the XRD peak for the (220) plane was around 0.2 degree for Ge-rich filmscomparable to a crystalline Ge substrate, indicating high crystallinity of Ge-rich films. On the other hand, the FWHM of Si-rich films were around 0.4 degree, which is inferior to that of Ge-rich films indeed, but still good enough compared with typical poly-Si films deposited by low-temperature CVD.

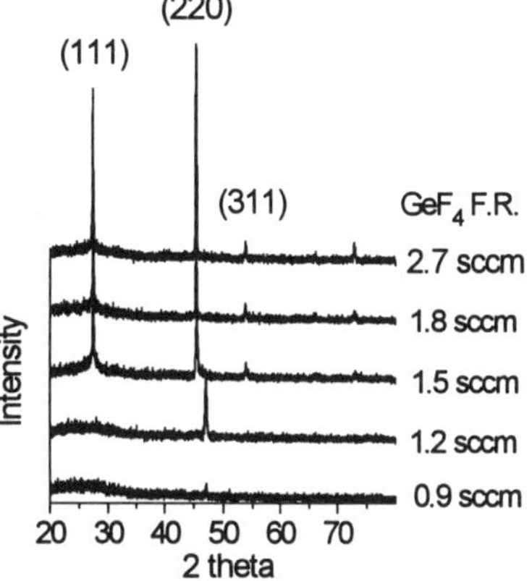

Fig.2 XRD patterns of poly-SiGe thin films deposited at different GeF_4 flow rates.

SEM observation of the film surface of poly-SiGe films indicated that the surface roughness decreases as the Si mole fraction was increased. This is probably because the grain size of polycrystals became smaller when the Si mole fraction is increased. The average grain size was estimated to be 50 to 60 nm for the most Si-rich film after Dash etching.

It is found that the crystallinity of Si-rich films was substantially affected by the growth rate of film. The FWHM of the Raman peak for the TO mode of crystalline Si was decreased from 12 cm^{-1} to 8 cm^{-1} when the growth rate was decreased from 9 nm/min. to 3.6 nm/min for poly-$Si_{0.95}Ge_{0.05}$ films.

Cross-sectional TEM observation of $Si_{0.95}Ge_{0.05}$ films prepared at different growth rates revealed a significant difference in the early stage of the film growth as shown in Fig. 4 and Fig. 5. For the film deposited at 9 nm/min as shown in Fig. 4, a 12 nm thick amorphous Si layer grew initially on the SiO_2 surface, followed by successive crystalline nucleation and grain growth. In contrast to this, no amorphous Si layer grew on the SiO_2 surface and the polycrystalline layer is formed directly on the SiO_2 surface for the film at 3.6 nm/min as shown in Fig. 5.

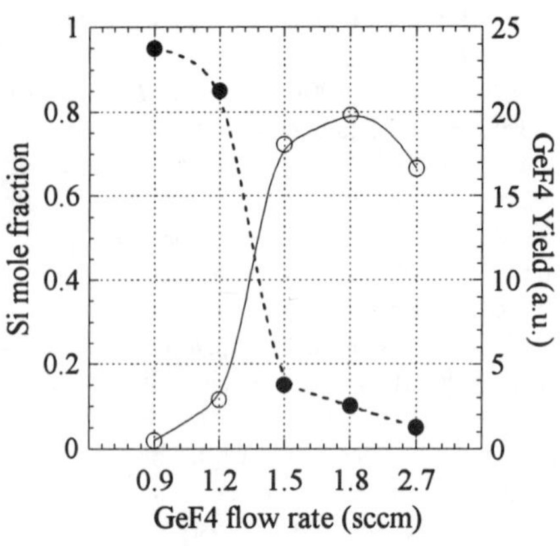

Fig. 3 Composition variety of poly-SiGe thin films deposited at different GeF_4 flow rates. Solid circles show Si mole fractions and open circles relative gas yields of GeF_4.

The present results are the first reported case of homogeneous growth of Si-rich poly-SiGe films on amorphous substrates by low-temperature CVD, although the direct nucleation took place on SiO_2 and glass substrates for Ge-rich films as reported previously[12],[13]. This is quite beneficial to

Fig. 4 Cross sectional TEM micrograph of poly-$Si_{0.95}Ge_{0.05}$ deposited at 9 nm/min (Film thickness: 130 nm).

Fig. 5 Cross sectional TEM micrograph of poly-$Si_{0.95}Ge_{0.05}$ deposited at 3 nm/min (Film thickness: 220 nm).

the establishment of a high crystallinity film in a very thin layer suitable for TFT application. The average grain size was about 50nm for both cases and consistent with the estimation from the SEM micrographs

The reason why the growth rate affects the structural properties of the film at a very early stage of growth has not been fully understood yet and could be answered after detailed studies of initial film growth at different growth conditions. However, a possible explanation might be given in terms of the competition between cluster growth and their structural relaxation into nuclei on the surface where the surface energy is relatively small. With the low surface energy, precursors for film growth can migrate on the surface within a certain diffusion length determined by the growth rate. At a given growth rate, the precursors migrate on the surface until they collide with each other to form a cluster. In the case of the higher growth rate, a short migration time causes insufficient structural relaxation of the resulting cluster because of a high clustering rate, resulting in the formation of amorphous Si tissue. In the case of the lower growth rate, however, the clustering rate is suppressed enough to give the precursur a sufficient time to relax into a nuclei.

CONCLUSIONS

In this work we have studied the composition variation of poly-SiGe thin films prepared at 450℃ by the reactive thermal CVD with GeF_4 and Si_2H_6 and characterized their structural properties. Controlling the gas flow ratio of Si_2H_6 to GeF_4 and their total flow rate, high crystallinity poly-Si_xGe_{1-x} thin films were prepared up to x=0.95.

The crystallinity in the Si-rich films was greatly improved by reducing the growth rate, resulting in no formation of amorphous tissue and direct polycrystalline growth on the substrate surface. These films are promising for TFT applications.

ACKNOWLEDGEMENTS

This work was supported in part by Grant-in-Aid for Scientific Research from Ministry of Education and Culture in Japan. We thank Central Glass Inc. for the donation of high purity GeF_4 and Mitsui Toatsu Chemicals Inc. for Si_2H_6.

REFERENCES

1. W. C. Omara, Liquid Crystal Displays, Manufacturing Science and Technology (Van Nostrand and Reinhold, New York 1993).

2. R. B. Iverson and R. Reif, J. Appl. Phys. **62** 1675 (1987).

3. T. Samejima, S. Usui, and M. Sekiya, IEEE Electron Dev. Lett. EDL-7, 276 (1986)

4. F. Okuyama, K. Sera, H. Tanabe, K. Yuda and H. Okumura, Mat.Res.Soc.Symp.Proc. **377** (1995) 877

5. S. W. Lee, Y. C. Jeon, and K. S. Joo, Appl. Phys. Lett. **66** (1996) 1671

6. T. Nagahara, K. Fujimoto, N. Kohno, Y. Kashiwagi, and H. Kakinoki, Jpn. J. Appl. Phys. **31** (1992) 4555

7. N. Shibata, K. Fukuda, H. Ohtoshi, J. Hanna, S. Oda, and I. Shimizu, Jpn. J. Appl. Phys. **26** (1987) L10

8. T. Komiya, A. Kamo, H. Kujirai, I. Shimizu, and J. Hanna, Mat. Res. Soc. Symp. Proc. **164** (1990) 63

9. K. Endo, M. Bunyo, I. Shimizu, and J. Hanna, Mat. Res. Soc. Symp. Proc. **283** (1993) 641

10. M. Yamamoto and J. Hanna: Proceedings of the 12th International Symp. on Chemical Vapor Deposition, (1993) 156

11. M. Yamamoto, M. Miyauchi and J. Hanna, Appl. Phys. Lett., **63** (1993) 641

12: J. Hanna, T. Ohuchi, and M. Yamamoto, Mat. Res. Soc. Symp. Proc. **358** (1995) 877

13. J. Hanna, T. Ohuchi, and M. Yamamoto, J. Non-cryst. Solids **198-200** (1996) 879

POLYCRYSTALLINE SILICON GROWN ON POROUS SILICON-ON-INSULATOR SUBSTRATES

Klaus Y.J. Hsu, C.H. Lee, and C.C. Yeh
Department of Electrical Engineering, National Tsing Hua University, Hsinchu, Taiwan 300, R.O.C., FAX: +886-3-571-5971, E-MAIL: yjhsu@ee.nthu.edu.tw

ABSTRACT

Inexpensive full-wafer SOI substrates are appealing for various applications such as ULSI. As an attempt to achieve this goal, low-temperature deposition of silicon on novel porous Si-on-insulator (PSOI) substrates was performed in this work. The bottom insulator was obtained by anodically oxidizing a pre-formed porous silicon film in HCl solution. The thickness, uniformity and quality of the resulted bottom oxide layer as well as the residual porous silicon layer above were well-controlled. Low-temperature PECVD growth of silicon on the PSOI wafer was conducted by using the residual porous silicon as the seed. Cross-sectional TEM pictures and electron diffraction patterns showed that poly-Si films were formed on PSOI substrates under the conditions of 98% hydrogen dilution ratio, 20 Watts RF power, and 300°C substrate temperature. Further thermal annealing at 1050°C for 30 minutes significantly enhanced the crystallinity of the deposited films. Combined with the excellent insulation ability of the bottom oxide, the technique is suitable for future inexpensive full-wafer SOI fabrication.

INTRODUCTION

In the area of advanced Si-related materials, silicon-on-insulator (SOI) and porous Si (PS) are two important subjects which contain both scientific interest and the potential for industrial application. The former presents many advantages over conventional Si substrates in the performance of ULSI. The latter creates the possibility for the birth of Si optoelectronic devices. In recent years, intensive studies in various aspects were conducted on the two materials and the control of materials' quality has been significantly improved.

Historically, PS had been used in the fabrication of SOI [1-3]. In these approaches, crystalline Si islands surrounded both at the four sides and at the bottom by PS were first formed on Si wafers. Because PS can be oxidized at a much faster rate than crystal Si, SOI structure was then formed at the Si island regions by applying thermal oxidation to the wafers. For VLSI design and fabrication, separate SOI islands are not practical. The demand for full-wafer scale SOI is obvious. Separation by ion implantation of oxygen (SIMOX) as well as wafer bonding and etch-back are now two commercialized methods to produce full-wafer SOI substrates. However, both of them involve sophisticated and expensive techniques. Therefore the price of these SOI substrates is high. Doubtless to say, inexpensive full-wafer SOI substrates will be appealing for various applications such as ULSI. If full-wafer scale SOI can be produced out of PS, it may play an important role in the field of SOI because PS is easy to form and the production cost is very low.

There are two possible ways to fabricate full-wafer SOI by using PS. The first is to oxidize the lower half of a PS layer to form a so-called porous Si-on-insulator (PSOI) structure and then thermally melt the upper PS region to let the Si recrystallize. The second method is to

Mat. Res. Soc. Symp. Proc. Vol. 452 © 1997 Materials Research Society

grow an epitaxial Si layer on top of PSOI by using the Si seeds at the surface. The feasibility of the two methods have not yet been ensured and therefore needs study. Since the first approach requires large amount of energy and the film quality is more difficult to control, we naturally paid our attention to the second method. MBE and low temperature (<800°C) LPCVD have been shown to be able to grow Si epi layers at least on p^+ PS [4-6]. In this work, the epitaxy approach is pushed one step further: We tried to deposit Si on PSOI at a very low temperature (300°C) by using PECVD and, so far, polycrystalline Si was obtained. In the following, the experimental details and the results are presented.

EXPERIMENTAL PROCEDURES

The formation of PSOI has been described in detail elsewhere [7]. Basically, it is a simple electrochemical anodization process in which HF solution was first used to form PS and then was replaced by HCl solution for the bottom oxide formation. In order to concentrate on PECVD growth, p^+-PS-on-p^--oxide samples were used as the substrates since they showed the best quality among the various PSOI forms.

After the formation of PSOI, it is inevitable that at the surface of each Si grain in the PS region, a thin native oxide layer exists. Therefore, it is essential to perform surface cleaning/etching prior to loading the samples into the PECVD chamber and/or before the deposition process to remove the surface oxide. However, protecting the bottom oxide is a basic constraint. This excludes the conventional HF dip method because the bottom oxide may be etched away as well. In this work, we adopted HF vapor etching to overcome this problem. The vapor etches the thin oxide at the surface and leaves the bottom oxide layer intact. The procedure of HF vapor etching is as following:

1. Samples cleaned in ultrasonic and rinsed by D.I. water;
2. Surface etching by the vapor of 49% HF solution for several minutes;
3. Removing residual HF vapor by nitrogen gas.

The HF etching established the hydrogen passivation at the PS surface which could prevent the Si from re-oxidation for several minutes in the air. So, the samples were loaded into the PECVD chamber within 5 minutes after the HF vapor etching. To reinforce the hydrogen passivation, H_2 gas kept flowing into the chamber for 15 minutes after the sample loading. The PECVD deposition process was commenced when the chamber pressure was pumped down to a base pressure of $5x10^{-5}$ mbar by a turbo pump. The reaction gases included SiH_4 and H_2 and the flow rate ratio of these two gases was chosen in the range between 2:98 and 5:95. The substrate temperature was 300°C. And the RF plasma power was varied between 20 W and 39 W. Scanning transmission electron microscopy (STEM) was used to investigate the deposited films.

RESULTS AND DISCUSSION

As a reference, PECVD deposition was also performed on pure PS samples. Fig. 1 shows the cross-sectional TEM micro-graph and the diffraction pattern of a Si film deposited on p-type PS under the conditions of 95% hydrogen dilution ratio, 0.88 mTorr pressure, 20 W RF power, and 300°C substrate temperature. The diffraction pattern suggests that the film is constituted by poly-Si and the film possesses good uniformity. Some bamboo shapes are visible in the film.

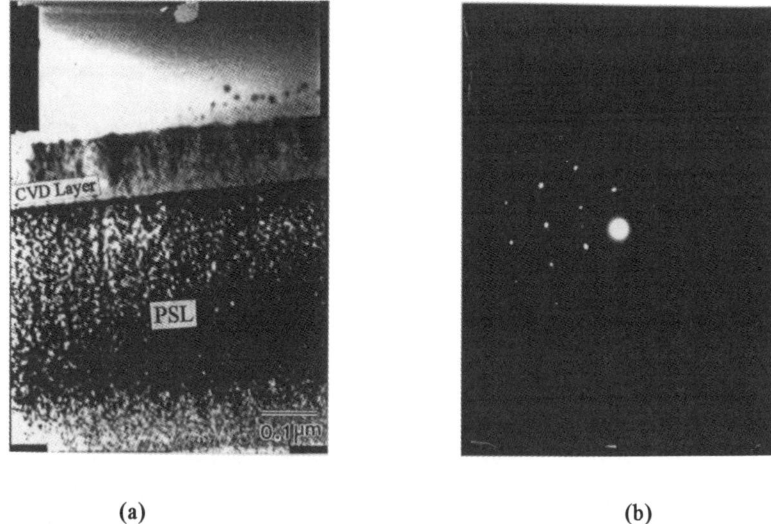

(a) (b)

Fig. 1 (a) The Bright-field cross-sectional TEM image of a PECVD Si film on PS; (b) The diffraction pattern of the deposited Si film.

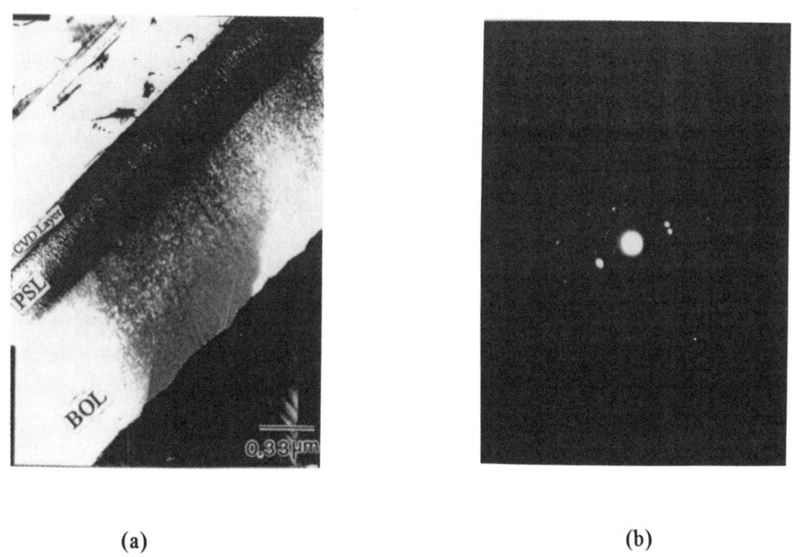

(a) (b)

Fig.2 (a) The Bright-field cross-sectional TEM image of a PECVD Si film on PSOI; (b) The diffraction pattern of the deposited Si film.

By applying the same deposition conditions to PSOI samples, similar results were obtained, as shown in Fig. 2. The Si grains were a little smaller. This could be due to that the surface oxide of PSOI samples is thicker and is more difficult to be removed completely by HF vapor etching. Therefore the number of seeds is decreased, the surface may be strained, and the crystallinity of the Si films is reduced.

The reduced crystallinity of the Si films on PSOI can be recovered by post-annealing. Fig. 3 shows that a thermal annealing at 1050°C for 30 minutes increased the Si grain size and the crystallinity revealed by the diffraction pattern was improved.

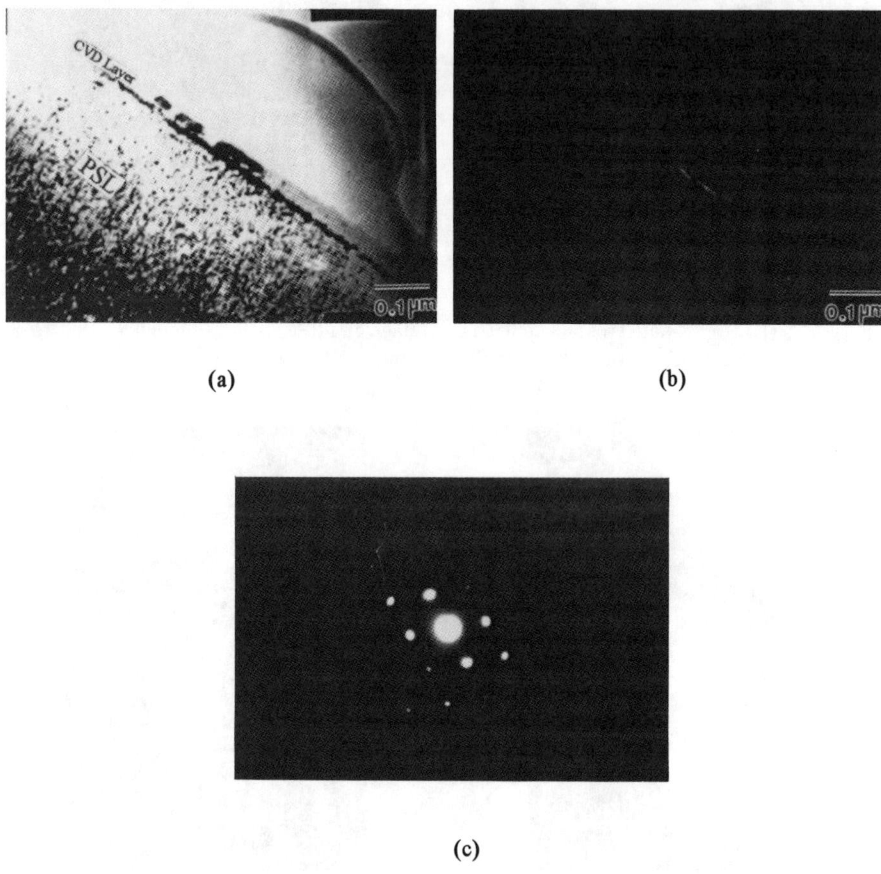

(a) (b)

(c)

Fig. 3 (a) The Bright-field and (b) the dark-field cross-sectional TEM images of a PECVD Si film on PSOI after an annealing in nitrogen environment at 1050°C for 30 minutes; (c) The diffraction pattern of the deposited Si film.

The growth of polycrystalline Si may be resulted from that the PS surface contains multiple seeds and the Si growth rate may be too high. If the growth rate is too high, the reactive precursors may have not enough time to find suitable sites. Reducing the RF power and surface roughness may help in lowering the multi-seed phenomenon. However, the deposition rate will be reduced as well. This is a trade-off which needs evaluation. Besides, the removal of PS surface oxide needs to be strengthened. After loading the sample into the chamber, some oxide may still grow during the pump-down and heating process. Hydrogen gas flow may not guarantee zero oxidation. As a consequence, some kind of *in-situ* cleaning/etching may be necessary. H_2-HCl mixture etching immediately before Si deposition can be a good way to ensure clean PS surfaces. Substrate temperature is another parameter which can be further adjusted for better results, as long as the PS structure is maintained stable. In addition, the bottom oxide in PSOI samples can contribute to maintaining the structure stability. So, varying the temperature should be of no problem. Although post-annealing was shown useful, it is not very desirable from the point of view of cost control.

Another possible way to improve the quality of low-temperature Si epitaxy on PSOI substrates is to reduce the pore number at the surface before the deposition process. Sato *et al.* [8] reported that no pores were observed at the PS surface after a $1040^{o}C$ prebake in hydrogen. If this method is applicable to PSOI wafers, better low-temperature PECVD Si films may be obtained. The feasibility of this approach is under investigation.

SUMMARY

As a summary, poly-Si on PSOI substrates was obtained by using low-temperature PECVD. Post-annealing enhanced the crystallinity of the deposited films. Refinement in surface cleaning and growth temperature adjustment could further improve the results and help reaching the goal of producing low-cost full-wafer SOI substrates.

ACKNOWLEDGMENT

Partial support from the National Science Council of the Republic of China under the contract number NSC84-0404-E007-009 is acknowledged by the authors.

REFERENCES

1. K. Imai *et al.*, IEEE Trans. Electron. Dev. ED**31** (1984), p.297-302.

2. S.S. Tsao *et al.*, J. Appl. Phys. **62** (1987), p.4182-4186.

3. R.P. Holmstrom *et al.*, Appl. Phys. Lett. **42** (1983), p.386-388.

4. F. Arnaud D'Avitaya *et al.*, 1st Intern. Symp. on MBE, Toronto, Canada, Extended Abstracts (1985), p.323-334.

5. M.I.J. Beale *et al.*, J. Vac. Sci. Tech. **B3** (1985), p.732-735.

6. G. Bomchil *et al.*, Appl. Surf. Sci. **41/42** (1989), p.604-613.

7. C.H. Lee, C.C. Yeh, and Klaus Y.J. Hsu, Appl. Surf. Sci. **92** (1996), p.621-625.

8. Nobuhiko Sato *et al.*, J. Electrochem. Soc. **142** (1995), p.3116.

CHARACTERIZATION AND METROLOGY OF LOW PRESSURE CHEMICAL VAPOR DEPOSITED (LPCVD) POLYSILICON

Leo Asinovsky *, Michael Schroth**, Fei Shen*, John Sweeney III**
* Rudolph Technologies Inc., Flanders, NJ 07836,USA
** Bruce Technologies Int'l, North Billerica, MA 01862,USA

ABSTRACT

Polycrystalline silicon (polySi) is widely used in the semiconductor industry as a gate electrode, interconnect material and for various other applications. Small variations in deposition conditions can significantly affect this material's properties, cause errors in metrology control of the film thicknesses and, ultimately, in device performance. In this work polySi was LPCVD deposited in a vertical reactor with multiple gas inlets and a flat temperature distribution. Deposition conditions were controllably changed to create a matrix of wafers. Characterization is done using production dual wavelength, multiple-angle-of-incidence and research grade spectroscopic ellipsometers. PolySi thickness and composition uniformity are analyzed.

INTRODUCTION

Low Pressure Chemical Vapor Deposition (LPCVD) is a dominant process for growth of the high-quality polySi used in semiconductor manufacturing, due to the high purity and uniformity of the deposited material as well as good process control. Uniformity of the polysilicon thickness across the reactor was historically one of the important characteristics. In the single gas inlet reactors, which are widely used in industry, thickness uniformity is achieved by biasing the temperature across the wafer load -- typically 20-30 ^{o}C -- to compensate for silane depletion. Physical properties (e.g., composition and surface morphology) are critically dependent on deposition temperature and, to a lesser extent, are sensitive to some other deposition parameters. Ignoring physical properties variation, which is common with use of the reflectometry and related techniques, can cause errors in metrology control of the film thickness. Both the optical constants and thickness of the polySi must be measured simultaneously, even when uniformity of the polySi composition is not by itself critical for device performance. In most cases surface roughness and native oxide layers must be also taken into account to achieve meaningful and reliable metrology control of the process.

Recent studies [1] done with the 100-150 mm wafers in a biased temperature furnace, have shown significant influence of the deposition parameters on the physical properties and thickness of the polySi. In this paper we use a state of the art vertical reactor with a 5-zone heater core to analyze the influence of the process parameters on uniformity of polySi deposition on 200 mm production wafers. To optimize and control the deposition process, it is essential to have metrology equipment that is capable of measuring simultaneously and independently physical properties and thickness of the film. We show that a production-oriented Dual Wavelength Multiple-Angle-of-Incidence (DW-MAI) ellipsometer meets all the requirements of a reliable metrology tool, and provides an excellent capability of tracking variation of both composition and thickness of the polySi. Measurements were found to be consistent with the results of a research grade spectroscopic ellipsometer (SE).

Mat. Res. Soc. Symp. Proc. Vol. 452 © 1997 Materials Research Society

EXPERIMENT

Polysilicon was deposited on 200 mm p-type prime grade Si <100> wafers with 1000 Å (nominal thickness) of a thermal oxide. Deposition was done in a BTI APOGEE vertical 5-zone LPCVD reactor with a 150 waferload size. The BTI APOGEE hot wall vertical reactor has its exhaust location at the base of the reactor chamber, a quartz liner for gas distribution, multi-hole distributed gas injection and a rotational tower mechanism. The matrix of the deposition conditions is summarized in Table 1. Total silane (SiH_4) gas flow and tower rotational speed were maintained at a constant value for all experiments. The runs were done with the full wafer load and eight wafers selected along the reactor -- from bottom to top -- were analyzed.

Measurements were taken with FEIV (Rudolph Research) fully automated DW-MAI production ellipsometer system. The system uses rotating compensator measurement system, patented Focused Beam Technology (FBT) that allows simultaneous measurement in an Angle-Of-Incidence (AOI) range of 40 to 70 degrees using 633 and 780 nm wavelengths. Standard 5-point measurements were taken on all wafers to analyze the uniformity across the reactor. Twenty point measurements along the wafer diameter were taken on wafers located at the bottom, center and top of the 5 zone heat core reactor to analyze wafer uniformity. One point measurements in the 250--850 nm range were taken using a research grade S2000 Spectroscopic Ellipsometer (Rudolph Research). Spectra at AOI 65 deg and 70 deg were taken in each case to confirm consistency of the model.

DATA ANALYSIS

Ellipsometry, like most optical methods, is an indirect measurement technique. It allows one to measure values of Δ and Ψ which are related to the optical properties of the analyzed sample as follows: $\rho(n_i\ d_i,\ \phi_i,\ \lambda_i) = r_p\ /\ r_s\ = \tan(\Psi)\exp(i\Delta)$, where ϕ is the AOI, λ is the wavelength of light, n_i and d_i are, respectively, the optical constants and thicknesses of the filmstack components. r_p and r_s are reflection coefficients for p- and s- polarized light and can be calculated using the Fresnel equations. A numerical procedure of minimizing the difference between measured and calculated values of Δ,Ψ, assuming a particular model of the filmstack, is used to determine parameters of the films.

Parameters of the assumed model are varied until the best fit between the measured and calculated values of Δ,Ψ is achieved. Corresponding parameters of the filmstack are then declared "measured". In multiple wavelength or/and multiple AOI measurements one determines Δ,Ψ pairs at each wavelength and/or each AOI, thus overdetermining the set of equations and allowing to find, potentially, more filmstack parameters or filmstack parameters with greater confidence.

Table 1. Matrix of Process Deposition Test Conditions.
All runs have been done with a full wafer load of 160 wafers and a tower rotation speed 2 rpm.

Run	Deposition T (°C)	Pressure (mTorr)	SiH_4 Flow (sccm)
1	540	360	400
2	600	360	400
Baseline	610	360	400
4	620	360	400
5	610	150	400
6	610	500	400

The number of parameters, which can be determined simultaneously is limited, however, due to the correlation between measured values [2,3]. In SE modeling one uses reference optical constants spectra, provided they are available and accurate, and determines the thicknesses of the films. When optical constants also have to be determined, a parametric representation of the spectra is normally used. This allows one to decrease the number of variable parameters and take advantage of the spectral information by indirectly varying the optical constants across the spectrum. In the case of multiple AOI ellipsometry, optical constants are independent of AOI and can be varied directly during the modeling. Thus spectroscopic and multiple AOI ellipsometry techniques constitute different approaches to multi-parametric measurements. Both FEIV and S2000 ellipsometers are combining the elements of these approaches: multiple AOI and dual wavelength capability in the former case, spectroscopic and several AOI measurement capability in the latter. This fact allows for modeling of the measured data in a fundamentally similar way. Polysilicon optical constants can be represented parametrically using the Bruggeman Effective Medium Approximation (BEMA) [4, 5]. BEMA assumes an isotropic mixture of scale smaller than the wavelength of light, with each constituent retaining its original optical constants. Variation of the volume fraction (VF) of constituents change the effective optical constants of the material. This approach reduces the number of independently determined parameters and puts reasonable constraints on the range of variation of the optical constants.

Interpretation of the data is done using 4 layer models of the filmstack: oxide/interface/polySi/oxide/Si. Bulk polySi optical functions are represented using BEMA as a mixture of a-Si + c-Si, interface/roughness layer as a BEMA of c-Si + a-Si + SiO_2 . The details of the modeling procedure are described in [4]. The thickness of the top oxide layer was fixed at 20Å, and the thickness of the bottom oxide at 1000 Å. Thicknesses of the bulk polySi, the interface layer and the composition of the bulk polySi are measured simultaneously.

RESULTS AND DISCUSSION

1. Polysilicon deposition.

The deposition rate and composition were averaged within the wafer and across the reactor for various runs. The deposition rate shows exponential increase with the temperature as expected for reaction rate-limited conditions. The rate is changing from 18.5 Å/min. for 540 °C to 105.8 Å/min for 620 °C. An Arrhenius type plot (Fig. 1) and the value of the activation energy (E_a=1.345 eV) confirms consistency of the results with the theoretical behavior and the insignificance of the silane depletion and other effects on the average deposition rate. Two points on Fig.1 corresponding to the different silane pressures (P=150 mT and P=500 mT) at T=610 °C are shown for comparison purposes. In fact, we found practically linear dependence between deposition rate and pressure in this range. The polysilicon composition changes also almost linearly with temperature in the 600-620 °C range: the degree of amorphization decreases from 17% to 12% (Fig. 2). At 540 °C the polysilicon was completely amorphous (100% amorphization). Increase of silane pressure in the 150mT-500mT range has an opposite effect -- increase of amorphization from 12% to 17% (Fig. 3).

2. Polysilicon Within-Wafer (WIW) and Within-Run (WIR) Uniformity.

WIW uniformity depends both on deposition conditions and reactor design. From this perspective variation of the WIW uniformity with the wafer position in the reactor can serve as a signature of a reactor. Indeed the shape of this dependence (Fig. 4) does not change qualitatively for various runs and shows general degradation of uniformity at the bottom and top of the

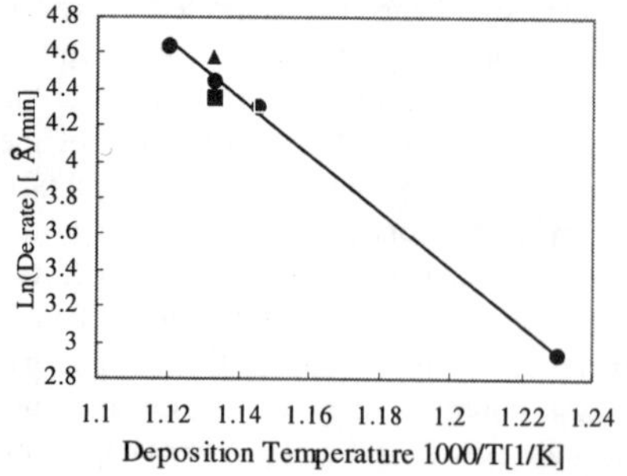

Figure 1. Deposition rate dependance on temperature (Arrhenius type plot). Temperature range 540 °C -- 620 °C, pressure 360 mT (•). Points corresponding to pressures of 150 mT (■) and 500 mT (▲) at 610 °C are given for comparison. The activation energy is E_a=1.345 eV.

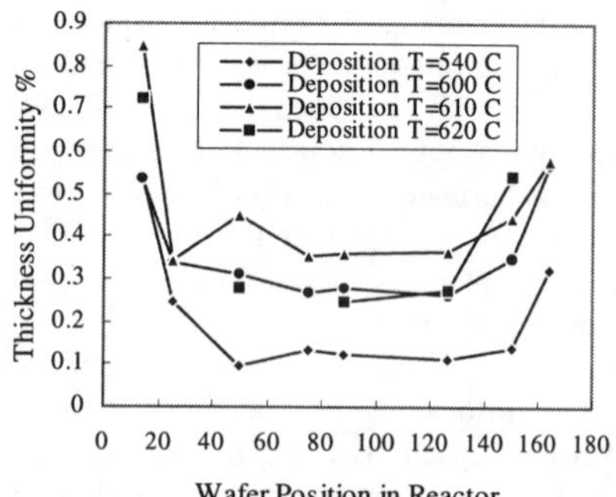

Figure 4. WIW thickness uniformity across the reactor.WIW uniformity dependence on wafer position is qualitatively the same for different runs and is a signature of the reactor design.

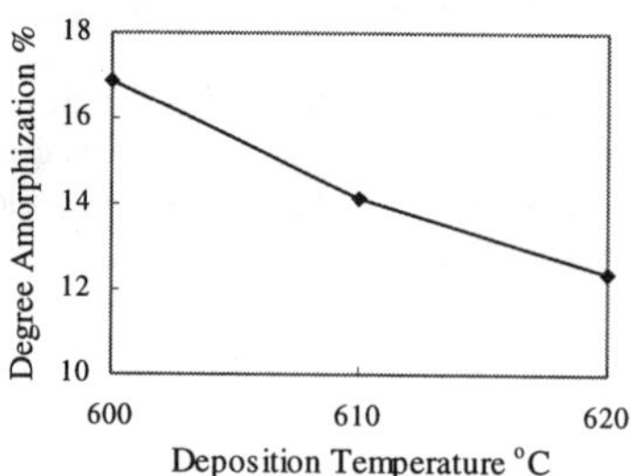

Figure 2. Change of polySi composition with temperature (silane pressure 360mT).

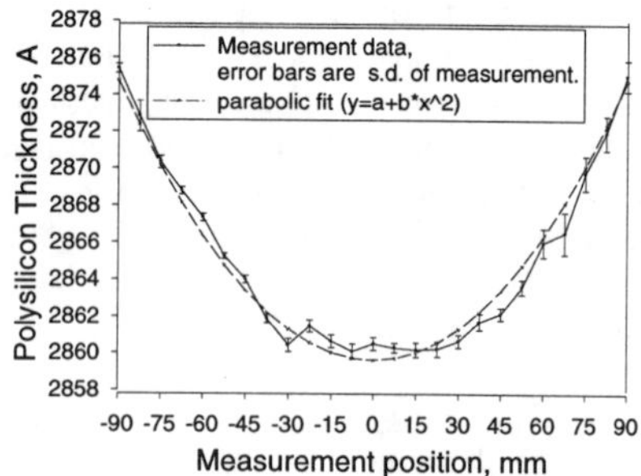

Figure 5. Thickness variation across the wafer (center of the reactor). These are the baseline conditions.

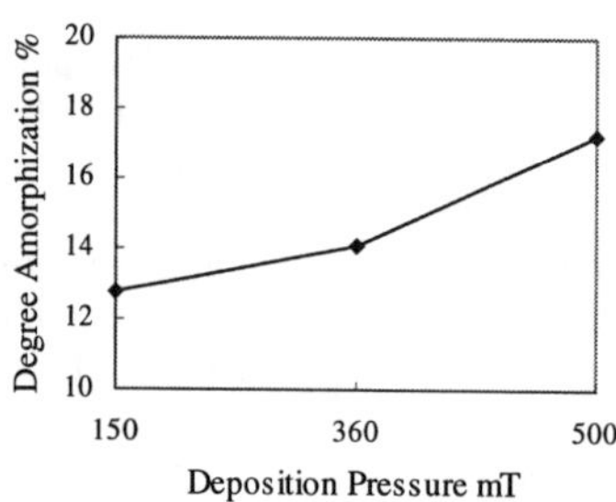

Figure 3. Change of polySi composition with silane pressure (deposition temperature 610 °C).

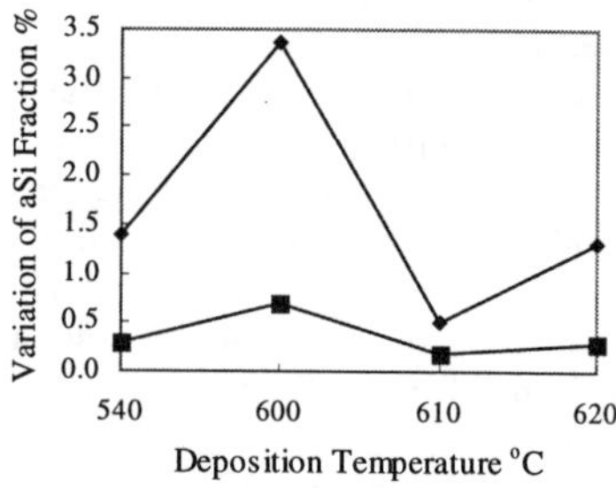

Figure 6. Composition uniformity vs. deposition temperature. Deposition pressure 360 mT . ■ - within wafer, ◆ - across the reactor. Optimum conditions correspond to 610 °C.

reactor. Both temperature and pressure have a weak influence on WIW uniformity. More detailed analysis of WIW uniformity was done using 20 point measurement across the wafer diameter taken from the central part of the reactor (Fig 5). The results of the 20 point measurements are summarized in Table 2. Composition variation is practically within the accuracy of the measurement (about 0.5% of constituent VF), which indicates good temperature stability across the wafer. The random character of the composition variation suggests that it may be primarily caused by the grain structure of the polysilicon. Thickness uniformity which reflects the effect of the silane depletion towards the center of the wafer is slightly increasing at 620 °C. The influence of the deposition parameters on the WIW and WIR uniformities, both compositional and thickness, are shown in Fig. 6-9. Clearly optimum conditions correspond to T=610 °C, P=360 mT.

Table 2. Within Wafer Uniformity. Twenty point measurements along the wafer diameter taken on wafers located in the center of the reactor to analyze WIW uniformity. Uniformity is calculated as $\sigma_X/\langle X\rangle$, where σ_X is the standard deviation and $\langle X\rangle$ is the mean value of the parameter X.

Parameters\Deposition Temperature		600 °C	610 °C	620 °C
Thick-ness	WIW variation (σ) Å	4.35 (0.35)	5.03 (0.45)	7.94 (1.5)
	Uniformity %	0.17	0.17	0.22
Compo-sition	WIW variation (σ) % VF	0.48 (0.31)	0.11 (0.11)	0.16 (0.13)
	Uniformity %	1.6	0.74	1.32

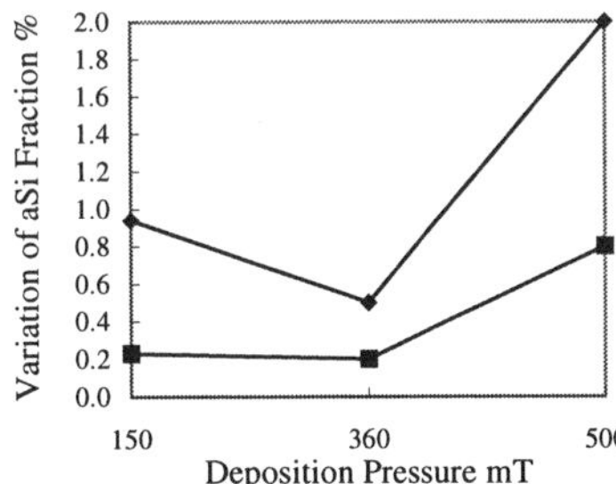

Figure 7. Composition uniformity vs. pressure. Deposition temperature 610 C ; ■ -- within wafer, ◆ -- across the reactor. Optimum conditions correspond to 360mT.

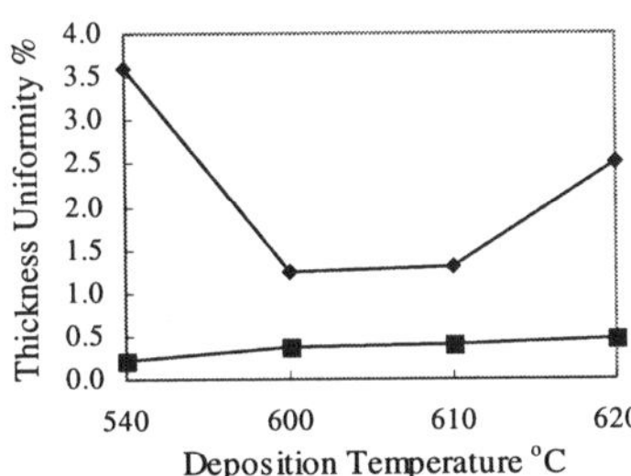

Figure 8. Thickness uniformity vs. deposition temperature.Deposition pressure 360 mT; ■ - within wafer, ◆ - across the reactor. Optimum conditions correspond to 610 °C.

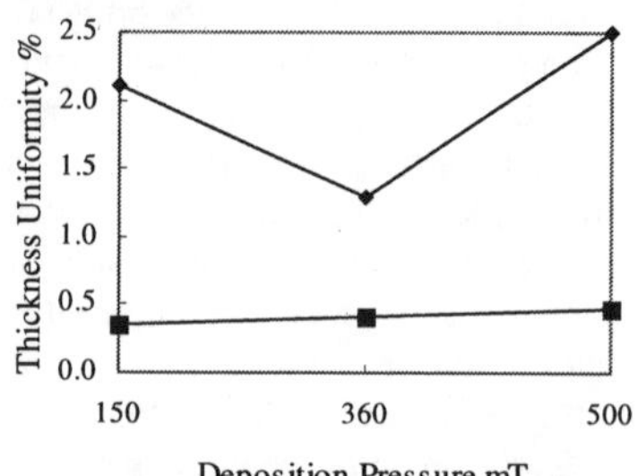

Figure 9. Thickness uniformity vs. deposition pressure. Deposition temperature 610 °C; ■ - within wafer, ◆ - across the reactor. Optimum conditions correspond to 360mT.

CONCLUSIONS

Dual wavelength multiple-angle-of-incidence ellipsometry is used for characterization of LPCVD polysilicon deposited in a state of the art vertical reactor. The effect of the reactor configuration and different deposition conditions on within wafer (WIW) and across reactor uniformity of thickness and composition of the polysilicon is analyzed. Application of the DW-MAI ellipsometer for optimization of the deposition process, qualification of the reactor and polysilicon metrology in a production environment is demonstrated. Simultaneous measurement of the optical constants and thickness of the polySi is a critical requirement for a reliable metrology tool. The results were found to be consistent with measurements on a research spectroscopic ellipsometer (SE).

REFERENCES.

[1] E. Ibok and S. Garg, *J. Electrochem. Soc.* **140**, 2927 (1993).

[2] R.M.A. Azzam and N.M. Bashara, Ellipsometry and Polarized Light, (Elsevier, New-York, 1979), p. 530.

[3] M.M. Ibrahim and N.M. Bashara, *J. Opt. Soc. Am.* **61**, 1622(1971).

[4] L. Asinovsky, S. Fox, E. Karagiannis, M. Schroth and J. Sweeney in Polycrystalline Semiconductors IV , edited by S. Pizzini, H. Strunk, J. Werner (Solid State Phenomena **51-52,** Scitec Publications, Switzerland, 1996) pp.179-185 .

[5] D.A.G. Bruggeman, *Ann.Phys.* (Leipzig), **24**, 636 (1935).

[6] L. Asinovsky, *Thin Solid Films* **233,** 210 (1993).

TEMPERATURE DEPENDENT LINE-SHAPE OF THE SILICON DANGLING BOND EPR-RESONANCE IN POLYCRYSTALLINE SILICON

N. H. NICKEL
Hahn-Meitner-Institut Berlin, Rudower Chaussee 5, 12489 Berlin, F. R. Germany.

E. A. SCHIFF
Department of Physics, Syracuse University, Syracuse, New York 13244, USA.

ABSTRACT

The temperature dependence of the silicon dangling-bond resonance in polycrystalline (poly-Si) and amorphous silicon (*a*-Si:H) was measured. At room temperature, electron paramagnetic resonance (EPR) measurements reveal an isotropic g-value of 2.0055 and a line width of 6.5 and 6.1 G for Si dangling-bonds in *a*-Si:H and poly-Si, respectively. In both materials spin density and g-value are independent of temperature. While in *a*-Si:H the width of the resonance did not change with temperature, poly-Si exhibits a remarkable T dependence of ΔH_{PP}. In unpassivated poly-Si a pronounced decrease of ΔH_{PP} is observed for temperatures above 300 K. At 384 K ΔH_{PP} reaches a minimum of 5.1 G, then increases to 6.1 G at 460 K, and eventually decreases to 4.6 G at 530 K. In hydrogenated poly-Si ΔH_{PP} decreases monotonically above 425 K. The decrease of ΔH_{PP} is attributed to electron hopping causing motional narrowing. An average hopping distance of 15 and 17.5 Å was estimated for unhydrogenated and H passivated poly-Si, respectively.

1. INTRODUCTION

Electrical and optical properties of polycrystalline silicon (poly-Si) are dominated by grain-boundary defects. These defects have been measured by electron paramagnetic resonance (EPR) and identified as silicon dangling-bonds with an average g-value of 2.0055.[1] In order to obtain technologically useful material, grain-boundary defects must be passivated. Commonly, this is achieved by exposing poly-Si to a hydrogen plasma at elevated temperatures. Hydrogen effectively passivates silicon dangling-bonds and thus improves the electrical properties of the material.[2] Defect passivation is diffusion limited and eventually the spin density saturates.[3]

The silicon dangling-bond defect has also been detected in amorphous silicon,[4] at Si-SiO_2 interfaces,[5] and divacencies.[6] In contrast to crystalline silicon, amorphous silicon samples are characterized by an isotropic distribution of the defect symmetry-axes with respect to the external magnetic field. This causes inhomogeneous broadening of the natural line-width.

In this paper we present a temperature dependent change of the width of the dangling-bond resonance in unpassivated and hydrogenated polycrystalline silicon. With increasing temperature the line width decreases which is indicative of motional narrowing. We propose that this effect is due to electron hopping between dangling-bond sites.

2. EXPERIMENTS

Undoped polycrystalline silicon films were prepared by two different methods using quartz wafers as substrates. A set of fine grain poly-Si films was deposited by low-pressure chemical

Mat. Res. Soc. Symp. Proc. Vol. 452

vapor deposition at 625 °C to a thickness of 0.55 μm. Cross-sectional transmission electron microscopy revealed that the films were composed of columnar grains extending from the substrate to the sample surface with an average diameter of 15 nm. A second set of samples was prepared by solid state crystallization of undoped LPCVD amorphous silicon at 600 °C. Solid-state crystallization produces material in which crystal grains with an average size of 150 nm are randomly distributed.

Prior to the passivation of grain-boundary defects the native oxide of the poly-Si films was removed with dilute HF to avoid a barrier for the incorporation of hydrogen. The poly-Si films were then hydrogenated through a sequence of 1-h exposures to monatomic H generated in an optically isolated remote plasma at 350 °C until the spin density was minimized.[3] According to secondary ion mass spectrometry measurements the hydrogenation produces a nearly depth independent H concentration of $\approx 2\times10^{20}$ cm^{-3}.

Electron-paramagnetic resonance (EPR) measurements of unpassivated and hydrogenated poly-Si films were performed in the X-band. The silicon dangling-bond resonance was measured as a function of microwave power and temperature. The temperature of the samples was controlled between 100 K and 550 K by a flow of either cooled or heated nitrogen gas. In order to avoid modulation induced broadening of the resonance a modulation amplitude of 0.4 G was chosen for all measurements.

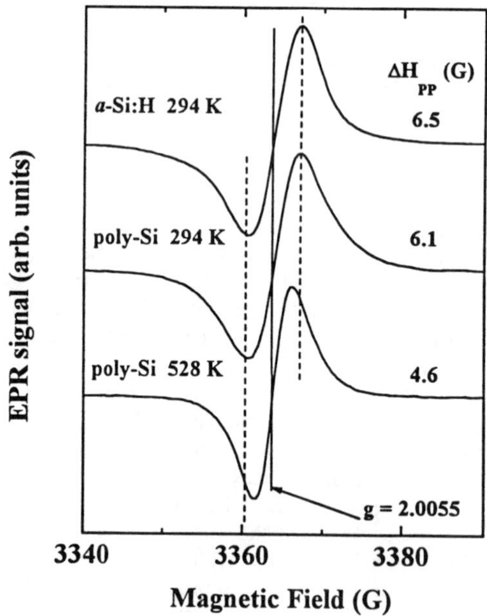

Fig. 1. EPR spectra of hydrogenated amorphous silicon (*a*-Si:H) recorded at 294 K and unpassivated polycrystalline silicon (poly-Si) measured at 294 and 528 K. The EPR spectra reveal a g-value of 2.0055 characteristic of a silicon dangling-bond defect.

3. RESULTS

In this section we present results taken on both sets of samples. Since the data are independent of the preparation method we use the notation poly-Si for both sets. Fig. 1 shows EPR spectra measured in hydrogenated amorphous silicon and unpassivated polycrystalline silicon. All EPR spectra exhibit a resonance with a central g-value of 2.0055 which is characteristic for silicon dangling-bond defects.[1,4,7] At room temperature the line width of the resonance in *a*-Si:H and poly-Si are ΔH_{PP}=6.5 G and 6.1 G, respectively. In both materials the Si dangling-bonds are randomly oriented with respect to the magnetic field. The somewhat narrower resonance found in poly-Si most likely is due to a smaller degree of structural disorder. In poly-Si dangling bonds are predominantly confined to grain boundaries. These two-dimensional boundaries display long-range order which is imposed by adjacent crystallites.

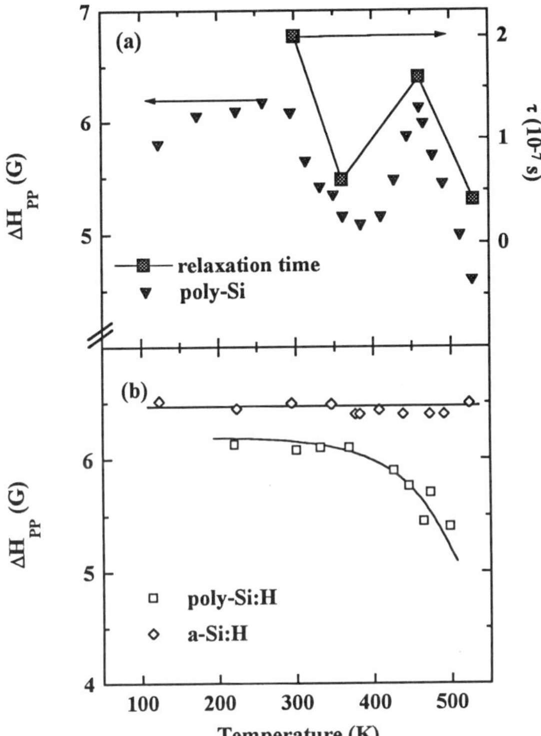

Fig. 2. Line width of the Si dangling-bond resonance as a function of the measurement temperature. (a) polycrystalline silicon and (b) hydrogenated amorphous silicon and hydrogen passivated poly-Si. Hydrogenation of the poly-Si samples was performed at 350 °C for 7 consecutive hours. The squares in (a) represent the characteristic relaxation time τ.

In poly-Si samples a striking decrease of the line width to ΔH_{PP}=4.6 G is observed when measuring the Si dangling bond resonance at T = 528 K. On the other hand, for *a*-Si:H the line width did not change with increasing temperature.

A detailed plot of the temperature dependence of the line width for unpassivated poly-Si, hydrogen passivated poly-Si, and *a*-Si:H is shown in Fig. 2. Unhydrogenated poly-Si reveals a striking temperature dependency of ΔH_{PP} [Fig. 2(a)]. For temperatures between 150 and 300 K the line width is independent of the measurement temperature (ΔH_{PP} = 6.1 G). However, above 300 K ΔH_{PP} decreases monotonically and reaches a minimum of 5.1 G at 384 K. With further increasing temperature the Si dangling-bond resonance broadens to ΔH_{PP} = 6.1 G at 460 K. Then, at even higher temperatures a second more pronounced decrease of ΔH_{PP} occurs. At the highest accessible measurement temperature of 528 K the line width decreased to ΔH_{PP} = 4.6 G. The same measurements were performed on hydrogen passivated polycrystalline silicon (poly-Si:H) and *a*-Si:H. In poly-Si:H samples the Si dangling-bond line-width is temperature independent up to T = 424 K. At higher temperatures a decrease of ΔH_{PP} similar to that for unhydrogenated specimens between 300 and 400 K is observed [open squares in Fig. 2(b)]. On the other hand, in *a*-Si:H the Si dangling-bond resonance has a width of ΔH_{PP} = 6.5 G and is independent of temperature [open diamonds in Fig. 2(b)].

The EPR spectra were further analyzed by calculating the spin concentrations and the g-values. The concentration of dangling bonds was obtained by double integration of the measured EPR spectra. At room temperature unpassivated poly-Si revealed a spin concentration of $N_S = 9.4\times10^{17}$ cm^{-3}. Exposing the specimens to monatomic H at 350 °C caused a reduction of the spin density to $N_S = 2\times10^{17}$ cm^{-3}. Assuming a Curie-law dependency of the spin density, N_S of poly-Si, poly-Si:H, and *a*-Si:H did not change with temperature. The temperature dependence of the g-value of unpassivated poly-Si is plotted in Fig. 3. The g-value is independent of temperature and hence is

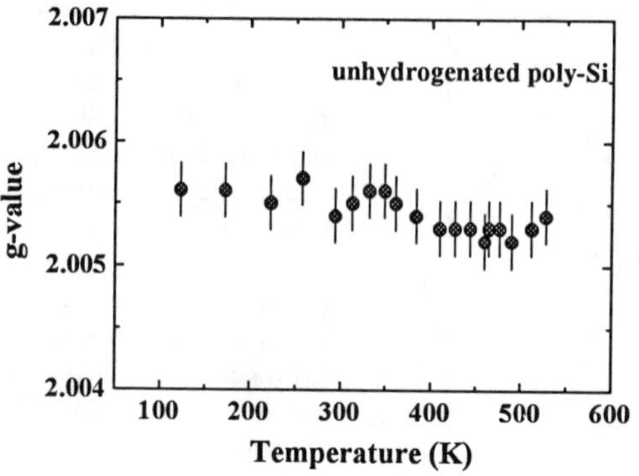

Fig. 3. Temperature dependence of the g-value in poly-Si.

unrelated to the narrowing and broadening of the line width. A similar result was obtained for hydrogenated poly-Si.

Previously, a similar decrease of the line width was reported in evaporated a-Si:H. The line width decreased from 10 to 5 G with increasing evaporation temperature. Simultaneously, the spin density decreased. Both effects were attributed to structural changes of the amorphous network.[8]

Unpassivated poly-Si specimens reveal a striking increase of the linewidth at temperatures between 400 and 460 K. This behavior suggests the presence of a second resonance line. However, a careful analysis of EPR spectra, taken in this temperature range with a variety of measurement parameters (microwave power, modulation frequency and amplitude), did not indicate the presence of a second resonance line suggesting that the observed temperature dependence of ΔH_{PP} is solely caused by the Si dangling-bond resonance.

Information on spin relaxation was obtained by measuring the saturation properties of the spectrometer signal as a function of the microwave field H_1 at room temperature, 360 K, 460 K, and 530 K, respectively. A characteristic relaxation time τ was estimated from the value for H_1 at which the signal had fallen 50% below its low-field linear dependence. For unpassivated poly-Si at room temperature we estimated $\tau = 2\times10^{-7}$ s. This value is compatible to results reported for poly-Si and a-Si:H.[9] The characteristic relaxation time was estimated at $T = 360$ K, 460 K, and 530 K and is represented by the solid squares in Fig. 2(a). It is important to note that in unhydrogenated poly-Si the relaxation time of the dangling bonds exhibits the same temperature dependence as the linewidth ΔH_{PP} [Fig. 2(a)].

3. DISCUSSION

The experimental results presented above have a number of important implications. The magnetic resonance spectra of disordered materials such as glasses, amorphous, and polycrystalline silicon are influenced by statistical distributions of crystal fields arising from the intrinsic randomness of the material. The field strength at a given dangling-bond site depends on the orientation of the spin with respect to the magnetic field. Hence, the measured resonance line consists of a spectral distribution of individual resonant lines.

With increasing temperature the relaxation time of the spins decreases which most likely is due to a resonance interaction between the spins. Comparing $1/\tau$ to the linewidth in frequency units measured in unpassivated poly-Si reveals that at 360 and 530 K the reverse relaxation time is comparable to the measured line width of 8.3×10^7 and 8×10^7 s^{-1}, respectively. A similar decrease of $1/\tau$ was observed in hydrogen passivated poly-Si.

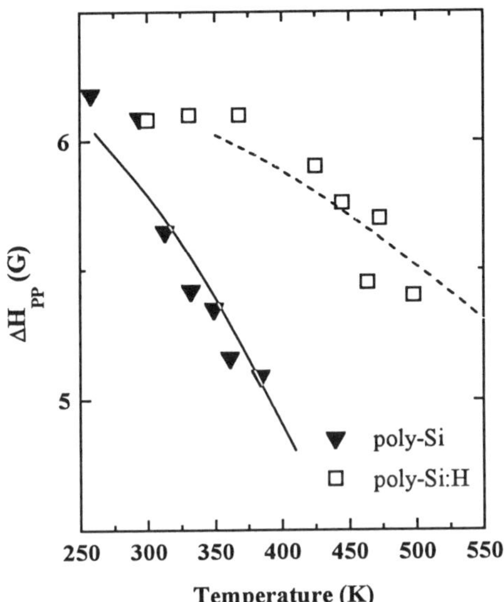

Fig. 4. Section of the line width as a function of temperature of unpassivated and hydrogenated poly-Si. The solid and dashed lines show the effect of motional narrowing of the Si dangling bond resonance assuming that the process is governed by electron hopping at the Fermi energy.

The fairly rapid decrease of the line width with temperature suggests some type of motional narrowing. One possibility is the following. We presume that the electrons on dangling bonds are to some extent free to hop to additional initially empty dangling bonds. If this hopping occurs slowly compared to the characteristic spin-lattice relaxation time, then each of the two dangling bonds contributes an independent „spin packet" to the line shape of the dangling-bond system. These packets reflect the orientations of the two defects as well as any structural differences, and occur at different g-values. When the hopping occurs rapidly, the pair contributes a single packet to the final line shape with a g-value corresponding to an average of the g-values.

Assuming that electron hopping occurs via deep defects the hopping rate is determined by the electron emission-rate[10]

$$\nu_H = \nu_0 \exp\left(\frac{-2R}{R_{db}} - \frac{E}{kT}\right) \quad (1)$$

where k is Boltzmann's constant, T the temperature, R is the distance and E the energy difference between two sites. The localization length of Si dangling-bonds was estimated by EPR measurements to $R_{db} \approx 3$ Å.[11] The resulting inhomogeneous line width can be estimated using the following relation[12]

$$\Delta H_{PP}(T) = \frac{\Delta H_0}{1 + \nu_H T_1} \quad , \quad (2)$$

where T_1 is the saturation life-time and $\Delta H_0 \approx 6.1$ G. Assuming an attempt frequency of $\nu_0 = 10^{12}$ s^{-1} the hopping distance and energy difference in unpassivated poly-Si can be estimated to $R \approx 15$ Å and $E \approx 135$ meV, respectively. On the other hand, in hydrogenated poly-Si the decrease of the line width occurs at a higher temperature and a hopping distance and energy difference of $R \approx 17.5$ Å and $E \approx 135$ meV were estimated, respectively. The fits are shown by the full and dashed lines in Fig. 4 and agree well with the data. However, the measured spin densities of $N_S = 9.4\times10^{17}$ and 2.1×10^{17} cm^{-3} for poly-Si and poly-Si:H, respectively, are too low to account for the estimated hopping distances. On the other hand, in poly-Si dangling-bond defects are predominantly confined to grain boundary regions. In poly-Si with an average grain size of 1000 Å the grain-boundary volume amounts to only 1% of the entire sample.[13] Thus, the grain-boundary spin concentration, N_S^{gb}, is consistent with the estimated hopping distances.

The hopping distance is proportional to $1/(N_S^{gb})$. Therefore, one would expect that the hopping distance increases by about 70 % for the observed decrease of the spin density. Thus, the observed increase of the hopping distance by only ≈3 Å suggests that Si dangling-bonds in poly-Si are clustered.

5. SUMMARY

In summary, we have shown a temperature dependent change of the line width of the silicon dangling-bond resonance in unpassivated and hydrogenated poly-Si. With increasing temperature the Si dangling-bond line width decreases for both, unpassivated and hydrogenated poly-Si samples. Moreover, in unpassivated poly-Si a pronounced increase of the line width was observed at 384 K. The decrease of the line width is accompanied by a decrease of the characteristic time constant. This is consistent with the idea that the decrease of the line width is due to motional narrowing. We propose that the underlying microscopic mechanism is electron hopping at the Fermi level. This model is consistent with both, unpassivated and hydrogenated poly-Si. An average hopping distance of 15 and 17.5 Å for poly-Si and poly-Si:H, respectively, was obtained from model calculations suggesting that Si dangling bond defects are clustered at grain boundaries.

ACKNOWLEDGMENTS

The authors are grateful to J. Boyce and R. A. Street for many stimulating discussions, and to J. Walker for technical support. This work was partially supported by NREL. One of the authors (N.H.N) also acknowledges support from the Alexander von Humboldt foundation, Germany.

REFERENCES

* Measurements were performed at Xerox PARC, 3333 Coyote Hill Road, Palo Alto, CA 94304
[1] N. M. Johnson, D. K. Biegelsen, and M. D. Moyer, Appl. Phys. Lett. **40**, 882 (1982).
[2] T. I. Kamins and P. J. Marcoux, IEEE Electron Devices Lett. **EDL-1**, 159 (1980).
[3] N. H. Nickel, N. M. Johnson, and W. B. Jackson, Appl. Phys. Lett. **62**, 3285 (1993).
[4] M. H. Brodsky and R. S. Title, Phys. Rev. Lett. **23**, 581 (1969).
[5] D. K. Biegelsen, N. M. Johnson, M. Stutzmann, E. H. Pointdexter, and P. J. Caplan, Appl. Surf. Science **22/23**, 879 (1985).
[6] G. D. Watkins and J. W. Corbett, Phys. Rev. **138**, A 543 (1965).
[7] H. Dersch, J. Stuke, and J. Beichler, Appl. Phys. Lett. **36**, 456 (1981).
[8] P. A. Thomas, M. H. Brodsky, D. Kaplan, and D. Lepine, Phys. Rev. B **18**, 3059 (1978).
[9] P. J. Gaczi and D. Booth, Solar Energy Materials **4**, 279 (1981).
[10] N. F. Mott, J. Non-Cryst. Solids **1**, 1 (1968).
[11] D. K. Biegelsen and M. Stutzmann, Phys. Rev. B **33**, 3006 (1986).
[12] C. P. Slichter, *Principles of Magnetic Resonance*, 3rd Edition (Springer, Berlin 1990).
[13] N. H. Nickel and E. A. Schiff, unpublished.

In-situ real time spectroscopic ellipsometry applied to the surface monitoring of semiconductors

P. BOHER and J.L. STEHLE,
SOPRA S.A., 26 rue Pierre Joigneaux, 92270 Bois-Colombes (France)

ABSTRACT

A new kind of Real Time Spectroscopic Ellipsometer (RTSE) system is presented in detail. A multichannel analyser with photointensifier allows one to get spectroscopic measurements over the 0.25-0.85μm wavelength range with resolution $\lambda/\Delta\lambda$ better than 500 and a good signal to noise ratio even for samples with poor reflectance. Precision and reproducibility of the system are around 0.3 and 0.1% for the two ellipsometric parameters. The measurement speed can be increased up to 15 spectra/s but must be adapted accurately to the process under control. An example of the heating process of a GaAs substrate measured in-situ by RTSE is presented and the temperature and native oxide thickness are deduced.

INTRODUCTION

Spectroscopic ellipsometry (SE) is widely used to characterize thin multilayer films for optical coatings or for the semiconductor industry. For the control of deposition or etching processes, SE can be applied with good success. In some cases scanning SE has been applied in-situ to control the surface state of the semiconductors see for example [1]. Nevertheless, the measurement speed was not sufficient to get spectroscopic information at each step of the process. For this reason, the use of multichannel detectors for SE was a great step in the development of the technique for in-situ applications. In 1986, SOPRA first developed a new instrument including a spectrograph and a silicon array photodetector [2]. Since this date different studies have been reported using the same kind of instrument [3-4]. Specific calibration procedures have also been developed [5]. Control of dynamic processes by ellipsometry has also been made for more than ten years. In-situ kinetic ellipsometry at the HeNe laser wavelength has been applied for example to the growth of nanometric multilayers [6]. Until very recently the use of SE for process control was limited moreover by the capacities of the computers to analyse in real time a huge number of data points.

With the rapid increase of computer capacities such large data sets can now be analyzed in real time. For this reason SOPRA began to develop a new kind of Real Time Spectroscopic Ellipsometer (RTSE) which provides not only raw measurement data but structural parameters in real time during the process, allowing process control and not only a posteriori analysis as was previously made [7]. In this paper we want to present this new system with special attention on its application to semiconductor process. Some characteristics of the multichannel detector will be shown and some details on the calibration procedures will be given. Specific problems of the semiconductor processes analysis by RTSE will be introduced through the practical example of a GaAs substrate heating experiment where the temperature and native oxide thickness have been extracted from RTSE measurements.

Mat. Res. Soc. Symp. Proc. Vol. 452 © 1997 Materials Research Society

DESCRIPTION and PERFORMANCES of the RTSE SYSTEM

General description of the system

The SOPRA RTSE instrument is a rotating polarizer instrument especially designed for easy adaptation to deposition or plasma treatment systems. The polariser and analyser arms can be attached directly to UHV flanges with easy adjustment by an optical collimator. The angle of incidence is fixed by the setup but can be determined accurately during the calibration procedure. A 5mm beam diameter is defined from an Xe lamp with a system of mirrors and passes through a quartz Rochon polarizer. It can be parallel or can be focused on the sample surface with a spot size lower than $1mm^2$. The reflected beam passes through another Rochon analyser and is focused onto the entrance of an optical fiber connected to a spectrograph.

Multichannel analyser detection

The optical fiber introduces the light inside a spectrograph with a silica prism for quasi linear dispersion versus energy. The dispersed light is focused on a low amplified intensifier and detected with a multichannel silicon photodiode aray. After amplication the signals are transmitted to a PC through a Digitial Signal Processor (DSP) card. The DSP allows control of all the steps of the measurement including non linearity corrections, averages and precalculations. During a RTSE measurement the processor of the PC is in fact only used for the transfer of the precalculated data (tan ψ and cos Δ or pseudo optical constants). In this way, data analysis can take place on the previous measurement during the current acquisition.

Signal/noise ratio of the intensified photodiode array detector

The intensifer in front of the silicon photodiode array is useful to increase the sensitivity in the UV range. Nevertheless, the amplification must be carefully adapted to optimize the signal/noise ratio of the detector. We have measured this parameter for all the pixels of the detector simply by making the same measurement thirty times using the emission spectra of the xenon lamp through the ellipsometer arms in straight line without sample. The noise evaluated by this method is reported versus the signal detected by each pixel in Figure 1. As shown in the figure, the signal/noise ratio is always higher than 25 even for the pixels with the lower signal level. It can reach values of 100 for the highest signals available (saturation at 65536 or 16 bits of precision of the photodiode array). These results have been obtained with 1s of integration time. If the speed of the measurement can be reduced, the signal/noise ratio can of course be improved by simple averaging.

Reproducibility and precision of the measurement

The precision and reproducibility of the RTSE instrument have been evaluated simply by making different measurements in straight line without sample. The precision is around 0.3% both for tan Psi and cos Delta parameters. The reproducibility is three times better around 0.1% for 1s of integration time as shown in Figure 2. The noise is lower than 0.1% in a great part of the wavelength range and only 0.2% in the UV part of the spectra where the number of photons is more limited.

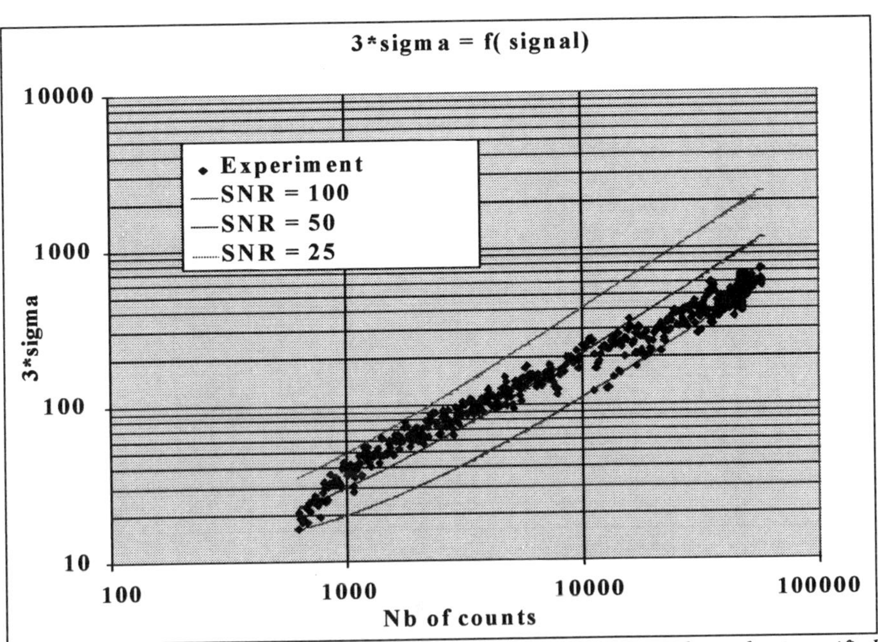

Figure 1: Signal to noise ratio versus number of counts measured on the intensified multichannel detector (integration time 1s, 30 successive measurements for SNR evaluation).

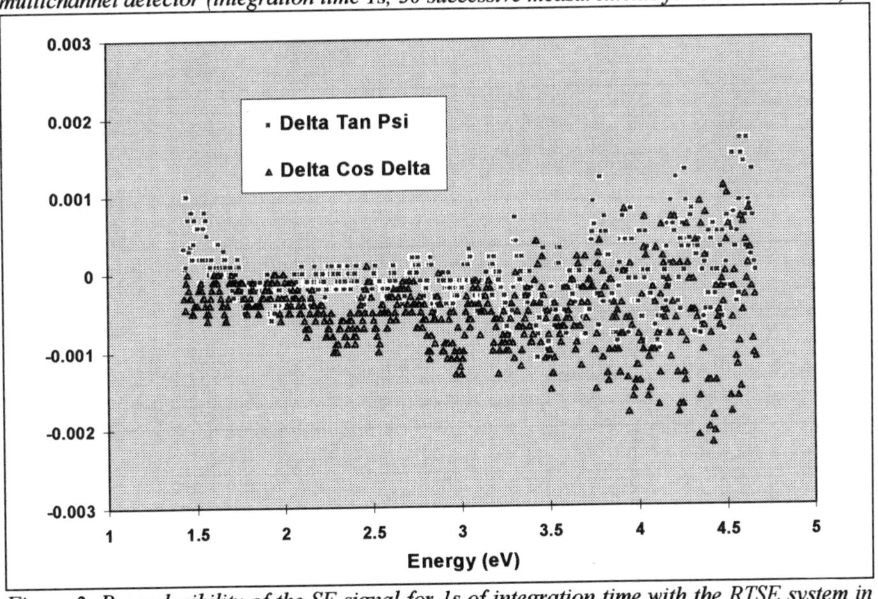

Figure 2: Reproducibility of the SE signal for 1s of integration time with the RTSE system in straight line.

GaAs HEATING EXPERIMENT

One two inch GaAs substrate was mounted on a heating system in air. The temperature of the sample holder was increased to 400°C after 300s. Then the heating current was cut off and the cooling process took place during 900s. During the entire procedure, the RTSE system measured the surface state of the sample every 3s. A 3D representation of the experimental curves is reported in Figure 3. The value reported is directly the optical absorption of the sample which presents two well defined absorption peaks around 3.5eV which correspond to the optical transition E_1 and $E_1+\Delta_1$ of GaAs. During the heating process, these transitions are shifted to lower energy in agreement with the literature [8].

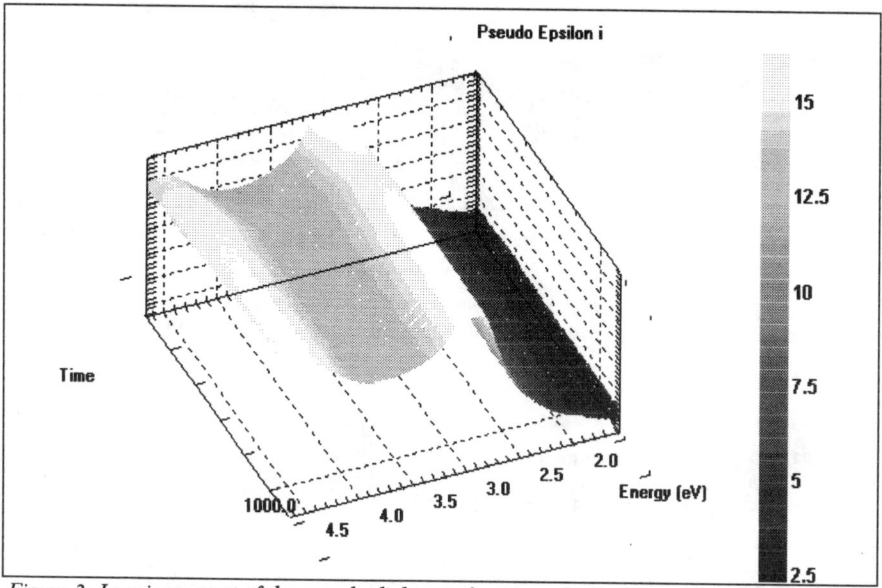

Figure 3: Imaginary part of the pseudo dielectric function of a GaAs sample measured during the heating experiment. (Heating up to 400 C was applied during the first 300s and then normal cooling occurred under air.)

To extract quantitative informations from these data, a simple model has been applied to each measurement. It includes a GaAs substrate with variable temperature and a native oxide layer on top. The optical indices of the GaAs substrate are calculated using an alloy model taking into account the known dependence of the optical transitions with temperature. At each step, the regression provides at the same time the temperature of the substrate and the thickness of the native oxide layer. These two quantities are reported in Figures 4.a and 4.b respectively. As expected the temperature of the substrate follows the same shape and that of the sample holder. The native oxide is reduced during the heating phase of the process and then remains constants at values around 27Å during the cooling phase.

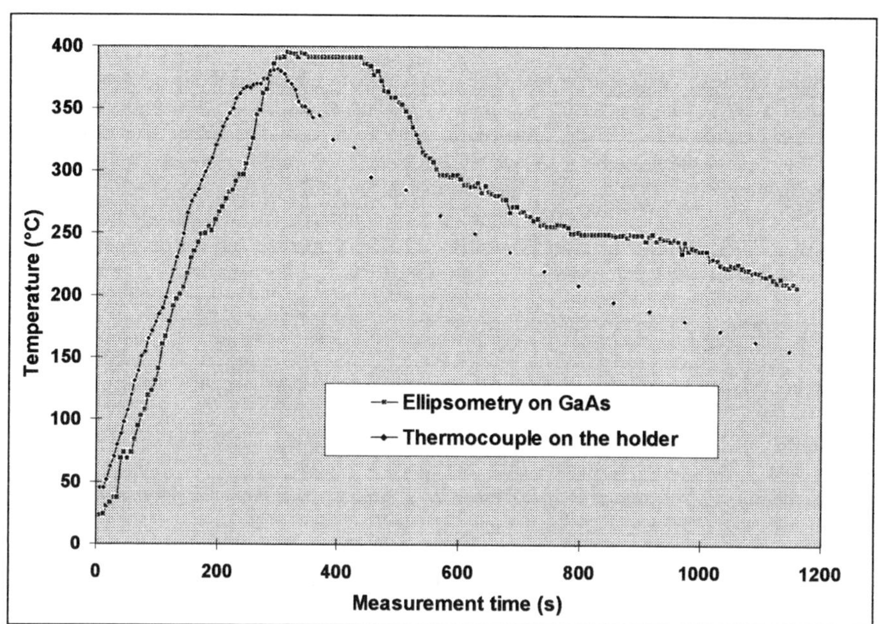

Figure 4.a: Temperature deduced by the RTSE system during the heating process. The temperature of the sample holder is also reported for comparison.

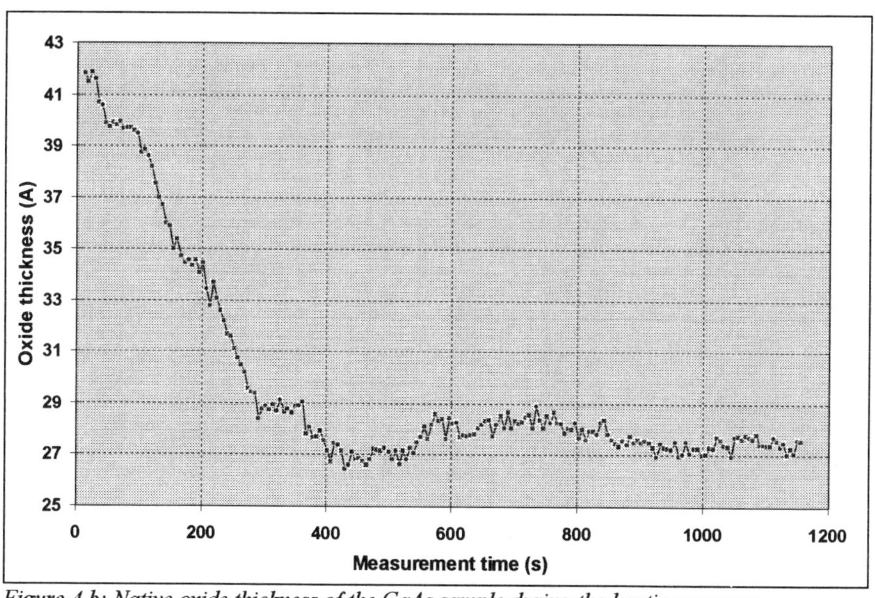

Figure 4.b: Native oxide thickness of the GaAs sample during the heating process.

CONCLUSION

First results obtained using a new real time spectroscopic ellipsometer are presented. The detection is a multichannel analyser with photointensifier which allows to get spectroscopic measurements in the 0.24-0.84μm wavelength range with resolution better than 500. The signal/noise ratio of the detector is always higher than 25 and can be increased by averaging automatically using the DSP. Precision and reproducibility of the system have been measured around 0.3 and 0.1% both for the two ellipsometric parameters. The example of the heating process of a GaAs substrate measured in-situ by RTSE is presented and the temperature and native oxide thickness are deduced

REFERENCES

(1) P. Boher, M. Renaud, J.Lopez-Villegas, J. Schneider, J.P. Chané, Applied Surface Science, 30, 100 (1987)
(2) F. Ferrieu, F. Bernoux, J.L. Stehle, Le Vide les couches Minces, suppl. 233, 17 (1986).
(3) Y.T. Kim, R.W. Collins, K.Vedam, Surf. Sci., 233, 341 (1990)
(4) C. Pickering, R.T. Carline, D.J. Robbins, W.Y. Leong, D.E.Gray, R.Greef, Thin Solid Films, 223, 126 (1993)
(5) I. An, R.W. Collins, Rev. Sci. Instrum., 62 , 1904 (1991)
(6) P. Boher, P. Houdy, C. Schiller, Thin Solid Films, 175 (1989)
(7) P. Boher, J.L Stehle, Materials. Science and Engineering, B37, 116 (1996)
(8) J.P. Walter, R.L. Zucca, M.L. Cohen, Y.R. Shen, Phys. Rev. Lett., 24, 102 (1970)

PHOSPHORUS DOPING OF BORON CARBON ALLOYS

SEONG-DON HWANG[1], P. A. DOWBEN[1,2], A. CHEESEMAN[2], J. T. SPENCER[2] AND D.N. MCILROY[3]
1) Department of Physics and the Center for Materials Research and Analysis, University of Nebraska-Lincoln, Lincoln, Nebraska 68588-0111, pdowben@unlinfo.unl.edu
2) Department of Chemistry and the W.M. Keck Center for Molecular Electronics, Syracuse University, Syracuse, New York 13244-4100, jtspence@mailbox.syr.edu
3) Department of Physics, University of Idaho, Moscow, Idaho 83844-0903, dmcilroy@uidaho.edu

ABSTRACT

Phosphorus doped boron carbon alloy films were made by chemical vapor deposition from a single source compound, dimeric chloro-phospha(III) carborane $((C_2B_{10}H_{10})_2(PCl)_2)$. Phosphorus doped B_5C materials exhibit increases in the band gap from 0.9 eV to 2.6 eV.

INTRODUCTION

Phosphorus will readily alloy with boron forming both boron phosphide (BP) [1] and $B_{12}P_2$, which is structurally similar to conventional icosahedral boron carbide [2,3]. The $B_{12}P_2$ structure is obtained by replacing the three-atom chain of the B_4C crystal with two phosphorus atoms; the icosahedra contain only boron in $B_{12}P_2$. Phosphorus is precluded from occupying an icosahedral site due to its relatively large size compared boron. Because of the structural similarity between $B_{12}P_2$ and boron carbides, it is anticipated that phosphorus will substitute at a chain position, and result in a larger band gap.

Since the molecular structure of dimeric chloro-phospha(III)carborane $([C_2B_{10}H_{10}PCl]_2)$ has similarities with orthocarborane (*closo*-1,2-dicarbadodecaborane or $C_2B_{10}H_{12}$), we anticipated that boron-carbon-phosphorus alloys could be fabricated by plasma enhanced chemical vapor deposition (PECVD) using this compound as a single source molecule in place of orthocarborane. Because of the similarity in molecular structures (as seen in Figure 1) between dimeric chloro-phospha(III)carborane and orthocarborane, the materials grown by PECVD from these source molecules should have similar structures. Orthocarborane has been successfully used to make boron-carbon alloy heterojunction diodes [4-7], homojunction diodes [5], tunnel diodes [5-6] and transistors [7] by chemical vapor deposition (CVD).

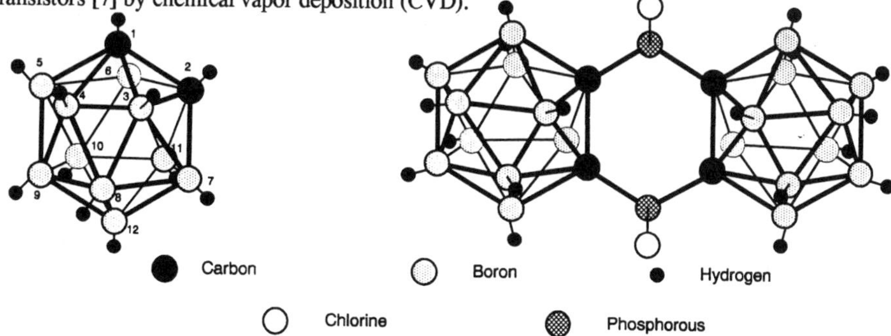

Fig. 1. Schematic diagrams of the source molecules orthocarborane $(C_2B_{10}H_{12})$ and dimeric chloro-phospha(III)carborane $([C_2B_{10}H_{10}PCl]_2)$ at the left and right, respectively.

Mat. Res. Soc. Symp. Proc. Vol. 452 © 1997 Materials Research Society

EXPERIMENTAL

Boron-carbon-phosphorus or phosphorus doped boron-carbon alloys were prepared by plasma enhanced chemical vapor deposition (PECVD) on Si(111) and glass substrates using the single source compound dimeric chloro-phospha(III)carborane (shown in Figure 1). The plasma reactor used in fabricating these materials has been described in detail elsewhere [8] and is operated at 13.56 MHz at 30 Watts. The general procedure for growing the films has been described in detail elsewhere [4-8].

The optical gap of the deposited materials was measured by a double beam, double monochromator, Perkin Elmer Lambda-9 series spectrophotometer. Both the transmittance (T) and reflectance (R) of the phosphorus doped boron carbon alloy thin films deposited on quartz glass substrates were measured. The absorption coefficient (α) of the films was assumed to be proportional to the absorption (A) determined from transmittance and reflectance measurements.

The source molecule chloro-phospha(III) carborane was synthesized using standard procedures described elsewhere [9].

RESULTS AND DISCUSSION

Films grown by PECVD under our deposition conditions from the phosphacarborane exhibit, in fact, very little phosphorus content as determined by X-ray emission spectroscopy. We conclude that there is considerable elimination of phosphorus from the parent molecule occurring during the deposition process. Thus, the resulting film is more akin to a phosphorus doped material rather than a ternary P-C-B alloy.

The optical band gap energy was determined by plotting $(\alpha E)^{1/2}$ versus energy (E), as seen in Figure 2. From the standard Tauc plot [10], the band gap of the doped boron carbon PECVD alloy was found to be 2.6 eV. This value is more consistent with the theoretical value for the band gap for $B_{12}P_2$ [3] than for the band gap of PECVD B_5C which is typically observed to be in the range from 0.7 to 0.9 eV [8,11].

The conductivity of phosphorus doped boron-carbon alloy films was also measured to be 1.47×10^{-11} $(\Omega cm)^{-1}$. This value is smaller than the typical conductivity values of PECVD B_5C films ($\sim 10^{-9}$ $(\Omega cm)^{-1}$) previously reported [8], although both materials exhibit generally high resistivities. This suggests that the PECVD grown phosphorus doped boron-carbon alloy is a near perfectly compensated material with few carriers.

X-ray diffraction data of thick (≥ 1 μm) phosphorus doped boron carbon alloy films deposited on Si (111) substrates using copper K_α radiation ($\lambda = 1.54$ Å) are displayed in Fig. 3A.

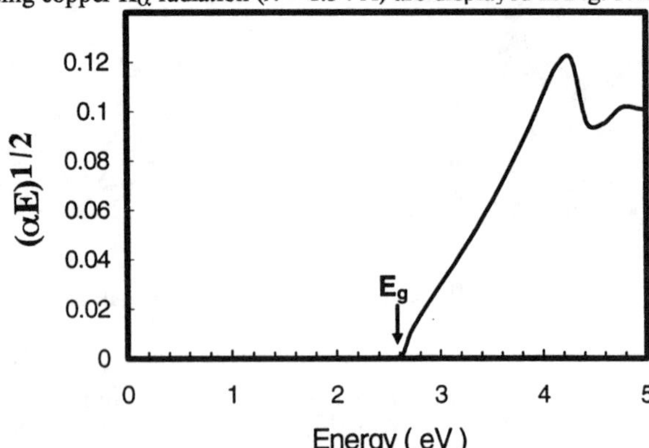

Fig.2 The Tauc plot of phosphorus doped PECVD boron carbon alloys, showing that the band gap is well over 2 eV (see text).

Broad diffraction lines are visible in the range of $42^\circ \leq 2\theta \leq 45^\circ$. The intensities of the features are relatively small due to the low scattering factors for this alloy. The X-ray diffraction data for the phosphorus doped boron carbon alloy phase produced by PECVD differs significantly from the undoped material. Unlike diffraction patterns of undoped B_5C (Fig. 3B [4,8]) and $B_{7.2}C$ [4,8] which exhibit the broad doublet peaks, diffraction of the phosphorus doped material shows only a single broad diffraction peak. We postulate that this result may be due to a new polytype of boron rich alloy as a result of doping with phosphorus. The calculated d-spacing value corresponding to the single broad peak is 2.10 Å. This value is somewhat larger than the value of $B_{7.2}C$ reported previously [8]. The average particle size of the phosphorus doped alloy is approximately 57 Å and

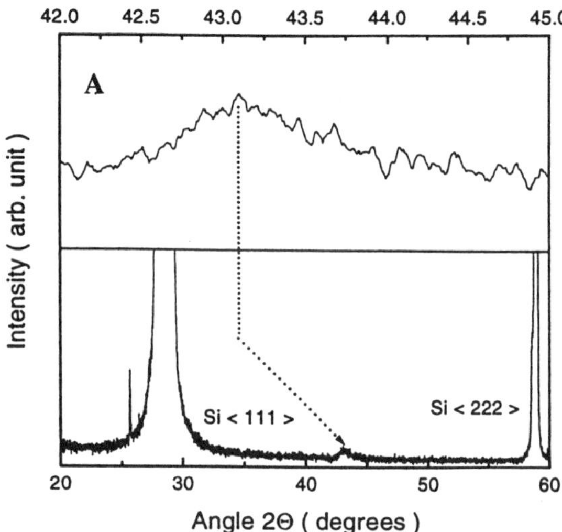

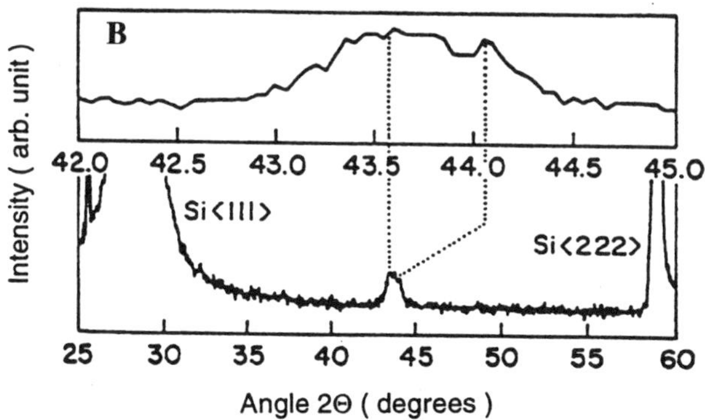

Fig. 3. Broad peak of diffraction pattern for (A) phosphorus doped boron carbon alloy and (B) B_5C on Si (111) are shown in the 2θ range from 20° to 60°. The large sharp peaks are identified as Si <111>, Si <222>, respectively.

the particle size of comparable undoped B_5C in Fig. 3B is about 100 Å. Boron-rich materials, particularly the α-rhombohedral polytype, are often regarded as natural structural models for amorphous semiconductors [12-13]. While much broader than that for crystalline boron carbide [2,14], nonetheless, the line width of the diffraction peak of this alloy film is seen to be smaller than that reported for amorphous materials. This is indicative of short range structural order in the doped film.

In Fig.4 the infrared absorption spectra of B_5C, nickel doped B_5C and phosphorus doped boron carbon alloys, are presented. The B_5C spectrum (Fig. 4(a)) and the phosphorus doped alloy spectrum (Fig. 4(b)) are very similar and show essentially the same peak positions. For B_5C and the alloy, strong vibration modes are observed between 900 and 1500 cm^{-1}. The modes around 942 cm^{-1} are attributed to the B-B vibrational modes of the boron network [15]. The continuum of states between 1100 - 1500 cm^{-1} have been tentatively assigned to the B-C vibrational modes. For $B_{1-x}C_x$ materials, these modes fall within a range of 1100 -1200 cm^{-1} [15]. For the B_5C films of this study, some C atoms reside within the boron icosahedral cage and therefore extend the vibrational states out to 1500 cm^{-1}, as observed in Fig. 4. A B-H mode is observed at 2535 cm^{-1} and is indicative of the inclusion of H into the films. There is also a C-H mode at 2916 cm^{-1} in the spectra of nickel and phosphorus doped B_5C.

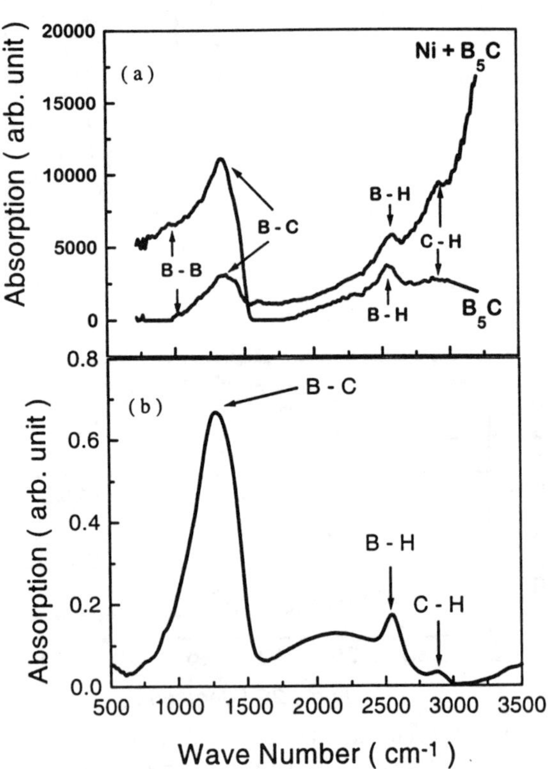

Fig.4. FTIR spectra of (a) B_5C and Ni doped B_5C and (b) phosphorous doped boron carbon alloys.

With nickel doping, as discussed elsewhere [5], there is an increasing background absorbance above 1500 to 200 cm^{-1}. This is possibly due to excitations from the donor states introduced with nickel doping roughly 200 meV below the conduction band edge to the conduction band continuum. This increase in the background absorbance is not observed in either the undoped B_5C or the phosphorus doped B_5C. While not by any means conclusive, this suggests that few carrier states are introduced with phosphorus doping. This postulate is supported by the very high resistivity of this material.

CONCLUSION

A phosphorus doped boron carbon alloys were fabricated by PECVD using a single source precursor dimeric chloro-phospha(III) carborane ($[C_2B_{10}H_{10}PCl]_2$). FTIR spectra from the boron carbon phosphorus alloys and B_5C materials show no significant local structural changes due to the presence of phosphorus in B_5C and provide no evidence for additional donor or acceptor states in the gap. This is supported by a decrease in the observed conductivity of this alloy by a few orders of magnitude (to a value of about ~ 10^{-11} $(\Omega cm)^{-1}$) relative to B_5C.

The optical properties of phosphorus doped boron carbon alloys are different from the undoped B_5C material. The energy band gap of the boron-carbon phosphorus alloy increased from 0.9 eV, for the undoped B_5C, to 2.6 eV. This large increase in band gap can be best understood if there is a change in the material polytype, or a large increase in the hydrogen content of the films [12], since the phosphorus content in these materials is not large, and the source molecule preserves the basic icosahedral building block.

The composition of this boron carbon phosphorus alloy has not yet been identified. The decomposition pathway of the precursor for this alloy is not known. The results of this study suggest that this alloy material has potential as a wide band gap semiconductor material.

REFERENCES

1. Y. Kumashiro, M. Hirabayashi and S. Takagi, Mater. Res. Symp. Proc. 162, 585 (1990); Y. Kumashiro, M. Hirabayashi, T. Koshiro and Y. Okada, J. Less-Common Metals, 143, 159 (1988); Y. Kumashiro, J. Mater. Res. 5, 2933 (1990); Y. Kumashiro, T. Yokohama, J. Nakamura, K. Matsuda, H. Yoshida and J. Takahashi, Mater. Res. Soc. Symp. Proc. 242, 629 (1992); H. Xia, Q. Xia and A. Ruoff, J. Appl. Phys. 74, 1660 (1993).

2. T. Aselage, D. Emin, G. Samara, D. Tallant, S. Van Deusen, M. Eatough, H. Tardy and E. Venturini, Phys. Rev. B 48, 11759 (1993); B. Morosin, A. W. Mullendore, D. Emin and G. Slack, in Boron-Rich Solids, edited by D. Emin, T. Aselage, C. Beckel, I. Howard and C. Wood, AIP Conf. Proc. No. 140 (AIP, New York, 1986), p. 70; H. Werheit, U. Kuhlmann, K. Shirai and Y. Kumashiro, J. Alloys Comp. 233, 121 (1996)

3. D. Li and W. Y. Ching, Phys. Rev. B 52, 17073 (1995).

4. Dongjin Byun, Seong-Don Hwang, P.A. Dowben, F. Keith Perkins, F. Filips and N.J. Ianno, Appl. Phys. Lett. 64, 1968 (1994); Dongjin Byun, B.R. Spady, N.J. Ianno, and P.A. Dowben, Nanostructured Materials 5, 465 (1995)

5. Seong-Don Hwang, Ken Yang, P.A. Dowben, A.A. Ahmad, N.J. Ianno, J.Z. Li, J.Y. Lin, H.X. Jiang and D.N. McIlroy, Appl. Phys. Lett. (1997), in press

6. Seong-Don Hwang, N.B. Remmes, P.A. Dowben, and D.N. McIlroy, J. Vac. Sci. Technol. B14, 2957 (1996)

7. Seong-Don Hwang, Dongjin Byun, N.J. Ianno, P.A. Dowben and H.R. Kim, Appl. Phys. Lett. 68, 1495 (1996)

8. S. Lee, J. Mazurowski, G. Ramseyer and P. A. Dowben, J. Appl. Phys. 72, 4925 (1992)

9. R.N. Grimes, in Carboranes, edited by P.M. Maitlis, F.G. Stone and R. West, (Academic Press, New York 1970), p. 55; R.Roy and H. Schroeder, Inorg. Chem. 2, 1107 (1963)

10. J. Tauc, in Optical Properties of Solids, edited by F. Abels, (American Elsevier Publishing Co. New York, 1972), p. 277.

11. Ahmad A. Ahmad, N.J. Ianno, P.G. Snyder, D. Welipitiya, Dongjin Byun and P.A. Dowben, J. Appl. Phys. **79**, 8643 (1996)

12. O. A. Golikova, Sov. Phys. Solid State 29, 1652 (1987).

13. O. A. Golikova, in Boron-Rich Solids, edited by D. Emin, T. Aselage, A. C. Switendick, B. Morosin and C. Beckel, AIP Conf. Proc. No. 231 (AIP, New York, 1990), p. 108.

14. T. L. Aselage and D. Emin, in Boron-Rich Solids, edited by D. Emin, T. Aselage, A. C. Switendick, B. Morosin and C. Beckel, AIP Conf. Proc. No. 231 (AIP, New York, 1990), p. 177.

15. K. Shirai, S. Emura, S. Gonda and Y. Kumashiro, J. Appl. Phys. 78, 3392 (1995).

Thermoelectric Properties of $RhSb_3$ Crystals and Thin Films

Baoxing Chen, Jun-Hao Xu, Siqing Hu, and Ctirad Uher
Department of Physics, University of Michigan, Ann Arbor, MI 48109-1120

ABSTRACT

$RhSb_3$ belongs to a class of solids called skutterudites which were recently identified as promising novel thermoelectric materials. Our intent was to explore thin film growth of $RhSb_3$ and to assess the thermoelectric properties of such films. $RhSb_3$ films were prepared by electron-beam deposition on silicon and sapphire substrates. Thermopower, electrical resistivity and the Hall effect were measured over the temperature range from 2K to 300K. The data are compared to measurements on single crystals of $RhSb_3$.

INTRODUCTION

Skutterudite compounds, the name originating from a town in Norway where mineral deposits of these compounds were first mined, were recently identified as promising thermoelectric materials [1,2]. The skutterudites are compounds of the form AB_3 where A is a metal such as Co, Rh, or Ir, and B is a pnicogen such as Sb, As, or P. The compounds crystallize with a cubic structure consisting of 32 atoms per unit cell (space group *Im3*). The pnictide atoms constitute four-member rings located at the center of the cubes formed by the metal atom [3]. The compounds show very high hole mobilities and large thermopower, the key ingredients for a good thermoelectric material. A relatively large thermal conductivity which is a distinct disadvantage from the perspective of thermoelectricity can be effectively suppressed by filling the voids and forming the so-called filled skutterudite structure [4,5]. The resulting figure of merit of filled skutterudites appears to be at least as high [6] as the figure of merit of the current state-of-the-art thermolectric materials. Low temperature transport measurements on a related compound $CoSb_3$ have been made recently [7]. $RhSb_3$ is another member of the skutterudite family and we are interested in exploring its thin film growth because the recent theoretical predictions point to a potential of obtaining very high figures of merit for 2-dimensional structures such as quantum wells [8]. The growth of high quality skutterudite films is a prerequisite for fabrication of quantum well structures in this material system.

EXPERIMENT

Films of $RhSb_3$ were grown in a UHV-compatible dual e-gun growth chamber. Antimony (99.9999% purity) and rhodium (99.9%) slugs were co-evaporated on either

Mat. Res. Soc. Symp. Proc. Vol. 452 © 1997 Materials Research Society

Si(111) or Al_2O_3 (110) substrates. Evaporation rates were controlled by two Inficon controllers and were typically set at 0.4A/s for Rh and 2.4A/s for Sb. The vacuum, normally on the order of $1x10^{-9}$ torr before the deposition, rises to $5x10^{-7}$ torr during the deposition, mostly on account of a high vapor pressure of antimony. X-ray measurements confirmed the resulting structure to be $RhSb_3$. Single crystals of $RhSb_3$, against which we compare our film results, were prepared at the Jet Propalsion Laboratory in Pasadena by the vertical gradient freeze technique [9]. Resistivity, thermopower and Hall effect measurements were made with the aid of a steady-state method in a liquid helium cryostat. Typically one end of the sample was thermally anchored to the cryostat using stycast epoxy, and a thin-film strain gauge was attached to the other end of the sample to generate a thermal gradient across the sample. Thermal gradients were monitored using chromel-constantant thermocouples.

RESULTS

Fig. 1 shows the results of x-ray 2-θ scans from a single crystal $RhSb_3$ powders and from four films (labelled #1, #2, #3, #4) of thickness about 900A grown and annealed at different temperatures. The top panel shows the intensity (count/sec) vs 2-θ for the single crystal powders and the next four panels display the same information for films #1, #2, #3, and #4 respectively. The lattice constant is 9.22A, similar to that reported by Zhuravlev et al. [10]. Film #1 was grown on a sapphire substrate at 300 C, then annealed at 500 C for an hour. Films #2 and #3 were grown on sapphire at 500 C and 300C respectively and no annealing was applied. Film #4 was grown and annealed together with film #1 but the substrate was Si(111). Annealing clearly results in a better quality of films. In the following we present data on films #1 and #4 (resistivity and thermopower) along with the results obtained on the single crystal of $RhSb_3$. More detailed studies of films on Si substrate are in progress.

Temperature dependence of resistivity is displayed in Fig. 2. While the single crystal shows a metallic behavior, the films have a semiconducting character. The hole concentrations and hole mobilities are presented in Figs. 3 and 4, respectively. Our thin film has a very large hole concentration on the order of 10^{20} cm^{-3} at room temperature and a correspondingly low mobility of only about 30 cm^2/V-s. This is to be compared to the room temperature mobility of the single crystal of about 3000 cm^2/V-s. For the crystal, the mobility obeys a $T^{-0.5}$ dependence at high temperatures which suggests the combination of phonon-limited mobility ($\propto T^{-1.5}$) and temperature independent neutral-impurity-limited mobility. For the thin film, the hole mobility increases as temperature increases, which indicates mixed conduction or ionized impurity scattering.

Fig. 5 shows the temperature dependence of thermopower in the single crystal and the two films. The single crystal has a considerably larger thermopower than the films and the approximately linear dependence on temperature indicates a degenerate behavior. The intrinsic conduction is expected to dominate the thermopower at higher temperatures.

From the expressions for the hole concentration and Seebeck coefficient of a partially degenerate conductor,

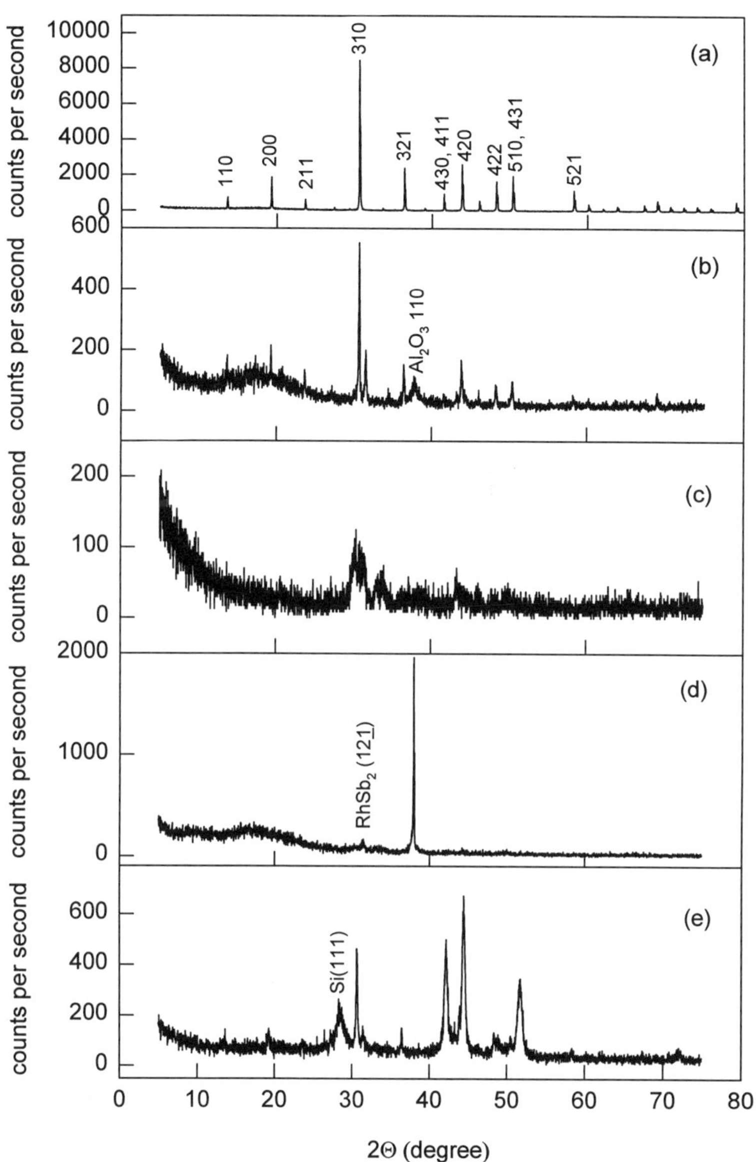

Fig. 1. X-ray diffraction pattern for $RhSb_3$ (a)single crystal, (b) thin film #1 grown on sapphire at 300 C, annealed at 500 C, (c) thin film #2, grown on sapphire at 500 C, (d) thin film #3, grown on sapphire at 300 C, and (e) thin film #4 grown and annealed together with film #1.

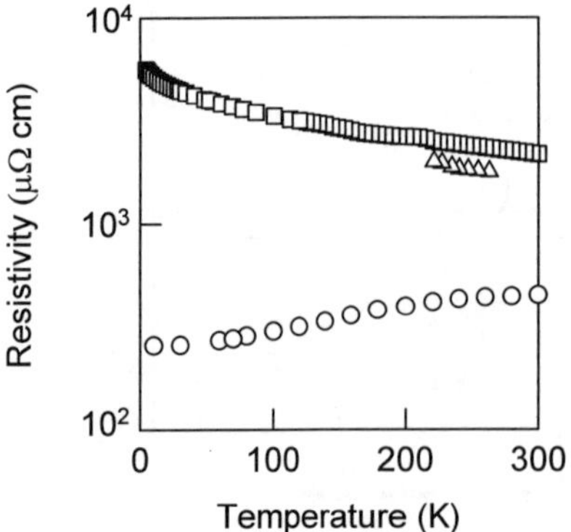

Fig. 2. Temperature dependences of the resistivity of the single crystal (circles), film #1 (squares), and film #4 (triangles).

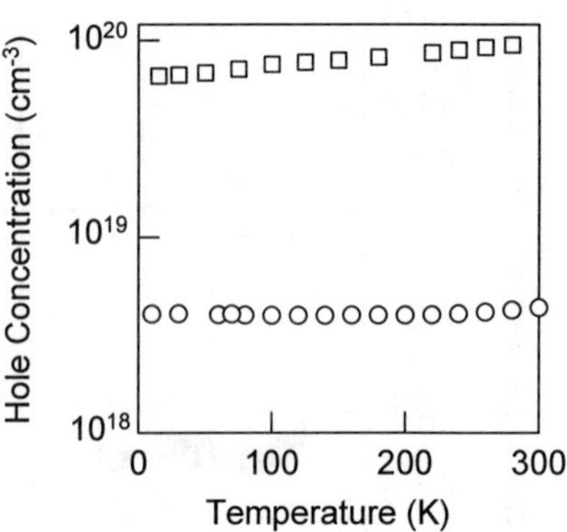

Fig. 3. Temperature dependences of the hole concentrations of the single crystal (circles), and film #1 (squares).

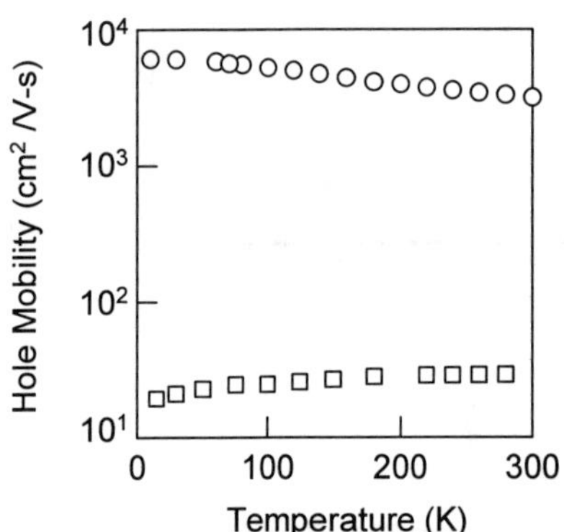

Fig. 4. Temperature dependences of the hole mobilities in the single crystal (circles), and in film #1 (squares).

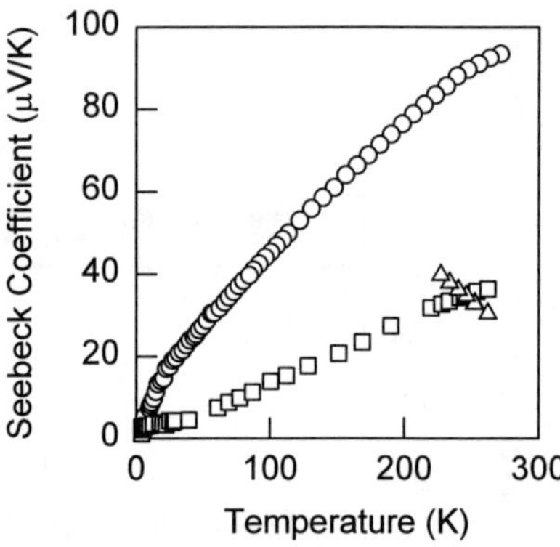

Fig. 5. Temperature dependences of the Seebeck coefficients in the single crystal (circles), in film #1 (squares), and in film #4 (triangles).

$$n_p = (4/\sqrt{\pi})\,[2\pi m^* k_B T/h^2]^{3/2} F_{1/2} \quad (1)$$

$$S = (k_B/e)\,[2F_1/F_0 - \eta] \quad (2)$$

where the F's are Fermi-Dirac integrals and $\eta = E_F/k_BT$, we can estimate the effective mass of the holes. For the single crystal of $RhSb_3$, the thermopower of $S \approx 95\ \mu V/K$ at 280 K is satisfied with $\eta \approx 3$. Using this value in Eq.1 yields $m^* \approx 0.11$, assuming $n_p \approx 5\times10^{24}\ m^{-3}$. For the data of film #1, we have $S \approx 40\ \mu V/K$ at 260 K in which case Eq.2 returns $\eta \approx 7.2$. For the hole density of $n_p \approx 1\times10^{26}\ m^{-3}$, Eq.1 gives the effective mass $m^* \approx 0.45$. Thus, the effective mass increases with the hole density. A similar trend has been seen recently in $CoSb_3$ [11].

For the single crystal we have also studied the thermal conductivity which is shown in Fig. 6. Similar to the findings for other unfilled skutterudites, $RhSb_3$ has a relatively large thermal conductivity. A sharp peak at about 23K indicates a competition between phonon-boundary scattering and Umklapp phonon-phonon processes. Using the Wiedemann-Franz law, we conclude that the thermal conductivity is essentially all due to phonons.

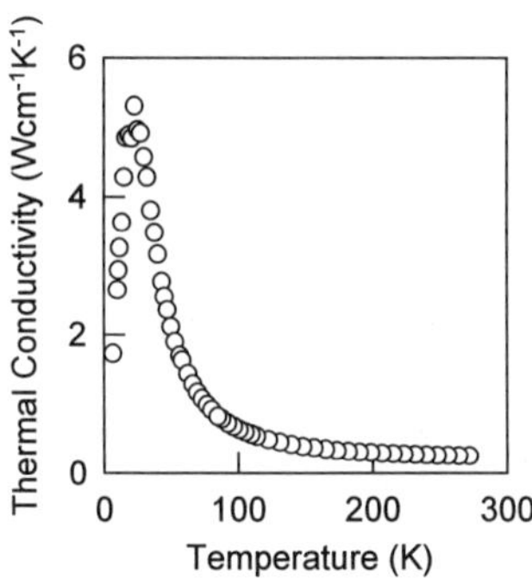

Fig. 6. Temperature dependence of the thermal conductivity in the single crystal.

CONCLUSIONS

We have studied thermoelectric properties of the $RhSb_3$ single crystals and thin films deposited on either sapphire or silicon substrates. While single crystals of $RhSb_3$ have large mobilities, thin films are not yet of the quality to replicate the promising behavior of the crystals. The large discrepancies between the properties of the single crystals and thin films seem to lie in the fact that the films of $RhSb_3$ have a tendency to be somewhat Sb deficient. Traces of $RhSb_2$ are detected in some of our $RhSb_3$ films from X-ray diffraction patterns and ways must be found to preserve Sb on the surface during the annealing process.

ACKNOWLEDGMENTS

We are grateful to Drs. T. Caillat and J.-P. Fleurial for making available their single crystals of $RhSb_3$ and to Dr. D. T. Morelli for stimulating discussions. The work was supported in part by the ONR Grant N00014-96-1-0181.

REFERENCES

[1] T. Caillat, A. Borshchevsky, and J.-P. Fleurial, in Proceedings of the XI-th International Conference on Thermoelectrics, University of Texas at Arlington, 1992, ed. by K. R. Rao (University of Texas at Arlington Press, p.98, 1993).
[2] G. A. Slack and V. G. Tsoukala, J. Appl. Phys. **76**, 1665 (1994).
[3] A. Kjekshus and T. Rakke, Acta Chem. Scand. **A28**, 99 (1974).
[4] D. T. Morelli and G. P. Meisner, J. Appl. Phys. **77**, 3777 (1995).
[5] G. S. Nolas, G. A. Slack, D. T. Morelli, T. M. Tritt, and A. C. Ehrlich, J. Appl. Phys. **79**, 4002 (1996).
[6] B. C. Sales, D. Mandrus, and R. K. Williams, Science **272**, 1325 (1996).
[7] D. T. Morelli, T. Caillat, J.-P. Fleurial, A. Borschevsky, J. Vandersande, B. Chen, and C. Uher, Phys. Rev. **B51**, 9622 (1995).
[8] J.-P. Fleurial, T. Caillat, and A. Borshchevsky, in Proceedings of the XIII-th International Conference on Thermoelectrics, Kansas City, 1994 , ed. by B. Mathiprakasam (AIP Conference Proceedings 316, p. 40, 1995).
[9] L. D. Hicks and M. S. Dresselhaus, Phys. Rev. B47, 12727 (1993).
[10] N. N. Zhuravlev and G. S. Zhdanov, Sov. Phys. Cryst. **1**, 404 (1956).
[11] T. Caillat, A. Borshchevsky, and J.-P. Fleurial, J. Appl. Phys. **80**, 4442 (1996).

AUTHOR INDEX

SUBJECT INDEX